The Properties of Gases and Liquids

ABOUT THE AUTHORS

J. Richard Elliott, PhD, is a professor emeritus of chemical engineering at the University of Akron in Ohio. He is co-author of the thermodynamics section in *Perry's Chemical Engineers' Handbook* and co-author of *Introductory Chemical Engineering Thermodynamics*.

Vladimir Diky, PhD, is the project leader of the ThermoData Engine at the NIST Thermodynamics Research Center Group, which supplies thermodynamic property data and data analysis tools for chemical engineering, research, public health, safety, and environmental protection.

Thomas A. Knotts IV, PhD, is a professor of chemical engineering at Brigham Young University. He is a co-PI of the AIChE/DIPPR 801 database project.

W. Vincent Wilding, PhD, is a professor of chemical engineering at Brigham Young University. He is co-author of the physical and chemical data section in *Perry's Chemical Engineers' Handbook* and a co-PI of the AIChE/DIPPR 801 database project.

The Properties of Gases and Liquids

J. Richard Elliott
Vladimir Diky
Thomas A. Knotts IV
W. Vincent Wilding

Sixth Edition

**New York Chicago San Francisco
Athens London Madrid
Mexico City Milan New Delhi
Singapore Sydney Toronto**

McGraw Hill books are available at special quantity discounts to use as premiums and sales promotions, or for use in corporate training programs. To contact a representative, please visit the Contact Us pages at www.mhprofessional.com.

The Properties of Gases and Liquids, Sixth Edition

2 3 4 5 6 LBC 29 28 27 26 25

Library of Congress Control Number: 2022051283

ISBN 978-1-260-11634-2
MHID 1-260-11634-4

Sponsoring Editor
Robin Najar

Editorial Supervisor
Janet Walden

Project Manager
Revathi Viswanathan,
KnowledgeWorks Global Ltd.

Acquisitions Coordinator
Elizabeth M. Houde

Copy Editor
Girish Sharma,
KnowledgeWorks Global Ltd.

Proofreader
Manish Tiwari,
KnowledgeWorks Global Ltd.

Indexer
Edwin Durbin

Production Supervisor
Lynn M. Messina

Composition
KnowledgeWorks Global Ltd.

Illustration
KnowledgeWorks Global Ltd.

Art Director, Cover
Jeff Weeks

Library of Congress Cataloging-in-Publication Data

Names: Elliott, J. Richard, author. | Poling, Bruce E. Properties of gases and liquids.
Title: The properties of gases and liquids / J. Richard Elliott, Vladimir Diky, Thomas A. Knotts IV, W. Vincent Wilding.
Description: Sixth edition. | New York : McGraw Hill, [2023] | Revised edition of: The properties of gases and liquids / Bruce E. Poling, John M. Prausnitz, John P. O'Connell. 2001. | Includes bibliographical references and index. | Summary: "Fully updated for the latest advances, this must-have chemical engineering guide serves as a single source for up-to-date physical data, chemical data, and predictive and estimation methods. The Properties of Gases and Liquids, Sixth Edition provides the latest curated data on over 480 compounds and includes a special section devoted to the interpretation of uncertainty in physical property estimation. You will get new coverage of advanced EOSs with correlated and predicted parameters, advanced computational methods, quantum density functional theory, and semi-empirical combinations. Clear explanations and sample calculations are provided throughout this all-inclusive resource."—Provided by publisher.
Identifiers: LCCN 2022051283 | ISBN 9781260116342
Classification: LCC TP242 .P62 2023 | DDC 660/.042—dc23/eng/20230111
LC record available at https://lccn.loc.gov/2022051283

CONTENTS

	Foreword	xi
	Preface	xiii
1	**Introduction**	**1**
1.1	Scope	1
1.2	Estimation of Physical Properties	1
1.3	Traditional Estimation Methods	2
1.4	Nontraditional Estimation Methods	5
1.5	Database Development and Method Evaluation	7
1.6	Organization of the Book	9
1.7	References	9
2	**Uncertainty**	**11**
2.1	Scope	11
2.2	Introduction: Why Are Uncertainties Important?	11
2.3	Historical Background	13
2.4	Key Documents: The GUM, NIST 1297, and VIM	13
2.5	Uncertainty Assessment for Thermophysical Properties: The GUM Approach	14
2.6	The Uncertainty Budget	16
2.7	Reference Materials and Standard Reference Materials	16
2.8	Challenges for the User of Experimental Data from the Literature	17
2.9	Examples of Common Problems in Articles Reporting Thermophysical Properties	19
2.10	Modern Uncertainty Assessment Procedures (Critical Evaluation)	22
2.11	Summary	31
2.12	Disclaimer	31
2.13	References	31
3	**Pure-Component Constants**	**35**
3.1	Scope	35
3.2	Vapor-Liquid Critical Properties	36
3.3	Acentric Factor	66
3.4	Melting and Boiling Points	66
3.5	Discussion of Estimation Methods for Pure Component Constants	85
3.6	Dipole Moments	92
3.7	Availability of Data and Computer Programs	93
3.8	Notation	93
3.9	References	94

4 Thermodynamic Properties of Ideal Gases 97

4.1 Scope and Definitions 97
4.2 Estimation Methods for the Ideal Gas Standard State 101
4.3 Method of Benson 102
4.4 Method of Domalski and Hearing 107
4.5 Modified Joback Method for Ideal Gas Heat Capacity 112
4.6 Quantum Mechanical Methods 114
4.7 Standard State Enthalpy of Formation and Enthalpy of Combustion 131
4.8 Discussion and Recommendations 133
4.9 Notation 141
4.10 References 144

5 Pure Fluid Thermodynamic Properties of the Single Variable Temperature 147

5.1 Scope 147
5.2 Saturated Liquid Density 147
5.3 Theory of Liquid Vapor Pressure and Enthalpy of Vaporization 156
5.4 Theory of Liquid Heat Capacity 157
5.5 Correlating Vapor-Pressure, Enthalpy of Vaporization, and Liquid Heat Capacity Data 158
5.6 Reliable Extrapolation of Vapor Pressure and Thermodynamic Consistency between Vapor Pressure, Enthalpy of Vaporization, and Liquid Heat Capacity 165
5.7 Prediction of Vapor Pressure 168
5.8 Extrapolation and Prediction of Enthalpy of Vaporization of Pure Compounds and Recommendations 174
5.9 Prediction of Liquid Heat Capacity 175
5.10 Discussion and Recommendations for Vapor-Pressure, Enthalpy of Vaporization, and Liquid Heat Capacity Estimation and Correlation 178
5.11 Enthalpy of Melting 180
5.12 Enthalpy of Sublimation 196
5.13 Solid Vapor Pressure (Sublimation Pressure) 199
5.14 Correlation and Estimation of Virial Coefficients 201
5.15 Notation 208
5.16 References 212

6 Thermodynamic Properties of Pure Gases and Liquids 217

6.1 Scope 217
6.2 Introduction to Equations of State 218
6.3 Theory of Equations of State 219
6.4 Fundamental Thermodynamic Relationships for Pure Compounds 230
6.5 Virial Equations of State 234
6.6 Cubic Equations of State 235
6.7 Multiparameter Equations of State 239

6.8 Perturbation Models with Customized Parameters 243
6.9 Perturbation Models with Transferable Parameters 256
6.10 Chemical Theory EOSs 267
6.11 Molecular Simulation Models 270
6.12 Residual Functions for Evaluated Models 272
6.13 Evaluations of Equations of State 273
6.14 Notation 281
6.15 References 282

7 Thermodynamic Properties of Mixtures 287

7.1 Scope 287
7.2 Mixture Properties—General Discussion 288
7.3 Theory of Mixture Modeling 291
7.4 Perturbation Models 294
7.5 Excess Gibbs Energy Mixing Rules 311
7.6 Mixing Rules for Multiparameter EOS 315
7.7 Virial Equations of State for Mixtures 318
7.8 Residual Functions for Evaluated Models 322
7.9 Empirical Correlations for Mixture Properties 324
7.10 Evaluations and Recommendations 324
7.11 Notation 328
7.12 References 330

8 Vapor-Liquid Equilibria in Mixtures 333

8.1 Scope 333
8.2 A Note about the Modeling of Temperature Effects 338
8.3 Thermodynamics of Vapor-Liquid Equilibria 338
8.4 Fugacity of a Pure Liquid 339
8.5 Simplifications in the Vapor-Liquid Equilibrium Relation 340
8.6 Activity Coefficients; Gibbs-Duhem Equation, and Excess Gibbs Energy 341
8.7 Theory of Activity Models 344
8.8 Correlating Low-Pressure Binary Vapor-Liquid Equilibria 363
8.9 Effect of Temperature on Low-Pressure Vapor-Liquid Equilibria 369
8.10 Multicomponent Vapor-Liquid Equilibria at Low Pressure 370
8.11 Predicting Activity Coefficients 374
8.12 Phase Equilibrium with Henry's Law 390
8.13 Vapor-Liquid Equilibria with Equations of State 396
8.14 Evaluations 404
8.15 Concluding Remarks 408
8.16 Acronyms 409
8.17 Notation 410
8.18 References 412

9 Specialized Phase Behavior in Mixtures 417

9.1 Scope 417
9.2 Infinite Dilution Activity Coefficients 418
9.3 Liquid-Liquid Equilibria 427
9.4 Solubilities of Solids in Liquids 436
9.5 Evaluations 443
9.6 Concluding Remarks 450
9.7 Notation New to Chapter 9 451
9.8 References 452

10 Viscosity 455

10.1 Scope 455
10.2 Definitions of Units of Viscosity 455
10.3 Theory of Gas Transport Properties 456
10.4 Estimation of Low-Pressure Gas Viscosity 457
10.5 Viscosities of Gas Mixtures at Low Pressures 467
10.6 Effect of Pressure on the Viscosity of Pure Gases 479
10.7 Viscosity of Gas Mixtures at High Pressures 495
10.8 Liquid Viscosity 499
10.9 Effect of High Pressure on Liquid Viscosity 502
10.10 Effect of Temperature on Liquid Viscosity 505
10.11 Estimation of Low-Temperature Liquid Viscosity 506
10.12 Estimation of Liquid Viscosity at High Temperatures 518
10.13 Liquid Mixture Viscosity 520
10.14 Notation 532
10.15 References 538

11 Thermal Conductivity 545

11.1 Scope 545
11.2 Theory of Thermal Conductivity 545
11.3 Thermal Conductivities of Polyatomic Gases 547
11.4 Effect of Temperature on the Low-Pressure Thermal Conductivities of Gases 553
11.5 Effect of Pressure on the Thermal Conductivities of Gases 553
11.6 Thermal Conductivities of Low-Pressure Gas Mixtures 562
11.7 Thermal Conductivities of Gas Mixtures at High Pressures 567
11.8 Thermal Conductivities of Liquids 574
11.9 Estimation of the Thermal Conductivities of Pure Liquids 575
11.10 Effect of Temperature on the Thermal Conductivities of Liquids 583
11.11 Effect of Pressure on the Thermal Conductivities of Liquids 585
11.12 Thermal Conductivities of Liquid Mixtures 588

11.13 Notation 596
11.14 References 600

12 Diffusion 605

12.1 Scope 605
12.2 Basic Concepts and Definitions 605
12.3 Progress in Self-Diffusivity Correlation 608
12.4 Diffusion Coefficients for Binary Gas Systems at Low Pressures: Prediction from Theory 613
12.5 Diffusion Coefficients for Binary Gas Systems at Low Pressures: Empirical Correlations 618
12.6 The Effect of Pressure on the Binary Diffusion Coefficients of Gases 620
12.7 The Effect of Temperature on Diffusion in Gases 626
12.8 Diffusion in Multicomponent Gas Mixtures 627
12.9 Diffusion in Liquids: Theory 628
12.10 Estimation of Binary Liquid Diffusion Coefficients at Infinite Dilution 629
12.11 Concentration Dependence of Binary Liquid Diffusion Coefficients 639
12.12 The Effects of Temperature and Pressure on Diffusion in Liquids 644
12.13 Diffusion in Multicomponent Liquid Mixtures 645
12.14 Diffusion in Electrolyte Solutions 648
12.15 Notation 652
12.16 References 654

13 Surface Tension 659

13.1 Scope 659
13.2 Introduction 659
13.3 Estimation of Pure-Liquid Surface Tension 660
13.4 Temperature Dependence of Pure-Liquid Surface Tension 669
13.5 Surface Tensions of Mixtures 671
13.6 Notation 682
13.7 References 685

Appendix A Property Data Bank 687

Appendix B Lennard-Jones Potentials as Determined from Viscosity Data 749

Index 753

FOREWORD

The Laws of Thermodynamics govern the physical and chemical behavior of our world. The conservation of mass and energy and the generation of entropy in real processes constrain human manipulations of nature. Yet, within these limitations, we have designed and built amazing systems and devices to create new substances, materials, and enterprises to enhance our quality of life.

While the laws are always true, they only provide algebraic and calculus relations among the measurable and conceptual properties. So, quantitative property descriptions are essential to the realization of facilities to produce desired products and actions. The advent of advanced computation has enabled examination and development of extremely complex characterizations from the molecular scale to enormous reactors and separators. Contemporary process simulators contain model codes and data banks for ready access and execution.

The series of *The Properties of Gases and Liquids*, initiated in 1957 by R. C. Reid and T. K. Sherwood, and guided by Bob Reid through four editions, has been utilized to an incredible extent in these efforts. J. M. Prausnitz joined for the third and succeeding editions and B. E. Poling started contributing at the fourth edition. The fifth edition published in 2000 was coauthored by Poling, Prausnitz, and O'Connell. This sixth edition has been coauthored by a team, led by J. R. Elliott, of talented and dedicated researchers and practitioners from a new generation.

Creating this sixth edition has been a monumental task because of the burgeoning literature in the area, for user expectations of electronic access to computer codes for many models and extensive values of property and phase equilibrium data, and the need to include treatments of molecular modeling and simulations.

The authors have responded with exceptional breadth, admirable depth, and optimal detail about the current literature in all the relevant areas. Further, while retaining much of the structure of prior editions, efforts have been made to not only update and expand the coverage, but also to develop reader understanding about the foundations and interconnections among classes of models.

The result is a treasure of methods, examples, data, and education for professionals and students in this vital discipline for producing traditional and novel chemical processes and products. The future demands greater efficiency in time, money, energy, and materials; more attention to health and safety; and minimal environmental impact. This book will be an invaluable resource for such endeavors.

–John P. O'Connell

PREFACE

It is an oversimplification to say that our goal in writing this sixth edition has been to compare all the theories in the world to all the data in the world, but it makes a good "elevator speech." Of course, our reach has exceeded our grasp in this regard, but it is a worthy goal, nonetheless. We encourage *The Properties of Gases and Liquids* community to let us know if we have overlooked important theories or data and to help us improve the next edition. Although the book is intended to serve primarily the practicing engineer, especially the process or chemical engineer, other engineers and scientists concerned with gases and liquids may find it useful and we welcome their contributions to the community as well.

We refer readers to Chap. 1 for an introduction to the content and organization of this edition. The scope is like that of previous editions. Some findings are new, however. Notably, methods based on molecular simulations and quantum mechanics are found to provide superior accuracy when compared to group contribution methods for predictions of vapor pressure and ideal gas properties like heats of formation. Quantum-based predictions of activity coefficients are found to provide competitive accuracy with fewer adjustable parameters and fewer gaps in coverage. Progress toward more fundamental molecular methods should be expected to advance even further in the near future.

As the methods become more sophisticated, their implementation becomes more challenging. To bridge that gap, we have constructed a website at PGL6ed.byu.edu where readers can find links to sample spreadsheets for hand calculations, tabulations of group contributions, scripts for implementing publicly available software, and open-source code to facilitate personal implementation. We gratefully acknowledge permission from ASTM to publish the Benson groups of Chap. 4 to the website and we hope this will minimize transcription errors and lead to quick fixes of any inconsistencies in our articulation of the method. We also acknowledge the contribution of source codes from Joachim Gross for the "polar perturbed chain statistical associating fluid theory" (PPC-SAFT) method, and from Jean-Noel Jaubert for the "translated consistent Peng-Robinson" (tcPR) method.

In some cases, we did our best to include methods in our evaluations that were either proprietary or beyond the scope of our implementation time. We thank Andreas Klamt for his help with the COSMO-RS/therm method, Jean-Charles DeHemptinne for help with his GC-PPC-SAFT method, Georgios Kontogeorgis and Xiaodong Liang for help with the CPA method, and Raphael Soares for help with his COSMO-RS/SAC methods.

Numerous other individuals have contributed their efforts to reviewing our coverage of the methods and even teaching us the details in some cases. Andrei Kazakov and Eugene Paulechka provided guidance in Chap. 4. Andreas Klamt, Juan Vera, Constantine Panayiotou, S.-T. Lin, Raphael Soares, and Andres Pina-Martinez provided guidance in Chap. 8. Ala Bazyleva, Marcia Huber, Sumnesh Gupta, John Shaw, James Jackson, Sasidhar Gumma, Allan Harvey, Ian Bell, and Yury Chernyak read preliminary drafts of various chapters and provided many helpful comments. John O'Connell provided support throughout the effort in multiple contexts. Alex Mansfield and Neil Giles, the assistant project and project coordinators, respectively, for the DIPPR® 801 project deserve special mention for their contributions. The former is primarily responsible for creating Appendix A, while the counsel from the latter was invaluable when evaluating the many pure-component prediction techniques found throughout the work. We also thank the DIPPR 801 project advisory committee for permission to use their database as the primary basis for Appendix A.

Family members have provided the least voluntary but most patient support for this multi-year project. JRE would like to thank Guliz, Serra, Eileen, and Brian for not prosecuting his thefts of family time too harshly. VD thanks Ala Bazyleva. TAK thanks Sharlene Knotts. WVW thanks Julie Wilding.

The Properties of Gases of Liquids has been a major resource for thermophysical property correlation and prediction since its inception. From the first edition to the fifth edition, Reid, Sherwood, Prausnitz, Poling, and O'Connell pioneered this effort and brought it to prominence. The present authors can only hope to have done justice to these previous achievements by continuing the tradition.

J. Richard Elliott
Vladimir Diky
Thomas A. Knotts IV
W. Vincent Wilding

1

Introduction

1.1 SCOPE

This chapter presents a brief outline of the entire sixth edition and its relation to the grand tradition of *The Properties of Gases and Liquids* ("PGL6ed") over the seven decades since the first edition. While several details have evolved, we remain committed to the spirit of "molecular engineering" expressed so long ago.

1.2 ESTIMATION OF PHYSICAL PROPERTIES

The structural engineer cannot design a bridge without knowing the properties of steel and concrete. Similarly, scientists and engineers often require the properties of gases and liquids. The chemical or process engineer finds knowledge of physical properties of fluids essential to the design of many kinds of products, processes, and industrial equipment. Even the theoretical physicist must occasionally compare theory with measured properties.

It is bittersweet to observe that the first page of the first edition of *The Properties of Gases and Liquids*[1] explains "The chemical or process engineer deals primarily with molecules, and his activities might be described as 'molecular engineering.'... Indeed, in the view of the modern physicist, 'everything is known about molecules.'... There is, unfortunately, a great gap between having the laws and having the numbers which the engineer and applied scientist need for design purposes." While it is remarkable how much relevance those words retain to this day, it is also frustrating, and a little embarrassing. Reid and Sherwood might well have expected that gap to be entirely closed by the year 2022.

Nevertheless, the gap is narrowing, and this monograph is broadly devoted to documenting that progress. We know that the physical properties of every substance depend directly on the nature of the molecules of the substance. Therefore, the ultimate generalization of physical properties of fluids must require quantitative characterization from the *ab initio* level to the macro scale. Many pieces of the puzzle have now fallen into place. For example, direct application of *ab initio* potentials for helium and neon can explain their peculiar acentric factors consistent with their behavior at conditions of elevated temperature and pressure.[2] Quantum density functional theory (QDFT) now provides a practical basis for estimating ideal gas formation energies and heat capacities that are more accurate than traditional methods.[3] Molecular simulation based on transferable intermolecular potentials now provides more accurate predictions of vapor pressure than traditional methods.[4–6]

QDFT combined with quasichemical theory (COSMO-RS/SAC) provides predictions of phase behavior that are competitive with traditional methods.[7,8] Instead of being frustrated with what we do not yet know, let us celebrate all that we have learned and the further advances that are just over the horizon.

We have focused on a singular mission in writing PGL6ed: to survey the field of physical property estimation in order to promote the best approaches and objectively characterize the quality of the available methods, such that current applications are reliable and future progress of the field is advanced. To promote objective characterization, we have brought together the experimental data compilation and curation efforts of BYU/DIPPR and NIST/TRC and excerpted data with broad coverage whenever feasible. We also identify cases where the breadth of experimental data appears lacking, in the hope that research efforts may be directed to fill those gaps.

We have tabulated observations of model quality and formulated all recommendations with reference to those tabulations. While it was not feasible to personally implement and evaluate every method that has been advanced since the fifth edition, we have solicited participation from a broad cross section of method developers. To the best of our ability, we have evaluated fundamental models, empirical models, and semiempirical models with equivalent metrics. We think we have compiled a reasonable representation of the available methods, but we would welcome the evaluation of more methods in the future. We hope that our efforts here will promote future research in directions that will be most productive to advancing the field.

1.3 TRADITIONAL ESTIMATION METHODS

Models that relate physical properties directly to characterizations of molecular forces do so through statistical mechanics. In this context, we refer to "statistical mechanics" as encompassing theories like those of Mayer and Mayer, Ornstein and Zernicke, Zwanzig, or Wertheim as well as molecular simulation methods like molecular dynamics (MD) or Monte Carlo (MC) simulation (cf. McQuarrie, 1976). Statistical mechanics is broader than statistical thermodynamics in that it encompasses transport properties like viscosity, thermal conductivity, and diffusivity as well as thermodynamic properties like pressure, enthalpy, heat capacity, and fugacity.

Unfortunately, we are not yet in a position where our molecular force models provide physical property estimates that supersede all empirical or semiempirical models. Statistical mechanical methods are only as good as their molecular force models, at best. Thus, despite impressive developments in molecular theory and information access, the engineer frequently finds a need for phenomenological means of physical property estimates. It has been a long-standing principle of *The Properties of Gases and Liquids* (PGL) that correlations are most reliable when they derive from molecular theory while retaining a (preferably small) number of empirical parameters to achieve accurate characterization of the property of interest. This semiempirical approach still predominates in providing the best accuracy in most cases. For example, corresponding states methods in the form of cubic equations of state with empirical mixing rules still dominate correlations of phase behavior, and group contribution methods still predominate as the most accurate methods for predicting critical properties.

1.3.1 Objectives of Estimation

An ideal system for the estimation of a physical property would (1) provide reliable physical and thermodynamic properties for pure substances and for mixtures at any temperature, pressure, and composition, (2) indicate the phase (solid, liquid, or gas), (3) require a minimum of input data, (4) choose the least-error route (i.e., the best estimation method), (5) indicate the probable error, and (6) minimize computation time. Few of the available methods approach this ideal, but some serve remarkably well. Thanks to modern technology, computation time is rarely a concern.

In numerous practical cases, the most accurate method may not be the best for the purpose. Many engineering applications properly require only approximate estimates, and a simple estimation method requiring little or no input data is often preferred over a complex, more accurate correlation. The ideal gas model is useful at low to modest pressures, although more accurate correlations are available. Unfortunately, it is often difficult to provide guidance on when to reject the simpler in favor of the more complex (but more accurate) method; the decision often depends on the problem, not the system.

We outline here several of the most well-known methods and attempt to put them into perspective. Our hope is that by understanding the fundamentals of various methods, the reader may guide themselves to the best possible decision, possessing expertise in their own problem supplemented by the insights that we may provide.

1.3.2 The Corresponding States Principle

Although a variety of molecular theories may be useful for data correlation, there is one theory that is particularly helpful. This theory, called the corresponding states principle, was originally based on macroscopic arguments, but its modern form has a molecular basis.

Proposed by van der Waals in 1873, the principle of corresponding states expresses the generalization that equilibrium properties that depend on common intermolecular forces are related to the critical properties in a universal way. In 1873, van der Waals showed corresponding states to be theoretically valid for all pure substances whose pressure-volume-temperature (PVT) properties could be expressed by a two-constant equation of state. Two-parameter corresponding states theory asserts that if pressure, volume, and temperature are divided by the corresponding critical properties, the function relating reduced pressure to reduced volume and reduced temperature becomes the same for all substances. The reduced property is commonly expressed as a fraction of the critical property: $P_r = P/P_c$; $V_r = V/V_c$; and $T_r = T/T_c$. As shown by Pitzer in 1939, it is similarly valid if the intermolecular potential function requires only two characteristic parameters.

To illustrate corresponding states, Fig. 1.1 shows reduced *PVT* data for methane and nitrogen. In effect, the critical point is taken as the origin. The data for saturated liquid and saturated vapor coincide well for the two substances. The isotherms (constant T_r), of which only one is shown, agree equally well.

Successful application of the principle of corresponding states for correlation of *PVT* data has encouraged similar correlations of other properties that depend primarily on intermolecular forces. Many of these have proved valuable to the practicing engineer. Modifications of the law are commonly made to improve accuracy or ease of use. For example, the TRAPP corresponding states method[9] is still a competitive method for predicting high-pressure transport properties of normal compounds. Good correlations of high-pressure gas viscosity have been obtained by expressing η^v/η_c^v as a function of P_r and T_r where η^v is the viscosity. But since η_c^v is seldom

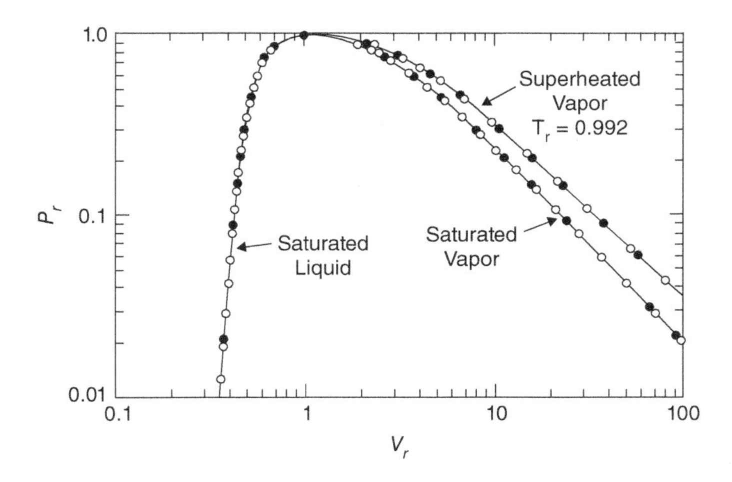

FIGURE 1.1
The law of corresponding states applied to the *PVT* properties of methane and nitrogen. Literature values (Din, 1961): ○ methane, • nitrogen.

known and not easily estimated, this quantity has been replaced in other correlations by other characteristics such as η_c°, η_T°, or the group $M^{1/2}P_c^{2/3}T_c^{1/6}$, where η_c° is the viscosity at T_c and low pressure, η_T° is the viscosity at the temperature of interest, again at low pressure, and the group containing M, P_c, and T_c is suggested by dimensional analysis. Other alternatives to the use of η_c might be proposed, each modeled on the principle of corresponding states but empirical as applied to transport properties.

1.3.3 Compound Classifications

Corresponding states holds well for fluids containing nonpolar molecules, and, upon semiempirical extension with a single additional parameter, it also holds for "normal" fluids where molecular orientation is not important, i.e., for molecules that are neither strongly polar, nor associating, nor heavy. For example, carbon dioxide is classified as a normal compound despite its quadrupole moment. Heavy compounds are defined relative to hydrocarbons with a molar mass greater than 212 g/mol. Thus n-hexadecane is "heavy." Polar compounds are esters, ketones, sulfones, carbonates, nitrites, and nitrates. Associating compounds are alcohols, acids, aldehydes, primary and secondary amines and amides, formates, nitriles, and thiols. Since many of the available models have their roots in corresponding states theory, we include discrimination of these four classes in our evaluations: normal, heavy, polar, and associating. Although crude, too detailed a classification scheme could confuse the analysis. We hope that this degree of resolution provides a reasonable tradeoff.

1.3.4 Models for Nonpolar and Weakly Polar Molecules

Small, spherically symmetric molecules (for example, Ar or CH_4) are well fitted by a two-constant law of corresponding states. However, nonspherical and polar molecules do not fit as well; deviations are often great enough to encourage development of correlations using a third parameter, e.g., the acentric factor, ω. The acentric factor is defined in terms of the experimental vapor pressure–temperature function such that it is near zero for nonpolar spherically-symmetric molecules (in the absence of quantum effects like helium, hydrogen, and neon). Typical corresponding-states correlations express a desired dimensionless property as a function of P_r, T_r, and acentric factor. We include cubic equations of state in this group, such as the Soave-Redlich-Kwong[10] and Peng-Robinson[11] models.

1.3.5 Models for Strongly Polar and Associating Molecules

Unfortunately, the properties of strongly polar or associating molecules are often unsatisfactorily represented by the two- or three-constant correlations which do so well for nonpolar molecules. An additional parameter based on the dipole moment has often been suggested but with limited success. Greater success has been achieved with equations of state that account explicitly for hydrogen bonding like the physical/chemical theory of Heidemann and Prausnitz[12] or the Statistical Associating Fluid Theory (SAFT)[13] adaptation of Wertheim's[14] initial conception. SAFT models can be shown to be equivalent to the Heidemann-Prausnitz model subject to details of the equation of state, emphasizing its chemical basis as well as its statistical mechanics. Unfortunately, these models introduce not one but two additional parameters, bonding volume, and bonding energy, that are highly coupled with each other and with the other parameters of molecular volume, shape, and dispersion energy. Resolving unique and optimal values for all these parameters remains a challenge, but association does fundamentally alter the modeling behavior. For example, the value of the critical compressibility factor, Z_c, decreases with increasing bonding energy, while cubic equations typically result in a single value for Z_c. This is one small but clear indication that association adds features to the model behavior. It should be noted, however, that most association models ignore the constraints of the critical point and focus directly on the molecular scale parameters. This approach deviates fundamentally from the traditional corresponding states approach.

Although accounting for association improves property estimation in some distinct ways, it is not the only way to achieve improved characterization of pure fluids and mixtures. It is feasible to introduce extra adjustable parameters into the cubic formalism, such that vapor pressure and mixture correlations are quite accurate. We have not yet reached a stage where association models or extended cubic models can be clearly recommended for all applications involving strongly polar and associating compounds. Once again, we can only provide insights for readers to make their own assessments relative to their own applications.

1.3.6 Structure, Bonding, and Group Additivity

All macroscopic properties are related to molecular structure and the bonds between atoms, which determine the magnitude and predominant type of the intermolecular forces. For example, structure and bonding determine the energy storage capacity of a molecule and thus the molecule's heat capacity.

This concept suggests that a macroscopic property can be calculated from group contributions. The relevant characteristics of structure are related to the atoms, atomic groups, bond type, etc.; to them we assign weighting factors and then determine the property, usually by an algebraic operation that sums the contributions from the molecule's parts. Sometimes the calculated sum of the contributions is not for the property itself but instead for a correction to the property as calculated by some simplified theory or empirical rule. For example, the methods for estimating T_c often start with the loose rule that the ratio of the normal boiling temperature to the critical temperature is about 2:3. Additive structural increments based on bond types are then used to obtain empirical corrections to that ratio.

Some of the better correlations of ideal-gas heat capacities employ theoretical values of C_p° (which are intimately related to structure) to obtain a polynomial expressing C_p° as a function of temperature; the constants in the polynomial are determined by contributions from the constituent atoms, atomic groups, and types of bonds.

"Transferable" potential models can be viewed from a perspective like group additivity, while lending themselves to the statistical mechanical approach for property estimation. For example, the force field of a CH_2 segment in *n*-pentadecane might be approximated by one that has been characterized using data for *n*-pentane. The force field is simply related to the potential model through taking a derivative. Like group contribution methods, first-order approximations may assume that, say, all CH_2 segments are identical while higher-order approximations distinguish between CH_2 segments in *n*-paraffins, napthenics, or attached to aromatic rings.

Other methods related to structure and bonding include quantum mechanical methods. QDFT and coupled cluster theory (CCT), for example, have become quite feasible computational techniques. These methods start with a characterization of a molecular structure that may include conformational variants and compute properties like heat of formation or ideal gas heat capacity directly from Schrödinger's equation, or its equivalent. QDFT methods have also been used to characterize the electron distributions around molecules in such a way that solution properties of mixtures can be estimated. These methods account for the roles of structure and bonding in ways that may be nonadditive, quite distinct from group contribution approaches.

1.4 NONTRADITIONAL ESTIMATION METHODS

In general, we include all the types of methods considered in previous editions of PGL, but we also seek to include methods that have become more practical and accurate in recent years. Molecular simulation and quantum mechanical methods, introduced in the context of group additivity and bonding, still qualify as nontraditional at present. Corresponding states, including cubic equations of state (EOSs), group contribution methods, and local composition activity models predominate the traditional methods. We occasionally refer the reader to previous editions of PGL for greater detail about traditional methods. We have not included quantitative structure property relationships (QSPR) and machine learning methods in PGL6ed, but we hope to learn more about them in the future.

1.4.1 Molecular Simulation

Shortness of coverage may be anticipated for methods like molecular simulation. Although we consider molecular simulation as a viable predictive tool, we do not perform molecular simulations as part of the PGL6ed evaluations. Therefore, we interpolated whatever simulation output was available for comparison to our databases. The narrower coverage for molecular simulation methods is reflected in the tabulated number of compounds for which they were applicable. We consider molecular simulation methods to be defined in terms of a clearly identifiable intermolecular potential function. For predictive applications, the potential functions should be transferable. Customized potential functions could be applicable in the same context as multiparameter correlations or models tuned to specific compounds, however.

1.4.2 Quantum Mechanical Methods

Programs like VASP, GAMESS, and Gaussian have become widely available in recent years. Properties like ideal gas heat capacity and dipole moment are naturally related to the principles of quantum mechanics, so it makes sense to include these methods for some degree of consideration. Our main concern is that the computations be reproducible by a broad range of readers. When this is clearly feasible, we include the methods.

1.4.3 QSPR and Machine Learning

In some instances, group contribution methods are like QSPR methods. When QSPR methods are in a closed form with few parameters, traditional regression procedures may be applied. In those cases, we include QSPR methods in the usual manner. Machine learning methods, on the other hand, rely on an arbitrary number of descriptors that have less apparent physical relation to the properties of interest. Increasingly, the list of parameters can number in the hundreds or thousands. Such methods rely on machine learning to correlate so many parameters and a large amount of training data. It is difficult to know the extent of overlap between the training set applied by the developers and the validation data set that is available to PGL6ed.

The bottleneck for evaluating QSPR and machine learning methods in PGL6ed is the generation of a sufficient number of descriptors to encompass all methods for all properties. Although descriptors are typically based on simple calculations for each descriptor, many descriptors are required. Generating such large databases of descriptors would delay PGL6ed excessively. We hope to address this issue to inform PGL7ed.

1.4.4 Method Consideration

It is difficult to keep track of every method developed for every type of property that might be of interest in chemistry and chemical engineering. We must acknowledge that we might have overlooked something important. To help, we attended many conferences and appealed to attendees to contribute methods of interest. We were able to incorporate direct contributions from Gross, Soares, Klamt, and their coworkers through this effort and we are grateful for their contributions. Additionally, the University of Akron, BYU, National Institute of Standards and Technology (NIST), and Thermodynamics Research Center (TRC) have volunteered methods, many complete with source code. Nevertheless, we must apologize in advance for oversights that might have occurred. We hope to begin work on PGL7ed soon, and we would welcome the opportunity to address these oversights. Method developers are encouraged to contact jelliott@uakron.edu if they would like to submit their method for consideration. Our aim has been to work with developers to facilitate interfacing with our efforts. We feel that PGL is stronger with the inclusion of as many methods as possible and we are committed to working diligently to achieve this.

1.5 DATABASE DEVELOPMENT AND METHOD EVALUATION

A lot has happened since the previous edition of this book was published. Despite the increased sophistication, nontraditional methods are not guaranteed to be more accurate than traditional methods, so rigorous evaluation of all available methods, both old and new, is essential for practical application. Doing so requires the specification of a common set of metrics against which the performance of all methods is compared. This set of metrics cannot simply be obtained from the original literature as the range of compounds, conditions, and comparative statistics vary from publication to publication. As in previous editions of PGL, care is taken in PGL6ed to define an objective and comprehensive set of metrics for proper evaluation of existing methods so that proper recommendations may be made about predictive methods.

Modern chemical databases and computer automation make evaluation possible in a manner not available previously, but the spirit of these objective comparisons is consistent with PGL editions of the past. Since its inception, PGL has adopted the working hypothesis that predictive methods are most reliable when they combine the fundamental principles of physics as a foundation with an appropriate degree of empiricism to ensure accurate reflection of experimental measurements. In this sense, experimental measurement comprises the final arbiter of any engineering evaluation. Therefore, the first order of business in evaluating physical property methods is the compilation of standard databases for comparing the methods on a consistent basis.

In the subsections below, we outline our guiding principles for database development and method comparisons in greater detail.

1.5.1 Database Development

Our efforts in database development reflect many years of tracing the provenance of literature data and detailed data curation. This experience informs our attention to uncertainty quantification as discussed in Chap. 2. Similarly, this experience teaches us that not all experimental measurements are equally reliable. Furthermore, confusion can exist about whether a particular data set strictly involves experimental measurements, or some form of extrapolation, or both. It is necessary to understand all the data comprehensively to compile a reliable data set. For example, suppose that chemical system X has been studied by several experimentalists, including author A, but system Y has only been studied by A. All the experimentalists report the same (low) uncertainty in their measurements, but the measurements of A deviate by 30% from all other measurements for system X. What uncertainty should be assigned to the measurements of A for system Y, and how would you know that without a comprehensive analysis of all data? Addressing issues like these may not be glamorous, but it is essential for systematic engineering progress. Reproducibility requires reliable metrics, and reproducibility is the cornerstone of the scientific method. It may be difficult for readers to imagine how much time goes into preparing the databases that we have used for the evaluations in PGL6ed, but describing the methods and running the evaluations represents a small portion of the time required, once the databases have been fully resolved. Note that full resolution of a database may require iteration as evaluations reveal inconsistencies or limitations.

In this context, we have compiled all experimental measurements known to us with careful attention to the reliability and uncertainty of the data. This compilation includes data for the 2371 pure compounds in the American Institute of Chemical Engineers (AIChE) DIPPR database and over 30,000 compounds in the TRC database. Each year since 2005, the TRC has conducted a survey of all experimental measurements reported in five journals, *J. Chem. Eng. Data, J. Chem. Thermo., Int. J. Thermo., Thermochim. Acta,* and *Fluid Phase Eq.*, with less comprehensive coverage of all other journals. All these measurements have been added to the TRC database. These measurements have been classified according to the chemical functionality scheme discussed in Sec. 1.3. Classification permits us to make comparisons on a targeted basis. For example, one method may be better for hydrocarbons, while another is better for alcohols. We prepare tables of comparisons that include metrics reflecting the breadth of coverage of chemical functionality and the coverage of compounds within each classification.

A recent policy of the US federal government has impacted the scope of some of our evaluation databases. The policy is that all publications including a federal resource must be fully reproducible by a knowledgeable reader. As one interpretation, this could mean that we would need to publish every data point in the TRC database if we used the entire database in our evaluations. Collecting and curating that database is highly resource intensive and requires significant sponsorship. This consideration motivates us to identify publicly available databases whenever possible. For example, the Danner-Gess[15] and Jaubert et al.[16] databases were very convenient for performing evaluations of vapor-liquid equilibrium (VLE) models. For evaluating models of infinite dilution activity coefficients (IDACs), the databases of Lazzaroni et al.[17] and Moine et al.[18] were adapted. In each case, the publicly available databases were checked for consistency with traceable literature sources through the NIST/TRC database. These consistency checks often resulted in significant reductions of the databases or substitutions. The final databases were then documented with sources when they were not available previously. We hope that these refined databases may promote advancement of the field.

1.5.2 The Implementation of Predictive Methods

Despite our best intentions, it would not be feasible for these authors to code and test every method that has been published since the previous edition. The advancing sophistication of predictive models has made it more difficult to reproduce and evaluate many methods. Therefore, a three-stage process was pursued for making comparisons. In the first stage, published methods were implemented to the best of our ability. Second, method developers were contacted to see if they would be interested to participate in our analysis. Several method developers submitted source code and facilitated its implementation. In limited cases, developers performed evaluations on their models using our datasets. The evaluations were tabulated, and a few methods were identified for the sample calculations. In the third stage, we performed sample calculations and documented them. Final recommendations include (subjective) consideration of how easily reproducible the calculations would be for our readers to perform, as well as performance on the tabulated metrics. Open-source codes are posted to the PGL6ed website (PGL6ed.byu.edu) whenever possible.

We attempted to include proprietary models for several properties in our evaluations. For example, Klamt et al. were kind enough to perform evaluations of the COSMO-RS/Therm model using our databases. In these cases, we did our best to verify the models with evaluations of publicly available similar models. In a few cases, we were provided with protected copies of proprietary models, but we were unable to facilitate adaptation of the software to the complete scale of computations required for all our evaluations. The CPA model[19] is an example of this case. We regret that we could not include its complete evaluation in PGL6ed.

1.5.3 Evaluation Metrics

We favor simple metrics whenever feasible. For example, many metrics have been suggested when evaluating VLE models, but we simply use the metric of percentage average absolute deviation (*%AAD*) in bubble pressure, defined by Eq. (1.5-1)

$$\%AAD = \frac{100\%}{NPTS} \sum \frac{|calc - expt|}{expt} \tag{1.5-1}$$

where *NPTS* is the number of points, *calc.* is the calculated value, and *expt.* is the experimentally measured value. A closely related metric is applied occasionally to provide an indication of how centered the distribution of deviations might be.

$$\%BIAS = \frac{100\%}{NPTS} \sum \frac{(calc - expt)}{expt} \tag{1.5-2}$$

Unfortunately, *%AAD* can lead to mischaracterizations of models when deviations are large. This occurs when properties vary on exponential scales as in the cases of vapor pressure and IDAC. Then we adopt a slightly different metric, the percentage absolute average logarithmic deviation (*%AALD*) defined by Eq. ((1.5-3).

$$\%AALD = \frac{100\%}{NPTS}\sum\left|\ln\left(\frac{calc}{expt}\right)\right| = \frac{100\%}{NPTS}\sum\left|\ln\left(1+\frac{calc-expt}{expt}\right)\right| \tag{1.5-3}$$

When deviations are small, *%AALD* ≈ *%AAD*. When deviations are large and negative, however, *%AALD* differs substantially from *%AAD*. Consider the case in which the calculated value is a factor of 5 less than the experimental value. The %LD for that point would be 100%ln(0.2) = −161%, but the %D would be −80%. On the other hand, the case in which the calculated value is a factor of 5 greater than the experimental value would lead to %LD = 161% whereas %D = 400%. A metric based on *%AAD* would favor negatively biased deviations, but a factor of 5 deviation is equally bad, whether positive or negative.

Another special case arises when properties can vary from positive to negative, like heats of mixing. In those cases, an offset metric is applicable, where the offset value is selected based on the uncertainty in the experimental measurement. This metric is defined by,

$$\%AADo = \frac{100\%}{NPTS}\sum\frac{|calc-expt|}{|expt|+o} \tag{1.5-4}$$

where o is the offset. For example, consider the case where *calc* = 1 kJ/mol, *expt.* = −0.01 kJ/mol, and o = 3 kJ/mol. Without the offset, the deviation would be 10,000%, but only 33% when the offset is included.

1.6 ORGANIZATION OF THE BOOK

The general layout for PGL6ed is much like the previous edition. Chapters are organized according to groupings of properties with conclusions and recommendations for each property group. The chapter headings are also the same except that the phase equilibrium chapter has been divided into two chapters and a new chapter has been inserted to discuss limitations inherent in experimental measurements and assessment of the uncertainty in these measurements. For phase equilibria, a single chapter is devoted to VLE while IDACs, liquid-liquid equilibria (LLE), and solid-liquid equilibria (SLE) are considered in a separate chapter. We expand coverage of SLE in this edition noting its importance in the growing pharmaceutical industry. Furthermore, its modeling is dominated by liquid activities that should be consistently related to VLE and LLE models. In general, pure component thermodynamic properties are considered in the early chapters, followed by thermodynamic properties for mixtures, and then transport properties for pure compounds and mixtures.

There are many obvious omissions in PGL6ed beyond the methods that have been overlooked. For example, electrolyte solutions, ionic liquids, and properties of polymers and their mixtures have not received their due attention. Entire books have been written about each of these topics, so substantial effort would be required to add a significant contribution. The effort required for PGL6ed with all new authors and revised databases has exhausted the available timeframe. We hope to return to these topics and more in future editions. We welcome your suggestions.

1.7 REFERENCES

1. R. C. Reid, A. E. Sherwood, *The Properties of Gases and Liquids*, McGraw-Hill, New York, NY, **1958**.
2. A. Aasen, M. Hammer, Å. Ervik, E. A. Müller, Ø. Wilhelmsen, *J Chem Phys* **2019**, *151*, 064508.

3. E. Paulechka, A. Kazakov, *J. Phys. Chem. A* **2021**, *125*, 8116–8131.
4. A. D. Sans, A. Vahid, J. R. Elliott, *J. Chem. Eng. Data* **2014**, *59*, 3069–3079.
5. J. R. Mick, M. S. Barhaghi, B. Jackman, L. Schwiebert, J. J. Potoff, *J. Chem. Eng. Data* **2017**, *62*, 1806–1818.
6. D. Weidler, J. Gross, *Ind. Eng. Chem. Res.* **2016**, *55*, 12123–12132.
7. A. Klamt, V. Jonas, T. Bu, J. C. W. Lohrenz, *J. Phys. Chem. A* **1998**, *102*, 5074–5085.
8. F. Ferrarini, G. B. Flôres, A. R. Muniz, R. P. de Soares, *AIChE Journal* **2018**, *64*, 3443–3455.
9. J. F. Ely, H. J. M. Hanley, *Ind. Eng. Chem. Fund.* **1981**, *20*, 323–332.
10. G. Soave, *Chem. Eng. Sci.* **1972**, *27*, 1197–1203.
11. D.-Y. Peng, D. B. Robinson, *Ind. Eng. Chem. Fund.* **1976**, *15*, 59–64.
12. R. A. Heidemann, J. M. Prausnitz, *Proc. Nat. Acad. Sci.* **1976**, *73*, 1773–1776.
13. W. G. Chapman, K. E. Gubbins, G. Jackson, M. Radosz, *Ind. Eng. Chem. Res.* **1990**, *29*, 1709–1721.
14. M. S. Wertheim, *J. Stat. Phys.* **1984**, *35*, 19–34.
15. R. P. Danner, M. A. Gess, *Fluid Phase Equilib.* **1990**, *56*, 285–301.
16. J.-N. Jaubert, Y. le Guennec, A. Piña-Martinez, N. Ramirez-Velez, S. Lasala, B. Schmid, I. K. Nikolaidis, I. G. Economou, R. Privat, *Ind. Eng. Chem. Res.* **2020**, *59*, 14981–15027.
17. M. J. Lazzaroni, D. Bush, C. A. Eckert, T. C. Frank, S. Gupta, J. D. Olson, *Ind. Eng. Chem. Res.* **2005,** *44*, 4075–4083.
18. E. Moine, R. Privat, B. Sirjean, J.-N. Jaubert, *J. Phys. Chem. Ref. Data* **2017,** *46*, 33102.
19. G. M. Kontogeorgis, G. K. Folas, *Thermodynamic Models for Industrial Applications. From Classical and Advanced Mixing Rules to Association Theories*, Wiley, West Sussex, U.K. **2010**.

2

Uncertainty

A contribution from:

Robert D. Chirico, Eugene Paulechka, Vladimir Diky, Ala Bazyleva

National Institute of Standards and Technology, Boulder Colorado

2.1 SCOPE

An old joke is that when a theorist gives a lecture, he is the only person in the room who believes 100% in what he is saying. When an experimentalist gives a lecture, he is the only person who does not believe 100% in what he is saying. These two perspectives are reflections on uncertainty. A careful experimentalist appreciates that each individual measurement includes its own degree of uncertainty: temperature, pressure, mass, volume, and so forth. The uncertainties in the measurements propagate to other properties like density and composition. This chapter is devoted to understanding these uncertainties and appreciating their proper documentation such that data are properly interpreted and problems are minimized. This understanding is essential to appreciating the compiled databases that form the basis of evaluations presented throughout the book, and teaches caution about interpretations of data and theory in general.

2.2 INTRODUCTION: WHY ARE UNCERTAINTIES IMPORTANT?

An important purpose of this book is to present and compare various methods for correlation and prediction of thermophysical properties of gases and liquids. These comparisons are made based on the relative degree of accord with critically evaluated experimental data. In this context, *critical evaluation* means assessment of the uncertainty for the experimental values based on an array of information, including experimental descriptions provided by the authors of the study, comparisons with well-established models and correlations, comparisons with other reports of the experimental values in the literature, the experience and knowledge of the particular evaluator, as well as knowledge codified in software systems for computer-aided evaluation.

Beyond the specific needs of this book, it can readily be argued that reliable uncertainties for experimental data are at the core of advancement in the practice and application of all scientific research. In the Foreword to the first edition of the *Guidelines for Evaluating and Expressing the Uncertainty of NIST Measurement Results (1993)*, the Director of NIST, John W. Lyons, summarized the essential nature of uncertainties: "It is generally agreed that the usefulness of measurement results, and thus much of the information that we provide as an institution, is to a large extent determined by the quality of the statements of uncertainty that accompany them."[1] Reliable uncertainties are a fundamental requirement for the advancement of science. Without reliable uncertainties, it is not possible to determine if a theory is supported or fails, if a process-design model is adequate or must be refined, or if better experimental data (i.e., data with smaller uncertainties) are required to resolve key questions, whether they arise in development of fundamental theory or in a practical application, such as equipment sizing or safety analysis. If uncertainties for experimental data are poorly known (i.e., we have no characterization of the reliability), the data are of no value for any of the tasks mentioned above.

In the context of this book, reliable uncertainties for experimental data are essential in determining if the data can be used as a meaningful discriminator between various published prediction and correlation methods. Figure 2.1 provides an analogy for this concept, where the center of the target represents the measured experimental value and the rings represent regions of probability or "levels of confidence" that the *true*—but unknowable—value lies within that region. (The exact *true value* is unknowable because it is impossible to make a measurement with zero uncertainty.)

For the sake of the example, we assume that the probability distribution for the measurement result is Gaussian or "normal."[2] The innermost black ring in the figure encloses a range of values for which there is a 68% probability (~1 standard deviation) that the true value is within that ring, while the second black ring encompasses the region of 95% probability (~2 standard deviations), and the third (outer) ring that of 99.7% probability (~3 standard deviations).

The three arrows in Fig. 2.1 indicate values estimated with three alternative predictions or models. Both the lower two arrows fall within the region of 95% probability, and although the middle arrow is slightly closer to the experimental value (i.e., closer to the center of the target), there remains a 5% probability that the *true value* lies outside of the second black ring. In contrast, the upper arrow has landed outside of the third (outer) ring, and there is less than a 0.3% probability that the true value is in this region. Although it is not possible to state unequivocally that a particular prediction is "right" or "wrong," it is an accepted practice in the field that values in agreement within the 95% range of probability are considered to be in accord. In this case, the experimental value serves as an adequate discriminator between the upper and middle or upper and lower estimates but does not allow us to distinguish the estimates represented by the middle and lower arrows. Details of model comparisons will be addressed in a separate chapter of this book.

If all experimental data reported in the literature included reliable uncertainties with well-defined levels of confidence, the task of comparing predictions and models would be quite straightforward. Unfortunately, this is not the case, and extensive critical evaluation of published experimental data is required to assign estimates of

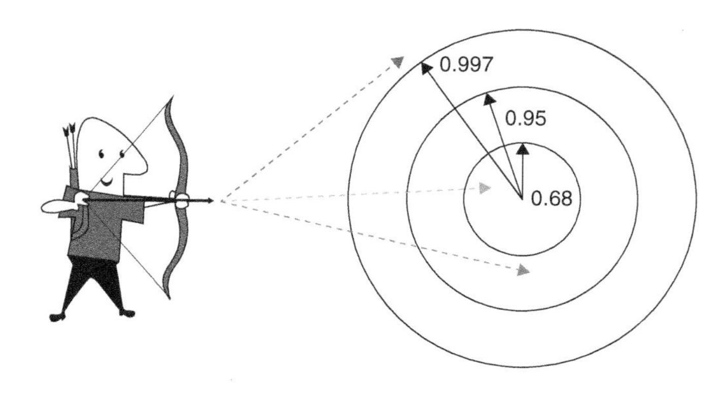

FIGURE 2.1
An analogy for comparison of data predictions and correlations (the arrows) with experimental property values (the target).

the uncertainties. All experimental data used in this book for comparison of models and predictions have been critically evaluated within the Thermodynamics Research Center (TRC) of the National Institute of Standards and Technology (NIST; mixtures and pure components) and the Design Institute for Physical Property Data (DIPPR; pure components).

The remainder of this chapter provides a historical perspective on the assessment of uncertainty and an overview of the philosophy and procedures that underlie modern methods of uncertainty assessment, as outlined in the *Guide to the Expression of Uncertainty in Measurement* (also known as the GUM).[3] This approach to uncertainty assessment was developed through international cooperation that included national metrology institutes from around the world, including NIST. Finally, some challenges that are specific to experimental data evaluation in the field of thermophysical and thermochemical properties are discussed.

2.3 HISTORICAL BACKGROUND

The word *uncertainty* is used widely in the English language and has multiple definitions that depend on context. For example, uncertainty is a characteristic of something that is not definitely ascertainable or fixed, as in time of occurrence, number, dimensions, or quality, but it can also indicate a lack of confidence or assurance (i.e., an *uncertain* smile), something that is dependent on chance or unpredictable factors (i.e., an *uncertain* outcome), a trait that is unreliable or undependable (i.e., as in *uncertain* loyalties), etc.[4] In fact, most definitions of uncertainty have little to do with numerical values, while for the present purposes we are concerned *only* with uncertainty as it applies to numerical results of experimental studies.

For centuries, scientists have used a wide variety of terms and phrases when describing uncertainty in measurements. These terms include *accuracy, absolute accuracy, error, absolute error, uncertainty, absolute uncertainty, maximum uncertainty, reproducibility, precision, known to within..., confidently given to..., measured to..., reliable to..., believed to be reproducible to..., personally guaranteed to be...*, the symbol +/–, and many more.[5] Unfortunately, none of these terms has a quantitative meaning.

To address the problem of poor definitions and quantification of uncertainties, the GUM was developed over a period of 17 years, culminating with its initial publication in 1993.[3] The GUM was prepared by a working group consisting of experts nominated by the Bureau International des Poids et Mesures (BIPM), the International Electrotechnical Commission (IEC), the International Organization for Standardization (ISO), and the International Organization of Legal Metrology (OIML). The GUM was developed through cooperation between 11 national metrology institutes.

The GUM describes a philosophy of uncertainty assessment put into mathematical language, without being overly prescriptive or formulaic. When followed, this approach yields well-defined uncertainties with specified "levels of confidence" (i.e., probabilities that the true value lies within the given range). Uncertainties are labeled as *standard* or *expanded*. For the *standard uncertainty*, the *level of confidence* (~68%) is part of the definition. For the *expanded uncertainty*, the *level of confidence* must be given explicitly. Without the adjectives *standard* or *expanded*, the word *uncertainty* retains its more general and non-quantitative meaning.

An overview of the GUM guidelines is given next. This overview is provided to give the reader sufficient background to understand the framework used in modern critical evaluations of experimental measurements of thermophysical properties reported in the literature. Readers are encouraged to consult the original publications for full details.[1,3]

2.4 KEY DOCUMENTS: THE GUM, NIST 1297, AND VIM

As noted previously, the GUM was first published in 1993. Subsequent modifications have been minor and largely for clarification and repair of typographical errors, with the most recent version released in 2008.[3] A closely related document is *NIST Technical Note 1297 (1994 Edition): Guidelines for Evaluating and Expressing the Uncertainty of NIST Measurement Results*.[1] The NIST document is terser than the GUM while providing, essentially, the same information. Last, the *International vocabulary of metrology—Basic and general concepts and associated terms* (the "*VIM3*") provides definitions for many common terms found in uncertainty analysis, as well as metrology.[6] The above documents are freely available online (see references), as are the VIM3 definitions with annotations.[7]

In the following section, the concepts and methods of the GUM are discussed within the context of application to uncertainty assessment for experimental thermophysical properties. This discussion is provided to aid the reader in understanding the general data-evaluation process used in assessing the experimental values employed to test the models described in this book, and it is not meant as a guide for the practicing experimentalist or data evaluator. The terms and definitions provided in this section are in accord with those provided in the GUM,[3] NIST Technical Note 1297,[1] and VIM3[6] to the extent practicable.

2.5 UNCERTAINTY ASSESSMENT FOR THERMOPHYSICAL PROPERTIES: THE GUM APPROACH

Some basic concepts and terms need to be defined at the outset. The result of an experimental property measurement is an approximation or estimate of the value of the quantity being sought. The result of the measurement is complete only when accompanied by a quantitative statement of its uncertainty. The quantity being sought is called the *measurand*, and it is assumed to have an exact value that is unknowable. The *value of the measurand* is also termed the *true value* of the property. Examples of measurands are the critical temperature of hexadecane, the vapor pressure of naphthalene at the temperature $T = 400$ K, or the gas-phase composition of a two-phase mixture of (ethanol + water) at $T = 350$ K and pressure $P = 0.1$ MPa. We presume that these quantities have single exact (i.e., "true") values, but it is impossible to make a measurement with zero uncertainty.

Uncertainty in a measurement result generally arises from multiple sources or components. In the GUM approach, these components are grouped into two types; "those which are evaluated by statistical methods" ("Type A") and "those evaluated by other means" ("Type B").[3] Although this categorization has led to much discussion of whether a particular component is type A or type B, in practice, the distinction is not of high importance, as all uncertainty components are combined without regard to type. In the field of thermophysical properties, the uncertainty components of type A are experimental repeatabilities, while other important components, such as sample purity, calibration data, uncertainties assigned to reference data, equipment manufacturer's specifications, etc., are type B. Repeatability is defined as "the closeness of the agreement between the results of successive measurements for the same measurand carried out under the same conditions of measurement (repeatability conditions) and may be expressed quantitatively in terms of the dispersion characteristics of the results."[3] Repeatability conditions include the same measurement procedure, observer, measuring instrument, location, and repetition over a short period of time. Generally, the type B components are far more important in the assessment of uncertainties for thermophysical properties than the simple repeatability. The key point here is that categorization of uncertainty components into types is not essential in the uncertainty evaluation process. Similarly, other common ill-defined categorizations of uncertainty components, such as *random* or *systematic*, do not alter the evaluation process in this context.

In essentially all determinations of thermophysical properties, the experimental value y for the measurand Y (i.e., true value of the property whose value is sought) is not obtained directly. The measurement result y (i.e., the *estimate* of Y) is determined from N other quantities or pieces of information $x_1, x_2, \ldots, x_N$ through a functional relation f:

$$y = f(x_1, x_2, \ldots, x_N) \tag{2.5-1}$$

The uncertainty in y is obtained from the law of propagation of uncertainty through combination of the uncertainties in each contributor x_i. It is important to note that the function f may include information that is difficult to quantify, such as the effect of sample purity or experience of the experimentalist. Nonetheless, these contributors must be considered. This point is amplified below.

In most practical measurements of thermophysical properties, the probability distribution characterized by a measurement result y is approximately normal (Gaussian), and the standard deviation of this normal distribution is assumed to be equal to the standard uncertainty $u(y)$. This implies that one standard uncertainty $u(y)$ corresponds to a level of confidence of ~68%, or if written $y \pm u(y)$, there is a level of confidence of ~68% that $y - u(y) \leq Y \leq y + u(y)$. In other words, there is a ~0.68 probability that the true value lies in the range $y \pm u(y)$. The quantity $u(y)$ is defined as the *standard uncertainty* in y. It is the convention in the field of thermophysical properties that values should be reported[8] with a 95% level of confidence; however, this level of confidence cannot be considered a default or assumed. For a normal distribution, an interval with a level of confidence of approximately 95% is obtained using $2 \cdot u(y)$, and $2 \cdot u(y)$ is termed the "expanded uncertainty for y which defines an interval with ~95% (or equivalently, ~0.95) level of confidence."[3] The level of confidence must be stated explicitly because the expanded uncertainty refers to any multiple of $u(y)$. For example, if a measurement result y was reported as $y \pm 3u(y)$, $3u(y)$ is still termed the *expanded uncertainty*, but this interval has ~0.997 *level of confidence*.

The standard uncertainty $u(y)$ for measurement result y (Eq. 2.5-1) is estimated with the law of propagation of uncertainty as follows.

$$u(y) = \left[\sum_{i=1}^{N} \left(\frac{\partial f}{\partial x_i} \right)^2 u^2(x_i) + 2\sum_{i=1}^{N-1} \sum_{j=i+1}^{N} \frac{\partial f}{\partial x_i} \frac{\partial f}{\partial x_j} u(x_i, x_j) \right]^{0.5} \tag{2.5-2}$$

The partial derivatives $\partial f/\partial x_i$ are termed *sensitivity coefficients*, as they reflect the sensitivity of y to changes in the values of the x_i. In the second term (i.e., the double summation) within the large brackets of Eq. (2.5-2), $u(x_i, x_j)$ represents the covariance between parameters x_i and x_j. In the experimental determination of thermophysical properties, this term often is not significant and Eq. (2.5-2) can be simplified.

$$u(y) = \left[\sum_{i=1}^{N} \left(\frac{\partial f}{\partial x_i} \right)^2 u^2(x_i) \right]^{0.5} \tag{2.5-3}$$

In the GUM,[3] a subscript c is added to the symbol $u(y)$ [i.e., $u_c(y)$], and $u_c(y)$ is termed the *combined* standard uncertainty to emphasize that this quantity is obtained through combination of the standard uncertainties in the parameters x_i. In the measurement of thermophysical properties, common variables, such as temperature and pressure, are themselves determined through combinations of other quantities (voltages, electrical currents, values of resistance and voltage standards, etc.) with contributing uncertainties, so distinction between a *combined standard uncertainty* and a *standard uncertainty* can be ambiguous and arbitrary. We use only the term *standard uncertainty* here, with the understanding that it necessarily is obtained through combination of the uncertainties for contributors $x_1, \ldots, x_N$ to the measurement process.

The NIST Guide for the Expression of Uncertainty (Technical Note 1297)[1] states, "...the function f [here, Eq. 2.5-1] should express not simply a physical law but a *measurement process* [our emphasis], and in particular, it should contain all quantities that can contribute a significant uncertainty to the measurement result." In measurement of thermophysical properties (and phase equilibria), typical input parameters x_i include temperature, pressure, and phase compositions, and possibly density, as one would expect based on the constraints of the Gibbs phase rule. However, in the GUM approach, *all* potential sources of uncertainty must be considered. These include sample purity (including decomposition during the experiment), as well as uncertainties in properties of reference materials and calibrants, laboratory conditions (humidity, temperature, atmospheric pressure, etc.), skill of the apparatus operator, appropriateness of calibration materials, etc. Though some of these are difficult to put into a mathematical form, they must be included in the uncertainty assessment.

A key contributor to uncertainty that is rarely considered in published uncertainty assessments for thermophysical properties is sample purity. The effect of impurities on measurement results can vary considerably depending upon the property and the nature of the impurity. For example, if densities are measured for an ionic liquid (IL) with mass fraction purity 0.98, results are very different if the impurity is water, which is a common impurity in hygroscopic ILs, or simply an isomer of the IL. Authors rarely consider sample purity in their uncertainty analyses. In 2005, a study was made of reporting practices for experimental thermophysical properties, using the determination of critical temperature as a case study.[5] Of 600 articles reviewed, involving ~2000 reports of a critical temperature, only one article[9] considered sample purity as part of the uncertainty assessment. Amongst the 2000 experimental values, some were made on samples with purities as low as 96% without consideration of sample quality. In most cases, only repeatability was considered in the uncertainty assessments which, as noted earlier, is a relatively small contributor to the uncertainty in measurement of thermophysical properties.

2.6 THE UNCERTAINTY BUDGET

The most informative approach to reporting an uncertainty analysis in an experimental report is through use of a tabular *uncertainty budget*. The benefits of an uncertainty budget include unambiguous listing of all considered uncertainty sources together with the associated standard uncertainties, sensitivity coefficients, and degrees of freedom. (Detailed information concerning tabular uncertainty budgets can be found in the *NIST/SEMATECH e-Handbook of Statistical Methods*.[10]) With this approach, the major contributors to the measurement uncertainty are clear, and the work of an independent reviewer in determining if important uncertainty sources were overlooked is eased considerably. At this time, uncertainty budgets are rarely seen in the literature outside of some reports from national metrology institutes. As stated in the GUM, the functional relationship for determination of a measurement result (Eq. 2.5-1) should include *all* sources of uncertainty. This ideal is hardly fully achievable, however; a high-quality uncertainty analysis is much closer to this ideal than a poor one, and explicit listing of all uncertainty sources considered and their estimated contributions within a concise tabular uncertainty budget is invaluable in judging the quality of the assessment. The value of the tabular uncertainty budget is gaining recognition in the field of thermophysical properties, and it is hoped that these will become a common feature in the literature. Although rare, examples of tabular uncertainty budgets can be found (cf. Ref. 11).

In the absence of an uncertainty budget or explicit consideration of key aspects by the authors of a given experimental report, the data evaluator is forced to make estimates for missing uncertainty contributions. These estimates are based on many factors, including the evaluator's knowledge base for the methods used, known chemical properties of the substance or family of substances studied, likely impurities and their effect, knowledge of calibration methods and calibrants, as well as less tangible considerations, such as the historical reliability of results from a particular research group or laboratory. In the absence of explicit and accurate information from the experimentalists, all estimates are necessarily conservative. Methods used by modern data evaluators are addressed more extensively later in this chapter.

2.7 REFERENCE MATERIALS AND STANDARD REFERENCE MATERIALS

In the measurement of thermophysical properties, the apparatus used is often calibrated or tested through measurements involving a substance with known properties. Such substances are known as *reference materials* or *standard reference materials*. A *reference material* is a substance—typically, a pure compound or mixture of a specified composition—for which property values have been carefully measured and critically evaluated (i.e., the uncertainty is well established). It is the responsibility of the experimentalist to obtain or synthesize the sample with an appropriate purity. Examples of pure compounds with well-established property values are liquid water,[12,13] used for calibration of viscometers, densimeters, and some vapor-pressure apparatus (particularly comparative ebulliometers), and sapphire,[14] used to calibrate differential scanning calorimeters (DSC) for the measurement of heat capacities.

In many cases, substances are used as reference materials even though the uncertainties for the reference values are not well established. It is simply assumed that the uncertainties in the reference values are much smaller than those generated with the apparatus being tested or calibrated. Examples include benzoic acid, copper, or heptane, used to test a calorimeter in determination of heat capacity, or naphthalene, used to test apparatus used for measurement of sublimation pressures.

In contrast to a *reference material*, a *standard reference material* is a physical material with a certified property value that can be used to test or calibrate a particular apparatus. Certification is done, typically, at a national metrology institute from whom the material can be purchased with a certificate stating the property value and its uncertainty determined in accord with the GUM. Examples include toluene, used for the measurement of liquid densities over wide ranges of temperature and pressure [see NIST Standard Reference Material (SRM) 211d];[15] molybdenum, used for determination of enthalpy and heat capacities for the temperature range $273.15 < (T/\mathrm{K}) < 2800$ [see NIST SRM 781D2];[16] and benzoic acid, used for calibration of combustion calorimeters [see NIST SRM 39j].[17]

It is essential that uncertainties in property values for calibrants be properly propagated to uncertainties of experimental results. Propagation of the uncertainty in calibrant property values in the field of combustion calorimetry (i.e., the enthalpy of combustion of benzoic acid) has been well established since the mid-20th century.[18] In other disciplines, application of appropriate uncertainty propagation procedures by authors is inconsistent, particularly in the measurement of viscosity and density with commercial instruments. Even for combustion calorimetry, some modern reports fail to use the well-established procedures or even mention that a calibrant was used.[19] This has led to development of a website (maintained at NIST) designed to aid experimentalists in providing complete results for publication, while simultaneously providing updates for some of the metadata used in typical data reduction.[20] Complete descriptions of the updated metadata are planned,[21] as well as reiterated and updated recommendations for the reporting requirements for combustion calorimetry.[22]

In general, proper propagation of uncertainties in calibrant values to uncertainties in the reported property values often falls to the data evaluator, who must be equipped with the necessary knowledge and tools to provide meaningful uncertainty estimates. The *NIST Uncertainty Machine* is a free web-based software application that is a practical tool to assess propagated uncertainty for measurement models provided by the user in the general form of Eq. (2.5-1) [i.e., $y = f(x_1, x_2, \ldots, x_N)$].[23]

2.8 CHALLENGES FOR THE USER OF EXPERIMENTAL DATA FROM THE LITERATURE

In the field of thermophysical-property and phase equilibrium measurements, the GUM was largely ignored by researchers prior to 2005, with the exception of some reports from national metrology institutes. This unfortunate state only started to change when key journals began to require use of the GUM language in reporting of uncertainties.[24] (In contrast, experimental procedures and comprehensive uncertainty assessment had been codified and used in the field of combustion calorimetry for more than 60 years,[18] and it is still valid today with minor modifications in language only.[21]) The *IUPAC Guidelines for Reporting of Phase Equilibrium Measurements (IUPAC Recommendations 2012)*[8] made application of the GUM procedures a requirement in all reports of experimental results for all types of phase equilibrium, as well as for thermophysical and thermochemical measurements on pure compounds. While many key journals have made implementation of these IUPAC guidelines a requirement for their authors, enforcement has been inconsistent. Overall, these developments have promoted awareness of the GUM in the field, but its general and appropriate application remains uncommon. It is our opinion that substantial progress will be made only when the GUM philosophy and procedures are embraced fully by the research community and become part of the core curriculum for all scientists and engineers working in this field.

Although the user of thermophysical property data might consider data quality (i.e., property values with well-defined uncertainties) to be of paramount importance, there are many factors that motivate scientific publication, leading to—at times—competing criteria used to judge quality by the parties involved. Stakeholders in the publication process include article authors (university professors, graduate students, postgraduate researchers, government and industrial laboratory scientists, etc.); journal editors, who are also commonly university professors; peer reviewers, who are like the authors by definition; journal publishers; and data users, such as process engineers, model and prediction developers, scientific policy developers, and regulatory agencies. For many of these, the act of publication itself is the primary goal, while subsequent use of the experimental data is secondary.

Publication allows degrees to be awarded, tenures granted, new issues published, research grants justified, etc. Major *users* of property data, such as those involved in process design and optimization or theory, modeling, and prediction development, as well as those involved in regulation and policy development, are not well represented in the publication or peer-review process. In addition, data users rarely have the detailed knowledge of key experimental details necessary to critique and improve the uncertainty assessments provided in most publications.

Exacerbating these problems is the fact that journal publishers are known to compete on the basis of submission-to-publication time, perhaps, in part, due to ease of measurement. In contrast, assessment of data quality is a difficult and time-consuming process. In some highly competitive fields, speed of publication is a significant concern, but this is rarely the case in the field of thermophysical properties, where most authors consider the professional standing of the journal to be paramount. Nonetheless, peer reviewers are strongly encouraged to work quickly and often do not have the time or resources to assess the relevant literature, much less to analyze large tables of numerical results for anomalies and inconsistencies. As peer reviewers are drawn from the same pool as authors and editors, they typically reside on the data-producer side of the process, rather than the data-user side, so there is little motivation to do labor-intensive assessments of numerical results.

Another major challenge for the modern data evaluator is the continued acceleration in the quantity of experimental data produced worldwide. Figure 2.2 shows the number of experimental values for thermophysical properties that were abstracted by NIST TRC from literature dating from ~1800 to the present. The acceleration in data production is ~6% per year, as shown by the line in the figure. This increase implies a doubling of the rate of data publication every 12 years. The growth rate in scientific publications for all fields was analyzed in 2010 and it was concluded that (1) the overall growth rate was near 4.7% per year and (2) the growth rate did not decrease in the previous 50 years.[25] The rate of accumulation for thermophysical-property data readily exceeds that of science in general.

The increase in productivity for thermophysical property measurements is supported, in part, by progress in measurement science, including development of high-precision commercial apparatus for measurement of temperature, pressure, and chemical composition, as well as fully automated data acquisition systems for key properties, such as density, viscosity, heat capacity, etc. Unfortunately, this increase in productivity can readily result in data of poor quality due, in part, to the fact that automated commercial equipment does not require involvement of personnel with high degrees of expertise and in-depth knowledge of the measurement principles involved. In addition, untenable uncertainty claims made by some equipment manufacturers seeking a competitive edge, when combined with a demonstrated[5] weak understanding within the scientific community of the meaning of uncertainty and its assessment, readily leads to grossly optimistic uncertainty assessments. If the uncertainties are of poor quality, meaningful application of the results in model assessment is lowered substantially.

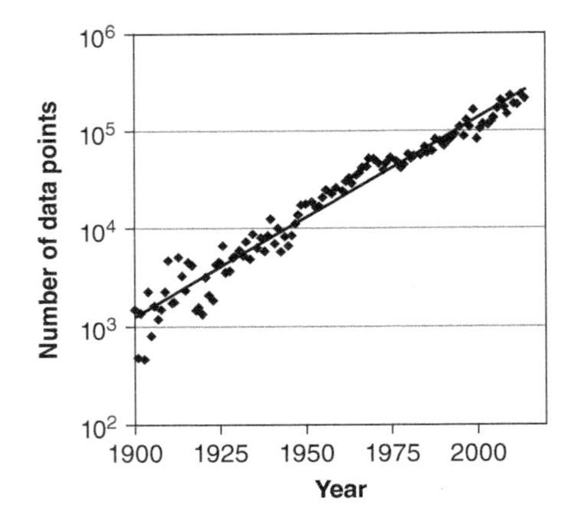

FIGURE 2.2
Black diamonds represent the number of experimental thermophysical-property values ("data points") abstracted by NIST/TRC in a given year for articles published since 1900. The line was calculated based on an increase in the rate of publication of 6% per year.

2.9 EXAMPLES OF COMMON PROBLEMS IN ARTICLES REPORTING THERMOPHYSICAL PROPERTIES

In this section, we provide examples of the sort of in-depth knowledge that is necessary to critically assess published reports of measured thermophysical properties. Due to recent efforts, the completeness and reporting quality for many articles have improved, but uncertainty assessment remains a problematic area.[26]

Example 2.1 Uncertainty contribution from sample impurity

In this example, we consider the contribution of sample impurity to the uncertainty in a measurement of liquid density for a single component. This contribution must be considered for any density measurement irrespective of the measurement principle. If the chemical identity and volumetric behavior of the impurity in mixtures with the main component are known, the effect of the impurity can be computed with a high degree of confidence; however, such information is seldom available, and approximate methods must be used.

The uncertainty due to sample impurity can be estimated approximately based on the following assumptions:

1. The impurity is a single component and the pressure is constant;
2. The main component and impurity have densities ρ differing by a fraction α at all temperatures, where $\alpha = [\,\rho(\text{impurity})/\rho(\text{main component}) - 1]$; and
3. The excess volume for the mixture (main component + impurity) is zero.

With these assumptions, the following expression is obtained for the relative standard uncertainty in the measured density $u_r(\rho)$ due to the impurity contribution:

$$u_r(\rho) = |\alpha| \cdot (1 - w_s)/(1 + \alpha \cdot w_s), \tag{2.9-1}$$

where w_s is the sample purity expressed as a mass fraction. This formula can be used if the impurity and its density are known or if the density of the impurity can be reasonably estimated.

In the majority of published density studies, commercial samples are employed, the sample purity is specified by the supplier, and the identity of the impurity is not known. In this case, a further simplification is needed, and the average of the two possible scenarios (i.e., a heavier or lighter impurity) can be used:

$$u_r(\rho) = |\alpha| \cdot (1 - w_s)/(1 - (\alpha \cdot w_s)^2), \tag{2.9-2}$$

and α can be assumed to be 0.1 (i.e., for all conditions, the densities of the main component and impurity differ by 10%, which is a conservative estimate). Application of Eq. (2.9-2) to a sample with 0.99 mass fraction purity yields $u_r(\rho) = 0.001$, which corresponds to roughly $u(\rho) \approx 1\ \text{kg} \cdot \text{m}^{-3}$ for many liquid substances well removed from the critical region (i.e., 0.1% of ~1000 $\text{kg} \cdot \text{m}^{-3}$). Reports of density measurements for single components with mass fraction purities as low as 0.95 are not uncommon in the literature.

Over the last 25 years or so, densities of pure compounds and mixtures are typically measured with commercial vibrating U-tube densimeters. These instruments require calibration with fluids of known density; typically, water and air. Uncertainty claims by equipment manufacturers are often remarkably small. For example, one commonly used instrument is accompanied by the following uncertainty information: "reproducibility" of 0.005 $\text{kg} \cdot \text{m}^{-3}$, and "accuracy (*under ideal conditions and for low densities/viscosities*)" of 0.007 $\text{kg} \cdot \text{m}^{-3}$.[27] The terms "reproducibility" and "accuracy" do not have well-defined numerical definitions, and the "ideal conditions" or the range of "low densities/viscosities" are not specified. If "accuracy" is assumed to mean "standard uncertainty," these values might be appropriate if measurements are made near the calibration point (i.e., on very dilute aqueous solutions prepared from degassed high-purity water), but are clearly inappropriate beyond that. Nevertheless, many researchers use statements of the manufacturers as uncertainty estimates for all of their measured values, which leads to the common problem in the literature of uncertainties for measured densities being underestimated by several orders of magnitude.

Example 2.2 Uncertainty associated with calibrant property values

In this example, we consider the contribution of uncertainty in calibrant values to the uncertainty in a measurement result. Specifically, we consider the measurement of viscosity with a commercial viscometer, but the concepts are also applicable to the measurement of heat capacities with a differential scanning calorimeter or densities with a vibrating tube densitometer.

Commercial viscometers are used in the large majority of viscosity studies and these always require calibration. Calibration is preferred to computations based on the precise determination of the physical characteristics of the instrument (e.g., diameter and length of a capillary, density and radius of a falling ball, etc.), which invariably deviate from the ideal shapes assumed in the equations used to model the experiment. Water is commonly used as a calibration fluid, as its viscosity is well studied.[12,28] The 2009 IAPWS reference equation gives the viscosity η of liquid water between 253.15 K and 383.15 K at 0.1 MPa with a relative expanded uncertainty (0.95 level of confidence) $U_r(\eta)$ equal to 1%.[12] (Values for temperatures below that of normal melting and above normal boiling are for metastable liquid water at 0.1 MPa.) Only the single value at $T = 293.15$ K and $P = 0.101325$ MPa has been assessed to have a smaller uncertainty [$U_r(\eta)$ equal to 0.17% with 0.95 level of confidence].[29] Toluene has been suggested recently as a reference material for viscometry in the temperature range from 263 K to 373 K and 0.1 MPa pressure with an expanded uncertainty of 0.7% (0.95 confidence level).[28]

Through careful technique, it is possible to obtain a repeatability of 0.1%—or even lower—in viscometric measurements. Unfortunately, this repeatability, which represents only a lower limit for the standard uncertainty, is often reported by researchers as the standard uncertainty for the measured viscosities. Clearly, the uncertainty in the viscosity values for the calibration fluids far outweigh such repeatabilities in determining the uncertainty in the measurement result (see Eq. 2.5-3).

The uncertainty associated with calibrant property values must be considered for any apparatus requiring calibration with a reference material. Consequently, proper assessment of literature values can only be done with an in-depth knowledge of the physical properties of the calibrants, as well as the evaluated uncertainties in their reference property values.

Example 2.3 Propagated uncertainty in the composition of a binary mixture

Uncertainty in the composition of a binary mixture is a contributor to the uncertainty in any property measurement for that mixture, and, in the case of phase equilibrium studies, may be the primary focus of the study. This example is illustrative for composition analyses made with analytical methods or for estimation of composition uncertainty due to sample impurities.

Here, we consider a binary mixture for which relative standard uncertainties in the amounts of components 1 and 2 are $u_r(n_1) = u(n_1)/n_1$ and $u_r(n_2) = u(n_2)/n_2$, respectively. Propagation of uncertainties for the mole fraction of the first component $x_1 = n_1/(n_1 + n_2)$ gives

$$u(x_1) \approx x_1 x_2 \sqrt{u_r(n_1)^2 + u_r(n_2)^2} \tag{2.9-3}$$

If $u_r(n_1) = u_r(n_2)$, then Eq. (2.5-1) simplifies to

$$u(x_1) \approx x_1 x_2 u_r(n) \sqrt{2} \tag{2.9-4}$$

This equation would be appropriate if, say, the amount of each component was determined with a relative standard uncertainty of 1% using gas-liquid chromatography. In this case, the relative standard uncertainty in the composition of the mixture would reach its maximum value [$u_r(n)\sqrt{2}/4$] at the equimolar composition.

Extension of Eq. (2.9-4) to a three-component system yields the following:

$$u(x_1) \approx x_1 u_r(n) \sqrt{(x_2 + x_3)^2 + x_2^2 + x_3^2} \tag{2.9-5}$$

In Eqs. (2.9-3) through (2.9-5), the component indices are interchangeable. When assessing the uncertainty of a property for any mixture, the composition uncertainty must be propagated into the uncertainty of the property value, as per Eq. (2.5-3). The magnitude of this contribution depends on the sensitivity of the property to changes in composition (i.e., $\partial f/\partial x_i$ of Eq. 2.5-2).

Figure 2.3 shows the results of the uncertainty calculation for measured vapor-liquid equilibrium (VLE) pressures P_{VLE} as a function of composition for a hypothetical binary mixture. For example, the relative standard uncertainty in the composition is taken as 2% [or $u_r(n_1) = 0.02$] and the relative standard uncertainty for the pressure sensor is $u_r(P_{sensor}) = 0.005$. Figure 2.3A shows the standard uncertainty in x_1, as per Eq. (2.9-3). Figure 2.3B shows the measured VLE pressures P_{VLE}, which are taken as linear with composition to simplify the calculations. The filled symbols in Fig. 2.3C show the standard uncertainty $u(P_{VLE})$ with propagation of $u(x_1)$ and $u(P_{sensor})$ into that of the

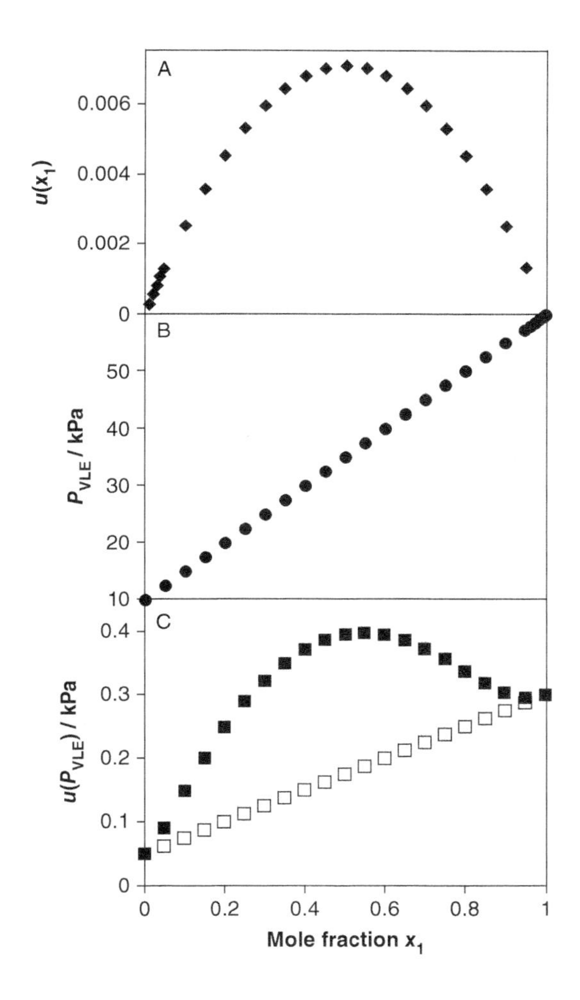

FIGURE 2.3
Example of propagation of uncertainty in composition for a binary mixture in the measurement of VLE pressure P_{VLE} for a binary system, as described in the text. All graphs are plotted as a function of mole fraction. Plots show (A) the standard uncertainty in composition $u(x_1)$, (B) measured VLE pressures P_{VLE}, (C) uncertainty in the measurement result $u(P_{VLE})$ computed through propagation of uncertainties for the pressure sensor and composition (■) and computed erroneously without consideration of the uncertainty in composition (□). As noted in the text, a complete analysis would include other uncertainty contributions, including those of temperature and sample purity.

measurement result $u(P_{VLE})$. The unfilled symbols at the bottom of the figure show the erroneous result obtained for $u(P_{VLE})$ that is obtained if the uncertainty in composition $u(x_1)$ is ignored, resulting in estimated uncertainties that are too low by more than a factor of 2 near $x_1 = 0.5$.

In this example (Fig. 2.3), only uncertainties associated with the composition and pressure sensor were considered. For a complete uncertainty analysis, all sources of uncertainty must be considered, including temperature and sample purity. It is common in the literature for authors to provide only the uncertainty associated with the pressure sensor, while ignoring all other contributions.

Example 2.4 Uncertainty for binodal composition determined by titration

A binodal composition [or liquid-liquid equilibrium (LLE) composition or liquid-liquid solubility] can be determined by titration of a transparent solution of known composition with another liquid or solution until turbidity is observed. The procedure can be reversed and, thus, the turbid system can be titrated to a clear solution. In many experiments, the volume of titrated solution is on the order of 10 to 50 mL. The volume of droplet depends, in part, on the composition of the solution, and for water, it is typically ~0.03 mL or somewhat larger. To estimate the uncertainty in the endpoint, one can assume that the densities of both solutions are $\sim 1000\ \mathrm{kg \cdot m^{-3}}$ and the volume of the titrated solution is 30 mL. Then, in a best-case scenario that includes reliable detection of the turbidity upon addition of a single droplet, a well-calibrated burette, an accurate mass of the initial solution, and good temperature control over the entire apparatus, the standard uncertainty for the mass fraction can be estimated $u(w) \approx 0.03/30 = 0.001$. This value is an estimate of the lower limit of uncertainty for these measurements. This value also represents an approximate lower limit of solubility that can be determined by titration. Considering all contributions, $u(w) \approx 0.005$ is a realistic limit.

2.10 MODERN UNCERTAINTY ASSESSMENT PROCEDURES (CRITICAL EVALUATION)

2.10.1 Introduction

The phrase "critically evaluated data" is often encountered in descriptions of collections of property values. Unfortunately, the term is poorly defined, and can mean as little as averaging a few values of ill-defined origin or can imply an extensive analysis of carefully curated experimental data. Any data collection that is claimed to be "critically evaluated" without a description of how the evaluation is done is necessarily suspect. A high-quality critical evaluation involves consideration of numerous factors, some of which cannot be readily quantified, but which can be used in weighing the results of one report relative to another. The degree of the relative weighting will depend on the knowledge, experience, and established procedures of the particular evaluator or evaluating organization. Because of such factors, it is extremely unlikely that any two critical evaluations will agree exactly in value and uncertainty, but if both are done well, the evaluations should be consistent within the specified uncertainties. This concept is depicted in cartoon form in Fig. 2.4.

In Fig. 2.4, the archer could represent, for example, an experimentalist wishing to use a critically evaluated property value to aid in calibrating an instrument; or—as in the case of the present book—a researcher seeking reference data to evaluate alternative models or predictions. Independent critical evaluations are represented in Fig. 2.4 by the two three-tiered targets (dashed lines represent evaluation A and solid lines represent evaluation B). Though the evaluations are based on the same experimental values (the **X**'s), the evaluating organizations used different weighting factors and arrived at slightly different, but consistent, results. The experimentalist/researcher will need to further weigh the results of the critical evaluations in carrying out their work. The "weighting" of the critical evaluations could be based on the quality of the uncertainty assessments provided with each evaluation, but clearly, there is no single correct answer. This example further reinforces the concept that the exact *true value* of a given property is unknowable.

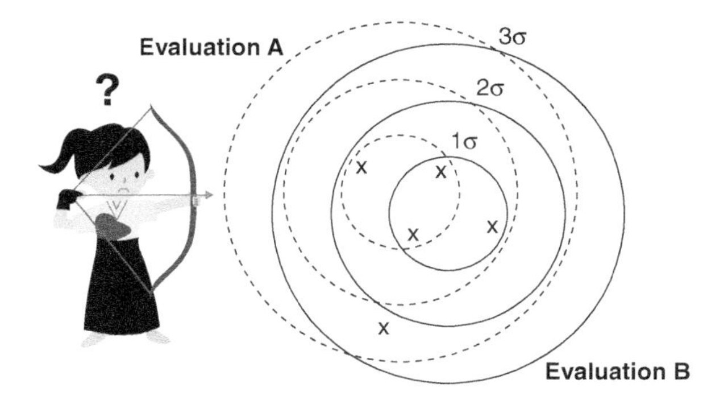

FIGURE 2.4
An analogy to reinforce the concept that critical evaluation does not produce a unique result. The two overlapping targets represent two independent critical evaluations for the same property that are different, but consistent.

The following sections (2.10.2–2.10.8) describe the various factors used in critical evaluation of the property values used in this book. Although details of the methods used by the organizations involved (NIST/TRC and DIPPR) differ, both are based on a thorough search and collection of the primary experimental literature and evaluation through consideration of the factors below. All evaluations by DIPPR are for pure compounds, while those at NIST/TRC are for binary and ternary mixtures, as well as pure compounds. Some factors discussed below are necessarily for mixtures only.

2.10.2 Article Content Analysis

The general quality of the content of an article reporting experimental property results can be judged against the *Guidelines for Reporting of Phase Equilibrium Measurements (IUPAC Recommendations 2012)*.[8] In spite of the title, these recommendations can be applied to all reports of thermophysical and thermochemical property measurements, and a simple checklist is provided that enumerates the key information that should be given (e.g., complete chemical sample information, appropriate detail of apparatus and experimental procedures, "stand-alone" tables with all quantities and uncertainties well defined, data validation through consistency checks and model fitting where possible, and assessment and comparison with previously published data). Full details and descriptions are given in the IUPAC recommendations.[8]

A report that is closely aligned with the IUPAC guidelines is strongly preferred to one that forces the evaluator do such things as guess the probable sample purity, assume calibrations were done appropriately, deduce the meaning of poorly defined uncertainties, or even guess key experimental conditions, such as temperature and pressure, when not given. Weightings of the experimental data based on this assessment are at the discretion of the evaluator. Each evaluating organization may have some general "rules of thumb" for this sort of general assessment, but any rigorous algorithm is not possible.

2.10.3 Literature Comparisons

Evaluation of experimental property values necessarily includes comparison with other experimental values existing in the literature. Such work requires extensive search, collection, and information extraction. High-quality measurements of thermophysical properties were first reported as early as the 19th century. An effective way to locate much historical literature is often through reference books and reviews. This, however, is extremely labor intensive because of the lack of electronic indexes in most cases, and simply locating and obtaining key reference books is a difficult search task in itself. The often-overlapping content of reference books also increases the work needed to execute a comprehensive data search.

Property evaluations for this book are based on two extensive and well-curated collections of experimental property values: (1) the DIPPR Database[30] of experimental properties and (2) the NIST/TRC SOURCE Database.[31] The DIPPR Database was created in support of property evaluations for specific pure compounds (49 properties for ~2500 compounds) and is available through the American Institute of Chemical Engineers (AIChE).[30] The NIST/TRC SOURCE Database is the experimental data archive associated with the NIST ThermoData Engine, which is available through NIST (Standard Reference Database 103a[32] and 103b[33]). The NIST/TRC SOURCE Database is not focused on specific chemical systems, but has the goal of coverage for as many pure, binary, and ternary chemical systems as is practical. At present this includes property data for 28,000 pures, 59,000 binary mixtures, 18,000 ternary mixtures, and 8000 chemical reactions.

2.10.4 Consistency with Endpoints

In the evaluation of properties for mixtures, an important consideration is the comparison of mixture values with independently evaluated endpoint properties, such as pure-component values for binary mixtures and binary endpoints for multicomponent mixtures. Specific examples are consistency of vapor-liquid equilibria with pure-component vapor pressures or solid-liquid equilibria with pure-component melting temperatures. If the coverage of the composition space is extensive, comparison with endpoints is a simple extrapolation task. If composition gaps are wide, the comparison will require data modeling.

In many cases, available experimental data for pure components is more extensive and of higher quality (i.e., lower uncertainty) than that for mixtures. This allows an added insight into the quality of the mixture results, which would not be possible through consideration of the mixture data in isolation. Multiple examples of the use of pure-component data to identify anomalous results for mixtures have been reported in articles authored by members and associates of the NIST Thermodynamics Research Center.[26,34] These methods have even been used to "repair" literature reports, where components were misidentified or experimental conditions (temperature or pressure) were incorrectly specified.[26,34]

As noted above with regard to article content analysis, uncertainties deduced through endpoint-consistency analysis in critical evaluation for a specific chemical system will necessarily vary between evaluator and evaluating organization. Nonetheless, endpoint-consistency analysis is an essential tool in assessment of data for multicomponent systems, and is practical only where an extensive database of experimental properties is available.

2.10.5 Property Prediction

The primary purpose of property predictions in critical evaluation of experimental data is the efficient identification of gross errors. Often, predicted (or computed) property values have only rough estimates of uncertainty themselves or none at all, which greatly reduces their value in critical analysis. Evaluation of uncertainties for prediction methods is a research area that has been largely ignored in the field of thermophysical properties, but recent advancements have been made, particularly in the area of prediction of enthalpies of formation for organic compounds.[35,36] Nonetheless, some property predictions have been established to be as good or better than experimental determinations, such as those for virial coefficients and heat capacities of simple dilute gases (cf., Refs. 37 and 38]), and gas-phase enthalpies of formation for many types of organic compounds.[35]

Property prediction methods vary considerably in the extent of their physical basis and reliance on experimental data used for parameterization. Well-known group-contribution methods include those for enthalpies of formation, heat capacities, and entropies of organic compounds in the ideal-gas state by Benson,[39] and UNIFAC predictions, first described by Fredenslund et al.[40] for prediction of phase equilibria involving vapors and liquids. The effectiveness of these predictions is evidenced by the continued interest in extending and improving their experiment-based parameterizations (cf., Refs. 41–43). The prediction of critical properties for pure compounds with the method of Joback[42] is another well-known example of such a method, with parameters published in the previous version of this book.[43] All group-contribution methods are strongly dependent on the quality of the underlying experimental values.

Other more recently developed prediction methods are COSMO-based models for phase equilibrium[44–46] and predictions for critical constants based on Quantitative Structure–Property Relationships (QSPR) methodology combined with the Support Vector Machines (SVM) regression.[45] Although all of these methods still require critically evaluated property values for parameterization, the underlying quantum chemical methods allow the parameters to be founded on a more physical basis than the earlier group methods. Of the results for these newer methods that are based, in part, on the results of computational chemistry, computed enthalpies of formation have been most extensively analyzed, and uncertainties near those for the best experimental studies have been established.[36]

2.10.6 Single-Property Regression and Visualization

Fitting of models (i.e., mathematical equations) to property data is an essential aspect of the critical evaluation process for any property, even including those that are invariant as per the Gibbs Phase Rule, such as the solid-liquid-gas triple point temperature or the critical temperature and pressure. Models are useful for many purposes, including comparison of multiple data sources, calculation of derivatives necessary for propagation of uncertainty in variables, extrapolations for end-point consistency assessment, data scatter detection (random deviations), and possibly, data bias (i.e., deviations correlated with state variables). Inadequate models or erroneous data may cause bias, and determining whether the model or the data are at fault is challenging, particularly with models involving a flexible number of parameters, where overfitting is a concern. Any model must be physically valid and in accord with validity rules, whether empirical or theoretical. Such rules define numerical ranges of valid model parameters and shapes of temperature, pressure, and composition dependence.

Visual analysis requires efficient generation of alternative graphical representations, and tools that allow transformation of plots into functions of any model variable are necessary. In recent articles concerning assessment of data quality for vapor-liquid equilibria, Mathias[46,47] has recommended use of "K values" (the ratio of the mole fraction of a component in the vapor to that in the liquid) and relative volatility α, where $\alpha = y_1{*}(1 - x_1)/[x_1{*}(1 - y_1)]$. VLE measurements often cover a broad range of compositions (from 0 to 1 mole fraction). When multiple data sources are plotted on a single xy-diagram, it is difficult to assess the relative quality of the datasets, particularly at the composition extremes (Fig. 2.5A). However, when the compositions are converted to relative volatility (Fig. 2.5B), the relative data scatter becomes clear. A general solution for revealing model and data behavior near composition endpoints has been proposed in the literature.[48] For binary mixtures, it consists of the transformation of the composition variable z (mole or mass fraction) to $\log[z/(1 - z)]$ for plotting.

The quality of a modeling result can be analyzed algorithmically or visually. Numerical criteria are used at NIST/TRC to determine the model sufficiency (termed "adequacy") and model bias. The adequacy A is calculated as

$$A = (1/n)\,\Sigma(d_i/U_i)^2, \qquad (2.10\text{-}1)$$

where n is the number of data points, d_i is the deviation from the model for the i-th data point, and U_i is the expanded uncertainty (0.95 level of confidence) associated with the data point. Bias is estimated from the absolute weighted deviations. Details have been published in Ref. 55. Although this chapter is concerned with critical evaluation of experimental data from the literature, it is apparent that if validated experimental data are used, the derived adequacy and bias can be used to judge the relative efficacy of alternative models. Consistency with experimental data, as well as general thermodynamic consistency, must be considered in assessment of any model, and this topic follows.

2.10.7 Thermodynamic Consistency

2.10.7.1 Equations of state (EOS) (pure compounds)

Implementation of an EOS allows, in principle, evaluation of all thermodynamic properties of the system simultaneously. In general, an EOS establishes a relationship between the thermodynamic variables of the system,

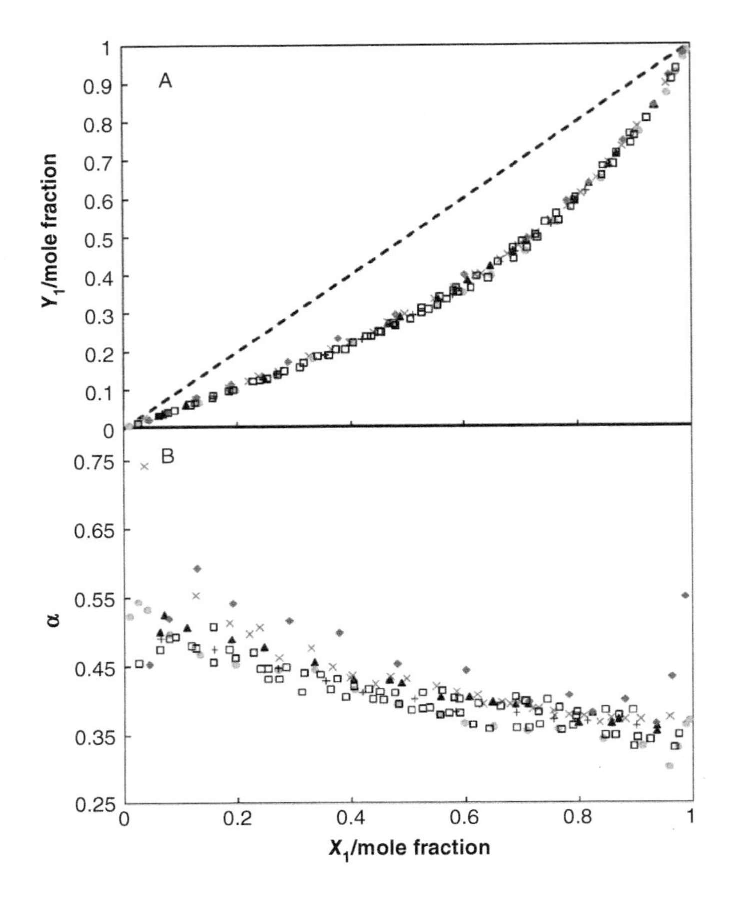

FIGURE 2.5
(A) Gas-phase composition against liquid phase composition. (B) Relative volatility α (defined in the text) against liquid phase composition. The chemical system is [toluene (1) + cyclohexane(2)] and values based on the following sources are shown: □,[49]; ◆,[50]; ●,[51]; ×,[52]; +,[53]; ▲.[54]

temperature T, volume V, and pressure P. For one mole of pure substance, the equation of state can generally be expressed as

$$PV = RT + f(T, P) \tag{2.10-2}$$

where R is the gas constant. EOS range considerably in complexity from the van der Waals EOS[56] to the 56-parameter Span-Wagner EOS that is used for water.[57] Selection of an EOS for a particular compound depends strongly on the particular data scenario (i.e., the variety, quality, and extent of the available experimental data). In spite of this complexity, even a relatively simple EOS can be helpful in identifying inconsistent or anomalous data.

At NIST/TRC, a range of EOS can be deployed on demand for various data scenarios. These include the Volume-Translated Peng-Robinson,[58] plus several subsequent variants,[59–62] the Modified Sanchez-Lacombe,[60,61] the PC-SAFT (with and without association),[62] and the 12-term Span Wagner formulation (polar and nonpolar).[63] For the types of compounds considered in this book, equations of state are limited to those modeling the liquid and gas phases. The solid state must also be considered, but alternative consistency tests must be applied, as described later in this chapter (Sec. 2.10.7.2).

A sample result obtained through application of the 12-term Span-Wagner EOS is shown in Fig. 2.6, where simultaneous consideration of measured values for vapor pressure, enthalpy of vaporization, liquid density, liquid heat capacity, and speed of sound allowed identification of anomalous property values in the literature for benzonitrile. The simultaneous consideration of all properties was particularly valuable for evaluation of enthalpy of vaporization, speed of sound, and liquid heat capacity, where large inconsistency is seen between the few existing datasets. References to the experimental data plotted in Fig. 2.6 can be found in Ref. 64.

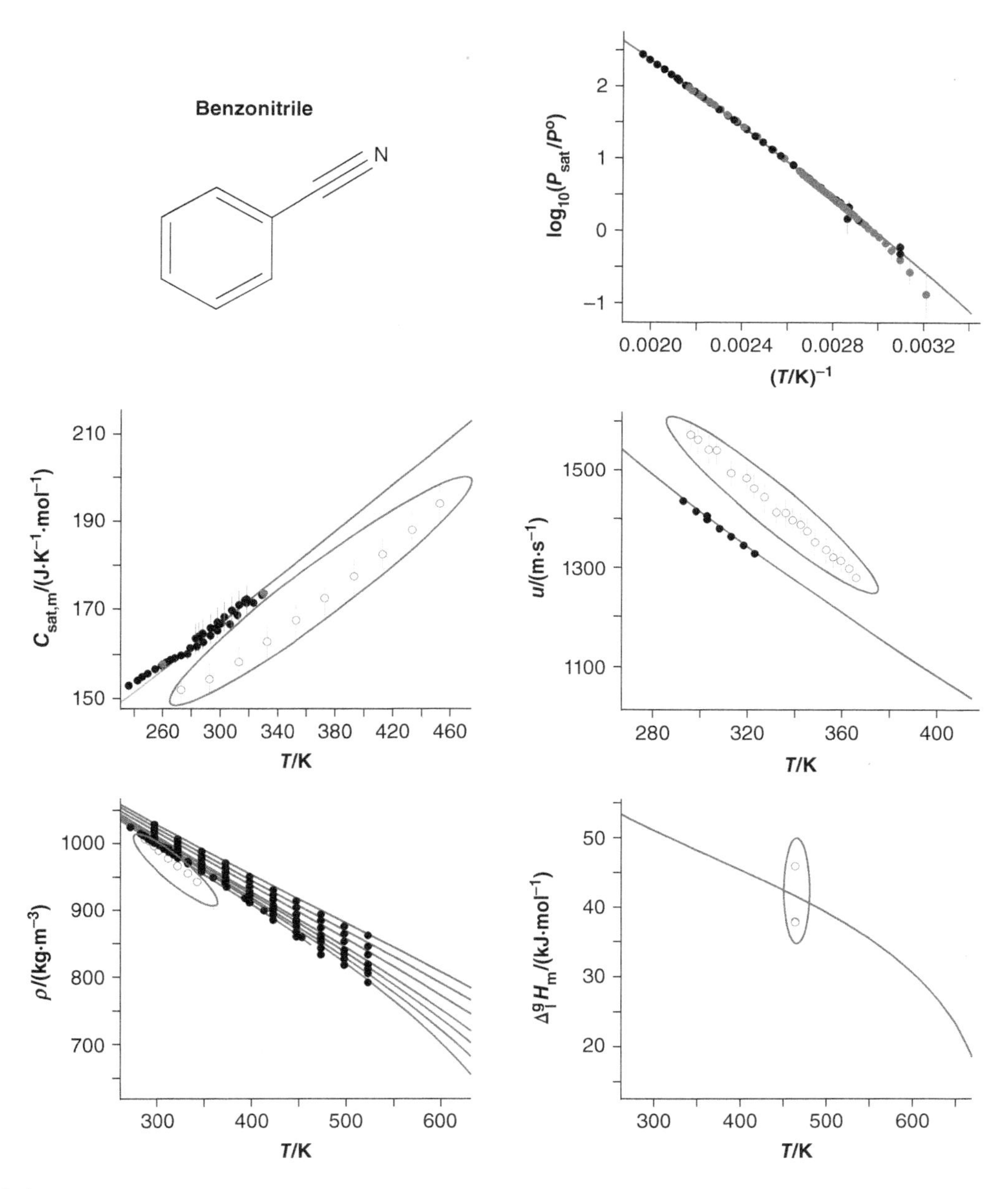

FIGURE 2.6
Results of application of the 12-term Span-Wagner EOS to experimental values of vapor pressures P_{sat}, liquid density ρ, enthalpy of vaporization ${}^{g}_{l}H_{m}$, liquid heat capacity $C_{sat,m}$, and speed of sound u for benzonitrile. The lines represent values calculated with the equation of state. Anomalous values are circled in the figures.

2.10.7.2 Properties represented by separate equations (pure compounds)

As noted above, equations of state for molecular compounds are typically limited to the fluid phases (gas and liquid), and consistency with solid phases cannot be addressed. However, thermodynamic consistency between properties of gas, liquid, and solid phases can be determined through consistency requirements that are checked and enforced for models associated with different phases.

Software implementation of extensible thermodynamic constraints in the NIST/TRC critical analysis tool *ThermoData Engine* (TDE) has been described in the literature.[65] The principles of consistent evaluation are similar to those for EOS fitting and are based on an objective function that combines the effect of deviations from fitted data and violation of thermodynamic constraints. Thermodynamic constraints implemented in TDE involving both fluid and solid phases are consistency between (1) vapor and sublimation pressure at the triple point, (2) enthalpy of fusion, vaporization, and sublimation at the triple point, (3) sublimation pressure and the gas-solid heat capacity difference, and (4) sublimation pressure and enthalpy of sublimation. Particular constraints on models for individual properties are also implemented. Examples are the Waring constraint,[66] which helps ensure a valid shape for the vapor pressure curve, and the compressibility-factor Z constraint, which ensures for a subcritical gas that Z is less than one and increases with temperature with a negative second derivative with respect to temperature along isobars (see Ref. 65 for details).

When different properties are represented by independent equations, inconsistencies can become apparent. In the absence of information that would allow rejection of some data, the uncertainty in the assessed property values is necessarily as large as the difference between the available data and property values derived through consistency enforcement. An example of such an assessment follows.

Figure 2.7A shows a plot of vapor and sublimation pressures for 3-methyoxy-4-hydroxybenzaldehyde (commonly known as vanillin). The normal melting temperature T_m and enthalpy of fusion $\Delta_{cr}^{l}H_m$ near ambient pressure have been reported many times in the literature, and the assessed values have the expanded uncertainty $U(T_m) \approx 1$ K and relative expanded uncertainty $U_r(\Delta_{cr}^{l}H_m) \approx 0.1$, respectively, with 0.95 level of confidence. T_m and the crystal-liquid-gas triple point temperature T_{tp} may be taken as equivalent due to measurement uncertainties, which are rarely less than 0.2 K. In Fig. 2.7A, it is apparent that extrapolations of the vapor pressures for the liquid and solid do not meet at T_{tp}. The curves represent the natural logarithm of the critically evaluated vapor pressures P with relative expanded uncertainty $U_r(P) \approx 0.2$ for both phases within 30 K of T_m. The large uncertainties in the assessed vapor and sublimation pressures reflect the poor thermodynamic consistency between the experimental values. This result is in spite of the small scatter (i.e., good repeatability) for the individual datasets.

Recently, new vapor and sublimation pressures were reported for vanillin that support the above analysis.[77] As seen in Fig. 2.7B, the new vapor and sublimation pressures are in excellent accord at $T = T_m$. In addition, the change in slope of the plot at $T = T_m$ is consistent with the evaluated enthalpy of fusion, adding further confidence in the new measurement results. Relative expanded uncertainties (0.95 level of confidence) for the vapor and sublimation pressures near $T = T_m$ are now reduced to $U_r(p) \approx 4\%$ for both phases, as a result of the thermodynamic consistency analysis.

2.10.7.3 Consistency tests for phase equilibrium studies on binary mixtures

Development and application of consistency tests for all types of phase equilibrium measurements continue to be an active area of research. The purpose of this section is not to exhaustively describe the variety of consistency tests used in the literature, but rather to provide an overview of how some common tests are used collectively in the assessment of phase equilibrium data at NIST/TRC.

For the assessment of VLE binary data, an algorithm is used that combines four widely used tests of consistency based on the requirements of the Gibbs-Duhem equation in combination with a check of consistency between the property values at the composition extremes of the VLE binary data and the corresponding pure-component vapor pressures.[34]

A VLE dataset must satisfy the constraint established by the Gibbs-Duhem equation,

$$\sum x_i \mathrm{d}\bar{M}_i - \left(\frac{\mathrm{d}M}{\mathrm{d}P}\right)_{x,T} \mathrm{d}P - \left(\frac{\mathrm{d}M}{\mathrm{d}T}\right)_{x,P} \mathrm{d}T = 0 \qquad (2.10\text{-}3)$$

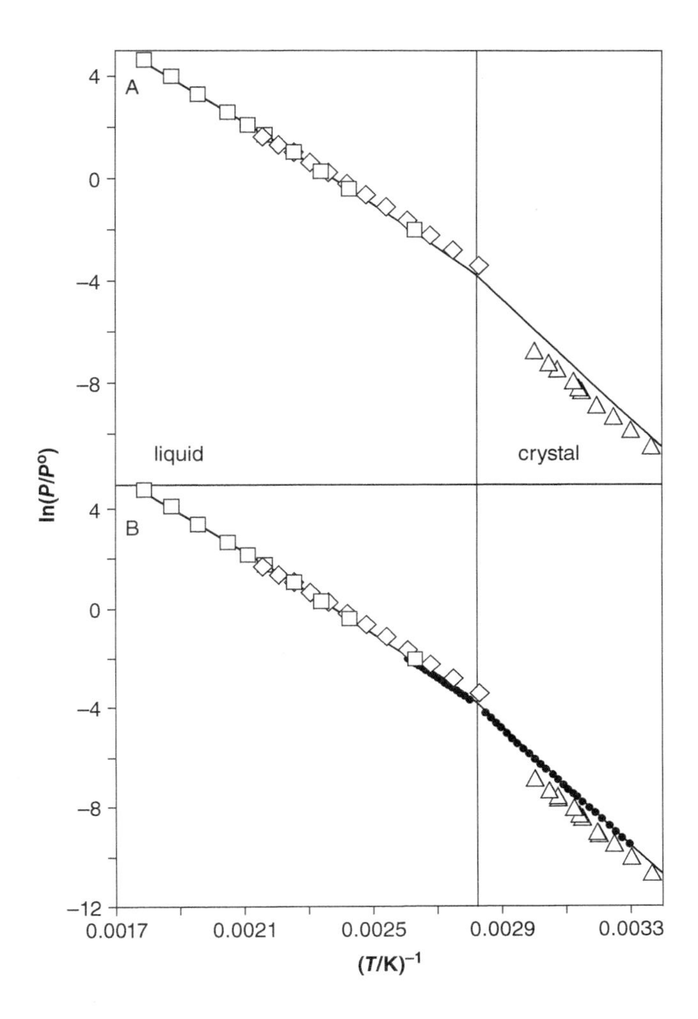

FIGURE 2.7
Vapor and sublimation pressures for vanillin. The vertical line indicates the triple-point temperature, and the curves represent the natural logarithm of the critically evaluated property values. The symbols □, Δ, ◊, and ●, represent experimental values reported in Refs. 67–70, respectively. (A) shows thermodynamic inconsistency of experimental vapor and sublimation pressures, and (B) shows consistency of newly reported values (●), as described in the text ($P^\circ = 1$ kPa).

where M is a molar thermodynamic property; $\bar{M}_i$ is a partial molar property; and T, P, and x are temperature, pressure, and liquid composition, respectively. The summation here and in the following equations is over the i components in the chemical system. If the property M is the excess Gibbs energy G^E divided by RT, where R is the gas constant, then

$$M \equiv \frac{G^E}{RT} = \sum x_i \ln \gamma_i, \tag{2.10-4}$$

and

$$\sum x_i \, d \ln \gamma_i - \frac{V^E}{RT} \, dP + \frac{H^E}{RT^2} \, dT = 0, \tag{2.10-5}$$

where γ is an activity coefficient; V^E is the excess volume; and H^E is the excess enthalpy. Equations 11 to 13 are the foundation for the VLE consistency tests used in the algorithm.

Specific consistency tests used in the algorithm are the *Herington Test*[68] that provides an indication of the compliance with the Gibbs-Duhem equation over the entire composition range, the *Van Ness Test*[69] that shows how well an activity coefficient model can reproduce the experimental data, the *Point Test*[70,71] that is used to assess the differential properties of G^E, and the *Infinite Dilution Test*[70,71] that provides a test of the limiting behavior of $G^E/(x_1x_2RT)$ and the activity coefficients γ_1 and γ_2. The effectiveness of each test depends upon the particular data scenario (i.e., the range of T, P, and x, and the data density to allow meaningful extrapolations and calculation of derivatives). Test results are combined with a measure of the consistency with pure-component vapor pressures to yield an overall numerical quality factor for the dataset. The quality factors are useful in weighting individual datasets when multiple sets are available for a given chemical system. Full details and many examples are provided in Ref. 34. A high-pressure equation-of-state test was subsequently added to expand the range of application.[72]

In the case of liquid-liquid equilibrium (LLE) data, the Gibbs-Duhem equation can also be applied; however, the experimental data needed to test for consistency are commonly lacking.[73] The large majority of experimental LLE data are obtained near atmospheric pressure, and are reported as LLE temperatures T as a function of composition. Such data are fitted with an empirical equation with an adjustable number of parameters. This approach is useful in identifying anomalous datasets, but only when numerous consistent measurements are available. Prediction of LLE data with group-contribution models (i.e., UNIFAC) or other methods based on computed molecular parameters (e.g., COSMO) do have the well-defined uncertainties needed to be useful in the critical evaluation process.

The primary consistency test for solid-liquid equilibria (SLE) at pressure $P \approx 0.1$ MPa is based on the equation

$$\ln\left(\frac{1}{\gamma_1 x_1}\right) = \frac{\Delta_{\mathrm{cr}}^{\mathrm{l}}H_{\mathrm{m}}}{RT_{\mathrm{m}}}\left(\frac{T_{\mathrm{m}}}{T} - 1\right) - \frac{\Delta C_p}{R}\left(\frac{T_{\mathrm{m}}}{T} - 1\right) + \frac{\Delta C_p}{R}\ln\left(\frac{T_{\mathrm{m}}}{T}\right) \tag{2.10-6}$$

where x_1 and γ_1, respectively, are the mole fraction and activity coefficient of component 1 in the liquid phase, T_{m} and $\Delta_{\mathrm{cr}}^{\mathrm{l}}H_{\mathrm{m}}$, respectively, are the temperature and molar enthalpy of melting of component 1, ΔC_p is the heat capacity of the liquid minus that of the solid at the melting temperature, R is the gas constant, and T is the SLE temperature. Two criteria are assessed in the limit of $x_1 \rightarrow 1$; (1) the limiting intercept must be in accord with the normal melting temperature of component 1, and (2), the limiting slope must be in accord with the enthalpy of fusion of component 1. Examples are provided in Ref. 73.

Due to the limited variety and amount of experimental VLE, LLE, and SLE data available for ternary mixtures, consistency checks are largely limited to comparisons with results for the component binary systems at the ternary composition extremes.[73] This is analogous to assessing the consistency of properties of binary systems with those of their pure constituents. Such assessments are only possible with extensive and well curated collections of experimental data, such as those that underlie the critically evaluated experimental data used in this book.

Analysis of thermodynamic consistency, particularly for vapor-liquid equilibrium experiments, continues to be an active area of research. Some recent examples include work by Olson,[74] Wisniak et al.,[75] and Carrero-Mantilla et al.[76]

2.10.8 Trends in a Series of Compounds (Pure Components)

A less rigorous but still powerful check can be based on variation of property values in families of compounds with systematic structural differences, such as incremental variation with addition of —CH_2— or functional groups. Less common structural differences can also be harnessed and constructing sets of related substances can be automated.[77] Variation of property values in such sets should be generally smooth, excluding the smallest molecules in the set, odd-even effects, and some property-specific effects, such as formation of plastic crystals and micellization. This approach is most effective in identifying anomalous data that can skew a critical assessment toward unphysical results.

2.11 SUMMARY

Evaluation of prediction and correlation methods can only be based on comparisons with critically evaluated property data founded on experimental measurements. To be of use in these comparisons, the critically evaluated data must have well-defined uncertainties with a stated level of confidence. An international standard (the GUM) has been established that outlines the philosophy and procedures to be used in the assessment of uncertainty in measurement. Although this standard has been in place for more than 25 years, it is only rarely well applied in studies of thermophysical properties and phase equilibria published in the scientific literature. In this chapter, we have summarized the GUM approach and described some of the challenges encountered when trying to apply the GUM procedures to uncertainty assessment for published experimental data. Some specific examples are given of common errors and omissions in uncertainty assessments encountered in the modern literature. Finally, an overview is given of the tools and procedures used in evaluation of uncertainties assigned to reference data used in this book.

2.12 DISCLAIMER

2.13 REFERENCES

1. Taylor, B. N., and C. E. Kuyatt: Guidelines for the evaluation and expression of uncertainty in NIST measurement results; NIST Technical Note 1297; NIST: Gaithersburg, MD, 1994. https://emtoolbox.nist.gov/Publications/NISTTechnicalNote1297s.pdf (accessed January, 2019).
2. See Annex C, subsection 2.14 of Ref. 3 for specific details.
3. Guide to the Expression of Uncertainty in Measurement; International Organization for Standardization: Geneva, Switzerland, 1993. This guide was prepared by ISO Technical Advisory Group 4 (TAG 4), Working Group 3 (WG 3). ISO/TAG 4 has as its sponsors the BIPM, IEC, International Federation of Clinical Chemistry (IFCC), ISO, International Union of Pure and Applied Chemistry (IUPAC), International Union of Pure and Applied Physics (IUPAP), and OIML. Although the individual members of WG 3 were nominated by the BIPM, IEC, ISO, or OIML, the guide is published by ISO in the name of all seven organizations. http://www.bipm.org/en/publications/guides/gum.html (accessed January, 2019).
4. https://www.dictionary.com/ (accessed January, 2019).
5. Dong, Q., R. D. Chirico, X. Yan; X. Hong, and M. Frenkel: *J. Chem. Eng. Data*, **50:** 546–550 (2005).
6. https://www.bipm.org/en/publications/guides/vim.html. The document can be accessed directly here: https://www.bipm.org/utils/common/documents/jcgm/JCGM_200_2012.pdf (accessed January, 2019).
7. https://jcgm.bipm.org/vim/en/index.html (accessed January, 2019).
8. Chirico, R. D., T. W. De Loos, J. Gmehling, A. R. H. Goodwin, S. Gupta, W.M. Haynes, K. N. Marsh, V. Rives, J. Olson, C. Spencer, J. F. Brennecke, and J. P. M. Trusler: *Pure Appl. Chem.*, **84:** 1785–1813 (2012).
9. Moldover, M. R.: *J. Chem. Phys.*, **61:** 1766–1778 (1974).
10. *NIST/SEMATECH e-Handbook of Statistical Methods*, http://www.itl.nist.gov/div898/handbook/ (accessed April 7, 2019).
11. Perkins, R. A., and M. O. McLinden: *J. Chem. Thermodyn.*, **91:** 43–61 (2015).
12. Huber, M. L., R. A. Perkins, A. Laesecke, D. G. Friend, J. V. Sengers, M. J. Assael, I. N. Metaxa, E. Vogel, R. Mareš, and K. Miyagawa: *J. Phys. Chem. Ref. Data*, **38:** 101–125 (2009).
13. Wagner, W., and A. Pruss: *J. Phys. Chem. Ref. Data*, **31:** 387–535 (2002).

14. Archer, D. G.: *J. Phys. Chem. Ref. Data*, **22:** 1441–1453 (1993).
15. SRM 211d—Toluene Liquid Density Extended Range (https://www.nist.gov/srm).
16. SRM 781D2—Molybdenum—Heat Capacity (https://www.nist.gov/srm).
17. SRM 39j—Benzoic Acid (Calorimetric Standard) (https://www.nist.gov/srm).
18. Hubbard, W. N., D. W. Scott, and G. Waddington: Standard States and Corrections for Combustions in a Bomb at Constant Volume. Chapter 5, *Experimental Thermochemistry*, Vol. I, 1956 (Ed. by F. D. Rossini).
19. Zhang, J. N., Z. C. Tan, Q. F. Meng, Q. Shi, B. Tong, and S. X. Wang: *J. Therm. Anal. Calorim.*, **95 (2):** 461–467 (2009).
20. https://trc.nist.gov/cctool/.
21. Paulechka, E., and D. Riccardi: *Corrections to Standard State in Combustion Calorimetry: An Update and a Web-based Tool*. To be published.
22. Bazyleva, A., and M. D. M. C. Ribeiro da Silva: *Oxygen Combustion Bomb Calorimetry: Good Measurement and Reporting Practice*. To be published.
23. NIST Uncertainty Machine. https://uncertainty.nist.gov/.
24. Cummings, P. T., T. de Loos, J. P. O'Connell, W. M. Haynes, D. G. Friend, A. Mandelis, K. N. Marsh, P. L. Brown, R. D. Chirico, A. R. H. Goodwin, J. Wu, R. D. Weir, J. P. M. Trusler, A. Pádua, V. Rives, C. Schick, S. Vyazovkin, and L. D. Hansen: *Fluid Phase Equilib.*, **276:** 165–166 (2009); *Int. J. Thermophys.*, **30:** 371–373 (2009); *J. Chem. Eng. Data*, **54:** 2–3 (2009); *J. Chem. Thermodyn.*, **41:** 575–576 (2009); *Thermochim. Acta*, **484:** vii–viii (2008).
25. Larsen, P. O., and M. von Ins: *Scientometrics*, **84:** 575–603 (2010).
26. Chirico, R. D., M. Frenkel, J. W. Magee, V. Diky, C. D. Muzny, A. F. Kazakov, K. Kroenlein, I. Abdulagatov, G. R. Hardin, W. E. Acree, Jr., J. F. Brenneke, P. L. Brown, P. T. Cummings, T. W. de Loos, D. G. Friend, A. R. H. Goodwin, L. D. Hansen, W. M. Haynes, N. Koga, A. Mandelis, K. N. Marsh, P. M. Mathias, C. McCabe, J. P. O'Connell, A. Pádua, V. Rives, C. Schick, J. P. M. Trusler, S. Vyazovkin, R. D. Weir, and J. Wu: *J. Chem. Eng. Data*, **58:** 2699–2716 (2013).
27. Anton Paar DMA 5000 M: https://www.anton-paar.com/us-en/products/details/dmatm-5000-m-density-meter/ (accessed January, 2019).
28. Assael, M. J., A. E. Kalyva, S. A. Monogenidou, M. L. Huber, R. A. Perkins, D. G. Friend, and E. F. May: *J. Phys. Chem. Ref. Data*, **47:** 021501-1–021501-10 (2018).
29. Viscosity of Water, ISO/TR Technical Report No. 3666: 1998(E), International Organization for Standardization (ISO), Geneva, 1998.
30. DIPPR Database: https://www.aiche.org/dippr/events-products/801-database.
31. Kazakov, A., C. D. Muzny, K. Kroenlein, V. Diky, R. D. Chirico, J. W. Magee, I. M. Abdulagatov, and M. Frenkel: *Int. J. Thermophys.*, **33:** 22–33 (2012).
32. https://www.nist.gov/mml/acmd/trc/thermodata-engine/srd-nist-tde-103a.
33. https://www.nist.gov/mml/acmd/trc/thermodata-engine/srd-nist-tde-103b.
34. Kang, J. W., V. Diky, R. D. Chirico, J.W. Magee, C. D. Muzny, I. Abdulagatov, A. F. Kazakov, and M. Frenkel: *J. Chem. Eng. Data*, **55:** 3631–3640 (2010).
35. Paulechka, E., and A. Kazakov: *J. Phys. Chem. A*, **121:** 4379–4387 (2017).
36. Paulechka, E., and A. Kazakov: *J. Chem. Theory Comput.*, **14:** 5920–5932 (2018).
37. Van, T. P., and U. K. Deiters: *Chem. Phys.*, **457:** 171–179 (2015).
38. Van, T. P., and U. K. Deiters: *Chem. Phys.*, **485–486:** 67–80 (2017).
39. Benson, S. W.: *Thermochemical Kinetics* (2nd Ed.). Wiley, New York, 1976.
40. Fredenslund, A., R. L. Jones, and J. M. Prausnitz: *AIChE J.*, **21:** 1086–1099 (1975).
41. Verevkin, S. P., V. N. Emel'yanenko, V. Diky, C. D. Muzny, R. D. Chirico, and M. Frenkel: *J. Phys. Chem. Ref. Data*, **42:** 033102-1–033102-33 (2013).
42. Joback, K. G. S. M.: Thesis. Massachusetts Institute of Technology, Cambridge, MA, 1984.
43. Poling, B. E., J. M. Prausnitz, and J. P. O'Connell: *The Properties of Gases and Liquids* (5th Ed.). McGraw-Hill, 2001.
44. Lin, S.-T., and S. I. Sandler: *Ind. Eng. Chem. Res.*, **41:** 899–913 (2002).
45. Kazakov, A., C. D. Muzny, V. Diky, R. D. Chirico, and M. Frenkel: *Fluid Phase Equilibria*, **298:** 131–142 (2010).
46. Mathias, P. M.: *J. Chem. Eng. Data*, **62:** 2231–2233 (2017).
47. Mathias, P. M.: *Fluid Phase Equilib.*, **408:** 265–272 (2016).
48. Diky, V.: *J. Chem. Eng. Data*, **62:** 2920–2926 (2017).
49. de Alfonso, C., A. A. Canovas, B. Llanas, M. Pintado, A. F. Saenz de la Torre: *An. Quim. A*, **82:** 320–330 (1986).
50. Chen, G., X. Yan, S. Han, Z. Ma, Q. Wang: *Huagong Xuebao*, **45:** 94–101 (1994).
51. Cholinski, J., M. Palczewska-Tulinska, A. Szafranska, and D. Wyrzykowska-Stankiewicz: *Chem. Eng. Sci.*, **36:** 173–181 (1981).
52. Myers, H. S.: *Ind. Eng. Chem.*, **48:** 1104–1108 (1956).

53. A. Delzenne: Bull. Soc. Chim. Fr., 295–298 (1961).
54. Sieg, L., *Chem.-Ing.-Tech.*, **22:** 322–326 (1950).
55. Diky, V., C. D. Muzny, E. W. Lemmon, R. D. Chirico, and M. Frenkel: *J. Chem. Inf. Model.*, **47:** 1713–1725 (2007).
56. van der Waals, J. D.: *On the Continuity of the Gaseous and Liquid States*, Doctoral Thesis, Universiteit Leiden, 1873.
57. Wagner, W., and A. Pruss: *J. Phys. Chem. Ref. Data*, **31:** 387–535 (2002).
58. Peng, D-Y., and D. B. Robinson: *Ind. Eng. Chem. Fundam.*, **15:** 59–64 (1976).
59. Twu, C. H., J. E. Coon, J. R. Cunningham: *Fluid Phase Equilib.*, **75:** 65–79 (1992).
60. Koak, N., R. A. Heidemann: *Ind. Eng. Chem. Res.* **35:** 4301–4309 (1996).
61. Krenz, R. A.: *Correlating the Fluid Phase Behavior of Polydisperse Polyethylene Solutions using the Modified Sanchez-Lacombe Equation of State*. Ph.D. Thesis, University of Calgary, 2005.
62. Gross, J., and G. Sadowski: *Ind. Eng. Chem. Res.*, **40:** 1244–1260 (2001).
63. Span, R., and W. Wagner: *Int. J. Thermophys.*, **24:** 1–39 (2003).
64. Diky, V., A. Bazyleva, E. Paulechka, J. W. Magee, V. Martinez, D. Riccardi, and K. Kroenlein: *J. Chem. Thermodyn.*, **133:** 208–222 (2019).
65. Diky, V., R. D. Chirico, C. D. Muzny, A. F. Kazakov, K. Kroenlein, J. W. Magee, I. Abdulagatov, and M. Frenkel: *J. Chem. Inf. Model.*, **53:** 3418–3430 (2013).
66. Waring, W.: *Ind. Eng. Chem.*, **46:** 762–763 (1954).
67. Anschütz, R., H. Reitter: Distillation under Reduced Pressure in the Laboratory, Friedrich Cohen (Bonn) 1895.
68. Herington, E. F. G.: *J. Inst. Pet.*, **37:** 457–470 (1951).
69. Van Ness, H. C., S. M. Byer, and R. E. Gibbs: *AIChE J.*, **19:** 238–244 (1973).
70. Kojima, K., H. M. Moon, and K. Ochi: *Fluid Phase Equilib.*, **56:** 269–284 (1990).
71. Kurihara, K., Y. Egawa, K. Ochi, and K. Kojima: *Fluid Phase Equilib.*, **219:** 75–85 (2004).
72. Diky, V., R. D. Chirico, C. D. Muzny, A. F. Kazakov, K. Kroenlein, J. W. Magee, I. Abdulagatov, J. W. Kang, and M. Frenkel: *J. Chem. Inf. Model.*, **52:** 260–276 (2012).
73. Kang, J.W., V. Diky, R. D. Chirico, J. W. Magee, C. D. Muzny, A. F. Kazakov, K. Kroenlein, and M. Frenkel: *J. Chem. Eng. Data*, **59:** 2283–2293 (2014).
74. Olson, J. D., *Fluid Phase Equilib.*, **418:** 50–56 (2016).
75. Wisniak, J., J. Ortega, and L. Fernández: *J. Chem. Thermodyn.*, **105:** 385–395 (2017).
76. Carrero-Mantilla, J. I., D. de Jesus Ramírez-Ramírez, and J. F. Súarez-Cifuentes: *Fluid Phase Equilib.*, **412:** 158–167 (2016).
77. Diky, V., A. Kazakov, and K. Kroenlein: Grid Evaluation of Pure-Compound Properties. AIChE Annual Meeting, Minneapolis, Paper 365d, Oct. 31, 2017. https://aiche.confex.com/aiche/2017/meetingapp.cgi/Paper/500012.

3

Pure-Component Constants

3.1 SCOPE*

The properties of pure components are important for all aspects of process design and optimization. Pure-component properties also underlie much of the observed behavior of mixtures. For example, property models intended for the whole range of composition must give pure-component properties at the pure-component limits. In addition, pure-component property constants are often used as the basis for models such as corresponding states correlations for PVT equations of state (Chap. 6). They are often used in composition-dependent mixing rules for the parameters to describe mixtures (Chap. 7).

As a result, we first study methods for obtaining *pure-component constants* of the more commonly used properties and show how they can be estimated if no experimental data are available. These include the vapor-liquid critical properties, normal ($P = 101325$ Pa) melting and boiling temperatures, acentric factors, and dipole moments. Others, such as the liquid molar volumes and heat capacities, are discussed in later chapters. Values for these properties for many substances are tabulated in Appendix A. Though the origins of current group contribution methods are over 50 years old, many of these methods are still used and are impressively accurate. These methods are also the foundation for improved methods that have been more recently developed.

In Secs. 3.2 (critical properties), 3.3 (acentric factor), and 3.4 (boiling and melting points), we illustrate several methods and compare each with the data tabulated in Appendix A and with each other. All the calculations were done with the program DIADEM which is a front-end interface for the DIPPR database that has the power to parse molecules into the chemical groups prescribed by group-contribution methods or perform the mathematical operations of other prediction approaches. This was done to maximize accuracy and consistency among the methods.

A primary objective of this edition was to establish the most extensive test possible of the available prediction methods to give the reader the broadest assessment of the accuracy of each. This was done by creating a test set database which was composed of evaluated, experimental data from the NIST TRC and DIPPR databases. This test database is the largest ever used to date to evaluate the accuracy of prediction methods. For example, the number of unique compounds with melting points in the test database exceeded 9500 and that for boiling points was more than 4900. Section 3.5 lists the results of the evaluations of the prediction methods and gives general recommendations for selecting a method. And Sec. 3.6 discusses dipole moments.

*With special contributions from V. Diky.

The molecular weights used in this edition, in both examples and Appendix A, are the 1999 International Union of Pure and Applied Chemistry (IUPAC) values (Coplen, 2001). IUPAC regularly reviews the relevant literature and updates values for atomic weights as warranted by experimental data, but it is common in industrial practice to update values with much less frequency for consistency. Changes for atoms in molecules of industrial relevance (e.g., C, H, N, O, S, P, halogens, common metals, etc.) are typically on the order of 0.001% or less. Uncertainties in the properties discussed in this book are typically no better than 0.1% in the best cases, so the error introduced by not using the most up-to-date IUPAC recommendations is negligible. Also, molecular weights only apply to a subset of the properties addressed in this book.

Most of the estimation methods presented in this chapter are of the *group, bond,* or *atom contribution* type. That is, the properties of a molecule are established from contributions from its fragments. The conceptual basis is that the intermolecular forces that determine the constants of interest depend mostly on the bonds between the atoms of the molecules. The fragment contributions are principally determined by the nature of the atoms involved (*atom contributions*), the bonds between pairs of atoms (*bond contributions* or equivalently *group interaction contributions*), or the bonds within and among small groups of atoms (*group contributions*). They all assume that the fragments can be treated independently of their arrangements or their neighbors. If this is not accurate enough, corrections for specific multigroup, conformational, or resonance effects can be included. Thus, there can be *levels* of contributions. The identity of the fragments to be considered (*group, bond,* or *atom*) are normally assumed in advance and their contributions obtained by fitting to data. Usually, applications to wide varieties of species start with saturated hydrocarbons and grow by sequentially adding different types of bonds, rings, heteroatoms, and resonance. The formulations for pure-component constants are quite similar to those of the ideal gas formation properties and heat capacities of Chap. 4.

Alternatives to *group/bond/atom* contribution methods have recently appeared. Most are based on adding weighted contributions of measured properties such as molecular weight and normal boiling point, etc. (*factor analysis*) or from "quantitative structure-property relationships" (QSPR) based on contributions from molecular properties such as electron or local charge densities, molecular surface area, etc. (*molecular descriptors*). Grigoras (1990), Horvath (1992), Katritzky et al. (1995; 1999), Egolf et al. (1994), Turner et al. (1998), Cholakov et al. (1999), and Kazakov et al. (2010) all describe the concepts and procedures. The descriptor values are computed from molecular mechanics or quantum mechanical descriptions of the substance of interest, and then property values are calculated as a function of the sum of contributions from the descriptors. The significant descriptors and their weighting factors are found by sophisticated regression techniques, and the resulting correlations are published in the literature. Unfortunately, values of the molecular descriptors can vary significantly depending on the methods and programs used in their calculation, and authors rarely publish the values used in the regression. This makes it difficult to reproduce or test the methods. This problem could be eliminated if tabulations of molecular descriptors for substances were available, but such do not currently exist. Rather, during creation of a QSPR-based prediction method, a molecular structure for each compound used in the training set is posed, the descriptors for each are computed, and the values are combined to create the correlation. Due to this customization, we have not been able to do any computations for these methods ourselves, but we quote the results from the literature when available.

The methods given here are not suitable for *pseudocomponent properties* such as for the poorly characterized mixtures often encountered with petroleum, coal, and natural products. These are usually based on measured properties such as average molecular weight, boiling point, and the specific gravity (at 20 °C) rather than molecular structure. We do not treat such systems here, but the reader is referred to the work of Tsonopoulos et al. (1986), Twu (1984), Twu and Coon (1996), and Zhang et al. (1998) for example. Older methods include those of Lin and Chao (1984) and Brule et al. (1982), Riazi and Daubert (1980), and Wilson et al. (1981).

3.2 VAPOR-LIQUID CRITICAL PROPERTIES

Vapor-liquid critical temperature, T_c, pressure, P_c, and volume, V_c are pure-component constants of great interest. They are used in many corresponding states correlations for volumetric (Chap. 6), thermodynamic (Chaps. 7–9), and transport (Chaps. 10–12) properties of gases and liquids. Experimental determination of their values can be challenging (Ambrose and Young, 1995). Historically, this has been especially true for larger components that can chemically degrade at elevated temperatures (Teja and Anselme, 1990), but recent experimental advances have improved this situation and yielded satisfactory results (Teja et al., 1989; Nikitin et al., 1993, 1999; Wilson et al., 1995; VonNiederhausern et al., 2000).

3.2.1 Estimation Techniques

One of the first successful *group contribution* methods to estimate critical properties was developed by Lydersen (1955). Since that time, more experimental values have been reported and efficient statistical techniques have been developed that allow determination of alternative group contributions and optimized parameters. We examine in detail the methods of Ambrose (1978; 1979; 1980), Joback (1984), Joback and Reid (1987), Constantinou and Gani (1994), Wilson and Jasperson (1996), and Nannoolal et al. (2007) for T_c, P_c, and V_c, the method of Lydersen (1955) for T_c and P_c, the method of Emami et al. (2009) for T_c, and that of Fedors (1979) for V_c. The Lydersen (1955) method for V_c is not described because it has proven inferior to the other methods discussed, as shown in the previous edition of this book.

Method of Lydersen (1955). The Lydersen (1955) method, though an early group contribution approach, has proven accurate for certain families of compounds and is still used by DIPPR and the American Petroleum Institute (API). The equations for T_c and P_c are

$$\frac{T_c}{\mathrm{K}} = \frac{T_b}{\mathrm{K}}\left[0.567 + \sum_i N_i\left(\Delta_{T_c}\right)_i - \left(\sum_i N_i\left(\Delta_{T_c}\right)_i\right)^2\right]^{-1} \tag{3.2-1}$$

$$\frac{P_c}{\mathrm{atm}} = \left(\frac{MW}{\mathrm{g\cdot mol^{-1}}}\right)\left(0.34 + \sum_i N_i\left(\Delta_{P_c}\right)_i\right)^{-2} \tag{3.2-2}$$

Here, T_b is the normal boiling point of the compound, MW is the molecular weight, the summations are over the unique groups in the molecule as defined in Table 3.1, $(\Delta_{T_c})_i$ and $(\Delta_{P_c})_i$ are the contributions of group i, and N_i is the number of times group i is found in the compound. Table 3.1 contains the original group values from Lydersen (1955) as well as updated and additional groups for siloxane and silane compounds from the work of Myers (1990). Example 3.1 demonstrates the Lydersen method.

Example 3.1 Estimate T_c and P_c for 2-methylphenol (*o*-cresol) by using the Lydersen group method.

Solution. 2-methylphenol contains one —CH_3, four =CrH—, two =Cr<, and one a—OH according to the group definitions outlined in Table 3.1. Note that the method does not distinguish between a =CrH— in aromatic and nonaromatic rings.

	Group	N	$N\Delta_{T_c}$	$N\Delta_{P_c}$
1	—CH_3	1	0.020	0.227
2	=CrH—	4	0.044	0.616
3	=Cr<	2	0.022	0.308
4	a—OH	1	0.031	−0.02
	$\sum_{k=1}^{4} N_k F_k$		0.117	1.131

The normal boiling point, T_b, and molecular weight, MW, of 2-methylphenol from Appendix A are 464.15 K and 108.138 g/mol, respectively, so the Lydersen estimates for T_c and P_c are as follows.

$$T_c = 464.15(0.567 + 0.117 - 0.117^2)^{-1} = 692.44 \text{ K}$$

$$P_c = 108.138(0.34 + 1.131)^{-2} = 49.98 \text{ atm} = 50.64 \text{ bar}$$

The experimental values recorded in Appendix A for the critical temperature and pressure are 697.55 K and 50.10 bar, so the differences are

$$T_c \text{ Difference} = 692.44 - 697.55 = -5.11 \text{ K or } -0.7\%$$

$$P_c \text{ Difference} = 50.64 - 50.10 = 0.54 \text{ bar or } 1.1\%.$$

TABLE 3.1 Group contributions for the Lydersen (1955) method for T_c and P_c to use in Eqs. (3.2-1) and (3.2-2). Groups with atoms in rings are identified by the letter "r" following the single-letter element symbol (e.g., Cr, Or, Sr, and Nr). Do not confuse these ring designations with symbols for elements such as chromium or strontium.

Group	Δ_{T_c}	Δ_{P_c}
Nonring Carbon Increments		
—CH_3	0.020	0.227
—CH_2—	0.020	0.227
>CH—	0.012	0.210
>C<	0.00	0.210
=CH_2—	0.018	0.198
=CH—	0.018	0.198
=C<	0.00	0.198
=C=	0.00	0.198
≡CH	0.005	0.153
≡C—	0.005	0.153
Ring Carbon Increments		
—CrH_2—	0.013	0.184
>CrH—	0.012	0.192
>Cr<	−0.007	0.154
=CrH—	0.011	0.154
=Cr<	0.011	0.154
=Cr=	0.011	0.154
Halogen Increments		
—F	0.018	0.224
—Cl	0.017	0.320
—Br	0.010	0.50
—I	0.012	0.83

(Continued)

TABLE 3.1 Group contributions for the Lydersen (1955) method for T_c and P_c to use in Eqs. (3.2-1) and (3.2-2). Groups with atoms in rings are identified by the letter "r" following the single-letter element symbol (e.g., Cr, Or, Sr, and Nr). Do not confuse these ring designations with symbols for elements such as chromium or strontium. (*Continued*)

Group	Δ_{T_c}	Δ_{P_c}
Oxygen Increments		
—OH (alcohol)	0.082	0.06
a—OH (aromatic)	0.031	−0.02
—O—	0.021	0.16
—Or—(ring)	0.014	0.12
>C=O (ketone)	0.040	0.29
>Cr=O (ring ketone)	0.033	0.2
—HC=O (aldehyde)	0.048	0.33
—COOH (acid)	0.085	0.4
—COO—(ester)	0.047	0.47
HCOO—(formate)	0.047	0.47
=O (except as above)	0.02	0.12
Nitrogen Increments		
—NH_2	0.031	0.095
—NH—	0.031	0.135
—NrH—(ring)	0.024	0.09
>N—	0.014	0.17
>Nr—(ring)	0.007	0.13
—C≡N	0.060	0.36
—NO_2	0.055	0.42
Sulfur Increments		
—SH	0.015	0.27
—S—	0.015	0.27
—Sr—(ring)	0.008	0.24
=S	0.003	0.24
Silicon Increments		
>Si<	0.026	0.468
>SiH—	0.04	0.513
—SiH_3	0.027	—
—SiO—	0.025	0.730
—SirOr—(ring)	0.027	0.668
Boron Increment		
B—	0.03	—

There is no increment for hydrogen. The bonds shown can be connected with any atom other than hydrogen. The silicon groups are those from Myers (1990).

Method of Joback. Joback (1984; Joback and Reid, 1987) reevaluated Lydersen's group contribution scheme, added several new functional groups, and determined new contribution values. His relations for the critical properties are

$$\frac{T_c}{\mathrm{K}} = \frac{T_b}{\mathrm{K}}\left[0.584 + 0.965\sum_i N_i(\Delta_{T_c})_i - \left(\sum_i N_i(\Delta_{T_c})_i\right)^2\right]^{-1} \tag{3.2-3}$$

$$\frac{P_c}{\mathrm{bar}} = \left[0.113 + 0.0032N_{\mathrm{atoms}} - \sum_i N_i(\Delta_{P_c})_i\right]^{-2} \tag{3.2-4}$$

$$\frac{V_c}{\mathrm{cm^3\cdot mol^{-1}}} = 17.5 + \sum_i N_i(\Delta_{V_c})_i \tag{3.2-5}$$

where, like the Lydersen method previously described, the summations are over the unique groups in the molecule, N_{atoms} is the total numbers of atoms in the compound, and the group values $(\Delta_{T_c})_i$ $(\Delta_{P_c})_i$, and $(\Delta_{V_c})_i$ are found in Table 3.2.

As with the Lydersen method, T_c, requires a value of the normal boiling point, T_b, which may be from experiment or by estimation from methods given in Sec. 3.4. An example of the use of Joback's groups is Example 3.2.

TABLE 3.2 Group contributions for the Joback (1984; Joback and Reid, 1987) method for T_c, P_c, V_c, T_m, and T_b. Use in Eqs. (3.2-3)–(3.2-5) for the critical properties and Eqs. (3.4-1) and (3.4-2) for T_m and T_b, respectively. Groups with atoms in rings are identified by the letter "r" following the single-letter element symbol (e.g., Cr, Or, Sr, and Nr). Do not confuse these ring designations with symbols for elements such as chromium or strontium.

Group	ΔT_c	Δ_{P_c}	Δ_{V_c}	Δ_{T_m}	Δ_{T_b}
		Nonring Carbon Increments			
—CH$_3$	0.0141	−0.0012	65	−5.10	23.58
—CH$_2$—	0.0189	0.0000	56	11.27	22.88
>CH—	0.0164	0.0020	41	12.64	21.74
>C<	0.0067	0.0043	27	46.43	18.25
=CH$_2$	0.0113	−0.0028	56	−4.32	18.18
=CH—	0.0129	−0.0006	46	8.73	24.96
=C<	0.0117	0.0011	38	11.14	24.14
=C=	0.0026	0.0028	36	17.78	26.15
≡CH	0.0027	−0.0008	46	−11.18	9.20
≡C—	0.0020	0.0016	37	64.32	27.38
		Ring Carbon Increments			
—CrH$_2$—	0.0100	0.0025	48	7.75	27.15
>CrH—	0.0122	0.0004	38	19.88	21.78
>Cr<	0.0042	0.0061	27	60.15	21.32
=CrH—	0.0082	0.0011	41	8.13	26.73
=Cr<	0.0143	0.0008	32	37.02	31.01

(Continued)

TABLE 3.2 Group contributions for the Joback (1984; Joback and Reid, 1987) method for T_c, P_c, V_c, T_m, and T_b. Use in Eqs. (3.2-3)–(3.2-5) for the critical properties and Eqs. (3.4-1) and (3.4-2) for T_m and T_b, respectively. Groups with atoms in rings are identified by the letter "r" following the single-letter element symbol (e.g., Cr, Or, Sr, and Nr). Do not confuse these ring designations with symbols for elements such as chromium or strontium. (*Continued*)

Group	ΔT_c	Δ_{P_c}	Δ_{V_c}	Δ_{T_m}	Δ_{T_b}
		Halogen Increments			
—F	0.0111	−0.0057	27	−15.78	−0.03
—Cl	0.0105	−0.0049	58	13.55	38.13
—Br	0.0133	0.0057	71	43.43	66.86
—I	0.0068	−0.0034	97	41.69	93.84
		Oxygen Increments			
—OH (alcohol)	0.0741	0.0112	28	44.45	92.88
a—OH (phenol)	0.0240	0.0184	−25	82.83	76.34
—O—	0.0168	0.0015	18	22.23	22.42
—Or—	0.0098	0.0048	13	23.05	31.22
>C═O (ketone)	0.0380	0.0031	62	61.20	76.75
>Cr═O (ring ketone)	0.0284	0.0028	55	75.97	94.97
—HC═O (aldehyde)	0.0379	0.0030	82	36.90	72.20
—COOH (acid)	0.0791	0.0077	89	155.50	169.09
—COO— (ester)	0.0481	0.0005	82	53.60	81.10
═O (except as above)	0.0143	0.0101	36	2.08	−10.50
		Nitrogen Increments			
—NH_2	0.0243	0.0109	38	66.89	73.23
—NH—	0.0295	0.0077	35	52.66	50.17
—NrH—	0.0130	0.0114	29	101.51	52.82
>N—	0.0169	0.0074	9	48.84	11.74
═N—	0.0255	−0.0099	—	—	74.60
═Nr—	0.0085	0.0076	34	68.40	57.55
═NH	—	—	—	68.91	83.08
—C≡N	0.0496	−0.0101	91	59.89	125.66
—NO_2	0.0437	0.0064	91	127.24	152.54
		Sulfur Increments			
—SH	0.0031	0.0084	63	20.09	63.56
—S—	0.0119	0.0049	54	34.40	68.78
—Sr—	0.0019	0.0051	38	79.93	52.10

Example 3.2 Estimate T_c, P_c, and V_c for 2-methylphenol (*o*-cresol) by using Joback's group method.

Solution. 2-methylphenol contains one —CH_3, four =CrH—, two =Cr<, and one a—OH according to the group definitions outlined in Table 3.2. Note that the method does not distinguish between a =CrH— in aromatic and non-aromatic rings.

	Group	N	$N\Delta_{T_c}$	$N\Delta_{P_c}$	$N\Delta_{V_c}$
1	—CH_3	1	0.0141	–0.0012	65
2	=CrH—	4	0.0328	0.0044	164
3	=Cr<	2	0.0286	0.0016	64
4	a—OH	1	0.0240	0.0184	–25
	$\sum_{k=1}^{4} N_k F_k$		0.0995	0.0232	268

The value of $N_{\text{atoms}} = 16$, while $T_b = 464.15$ K, so the Joback estimates of the critical constants are

$$T_c = 464.15[0.584 + 0.965(0.0995) - 0.0995^2]^{-1} = 692.64 \text{ K}$$
$$P_c = [0.113 + 0.0032(16) - 0.0232]^{-2} = 50.30 \text{ bar}$$
$$V_c = 17.5 + 268 = 285.50 \text{ cm}^3/\text{mol}$$

Appendix A values for the critical temperature, pressure, and volume are 697.55 K, 50.10 bar, and 282 cm^3/mol. Thus the differences are

$$T_c \text{ Difference} = 692.64 - 697.55 = -4.91 \text{ K or } -0.7\%$$
$$P_c \text{ Difference} = 50.30 - 50.10 = 0.2 \text{ bar or } 0.4\%$$
$$V_c \text{ Difference} = 285.50 - 282 = 3.5 \text{ cm}^3/\text{mol or } 1.2\%.$$

Method of Constantinou and Gani (CG). Constantinou and Gani (1994) developed an advanced *group contribution* method based on the UNIFAC (UNIQUAC Functional-group Activity Coefficients) groups (see Chap. 8) but they allow for more sophisticated functions of the desired properties and for contributions at a "second order" level. The functions give more flexibility to the correlation while the second order partially overcomes the limitation of UNIFAC, which cannot distinguish special configurations such as isomers, multiple groups located close together, resonance structures, etc., at the "first order." The general CG formulation of a function f of a property F (e.g., T_c, P_c, V_c) is

$$F = f\left[\sum_i N_i \left(\Delta_F^{(1)}\right)_i + W\sum_j M_j \left(\Delta_F^{(2)}\right)_j\right] \tag{3.2-6}$$

where f can be a linear or nonlinear function [see Eqs. (3.2-7)–(3.2-9)], N_i is the number of first-order groups of type i in the molecule; $\left(\Delta_F^{(1)}\right)_i$ is the contribution for the first-order group to the specified property, F; M_j is the number of second-order groups of type j in the molecule; and $\left(\Delta_F^{(2)}\right)_j$ is the contribution for the second-order group to the specified property, F. The value of W is set to zero for First-Order calculations and set to unity for second-order calculations. For the critical properties, the CG formulations are

$$\frac{T_c}{\text{K}} = 181.128 \ln\left[\sum_i N_i \left(\Delta_{T_c}^{(1)}\right)_i + W\sum_j M_j \left(\Delta_{T_c}^{(2)}\right)_j\right] \tag{3.2-7}$$

$$\frac{P_c}{\text{bar}} = \left[\sum_i N_i \left(\Delta_{P_c}^{(1)}\right)_i + W\sum_j M_j \left(\Delta_{P_c}^{(2)}\right)_j + 0.10022\right]^{-2} + 1.3705 \tag{3.2-8}$$

$$\frac{V_c}{\text{m}^3 \cdot \text{kmol}^{-1}} = \left[\sum_i N_i \left(\Delta_{V_c}^{(1)}\right)_i + W \sum_j M_j \left(\Delta_{V_c}^{(2)}\right)_j\right] - 0.004350 \qquad (3.2\text{-}9)$$

Note that T_c does not require a value for T_b. The group values for Eqs. (3.2-7)–(3.2-9) are given in Table 3.3 for the first-order groups and Table 3.4 for the second-order groups. The latter has example molecules to aid in group identification.

One difficulty with the method is that group definitions can be ambiguous. Sometimes, more than one overlapping second-order contribution can apply to a molecule, and the literature does not specify how to resolve the conflict. More explicitly, it is unknown whether all applicable groups should be used or if some should be dropped.

Example 3.3 Estimate T_c, P_c, and V_c for 2-methylphenol (*o*-cresol) by using Constantinou and Gani's group method.

Solution. The first-order groups for 2-methylphenol are one >$ACCH_3$, four >ACH, and one >ACOH. There are no second-order groups (even though the ortho proximity effect might suggest it), so the first-order and second-order calculations are the same. From Table 3.3

	Group	N	$N\Delta_{T_c}^{(1)}$	$N\Delta_{P_c}^{(1)}$	$N\Delta_{V_c}^{(1)}$
1	>$ACCH_3$	1	8.2130	0.0194	0.1036
2	>ACH	4	14.9348	0.0300	0.1688
3	>ACOH	1	25.9145	−0.0074	0.0316
	$\sum_{k=1}^{3} N_k F_k$		49.0623	0.04200	0.3040

$$T_c = 181.128 \ln[49.0623 + W(0)] = 705.15 \text{ K}$$

$$P_c = [0.042084 + W(0) + 0.10022]^{-2} + 1.3705 = 50.81 \text{ bar}$$

$$V_c = [(0.30386 + W\{0\}) - 0.004350]1000 = 299.65 \text{ cm}^3/\text{mol}$$

Appendix A values for the critical temperature, pressure, and volume are 697.55 K, 50.10 bar, and 282 cm^3/mol. Thus, the differences are

$$T_c \text{ Difference} = 705.15 - 697.55 = 7.60 \text{ K or } 1.09\%$$

$$P_c \text{ Difference} = 50.75 - 50.10 = 0.2 \text{ bar or } 1.3\%$$

$$V_c \text{ Difference} = 299.51 - 282 = 17.5 \text{ cm}^3/\text{mol or } 6.2\%.$$

Example 3.4 Estimate T_c, P_c, and V_c for the four butanols using Constantinou and Gani's group method.

Solution. The first- and second-order groups for the butanols are:

Groups/Butanol	1-butanol	2-methyl-1-propanol	2-methyl-2-propanol	2-butanol
# First-Order groups, N_i	—	—	—	—
—CH_3	1	2	3	2
>CH_2	3	1	0	1
>CH—	0	1	0	1

(Continued)

Groups/Butanol	1-butanol	2-methyl-1-propanol	2-methyl-2-propanol	2-butanol
>C<	0	0	1	0
—OH	1	1	1	1
Second-Order groups, M_j	—	—	—	—
$(CH_3)_2CH$—	0	1	0	0
$(CH_3)_3C$—	0	0	1	0
>CHOH	0	0	0	1
≥COH	0	0	1	0

Since 1-butanol has no second-order group, its calculated results are the same for both orders. Using values of group contributions from Table 3.3 and Table 3.4, and experimental values from Appendix A, the results are:

Property/Butanol	1-butanol	2-methyl-1-propanol	2-methyl-2-propanol	2-butanol
T_c, K				
Experimental	563.1	547.8	506.2	535.9
Calculated (first order)	558.91	548.06	539.37	548.06
Abs. percent Err. (first order)	0.74	0.05	6.55	2.27
Calculated (second order)	558.91	543.31	497.46	521.57
Abs. percent Err. (second order)	0.74	0.82	1.73	2.67
P_c, bar				
Experimental	44.14	42.95	39.72	41.89
Calculated (first order)	41.97	41.91	43.17	41.91
Abs. percent Err. (first order)	4.92	2.41	8.68	0.07
Calculated (second order)	41.97	41.66	42.32	44.28
Abs. percent Err. (second order)	4.92	2.99	6.55	5.72
V_c, cm^3/mol				
Experimental	273	274	275	270
Calculated (first order)	276.94	272.0	259.4	272.0
Abs. percent Err. (first order)	1.44	0.73	5.67	0.74
Calculated (second order)	276.94	276.0	280.2	264.2
Abs. percent Err. (second order)	1.44	0.73	1.90	2.14

The first-order results are generally good except for 2-methyl-2-propanol (*t*-butanol). The steric effects of its crowded methyl groups make its experimental value quite different from the others; most of this is taken into account by the first-order groups, but the second-order contribution is significant. Notice that the second-order contributions for the other species are small and may change the results in the wrong direction so that the second-order estimate can be slightly worse than the first-order estimate. This problem occurs often, but its effect is normally small including second-order effects usually helps and rarely hurts much. The performance analysis of the methods found later includes second-order contributions.

TABLE 3.3 First-order group contributions for the Constantinou and Gani's (1994) method for T_c, P_c, V_c, T_m, and T_b. Use in Eqs. (3.2-7)–(3.2-9) for the critical properties and Eqs. (3.4-3) and (3.4-4) for T_m and T_b, respectively.

Group	$\Delta_{T_c}^{(1)}$	$\Delta_{P_c}^{(1)}$	$\Delta_{V_c}^{(1)}$	$\Delta_{T_m}^{(1)}$	$\Delta_{T_b}^{(1)}$
—CH_3	1.6781	0.019904	0.07504	0.4640	0.8894
—CH_2—	3.4920	0.010558	0.05576	0.9246	0.9225
>CH—	4.0330	0.001315	0.03153	0.3557	0.6033
>C<	4.8823	−0.010404	−0.00034	1.6479	0.2878
CH_2=CH—	5.0146	0.025014	0.11648	1.6472	1.7827
—CH=CH—	7.3691	0.017865	0.09541	1.6322	1.8433
CH_2=C<	6.5081	0.022319	0.09183	1.7899	1.7117
—CH=C<	8.9582	0.012590	0.07327	2.0018	1.7957
>C=C<	11.3764	0.002044	0.07618	5.1175	1.8881
CH_2=C=CH—	9.9318	0.031270	0.14831	3.3439	3.1243
>ACH	3.7337	0.007542	0.04215	1.4669	0.9297
>AC—	14.6409	0.002136	0.03985	0.2098	1.6254
>$ACCH_3$	8.2130	0.019360	0.10364	1.8635	1.9669
>$ACCH_2$—	10.3239	0.012200	0.10099	0.4177	1.9478
>ACCH<	10.4664	0.002769	0.07120	−1.7567	1.7444
—OH	9.7292	0.005148	0.03897	3.5979	3.2152
>ACOH	25.9145	−0.007444	0.03162	13.7349	4.4014
CH_3CO—	13.2896	0.025073	0.13396	4.8776	3.5668
—CH_2CO—	14.6273	0.017841	0.11195	5.6622	3.8967
—CHO	10.1986	0.014091	0.08635	4.2927	2.8526
CH_3COO—	12.5965	0.029020	0.15890	4.0823	3.6360
—CH_2COO—	13.8116	0.021836	0.13649	3.5572	3.3953
HCOO—	11.6057	0.013797	0.10565	4.2250	3.1459
CH_3O—	6.4737	0.020440	0.08746	2.9248	2.2536
—CH_2O—	6.0723	0.015135	0.07286	2.0695	1.6249
>CH-O—	5.0663	0.009875	0.05865	4.0352	1.1557
—CH_2O— (substituted furan)	9.5059	0.009011	0.06858	−4.5047	2.5892
—CH_2NH_2	12.1726	0.012558	0.13128	6.7684	3.1656
>$CHNH_2$	10.2075	0.010694	0.07527	4.1187	2.5983
CH_3NH<	9.8544	0.012589	0.12152	4.5341	3.1376
—CH_2NH<	10.4677	0.010390	0.09956	6.0609	2.6127
>CHNH<	7.2121	−0.000462	0.09165	3.4100	1.5780
CH_3N<	7.6924	0.015874	0.12598	4.0580	2.1647
—CH_2N<	5.5172	0.004917	0.06705	0.9544	1.2171
>$ACNH_2$	28.7570	0.001120	0.06358	10.1031	5.4736
C_5H_4N (pyridine, one substitution)	29.1528	0.029565	0.24831	—	6.2800
C_5H_3N (pyridine, two substitutions)	27.9464	0.025653	0.17027	12.6275	5.9234
—CH_2CN	20.3781	0.036133	0.15831	4.1859	5.0525

(*Continued*)

TABLE 3.3 First-order group contributions for the Constantinou and Gani's (1994) method for T_c, P_c, V_c, T_m, and T_b. Use in Eqs. (3.2-7)–(3.2-9) for the critical properties and Eqs. (3.4-3) and (3.4-4) for T_m and T_b, respectively. ***(Continued)***

Group	$\Delta_{T_c}^{(1)}$	$\Delta_{P_c}^{(1)}$	$\Delta_{V_c}^{(1)}$	$\Delta_{T_m}^{(1)}$	$\Delta_{T_b}^{(1)}$
—COOH	23.7593	0.011507	0.10188	11.5630	5.8337
>CH_2Cl	11.0752	0.019789	0.11564	3.3376	2.9637
>CHCl	10.8632	0.011360	0.10350	2.9933	2.6948
>CCl—	11.3959	0.003086	0.07922	9.8409	2.2073
—$CHCl_2$	16.3945	0.026808	0.16951	5.1638	3.9300
—CCl_3	—	—	—	—	3.5600
>CCl_2	18.5875	0.034935	0.21031	10.2337	4.5797
>ACCl	14.1565	0.013135	0.10158	2.7336	2.6293
—CH_2NO_2	24.7369	0.020974	0.16531	5.5424	5.7619
>$CHNO_2$	23.2050	0.012241	0.14227	4.9738	5.0767
>$ACNO_2$	34.5870	0.015050	0.14258	8.4724	6.0837
—CH_2SH	13.8058	0.013572	0.10252	3.0044	3.2914
—I	17.3947	0.002753	0.10814	4.6089	3.6650
—Br	10.5371	−0.001771	0.08281	3.7442	2.6495
CH≡C—	7.5433	0.014827	0.09331	3.9106	2.3678
—C≡C—	11.4501	0.004115	0.07627	9.5793	2.5645
—C(Cl)═C<	5.4334	0.016004	0.05687	1.5598	1.7824
>ACF	2.8977	0.013027	0.05672	2.5015	0.9442
—$HCON(CH_2)_2^-$	—	—	—	—	7.2644
—CF_3	2.4778	0.044232	0.11480	3.2411	1.2880
>CF_2	1.7399	0.012884	0.09519	—	0.6115
≥CF	3.5192	0.004673	—	—	1.1739
—COO—	12.1084	0.011294	0.08588	3.4448	2.6446
—CCl_2F	9.8408	0.035446	0.18212	7.4756	2.8881
—HCClF	—	—	—	—	2.3086
—$CClF_2$	4.8923	0.039004	0.14753	2.7523	1.9163
—F (except as above)	1.5974	0.014434	0.03783	1.9623	1.0081
—$CONH_2$	65.1053	0.004266	0.14431	31.2786	10.3428
—$CON(CH_3)_2$	36.1403	0.040149	0.25031	11.3770	7.6904
—$CON(CH_2-)_2$	—	—	—	—	6.7822
—$C_2H_5O_2$	17.9668	0.025435	0.16754	—	5.5566
—$C_2H_4O_2$–	—	—	—	—	5.4248
CH_3S—	14.3969	0.016048	0.13021	5.0506	3.6796
—CH_2S—	17.7916	0.011105	0.11650	3.1468	3.6763
>CHS—	—	—	—	—	2.6812
C_4H_3S (thiophene, one substitution)	—	—	—	—	5.7093
C_4H_2S (thiophene, two substitutions)	—	—	—	—	5.8260

TABLE 3.4 Second-order group contributions for the Constantinou and Gani's (1994) method for T_c, P_c, V_c, T_m, and T_b. Use in Eqs. (3.2-7)–(3.2-9) for the critical properties and Eqs. (3.4-3) and (3.4-4) for T_m and T_b, respectively.

Group	$\Delta_{T_c}^{(2)}$	$\Delta_{P_c}^{(2)}$	$\Delta_{V_c}^{(2)}$	$\Delta_{T_m}^{(2)}$	$\Delta_{T_b}^{(2)}$	Example (# of Groups)
$(CH_3)_2CH$—	−0.5334	0.000488	0.00400	0.0381	−0.1157	2-methylpentane (1)
$(CH_3)_3C$—	−0.5143	0.001410	0.00572	−0.2355	−0.0489	2,2,4,4-tetramethylpentane (2)
—$CH(CH_3)CH(CH_3)$—	1.0699	−0.001849	−0.00398	0.4401	0.1798	2,2,3,4-tetramethylpentane (1)
—$CH(CH_3)C(CH_3)_2$—	1.9886	−0.005198	−0.01081	−0.4923	0.3189	2,2,3,4,4-pentamethylpentane (2)
—$C(CH_3)_2C(CH_3)_2$—	5.8254	−0.013230	−0.02300	6.0650	0.7273	2,2,3,3,4,4-hexamethylpentane (2)
3-membered ring	−2.3305	0.003714	−0.00014	1.3772	0.4745	cyclopropane (1)
4-membered ring	−1.2978	0.001171	−0.00851	—	0.3563	cyclobutane (1)
5-membered ring	−0.6785	0.000424	−0.00866	0.6824	0.1919	cyclopentane (1)
6-membered ring	0.8479	0.002257	0.01636	1.5656	0.1957	cyclohexane (1)
7-membered ring	3.6714	−0.009799	−0.02700	6.9709	0.3489	cycloheptane (1)
$CH_n{=}CH_m{-}CH_p{=}CH_k$ $m, p \in (0,1)$, $k, n \in (0,1,2)$	0.4402	0.004186	−0.00781	1.9913	0.1589	1,3-butadiene (1)
$CH_3{-}CH_m{=}CH_n$ $m \in (0,1)$, $n \in (0,1,2)$	0.0167	−0.000183	−0.00098	0.2476	0.0668	2-methyl-2-butene (3)
$-CH_2{-}CH_m{=}CH_n$ $m \in (0,1)$, $n \in (0,1,2)$	−0.5231	0.003538	0.00281	−0.5870	−0.1406	1,4-pentadiene (2)
$>CH{-}CH_m{=}CH_n$ or $\geq C{-}CH_m{=}CH_n$ $m \in (0,1)$, $n \in (0,1,2)$	−0.3850	0.005675	0.00826	−0.2361	−0.0900	4-methyl-2-pentene (1)
Alicyclic side-chain $C_{cyclic}C_m$ $m > 1$	2.1160	−0.002546	−0.01755	−2.8298	0.0511	propylcycloheptane (1)
CH_3CH_3	2.0427	0.005175	0.00227	1.4880	0.6884	ethane (only)
>CHCHO or ≥CCHO	−1.5826	0.003659	−0.00664	2.0547	−0.1074	2-methylbutanal (1)
CH_3COCH_2—	0.2996	0.001474	−0.00510	−0.2951	0.0224	2-pentanone (1)
CH_3COCH< or CH_3COC≤	0.5018	−0.002300	−0.00122	−0.2986	0.0920	3-methyl-2-pentanone (1)
$>C_{cyclic}{=}O$	2.9571	0.003818	−0.01966	0.7143	0.5580	cyclohexanone (1)
>ACCHO	1.1696	−0.002481	0.00664	−0.6697	0.0735	benzaldehyde (1)
>CHCOOH or ≥CCOOH	−1.7493	0.004920	0.00559	−3.1034	−0.1552	2-methylbutanoic acid (1)
>ACCOOH	6.1279	0.000344	−0.00415	28.4324	0.7801	benzoic acid (1)
CH_3COOCH< or CH_3COOC≤	−1.3406	0.000659	−0.00293	0.4838	−0.2383	1-methylethyl ethanoate (1)
$-COCH_2COO-$ or —COC(—)HCOO— (monosubstitution on middle C) or —COC(<)COO— (disubstitution on middle C)	2.5413	0.001067	−0.00591	0.0127	0.4456	ethyl acetoethanoate (1)
—CO—O—CO—	−2.7617	−0.004877	−0.00144	−2.3598	−0.1977	acetic anhydride (1)
>ACCOO—	−3.4235	−0.000541	0.02605	−2.0198	0.0835	ethyl benzoate (1)

(Continued)

TABLE 3.4 Second-order group contributions for the Constantinou and Gani's (1994) method for T_c, P_c, V_c, T_m, and T_b. Use in Eqs. (3.2-7)–(3.2-9) for the critical properties and Eqs. (3.4-3) and (3.4-4) for T_m and T_b, respectively. *(Continued)*

Group	$\Delta_{T_c}^{(2)}$	$\Delta_{P_c}^{(2)}$	$\Delta_{V_c}^{(2)}$	$\Delta_{T_m}^{(2)}$	$\Delta_{T_b}^{(2)}$	Example (# of Groups)
>CHOH	−2.8035	−0.004393	−0.00777	−0.5480	−0.5385	2-butanol (1)
≥COH	−3.5442	0.000178	0.01511	0.3189	−0.6331	2-methyl-2-butanol (1)
$CH_m(OH)CH_n(OH)$ $m, n \in (0,1,2)$	5.4941	0.005052	0.00397	0.9124	1.4108	1,2,3-propanetriol (2)
$C_{cyclic}H_mOH$ $m \in (0,1)$	0.3233	0.006917	−0.02297	9.5209	−0.0690	cyclopentanol (1)
$CH_n(OH)CH_m(NH_p)$ $m \in (0,1)\ n, p \in (0,1,2)$	5.4864	0.001408	0.00433	2.7826	1.0682	2-amino-1-butanol (1)
$CH_m(NH_2)CH_n(NH_2)$ $m, n \in (0,1,2)$	2.0699	0.002148	0.00580	2.5114	0.4247	1,2-diaminopropane (1)
$C_{cyclic}H_m$—NH_p—$C_{cyclic}H_n$ $m, n \in (0,1,2), p \in (0,1)$	2.1345	−0.005947	−0.01380	1.0729	0.2499	pyrrolidine (1)
CH_n—O—CH_m=CH_p $m \in (0,1), n, p \in (0,1,2)$	1.0159	−0.000878	0.00297	0.2476	0.1134	ethyl vinyl ether (1)
>AC—O—CH_m $m \in (0,1,2,3)$	−5.3307	−0.002249	−0.00045	0.1175	−0.2596	ethyl phenyl ether (1)
$C_{cyclic}H_m$—S—$C_{cyclic}H_n$ $m, n \in (0,1,2)$	4.4847	—	—	−0.2914	0.4408	tetrahydrothiophene (1)
CH_n=CH_m—F $m \in (0,1), n \in (0,1,2)$	−0.4996	0.000319	−0.00596	−0.0514	−0.1168	1-fluoro-1-propene (1)
CH_n=CH_m—Br $m \in (0,1), n \in (0,1,2)$	−1.9334	−0.004305	0.00507	−1.6425	−0.3201	1-bromo-1-propene (1)
CH_n=CH_m—I $m \in (0,1), n \in (0,1,2)$	—	—	—	—	−0.4453	1-iodo-1-propene (1)
>ACBr	−2.2974	0.009027	−0.00832	2.5832	−0.6776	bromobenzene (1)
>ACI	2.8907	0.008247	−0.00341	−1.5511	−0.3678	iodobenzene (1)

Note: C_{cyclic} means carbons in nonaromatic rings.

Method of Nannoolal et al. To improve on previous methods, Nannoolal et al. (2007) expanded the number of groups to account for attached neighbors to a greater extent than was done before. They also included "second order" groups and "group interaction" effects. The equations for the critical properties are

$$\frac{T_c}{T_b} = \left(0.6990 + \frac{1}{0.9889 + \left(\sum_i N_i(\Delta_{T_c})_i + GI_{T_c}\right)^{0.8607}}\right) \tag{3.2-10}$$

$$\frac{P_c}{\text{kPa}} = \frac{\left(\dfrac{MW}{\text{g mol}^{-1}}\right)^{-0.14041}}{\left(0.00939 + \sum_i N_i(\Delta_{P_c})_i + GI_{P_c}\right)^2} \tag{3.2-11}$$

$$\frac{V_c}{\text{cm}^3\ \text{mol}^{-1}} = \frac{\sum_i N_i(\Delta_{V_c})_i + GI_{V_c}}{n^{-0.2266}} + 86.1539 \tag{3.2-12}$$

where the summations are over the unique groups in the molecule, n is the total numbers of *non-hydrogen* atoms in the compound, the group values $(\Delta_{T_c})_i$, $(\Delta_{P_c})_i$, and $(\Delta_{V_c})_i$ are found in Table 3.5 and Table 3.6, and GI_p is the *total group interaction* contribution for property p—either T_c, P_c, or V_c. GI_p is used to improve the prediction of compounds with multiple functional groups and is calculated according to

$$GI_p = \frac{1}{n}\sum_{i=1}^{m}\sum_{j=1}^{m}\frac{(\Delta_{GI_{ij}})_p}{m-1} \tag{3.2-13}$$

Here, Δ_{GIij} is the group interaction contribution between groups i and j as defined in Table 3.7 and m is the total number of interaction groups in the molecule. Also note that $\Delta_{GI_{ij}} = \Delta_{GI_{ji}}$ and $\Delta_{GI_{ii}} = \Delta_{GI_{jj}} = 0$. The authors explain that the GI_p contribution is needed to fix the assumption that groups are additive, a supposition that breaks down for groups which associate (e.g., hydrogen bond). Example 3.5 shows an application of the Nannoolal et al. method with first- and second-order groups (Table 3.5) and correction groups (Table 3.6). Example 3.6 demonstrates how to calculate GI_p using Eq. (3.2-13) and Table 3.7.

Example 3.5 Estimate T_c, P_c, and V_c for 2-methylphenol (*o*-cresol) using the method of Nannoolal et al. (2007).

Solution. The first-order groups of 2-methylphenol are one CH_3—(a), four =C(a)H—, one =C(a)<(e), one =C(a)<(ne), and one HO—(C(a)) according to Table 3.5. The molecule also has one Ortho pair(s) as found in Table 3.6. There are no interacting groups in the molecule according to Table 3.7, so $GI_p = 0$. The resulting group summations are

	Group	N	$N\Delta_{T_c}10^3$	$N\Delta_{P_c}10^4$	$N\Delta_{V_c}$
1	CH_3—(a)	1	−1.071	4.166	26.7237
2	=C(a)H—	4	64.4616	8.4256	77.608
3	=C(a)<(e)	1	68.1923	3.55	5.6704
4	=C(a)<(ne)	1	68.2045	4.1826	25.0434
5	HO—(C(a))	1	14.0159	−12.1664	25.6584
6	Ortho pair(s)	1	1.2823	0.7061	−3.5964
	$\sum_{k=1}^{6} N_k F_k$		215.0856	8.8639	157.1075

The number of non-hydrogen atoms is $n = 8$. With $MW = 108.138$ g/mol and $T_b = 464.15$ K, the Nannoolal estimates of the critical constants are

$$T_c = 464.15[0.6990 + (0.9889 + \{215.0856/1000\}^{0.8607})^{-1}] = 694.19 \text{ K}$$
$$P_c = 108.138^{-0.14041}(0.00939 + 8.8639/10000)^{-2} = 4906 \text{ kPa} = 49.06 \text{ bar}$$
$$V_c = 157.1075/8^{-0.2266} + 86.1539 = 337.83 \text{ cm}^3 \text{ mol}^{-1}$$

Appendix A values for the critical temperature, pressure, and volume are 697.55 K, 50.10 bar, and 282 cm^3/mol. Thus, the differences are

$$T_c \text{ Difference} = 694.19 - 697.55 = -3.36 \text{ K or } -0.48\%$$
$$P_c \text{ Difference} = 49.06 - 50.10 = -1.04 \text{ bar or } -2.08\%$$
$$V_c \text{ Difference} = 337.83 - 282 = 55.83 \text{ cm}^3\text{/mol or } 19.8\%.$$

TABLE 3.5 Primary and secondary group contributions for the Nannoolal et al. (2007) method for T_c, P_c, and V_c to use in Eqs. (3.2-10)–(3.2-12). The ID column is a numerical identifier for the group. The Pr column is for "Priority." A group with a lower priority number should be used over a group with higher priority number when two may be chosen. The Example(s) column contains representative molecule(s) where the group may be found.

Group	Description	ID	Pr	$10^3\Delta_{T_c}$	$10^4\Delta_{P_c}$	Δ_{V_c}	Example (# of Groups)
		Carbon Groups					
CH_3—(ne)	CH_3— connected to atoms other than N, O, F, or Cl	1	105	41.8682	8.1620	28.7855	decane (2)
CH_3—(e)	CH_3— connected to either N, O, F, or Cl	2	103	33.1371	5.5262	28.8811	dimethoxymethane (2)
CH_3—(a)	CH_3— connected to an aromatic atom (not necessarily C)	3	104	−1.071	4.1660	26.7237	toluene (1)
—C(c)H_2—	—CH_2— in a chain	4	112	40.0977	5.2623	32.0493	butane (2)
>C(c)H—	>CH— in a chain	5	119	30.2069	2.3009	32.1108	2-methylpentane (1)
>C(c)<	>C< in a chain	6	121	−3.8778	−2.9925	28.0534	neopentane (1)
>C(c)<(e)	>C< in a chain connected to one or more F, Cl, N, or O; other bonds may be to any atoms	7	108	52.8003	3.4310	33.7577	ethanol (1)
>C(c)<(a)	>C< in a chain connected to at least one aromatic carbon; other bonds may be to any atoms	8	109	9.4422	2.3665	28.8792	ethylbenzene (1)
—C(r)H_2—	—CH_2— in a ring; connected to atoms other than N, O, F, or Cl	9	113	21.2898	3.4027	24.8517	cyclopentane (5)
>C(r)H—	>CH— in a ring; connected to atoms other than N, O, F, or Cl	10	118	26.3513	3.6162	30.9323	methylcyclohexane (1)
>C(r)<	>C< in a ring; connected to atoms other than N, O, F, or Cl	11	120	−17.0459	−5.1299	5.9550	beta-pinene (1)
>C(r)<(e,c)	>C< in a ring connected to one or more N or O which are not part of the ring, or one Cl or F; other bonds are not to N, O, Cl, or F	12	110	51.7974	4.1421	29.5901	cyclopentanol (1)
>C(r)<(e,r)	>C< in a ring connected to one or more N or O which are part of the ring; other bonds are not to N, O, Cl, or F	13	111	18.9549	0.8765	20.2325	morpholine (4)

>C(r)<(C(a))	>C< in a ring connected to one or more aromatic carbon; other bonds are not to N, O, Cl, or F	14	107	−29.1568	−0.1320	10.5669	2-methyltetraline (2)
=C(a)H—	Aromatic =CH—	15	106	16.1154	2.1064	19.4020	benzene (6)
=C(a)<(ne)	Aromatic C not connected to either O, N, Cl, or F	16	117	68.2045	4.1826	25.0434	ethylbenzene (1)
=C(a)<(e)	Aromatic C connected to either O, N, Cl, or F	17	114	68.1923	3.5500	5.6704	phenol (1)
(a)=C(a)<2(a)	Aromatic C with three aromatic neighbors and three aromatic bonds	18	115	29.8039	1.0997	16.4118	naphthalene (2)
C(a)=C(a)<C_2(a) (bridge)	Aromatic C with three aromatic neighbors and two aromatic bonds (aliphatic bridge bond between aromatic rings)	214	116	48.168	1.5574	16.3122	*m*-terphenyl (4)
>C(c)=C(c)<	>C=C< (both C have at least one non-H neighbor)	58	62	45.1531	7.1581	—	2-heptene (1)
—(e)>C(c)=C(c)<	Non-cyclic >C=C< substituted with at least one F, Cl, N, or O	60	58	67.9821	−6.2791	51.071	*trans*-1,2-dichloroethylene (1)
H_2C(c)=C<	H_2C=C< (1-ene)	61	57	45.4406	9.6413	48.1957	1-hexene (1)
>C(r)=C(r)<	Cyclic >C=C<	62	60	56.4059	3.4731	34.1240	cyclopentene (1)
>C=C=C<	Two cumulated double bonds	87	5	53.6350	12.6128	—	1,2-butadiene (1)
>C=C—C=C<	Two conjugated double bonds in a ring	88	6	24.7302	−10.2451	64.4616	cyclopentadiene (1)
—C≡C—	—C≡C— (-yne-) with two non-H neighbors	63	61	−19.9737	−2.2718	40.9263	2-octyne (1)
HC≡C—	HC≡C— (1-yne) with one non-H neighbor	64	56	36.0883	2.4489	29.8612	1-heptyne (1)
		Halogen Groups					
F—(C,Si)	F— connected to nonaromatic C or Si not already substituted with F or Cl	19	87	15.6068	0.7328	−5.0331	2-fluoropropane (1)

(*Continued*)

TABLE 3.5 Primary and secondary group contributions for the Nannoolal et al. (2007) method for T_c, P_c, and V_c to use in Eqs. (3.2-10) to (3.2-12). The ID column is a numerical identifier for the group. The Pr column is for "Priority." A group with a lower priority number should be used over a group with higher priority number when two may be chosen. The Example(s) column contains representative molecule(s) where the group may be found. (*Continued*)

Group	Description	ID	Pr	$10^3\Delta_{T_c}$	$10^4\Delta_{P_c}$	Δ_{V_c}	Example (# of Groups)
F—(C(—)=C)<	F— on a nonaromatic C=C; used once for each F	20	85	11.0757	4.3757	1.5646	trifluoroethylene (3)
F—((C,Si)—([F]))—b	F— connected to C or Si already substituted with one F and two other non-halogen atoms (the C is fully substituted with no H's); used once for each F	21	81	18.1302	3.4933	3.3646	2,2,3,3-tetrafluorobutane (4)
F—((C,Si)—([Cl]))—b	F— connected to C or Si already substituted with at least one Cl and two other non-F atoms (the C is fully substituted with no H's)	102	82	1.3231	3.3971	1.3597	1,1-dichloro-1-fluoroethane (1)
F—((C,Si)—([F,Cl]))—a	F— connected to C or Si already substituted with at least one F or Cl and one other non-halogen atom (the C is trisubstituted with one H remaining)	22	84	19.1772	2.6558	1.0897	1-chloro-1,2,2,2-tetrafluoroethane (1)
F—((C,Si)—([F,Cl]$_2$))	F— connected to C or Si already substituted with two F or Cl; used once for each F	23	83	20.8519	1.6547	1.1084	1,1,1-trifluoroethane (3)
F—(C(a))—	F— connected to an aromatic C	24	86	−24.0220	0.5236	19.3190	fluorobenzene (1)
Cl—(C,Si)	Cl— connected to C or Si not already substituted with F or Cl	25	71	−1.3329	−2.2611	22.0457	1-chlorobutane (1)
Cl—((C,Si)—([F,Cl]))	Cl— connected to C or Si already substituted with one F or Cl; used once for each Cl	26	70	2.6113	−1.4992	23.9279	dichloroacetic acid (2)
Cl—((C,Si)—([F,Cl]$_2$))	Cl— connected to C or Si already substituted with at least two F or Cl; used once for each Cl	27	68	15.5010	0.4883	26.2582	1,1,1-trichloroethane (3)
Cl—(C(a))	Cl— connected to an aromatic C	28	72	−16.1905	−0.9280	36.7624	chlorobenzene (1)

Cl—(C(—)═C)<	Cl— on a C═C; used once for each Cl	29	69	60.1907	11.8687	34.4110	trichloroethylene (3)
Br—(C,Si(na))	Br— connected to a non-aromatic C or Si	30	65	5.2621	−4.317	36.0223	ethyl bromide (1)
Br—(C(a))	Br— connected to an aromatic C or Si	31	66	−21.5199	−2.2409	30.7004	bromobenzene (1)
I—(C,Si)	I— connected to C or Si	32	63	−8.6881	−4.7841	48.2989	ethyl iodide (1)
		Oxygen Groups					
—OH tert	—OH connected to C which has four non-H neighbors (tertiary alkanols)	33	91	84.8567	−7.4244	10.6790	*tert*-butanol (1)
HO—((C,Si)$_2$H—(C,Si)—(C,Si)—)	—OH connected to a C or Si substituted with two C or Si in at least three C or Si containing chain (secondary alkanols)	34	90	79.3047	−4.4735	5.6645	2-butanol (1)
—OH (>C_4)	—OH connected to C or Si substituted with one C or Si in an at least five C or Si containing chain (primary alkanols)	35	88	49.5968	−1.8153	2.0869	1-nonanol (1)
—OH (<C_5)	—OH for aliphatic chains with less than five C (cannot be connected to aromatic groups)	36	92	130.132	−6.8991	3.7778	ethanol (1)
—OH (C(a))	—OH connected to an aromatic C (phenols)	37	89	14.0159	−12.1664	25.6584	phenol (1)
(C,Si)—O—(C,Si)	—O— connected to two neighbors which are each either C or Si (ethers)	38	94	12.5082	2.0592	11.6284	diethyl ether (1)
>(OC_2)<	>(OC_2)< (epoxide)	39	50	41.3490	0.1759	46.7680	propylene oxide (1)
(C(a))—O(a)—(C(a))	—O— in an aromatic ring with aromatic C as neighbors	65	93	10.4146	−0.5403	4.7476	furan (1)
CHO—(C(na))	H(C═O)— connected to non-aromatic C (aldehydes)	52	53	44.2000	−2.3615	25.5034	acetaldehyde (1)
CHO—(C(a))	H(C═O)— connected to aromatic C (aldehydes)	90	52	38.4681	−4.0133	20.0440	benzaldehyde (1)

(Continued)

TABLE 3.5 Primary and secondary group contributions for the Nannoolal et al. (2007) method for T_c, P_c, and V_c to use in Eqs. (3.2-10)–(3.2-12). The ID column is a numerical identifier for the group. The Pr column is for "Priority." A group with a lower priority number should be used over a group with higher priority number when two may be chosen. The Example(s) column contains representative molecule(s) where the group may be found. (*Continued*)

Group	Description	ID	Pr	$10^3\Delta_{T_c}$	$10^4\Delta_{P_c}$	Δ_{V_c}	Example (# of Groups)
COOH—(C)	—(C=O)OH connected to C (carboxylic acid)	44	23	199.9042	3.9873	40.3909	acetic acid (1)
(C)—COO—(C)	—(C=O)O— connected to two C (ester) in a chain	45	24	75.7089	4.3592	42.6733	ethyl acetate (1)
HCOO—(C)	H(C=O)O— connected to C (formic acid ester)	46	26	58.0782	1.0266	36.1286	ethyl formate (1)
—C(r)OO—	—(C=O)O— in a ring, C is connected to C (lactone)	47	25	109.1930	0.4329	—	ε-caprolactone (1)
O=C<(C(na))$_2$	—(C=O)— connected to two non-aromatic C (ketones)	51	4	56.1572	0.1190	30.9229	acetone (1)
(O=C<(C)$_2$)(a)	—(C=O)— connected to two C with at least one aromatic C	92	54	63.6504	−5.0403	28.7127	acetophenone (1)
O=C(—O—)$_2$	Non-cyclic carbonate diester	79	14	97.2830	0.2822	52.8789	dimethyl carbonate (1)
—O—(C=O)—O—	C=O connected to two O (carbonates)	103	33	764.9595	58.9190	—	propylene carbonate (1)
		Nitrogen Groups					
NH_2—(C,Si)	NH_2— connected to either C or Si (primary amine)	40	96	18.3404	−4.4164	13.2571	1-hexylamine (1)
NH_2—(C(a))	NH_2— connected to an aromatic C (aromatic primary amine)	41	95	−50.6419	−9.0065	73.7444	aniline (1)
(C,Si)—NH—(C,Si)	—NH— connected to two C or Si neighbors (secondary amine)	42	100	17.1780	−0.4086	20.5722	diethylamine (1)
(C,Si)(r)—NH—(C,Si)(r)	—NH— connected to two C or Si neighbors in a ring (cyclic secondary amine)	97	99	27.3441	−4.3834	29.3068	pyrrolidine (1)
(C,Si)$_2$>N—(C,Si)	>N— connected to three C or Si neighbors (tertiary amine); neighbors can be aromatic or nonaromatic	43	101	−0.5820	2.3625	6.0178	N,N-dimethylaniline (1)

=N(a)—(r5)	Aromatic —N— in a five-membered ring, free electron pair	66	98	18.9903	8.3052	−25.3680	pyrrole (1)
=N(a)—(r6)	Aromatic =N— in a six-membered ring	67	97	10.9495	−4.7101	23.6094	pyridine (1)
N≡C—(C)	—C≡N (cyano-group) connected to C (cyanide)	57	55	117.1330	5.1666	43.7983	acetonitrile (1)
—ON=(C,Si)	—ON= connected to C or Si (isoazole)	115	45	36.0361	−5.1116	16.2688	5-phenylisoxazole (1)
NO_2—(C)	NO_2— connected to aliphatic C	68	20	82.6239	−5.0929	34.8472	1-nitropropane (1)
OCN—	O=C=N— connected to C or Si (isocyanate)	80	28	153.7225	—	27.1026	butyl isocyanate (1)
—CON<	—(C=O)N< (disubstituted amide)	48	49	102.1024	0.5172	64.3506	N,N-dimethylformamide (1)
		Silicon Groups					
>Si<	>Si<	70	80	25.4209	5.727	75.7193	butylsilane (1)
>Si<$(C,H)_3$	>Si< attached to three carbons or hydrogens	71	79	72.5587	2.7602	69.5645	hexamethyldisiloxane (2)
>Si<$(C,H)_2$	>Si< attached to two carbons or hydrogens	93	78	34.2058	3.2023	55.3822	dichlorodimethylsilane (1)
>Si<$(C,H)_1$	>Si< attached to one carbon or hydrogen	215	77	0.2842	3.8751	37.0423	trichlorosilane (1)
>Si<$(C,H)_0$	>Si< attached to no carbons or hydrogens	216	76	−0.6536	4.4882	55.7432	tetrachlorosilane (1)
Sulfur Groups							
SH—(C)	—SH connected to C (thiols, mercaptanes)	53	73	−7.1070	−9.4154	34.7699	1-propanethiol (1)
(C)—S—(C)	—S— connected to two C (thioether)	54	74	0.5887	−8.2595	38.0185	methyl ethyl sulfide (1)
—S(a)—	—S— in an aromatic ring (aromatic thioether)	56	75	−7.7181	−4.9259	20.3127	thiophene (1)
(C)—SO_2—(C)	Non-cyclic sulfone connected to two C (sulfone)	82	17	90.9726	−23.9221	68.0701	divinylsulfone (1)
		Other Groups					
$(C)_2$>Sn<$(C)_2$	>Sn< connected to four carbons	83	64	62.3642	0.7043	—	tetramethylstannane (1)
B(O—$)_3$	Non-cyclic boric acid ester	78	15	157.3401	12.6786	—	triethyl borate (1)

TABLE 3.6 Corrections for the Nannoolal et al. (2007) method for T_c, P_c, and V_c to use in Eqs. (3.2-10)–(3.2-12). These corrections are added into the summation of the groups found in Table 3.5; they do not replace the groups. See Example 3.5.

Name	Description	ID	$10^3\Delta_{T_c}$	$10^4\Delta_{P_c}$	Δ_{V_c}	Example (# of Corrections)
—C=C—C=O—	—C=O connected to sp² carbon	134	−35.6113	1.0934	2.8889	benzaldehyde (1)
(C=O)—C([F,Cl]$_{2,3}$)	Carbonyl connected to C with two or more halogens	119	32.1829	7.3149	−3.8033	dichloroacetyl chloride (1)
(C=O)—(C([F,Cl]$_{2,3}$)$_2$	Carbonyl connected to two C with two or more halogens each	120	11.4437	4.1439	27.5326	1,1,1,3,3,3-hexafluoro-2-propanone (1)
C—[F,Cl]$_3$	C with three halogens	121	−1.3023	0.4387	1.5807	α,α,α-trifluorotoluene (1)
(C)$_2$—C—[F,Cl]$_2$	Secondary C with two halogens	122	−34.3037	−4.2678	−2.6235	2,2-dichloropropane (1)
no hydrogen	Component has no H	123	−1.3798	4.8944	−5.3091	perfluoro compounds (1)
one hydrogen	Component has one H	124	−2.7180	2.8103	−6.1909	1H-perfluorobutane (1)
3/4 ring	A three or four-membered non-aromatic ring	125	11.3251	−0.3035	3.2219	cyclobutene (1)
five ring	A five-membered non-aromatic ring	126	−4.7516	0.0930	−6.3900	cyclopentane (1)
Ortho	Ortho position—counted only once and only if there are no meta or para pairs	127	1.2823	0.7061	−3.5964	*o*-xylene (1)
Meta	Meta position—counted only once and only if there are no para or ortho pairs	128	6.7099	−0.7246	1.5196	*m*-xylene (1)
((C=)(C)C—CC$_3$)	C—C bond with four single bonded and one double bonded C neighbor	130	−33.8201	−8.8457	−4.6483	*tert*-butylbenzene (1)
C$_2$C—CC$_2$	C—C bond with four C neighbors, two on each side	131	−18.4815	−2.2542	−5.0563	bicyclohexyl (1)
C$_3$C—CC$_2$	C—C bond with five C neighbors	132	−23.6024	−3.246	−6.3267	2,2,3-trimethylbutane (1)
C$_3$C—CC$_3$	C—C bond with six C neighbors	133	−24.5802	−5.3113	4.9392	2,2,3,3-tetramethylbutane (1)
≥Si—(F,Cl,Br,I)	Si attached to a halogen atom	217	62.0286	8.6126	19.4348	trichloroethylsilane (1)

TABLE 3.7 Group interaction contributions for the Nannoolal et al. (2007) method for T_c, P_c, and V_c to use in Eq. (3.2-13). See Example 3.6.

Group ID	Interacting Groups (i—j)	$10^3\Delta_{GIijT_c}$	$10^4\Delta_{GIijP_c}$	Δ_{GIijV_c}
135	OH—OH	–434.8568	–5.6023	—
136	OH—NH_2	120.9166	69.8200	—
137	OH—NH	–30.4354	6.1331	–8.0423
140	OH—EtherO	–146.7881	7.3373	19.7707
148	OH(a)—OH(a)	144.4697	57.8350	97.5425
157	NH_2—NH_2	–60.9217	–0.6754	—
159	NH_2—EtherO	–738.0515	–125.5983	—
165	NH—NH	–49.7641	22.1871	–57.1233
176	OCN—OCN	–1866.0970	—	44.1062
178	EtherO—EtherO	162.6878	2.6751	–23.6366
179	EtherO—Epox	707.4116	88.8752	–329.5074
180	EtherO—Ester	128.2740	–1.0295	–55.5112
183	EtherO—Teth	–654.1363	25.8246	–37.2468
185	EtherO—CN	741.8565	—	—
189	Ester—Ester	366.2663	0.5195	–74.8680
194	Ketone—Ketone	1605.5640	–78.2743	–413.3976
205	Teth—Teth	–861.1528	43.9001	–403.1196
208	AtS—AN_5	131.7924	–19.7033	164.293
210	AO—AN_5	24.0243	–35.1998	217.9243
212	AN_6—AN_6	–32.3208	12.5371	–26.4556
218	COOH—NH_2	Do not estimate	Do not estimate	Do not estimate

Abbreviation for Group *i* from Above	Group Description	Group ID(s) from Table 3.5
OH	Alkanol (—OH)	34, 35, 36
OH(a)	Phenol (—OH (C(a)))	37
COOH	Carboxylic Acid (—COOH)	44
EtherO	Ether (—O—)	38
Epox	Epoxide (>(OC_2)<)	39
Ester	Ester (—OOC—)	45, 46, 47
Ketone	Ketone (—CO—)	51, 92
Alde	Aldehyde (—CHO)	52, 90
AO	Aromatic oxygen (—O(a)—)	65
Teth	Sulfide (thioester) (—S(na)—)	54
Ats	Aromatic sulfur (—S(a)—)	56
NH_2	Primary amine (—NH_2)	40, 41
NH	Secondary amine (>NH)	42, 97
OCN	Isocyanate (—OCN)	80
CN	Cyanide (—CN)	57
AN_5	Aromatic N in 5-ring (=N(a)—(r5))	66
AN_6	Aromatic N in 6-ring (=N(a)—(r6))	67

Example 3.6 Estimate T_c, P_c, and V_c for 1,2-dimethoxyethane using the method of Nannoolal et al.

Solution. The first-order groups of 1,2-dimethoxyethane are two CH3—(e), two >C(c)<(e), and two (C,Si)—O—(C,Si) according to Table 3.5. The molecule has no second-order group or corrections. The resulting group summations are

	Group	N	$N\Delta_{T_c}10^3$	$N\Delta_{P_c}10^4$	$N\Delta_{V_c}$
1	CH_3—(e)	2	66.2742	11.0524	57.7622
2	>C(c)<(e),	2	105.6006	6.862	67.5154
3	(C,Si)—O—(C,Si)	2	25.0164	4.1184	23.2568
	$\sum_{k=1}^{3} N_k F_k$		196.8912	22.0328	148.5344

The two interacting groups are both EtherO (arbitrarily called EtherO 1 and EtherO 2). Determination of GI_p from Eq. (3.2-13) requires setting up three matrices as outlined below. The entries of the matrix are the $\Delta_{GI_{ij_p}}$ values for property p, as indicated in the upper left corner, needed for Eq. (3.2-13). The values are found in Table 3.7. Notice that the ij combinations cover 11, 12, 21, and 22 because there are two interacting groups.

$10^3\Delta_{GI_{ijT_c}}$	$j \rightarrow$	1	2
$i \downarrow$	INTERACTING GROUPS	EtherO 1	EtherO 2
1	EtherO 1	0	162.6878
2	EtherO 2	162.6878	0

$10^4\Delta_{GI_{ijP_c}}$	$j \rightarrow$	1	2
$i \downarrow$	INTERACTING GROUPS	EtherO 1	EtherO 2
1	EtherO 1	0	2.6751
2	EtherO 2	2.6751	0

$\Delta_{GI_{ijV_c}}$	$j \rightarrow$	1	2
$i \downarrow$	INTERACTING GROUPS	EtherO 1	EtherO 2
1	EtherO 1	0	−23.6366
2	EtherO 2	−23.6366	0

The number of non-hydrogen atoms is $n = 6$, so GI_p is calculated for each critical constant, based on the above matrices and Eq. (3.2-13), as

$$10^3 GI_{T_c} = 1/6(0 + 162.6878 + 0 + 162.6878) = 54.22927$$

$$10^4 GI_{P_c} = 1/6(0 + 2.6751 + 0 + 2.6751) = 0.8917$$

$$GI_{V_c} = 1/6(0 - 23.6366 + 0 - 23.6366) = -7.878867$$

With $MW = 90.121$ g/mol and $T_b = 357.75$ K, the Nannoolal estimates of the critical constants are

$$T_c = 357.75[0.6990 + (0.9889 + \{(196.8912 + 54.22927)/1000\}^{0.8607})^{-1}] = 526.68 \text{ K}$$

$$P_c = 90.121^{-0.14041}[0.00939 + (22.0328 + 0.8917)/10000]^{-2} = 3894.5 \text{ kPa} = 38.95 \text{ bar}$$

$$V_c = (148.5344 - 7.878867)/6^{-0.2266} + 86.1539 = 297.25 \text{ cm}^3/\text{mol}$$

Wilson et al. (2018) report values of 535.32 K, 38.3 bar, and 310 cm^3/mol for the critical temperature, pressure, and volume, respectively. Thus, the differences are

$$T_c \text{ Difference} = 526.7 - 535.32 = -8.64 \text{ K or } -1.61\%$$

$$P_c \text{ Difference} = 38.95 - 38.3 = 0.65 \text{ bar or } 1.70\%$$

$$V_c \text{ Difference} = 297.25 - 310 = -12.75 \text{ cm}^3/\text{mol or } -4.11\%.$$

Method of Emami et al. Motivated by a desire to create a UNIFAC-based prediction of boiling temperature at multiple pressures, Emami et al. (2009) created a group contribution method to predict T_c. This method differs from most others because it requires the boiling temperature at 10 mmHg rather than the normal boiling point. This was done with the thought that such a temperature would be more applicable to compounds of high molecular weight for which experimental T_c and normal T_b data are difficult to obtain due to decomposition or purity issues. The equation for the critical temperature is

$$\frac{T_c}{T_b^{10\text{ mmHg}}} = \left(1 + \frac{1}{0.596 + \sum_i N_i (\Delta_{T_c})_i}\right) \qquad (3.2\text{-}14)$$

where the summations are over the unique groups in the molecule, $T_b^{10\text{ mmHg}}$ is the boiling temperature at 10 mmHg (the temperature at which the vapor pressure of the compound is equal to 10 mmHg), the values for $(\Delta_{T_c})_i$ are found in Table 3.8, and N_i is the number of times group i is found in the compound. As mentioned, the method relies on $T_b^{10\text{ mmHg}}$ rather than T_b. If the former is not available, it may be predicted from the following equation.

$$\frac{T_b^{10\text{ mmHg}}}{\text{K}} = \left(\frac{1000}{0.5 + \dfrac{59.15}{\sqrt{\sum_i N_i (\Delta_{T_{b10}})_i}} + \dfrac{1000}{274.22 + \sum_i N_i (\Delta_{T_{b10}})_i}}\right) \qquad (3.2\text{-}15)$$

Here, the summations are over the unique groups in the molecule, the values for $(\Delta_{T_{b10}})_i$ are found in Table 3.8, and N_i is the number of times group i is found in the compound. An additional motivation of Emami et al. is apparent in the form of Eq. (3.2-15); $T_b^{10\text{ mmHg}}$ and by inference T_c approach finite values even when the molecular weight approaches infinity. As discussed by Emami et al., this observation has been noted by several authors previously, but other methods in this chapter do not comply with this requirement.

Because experimental values for $T_b^{10\text{ mmHg}}$ are not as common in the literature compared to T_b, two steps are generally needed to use this method to calculate T_c.

1. Calculate $T_b^{10\text{ mmHg}}$ from Eq. (3.2-15).
2. Calculate T_c from Eq. (3.2-14).

This two-step approach was followed when the method was evaluated against the test database as discussed in Sec. 3.5. Application of the Emami et al. method is shown in Example 3.7.

Example 3.7 Estimate T_c for 2-methylphenol (*o*-cresol) using the method of Emami et al. (2009).

Solution. The groups of 2-methylphenol are one >ACCH3, four >ACH, and one >ACOH according to Table 3.8. The resulting group summations are:

	Group	N	$N\Delta_{T_{b10}}$	$N\Delta_{T_c}$
1	$>ACCH_3$	1	243.180	0.099
2	>ACH	4	432.000	0.148
3	>ACOH	1	571.449	0.163
	$\sum_{k=1}^{3} N_k F_k$		1246.629	0.410

The Emami estimate of the boiling temperature at 10 mmHg is

$$T_b^{10\text{ mmHg}} = 1000/[0.5 + 59.15/1246.629^{0.5} + 1000/(274.22 + 1246.629)] = 353.007\text{ K}$$

The vapor pressure correlation for 2-methylphenol found in Appendix A, which is regressed from experimental data, gives $T_b^{10\text{ mmHg}} = 350.099$ K. The Emami estimates of the critical temperature using both the predicted and experimental $T_b^{10\text{ mmHg}}$ values are

$$T_c\text{ (with predicted } T_b^{10\text{ mmHg}}) = 353.007 \cdot [1 + 1/(0.596 + 0.410)] = 703.91\text{ K}$$

$$T_c\text{ (with experimental } T_b^{10\text{ mmHg}}) = 350.099 \cdot [1 + 1/(0.596 + 0.410)] = 698.11\text{ K}$$

The Appendix A value for the critical temperature is 697.55 K, so the differences are

$$T_c\text{ Difference (with predicted } T_b^{10\text{ mmHg}}) = 703.91 - 697.55 = -6.36\text{ K or } -0.91\%$$

$$T_c\text{ Difference (with experimental } T_b^{10\text{ mmHg}}) = 698.11 - 697.55 = -0.56\text{ K or } -0.08\%$$

As will be discussed in Sec. 3.5, the Emami method performs well in general, but groups with fluorine seem to produce incorrect results. Thus, the method is not recommended for such compounds.

Method of Wilson and Jasperson. Wilson and Jasperson (1996) reported three methods for T_c and P_c that apply to both organic and inorganic species. The zero-order method uses *factor analysis* with boiling point, liquid density, and molecular weight as the descriptors. At the first order, the method uses *atomic contributions* along with boiling point and number of rings, while the second-order method also includes *group contributions*. The zero-order has not been tested here; it is iterative and the authors report that it is less accurate by as much as a factor of two or three than the others, especially for P_c. The first-order and second-order methods use the following equations:

$$\frac{T_c}{T_b} = \left(0.048271 - 0.019846 N_r + \sum_i N_i \left(\Delta_{T_c}^{(1)}\right)_i + \sum_j M_j \left(\Delta_{T_c}^{(2)}\right)_j\right)^{-0.2} \tag{3.2-16}$$

$$\frac{P_c}{\text{bar}} = \frac{0.0186233 \frac{T_c}{\text{K}}}{-0.96601 + \exp(Y)} \tag{3.2-17}$$

$$Y = -0.00922295 - 0.0290403 N_r + 0.041\left(\sum_i N_i \left(\Delta_{P_c}^{(1)}\right)_i + \sum_j M_j \left(\Delta_{P_c}^{(2)}\right)_j\right) \tag{3.2-18}$$

TABLE 3.8 Group contributions for the Emami et al. (2009) method for T_c and T_b use in Eqs. (3.2-14) and (3.2-15) for T_c and $T_b^{10\ \mathrm{mmHg}}$, respectively. Use in Eq. (3.4-7) for T_b. Groups with fluorine (F) do not perform well.

Group	Δ_{T_c}	$\Delta_{T_{b10}}$	Δ_{T_b}
—CH_3	0.086	76.569	116.284
—CH_2—	0.041	111.400	123.631
—RCH_2—	0.036	110.859	133.831
>CH—	−0.009	111.000	95.944
>RCH—	0.015	101.100	111.630
>C<	−0.086	90.711	58.450
>RC<	0.038	72.857	36.300
CH_2═CH—	0.112	178.000	229.791
—CH═CH—	0.079	225.819	258.038
CH_2═C<	0.050	208.545	240.287
—CH═C<	0.025	264.290	266.852
>C═C<	−0.176	180.000	188.120
>ACH	0.037	108.000	132.609
>AC—	0.004	192.740	204.558
>$ACCH_3$	0.099	243.180	282.512
>$ACCH_2$—	0.054	267.900	273.625
>ACCH<	−0.005	248.000	226.530
—OH	0.225	397.860	412.000
>ACOH	0.163	571.449	525.221
CH_3CO—	0.193	447.547	518.749
—CH_2CO—	0.167	468.100	493.090
—CHO	0.180	332.500	384.480
CH_3COO—	0.222	462.720	520.791
—CH_2COO—	0.204	472.520	497.163
HCOO—	0.188	360.579	409.000
CH_3O—	0.145	207.642	250.000
—CH_2O—	0.108	234.651	245.957
>CHO—	0.046	188.000	213.800
—CH_2O— (substituted furan)	0.112	256.000	337.000
—CH_2NH_2	0.189	367.300	409.408
>$CHNH_2$	0.105	323.100	351.959
CH_3NH—	0.148	308.000	325.300
—CH_2NH—	0.105	308.000	325.300
>CHNH—	0.137	308.000	325.300
CH_3N<	0.144	217.681	142.000
—CH_2N<	0.008	217.681	142.000

(Continued)

TABLE 3.8 Group contributions for the Emami et al. (2009) method for T_c and T_b use in Eqs. (3.2-14) and (3.2-15) for T_c and $T_b^{10\,\mathrm{mmHg}}$, respectively. Use in Eq. (3.4-7) for T_b. Groups with fluorine (F) do not perform well. (*Continued*)

Group	Δ_{T_c}	$\Delta_{T_{b10}}$	Δ_{T_b}
$>ACNH_2$	0.112	546.553	611.200
—C_5H_4N	0.196	828.662	925.245
—$CH_2C{\equiv}N$	0.179	628.005	734.001
—COOH	0.317	797.129	800.052
—CH_2Cl	0.129	346.591	399.999
>CHCl	0.074	265.517	312.000
≥CCl	0.066	236.610	275.600
—$CHCl_2$	0.161	474.200	568.324
$>CCl_2$	0.087	375.331	469.500
—CCl_3	0.148	519.421	647.410
>ACCl	0.080	310.470	366.305
—CH_2NO_2	0.220	696.000	797.000
$>CHNO_2$	0.174	672.000	744.000
$>ACNO_2$	0.164	671.000	733.000
—CH_2SH	0.129	418.277	501.000
—I	0.072	424.420	509.000
—Br	−0.023	125.556	201.724
CH≡C—	0.106	210.342	263.387
—C≡C—	−0.185	97.538	309.648
—C(Cl)=C<	0.092	140.000	183.171
>ACF	0.071	112.323	131.000
—CF_3	−0.122	0.001	0.001
—CF_2—	0.088	10.000	90.000
>CF—	0.107	92.000	110.000
—COO—	0.150	361.331	385.893
—SiH_3	0.071	137.000	167.864
—SiH_2—	0.029	153.000	160.000
>SiH—	−0.026	144.000	133.000
>Si<	−0.051	135.000	94.000
≥SiO—	0.026	138.000	108.481
>N—	−0.041	140.000	94.000
—CCl_2F	0.161	344.181	408.557
—$CClF_2$	−0.145	37.700	14.500
—$CONH_2$	0.125	1066.585	1371.950
CH_3S—	0.119	411.035	464.600
—CH_2S—	0.091	411.035	464.600

where N_r is the number of rings in the compound, N_i is the number of atoms of type i with first-order atomic contributions $(\Delta_{T_c}^{(1)})_i$ and $(\Delta_{P_c}^{(1)})_i$ while M_j is the number of groups of type j with second-order group contributions $(\Delta_{T_c}^{(2)})_j$ and $(\Delta_{P_c}^{(2)})_j$. Values of the first-order atomic contributions are given in Table 3.9 and those for the second-order group contributions in Table 3.10. Note that T_c requires T_b. Application of the Wilson and Jasperson method is shown in Example 3.8.

One area of confusion when using the method is determining N_r in a molecule with fused rings. Figure 3.1 gives several examples. For compounds with fused rings *without* bridging, such as naphthalene, counting the number of rings is straightforward as indicated in the figure. The challenge comes with compounds with bridges such as bicyclo[2.2.1]heptane. Wilson and Jasperson do not discuss how to count the number of rings in these instances, and likely did not have experimental data for such compounds when the approach was created, so the method should be avoided in these cases.

TABLE 3.9 Wilson-Jasperson (1996) first-order atomic contributions for T_c and P_c for Eqs. (3.2-16) and (3.2-18).

Atom	$\Delta_{T_c}^{(1)}$	$\Delta_{P_c}^{(1)}$
H	0.002793	0.12660
D	0.002793	0.12660
T	0.002793	0.12660
He	0.320000	0.43400
B	0.019000	0.91000
C	0.008532	0.72983
N	0.019181	0.44805
O	0.020341	0.43360
F	0.008810	0.32868
Ne	0.036400	0.12600
Al	0.088000	6.05000
Si	0.020000	1.34000
P	0.012000	1.22000
S	0.007271	1.04713
Cl	0.011151	0.97711
Ar	0.016800	0.79600
Ti	0.014000	1.19000
V	0.018600	—
Ga	0.059000	—
Ge	0.031000	1.42000
As	0.007000	2.68000
Se	0.010300	1.20000
Br	0.012447	0.97151
Kr	0.013300	1.11000
Rb	−0.027000	—
Zr	0.175000	1.11000

(Continued)

TABLE 3.9 Wilson-Jasperson (1996) first-order atomic contributions for T_c and P_c for Eqs. (3.2-16) and (3.2-18). (*Continued*)

Atom	$\Delta_{T_c}^{(1)}$	$\Delta_{P_c}^{(1)}$
Nb	0.017600	2.71000
Mo	0.007000	1.69000
Sn	0.020000	1.95000
Sb	0.010000	—
Te	0.000000	0.43000
I	0.005900	1.315930
Xe	0.017000	1.66000
Cs	–0.027500	6.33000
Hf	0.219000	1.07000
Ta	0.013000	—
W	0.011000	1.08000
Re	0.014000	—
Os	–0.050000	—
Hg	0.000000	–0.08000
Bi	0.000000	0.69000
Rn	0.007000	2.05000
U	0.015000	2.04000

TABLE 3.10 Wilson-Jasperson (1996) second-order group contributions for T_c and P_c for Eqs. (3.2-16) and (3.2-18).

Group	$\Delta_{T_c}^{(2)}$	$\Delta_{P_c}^{(2)}$
—OH, C_4 or less	0.0350	0.00
—OH, C_5 or more	0.0100	0.00
—O—	–0.0075	0.00
—NH_2, >NH, >N—	–0.0040	0.00
—CHO	0.0000	0.50
>CO	–0.0550	0.00
—COOH	0.0170	0.50
—COO—	–0.0150	0.00
—CN	0.0170	1.50
—NO_2	–0.0200	1.00
Organic Halides (once/molecule)	0.0020	0.00
—SH, —S—, —SS—	0.0000	0.00
Siloxane bond	–0.0250	–0.50

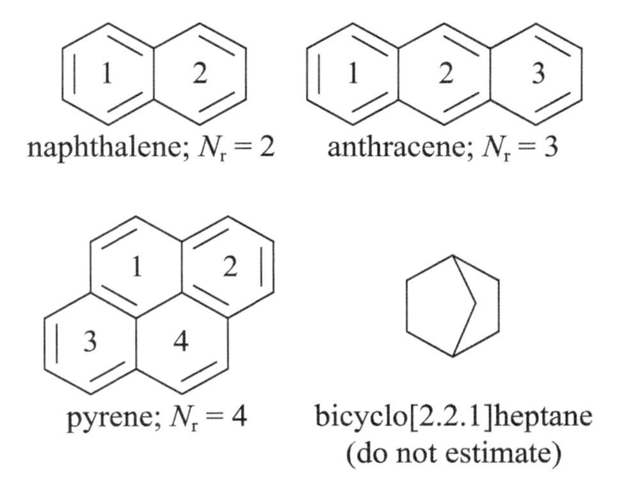

FIGURE 3.1
Examples of determining the number of rings (N_r) for fused compounds for the method of Wilson and Japserson (1996) for T_c and P_c.

Example 3.8 Estimate T_c and P_c for 2-methylphenol (*o*-cresol) using the method of Wilson and Jasperson.

Solution. The atoms of 2-methylphenol are 7 C, 8 H, 1 O, and there is 1 ring. For groups, there is 1 —OH for "C5 or more." The values of T_b and T_c from Appendix A are 464.15 K and 697.55 K, respectively. From Table 3.9

Atom		N	$N\Delta_{T_c}^{(1)}$	$N\Delta_{P_c}^{(1)}$
1	C	7	0.059724	5.10881
2	H	8	0.022344	1.01280
3	O	1	0.020341	0.43360
$\sum_{k=1}^{3} N_k F_k$			0.102409	6.55521

Thus, the first-order estimates are

$$T_c = 464.15(0.048271 - 0.019846 \cdot 1 + 0.102409)^{-0.2} = 697.13 \text{ K}$$
$$Y = -0.00922295 - 0.0290403 \cdot 1 + 0.041 \cdot 6.55521 = 0.2305004$$
$$P_c = 0.0186233 \cdot 697.55/[-0.96601 + \exp(0.2305004)] = 44.30 \text{ bar}$$

From Table 3.10, there is the "—OH, C5 or more" contribution of $N\Delta_{T_c}^{(2)} = 0.01$ though for P_c there is no contribution. Thus, only the second-order estimate for T_c is changed to

$$T_c = 464.15(0.048271 - 0.019846 \cdot 1 + 0.102409 + 0.01)^{-0.2} = 686.94 \text{ K}$$

The Appendix A values for the critical properties are 697.55 K and 50.10 bar, respectively. Thus, the differences are

First Order

T_c Difference = 697.13 – 697.55 = –0.42 K or –0.06%

P_c Difference = 44.30 – 50.10 = –5.80 bar or –11.6%

Second Order

T_c Difference = 686.94 – 697.55 = –10.61 K or –1.52%

The first-order estimate for T_c in this example is more accurate than the second-order estimate which occasionally occurs.

3.3 ACENTRIC FACTOR

Along with the critical properties of Sec. 3.2, a commonly used pure-component constant for property estimation is the acentric factor which was originally defined by Pitzer et al. (1955) as

$$\omega \equiv -\log_{10}\left[\frac{P^{vp}(T = 0.7T_c)}{P_c}\right] - 1.0 \tag{3.3-1}$$

where $P^{vp}(T = 0.7T_c)$ is the saturated liquid vapor pressure of the compound at a temperature of $0.70T_c$. The particular definition of Eq. (3.3-1) arose because the monatomic gases (Ar, Kr, Xe) have $\omega \approx 0$. Most species have positive values in the range $0 < \omega < 2.5$. Exceptions, where $\omega < 0$, include the quantum gases (H_2, He, Ne), some elements (e.g., K, Al, Au, Rn, etc.), and some inorganic molecules (e.g., NaF, KI, etc.). To obtain values of ω from its definition, one must know the constants T_c, P_c, and the property P^{vp} at the reduced temperature, $\dfrac{T}{T_c} = 0.7$.

The most accurate technique to estimate an unknown acentric factor is to obtain (or estimate) the critical constants T_c and P_c and use the vapor pressure correlation regressed from experimental data or from estimation techniques recommended in Chap. 5. The accuracy of the acentric factor is dependent on the accuracy of the properties involved. Previous editions of this book presented other methods for estimating the acentric factor, but these are no longer recommended.

3.4 MELTING AND BOILING POINTS

Melting point is a term used to describe the temperature at which a solid to liquid phase transition occurs. It is weakly dependent on pressure, so a curve of melting points can be reported. The melting point values found in Appendix A are the transitions that occur at 101325 Pa and are given the symbol T_m.

The triple point is the temperature and pressure at which solid, liquid, and vapor phases coexist in thermodynamic equilibrium. The triple point temperature is close to the value of T_m, with the latter usually being about 0.01 K greater than the latter. (However, if T_m is measured in air, the air dissolved in the liquid may cause the opposite effect.) Because the two properties are so close in value, the triple point temperature is assumed to be equal to T_m in the absence of experimental data.

Boiling point is a term used to describe the temperature at which a substance undergoes a phase transition from liquid to vapor. The vapor pressure curve is composed of all the boiling points from the triple point temperature to the critical temperature. The "normal" boiling point (T_b) has historically been defined to be the temperature at which the vapor pressure is equal to 101325 Pa. This convention is used in this (and previous) editions of this book, and values for T_b are tabulated in Appendix A. From a thermodynamic standpoint, there is nothing special about tabulating the "normal" boiling point over temperatures at other vapor pressures. However, from a practical standpoint, knowing the point at sea-level atmospheric pressure serves as a useful metric for process, safety, environmental, and other design activities. Note that several of the estimation methods in Sec. 3.2 use T_b as input information for T_c.

Several methods to estimate the melting point and normal boiling point have been proposed. Some were reviewed in the previous editions. Several of *group/bond/atom* methods described in Sec. 3.2 have been applied to T_m and T_b. We describe the application of these in a similar manner to that used above for critical properties.

Method of Joback for T_m and T_b. The Joback (1984; Joback and Reid, 1987) relations for T_m and T_b are

$$\frac{T_m}{K} = 122.5 + \sum_i N_i(\Delta_{T_m})_i \tag{3.4-1}$$

$$\frac{T_b}{K} = 198.2 + \sum_i N_i(\Delta_{T_b})_i \tag{3.4-2}$$

where the identity and values of the group contributions are indicated as $(\Delta_{T_m})_i$ and $(\Delta_{T_b})_i$ and are found in Table 3.2. The linear form of Eq. (3.4-2) is adequate for a narrow range of molecular sizes but has been shown to be insufficient for many compounds. Stein and Brown (1994) recognized this limitation and presented an improved method discussed later [Eqs. (3.4-8)–(3.4-10)]. Example 3.9 demonstrates the method of Joback.

Example 3.9 Estimate T_m and T_b for 2,4-dimethylphenol (2,4-xylenol) by using Joback's group method.

Solution. 2,4-dimethylphenol contains two —CH_3, three =CrH—, three =Cr<, and one a-OH (phenol). From Table 3.2

	Group	N	$N\Delta_{T_m}$	$N\Delta_{T_b}$
1	—CH_3	2	−10.20	47.16
2	=CrH—	3	24.39	80.19
3	=Cr<	3	111.06	93.03
4	a—OH (phenol)	1	82.83	76.34
	$\sum_{k=1}^{4} N_k F_k$		208.08	296.72

The estimates are:

$$T_m = 122.5 + 208.08 = 330.58 \text{ K}$$

$$T_b = 198.2 + 296.72 = 494.92 \text{ K}$$

Appendix A values for these properties are $T_m = 297.68$ K and $T_b = 484.13$ K. Thus, the differences are:

$$T_m \text{ Difference} = 330.58 - 297.68 = 32.90 \text{ K or } 11.1\%$$

$$T_b \text{ Difference} = 494.92 - 484.13 = 10.79 \text{ K or } 2.2\%$$

Method of Constantinou and Gani (CG) for T_m and T_b. The Constantinou and Gani (1994, 1995) equations for T_m and T_b are

$$\frac{T_m}{K} = 102.425 \ln\left[\sum_i N_i(\Delta^{(1)}_{T_m})_i + W\sum_j M_j(\Delta^{(2)}_{T_m})_j\right] \tag{3.4-3}$$

$$\frac{T_b}{K} = 204.359 \ln\left[\sum_i N_i(\Delta^{(1)}_{T_b})_i + W\sum_j M_j(\Delta^{(2)}_{T_b})_j\right] \tag{3.4-4}$$

The group values $(\Delta_{T_m}^{(1)})_i$, $(\Delta_{T_m}^{(2)})_j$, $(\Delta_{T_b}^{(1)})_i$, and $(\Delta_{T_b}^{(2)})_j$ for Eqs. (3.4-3) and (3.4-4) and are given in Table 3.3 and Table 3.4. Example 3.10 illustrates this method for a compound with only first-order groups. Example 3.11 demonstrates the technique for compounds with first- and second-order groups.

Example 3.10 Estimate T_m and T_b for 2,4-dimethylphenol (2,4-xylenol) using Constantinou and Gani's group method.

Solution. The first-order groups for 2,4-dimethylphenol are three >ACH, two $>ACCH_3$, and one >ACOH. There are no second-order groups, so the first-order and second-order calculations are the same. From Table 3.3

	Group	N	$N\Delta_{T_m}^{(1)}$	$N\Delta_{T_b}^{(1)}$
1	>ACH	3	4.4007	2.7891
2	$>ACCH_3$	2	3.7270	3.9338
3	>ACOH	1	13.7349	4.4014
	$\sum_{k=1}^{3} N_k F_k$		21.8626	11.1243

$$T_m = 102.425 \ln(21.8626) = 315.96 \text{ K}$$

$$T_b = 204.359 \ln(11.1243) = 492.33 \text{ K}$$

Appendix A values for these properties are $T_m = 297.68$ K and $T_b = 484.13$ K. Thus, the differences are

$$T_m \text{ Difference} = 315.96 - 297.68 = 18.28 \text{ K or } 6.1\%$$

$$T_b \text{ Difference} = 492.33 - 484.13 = 8.20 \text{ K or } 1.7\%$$

Example 3.11 Estimate T_m and T_b for five cycloalkanes with formula C_7H_{14} using the group method of Constantinou and Gani.

Solution. The first- and second-order groups for the C_7H_{14} cycloalkanes are:

	cycloheptane	methyl cyclohexane	ethyl cyclopentane	cis-1,3- dimethyl cyclopentane	trans-1,3- dimethyl cyclopentane
First-Order groups, N_i					
$—CH_3$	0	1	1	2	2
$—CH_2—$	7	5	5	3	3
>CH—	0	1	1	2	2
Second-Order groups, M_j					
7-ring	1	0	0	0	0
6-ring	0	1	0	0	0
5-ring	0	0	1	1	1
Alicyclic Side Chain	0	0	1	0	0

All the substances have one or more second-order groups. Using first-order group contribution values from Table 3.3, the second-order values from Table 3.4, and experimental values from Appendix A, the results are

Property	cycloheptane	methyl cyclohexane	ethyl cyclopentane	cis-1,3- dimethyl cyclopentane	trans-1,3- dimethyl cyclopentane
T_m, K					
Experimental	265.12	146.58	134.71	139.45	139.18
Calculated (first order)	191.28	173.54	173.54	152.06	152.06
Abs. percent Err. (first order)	27.85	18.39	28.82	9.04	9.25
Calculated (second order)	266.15	199.43	122.14	166.79	166.79
Abs. percent Err. (second order)	0.39	36.06	9.33	19.64	19.84
T_b, K					
Experimental	391.94	374.08	376.62	364.88	363.92
Calculated (first order)	381.18	369.71	369.71	357.57	357.57
Abs. percent Err. (first order)	2.75	1.17	1.83	2.00	1.74
Calculated (second order)	391.93	376.16	377.69	364.27	364.27
Abs. percent Err. (second order)	0.00	0.56	0.28	0.17	0.10

The first-order results are generally good for boiling but not melting. For T_m, the second-order contributions improve the agreement for only cycloheptane and ethylcyclopentane. The correction goes in the wrong direction for the other compounds. Overall, the performance of the method for T_m is poor for these compounds.

Method of Nannoolal et al. for T_b. As explained in Sec. 3.2, Nannoolal et al. (2007) created a group contribution method for the critical constants that accounts for attached neighbors. Prior to this, they reported a method to predict T_b (Nannoolal et al., 2004). The equation is

$$\frac{T_b}{K} = \frac{\sum_i N_i \left(\Delta_{T_b}\right)_i + GI_{T_b}}{n^{0.6583} + 1.6868} + 84.3395 \tag{3.4-5}$$

where the summation is over the unique groups in the molecule, n is the total numbers of *non-hydrogen* atoms in the compound, the group values $(\Delta_{T_b})_i$ are found in Table 3.11 and Table 3.12, and GI_{T_b} is the *total group interaction* contribution. Note that some group definitions are different between the method for T_b and that for the critical constants in Sec. 3.2. Also, as explained for the critical constants method, GI_{T_b} is used to improve the prediction of compounds with multiple functional groups. It is calculated according to

$$GI_{T_b} = \frac{1}{n} \sum_{i=1}^{m} \sum_{j=1}^{m} \frac{(\Delta_{GIij})_{T_b}}{m-1} \tag{3.4-6}$$

TABLE 3.11 Primary and secondary group contributions for the Nannoolal et al. (2004) method for T_b to use in Eq. (3.4-5). The ID column is a numerical identifier for the group. The Pr column is for "Priority." A group with a lower priority number should be used over a group with higher priority number when two may be chosen. The Example(s) column contains representative molecule(s) where the group may be found.

Group	Description	ID	Pr	Δ_{T_b}	Example (# of Groups)
	Carbon Groups				
CH_3—(ne)	CH_3— connected to atoms other than N, O, F, or Cl	1	104	177.3066	decane (2)
CH_3—(e)	CH3— connected to either N, O, F, or Cl	2	102	251.8338	dimethoxymethane (2)
CH_3—(a)	CH3— connected to an aromatic atom (not necessarily C)	3	103	157.9527	toluene (1)
—C(c)H_2—	—CH_2— in a chain	4	111	239.4531	butane (2)
>C(c)H—	>CH— in a chain	5	117	240.6785	2-methylpentane (1)
>C(c)<	>C< in a chain	6	119	249.5809	neopentane (1)
>C(c)<(e)	>C< in a chain connected to at least one F, Cl, N, or O; other bonds may be to any atoms	7	107	266.8769	ethanol (1)
>C(c)<(a)	>C< in a chain connected to at least one aromatic carbon; other bonds may be to any atoms	8	108	201.0115	ethylbenzene (1)
—C(r)H_2—	—CH_2— in a ring; connected to atoms other than N, O, F, or Cl	9	112	239.4957	cyclopentane (5)
>C(r)H—	>CH— in a ring; connected to atoms other than N, O, F, or Cl	10	116	222.1163	methylcyclohexane (1)
>C(r)<	>C< in a ring; connected to atoms other than N, O, F, or Cl	11	118	209.9749	beta-pinene (1)
>C(r)<(e,c)	>C< in a ring connected to one or more N or O which are not part of the ring, or one Cl or F; other bonds are not to N, O, Cl, or F	12	109	492.0707	cyclopentanol (1)
>C(r)<(e,r)	>C< in a ring connected to one or more N or O which are part of the ring; other bonds are not to N, O, Cl, or F	13	110	291.2291	morpholine (4)
>C(r)<(C(a))	>C< in a ring connected to one or more aromatic carbon; other bonds are not to N, O, Cl, or F	14	106	244.3581	2-methyltetraline (2)
=C(a)H—	Aromatic =CH—	15	105	235.3462	benzene (6)
=C(a)<(ne)	Aromatic =C< not connected to either O, N, Cl, or F	16	115	315.4128	ethylbenzene (1)
=C(a)<(e)	Aromatic =C< connected to either O, N, Cl, or F	17	113	348.2779	phenol (1)

(a)=C(a)<2(a)	Aromatic =C< with three aromatic neighbors and three aromatic bonds	18	114	367.9649	naphthalene (2)
>C(c)=C(c)<	>C=C< (both C have at least one non-H neighbor)	58	63	475.7958	2-heptene (1)
>C(c)=C(c)<(C(a))	Non-cyclic >C=C< connected to at least one aromatic C	59	60	586.1413	*trans*-cinnamic alcohol (1)
—(e)>C(c)=C(c)<	Non-cyclic >C=C< substituted with at least one F, Cl, N, or O	60	59	500.2434	*trans*-1,2-dichloroethylene (1)
H2C(c)=C<	H2C=C< (1-ene)	61	58	412.6276	1-hexene (1)
>C(r)=C(r)<	Cyclic >C=C<	62	61	475.9623	cyclopentene (1)
>C=C=C<	Two cumulated double bonds	87	6	664.0903	1,2-butadiene (1)
>C(r)=C(r) —C(r)=C(r)<	Two conjugated double bonds in a ring	88	7	957.6388	cyclopentadiene (1)
>C=C—C=C<	Two conjugated double bonds not in a ring	89	8	928.9954	1,3-hexadiene (1)
—C≡C—C≡C—	Conjugated triple bond	95	9	1218.1878	2,4-hexadiyne (1)
—C≡C—	—C≡C— (-yne-) with two non-H neighbors	63	62	512.2893	2-octyne (2)
HC≡C—	HC≡C— (1-yne) with one non-H neighbor	64	57	422.2307	1-heptyne (1)
	Halogen Groups				
F—(C,Si)	F— connected to nonaromatic C or Si not already substituted with F or Cl	19	86	106.5492	2-fluoropropane (1)
F—(C(—)=C)<	F— on a C=C; used once for each F	20	84	49.2701	trifluoroethylene (3)
F—((C,Si)—([F]))—b	F— connected to C or Si already substituted with at one F and two other non-halogen atoms (the C is fully substituted with no H's); used once for each F	21	80	53.1871	2,2,3,3-tetrafluorobutane (4)
F—((C,Si)—([Cl]))—b	F— connected to C or Si already substituted with at least one Cl and two other non-F atoms (the C is fully substituted with no H's)	102	81	111.0590	1,1-dichloro-1-fluoroethane (1)

(Continued)

TABLE 3.11 Primary and secondary group contributions for the Nannoolal et al. (2004) method for T_b to use in Eq. (3.4-5). The ID column is a numerical identifier for the group. The Pr column is for "Priority." A group with a lower priority number should be used over a group with higher priority number when two may be chosen. The Example(s) column contains representative molecule(s) where the group may be found. (*Continued*)

Group	Description	ID	Pr	Δ_{T_b}	Example (# of Groups)
F-((C,Si)—([F,Cl]))—a	F— connected to C or Si already substituted with at least one F or Cl and one other non-halogen atom (the C is trisubstituted with one H remaining)	22	83	78.7578	1-chloro-1,2,2,2-tetrafluoroethane (1)
F-((C,Si)—([F,Cl]$_2$))	F— connected to C or Si already substituted with two F or Cl; used once for each F	23	82	103.5672	1,1,1-trifluoroethane (3)
F—(C(a))—	F— connected to an aromatic C	24	85	−19.5575	fluorobenzene (1)
Cl—(C,Si)	Cl— connected to C or Si not already substituted with F or Cl	25	72	330.9117	1-chlorobutane (1)
Cl—((C,Si) —([F,Cl]))	Cl— connected to C or Si already substituted with one F or Cl; used once for each Cl	26	71	287.1863	dichloroacetic acid (2)
Cl—((C,Si)—([F,Cl]$_2$))	Cl— connected to C or Si already substituted with at least two F or Cl; used once for each Cl	27	69	267.417	1,1,1-trichloroethane (3)
Cl—(C(a))	Cl— connected to an aromatic C	28	73	205.7363	chlorobenzene (1)
Cl—(C(—)=C)<	Cl— on a C=C; used once for each Cl	29	70	292.5816	trichloroethylene (3)
—COCl	—(C=O)Cl connected to C (acid chloride)	77	19	778.9151	acetyl chloride (1)
Br—C,Si(na))	Br— connected to a non-aromatic C or Si	30	66	419.4959	ethyl bromide (1)
Br—(C(a))	Br— connected to an aromatic C or Si	31	67	377.6775	bromobenzene (1)
I—(C,Si)	I— connected to C or Si	32	64	556.3944	ethyl iodide (1)
	Oxygen Groups				
—OH tert	—OH connected to C which has four non-H neighbors (tertiary alcohols)	33	90	349.9409	*tert*-butanol (1)
HO—((C,Si)$_2$H—(C,Si)-(C,Si) —)	—OH connected to a C or Si substituted with two C or Si in at least three C or Si containing chain (secondary alcohols)	34	89	390.2446	2-butanol (1)
—OH (>C$_4$)	—OH connected to C or Si substituted with one C or Si in an at least five C or Si containing chain (primary alcohols)	35	87	443.8712	1-nonanol (1)

—OH (<C_5)	—OH for aliphatic chains with less than five C (cannot be connected to aromatic groups)	36	91	488.0819	ethanol (1)
—OH (C(a))	—OH connected to an aromatic C (phenols)	37	88	361.4775	phenol (1)
(C,Si)—O—(C,Si)	—O— connected to two neighbors which are each either C or Si (ethers)	38	93	146.4836	diethyl ether (1)
>(OC_2)<	>(OC_2)< (epoxide)	39	50	820.7118	propylene oxide (1)
(C(a))—O(a)—(C(a))	—O in an aromatic ring with aromatic C as neighbors	65	92	37.1936	furan (1)
CHO—(C(na))	H(C=O)— connected to non-aromatic C (aldehydes)	52	53	553.809	acetaldehyde (1)
CHO—(C(a))	H(C=O)— connected to aromatic C (aldehydes)	90	52	560.1024	benzaldehyde (1)
COOH—(C)	—(C=O)OH connected to C (carboxylic acid)	44	24	1080.3139	acetic acid (1)
(C) —COO—(C)	—(C=O)O— connected to two C (ester) in a chain	45	25	636.2020	ethyl acetate (1)
HCOO—(C)	H(C=O)O— connected to C (formic acid ester)	46	27	642.0427	ethyl formate (1)
—C(r)OO—	—(C=O)O— in a ring, C is connected to C (lactone)	47	26	1142.6119	ε-caprolactone (1)
O=C<$(C(na))_2$	—(C=O)— connected to two non-aromatic C (ketones)	51	55	618.9782	acetone (1)
(O=C<$(C)_2$)(a)	—(C=O)— connected to two C with at least one aromatic C	92	54	606.1797	acetophenone (1)
O=C$(—O—)_2$	Non-cyclic carbonate diester	79	15	879.7062	dimethyl carbonate (1)
—O—(C=O)—O—	—O—(C=O) —O— connected to two O (carbonates)	103	34	1573.3769	propylene carbonate (1)
—S—(C=O)—	C=O connected to S	109	39	492.0707	S-methyl thioacetate
>N—(C=O)—N<	C=O connected to two N	100	2	1045.0343	1,1,3,3-tetramethylurea (1)
—O—(C=O)—N<	C=O connected to O and N (carbamate)	99	1	886.7613	ethyl carbamate (1)
—(C=O)—O—(C=O)—	Anhydride connected to two C	76	12	1251.2675	butyric anhydride (1)
(—(C=O)—O—(C=O)—)r	Cyclic anhydride connected to two C	96	11	2082.3288	maleic anhydride (1)
—O—O—	Peroxide	94	32	273.1755	di-*tert*-butyl peroxide (1)

(*Continued*)

TABLE 3.11 Primary and secondary group contributions for the Nannoolal et al. (2004) method for T_b to use in Eq. (3.4-5). The ID column is a numerical identifier for the group. The Pr column is for "Priority." A group with a lower priority number should be used over a group with higher priority number when two may be chosen. The Example(s) column contains representative molecule(s) where the group may be found. (*Continued*)

Group	Description	ID	Pr	Δ_{T_b}	Example (# of Groups)
	Nitrogen Groups				
NH_2—(C,Si)	NH_2— connected to either C or Si (primary amine)	40	95	321.1759	1-hexylamine (1)
NH_2—(C(a))	NH_2— connected to an aromatic C (aromatic primary amine)	41	94	441.4388	aniline (1)
(C,Si)—NH—(C,Si)	—NH— connected to two C or Si neighbors (secondary amine)	42	99	223.0992	diethylamine (1)
(C,Si)r—NH—(C,Si)r	—NH— connected to two C or Si neighbors in a ring (cyclic secondary amine)	97	98	201.3224	pyrrolidine (1)
$(C,Si)_2$>N—(C,Si)	>N— connected to three C or Si neighbors (tertiary amine); neighbors can be aromatic or nonaromatic	43	100	126.2952	N,N-dimethylaniline (1)
(C,Si)═N—	Double-bonded amine connected to at least on C or Si	91	101	229.228	acetonin (1)
═N(a)—(r5)	Aromatic —N— in a five-membered ring, free electron pair	66	97	453.3397	pyrrole (1)
═N(a)—(r6)	Aromatic ═N— in a six-membered ring	67	96	306.7139	pyridine (1)
N≡C—(C)	—C≡N (cyano-group) connected to C	57	56	719.2462	acetonitrile (1)
N≡C—(N)	—C≡N (cyano-group) connected to N	111	41	971.0365	dimethylcyanamide (1)
N≡C—(S)	—C≡N (cyano-group) connected to S	108	38	659.7336	methyl thiocyanate (1)
CNCNC-ring	…═CNC═NC═… (imidazole ring)	106	3	484.6371	1-methyl-1H-imidazole (1)
—$CONH_2$	—(C═O)NH_2 (amide)	50	28	1487.4109	acetamide (1)
—CONH—	—(C═O)NH— (monosubstituted amide)	49	48	1364.5333	N-methylformamide (1)
—CON<	—(C═O)N< (disubstituted amide)	48	49	1052.6072	N,N-dimethylformamide (1)
OCN—	O═C═N— connected to C or Si (isocyanate)	80	29	660.4645	butyl isocyanate (1)
>C═NOH	>C═NOH (oxime)	75	30	1041.0851	methyl ethyl ketoxime (1)
—ON═(C,Si)	—ON═ connected to C or Si (isoazole)	115	45	612.9506	5-phenylisoxazole (1)
O═N—O—(C)	nitrites (esters of nitrous acid)	74	23	494.2668	ethyl nitrite (1)

NO_2—(C)	NO_2— connected to aliphatic C	68	21	866.5843	1-nitropropane (1)
NO_2—(C(a))	NO_2— connected to aromatic C	69	22	821.4141	nitrobenzene (1)
NO_3—(C)	nitrate (esters of nitric acid)	72	14	920.3617	*n*-butyl nitrate (1)
		Silicon Groups			
>Si<	>Si< not attached to N, O, F, or Cl	70	79	282.0181	triethylsilane (1)
>Si<(O)	>Si< connected to at least one O	71	77	207.9312	hexamethyldisiloxane (2)
>Si<(F,Cl)	>Si< connected to at least one F or Cl	78	16	540.0895	hexachlorodisilane (2)
		Sulfur Groups			
(C)—S—S—(C)	—S—S— (disulfide) connected to two C	55	51	864.5074	dimethyl disulfide (1)
SH—(C)	—SH connected to C (thiols, mercaptans)	53	74	434.0811	1-propanethiol (1)
(C)—S—(C)	—S— connected to two C (thioether)	54	75	461.5784	methyl ethyl sulfide (1)
—S(a)—	—S— in an aromatic ring (aromatic thioether)	56	76	304.3321	thiophene (1)
(C)—SO_2—(C)	Non-cyclic sulfone connected to two C (sulfone)	82	18	1559.9840	divinylsulfone (1)
—O—S(=O)$_2$—O—	S(=O)$_2$ connected to two O (sulfates)	104	35	1483.1290	dimethyl sulfate (1)
—S(=O)$_2$N<	>S(=O)$_2$ connected to N and another atom	105	36	1506.8136	N,N-diethylmethanesulfonamide (1)
>S=O	sulfoxide	107	37	1379.4485	dimethyl sulfoxide (1)
S=C=N—(C)	S=C=N— (thiocyanate) connected to C	81	20	1018.4865	ethyl isothiocyanate (1)
		Other Groups			
(C)$_2$>Sn<(C)$_2$	Sn connected to four carbons	83	65	510.4223	tetramethylstannane (1)
B(O—)$_3$	Non-cyclic boric acid ester	78	16	540.0895	triethyl borate (1)
—Se—, >Se<	Se connected to at least one C or Si	116	26	562.1791	dimethyl selenide (1)
PO(O—)$_3$	O=P(O—)$_3$ (phosphates)	73	10	1153.1344	triethyl phosphate (1)
>P—	Phosphorus connected to at least 1 C or S (phosphine)	113	43	428.8911	triphenylphosphine (1)
—$AsCl_2$	$AsCl_2$ connected to C	84	17	1149.9670	ethyldichloroarsine (1)
(C)$_2$>Ge<(C)$_2$	>Ge< connected to four carbons	86	68	347.7717	tetramethylgermane (1)
—$GeCl_3$	—$GeCl_3$ connected to carbons	85	13	1209.2972	phenyltrichlorogermane (1)
>Al—	>Al< connected to at least one C or Si	117	47	761.6006	triethylaluminum (1)

TABLE 3.12 Corrections for the Nannoolal et al. (2004) method for T_b to use in Eq. (3.4-6). These corrections are added into the summation of the groups found in Table 3.11; they do not replace the groups. See Example 3.12.

Name	Description	ID	Δ_{T_b}	Example (# of corrections)
—C=C—C=O	—C=O connected to sp^2 carbon	118	40.4205	benzaldehyde (1)
(C=O)—C([F,Cl]$_{2,3}$)	Carbonyl connected to C with two or more halogens	119	−82.2328	dichloroacetyl chloride (1)
(C=O)—(C([F,Cl]$_{2,3}$)$_2$	Carbonyl connected to two C with two or more halogens each	120	−247.8893	1,1,1,3,3,3-hexafluoro-2-propanone (1)
C—[F,Cl]$_3$	C with three halogens	121	−20.3996	α,α,α-trifluorotoluene (1)
(C)$_2$—C—[F,Cl]$_2$	Secondary C with two halogens	122	15.4720	2,2-dichloropropane (1)
no hydrogen	Component has no H	123	−172.4201	perfluoro compounds (1)
one hydrogen	Component has one H	124	−99.8035	1H-perfluorobutane (1)
3/4 ring	A three- or four-membered non-aromatic ring	125	−62.3740	cyclobutene (1)
five ring	A five-membered non-aromatic ring	126	−40.0058	cyclopentane (1)
ortho	Ortho position—counted only once and only if there are no meta or para pairs	127	−27.2705	*o*-xylene (1)
meta	Meta position—counted only once and only if there are no para or ortho pairs	128	−3.5075	*m*-xylene (1)
para	Para position—counted only once and only if there are no meta or ortho pairs	129	16.1061	*p*-xylene (1)
((C=)(C)C—CC$_3$)	C—C bond with four single bonded and one double bonded C neighbor	130	25.8348	*tert*-butylbenzene (1)
C$_2$C—CC$_2$	C—C bond with four C neighbors, two on each side	131	35.8330	bicyclohexyl (1)
C$_3$C—CC$_2$	C—C bond with five C neighbors	132	51.9098	2,2,3-trimethylbutane (1)
C$_3$C—CC$_3$	C—C bond with six C neighbors	133	111.8372	2,2,3,3-tetrametylbutane (1)
$(C,Si)_2{>}NCN{<}(C,Si)_2$	Quaternary diamine connected to four C or Si	101*	−109.6269	N,N,N',N'-tetramethylmethylenediamine (1)

*Erroneously identified as a group in the original publication

Here, $(\Delta_{GI_{ij}})_{T_b}$ is the group interaction contribution between groups i and j as defined in Table 3.13 and m is the total number of interaction groups in the molecule. Also note that $\Delta_{GI_{ij}} = \Delta_{GI_{ji}}$ and $\Delta_{GI_{ii}} = \Delta_{GI_{jj}} = 0$. The authors explain that the GI_{T_b} contribution is needed to fix the assumption that groups are additive, a supposition that breaks down for groups which associate (e.g., hydrogen bond). Application of the Nannoolal et al. (2004) method for T_b is shown in Examples 3.12 and 3.13. Example 3.12 does so for a molecule without interacting groups ($GI_{T_b} = 0$), and Example 3.13 is for a compound with interacting groups ($GI_{T_b} \neq 0$).

Example 3.12 Estimate T_b for 2-methylphenol (*o*-cresol) using the method of Nannoolal et al. (2004).

Solution. The first- and second-order groups of 2-methylphenol are one CH_3—(a), four =C(a)H—, one =C(a)<(e), one —C(a)<(ne), and one —OH (C(a)) according to Table 3.11. There is one Ortho pair(s) correction according to Table 3.12. There are no interacting groups in the molecule according to Table 3.13, so $GI_{T_b} = 0$. The resulting group summations are

	Group	N	$N\Delta_{T_b}$
1	CH_3—(a)	1	157.9527
2	=C(a)H—	4	941.3848
3	=C(a)<(e)	1	348.2779
4	=C(a)<(ne)	1	315.4128
5	—OH (C(a))	1	361.4775
6	Ortho pair(s)	1	–27.2705
	$\sum_{k=1}^{6} N_k F_k$		2097.2352

TABLE 3.13 Group interactions for the Nannoolal et al. (2004) method for T_b to use in Eqs. (3.4-5) and (3.4-6). See Example 3.13.

Group ID	Interacting Groups (i—j)	Δ_{GIijT_b}
135	OH—OH	291.7985
136	OH—NH_2	314.6126
137	OH—NH	286.9698
138	OH—SH	38.6974
139	OH—COOH	146.7286
140	OH—EtherO	135.3991
141	OH—Epox	226.4980
142	OH—Ester	211.6814
143	OH—Ketone	46.3754
144	OH—Teth	–74.0193
145	OH—CN	306.3979
146	OH—AO	435.0923
147	OH—AN6	1334.6747
148	OH(a)—OH(a)	288.6155

(*Continued*)

TABLE 3.13 Group interactions for the Nannoolal et al. (2004) method for T_b to use in Eqs. (3.4-5) and (3.4-6). See Example 3.13. (*Continued*)

Group ID	Interacting Groups (i—j)	$\Delta_{GIij_{T_b}}$
149	OH(a)—NH_2	797.4327
150	OH(a)—COOH	−1477.9671
151	OH(a)—EtherO	130.3742
152	OH(a)—Ester	−1184.9784
153	OH(a)—Ketone	—
154	OH(a)—Alde	43.9722
155	OH(a)—Nitro	−1048.1236
156	OH(a)—AN6	−614.3624
157	NH_2—NH_2	174.0258
158	NH_2—NH	510.3473
159	NH_2—EtherO	124.3549
160	NH_2—Ester	182.6291
161	NH_2—Teth	−562.3061
162	NH_2—Nitro	663.8009
163	NH_2—AO	395.4093
164	NH_2—AN6	27.2735
165	NH—NH	239.8076
166	NH—EtherO	101.8475
167	NH—Ester	317.0200
168	NH—Ketone	−215.3532
169	NH—AN6	758.9855
170	SH—SH	217.6360
171	SH—OH—Ester	501.2778
172	COOH—COOH	117.2044
173	COOH—EtherO	612.8821
174	COOH—Ester	−183.2986
175	COOH—Ketone	−55.9871
176	OCN—OCN	−356.5017
177	OCN—Nitro	−263.0807
178	EtherO—EtherO	91.4997
179	EtherO—Epox	178.7845
180	EtherO—Ester	322.5671
181	EtherO—Ketone	15.6980
182	EtherO—Alde	17.0400
183	EtherO—Teth	394.5505
184	EtherO—Nitro	963.6518
185	EtherO—CN	293.5974
186	EtherO—AO	329.0050
187	Epox—Epox	1006.3880
188	Epox—Alde	163.5475
189	Ester—Ester	431.0990
190	Ester—Ketone	22.5208

(*Continued*)

TABLE 3.13 Group interactions for the Nannoolal et al. (2004) method for T_b to use in Eqs. (3.4-5) and (3.4-6). See Example 3.13. (*Continued*)

Group ID	Interacting Groups (i—j)	$\Delta_{GIij_{T_b}}$
191	Ester—Nitro	−205.6165
192	Ester—CN	517.0677
193	Ester—AO	707.9404
194	Ketone—Ketone	−303.9653
195	Ketone—Alde	−391.3690
196	Ketone—Nitro	−3628.9026
197	Ketone—Ats	381.0107
198	Ketone—CN	−574.2230
199	Ketone—AO	176.5481
200	Ketone—AN6	124.1943
201	Alde—Alde	582.1763
202	Alde—Nitro	140.9644
203	Alde—Ats	397.5750
204	Alde—AO	674.6858
205	Teth—Teth	−11.9406
206	Nitro—Nitro	65.1432
207	Ats—CN	−101.2319
208	Ats—AN5	−348.7400
209	CN—AN6	−370.9729
210	AO—AN5	−888.6123
211	AN5—AN5	—
212	AN6—AN6	−271.9449

Abbreviation for Group *i* from Above	Group Description	Group ID(s) from Table 3.11
OH	Alcohol (—OH)	34, 35, 36
OH(a)	Phenol (—OH (C(a)))	37
COOH	Carboxylic Acid (—COOH)	44
EtherO	Ether (—O—)	38
Epox	Epoxide (>(OC_2)<)	39
Ester	Ester (—OOC—)	45, 46, 47
Ketone	Ketone (—CO—)	51, 92
Alde	Aldehyde (—CHO)	52, 90
AO	Aromatic oxygen (—O(a)—)	65
Teth	Sulfide (thioester) (—S(na)—)	54
Ats	Aromatic sulfur (—S(a)—)	56
SH	Thiol (—SH)	53
NH2	Primary amine (—NH_2)	40, 41
NH	Secondary amine (>NH)	42, 97
OCN	Isocyanate (—OCN)	80
CN	Cyanide (—CN)	57
Nitro	NO_2— connected to aromatic C	69
AN5	Aromatic N in 5-ring (=N(a)—(r5))	66
AN6	Aromatic N in 6-ring (=N(a)—(r6))	67

The number of non-hydrogen atoms is $n = 8$, so the Nannoolal estimate of the normal boiling point is

$$T_b = 2097.2352/(8^{0.6583} + 1.6868) + 84.3395 = 457.66 \text{ K}$$

The Appendix A value for the normal boiling point is 464.15 K. Thus, the difference is

$$T_b \text{ Difference} = 457.66 - 464.15 = -6.49 \text{ K or } -1.4\%$$

Example 3.13 Estimate T_b for 1,2-dimethoxyethane using the method of Nannoolal et al. (2004).

Solution. The first- and second-order groups of 1,2-dimethoxyethane are two CH_3—(e), two >C(c)<(e), and two (C,Si) —O—(C,Si) according to Table 3.11. The molecule has no second-order group or corrections (see Table 3.12). The resulting group summations are

	Group	N	$N\Delta_{T_b}$
1	CH_3—(e)	2	503.6676
2	>C(c)<(e),	2	533.7538
3	(C,Si)—O—(C,Si)	2	292.9672
	$\sum_{k=1}^{3} N_k F_k$		1330.3886

The two interacting groups are both EtherO (arbitrarily called EtherO 1 and EtherO 2). Determination of GI_{T_b} from Eq. (3.4-6) requires setting up the matrix as outlined below. The values for $(\Delta_{GI_{ij}})_{T_b}$ are found in Table 3.13. Notice that the ij combinations cover 11, 12, 21, and 22, because there are two interacting groups.

$\Delta_{GI_{ij}T_b}$	$j \rightarrow$	1	2
$i \downarrow$	INTERACTING GROUPS	EtherO 1	EtherO 2
1	EtherO 1	0	91.4997
2	EtherO 2	91.4997	0

The number of non-hydrogen atoms is $n = 6$, so GI_{T_b} from the matrix above and Eq. (3.4-6) is

$$GI_{T_b} = 1/6(0 + 91.4997 + 0 + 91.4997) = 30.4999$$

The Nannoolal estimate of the normal boiling point is thus

$$T_b = (1330.3886 + 30.4999)/(6^{0.6583} + 1.6868) + 84.3395 = 359.85 \text{ K}$$

The Appendix A value for the normal boiling point is 357.75 K. Thus, the difference is

$$T_b \text{ Difference} = 359.85 - 357.75 = 2.10 \text{ K or } 0.59\%$$

Method of Emami et al. (2009) for T_b. As explained in Sec. 3.2, Emami et al. (2009) created a group contribution method for T_c based on UNIFAC groups. They also created a correlation for T_b at the same time. The equation is

$$\frac{T_b}{\text{K}} = \left(\frac{1000}{0.5 + \dfrac{38.36}{\sqrt{\sum_i N_i (\Delta_{T_b})_i}} + \dfrac{1000}{256.6 + \sum_i N_i (\Delta_{T_b})_i}} \right) \tag{3.4-7}$$

where the summation is over the unique groups in the molecule, and the group values $(\Delta_{T_b})_i$ are found in Table 3.8. Application of the Emami et al. method for T_b is shown in Example 3.14.

Example 3.14 Estimate T_b for 2-methylphenol (*o*-cresol) using the method of Emami et al. (2009).

Solution. The groups of 2-methylphenol are one >ACCH3, four >ACH, and one >ACOH according to Table 3.8. The resulting group summations are

	Group	N	$N\Delta_{T_b}$
1	>ACCH$_3$	1	282.512
2	>ACH	4	530.436
3	>ACOH	1	525.221
	$\sum_{k=1}^{3} N_k F_k$		1338.169

The Emami estimate of the normal boiling point is

$$T_b = 1000/[0.5 + 38.36/(1338.169)^{0.5} + 1000/(256.6+1338.169)] = 459.63 \text{ K}$$

Appendix A value for the normal boiling point is 464.15 K. Thus, the differences are

$$T_b \text{ Difference} = 459.63 - 464.15 = -4.52 \text{ K or } -0.97\%$$

Method of Stein and Brown. Stein and Brown (1994) built on the method of Joback to expand the number of groups and overcome the limitations of the linear relation between the numbers of groups and boiling temperature. Their work also addressed the fact that the Joback T_b method systematically overpredicts for normal boiling points above 500 K. The method uses an equation similar to Joback to obtain the initial estimate of the normal boiling point, $T_{b,1}$, and then *corrects* this value to give the final result for T_b. The equations are

$$\frac{T_{b,1}}{\text{K}} = 198.2 + \sum_i N_i (\Delta_{T_b})_i \tag{3.4-8}$$

$$\frac{T_b}{\text{K}} = T_{b,1} - 94.84 + 0.5577 T_{b,1} - 0.0007705 T_{b,1}^2 \qquad T_{b,1} \leq 700 \text{ K} \tag{3.4-9}$$

$$\frac{T_b}{\text{K}} = T_{b,1} + 287.7 - 0.5209 T_{b,1} \qquad T_{b,1} > 700 \text{ K} \tag{3.4-10}$$

where the summation is over the unique groups in the molecule, and the group values $(\Delta_{T_b})_i$ are found in Table 3.14. Application of the method of Stein and Brown for T_b is shown in Example 3.15.

TABLE 3.14 Group contributions for method of Stein and Brown (1994) for T_b for use in Eq. (3.4-8).

Group	Δ_{T_b}
Nonring Carbon Increments	
—CH_3	21.98
—CH_2—	24.22
>CH—	11.86
>C<	4.5
═CH_2	16.44
═CH—	27.95
═C<	23.58
≡CH	21.71
≡C—	32.99
Nonaromatic Ring Carbon Increments	
—CrH_2—	26.44
>CrH—	21.66
>Cr<	11.12
═Cr<	28.19
═CrH—	28.03
Aromatic Ring Carbon Increments	
aaCH	28.53
aaC—	30.76
aaaC	45.46
Halogen Increments	
—Br	76.28
ϕ—Br	61.85
—F	0.13
ϕ—F	−7.81
—I	111.67
ϕ—I	99.93
—Cl	34.08
1—Cl	62.63
2—Cl	49.41
3—Cl	36.23
ϕ—Cl	36.79
Oxygen Increments	
—OH	106.27
1—OH	88.46
2—OH	80.63
3—OH	69.32
ϕ—OH	70.48
—O—	25.16
—Or—	32.98

(*Continued*)

TABLE 3.14 Group contributions for method of Stein and Brown (1994) for T_b for use in Eq. (3.4-8). ***(Continued)***

Group	Δ_{T_b}
—O—OH	72.92
—HC═O	83.38
>C═O	71.53
>Cr═O	94.76
—(C═O)O—	78.85
—(Cr═O)Or—	172.49
—(C═O)OH	169.83
—(C═O)NH2	230.39
—(C═O)NH—	225.09
—(Cr═O)NrH—	246.13
—(C═O)N<	142.77
—(Cr═O)Nr<	180.22
Nitrogen Increments	
—NH2	61.98
ϕ—NH2	86.63
—NH—	45.28
—NrH—	65.50
>N—	25.78
>Nr—	32.77
>NOH	104.87
>NN═O	184.68
aaN	39.88
═NH	73.40
═N—	31.32
═Nr—	43.54
═NrNrH—	179.43
—Nr═Cr(R)NrH—	284.16
—N═NNH—	257.29
—N═N—	90.87
—N═O	30.91
—NO_2	113.99
—C≡N	119.16
ϕ—C≡N	95.43
Sulfur Increments	
—SH	81.71
ϕ—SH	77.49
—S—	69.42
—Sr—	69.00
>SO	154.5
>SO_2	171.58
>C═S	106.20
>Cr═S	179.26

(*Continued*)

TABLE 3.14 Group contributions for method of Stein and Brown (1994) for T_b for use in Eq. (3.4-8). (*Continued*)

Group	Δ_{T_b}
Phosphorus Increments	
—PH_2	59.11
—PH—	40.54
>P—	43.75
>PO—	107.23
Silicon Increments	
>SiH—	27.15
>Si<	8.21
>Sir<	−12.16
Miscellaneous Increments	
>B—	−27.27
—Se—	92.06
>Sn<	62.89

Notes on Symbols:
- Atoms with an "r" are in rings.
- The symbol "a" denotes an aromatic bond.
- The symbol "ϕ" denotes an aromatic system.
- Numbers 1, 2, and 3 denote attachment to primary, secondary, and tertiary carbon atoms, respectively.

Example 3.15 Estimate T_b for 2-methylphenol (*o*-cresol) using the method of Stein and Brown (1994).

Solution. The groups of 2-methylphenol (o-cresol) are one —CH3, four aaCH, two aaC—, and one ϕ—OH according to Table 3.14. The resulting group summations are

	Group	N	$N\Delta_{T_b}$
1	—CH_3	1	21.98
2	aaCH	4	114.12
3	aaC—	2	61.52
4	ϕ—OH	1	70.48
	$\sum_{k=1}^{4} N_k F_k$		268.10

The Stein and Brown *initial* estimate of the normal boiling point, $T_{b,1}$ from Eq. (3.4-8), is

$$T_{b,1} = 198.2 + 268.10 = 466.30 \text{ K}$$

This value suggests using Eq. (3.4-9) to correct the initial estimate. Doing so yields

$$T_b = 466.30 - 94.84 + 0.5577(466.30) - 0.0007705(466.30)^2 = 463.98 \text{ K}$$

The Appendix A value for the normal boiling point is 464.15 K. Thus, the difference is

$$T_b \text{ Difference} = 463.98 - 464.15 = -0.17 \text{ K or } -0.04\%$$

3.5 DISCUSSION OF ESTIMATION METHODS FOR PURE COMPONENT CONSTANTS

Philosophy. As mentioned in Sec. 3.1, a major focus of this edition is to establish the most extensive test possible of the available prediction methods in order to give the reader the broadest assessment of the accuracy of each. The first step in this endeavor was the creation of a test database with experimental values for T_c, P_c, V_c, T_m, and T_b. This was done by combining and reconciling data found in the NIST TRC and DIPPR databases. Both databases are *evaluated* in that they have assessed the available literature to give a recommendation for the property values for each compound. The test database contains only values that (1) came from experiments and (2) agreed between the two source databases. This test database is the largest ever used to date to evaluate the accuracy of prediction methods for the properties in this chapter. Table 3.15 lists the number of unique compounds with experimental values found in the test database for each of the properties discussed in Secs. 3.2 through 3.4.

The second step in establishing the most extensive test of available prediction methods was to obtain or develop computer programs capable of performing the predictions. To remove bias, the computer tools required the capability to be run by the authors of this edition and not the developers of the method. The computer program also had to have *batching* capabilities so that the predictions could be done in a timely manner. Prediction techniques that did not satisfy the above two criteria were not considered for inclusion in this edition.

Solicitation for prediction methods was done over two years at multiple professional meetings, through the establishment and advertising of a website, and through personal contact to potentially interested parties. Every reasonable effort was made to accommodate contributions, but some groups chose not to supply the required computer programs or web interfaces. In other cases, the computer programs supplied did not function in a batch mode or experienced crashes that could not be resolved. The methods discussed below are only those that could be automated as discussed above.

TABLE 3.15 Size of the test database for each of the properties discussed in Secs. 3.2 and 3.4.

Property	Number of Unique Compounds with Experimental Values in the Test Database
T_c	1041
P_c	720
V_c	428
T_m	9667
T_b	4950

To identify if a method was better at one class of compounds than another, each chemical in the test database was assigned a *class* according to the scheme below.

- **Normal:** Compounds containing only C, H, F, Br, and/or I with molecular weights below 213 g/mol.
- **Heavy:** Compounds containing only C, H, F, Br, and/or I with molecular weights greater than or equal to 213 g/mol. Also compounds containing only C, H, F, Br, I and/or metal atoms (e.g., Fe, As, Al, etc.) regardless of MW.
- **Polar:** Compounds containing atoms other than C, H, F, Br, metals (e.g., O, N, P, S, etc.) that cannot form hydrogen bonds (e.g., esters, ketones, aldehydes, thioesters, sulfonyl, etc.).
- **Associating:** Compounds containing atoms other than C, H, F, Br, metals (e.g., O, N, P, S, etc.) that can form hydrogen bond (e.g., acids, amines, alcohols, etc.).

Results are presented in Tables 3.17 to 3.20 and 3.22 for each class and prediction method for the various properties. Each case shows the number of compounds tested with the method (N) and the average absolute relative deviation ($AARD$). The latter is defined as

$$AARD = \frac{1}{N}\sum_{i=1}^{N}\left|\frac{X_{\mathrm{pred}_i} - X_{\mathrm{exp}_i}}{X_{\mathrm{exp}_i}}\right| \tag{3.5-1}$$

where the index is for compound i in the test set, X_{pred_i} is the predicted value for property X for compound i, and X_{exp_i} is the experimental value for property X for compound i. Table 3.16 contains abbreviations used to delineate each prediction method.

3.5.1 Critical Temperature

Table 3.17 shows the performance of the prediction methods against the test database for T_c. Because most T_c prediction methods require T_b, only compounds with an experimental value for T_b were used to test the T_c methods. As mentioned above, the results are listed according to compound class. The following are listed for each method/class pair: number of compounds in the test set (N) and average absolute relative deviation ($AARD$). The results for Emami et al. (2009) exclude all molecules containing F because testing showed these groups produced very poor agreement with experiment. The number of compounds in each test set is different because each method has different groups that may or may not be able to completely describe the molecule.

TABLE 3.16 Abbreviations for the predictions methods evaluated against the test database.

Abbreviation	Method, Property(ies)
LY	Lydersen (1955), T_c, P_c
AM	Ambrose (1979), T_c, P_c, V_c
FE	Fedors (1979), V_c
JO	Joback (1984; Joback and Reid, 1987), T_c, P_c, V_c, T_m, T_b
ST	Stein and Brown (1994), T_b
CO	Constantinou and Gani (1994), T_c, P_c, V_c, T_m, T_b
WJ1	Wilson and Jasperson (1996), T_c, P_c, 1st order
WJ2	Wilson and Jasperson (1996), T_c, P_c, 2nd order
NA	Nannoolal et al. (2004), T_b; Nannoolal et al. (2007) T_c, P_c, V_c
EM	Emami et al. (2009), T_c, T_b

TABLE 3.17 Performance of prediction techniques for T_c against the test database.

	Compound Class							
	Normal		Heavy		Polar		Associating	
Method	N	AARD	N	AARD	N	AARD	N	AARD
LY	408	1.0%	67	5.8%	134	1.6%	165	2.3%
AM	390	1.0%	43	1.1%	132	0.8%	164	1.3%
JO	377	1.0%	67	4.0%	133	1.4%	163	2.2%
CO	332	4.7%	61	5.3%	129	6.0%	153	2.4%
WJ1	422	1.4%	68	1.4%	135	1.9%	167	1.8%
WJ2	422	1.4%	68	1.4%	135	3.1%	167	1.6%
NA	397	1.0%	68	1.4%	131	1.0%	163	1.3%
EM	277	2.8%	33	1.0%	148	1.7%	221	3.5%

Across all classes of compounds, the method of Constantinou and Gani (1994) consistently performs poorly while the Ambrose (1979) method performs consistently well. Both the first- and second-order methods of Wilson and Jasperson also perform consistently well—especially considering these methods are atom-based and apply to the largest number of molecules. The method of Emami et al. (2009) is accurate for heavy and polar molecules but suffers for normal (lighter) compounds and those which hydrogen bond. Also, as mentioned above, this method should never be used to predict properties of compounds containing fluorine (F). The methods of Lydersen (1955), Joback (1984; Joback and Reid, 1987), and Constantinou and Gani (1994) suffer significantly for heavy molecules and should not be used to predict T_c of such compounds. More discussion is found later explaining how to further narrow the selection of prediction method for an unknown value of T_c.

3.5.2 Critical Pressure

Table 3.18 illustrates the accuracy of the prediction methods for P_c. Fewer experimental values exist for this property than for T_c. For compounds that fall in the normal class, each of the methods appear accurate with the Lydersen (1955) method showing the best average accuracy and the Nannoolal (2007) technique the least. The Wilson and Jasperson methods predict the best for heavy compounds but suffer for polar molecules. All methods

TABLE 3.18 Performance of prediction techniques for P_c against the test database.

	Compound Class							
	Normal		Heavy		Polar		Associating	
Method	N	AARD	N	AARD	N	AARD	N	AARD
LY	298	5.2%	73	14.3%	103	5.4%	185	6.3%
AM	284	5.7%	49	13.1%	102	5.1%	184	8.8%
JO	263	5.6%	73	10.6%	102	6.6%	185	6.4%
CO	242	6.0%	65	12.6%	100	7.1%	178	7.4%
WJ1	322	5.5%	82	7.9%	105	7.9%	187	6.6%
WJ2	322	5.4%	82	7.9%	105	7.5%	189	6.4%
NA	297	6.9%	74	10.8%	101	6.4%	183	6.7%

TABLE 3.19 Performance of prediction techniques for V_c against the test database.

	Compound Class							
	Normal		Heavy		Polar		Associating	
Method	N	AARD	N	AARD	N	AARD	N	AARD
AM	180	2.8%	17	5.2%	70	4.4%	78	3.4%
FE	104	3.1%	28	3.6%	72	5.9%	79	4.0%
JO	194	2.8%	28	3.2%	69	3.8%	79	3.7%
CO	173	4.0%	24	8.0%	70	8.3%	74	3.7%
NA	201	6.4%	28	8.9%	69	2.0%	79	3.1%

appear to work equally well for associating compounds except those by Ambrose (1979) and Constantinou and Gani (1994), which perform worse than the others.

3.5.3 Critical Volume

Table 3.19 contains the performance data of the prediction techniques for V_c broken down according to compound class. Notice that there are less experimental data for V_c than for the other critical properties. The method of Nannoolal et al. (2007) appears to work better for compounds with groups other than carbon, hydrogen, and halogens (polar and associating compounds), while the early methods (AM, FE, and JO) work well for normal and heavy classes of molecules. Constantinou and Gani's method (1994) performs poorly as with T_c and P_c.

3.5.4 Melting Point

Table 3.20 displays the accuracies of the prediction methods for T_m according to compound class. These data demonstrate the prediction of melting temperature is very difficult with the average accuracy never less than 14% for any class of molecules for either method. The method of Constantinou and Gani (1994) is more accurate than that of Joback (1984; Joback and Reid, 1987) and is the recommended method for this property for all classes of compounds. Of note, neither contains groups for silicon compounds.

The difficulty of predicting melting points compared to critical constants likely lies in the fact that the energy of intermolecular interactions in oriented crystal states strongly depends on the shape and packing of the molecules. Take 1,1,2-trimethylcyclohexane and 1,1,3-trimethylcyclohexane. These isomers differ only in the position of the monosubstituted ring carbon and would be parsed into the same groups for all the group contribution methods discussed here. Reliable experimental data (see Streiff et al., 1947; ASTM, 1963; Forziati et al., 1949) for T_m, T_c, and T_b are available for both compounds and are summarized in Table 3.21. Notice the

TABLE 3.20 Performance of prediction techniques for T_m against the test database.

	Compound Class							
	Normal		Heavy		Polar		Associating	
Method	N	AARD	N	AARD	N	AARD	N	AARD
JO	1520	16.7%	1220	28.3%	1674	36.8%	3198	25.5%
CO	1304	17.8%	1045	17.6%	1044	14.2%	1913	14.7%

TABLE 3.21 T_m, T_c, and T_b for 1,1,2-trimethylcyclopentane and 1,1,3-trimethylcyclopentane.

Compound	T_m (K)	T_c (K)	T_b (K)
1,1,2-trimethylcyclopentane	252	580	387
1,1,3-trimethylcyclopentane	131	570	378
Absolute difference	121	10	9

large difference in T_m compared to the differences in T_c and T_b. At the melting/freezing transition, molecules must align into the crystalline form, and this process is apparently very different between the two isomers. The critical temperature and normal boiling point describe transitions between two states characterized by fairly random molecular positions that are much more mobile than the solid phase and they do not depend on static positions and orientations of molecules. Prediction methods based on groups of atoms are based on the idea that such groups have similar thermodynamic properties regardless of the exact location in the molecule. Improvements at predicting melting temperature will likely need to account for crystal structures and directly computed intermolecular interactions (Hylton et al., 2015).

3.5.5 Normal Boiling Point

Table 3.22 lists the performance of the T_b prediction techniques against the test database according to compound class. Stein and Brown's method (1994) produces more accurate values for associating compounds than the other methods and applies to a wide array of structural groups. The Emami et al. (2009) method works better for the remaining classes of compounds, but also has a limited number of groups compared to the other methods. Also, as mentioned for T_c, this method should not be used for compounds containing fluorine (F).

3.5.6 Selecting a Prediction Method Best Practice

As evidenced above, multiple methods perform well when averaging is done across a large number of compounds. When selecting a prediction method for an unknown property for a compound, DIPPR and NIST-TRC best practice, and the recommendation of this edition, is therefore to evaluate the performance of the methods for compounds similar in structure to that of the unknown value. One way to do this is by assigning each compound to a family or class and finding the prediction method that best reproduces the properties in question for other compounds in the same category that have experimental data. Example 3.16 demonstrates the process.

TABLE 3.22 Performance of prediction techniques for T_b against the test database.

	Compound Class							
	Normal		Heavy		Polar		Associating	
Method	N	AARD	N	AARD	N	AARD	N	AARD
JO	2489	3.7%	279	7.1%	703	6.5%	977	5.7%
CO	2489	4.6%	279	6.0%	706	5.2%	982	6.6%
ST	2499	3.3%	279	8.8%	708	3.6%	987	2.9%
NA	2499	4.0%	279	8.3%	707	5.0%	987	3.7%
EM	2153	2.6%	165	3.4%	515	2.2%	887	3.0%

Example 3.16 Determine which prediction method will yield the most accurate prediction for T_c of 3-ethylthiophene.

Solution. 3-ethylthiophene is a member of the *Sulfides/Thiophenes* family in the DIPPR database. The database contains 62 compounds in this family, but only 15 of these have experimental values for T_c. Determining which of the prediction methods in this chapter is likely to give the best estimate of T_c for 3-ethylthiophene was done automatically using DIPPR's DIADEM program in a process described by Rowley et al. (2007). The general procedure is as follows:

1. Calculate a predicted value for T_c for each of the 15 compounds with experimental T_c data using each prediction technique discussed in this chapter. (Note that some techniques cannot be applied to all the chemicals due to missing groups.)
2. Calculate the minimum, maximum, and average [see Eq. (3.5-1)] relative deviation between the experimental data and the values produced by each prediction technique.
3. Compare the accuracy of each prediction technique based on the metrics calculated in Step 2 to identify which method performs best for the family.

The results for this approach are found in the following table:

Method	Number of Compounds	Ave. Abs. Rel. Dev. (%)	Min. Abs. Rel. Dev. (%)	Max. Abs. Rel. Dev. (%)
LY	15	1.09	0.06	6.97
NA	12	1.33	0.04	5.99
JO	15	1.41	0.01	5.74
WJ1*	15	1.54	0.05	5.82
AM	15	1.57	0.05	6.54
EM	6	1.87	0.49	6.72
CO	10	1.98	0.00	5.02

*WJ2 = WJ1 for this family.

Notice that groups are not available in some methods to predict a value for all 15 compounds in the family. The maximum and minimum deviations for each method are comparable, but the Lydersen method performs the best on average. The Lydersen method is thus the preferred approach to predict the critical temperature of 3-ethylthiophene.

3.5.7 QSPR and Other Methods

There is a great variety of other types of estimation methods for pure-component properties besides the above *group/bond/atom* approaches. The techniques generally fall into three classes. The first is based on *factor analysis* that builds correlation equations from data of other measurable, macroscopic properties such as densities, molecular weight, boiling temperature, etc. Such methods include those of Klincewicz and Reid (1984) and of Vetere (1995) for many types of substances. Somayajulu (1991) treats only alkanes but also suggests ways to approach other homologous series. However, the results of these methods are either of reduced accuracy or extra complexity. The way the parameters depend upon the type of substance and their need for other input information prevents automation for evaluation against the test database. We have thus not given any results for these methods.

The second type of "other" techniques are based on the concept of *molecular descriptors*. These molecular properties are calculated mathematically via specified algorithms and attempt to describe the physical and chemical information of the molecule. These descriptors are then correlated to macroscopic properties to create so-called "Quantitative Structure-Property Relationships" (QSPR).

Molecular descriptors are usually obtained in a two-step fashion. First, quantum mechanical calculations are done to obtain the optimized geometry of the molecule. This geometry is then input into a second computer program which calculates the desired molecular descriptors.

The DIPPR 801 Database includes hundreds of molecular descriptors for each compound. The computer program Gaussian from Gaussian Inc. (Gaussian 16, 2016) is used by the project for geometry optimization, and the program Dragon 7 from Kode Chemoinformatics (Dragon, 2021) is used to obtain the molecular descriptors. The latter can calculate over 5000 descriptors describing different aspects of the molecules such as topology, connectivity, geometry, constitution, ring structures, charge distribution, and many more. A free database of molecular descriptors is also maintained by the Milano Chemometrics and QSAR Research Group (Milano, 2021; Todeschini and Consonni, 2009).

Once molecular descriptors for a set of molecules are obtained, developing QSPR models is usually done using multiple linear regressions. For example, Kier and Hall (Kier, 1976; Kier and Hall, 1976) proposed multiple *valence connectivity indices* to characterize the molecular size and branching of a molecule. The zero-order index (${}^0\chi^V$) parses the molecules into individual atoms, the first-order index (${}^1\chi^V$) into bonded pairs, the second-order (${}^2\chi^V$) index into all fragments with two bonds (three atoms), etc. Once the valence connectivity indices are calculated for the molecule, they can be used to predict properties. One of the many correlations given by the authors is an equation for the critical volume for normal alcohols:

$$\frac{V_c}{\text{cm}^3\ \text{mol}^{-1}} = 50.76 + 109.8({}^1\chi^V) \tag{3.5-2}$$

This example shows one of the weaknesses of QSPR correlations. Specifically, the correlations are usually only applicable to a specific family of compounds and are not transferrable to other types of molecules.

Because multiple linear regression is relatively easy to perform, many QSPR methods have been proposed for pure-component properties. But one of the problems with using these methods is that the descriptors must be calculated using the same methods and programs/algorithms used by the author, and sometimes needed information is missing in the literature or the program can no longer be obtained. For example, CODESSA Pro (2013) was a molecular descriptor program used by many groups for several years, but it has been sold to various companies which are now defunct, is no longer being upgraded, and is difficult to obtain. This is problematic because the molecular descriptor programs that are currently available may not use the same algorithms to determine the molecular descriptors. Two such descriptors are van der Waals volume and van der Waals surface area. Several algorithms have been proposed to calculate these values, but the methods used in CODESSA Pro do not produce the same values as the methods from Dragon 7. Another problem is the level of theory and basis set used to generate the optimized structures. The QSPR methods found in the literature rarely use the same quantum mechanical approach, so applying the techniques in a general way becomes cumbersome. The QSPR method of Turner et al. (1998) is now described to illustrate the difficulties.

Turner et al. (1998) correlated critical temperature and pressure to molecular descriptors using a set of 165 compounds. The optimized geometries of each molecule were obtained using the PM3 (Stewart, 1989) method and the computer program MOPAC (MOPAC2016, 2016). The paper does not state which program was used to obtain the molecular descriptors, but the model developed for T_c used eight descriptors: dipole moment μ, area A_+, a connectivity index ("simple path 3 connectivity index"), number of oxygens, number of secondary carbon bonds of the sp^3 type, gravitation index, a function of acceptor atom charge, and average positive charge on carbons. A completely different descriptor set was used for P_c.

We see here why it is often difficult to use QSPR methods. First, one must optimize the geometry of the molecule using the PM3 method. The hope is that doing so with the quantum mechanical program of choice at your institution will produce the same geometry as that of MOPAC. Then, this geometry must be used to obtain descriptors. There are two further concerns in this regard. First, the "simple path 3 connectivity index" could be one of many such descriptors available. Second, the paper does not state the program needed to calculate the descriptors to ensure they are consistent with the model.

The third class of non-group-contribution prediction methods is machine learning (ML) or artificial intelligence (AI) techniques. To date, this is an extension of QSPR methods in that molecular descriptors are correlated to macroscopic properties. But rather than doing so using linear or nonlinear regression, the relationships are determined by, and stored in, the computer, so a closed-form equation is not available to be shared. These methods suffer from many of the same problems as those discussed for QSPR methods from linear regression—such as reproducing the quantum mechanical calculations used to optimize geometry and ambiguous or irreproducible determination of the descriptors—but they include an additional difficulty. Using the techniques requires the exact database of values and the exact algorithms used to train the method, and these are notoriously difficult to reproduce among different groups.

QSPR methods—whether obtained from linear regression, ML, or AI—show great promise in improving prediction of pure-component thermophysical properties, but changes are needed to allow verification of the method by outside groups and to make them useful to the scientific community. Unfortunately, those methods are frequently published as "black boxes" without the information needed to reproduce the results and use for additional predictions. Ideally, the developers of such methods would set up a web interface that allows users to upload a list of chemical structures [e.g., SMILES (Weininger, 1988)] and return the list of corresponding predicted property values. Chemical structures are needed, rather than chemical names, CAS RNs, or other identifiers, to maintain the integrity of the test and prevent simple database searches of experimental values. Uploading a list is also required for computational expediency as a tool that only allows one molecule to be processed at a time would be prohibitively time consuming.

Kazakov et al. (2010) outline additional criteria for constructing rigorous QSPR methods and applied these criteria to create a prediction for T_c and P_c. The recommendations focus on creating sound training sets composed of *experimental* data, ensuring the uncertainties in the training set data points are quantified and used to weight the regression, and specifying the exact algorithms used in generating the descriptors and performing the regression. Their method performed better than multiple group contribution techniques discussed in this chapter. However, use of the method by outside groups remains difficult as users must still construct their own database of descriptors and perform their own regression. As discussed above for QSPR techniques in general, producing the exact results as Kazakov et al. would prove difficult because the database of descriptors and experimental values are not provided, and the researchers used a customized support vector machine algorithm package for the regression.

It will be important that users follow the developments in this area so that the most prudent decisions about investment and commitment can be made.

3.6 DIPOLE MOMENTS

Dipole moments of molecules are often required in property correlations for polar materials such as for virial coefficients (Chap. 5) and viscosities (Chap. 10). The best sources of this constant are the compilations by McClellan (1963), Nelson et al. (1967), and Computational Chemistry Comparison and Benchmark DataBase at NIST (CCCBD, 2020). All the programs used for *molecular descriptors* and *quantum mechanics* yield molecular dipole moments as a part of the analysis when an estimated value is needed.

Dipole moments for many materials are listed in Appendix A. Their variations with temperature and phase (gas or liquid) are not considered. Such variations are usually not large enough to affect the estimation result. It should be noted that the dipole is only the lowest of a series of electrostatic effects on intermolecular forces; higher-order terms such as quadrupoles can also be important, as it is in the case of CO_2.

It is often of interest to determine whether electrostatic contributions are significant compared to van der Waals attraction (dispersion). The importance of the dipolar forces depends on the ratio of electrostatic to van der Waals energies. This ratio can be estimated in dimensionless fashion by

$$\mu^* = \frac{4362\mu^2}{T_c V_c} \tag{3.6-1}$$

where μ is the dipole moment in Debye (D), T_c is critical temperature in K, and V_c is critical volume in cm^3/mol. It is estimated that if μ^* of Eq. (3.6-1) is less than 0.03, dipolar effects can be neglected. This criterion is based on Eq. (5-71) from Prausnitz et al. (1999) where the magnitudes of the dispersion and dipole contributions to estimated second virial coefficients are compared.

Another way to determine if polar forces are important was discussed by Pitzer (1995) who was seeking to identify if a compound could be described by the three-parameter corresponding states principle (CSP) and its acentric factor (see Sec. 3.3). As originally envisioned, ω describes only "normal" fluids which are not strongly polar or associating. Pitzer (1995) suggested that the surface tension (σ) was very sensitive to such effects and could thus be used to separate "normal" compounds from those which are polar or associating. The process involves estimating the surface tension using the CSP-based method of Eq. (13.3-7) and comparing this estimate to experimental data. Pitzer (1995) states that estimates deviating more than 5% from the data "indicate significant abnormality" and should not be considered "normal." If the substance is found to be "normal" using this test, then polar forces are not important.

3.7 AVAILABILITY OF DATA AND COMPUTER PROGRAMS

Several commercial products for obtaining pure-component constants are readily available. These include data and correlation-based tabulations and computer-based group contribution methods. The data for Appendix A were obtained from the DIPPR 801 database (Wilding et al., 2017; DIPPR 2021), and the automated evaluation of the prediction methods from molecular structures was done with DIPPR's DIADEM software (Rowley et al., 2007). There is a similar tabulation as well as automated prediction methods based on molecular structure available in ThermoData Engine from the Thermodynamics Research Center (Diky et al., 2021). Joback has established a program (Cranium, 2021) for computing many properties by various group contribution methods. Gani and coworkers at PSE for SPEED also have a program available (ProCAPE, 2021) for many properties.

3.8 NOTATION

$AARD$	average absolute relative deviation; see Eq. (3.5-1)
f	general function in the prediction method of Constantinou and Gani (1994); see Eq. (3.2-6)
F	general symbol for one of the properties (T_c, P_c, V_c, T_b, or T_m) in the prediction method of Constantinou and Gani (1994)
GI_p	total group interaction contribution for property p in the prediction method of Nannoolal et al. (2007)
M_j	number of times group j is found in a compound
MW	molecular weight
n	number of non-hydrogen atoms in a compound in the prediction method Nannoolal et al. (2004) for T_b
N_A	Avogadro's number ($6.02214076 \times 10^{23}$ mol^{-1})
N_{atoms}	number of atoms in a molecule
N_i	number of times group i is found in a compound
N_r	number of rings in a molecule
P	pressure
P_c	critical pressure
P^{vp}	saturated liquid vapor pressure
R	gas constant (8.314462618 J/(mol·K))
T	temperature
T_b	normal (P = 101325 Pa) boiling point
$T_{b,1}$	initial estimate of the normal (P = 101325 Pa) boiling point in the prediction method of Stein and Brown (1994) for T_b

$T_b^{10\text{ mmHg}}$	boiling temperature at a pressure of 10 mmHg; used in the prediction method of Emami et al. (2009)
T_c	critical temperature
T_m	normal ($P = 101325$ Pa) melting temperature
V_c	critical volume
W	weighting factor in the prediction method of Constantinou and Gani (1994); see Eq. (3.2-6)
Y	function in the prediction method of Wilson and Jasperson (1996) for P_c

Greek

$(\Delta_F^{(1)})_i$	contribution for first-order group i for property F (T_c, P_c, V_c, T_b, or T_m) in the prediction method of Constantinou and Gani (1994)
$(\Delta_F^{(2)})_j$	contribution for second-order group j for property F (T_c, P_c, V_c, T_b, or T_m) in the prediction method of Constantinou and Gani (1994)
$(\Delta_p^{(1)})_i$	contribution for first-order group i for property p (T_c or P_c) in the prediction method of Wilson and Jasperson (1996)
$(\Delta_p^{(2)})_j$	contribution for second-order group j for property p (T_c or P_c) in the prediction method of Wilson and Jasperson (1996)
$(\Delta_{P_c})_i$	contribution for group i for a prediction method for P_c
$(\Delta_{T_b})_i$	contribution for group i for a prediction method for T_b
$(\Delta_{T_{b10}})_i$	contribution for group i for the prediction method of Emami et al. (2009) for $T_b^{10\text{ mmHg}}$
$(\Delta_{T_c})_i$	contribution for group i for a prediction method for T_c
$(\Delta_{T_m})_i$	contribution for group i for a prediction method for T_m
$(\Delta_{V_c})_i$	contribution for group i for a prediction method for V_c
$(\Delta_{GIij})_p$	group interaction contribution between groups i and j for property p in the prediction method of Nannoolal et al. (2007)
μ	dipole moment
μ^*	dimensionless dipole moment
ω	acentric factor
${}^z\chi^V$	valence connectivity index of order z of Kier and Hall (1976)

Disclaimer. Commercial equipment, instruments, or materials are identified only in order to adequately specify certain procedures. In no case does such identification imply recommendation or endorsement by the National Institute of Standards and Technology, nor does it imply that the products identified are necessarily the best available for the purpose.

3.9 REFERENCES

Ambrose, D.: "Correlation and Estimation of Vapour-Liquid Critical Properties. I. Critical Temperatures of Organic Compounds," National Physical Laboratory, Teddington, *NPL Rep. Chem.* 92: 1978 (1980).

Ambrose, D.: "Correlation and Estimation of Vapour-Liquid Critical Properties. I. Critical Temperatures of Organic Compounds," National Physical Laboratory, Teddington, *NPL Rep. Chem.* 98: (1979).

Ambrose, D., and C. L. Young: *J. Chem. Eng. Data*, **40:** 345 (1995).

ASTM Committee D-2 on Petroleum Products and Lubricants: *Physical Constants of Hydrocarbons C1 to C10* (1963).

Brulé, M. R., C. T. Lin, L. L. Lee, and K. E. Starling: *AIChE J.*, **28:** 616 (1982).

CCCBD, NIST Computational Chemistry Comparison and Benchmark Database, NIST Standard Reference Database Number 101, Release 21, Editor: R. D. Johnson III, http://cccbdb.nist.gov/, (2020).

Cholakov, G. St., W. A. Wakeham, and R. P. Stateva: *Fluid Phase Equil.*, **163:** 21 (1999).
CODESSA Pro, CompuDrug International, https://www.compudrug.com/, (2013).
Constantinou, L., and R. Gani: *AIChE J.,* **40:** 1697 (1994).
Constantinou, L., R. Gani, and J. P. O'Connell: *Fluid Phase Equil.*, **104:** 11 (1995).
Coplen, T. B.: *J. Phys. Chem. Ref. Data*, **30:** 701–712 (2001).
Cranium Version 4.2: Molecular Knowledge Systems, Inc., PO Box 10755, Bedford, NH 03110-0755 USA; https://www.molecularknowledge.com/, (2021).
Diky, V., R. D. Chirico, M. Frenkel, A. Bazyleva, J. W. Magee, E. Paulechka, A. Kazakov, E. W. Lemmon, C. D. Muzny, A. Y. Smolyanitsky, S. Townsend, and K. Kroenlein. *NIST ThermoData Engine, NIST Standard Reference Database 103 a/b*, National Institute of Standards and Technology, USA (2021): https://www.nist.gov/mml/acmd/trc/thermodata-engine
DIPPR, Design Institute for Physical Properties, American Institute of Chemical Engineers, 120 Wall St. FL 23, New York, NY 10005-4020 USA; https://www.aiche.org/dippr, (2021).
Dragon (Software for Molecular Descriptor Calculation) Version 7; https://chm.kode-solutions.net/pf/dragon-7-0/, (2021).
Egolf, L. M., M. D. Wessel, and P. C. Jurs: *J. Chem. Inf. Comput. Sci.*, **34:** 947 (1994).
Emami, F. S., A. Vahid, and J. R. Elliot: *J. Chem. Thermodyn.*, **41:** 530–537 (2009).
Fedors, R. F.: *AIChE J.,* **25:** 202 (1979).
Forziati, A.F., W. R. Norris, and F. D. Rossini: *J. Res. Natl. Bur. Stand. A*, **43:** 555–563 (1949).
Gaussian 16, Revision D.01, M. J. Frisch, G. W. Trucks, H. B. Schlegel, G. E. Scuseria, M. A. Robb, J. R. Cheeseman, G. Scalmani, V. Barone, G. A. Petersson, H. Nakatsuji, X. Li, M. Caricato, A. V. Marenich, J. Bloino, B. G. Janesko, R. Gomperts, B. Mennucci, H. P. Hratchian, J. V. Ortiz, A. F. Izmaylov, J. L. Sonnenberg, D. Williams-Young, F. Ding, F. Lipparini, F. Egidi, J. Goings, B. Peng, A. Petrone, T. Henderson, D. Ranasinghe, V. G. Zakrzewski, J. Gao, N. Rega, G. Zheng, W. Liang, M. Hada, M. Ehara, K. Toyota, R. Fukuda, J. Hasegawa, M. Ishida, T. Nakajima, Y. Honda, O. Kitao, H. Nakai, T. Vreven, K. Throssell, J. A. Montgomery, Jr., J. E. Peralta, F. Ogliaro, M. J. Bearpark, J. J. Heyd, E. N. Brothers, K. N. Kudin, V. N. Staroverov, T. A. Keith, R. Kobayashi, J. Normand, K. Raghavachari, A. P. Rendell, J. C. Burant, S. S. Iyengar, J. Tomasi, M. Cossi, J. M. Millam, M. Klene, C. Adamo, R. Cammi, J. W. Ochterski, R. L. Martin, K. Morokuma, O. Farkas, J. B. Foresman, and D. J. Fox, Gaussian, Inc., Wallingford CT, 2016.
Grigoras, S.: *J. Comp. Chem.*, **11:** 493 (1990).
Horvath, A. L.: '*Molecular Design: Chemical Structure Generation from the Properties of Pure Organic Compounds*, Elsevier, 1992.
Hylton, R. K., G. J. Tizzard, T. L. Threlfall, A. L. Ellis, S. J. Coles, C. C. Seaton, E. Schulze, H. Lorenz, A. Seidel-Morgenstern, M. Stein, and S. L Price: *J. Am. Chem. Soc.*, **137:** 11095–11104 (2015).
Joback, K. G.: "A Unified Approach to Physical Property Estimation Using Multivariate Statistical Techniques," S. M. Thesis, Department of Chemical Engineering, Massachusetts Institute of Technology, Cambridge, MA, 1984.
Joback, K. G., and R. C. Reid: *Chem. Eng. Comm.*, **57:** 233 (1987).
Katritzky, A. R., V. S. Lobanov, and M. Karelson: *Chem Soc. Rev.*, **24:** 279 (1995).
Katritzky, A. R., T. Tamm, Y. Wang, S. Sild, and M. Karelson: *J. Chem. Inf. Comput. Sci.*, **39:** 684 (1999).
Kazakov, A., C. D. Munzy, V. Diky, R. D. Chirico, and M. Frenkel: *Fluid Phase Equil.*, **298:** 131–142 (2010).
Kier, L. B: *Molecular Connectivity in Chemistry and Drug Research, Medical Chemistry*, Vol. 14, New York, Academic Press, 1976.
Kier, L. B., and L. H. Hall: *J. Pharm. Sci.*, **65:** 1806–1809 (1976).
Klincewicz, K. M., and R. C. Reid: *AIChE J.*, **30:** 137 (1984).
Lin, H.-M., and K. C. Chao: *AIChE J.*, **30:** 981 (1984).
Lydersen, A. L.: "Estimation of Critical Properties of Organic Compounds," *Univ. Wisconsin Coll. Eng., Eng. Exp. Stn. Rept. 3*, Madison, WI, April, 1955.
McClellan, A. L., *Tables of Experimental Dipole Moments*, Vol. 1, San Francisco, W. H. Freeman Pub., 1963.
Milano Chemometrics and QSAR Research Group: University of Milano-Bicocca, Milano, Italy; https://michem.unimib.it/, (2021).
MOPAC2016, J. J. P. Stewart, Stewart Computational Chemistry, Colorado Springs, CO, USA, http://OpenMOPAC.net, (2016).
Myers, K. H.: "Thermodynamic and Transport Property Prediction Methods for Organometallic Compounds," M. S. Thesis, University Park, Pennsylvania, The Pennsylvania State University, 1990.
Nannoolal, Y., J. Rarey, D. Ramjugernath, and W. Cordes: *Fluid Phase Equilib.*, **226:** 45–63 (2004).
Nannoolal, Y., J. Rarey, D. Ramjugernath: "Estimation of Pure Component Properties Part 2. Estimation of Critical Property Data by Group Contribution," *Fluid Phase Equilib.*, **252:** 1–27 (2007).
Nelson Jr., R. D., D. R. Lide Jr., and A. A. Maryott: "Selected Values of Electric Dipole Moments for Molecules in the Gas Phase," NSRDS 10, National Bureau of Standards, Washington, D.C., 1967.

Nikitin, E. D., P. A. Pavlov, and P. V. Skripov: *J. Chem. Thermodyn.*, **25:** 869–880 (1993).
Nikitin, E. D., P. A. Pavlov, and M. G. Skutin: *Fluid Phase Equilib.*, **161:** 119–134 (1999).
Pitzer, K. S.: *Thermodynamics*, 3rd ed., New York, McGraw-Hill, p. 521, 1995.
Pitzer, K. S., D. Z. Lippmann, R. F. Curl, C. M. Huggins, and D. E. Petersen: *J. Am. Chem. Soc.*, **77:** 3433 (1955).
Prausnitz, J. M., R. N. Lichtenthaler, and E. G. de Azevedo: *Molecular Thermodynamics of Fluid-Phase Equilibria*, 3rd ed., Prentice-Hall, 1999.
ProCAPE, PSE for SPEED Company Limited, Ordrup Jagtvej 42D, DK-2920 Charlottenlund, Denmark; https://www.pseforspeed.com/ (2021).
Riazi, M. R., and T. E. Daubert: *Hydrocarbon Process. Petrol. Refiner*, **59**(3): 115 (1980).
Rowley, J. R., W. V. Wilding, J. L. Oscarson, and R. L. Rowley: *Int. J. Thermophys.*, **28:** 824–834 (2007).
Somayajulu, G. R.: *Int. J. Thermophys.*, **12:** 1039 (1991).
Stein, S. E., and R. L. Brown, R. L.: *J. Chem. Inf. Comput. Sci.*, **34:** 581–587 (1994).
Stewart, J. J. P.: *J. Comput. Chem.*, **10:** 209–220 (1989).
Streiff, A. J., E. T. Murphy, J. C. Cahill, H. F. Flanagan, V. A. Sedlak, C. B. Willingham, and F. D. Rossini: *J. Res. Natl. Bur. Stand.*, **38:**, 53–94 (1947).
Teja, A. S., M. Gude, and D. J. Rosenthal: *Fluid Phase Equil.*, **52:** 193–200 (1989).
Teja, A. S. and M. J. Anselme: *AIChE Symp. Ser.*, **279:** 115, 122 (1990).
Todeschini, R., and V. Consonni: *Molecular Descriptors for Chemoinformatics*, 2 vols, Wiley-VCH, 2009.
Tsonopoulos, C., J. L. Heidman, and S. C. Hwang: *Thermodynamic and Transport Properties of Coal Liquids*, New York, Wiley, 1986.
Turner, B. E., C. L. Costello, and P. C. Jurs: *J. Chem. Inf. Comput. Sci.*, **38:** 639 (1998).
Twu, C.: *Fluid Phase Equil.*, **16:** 137 (1984).
Twu, C. H., and J. E. Coon: *Fluid Phase Equil.*, **117:** 233 (1996).
Vetere, A.: *Fluid Phase Equil.*, **109:** 17 (1995).
VonNiederhausern, D., G. M. Wilson, and N. F. Giles: *J. Chem. Eng. Data,* **45**: 157–160 (2000).
Weininger, D.: *J. Chem. Inf. Comput. Sci.*, **28:** 31–36 (1988).
Wilding, W. V., T. A. Knotts, N. F. Giles, R. L. Rowley, and J. L. Oscarson, *DIPPR® Data Compilation of Pure Chemical Properties*, Design Institute for Physical Properties, AIChE, New York, NY (2017). (https://dippr.aiche.org/)
Wilson, G. M., R. H. Johnston, S. C. Hwang, and C. Tsonopoulos: *Ind. Eng. Chem. Proc. Des. Dev.*, **20:** 94 (1981).
Wilson, L. C., W. V. Wilding, H. L. Wilson, and G. M. Wilson: *J. Chem. Eng. Data*, **40:** 765–768 (1995).
Wilson, G. M., and L. V. Jasperson: "Critical Constants T_c, P_c, Estimation Based on Zero, First and Second Order Methods," AIChE Spring Meeting, New Orleans, LA, (1996).
Wilson, L. C., L. V. Jasperson, D. VonNiederhausern, N. F. Giles, and C. Ihmels: *J. Chem. Eng. Data*, **63:** 3408–3417 (2018).
Zhang, J., B. Zhang, S. Zhao, R. Wang, and G. Yang: *Ind. Eng. Chem. Res.,* **37:** 2059 (1998).

4

Thermodynamic Properties of Ideal Gases

4.1 SCOPE AND DEFINITIONS*

This chapter describes methods to estimate the standard state enthalpy, $\Delta_f H^\circ(T)$, Gibbs energy of formation, $\Delta_f G^\circ(T)$, and absolute entropy, $S^\circ(T)$, of a compound. The *reference* temperature is 298.15 K, and the standard pressure is 1 *atm* (1.01325×10^5 Pa). As compounds can exist in multiple phases (e.g., a reaction produces gaseous versus liquid water), the state of aggregation of the species must also be identified. This chapter focuses mainly on the *ideal gas* state. In addition, techniques are given for estimating the ideal gas heat capacity, $C_p^\circ(T)$, as a function of temperature.

The standard enthalpy of formation of a compound in any aggregation state is defined as the enthalpy change to form the species from chemical elements in their *standard* states by an isothermal reaction. In such a reaction scheme, the elements are assumed initially to be at the reaction temperature, at 1 atm, and in their most stable phase, e.g., diatomic oxygen as an ideal gas, carbon as a solid in the form of β-graphite, bromine as a pure liquid at 1 atm, etc. Numerical values of properties of the constituent elements are not of concern, since, when the standard enthalpy of a reaction with several species is calculated, all the enthalpies of formation of the *elements* cancel. For a reaction at other than *standard* conditions, corrections must be made such as for fluid nonidealities.

Any reaction can be written in mathematical notation as

$$\sum_i v_i A_i = 0 \tag{4.1-1}$$

where the species (reactants and products) are identified by the subscript i and are named A_i. The stoichiometric coefficients v_i are positive for products and negative for reactants. An example of this notation is the steam oxidation of propane which is usually written as

$$-1\ C_3H_8\ (g) - 3\ H_2O\ (g) + 3\ CO\ (g) + 7\ H_2\ (g) = 0$$

*With special contributions from V. Diky.

where the stoichiometric coefficients for propane (C_3H_8), water (H_2O), carbon monoxide (CO), and hydrogen (H_2) are −1, −3, 3, and 7, respectively. In more familiar form, this would be

$$C_3H_8\ (g) + 3\ H_2O\ (g) = 3\ CO\ (g) + 7\ H_2\ (g)$$

The purpose of the notation of Eq. (4.1-1) is to express the properties associated with the reaction more compactly. Thus, the standard enthalpy of reaction ($\Delta_r H^\circ(T)$)—the enthalpy change when stoichiometric amounts of reactants, at 1 atm and in their states of aggregation specified in the reaction, are reacted to completion with the products at 1 atm and in their states of aggregation specified in the reaction—is obtained from the standard state enthalpies of formation of the species at the same temperature, $\Delta_f H_i^\circ(T)$. Specifically,

$$\Delta_r H^\circ(T) = \sum_i \nu_i \Delta_f H_i^\circ(T) \tag{4.1-2}$$

where

$$\Delta_f H_i^\circ(T) = H_i^\circ(T) - \sum_e \nu_e^{(i)} H_e^\circ(T) \tag{4.1-3}$$

Here, the summation runs over all the elements found in species i, $H_i^\circ(T)$ is the enthalpy of species i at the standard pressure of 1 atm, state of aggregation specified in the reaction, and temperature T, $\nu_e^{(i)}$ is the number of atoms of an element of type e in species i, and $H_e^\circ(T)$ is the enthalpy of element e in its natural state of aggregation at the standard pressure of 1 atm and temperature T. Note that some elements are diatomic in their most abundant form found in nature such as molecular hydrogen (H_2), molecular nitrogen (N_2), and molecular oxygen (O_2), and these are used instead of their atomic counterparts (e.g., H, N, or O). Thus, for propane, $\nu_C^{(C_3H_8)} = 3$ and $\nu_{H_2}^{(C_3H_8)} = 4$ while for carbon monoxide $\nu_C^{(CO)} = 1$ and $\nu_{O_2}^{(CO)} = \frac{1}{2}$. Since all the values of $H_e^\circ(T)$ cancel when Eq. (4.1-3) is used in Eq. (4.1-2), their values are never obtained explicitly. For our steam oxidation example, Eq. (4.1-2) becomes

$$\Delta_r H^\circ(T) = 3\Delta_f H_{CO}^\circ(T) + 7\Delta_f H_{H_2}^\circ(T) - \Delta_f H_{C_3H_8}^\circ(T) - 3\Delta_f H_{H_2O}^\circ(T) \tag{4.1-4}$$

where, because the state of aggregation of each species was specified in the reaction to be gaseous, the enthalpies are for each species in the ideal gas state at 1 atm and temperature T. Enthalpies of formation, $\Delta_f H^\circ(T)$, are normally tabulated only for the *reference* state of 298.15 K and 1 atm with enthalpy values for all elements in the *standard* state set to zero. The enthalpies of formation at other temperatures may be obtained from

$$\Delta_f H_i^\circ(T) = \Delta_f H_i^\circ(298.15\text{ K}) + \int_{298.15\text{ K}}^{T} \Delta C_{p_i}^\circ(T)\, dT + \sum_e \Delta_t H_e^{(i)} \tag{4.1-5}$$

where the temperature effects of the elements e in species i are taken into account by

$$\Delta C_{p_i}^\circ(T) = C_{p_i}^\circ(T) - \sum_e \nu_e^{(i)} C_{p_e}^\circ(T) \tag{4.1-6}$$

Here, $\Delta C_{p_i}^\circ(T)$ is the difference between the standard state heat capacity of species i and the sum of the standard state heat capacities of elements e, $C_{p_e}^\circ(T)$, in their stoichiometric proportions. Moreover, the indices of the summation in Eq. (4.1-6) are as those described for Eq. (4.1-3), and $C_{p_e}^\circ(T)$ is the standard heat capacity of element e at T. Finally, $\Delta_t H_e^{(i)}$ is the change in enthalpy that occurs when element e of species i undergoes a phase or structural change, such as melting and crystal habit transitions, in the temperature range from 298.15 K to T. If transitions are present, the $C_{p_i}^\circ$ value of the integral must be consistent with the physical state of the species and will be different in different T ranges. In our example, to find $\Delta_f H_{C_3H_8}^\circ(T)$ for C_3H_8 from carbon and hydrogen, $C_{p_C}^\circ(T)$ would be for carbon (β-graphite), $C_{p_{H_2}}^\circ(T)$ would be for diatomic hydrogen ideal gas, and there are no transitions. If, however, the elements change phase between 298.15 K and T, the enthalpy change for this process must be included. Consider obtaining $\Delta_f H^\circ$ for bromobenzene at $T = 350$ K. The elements have $\nu_C^{(C_6H_5Br)} = 6$, $\nu_{H_2}^{(C_6H_5Br)} = \frac{5}{2}$, and $\nu_{Br_2}^{(C_6H_5Br)} = \frac{1}{2}$. Since the standard-state pressure is 1 atm, and T is greater than the normal boiling temperature of bromine, $T_b = 332$ K, one must use the liquid $C^{\text{liq}}_{p\text{Br}_2}(T)$ up to T_b, subtract ½ of the enthalpy of vaporization of Br_2 at T_b, and then use the vapor $C^\circ_{p\text{Br}_2}(T)$ between T_b and 350 K. That is,

$$\Delta_f H^\circ_{C_6H_5Br}(350\text{ K}) = \Delta_f H^\circ_{C_6H_5Br}(298.15\text{ K}) + \int_{298.15\text{ K}}^{332\text{ K}} \Delta C^\circ_{pC_6H_5Br}(T)dT - \frac{1}{2}\Delta_v H_{Br_2} - \frac{1}{2}\Delta^{ig}_g H_{Br_2} + \int_{332\text{ K}}^{350\text{ K}} \Delta C'^\circ_{pC_6H_5Br}(T)dT \tag{4.1-7}$$

where $C^{\text{liq}}_{p\text{Br}_2}(T)$ is used in Eqs. (4.1-6) and (4.1-7) for $\Delta C^\circ_{pC_6H_5Br}(T)$ and ideal gas $C^\circ_{p\text{Br}_2}(T)$ is used in Eqs. (4.1-6) and (4.1-7) for $\Delta C'^\circ_{pC_6H_5Br}(T)$. $\Delta^{ig}_g H_{Br_2}$ is a nonideality correction described in Chap. 8. It is needed because vaporization enthalpies are available between liquid and real gas states, while the reference state for formation enthalpies is ideal gas rather than real gas.

A similar analysis can be done for standard-state entropies, and there are equivalent relations to Eqs. (4.1-2) and (4.1-5). Thus, for our example gas-phase reaction,

$$\Delta_r S^\circ(T) = 3S^\circ_{CO}(T) + 7S^\circ_{H_2}(T) - S^\circ_{C_3H_8}(T) - 3S^\circ_{H_2O}(T) \tag{4.1-8}$$

where each species' standard entropy would be for the ideal gas state at 1 atm and temperature T. However, there is one apparent difference for entropy when obtaining and tabulating values in practice. Unlike energy and enthalpy, the *absolute entropy*, $S^\circ(T)$, may be found which has a zero value when the species and the elements are at $T = 0$ K in a perfectly ordered solid state. This means that the entropy of formation is not normally used explicitly; however, the value may be obtained using the expression

$$\Delta_f S^\circ(T) = \sum_i \nu_i \left[S^\circ_i(T) - \sum_e \nu^{(i)}_e S^\circ_e(T) \right] \tag{4.1-9}$$

where the indices of the summation are as those described for Eq. (4.1-3), S°_i is the *absolute* entropy of species i, and S°_e is the *absolute* entropy of the elements comprising compound i. Though all the S°_e values cancel out in Eq. (4.1-9), they are tabulated separately because they, like the values for all species, can be found experimentally from

$$S^\circ_i(T) = \int_0^T \frac{C^\circ_{p_i}(T)}{T} dT + \sum_j \frac{\Delta_t H^{(i)}_j}{T_{t_j}} \tag{4.1-10}$$

where the summation is over transitions j that occur in species i between 0 and T, and T_{t_j} is the temperature of transition j. [For the gas state, the nonideality correction, $\Delta^{ig}_g H_i$, is applied as in Eq. (4.1-7)]. Evaluating the integral in Eq. (4.1-10) must follow the same procedure as described for Eq. (4.1-5). The NIST JANAF Thermochemical Tables (Chase, 1998) contain the values of absolute entropy for many compounds including elements, and Table 4.1 lists the values for $S^\circ_e(298.15\text{ K})$ from this reference for several common species.

The standard Gibbs energy change of reaction, $\Delta_r G^\circ(T)$, is defined analogously to $\Delta_r H^\circ(T)$ and $\Delta_r S^\circ(T)$. It is especially useful because it is related to chemical equilibrium constants by

$$\ln K = -\frac{\Delta_r G^\circ(T)}{RT} \tag{4.1-11}$$

There is a variety of routes to determine $\Delta_r G^\circ(T)$. The first is to compute enthalpy and entropy changes individually from Eqs. (4.1-2) and (4.1-9) and then use

$$\Delta_r G^\circ(T) = \Delta_r H^\circ(T) - T\Delta_r S^\circ(T) \tag{4.1-12}$$

Another way to obtain $\Delta_r G^\circ(T)$ is to use tabulated values of $\Delta_f G^\circ(298.15\text{ K})$ in a manner similar to Eq. (4.1-5).

$$\frac{\Delta_r G^\circ(T)}{T} = \frac{1}{298.15\text{ K}} \sum_i \nu_i \Delta_f G^\circ_i(298.15\text{ K}) + \Delta_r H^\circ(298.15\text{ K})\left(\frac{1}{T} - \frac{1}{298.15\text{ K}}\right) + \frac{1}{T}\int_{298.15\text{ K}}^{T} \sum_i \nu_i \Delta C^\circ_{p_i}(T)\,dT - \int_{298.15\text{ K}}^{T} \sum_i \nu_i \frac{\Delta C^\circ_{p_i}(T)}{T} dT \tag{4.1-13}$$

TABLE 4.1 Absolute entropies (S_e°) for several common elements in their natural state at 298.15 K as reported in the NIST-JANAF Thermodynamic Tables (Chase, 1998). Note: Gas means ideal gas here.

Element	State	S°(298.15 K) (J/(mol·K))	Element	State	S°(298.15 K) (J/(mol·K))
Al	crystal	28.275	I_2	crystal	116.142
Ar	gas	154.845	K	crystal	64.670
B	crystal	5.834	Kr	gas	164.084
Ba	crystal, α	62.475	Li	crystal	29.085
Be	crystal, α	9.440	Mg	crystal	32.671
Br_2	liquid	152.206	Mn	crystal, α	32.010
C	crystal, graphite	5.740	N_2	gas	191.609
Ca	crystal	41.588	Na	crystal	51.455
Cl_2	gas	223.079	Ne	gas	146.327
Co	crystal, α	30.067	Ni	crystal	29.870
Cr	crystal	23.618	O_2	gas	205.147
Cs	crystal	85.147	P	crystal, white, α	41.077
Cu	crystal	33.164	Pb	crystal	64.785
F_2	gas	202.789	S	crystal,	32.056
Fe	crystal, α	27.321	Si	crystal	18.820
Ga	crystal	40.838	Ti	crystal, α	30.759
H_2	gas	130.680	V	crystal	28.936
He	gas	126.152	Xe	gas	169.684
Hg	liquid	76.028	Zn	crystal	41.717

In this case, there are no explicit terms for transitions since $\Delta_t G_i = 0$. However, appropriate ranges of T and values of C_p° must still be used.

If tabulated property values are all consistent, results from the different treatments will be equal. When estimation methods for different properties are employed or errors occur in doing the calculations, inconsistencies can result, and it is best to check important values by using different routes.

With multiproperty equations of state (see Sec. 6.7), it is common to select a zero-value reference state for a substance's H° and S°, such as at 298.15 K and 1 atm, instead of using pure-component ideal gas properties of formation. These so-called Helmholtz energy equations of state are constructed to give values for the ideal gas Helmholtz energy, $\frac{A^\circ}{RT}$, as a function of V and T. An example is given by Setzmann and Wagner (1991) for methane.

$$\frac{A^\circ}{RT} = \ln\left(\frac{V_c}{V}\right) + \sum_{i=1}^{8} a_i f_i\left(\frac{\theta_i}{T}\right) \tag{4.1-14}$$

where a_i and θ_i are fitted parameters and the functions f_i are either simple or of the form $\ln[1 - \exp(\frac{\theta_i}{T})]$. All other properties relative to the chosen reference states can be obtained by differentiation of Eq. (4.1-14).

In the case of reaction equilibrium constants, the exponential character of Eq. (4.1-11) for K amplifies small errors in $\Delta_r G^\circ(T)$ since the percentage error in K is exponentially related to the error in the value of $\frac{\Delta_r G^\circ(T)}{RT}$. Thus,

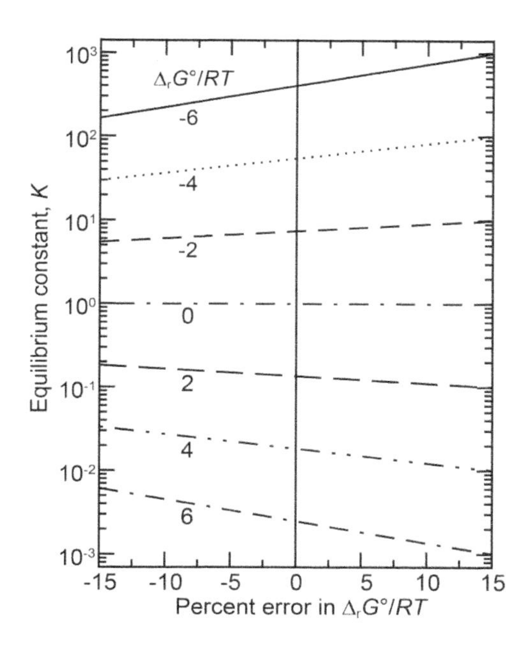

FIGURE 4.1

Effect of errors in $\frac{\Delta_r G°(T)}{RT}$ on the equilibrium constant K.

percentage errors in $\frac{\Delta_r G°(T)}{RT}$ are not indicative. We illustrate this in Fig. 4.1 where values of K are plotted versus percentage error in $\frac{\Delta_r G°(T)}{RT}$ for different values of $\frac{\Delta_r G°(T)}{RT}$. The correct values for K are the intersections of the vertical line with the lines for the computed values of $\frac{\Delta_r G°(T)}{RT}$. If $\frac{\Delta_r G°(T)}{RT}$ is 6 and too small by 15%, the computed value of K is 0.0061 rather than 0.0025; it is too large by a factor of almost 2.5!

The measurement of properties of formation is difficult due to many problems. Impurities and instrument errors can give results that are in error by as much as a few kJ/mol. Evidence of this can be found by examining values from different sources. The result of these uncertainties is that estimation methods may be more accurate than experimental data, and it is now becoming common to compute properties of formation from quantum mechanical (QM) methods (see, for example, DeYonker et al., 2006; Somers and Simmie, 2015; Chirico et al., 2017; Paulechka and Kazakov, 2018). Not only are values obtained much more rapidly, but better reliability and self-consistency are often found from modern chemistry and powerful computers. These methods are discussed later.

Disclaimer. Commercial equipment, instruments, or materials are identified only to adequately specify certain procedures. In no case does such identification imply recommendation or endorsement by the National Institute of Standards and Technology, DIPPR, or the authors, nor does it imply that the products identified are necessarily the best available for the purpose.

4.2 ESTIMATION METHODS FOR THE IDEAL GAS STANDARD STATE

Since the properties of most of the species treated in this chapter are for ideal gases, intermolecular forces play no role in their estimation and, as a result, the law of corresponding states is inapplicable. Rather, two different approaches may be taken to estimate ideal gas $\Delta_f H°(T)$, $S°(T)$, $\Delta_f G°(T)$, and $C_p^\circ(T)$: schemes based on molecular structure [e.g., group contribution (GC) methods] and QM approaches (e.g., ab initio, density functional theory, etc.).

Concerning the former approaches, earlier editions of this book describe limited methods, such as those of Yoneda (1979), Thinh et al. (1971), Thinh and Trong (1976), and Cardozo (1986), for some of the properties discussed in this chapter, but they are not repeated here. Neither are the methods of Joback (1984) and Constantinou and Gani (1994), which have proven to be less accurate than other methods (see Sec. 4.8), but the reader may find these approaches described in previous editions of this book.

In this chapter, we present details of the estimation methods for ideal gas properties that have proven superior to others. These are the GC methods of Benson et al. (Sec. 4.3), Domalski and Hearing (Sec. 4.4), and modified Joback (Sec. 4.5) and QM approaches (Sec. 4.6). Most of the methods discussed in this edition (both GC and QM) provide estimates of ideal gas $\Delta_f H°(298.15 \text{ K})$ and $S°(298.15 \text{ K})$ from which $\Delta_f G°(298.15 \text{ K})$ may be determined from Eq. (4.3-5). The method of Benson et al., modified Joback, and QM approaches provide estimates of ideal gas $C_p^\circ(T)$ at varied temperatures whereas that of Domalski and Hearing gives the value only at 298.15 K. Evaluation of the prediction methods and recommendations are discussed in Sec. 4.8. The group contributions were evaluated against experimental data found in the DIPPR database that were validated by NIST as will be described later. Computing standard state enthalpies of formation and standard enthalpies of combustion from ideal gas standard enthalpies for formation is described in Sec. 4.7.

4.3 METHOD OF BENSON

Benson and coworkers have developed extensive techniques for estimating ideal gas $\Delta_f H°(298.15 \text{ K})$, $S°(298.15 \text{ K})$, and C_p°, which then allow one to obtain ideal gas $\Delta_f G°(298.15 \text{ K})$, energy release information, enthalpies of combustion (see Sec. 4.7), and lower flammability limits. There are several references to Benson's work (1968; Benson and Buss, 1969; Benson et al., 1969; O'Neal and Benson, 1970; Eigenmann et al., 1973; Stein et al., 1977; the CHETAH program, 2020). Because group values have been developed by different people at different times, there are some inconsistencies among the different implementations. Here, we adopt the notation of the CHETAH (version 11.0) program from ASTM International. This differs from Benson's original and also from that of previous editions of this book because it makes clearer the distinction between the *structural* groups and the *neighbor* groups. There are contributions from all the bonding arrangements ("type") that the chosen groups can have with every other type of group or atom (except hydrogen). Thus, the method involves next-nearest neighbor interactions.

Table 4.2 shows some of the many distinct groups of the elements C, N, O, and S that bond to more than one neighbor. The column "valence" contains the number of single-bonded groups, such as H or a halogen, that can be attached to the group. Thus, for C, four single-bonded groups can be attached, for Ct, only one, and for =C=, no single-bonded groups (only double-bonded groups) can be attached. There is also a word description of the group. In addition to the above elements, the method can treat many other atoms and groups containing F, Cl, Br, I, P, B, Si, and 10 different metals. For each type, the notation gives the key group or atom followed in parentheses by the groups or atoms it is bonded to. Thus, the repeating CH_2 group in polyethylene is CH_2—(2C) since each CH_2 is bonded to two C atoms. The methylene group attached to the oxygen in methyl ethyl ether (methoxyethane) is CH_2—(C,O). The carbon which is bonded to the ring and to the two methyl groups of the side group in 1-(1-methylethyl)-4-methylbenzene (*p*-cymene) is CH—(2C,Cb).

The CHETAH compilation group values for the Benson method are found in the file CH04Benson-CHETAHdHSCp.xlsx on the PGLed6 website. (Reprinted, with permission from The ASTM Computer Program for Chemical Thermodynamic and Energy Release Evaluation—Chetah® Version 5.0, copyright ASTM International, 100 Barr Harbor Drive, West Conshohocken, PA 19428. A copy of the complete current program may be obtained from ASTM, https://www.astm.org/ds51hol-eb.html.) This file lists the contributions from the 623 groups that are organized into 16 categories along with the 98 different ring configurations, 13 gauche and 1,5 repulsion types, and 22 *cis* and *ortho*/*para* interactions. The CHETAH program allows computation of all the relevant properties for species made of these groups and includes a sizable database of values for molecules as obtained from the literature.

TABLE 4.2 Some multivalent groups in Benson's method for ideal gas properties.

Group	Valence	Definition
C	4	tetravalent carbon (alkanes)
=C	2	double bonded carbon (alkenes), note that Cd represents cadmium
Cb	1	benzene-type carbon (aromatic)
Cp	3	aromatic carbon of fused ring in polyaromatics, e.g. the two fused carbons in naphthalene
Ct	1	triple bonded carbon (alkynes)
=C=	0	allene carbon
=Cim	2	carbon double bonded to nitrogen (C in >C=N—)
CO	2	carbonyl group (aldehydes, ketones, esters, carboxylic acids)
O	2	oxygen (non-carbonyl oxygen atom in ethers, esters, acids, alcohols)
N	3	trivalent nitrogen (amines)
=Nim	1	imino nitrogen (N in >C=N—)
=Naz	1	azo, nitrogen (N in —N=N—)
Nb	0	aromatic nitrogen not next to another aromatic nitrogen (e.g. pyridine, pyrazine and pyrimidine, but not pyridazine)
CS	2	thiocarbonyl
S	2	divalent sulfur (sulfides)
SO_2	2	sulfoxide group
SO	2	sulfone group

Example 4.1 shows other examples of Benson groups to construct species. When adding the contributions, there should be terms from both the group before a dash (—) and from the group (or groups) in the parenthesis following the dash (—). Thus, the group Cb—(C) would need to be accompanied by a group such as CH_3—(Cb) to complete the side group contributions for toluene. If a parenthetical group exists without a dash, only the contribution from the group listed is included. Thus, the groups CbBr and Cb(CN) are for both the aromatic carbons and their side groups for bromobenzene and cyanobenzene, respectively. Finally, if multiple bonds are indicated for a group, it must be the parenthetical group to another group. For example, the group =CHI must be accompanied by a group such as =CHF which would complete the species 1-fluoro-2-iodoethene or by a group such as =CCl-(C), which when additionally accompanied by CH_3—(=C) would give the species 1-iodo-2-chloropropene. Example 4.1 shows some other examples of Benson groups to construct species.

Example 4.1 Examples of Benson groups (CHETAH, 2020)

Name	Formula	Group	# Groups	Name	Formula	Group	# Groups
4-hydroxy-2-heptanone	$C_7H_{14}O_2$	CH_3—(CO)	1	propene	C_3H_6	CH_3—(=C)	1
		CO—(2C)	1			=CH—(C)	1
		CH_2—(C,CO)	1			$=CH_2$	1
		CH—(2C,O)	1	3-chloropropanoic acid	$C_3H_5O_2Cl$	CH_2Cl—(C)	1
		OH—(C)	1			CH_2—(C,CO)	1
		CH_2—(2C)	2			CO—(C,O)	1
		CH_3—(C)	1			OH—(CO)	1

(Continued)

Name	Formula	Group	# Groups	Name	Formula	Group	# Groups
benzylideneaniline	$C_{13}H_{11}N$	CbH	10	anthracene	$C_{14}H_{10}$	CbH	10
		=Nim—(Cb)	1			Cp—(2Cb,Cp)	4
		=CimH—(Cb)	1	phenanthrene	$C_{14}H_{10}$	CbH	10
		Cb—(=C)	1			Cp—(2Cb,Cp)	2
		Cb—(N)	1			Cp—(Cb,2Cp)	2
1-butanol	$C_4H_{10}O$	CH_3—(C)	1	2-butanol	$C_4H_{10}O$	CH_3—(C)	2
		CH_2—(2C)	2			CH_2—(2C)	1
		CH_2—(C,O)	1			CH—(2C,O)	1
		OH—(C)	1			OH—(C)	1
2-methyl-1-propanol	$C_4H_{10}O$	CH_3—(C)	2	2-methyl-2-propanol	$C_4H_{10}O$	CH_3—(C)	3
		CH—(3C)—	1			C—(3C,O)	1
		CH_2—(C,O)	1			OH—(C)	1
		OH—(C)	1				

Values from the Benson groups can be summed directly to obtain ideal gas $\Delta_f H°(298.15\text{ K})$ and $C_p^\circ(T)$ values. However, obtaining ideal gas $S°(298.15\text{ K})$ also requires taking molecular symmetry into account. Finally, obtaining ideal gas $\Delta_f G°(298.15\text{ K})$ requires subtracting the entropy of the elements. The relations are

$$\Delta_f H°(298.15\text{ K}) = \sum_i N_i\left(\Delta_{\Delta_f H°(298.15\text{ K})}\right)_i \tag{4.3-1}$$

$$C_p^\circ(T) = \sum_i N_i\left(\Delta_{C_p^\circ(T)}\right)_i \tag{4.3-2}$$

$$S°(298.15\text{ K}) = \sum_i N_i \Delta_{S°(298.15\text{ K})} + S_s^\circ \tag{4.3-3}$$

$$S_{el}^\circ(298.15\text{ K}) = \sum_e N_e S_e^\circ \tag{4.3-4}$$

$$\Delta_f G°(298.15\text{ K}) = \Delta_f H°(298.15\text{ K}) - 298.15\text{ K}[S°(298.15\text{ K}) - S_{el}^\circ(298.15\text{ K})] \tag{4.3-5}$$

where the GC values are in the file CH04BensonCHETAHdHSCp.xlsx on the PGLed6 website, values of S_e° are found in Table 4.1, and the symmetry entropy, S_s°, is given in Eq. (4.3-6). Though group values for ideal gas C_p° are only reported at a few temperatures, the CHETAH program can provide values at any specified T. Though apparently complicated, the rules for these adjustments are straightforward, and the CHETAH (2020) program performs all the necessary calculations.

A stepwise procedure for obtaining symmetry numbers is described in Appendix B, which was written by Davies et al. (2020) from the Dow Chemical Company, of the CHETAH 11.0 User Guide (Harrison, 2020). Statistical mechanics shows that entropy varies as $R \ln W$, where W is the number of distinguishable configurations of a compound. If one can find indistinguishable configurations by rotating a molecule either totally as if it were rigid or along bonds between atoms, the result will be an overcounting, and W must be reduced by division. General rules and examples based on the CHETAH program manual are given here; the reader is referred to Benson et al. (1969) and CHETAH (2020) for more complete treatments.

The symmetry entropy, S_s°, is independent of T and is given by

$$S_s^\circ = R \ln N_{iso} - R \ln N_{ts} \tag{4.3-6}$$

The first term on the right-hand side accounts for entropy contributions for certain types of stereoisomerism in compounds with chiral centers or the number of indistinguishable arrangements in molecules whose structures are "frozen" by steric effects. The second term is related to different ways the molecule or parts of the molecule can be rotated into indistinguishable structures. Both are now described.

For achiral compounds, mixtures of diastereomers, unequal mixtures of enantiomers, or when the entropy of a specific stereoisomer of a compound with chiral centers is sought, $N_{iso} = 1$, so the first term makes no contribution in Eq. (4.3-6). Two cases will lead to nonunit values. The first is for racemic mixtures of enantiomers. Enantiomers are stereoisomers that are mirror images of each other. For example, the four atoms (H, F, Cl, I) bonded to the carbon in CHFClI can be arranged in two distinct ways, each being a mirror image of the other. Each chiral center in a racemic mixture of such a compound contributes to N_{iso} yielding $N_{iso} = 2$. The second way for N_{iso} to be different from unity is if an otherwise symmetrical molecule is frozen by steric effects into an asymmetrical conformation. For example, 2,2′,6,6′-tetramethylbiphenyl cannot rotate about the bond between the two benzene rings due to its 2,2′ steric effects. Therefore, the plane of the rings can have two distinct arrangements ($N_{iso} = 2$) that must be included in the entropy calculation.

To obtain N_{ts}, one multiplies the two distinct types of indistinguishability that can occur: "internal," designated N_{is}, and "external," designated N_{es}. The value of N_{is} can be found by rotating terminal groups about their bonds to interior groups. An example is methyl (—CH_3), which has three indistinguishable conformations ($N_{is} = 3$), and phenyl, which has $N_{is} = 2$. Other examples are given in Table 4.3. (Note that in the 2,2′,6,6′-tetramethylbiphenyl example above, the expected indistinguishability about the bond connecting the two benzene rings cannot occur, but each methyl can be rotated, so $N_{is} = 3^4 = 81$.) The value of N_{es} comes from indistinguishability when the whole molecule is rotated as if it were rigid. Thus, homodiatomics have $N_{es} = 2$ from rotation about their bond axis, heterodiatomics have $N_{es} = 1$, and benzene has six from rotation about its ring center and two from flipping the plane to give $N_{es} = 12$, etc.

Finally N_{ts} is found from

$$N_{ts} = N_{es} \prod_k (N_{is})_k \tag{4.3-7}$$

TABLE 4.3 Examples of Benson group indistinguishabilities (CHETAH, 2020).

Molecule	Formula	N_{is}	N_{es}	N_{ts}	N_{iso}
methane	CH_4	1	$4 \times 3 = 12$	12	1
oxygen	O_2	1	2	2	1
carbon monoxide	CO	1	1	1	1
benzene	C_6H_6	1	$6 \times 2 = 12$	12	1
phosphorus pentafluoride	PF_5	1	$3 \times 2 = 6$	6	1
1,1-dichloroethene	$C_2H_2Cl_2$	1	$1 \times 2 = 2$	2	1
hydrogen peroxide	H_2O_2	1	$1 \times 2 = 2$	2	1
dimethylamine	C_2H_7N	$3^2 = 9$	1	9	1
2,2-dimethylpropane	C_5H_{12}	$3^4 = 81$	$4 \times 3 = 12$	972	1
1,4-di-*tert*-butylbenzene	$C_{14}H_{22}$	$3^6 \times 3^2 = 6561$	$2 \times 2 = 4$	26244	1
2,2-dimethyl-4-nitro-3-(4-nitrophenyl)-butane	$C_{12}H_{16}N_2O_4$	$3^3 \times 2^2 \times 3^1 \times 2^1 = 648$	1	648	2
2-(3,5-di-(3-trichloromethylphenyl)-phenyl)-butane	$C_{24}H_{20}Cl_6$	$3^2 \times 3^2 \times 2^1 = 162$	1	162	2

TABLE 4.4 Benson (CHETAH, 2020) values for groups in 2-methylphenol. Taken from the file CH04BensonCHETAHdHSCp.xlsx on the PGLed6 website.

Group	$\Delta_{\Delta_f H°(T)}$	$\Delta_{S°(T)}$	$\Delta_{C_p^\circ(T)}$
Temperature (T) of Group	298.15 K	298.15 K	800 K
Units of Group	kJ/mol	J/(mol·K)	J/(mol·K)
CH_3—(Cb)	−42.19	127.29	54.5
CbH	13.81	48.26	31.56
Cb—(C)	23.06	−32.19	20.76
Cb—(O)	−3.77	−42.7	28.88
OH—(Cb)	−158.64	121.81	25.12
other ortho-(nonpolar-polar)	1.42	—	—

Table 4.4 shows examples of the analysis.

Example 4.2 Estimate ideal gas $\Delta_f H°(298.15\text{ K})$, $\Delta_f G°(298.15\text{ K})$, and $C_p^\circ(800\text{ K})$ for 2-methylphenol (*o*-cresol) by using Benson's group method.

Solution. The Benson groups for 2-methylphenol are one CH_3—(Cb), four CbH, one Cb—(C), one Cb—(O), one OH—(Cb), and one other ortho-(nonpolar-polar) ring effect. The groups values, from the file CH04BensonCHETAHdHSCp.xlsx on the PGLed6 website, are summarized below.

The methyl group makes $N_{is} = 3$, and there are no optical isomer corrections, so $N_{ts} = 3$. The elements are 7 C, 4 H_2, and ½ O_2. Using Eqs. (4.3-1)–(4.3-5) with the values in Tables 4.1 and 4.4, the property predictions are

$$\Delta_f H°(298.15\text{ K}) = \sum_{i=1}^{6} N_i\left(\Delta_{\Delta_f H°(298.15\text{ K})}\right)_i = -124.88\text{ kJ/mol}$$

$$S°(298.15\text{ K}) = \sum_{i=1}^{6} N_i\left(\Delta_{S°(298.15\text{ K})}\right)_i + S_s^\circ = 0.35812\text{ kJ/(mol·K)}$$

$$S_{el}^\circ(298.15\text{ K}) = \sum_{e=1}^{3} N_e S_e^\circ = 0.66547\text{ kJ/(mol·K)}$$

$$\Delta_f G°(298.15\text{ K}) = \Delta_f H°(298.15\text{ K}) - 298.15\text{ K}[S°(298.15\text{ K}) - S_{el}^\circ(298.15\text{ K})] = -33.24\text{ kJ/mol}$$

$$C_p^\circ(800\text{ K}) = \sum_{i=1}^{6} N_i\left(\Delta_{C_p^\circ(800\text{ K})}\right)_i 255.5\text{ J/(mol·K)}$$

The Appendix A values for the ideal gas enthalpy and Gibbs energy formation properties are −128.57 and −35.43 kJ/mol, respectively, which yield an absolute entropy value of 0.35308 J/(mol·K). The ideal gas heat capacity at 800 K calculated from the coefficients of Appendix A (not experimental data) is 257.5 J/(mol·K). Thus, the differences are

$$\Delta_f H°(298.15\text{ K})\text{ Difference} = -124.88 - (-128.57) = 3.69\text{ kJ/mol}$$

$$\Delta_f G°(298.15\text{ K})\text{ Difference} = -33.24 - (-35.43) = 2.19\text{ kJ/mol}$$

$$S°(298.15\text{ K})\text{ Difference} = 0.35812 - 0.35308 = 0.00504\text{ J/(mol·K) or } 1.43\%$$

$$C_p^\circ(800\text{ K})\text{ Difference} = 255.5 - 257.5 = 2.0\text{ J/(mol·K) or } 0.78\%$$

Results computed with the CHETAH program are the same as those given here.

Example 4.3 Estimate ideal gas $\Delta_f H°$(298.15 K), $\Delta_f G°$(298.15 K), and C_p°(298.15 K) for the four butanols using Benson's group method.

Solution. The groups are given in Example 4.1, and their values are from the file CH04BensonCHETAHdHSCp.xlsx on the PGLed6 website. The absolute entropies of the elements, S_{el}° (298.15 K), are given in Table 4.1. The symmetry numbers and optical isomer numbers are listed in the table below. The predicted values are compared to experimental data where possible.

Property	1-butanol	2-methyl-1-propanol	2-methyl-2-propanol	2-butanol
N_{is}	3	9	81	9
N_{es}	1	1	1	1
N_{ts}	3	9	81	9
N_{iso}	1	1	1	2
		$\Delta_f H°$(298.15 K), kJ/mol		
Experimental	−275.1	−283.2	−312.4	−292.9
Calculated	−275.94	−284.80	−312.76	−293.72
Abs. % Err.	0.30	0.56	0.12	0.24
		$S°$(298.15 K), J/(mol·K)		
Experimental	361.69	—	326.81	356.66
Calculated	359.72	348.49	326.53	357.14
Abs. % Err.	0.55	—	0.09	0.13
S_{el}°(298.15 K), J/(mol·K)	778.934	778.934	778.934	778.934
		$\Delta_f G°$(298.15 K), kJ/mol		
Experimental	−150.7	—	−177.6	−167.0
Calculated	−150.95	−156.46	−177.88	−167.96
Abs. % Err.	0.17	—	0.16	0.58
		C_p°(298 K), J/(mol·K)		
Calculated	111	110	114	113

4.4 METHOD OF DOMALSKI AND HEARING

Domalski and Hearing (1993; 1994) expanded the work of Benson using 1512 compounds and over 3700 experimental data points to develop a method to predict $\Delta_f H°$, C_p°, and $S°$ at 298.15 K for ideal gas, liquid, and solid phases. The approach follows the scheme proposed by Benson in that the groups are similar, estimates for enthalpy and heat capacity are direct summations of the values for the groups comprising the molecule, and entropy is the direct summation minus the contribution due to molecular symmetry. Specifically, Eqs. (4.3-1)–(4.3-5) are used with the groups found in the file CH04DomalskiHearingdHSCp.xlsx on the PGL6ed website. A sample of the groups is found in Table 4.5 for 2-methylphenol. Notice that each group specifies the bonding neighbors in parentheses. For example, C—$(H)_3(C_B)$ designates a nonaromatic carbon, C, bonded to three hydrogens, $(H)_3$,

TABLE 4.5 Domalski and Hearing (1993) values for groups in 2-methylphenol. Taken from CH04DomalskiHearingdHSCp.xlsx on the PGL6ed website with equivalent groups as found in Table 4.6.

Group	$\Delta_{\Delta_f H^\circ(T)}$	$\Delta_{C_p^\circ(T)}$	$\Delta_{S^\circ(T)}$
Temperature (T) of group	298.15 K	298.15 K	298.15 K
Units of group	kJ/mol	J/(mol·K)	J/(mol·K)
$C—(H)_3(C_B)$; equivalent to $C—(H)_3(C)$	–42.26	25.73	127.32
$C_B—(H)(C_B)_2$	13.81	13.61	48.31
$C_B—(C)(C_B)_2$	23.64	9.75	–35.61
$C_B—(O)(C_B)_2$	–4.75	15.86	–43.72
$O—(H)(C_B)$	–160.3	18.16	121.5
ortho corr, hydrocarbons	1.26	6.4	–2.5

and one aromatic carbon, (C_B), and $C_B—(H)(C_B)_2$ is an aromatic carbon, C_B, bonded to one hydrogen, (H), and two other aromatic carbons, $(C_B)_2$. The groups in parentheses are not counted when parsing the molecule, so $C—(H)_3(C_B)$ needs another C_B group to complete the molecule, and $C_B—(H)(C_B)_2$ would need two other C_B groups. There are no groups for hydrogen atoms. Table 4.6 lists several equivalent groups and should be consulted when a group appears to be missing from the method when parsing a molecule. Table 4.7 explains the nomenclature behind the group symbols. Example 4.4 shows an example of the Domalski and Hearing (1993) method. The reference itself contains over 1000 examples.

TABLE 4.6 Equivalent groups in the method of Domalski and Hearing (1993). See CH04DomalskiHearing-dHSCp.xlsx on the PGL6ed website.

Group Listed in the File on the PGL6ed Website	Equivalent Groups Possibly Encountered in Molecules			
$C—(H)_3(C)$	$C—(H)_3(C_d)$	$C—(H)_3(C_t)$	$C—(H)_3(C_B)$	$C—(H)_3(O)$
	$C—(H)_3(CO)$	$C—(H)_3(N)$	$C—(H)_3(N_A)$	$C—(H)_3(N_I)$
	$C—(H)_3(S)$	$C—(H)_3(SO_2)$		
$C—(H)(C)_3$	$C—(H)(C)(C_d)_2$			
$C_d—(H)$	$C_d—(H)(O)$	$C_d—(H)(S)$		
$C_d—(H)(C_d)$	$C_d—(H)(C_t)$	$C_d—(H)(C_B)$	$C_d—(O)(C_d)$	
$C_B—(H)(C_B)_2$	$C_B—(H)(O)(C_B)$	$C_B—(H)(N)(C_B)$	$C_B—(H)(N_I)(C_B)$	$C_B—(H)(S)(C_B)$
	$C_B—(C)(O)(C_B)$	$C_B—(C)(N)(C_B)$	$C_B—(C)(S)(C_B)$	
$C_B—(C_d)(C_B)_2$	$C_B—(C_t)(C_B)_2$			
$C_B—(SO)(C_B)_2$	$C_B—(SO_2)(C_B)_2$			
$S—(C_d)_2$	$S—(C_B)_2$			

TABLE 4.7 Group definitions and explanations for the method of Domalski and Hearing (1993).

Group	Explanation
$C-(H)_3(C)$	A carbon atom with three bonds to hydrogen atoms and the fourth bond to a carbon atom. (e.g., ethane)
$C-(H)_2(C)_2$	A carbon atom with two bonds to hydrogen atoms and two bonds to carbon atoms. (e.g., *n*-hexane)
$C-(H)(C)_3$	A carbon atom with one bond to a hydrogen atom and three bonds to carbon atoms. (e.g., 2-methylpropane)
$C-(C)_4$	A carbon atom with four bonds to carbon atoms. (e.g., 2,2,dimethylpropane)
$C_d-(H)_2$	A doubly bonded carbon atom attached to two hydrogen atoms. (e.g., ethylene)
$C_d-(C)_2$	A doubly bonded carbon atom attached to two carbon atoms. (e.g., propene)
$C_t-(H)$	A triply bonded carbon atom attached to a hydrogen atom. (e.g., ethyne)
$C_t-(C)$	A triply bonded carbon atom attached to a carbon atom. (e.g., propyne)
$C_B-(H)(C_B)_2$	An aromatic ring (benzene) carbon atom bonded to a hydrogen atom and two other aromatic ring carbon atoms. (e.g., benzene)
$C_B-(C_B)_3$	An aromatic ring (benzene) carbon atom bonded to three aromatic ring carbon atoms. (e.g., biphenyl)
$C_{BF}-(C_{BF})(C_B)_2$	A fused aromatic ring carbon atom (such as the two fused ring carbon atoms in naphthalene) bonded to one other fused aromatic ring carbon atom and aromatic ring carbon atoms. (e.g., naphthalene)
$C_{BF}-(C_{BF})_3$	A fused aromatic ring carbon atom bonded to three other fused aromatic ring carbon atoms. (e.g., pyrene)
C_a	An allenic carbon atom. When allene is unsubstituted, the group values are equal to allene itself. (e.g., allene)
$-CH_3$ corr (tertiary)	A correction for the attachment of each methyl group to a tertiary carbon atom. (e.g., 2-methylpropane)
$-CH_3$ corr (quaternary)	A correction for the attachment of each methyl group to a quaternary carbon atom. (e.g., 2,2-dimethylpropane)
$-CH_3$ corr (tert/quat)	A correction for the attachment of each methyl group when there is both a tertiary and a quaternary carbon atom present in the longest chain of a hydrocarbon. (e.g., 2,2,3-trimethylpentane)
$-CH_3$ corr (quat/quat)	A correction for the attachment of each methyl group when there are two quaternary carbon atoms present in the longest chain of a hydrocarbon. (e.g., 2,2,4,4-tetramethylpentane)
ortho corr, hydrocarbons	An aromatic ring correction for *ortho* substitution in hydrocarbon compounds. (e.g. *o*-xylene)
meta corr, hydrocarbons	An aromatic ring correction for *meta* substitution in hydrocarbon compounds. (e.g., *m*-xylene)
rsc	Ring strain correction, rsc, for a cyclic non-aromatic compound. (e.g., cyclopropane)
rsc (unsub)	Ring strain correction, rsc, for a cyclic non-aromatic unsubstituted compound. (e.g., cyclopentane)
rsc (sub)	Ring strain correction, rsc, for a cyclic non-aromatic substituted compound. (e.g., methylcyclopentane)
$C-(H)_2(C)(O)$	A carbon atom bonded to two hydrogen atoms, a carbon atom, and an oxygen atom. (e.g., ethanol)
$O-(C)_2$	An oxygen atom bonded to two carbon atoms. (e.g. dimethyl ether)

(*Continued*)

TABLE 4.7 Group definitions and explanations for the method of Domalski and Hearing (1993). (*Continued*)

Group	Explanation
C—(H)(O)$(C)_2$ (alcohols, peroxides)	Tertiary carbon atom group in alcohols and peroxides. (e.g., 2-propanol, *n*-heptyl-2-hydroperoxide)
C—(H)(O)$(C)_2$ (esters, ethers)	Tertiary carbon atom group in ethers and esters. (e.g., methylisopropyl ether, isopropyl acetate)
C—(O)$(C)_3$ (alcohols, peroxides)	Quaternary carbon atom group in alcohols and peroxides. (e.g., *tert*-butyl alcohol, di-*tert*-butyl peroxide)
C—(O)$(C)_3$ (esters, ethers)	Quaternary carbon atom group in ethers and esters. (e.g., di-*tert*-butyl ether, *tert*-butyl acetate)
C—$(H)_2$(C)(CN)	A carbon atom bonded to two hydrogen atoms, a carbon atom, and a nitrile (cyano) group. (e.g., propanenitrile)
C_B—$(NO_2)(C_B)_2$	An aromatic ring carbon atom bonded to a nitro group and two other aromatic ring carbon atoms. (e.g., nitrobenzene)
NO_2—NO_2 (*ortho* corr)	A correction for adjacent *(ortho)* substitution of NO_2 groups on an aromatic ring. (e.g., *o*-dinitrobenzene)
NO_2—COOH (*ortho* corr)	A correction for substitution of an NO_2 group adjacent to a COOH group on an aromatic ring. (e.g., *o*-nitrobenzoic acid)
N—$(H)_2$(C) (first, amino acids)	The first (and only) NH_2 group bonded to a carbon atom in an amino acid. (e.g., glycine)
N—$(H)_2$(C) (second, amino acids)	The second NH_2 group bonded to a carbon atom in an amino acid. (e.g., lysine)
N—$(H)_2$(CO) (amides, ureas)	An NH_2 group bonded to a carbonyl group, CO, in amides and ureas. (e.g., acetamide, urea)
N—$(H)_2$(CO) (amino acids)	An NH_2 group bonded to a carbonyl group, CO, in amino acids. (e.g., asparagine)
N—(H)(C)(CO) (amides, ureas)	An NH group bonded to a hydrogen atom, carbon atom, and a carbonyl group in amides and ureas. (e.g., *N*-methylformamide, methylurea)
N—(H)(C)(CO) (amino acids)	An NH group bonded to a hydrogen atom, carbon atom, and a carbonyl group in amino acids. (e.g., glycylglycine)
Zwitterion energy, aliphatic	A correction for the conversion of an amino acid or to a zwitterion in amino acids and peptides with aliphatic moieties. (e.g., glycine, glycylalanine)
Zwitterion energy, aromatic I	A correction for the conversion of an aromatic amino acid or peptide to a zwitterion containing an aromatic ring attached directly to a conjugation deterring group (such as a —CH_2— group). (e.g., phenylalanine, glycylphenylalanine)
Zwitterion energy, Aromatic II	A correction for the conversion of an aromatic amino acid or peptide to a zwitterion containing an aromatic ring attached directly to a conjugation enhancing group (such as a >C=O group). (e.g., hippuric acid, hippurylglycine.)
N_A—(C)	A doubly bonded (azo) nitrogen atom bonded to a carbon atom. (e.g., azomethane)
N_A—(C_B)	A doubly bonded (azo) nitrogen atom bonded to an aromatic ring carbon atom. (e.g., *trans*-azobenzene)
N_A—(oxide)(C)	A doubly bonded (azoxy) nitrogen atom bonded to a carbon atom. (e.g., di-*tert*-butyldiazene *N*-oxide)
N_I—(C)	A doubly bonded (imino) nitrogen atom bonded to a carbon atom. (e.g., *N*-butylisobutyleneimine)
N_I—(C_B)	A doubly bonded (pyridine-type) nitrogen atom bonded to an aromatic ring carbon atom. (e.g., pyridine)

(*Continued*)

TABLE 4.7 Group definitions and explanations for the method of Domalski and Hearing (1993). (*Continued*)

Group	Explanation
N_I—(CH_3) (*ortho* corr)	A doubly bonded (pyridine-type) nitrogen atom in an aromatic ring adjacent to a substituted methyl group. (e.g., 2-picoline)
N_I—N_I (*ortho* corr)	A doubly bonded (pyridine-type) nitrogen atom adjacent to an identical (pyridine-type) nitrogen atom in an aromatic ring. (e.g., pyridazine)
C—$(H)_2$(C)(S)	A carbon atom bonded to two hydrogen atoms, a carbon atom, and a sulfur atom. (e.g., methanethiol)
S—(C)(S)	A sulfur atom bonded to a carbon atom and another sulfur atom. (e.g. dimethyl sulfide)
C—$(H)_2$(C)(F)	A carbon atom bonded to two hydrogen atoms, a carbon atom, and a fluorine atom. (e.g., fluoroethane)
ortho corr, (F)(F)	A correction for the adjacent *(ortho)* substitution of two fluorine atoms on an aromatic ring. (e.g., *o*-difluorobenzene)
ortho corr, (I)(COOH)	A correction for the substitution of an iodine atom adjacent *(ortho)* to a COOH group on an aromatic ring. (e.g., 2-iodobenzoic acid)
ortho corr, (Cl)(Cl')	A correction for the substitution of a chlorine atom in an aromatic ring in the near proximity of another chlorine atom in a different aromatic ring which is bonded to the first ring. (e.g., 2,2′-dichlorobiphenyl)

Example 4.4 Estimate ideal gas $\Delta_f H°$(298.15 K), $\Delta_f G°$(298.15 K), and C_p°(298.15 K) for 2-methylphenol (*o*-cresol) by using Domalski's group method.

Solution. The Domalski groups for 2-methylphenol are one C—$(H)_3(C_B)$, four C_B—$(H)(C_B)_2$, one C_B—$(C)(C_B)_2$, one C_B—$(O)(C_B)_2$, one O—$(H)(C_B)$, and one "*ortho* corr, hydrocarbons." The methyl group makes $N_{is} = 3$ and there are no optical isomer corrections, so $N_{ts} = 3$. The elements are 7 C, 4 H_2, and ½ O_2. Using Eqs. (4.3-1)–(4.3-5) with the values in Table 4.5, the results are

$$\Delta_f H°(298.15\text{ K}) = \sum_{i=1}^{6} N_i \left(\Delta_{\Delta_f H°}\right)_i = -127.17\text{ kJ/mol}$$

$$S°(298.15\text{ K}) = \sum_{i=1}^{6} N_i (\Delta_{S°})_i + S_s^\circ = 0.35110\text{ kJ/(mol·K)}$$

$$S_{el}^\circ(298.15\text{ K}) = \sum_{e=1}^{3} N_e S_e^\circ = 0.66547\text{ kJ/(mol·K)}$$

$$\Delta_f G°(298.15\text{ K}) = \Delta_f H°(298.15\text{ K}) - 298.15\text{ K}[S°(298.15\text{ K}) - S_{el}^\circ(298.15\text{ K})] = -33.44\text{ kJ/mol}$$

$$C_p^\circ(298.15\text{ K}) = \sum_{i=1}^{6}\sum_{i=1}^{6} N_i \left(\Delta_{C_p^\circ}\right)_i = 130.34\text{ J/(mol·K)}$$

The Appendix A values for the ideal gas enthalpy and Gibbs energy formation properties are −128.57 and −35.43 kJ/mol, respectively, while the ideal gas heat capacity at 298.15 K calculated from the coefficients of Appendix A (not experimental data) is 127.2 J/(mol·K). Thus the differences are

$$\Delta_f H°(298.15\text{ K})\text{ Difference} = -127.17 - (-128.57) = 1.4\text{ kJ/mol or } 1.09\%$$

$$\Delta_f G°(298.15\text{ K})\text{ Difference} = -33.44 - (-35.43) = 2.0\text{ kJ/mol or } 5.62\%$$

$$C_p^\circ(298.15\text{ K})\text{ Difference} = 130.34 - 127.2 = 3.14\text{ J/(mol·K) or } 2.47\%$$

4.5 MODIFIED JOBACK METHOD FOR IDEAL GAS HEAT CAPACITY

As discussed in Chap. 3, the original Joback (1984; Joback and Reid, 1987) GC approach for several properties is limited to small ranges because the form of the correlating equation is linear. One of the authors of this book (Diky) has modified the method for ideal gas heat capacity to refine the predictions. The Joback group scheme remains largely the same (Table 4.8), but the parameters have been refitted. Specifically, the ideal gas heat capacity in the modified scheme is given by

$$\frac{C_p^\circ}{\mathrm{J/(mol{\cdot}K)}} = \sum_{i=0}^{3}\sum_j N_j A_{ij}\left(\frac{T}{\mathrm{K}}\right)^i \tag{4.5-1}$$

where the summation for j is over the number of groups in the molecule as defined in Table 4.8, N_j is the number of the groups of type j in the molecule, and A_{ij} is parameter A_i for group type j. Note that each molecule must include the second-order "entity term," one per molecule, as indicated in the last section (Second-order corrections) of Table 4.8, and the other second-order corrections differ from those for the properties found in Chap. 3. Example 4.5 demonstrates the technique.

TABLE 4.8 Parameters for the modified Joback method for prediction of ideal gas C_p°. Use in Eq. (4.5-1). Groups with atoms in rings are identified by the letter "r" following the single-letter element symbol (e.g., Cr, Or, Sr, and Nr). Do not confuse these ring designations with symbols for elements such as chromium or strontium.

Group	A_0	$10^2{\cdot}A_1$	$10^5{\cdot}A_2$	$10^8{\cdot}A_3$
	Non-ring Carbon Increments			
—CH3	2.883	11.622	−6.401	1.503
—CH2—	1.300	8.192	−2.691	−0.447
>CH—	−20.574	15.684	−16.190	6.298
>C<	−26.584	14.800	−16.106	6.373
=CH2	−0.328	12.425	−10.356	3.531
=CH—	−0.798	7.427	−4.585	1.006
=C<	−10.284	9.240	−10.108	4.069
=C=	14.155	0.737	0.187	−0.271
≡CH	0.328	13.787	−17.018	7.455
≡C—	4.936	4.755	−5.159	2.097
	Ring Carbon Increments			
—CrH2—	−3.203	8.802	−2.277	−0.794
>CrH—	−8.253	7.841	−3.441	0.015
>Cr<	−14.227	7.466	−5.183	1.108
=CrH—	−4.846	8.941	−6.342	1.675
=Cr<	−5.445	6.101	−4.998	1.558

(Continued)

TABLE 4.8 Parameters for the modified Joback method for prediction of ideal gas C_p°. Use in Eq. (4.5-1). Groups with atoms in rings are identified by the letter "r" following the single-letter element symbol (e.g., Cr, Or, Sr, and Nr). Do not confuse these ring designations with symbols for elements such as chromium or strontium. (*Continued*)

Group	A_0	$10^2 \cdot A_1$	$10^5 \cdot A_2$	$10^8 \cdot A_3$
Halogen Increments				
—F	9.179	4.446	−4.952	1.929
—Cl	12.562	5.221	−7.090	3.117
—Br	16.135	4.491	−6.591	3.005
—I	14.598	6.211	−9.696	4.624
Oxygen Increments				
—OH (alcohol)	12.212	6.095	−5.274	1.813
a—OH (phenol)	7.585	9.557	−10.256	3.931
—O—	30.820	−8.548	14.639	−7.613
—Or—	2.682	4.430	−3.634	1.029
>C═O (ketone)	17.426	4.941	−2.812	0.469
>Cr═O (ring ketone)	−9.621	17.654	−21.400	9.217
—HC═O (aldehyde)	10.873	10.913	−7.941	2.287
—COOH (acid)	17.190	11.076	−3.829	−0.909
—COO— (ester)	−10.024	20.582	−21.855	8.588
═O in —N═O	−21.998	18.439	−22.067	9.287
Nitrogen Increments				
—NH_2	8.810	8.952	−6.369	1.800
—NH—	−16.492	14.229	−14.854	5.850
—NrH—	2.908	5.369	−1.186	−0.845
>N—	−22.180	15.386	−18.387	7.228
═N—	−5.019	9.540	−13.522	6.531
═Nr—	−1.985	6.065	−5.521	1.867
═NH	10.222	6.030	−5.506	2.310
—C≡N	14.326	10.015	−11.940	5.109
—NO_2	3.705	16.500	−15.863	5.556
Sulfur Increments				
—SH	17.513	6.800	−7.410	3.117
—S—	8.896	7.209	−10.611	4.921
—Sr—	7.568	6.729	−8.261	3.338
Second-order Corrections				
entity term, one per molecule	5.675	−10.825	15.933	−7.407
3-membered ring	5.523	3.924	−8.793	4.542
4-membered ring	9.084	−1.774	1.903	−1.011

Example 4.5 Estimate ideal gas C_p° C at 475.25 K of 2-methyl-2-butanol using the modified Joback method.

Solution. 2-methyl-2-butanol contains three —CH_3, one —CH_2—, one >C<, and one —OH (alcohol) according to the group definitions outlined in Table 4.8. Recall also that all compound descriptions must include the constant term.

Group		N	NA_0	10^2NA_1	10^5NA_2	10^8NA_3
1	—CH_3	3	8.649	34.866	−19.203	4.509
2	—CH_2—	1	1.300	8.192	−2.691	−0.447
3	>C<	1	−26.584	14.800	−16.106	6.373
4	—OH (alcohol)	1	12.212	6.095	−5.274	1.813
5	Constant term	1	5.675	−10.825	15.933	−7.407
$\sum_{k=1}^{5} N_k F_k$			1.252	53.128	−27.341	4.841

With the group summations indicated, the modified Joback estimate of ideal gas heat capacity at 475.25 K is

$$C_p^\circ(475.25\ \text{K}) = 1.252 + 10^{-2}(53.128)(475.25) + 10^{-5}(-27.341)(475.25^2) + 10^{-8}(4.841)(475.25^3) = 197.19\ \text{J/(mol·K)}$$

The experimental value from Strömsöe et al. (1970) is 201.4 J/(mol·K). Thus, the difference is

$$C_p^\circ(500\ \text{K})\ \text{Difference} = 197.19 - 201.4 = -4.21\ \text{J/(mol·K) or } -2.1\%$$

4.6 QUANTUM MECHANICAL METHODS

Quantum mechanics and modern computational power have progressed to a state that many ab initio and density functional theory methods for determining ideal gas properties are considered as accurate as experimental methods. A stated goal of the field is to predict the properties within ±1 kcal/mol which is considered *experimental accuracy*, and many methods perform better than this (DeYonker, 2006). Two main issues must be considered when using quantum chemical methods: the level of theory (LT) and the basis set (BS). The level of theory refers to how electron-electron correlation is taken into account when modeling the energy of the molecule. The basis set concerns the number and types of functions used to approximate the electronic wave function of the molecule. An LT/BS pair is called the *model chemistry*, and the exact pair chosen when performing the calculations affects the accuracy of some properties differently than others. Determining accurate *energies* (e.g., internal energies, enthalpies, Gibbs energies) requires careful attention to detail to account for biases introduced by the different approximations used to solve Schrödinger's equation for each LT/BS. However, entropies and heat capacities can be determined at accuracies sufficient for most applications with relatively simple levels of theory and small basis sets. This section describes approaches for using quantum mechanics to predict ideal gas $\Delta_f H^\circ(298.15\ \text{K})$, $S^\circ(298.15\ \text{K})$, and $C_p^\circ(T)$.

Many software packages, both commercial and open-source, are available for performing quantum chemical calculations. Generally, commercial packages are easy to use, remain current with the field (both theoretically and computationally), and are computationally fast but come at a large up-front cost. Many open-source software options exist, but usability, speed, and/or currency may be more limited than with commercial options.

Regardless of the choice of software, using quantum chemical calculations to predict thermodynamic properties of ideal gases generally consists of two fundamental steps. The first is *geometry optimization* and the second is *frequency calculation*. During the first step, the software searches for the configuration of the atomic nuclei that gives the lowest energy for the system (calculated using Schrödinger's equation) for the given model chemistry. During the frequency calculation, the software calculates the second derivatives of the energy with respect to the nuclear positions to determine the vibrational (ν_{v_i}) and rotational (ν_{r_i}) frequencies of the molecule. Linear molecules have two rotational modes while nonlinear molecules have three. Also, linear molecules with n atoms have $3n - 5$ vibrational modes, while nonlinear molecules have $3n - 6$.

The frequencies are related to thermodynamic properties through statistical mechanics as described later and can also be reported as wavenumbers ($\tilde{\nu}_{v_i}$, $\tilde{\nu}_{r_i}$) or characteristic temperatures (Θ_{v_i}, Θ_{r_i}). The relationship between frequencies, wavenumbers, and characteristic temperatures is given by the standard equations from physics as found below where h is Planck's constant, k is Boltzmann's constant, c is the speed of light in vacuum, and λ is wavelength.

$$\lambda = \frac{1}{\tilde{\nu}} = \frac{c}{\nu} \tag{4.6-1}$$

$$\Theta = \frac{h\nu}{k} = \frac{hc\tilde{\nu}}{k} \tag{4.6-2}$$

One important value that the frequency calculation provides is the *zero-point energy* (*ZPE*) or *zero-point vibrational energy*. *ZPE* refers to the energy of the system at 0 K and is needed when determining accurate energies. Due to the Heisenberg Uncertainty Principle, subatomic particles can never be "at rest," even at 0 K but always have some *residual energy* that cannot be removed. The geometry optimization step done by the quantum mechanical program models the system without regard to the Uncertainty Principle as it fixes the locations of the atomic nuclei at each step in its search for the lowest energy of the system and culminates in an "at rest" final state of the molecule. The electronic energy of this state (E_0) is often termed the "bottom of the well" energy. A frequency calculation must be done to determine the residual energy, the *ZPE*, that remains so that it can be added to E_0 to bring the system into agreement with the Uncertainty Principle. The *ZPE* is found from

$$ZPE = \frac{1}{2}\sum_i h\nu_i = \frac{1}{2}\sum_i hc\tilde{\nu}_i \tag{4.6-3}$$

where the summation is over all vibrational modes of the molecule.

Additionally, quantum mechanical calculations are done at 0 K, and corrections must be made to determine values at other temperatures. From statistical mechanics (McQuarrie, 2000), the thermal correction that must be added to the 0 K, bottom-of-the-well electronic energy (E_0) to determine the internal energy of the system for a linear molecule at temperature T is given by

$$\frac{U_{\text{corr}}(T)}{RT} = \frac{3}{2} + 1 + \sum_{i=1}^{3n-5} \frac{\frac{\Theta_{v_i}}{T}}{e^{\frac{\Theta_{v_i}}{T}} - 1} \tag{4.6-4}$$

and that for a nonlinear molecule is

$$\frac{U_{\text{corr}}(T)}{RT} = \frac{3}{2} + \frac{3}{2} + \sum_{i=1}^{3n-6} \frac{\frac{\Theta_{v_i}}{T}}{e^{\frac{\Theta_{v_i}}{T}} - 1} \tag{4.6-5}$$

In each of these expressions, the first term on the right-hand side of the equation is the contribution due to translational modes, the second is those due to rotational modes, and the third is those due to vibrational modes. (In both cases, the electronic contribution is identically zero because the excited electronic states are assumed to be inaccessible at relevant temperatures, and the zero in energy is arbitrarily set equal to the energy of the ground state of the molecule. See McQuarrie (2000) for more discussion.) Corrections can also be derived to calculate the enthalpy from E_0 by adding PV to $U_{\text{corr}}(T)$. Because $PV = RT$ for the ideal gas conditions considered in this chapter, the expressions for the thermal correction to the enthalpy, $H_{corr}(T)$, for linear and nonlinear molecules, respectively, are

$$\frac{H_{corr}(T)}{RT} = \frac{7}{2} + \sum_{i=1}^{3n-5} \frac{\frac{\Theta_{v_i}}{T}}{e^{\frac{\Theta_{v_i}}{T}} - 1} \tag{4.6-6}$$

$$\frac{H_{corr}(T)}{RT} = 4 + \sum_{i=1}^{3n-6} \frac{\frac{\Theta_{v_i}}{T}}{e^{\frac{\Theta_{v_i}}{T}} - 1} \tag{4.6-7}$$

As will be discussed below, ZPE must be added to E_0 in almost every practical use of quantum mechanics to predict energies, and thermal corrections (e.g., $U_{corr}(T)$, $H_{corr}(T)$) must be added for temperatures other than 0 K (most cases). Consider determining an enthalpy of a reaction at 298.15 K using quantum mechanics to calculate the enthalpies of each species. By not adding ZPE_i to E_{0_i}, the compounds involved are lower in energy than found in nature, so the resulting enthalpy of reaction value will be incorrect. By not adding $H_{corr_i}(T)$, the enthalpy of reaction determined would be at 0 K rather than 298.15 K. The proper procedure to follow when calculating changes in enthalpy for the reaction is to add both ZPE_i and $H_{corr_i}(T)$ to E_{0_i} before taking the difference between the products and reactants.

Several assumptions are made to arrive at Eqs. (4.5-4)–(4.5-7) including the Born-Oppenheimer approximation, the assumption that the internal modes of the molecules are independent, modeling the rotational motion as that of a rigid rotor, and modeling the vibrational motion as that of a harmonic oscillator. Deviations from these approximations can become significant at under certain conditions with the degree of error being individual for each molecule, property, and assumption. Later sections explain cases pertinent to the properties covered in this chapter in more detail.

As mentioned previously, the model chemistry (level of theory and basis set) chosen when performing the calculation affects the accuracy of the results. It has been found that the errors introduced are dependent on the specific LT/BS pair, and that an effective way to account for the errors in thermodynamic properties is to scale the frequencies and ZPE produced by the software. These scaling factors are determined by matching the quantum mechanical vibrational frequencies and ZPEs to their experimental counterparts (Alecu, 2010). Determining scaling factors is an area of active research, so many articles on the subject may be found in the literature. NIST maintains a large list of frequency scaling factors in the Computational Chemistry Comparison and Benchmark DataBase (CCCBDB), which can be found by searching "CCCBDB scaling factors" on the internet. A few of these are listed in Table 4.9. They are applied by multiplying each vibrational frequency produced by the computation by the appropriate scaling factor. The frequency scaling approach leads to reasonably accurate values for $S°$ and $C_p°$ but more complicated methods are needed to determine accurate values for $\Delta_f H°$ as will be described in the next sections. Scaling factors for ZPE are also found in the literature.

Traditionally, scaling factors like those found in Table 4.9 from the CCCBDB database are applied to all vibrational frequencies in the molecule. Recently, researchers have improved agreement with experimental data by applying different scaling factors to different frequencies produced by the same LT/BS pair. For example,

TABLE 4.9 Example vibration frequency scaling factors from the CCCBDB at NIST.

Method	Frequency Scaling Factor
HF/3-21G*	0.903
HF/6-31G*	0.899
HF/6-311G**	0.909
B3LYP/6-311+G(3df,2p)	0.967
MP2/6-311+G(3df,2pd)	0.950

Paulechka and Kazakov (2018) scale all frequencies by 0.990 when calculating *ZPE*, but for $H_{corr}(298.15\ K)$, frequencies corresponding to hydrogen stretches (>2800 cm^{-1}) are scaled by 0.960 while all other frequencies are scaled by 0.985.

A. Ideal Gas Standard State Enthalpy of Formation

Accurate ideal gas enthalpies of formation from quantum mechanical calculations require computationally intensive high levels of theory and very large basis sets, but procedures can be used to achieve reliable values with more moderate effort. One approach is to use an *isodesmic* reaction (Hehre et al., 1970). An isodesmic reaction is one where the number and types of bonds in the reactants are the same as those in the products. For example (Hehre et al., 1970), the reaction

$$CH_3{-}CH{=}C{=}O + 2CH_4 = H_3C{-}CH_3 + H_2C{=}CH_2 + H_2C{=}O$$

is isodesmic because the reactants and the products each have the following bond types: 12 C—H, 1 C—C, 1 C═C, and 1 C═O. For such reactions, the errors in the energies determined from quantum mechanical calculations for the products are expected to be the same as those for the reactants, so the computational enthalpy of reaction benefits from fortuitous cancelation of errors and is considered accurate. Once the computational enthalpy of reaction is determined, an enthalpy of formation for one of the compounds in the reaction can be determined if the enthalpies of formation for the other compounds are known from experiment. The reaction used can be hypothetical and does not have to correspond to one that physically occurs, which gives the method significant flexibility to select compounds that already have experimental enthalpies of formation. Example 4.6 gives an example of this approach.

Example 4.6 Estimate the ideal gas enthalpy of formation of 2-methylphenol (*o*-cresol) at 298.15 K using computational chemistry techniques and an isodesmic reaction.

Solution. Many isodesmic reactions are possible, and the selected reaction can be hypothetical as long as the enthalpies of formation of the products and reactants, other than the compound under investigation, are known from experiment. For this example, the following reaction is suitable for the calculations but is not a known pathway to synthesize 2-methylphenol (which is typically done through alkylation of phenol with methanol).

$$\begin{aligned} &1 \text{ toluene} \;+\; 1 \text{ methanol} = 1 \text{ 2-methylphenol} + 1 \text{ methane} \\ &1\,(C_6H_5)CH_3 + 1\,CH_3OH = 1\,CH_3(C_6H_4)OH + 1\,CH_4 \end{aligned}$$

The reactants are composed of 11 C—H bonds, 4 C—C bonds, 3 C═C bonds, 1 C—O bond, and 1 O—H bond. Likewise, the products have a total of 11 C—H bonds, 4 C—C bonds, 3 C═C bonds, 1 C—O bond, and 1 O—H bond. Thus, the reaction is isodesmic in nature.

Determining the enthalpy of formation of 2-methylphenol requires running a geometry optimization followed by a frequency calculation on all compounds in the reaction separately in order to determine the energies of the molecules and from these the enthalpy of reaction. The same level of theory and basis set must be used for each molecule, and the *ZPE* must be included. For this example, each molecule was built using WebMO Basic (WebMO, 2021). Quantum mechanical geometry optimization and frequency calculations were done using Version 1.4rc1 of Psi4 (Smith et al., 2020) with B3LYP/6-311+G(3df,2p) model chemistry. Defaults for other parameters were used in the calculation except the geometry optimization criteria (`g_convergence`) were set equal to `gau_tight`, the grid for the radial integration was set to 99 points (`dft_radial_points 99`), and the grid for the spherical integration was set to the Lebedev Points number of 590 (`dft_spherical_points 590`). The job was run using 12 cores on a modestly sized Dell Precision Workstation fitted with 64 GB RAM and four Intel Xeon E5-2650 v2 processors, which have eight cores each running at 2.60 GHz. The calculation for all four compounds combined took about 3.2 wall-clock hours and 34.7 hours of CPU time.

The Psi4 output files contain much data, but only a few values are needed for the problem at hand. These are: (1) the electronic energy of each optimized geometry at the bottom of the well (E_0), (2) the enthalpy correction from 0 K to 298.15 K [H_{corr}(298.15 K)], and (3) the *ZPE*. Adding the latter two to the first yields the enthalpy of each molecule relative to the atoms separated at infinite distances. These enthalpies can then be used to obtain the enthalpy of the isodesmic reaction.

The labels used by Psi4 for each of the needed quantities are listed below. Notice that the enthalpy correction and the *ZPE* are reported together. This is because the former is rarely used without the latter, so Psi4 reports the two values combined. It also reports the summation of the three quantities (see last line of the list).

Variable	Psi4 Label in Output File
E_0	`Total E0, Electronic energy at well bottom at 0 [K]`
$H_{corr}(298.15\text{ K}) + ZPE$	`Correction H`
$E_0 + H_{corr}(298.15\text{ K}) + ZPE$	`Total H, Enthalpy at 298.15 [K]`

Table 4.10 contains the numerical values for E_0, H_{corr} (298.15 K) + *ZPE*, and the total sum of the three quantities for each species in the reaction. The experimental ideal gas enthalpy of formation for the compounds other than 2-methylphenol (the value being determined) are also listed. The base energy unit in Psi4 is the hartree [1 hartree = $4.3597447222071 \times 10^{-18}$ J (Tiesinga et al., 2021)], so the table lists the *calculated* energies in units of hartree and the *experimental* ΔH_f° values in units of kJ/mol. The experimental enthalpies of formation were obtained from Appendix A.

The predicted enthalpy of reaction ($\Delta_r H$) from the computational chemistry is

$$-346.77145898 + (-40.48842903) - (-271.52340332) - (-115.71769366) = -0.01879103 \text{ hartree}$$
$$= -49.33585 \text{ kJ/mol}$$

The unknown enthalpy of formation is obtained from

$$\Delta_f H^\circ_{2\text{-methylphenol}} = \Delta_r H + \Delta_f H^\circ_{\text{toluene}} + \Delta_f H^\circ_{\text{methanol}} - \Delta_f H^\circ_{\text{methane}}$$

Inserting the numbers yields

$$-49.33582 + 50.17 + (-200.94) - (-74.52) = -125.59 \text{ kJ/mol}$$

Appendix A lists the ideal gas enthalpy of formation as −128.57 kJ/mol. The error in the predicted value is thus 2.3% or 3.0 kJ/mol.

TABLE 4.10 Computational chemistry results with model chemistry B3LYP/6-311G+(3df,2p) and experimental ideal gas enthalpies of formation for the isodesmic reaction of Example 4.6 needed to determine the ideal gas enthalpy of formation of 2-methylphenol (1 hartree = $4.3597447222071 \times 10^{-18}$ J).

	Computational (B3LYP/6-311G+(3df,2p))			Experimental
	E_0	H_{corr} (298.15 K) + *ZPE*	Sum	$\Delta_f H^\circ$(298.15 K)
Compound	hartree	hartree	hartree	kJ/mol
toluene	−271.65790077	0.13449745	−271.52340332	50.17
methanol	−115.77311018	0.05541652	−115.71769366	−200.94
methane	−40.53680086	0.04837183	−40.48842903	−74.52
2-methylphenol	−346.91150308	0.14004410	−346.77145898	—

Subsets of isodesmic reactions have also been proposed to ensure that the products and reactants have even more in common with each other than what is required for an isodesmic reaction. This increased similarity theoretically leads to better accuracy when calculating $\Delta_f H°$. The terms *homodesmic* and *homodesmotic* are found in widespread use, but the definitions have been ambiguous. Wheeler et al. (2009) discuss the different definitions found in the literature and give a hierarchy to identify different types. Their definition of a *homodesmotic* reaction is one in which the products and reactants have equal numbers of (a) each type of carbon—carbon bonds (C_{sp3}—C_{sp3}, C_{sp3}—C_{sp}, C_{sp3}—C_{sp}, C_{sp2}—C_{sp2}, C_{sp2}—C_{sp}, C_{sp}—C_{sp}, C_{sp2}=C_{sp2}, C_{sp2}=C_{sp}, C_{sp}=C_{sp}, C_{sp}≡C_{sp}) and (b) each type of carbon atom (sp^3, sp^2, sp) with zero, one, two, and three hydrogens attached. Hyperhomodesmotic reactions and hypohomodesmotic reactions have also been defined. The point with each of these is to classify the similarity between the types of bonds in the products and reactants. Using isodesmic or more restrictive reactions can be done following the procedure in Example 4.6.

Other approaches to determine accurate energies from quantum mechanical calculations are so-called *composite* and *focal point* methods. These techniques achieve accuracy by performing a set of calculations to remove or account for biases and limitations known to occur when using single computations with one level of theory and basis set selection. They also attempt to overcome the shortcomings of using a *finite* number of basis set functions to mathematically describe the wave function of the molecule—a process which requires an infinite number of functions for perfect accuracy.

The idea for such approaches was first proposed by Pople et al. (1989) with the Gaussian-1 (G1) method but has quickly spread to include many recipes including updates to the G1 method (collectively known as Gn methods) by Curtiss et al. (1991, 1998, 2007) and new methods such as various complete basis set (CBS) approaches (Ochterski et al., 1996; Montgomery et al., 1999; Montgomery et al., 2000), the W (Weizmann) methods (Martin and de Oliveira, 1999; Parthiban, 2001; Barnes, 2009), HEAT (Tajti et al., 2004; Bomble et al., 2006), ATOMIC (Bakowies, 2009), the correlation consistent composite approach (ccCA) (DeYonker, 2006), and the various coupled-cluster methods (Riplinger et al., 2013). Each of these methods has been shown to give accurate energies, but most require expertise in quantum chemical calculations. The Gn methods were intentionally designed to be "black box," and an example of the G4 method is presented below because it is straightforward to implement for the nonspecialist. However, it is computationally demanding and, as will be discussed in Sec. 4.8, is not as accurate as newer approaches. For this reason, an example using the local coupled-cluster approach of Paulechka and Kazakov (2018) is also given. This method, though requiring more training in quantum mechanics to implement, has been shown be more accurate than black box approaches even though it requires less computational time. This method is also one of those that is evaluated later against a test database.

Example 4.7 Estimate the ideal gas enthalpy of formation of 2-methylphenol (*o*-cresol) at 298.15 K using the G4 method of Curtiss et al. (2007).

Solution. The molecule was built using GaussView 5.0 from Gaussian Inc., and the calculation was done using Gaussian 09. The job was run using 24 cores on a modestly sized Dell Precision Workstation fitted with 64 GB RAM and four Intel Xeon E5-2650 v2 processors which have eight cores each running at 2.60 GHz. The pertinent energies found in the log file are listed in Table 4.11. The first column is the name of the energy in the log file—a string that is consistent regardless of the system of molecules on which calculations are performed. The second column is the value of the energy. The third column is a description of the energy.

TABLE 4.11 Relevant energies from the log file for the G4 calculation of 2-methylphenol (*o*-cresol).

Log File Entry	Value in Hartree	Description
G4(0 K)	−346.583701	The bottom of the well electronic energy of the molecule plus the zero-point energy ($E_{0+ZPE} = E_0 + ZPE$).
G4 enthalpy	−346.575516	E_{0+ZPE} plus the translational, vibrational, rotational, energies plus $PV = RT$.

The reference state for the energies in Gaussian are the individual atomic particles at 0 K. More specifically, the energies found by solving Schrödinger's equation represent the change in energy of bringing all the atomic particles in the molecules, initially separated at infinite distance, together and forming the molecule at 0 K. This reference state is different than that used for the standard ideal gas enthalpy of formation discussed in this chapter, so determining $\Delta_f H°(298.15 \text{ K})$ from the quantum mechanical energies involves moving to the reference state suggested by Eq. (4.1-3). As mentioned at the start of this section, the influence of *ZPE* must also be incorporated to ensure the correct dissociation energy is used. The procedure described below consists of three steps.

1. **Calculate the atomization energy of the molecule (D_0).** This value is needed to set the reference state in accordance with Eq. (4.1-3). E_0 of the compound is subtracted from the electronic energies of the atoms in the molecule to accomplish this. Specifically

$$D_0 = \sum_e N_e E_{0_e} - E_{0+ZPE} \tag{4.6-8}$$

where N_e is the number of atoms of element e in the molecule and E_{0_e} is the electronic energy of atom e. The electronic energies of the atoms are determined from G4 calculations and include the *ZPE*. They are the "G4 (0K)" energies in the Gaussian log file for a G4 calculation of an atom. Curtiss et al. (2007) list many of these atomic G4 energies, and the values for some of the most common atoms are found in Table 4.12.

2. **Calculate the enthalpy of formation of the molecule at 0 K [$\Delta_f H°(0 \text{ K})$].** The enthalpy of formation of the molecule at 0 K is determined by

$$\Delta_f H°(0 \text{ K}) = \sum_e N_e \Delta_f H_e^\circ(0 \text{ K}) - D_0 \tag{4.6-9}$$

where $\Delta_f H_e^\circ(0 \text{ K})$ is the enthalpy of formation of atom e at 0 K, and D_0 is found from Eq. (4.6-8). The atomic enthalpies of formation are found from experiments and are listed in Curtiss et al. (1997). Some of the pertinent values are found in Table 4.13.

3. **Calculate the enthalpy of formation of the molecule at 298.15 K [$\Delta_f H°(298.15 \text{ K})$].** Once $\Delta_f H°(0 \text{ K})$ is known, the standard enthalpy of formation at 298.15 K is determined by

$$\Delta_f H°(298.15 \text{ K}) = \Delta_f H°(0 \text{ K}) + [H°(298.15 \text{ K}) - H°(0 \text{ K})] - \sum_e N_e \left[H_e^\circ(298.15 \text{ K}) - H_e^\circ(0 \text{ K})\right] \tag{4.6-10}$$

where $[H°(298.15 \text{ K}) - H°(0 \text{ K}]$ is the change in enthalpy of the molecule in the ideal gas state that occurs when moving from 0 K to 298.15 K, and $[H_e^\circ(298.15 \text{ K}) - H_e^\circ(0 \text{ K})]$ is the change in enthalpy of

TABLE 4.12 Total G4 energies for common atoms as reported by Curtiss et al. (2007).

Atom	E_{0+ZPE} [*G*4(0 K)] (hartree)
H	−0.50142
C	−37.83417
O	−75.04550
N	−54.57367
S	−397.98018
Si	−289.23704
F	−99.70498
Cl	−460.01505
P	−341.13463

TABLE 4.13 Enthalpies of formation of gaseous atoms at 0 K as reported in Curtiss et al. (1997). (1 kcal = 4.184 J).

Atom	$\Delta_f H°(0 \text{ K})$(kcal/mol)
H	51.63
C	169.98
O	58.99
N	112.53
S	65.66
Si	106.6
F	18.87
Cl	28.59
P	75.42

TABLE 4.14 Atomic ideal gas thermal corrections to enthalpy from 0 to 298.15 K for the G4 method of Curtiss et al. (2007). (1 kcal = 4.184 J).

Atom	$[H_e^\circ(298.15\text{ K}) - H_e^\circ(0\text{ K})]$ (kcal/mol)
H	1.01
C	0.25
O	1.04
N	1.04
S	1.05
Si	0.76
F	1.05
Cl	1.10
P	1.28

ideal gas atom e that occurs when moving from 0 K to 298.15 K. The bracket notation used here is that used by the Gaussian literature. In terms of quantities discussed previously, they are thermal corrections that are calculated via the process suggested by Eqs. (4.5-6) and (4.5-7) except that for atoms the rotational and vibrational modes are identically zero. However, starting at Eq. (4.5-6) or (4.5-7) is not done by the user. Rather, the thermal correction for the molecule can be obtained from the G4 energy and enthalpy in the Gaussian log file (see Table 4.11) as

$$[H^\circ(298.15\text{ K}) - H^\circ(0\text{ K})] = \text{G4 Enthalpy} - \text{G4}(0\text{ K}) \tag{4.6-11}$$

The thermal corrections for relevant atoms are listed by Curtiss et al. (1997), and a few are found in Table 4.14.

2-Methylphenol is composed of three elements—C, H, and O—with $N_C = 7$, $N_H = 8$, and $N_O = 1$. With these values, and those in Table 4.11 and Table 4.12, Eq. (4.6-8) gives

$$D_0 = 7(-37.83417) + 8(-0.50142) + 1(-75.04550) - (-346.583701) = 2.687651\text{ hartree} = 1686.526\text{ kcal/mol}$$

This value, along with Eq. (4.6-9) and the values in Table 4.13 give

$$\Delta_f H^\circ(0\text{ K}) = 7(169.98) + 8(51.63) + 1(58.99) - 1686.526 = -24.636\text{ kcal/mol}$$

Finally, converting to 298.15 K using Eqs. (4.5-10) and (4.5-11) and the values in Table 4.14 yields

$$\Delta_f H^\circ(298.15\text{ K}) = -24.636 + (6.02214\times10^{23})(1.0420039\times10^{-21})[-346.575516 - (-346.583701)] - 7(0.25) - 8(1.01) - 1(1.04) = -30.37\text{ kcal/mol} = -127.07\text{ kJ/mol.}$$

Appendix A lists the ideal gas enthalpy of formation as −128.57 kJ/mol. The error in the predicted value is thus 1.2% or 1.5 kJ/mol.

Example 4.8 Estimate the ideal gas enthalpy of formation of 2-methylphenol (*o*-cresol) at 298.15 K using the local coupled-cluster "aLL5" method of Paulechka and Kazakov (2018).

Solution. As described in Example 4.7, care must be taken when using quantum mechanics to calculate ideal gas $\Delta_f H^\circ(298.15\text{ K})$ to move from the conditions of the calculation (0 K, zero in energy is isolated atomic particles) to the standard state commonly used to tabulate the property (298.15 K, 101325 Pa, zero in energy is compounds in their natural state at 298.15 K). In the G4 method, moving from defining the zero in energy as isolated atomic particles to the macroscopic definition using naturally occurring compounds required the electronic energies of the atoms (see Table 4.12), the enthalpy of formation of the atoms comprising the molecule at 0 K (see Table 4.13) and the change in enthalpy of each atom when moving from 0 to 298.15 K (see Table 4.14). Each of these contributions to the final energy are additive, so recent techniques, like that of Paulechka and Kazakov (2018), combine all of these contributions into one empirical parameter that is fit to critically evaluated experimental data. With such an approach, the standard state enthalpy of formation of a compound at 298.15 K is obtained from

$$\Delta_f H^\circ(298.15\text{ K}) = E_0 + ZPE + H_{corr}(298.15\text{ K}) - \sum_{\text{atom types}} n_i h_i \tag{4.6-12}$$

Here, E_0 is the "bottom of the well" total electronic energy, ZPE is the zero-point energy calculated from Eq. (4.5-3), and $H_{corr}(298.15\text{ K})$ is the change in enthalpy when moving the compound from 0 to 298.15 K (the thermal correction to the compound) as given by Eq. (4.5-6) or (4.5-7). The summation is over all atom types in the compound, n_i is the number of atoms of type i in the compound, and h_i are the empirical parameters for each atom type previously mentioned. For the "aLL5" method, the frequencies (ν_i) are scaled by 0.990 when calculating ZPE. And, for $H_{corr}(298.15\text{ K})$, frequencies corresponding to hydrogen stretches (>2800 cm^{-1}) are scaled by 0.960 while all other frequencies are scaled by 0.985.

Table 4.15 lists the value of h_i for the "aLL5" method as reported in Paulechka and Kazakov (2018, 2019) for compounds containing C, H, O, N, and F. Sulfur-containing compounds have also been examined (Paulechka and Kazakov, 2021), but the parameter for each sulfur atom in the molecule, $h_{S\text{ type}}$, depends on the oxidation state and bonding configuration of the atom. It is determined by defining a *base value* and then adding *corrections*. Specifically, the coefficient for Eq. (4.6-12) for each sulfur type i (multiple sulfur *types* may be present in the molecule) is obtained by

$$h_{S_i} = h(S) + \sum_j N_j \Delta h(j) \tag{4.6-13}$$

where $h(S)$ is the base value for a sulfur, N_j is the number of corrections of type j for sulfur type i, and $\Delta h(j)$ is the correction of type j that accounts for the oxidation state and bonding configuration of sulfur type i in the molecule. Table 4.16 contains the values of the parameters (the base and the corrections) needed to use Eq. (4.6-13). For example (Paulechka and Kazakov, 2021), the coefficient to use in Eq. (4.6-12) for the one sulfur type ($n_S = 1$) in sulfuric acid (H_2SO_4) would be $h_S = h(S) + \Delta h[S(VI)] + 2\Delta h(S{=})$ where $N_{S(VI)} = 2$ and $N_{S=} = 2$. In hydrogen disulfide (H_2S_2), the two sulfur atoms are the same type, so $n_S = 2$ and $h_S = h(S) + \Delta h\left(\frac{1}{2}\text{SS}\right)$ where $N_{\frac{1}{2}SS} = 1$. There is no correction for an S-H bond, so the coefficient to use in Eq. (4.6-12) for the sulfur in ethanethiol (CH_3CH_2SH) would be $h_S = h(S)$ with $n_S = 1$.

TABLE 4.15 Coefficients in Eq. (4.6-12) to determine ideal gas $\Delta_f H^\circ$(298.15 K) using the "aLL5" quantum mechanical technique of Paulechka and Kazakov (2018). Values for C, H, O, and N are from Paulechka and Kazakov (2018). The value for F is from Paulechka and Kazakov (2019).

Atom	$-h_i$ (kJ/mol)
C (saturated or aromatic)	99910.32
C (unsaturated)	99909.44
H	1524.23
O	197138.05
N	143612.32
F	261711.75
S	See Table 4.16

TABLE 4.16 Base value and corrections to use in Eq. (4.6-13) to calculate the coefficient for each sulfur type i in a molecule (h_{si}) to use Eq. (4.6-12) to determine ideal gas $\Delta_f H°$(298.15 K) using the "aLL5" quantum mechanical technique of Paulechka and Kazakov (2018, 2021).

Quantity	Description	Value (kJ/mol)
h(S)	sulfur base value	−1044348.68
Δh(S=)	correction of each nonaromatic double bond connected to the sulfur atom	1.76
$\Delta h\left(\frac{1}{2}\text{SS}\right)$	correction for a (one) sulfur participating in an S—S bond	1.76
Δh[S(arom)]	correction for a sulfur in an aromatic ring	1.76
Δh[S(IV)]	correction for each group (IV) atom bonded to the sulfur	5.03
$\frac{1}{2}\Delta h$[S(VI)]	correction for each group (VI) atom bonded to the sulfur	5.03

To summarize the method of Paulechka and Kazakov (2018), the steps to obtain $\Delta_f H°$(298.15 K) from Eq. (4.5-12) are:

1. Run a geometry optimization and frequency calculation on the compound using B3LYP-D3(BJ)/def2-TZVP model chemistry. Paulechka and Kazakov (2018) use Gaussian for this step with "tight" optimization convergence criteria and "ultrafine" grid spacing for integrations.
   a. Scale the raw frequencies found in the log file by 0.990 and use these values to obtain *ZPE* from Eq. (4.5-3).
   b. Scale the raw frequencies found in the log file that are >2800 cm^{-1} (hydrogen stretching modes) by 0.960. Scale all other raw frequencies by 0.985. Use these scaled frequencies to determine H_{corr}(298.15 K) from Eq. (4.5-6) for linear molecules or (4.5-7) for nonlinear molecules.
2. Run a geometry optimization on the compound using DF-MP2/aug-cc-pVQZ model chemistry. Paulechka and Kazakov (2018) use Psi4 for this step with "gau_tight" optimization convergence criteria.
3. Run an energy calculation on the geometry from the previous step using LCCSD(T)/aug-cc-pVQZ model chemistry to obtain E_0, the "bottom of the well" electronic energy. Paulechka and Kazakov (2018) use MRCC with the 2016 local correlation calculation (localcc=2016) and "tight" accuracy of the local correlation calculations (lcorthr=tight) for this step.
4. Calculate $\Delta_f H°$(298.15 K) using Eq. (4.5-12) with the values for E_0, *ZPE*, and H_{corr}(298.15 K) obtained from the previous steps and the values of h_i from Table 4.15.

The frequencies produced by Step 1 are found in Table 4.17. The "bottom of the well" electronic energy (E_0) produced from Step 3 is −346.293858601249 hartree = −909194.401 kJ/mol. The *ZPE*, calculated as outlined in Step 1A from the frequencies found in Table 4.17, is 0.13058898 hartree = 342.861 kJ/mol, and the thermal correction calculated as outlined in Step 1B from the frequencies found in Table 4.17 is 21.584 kJ/mol. 2-Methylphenol is composed of 8 H, 7 saturated or aromatic C, and 1 O, so and $\sum_{\text{atom types}} n_i h_i = -908704.130$ kJ/mol.

With these values, the ideal gas enthalpy of formation from Eq. (4.5-12) is

$$-909194.401 + 342.861 + 21.584 - (-908704.130) = -125.83 \text{ kJ/mol}$$

Appendix A lists the ideal gas enthalpy of formation as −128.57 kJ/mol. The error in the predicted value is thus 2.1% or 2.7 kJ/mol.

TABLE 4.17 Unscaled vibrational wave numbers for 2-methylphenol in units of cm^{-1}. The values were determined from quantum mechanical calculations with B3LYP-D3(BJ)/def2-TZVP model chemistry.

133.6148	756.0442	1189.737	1654.126
175.8666	766.0807	1239.58	3033.847
263.9641	843.2582	1285.742	3078.95
293.8025	856.1972	1335.578	3112.462
351.0306	916.2731	1362.737	3150.505
440.4393	942.094	1420.283	3164.72
449.9456	1009.621	1474.931	3180.707
534.8003	1065.246	1480.062	3197.062
554.6771	1071.492	1507.934	3804.341
595.0808	1125.872	1542.715	
718.2388	1183.081	1638.206	

B. Ideal Gas Absolute Entropy

From statistical mechanics (McQuarrie, 2000), the ideal gas absolute entropy of a *linear* molecule in the rigid rotator-harmonic oscillator (RRHO) approximation is

$$\frac{S^\circ(T)}{R} = \ln\left[\left(\frac{2\pi mkT}{h^2}\right)^{\frac{3}{2}}\frac{Ve^{\frac{5}{2}}}{N}\right] + \ln\left(\frac{Te}{\sigma\Theta_r}\right) + \sum_{i=1}^{3n-5}\frac{\frac{\Theta_{v_i}}{T}}{e^{\frac{\Theta_{v_i}}{T}}-1} - \ln\left(1-e^{\frac{-\Theta_{v_i}}{T}}\right) + \ln W_0 \tag{4.6-14}$$

For a *nonlinear* molecule, the expression in the RRHO approximation is

$$\frac{S^\circ(T)}{R} = \ln\left[\left(\frac{2\pi mkT}{h^2}\right)^{\frac{3}{2}}\frac{Ve^{\frac{5}{2}}}{N}\right] + \ln\left[\frac{1}{\sigma}\left(\frac{\pi T^3 e^3}{\Theta_{rA}\Theta_{rB}\Theta_{rC}}\right)^{\frac{1}{2}}\right] + \sum_{i=1}^{3n-6}\frac{\frac{\Theta_{v_i}}{T}}{e^{\frac{\Theta_{v_i}}{T}}-1} - \ln\left(1-e^{\frac{-\Theta_{v_i}}{T}}\right) + \ln W_0 \tag{4.6-15}$$

The first term in these expressions is the translational contribution to the entropy, the second is the rotational, the third the vibrational, and the fourth the electronic. Also, h is Planck's constant, k is Boltzmann's constant, n is the number of atoms in the molecule, σ is the rotational symmetry number, Θ_r or $\Theta_{rA}/\Theta_{rB}/\Theta_{rC}$ is/are the characteristic rotational temperature(s) of the molecule, Θ_{v_i} are the characteristic vibrational temperatures of the molecule, W_0 is the degeneracy of the ground electronic state of the molecule, m is the mass of the molecule, V is the volume of the system, and N is the number of molecules in the system. The assumptions made in deriving these equations are found in the discussion surrounding Eqs. (4.5-4)–(4.5-7).

Care must be taken when applying Eqs. (4.5-13) and (4.5-14) to ensure dimensional consistency. Toward this end, m can be obtained in units of kg by dividing the molecular weight (in units of kg/mol) of the molecule by Avogadro's number. Also, because the expressions were derived for the ideal gas, $V = \frac{kT}{P}$ (where P is the pressure of the system; 1 atm (101325 Pa) for the standard conditions) and $N = 1$. The characteristic rotational temperature(s), Θ_r or $\Theta_{rA}/\Theta_{rB}/\Theta_{rC}$, and the characteristic vibrational temperatures Θ_{v_i} are also needed. The former can be obtained experimentally from microwave spectroscopy and the latter from IR spectroscopy, but for prediction purposes are found by performing quantum mechanical calculations. Specifically, one first performs a geometry optimization on the molecule followed by a frequency calculation to determine the characteristic temperatures. The electronic ground state degeneracy, W_0, is also obtained during the calculations.

Most quantum mechanical software performs the mathematical operations outlined in Eqs. (4.5-13) and (4.5-14) to determine $S^\circ(298.15 \text{ K})$ automatically when frequency calculations are requested, and this precomputed value may be used. However, the software does not take into account frequency scaling factors by default during this determination. Thus, if including the scaling factors is desired, the user will need to perform the calculation of $S°(298.15\ K)$ manually after obtaining the Θ's and W_0 from the log file. Example 4.9 demonstrates the process.

Example 4.9 Estimate ideal gas $S°(298.15\ K)$ of 2-methyl-2-butanol using computational chemistry techniques.

Solution. The molecule was built using WebMO Basic (WebMO, 2021). Quantum mechanical geometry optimization and frequency calculations were done using Version 1.4rc1 of Psi4 (Smith et al., 2020) with B3LYP/6-311+G(3df,2p) model chemistry. Defaults for other parameters were used in the calculation except the geometry optimization criteria (`g_convergence`) were set equal to `gau_tight`, the grid for the radial integration was set to 99 points (`dft_radial_points 99`), and the grid for the spherical integration was set to the Lebedev Points number of 590 (`dft_spherical_points 590`). The job was run using 12 cores on a modestly sized Dell Precision Workstation fitted with 64 GB RAM and four Intel Xeon E5-2650 v2 processors which have eight cores each running at 2.60 GHz. The calculation took 114 wall-clock minutes and 20.6 CPU hours.

The resulting rotational temperatures are listed in Table 4.18, and vibrational temperatures are found in Table 4.19. The rotational symmetry number reported in the output file was $\sigma = 1$, the ground state degeneracy, given by the program as "multiplicity," is $W_0 = 1$, and the mass of the molecule is 88.08882 amu.

Table 4.20 contains the estimates of ideal gas $S°(298.15\ K)$ obtained using Eq. (4.5-14) and the values as just described. The results are broken down into the individual contributions to the total, and two different estimates are given. The first uses unscaled frequencies (the values in Table 4.19), and the second uses scaled frequencies by multiplying the values in Table 4.19 by 0.967 (see Table 4.9). Also shown are the differences (and % difference) from the experimental value of 362.8 J/(mol·K) reported by Wilholt and Zwolinski (1973). Notice that scaling the frequencies improves the agreement between the prediction and the experiment.

TABLE 4.18 Characteristic rotational temperatures—Θ_{r_A}, Θ_{r_B}, Θ_{r_C},—for 2-methyl-2-butanol in units of K. The values were determined from quantum mechanical calculations with B3LYP/6-311+G(3df,2p).

0.21888	0.12697	0.12543

TABLE 4.19 Characteristic vibrational temperatures, Θ_{v_i}, for 2-methyl-2-butanol in units of K. The values were determined from quantum mechanical calculations with B3LYP/6-311+G(3df,2p).

141.0	751.4	1659.8	2137.2	4386.7
304.0	1042.3	1741.9	2145.8	4419.2
330.7	1128.7	1756.4	2161.4	4430.7
365.5	1269.7	1895.4	2163.2	4436.3
376.5	1323.4	1942.1	2174.0	4455.7
393.2	1352.5	1994.6	2181.7	4468.1
492.3	1438.6	2012.5	4336.0	4474.0
528.9	1453.2	2031.1	4338.9	5506.6
594.3	1531.7	2041.3	4349.4	
656.2	1543.2	2125.1	4360.8	

TABLE 4.20 Ideal gas absolute entropy at 298.15 K [$S°$(298.15 K)] of 2-methyl-2-butanol from quantum mechanical calculations. One estimate uses unscaled vibrational frequencies and the other scaled vibrational frequencies.

	$S°$(298.15 K) [J/(mol·K)]	
Contribution	**Unscaled Frequencies**	**Scaled Frequencies**
Translational	164.599	164.599
Rotational	111.815	111.815
Vibrational	77.304	80.736
Electronic	0.000	0.000
Total	353.717	357.150
Difference from experimental value of 362.8 J/(mol·K)	−9.08 (−2.5%)	−5.65 (−1.6%)

C. Ideal Gas Heat Capacity

From statistical mechanics (McQuarrie, 2000), the isobaric ideal gas heat capacity as a function of temperature ($C_p^\circ(T)$) for a *linear* polyatomic molecule in the RRHO approximation is given by

$$\frac{C_p^\circ(T)}{R} = \frac{7}{2} + \sum_{i=1}^{3n-5} \left(\frac{\Theta_{v_i}}{T}\right)^2 \frac{e^{\frac{\Theta_{v_i}}{T}}}{\left(e^{\frac{\Theta_{v_i}}{T}} - 1\right)^2} \tag{4.6-16}$$

For a *nonlinear* molecule, the expression in the RRHO approximation is

$$\frac{C_p^\circ(T)}{R} = 4 + \sum_{i=1}^{3n-6} \left(\frac{\Theta_{v_i}}{T}\right)^2 \frac{e^{\frac{\Theta_{v_i}}{T}}}{\left(e^{\frac{\Theta_{v_i}}{T}} - 1\right)^2} \tag{4.6-17}$$

The assumptions made in deriving these equations, and the definition of the variables, are found in the discussion surrounding Eqs. (4.5-4)–(4.5-7). As with those expressions, the characteristic vibrational temperatures Θ_{v_i} are found from first performing a geometry optimization on the molecule followed by a frequency calculation. An example of the process is found below.

Example 4.10 Estimate the ideal gas heat capacity of 2-methyl-2-butanol from 350 to 550 K using computational chemistry techniques.

Solution. The molecule was built using WebMO Basic (WebMO, 2021). Quantum mechanical geometry optimization and frequency calculations were done using Version 1.4rc1 of Psi4 (Smith et al., 2020) with B3LYP/6-311+G(3df,2p) model chemistry. Defaults for other parameters were used in the calculation except the geometry optimization criteria (`g_convergence`) were set equal to `gau_tight`, the grid for the radial integration was set to 99 points (`dft_radial_points 99`), and the grid for the spherical integration was set to the Lebedev Points number of 590 (`dft_spherical_points 590`). The job was run using 12 cores on a modestly sized Dell Precision Workstation fitted with 64 GB RAM and four Intel Xeon E5-2650 v2 processors which have eight cores each running at 2.60 GHz. The calculation took 114 wall-clock minutes and 20.6 CPU hours.

The resulting vibrational temperatures are found in Table 4.17. Recall that the frequencies in Table 4.17 are unscaled and were read directly from the log file produced by the quantum chemical software. Determining $C_p^\circ(T)$ was done by multiplying the values in Table 4.17 by 0.967 and then using these scaled values in Eq. (4.5-16).

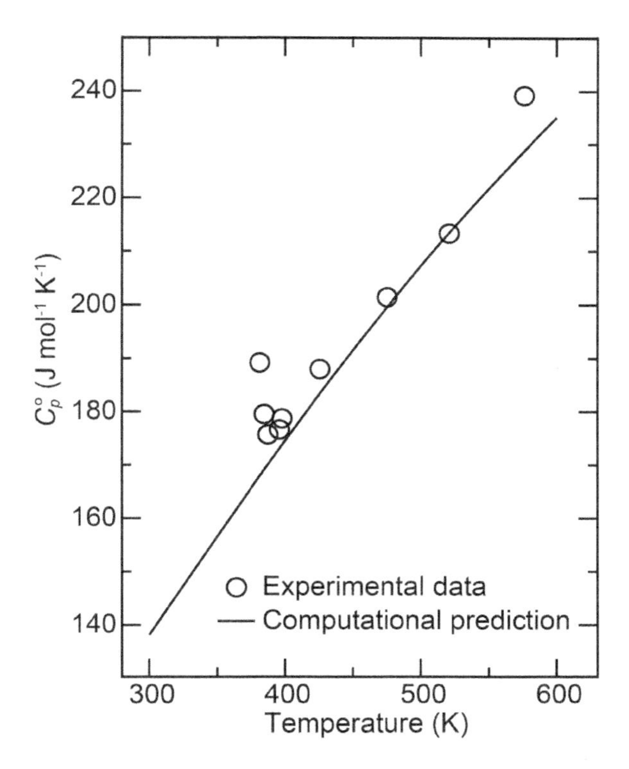

FIGURE 4.2
Ideal gas heat capacity of 2-methyl-2-butanol as a function of temperature predicted from quantum mechanical calculations and measured experimentally by Strömsöe et al. (1970).

Figure 4.2 shows the results of this approach. The solid line is the quantum mechanical prediction as just outlined. The points are experimental data from Strömsöe et al. (1970). The prediction agrees well with the data especially considering the scatter in the latter. The average absolute deviation between the experimental points and the quantum mechanical prediction is 3.6%.

D. Flexible Molecules and Quantum Mechanical Approaches

If a molecule has rotatable bonds, several configurations of the molecule may exist that have comparable energies. This means that many different conformations of the same ideal gas molecule occur in equilibrium in nature. Geometry optimization finds only one conformation, a local electronic energy minimum, so the subsequent properties calculated from this geometry, like the examples presented above, are only for a single conformer. For compounds with many rotatable bonds, the energies among the different conformers can vary more than a few kJ/mol, so improved accuracy is obtained by performing several calculations, using the same model chemistry, for different conformations of the molecules and properly averaging the subsequent property values from each.

For ideal gas enthalpy of formation, the averaging among the difference conformers is the mole fraction average

$$\Delta_f H^\circ(T) = \sum_{i=1}^{N_{\text{conformers}}} x_i(T)\Delta_f H_i^\circ(T) \tag{4.6-18}$$

where the summation is over the number of conformers $N_{\text{conformers}}$, $\Delta_f H_i^\circ(T)$ is the value of the property for the ith conformer of the compound at temperature T, and $x_i(T)$ is the mole fraction of the ith conformer at temperature T. The mole fractions are computed from the Gibbs energy of each conformer $G_i^\circ(T)$ as

$$x_i(T) = \frac{e^{-\frac{G_i^\circ(T)}{RT}}}{\sum_{i=1}^{N_{\text{conformers}}} e^{-\frac{G_i^\circ(T)}{RT}}} \quad (4.6\text{-}19)$$

The Gibbs energy of each conformer is obtained from the quantum mechanics results and the definition $G = H - TS$, namely

$$G_i^\circ(T) = E_{0_i} + H_{\text{corr}_i}(T) + ZPE_i - TS_i^\circ(T) \quad (4.6\text{-}20)$$

Here, $H_{\text{corr}_i}(T)$ is obtained from either Eq. (4.5-6) or (4.5-7), $S_i^\circ(T)$ is found from either Eq. (4.5-13) or (4.5-14), and ZPE_i is calculated using Eq. (4.5-3). The ZPE must be added to ensure all conformations are set to the correct reference state, and frequencies should be scaled according to the model chemistry used in the calculations.

The average ideal gas absolute entropy of different conformers of a molecule is obtained from

$$S^\circ(T) = \sum_{i=1}^{N_{\text{conformers}}} x_i(T) S_i^\circ(T) - R \sum_{i=1}^{N_{\text{conformers}}} x_i \ln x_i \quad (4.6\text{-}21)$$

The average ideal gas heat capacity of different conformers of a molecules is

$$C_p^\circ(T) = \sum_{i=1}^{N_{\text{conformers}}} \left[x_i(T) C_{p_i}^\circ(T) + H_i^\circ(T) \left(\frac{dx_i(T)}{dT} \right) \right] \quad (4.6\text{-}22)$$

where $H_i(T) = E_{0_i} + H_{\text{corr}_i}(T) + ZPE_i$ with the terms on the right hand side of the equation coming from the quantum mechanical calculations as described previously. The generation of different conformers can be done in a variety of ways. For a small number of rotatable bonds, it is feasible to directly rotate each applicable bond through 360° and scan through all the possible configurations. For larger molecules, one approach is to use molecular simulations techniques, such as Monte Carlo (MC) or molecular dynamics (MD), to generate a number of candidates. Simulated annealing can be done to help overcome rotational barriers, and the random number seed can be changed among different runs to generate different trajectories. Because the system consists of a single molecule, the computational costs are low, and a large number of candidate conformers (hundreds) can be generated using either MD or MC. The structure of the molecule produced at the end of each simulation run can be minimized, and the root mean squared deviation between the different minimized structures can be calculated to eliminate replicate structures. Geometry optimization can then be done on the unique structures of the molecule to find the individual property values for the ith conformer needed for Eqs. (4.5-17)–(4.5-21).

Many variations on the theme can be envisioned to creating the suite of unique conformers. Pracht et al. (2020) have proposed an automated scheme combining tight-binding methods with meta-dynamics approaches to map out the potential energy landscape of the molecule and thus identify each of the local minimum containing unique structures of the molecule. The tool is called Conformer-Rotamer Ensemble Sampling Tool (CREST), and it systematically samples the possible conformations of the molecule to find the low-energy structures. The methods used to accomplish this include MD, meta-dynamics simulation, and genetic cross algorithms. Moreover, the code is freely available for use.

The examples presented up to this point in this section used only a single conformer of the molecule. Example 4.11 shows how consideration of multiple conformers changes the prediction for 2-methylphenol (o-cresol). This compound only has one rotatable bond of interest, so the effect of multiple conformations on properties is minimal. However, this simplicity is intentional to allow clear understanding of the process.

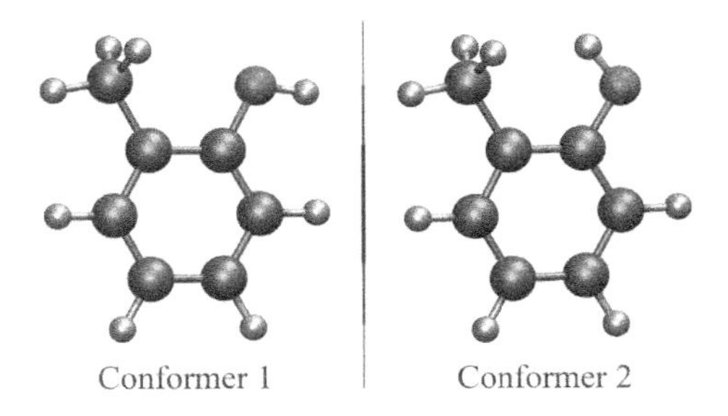

FIGURE 4.3
Conformers of 2-methylphenol (o-cresol). Conformer 1 is the lowest energy conformer and is the state used in Example 4-8. Conformer 2 was found by systematically rotating the C—O bond through 360° and optimizing the structure to find other possible energy minima. The figures were created using VMD (Humphrey et al., 1996).

Example 4.11 Estimate the ideal gas enthalpy of formation of 2-methylphenol (*o*-cresol) at 298.15 K using the local coupled-cluster "aLL5" method of Paulechka and Kazakov (2018), as was done in Example 4.8, but take different conformations into account.

Solution. The details of the quantum mechanical calculations are the same as those of Example 4.8. That process is repeated for each low-energy-minimum conformer of the molecule, and the results are averaged as per Eq. (4.5-17). 2-methylphenol has only one rotatable bond of interest, the C—O bond, so generation of different conformers consisted of systematically rotating the bond through 360° and performing geometry optimizations for each configuration to identify low energy minima. This process produced two unique conformers to consider when calculating properties. Figure 4.3 depicts the structure of each conformer. Notice that the two states differ as to whether the hydrogen of the alcohol group points toward or away from the methyl group. Conformer 1 is the lowest energy configuration and is the isomer of Example 4-8. Conformer 2 is another low-energy structure whose effect on the average property is now considered.

Determining $\Delta_f H°(298.15\text{ K})$ using Eq. (4.5-17) requires the standard enthalpies of formation of each conformer and the mole fractions of each conformer at equilibrium. The process for determining $\Delta_f H_i^\circ(T)$ from the quantum chemical calculations is demonstrated in Example 4.8. Calculating $x_i(T)$ is accomplished using Eqs. (4.5-18) and (4.5-19). All the information needed comes from the quantum mechanical calculations as outlined previously. The pertinent quantities are outlined in Table 4.21 for each conformer with the data for Conformer 1 being the same as that in Example 4.8. All values are scaled as per the "aLL5" method of Paulechka and Kazakov (2018).

TABLE 4.21 Values to calculate ideal gas $\Delta_f H°(298.15\text{ K})$ of 2-methylphenol using quantum chemical calculations on its different conformers and the "aLL5" method of Paulechka and Kazakov (2018). All values are evaluated at 298.15 K.

		Conformer 1	Conformer 2	Origin
E_{0_i}	(hartree)	−346.293858601249	−346.293519007516	Log file
H_{corr_i}	(kJ/mol)	21.584	21.634	Eq. (4.5-7)
ZPE_i	(kJ/mol)	342.861	342.798	Eq. (4.5-3)
TS_i	(kJ/mol)	103.294	103.287	Eq. (4.5-14)
G_i	(kJ/mol)	−908933.294	−908932.363	Eq. (4.5-19)
x_i		0.5884	0.4116	Eq. (4.5-18)
$\Delta_f H_i^\circ$	(kJ/mol)	−125.825	−124.947	Process of Example 4.8

TABLE 4.22 Unscaled vibrational wave numbers for Conformer 2 of 2-methylphenol in units of cm^{-1}. The values were determined from quantum mechanical calculations with B3LYP-D3(BJ)/def2-TZVP model chemistry.

154.7031	762.1828	1190.861	1662.432
171.3179	763.3645	1235.842	3004.979
260.8848	855.5165	1288.379	3042.727
294.5835	865.7026	1331.705	3113.146
306.173	930.8655	1367.808	3162.984
439.2671	952.0211	1416.523	3176.413
451.0367	1001.782	1483.31	3192.181
533.6645	1059.467	1489.999	3200.352
551.3326	1064.034	1516.788	3810.224
595.8984	1128.534	1530.523	
720.7815	1179.324	1630.715	

Determining the values listed in Table 4.21 required the vibrational and rotational constants of each conformer. The vibrational wave numbers for Conformer 1 are listed in Table 4.17. Those for Conformer 2 are found in Table 4.22. The rotational frequencies—ν_{r_A}, ν_{r_B}, ν_{r_C}—and symmetry numbers (σ) for both conformers are found in Table 4.23. Also, electronic degeneracy (W_0), needed when determining TS_i° from Eq. (4.5-14), is 1. Notice that the lowest-energy conformer is present about 59% of the time at equilibrium and the other conformer is found about 41% of the time.

With x_i and $\Delta_f H_i^\circ$ from Table 4.21, the standard enthalpy of formation for 2-methylphenol becomes

$$(0.5884)(-125.825) + (0.4116)(-124.947) = -125.46 \text{ kJ/mol}$$

The values of $\Delta_f H_i^\circ(298.15 \text{ K})$ for each conformer are within 1 kJ/mol of each other for this molecule, so considering multiple conformers only changes the standard enthalpy of formation by about 0.3%. This is because 2-methylphenol contains a small number of rotatable bonds. The difference between single and conformer-averaged properties becomes significant as the number of rotatable bonds increases. For example, Paulechka and Kazakov (2018) found that taking conformers into account for 3-morpholinopropan-1-amine gave $\Delta_f H^\circ(298.15 \text{ K}) = -173.8$ kJ/mol whereas using only the lowest-energy conformer yielded a value of -179.3 kJ/mol—a difference exceeding chemical accuracy of ~4 kJ/mol.

TABLE 4.23 Rotational constants for conformers of 2-methylphenol. The values were determined from quantum mechanical calculations with B3LYP-D3(BJ)/def2-TZVP model chemistry.

	Conformer 1	Conformer 2
ν_{r_A} (GHz)	1.3327859	1.3311897
ν_{r_B} (GHz)	2.2096067	2.2137766
ν_{r_C} (GHz)	3.2910473	3.2709651
σ	1	1

4.7 STANDARD STATE ENTHALPY OF FORMATION AND ENTHALPY OF COMBUSTION

This chapter has examined *ideal gas* enthalpies of formation where the product of the formation reaction is in the ideal gas state regardless of the compound's natural state at T and 1 atm. It is useful to define a standard state enthalpy of formation where the elements in their standard states as found in nature at 298.15 K and 1 atm combine to form the product species in its natural state of aggregation at 298.15 K and 1 atm ($\Delta_f H^\circ_{std}$). For compounds whose standard state at 298.15 K and 1 atm is gaseous, ideal gas $\Delta_f H^\circ = \Delta_f H^\circ_{std}$. The relationship between $\Delta_f H^\circ$ and $\Delta_f H^\circ_{std}$ is explained later for other situations.

Another common enthalpy change is the *enthalpy of combustion* which, from the perspective of a reader of this book, may be input or output values. Experimentally measured enthalpies of combustion are used for deriving standard enthalpies of formation, while enthalpies of combustion used in chemical engineering are usually derived from well-established standard enthalpies of formation. The engineering values (traditionally called net and gross heats of combustion, which are negative enthalpies of combustion) are for combustion in oxygen and associated with reaction products defined for each element in the context of each combustion reaction (gross or net). The reported standard enthalpies of combustion are associated with certain reactions and physical states of their participants (including solution of certain concentration). As actual combustion experiments typically give mixtures of products, corrections to the nominal reactions and states are applied. When reported, combustion enthalpies must be accompanied by the complete specification of the states of all participants. Depending on the experiment design, the final states of the same elements may be different, for example, SO_2, $H_2SO_4 \cdot 116\ H_2O$, or even SF_6 for sulfur. Standard enthalpies of formation of all reaction participants other than the tested compounds are used to derive the formation enthalpy of the tested compound.

In engineering practice, particularly in the energy, petroleum, and chemical process industries, it is common to define the so-called standard state heat of combustion as the difference in enthalpy of a compound and that of its products of combustion in which both products and reactants are at 298.15 K and 1 atm. Applying the label "standard" to any combustion reaction is somewhat of a misnomer because there is no universal consensus on combustion products. However, the above-mentioned industries have found it useful to define the following as "standard" products: $CO_2(g)$, $SO_2(g)$, $N_2(g)$, HX(g) (where X is a halogen atom), and either $H_2O(l)$ or $H_2O(g)$. If the product water is designated to be in gas phase, the enthalpy of combustion value is termed *net*. If the product water is in the liquid state, then the value is termed *gross*. In this book, this industrial standard enthalpy of combustion is given the symbol $\Delta_c H^\varnothing$. The superscript is different than the symbol used for strict standard quantities to remind the reader of the distinction.

Multiple rigorous thermodynamic relationships can be created to relate $\Delta_f H^\circ$ for the ideal gas state, $\Delta_f H^\circ_{std}$, and $\Delta_c H^\varnothing$. Figure 4.4 depicts a four-step path to obtain $\Delta_f H^\circ_{std}$ from ideal gas $\Delta_f H^\circ$ for a compound that is a *liquid* at standard conditions. T_b in Fig. 4.4 is the normal boiling point of the compound as discussed in Chap. 3. This path uses properties that are commonly found in data tabulations, such as those in Appendix A, but others could be constructed. The dashed line represents the overall change in enthalpy going from ideal gas $\Delta_f H^\circ$ to $\Delta_f H^\circ_{std}$ that is determined by adding the individual enthalpy changes denoted by Steps 1–4. Step 2 is discussed in Chap. 8 and consists of moving from an *ideal* gas to a *real* gas state at the normal boiling point. The contribution of this

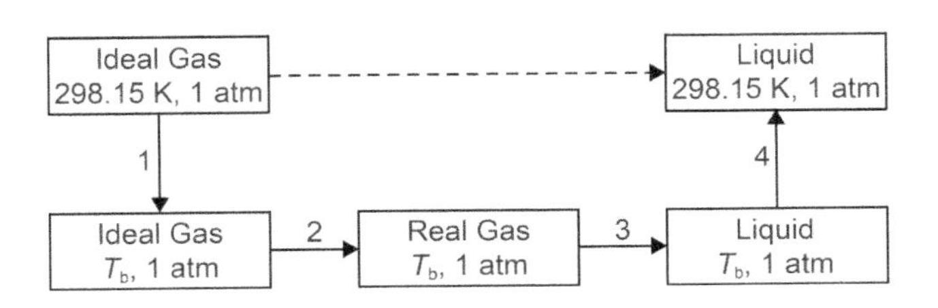

FIGURE 4.4
Thermodynamic path to obtain the standard state enthalpy of formation from the ideal gas enthalpy of formation for a compound whose standard state is a liquid.

step to the overall change in enthalpy is often negligible except for extremely low-boiling compounds as most gases behave ideally at 1 atm and temperatures above 200 K.

As suggested by the four-step path in Fig. 4.4, the standard enthalpy of formation for a liquid is obtained from the ideal gas value through

$$\Delta_f H^\circ_{\text{std}} = \Delta_f H^\circ + \int_{298.15\text{ K}}^{T_b} C^\circ_p(T)\,dT - \Delta^{\text{ig}}_{\text{g}} H(T_b) - \Delta_v H(T_b) + \int_{T_b}^{298.15\text{ K}} C^{\text{liq}}_p(T)\,dT \tag{4.7-1}$$

where $\Delta_f H^\circ$ is the standard enthalpy of formation for the ideal gas, $\Delta_v H$ is its enthalpy of vaporization of the compound, $\Delta^{\text{ig}}_{\text{g}} H\ (T_b)$ is the enthalpy change when moving from a gas to an ideal gas state at T_b (see Chap. 8), and $C^{\text{liq}}_p\ (T)$ is the compound's liquid heat capacity.

A four-step path similar to Fig. 4.4 can also be used to obtain $\Delta_f H^\circ_{\text{std}}$ from ideal gas $\Delta_f H^\circ$ if the stable state at 298.15 K and 1 atm is a crystal instead of a liquid. In this case, the enthalpy of sublimation ($\Delta_{\text{sub}} H$, see Chap. 5) must be known at some temperature (T_{sub}) instead of $\Delta_v H(T_b)$, and Step 4 uses the heat capacity of the solid (C^{sol}_p) and assumes no solid–solid transitions occur. For such a case, the expression becomes

$$\Delta_f H^\circ_{\text{std}} = \Delta_f H^\circ + \int_{298.15\text{ K}}^{T_{\text{sub}}} C^\circ_p(T)\,dT - \Delta^{\text{ig}}_{\text{g}} H(T_{\text{sub}}) - \Delta_{\text{sub}} H(T_{\text{sub}}) + \int_{T_{\text{sub}}}^{298.15\text{ K}} C^{\text{sol}}_p(T)\,dT \tag{4.7-2}$$

The normal melting point (T_m) or the triple point (T_{tp}) are common values to use for T_{sub} as such are frequently found in data tabulations.

Thermodynamic paths may also be constructed to determine the standard state enthalpy from the ideal gas standard enthalpy of a compound that is a solid at standard conditions even though the enthalpy of sublimation is not known. One such path (others could be constructed) is depicted in Fig. 4.5 where T_v is the temperature at which the vaporization is known. (T_{tp} or T_m are commonly used for T_v as the triple point and normal melting point usually only differ by a fraction of a kelvin and one or both are commonly found in data tabulations such as those in Appendix A.) Note, again, that the transition from the ideal gas to real gas state accounts for vapor phase nonidealities (see Chap. 8).

Step 2 of Fig. 4.5 is equal to zero because the operation is for an ideal gas. Step 5 is a very small contribution to the overall change in enthalpy represented by the path, usually only hundredths of a percent, and is commonly neglected. If T_v is chosen to be T_{tp}, then Step 6 is not needed. With these assumptions, and selecting T_{tp} for T_v, the standard enthalpy of formation for a solid is obtained from the ideal gas value through

$$\begin{aligned}\Delta_f H^\circ_{\text{std}} = {} & \Delta_f H^\circ + \int_{298.15\text{ K}}^{T_{tp}} C^\circ_p(T)dT - \Delta^{\text{ig}}_{\text{g}} H(T_{tp}) - \Delta_v H(T_{tp}) \\ & - \Delta_{\text{fus}} H(T_{tp}, 1\text{ atm}) + \int_{T_{tp}}^{298.15\text{ K}} C^{\text{sol}}_p(T)dT \end{aligned} \tag{4.7-3}$$

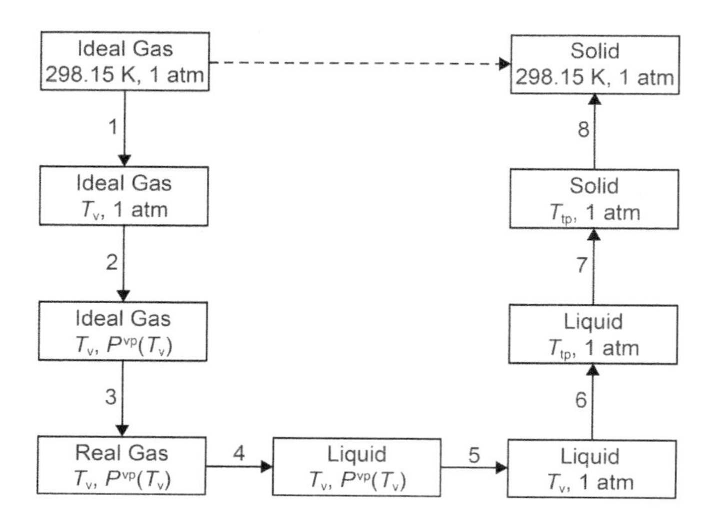

FIGURE 4.5
Thermodynamic path to obtain the standard state enthalpy of formation from the ideal gas enthalpy of formation for a compound whose standard state is a solid. In practice, T_m may be used instead of T_{tp} and either for T_v.

where $\Delta_f H^\circ$ is the standard enthalpy of formation for the ideal gas, $\Delta_{fus}H\ (T_{tp}, 1\ \text{atm})$ is the change in enthalpy that occurs as the compound changes from a liquid to a solid at 1 atm, and $C_p^{sol}(T)$ is the heat capacity of the solid compound. The latter quantity is technically also dependent on the pressure of the system and changes for different solid phases. If solid–solid phase changes occur between T_{tp} and 298.15 K, the changes in enthalpy of these transitions should be included in the expression, and the integrations should use the heat capacities of the relevant solid phases.

There is a direct relationship between the enthalpy of combustion ($\Delta_c H^{\varnothing}$) and the standard-state enthalpy of formation ($\Delta_f H^\circ_{std}$ (298.15 K). As mentioned above, the expression is dependent on the products specified. For all reactants and products at 298.15 K and 1 atm, and for products of H_2O(g), CO_2(g), SO_2(g), Cl_2(g), Br_2(g), I_2(g), P_4O_{10}(crystal), SiO_2(cristobalite), and Al_2O_3(crystal, alpha), the equation for the enthalpy of combustion of component j is

$$\begin{aligned}\Delta_c H^{\varnothing}_{net_j}(298.15\ \text{K}) = {} & \frac{N_H}{2}\Delta_f H^\circ_{H_2O}(298.15\ \text{K}) + N_C \Delta_f H^\circ_{CO_2}(298.15\ \text{K}) \\ & + N_S \Delta_f H^\circ_{SO_2}(298.15\ \text{K}) + \frac{N_P}{4}\Delta_f H^\circ_{std_{P_4O_{10}}}(298.15\ \text{K}) \\ & + N_{Si}\Delta_f H^\circ_{std_{SiO_2}}(298.15\ \text{K}) + \frac{N_{Al}}{2}\Delta_f H^\circ_{std_{Al_2O_3}}(298.15\ \text{K}) \\ & - \Delta_f H^\circ_{std_j}(298.15\ \text{K})\end{aligned} \tag{4.7-4}$$

where N_H, N_C, N_S, N_P, N_{Si}, and N_{Al} are the number of atoms of hydrogen, carbon, sulfur, phosphorus, silicon, and aluminum in the compound, respectively, $\Delta_f H_i^\circ(298.15\ \text{K})$ is the ideal gas enthalpy of formation of product compound i at 298.15 K and 1 atm, $\Delta_f H^\circ_{std_i}$ (298.15 K) is the standard state enthalpy of formation of product compound i at 298.15 K and 1 atm, and $\Delta_f H^\circ_{std_j}$(298.15 K) is the standard state enthalpy of formation of reactant compound j. The enthalpies of formation for each product compound may be found in Appendix A, and because the water product is in the gaseous state the enthalpy of combustion found by Eq. (4.7-4) is the *net* value. If the combustion products are HCl(g), HF(g), HBr(g), and HI(g) instead of the molecular halogen gases, then the standard enthalpy of combustion is related to the standard enthalpy of formation through

$$\begin{aligned}\Delta_c H^{\varnothing}_{net_j}(298.15\ \text{K}) = {} & \frac{(N_H - N_X)}{2}\Delta_f H^\circ_{H_2O}(298.15\ \text{K}) \\ & + N_C \Delta_f H^\circ_{CO_2}(298.15\ \text{K}) + N_S \Delta_f H^\circ_{SO_2}(298.15\ \text{K}) \\ & + \frac{N_P}{4}\Delta_f H^\circ_{std_{P_4O_{10}}}(298.15\ \text{K}) + N_{Si}\Delta_f H^\circ_{std_{SiO_2}}(298.15\ \text{K}) \\ & + \frac{N_{Al}}{2}\Delta_f H^\circ_{std_{Al_2O_3}}(298.15\ \text{K}) + N_{Cl}\Delta_f H^\circ_{HCl}(298.15\ \text{K}) \\ & + N_F \Delta_f H^\circ_{HF}(298.15\ \text{K}) + N_{Br}\Delta_f H^\circ_{HBr}(298.15\ \text{K}) \\ & + N_I \Delta_f H^\circ_{HI}(298.15\ \text{K}) - \Delta_f H^\circ_{std_j}(298.15\ \text{K})\end{aligned} \tag{4.7-5}$$

where N_{Cl}, N_F, N_{Br}, and N_I are the number of atoms of chlorine, fluorine, bromine, and iodine in the reactants, $N_X = N_{Cl} + N_F + N_{Br} + N_I$, and the other variables are as defined for Eq. (4.7-4). Other expressions may be derived if different products are specified following the patterns outlined above.

4.8 DISCUSSION AND RECOMMENDATIONS

The methods discussed in this chapter were evaluated against a test database of reliable experimental data for ideal gas $\Delta_f H^\circ$(298.15 K), S°(298.15 K), and $C_p^\circ(T)$. The creation of each database, and the performance of each method against the compilation, is discussed below for each property. To identify if a method was better at one class of compounds than another, each chemical in the test database was assigned a *class* according to the scheme below.

- **Normal:** Compounds containing only C, H, F, Br, and/or I with molecular weights below 213 g/mol.
- **Heavy:** Compounds containing only C, H, F, Br, and/or I with molecular weights greater than or equal to 213 g/mol. Also compounds containing only C, H, F, Br, I and/or metal atoms (e.g., Fe, As, Al, etc.) regardless of MW.
- **Polar:** Compounds containing atoms other than C, H, F, Br, metals (e.g., O, N, P, S, etc.) that cannot form hydrogen bonds (e.g., esters, ketones, aldehydes, thioesters, sulfonyl, etc.).
- **Associating:** Compounds containing atoms other than C, H, F, Br, metals (e.g., O, N, P, S, etc.) that can form hydrogen bond (e.g., acids, amines, alcohols, etc.)

Table 4.24 contains abbreviations used to delineate each prediction method. Results are presented for each class and prediction method for each property. For ideal gas $S°$ and $C_p^\circ(T)$, each case shows the number of compounds tested with the method (N) and the average absolute relative deviation ($AARD$). The latter is defined as

$$AARD = \frac{1}{N}\sum_{i=1}^{N}\left|\frac{X_{\text{pred}i} - X_{\text{exp}i}}{X_{\text{exp}i}}\right| \tag{4.8-1}$$

where the index is for compound i in the test set, X_{pred_i} is the predicted value for property X for compound i, and X_{exp_i} is the experimental value for property X for compound i.

Because $\Delta_f H°(298.15\text{ K})$ is a combination of positive and negative deviations from the enthalpies of the elements, percent or relative deviations between the calculated and experimental values are not adequate criteria to evaluate the prediction methods. This is because such error definitions are artificially amplified when the positive and negative contributions are similar in magnitude such that their sum approaches 0. As such, the error analysis for this property is done using a metric that removes this artifact. Specifically, prediction methods for ideal gas $\Delta_f H°(298.15\text{ K})$ are evaluated on an absolute basis defined as

$$\Delta\Delta_f H° = \frac{1}{N}\sum_{i=1}^{N}\frac{\left|\Delta_f H°(298.15\text{ K})_{\text{pred}_i} - \Delta_f H°(298.15\text{ K})_{\text{exp}_i}\right|}{N_{\text{atoms}_i}} \tag{4.8-2}$$

where N is the number of compounds in the test set, N_{atoms_i} is the number of atoms in compound i, and the other quantities are as just defined. The number of atoms is used for normalization as a surrogate for the size of the molecule. Overall, this definition gives an idea of the number of J/mol each atom will contribute to the error of the prediction method.

TABLE 4.24 Abbreviations for the predictions methods evaluated against the test databases for ideal gas properties $\Delta_f H°(298.15\text{ K})$, $S°(298.15\text{ K})$, and $C_p^\circ(T)$.

Abbreviation	Method, Property(ies)
JO	Joback (1984; Joback and Reid, 1987), $\Delta_f H°(298.15\text{ K})$, $C_p^\circ(T)$
CO	Constantinou and Gani (1994), $\Delta_f H°(298.15\text{ K})$, $C_p^\circ(T)$
BE	Benson (CHETAH, 2021), $\Delta_f H°(298.15\text{ K})$, $S°(298.15\text{ K})$, $C_p^\circ(T)$
DO	Domalski and Hearing (1993; 1994), $\Delta_f H°(298.15\text{ K})$, $S°(298.15\text{ K})$, $C_p^\circ(298.15\text{ K})$
mJO	Modified Joback (Sec. 4.5), $C_p^\circ(T)$
QM	Quantum mechanical approach of Paulechka and Kazakov (2018), $S°(298.15\text{ K})$, $C_p^\circ(T)$
QM-LCC	Quantum mechanical local coupled-cluster method of Paulechka and Kazakov (2017), $\Delta_f H°(298.15\text{ K})$
QM-ATOMIC	Quantum mechanical ATOMIC/ATOMIC(hc) technique of Bakowies (2009, 2020), $\Delta_f H°(298.15\text{ K})$

A. Ideal Gas Standard State Enthalpy of Formation, $\Delta_f H°(298.15\text{ K})$

As explained in this chapter, multiple methods are available to obtain values of ideal gas $\Delta_f H°(298.15\text{ K})$ when no experimental data are available for the property. The list below is presented in order of recommendation.

1. Use Eqs. (4.7-4)–(4.7-5) if *reliable* experimental enthalpy of combustion ($\Delta_c H^{\varnothing}(298.15\text{ K})$) data are available.
2. Use Eqs. (4.7-2)–(4.7-3) if *reliable* experimental standard enthalpy of formation ($\Delta_f H^{\circ}_{std}(298.15\text{ K})$) data are available.
3. Use an appropriate quantum mechanical approach.
4. Use an appropriate GC approach which is likely either the method of Benson (CHETAH, 2020) or Domalski and Hearing (1993).

Using rigorous thermodynamic relationships to derive a value for ideal gas $\Delta_f H°(298.15\text{ K})$ from experimental $\Delta_c H^{\varnothing}(298.15\text{ K})$ or $\Delta_f H^{\circ}_{std}(298.15\text{ K})$ data (Recommendations 1 and 2) provides the most reliable and thermodynamically consistent estimate. In the absence of any reliable experimental data, ideal gas $\Delta_f H°(298.15\text{ K})$ must be estimated first and then $\Delta_c H^{\varnothing}(298.15\text{ K})$ and $\Delta_f H^{\circ}_{std}(298.15\text{ K})$ may be obtained.

GC and QM methods for ideal gas $\Delta_f H°(298.15\text{ K})$ were evaluated against a test database consisting of accepted experimental data for compounds in the October 2020 sponsor release of the DIPPR database which were cross-validated against values from NIST. Over 900 compounds were included in the analysis. The GC methods of Joback (1984), Constantinou and Gani (1994), Benson (as encoded in CHETAH, 2020), and Domalski and Hearing (1993) (abbreviated below JO, CO, BE, and DO, respectively) were assessed. Due to computational limitations, testing all the available QM approaches is not feasible, but the local coupled-cluster method of Paulechka and Kazakov (2017) and the ATOMIC/ATOMIC(hc) technique of Backowies (2009, 2020) (abbreviated below QM-LCC AND QM-ATOMIC, respectively) were examined.

As depicted in Table 4.25, the number of compounds tested for each prediction method differ. For the GC methods, this occurs from excluding a compound in the test set from the evaluation of the method if groups were not available for the technique. The lower number of compounds for QM compared to the GC methods is due to the computational requirements of the former. The design taken in this table was to evaluate the maximum number of compounds possible for each approach; however, this tactic can mask problems with a method if excluded compounds for one technique are difficult to predict in general but were included in the compounds tested for another method. To account for this possibility, a subset of the test set was created to evaluate each method on the exact same compounds. This subset consisted of compounds for which QM-LCC values were available and which could be completely described by all four GC methods. Table 4.26 has the results.

TABLE 4.25 Performance of prediction techniques for ideal gas $\Delta_f H°(298.15\text{ K})$ against the test database in terms of absolute errors as defined by Eq (4.8-2). For the group contribution methods, all possible compounds in the test database are included.

	Compound Class†							
	Normal		Heavy		Polar		Associating	
Method	N	$\Delta\Delta_f H°$	N	$\Delta\Delta_f H°$	N	$\Delta\Delta_f H°$	N	$\Delta\Delta_f H°$
JO	399	1.305	21	1.368	91	1.932	218	1.533
CO	384	2.550	20	8.789	79	1.621	190	1.120
BE	353	0.805	19	1.453	90	1.066	174	0.655
DO	410	0.879	20	0.373	90	1.692	222	0.819
QM-LCC	170	0.090	3	0.050	39	0.197	106	0.294
QM-ATOMIC	82	0.332	3	0.125	13	0.545	38	0.360

†$\Delta\Delta_f H°$ has units of kJ/(mol·atom) and is defined in Eq. (4.8-2).

TABLE 4.26 Performance of prediction techniques for ideal gas $\Delta_f H°$(298.15 K) against a subset of the test database in terms of absolute errors as defined by Eq. (4.8-2). The subset was selected so that all methods were applicable to all compounds involved.

	Compound Class†							
	Normal		Heavy		Polar		Associating	
Method	N	$\Delta\Delta_f H°$	N	$\Delta\Delta_f H°$	N	$\Delta\Delta_f H°$	N	$\Delta\Delta_f H°$
JO	154	1.110	3	4.588	31	1.436	83	1.238
CO	154	0.714	3	2.036	31	1.188	83	1.048
BE	154	0.252	3	0.507	31	0.432	83	0.350
DO	154	0.627	3	0.328	31	0.637	83	0.658
QM-LCC	154	0.086	3	0.050	31	0.184	83	0.261

†$\Delta\Delta_f H°$ has units of kJ/(mol·atom) and is defined in Eq. (4.8-2).

Notice first that, regardless of the metric or the set of compounds examined, the QM methods perform significantly better than any of the GC methods, which is reflected in the recommendations previously given. The QM-LCC was tested on more compounds than the QM-ATOMIC approach and performed better in each metric and class of compounds except in two cases which is also reflected in the recommendations. Regarding the GC approaches, the method of Joback performs much worse than other methods and is not recommended. The Benson and Domalski and Hearing methods perform consistently better than the other GC techniques though the Constantiou and Gani technique seems equally valid for polar and associating compounds. Recommendation 4 reflects the ordering just described; however, to ensure the best GC method is selected one can follow the family analysis approach discussed at the end of Sec. 3.5. This is particularly important as all the group techniques listed in Table 4.24 can produce large errors for compound classes that were not considered when the methods were parameterized. (QM approaches are preferred to GC methods for this reason.) An example of the family analysis selection method for GC approaches is found in Example 4.12.

Example 4.12 Determine which group contribution prediction method will yield the most accurate value for ideal gas $\Delta_f H°$(298.15 K) of 3-ethylthiophene. No experimental data for ideal gas $\Delta_f H°$(298.15 K) exist for this compound.

Solution. 3-Ethylthiophene is a member of the *Sulfides/Thiophenes* family in the DIPPR database. The database contains 62 compounds in this family, but only 23 have experimental values for ideal gas $\Delta_f H°$(298.15 K). Determining which of the prediction methods in this chapter is likely to give the best estimate of ideal gas $\Delta_f H°$(298.15 K) for 3-ethylthiophene was done automatically using DIPPR's DIADEM program in a process described by Rowley et al. (2007) and summarized in Example 3.16. Using this process, the GC prediction methods discussed in this chapter reproduce these 26 experimental values according to the statistics found in the table below.

Method	Number of Compounds	Ave. Abs. $\Delta\Delta_f H°$ [kJ/(mol·K)]	Min. Abs. $\Delta\Delta_f H°$ [kJ/(mol·K)]	Max. Abs. $\Delta\Delta_f H°$ [kJ/(mol·K)]
JO	26	0.939	0.000	4.030
CO	19	0.910	0.001	4.366
BE	24	0.798	0.004	8.009
DO	26	0.344	0.027	1.634

Notice that groups are not available in some methods to predict a value for all 26 compounds. The method of Domalski and Hearing performs the best on average and has the lowest maximum $\Delta\Delta_f H°$ for the family. The Joback and Constantinou and Gani methods have average errors more than triple that of Domalski and Hearing. The Benson method has a comparable average error to that of Domalski and Hearing but also has a much larger max deviation. For these reasons, the Domalski and Hearing approach is the preferred GC method to predict the ideal gas enthalpy of formation of 3-ethylthiophene.

The data in Table 4.25 indicate that QM methods have become superior to GC approaches for predicting ideal gas $\Delta_f H°(298.15\ \text{K})$. However, the reader should remember that a simple quantum mechanical geometry optimization and frequency calculation using typical model chemistries on the compound of interest is not sufficient. In the very least, the isodesmic reaction approach should be used, but more advanced composite or focal point methods are recommended. Recall the stated goal of these techniques to yield computed values of *experimental accuracy* (DeYonker, 2006). This is typically defined this to be ±1 kcal/mol, but many methods are now much better than this. As explained previously, all the possible QM methods for ideal gas $\Delta_f H°(298.15\ \text{K})$ could not be tested large-scale by the authors due to the computational requirements of such approaches; however, others (Simmie and Somers, 2017; Paulechka and Kazakov, 2017) have done this analysis, so some general statements may be made.

The black-box G4 composite approach is expected to yield accuracies of ±1 kcal/mol for small to moderately sized molecules but has been shown to systematically overpredict the true value (Simmie and Somers, 2015; Paulechka and Kazakov, 2017). This may be an acceptable tradeoff in most cases of practical importance due to its ease of use, and some work has indicated that the G3 or G4 implementations (Curtiss et al., 1998, 2007) are superior to the various Weizmann (W) methods (Martin and de Oliveira, 1999; Parthiban, 2001; Barnes, 2009) or CBS approaches (Ochterski et al., 1996; Montgomery et al., 1999; Montgomery et al., 2000) despite the bias (Somers and Simmie, 2015). To alleviate this bias and achieve accuracies better than ±1 kcal/mol, more recent focal point methods can be used. The various implementations of the local coupled-cluster approaches have been rigorously evaluated and shown to give enthalpies of formation within ±0.5 kcal/mol with no bias (Paulechka and Kazakov, 2017, 2018, 2019; Chirico et al., 2017), though more expertise in computational chemistry is needed to implement these solutions. The ccCA has been proposed as an alternative "black box" method to the Gn techniques with an expected accuracy of ±1 kcal/mol. (DeYonker et al., 2006; Wilson et al., 2012). Finally, recall that the effect of different conformers should be considered for molecules containing a large number of rotatable bonds. Paulechka and Kazakov (2018) found that the difference between single conformer and multiple conformer enthalpies of formation differ by less than 1 kJ/mol for small to medium compounds, but the difference becomes significant (>4 kJ/mol) for flexible molecules. This is particularly problematic for long-chain compounds where it can become computationally intractable to sample all the possible configurations.

B. Ideal Gas Absolute Entropy, $S°(T)$

GC and QM methods for ideal gas $S°(298.15\ K)$ were evaluated against a test database consisting of accepted experimental data for compounds in the October 2020 sponsor release of the DIPPR database which were cross-validated against values from NIST. To increase the size of the test set, additional compounds were added from the Thermodynamic Research Center (TRC) database when they were deemed to come from experiments. All values were also thermodynamically consistent with other relevant properties such as condensed-phase heat capacities and enthalpies of transitions. Five hundred sixty-nine compounds ultimately comprised the test set. The GC methods of Benson (CHETAH, 2020) and Domalski and Hearing (1993) (abbreviated below BE and DO, respectively) were assessed. A quantum mechanical value was also determined using Eq. (4.5-13) or (4.5-14) per the scheme of Paulechka and Kazakov (2018). Specifically, vibrational frequencies were obtained from B3LYP-D3(BJ)/def2-TZVP calculations done in Gaussian® with tight optimization and an ultrafine grid for integrals, frequencies corresponding to hydrogen stretching (>2800 cm^{-1}) were scaled by 0.96, and all other modes were scaled by 0.985. The performance reported in Table 4.27 is the absolute average percent deviation defined by Eq. (4.8-1).

TABLE 4.27 Performance of prediction techniques for ideal gas $S°(298.15\text{ K})$ against the test database in terms of average absolute relative percent deviation [see Eq. (4.8-1)].

	Compound Class							
	Normal		Heavy		Polar		Associating	
Method	N	AARD	N	AARD	N	AARD	N	AARD
BE	287	7.9%	15	11.4%	22	2.9%	93	4.7%
DO	335	5.6%	15	9.9%	29	4.7%	97	4.5%
QM	141	1.1%	5	0.9%	27	1.7%	61	1.5%

As with ideal gas $\Delta_f H°(298.15\text{ K})$, a subset of the test set for ideal gas $S°(298.15\text{ K})$ was created to remove possible anomalies that could be introduced into the analysis by having a different number of compounds tested for each technique. All three methods (one QM, two GC) were applicable to all compounds in this subset, and the results are found in Table 4.28.

From the results in Table 4.27 and Table 4.28, the recommended prediction approaches for ideal gas absolute entropy ($S°(298.15\text{ K})$) follow that for the ideal gas standard enthalpy of formation, namely, quantum mechanics (Sec. 4.5) should be the first choice followed by an appropriate GC method. In contrast to predicting ideal gas $\Delta_f H°(298.15\text{ K})$ from quantum mechanics, which requires large computational costs to achieve reasonable accuracy as previously described, ideal gas $S°(298.15\text{ K})$ may be reliably predicted with more modest numerical efforts. A model chemistry of B3LYP/6-311+G(3df,2p) or comparable is recommended. With appropriate scaling factors, such approaches can be expected to give ideal gas $S°(298.15\text{ K})$ accuracies within 3 to 5%. Higher levels of theory and larger basis sets will produce even better results. Lower model chemistries are not recommended.

Equations (4.6-14) and (4.6-15) were derived assuming the rotational and vibrational movements of the molecule are independent of each other—the so-called RRHO approximation. This assumption becomes more questionable for very small frequencies such as those present in flexible molecules (Frenkel et al., 1994). Increased accuracy can be achieved by taking the coupling between the rotations and vibrations into account. Different approaches have been proposed to do so, and the reader is referred to the literature for more discussion (e.g., Ribeiro et al., 2011; Grimme, 2012). Common quantum mechanical programs often provide options to account for mode coupling.

If QM approaches are infeasible, the GC approach of Domalski and Hearing followed by that of Benson may be used with an average error of 5%. However, caution should be used. GC methods can work well, but they can also fail spectacularly for compounds with unusual structures that are different than the types of compounds used when parameterizing the techniques. For many types of compounds, the GC methods work much

TABLE 4.28 Performance of prediction techniques for ideal gas $S°(298.15\text{ K})$ against a subset of the test database in terms of average absolute relative percent deviation [see Eq. (4.8-1)]. The subset was selected so that all methods were applicable to all compounds involved.

	Compound Class							
	Normal		Heavy		Polar		Associating	
Method	N	AARD	N	AARD	N	AARD	N	AARD
BE	121	3.8%	3	4.9%	16	3.1%	49	5.0%
DO	121	3.4%	3	3.5%	16	3.7%	49	4.7%
QM	121	1.1%	3	0.9%	16	1.3%	49	1.6%

better than the numbers in Table 4.27 indicate and for others worse. For example, several halogen-containing compounds were found to be problematic for the GC methods as were fused ring compounds. It is therefore recommended to follow the family analysis approach discussed at the end of Sec. 3.5. An example of the family analysis selection method for GC approaches for ideal gas $S°(298.15\ K)$ is found in Example 4.13.

Example 4.13 Determine which group contribution prediction method will yield the most accurate value for ideal gas $S°(298.15\ K)$ of 3-ethylthiophene. No experimental data for $S°(298.15\ K)$ exist for this compound.

Solution. 3-ethylthiophene is a member of the *Sulfides/Thiophenes* family in the DIPPR database. The database contains 62 compounds in this family, but only 23 have experimental values for ideal gas $S°(298.15\ K)$. Determining which of the prediction methods in this chapter is likely to give the best estimate of ideal gas $S°(298.15\ K)$ for 3-ethylthiophene was done automatically using DIPPR's DIADEM program in a process described by Rowley et al. (2007) and summarized in Example 3.16. Using this process, the group contribution prediction methods discussed in this chapter reproduce the 23 experimental values according to the statistics found in the table below.

Method	Number of Compounds	Ave. Abs. Rel. Dev. (%)	Min. Abs. Rel. Dev. (%)	Max. Abs. Rel. Dev. (%)
BE	22	5.1	1.2	13.1
DO	20	7.4	1.4	30.0

Notice that groups are not available in the methods to predict a value for all 23 compounds. The Benson method performs better than the technique of Domalski and Hearing in terms of average and maximum error. The former is therefore the preferred GC method to predict the ideal gas absolute entropy of 3-ethylthiophene.

C. Ideal Gas Standard State Gibbs Energy of Formation, $\Delta_f G°(298.15\ K)$

The preferred approach to predicting ideal gas $\Delta_f G°(298.15\ K)$ is to first obtain $\Delta_f H°(298.15\ K)$ and $S°(298.15\ K)$ for the ideal gas state, either from experiment or one of the prediction methods outlined above, and then use Eq. (4.3-5) to calculate $\Delta_f G°(298.15\ K)$. This approach ensures thermodynamic consistency among the properties. Previous editions of this book explained that the method of Joback (1984) or Constantinou and Gani (1994) could be used to obtain ideal gas $\Delta_f G°(298.15\ K)$, but these are no longer recommended as their accuracies are inferior to the GC methods of Benson (CHETAH, 2020) or Domalski and Hearing (1993) as evidenced by the results in Table 4.25. Illustrations of obtaining $\Delta_f G°(298.15\ K)$ from $\Delta_f H°(298.15\ K)$ and $S°(298.15\ K)$ can be found in Examples 4.2 through 4.4.

D. Ideal Gas Heat Capacity, $C_p^\circ(T)$

GC and QM methods for ideal gas $C_p^\circ(T)$ were evaluated against a test database consisting of accepted experimental data for compounds in the October 2020 sponsor release of the DIPPR database which were cross-validated against values from NIST. Data for a few compounds from REFPROP (Lemmon et al., 2018) were also included to increase the size of the test database. In total, 178 compounds were available for comparison.

The GC methods of Joback (1984), Constantinou and Gani (1994), Benson (as encoded in CHETAH, 2020), Domalski and Hearing (1993), and modified Joback (see Sec. 4.5)—abbreviated below JO, CO, BE, DO, and mJO, respectively—were assessed as was the QM method of Eqs. (4.5-15) and (4.5-16) per the computational scheme of Paulechka and Kazakov (2018). Specifically for the QM method, vibrational frequencies were obtained from B3LYP-D3(BJ)/def2-TZVP calculations done in Gaussian with tight optimization and an

TABLE 4.29 Performance of prediction techniques for ideal gas $C_p^\circ(T)$ against the test database in terms of average absolute relative percent deviation [see Eq. (4.8-1)]. For the DO method, only values at 298 K were tested. For the other methods, values at 298 K, 400 K, and 500 K were tested for each compound.

	Compound Class							
	Normal		Heavy		Polar		Associating	
Method	N	AARD	N	AARD	N	AARD	N	AARD
JO	102	11.0%	7	3.6%	22	2.1%	18	5.5%
CO	92	8.6%	7	11.1%	21	11.4%	16	7.1%
BE	78	5.9%	7	8.5%	12	16.5%	16	3.4%
DO	91	6.2%	7	11.5%	20	6.9%	16	6.2%
mJO	97	1.8%	14	1.8%	21	1.9%	18	2.3%
QM	39	2.3%	4	1.8%	11	1.2%	15	3.1%

ultrafine grid for integrals, frequencies corresponding to hydrogen stretching (>2800 cm^{-1}) were scaled by 0.96, and all other modes were scaled by 0.985. The method of Domalski and Hearing only provides a prediction at 298 K, so only comparisons at this temperature were made for this approach. The performance reported in Table 4.29 for this method is the absolute average percent deviation defined by Eq. (4.8-1). For the other techniques, both GC and QM, values at 298 K, 400 K, and 500 K were predicted and compared to the test database. In these cases, three absolute percent deviations (one for each temperature) for each compound contribute to the absolute average percent deviation.

As with $\Delta_f H^\circ(298.15\text{ K})$ and $S^\circ(298.15\text{ K})$, a subset of the test set for C_p° was created to remove possible anomalies that could be introduced into the analysis by having a different number of compounds tested for each technique. All five methods (one QM, four GC) were applicable to all compounds in this subset, and the results are found in Table 4.30.

From the results in Table 4.29 and Table 4.30, the primary recommendation for predicting ideal gas heat capacity is to use QM methods as described in Sec. 4.5 followed by the modified Joback GC method explained

TABLE 4.30 Performance of prediction techniques for ideal gas $C_p^\circ(T)$ against a subset of the test database in terms of average absolute relative percent deviation [see Eq. (4.8-1)]. The subset was selected so that all methods were applicable to all compounds involved. For the DO method, only values at 298 K were tested. For the other methods, values at 298 K, 400 K, and 500 K were tested for each compound.

	Compound Class							
	Normal		Heavy		Polar		Associating	
Method	N	AARD	N	AARD	N	AARD	N	AARD
JO	28	21.9%	4	5.8%	9	2.1%	10	7.4%
CO	28	3.2%	4	18.4%	9	9.3%	10	8.7%
BE	28	1.6%	4	13.7%	9	18.3%	10	4.5%
DO	28	1.7%	4	12.4%	9	10.3%	10	9.6%
DI	28	1.6%	4	2.2%	9	1.8%	10	3.2%
QM	28	2.7%	4	1.8%	9	1.3%	10	3.9%

in Sec. 4.5. The other GC methods can work well in some cases but fail spectacularly in others (e.g., Joback error is 11.0% for normal vs 2.1% for polar compound in Table 4.29). Previous editions of this book also showed this—especially for compounds with unusual structures that are different than the types of compounds used when parameterizing the techniques. The data in Table 4.29 demonstrate this as the average error for the GC methods for some classes of compounds can be up to 15 times larger than the QM methods. QM methods do not suffer from the same unique structural problems encountered by GC method and are thus more robust. GC methods are also limited in the applicable temperature range whereas quantum chemical approaches can reach to very low and high temperatures.

In contrast to accurately predicting ideal gas $\Delta_f H°(298.15\ \text{K})$ from quantum mechanics, obtaining reliable values for ideal gas $C_p^\circ(T)$ can be done with relatively little computational cost. A model chemistry of B3LYP/6-311+G(3df,2p) or comparable is recommended. With appropriate scaling factors, such can be expected to give ideal gas $C_p^\circ(T)$ accuracies within 5% depending on the class of the molecules. Higher levels of theory and larger basis sets will produce even better results. Lower model chemistries may be used for larger compounds that may be computationally intractable, but the errors will be larger. HF/6-31G* should be able to handle almost any situation with reasonable computer requirements and produce accuracies within 10%.

Equations (4.6-16) and (4.6-17) for ideal gas C_p°, like Eqs. (4.6-14) and (4.6-15) for ideal gas $S°$, were derived using the RRHO approximation. As explained in the recommendations for $S°$ (discussed previously), this assumption becomes more questionable for very small frequencies such as those present in flexible molecules (Frenkel et al., 1994). Accuracy can be improved in such cases by taking the coupling between the rotations and vibrations into account using methods such as those proposed by Ribeiro et al. (2011) and Grimme (2012). The reader is referred to the literature for more discussion on this topic.

E. Availability of Data and Computer Software

There are several readily available commercial products for obtaining ideal-gas properties. These include data and correlation-based tabulations and computer-based GC methods. The data for Appendix A were obtained from the DIPPR 801 database (Wilding et al., 2017; DIPPR, 2021), and the automated evaluation of the prediction methods from molecular structures was done with DIPPR's DIADEM software (Rowley et al., 2007). There is a similar tabulation as well as automated prediction methods based on molecular structure available in ThermoData Engine from the Thermodynamics Research Center (Diky et al., 2021). Frenkel et al. (1994) has a tabulation of ideal gas properties of organic molecules as does Cox and Pilcher (1970) and Burcat and Ruscic (2005). Joback has established a program (Cranium, 2021) for computing many properties by various GC methods. Gani and coworkers at PSE for SPEED also have a program available (ProCAPE, 2021) for many properties. Other programs for the Benson method are available from ASTM (CHETAH, 2020) and NIST Chemistry WebBook (Linstrom, 2022). Karton (2016) provides a review and statistical analysis of computationally intensive QM methods for roughly 430 compounds to give recommendations on which approaches to use when balancing accuracy versus computational cost.

4.9 NOTATION

a_i	parameter in a multiproperty equation of state
$A°$	standard state ($P = 1$ atm) Helmholtz energy
A_i	species i in a chemical reaction
A_{ij}	parameter in the modified Joback prediction method for ideal gas C_p°
$AARD$	average absolute relative deviation; see Eq. (4.8-1)
c	speed of light in vacuum (2.99792458×10^8 m/s)
$C_p^\circ(T)$	standard state ($P = 101325$ Pa) heat capacity at temperature T in a specified state of aggregation (e.g., crystal, liquid, ideal gas)

$C^{\circ}_{P_e}(T)$	standard state ($P = 101325$ Pa) heat capacity of element e at temperature T in a specified state of aggregation (e.g., crystal, liquid, ideal gas)
$C^{\circ}_{P_i}$	standard state ($P = 101325$ Pa) heat capacity of species i at temperature T in a specified state of aggregation (e.g., crystal, liquid, ideal gas)
$C^{\text{liq}}_{P_i}(T)$	liquid heat capacity of species i at temperature T
$C^{\text{sol}}_{P_i}(T)$	solid heat capacity of species i at temperature T
D_0	atomization energy of a molecule
E_0	bottom-of-the-well electronic energy of a molecule
E_{0+ZPE}	$E_0 + ZPE$
$G^{\circ}_i(T)$	Gibbs energy of species i at temperature T
f_i	parameter in a multiproperty equation of state
h	Planck's constant ($6.62607015 \times 10^{-34}$ J/Hz)
h_i	empirical parameter for atom type i in prediction method of Paulechka and Kazakov (2018) for ideal gas $\Delta_f H^{\circ}$
h_{S_i}	empirical parameter for sulfur atom type i in prediction method of Paulechka and Kazakov (2018) for ideal gas $\Delta_f H^{\circ}$
$h(\text{S})$	base value for h_{S_i}
H°	standard state ($P = 1$ atm) enthalpy
$H_{\text{corr}}(T)$; $[H^{\circ}(298.15\text{ K}) - H^{\circ}(0\text{ K})]$	thermal correction added to the 0 K, bottom-of-the-well electronic energy to determine the enthalpy of a molecule at temperature T
$H^{\circ}_e(T)$	standard state ($P = 1$ atm) enthalpy of element e at temperature T in a specified state of aggregation (e.g., crystal, liquid, ideal gas)
$H^{\circ}_i(T)$	standard state ($P = 1$ atm) enthalpy of species i at temperature T in a specified state of aggregation (e.g., crystal, liquid, ideal gas)
k	Boltzmann's constant (1.380649×10^{-23} J/K)
K	equilibrium constant of a reaction
m	mass of a molecule
n	number of atoms in a molecule
N	number of molecules in a system
N_A	Avogadro's number ($6.02214076 \times 10^{23}$ mol^{-1})
N_{atoms}	number of atoms in a molecule
$N_{\text{conformers}}$	number of conformers of a molecule
N_e	number of atoms of element e
N_i	number of times group i is found in a compound
N_{is}	number of indistinguishable conformations observed when rotating the terminal groups of a molecule
N_{es}	number of indistinguishable conformations observed with rotating an entire molecule
N_{iso}	number of stereoisomers in a molecule [see Eq. (4.3-6)]
N_{ts}	number of indistinguishable structures in a molecule
P	pressure
P^{vp}	saturated liquid vapor pressure
R	gas constant [8.314462618 J/(mol·K)]

$S^\circ(T)$	standard state ($P = 1$ atm) absolute entropy at temperature T in a specified state of aggregation (e.g., crystal, liquid, ideal gas)
$S_e^\circ(T)$	standard state ($P = 1$ atm) absolute entropy of element e at temperature T in a specified state of aggregation (e.g., crystal, liquid, ideal gas)
$S_i^\circ(T)$	standard state ($P = 1$ atm) absolute entropy of species i at temperature T in a specified state of aggregation (e.g., crystal, liquid, ideal gas)
S_s°	symmetry entropy needed for the methods of Benson (CHETAH, 2020) and Domalski and Hearing (1993)
T	temperature
T_b	normal ($P = 101325$ Pa) boiling point
T_v	vaporization temperature (temperature along the vapor pressure curve)
T_m	normal ($P = 101325$ Pa) melting temperature
T_{sub}	sublimation temperature
T_{ti}	temperature at phase transition i
T_{tp}	temperature at the triple point
$U_{corr}(T)$	thermal correction added to the 0 K, bottom-of-the-well electronic energy to determine the internal energy of a molecule at temperature T
V	molar volume
V_c	critical volume
W	number of distinguishable configurations of a compound
W_0	degeneracy of the ground electronic state of the molecule
x_i	mole fraction of component i
ZPE	zero-point energy (zero-point vibrational energy)

Greek

$\Delta C_{p_i}^\circ(T)$	change is ideal gas isobaric heat capacity during the formation of species i from its elements [see Eq. (4.1-6)]
$\Delta_f G^\circ(T)$	standard state ($P = 1$ atm) Gibbs energy of formation at temperature T in a specified state of aggregation (e.g., crystal, liquid, ideal gas)
$\Delta_f G_i^\circ(T)$	standard state ($P = 1$ atm) Gibbs energy of formation of species i at temperature T in a specified state of aggregation (e.g., crystal, liquid, ideal gas)
$\Delta_r G^\circ(T)$	change in standard state ($P = 1$ atm) Gibbs energy that occurs for a reaction happening at temperature T
$\Delta h(j)$	correction of type j for determining the parameter for sulfur type i, h_{S_i}
$\Delta_c H^\varnothing$	"standard" heat of combustion
$\Delta_f H^\circ(T)$	standard state ($P = 1$ atm) enthalpy of formation at temperature T in a specified state of aggregation (e.g., crystal, liquid, ideal gas)
$\Delta_f H_i^\circ(T)$	standard state ($P = 1$ atm) enthalpy of formation for species i at temperature T in a specified state of aggregation (e.g., crystal, liquid, ideal gas)

$\Delta_f H^\circ_{std}(T)$	standard state ($P = 1$ atm) enthalpy of formation for species i at temperature T in its natural state of aggregation
$\Delta_{fus} H$	enthalpy of fusion (change in enthalpy that occurs when a substance changes from a liquid to a solid
$\Delta_g^{ig} H_i$	change in enthalpy in species i that occurs when the substance moves from a real gas to an ideal gas state
$\Delta_r H$	enthalpy of reaction at temperature (change in enthalpy that occurs for a reaction happening at temperature T)
$\Delta_r H^\circ(T)$	standard enthalpy of reaction at temperature (change in standard state ($P = 1$ atm) enthalpy that occurs for a reaction happening at temperature T)
$\Delta_{sub} H$	enthalpy of sublimation (change in enthalpy that occurs when a substance changes from a saturated solid to a saturated vapor)
$\Delta_t H_j^{(i)}$	enthalpy change occurring at phase transition j of species i
$\Delta_v H_i$	enthalpy of vaporization of species i (change in enthalpy that occurs when a substance changes from a saturated liquid to a saturated vapor)
$\Delta_{sub} H_i$	enthalpy of sublimation of species i (change in enthalpy that occurs when a substance changes from a saturated solid to a saturated vapor)
$\Delta_f S^\circ(T)$	standard state ($P = 1$ atm) entropy of formation at temperature T in a specified state of aggregation (e.g., crystal, liquid, ideal gas)
$\Delta_r S^\circ(T)$	entropy of reaction at temperature T (change in standard state (P = 1 atm) absolute entropy that occurs during a reaction at temperature T)
$\left(\Delta_{C_p^\circ(T)}\right)_i$	contribution for group i for a prediction method for $C_p^\circ(T)$
$\left(\Delta_{\Delta_f H^\circ(T)}\right)_i$	contribution for group i for a prediction method for $\Delta_f H^\circ(T)$
$\left(\Delta_{S^\circ(T)}\right)_i$	contribution for group i for a prediction method for $S^\circ(T)$
$\Delta\Delta_f H^\circ$	per atom absolute deviation between predicted and experimental values for $\Delta_f H^\circ$ [see Eq. (4.8-2)]
θ_i	parameter in a multiproperty equation of state
Θ_{r_i}	rotational temperature i of a molecule
Θ_{v_i}	vibrational temperature i of a molecule
λ	wavelength
$\nu_e^{(i)}$	number of atoms of element e in species i
ν_i	stoichiometric coefficient of species i in a reaction
ν_{r_i}	rotational frequency i of a molecule
$\tilde{\nu}_{r_i}$	rotational wave number i of a molecule
ν_{v_i}	vibrational frequency i of a molecule
$\tilde{\nu}_{v_i}$	vibrational wave number i of a molecule
σ	rotational symmetry number

4.10 REFERENCES

Alecu, I. M., J. Zheng, Y. Zhao, and D. G. Truhlar: *J. Chem. Theory Comput.*, **6:** 2872–2887 (2010).
Bakowies, D.: *J. Chem. Phys.*, **130:** 144113 (2009).

Bakowies, D.: *J. Phys. Chem. A*, **113:** 11517–11534 (2009).
Bakowies, D.: *J. Chem. Theory Comput.*, **16:** 399–426 (2020).
Barnes, E. C., G. A. Petersson, J. A. Montgomery Jr., M. J. Frisch, and J. M. L. Martin: *J. Chem. Theor. Comput.*, **5:** 2687 (2009).
Benson, S. W.: *Thermochemical Kinetics*, Wiley, New York, 1968, Chap. 2.
Benson, S. W., and J. H. Buss: *J. Chem. Phys.*, **29:** 279 (1969).
Benson, S. W., F. R. Cruickshank, D. M. Golden, G. R. Haugen, H. E. O'Neal, A.S. Rodgers, R. Shaw, and R. Walsh: *Chem. Rev.*, **69:** 279 (1969).
Bomble, Y. J., J. Vazquez, M. Kállay, C. Michauk, P. G. Szalay, A. G. Csaszár, J. Gauss, J. F. Stanton: *J. Chem. Phys.*, **125:** 064108 (2006).
Burcat, A. and B. Ruscic: *Third Millennium Ideal Gas and Condensed Phase Thermochemical Database for Combustion with Updates from Active Thermochemical Tables*, Argonne National Laboratory, Argonne, IL (2005). https://www.osti.gov/biblio/925269; http://garfield.chem.elte.hu/Burcat/burcat.html
Cardozo, R. L.: *AIChE J.*, **32:** 844 (1986).
Chase Jr., M. W.: *J. Phys. Chem. Ref. Data*, Monograph No. 9 NIST-JANAF Thermochemical Tables Part 1, Al-Co (1998).
Chase Jr., M. W.: *J. Phys. Chem. Ref. Data*, Monograph No. 9 NIST-JANAF Thermochemical Tables Part 2, Cr-Zr (1998).
CHETAH Version 11.0: *The ASTM Computer Program for Chemical Thermodynamic and Energy Release Evaluation.* ASTM International, West Conshohocken, PA (2020).
Chirico, R. D., A. Kazakov, A. Bazyleva, V. Diky, K. Kroenlein, V. N. Emel'yanenko, and S. P. Verevkin: *J. Phys. Chem. Ref. Data*, **46:** 023105 (2017).
Constantinou, L. and R. Gani: *AIChE J.*, **40:** 1697–1710 (1994).
Cox, J. D., and G. Pilcher: *Thermochemistry of Organic and Organometallic Compounds*. Academic Press, London and New York (1970).
Cranium Version 4.2: Molecular Knowledge Systems, Inc., PO Box 10755, Bedford, NH 03110-0755 USA (2021). https://www.molecularknowledge.com/
Curtiss, L. A., K. Raghavachari, G. W. Trucks, and J. A. Pople: *J. Chem. Phys.*, **94:** 7221–7230 (1991).
Curtiss, L. A., K. Raghavachari, P. C. Redfern, and J. A. Pople: *J. Chem. Phys.*, **106:** 1063–1079 (1997).
Curtiss, L. A., K. Raghavachari, P. C. Redfern, V. Rassolov, and J. A. Pople: *J. Chem. Phys.*, **109:** 7764–7776 (1998).
Curtiss, L. A., P. C. Redfern, and K. Raghavachari: *J. Chem. Phys.*, **126:** 084108 (2007).
Davies, C. A., A. N. Syverud, and E. C. Steiner: "Appendix B: Definition of Symmetry Numbers and Optical Isomers Used in CHETAH Calculations and a Simple Method for Choosing These Numbers," *The ASTM Computer Program for Chemical Thermodynamic and Energy Release Evaluation: CHETAH® Version 11.0 User Guide,* ASTM International, West Conshohocken, PA (2020).
DeYonker, N. J., T. R. Cundari, and A. K. Wilson: *J. Chem. Phys.*, **124:** 114104 (2006).
DeYonker, N. J., T. Grimes, S. Yockel, A. Dinescu, B. Mintz, T. R. Cundari, and A. K. Wilson: *J. Chem. Phys.*, **125:** 104111 (2006).
Diky, V., R. D. Chirico, M. Frenkel, A. Bazyleva, J. W. Magee, E. Paulechka, A. Kazakov, E. W. Lemmon, C. D. Muzny, A. Y. Smolyanitsky, S. Townsend, and K. Kroenlein. *NIST ThermoData Engine, NIST Standard Reference Database 103 a/b*, National Institute of Standards and Technology, USA (2021). https://www.nist.gov/mml/acmd/trc/thermodata-engine
DIPPR, Design Institute for Physical Properties, American Institute of Chemical Engineers, 120 Wall St. FL 23, New York, NY 10005-4020 USA (2021). https://www.aiche.org/dippr
Domalski, E. S., and E. D. Hearing: *J. Phys. Chem. Ref. Data*, **22:** 805 (1993); **23:** 157 (1994).
Eigenmann, H. K., D. M. Golden, and S. W. Benson: *J. Phys. Chem.*, **77:** 1687 (1973).
Frenkel, M., K. N. Marsh, G. J. Kabo, R. C. Wilhoit, and G. N. Roganov: *Thermodynamics of Organic Compounds in the Gas State,* Vols. I and II, Thermodynamics Research Center, College Station, TX (1994).
Grimme, S.: *Chem. Eur. J.*, **18:** 9955–9964 (2012).
Harrison, B. K.: *The ASTM Computer Program for Chemical Thermodynamic and Energy Release Evaluation: CHETAH® Version 11.0 User Guide,* ASTM International, West Conshohocken, PA (2020).
Hehre, W. J., R. Ditchfield, L. Radom, and J. A. People: *J. Am. Chem. Soc.*, **92:** 4796–4801 (1970).
Humphrey, W., A. Dalke, and K. Schulten, "VMD—Visual Molecular Dynamics,": *J. Molec. Graphics*, **14:** 33–38 (1996).
Joback, K. G.: *A Unified Approach to Physical Property Estimation Using Multivariate Statistical Techniques*, S.M. Thesis, Department of Chemical Engineering, Massachusetts Institute of Technology, Cambridge, MA (1984).
Joback, K. G., and R. C. Reid: *Chem. Eng. Comm.*, **57:** 233–243 (1987).
Karton, A.: *WIREs Comput. Mol. Sci.*, **6:** 292–310 (2016).
Lemmon, E. W., I. H. Bell, M. L. Huber, and M. O. McLinden: NIST Standard Reference Database 23: Reference Fluid Thermodynamic and Transport Properties-REFPROP, Version 10.0, National Institute of Standards and Technology, Standard Reference Data Program, Gaithersburg (2018).

Linstrom, P. J.: NIST Standard Reference Database 69: NIST Chemistry WebBook, Group Additivity Based Estimates, National Institute of Standards and Technology, Standard Reference Data Program, Gaithersburg (2022). https://webbook.nist.gov/chemistry/grp-add/

Martin, J. M. L., and G. de Oliveira: *J. Chem. Phys.*, **111:** 1843–1856 (1999).

McQuarrie, D. A.: *Statistical Mechanics*, University Science Books, Sausalito, CA, 2000.

Montgomery Jr., J. A., M. J. Frisch, J. W. Ochterski, and G. A. Petersson: *J. Chem. Phys.*, **110:** 2822–2827 (1999); **112:** 6532–6542 (2000).

Ochterski, J. W., G. A. Petersson, and J. A. Montgomery Jr.: *J. Chem. Phys.*, **104:** 2598–2619 (1996).

O'Neal, H. E., and S. W. Benson: *J. Chem. Eng. Data*, **15:** 266 (1970).

Paulechka, E., and A. Kazakov: *J. Phys. Chem. A*, **121:** 4379–4387 (2017).

Paulechka, E., and A. Kazakov: *J. Chem. Theory. Comput.*, **14:** 5920–5932 (2018).

Paulechka, E., and A. Kazakov: *J. Chem. Eng. Data*, **64:** 4863–4874 (2019).

Paulechka, E., and A. Kazakov: *J. Phys. Chem. A*, **125:** 8116–8131 (2021).

Parthiban, S., and J. M. L. Martin: *J. Chem. Phys.*, **114:** 6014–6029 (2001).

Pople, J. A., M. Head-Gordon, and D. J. Fox: *J. Chem. Phys.*, **90:** 5622 (1989).

Pracht, P., F. Bohle, and S. Grimme: *Phys. Chem. Chem. Phys.*, **22:** 7169 (2020).

Ribeiro, R. F., A. V. Marenich, C. J. Cramer, and D. G. Truhlar: *J. Phys. Chem. B*, **115:** 14556–14562 (2011).

Riplinger, C., and F. Neese: *J. Chem. Phys.*, **138:** 034106 (2013).

Riplinger, C., B. Sandhoefer, A. Hansen, and F. Neese: *J. Chem. Phys.*, **139:** 134101 (2013).

Rowley, J. R., W. V. Wilding, J. L. Oscarson, and R. L. Rowley: *Int. J. Thermophys.*, **28:** 824–834 (2007).

Setzmann, U., and W. Wagner: *J. Phys. Chem. Ref. Data,* **20:** 1061 (1991).

Simmie, J. M., and K. P. Somers: *J. Phys. Chem. A*, **119:** 7235–7246 (2015).

Smith, D. G. A., L. A. Burns, A. C. Simmonett, R. M. Parrish, M. C. Schieber, R. Galvelis, P. Kraus, H. Kruse, R. Di Remigio, A. Alenaizan, A. M. James, S. Lehtola, J. P. Misiewicz, M. Scheurer, R. A. Shaw, J. B. Schriber, Y. Xie, Z. L. Glick, D. A. Sirianni, J. S. O'Brien, J. M. Waldrop, A. Kumar, E. G. Hohenstein, B. P. Pritchard, B. R. Brooks, H. F. Schaefer III, A. Yu. Sokolov, K. Patkowski, A. E. DePrince III, U. Bozkaya, R. A. King, F. A. Evangelista, J. M. Turney, T. D. Crawford, and C. D. Sherrill: *J. Chem. Phys.*, **152:** 184108 (2020).

Stein, S. E., D. M. Golden, and S. W. Benson: *J. Phys. Chem.*, **81:** 314 (1977).

Strömsöe, E., H. G. Ronne, and A. L. Lydersen: *J. Chem. Eng. Data*, **15:** 286–290 (1970).

Tajti, A., P. G. Szalay, A. G. Csaszár, M. Kállay, J. Gauss, E. F. Valeev, B. A. Flowers, J. Vazquez, and J. F. Stanton: *J. Chem. Phys.*, **121:** 11599–11613 (2004).

Thinh, T.-P., J.-L. Duran, and R. S. Ramalho: *Ind. Eng. Chem. Process Design Develop.*, **10:** 576 (1971).

Thinh, T.-P., and T. K. Trong: *Can. J. Chem. Eng.*, **54:** 344 (1976).

Tiesinga, E. P. J. Mohr, D. B. Newell, and B. N. Taylor: *J. Phys. Chem. Ref. Data*, **50:** 033105 (2021).

WebMO, Basic Version 20.0.012(2021). https://www.webmo.net/

Wheeler, S. E., K. N. Houlk, P. v. R. Schleyer, and W. D. Allen: *J. Am. Chem. Soc.*, **131:** 2547–2560 (2009).

Wilding, W. V., T. A. Knotts, N. F. Giles, R. L. Rowley, and J. L. Oscarson, *DIPPR® Data Compilation of Pure Chemical Properties*, Design Institute for Physical Properties, AIChE, New York, NY (2017).

Wilholt, R. C., and B. J. Zwolinski: *J. Phys. Chem. Ref. Data*, **2:** Supplement 1 (1973).

Wilson, B. R., N. J. DeYonker, and A. K. Wilson: *J. Compt. Chem.*, **33:** 2032–2042 (2012).

Yoneda, Y.: *Bull. Chem. Soc. Japan,* **52:** 1297 (1979).

5

Pure Fluid Thermodynamic Properties of the Single Variable Temperature

5.1 SCOPE

This chapter covers methods for predicting and correlating pure compound thermodynamic properties that depend on temperature. These properties include saturated liquid density, liquid vapor pressure, enthalpy of vaporization, constant pressure liquid heat capacity, solid vapor pressure, enthalpy of sublimation, enthalpy of melting, and virial coefficients. Although surface tension could also be considered under this heading, we reserve discussion of surface tension for a later chapter. Since saturated liquid densities, enthalpies of vaporization, and saturated liquid heat capacity are related to vapor pressure, we discuss those first. The enthalpy of melting, enthalpy of sublimation, and solid vapor pressure are discussed next as these are related to the liquid vapor pressure at the triple point. The chapter ends with a discussion of virial coefficients as these share a less direct relationship with vapor pressure.

Methods for predicting these properties are generally less reliable than for many other properties, especially in the case of vapor pressure. Vapor pressure varies on an exponential scale with temperature, meaning that small deviations in the log-linear relation of vapor pressure to reciprocal temperature result in large relative deviations. For example, vapor pressures predicted by group-contribution methods typically generate deviations of 40–80%. Experimental databases for some properties considered in this chapter are also relatively small. For example, experimentally determined second virial coefficients are available for less than 300 compounds. On the other hand, liquid density data are available for many compounds, but high accuracy is often required for this property due to its importance in custody transfer operations. Thus, each property in this chapter presents distinct challenges and significance.

5.2 SATURATED LIQUID DENSITY

Several techniques are available to estimate pure saturated liquid molar or specific volumes or densities as a function of temperature. Here, one group-contribution technique and several corresponding states methods are presented to estimate saturated liquid densities.

5.2.1 Rackett Equation and Other Corresponding States Approaches

Rackett (1970) proposed that saturated liquid volumes (the reciprocal of saturated liquid density, $\rho_s = \frac{1}{V_s}$) at temperature T be calculated by

$$V_s = V_c Z_c^{\left(1-\frac{T}{T_c}\right)^{\frac{2}{7}}} \tag{5.2-1}$$

where V_s is the saturated liquid volume, V_c is the critical volume, Z_c is the critical compressibility factor, and T_c is the critical temperature. Equation (5.2-1) is often written in the equivalent form

$$V_s = \frac{RT_c}{P_c} Z_c^{\left[1+\left(1-\frac{T}{T_c}\right)^{\frac{2}{7}}\right]} \tag{5.2-2}$$

While Eq. (5.2-1) is remarkably accurate for many substances, it underpredicts V_s when $Z_c < 0.22$.

Yamada and Gunn (1973) proposed that Z_c in Eq. (5.2-1) be correlated with the acentric factor:

$$V_s = V_c(0.29056 - 0.08775\omega)^{\left(1-\frac{T}{T_c}\right)^{\frac{2}{7}}} \tag{5.2-3}$$

If one experimental density/volume, V_s^R, is available at a reference temperature, T^R, Eqs. (5.2-1) and (5.2-3) can be modified to give

$$V_s = V_s^R(0.29056 - 0.08775\omega)^{\phi} \tag{5.2-4}$$

$$V_s = V_s^R Z_c^{\phi} \tag{5.2-5}$$

where

$$\phi = \left(1 - \frac{T}{T_c}\right)^{\frac{2}{7}} - \left(1 - \frac{T^R}{T_c}\right)^{\frac{2}{7}} \tag{5.2-6}$$

Equation (5.2-4) is obtained from Eq. (5.2-3) by using the known reference volume to eliminate V_c. The same approach is used to obtain Eq. (5.2-5) from Eq. (5.2-1). It is also possible to eliminate Z_c from Eq. (5.2-1), but then V_c appears in the final equation and it is generally known less accurately than the quantities that appear in Eqs. (5.2-4) and (5.2-5).

An often-used variation of Eq. (5.2-3) is

$$V_s = \frac{RT_c}{P_c}(0.29056 - 0.08775\omega)^{\left[1+\left(1-\frac{T}{T_c}\right)^{\frac{2}{7}}\right]} \tag{5.2-7}$$

However, this form does not predict V_c correctly unless the actual Z_c is equal to $0.29056 - 0.08775\omega$, in which case it is identical to Eq. (5.2-2).

Equation (5.2-1) has been used as the starting point to develop a variety of equations for correlating liquid densities. For example, Spencer and Danner (1972) replaced Z_c with an adjustable parameter, Z_{RA}, values of which are tabulated in Spencer and Danner (1972), Spencer and Adler (1978), and the 4th Edition of this book. Daubert et al. (1997) changed the physical quantities and constants of Eq. (5.2-1) into four adjustable parameters to give

$$V_s = \frac{B^{\left[1+\left(1-\frac{T}{C}\right)^{D}\right]}}{A} \tag{5.2-8}$$

Values of the four constants A through D are tabulated in Daubert et al. (1997) for approximately 1200 compounds. The value of C is generally equal to T_c while A, B, and D are generally close to the values used in Eq. (5.2-3). DIPPR (Wilding et al., 2017) uses the reciprocal of Eq. (5.2-8) to correlate liquid density values, so Appendix A contains values for these constants for the compounds found therein.

When reliable experimental points are available, values of saturated liquid density at temperatures from T_m to T_c can be reliably extrapolated using a suitable form of the Rackett equation. For example, Eq. (5.2-4) uses a known reference value, V_s^R and does not require V_c or P_c. It is more accurate when Z_c is low. Errors associated with the assumption that a correlation in ω applies to all substances is mitigated by use of the reference value. As mentioned above, DIPPR uses the reciprocal of Eq. (5.2-8) to correlate saturated liquid densities ($\rho_s = \frac{1}{V_s}$). Specifically, DIPPR follows a hierarchical approach when fitting the A, B, and D coefficients depending on the amount and temperature range of the available, reliable experimental V_s data, and the reliability of V_c, while setting $C = T_c$. This technique has the advantage that it does not rely on ω as does Eq. (5.2-4).

- Case 1:
  - **Situation**: V_s data are available over a *large* temperature range; reliable value for V_c is available
  - **Regression Procedure**: B and D are fit to data; function forced through V_c by setting $A = \frac{B}{V_c}$
- Case 2:
  - **Situation**: V_s data are available over a *large* temperature range; value for V_c is questionable
  - **Regression Procedure**: A is set equal to $\frac{RT_c}{P_c}$; B and D are fit to data; V_c is obtained from the resulting correlation (i.e., $V_c = V_s(T_c)$)
- Case 3:
  - **Situation**: V_s data are available over a *narrow* temperature range; reliable value for V_c is available
  - **Regression Procedure**: D is set equal to $\frac{2}{7}$; B is fit to data; function forced through V_c by setting $A = \frac{B}{V_c}$
- Case 4:
  - **Situation**: V_s data are available over a *narrow* temperature range; value for V_c is questionable
  - **Regression Procedure**: A is set equal to $\frac{RT_c}{P_c}$; D is set equal to $\frac{2}{7}$; B is fit to data; V_c is obtained from the resulting correlation (i.e., $V_c = V_s(T_c)$)

Another liquid volume correlation was proposed by Hankinson and Thomson (1979) and further developed in Thomson et al. (1982). This correlation, herein referred to as the HBT correlation, is

$$V_s = V^* V^{(0)}\left[1 - \omega_{\mathrm{SRK}} V^{(\delta)}\right] \tag{5.2-9}$$

$$V^{(0)} = 1 + a(1 - T_r)^{\frac{1}{3}} + b(1 - T_r)^{\frac{2}{3}} + c(1 - T_r) + d(1 - T_r)^{\frac{4}{3}} \tag{5.2-10}$$

$$V^{(\delta)} = \frac{e + fT_r + gT_r^2 + hT_r^3}{T_r - 1.00001} \tag{5.2-11}$$

where ω_{SRK} is that value of the acentric factor that causes the Soave equation of state to give the best fit to pure component vapor pressures, V^* is a parameter whose value is close to the critical volume, and constants a through h are given by

a	−1.52816	b	1.43907
c	−0.81446	d	0.190454
e	−0.296123	f	0.386914
g	−0.0427258	h	−0.0480645

Values of ω_{SRK} and V^* are tabulated for a number of compounds in Hankinson and Thomson (1979) and in the 4th edition of this book. The 5th edition explained that ω_{SRK} and V^* can be replaced with ω and V_c with little loss in accuracy. This approach was used when performing the evaluations discussed later. Equation (5.2-10) may be used in the range $0.25 < T_r < 0.95$ and Eq. (5.2-11) may be used when $0.25 < T_r < 1.0$.

Example 5.1

Predict the saturated liquid volume of 1,1,1-trifluoroethane (R143a) at 300 K using various Rackett-based methods. For those requiring a reference density at another temperature, use $V_s^R = 75.38$ cm³/mol at $T^R = 245$ K from Defibaugh and Moldover (1997). The value for the liquid volume at 300 K from the correlation in Appendix A, which is regressed from multiple sets of experimental data, is 91.32 cm³/mol.

Solution. From Appendix A for 1,1,1-trifluoroethane, $T_c = 345.88$ K, $P_c = 37.64$ bar, $V_c = 195$ cm³/mol, $Z_c = 0.255$, and $\omega = 0.261421$. The reduced temperature is $T_r = 300/345.88 = 0.8674$, so $1 - T_r = 0.1326$ and $(1 - T_r)^{\frac{2}{7}} = (0.1326)^{0.2857} = 0.5614$.

RACKETT PREDICTION WITHOUT REFERENCE POINT: As per Eq. (5.2-1)

$$V_s = (195)(0.255)^{0.5614} = 90.54 \text{ cm}^3/\text{mol}$$

$$\text{Error} = 90.54 - 91.32 = -0.78 \text{ cm}^3/\text{mol or } -0.85\%$$

YAMADA-GUNN PREDICTION WITHOUT REFERENCE POINT: From Eq. (5.2-3)

$$V_s = (195)[0.29056 - (0.08775)(0.261421)]^{0.5614} = 93.03 \text{ cm}^3/\text{mol}$$

$$\text{Error} = 93.03 - 91.32 = 1.71 \text{ cm}^3/\text{mol or } 1.9\%$$

RACKETT PREDICTION WITH REFERENCE POINT: With one reference point and a reliable value for V_c, the constants in Eq. (5.2-8) should be obtained with the procedures listed in Case 3.

$$C = 345.88 \text{ K}$$

$$D = 0.2857$$

With only one data point, B is obtained by substituting $A = \dfrac{B}{V_c}$ into Eq. (5.2-8) to give

$$V_s = \frac{B^{\left[1+\left(1-\frac{T}{C}\right)^D\right]}}{\left(\dfrac{B}{V_c}\right)} = V_c B^{\left(1-\frac{T}{C}\right)^D}$$

With T^R, $\left(1 - \dfrac{T}{C}\right)^D = (1 - 245/345.88)^{0.2857} = 0.7033$, so B is obtained by solving

$$75.38 = 195B^{0.7033}$$

which yields $B = 0.2589$. Thus,

$$A = 0.2589/195 = 0.001328 \text{ mol/cm}^3$$

and the complete correlation becomes

$$V_s = \frac{0.2589^{\left[1+\left(1-\frac{T}{345.88}\right)^{0.2857}\right]}}{0.001328}$$

where V_s is in cm^3/mol and T is in K. For T = 300 K, V_s = 91.29 cm^3/mol.

$$\text{Error} = 91.29 - 91.32 = -0.03 \text{ cm}^3/\text{mol or } -0.033\%$$

YAMADA-GUNN PREDICTION WITH REFERENCE POINT: With Eqs. (5.2-4) and (5.2-6)

$$\phi = (0.1326)^{0.2857} - (1 - 245/345.88)^{0.2857} = -0.1418$$

$$V_s = (75.38)[0.29056 - (0.08775)(0.261421)]^{-0.1418} = 90.87 \text{ cm}^3/\text{mol}$$

$$\text{Error} = 90.87 - 91.32 = -0.45 \text{ cm}^3/\text{mol or } -0.49\%$$

HBT PREDICTION (NO REFERENCE POINT): With Eqs. (5.2-9)–(5.2-11)

$$V^{(0)} = 1 + (-1.52816)(0.1326)^{0.3333} + (1.43907)(0.1326)^{0.6667} + (-0.81446)(0.1326) + (0.190454)(0.1326)^{1.3333} = 0.49975$$

$$V^{(\delta)} = [-0.296123 + (0.386914)(0.8674) + (-0.0427258)(0.8674)^2 + (-0.0480645)(0.8674)^3]/(0.8674 - 1.00001) = 0.18119$$

$$V_s = (195)(0.49975)[1 - (0.261421)(0.18119)] = 92.84 \text{ cm}^3/\text{mol}$$

$$\text{Error} = 92.84 - 91.32 = 1.52 \text{ cm}^3/\text{mol or } 1.7\%$$

Figure 5.1 shows the percent deviation in the liquid density of 1,1-difluoroethane (R152a) when calculated by the different equations to show the performance as a function of temperature. The errors are the deviation from the correlation in Appendix A which was regressed from multiple data sets from different authors at temperatures ranging from 231.74 K to 382.71 K. The methods used are: (1) the purely predictive Rackett technique of Eq. (5.2-1), (2) the fully predictive Yamada-Gunn approach with Eq. (5.2-3), (3) Eq. (5.2-8) from Daubert et al. with the B coefficient obtained using a reference point (DIPPR Case 3 procedure), (4) the Yamada-Gunn approach with a reference point Eq. (5.2-4), and (5) the fully predictive HBT of Eq. (5.2-9). For the two methods requiring a reference point, the value was T^R = 300 K, V_s^R = 73.85 cm^3/mol. Notice, as indicated above, that the

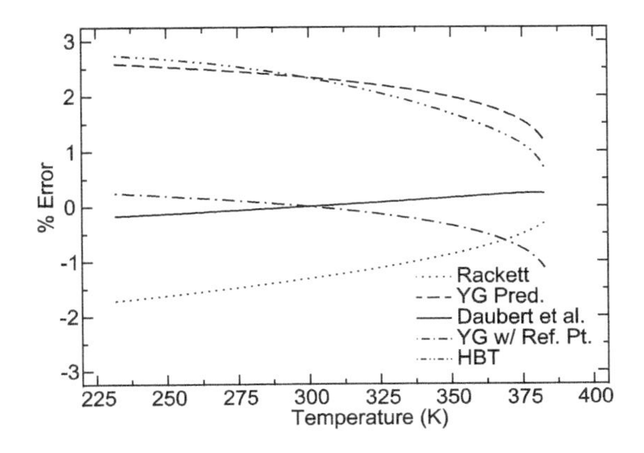

FIGURE 5.1
Percent error of saturated liquid density predictions and extrapolations as a function of temperature for 1,1-difluoroethane. The temperature range of each line covers the extent of the experimental data. Rackett is from Eq. (5.2-1). YG Pred. is from Eq. (5.2-3). Daubert et al. is from (5.2-8) with the DIPPR Case 3 procedure which requires a reference point. YG w/Ref. Pt. is from Eq. (5.2-4). HBT is from Eq. (5.2-9).

methods using a reference point perform better than the purely predictive approaches. The DIPPR method using the equation of Daubert et al. (1997), which is closely related to the Rackett equation, has the best overall agreement with the experimental data with the Yamada-Gunn reference point method a close second. The average absolute relative deviation for each plot in Figure 5.1 are as follows:

Rackett	1.2%
YG Pred.	2.2%
Daubert et al.	0.1%
YG w/Ref. Pt.	0.3%
HBT	2.1%

For this compound, the Rackett method performed best among those which are purely predictive while the HBT and Yamada-Gunn predictions performed about the same.

5.2.2 Method of Elbro et al. (1991)

Elbro et al. (1991) have presented a group-contribution method for the prediction of liquid densities as a function of temperature from the triple point to the normal boiling point. The groups are like those used by UNIFAC. In addition to being applicable to simple organic compounds, the method can also be used for amorphous polymers from the glass transition temperature to the degradation temperature. The method should not be used for cycloalkanes. To use the technique, the volume is calculated by

$$V_s = \sum_i N_i(\Delta_{V_s})_i \tag{5.2-12}$$

where N_i is the number of group i in the substance and $(\Delta_{V_s})_i$ is a temperature-dependent group molar volume given by

$$(\Delta_{V_s})_i = A_i + B_iT + C_iT^2 \tag{5.2-13}$$

Values for the group volume temperature constants are given in Table 5.1. To calculate the density of a polymer, only groups present in the repeat unit need be considered. The technique first obtains the molar volume of the repeat unit and then divides this into the repeat unit molar mass to obtain the polymer density. The method is illustrated in Example 5.2 and Example 5.3.

Example 5.2

Estimate V_s of hexadecane at 298.15 K with the method of Elbro et al. (1991). From the correlation in Appendix A, which is regressed from experimental data from multiple authors, $V_s = 294.30$, cm^3/mol at 298.15 K.

Solution. Hexadecane is composed of 14 >CH_2 groups and 2 —CH_3 groups. Using values from Table 5.1 in Eq. (5.2-13) with $T = 298.15$ K yields:

Group i	A_i cm^3/mol	10^3B_i cm^3/(mol·K)	10^5C_i cm^3/(mol·K^2)	$(\Delta_{V_s})_i$ cm^3/mol
—CH_3	18.960	45.58	0	32.550
>CH_2	12.520	12.94	0	16.378

TABLE 5.1 Group-contributions for the Elbro et al. (1991) method for V_s to use in Eq. (5.2-13).

No.	Group	A cm³/mol	10^3B cm³/(mol·K)	10^5C cm³/(mol·K²)
1	—CH_3	18.960	45.58	0
2	—CH_2—	12.520	12.94	0
3	>CH—	6.297	−21.92	0
4	>C<	1.296	−59.66	0
5	>ACH	10.090	17.37	0
6	>$ACCH_3$	23.580	24.43	0
7	>$ACCH_2$—	18.160	−8.589	0
8	>ACCH<	8.925	−31.86	0
9	>ACC≤	7.369	−83.60	0
10	=CH_2	20.630	31.43	0
11	=CH—	6.761	23.97	0
12	=C<	−0.3971	−14.10	0
13	—CH_2OH	39.460	−110.60	23.31
14	>CHOH	40.920	−193.20	32.21
15	>ACOH	41.20	−164.20	22.78
16	CH_3CO—	42.180	−67.17	22.58
17	—CH_2CO—	48.560	−170.40	32.15
18	>CHCO—	25.170	−185.60	28.59
19	—CHO	12.090	45.25	0
20	CH_3COO—	42.820	−20.50	16.42
21	—CH_2COO—	49.730	−154.10	33.19
22	>CHCOO—	43.280	−168.70	33.25
23	—COO—	14.230	11.93	0
24	>ACCOO—	43.060	−147.20	20.93
25	CH_3O—	16.660	74.31	0
26	—CH_2O—	14.410	28.54	0
27	>CHO—	35.070	−199.70	40.93
28	≥CO—	30.120	−247.30	40.69
29	—CH_2Cl	25.29	49.11	0
30	>CHCl	17.40	27.24	0
31	>CCl—	37.62	−179.1	32.47
32	—$CHCl_2$	36.45	54.31	0
33	—CCl_3	48.74	65.53	0
34	>ACCl	23.51	9.303	0
35	>Si<	86.71	−555.5	97.90
36	≥SiO—	17.41	−22.18	0

With Eq. (5.2-12)

$$V_s = 2(32.550) + 14(16.378) = 294.39 \text{ cm}^3/\text{mol}$$

$$\text{Error} = 294.39 - 294.30 = 0.09 \text{ cm}^3/\text{mol or } 0.031\%$$

Example 5.3

Estimate the density of poly (methyl acrylate) at 298.15 K with the method of Elbro et al. (1991). The value given in van Krevelen and Hoftyzer (1972) is 1.220 g/cm^3.

Solution. For poly (methyl acrylate), the repeat unit is

$$\text{—(—H}_2\text{C—CH—)}_n\text{—with } M = 86.09$$
$$\text{(CH substituent: COOCH}_3\text{)}$$

which is composed of the following groups: 1 —CH_3, 1 >CH_2, and 1 >CHCOO—. Using values from Table 5.1 in Eq. (5.2-13) with $T = 298.15$ K gives:

Group i	A_i cm^3/mol	10^3B_i cm^3/(mol·K)	10^5C_i cm^3/(mol·K^2)	$(\Delta_{V_s})_i$ cm^3/mol
—CH_3	18.960	45.58	0	32.550
>CH_2	12.520	12.94	0	16.378
>CHCOO—	43.280	−168.70	33.25	22.539

From Eq. (5.2-12)

$$V_s = 1(32.550) + 1(16.378) + 1(22.539) = 71.47 \text{ cm}^3/\text{mol}$$

Taking the reciprocal to determine the density and using the molar mass to convert to a mass basis yields

$$\rho_s = \frac{M}{V_s} = 86.09/71.47 = 1.205 \text{ g/cm}^3$$

$$\text{Error} = 1.205 - 1.220 = -0.015 \text{ g/cm}^3 \text{ or } -1.2\%$$

5.2.3 Discussion and Recommendations

The best source of liquid density values are correlations provided in evaluated databases or compilations such as those of Daubert et al. (1997), NIST TRC, or DIPPR (Wilding et al., 2017). In the absence of such resources, the best course of action to predict saturated liquid densities is to use a reference point available at another temperature. The 4-case DIPPR scheme utilizing Eq. (5.2-8) and fitting the coefficients to the reference data is recommended first followed by the reference point method of Yamada and Gunn outlined in Eq. (5.2-4). The former appears to work slightly better across a wider range of temperature (see Figure 5.1) and can handle cases with single or multiple reference data points.

A database was created from the May 2021 release of the DIPPR database to test the different prediction methods. Only reliable liquid density values from experiment were included in this test database. This gave 32000+ total points (N_{pt}) from over 1900 unique compounds (N_c) spanning temperature ranges from T_m to T_c. Four prediction techniques were tested. Three were corresponding states method: Rackett Eq. (5.2-1) (abbreviated RA below), Yamada-Gunn Eq. (5.2-3) (YG), and HBT Eq. (5.2-9). The other was the group-contribution approach of Elbro et al., Eq. (5.2-12) (EL).

To identify if a method was better at one class of compounds than another, each chemical in the test database was assigned a *class* according to the scheme below.

- **Normal**: Compounds containing only C, H, F, Br, and/or I with molar mass below 213 g/mol.
- **Heavy**: Compounds containing only C, H, F, Br, and/or I with molar mass greater than or equal to 213 g/mol. Also, compounds containing only C, H, F, Br, I and/or metal atoms (e.g., Fe, As, Al, etc.), regardless of MW.
- **Polar**: Compounds containing atoms other than C, H, F, Br, metals (e.g., O, N, P, S, etc.) that cannot form hydrogen bonds (e.g., esters, ketones, aldehydes, thioesters, sulfonyl, etc.)
- **Associating**: Compounds containing atoms other than C, H, F, Br, metals (e.g., O, N, P, S, etc.) that can form hydrogen bond (e.g., acids, amines, alcohols, etc.)

Results were quantified using the average absolute relative deviation (AARD) defined as

$$\mathrm{AARD} = \frac{1}{N_{pt}} \sum_{i=1}^{N_{pt}} \left| \frac{X_{\mathrm{pred}_i} - X_{\exp_i}}{X_{\exp_i}} \right| \tag{5.2-14}$$

where the index is for point i in the test set, X_{pred_i} is the predicted value for property X for point i, and $X_{\exp_i}$ is the experimental value for property X for point i.

Table 5.2 contains the absolute average relative deviation [see Eq. (5.2-14)] of each method according to compound class. Also listed are the number of points tested in each class (N_{pt}) and number of compounds (N_c). The corresponding states methods all perform about the same with each AARD for each compound class being within ~0.5% of each other. Thus, any of the methods will likely provide a reliable estimate of liquid density with an error below about 10%. The group-contribution method of Elbro et al. performed significantly worse in the normal category, better in the heavy category, and about equal for the other classes. However, notice that the method was tested against fewer compounds and points than the corresponding states approaches. This is because the method has a limited number of groups. For example, the only halogen represented is chlorine, and there are no alkyne groups.

The Elbro method suffers from other weaknesses. The results in Table 5.2 for the technique exclude silicon-containing compounds. This is because the predictions for these chemicals were several orders of magnitude in error despite the availability of silicon groups. Even excluding these compounds, the method had very high maximum deviations regularly exceeding 200%. These issues make it difficult to recommend the technique.

TABLE 5.2 Performance of prediction techniques for V_s against the test database.

	Compound Class											
	Normal			Heavy			Polar			Associating		
Method	N_{pt}	N_c	**AARD**	N_{pt}	N_c	**AARD**	N_{pt}	N_c	**AARD**	N_{pt}	N_c	**AARD**
RA	15841	859	3.1%	1113	90	6.9%	6329	378	4.4%	8801	579	6.1%
YG	15841	859	2.8%	1113	90	7.1%	6329	378	5.0%	8794	577	5.9%
HBT	15841	859	2.7%	1113	90	6.2%	6329	378	4.6%	8794	577	5.5%
EL	9785	552	7.1%	740	59	1.6%	4293	219	5.6%	5506	314	6.1%

5.3 THEORY OF LIQUID VAPOR PRESSURE AND ENTHALPY OF VAPORIZATION

When the vapor phase of a pure fluid is in equilibrium with its liquid phase, the equality of chemical potential, temperature, and pressure in both phases leads to the Clapeyron equation (Clapeyron 1834; Smith et al., 2018)

$$\frac{\mathrm{d}P^{\mathrm{vp}}}{\mathrm{d}T} = \frac{\Delta_{\mathrm{v}}H}{T\Delta_{\mathrm{v}}V} \tag{5.3-1}$$

In this equation, P^{vp} is the saturated liquid vapor pressure at temperature T and $\Delta_{\mathrm{v}}H$ and $\Delta_{\mathrm{v}}V$ are differences in the enthalpies and volumes of the compound as the system moves from a saturated liquid to a saturated vapor. This equation is exact, and with accurate information about the volume change of vaporization, gives a reliable relationship between vapor pressure and enthalpy of vaporization. A commonly used simplification is obtained by assuming that the vapor phase is an ideal gas and the molar volume of the liquid is negligible compared to the molar volume of the vapor. This gives the Clausius-Clapeyron equation

$$\frac{d \ln P^{\mathrm{vp}}}{d\left(\frac{1}{T}\right)} = -\frac{\Delta_{\mathrm{v}}H}{R} \tag{5.3-2}$$

For most substances, a plot of $\ln P^{\mathrm{vp}}$ vs. $1/T$ gives a nearly straight line at temperatures near the normal boiling point and higher. Below the normal boiling point, the plot typically breaks slightly then resumes a new line with different slope. If the line is assumed to be straight, then the Clausius-Clapeyron equation suggests that the enthalpy of vaporization is a constant value with temperature. This is not strictly true but can be a valuable approximation over limited temperature ranges.

The Clapeyron equation is useful in developing appropriate equation forms to correlate experimental vapor pressure data. Substituting the compressibility factor $\left(Z = \frac{PV}{RT}\right)$ for volume in Eq. (5.3-1) gives

$$\frac{dP^{\mathrm{vp}}}{dT} = \frac{\Delta_{\mathrm{v}}H}{T\Delta_{\mathrm{v}}V} = \frac{\Delta_{\mathrm{v}}H}{\left(\frac{RT^2}{P^{\mathrm{vp}}}\right)\Delta_{\mathrm{v}}Z} \tag{5.3-3}$$

$$\frac{d \ln P^{\mathrm{vp}}}{d\left(\frac{1}{T}\right)} = -\frac{\Delta_{\mathrm{v}}H}{R\Delta_{\mathrm{v}}Z} \tag{5.3-4}$$

The enthalpy of vaporization $\Delta_{\mathrm{v}}H$ is sometimes referred to as the latent enthalpy of vaporization. It is the difference between the enthalpy of the saturated vapor and that of the saturated liquid at a given temperature and originates from the difference in molecular attractions among molecules when in the different states. Molecules in the vapor phase are much farther apart than when in a liquid state. This means the energy of attraction in the vapor phase is less than that of the liquid, so energy must be supplied for vaporization to occur. This is the internal energy of vaporization $\Delta_{\mathrm{v}}U$. PV work is also done on the vapor phase as vaporization proceeds since the vapor volume increases if the pressure is maintained constant at P^{vp}. This work is $P^{\mathrm{vp}}(V^{\mathrm{vap}} - V^{\mathrm{liq}})$. Thus

$$\Delta_{\mathrm{v}}H = \Delta_{\mathrm{v}}U + P^{\mathrm{vp}}(V^{\mathrm{vap}} - V^{\mathrm{liq}}) = \Delta_{\mathrm{v}}U + RT(Z^{\mathrm{vap}} - Z^{\mathrm{liq}}) = \Delta_{\mathrm{v}}U + RT\Delta_{\mathrm{v}}Z \tag{5.3-5}$$

where Z is the compressibility factor and superscripts vap and liq refer to vapor and liquid states, respectively.

Directly measuring $\Delta_{\mathrm{v}}H$ through experiment is difficult in terms of both accuracy and repeatability. Obtaining reliable vapor pressure data is easier from an experimental standpoint, so values of $\Delta_{\mathrm{v}}H$ are commonly obtained through Eq. (5.3-4) using a fit of P^{vp} data to one of the correlations discussed in Section 5.5. Many data for $\Delta_{\mathrm{v}}H$ found in the literature and classified as "experimental" were *derived* in this manner.

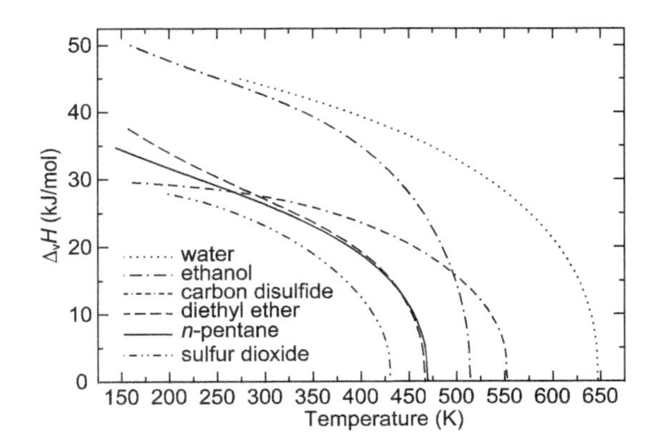

FIGURE 5.2
Enthalpy of vaporization as a function of temperature for multiple compounds.

Despite the difficulty, many experimentally determined values for $\Delta_v H$ are available. Majer and Svoboda (1985) provide a comprehensive and critical compilation of experimental values of $\Delta_v H$ measured since 1932 for approximately 600 organic compounds. Additional compilations of enthalpy of vaporization information can be found in Tamir et al. (1983) and DIPPR (Wilding et al. 2017). Most recently, Acree and Chickos (2016, 2017) have compiled data on organic and organometallic compounds from 1885 to 2015 for C_1 to C_{192}.

The latent enthalpy of vaporization decreases steadily with temperature and is zero at the critical point. Typical data are shown in Figure 5.2. The shapes of these curves are consistent with most other enthalpy of vaporization data.

5.4 THEORY OF LIQUID HEAT CAPACITY

Three liquid heat capacities are in common use: C_p^{liq}, C_σ^{liq}, and $C_{\text{sat}}^{\text{liq}}$. The first represents the change in *enthalpy* with temperature at constant pressure; the second shows the change in *enthalpy* of a saturated liquid with temperature; the third indicates the *heat* required to cause a temperature change while maintaining the liquid in a saturated state. Mathematically, the three heat capacities are

$$C_p^{\text{liq}} = \left(\frac{\mathrm{d}H^{\text{liq}}}{\mathrm{d}T}\right)_P \tag{5.4-1}$$

$$C_\sigma^{\text{liq}} = \left(\frac{\mathrm{d}H^{\text{liq}}}{\mathrm{d}T}\right)_\sigma \tag{5.4-2}$$

$$C_{\text{sat}}^{\text{liq}} = \left(\frac{\mathrm{d}Q}{\mathrm{d}T}\right)_\sigma \tag{5.4-3}$$

where H^{liq} is the enthalpy of the liquid phase, Q is heat, and the subscript σ indicates the process occurs along the saturation curve. The three quantities are related as follows

$$C_\sigma^{\text{liq}} = C_p^{\text{liq}} + \left[V_\sigma^{\text{liq}} - T\left(\frac{\partial V}{\partial T}\right)_P\right]\left(\frac{\mathrm{d}P}{\mathrm{d}T}\right)_\sigma = C_{\text{sat}}^{\text{liq}} + V_\sigma^{\text{liq}}\left(\frac{\mathrm{d}P}{\mathrm{d}T}\right)_\sigma \tag{5.4-4}$$

where $\left(\frac{dP}{dT}\right)_\sigma$ is the change of P^{vp} with T and V_σ^{liq} is the molar volume of the liquid at saturation ($V_\sigma^{\text{liq}} = V_s$). Except at high reduced temperatures, all three forms of the liquid heat capacity are in close numerical agreement. Most estimation techniques yield either C_p^{liq} or C_σ^{liq}, although $C_{\text{sat}}^{\text{liq}}$ is often the quantity measured experimentally.

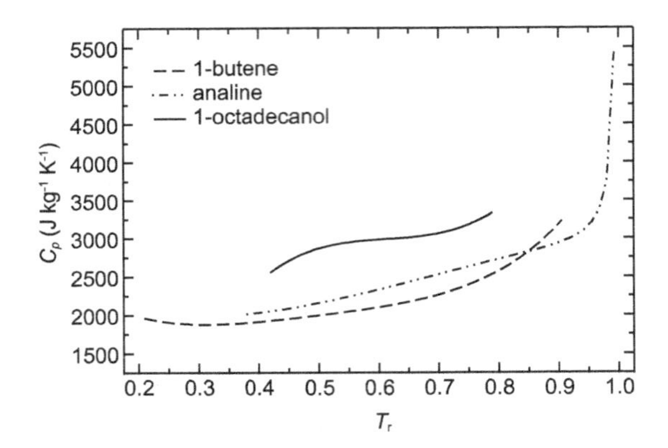

FIGURE 5.3
Liquid heat capacity as a function of temperature for multiple compounds illustrating the different possible behaviors for the property.

Liquid heat capacities can exhibit different behaviors depending on the compound. In general, they are weak functions of temperature below $T_r = 0.7$ to 0.8 with most compounds displaying a modest monotonic increase in the property with temperature. Some compounds exhibit a shallow minimum at temperatures below the normal boiling point. For other compounds, the heat capacity drops at low temperatures near the melting point giving the curve an S shape. At high reduced temperatures, C_p^{liq} values for all compounds are large and strong functions of temperature as they approach infinity at the critical point. Examples of these trends are illustrated in Figure 5.3 for 1-butene, aniline, and 1-octadecanol. Each of the curves was obtained from reliable experimental data across the entire temperature range depicted for the curve (1-butene: Schlinger and Sage, 1949; Takeda et al., 1991; Aston et al., 1946; Todd and Parks, 1936; analine: Steele et al., 2002; 1-octadecanol: van Miltenburg and Oonk, 2001; Zábranský et al., 1990; Khasashin and Zykova, 1989).

Near the normal boiling point, most liquid organic compounds have heat capacities between 1.2 and 2.0 J/(g·K). In this temperature range, there is essentially no effect of pressure (Gambill, 1957).

Experimentally reported liquid heat capacities for over 1600 substances have been compiled and evaluated by Zábranský et al. (1996), and values at 298.15 K for over 2500 compounds are given by Domalski and Hearing (1996) with two supplements from Zábranský et al. (2001) and Zábranský et al. (2010). Constants for equations that may be used to calculate liquid heat capacities are presented by Daubert et al. (1997). NIST TRC and DIPPR contain experimental and predicted data and correlations with the equations found in Appendix A coming from the latter. Most of the available heat capacity data are for temperatures below the normal boiling point as data for higher temperatures are more difficult to measure and thus far less plentiful.

5.5 CORRELATING VAPOR-PRESSURE, ENTHALPY OF VAPORIZATION, AND LIQUID HEAT CAPACITY DATA

5.5.1 Vapor Pressure Correlations

Vapor pressures have been measured for many substances. When reliable measurements are available, they are preferred over results from the estimation methods presented later in this chapter. Boublik (1984) presents tabulations of experimental data that have been judged to be of high quality for approximately 1000 substances. Numerous additional tabulations of "experimental" vapor pressure exist. However, sometimes these vapor pressures are calculated rather than original data, and therefore the possibility exists that errors have been introduced in fitting, interpolation, or extrapolation of the values. Literature references to experimental vapor pressure data can be found in Dykyj and Repá (1979), Dykyj et al. (1984), Rumble (2021), Majer et al. (1989), Ohe (1976), and Green and Southard (2019). Data for environmentally significant solids and liquids, including polycyclic aromatics, polychlorinated biphenyls, dioxins, furans, and selected pesticides, are compiled in Dellesite (1997). NIST TRC and DIPPR (Wilding et al. 2017) have both experimental and predicted values for vapor pressure. These databases include both the raw data and fits of the data to equations presented below.

Most vapor-pressure estimation and correlation equations stem from integration of Eq. (5.3-4). Doing so requires consideration of the temperature dependence of $\Delta_v H$ and $\Delta_v Z$, and a vapor pressure-temperature point must be used to evaluate the constant of integration. The simplest approach is to assume that the group $\frac{\Delta_v H}{\Delta_v Z}$ is constant and independent of temperature. Then, with the constant of integration denoted as A, integration of Eq. (5.3-4) leads to

$$\ln P^{vp} = A - \frac{B}{T} \tag{5.5-1}$$

where $B = \frac{\Delta_v H}{R\Delta_v Z}$. This equation is a fairly good relation for approximating vapor pressure over *small* temperature intervals due to a fortuitous relationship. Except near the critical point, $\Delta_v H$ and $\Delta_v Z$ are both weak functions of temperature, and since both decrease with rising temperature, they provide a compensatory effect. However, over large temperature ranges, especially when extrapolated below the normal boiling point, Eq. (5.5-1) represents vapor pressure data poorly. This is demonstrated in Figure 5.4 which is a plot of the deviation of the

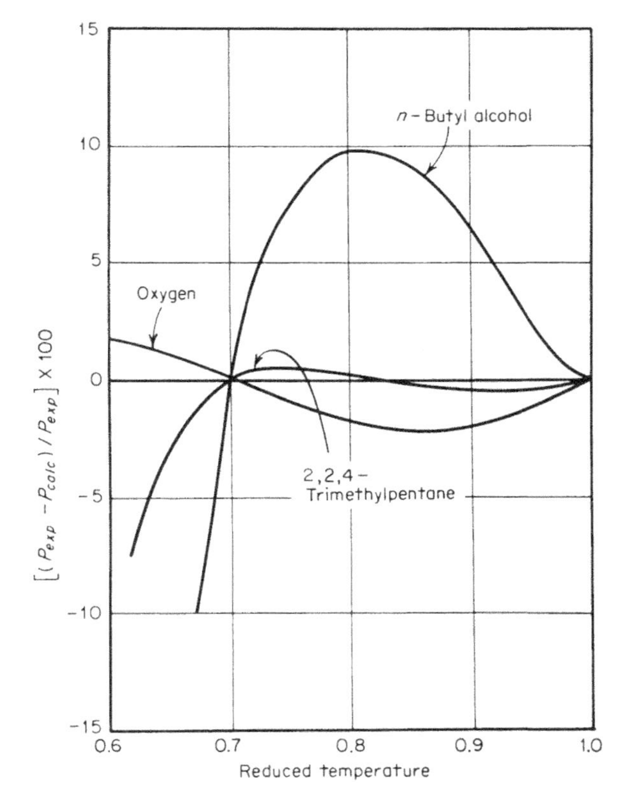

FIGURE 5.4
Comparison of experimental vapor pressure data with Eq. (5.5-1) [Ambrose (1972)].

true vapor pressure from that suggested by Eq. (5.5-1). The ordinate in this plot is the ratio $\frac{[P_{exp} - P_{calc}]}{P_{exp}}$ and the abscissa the reduced temperature $T_r = \frac{T}{T_c}$. P_{calc} is obtained from Eq. (5.5-1) where constants A and B are set by the value of P^{vp} at $T = 0.7T_c$ and $P^{vp} = P_c$ at T_c. Thus, P_{calc} is obtained from

$$\ln\left(\frac{P_{calc}}{P_c}\right) = -\beta\left(1 - \frac{1}{T_r}\right) \tag{5.5-2}$$

where $\beta = \frac{7}{3}\ln\left(P_r^{vp}(T_r = 0.7)\right)$. The figure shows that at high reduced temperatures, the fit of Eq. (5.5-2) (and thus Eq. 5.5-1) is reasonably good for oxygen and a typical hydrocarbon, 2,2,4-trimethylpentane, but for the associating liquid, *n*-butanol, errors as high as 10% can result for reduced temperatures between 0.7 and 1.0. When Eq. (5.5-2) is used to extrapolate to lower temperatures, much larger errors result. For ethanol, for example, Eq. (5.5-2) predicts a vapor pressure at the melting temperature that is 24 times too high.

Extending our consideration of Eq. (5.5-1) one step further, a common practice is to use both the normal boiling point [rather than the vapor pressure at $T_r = 0.7$ as in Eq. (5.5-2)] and the critical point to obtain generalized constants. Expressing pressure in bars and temperature on the absolute scale (kelvins or degrees Rankine), with $P^{vp} = P_c$ at $T = T_c$ and $P^{vp} = 1.01325$ bar at $T = T_b$, the normal boiling temperature at 1 atm = 1.01325 bar, Eq. (5.5-1) becomes

$$\ln P_r^{vp} = h\left(1 - \frac{1}{T_r}\right) \tag{5.5-3}$$

$$h = T_{b_r}\frac{\ln\left(\frac{P_c}{1.01325\ \text{bar}}\right)}{1 - T_{b_r}} \tag{5.5-4}$$

The behavior of Eq. (5.5-3) is similar to that of Eq. (5.5-2) i.e., the equation is satisfactory for describing vapor-pressure behavior over *small* temperature ranges. Over large temperature ranges, or when used to extrapolate data, Eq. (5.5-3) can lead to unacceptably large errors.

Many different equations have been presented to correlate vapor pressures as a function of temperature over wider temperature ranges. Three of these, the Antoine, Riedel, and Wagner equations are discussed below.

5.5.1.1 Antoine Equation

Antoine (1888) proposed a simple modification of Eq. (5.5-1) which has been widely and successfully used over limited temperature ranges.

$$\log_{10} P^{vp} = A - \frac{B}{T + C} \tag{5.5-5}$$

where T is in kelvin, and A, B, and C are constants. Simple rules have been proposed (Fishtine, 1963; Thomson, 1959) to relate C to the normal boiling point for certain classes of materials, but these rules are not dependable, and the only reliable way to obtain values of the constants A, B, and C is to regress experimental data. Values of A, B, and C are tabulated for a number of materials in previous editions of this book. Additional tabulations of Antoine constants may be found in Boublik et al. (1984), Speight (2017), and Yaws (1992). The applicable temperature range is not large and, in most instances, corresponds to a pressure interval of about 0.01 to 2 bars. The Antoine equation should never be used outside the stated temperature limits. Extrapolation beyond these limits may lead to unreliable values. The constants A, B, and C form a set. Never use one constant from one tabulation and the other constants from a different tabulation.

Cox (1923) suggested a graphical correlation in which the ordinate, representing P^{vp}, is a log scale, and a straight line (with a positive slope) is drawn. The sloping line is taken to represent the vapor pressure of water (or some other reference substance). Since the vapor pressure of water is accurately known as a function of temperature, the abscissa scale can be marked in temperature units. When the vapor pressure and temperature scales are prepared in this way, vapor pressures for other compounds are often found to be nearly straight lines, especially for homologous series. Calingaert and Davis (1925) have shown that the temperature scale on this Cox chart is nearly equivalent to the function $(T + C)^{-1}$, where C is approximately −43 K for many materials boiling between 273 and 373 K. Thus, the Cox chart closely resembles a plot of the Antoine vapor pressure equation. Also, for homologous series, a useful phenomenon is often noted on Cox charts. The straight lines for different members of the homologous series often converge to a single point when extrapolated. This point, called the infinite point, is useful for providing one value of vapor pressure for a new member of the series. Dreisbach (1952) presents a tabulation of these infinite points for several homologous series.

5.5.1.2 Riedel and Modified Riedel Equations

Riedel (1954) proposed the following equation for the correlation of vapor pressure.

$$\ln P^{vp} = A + \frac{B}{T} + C \ln T + DT^6 \tag{5.5-6}$$

The T^6 term provides flexibility to the equation which is particularly important near the freezing and critical points. As will be described later in this chapter, the value of 6 was selected because it seemed to provide the best shape near the critical point for several compounds investigated when the equation was first proposed; however, it has been found that the utility of this equation is greatly improved by allowing this exponent to vary. This leads to the *modified* Riedel equation

$$\ln P^{vp} = A + \frac{B}{T} + C \ln T + DT^E \tag{5.5-7}$$

Table 5.3 contains the values of A, B, C, D, and E for Eq. (5.5-7) for a few compounds. Appendix A has the values for 473 compounds.

5.5.1.3 Wagner Equation

Wagner (1973, 1977) used an elaborate statistical method to develop an equation for representing the vapor pressure behavior of nitrogen and argon over the entire temperature range for which experimental data were available. In this method, the actual terms as well as their coefficients were variables, i.e., a superfluity of terms was available and the most significant ones were chosen according to statistical criteria. The resulting equation is

$$\ln P_r^{vp} = \frac{(a\tau + b\tau^{1.5} + c\tau^3 + d\tau^6)}{T_r} \tag{5.5-8}$$

TABLE 5.3 Constants for the modified Riedel expression for vapor pressure, Eq. (5.5-7), for some compounds. Use with units of K and Pa for temperature and pressure, respectively.

Compound	*A*	*B*	*C*	*D*	*E*
2-pentanol	116.828	−10453.0	−13.1768	1.07123×10^{-17}	6
n-octane	96.084	−7900.2	−11.003	7.18020×10^{-6}	2
ammonia	90.483	−4669.7	−11.607	0.017194	1
n-octanoic acid	116.477	−13300.4	−12.6746	3.98338×10^{-18}	6

where P_r^{vp} is the reduced vapor pressure, T_r is the reduced temperature, and $\tau = 1 - T_r$. However, since Eq. (5.5-8) was first presented, the following form has come to be preferred (Ambrose, 1986; Ambrose and Ghiassee, 1987a):

$$\ln P_r^{vp} = \frac{(a\tau + b\tau^{1.5} + c\tau^{2.5} + d\tau^5)}{T_r} \tag{5.5-9}$$

Both Eqs. (5.5-8) and (5.5-9) can represent the vapor pressure behavior of most substances over the entire liquid range. Various forms of the Wagner equation that employ a fifth term have been presented; for some substances, e.g., water (Wagner, 1977; Wagner and Pruss, 1993), oxygen (Wagner et al., 1976), and some alcohols (Poling, 1996), this fifth term can be justified. However, Ambrose (1986) points out that except in such cases, a fifth term cannot be justified and is not necessary. The constants in Eq. (5.5-8) have been given by McGarry (1983) for 250 fluids. More recently, constants for Eq. (5.5-9) have been given by Ambrose and Ghiassee (1987a, 1987b, 1987c, 1988a, 1988b, 1990), Ambrose et al. (1988, 1990), and Ambrose and Walton (1989) for 92 fluids. The values of the constants a, b, c, d as well as the values of T_c and P_c to be used in Eq. (5.5-9) for all 92 fluids are listed in Appendix A of the 5th edition of this book.

5.5.1.4 Fitting Vapor Pressure Data

Extensive research over many years has shown that vapor pressure curves for nearly all compounds have similar characteristics. The Antoine Equation, Eq. (5.5-5), has difficulty reproducing these features and is avoided in modern engineering practice. Regardless of the form of the correlating equation chosen from those discussed above (e.g., Riedel, Wagner), fitting of vapor pressure data should be constrained to satisfy these common attributes. Three constraints are routinely used in this regard. The first of these features is a minimum in the $\dfrac{\Delta_v H}{\Delta_v Z}$ vs. T_r curve at some reduced temperature between 0.8 and 1.0. This minimum was first observed by Waring (1954) for water. Ambrose and Ghiassee (1987a) point out that this constraint causes b and c in Eqs. (5.5-8) and (5.5-9) to have different signs, and DIPPR (Wilding et al. 2017) satisfies this same criterion by requiring $A > 0$, $B < 0, C < 0$, and $D > 0$ in Eqs. (5.5-6) and (5.5-7).

The second characteristic feature, first identified by Thodos (1950), requires that there be an inflection point in the $\ln P^{vp}$ vs. $\dfrac{1}{T}$ curve—a feature that gives rise to the "elongated S-shape" characteristic of such plots. One way to ensure the presence of this inflection point is by requiring that the quantity $\ln \dfrac{P^{vp}}{P_{calc}}$ at $T_r = 0.95$ take on a value that falls within some specified range, where P_{calc} is determined from Eq. (5.5-2). For example, Ambrose et al. (1978) impose this constraint by requiring that the selected constants generate a P^{vp} value at $T_r = 0.95$ such that $-0.010 < \ln \dfrac{P^{vp}}{P_{calc}} < -0.002$ for non-associated compounds. Another way to check for the presence of the inflection point or proper S-shape is by transforming the data and correlation into a new coordinate system which emphasizes deviations from the proper shape. This approach, described by Wilsak and Thodos (1984) and explained below, produces a so-called *shifted-rotated* plot which should be parabolic in nature if the correlation has the proper shape.

The third constraint employs the Watson relation (see the next section) to ensure that the temperature behavior of the vapor-pressure equation matches the temperature dependence of the enthalpy of vaporization predicted by Eq. (5.5-16). To do this, Ambrose et al. (1978) calculate the quantity $g = \dfrac{\Delta_v H}{(1 - T_r)^{0.375}}$ at several reduced temperatures between 0.5 and 0.6, where it is supposed to be approximately constant. The constraint is satisfied if the standard deviation of g from its mean value g', over this range, is less than 5%. DIPPR uses a wider temperature range, from T_b to $T_r = 0.99$, by fitting $\Delta_v H$ vs T over this range to Eq. (5.5-16) by varying the exponent n and requiring $n = 0.38 \pm 30\%$.

A shifted-rotated plot, mentioned previously, can be used to ensure the presence of an inflection point in the vapor pressure correlation (the second constraint explained above). It can also discriminate between reliable and unreliable data when determining which sources to trust when regressing coefficients from published values. Such a plot is created by transforming the data onto a coordinate system that emphasizes deviations from the expected approximately linear trend for $\ln P^{\mathrm{vp}}$ vs $\frac{1}{T}$. The new coordinate system is created by selecting two points to define a straight line along the expected linear behavior. The critical point (T_c, P_c) and triple point (T_{tp}, P_{tp}) are convenient selections in this regard. Selecting the critical point as the origin and rotating the coordinate system so that its abscissa is found on the line between the critical and triple points defines the new coordinate system in terms of the original. Specifically, the following equations may be used to transform P^{vp} vs T data into the new coordinate system.

$$\tan\phi = \frac{\ln P_{\mathrm{tp_r}}}{\frac{1}{T_{\mathrm{tp_r}}} - 1} \tag{5.5-10}$$

$$x = (\sin\phi)\ln P_r + (\cos\phi)\left(\frac{1}{T_r} - 1\right) \tag{5.5-11}$$

$$y = (\cos\phi)\ln P_r - (\sin\phi)\left(\frac{1}{T_r} - 1\right) \tag{5.5-12}$$

$$X = \frac{x - x_{\min}}{x_{\max} - x_{\min}} \tag{5.5-13}$$

$$Y = \frac{y - y_{\min}}{y_{\max} - y_{\min}} \tag{5.5-14}$$

Here, $P_{\mathrm{tp_r}} = \frac{P_{\mathrm{tp}}}{P_c}$, $T_{\mathrm{tp_r}} = \frac{T_{\mathrm{tp}}}{T_c}$, and subscripts min and max refer to the minimum and maximum values of x and y in the set of data. The last two expressions translate the points in the new coordinate system to all lie in the first quadrant for graphing purposes.

Figure 5.5 is the *shifted-rotated plot* for n-octanoic acid obtained by transforming P^{vp} vs T pairs of values with Eqs. (5.5-13) and (5.5-14). The points are from various literature sources with values from the same source having the same symbol. The line is created from the modified Riedel correlation and the coefficients found in Table 5.3. Correlations for the vapor pressure, or vapor pressure data, possessing the characteristic inflection

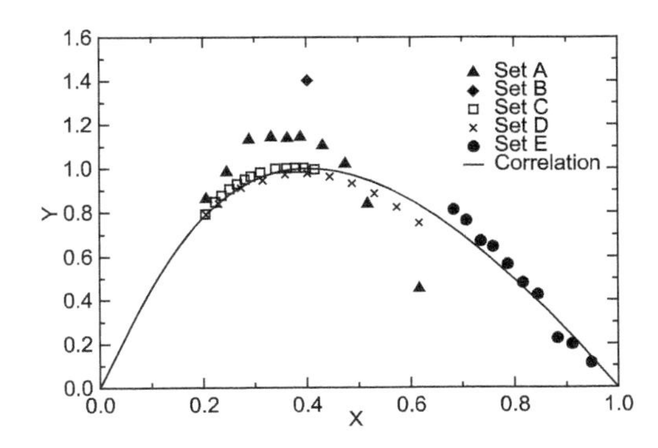

FIGURE 5.5
Shifted-rotated plot for n-octanoic acid. The axes are defined by Eqs. (5.5-13) to (5.5-14). The line is from the modified Riedel equation for the compound. The points are sets of values available in the literature.

point (elongated S-shape) in the $\ln P^{\mathrm{vp}}$ vs $\frac{1}{T}$ curve will produce shifted-rotated plots with a parabolic shape opening downwards. Correlations or data which do not follow this parabolic pattern (multiple maxima and minima) do not possess the characteristic elongated S-shape with its associated inflection point. The shifted-rotated plot can therefore be used to ensure the second constraint on fitting mentioned above by requiring the correlation produce the correct downward parabola.

In addition, the shifted-rotated plot can be used to identify vapor pressure data that may be less reliable. Notice Sets A and B in Figure 5.5 do not follow the expected shape. Set A cuts through the other data and correlation while Set B (a single point) is far above the expected trend. Because of the disagreement between these two sets and the others available, they should be omitted when regressing coefficients for vapor pressure correlations in favor of Sets C–E.

5.5.2 Enthalpy of Vaporization Correlation

As pictured in Figure 5.2, the enthalpy of vaporization for a compound is at a maximum at the triple point temperature and monotonically decreases with temperature. Near the critical temperature of the compound, the decrease is very rapid until $\Delta_{\mathrm{v}}H = 0$ at T_{c}. This behavior across the entire temperature range (T_{m} to T_{c}) can be correlated using an equation of the form

$$\Delta_{\mathrm{v}}H = A(1 - T_{\mathrm{r}})^{\left(B + CT_{\mathrm{r}} + DT_{\mathrm{r}}^2 + ET_{\mathrm{r}}^3\right)} \tag{5.5-15}$$

where A–E are coefficients determined by fitting the equation to enthalpy of vaporization values. As with the correlation equations discussed above for vapor pressure, correlations for enthalpy of vaporization should not be used outside the temperature range over which the data were fit, and the coefficients for one compound should not be used for another. Additional terms in the series can be added to the exponent if there are enough reliable data to warrant this.

A correlation between $\Delta_{\mathrm{v}}H$ and T_{r}, used for many years, is the Watson relation (Thek and Stiel, 1967)

$$\Delta_{\mathrm{v}}H_2 = \Delta_{\mathrm{v}}H_1\left(\frac{1 - T_{\mathrm{r}_2}}{1 - T_{\mathrm{r}_1}}\right)^n \tag{5.5-16}$$

where subscripts 1 and 2 refer to reduced temperatures T_{r_1} and T_{r_2}. A common choice for n is 0.375 or 0.38 (Thodos, 1950). This equation can correlate enthalpy of vaporization data over the temperature range T_{b} to T_{c}; however, it is no longer recommended because its functional form does not allow it to reproduce the temperature behavior of $\Delta_{\mathrm{v}}H$ across the temperature range from T_{m} to T_{c} as does Eq. (5.5-15). Despite this, the expression remains useful as a constraint when fitting vapor pressure data as explained in the previous section.

5.5.3 Liquid Heat Capacity Correlations

As depicted in Figure 5.3, liquid heat capacity behavior with temperature can be diverse below the critical temperature, but it rises very sharply near T_{c}. For T_{r} below approximately 0.8, correlating liquid heat capacity data can be adequately done using a simple polynomial. To cover the entire range of temperatures, including the sharp rise near T_{c}, an equation of the form found below may be used

$$C_p^{\mathrm{liq}} = A + \frac{B}{\tau} + C\tau + D\tau^2 + E\tau^3 \tag{5.5-17}$$

where $\tau = 1 - T_{\mathrm{r}}$.

5.6 RELIABLE EXTRAPOLATION OF VAPOR PRESSURE AND THERMODYNAMIC CONSISTENCY BETWEEN VAPOR PRESSURE, ENTHALPY OF VAPORIZATION, AND LIQUID HEAT CAPACITY

The three properties discussed up to this point in this chapter are related through rigorous thermodynamic relationships. As already discussed, the Clapeyron Equation, Eq. (5.3-1), relates vapor pressure and enthalpy of vaporization. Solving this equation for ΔH_v and taking the temperature derivative of the results yields

$$\frac{\mathrm{d}\Delta_v H}{\mathrm{d}T} = \frac{\mathrm{d}}{\mathrm{d}T}(H^{\mathrm{vap}} - H^{\mathrm{liq}})_\sigma = C_\sigma^{\mathrm{vap}} - C_\sigma^{\mathrm{liq}} \tag{5.6-1}$$

where H^{vap} and H^{liq} are the enthalpy of the vapor and liquid. Through Eq. (5.4-4), this becomes

$$\frac{\mathrm{d}\Delta_v H}{\mathrm{d}T} = C_p^{\mathrm{vap}} + \left[V_\sigma^{\mathrm{vap}} - T\left(\frac{\partial V_\sigma^{\mathrm{vap}}}{\partial T}\right)_P\right]\left(\frac{\partial P}{\partial T}\right)_\sigma - C_p^{\mathrm{liq}} - \left[V_\sigma^{\mathrm{liq}} - T\left(\frac{\partial V_\sigma^{\mathrm{liq}}}{\partial T}\right)_P\right]\left(\frac{\partial P}{\partial T}\right)_\sigma \tag{5.6-2}$$

Here, C_p^{vap} is the isobaric heat capacity of the vapor at the saturation pressure, C_p^{liq} is the isobaric heat capacity of the liquid phase, and $\left(\frac{\partial P}{\partial T}\right)_\sigma$ is the temperature derivative of the vapor pressure.

Solving for C_p^{liq} and rearranging gives

$$C_p^{\mathrm{liq}} = C_p^{\mathrm{vap}} - \frac{\mathrm{d}\Delta_v H}{\mathrm{d}T} + \left\{\left[V_\sigma^{\mathrm{vap}} - T\left(\frac{\partial V_\sigma^{\mathrm{vap}}}{\partial T}\right)_P\right] - \left[V_\sigma^{\mathrm{liq}} - T\left(\frac{\partial V_\sigma^{\mathrm{liq}}}{\partial T}\right)_P\right]\right\}\left(\frac{\partial P}{\partial T}\right)_\sigma \tag{5.6-3}$$

As will be discussed in Chap. 6, C_p^{vap} can be determined from the ideal gas heat capacity (C_p°), which was discussed in Chap. 4, and the residual heat capacity $C_p^{\mathrm{r}} = -T\int_0^P \left(\frac{\partial^2 V^{\mathrm{vap}}}{\partial T^2}\right)\mathrm{d}P$ through $C_p^{\mathrm{vap}} = C_p^\circ + C_p^{\mathrm{r}}$. The pressure of the upper-bound of the integral is the vapor pressure of the compound at the temperature of the system. Substitution for C_p^{vap} thus gives

$$C_p^{\mathrm{liq}} = C_p^\circ - T\int_0^{P^{\mathrm{vp}}}\left(\frac{\partial^2 V^{\mathrm{vap}}}{\partial T^2}\right)dP - \frac{\mathrm{d}\Delta_v H}{\mathrm{d}T} + \left\{\left[V_\sigma^{\mathrm{vap}} - T\left(\frac{\partial V_\sigma^{\mathrm{vap}}}{\partial T}\right)_P\right] - \left[V_\sigma^{\mathrm{liq}} - T\left(\frac{\partial V_\sigma^{\mathrm{liq}}}{\partial T}\right)_P\right]\right\}\left(\frac{\partial P}{\partial T}\right)_\sigma \tag{5.6-4}$$

The power of this expression is it can be used to ensure the thermodynamic consistency of several important properties. These are the ideal gas heat capacity, the liquid heat capacity, the enthalpy of vaporization, the saturated liquid density (through V_σ^{liq}), and the vapor pressure. This can help identify problems in data and/or correlations for any of these properties. Moreover, as will be described below, Eq. (5.6-4) can be used to *predict* liquid heat capacity from the other properties.

When applying Eq. (5.6-4), C_p° is obtained using the methods discussed in Chap. 4. Vapor volumes and derivatives of vapor volumes are obtained using an appropriate equation of state (see Chap. 6). The saturated liquid density, and its derivative, are calculated from an equation such as (5.2-8). The temperature derivative of the vapor pressure is obtained through straightforward differentiation of the selected correlation [one of Eq. (5.3-2)–(5.3-4)]. The temperature derivative of enthalpy of vaporization comes from differentiation of Eq. (5.5-15).

Vapor pressure data are most abundant at temperatures near the normal boiling point. Low and high temperature vapor pressure, those approaching the melting and critical points, respectively, are less frequent in the literature. As explained above, none of the correlating equations for vapor pressure extrapolate well outside the temperature range of the data used to obtain the coefficients. Extrapolation to higher temperatures can reliably be done by ensuring the correlation accurately reproduces the critical point, whether predicted or experimental, when regressing the parameters. DIPPR (Wilding et al. 2017) policy in this regard is to ensure that $P^{vp}_{corr}(T_c) = P_c$ to within 0.5%. Extrapolating to lower temperatures is best done using liquid heat capacity data and Eq. (5.6-4). Specifically, if liquid heat capacity data are available, regression of the parameters for vapor pressure should be done so that the selected vapor pressure correlation replicates the vapor pressure data while simultaneously replicating available liquid heat capacity data through Eq. (5.6-4)—a process which can produce reliable extrapolation of vapor pressures down to the melting temperature of the compound. DIPPR procedures accomplish this by requiring that liquid heat capacities predicted by Eq. (5.6-4) agree to within an average absolute deviation of 3% of reliable liquid heat capacity data.

Vapor pressure and thermal data have been simultaneously used by several authors (Ambrose and Davies, 1980; King and Al-Najjar, 1974; Majer et al., 1989; Moelwyn-Hughes, 1961; Poling, 1996; Růžička and Majer, 1994; Myrdal and Yalkowsky, 1997; Huber et al., 2006) to generate more reliable low-temperature vapor-pressure equations. Many of these used approximations to describe what Eq. (5.6-4) does rigorously. Hogge et al. (2016a, 2016b) have described the rigorous process in detail—an approach they term *multiproperty optimization*. An example of the technique appears below.

Example 5.4

Correlate the vapor pressure of acetic anhydride for accurate extrapolation down to the melting point of the compound.

Solution. Experimental vapor pressure data for the compound are reported by several authors. The values span from 273 to 429 K. Experimental liquid heat capacity data have been measured by Dai et al. (2010) with a reported error of < 0.2%. The critical temperature and pressure of the compound, reported by Ambrose and Ghiasse (1987), are 606.0 K and 40.0 bar, respectively. Figure 5.6 displays the reliable experimental vapor pressure data for the compound along with fits of these data to the modified Riedel equation [Eq. (5.5-7)] with values of E from 1 to 6. The plot is presented as $\ln P^{vp}$ vs $\frac{1}{T}$ with vapor pressures in units of Pa and temperatures in K. Notice that the value of E does not significantly change how effectively Eq. (5.5-7) fits the data but does affect the extrapolation to low temperatures where no data exist. Evidence of this is found in Table 5.4 which lists the statistics of the fits to the data for various values of E. The average absolute relative deviation (AARD) between the correlation and the data [see Eq. (3.5-1)], the maximum absolute relative deviation (MARD), and the square of the correlation coefficient (R^2) are listed. Notice that the variation in R^2 between the different value of E is on the order of 10^{-5} and that AARD ranges from 1.28% to 1.92%. The only meaningful difference in the statistics of the different fits is MARD = 8.83%

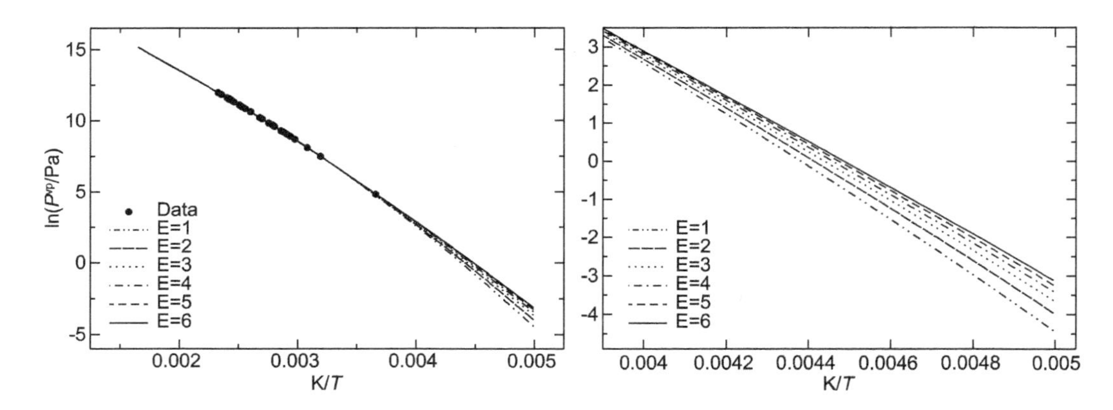

FIGURE 5.6
Experimental vapor pressure data (points) for acetic anhydride and fits of these data (lines) to the modified Riedel equation [Eq. (5.5-7)] for various values of E. Left: Temperature range from T_{tp} to T_c. Right: Low temperature range.

TABLE 5.4 Statistics on the fits of the experimental data for the vapor pressure of acetic anhydride to the modified Riedel equation [Eq. (5.5-7)] for various values of *E*.

	E						
	1	**2**	**3**	**4**	**5**	**6**	**2.75**
AARD	1.92%	1.51%	1.34%	1.28%	1.31%	1.44%	1.37%
MARD	8.83%	4.77%	2.74%	3.34%	3.93%	4.40%	2.75%
R^2	0.99943	0.99944	0.99944	0.99943	0.99942	0.99940	0.99944

for $E = 1$ which is almost double this highest MARD of the other correlations which range from 2.74% to 4.77%. In short, a value of E from 2-6 could be chosen to produce a reliable correlation of the vapor pressure of the compound across the temperature range of the experimental data, and the difference between the assorted options is found only in the extrapolated region at low temperatures (high values of $\frac{1}{T}$).

Because the vapor pressure data themselves do not discriminate between different values of E, determining the best value for the parameter is done by selecting which best reproduces the experimental *liquid heat capacity data.* Figure 5.7 shows the liquid heat capacity of acetic anhydride from various sources. The points are experimental data from Dai et al. (2010) and the curves are predictions using Eq. (5.6-4). The values of $\left(\frac{\partial P}{\partial T}\right)_\sigma$ and $\frac{\mathrm{d}\Delta_v H}{\mathrm{d}T}$ needed to evaluate the expression are found from differentiation of the vapor pressure correlations depicted in Figure 5.6 (those with different E values) and one additional correlation for $E = 2.75$. The prediction of liquid heat capacity, especially at temperatures below the normal boiling point, is very sensitive to E indicating the importance of selecting the correct value when regressing the vapor pressure data. Notice $E = 1$ gives a heat capacity curve with excessive curvature, and $E = 5$ and 6 produce curves that fall off at low temperature more than is commonly expected. If the vapor pressure data are regressed with $E = 2.75$, the liquid heat capacity predicted from Eq. (5.6-4) passes through the experimental points (0.29% AARD). This means the vapor pressure correlation is thermodynamically consistent with the liquid heat capacity yielding a more reliable extrapolation of vapor pressures at temperatures below those of the experimental values.

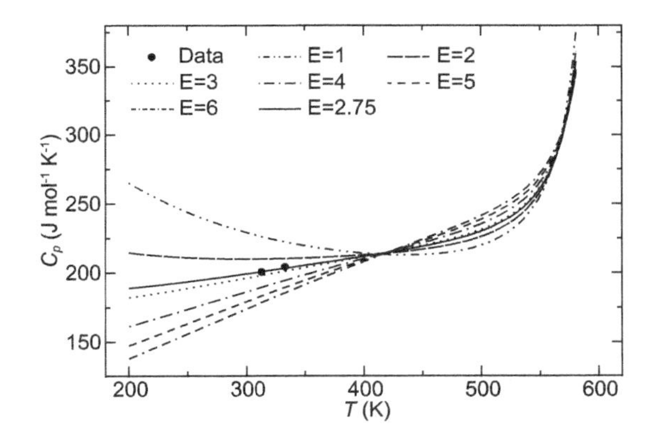

FIGURE 5.7
Liquid heat capacity of acetic anhydride. The points are experimental data. The lines are predictions using Eq. (5.6-4) with the vapor pressure data regressed with different value of E (see Figure 5.6).

5.7 PREDICTION OF VAPOR PRESSURE

Several methods have been proposed to predict vapor pressure. Multiple of these approaches are based on the principle of corresponding states (CSP). Previous editions of this book have outlined these methods, but these are not repeated here because these techniques require the acentric factor, ω, of the compound of interest. As described in Chap. 3, ω is *derived* from the vapor pressure curve, so CSP prediction approaches for vapor pressure are based on circular reasoning. Methods have been published to predict ω independent of vapor pressure, but this can lead to thermodynamic inconsistency between the vapor pressure curve and the value of ω and/or T_c. Both NIST TRC and DIPPR (Wilding et al. 2017) policy concerning this matter is to first obtain the vapor pressure curve and then use this curve, and the accepted value for T_c, to *derive* ω.

One exception to the above rationale is the CSP method of Lee and Kesler (1975). This approach is different from other CSP techniques because its derivation includes an associated expression for the acentric factor in terms of T_b and T_c thus removing the need to obtain ω using other techniques. This yields a consistent prediction approach. This method is discussed below, but readers are referred to previous editions of the book if other corresponding states methods for predicting vapor pressure are desired.

5.7.1 Riedel and Vetere Methods

As mentioned previously in this chapter, Riedel (1954) proposed a vapor pressure equation of the form

$$\ln P^{\text{vp}} = A + \frac{B}{T} + C \ln T + DT^6 \tag{5.5-6}$$

The T^6 term allows description of the inflection point of the vapor pressure curve in the high-pressure region. To determine the constants in Eq. (5.5-6), Riedel defined a parameter α

$$\alpha = \frac{\text{d} \ln P_{\text{r}}^{\text{vp}}}{\text{d} \ln T_{\text{r}}} \tag{5.7-1}$$

where $P_{\text{r}}^{\text{vp}} = \dfrac{P^{\text{vp}}}{P_{\text{c}}}$ is the reduced vapor pressure.

From a study of experimental vapor pressure data, Plank and Riedel (1948) showed that

$$\left.\frac{\text{d}\alpha}{\text{d}T_{\text{r}}}\right|_{T_{\text{r}}=1} = 0 \tag{5.7-2}$$

Using Eq. (5.7-2) as a constraint on (5.5-6), Riedel (1954) used experimental vapor pressure data for 20 compounds to show that

$$\ln P_{\text{r}}^{\text{vp}} = A^+ - \frac{B^+}{T_{\text{r}}} + C^+ \ln T_{\text{r}} + D^+ T_{\text{r}}^6 \tag{5.7-3}$$

where

$$\begin{aligned} A^+ &= -35Q \qquad & B^+ &= -36Q \qquad & C^+ &= 42Q + \alpha_{\text{c}} \\ D^+ &= -Q \qquad & Q &= K\left(X_{\text{c}} - \alpha_{\text{c}}\right) \end{aligned} \tag{5.7-4}$$

Here, α_{c} is α at the critical point, $K = 0.0838$, and $X_{\text{c}} = 3.758$. Since it is not easy (or desirable) to determine α_{c} by its defining equation at the critical point, α_{c} is usually found from Eqs. (5.7-3) and (5.7-4) by inserting $P = 1.01325$ bar at $T = T_{\text{b}}$ and calculating α_{c}. The equations that result from this process are:

TABLE 5.5 Vetere (2006) rules for Riedel Constant *K*.

Class	Relationship
saturated and branched hydrocarbons	$K = 0.075 + 0.0014h - 0.023(T_r - 0.4)$
olefins	$K = 0.089 - 0.033(T_r - 0.35)$
aromatics	$K = 0.10 - 0.0004\omega_p - 0.037(T_r - 0.30)$
alcohols	$K = 0.24 - 0.029 \ln \omega_p$
ketones	$K = 0.138 - 0.014 \ln \omega_p$
esters	$K = 0.09 - 0.025(T_r - 0.35)$
acids	$K = 0.12 - 0.00025\omega_p$
refrigerants and organic polar compounds	$K = 0.088 + 0.035(0.35 - T_r)$

$$\alpha_c = \frac{3.758\,K\psi_b - \ln P_{atm_r}}{K\psi_b - \ln T_{b_r}} \tag{5.7-5}$$

$$\psi_b = -35 + \frac{36}{T_{b_r}} + 42 \ln T_{b_r} - T_{b_r}^6 \tag{5.7-6}$$

where $T_{b_r} = \dfrac{T_b}{T_c}$ and $P_{atm_r} = \dfrac{1.01325 \text{ bar}}{P_c}$.

Vetere (1991), seeking to improve the predictive capabilities of Riedel's equation, used not only vapor pressure data to determine K (as Riedel did) but also normal boiling point, critical point, acentric factor, reduced temperature, and chemical family. He proposed equations for K for five classes of compounds—nonpolar, acids, alcohols, glycols, other polar compounds; however, only the acids and alcohol classes improved upon the original method Riedel method. More recently, Vetere (2006) updated his previous efforts to include more classes of compounds and make K dependent on temperature. The effort results in a method that was superior to the original Riedel method for all classes of compounds examined. Table 5.5 contains the equations to calculate K from Vetere (2006) for each family examined. In these expressions, h, is defined by Eq. (5.5-4) and $\omega_p = \dfrac{\left(\dfrac{T_b}{K}\right)^{1.72}}{\dfrac{M}{\text{g}\cdot\text{mol}^{-1}}} - 263$ where M is molar mass and T_b is the normal boiling point of the compound. Also, $X_c = 3.758$ for Eq. (5.7-4) as proposed by Riedel.

Example 5.5

Estimate the vapor pressure of ethylbenzene at 350 and 450 K using the methods of Riedel (1954) and Vetere (2006). For this compound, the correlation in Appendix A comes from experimental data and gives values of 0.1487 and 2.6935 bar at 350 and 450 K, respectively.

Solution. For ethylbenzene from Appendix A, $T_b = 409.31$ K, $T_c = 617.26$ K, and $P_c = 36.16$ bar. Thus, $T_{b_r} = 409.31/617.26 = 0.6631$ and $P_{atm_r} = 1.01325/36.16 = 0.02802$. From Eqs. (5.7-6)

$$\psi_b = -35 + 36/0.6631 + 42 \ln 0.6631 - 0.6631^6 = 1.9506$$

Riedel Method With Eq. (5.7-5) and $K = 0.0838$,

$$\alpha_c = [(3.758)(0.0838)(1.9506) - \ln(0.02802)] / [(0.0838)(1.9506) - \ln(0.6631)] = 7.2944$$

The constants in Eq. (5.7-4) become

$$Q = (0.0838)(3.758 - 7.2944) = -0.29635$$
$$A^+ = (-35)(-0.29635) = 10.372$$
$$B^+ = (-36)(-0.29635) = 10.669$$
$$C^+ = (42)(-0.29635) + 7.2944 = -5.152$$
$$D^+ = -(-0.29635) = 0.2964$$

and Eq. (5.7-3) becomes

$$\ln P_r^{vp} = 10.372 - \frac{10.669}{T_r} - 5.152T_r + 0.2964T_r^6$$

The predicted values of vapor pressure for the compound at 350 and 450 K, along with the experimental values from Appendix A are below

		P^{vp} (bar)		
T (K)	**T_r**	**Pred.**	**Exp.**	$\frac{P_{pred} - P_{exp}}{P_{exp}} \times 100$
350	0.5670	0.1462	0.1487	−1.71
450	0.7290	2.713	2.694	0.72

Vetere Method The values for K in Eqs. (5.7-4) and (5.7-5), and thus the coefficients of Eq. (5.7-3), are dependent on the class of the compound under consideration and the temperature at which the vapor pressure is being predicted. Ethylbenzene (M = 106.165 g/mol) belongs to the aromatics class of Table 5.5, so ω_p is needed to calculate the needed values for K. The following are obtained at 350 K (T_r = 0.5670) and 450 K (T_r = 0.7290) with the last two quantities coming from to Eq. (5.7-5).

$$\omega_p = 409.31^{1.72}/106.165 - 263 = 29.9198$$
$$K\,(T_r = 0.5670) = 0.10 - (0.0004)(29.9198) - 0.037(0.5670 - 0.30) = 0.07815$$
$$K\,(T_r = 0.7290) = 0.10 - (0.0004)(29.9198) - 0.037(0.7290 - 0.30) = 0.07216$$
$$\alpha_c\,(T_r = 0.5670) = [(3.758)(0.07815)(1.9506) - \ln(0.02802)] / [(0.07815)(1.9506) - \ln(0.6631)] = 7.3636$$
$$\alpha_c\,(T_r = 0.7290) = [(3.758)(0.07216)(1.9506) - \ln(0.02802)] / [(0.07216)(1.9506) - \ln(0.6631)] = 7.4400$$

Using these values in Eq. (5.7-4) gives the following for Q and coefficients to use in Eq. (5.7-3).

	350 K	**450 K**
Q	−0.28178	−0.26569
A^+	9.8622	9.2993
B^+	10.1440	9.5650
C^+	−4.4711	−3.7191
D^+	0.28178	0.26569
$\ln P_r^{vp}$	−5.4822	−2.606

The predicted values of vapor pressure using these coefficients along with the experimental values from Appendix A are below.

T (K)	T_r	P^{vp} (bar) Pred.	P^{vp} (bar) Exp.	$\frac{P_{pred} - P_{exp}}{P_{exp}} \times 100$
350	0.5670	0.1504	0.1487	1.14
450	0.7290	2.6697	2.694	−0.90

5.7.2 Hogge et al. Method

Hogge et al. (2017) examined the performance of Eq. (5.5-7) by letting E vary between 1 and 6 in half integer increments. The work was prompted from an examination of the ability of Eq. (5.6-4) to predict liquid heat capacity when reliable experimental data for vapor pressure were available and correlated to Eq. (5.5-7). The results showed that $E = 6$ was rarely the best value to ensure thermodynamic consistency between vapor pressure and liquid heat capacity through Eq. (5.6-4); however, with judicious selection of this exponent, the modified Riedel equation can simultaneously represent both properties. Table 5.6 gives the recommendations for E for various families in this regard. In general, a value a 3 or 4 is best for all compounds but amines and gases.

The Riedel and Vetere prediction methods were optimized to obtain values for D in Eq. (5.6-4) [through K and X_c of Eq. (5.7-4)] assuming $E = 6$. A different scheme is required for other values of E. In such an approach, Eq. (5.6-3) must be generalized for arbitrary values of E and becomes

$$\ln P_r^{vp} = A^+ - \frac{B^+}{T_r} + C^+ \ln T_r + D^+ T_r^E \tag{5.7-7}$$

TABLE 5.6 Recommended value of E in Eq. (5.5-7) by family (see Hogge et al. 2017).

Family	E
aldehydes	4
alkane	3
alkene	3
alkyne	3
amine	2
aromatic	4
ester	4
ether	3
gas	5
halogenated	4
ketone	3
multifunctional	3–4
ring alkane	4
silane	3
sulfide	3

Using Eq. (5.7-2) as a constraint on (5.5-7), Hogge et al. (2017) simultaneously fit reliable experimental vapor pressure and liquid heat capacity data to Eqs. (5.5-7) and (5.6-4), respectively, for 37 compounds from multiple families to determine the following relationships for K and X_c as a function of E.

$$K = 4.96465E^{-2.29195} \tag{5.7-8}$$

$$X_c = 4.14524 - 0.0818433E + 0.00310685E^2 \tag{5.7-9}$$

The coefficients of Eq. (5.7-7) are obtained through

$$A^+ = (E^2 - 1)D^+ \tag{5.7-10}$$

$$B^+ = E^2D^+ \tag{5.7-11}$$

$$C^+ = \alpha_c - E(E+1)D^+ \tag{5.7-12}$$

$$D^+ = K[\alpha_c - X_c] \tag{5.7-13}$$

$$\alpha_c = \frac{\ln P_{\text{atm}_r} + KX_c\phi}{\ln T_{b_r} + K\phi} \tag{5.7-14}$$

$$\phi = E^2 - 1 - \frac{E^2}{T_{b_r}} - E(E+1)\ln T_{b_r} + T_{b_r}^E \tag{5.7-15}$$

To summarize predicting vapor pressure using the method of Hogge et al. (2017), first select the desired value of E then calculate K, X_c, ϕ, and α_c from which A^+ through D^+ may be obtained.

Example 5.6

Estimate the vapor pressure of ethylbenzene at 350 and 450 K using the method of Hogge et al. (2017). For this compound, the correlation in Appendix A comes from experimental data and gives values of 0.1487 and 2.6935 bar at 350 and 450 K, respectively.

Solution. For ethylbenzene from Appendix A, $T_b = 409.31$ K, $T_c = 617.26$ K, and $P_c = 36.16$ bar. Thus, $T_{b_r} = 409.31/617.26 = 0.6631$ and $P_{\text{atm}_r} = 1.01325/36.16 = 0.02802$. Ethylbenzene belongs to the aromatic family of Table 5.3, so $E = 4$ is selected for the compound. With this value, Eqs. (5.7-5), (5.7-6), (5.7-15), and (5.7-14) give the following for K, X_c, ϕ, and α_c, respectively.

$$K = (4.96465)(4)^{-2.2195} = 0.22888$$

$$X_c = 4.14524 - (0.0818433)(4) + (0.00310685)(4)^2 = 3.86758$$

$$\phi = 4^2 - 1 - 4^2/0.6631 - (4)(5)\ln(0.6631) + 0.6631^4 = -0.71916$$

$$\alpha_c = [\ln(0.\ 02802) + (0.22888)(3.86758)(-0.71916)]/[\ln(0.6631) + (0.22888)(-0.71916)] = 7.31877$$

Using these quantities in Eqs. (5.7-10)–(5.7-13) yields the following coefficients for Eq. (5.7-7).

$$D^+ = 0.22888(7.31877 - 3.86758) = 0.78991$$

$$A^+ = (4^2 - 1)(0.78991) = 11.84865$$

$$B^+ = (4^2)(0.78991) = 12.63856$$

$$C^+ = 7.31877 - (4)(5)(0.78991) = -8.47943$$

The predicted values of vapor pressure using these coefficients along with the experimental values from Appendix A are below.

T (K)	T_r	P^{vp} (bar) Pred.	P^{vp} (bar) Exp.	$\frac{P_{pred} - P_{exp}}{P_{exp}} \times 100$
350	0.5670	0.1408	0.1487	−5.30
450	0.7290	2.7276	2.694	1.25

5.7.3 Lee and Kesler Method

The Lee and Kesler (1975) method for predicting vapor pressures is a corresponding states approach where the reduced vapor pressure (P_r^{vp}) is given by

$$\ln P_r^{vp} = f^{(0)} + \omega_{lk} f^{(1)} \tag{5.7-16}$$

where

$$f^{(0)} = 5.92714 - \frac{6.09648}{T_r} - 1.28862 \ln T_r + 0.169347 T_r^6 \tag{5.7-17}$$

$$f^{(1)} = 15.2518 - \frac{15.6875}{T_r} - 13.4721 \ln T_r + 0.43577 T_r^6 \tag{5.7-18}$$

$$\omega_{lk} = \frac{\ln P_{atm_r} - 5.92714 + \dfrac{6.09648}{T_{b_r}} + 1.28862 \ln T_{b_r} - 0.169347 T_{b_r}^6}{15.2518 - \dfrac{15.6875}{T_{b_r}} - 13.4721 \ln T_{b_r} + 0.43577 T_{b_r}^6} \tag{5.7-19}$$

Here, $T_{b_r} = \frac{T_b}{T_c}$, $P_{atm_r} = \frac{1.01325 \text{ bar}}{P_c}$, and ω_{lk} is the acentric factor for the Lee and Kesler method which would be obtained from the equation indicated and not from other sources to ensure consistency among the properties. By combining Eqs. (5.7-16)–(5.7-19) in the form suggested by Eq. (5.7-3), the following coefficients for the Lee and Kelser vapor pressure correlation using the Riedel equation are obtained.

$$A^+ = 5.92714 + 1.28862 \ln T_c + 13.4721\,\omega_{lk} \ln T_c + 15.2518\,\omega_{lk} \tag{5.7-20}$$

$$B^+ = -\left(-6.09648 - 15.6875\,\omega_{lk}\right) T_c \tag{5.7-21}$$

$$C^+ = -1.28862 - 13.4721\,\omega_{lk} \tag{5.7-22}$$

$$D^+ = \left(0.169347 + 0.43577\,\omega_{lk}\right) T_c^{-6} \tag{5.7-23}$$

The units of T_c in these equations should be on the absolute scale, and the resulting correlation should be evaluated using temperatures in these same units.

Example 5.7

Estimate the vapor pressure of ethylbenzene at 350 and 450 K using the method of Lee and Kesler (1975). For this compound, the correlation in Appendix A comes from experimental data and gives values of 0.1487 and 2.6935 bar at 350 and 450 K, respectively.

Solution. For ethylbenzene from Appendix A, T_b = 409.31 K, T_c = 617.26 K, and P_c = 36.16 bar. Thus, T_{b_r} = 409.31/617.26 = 0.6631 and P_{atm_r} = 1.01325/36.16 = 0.02802. With these values, the Lee-Kesler acentric factor is

$$\omega_{lk} = [\ln(0.02802) - 5.92714 + 6.09648/0.6631 + 1.28862\ln(0.6631) - (0.169347)(0.6631)^6]/[15.2518 - 15.6875/0.6631 - 13.4721\ln(0.6631) + 0.43577(0.6631)^6] = 0.30056$$

and Eqs. (5.7-20)–(5.7-23) yield the following coefficients for Eq. (5.7-3).

$$A^+ = 5.92714 + 1.28862\ln(617.26) + (13.4721)(0.30056)\ln(617.26) + (15.2518)(0.30056) = 44.080$$

$$B^+ = -[(-6.09648) - (15.6875)(0.30056)]617.26 = 6673.5$$

$$C^+ = -1.28862 - (13.4721)(0.30056) = -5.3378$$

$$D^+ = [0.169347 + (0.43577)(0.30056)]617.26^{-6} = 5.4297 \times 10^{-18}$$

The predicted values of vapor pressure using these coefficients along with the experimental values from Appendix A are below.

T (K)	T_r	P^{vp} (bar) Pred.	P^{vp} (bar) Exp.	$\frac{P_{pred} - P_{exp}}{P_{exp}} \times 100$
350	0.5670	0.1452	0.1487	–2.36
450	0.7290	2.721	2.694	1.01

5.8 EXTRAPOLATION AND PREDICTION OF ENTHALPY OF VAPORIZATION OF PURE COMPOUNDS AND RECOMMENDATIONS

As mentioned in Section 5.2, enthalpy of vaporization measurements are difficult, and the amount of data in the literature is relatively small compared to that of other properties. Even when experimental values for $\Delta_v H$ are available, they are usually only found at one or a few temperatures, so extrapolation is often needed. Both prediction and extrapolation are accomplished using Eq. (5.3-4). The following is the recommended procedure for extrapolation, prediction, and correlation of $\Delta_v H$.

1. Determine the vapor pressure and liquid density correlations for the compound using experimental data or a suitable prediction method and fitting to an appropriate correlating equation as discussed in this chapter.
2. Select an appropriate equation of state (see Chap. 6) to represent the vapor volume.
3. Determine values for $\Delta_v H$ using the quantities from the previous two steps and Eq. (5.3-4) at several temperatures from T_m to T_c.
4. Fit the values from Step 3, along with any reliable data for $\Delta_v H$, to Eq. (5.5-15).

This process produces a correlation for $\Delta_v H$ with average errors <10% which is thermodynamically consistent with P^{vp}. This accuracy is possible even with predicted vapor pressures because the slopes produced by the vapor pressure prediction methods discussed in this chapter are generally accurate even if the estimated absolute values are less reliable.

Previous editions of this book list several methods to predict or extrapolate enthalpy of vaporization values. These approaches should be used with caution as they do not ensure thermodynamic consistency between $\Delta_v H$ and P^{vp} and are usually less accurate than using the vapor pressure correlation with the Clapeyron relationship [Eq. (5.3-4)]. Both NIST TRC and DIPPR (Wilding et al. 2017) follow the 4-step procedure outlined above when reporting heats of vaporization.

5.9 PREDICTION OF LIQUID HEAT CAPACITY

5.9.1 Group-contribution Methods for Liquid Heat Capacity

Several different group-contribution methods have been proposed to estimate liquid heat capacities. In some of these, the assumption is made that various groups in a molecule contribute a definite value to the total molar heat capacity that is independent of other groups present. Such methods include those of Chueh and Swanson (1973a, 1973b) and Missenard (1965) which are described in the 4th edition of this book.

More recently, methods have been presented that account for differing contributions depending on what a particular atom is bonded to in the same way as Benson's method described in Chap. 4. Domalski and Hearing (1993) present such a method to estimate liquid heat capacities at 298.15 K. The method employs over 600 groups and energy corrections and covers 1512 compounds. Another Benson-type method, that of Růžička and Domalski (1993a, 1993b), is applicable across a range of temperatures and is described below.

5.9.1.1 Method of Růžička and Domalski (1993)

Růžička and Domalski (1993a, 1993b) developed a second-order group-contribution method to predict liquid heat capacity as a function of temperature from T_m to T_b. The method has been amended by Zábranský and Růžička (2004, 2005). The equation to calculate C_p^{liq} at temperature T is

$$\frac{C_p^{liq}(T)}{R} = A + B\frac{T}{100} + D\left(\frac{T}{100}\right)^2 \tag{5.9-1}$$

$$A = \sum_i N_i\left(\Delta_{C_p^{liq}}^{A}\right)_i \qquad B = \sum_i N_i\left(\Delta_{C_p^{liq}}^{B}\right)_i \qquad D = \sum_i N_i\left(\Delta_{C_p^{liq}}^{D}\right)_i \tag{5.9-2}$$

where R is the gas constant; T is in units of Kelvin; A, B and D are the coefficieints of the temperature-dependent terms; $\left(\Delta_{C_p^{liq}}^{A}\right)_i$, $\left(\Delta_{C_p^{liq}}^{B}\right)_i$, and $\left(\Delta_{C_p^{liq}}^{D}\right)_i$ are the values of group i contributing to the coefficients; and N_i is the number of times group i is found in the compound.

The file CH05RuzickaDomalskiCp.xlsx on the PGLed6 website lists the values for 114 different groups and 36 different ring strain corrections (rsc) for the method. Twenty-one more groups can be accommodated with the group equivalency table, Table 5.7. Table 5.8 contains example groups for the technique for the compound 1,3-cyclohexadiene. See Chap. 4 (or Růžička and Domalski 1993a, 1993b) for discussion and examples of group assignments in a molecule.

TABLE 5.7 Equivalent groups for the Růžička-Domalski (1993a, 1993b) method to predict liquid heat capacity. The symbol ≡ is placed between equivalent groups.

C—(3H,C) ≡ C—(3H,=C) ≡ C—(3H,Ct) ≡ C—(3H,Cb)
C—(2H,C,Ct) ≡ C—(2H,C,=C)
Cb—(Ct) ≡ Cb—(=C)
=C—(H,Cb) ≡ =C—(H, =C)
=C—(C,Cb) ≡ =C—(C, =C)
C—(3H,C) ≡ C—(3H,N) ≡ C—(3H,O) ≡ C—(3H,CO) ≡ C—(3H,S)
N—(2H,Cb) ≡ N—(2H,C)
S—(H,Cb) ≡ S—(H,C)
O—(H,Cb) (diol) ≡ O—(H,C) (diol)
CO—(H,Cb) ≡ CO—(H, =C)
C—(2H, =C,Cl) ≡C—(2H,C,Cl)
C—(2H,Cb,N) ≡ C—(2H,C,N)
N—(C,2Cb) ≡ N—(3C)
C—(2H, =C,O) ≡ C—(2H,Cb,O)
S—(Cb,S) ≡ S—(2C)
S—(2Cb) ≡ S—(2C)

TABLE 5.8 Group-contribution parameters for the Růžička-Domalski (1993a, 1993b) method for 1,3-cyclohexadiene. Taken from the file CH05RuzickaDomalskiCp.xlsx on the PGLed6 website.

Group	$\Delta^A_{C_p^{\text{liq}}}$	$\Delta^B_{C_p^{\text{liq}}}$	$\Delta^D_{C_p^{\text{liq}}}$	T range (K)
Cd—(H)(C)	4.0749	−1.0735	0.21413	90–355
Cd—(H)(Cd)	3.6968	−1.6037	0.55022	130–305
C—(H)2(C)(Cd)	2.0268	0.20137	0.11624	90–355
cyclohexadiene rsc	−8.9683	6.4959	−1.5272	170–300

Example 5.8

Estimate the liquid heat capacity of 1,3-cyclohexadiene at 300 K by using the Růžička-Domalski group-contribution method. The recommended value given by Zábranský et al. (1996) is 142 J/(mol·K).

Solution. Since 300 K is less than the boiling point temperature of 353.49 K (Daubert et al., 1997), the Rùzicka-Domalski method can be used. The six groups of 1,3-cyclohexadiene are two each of: Cd—(H)(C), Cd—(H)(Cd), and C—(H)2(C)(Cd). There is also a cyclohexadiene rsc. Using the group values and ring strain correction listed in Table 5.8 gives the following coefficients according to Eq. (5.9-2).

$$A = 2(4.0749) + 2(3.6968) + 2(2.0268) - 8.9683 = 10.6287$$
$$B = 2(-1.0735) + 2(-1.6037) + 2(0.20137) + 6.4959 = 1.54424$$
$$D = 2(0.21413) + 2(0.55022) + 2(0.11624) - 1.5272 = 0.23398$$

When a value of 300 is used for T in Eq. (5.9-1), the result is $C_p^{\text{liq}} = 144.4$ J/(mol·K) for an error of 1.7%.

5.9.2 Corresponding States Methods (CSP) for Liquid Heat Capacity

Several CSP methods for liquid C_p estimation have been developed using the residual C_p^{r} of Eq. (6.4-22). One such equation is

$$\frac{C_p^{\text{r}}}{R} = \frac{C_p - C_p^{\circ}}{R} = 1.586 + \frac{0.49}{1 - T_{\text{r}}} + \omega\left[4.2775 + \frac{6.3(1 - T_{\text{r}})^{\frac{1}{3}}}{T_{\text{r}}} + \frac{0.4355}{1 - T_{\text{r}}}\right] \tag{5.9-3}$$

where $C_p = C_p^{\text{liq}}$ for liquid heat capacity, C_p° is the ideal gas heat capacity of Chap. 4, R is the gas constant, T_{r} is reduced temperature, and ω is the acentric factor. Equation (5.9-3) is similar to one given by Bondi (1968), but the first two constants were refitted to more accurately describe liquid argon behavior than Bondi's form as described in the 5th edition of this book.

If the substance follows CSP behavior, C_p^{liq}, C_σ^{liq}, and $C_{\text{sat}}^{\text{liq}}$ can also be related to each other by CSP relations or the EoS quantities of Eq. (5.4-4).

$$\frac{C_p - C_\sigma}{R} = \exp(20.1T_{\text{r}} - 17.9) \tag{5.9-4}$$

$$\frac{C_\sigma - C_{\text{sat}}}{R} = \exp(8.655T_{\text{r}} - 8.385) \tag{5.9-5}$$

Equations (5.9-4) and (5.9-5) are valid for $T_{\text{r}} < 0.99$. Below $T_{\text{r}} \sim 0.8$, C_p^{liq}, C_σ^{liq}, and $C_{\text{sat}}^{\text{liq}}$ may be considered to have the same value.

Example 5.9

Estimate the liquid heat capacity of *cis*-2-butene at 350 K using the corresponding states method of Eq. (5.9-3). The value from the correlations in Appendix A is 150.2 J/(mol·K).

Solution. From Appendix A, $T_{\text{c}} = 435.5$ K and $\omega = 0.201877$. The ideal gas heat capacity constants from Appendix A give $C_p^{\circ} = 91.30$ J/(mol·K). The reduced temperature, $T_{\text{r}} = 350/435.5 = 0.8037$. Equation (5.9-3) gives

$$\frac{C_p^{\text{liq}} - C_p^{\circ}}{R} = 1.586 + \frac{0.49}{1 - 0.8037} + 0.201877\left[4.2775 + \frac{6.3(1 - 0.8037)^{\frac{1}{3}}}{0.8037} + \frac{0.4355}{1 - 0.8037}\right] = 6.314$$

With this value for the residual heat capacity, the estimate of the liquid heat capacity becomes

$$C_p^{\text{liq}} = 91.30 + (8.3145)(6.314) = 143.8 \text{ J/(mol·K)}$$

for an error of $\dfrac{143.8 - 150.2}{150.2} \times 100 = -4.3\%$.

5.10 DISCUSSION AND RECOMMENDATIONS FOR VAPOR-PRESSURE, ENTHALPY OF VAPORIZATION, AND LIQUID HEAT CAPACITY ESTIMATION AND CORRELATION

Historically, correlating or estimating values for P^{vp}, $\Delta_v H$, and C_p^{liq} has been done independent of the other properties. For example, vapor pressure data have frequently been fit to the Antoine equation without regard to how the resulting correlation predicted the enthalpy of vaporization or the heat capacity of the compound. Within this context, any of the correlations presented in this chapter can adequately reproduce values for the property within the temperature range of the data used in the regression. However, the accuracy of the underlying data may be questionable if care has not been taken to ensure the correlations yield thermodynamic consistency. Towards this end, the Antione equation does not have the flexibility needed to reproduce vapor pressures from the melting point to the critical point and either the Wagner or Modified Riedel equations are recommended.

The estimation methods presented in this chapter require the critical properties of a compound and the normal boiling point. Thus, the first task is to obtain values for these properties. When such are not known from experiment, they should be estimated using the approaches outlined in Chap. 3 prior to using the prediction methods in this chapter to obtain the correlations for P^{vp}, $\Delta_v H$, and C_p^{liq}.

Some methods found in the literature for P^{vp} do not require the critical constants. For example, Li et al. (1994) developed a combination group-contribution, corresponding states method that requires only the normal boiling point and the chemical structure of the compound. The authors claim more accurate results than those obtained with the Riedel method, but this approach could not be tested for this book. The group-contribution method of Tu (1994) doesn't require any other properties. However, testing has found it to be far less reliable (AARD > 50%) than the methods presented in this chapter. Also, because it does not use the normal boiling point as an input, the resulting equation will likely not be consistent with available T_b values which are often found in the literature even when vapor pressures at other conditions are not. DIPPR has used the methods of Maxwell and Bonnel (1955) and Othmer and Yu (1968) in the past, and still uses them for comparison purposes, but currently relies on the method of Hogge et al. (2017) followed by those of Riedel (1954) and Lee and Kesler (1975). Other correlations that have been published may be found in Smith et al. (1976), Edwards et al. (1981), Macknick et al. (1978), Bloomer (1990), Xiang and Tan (1994), Campanella (1995), and Ledanois et al. (1997).

5.10.1 Recommendations

As discussed extensively in this chapter, P^{vp}, $\Delta_v H$, and C_p^{liq} are related through rigorous thermodynamic relationships. As such, these properties should be considered simultaneously to ensure consistency and increase accuracy. Before doing this, the correlation for liquid density should be developed so that it can be used in the process. (See Section 5.2). Then, if experimental vapor pressure and liquid heat capacity data are available, the vapor pressure should be fit to an appropriate correlation (See Section 5.5) so that it also replicates the enthalpy of vaporization correlation through the Clapeyron equation [Eq. (5.3-1)], and the liquid heat capacity data through Eq. (5.6-4). Any reliable experimental data for enthalpy of vaporization that is available can be included with the Clapeyron values when regressing the correlation for $\Delta_v H$ (see Section 5.8).

Constants in Appendix A should be used first if available for the compound of interest as the correlations presented therein have been evaluated by DIPPR (Wilding et al. 2017) using the methods in this chapter. The Antoine equation, found in many textbooks and handbooks, should not be used for vapor pressures at temperatures outside the range of the data used in the fitting of the coefficients and never at temperatures approaching the melting point of the compound. The Riedel, Modified Riedel, and Wagner equations may be used to extrapolate from T_m to T_c if the fitting was done to satisfy the criteria listed in Section 5.6 and the low temperature vapor pressure behavior was set using multi-property optimization (Example 5.4) or the method of Hogge et al. (2017).

A test database was created from the May 2021 release of the DIPPR database to evaluate the prediction methods of P^{vp}. Only liquid vapor pressure values from experiment that were reliable were included in this test database. This gave nearly 374,000 total points (N_{pt}) from almost 2000 unique compounds (N_c) spanning temperature ranges from T_m to T_c. Four prediction techniques were tested: Riedel (abbreviated RI below), Hogge et al. (2017) (HO), Vetere (2006) (VE), and Lee-Kesler (1975) (LK). The methods of Riedel and Lee-Kesler are more general than those of Hogge et al. and Vetere and could thus be tested against a greater number of compounds. Compounds used to test the Vetere and Hogge et al. techniques are those in the test database that are also members of one of the families found in Table 5.5 and Table 5.6, respectively.

To identify if a method was better at one class of compounds than another, each chemical in the test database was assigned a *class* according to the scheme described at the end of Sec. 5.2 where the prediction methods for ρ_s were evaluated. Results were quantified using the average absolute relative deviation (AARD) defined by Eq. (5.2-14).

Table 5.9 contains the absolute average deviation [see Eq. (5.2-14)] of each method according to compound class. Also listed are the number of points tested in each class (N_{pt}) and number of compounds (N_c). The Vetere method performs the worst of all the methods tested. Riedel and Lee-Kesler methods perform about the same across all classes with the latter being marginally better for the heavy class of compounds. The method of Hogge et al. performed slightly worse than the Riedel and Lee-Kelser techniques for the normal and polar classes but slightly better for heavy and associating compounds. However, it has the advantage of variable E values over these other approaches. Overall, the Riedel, Lee-Kesler, and Hogge et al. methods can be recommended. As discussed above, selecting a method should be done to ensure thermodynamic consistency with the liquid heat capacity, enthalpy of vaporization, and liquid density of the compound.

A test database was created from the May 2021 release of the DIPPR database to evaluate the prediction methods for C_p^{liq}. Only liquid heat capacity values from experiment that were reliable were included in this test database. This gave almost 15200 total points from over 870 unique compounds spanning temperature ranges from T_m to T_c. Two prediction techniques were tested: the group-contribution method of Růžička and Domalski (1993) (abbreviated RD) and the corresponding states method given by Eq. (5.9-3) (abbreviated CS). Compounds were organized into the same classes as those used to evaluate ρ_s to determine how each method worked for different types of compounds (see the end of Sec. 5.2). Results are quantified using the average absolute relative deviation given by Eq. (5.2-14).

Table 5.10 contains the AARD [see Eq. (5.2-14)] of each method according to compound class. Also listed are the number of points tested in each class (N_{pt}) and number of compounds (N_c). The two methods perform

TABLE 5.9 Performance of prediction techniques for p^{vp} against the test database.

	Compound Class											
	Normal			Heavy			Polar			Associating		
Method	N_{pt}	N_c	AARD	N_{pt}	N_c	AARD	N_{pt}	N_c	AARD	N_{pt}	N_c	AARD
RI	33790	873	3.3%	2059	93	14.1%	11786	396	9.8%	16241	634	13.8%
LK	33790	873	3.6%	2059	93	12.8%	11786	396	9.7%	16241	634	13.3%
HO	31874	783	6.2%	2091	88	10.3%	10061	343	12.3%	6676	337	12.7%
VE	25118	605	4.9%	1941	79	19.5%	9076	284	9.8%	9291	325	15.6%

TABLE 5.10 Performance of prediction techniques for C_p^{liq} against the test database.

	Compound Class											
	Normal			Heavy			Polar			Associating		
Method	N_{pt}	N_c	AARD	N_{pt}	N_c	AARD	N_{pt}	N_c	AARD	N_{pt}	N_c	AARD
RD	6328	348	3.3%	447	29	5.6%	1358	109	3.7%	4086	218	8.5%
CS	7750	429	5.4%	685	40	5.3%	2083	150	5.9%	4642	252	12.2%

approximately the same. Růžička and Domalski (as amended by Zábranský and Růžička, see above) have better performance for three of the four classes, but the corresponding states method can be used to predict more compounds. Also, as recommended above, selecting a method should be done to ensure thermodynamic consistency with the liquid vapor pressure, enthalpy of vaporization, and liquid density of the compound.

5.11 ENTHALPY OF MELTING

The enthalpy of melting ($\Delta_m H$) is the change in enthalpy that occurs when a substance transitions from the solid to the liquid phase. Technically, the enthalpy of melting is a function of temperature, but because the relationship is very weak, the property is commonly treated as a constant and referenced as the value at the normal melting temperature T_m—a practice that is followed here. The literature commonly uses the terms "enthalpy of fusion" and "enthalpy of melting" interchangeably. The reader should understand that regardless of the sign on the value listed in a table in this book or other sources, the process of moving from a solid to a liquid phase (melting) increases the enthalpy of the system yielding a positive *change*, while that proceeding from the liquid to the solid phase decreases then enthalpy of the system yielding a negative *change*. References for literature values are listed in Tamir et al. (1983). Domalski and Hearing (1996), Chickos et al. (1999, 2003, 2009), and Acree and Chickos (2016, 2017) provide extensive tabulations of experimental values. The enthalpy of melting is related to the entropy change on melting $\Delta_m S$ and the melting point temperature by

$$\Delta_m H = T_m \Delta_m S \qquad (5.11\text{-}1)$$

Reliable methods to estimate the enthalpy or entropy of melting at the melting point have not been developed. However, as first suggested by Bondi (1963), methods have been developed to estimate the total change in entropy, ΔS_{tot}, due to phase changes when a substance goes from the solid state at 0 K to a liquid at its melting point. For substances that do not have solid-solid transitions, ΔS_{tot} and $\Delta_m S$ are the same. For these substances, multiple methods can be used to estimate ΔS_{tot} which can then be used with Eq. (5.11-1) to predict $\Delta_m H$. Four methods using this approach, two by Bondi and one by Chickos et al. (1990, 1991, 1998, 1999, 2003, 2009) are described below.

For substances that demonstrate solid-solid transitions, ΔS_{tot} can be much greater than $\Delta_m S$. This can be seen in Table 5.11 which lists ΔS_{tot} and $\Delta_m S$ for 44 hydrocarbons. For 14 of the 44 hydrocarbons listed, ΔS_{tot} and $\Delta_m S$ are different due to solid-solid transitions below the melting point. This difference, ΔS_{sol}, is also tabulated. $\Delta_m S$ can be much less than ΔS_{tot} when a solid-solid transitions exists. For example, 2,2-dimethylbutane has two solid-solid transitions, one at 127 K for which ΔS = 42.66 J/(mol·K) and one at 141 K for which ΔS = 2.03 J/(mol·K). For this compound, ΔS_{tot} = 47.9 J/(mol·K) while $\Delta_m S$ = 3.3 J/(mol·K). Also notice that Eq. (5.11-1) is satisfied regardless of whether a solid-solid transition occurs.

For substances with solid-solid transitions, no reliable method exists for the estimation of $\Delta_m S$ because there is no way to predict whether solid-solid transitions occur. Still the methods to estimate ΔS_{tot} represent a significant development.

5.11.1 Bondi Methods (1968)

Bondi (1968) gives several approaches to calculate ΔS_{tot}. Two of these used by DIPPR are presented. The first is a family approach where ΔS_{tot} is calculated from the number of carbons (N_C) in the compound according to the equations in Table 5.12. Once a value of ΔS_{tot} is determined, Eq. (5.11-1) may be used to calculate $\Delta_m H$.

The second Bondi approach is for compounds that are not hydrocarbons nor members of one of the families listed in Table 5.12 and uses the idea of a homomorph. A homomorph of a compound is created by replacing all heavy atoms in the molecule with carbons. For example, the homomorph of urea is 2-methylpropene and that of diphenylamine is diphenylmethane. The technique requires experimental values of the enthalpy of melting

TABLE 5.11 Entropies and enthalpies of melting for various hydrocarbons. The experimental (Exp.) values are from Domalski and Hearing (1996). T_m values are from Dreisbach (1995, 1959). Predicted (Pred.) is from Eq. (5.11-13). The difference is Predicted – Experimental.

	Exp. $\Delta_m S$	Exp. ΔS_{tot}	Exp. ΔS_{sol}	Pred. ΔS_{tot}	diff	T_m	Exp. $\Delta_m H$
	J/(mol·K)	J/(mol·K)	J/(mol·K)	J/(mol·K)	J/(mol·K)	K	J/mol
Methane	10.4	14.9	4.5	11.7	–3.2	90.7	943
Ethane	6.5	31.9	25.4	25.1	–6.8	89.9	584
Propane	41.2	41.2		44.2	3	85.5	3523
n-Butane	34.6	53.8	19.2	58.7	4.9	134.8	4664
Isobutane	39.9	39.9		40.9	1	113.6	4533
n-Pentane	58.6	58.6		67.4	8.8	143.4	8403
Isopentane	45.4	45.4		58.7	13.3	113.3	5144
Neopentane	12.4	30.9	18.5	29.3	–1.6	256.6	3182
n-Hexane	71	71		76.1	5.1	177.8	12624
2-Methylpentane	52.4	52.4		67.4	15	119.5	6262
2,2-Dimethylbutane	3.3	47.9	44.6	50	2.1	173.3	572
2,3-Dimethylbutane	5.5	53	47.5	58.7	5.7	144.6	795
n-Heptane	78	78		84.8	6.8	182.6	14243
2-Methylhexane	59.3	59.3		76.1	16.8	154.9	9186
3-Ethylpentane	61.8	61.8		76.1	14.3	154.6	9554
2,2-Dimethylpentane	39	39		58.7	19.7	149.4	5827
2,4-Dimethylpentane	44.5	44.5		67.4	22.9	154	6853
3,3-Dimethylpentane	49.3	55.3	6	67.4	12.1	138.7	6838
2,2,3-Trimethylbutane	8.9	28.5	19.6	50	21.5	248.3	2210
n-Octane	95.8	95.8		93.5	–2.3	216.4	20731
2-Methylheptane	72.6	72.6		84.8	12.2	164.2	11921
n-Nonane	70.4	99.3	28.9	102.2	2.9	219.7	15467
n-Decane	117.9	117.9		111	–6.9	243.5	28709
n-Dodecane	139.7	139.7		128.4	–11.3	263.6	36825
n-Octadecane	203	203		180.6	–22.4	301.3	61164
n-Nonadecane	153	199	46	189.3	–9.7	305	46665
n-tetratriacontane	231.2	371	139.8	319.9	–51.1	346	79995
Benzene	35.4	35.4		29.3	–6.1	278.7	9866
Toluene	37.2	37.2		44.2	7	178	6622
Ethylbenzene	51.4	51.4		54.4	3	178.2	9159
o-Xylene	54.9	54.9		44.2	–10.7	248	3615
m-Xylene	51.4	51.4		44.2	–7.2	225.3	11580
p-Xylene	59.8	59.8		38.5	–21.3	286.4	17127
n-Propylbenzene	53.4	53.4		63.1	9.7	173.7	9276
Isopropylbenzene	41.4	41.4		54.4	13	177.1	7332
1,2,3-Trimethylbenzene	33	41.8	8.8	44.2	2.4	247.1	8154
1,2,4-Trimethylbenzene	57.5	57.5		50	–7.5	227	13053
1,3,5-Trimethylbenzene	41	7 41.7		35.1	–6.6	228.4	9524

(Continued)

TABLE 5.11 Entropies and enthalpies of melting for various hydrocarbons. The experimental (Exp.) values are from Domalski and Hearing (1996). T_m values are from Dreisbach (1995, 1959). Predicted (Pred.) is from Eq. (5.11-13). The difference is Predicted – Experimental. (*Continued*)

	Exp. $\Delta_m S$	Exp. ΔS_{tot}	Exp. ΔS_{sol}	Pred. ΔS_{tot}	diff	T_m	Exp. $\Delta_m H$
	J/(mol·K)	J/(mol·K)	J/(mol·K)	J/(mol·K)	J/(mol·K)	K	J/mol
Cyclohexane	9.5	45.5	36	35.1	−10.4	279.6	2656
Methylcyclohexane	46.1	46.1		50	3.9	146.6	6758
Ethylcyclohexane	51.5	51.5		54.4	2.9	161.8	8333
1,1-Dimethylcyclohexane	8.4	47.4	39	50	2.6	239.7	2013
1,cis-2-Dimethylcyclohexane	7.4	55.2	47.8	50	−5.2	223.1	1651
1,trans-2-Dimethylcyclohexane	56.7	56.7		50	−6.7	185	10490

TABLE 5.12 Bondi (1968) equations to calculate ΔS_{tot} for select families to predict $\Delta_m H$ using Eq. (5.11-1).

Family	$\dfrac{\Delta S_{tot}}{R}$
n-paraffins (N_C = even)	$0.80 + 1.33N_C$
n-paraffins (N_C = odd)	$1.10 + 1.18N_C$
2-methyl-n-alkanes	$-1.24 + 1.2N_C + \dfrac{20}{N_C^2}$
2,2-dimethyl-n-alkanes	$-6.26 + 1.33N_C + \dfrac{83}{N_C^2}$
n-alkylcyclopentanes	$-5.1 + 1.30N_C + \dfrac{131}{N_C^2}$
n-alkylcyclohexanes	$-6.3 + 1.45N_C + \dfrac{56}{N_C^2}$
n-alkylbenzenes	$-5.6 + 1.18N_C + \dfrac{100}{N_C^2}$
n-alkanethiols (N_C = even)	$3.3 + 1.2N_C$
n-alkanethiols (N_C = odd)	$3.90 + 1.33N_C$
n-alkylbromide (N_C = odd)	$2.4 + 1.38N_C + \dfrac{0.54}{N_C^2}$
n-alkanoic acid (N_C = even)	$-2.56 + 1.33N_C + \dfrac{36}{N_C^2}$
n-alkanoic acid (N_C = odd, > 5)	$-2.7 + 1.25N_C$
n-alkanoates (N_C = even)	$-6.6 + 1.35N_C$

and melting temperature of the homomorph ($\Delta_m H_{homo}$ and $T_{m,homo}$, respectively), the melting temperature of the compound of interest (T_m), and corrections for polar functional groups as found in Table 5.13. The enthalpy of melting ($\Delta_m H$) is predicted from these values using the following equations.

$$\Delta_m S_{homo} = \frac{\Delta_m H_{homo}}{T_{m,homo}} \tag{5.11-2}$$

$$\delta_{bondi} = \sum_i N_i \left(\Delta_{S_{tot}}\right)_i \tag{5.11-3}$$

$$\Delta S_{tot} = \Delta_m S_{homo} + \delta_{bondi} \tag{5.11-4}$$

$$\Delta_m H = T_m \Delta S_{tot} \tag{5.11-5}$$

Here, N_i is the number of groups i in the chemical, $\left(\Delta_{S_{tot}}\right)_i$ is the value of group i, and T_m is the meting temperature of the compound whose enthalpy of melting is being estimated. Eq. (5.11-2) calculates the entropy of melting for the homomorph using the relationship in Eq. (5.11-1). Eq. (5.11-3) accounts for difference in

TABLE 5.13 Group-contributions for the Bondi (1968) method for $\Delta_m H$ to use in Eq. (5.11-3). Notice that the group values are divided by R in the table.

Polar Group	Functional Group	$\frac{\Delta_{S_{tot}}}{R}$
Aliphatic ether	—O—	–0.5
Aliphatic thioether (*n*-alkyl)	—S—	–1.0
Aliphatic thioether (*sec*-alkyl)	—S—	0.85
Aliphatic thioether (*t*-alkyl)	—S—	1.0
Aliphatic and diaromatic ketone	>C═O	–1.5
Quinone carbonyl	>C═O	–0.7
Aliphatic aldehyde	—CH═O	0.8
Aliphatic ester	—O(C═O)—	0
Aliphatic alcohol (one —OH per molecule)	—OH	–1.2
Aliphatic primary amine	$—NH_2$	0.9
Aromatic amine (unhindered)	$—NH_2$	0.3
Aliphatic and aromatic secondary amine	>NH	–0.9
Monocyano alkane	—C≡N	0.7
Dicyano alkane	—C≡N	–1.8
Alkane amide	$—(C{=}O)NH_2$	–0.5
Alkane thiols (N_C = even)	—SH	1.0
Alkane thiols (N_C = odd)	—SH	1.8
Secondary alkane thiols	—SH	0.25
Teritary alkane thiols	—SH	1.2
Dialkyl disulfide	—SS—	–1.2
Dialkyl or diaryl sulfone	$—SO_2—$	3.3
Heterocyclic sulfone	$—SO_2—$	–1.4

polarity between the homomorph and the compound under consideration. Eq. (5.11-4) is the estimated entropy of melting of the compound under consideration, and Eq. (5.11-5) is the estimate of the enthalpy of melting from the estimated entropy of melting and the melting temperature as per Eq. (5.11-1). Example 5.15 demonstrates both Bondi methods described above.

Example 5.10

Estimate the enthalpy of melting at the melting point of *n*-propylbenzene and propanal using an appropriate method from Bondi (1968). The values in Appendix A for *n*-propylbenzene and propanal are 9.268 kJ/mol and 8.590 kJ/mol, respectively.

Solution

Ethylbenzene

From Appendix A, the melting temperature of *n*-propylbenzene is 173.55 K. This compound is a member of one of the families of Table 5.12, so the ΔS_{tot} Bondi approach should be used over the homomorph technique. The molecule has 9 carbons so $N_C = 9$. Using the appropriate equation from Table 5.12 gives

$$\Delta S_{tot} = (8.31451)[-5.6 + (1.18)(9) + 100/9^2] = 52.004 \text{ J/(mol} \cdot \text{K)}$$

which through Eq. (5.11-1) yields

$$\Delta_m H = (52.004)(173.55) = 9.025 \text{ kJ/mol}$$

With the experimental value from Appendix A,

$$\text{Error} = 9.025 - 9.268 = -0.243 \text{ kJ/mol or } -2.62\%.$$

Propanal

From Appendix A, the melting temperature of propanal is 165.0 K. This compound is not a member of one of the families of Table 5.12, so the homomorph technique is required. The homomorph of propanal is 1-butene. From Appendix A, the melting temperature and enthalpy of melting of 1-butene are 87.8 K and 3.848 kJ/mol, respectively. As listed in Table 5.13, propanal has one aliphatic aldehyde polar group. With these values, Eqs. (5.11-2)–(5.11-5) give the following.

$$\Delta_m S_{homo} = 3848/87.8 = 43.827 \text{ J/(mol} \cdot \text{K)}$$

$$\delta_{bondi} = (8.31451)(0.8) = 6.652 \text{ J/(mol} \cdot \text{K)}$$

$$\Delta S_{tot} = 43.827 + 6.652 = 50.478 \text{ J/(mol} \cdot \text{K)}$$

$$\Delta_m H = (50.478)(165.0) = 8329 \text{ J/mol} = 8.329 \text{ kJ/mol}$$

With the experimental value from Appendix A,

$$\text{Error} = 8.329 - 8.590 = -0.261 \text{ kJ/mol or } -3.04\%.$$

5.11.2 Chickos et al. Method (1990, 1991, 1998, 1999, 2003, 2009)

Chickos et al. (1990, 1991, 1998, 1999, 2003, 2009) created a group-contribution approach to estimating ΔS_{tot} and have refined it multiple times as new experimental values have been reported in the literature. The most recent implementation (1999, 2003, 2009) of the method estimates the total entropy of melting as a combination of several contributions depending on whether nonaromatic rings and atoms other than C and H are present. For *acyclic and aromatic hydrocarbons* (aah), the total entropy of melting is given by:

$$\Delta S_{\text{tot,aah}} = \sum_{i \neq \text{CH}_2} N_i \left(\Delta_{\Delta S_{\text{tot,aah}}}\right)_i + \left(N_{\text{CH}_2\text{,non}} + N_{\text{CH}_2\text{,con}} c_{\text{CH}_2}\right)\left(\Delta_{\Delta S_{\text{tot,aah}}}\right)_{\text{CH}_2} \tag{5.11-6}$$

$$c_{\text{CH}_2} = \begin{cases} 1.31 & \text{if} \quad N_{\text{CH}_2\text{,con}} \geq \sum\limits_{i \neq \text{CH}_2} N_i \\ 1 & \text{otherwise} \end{cases} \tag{5.11-7}$$

In this equation, i runs over all *non-methylene* groups (all groups other than $>CH_2$), N_i is the number of times groups i is found in the molecule, $\left(\Delta_{\Delta S_{\text{tot,aah}}}\right)_i$ is the contribution of group i to the summation (see Table 5.14), $N_{\text{CH}_2\text{,non}}$ is the total number of *nonconsecutive* methylene groups, $N_{\text{CH}_2\text{,con}}$ is the total number of *consecutive* methylene groups, and c_{CH_2} is a coefficient applied to *consecutive* methylene groups. The coefficient takes on a value of 1.31 whenever the number of consecutive methylene groups is greater than or equal to the number of non-methylene groups ($N_{\text{CH}_2\text{,con}} \geq \sum_i N_i$) and is equal to 1 otherwise. The introduction of c_{CH_2} distinguishes long chain methylene groups from those found in short chains.

Correctly identifying methylene groups is important to accurately applying the method, and Chikos et al. (1999, 2003, 2009) provide many examples in the articles and supplementary information. A few are discussed

TABLE 5.14 Aliphatic and aromatic group-contributions for the hydrocarbon portion of the molecule for the Chickos (2009) method for ΔS_{tot}.

		$\Delta_{\Delta S_{\text{tot,aah}}}$ [a]		$c_{\text{C(X)}_k}$ [b]	
Description	**Formula**[f]	**Value [J/(mol·K)]**	**Label**	**Value**	**Label**
Primary sp^3 C	$—CH_3$	17.6	(A1)		
Secondary sp^3 C	$>CH_2$	7.1	(A2)	1.31	(B2)
Tertiary sp^3 C	$>CH—$	−16.4	(A3)	0.60	(B3)
Quaternary sp^3 C	$>C<$	−34.8	(A4)	0.66	(B4)
Secondary sp^2 C	$=CH_2$	17.3	(A5)		
Tertiary sp^2 C	$=CH—$	5.3	(A6)	0.75	(B6)
Quaternary sp^2 C	$=C<$	−10.7	(A7)		
Tertiary sp C	$\equiv CH$	17.5	(A8)		
Quaternary sp C	$\equiv C—$	−4.3	(A9)		
Tertiary benzenoid sp^2 C	$=CH—$	7.4	(A10)		
Quaternary benzenoid sp^2 C adjacent to an sp^3 atom[c]	$=C<$	−9.6	(A11)		
Quaternary benzenoid sp^2 C adjacent to an sp^2 atom[d]	$=C<$	−7.5	(A12)		
Internal benzenoid quaternary sp^2 C adjacent to an sp^2 atom[e]	$=C<$	−0.7	(A13)		

[a]Group values for Eq. (5.11-6); group (A2) is $\Delta_{\Delta S_{\text{tot,aah}_{\text{CH}_2}}}$ of Eq. (5.11-6)

[b]Coefficients used in Eq. (5.11-10) except for (B2) which is c_{CH_2} of Eqs. (5.11-6) for consecutive methylene groups for the criteria found in Eq. (5.11-7); unlisted values are equal to 1.0

[c]Quaternary benzenoid carbon adjacent to an sp^3 hybridized atom with no lone pair of electrons

[d]Quaternary benzenoid carbon adjacent to any sp^2 hybridized atom and to sp^3 hybridized atoms with non-bonding electrons except for internal quaternary carbon atoms (see Footnote e)

[e]Internal quaternary benzenoid carbon that is not at the periphery of a molecule; e.g. the six internal quaternary benzenoid carbon atoms of coronene

[f]Indicated bonds may be with all atoms other than H

for illustrative purposes. 1-Heptene contains 4 consecutive methylene groups and this number exceeds the sum of the other groups which is 3 (e.g. 1, $=CH_2$; 1, $=CH-$; 1, $-CH_3$). The coefficient, c_{CH_2}, for the 4 methylene groups in the molecule would be 1.31 [see Eq. (5.11-7)]. 3-Heptene has 2 consecutive methylene groups, 1 non-consecutive methylene group, 2 $=CH-$ groups, and 2 $-CH_3$ groups. For this compound, $N_{CH_2,con} = 2$ and the sum of the non-methylene groups is 4, so $c_{CH_2} = 1$ according to Eq. 5.11-7. For 3-decene, $N_{CH_2,con} = 5$, $N_{CH_2,non} = 1$, and $\sum_{i \neq CH_2} N_i = 4$. In this case, the number of consecutive methylene groups is greater than the number of non-methylene groups, so $c_{CH_2} = 1.31$. To emphasize, this coefficieint is not applied to the *nonconsecutive* methylene group in the molecule as outlined in Eq. (5.11-6).

For molecules with nonaromatic rings and polycyclic hydrocarbons, the contributions to the total entropy of melting is split into three contributions: the portion for the *parent* ring structure ($\Delta S_{tot,ring}$) where all the atoms in the ring are treated as *sp*^{3} hybridized that are bonded to other ring carbon or hydrogens, corrections to the parent ring structure for substition or non-*sp*^{3} orbitals ($\Delta S_{tot,corr}$), and the acyclic and aromatic hydrocarbons (aah) portion of the remaining parts of the molecule ($\Delta S_{tot,aah}$) which is calculated using Eqs. (5.11-6)–(5.11-7). The parent ring and ring correction contributions are determined using

$$\Delta S_{tot,ring} = 33.4N_{rings} + 3.7(N_{ring\ atoms} - 3N_{rings}) \tag{5.11-8}$$

$$\Delta S_{tot,corr} = \sum_j N_j \left(\Delta_{\Delta S_{tot,corr}}\right)_j \tag{5.11-9}$$

where N_{rings} is the number of rings in the molecule, $N_{ring\ atoms}$ is the number of ring atoms in the molecule, N_j is the number of ring correction groups of type j (see Table 5.15), and $\left(\Delta_{\Delta S_{tot,corr}}\right)_j$ is the contribution of ring correction group type j. Chickos et al. (1999) give 10,10,13,13-tetramethyl-1,5-cyclohexadecadiyne as an example of parsing a single ring molecule. The parent ring is hexadecane and its contribution is calculated using Eq. (5.11-8) with $N_{ring\ atoms} = 16$ and $N_{rings} = 1$ giving 33.4 + (3.7)(13). The corrections to the parent ring are 2 cyclic quaternary *sp*^{3} ($>C_c(R)_2$) groups, and 4 cyclic quaternary *sp* ($R\text{-}C_c\equiv$) groups used in Eq. (5.11-9). Finally, the aah contribution consists of 4 methyl groups ($-CH_3$) used in Eq. (5.11-6).

Norborene (bicyclo[2.2.1]hept-2-ene) is an example of a multiring compound. The number of rings (N_{rings}) is found by determing the minimum number of bonds that need to be broken to form a completely acyclic molecule. This is two in the case of norborene. The parent ring contribution is thus determined using $N_{rings} = 2$ and

TABLE 5.15 Corrections for cyclic hydrocarbon portions of the molecule for the Chickos and Acree (2009) method for ΔS_{tot}.

		$\Delta_{\Delta S_{tot,corr}}$ [a]		$c_{C(X)_k}$ [b]	
Description	**Formula**[c]	**Value [J(mol·K)]**	**Label**	**Value**	**Label**
Number of rings coefficient[d]	N_{rings}	33.4	(A14)		
Number of ring atoms[e]	$N_{ring\ atoms}$	3.7	(A15)		
Cyclic tertiary sp^3 carbon	$>CH-$	−14.7	(A16)		
Cyclic quaternary sp^3 carbon	$>C<$	−34.6	(A17)		
Cyclic tertiary sp^2 carbon	$=CH-$	−1.6	(A18)	1.92	(B18)
Cyclic quaternary sp^2 carbon	$=C<$	−12.3	(A19)		
Cyclic quaternary sp carbon	$=C=$; $\equiv C-$	−4.7	(A20)		

[a]Group values for Eq. (5.11-9)
[b]Coefficients used in Eq. (5.11-10); unlisted values are equal to 1.0
[c]Indicated bonds may be with all atoms other than H
[d]The coefficient on N_{rings} of Eq. (5.11-8) is label A14
[e]The coefficient on $N_{ring\ atoms}$ of Eq. (5.11-8) is label A15

$N_{\text{ring atoms}} = 7$ to give (33.4)(2) + (3.7)(1) using Eq. (5.11-8). The corrections to the parent ring are 2 cyclic tertiary sp^3 [$>C_cH(R)$] and 2 cyclic tertiary sp^2 ($=C_c(R)—$). There is no aah contribution for this molecule.

Estimations of the total entropy of melting for acyclic molecules with atoms other than C and H consist of three contributions: (1) The hydrocarbon portion calculated using Eqs. (5.11-6)–(5.11-9) as already described, (2) a portion for the carbon(s) bonded to the heteroatom (denoted X) functional groups ($\Delta S_{\text{tot,C(X)}}$), and (3) a portion for the contribution of the functional groups themselves ($\Delta S_{\text{tot,X}}$). (The term "functional group" is used by Chickos and Acree (2009) to mean non-carbon (heteroatom) groups such as those involving O, S, P, F, etc.) The last two contributions are determined by

$$\Delta S_{\text{tot,C(X)}} = \sum_k N_k c_{\text{C(X)}k} \left(\Delta_{\Delta S_{\text{tot,C(X)}}}\right)_k \tag{5.11-10}$$

$$\Delta S_{\text{tot, X}} = \sum_l N_l c_{\text{X}_m} \left(\Delta_{\Delta S_{\text{tot,X}}}\right)_l \tag{5.11-11}$$

where N_k is the number of C(X) groups of type k (see Table 5.14), $c_{\text{C(X)}_k}$ is the coefficient for C(X) group k, $\left(\Delta_{\Delta S_{\text{tot,C(X)}}}\right)_k$ is the contribution of C(X) group k, N_l is the number of functional groups l (see Table 5.17), c_{X_m} is the coefficient for functional group l, and $\left(\Delta_{\Delta S_{\text{tot,X}}}\right)_l$ is the contribution of functional group l. Also, m is the total number of functional groups in the molecule. Two examples of non-hydrocarbon molecules are given below. See Chickos et al (1999, 2002, 2009) for many more examples.

Consider 3-decanol as an example of a compound with a functional group. The hydrocarbon portion of the molecule is calculated using Eqs. (5.11-6) and (5.11-7). This molecule has 1 nonconsecutive methylene group ($N_{\text{CH}_2,\text{ non}} = 1$), 6 consecutive methylene groups ($N_{\text{CH}_2,\text{ con}} = 6$) and 4 non-methylene groups (2 $—CH_3$, 1 $>CH—$, 1 $—OH$). Because $6 > 4$, $c_{\text{CH}_2} = 1.31$ in Eq. (5.11-6). Only one functional (heteroatom) group is present in the molecule, so $m = 1$. The final entropy consists of the following group operations: 1(A2) + 6(A2)(B2) + 2(A1) + 1(A3) + 1(A30).

An example of a molecule with multiple functional groups is trichloroacetic acid. There are no carbon atoms in the molecule that are not connected to functional groups, so there is no aah contribution from Eq. (5.11-6). There are also no rings, so Eq. (5.11-8) does not come into play. The carbon connected to the three chlorines contributes to the total entropy through Eq. (5.11-10) and requires the group value $\left(\Delta_{\Delta S_{\text{tot,C(X)}}}\right)_k$ and the associated coefficient $c_{\text{C(X)}_k}$ from Table 5.14. The carboxyl group and the three chlorines are functions groups and are taken into account using Eq. (5.11-11) with $m = 4$. The group values $\left(\Delta_{\Delta S_{\text{tot,X}}}\right)_l$ and their coefficients c_{X_m} are obtained from Table 5.16. The final entropy consists of the following group operations: (A4)(B4) + 3(A22)(D22) + (A36)(D36).

TABLE 5.16 Contributions for the functional group (heteroatom) portions of the molecule for the Chickos and Acree (2009) method for ΔS_{tot}. Groups in this table are acyclic and are dependent on the substitution pattern and thus have non-unity c_{X_m} values.

		$\Delta_{\Delta S_{\text{tot,X}}}$ [a]		c_{X_m} [b]					
				$m = 2$		$m = 3$		$m > 3$	
Description	**Formula** [c]	**Value [J/(mol·K)]**	**Label**	**Value**	**Label**	**Value**	**Label**	**Value**	**Label**
Chlorine	$—Cl$	10.8	(A22)					1.5	(D22)
2-Fluorines on an acyclic sp^3 carbon	$>CF_2$	13.2	(A26)	1.06	(B26)	1.06	(C26)	1.15	(D26)[d]
Hydroxyl group	$—OH$	1.7	(A30)	10.4	(B30)	9.7	(C30)	13.1	(D30)
Carboxylic group	$—C(=O)OH$	13.4	(A36)	1.21	(B36)			2.25	(D36)

[a]Group values for Eq. (5.11-14)
[b]Coefficients used in Eq. (5.11-14); use the $m > 3$ column for unlisted values
[c]Indicated bonds may be with all atoms other than H
[d]Used in acyclic perfluorinated hydrocarbons

TABLE 5.17 Contributions for the functional group (heteroatom) portions of the molecule for the Chickos and Acree (2009) method for ΔS_{tot}. Groups in this table are acyclic but are not dependent on substitution patterns and thus have c_{X_m} values equal to 1.

		$\Delta_{\Delta S_{tot},X}$ [a,b]	
Description	Formula[c,d]	Value [J/(mol·K)]	Label
Bromine	—Br	17.5	(A21)
Fluorine on an sp^2 carbon	=CF—	19.5	(A23)
Fluorine on an aromatic carbon	=CF—	16.6	(A24)
3 Fluorines on an sp^3 carbon	—CF_3	13.2	(A25)
1 Fluorine on an sp^3 carbon	>CF—	12.7	(A27)
1 Fluorine on a ring carbon	—CHF—	17.5	(A28)
2 Fluorines on a ring carbon	—CF_2—	17.5	(A28)
Iodine	—I	19.4	(A29)
Phenol	=C(OH)—	20.3	(A31)
Hydroperoxide	—OOH	31.8	(A158)
Ether	—O—	4.71	(A32)
Peroxide	—O—O—	10.6	(A33)
Aldehyde	—C(=O)H	21.5	(A34)
Ketone	—C(=O)—	4.6	(A35)
Formate ester	—OC(=O)H	22.3	(A37)
Ester	—OC(=O)—	7.7	(A38)
Carbonate	—OC(=O)O—	7.1	(A149)
Anhydride	—C(=O)OC(=O)—	10.0	(A39)
Acyl chloride	—C(=O)Cl	25.8	(A40)
Aromatic heterocyclic nitrogen	=N—	10.9	(A41)
Acyclic sp^2 nitrogen	=N—	−1.8	(A42)
Tertiary amine	>N—	−22.2	(A43)
Secondary amine	—NH—	−5.3	(A44)
Primary amine	—NH_2	21.4	(A45)
Azide	—N=N=N	−23	(A46)
Tertiary amine N-nitro	>N—$\underline{NO_2}$	−21	(A47)
1 Fluorine on a ring nitrogen	>NF	39.3	(A48)
2 Fluorines on a nitrogen	>NF_2	39.3	(A48)
Diazo nitrogen	>C=N=N	9.2	(A49)
Nitro group	—NO_2	17.7	(A50)
N-Nitro	>$\underline{N}$—NO_2	31.3	(A51)
N-Nitroso	>$\underline{N}$—N=O	25.6	(A52)
Oxime	=N—OH	13.6	(A53)
Azoxy nitrogen	—$\underline{N}$=$N^+(O^-)$—	6.8	(A54)

(Continued)

TABLE 5.17 Contributions for the functional group (heteroatom) portions of the molecule for the Chickos and Acree (2009) method for ΔS_{tot}. Groups in this table are acyclic but are not dependent on substitution patterns and thus have c_{X_m} values equal to 1. (*Continued*)

		$\Delta_{\Delta S_{tot,X}}$ [a,b]	
Description	**Formula**[c,d]	**Value [J/(mol·K)]**	**Label**
Nitrate ester	—ONO_2	24.4	(A55)
Nitrile	—C≡N	17.7	(A56)
Isocyanide	—N≡C	17.5	(A57)
Isocyanate	—N=C=O	23.1	(A58)
Tertiary amides	—C(=O)N<	−11.2	(A59)
Secondary amides	—C(=O)NH—	1.5	(A60)
Primary amide	—C(=O)NH_2	27.9	(A61)
N,N-Dialkylformamide	HC(=O)N<	6.9	(A62)
N-Alkylformamide	HC(=O)NH—	27.0	(A162)
Hydrazide	—C(=O)$NHNH_2$	26.0	(A147)
Iminohydrazide	—C(=O)NHN=$\underline{CH}$—	18.6	(A159)
Tetra-substituted urea	>NC(=O)N<	−19.3	(A63)
1,1,3-Trisubstituted urea	>NC(=O)NH—	0	(A64)
1,1-Disubstituted urea	>NC(=O)NH_2	19.5	(A65)
1,3-Disubstituted urea	—NHC(=O)NH—	−8.1	(A66)
Monosubstituted urea	—NHC(=O)NH_2	14.1	(A67)
N,N-Disubstituted carbamate	—OC(=O)N<	−23	(A68)
N-Substituted carbamate	—OC(=O)NH—	7.8	(A69)
Carbamate	—OC(=O)NH_2	27.1	(A70)
Carbamic acid	—NHC(=O)OH	11	(A178)
Imide	—C(=O)NHC(=O)—	10.4	(A71)
Phosphine	>P—	−20.7	(A72)
Phosphine oxide	≥P=O	−32.7	(A73)
Phosphate ester	O=P(O—$)_4$	−10	(A74)
Phosphonate ester	—P(=O)(O—$)_2$	−11.4	(A75)
Phosphonic acid	—P(=O)$(OH)_2$	−12.1	(A76)
Phosphinic acid	>P(=O)OH	−12	(A173)
Phosphonyl halide (X = any halide)	—P(=O)$\underline{X}_2$	4.8	(A77)
Phosphoramidate ester	P(=O)(O—$)_2$NH—	−0.7	(A78)
Phosphorothioate ester	P(=S)(O—$)_3$	1.1	(A79)
Phosphorodithioate ester	—S—P(=S)(O—$)_2$	−9.6	(A80)
Phosphonothioate ester	—P(=S)(O—$)_2$	5.2	(A81)
Phosphoroamidothioate ester	—NHP(=S)(O—$)_2$	16.0	(A82)
Phosphoroamidodithioate ester	NH_2P(=S)(S—)(O—)	6.9	(A83)

(*Continued*)

TABLE 5.17 Contributions for the functional group (heteroatom) portions of the molecule for the Chickos and Acree (2009) method for ΔS_{tot}. Groups in this table are acyclic but are not dependent on substitution patterns and thus have c_{X_m} values equal to 1. (*Continued*)

		$\Delta_{\Delta S_{tot,X}}$ [a,b]	
Description	**Formula**[c,d]	**Value [J/(mol·K)]**	**Label**
Sulfides	—S—	2.1	(A84)
Disulfides	—SS—	9.6	(A85)
Thiols	—SH	23.0	(A86)
Sulfoxide	—S(=O)—	8.0	(A87)
Sulfones	—S(=O)$_2$—	0.6	(A88)
Sulfonate ester	—S(=O)$_2$O—	7.3	(A89)
1,3-Disubstituted thiourea	—NHC(=S)NH—	7.8	(A90)
Tetrasubstituted thiourea	>NC(=S)N<	−7.2	(A148)
Isothiourea	—S—C(=NH)NH—	23.8	(A160)
S,N,N′-Trisubstituted isothiourea	—S—C(=N—)NH—	0.7	(A161)
Monosubstituted thiourea	—NHC(=S)NH$_2$	30.1	(A91)
Thioamide	—C(=S)NH$_2$	15.0	(A92)
N-Substituted thioamide	—C(=S)NH—	4.07	(A174)
N,N-Disubstituted thioamide	—C(=S)N<	−13.5	(A175)
N,N disubstituted thiocarbamate	—S(C=O)N<	5.6	(A93)
Thiocarbamic ester	—NHC(=S)O—	4.8	(A176)
Sulfonic acid	—S(=O)$_2$OH	1.8	(A145)
N,N-Disubstituted sulfonamide	—S(=O)$_2$N<	−11.3	(A94)
N-Substituted sulfonamide	—S(=O)$_2$NH—	6.6	(A95)
Sulfonamide	—S(=O)$_2$NH$_2$	25.2	(A96)
N-Acyl sulfonamide	—S(=O)$_2$NH(C=O)—	11.9	(A177)
Sulfonyl chloride	—S(=O)$_2$Cl	23.4	(A157)
Trisubstituted aluminum	>Al—	−24.7	(A97)
Trisubstituted arsenic	>As—	3.1	(A98)
Trisubstituted boron	>B—	−17.2	(A99)
Trisubstituted bismuth	>Bi—	−14.5	(A100)
Aryltricarbonyl chromium	—Cr(C=O)$_3$	20.8	(A182)
Trisubstituted gallium	>Ga—	−11.3	(A101)
Tetrasubstituted germanium	>Ge<	−35.2	(A102)
Disubstituted germanium	>GeH$_2$	−14.7	(A103)
Disubstituted mercury	>Hg	8.4	(A104)
Trisubstituted indium	>In—	−19.3	(A105)
Ferrocenyl iron	>Fe	−5	(A179)
Tetrasubstituted lead	>Pb<	−30.2	(A106)
Trisubstituted antimony	>Sb—	−12.7	(A107)

(*Continued*)

TABLE 5.17 Contributions for the functional group (heteroatom) portions of the molecule for the Chickos and Acree (2009) method for ΔS_{tot}. Groups in this table are acyclic but are not dependent on substitution patterns and thus have c_{X_m} values equal to 1. (*Continued*)

		$\Delta_{\Delta S_{tot,X}}$ [a,b]	
Description	**Formula**[c,d]	**Value [J/(mol·K)]**	**Label**
Disubstituted selenium	>Se	6.0	(A108)
Alkyl arsonic acid	—(As=O)(OH)$_2$	−2.9	(A181)
Quaternary silicon	>Si<	−27.1	(A109)
Quaternary tin	>Sn<	−24.2	(A110)
Disubstituted zinc	>Zn	11.1	(A111)
Disubstituted telluride	>Te	5.1	(A140)
Trisubstituted germanium	>GeH—	−27.8	(A141)
Disubstituted arsinic acid	>(As=O)(OH)	−24	(A142)
Trisubstituted thallium	>Th—	1	(A143)
Disubstituted cadmium	>Cd	−2	(A144)
Dialkyl ammonium carboxylate	—CO_2^- $^+NH_2$—	4.3	(A180)

[a]Group values for Eq. (5.11-14)
[b]c_{X_m} = 1 in Eq. (5.11-14) for groups in this table
[c]Indicated bonds may be with all atoms other than H
[d]Underlined atoms are ***not*** part of the atoms in the group but indicate bonding patterns. Another group must be included to fully describe the molecule in such cases. For example, group (A51) is used together with another group such as (A47), (A120), etc. (A51) is the NO_2 contribution, and (A47) or (A120) is the N contribution.

Cyclic molecules with heteroatoms as part of the ring are treated similarly to cyclic hydrocarbons. The *parent* ring structure is calculated using Eq. (5.11-8) and a ring correction is added to this value. Specifically, the cyclic heteroatom corrections are calculated as

$$\Delta S_{tot,\ corr,X} = \sum_j N_j \left(\Delta_{\Delta S_{tot,corr,X}}\right)_j \tag{5.11-12}$$

where N_j is the number of times heteroatom ring group j is found in the molecule (see Table 5.18) and $\left(\Delta_{\Delta S_{tot,corr,X}}\right)_j$ is the contribution of this group. Eqs. (5.11-9) and (5.11-12) function the same way, and the delineation between the hydrocarbon corrections in the former and the heterocarbon corrections in the latter is artificial. However, this delineation is kept for consistency with the presentation of the method in Chickos and Acree (2009).

Furan is a cyclic molecule with one heteroatom. The parent ring is cyclopentane which has $N_{rings} = 1$ and $N_{ring\ atoms} = 5$ to use to calculate $\Delta S_{tot,ring}$ using Eq. (5.11-8). The ring corrections of the molecule, for $\Delta S_{tot,corr}$ of Eq. (5.11-9) consists of 2 cyclic tertiary sp^2 (=C_cH—) groups. (Note: these groups are counted for Eq. (5.11-9) which does *not* include an associated coefficient). The molecule does not have any aah portion, so the contribution of Eq. (5.11-6) to the entropy is zero. The other two carbons in the ring are attached to a heteroatom. They are also cyclic tertiary sp^2 (=C_cH—) groups, but since they are included in calculation as part of Eq. (5.11-10) the group value is multiplied by the coefficient found in the table. The last atom is the oxygen which constitutes the only group (i.e., $m = 1$) for $\Delta S_{tot.corr,\ X}$ in Eq. (5.11-12). The specific group is cyclic ether (*R—O—R*). With each of these contributions, the entropy is calculated with the following group operations: 33.4 + 2(3.7) + A112 + 2(A18)(B18) + 2(A18).

TABLE 5.18 Corrections for cyclic function group (heteroatom) portions of the molecule for the Chickos and Acree (2009) method for ΔS_{tot}.

		$\Delta_{\Delta S_{tot,corr,X}}$[a]	
Description	Formula[b,c,d]	Value [J/(mol·K)]	Label
Cyclic ether	$R{—}O{—}R$	1.2	(A112)
Cyclic peroxide	$R{—}O{—}R$	27.7	(A113)
Cyclic ketone	$R{—}C({=}O){—}R$	−1.4	(A114)
Cyclic ester	$R{—}C({=}O){—}O{—}R$	3.1	(A115)
Cyclic carbonate	$R{—}O{—}C({=}O){—}O{—}R$	1.3	(A116)
Cyclic anhydride	$R{—}C({=}O){—}O{—}C({=}O){—}R$	2.3	(A117)
Cyclic sp² nitrogen	$R{=}N{—}R$	0.5	(A118)
Cyclic tertiary amine	$R_2{>}N{—}R$	−19.3	(A119)
Cyclic hydrazine	$R_2{>}NNH_2$	21.7	(A153)
Cyclic tertiary amine-N-nitro	$R_2{>}N{—}(NO_2)$	−20.6	(A120)
Cyclic tertiary amine-N-nitroso	$R_2{>}N{—}(N{=}O)$	−20.6	(A120)
Cyclic secondary amine	$R{—}N(H){—}R$	2.2	(A121)
Cyclic tertiary amine-N-oxide	$R_2{>}N^+(O^-){—}R$	−22.2	(A122)
Cyclic azoxy group	$R{=}N^+(O^-){—}R$	2.9	(A123)
Cyclic sec amide	$R{—}C({=}O)N(H){—}R$	2.7	(A124)
Cyclic tertiary amide	$R{—}C({=}O)N{<}(R,R)$	−21.7	(A125)
Cyclic tertiary amide	$R{—}C({=}O)N{<}R_2$	−16.7	(A146)
N Substituted cyclic carbamate	$R{—}OC({=}O)N{<}(R,R)$	−5.2	(A126)
N,N Substituted cyclic carbamate	$R{—}OC({=}O)N{<}R_2$	−22.2	(A169)
Cyclic carbamate	$R{—}OC({=}O)N(H){—}R$	15.3	(A154)
Cyclic N,N′-disubstituted urea	$(R,R){>}NC({=}O)N{<}(R,R)$	−34.8	(A127)
Cyclic N,N-disubstituted urea	$R_2{>}NC({=}O)NH_2$	0.3	(A170)
N-Substituted cyclic imide	$R{—}C({=}O)N(R)C({=}O){—}R$	−13.6	(A128)
Cyclic imide	$R{—}C({=}O)N(H)C({=}O){—}R$	2.8	(A129)
Cyclic phosphorothioate	$R{—}O{—}P({=}S){<}(O{—}R)(O{—}R)$	−15.6	(A130)
Cyclic phosphazene (X is any halogen)	${—}N{=}P(X_2){—}N{=}$	−26.7	(A155)
Cyclic sulfide	$R{—}S{—}R$	2.9	(A131)
Cyclic disulfide	$R{—}SS{—}R$	−6.4	(A132)
Cyclic disulfide S-oxide	$R{—}SS({=}O){—}R$	4.0	(A133)
Cyclic sulfoxide	$R{—}S({=}O){—}R$	−2.2	(A134)
Cyclic sulfone	$R{—}S({=}O)_2{—}R$	15.1	(A164)
Cyclic thiocarbonate	$R{—}OC({=}O)S{—}R$	14.3	(A135)
Cyclic sulfite	$R{—}OS({=}O)O{—}R$	−5.8	(A150)
Cyclic thioester	$R{—}C({=}O)S{—}R$	17.0	(A167)
Cyclic dithioester	$R{—}C({=}S)S{—}R$	11.0	(A151)

(Continued)

TABLE 5.18 Corrections for cyclic function group (heteroatom) portions of the molecule for the Chickos and Acree (2009) method for ΔS_{tot}. (*Continued*)

		$\Delta_{\Delta S_{tot,corr,X}}$ [a]	
Description	**Formula** [b,c,d]	**Value [J/(mol·K)]**	**Label**
Cyclic sulfate	*R*—*OS*(=O)$_2$*O*—*R*	0.9	(A136)
Cyclic N–substituted sulphonamide	*R*—*S*(=O)$_2$*N*(H)—*R*	–0.4	(A137)
Cyclic tertiary sulfonamide	*R*—*S*(=O)$_2$*N*<(*R*,R)	–27.1	(A152)
Cyclic carboxyl sulfimide	*R*—*S*(=O)$_2$*N*(H)*C*(=O)—*R*	13.9	(A166)
Cyclic thiocarbamate	*R*—*SC*(=O)*N*(H)—*R*	13.9	(A138)
Cyclic isothiocarbamate	*R*—*OC*(=S)*N*(H)—*R*	2.6	(A163)
Cyclic dithiocarbamate	*R*—*SC*(=S)*N*(H)—*R*	3.8	(A165)
Cyclic thiourea	*R*—*N*(H)*C*(=S)*N*(H)—*R*	4.9	(A168)
Cyclic isothiourea	*R*—*N*=*C*(N(H)R)*S*—*R*	–49.7	(A171)
Cyclic alkylboronate ester	R—*B*<(*OR*)$_2$	–54.2	(A172)
Cyclic quaternary silicon	R$_2$>*Si*<*R*$_2$	–34.7	(A139)

[a] Group values for Eq. (5.11-12)
[b] Italicized letters indicate atoms or groups that are part of the ring structure
[c] The letter R is used to denote bonding to any alkyl or aryl group; R's may (italicized) or may not be (un-italiziced) part of the ring.
[d] Underlined atoms are ***not*** part of the atoms in the group but indicate bonding patterns. Another group must be included to fully describe the molecule in such cases. For example, group (A120) is used together with another group such as (A51). (A51) is the NO_2 contribution, and (A120) is the N contribution. R's are ***not*** part of the atoms in the group. The underlines are implied for all R's.

Example 5.11

Estimate the enthalpy of melting at the melting point of *n*-propylbenzene, propanal, and L-galactono-1,4-lactone using the Chickos and Acree (2009) method. The values in Appendix A for *n*-propylbenzene and propanal are 9.268 kJ/mol and 8.590 kJ/mol, respectively. Flores and Amador (2004) report a value of 35.98 kJ/mol for L-galactono-1,4-lactone.

Solution

Ethylbenzene

From Appendix A, the melting temperature of *n*-propylbenzene is 173.55 K. This compound is composed of only aah groups, specifically: 1 —CH_3, 1 >CH_2, 5 aromatic =CH—, and 1 aromatic =C<. There are no *consecutive* methylene groups ($N_{CH_2,con} = 0$), so by Eq. (5.11-6) and the definitions in Table 5.14, the group operations to obtain the total entropy are A1 + A2 + 5(A10) + A11 to give

$$\Delta S_{tot} = 17.6 + 7.1 + (5)(7.4) + (-9.6) = 52.1 \text{ J/(mol}\cdot\text{K)}$$

which through Eq. (5.11-1) yields

$$\Delta_m H = (52.1)(173.55) = 9042 \text{ J/mol} = 9.042 \text{ kJ/mol}$$

With the experimental value from Appendix A,

$$\text{Error} = 9.042 - 9.268 = -0.226 \text{ kJ/mol or } -2.44\%.$$

Propanal

From Appendix A, the melting temperature of propanal is 165.0 K. This compound is composed of one aah group (—CH_3) contributing to Eq. (5.11-6), one C(X) group (>CH_2) with $c_{C(X)} = 1$ from Table 5.14 contributing to Eq. (5.11-13) and one ($m = 1$) functional group (—C(=O)H) with $c_{X_1} = 1$ from Table 5.17 contributing to Eq. (5.11-11). [$c_{C(X)} = 1$ for the methylene group because group B2 in Table 5.14 is for *consecutive* methylene groups if the proper criterion is met in Eq. (5.11-11). $c_{X_1} = 1$ because the coefficieints of all the groups found in Table 5.17 are equal to 1.] The group operations to calculate the total entropy are thus A1 + A2 + A34 to give

$$\Delta S_{\text{tot}} = 17.6 + 7.1 + 21.5 = 46.2 \text{ J/(mol}\cdot\text{K)}$$

$$\Delta_{\text{m}} H = (46.2)(165.0) = 7623 \text{ J/mol} = 7.623 \text{ kJ/mol}$$

With the experimental value from Appendix A,

$$\text{Error} = 7.623 - 8.590 = -0.967 \text{ kJ/mol or } -11.3\%.$$

L-galactono-1,4-lactone

Flores and Amador (2004) report the melting temperature of L-galactono-1,4-lactone to be 409.82 K. This compound, pictured in Figure 5.8, has several elements contributing to the total entropy of melting. There are no pure aah contributions, rather, all the carbons are connected to functions groups and thus comprise C(X) groups. The parent ring structure is cyclopentane with $N_{\text{rings}} = 1$ and $N_{\text{ring atoms}} = 5$. The carbon ring corrections are accounted for through Eq. (5.11-10) rather than Eq. (5.11-9) since each is bonded to a heteroatom function group. The group values and coefficients for these ring carbons are found Table 5.15 and are 3 cyclic tertiary sp^3 carbon (>CH—; A16). The ester group ring adjustment (*R*—*C*(=O)*O*—*R*) is added through Eq. (5.11-12) with the group value A115 from Table 5.18. The tertiary sp^3 >CH— in the ring is added through Eq. (5.11-10) with group A3 and group coefficieint ($c_{C(X)}$) B3. The methylene group is added through Eq. (5.11-10) with group A2. Group coefficient B2 is not applied because it is only used for *consecutive* methylene groups if the the proper criterion is met in Eq. (5.11-11). Finally, the four ($m = 4$) hydroxyl groups are added using Eq. (5.11-11) with group A30 and coefficient (c_{X_m}) D30. The group operations to calculate the total entropy are thus A14 + 2(A15) + 3(A16) + A115 + (A3)(B3) + A2 + 4(A30)(D30) to give

$$\Delta S_{\text{tot}} = 33.4 + 2(3.7) + 3(-14.7) + 3.1 + (-16.4)(0.60) + 7.1 + 4(1.7)(13.1) = 86.1 \text{ J/(mol}\cdot\text{K)}$$

$$\Delta_{\text{m}} H = (86.1)(409.82) = 35284 \text{ J/mol} = 35.28 \text{ kJ/mol}$$

With the experimental value from Flores and Amador (2004),

$$\text{Error} = 35.28 - 35.98 = -0.70 \text{ kJ/mol or } -1.95\%.$$

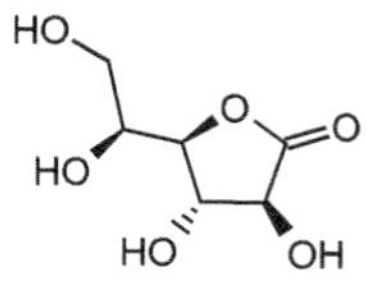

FIGURE 5.8
Structure of L-galactono-1,4-lactone.

5.11.3 Dannenfelser-Yalkowski Method (Dannenfelser and Yalkowsky, 1996)

In this method, ΔS_{tot} is calculated by

$$\Delta S_{tot} = 50 - R \ln \sigma + 1.047 R\tau \quad (5.11\text{-}13)$$

where R is the gas constant in J/(mol·K), ΔS_{tot} is in J/(mol·K), σ is a symmetry number, and τ is the number of torsional angles in the compound. The symmetry number is the number of ways a molecule can be *rigidly* rotated so that atoms are in a different position but the molecule occupies the same space. Thus, σ for benzene and carbon tetrafluoride is 12, and for fluorobenzene it is 2. In the assignment of a value to σ, the structure is hydrogen suppressed and the following groups are assumed to be radially symmetrical and/or freely rotating: halogens, methyl, hydroxyl, mercapto, amine, and cyano. This is different than in Benson's method for ideal gas properties (see Chap. 4). Thus, for propane, $\sigma = 2$; for all higher *n*-alkanes, $\sigma = 1$. For spherical molecules, such as methane or neon, $\sigma = 100$. Molecules that are conical (e.g., hydrogen cyanide and chloromethane) or cylindrical (e.g., carbon dioxide and ethane) have one infinite rotational axis and are empirically assigned σ values 10 and 20 respectively. Dannenfelser and Yalkowsky (1996) gives additional examples, and the value of σ for 949 compounds is found in the supplementary material of Dannenfelser and Yalkowsky (1996).

The number of torsional angles, τ, is calculated by

$$\tau = SP3 + 0.5SP2 + 0.5RING - 1 \quad (5.11\text{-}14)$$

$SP3$ is the number of sp^3 chain atoms, $SP2$ is the number of sp^2 chain atoms, and $RING$ is the number of fused-ring systems. τ cannot be less than zero. Note that the radially symmetrical end groups mentioned above, as well as carbonyl oxygen and *tert*-butyl groups are not included in the number of chain atoms. This method is illustrated in Example 5.12, and results for 949 compounds (as well as the value of τ) are found in Dannenfelser and Yalkowsky (1996).

Example 5.12

Calculate ΔS_{tot} and $\Delta_m H$ for isobutane with Eqs. (5.11-13) and (5.11-14). Literature values (Domalski and Hearing, 1996) are 39.92 J/(mol·K) and 4540 J/mol respectively. The melting point temperature is 113.2 K.

Solution. For isobutane, $SP3 = 1$ so $\tau = 0$ and $\sigma = 3$. With Eq. (5.11-14) first and then (5.11-13)

$$\Delta S_{tot} = 50 - 8.3145 \ln(3) = 40.866 \text{ J/(mol·K)}$$

$$\Delta_m H = (113.2)(40.866) = 4626 \text{ J/mol}$$

$$\text{Error} = 4626 - 4540 = 86 \text{ J/mol or } 1.89\%$$

5.11.4 Discussion and Recommendations

The enthalpy of melting ($\Delta_m H$) is related to the enthalpies of vaporization ($\Delta_v H$) and sublimation ($\Delta_{sub} H$; see Sec. 5.12) *at the triple point* through

$$\Delta_m H = \Delta_{sub} H - \Delta_v H \quad (5.11\text{-}15)$$

Thus, in the absence of experimental data for $\Delta_m H$, an estimate may be obtained if reliable experimental data are available for the sublimation and vaporization phase changes, and this should be the first choice for prediction of the property. Because Eq. (5.11-15) is subtracting two relatively large numbers to obtain a small number, the accuracy of the resulting $\Delta_m H$ value is highly sensitive to the accuracy of the other two enthalpies.

Automated algorithms for the methods presented in this section to predict $\Delta_m H$ are not available, so the techniques could not be tested independently. The 5th edition of this book tested the Dannenfelser and Yalkowsky method for 43 compounds and found an average error of 18% for this approach. DIPPR (Wilding et al. 2017) considers the uncertainty in the Bondi method to be approximately 25%. Chickos et al. (1999, 2003, 2009) reported errors of approximately 20% on tests for over 2000 compounds with experimental data. Any of these methods can be used when Eq. (5.11-15) cannot be applied.

5.12 ENTHALPY OF SUBLIMATION

Solids vaporize without melting (sublime) at temperatures below the triple-point temperature. Sublimation is accompanied by an enthalpy increase, or latent enthalpy of sublimation ($\Delta_{sub}H$). This may be taken as the sum of a latent enthalpy of melting and a hypothetical latent enthalpy of vaporization, even though liquid cannot exist at the pressure and temperature in question.

The latent enthalpy of sublimation $\Delta_{sub}H$ is best obtained from solid vapor pressure (P^{svp}) data. For this purpose, the Clausius-Clapeyron equation is applicable, namely,

$$\frac{d \ln P^{svp}}{d\left(\frac{1}{T}\right)} = -\frac{\Delta_{sub}H}{R} \tag{5.12-1}$$

Solid vapor pressure data have become more common in the literature in the last two decades. When such are not available, one sublimation pressure may be obtained from the liquid vapor pressure correlation. At the triple point temperature (T_{tp}) the sublimation pressure is the same as the vapor pressure of the liquid ($P^{svp}(T_{tp}) = P^{vp}(T_{tp})$) of which the latter can be determined directly through experimental liquid vapor pressure measurements at temperatures near the triple point or by appropriate extrapolation of the vapor pressure correlation using thermal data as discussed in Sec. 5.8. However, even if P^{vp} at T_{tp} is known, at least one other value of the solid vapor pressure is necessary to calculate $\Delta_{sub}H$ from the integrated form of the Clausius-Clapeyron equation. The enthalpy of sublimation can also be determined by adding the enthalpy of vaporization to the enthalpy of fusion at the melting point. Since P^{svp} is often difficult to obtain but $\Delta_m H$ can be measured with calorimetry, this may be the most reliable approach. Vapor-pressure data for solids may be found in compilations by Dellesite (1997), Oja and Suuberg (1998, 1999), Pouillot et al. (1996), Acree and Chickos (2016), and Acree and Chickos (2017).

The enthalpy of sublimation is technically temperature dependent, but the dependence is weak, and it is common to consider the property a constant. Assuming the applicability of the Clausius-Clapeyron equation, having a constant enthalpy of sublimation means that the logarithm of the solid vapor pressure is directly proportional to the inverse temperature (i.e., $\ln P^{vp} \propto \frac{1}{T}$). The temperature dependence of the enthalpy of sublimation may be determined if reliable solid vapor pressure data are available over a wide temperature range, but the temperature variation is less than the error in the value for only a few compounds where extensive experimentation has been done. In short, the enthalpy of sublimation may be considered constant in most practical situations.

5.12.1 Goodman et al. (2004) Method

Goodman et al. (2004) developed a group-contribution method to predict $\Delta_{sub}H(T_{tp})$. The expression is

$$\frac{\Delta_{sub}H(T_{tp})}{R} = 698.04\text{K} + \left(3.83798 \times 10^{12}\,\frac{\text{K}}{\text{m}}\right) R_g + \sum_i N_i(\Delta_a)_i + \sum_j N_j^2(\Delta_b)_j + \sum_k \frac{N_k}{N_x}(\Delta_c)_k \tag{5.12-2}$$

where R_g is the radius of gyration of the compound, N_i is the number of times group i is found in the molecule, N_x is the total number of halogens and hydrogens attached to either C or Si in the molecule, $(\Delta_a)_i$ are contributions with linear dependence (see Table 5.19), $(\Delta_b)_j$ are contributions with nonlinear dependence (see Table 5.20), $(\Delta_c)_k$ the contributions for halogen groups (see Table 5.20). The radius of gyration can be found in many compilations. Accurate values may also be obtained from standard quantum chemical geometry optimization calculations using relatively modest levels of theory and basis sets such as B3LYP/6-311+G(3df,2p) or even HF/6-31G* (see Ch. 4).

TABLE 5.19 Contributions for the linear groups $(\Delta_a)_i$ for the Goodman et al. (2004) method for $\Delta_{sub}H(T_{tp})$ of Eq. (5.12-2).

Description	Formula	Δ_a(K)
Methyl	$—CH_3$	736.5889
Methylene	$>CH_2$	561.3543
Secondary carbon	>CH—	111.0344
Tertiary carbon	>C<	–800.517
Terminal alkene	$=CH_2$	572.6245
Alkene	=CH—	541.2918
Substituted alkene	=C<	117.9504
Aromatic carbon	$=C_aH—$	626.7621
Substituted aromatic arbon	$=C_a<$	348.8092
Furan oxygen (aromatic oxygen)	$—O_a—$	763.284
Pyridine nitrogen (aromatic nitrogen)	$=N_a—$	1317.056
Thiophene sulfur (aromatic sulfur)	$—S_a—$	911.2903
Ether	—O—	970.4474
Hydroxyl	—OH	3278.446
Aldehyde	—C(=O)H	2402.093
Ketone	>C=O	1816.093
Ester	—C(=O)O—	2674.525
Acid	—C(=O)OH	5006.188
Primary amine	$—NH_2$	2219.148
Secondary amine	>NH	1561.222
Tertiary amine	>N—	325.9442
Nitro	$—NO_2$	3661.233
Thiol/mercaptan	—SH	1921.097
Sulfide	—S—	1930.84
Disulfide	—SS—	2782.054
Fluoride	—F	626.4494
Chloride	—Cl	1243.445
Bromide	—Br	669.9302
Silane	>Si<	–83.7034
Siloxane	>Si(O—)—	–16.0597

TABLE 5.20 Contributions for the nonlinear $(\Delta_b)_j$ and halogen fraction $(\Delta_c)_k$ groups for the Goodman et al. (2004) method for $\Delta_{sub}H(T_{tp})$ of Eq. (5.12-2).

Description	Formula	Value
Nonlinear Groups		
		Δ_b (K)
Methylene	$>CH_2$	9.5553
Aromatic carbon	$=C_aH—$	–2.21614
Halogen Fraction Groups		
		Δ_c (K)
Chloride fraction	—Cl	−1543.66
Fluoride fraction	—F	−1397.4
Bromide fraction	—Br	5812.49

Example 5.13

Estimate the enthalpy of sublimation at the melting point of 1,2-dibromoethane, 1-octadecanol, and benzophenone using the Goodman et al. (2004) method. Nitta (1948) found the enthalpy of sublimation to be 54.85 kJ/mol for 1,2-dibromoethane, Davies and Kybett (1965) give a value of 18.8 kJ/mol for 1-octadecanol, and de Kruif et al. (1983) report a value of 94.67 kJ/mol for benzophenone.

Solution

1,2 dibromoethane

The radius of gyration of 1,2-dibromoethane, reported by Stuper et al. (1979) is 2.833×10^{-10} m. The molecule is composed of two $>CH_2$ groups and two —Br groups. The former group has both linear (through $(\Delta_a)_i$) and nonlinear (through $(\Delta_b)_j$) contributions, while the latter has both linear (through $(\Delta_a)_i$) and halogen fraction (through $(\Delta_c)_k$) contributions. With $N_x = 6$, Eq. (5.12-2) yields

$$\Delta_{sub}H = (8.3145)[698.04 + (3.83798 \times 10^{12})(2.833 \times 10^{-10}) + 2(561.3543) + 2(669.9302) + 2^2(9.5553) + (2/6)(5812.49)] = 5.175 \times 10^4 \text{ J/mol} = 51.75 \text{ kJ/mol}$$

$$\text{Error} = 51.75 - 54.85 = -3.1 \text{ kJ/mol or } -5.65\%$$

1-octadecanol

The radius of gyration of 1-octadecanol, reported by Stuper et al. (1979) is 7.930×10^{-10} m. The molecule is composed of one $—CH_3$ group, seventeen $>CH_2$ groups and one —OH group. The methylene group has both linear (through $(\Delta_a)_i$) and nonlinear (through $(\Delta_b)_j$) contributions. The other groups only have linear (through $(\Delta_a)_i$) contributions. Eq. (5.12-2) yields

$$\Delta_{sub}H = (8.3145)[698.04 + (3.83798 \times 10^{12})(7.930 \times 10^{-10}) + 1(736.5889) + 17(561.3543) + 17^2(9.5553) + 1(3278.446)] = 1.668 \times 10^4 \text{ J/mol} = 16.68 \text{ kJ/mol}$$

$$\text{Error} = 16.68 - 18.8 = 2.12 \text{ kJ/mol or } 11.3\%$$

benzophenone

The radius of gyration of benzophenone, reported by Stuper et al. (1979) is 5.302×10^{-10} m. The molecule is composed of ten $=C_aH—$, two $=C_a<$, and one $>C=O$ group. The unsubstituted aromatic carbon group has both linear (through $(\Delta_a)_i$) and nonlinear (through $(\Delta_b)_j$) contributions. The other groups only have linear (through $(\Delta_a)_i$) contributions. Eq. (5.12-2) yields

$$\Delta_{sub}H = (8.3145)[698.04 + (3.83798 \times 10^{12})(5.302 \times 10^{-10}) + 10(626.7621) + 2(348.8092) + 10^2(-2.21614) + 1(1816.093)] = 9.389 \times 10^4 \text{ J/mol} = 93.89 \text{ kJ/mol}$$

$$\text{Error} = 93.89 - 94.67 = -0.78 \text{ kJ/mol or } -0.82\%$$

5.12.2 Discussion and Recommendations

In some cases, it is possible to obtain $\Delta_{sub}H$ from thermochemical data by subtracting known values of the enthalpies of formation of solid and vapor. This is hardly a basis for estimation of an unknown $\Delta_{sub}H$; however, since the enthalpies of formation tabulated in the standard references are often based in part on measured values of $\Delta_{sub}H$. If the enthalpies of dissociation of both solid and gas phases are known, it is possible to formulate a cycle including the sublimation of the solid, the dissociation of the vapor, and the recombination of the elements to form the solid compound.

The first approach that should be taken to obtain enthalpy of sublimation values is to use experimental solid vapor pressure data along with Eq. (5.12-1)—the Clausius-Clapeyron equation. The second approach is to obtain the value by adding $\Delta_m H$ and $\Delta_v H(T_{tp})$ when reliable experimental data are available. This approach may be used, and be sufficiently accurate, even if $\Delta_m H$ is estimated using one of the techniques found in Sec. 5.11, since the latent enthalpy of melting is usually less than one-quarter of the sum.

In situations where these two methods are not applicable, when reliable values for $\Delta_m H$ and $\Delta_v H(T_{tp})$ cannot be found in the literature or estimated, the technique of Goodman et al. (2004) may be used to obtain $\Delta_{sub}H$. To evaluate the performance of the method, a test database was created from the October 2021 release of the DIPPR database which included compounds with $\Delta_{sub}H$ data that were measured experimentally, derived from experimental solid vapor pressure using the Clausius-Clapeyron equation, or derived from reliable values of $\Delta_m H$ and $\Delta_v H$ and Eq. (5.12-2). A total of 584 compounds were included in the evaluation. Compounds were organized into the same classes as those used to evaluate ρ_s (see the end of Sec. 5.2) to determine how the method worked for different types of compounds. Results are quantified using the average absolute relative deviation given by Eq. (5.2-14), and the results are found in Table 5.21. The data indicate that the Goodman et al. (2004) method may be expected to give errors on the order 10% regardless of compound class.

5.13 SOLID VAPOR PRESSURE (SUBLIMATION PRESSURE)

The vapor pressure of a solid (P^{svp}), also called the sublimation pressure, is the pressure at which the solid is in equilibrium with its vapor. It is commonly correlated with a form of Eq. (5.5-6). When data are available over only a limited temperature range, just two terms are used in the correlation resulting in Eq. (5.5-1). When written in this form, $B = -\dfrac{\Delta_{sub}H}{R}$ due to the Clausius-Clapeyron equation [Eq. (5.12-1)].

TABLE 5.21 Performance of the Goodman et al. (2004) method for estimating $\Delta_{sub}H$ in terms of average absolute relative deviation [see Eq. (5.2-14)].

Compound Class							
Normal		Heavy		Polar		Associating	
N	AARD	N	AARD	N	AARD	N	AARD
293	6.3%	38	8.2%	88	7.2%	165	10.9%

Multiple methods for estimating P^{svp} have been described in the literature. Most are dependent on knowing either $\Delta_m H$ or $\Delta_{sub} H$—both of which may be obtained experimentally or using one of the estimation techniques described in this chapter. They then obtain the solid vapor pressure by integrating either the Clapeyron or Clausius-Clapeyron equation from the triple point to the lower pressures. These approaches thus require the triple-point temperature and pressure.

For example, integration of Eq. (5.12-1) from the triple point to an arbitrary solid vapor pressure P^{svp} at arbitrary temperature (T) yields

$$\ln \frac{P^{svp}}{P_{tp}} = -\frac{\Delta_{sub} H}{R}\left(\frac{1}{T} - \frac{1}{T_{tp}}\right) \tag{5.13-1}$$

where T_{tp} and P_{tp} are the triple point temperature and pressure, respectively. In this equation, $T < T_{tp}$. This integration assumes $\Delta_{sub} H$ is independent of temperature over the range T to T_{tp}, a valid approximation if the range is less than 80 K.

Goodman et al. (2004) tested this approach for 87 compounds with 1103 experimental data points for P^{svp}. The enthalpies of sublimation $\Delta_{sub} H$ were predicted using Eq. (5.11-15). This predicted value was then used with Eq. (5.13-1) to estimate values for P^{svp}. They found their method gave deviations of < 50% down to vapor pressure values of 0.01 Pa. Such deviations are common for P^{svp} as the experimental values are commonly much less than 0.13 kPa. The literature commonly reports uncertainties of 25% or greater at such low pressure even for experimental data.

Prausnitz et al. (1999) describe using the full Clapeyron equation, Eq. (5.3-1), to estimate vapor pressures of solids when reliable enthalpy of melting, liquid heat capacity, and solid heat capacity information are available. The technique, along with its limitations, is illustrated with Example 7.9 (Prausnitz et al., 1999).

Example 5.14

Use information at the triple point and Eq. (5.3-1) to estimate the vapor pressure of ice at 263 K.

Solution. Eq. (5.3-1) may be written for the solid in equilibrium with vapor

$$\frac{dP^{svp}}{dT} = \frac{\Delta_{sub} H}{T\Delta_{sub} V} \tag{i}$$

and for hypothetical subcooled liquid in equilibrium with vapor

$$\frac{dP^{vp}}{dT} = \frac{\Delta_v H}{T\Delta_v V} \tag{ii}$$

In these equations, $\Delta_{sub} V$ is the change in molar volume that occurs upon sublimation and $\Delta_v V$ that which happens upon vaporization. In both cases, the volume change happens when moving from a condensed phase (either liquid or vapor) to the gas phase. In each instance, the condensed phase volume may be considered negligible compared to that of the gas phase to give:

$$\Delta_{sub} V = V^{vap} - V^{sol} \cong V^{vap} = \frac{RT}{P} \tag{i}$$

and

$$\Delta_v V = V^{vap} - V^{liq} \cong V^{vap} = \frac{RT}{P} \tag{ii}$$

where the ideal gas law (applicable at the low pressures involved) is invoked. Subtracting Eq. (i) from (ii), using $\Delta_v H - \Delta_{sub} H = -\Delta_m H$ from Eq. (5.11-15), and integrating from 273.16 to 263 K gives

$$\ln\left(\frac{P^{vp}(263\text{ K})}{P^{vp}(273.16\text{ K})}\right) - \ln\left(\frac{P^{svp}(263\text{ K})}{P^{svp}(273.16\text{ K})}\right) = -\int_{273.16}^{263} \frac{\Delta_m H}{RT^2}\, dT \tag{iii}$$

A similar derivation to that used to obtain Eq. (5.6-1) can be done to give $\Delta_m H$ as a function of temperature T as

$$\Delta_m H(T) = \Delta_m H(T_1) + \int_{T_1}^{T} \left(C_\sigma^{\text{liq}} - C_\sigma^{\text{sol}}\right) dT = \Delta H_m(T_1) + \int_{T_1}^{T} \left(C_p^{\text{liq}} - C_p^{\text{sol}}\right) dT \qquad \text{(iv)}$$

where the heat capacities of the liquid and solid at saturation (C_σ^{liq} and C_σ^{sol} respectively) can be replaced by their isobaric counterparts (C_p^{liq} and C_p^{sol}) at the conditions involved.

For H_2O, $\Delta_m H$ at 273 K is 6008 J/mol; $C_p^{\text{liq}} = 75.3$ J/(mol·K), and C_p^{sol} 37.7 J/(mol·K). This means

$$\Delta_m H(T) = 6008 + 37.6(T - 273) = 37.6T - 4284 \qquad \text{(v)}$$

Substitution of (v) into (iii) and integration gives

$$\ln \frac{P^{\text{svp}}(263\text{ K})}{P^{\text{svp}}(273.16\text{ K})} - \ln \frac{P^{\text{vp}}(263\text{ K})}{P^{\text{vp}}(273.16\text{ K})} = \frac{37.6}{8.314} \ln \frac{263}{273.16} + \frac{4257}{8.314}\left(\frac{1}{263} - \frac{1}{273.16}\right) \qquad \text{(vi)}$$

$P^{\text{svp}}(273.16\text{ K}) = P^{\text{vp}}(273.16\text{ K}) = 0.611657$ kPa (Guildner et al., 1976). $P^{\text{vp}}(263\text{ K})$ is the vapor pressure of subcooled liquid at 263 K. An extrapolation based on the assumption that $\ln P^{\text{vp}}$ vs $\frac{1}{T}$ is linear over the small temperature range gives $P^{\text{vp}}(263\text{ K}) = 0.288$ kPa. Solving Eq. (vi) gives $P^{\text{svp}}(263\text{ K}) = 0.261$ kPa, which is close to the IAPWS value of 0.256 kPa. The current standards for P^{svp} are given by Wagner and Riethmann (2011).

The technique used in Example 5.14 requires care. Discontinuities can occur in $\Delta_m H$ because of possible solid-solid transitions as discussed in Sec. 5.11. Unless these are accounted for, the integration as shown above in (iii) will not be correct. Also, the extrapolation of vapor pressures for hypothetical subcooled liquid over large temperature ranges is uncertain (see Sec. 5.8).

5.14 CORRELATION AND ESTIMATION OF VIRIAL COEFFICIENTS

5.14.1 Second Virial Coefficients

The second coefficient of the virial equation of state (B_2) is a function of temperature and is often correlated with an expression of the form

$$B_2 = \sum_{n=0}^{9} \frac{B_{2,n}}{T^n} \qquad (5.14\text{-}1)$$

Experimental values for second virial coefficients are relatively scarce in the literature compared to the other properties listed in this chapter. This is likely due to the diminished importance of the virial equation of state in industrial applications. When truncated at the second term, the equation is only accurate over limited pressure ranges compared to the cubic equations of state described in Chap. 6. Adding the third coefficient increases accuracy, but at this point it is easier to simply use a cubic equation of state.

Several techniques have been developed to predict second virial coefficieints of pure substances (Tsonopoulos, 1974; Hayden and O'Connell, 1975; Tarakad and Danner, 1977; McCann and Danner, 1984; Orbey, 1988, Orbey and Vera, 1983; Kis and Orbey, 1989; Abusleme and Vera, 1989; Olf et al., 1989; Lee and Chen, 1998; Vetere, 1999; Iglesias-Silva and Hall, 2001). Unlike for empirical EOSs, there is direct theoretical justification for extending simple CSP for B_2 to complex substances by merely adding terms to those for simple substances. Thus, most of the methods referenced above can be written in the form

$$\frac{B_2(T)}{V^*} = \sum_i a_i f^{(i)}\left(\frac{T}{T^*}\right) \qquad (5.14\text{-}2)$$

where V^* is a characteristic volume, such as V_c or $\frac{RT_c}{P_c}$, the a_i are strength parameters for various intermolecular forces, and the $f^{(i)}$ are sets of universal functions of reduced temperature, $\frac{T}{T^*}$, with T^* typically being T_c. Then, $f^{(0)}$ is for simple substances with a_0 being unity, $f^{(1)}$ corrects for nonspherical shape and globularity of normal fluids with a_1 commonly being ω, $f^{(2)}$ takes account of polarity with a_2 being a function of the dipole moment, μ (see Sec. 3.6), and $f^{(3)}$ takes account of association with a_3 an empirical parameter. In methods such as those of Tsonopoulos (1974) and Tarakad and Danner (1977), the terms in the various $f^{(i)}(T)$ are obvious; in the Hayden-O'Connell (1975) and Abusleme-Vera (1989) methods, the derivation and final expressions might disguise this simple division, but the correlations can be expressed this way.

Discussion about some methods is given below. Because no single technique is significantly better than the others, only two approaches are presented in detail. One is the Tsonoplous method because it is one of the most popular. The Iglesias-Silva and Hall technique is the other as it takes a different approach than that of Eq. (5.14-2).

5.14.1.1 Tsonopoulos Method

The Tsonopoulos (1974) correlation uses $V^* = \frac{RT_c}{P_c}$ and $T^* = T_c$. The substance-dependent strength coefficients are $a_1 = \omega$, $a_2 = a$, and $a_3 = b$. The last two parameters are empirical constants set with fitting to experimental data and can be constant parameters or variable functions of the dipole moment, μ, (see Sec. 3.6). They may depend on the "family" of the substance or the substance itself (see Table 5.22). The full form is

$$\frac{BP_c}{RT_c} = f^{(0)} + \omega f^{(1)} + af^{(2)} + bf^{(3)} \tag{5.14-3}$$

$$f^{(0)} = 0.1445 - \frac{0.330}{T_r} - \frac{0.1385}{T_r^2} - \frac{0.0121}{T_r^3} - \frac{0.000607}{T_r^8} \tag{5.14-4}$$

$$f^{(1)} = 0.0637 + \frac{0.331}{T_r^2} - \frac{0.423}{T_r^3} - \frac{0.008}{T_r^8} \tag{5.14-5}$$

$$f^{(2)} = \frac{1}{T_r^6} \tag{5.14-6}$$

$$f^{(3)} = -\frac{1}{T_r^8} \tag{5.14-7}$$

TABLE 5.22 Parameters for Eq. (5.14-3) for estimation of second virial coefficients using the Tsonopolous method (Tsonopoulos et al., 1975, 1978, 1979, 1989, 1990, 1997).

Compound Class	a	b
Nonpolar, Haloalkanes (2+ halogens)	0	0
Monohaloalkanes, Mercaptans, Sulfides, Disulfides	$-2.188 \times 10^{-11}\mu_r^4 - 7.831 \times 10^{-21}\mu_r^8$	0
Nonhydrogen Bonding Polar (Ketones, Aldehydes, Nitriles, Ethers, Esters, Amines)	$-2.14 \times 10^{-4}\mu_r - 4.308 \times 10^{-21}\mu_r^8$	0
1-Alkanols (except methanol)	0.0878	$0.00908 + 0.0006957\mu_r$
Other Alcohols (branched)	0.0878	0.05

$\mu_r = 10^5 \frac{\mu^2 P_c}{T_c^2}$ where μ is in Debye, P_c is in atm (1.01325 bar), and T_c is in K.

where $T_r = \dfrac{T}{T_c}$. Equations (5.14-4) and (5.14-5) are modifications of the early correlation of Pitzer and Curl (1955). Several revisions and extensions have appeared for the Tsonopoulos model (Tsonopoulos et al., 1975, 1978, 1979, 1989, 1990, 1997) mainly treating new data for alkanes and alcohols with revised parameters and making comparisons with other models. Table 5.22 summarizes current recommendations for a and b in Eq. (5.14-3).

There is considerable sensitivity to the values of a and b in this model because of the large powers on T_r in $f^{(2)}$ and $f^{(3)}$. As a result, for highest accuracy, fitting one of the parameters should be done when experimental data are available. The parameter b would be the first choice in that case.

Example 5.15

Estimate the second virial coefficient of ethanol at 400 K using the Tsonopoulos method.

Solution. From Appendix A, $T_c = 514.0$ K, $P_c = 61.37$ bar $= 60.567$ atm, $\omega = 0.643558$, and $\mu = 1.69087$ Debye. With these values,

$$\mu_r = (10^5)(1.69087)^2(60.567)/514.0^2 = 65.544$$

From Table 5.22

$$a = 0.0878$$

$$b = 0.00908 + (0.0006957)(65.544) = 0.05467$$

With $T_r = 400/514.0 = 0.7782$, Eqs. (5.14-4)–(5.14-7) give

$$f^{(0)} = 0.1445 - 0.330/0.7782 - 0.1385/0.7782^2 - 0.0121/0.7782^3 - 0.000607/0.7782^8 = -0.5384$$

$$f^{(1)} = 0.0637 + 0.331/0.7782^2 - 0.423/0.7782^3 - 0.008/0.7782^8 = -0.3468$$

$$f^{(2)} = 1/0.7782^6 = 4.5025$$

$$f^{(3)} = -1/0.7782^8 = -7.4348$$

From Eq. (5.14-3)

$$\frac{BP_c}{RT_c} = -0.5384 + (0.643558)(-0.3468) + (0.0878)(4.5025) + (0.05467)(-7.4348) = -0.7727$$

giving

$$B = (-0.7727)(8.3145)(514.0)/(61.37 \times 10^5) = -5.38 \times 10^5 \text{ m}^3/\text{mol} = -538 \text{ cm}^3/\text{mol}.$$

The recommended experimental value is -535 cm^3/mol (Tsonopoulos et al., 1989). Thus the difference is

$$-538 - (-535) = -3 \text{ cm}^3/\text{mol or } 0.56\%$$

which is within the experimental uncertainty of ± 40 cm^3/mol.

5.14.1.2 Iglesias-Silva and Hall Method

Iglesias-Silva and Hall (2001) took an approach that is different than that suggested by Eq. (5.14-2). Rather than fit to an equation where each term accounts for perturbations from the base case of nonpolar, spherical molecules (e.g., shape, polarity, association), they selected a model that satisfies several physical features that second virial coefficients are known to follow. These characteristics are:

- $B_2 = 0$ when $T = T_B$ (the Boyle temperature)
- $B_2 \to \infty$ when $T \to 0$
- $B_2 = \max$ at $T = T_{\max}$ where $T_{\max} > T_B$
- $B_2 = 0$ when $T \to \infty$

Because of the first constraint, the approach uses the Boyle temperature (T_B) as the normalization factor when calculating reduced temperatures rather than the more typical T_c. This gives the following equation for B_2 which satisfies all the constraints.

$$\frac{B(T)}{b_0} = \left(\frac{T_B}{T}\right)^{0.2}\left[1-\left(\frac{T_B}{T}\right)^{0.8}\right]\left\{\frac{B_c}{b_0\left[\left(\frac{T_B}{T_c}\right)^{0.2}-\left(\frac{T_B}{T_c}\right)\right]}\right\}^{\left(\frac{T_c}{T}\right)^n} \tag{5.14-8}$$

where B_c, b_0, and n are adjustable parameters set through fitting to experimental data. The authors give the following equations for determining these quantities using the critical constants (i.e., T_c and V_c), the acentric factor ω, and the reduced dipole moment defined in Table 5.22.

$$\frac{B_c}{V_c} = -1.1747 - 0.3668\omega - 0.00061\mu_r \tag{5.14-9}$$

$$n = 1.4187 + 1.2058\omega \tag{5.14-10}$$

$$\frac{b_0}{V_c} = 0.1368 - 0.4791\omega + 13.81\left(\frac{T_B}{T_c}\right)^2 \exp\left[-1.95\left(\frac{T_B}{T_c}\right)\right] \tag{5.14-11}$$

When experimental Boyle temperatures are not available, the value may be predicted using

$$\frac{T_B}{T_c} = 2.0525 + 0.6428 \exp[-3.6167\omega] \tag{5.14-12}$$

Example 5.16

Estimate the second virial coefficient of ethanol at 400 K using the Iglesias-Silva and Hall (2001) method.

Solution. From Appendix A, $T_c = 514.0$ K, $P_c = 61.37$ bar $= 60.567$ atm, $V_c = 0.000168$ m^3/mol, $\omega = 0.643558$, and $\mu = 1.69087$ Debye. With these values,

$$\mu_r = (10^5)(1.69087)^2(60.567)/514.0^2 = 65.544$$

Eqs. (5.14-9)–(5.14-12) give the following.

$T_B = (514.0)(2.0525 + 0.6428 \exp[(-3.6167)(0.643558)] = 1087.2$ K

$B_c = 0.000168[-1.1747 - (0.3668)(0.643558) - (0.00061)(65.544)] = -0.0002437$ m^3/mol

$n = 1.4187 + (1.2058)(0.643558) = 2.1947$

$b_0 = 0.000168\{0.1368 - (0.4791)(0.643558) + 13.81(1087.2/514)^2 \exp[-(1.95)(1087.2/514)]\} = 0.0001390$ m^3/mol.

With these values, and the conditions specified (i.e. 400 K), B_2 is found by Eq. (5.14-8) through

$$\frac{T_B}{T} = 1087.2/400 = 2.7180$$

$$\frac{T_B}{T_c} = 1087.2/514 = 2.1152$$

$$\left(\frac{T_c}{T}\right)^n = (514/400)^{2.1947} = 1.7338$$

$$B_2 = (0.0001390)(2.7180)^{0.2}\,[1 - (2.7180)^{0.8}]\{-0.0002437/[(0.0001390)(2.1152^{0.2} - 2.1152)]\}^{1.7338}$$
$$= -0.000598\ \text{m}^3/\text{mol} = -598\ \text{cm}^3/\text{mol}$$

The recommended experimental value is -535 cm^3/mol (Tsonopoulos et al., 1989). Thus the difference is

$$-598 - (-535) = 63\ \text{cm}^3/\text{mol or } 11.7\%.$$

5.14.1.3 Molecular-based Methods

The virial equation of state, discussed more in Section 6.3, can be derived from statistical mechanics and thus has a rigorous theoretical basis not shared by other equations of state (except ideal gas). Specifically, for spherically symmetric models, B_2 is related to interactions between pairs of molecules according to (McQuarrie 2000)

$$B_2(T) = -2\pi\int_0^{\infty}\left[e^{-\beta u(r)} - 1\right]r^2 dr \tag{5.14-13}$$

where $\beta = \dfrac{1}{kT}$, k is Boltzmann's constant, r is the distance between two molecules, and $u(r)$ is the potential between the two molecules at distance r. Common intermolecular potentials include the hard sphere, Lennard-Jones, and Mie models. For the hard-sphere (HS) potential,

$$u(r) = \begin{cases} \infty & r \le \sigma_{HS} \\ 0 & r > \sigma_{HS} \end{cases} \tag{5.14-14}$$

where σ_{HS} is the hard-sphere diameter of the molecule. When this expression is inserted into Eq. (5.14-13), and the results simplified, the following is obtained for the second virial coefficient for the hard sphere model ($B_{2,HS}$)

$$B_{2,HS}(T) = \frac{2}{3}\pi\sigma_{HS}^3 \tag{5.14-15}$$

This solution is often used as a normalization factor for other potentials and is given the symbol b_0 (i.e., $b_0 = \frac{2}{3}\pi\sigma_{HS}^3$).

An exact solution is also available for square well spheres,

$$B_{2,SW}(T) = \frac{2}{3}\pi\sigma_{SW}^3\left\{1 + \left(\lambda_{SW}^3 - 1\right)\left[\exp\left(\beta\varepsilon_{SW}\right) - 1\right]\right\} \tag{5.14-16}$$

where λ_{SW} is the width of the square well model. A typical value is 1.7.

More complex potentials can also be solved analytically. The well-known Lennard-Jones (LJ) potential is

$$u(r) = 4\epsilon\left[\left(\frac{\sigma}{r}\right)^{12} - \left(\frac{\sigma}{r}\right)^{6}\right] \tag{5.14-17}$$

where ϵ is the parameter describing the depth of the attractive well between the two molecules and σ is the parameter related to the size of the molecule and the distance overwhich interactions are experienced between the two molecules. Combining this potential with Eq. (5.14-13) gives the expression below for the second virial coefficient for the Lennard-Jones potential ($B_{2,LJ}$).

$$\frac{B_{2,\mathrm{LJ}}(T)}{b_0} = -\sum_{i=0}^{\infty} \frac{2^{\frac{2n+1}{2}}}{4n!}\left(\frac{\epsilon}{kT}\right)^{\frac{2n+1}{4}} \Gamma\left(\frac{2n-1}{4}\right) \tag{5.14-18}$$

Here, $\Gamma(x)$ is the gamma function of x. The series converges quickly—within 100 terms and usually much fewer. Readers should determine the accuracy they need then add terms until they achieve the desired accuracy. Twenty terms usually suffice for four-digit precision.

The Mie potential, Eq. 6.8-59, is a generalization of the LJ formalism which allows the exponents in Eq. (5.14-14) to vary (Mie, 1903). Sadus (2018) gives the following exact solution for the second virial coefficient of this model ($B_{2,\mathrm{Mie}}$)

$$\frac{B_{2,\mathrm{Mie}}(T)}{b_0} = y^{\left[\frac{3}{n-m}\right]}\left[\Gamma\left(\frac{n-3}{n}\right) - \frac{3}{n}\sum_{i=1}^{\infty}\Gamma\left(\frac{im-3}{n}\right)\frac{y^i}{i!}\right] \tag{5.14-19}$$

where $y^n = \left(\frac{n}{n-m}\right)^n\left(\frac{n-m}{n}\right)^m\left(\frac{\epsilon}{kT}\right)^{n-m}$ and n and m are the Mie exponents. Eq. (5.14-19) reduces to Eq. (5.14-18) when $n = 12$ and $m = 6$.

Second virial coefficients for most nearly spherical molecules like noble gases, methane, and carbon monoxide can be easily correlated using either Eq. (5.14-16), Eq. (5.14-18), or Eq. (5.14-19). Exceptions include helium, hydrogen, and neon, in which cases which quantum effects must be considered. Recent work by Garberoglio et al. (2017), Schultz et al. (2019), and Aasen et al. (2019) shows how potential functions accounting for quantum effects can provide good accuracy in these cases.

Non-spherical molecules typically require numerical integration, but the procedure is relatively simple. The key modification is that the orientational effect must be averaged before performing the integration of Eq. (5.14-13). This orientational average can be achieved by summing over all molecular pairs from a single molecular configuration of a Monte Carlo or molecular dynamics simulation (Elliott, 2021). The procedure is remarkably simple and provides coefficients of temperature from a single simulation sample. The averaged potential function can also provide useful insights into the ways in which large molecules interact at low density. Their repulsions are relatively "soft," for example, when viewed from the center of mass perspective, because many combinations of orientation and conformation result in interaction energies approaching zero even when the centers of mass are closer than the radius of gyration. Impacts of density on molecular conformation are also evident through the analysis.

5.14.1.4 Discussion and Recommendations

Second virial coefficients from molecular methods can be calculated if the parameters of the intermolecular potential model are known and sample configurations are available. The potential model can be estimated using a transferable force field or by using quantum mechanical calculations to characterize a multiparameter force field. Unfortunately, the unavailability of sample configurations and a lack of time precluded evaluation of this method.

A test database was created to evaluate the performance of several prediction methods for second virial coefficient other than molecular approaches. This test database consisted of experimental data for the property

TABLE 5.23 Performance of prediction techniques for B_2 against the test database.

	Compound Class											
	Normal			Heavy			Polar			Associating		
Method	N_{pt}	N_c	AARD	N_{pt}	N_c	AARD	N_{pt}	N_c	AARD	N_{pt}	N_c	AARD
TS	4231	185	14.3%	92	5	10.3%	505	24	16.8%	1031	57	16.8%
SH	4231	185	11.1%	92	5	16.8%	505	24	15.1%	1031	57	18.5%
TD	4206	183	16.1%	92	5	27.1%	505	24	18.8%	1031	47	23.3%
MD	2436	108	15.3%	0	0	—	243	14	9.6%	855	48	20.5%

from the October 2021 release of the DIPPR database. A similar compilation of virial coefficients is given by Frenkel and Marsh (2002). A total of 5859 temperature points from 271 compounds were used in the evaluation, and the Tsonopoulos et al. (1975, 1978, 1979, 1989, 1990, 1997), Iglesias-Silva and Hall (2001), Tarakad and Danner (1977), and McCann and Danner (1984) (abbreviated below as TS, SH, TD, and MD, respectively) methods were assessed. Compounds were organized into the same classes as those used to evaluate ρ_s (see the end of Sec. 5.2) to determine how each method worked for different types of compounds. Results are quantified using the average absolute relative deviation given by Eq. (5.2-14). Table 5.23 contains the AARD of each method according to compound class. Also listed are the number of points tested in each class (N_{pt}) and number of compounds (N_c).

When at least one reliable experiment point for B_2 is available, the best approach to predicting second virial coefficients at other temperatures is to fit the parameters found in a method cited above to these data. For example, the a and/or b parameters for the Tsonopolous method can be varied to give the best agreement with the available experimental data, and then B_2 values at other temperatures can be reliably obtained. In the absence of data, most of the recent prediction methods perform equally well as indicated in Table 5.23. Of those surveyed, Tarakad and Danner (1977) should be avoided as it performs consistently worse than other approaches. Iglesias-Silva and Hall (2001) is slightly better for compounds in the normal class than other methods. Overall, errors between 10 and 20% can be expected when using the Tsonopolous, Iglesias-Silva and Hall, or McCann and Danner formalisms.

All the methods can produce wild errors, and no consistent patterns could be found for such occurrences. Due to this, and the fact that several of the methods perform equally well, the family analysis approach discussed in Chap. 3 can be followed to help identify which method to use. Comparing the predictions from the various methods, and ensuring that they agree to within 10–20%, can also increase the confidence in the values obtained and identify outliers.

5.14.2 Third Virial Coefficients

As with second virial coefficients, it is possible to derive third virial coefficient (B_3) correlations from molecular theory, but these are not generally accurate. A significant theoretical problem is that the three-body intermolecular potential includes significant contributions that cannot be determined from the pairwise potentials that describe second virial coefficients. Corresponding states approaches have been proposed to estimate B_3, but the range of substances considered is very limited. From a practical standpoint, B_3 is not very useful. A brief comparison of the virial equation of state is given in Chap. 6, including the exact solution for the third virial coefficient of square well spheres. The benefit of the virial expansion is that truncation after the second term (B_2) yields an equation for pressure that is quadratic in density and easily solvable. The reason for adding B_3 is to model the PVT behavior of the vapor more accurately than is possible with B_2 alone, but the linear-in-volume advantage is removed when this is done as the equation then becomes cubic in volume. Other cubic equations of state, discussed in Chap. 6, have been developed to a much greater extent than B_3, so the former are usually selected in practice over than the latter. However, the virial equation of state, including B_3, is important from theoretical and pedagogical standpoints, and is worthwhile in some practical cases such as supercritical mixtures, so a few estimation methods have been developed for normal fluids.

The principal techniques for B_3 are the corresponding states methods of Chueh and Prausnitz (1967), De Santis and Grande (1979), and Orbey and Vera (1983). All use $T^* = T_c$ in the equation

$$\frac{B_3(T)}{V^{*2}} = \sum_i a_i g^{(i)}\left(\frac{T}{T^*}\right) \tag{5.14-20}$$

but they differ in the choice of V^* and of the third parameter. Chueh and Prausnitz select $V^* = V_c$ and use a special third parameter that must be found from B_3 data. De Santis and Grande use $V^* = V_c$ while reformulating Chueh and Prausnitz' expressions for the $g^{(i)}\left(\frac{T}{T^*}\right)$ and choose to correlate their special third parameter with ω and other molecular properties. The correlation of Orbey and Vera uses the more accessible $V^* = \frac{RT_c}{P_c}$ and ω directly. They take two terms in the series of Eq. (5.14-20) with $a_0 = 1$ and $a_1 = \omega$ and

$$g^{(0)} = 0.01407 + \frac{0.02432}{T_r^{2.8}} - \frac{0.00313}{T_r^{10.5}} \tag{5.14-21}$$

$$g^{(1)} = -0.02676 + \frac{0.01770}{T_r^{2.8}} + \frac{0.04}{T_r^{3}} - \frac{0.003}{T_r^{6}} - \frac{0.00228}{T_r^{10.5}}$$

The correlation is the best available, and its estimates should be adequate for simple and normal substances over the range of conditions that Table 6.3 indicates that Eq. (6.5-1) should be used. There is no estimation method for third virial coefficients of polar and associating substances.

5.14.3 Higher Order Virial Coefficients

For B_4 and higher virial coefficients, a molecular method is the only reliable means to make predictions. Open-source programs (e.g., Etomica, 2022; Schultz and Kofke, 2015) for computing these rigorous virial coefficients are available and ready for implementation. For example, Kofke and coworkers have explored virial coefficients for many molecular models, including n-alkanes (Schultz and Kofke, 2010), methanol (Shaul et al., 2010), water (Benjamin et al. 2007, 2009), and many other compounds. Virial coefficients are mostly available through B_5, but higher-order coefficients are available in a few cases. For example, up to B_{16} is available for the Lennard-Jones model (Feng et al., 2015). The perturbative virial coefficients mentioned in Chap. 6 are available for square-well spheres through B_9 and n-alkane chains through B_5.

5.15 NOTATION

$A, a, B, b, C, c, D, d, \ldots$	parameters in correlating equations or prediction methods
A^+, B^+, C^+, D^+	dimensionless parameters Riedel-based prediction methods for P^{vp}
a_i	parameter in predictions of B_2 or B_3
$AARD$	average absolute relative deviation
b_0	parameter in Iglesias-Silva and Hall prediction method for B_2

B_c	parameter in Iglesias-Silva and Hall prediction method for B_2
B_2	second virial coefficient
$B_3, B_4, \ldots$	third virial coefficient, fourth virial coefficieint, …
$c_{CH_2}, c_{C(X)_k}, c_{X_m}$	coefficients in Chickos and Acree prediction method of ΔS_{tot} and $\Delta_m H$
C_p	isobaric heat capacity
$C_p^{liq}, C_\sigma^{liq}, C_{sat}^{liq}$	liquid heat capacities commonly encountered
$C_p^{sol}, C_\sigma^{sol}$	solid heat capacities commonly encountered
$C_p^{vap}, C_\sigma^{vap}$	vapor heat capacities commonly encountered
C_p^{r}	residual isobaric heat capacity
C_p°	ideal gas isobaric heat capacity
$f^{(0)}, f^{(1)}, \ldots, f^{(i)}$	1) parameters in Lee-Kesler prediction method for P_{vp}, 2) parameters in prediction of B_2
$g^{(i)}$	parameters in prediction of B_3
H^{liq}	enthalpy of the liquid phase
H^{vap}	enthalpy of the vapor phase
K	parameter in Riedel-based prediction methods for P^{vp}
k	Boltzmann's constant
M	molar mass
$MARD$	maximum absolute relative deviation
N_c	number of compounds in test set database
N_C	number of carbon atoms in the Bondi prediction method for ΔS_{tot}
$N_{CH_2,con}$	number of consecutive methylene groups in Chickos and Acree prediction of ΔS_{tot} and $\Delta_m H$
$N_{CH_2,non}$	number of non-consecutive methylene groups in Chickos and Acree prediction of ΔS_{tot} and $\Delta_m H$
n	parameter in Iglesias-Silva and Hall prediction method for B_2
N_i	number of times group i is found in a compound
N_{pt}	number of points in test set database
N_{rings}	number of rings in a compound
$N_{ring\ atoms}$	number of ring atoms in a compound
N_x	number of halogen atoms in compound
P	system pressure
P_{atm_r}	reduced atmospheric pressure; $\frac{1.01325\ \text{bar}}{P_c}$
P_c	critical pressure
P^{svp}	solid vapor pressure (sublimation pressure)
P_{tp}	triple point pressure
P_{tp_r}	reduced triple point pressure; $\frac{P_{tp}}{P_c}$
P^{vp}	saturated liquid vapor pressure
P_{corr}^{vp}	saturated liquid vapor pressure calculated from a correlation
P_r^{vp}	reduced saturated liquid vapor pressure; $\frac{P^{vp}}{P_c}$
Q	1) heat; 2) parameter in Riedel-based prediction methods for P^{vp}
R	gas constant
r	distance between two molecules
R_g	radius of gyration

$RINGS$	number of fused-ring systems in compound
$SP2$	number of sp^2 chain atoms
$SP3$	number of sp^3 chain atoms
T	system temperature
T_B	Boyle temperature
T_b	normal ($P = 101325$ Pa) boiling temperature
T_{b_r}	reduced normal boiling temperature; $\frac{T_b}{T_c}$
T_c	critical temperature
T_r	reduced temperature; $\frac{T}{T_c}$
T^R	reference temperature
T_m	melting temperature
T_{tp}	triple point temperature
T_{tp_r}	reduced triple point temperature; $\frac{T_{tp}}{T_c}$
T^*	characteristic temperature
$u(r)$	potential between two molecules
$V^{(0)}, V^{(\delta)}$	parameters in Yamada and Gunn (1973) prediction of V_s
V^*	(1) parameter in Yamada and Gunn (1973) prediction of V_s; (2) characteristic volume
V_c	critical volume
V^{vap}	volume of vapor phase
V^{liq}	volume of liquid phase
V_s^R	experimental saturated volume at reference temperature T^R in Yamada and Gunn (1973) prediction of V_s
V_s, V_σ^{liq}	saturated liquid volume
V_σ^{vap}	saturated vapor volume
x	x coordinate in a shifted-rotated plot
X	scaled x coordinate in a shifted-rotated plot
X_c	parameter in Riedel-based prediction methods for P^{vp}
y	y coordinate in a shifted-rotated plot
Y	scaled y coordinate in a shifted-rotated plot
Z	compressibility factor
Z_c	critical compressibility factor
Z^{vap}	compressibility factor of vapor phase
Z^{liq}	compressibility factor of liquid phase

Greek

α, α_c	parameters in Riedel-based predictions for P^{vp}
β	$\frac{1}{kT}$
δ_{bondi}	parameter in Bondi method to predict $\Delta_m H$
ϵ	parameter in Lennard Jones potential
$\Delta_m H$	enthalpy (heat) of melting; change in enthalpy as system moves from a solid to a liquid
$\Delta_m H_{homo}$	enthalpy (heat) of melting of homomorph used in the Bondi prediction method for ΔS_{tot}
$\Delta_{sub} H$	enthalpy (heat) of sublimation; change in enthalpy as system moves from a saturated solid to a saturated vapor

$\Delta_v H$	enthalpy (heat) of vaporization; change in enthalpy as system moves from a saturated liquid to a saturated vapor
$\Delta_m S_{homo}$	entropy of melting of homomorph used in the Bondi prediction method for ΔS_{tot}
$\Delta_m S$	entropy of melting; change in entropy as system moves from a saturated solid to a saturated liquid
ΔS_{tot}	total change in entropy as a substance moves from a solid at 0 K to a saturated liquid at T_m
$\Delta S_{tot,aah}$	acyclic and aromatic hydrocarbon contribution to total change in entropy for the Chickos and Acree method for prediction of ΔS_{tot}
$\Delta S_{tot,corr}$	correction to parent ring contribution to total change in entropy for the Chickos and Acree method for prediction of ΔS_{tot}
$\Delta S_{tot,\ corr,X}$	functional group correction to parent ring contribution to total change in entropy for the Chickos and Acree method for prediction of ΔS_{tot}
$\Delta S_{tot,C(X)}$	carbon attached to a functional group-contribution to total change in entropy for the Chickos and Acree method for prediction of ΔS_{tot}
$\Delta S_{tot,ring}$	parent ring contribution to total change in entropy for the Chickos and Acree method for prediction of ΔS_{tot}
$\Delta S_{tot,\ X}$	functional group-contribution to total change in entropy for the Chickos and Acree method for prediction of ΔS_{tot}
$\Delta_v U$	change in internal energy as system moves from a saturated liquid to a saturated vapor
$\Delta_{sub} V$	change in volume as system moves from a saturated solid to a saturated vapor
$\Delta_v V$	change in volume as system moves from a saturated liquid to a saturated vapor
$(\Delta_a)_i, (\Delta_b)_j, (\Delta_c)_k$	contribution for group i, j, or k for Goodman et al. prediction method for $\Delta_{sub} H$
$\left(\Delta_{\Delta S_{tot,aah}}\right)_i$	contribution for acyclic and aromatic hydrocarbon group i for Chickos and Acree prediction of ΔS_{tot}
$\left(\Delta^A_{C_p^{liq}}\right)_i, \left(\Delta^B_{C_p^{liq}}\right)_i, \left(\Delta^C_{C_p^{liq}}\right)_i$	contributions for group i for Růžička and Domalski prediction of C_p^{liq}
$\left(\Delta_{V_s}\right)_i$	contribution for group i for prediction of V_s
$\left(\Delta_{S_{tot}}\right)_i, \left(\Delta_{S_{tot,corr}}\right)_j, \left(\Delta_{S_{tot,C(X)}}\right)_k, \left(\Delta_{S_{tot,X}}\right)_l, \left(\Delta_{S_{tot,corr,X}}\right)_j$	contribution for group i, j, k, or l for prediction of ΔS_{tot}
$\Delta_v Z$	change in compressibility factor as system moves from a saturated liquid to a saturated vapor
ϕ	(1) exponent in Yamada and Gunn (1973) prediction of V_s; (2) parameter in creating a shifted-rotated plot; (3) parameter in Riedel-based predictions for P^{vp}
$\Gamma(x)$	gamma function of x
μ	diplole moment
μ_r	reduced dipole moment
ψ_b	parameter in Riedel-based prediction methods for P^{vp}
ρ_s	saturated liquid density

σ	(1) symmetry number in Dannenfelser and Yalkowsky prediction method for ΔS_{tot}; (2) parameter in Lennard Jones potential
σ (subscript)	indicates process occur while maintaining the system at saturation
τ	(1) $(1 - T_r)$; (2) number of torsion angles in compound
ω	acentric factor
ω_{lk}	parameters in Lee-Kesler prediction method for P^{vp}
ω_p	parameter in Vetere prediction for P^{vp}
ω_{SRK}	value of acentric factor that causes the Soave (1972) equation of state to best fit pure component vapor pressures in Yamada and Gunn (1973) prediction of V_s

5.16 REFERENCES

Aasen, A., M. Hammer, A. Ervik,2 E.A. Müller, O. Wilhelmsen, *J. Chem. Phys.,* **151:** 064508 (2019).

Abusleme, J. A., and J. H. Vera: *AIChE J.,* **35:** 481 (1989).

Acree, W., J.S. Chickos, *J. Phys. Chem. Ref. Data,* **45:** 033101 (2016).

Acree, W., J.S. Chickos, *J. Phys. Chem. Ref. Data,* **46:** 013104 (2017).

Ambrose, D.: *J. Chem. Thermodynamics,* **18:** 45 (1986).

Ambrose, D., and R. H. Davies: *J. Chem. Thermodynamics,* **12:** 871 (1980).

Ambrose, D., and N. B. Ghiassee: *J. Chem. Thermodynamics,* **19:** 505 (1987a).

Ambrose, D., and N. B. Ghiassee: *J. Chem. Thermodynamics,* **19:** 903 (1987b).

Ambrose, D., and N. B. Ghiassee: *J. Chem. Thermodynamics,* **19:** 911 (1987c).

Ambrose, D., and N. B. Ghiassee: *J. Chem. Thermodynamics,* **20:** 765 (1988a).

Ambrose, D., and N. B. Ghiassee: *J. Chem. Thermodynamics,* **20:** 1231 (1988b).

Ambrose, D., and N. B. Ghiassee: *J. Chem. Thermodynamics,* **22:** 307 (1990).

Ambrose, D., and J. Walton: *Pure & Appl. Chem.,* **61:** 1395 (1989).

Ambrose, D., J. F. Counsell, and C. P. Hicks: *J. Chem. Thermodynamics,* **10:** 771 (1978).

Ambrose, D., N. B. Ghiassee, and R. Tuckerman: *J. Chem. Thermodynamics,* **20:** 767 (1988).

Ambrose, D., M. B. Ewing, N. B. Ghiassee, and J. C. Sanchez Ochoa: *J. Chem. Thermodynamics,* **22:** 589 (1990).

Antoine, C.: *C.R.,* **107:** 681, 836 (1888).

Aston, J. G., H. L. Fink, A. B. Bestul, E. L. Page, and G. J. Szasz: *J. Am. Chem. Soc.,* **68:** 52 (1946).

Benjamin, K. M., A. J. Schultz, and D. A. Kofke: *J. Phys. Chem. C,* **43:** 16021–16027 (2007). *J. Phys. Chem. B,* **113:** 7810–7815 (2009).

Bloomer, O. T.: *Ind. Eng. Chem. Res.,* **29:** 128 (1990).

Bondi, A.: *J. Chem. Eng. Data,* **8:** 371 (1963).

Bondi, A.: "Physical Properties of Molecular Crystals, Liquids and Glasses," Wiley, New York, 1968.

Boublik, T., V. Fried, and E. Hala: *The Vapor Pressures of Pure Substances,* 2d rev. ed., Elsevier, New York, 1984.

Calingaert, G., and D. S. Davis: *Ind. Eng. Chem.,* **17:** 1287 (1925).

Campanella, E. A.: *J. Chem. Eng. Japan,* **28:** 234 (1995).

Chueh, C. F., and A. C. Swanson: *Chem Eng. Progr.,* **69**(7): 83 (1973a).

Chueh, C. F., and A. C. Swanson: *Can. J. Chem. Eng.,* **51:** 596 (1973b).

Chueh, P. L., and J. M. Prausnitz: *AIChE J.,* **13:** 896 (1967).

Chickos, J. S. and D. G. Hesse: *J. Org. Chem.,* **55:** 3833 (1990).

Chickos, J. S., C. M. Braton, D. G. Hesse, and J. F. Liebman: *J. Org. Chem.,* **56:** 927 (1991).

Chickos, J. S., W. E. Acree Jr., and J. F. Liebman: *Estimating Phase Change Enthalpies and Entropies, in Computational Thermochemistry, Prediction and Estimation of Molecular Thermodynamics.* D. Frurip and K. Irikura, (eds.), ACS Symp. Ser. 677, p. 63, ACS, Washington, D. C., 1998.

Chickos, J. S., W. E. Acree Jr., and J. F. Liebman: *J. Phys. Chem. Ref. Data,* **28**(6): 1535 (1999).

Chickos, J. S. and W. E. Acree Jr.: *Thermochim. Acta,* **395:** 59 (2003).

Chickos, J. S. and W. E. Acree Jr.: *Thermochim. Acta,* **495:** 5 (2009).

Clapeyron, E. J.: *J. de l'Ecole Polytech.,* **14:** 153–190 (1834).

Cox, E. R.: *Ind. Eng. Chem.,* **15:** 592 (1923).

Dai, L.-Y., Q. Li, M. Lei, and Y.-Q. Chen: *J. Chem. Eng. Data*, **55:** 1704–1707 (2010).
Dannenfelser, R.-M., and S. H. Yalkowsky: *Ind. Eng. Chem. Res.,* **35:** 1483 (1996).
Daubert, T. E., R. P. Danner, H. M. Sibel, and C. C. Stebbins: *Physical and Thermodynamic Properties of Pure* Chemicals*: Data Compilation,* Taylor & Francis, Washington, D. C., 1997.
Davies, M. and B. Kybett: *Trans. Faraday Soc.*, **61:** 1608 (1965).
de Kruif, C. G., J. C. van Miltenburg, and J. G. Blok: *J. Chem. Thermodyn.*, **15:** 129 (1983).
Defibaugh, D. R., and M. R. Moldover: *J. Chem. Eng. Data,* **42:** 160 (1997).
Dellesite, A.: *J. Phys. Chem. Ref. Data,* **26:** 157 (1997).
DeSantis, R., and B. Grande: *AIChE J.,* **25:** 931 (1979).
Domalski, E. S., and E. D. Hearing: *J. Phys. Chem. Ref. Data,* **22:** 805 (1993).
Domalski, E. S., and E. D. Hearing: *J. Phys. Chem. Ref. Data,* **25:** 1 (1996).
Dreisbach, R. R.: *Pressure-Volume-Temperature Relationships of Organic Compounds,* 3d ed., McGraw-Hill, New York, 1952.
Dreisbach, R. R.: Physical Properties of Chemical Compounds, *Advan. Chem. Ser., ACS Monogr. 15 and 22,* Washington, D.C., 1955, 1959.
Dykyj, J., and M. Repa´: *The Vapour Pressures of Organic Compounds* (in Slovak), Veda, Bratislava, 1979.
Dykyj, J., M. Repa´ and J. Svoboda: *The Vapour Pressures of Organic Compounds* (in Slovak), Veda, Bratislava, 1984.
Edwards, D., Van de Rostyne, C. G., Winnick, J., and J. M. Prausnitz: *Ind. Eng. Chem. Process Des. Dev.,* **20:** 138 (1981).
Elbro, H. S., A. Fredenslund, and P. Rasmussen: *Ind. Eng. Chem. Res.,* **30:** 2576 (1991).
Elliott, J.R., *J. Phys. Chem. B,* 125: 4494–4500 (2021).
Etomica, https://github.com/etomica, (2022).
Feng, C. A. J. Schultz, V. Chaudhary, and D. A. Kofke: *J. Chem. Phys.*, **143:** 044504 (2015).
Fishtine, S. H.: *Ind. Eng. Chem.,* **55**(4): 20, **55**(5): 49, **55**(6): 47(1963); *Hydrocarbon Process. Pet. Refiner,* **42**(10): 143(1963).
Flores, H. and P. Amador: *J. Chem. Thermodyn.*, **36**(11): 1019 (2004).
Frenkel, M. and K.N. Marsh, Eds. "Virial Coefficients of Pure Gases and Mixtures," Landolt-Börnstein—Group IV Physical Chemistry Volume 21A, Springer-Materials, Springer-Nature, Switzerland AG (2002).
Gambill, W. R.: *Chem. Eng.,* **64**(5): 263, **64**(6): 243, **64**(7): 263; **64**(8): 257 (1957).
Garberoglio, G., P. Jankowski, K. Szalewicz, A.H. Harvey, J. Chem. Phys., **146:** 054304 (2017).
Goodman, B. T., W. V. Wilding, J. L. Oscarson, and R. L. Rowley: *Int. J. Thermophys.*, **25**(2): 337 (2004).
Guildner, L., D.P. Johnson, and F.E. Jones, *J. Res. Nat. Bur. Stand.* **80A:** 505 (1976).
Green, D. W. and M. L. Southard, eds.: *Perry's Chemical Engineers' Handbook,* 9th ed., McGraw-Hill, New York, 2019.
Hankinson, R. W., and G. H. Thomson: *AIChE J.,* **25:** 653 (1979).
Hayden, J. G., and J. P. O'Connell: *Ind. Eng. Chem. Process Des. Dev.,* **14:** 209 (1975).
Hogge, J. W., N. F. Giles, R. L. Rowley, T. A. Knotts IV, and W. V. Wilding: *Ind. Eng. Chem. Res.*, **56:** 14678–14685 (2017).
Hogge, J. W., N. F. Giles, T. A. Knotts IV, R. L. Rowley, and W. V. Wilding: *Fluid Phase Equil.*, **429:** 149–165 (2016a).
Hogge, J. W., R. Messerly, N. Giles, T. Knotts, R. L. Rowley, and W. V. Wilding: *Fluid Phase Equil.*, **418:** 37–43 (2016b).
Huber, M.L., A. Laesecke, and D.G. Friend, *Ind. Eng. Chem. Res.*, **45:** 7351–7361 (2006).
Iglesias-Silva, G. A. and K. R. Hall: *Ind. Eng. Chem. Res.*, **40:** 1968 (2001).
Khasashin, T. S. and T. B. Zykova: *Inzh.-Fiz. Zh.*, **56:** 991–994 (1989).
King, M. G., and H. Al-Najjar: *Chem. Eng. Sci.,* **29:** 1003 (1974).
Kis, K. and H. Orbey: *Chem Eng. J.,* **41:** 149 (1989).
Ledanois, J.-M., C. M. Colina, J. W. Santos, D. González-Mendizabal, and C. Olivera-Fuentes:Ind. *Eng. Chem. Res.,* **36:** 2505 (1997).
Lee, B. I., and M. G. Kesler: *AIChE J.,* **21:** 510 (1975).
Lee, M.-J., and J.-T. Chen: *J. Chem. Eng. Japan,* **31:** 518 (1998).
Li, P., P.-S. Ma, S.-Z. Yi, Z.-G. Zhao, and L.-Z. Cong: *Fluid Phase Equil,* **101:** 101 (1994).
Rumble, J.: *CRC Handbook of Chemistry and Physics,* 102nd ed., CRC Press, Taylor and Francis, 2021.
Macknick, A. B., J. Winnick, and J. M. Prausnitz: *AIChE J.,* **24:** 731 (1978).
Majer, V., and V. Svoboda: Enthalpies of Vaporization of Organic Compounds, A Critical Review and Data Compilation, *IUPAC Chem. Data Ser No.* 32, Blackwell Sci. Pub., Oxford, 1985.
Majer, V., V. Svoboda, and J. Pick: Heats of Vaporization of Fluids, *Studies in Modern* Thermodynamics **9,** Elsevier, Amsterdam, 1989.
Maxwell, J. B. and L. S. Bonnell: "Vapor Pressure Charts for Petroleum Engineers," Esso Research and Engineering Company, Linden, New Jersey (1955).
McCann, D. W., and R. P. Danner: *Ind. Eng. Chem. Process Des. Dev.,* **23:** 529 (1984).
McGarry, J.: *Ind. Eng. Chem. Process Des. Dev.,* **22:** 313 (1983).

McQuarrie, D. A.: *Statistical Mechanics*, University Science Books, Sausalito, CA, 2000.
Mie, G.: *Ann. Phys.*, **11:** 657 (1903).
Missenard, F.-A.: *Compte Rend.,* **260:** 5521 (1965).
Moelwyn-Hughes, E. A.: *Physical Chemistry,* 2d ed., pp. 699–701, Pergamon Press, New York, 1961.
Myrdal, P. B., and S. H. Yalkowsky: *Ind. Eng. Chem. Res.,* **36:** 2494 (1997).
Nitta: *Nippon Kagaku Zasshi*, **69:** 85 (1948).
Ohe, S.: *Computer Aided Data Book of Vapor Pressure, Data Book,* Tokyo, 1976.
Oja, V., and E. M. Suuberg: *J. Chem Eng. Data,* **43:** 486 (1998).
Oja, V., and E. M. Suuberg: *J. Chem Eng. Data,* **44:** 26 (1999).
Olf, G., J. Spiske, and J. Gaube: *Fluid Phase Equil.,* **51:** 209 (1989).
Orbey, H., and J. H. Vera: *AIChE J.,* **29:** 107 (1983).
Orbey, H.: *Chem. Eng. Comm.,* **65:** 1 (1988).
Othmer, D. F. and E. Yu: *Ind. Eng. Chem.*, **60:** 22 (1968).
Pitzer, K. S. and R. F. Curl: *J. Am. Chem. Soc.,* **77:** 3427 (1955).
Plank, R. and L. Riedel: *Ing. Arch.,* **16:** 255 (1948).
Poling, B. E.: *Fluid Phase Equil.,* **116:** 102 (1996).
Pouillot, F. L. L., D. Chandler, and C. A. Eckert: *Ind. Eng. Chem. Res.,* **35:** 2408 (1996).
Prausnitz, J. M., R. N. Lichtenthaler, and E. G. de Azevedo: *Molecular Thermodynamics of Fluid-Phase* Equilibria*,* 3d ed. Prentice Hall, Englewood Cliffs, New Jersey, 1999, pp. 638–641.
Rackett, H. G.: *J. Chem. Eng. Data,* **15:** 514 (1970).
Riedel, L: *Chemie Ingenieur Technik*, **26:** 83–39 (1954).
Růžička, V. and E. S. Domalski: *J. Phys. Chem. Ref. Data*, **22:** 597–618 (1993a).
Růžička, V. and E. S. Domalski: *J. Phys. Chem. Ref. Data*, **22:** 619–657 (1993b).
Růžička, K. and V. Majer: *J. Phys. Chem. Ref. Data,* **23:** 1 (1994).
Sadus, R. J.: *J. Chem. Phys.*, **149:** 074504 (2018).
Schlinger, W. G. and B. H. Sage: *Ind. Eng. Chem.*, **41:** 1779–1782 (1949).
Schultz, A. J. and D. A. Kofke: *J. Chem. Phys.*, 1**33:** 104101 (2010).
Schultz, A. J. and D. A. Kofke: *J. Comput. Chem.*, **36:** 573–583 (2015).
Schultz, A. J. and D. A. Kofke: *J. Chem. Eng. Data,* **64**, 3742–3754 (2019).
Shaul, K. R. S., A. J. Schultz, and D. A. Kofke: *Mol. Simulat.* **36:** 1282–1288 (2010).
Smith, G., J. Winnick, D. S. Abrams, and J. M. Prausnitz: *Can. J. Chem. Eng.,* **54:** 337 (1976).
Smith, J. M., H. C. Van Ness, M. M. Abbott, and M. T. Swihart: *Introduction to Chemical Engineering Thermodynamics,* 8th ed., McGraw-Hill, New York, 2018.
Speight, J. G.: *Lange's Handbook of Chemistry,* 17th ed., McGraw-Hill, New York, 2017.
Spencer, C. F. and R. P. Danner: *J. Chem. Eng. Data,* **17:** 236 (1972).
Spencer, C. F. and S. B. Adler: *J. Chem. Eng. Data,* **23:** 82 (1978).
Steele, W. V., R. D. Chirico, S. E. Knipmeyer, and A. Nguyen: *J. Chem. Eng. Data*, **47:** 648–666 (2002).
Stuper, A. J., W. E. Brugger, and P. C. Jurs: *Computer Assisted Studies of Chemical Structure and Biological Function*, John Wiley, New York, 1979.
Takeda, K., O. Yamamuro, and H. Suga: *J. Phys. Chem. Solids*, **52:** 607–615 (1991).
Tamir, A., E. Tamir, and K. Stephan: *Heats of Phase Change of Pure Components and Mixtures,* Elsevier, Amsterdam, 1983.
Tarakad, R. R., and R. P. Danner: *AIChE J.,* **23:** 685 (1977).
Thek, R. E., and L. I. Stiel: *AIChE J.,* **12:** 599 (1966), **13:** 626 (1967).
Thodos, G.: *Ind. Eng. Chem.* **42:** 1514 (1950).
Thomson, G. W.: *Techniques of Organic Chemistry,* A. Weissberger (ed.), 3d. ed., vol. I, pt. I, p. 473, Interscience, New York, 1959.
Thompson, G. H., K. R. Brobst, and R. W. Hankinson: *AIChE J.*, **28:** 671 (1982).
Todd, S. S. and G. S. Parks: *J. Am. Chem. Soc.*, **58:** 134–137 (1936).
Tsonopoulos, C.: *AIChE J.,* **20:** 263 (1974).
Tsonopoulos, C.: *AIChE J.,* **21:** 827 (1975).
Tsonopoulos, C.: *AIChE J.,* **24:** 1112 (1978).
Tsonopoulos, C.: *Adv. in Chem. Ser.,* **182:** 143 (1979).
Tsonopoulos, C., J. H. Dymond, and A. M. Szafranski: *Pure. Appl. Chem.* **61:** 1387 (1989).
Tsonopoulos, C., and J. L. Heidman: *Fluid Phase Equil.,* **57:** 261 (1990).
Tsonopoulos, C., and J. H. Dymond: *Fluid Phase Equil.,* **133:** 11 (1997).
Tu, C.-H.: *Fluid Phase Equil.*, **99:** 105 (1994).

van Krevelen, D. W. and P. J. Hoftyzer: *Properties of Polymers, Correlations with Chemical Structure,* Elsevier, Amsterdam, 1972.
van Miltenburg, J. C. and H. A. J. Oonk: *J. Chem. Eng. Data*, **46:** 90–97 (2001).
Vetere, A.: *Ind. Eng. Chem Res.,* **30:** 2487 (1991).
Vetere, A.: *Fluid Phase Equil.,* **164:** 49 (1999).
Vetere, A.: *Fluid Phase Equil.,* **240:** 155–160 (2006).
Wagner, W.: *Cyrogenics,* **13:** 470 (1973).
Wagner, W.: *A New Correlation Method for Thermodynamic Data Applied to the Vapor-pressure Curve of* Argon, *Nitrogen, and Water*, J. T. R. Watson (trans. and ed.). IUPAC Thermodynamic Tables Project Centre, London, 1977.
Wagner, W., J. Evers, and W. Pentermann: *J. Chem. Thermodynamics,* **8:** 1049 (1976).
Wagner, W., A. Pruss, *J. Phys. Chem. Ref. Data,* **22**, 783 (1993).
Wagner, W., T. Riethmann, *J. Phys. Chem. Ref. Data,* **40:** 043103 (2011).
Waring, W.: *Ind.* Eng. *Chem.,* **46:** 762 (1954).
Wilding, W. V., T. A. Knotts, N. F. Giles, R. L. Rowley, DIPPR® Data Compilation of Pure Chemical Properties, Design Institute for Physical Properties, AIChE, New York, NY (2017).
Wilsak R. A. and G. Thodos: *Ind. Eng. Chem. Fund.* **23:** 75–82 (1984).
Xiang, H. W. and L. C. Tan: *Int. J. Thermophys.,* **15:** 711 (1994).
Yamada, T. and R. D. Gunn: *J. Chem. Eng. Data,* **18:** 234 (1973).
Yaws, C. L.: *Thermodynamic and Physical Property Data,* Gulf Pub. Co., Houston, 1992.
Zábranský, M., V. Růžička, and V. Majer: *J. Phys. Chem. Ref. Data*, **19:** 719 (1990).
Zábranský, M., V. Růžička, V. Majer, and E. S. Domalski: *Heat Capacity of Liquids: Critical Review and Recommended Values,* Amer. Chem. Soc. and Amer. Inst. Phys. For NIST, Washington D. C., 1996.
Zábranský, M., V. Růžička, E.S. Domalski: *J. Phys. Chem. Ref. Data,* **30:** 1199 (2001).
Zábranský, M., V. Růžička: *J. Phys. Chem. Ref. Data,* **33:** 1071 (2004).
Zábranský, M., V. Růžička: *J. Phys. Chem. Ref. Data,* **34:** 39 (2005).
Zábranský, M., Z. Kolska, V. Růžička, E.S. Domalski: *J. Phys. Chem. Ref. Data,* **39:** 013103 (2010).

6

Thermodynamic Properties of Pure Gases and Liquids

6.1 SCOPE*

Methods are presented in this chapter foremost for estimating the molar volume, v, of a pure gas or liquid as a function of temperature, T, and pressure, P. Relationships between P, v, and T are formally known as equations of state (EOSs). Through the relations of classical thermodynamics, however, thermodynamic properties like enthalpy, H, Gibbs energy, G, and isobaric heat capacity, C_p, are intrinsically bound to the expression of the PvT relation. Therefore, it is reasonable to assess EOSs holistically, evaluating their volumetric accuracy simultaneously with the accuracy of thermodynamic properties. Sections 6.2 to 6.4 introduce the framework and a brief historical review of PvT relations and the evolution of the physical basis for their modeling. The fundamental models include the virial series and perturbation theory, from which modern EOSs derive: virial EOSs, multiparameter EOSs (e.g., Wagner-Pruss[1]), cubic EOSs (e.g., Soave[2], and Peng-Robinson[3]), and perturbative and chemical EOSs (e.g., PC-SAFT[4]), and transferable and customized molecular simulation EOSs (e.g. SPEADMD). Overviews for each of these EOS classes and their computational implementations are given in Secs. 6.5 to 6.11. Section 6.12 summarizes the residual functions for the internal energy of several models that are important for our evaluations. Section 6.13 summarizes our evaluation of EOS methods.

Extension of this chapter's methods to thermodynamic properties of mixtures is given in Chap. 7. Like the thermodynamic consistency that intrinsically binds thermodynamic properties such as H and G to PvT relations, details of the composition dependence of mixture properties are bound to phase equilibrium estimation. The great challenge of accurate phase equilibrium estimation leads to "adjustments" in some EOSs that may seem unexpected when considered in isolation, volume translation of a cubic EOS, for example. These adjustments become more comprehensible in the context of constructing an expedient framework that can accurately characterize phase behavior while characterizing other properties as accurately as possible. For example, it might be useful to accept less accuracy in PvT properties if computationally efficient and accurate phase equilibrium estimation is the primary goal. Nevertheless, one should know how much accuracy is being sacrificed. The accuracy of pure compound properties is assessed in this chapter and mixture properties in Chap. 7. Accuracy of phase equilibrium models is assessed in Chaps. 8 and 9.

*With special contributions from V. Diky.

We reiterate here that important aspects of our mission are to promote reliable applications and advancement of the field. Both aspects are served by descriptions of the physical basis of the model equations. Readers interested in applications are likely to be interfacing with the models through a process simulator. A key step in process simulation is the specification of the thermodynamic model(s). At first glance, this specification might seem simple, as described in appendix D of Elliott and Lira.[5] For example, preliminary analysis of a natural gas pipeline might focus on high pressure hydrocarbons, but the specification gets more complicated if water or a hydrate inhibitor like methanol must be added to the mix. Corrosion inhibitors complicate matters further. In other applications, carboxylic acids or formaldehyde may require special consideration. "Tried and true" methods may suddenly be rendered unreliable as the applications evolve. Understanding the entirety of what can be done is essential to knowing when it is time to reassess the "best" specification for any given application.

6.2 INTRODUCTION TO EQUATIONS OF STATE

The volumetric properties of a pure fluid in each state are commonly expressed with the compressibility factor, Z, which can be written as a function of T and P or of T and v

$$Z \equiv \frac{Pv}{RT} = f_v(T, v) \tag{6.2-1}$$

$$Z = f_P(T, P) \tag{6.2-2}$$

where R is called the universal gas constant. The value of R depends upon the units of the variables used. Common values are shown in Table 6.1. In this chapter, unless otherwise noted, P is in MPa, v in cm^3 mol^{-1}. Note that 1 MPa = 10 bar = 10^6 N·m^{-1} and 1 atm = 0.101325 MPa.

The choice of independent variables, T and V in Eq. (6.2-1) and T and P in Eq. (6.2-2), depends upon the application. Commonly, a user specifies the state with T and P since these are most easily measured and so Eq. (6.2-2) is considered most convenient. However, if one seeks an equation that can describe both gaseous and liquid phases with the same parameters, the needed multiplicity of volume or density roots demands a function of the form of Eq. (6.2-1). Thus, most applied EOSs are in the form of Eq. (6.2-1).

TABLE 6.1 Values of the gas constant, R[6].

The gas constant is defined as: $R = k_B \cdot N_A$.*
where $k_B = 0.01380649(10^{-21})$ J·K^{-1}, $N_A = 602.214076(10^{21})$ mol^{-1}.
In this manner values of R can be inferred in various units.

Value of R	Units on R
$k_B \cdot N_A = 8.31446261815324$	J mol^{-1} K^{-1}
$k_B \cdot N_A$	MPa-cm^3 mol^{-1} K^{-1}
$k_B \cdot N_A$	Pa-m^3 mol^{-1} K^{-1}
$k_B \cdot N_A$	kPa-L mol^{-1} K^{-1}
$10 \cdot k_B \cdot N_A$	bar cm^3 mol^{-1} K^{-1}
$10 \cdot k_B \cdot N_A/\mathbf{1.01325}^+ \approx 82.0574$	atm cm^3 mol^{-1} K^{-1}
$k_B \cdot N_A \cdot 1.29071^+ \approx 10.7316$	psia ft^3 lb-mol^{-1} R^{-1}
$k_B \cdot N_A/\mathbf{4.184}^+ \approx 1.98720$	cal$_{th}$ mol^{-1} K^{-1}
$k_B \cdot N_A/\mathbf{4.184}^+ \approx 1.98720$	Btu$_{th}$ lb-mol^{-1} R^{-1}

*https://www.nist.gov/si-redefinition;
+https://www.nist.gov/pml/special-publication-811 defines **boldface** conversion factors "exactly," so precision is not constrained by the number of significant figures in the conversion factor. cal$_{th}$ is the thermochemical calorie defined exactly as 4.184 J. Btu$_{th}$ is the analogous quantity. Multiple definitions are in use for the calorie and Btu; avoid these units if possible.

For an ideal gas, $Z^{ig} = 1.0$. For real gases, Z is somewhat less than 1 except at high reduced temperatures and pressures. For liquids that are saturated between the triple or melting point and the boiling point or under low applied pressure, Z is normally quite small. Near the vapor-liquid critical point, Z is usually between 0.15 and 0.4, but it varies rapidly with changes in T and P.

Since the compressibility factor is dimensionless, it is often represented by a function of dimensionless (reduced) temperature, $T_r = T/T_c$, and dimensionless (reduced) pressure, $P_r = P/P_c$, where T_c and P_c are the component's vapor-liquid critical properties. Noting that most compounds exhibit a vapor-liquid critical point, with qualitatively similar behavior in reduced coordinates, the corresponding states principle allows many substances to be represented graphically in generalized form. For example, $f_P(T_r, P_r)$ of the Lee-Kesler correlation can be illustrated in the form of Fig. 6.1, where $Z = Z^0 + \omega Z^1$ and ω is the acentric factor.[7] This qualitative similarity of compounds when expressed in reduced coordinates is known as the corresponding states principle. Appendix A lists values of T_c, P_c, and ω and methods to estimate them are described in Chap. 3.

Modern computers obviate the need to manually obtain volumetric behavior such as from graphs and allow more accurate results by using equations with more sophisticated correlations depending on molecular polarity, size, and shape. There have been countless numbers of EOS functions generated.[8,9] It is not possible to evaluate all such models here, but the discussion of this chapter is intended to provide guidance about ranges of accuracy, reliability and computational difficulties encountered when doing pure-component analysis with literature models.

6.3 THEORY OF EQUATIONS OF STATE

Our current understanding of EOSs is best related in terms of the fundamentals underlying their development. These fundamentals have their historical roots in the van der Waals EOS[10] and the virial series (cf. Kammerlingh-Onnes[11] but significant work was done previously, e.g., Boltzmann obtained and exact solution of the fourth virial coefficient for hard spheres[12]). These fundamentals evolved with better articulation in works by Mayer and Mayer[13] for the virial series and Zwanzig[14] for thermodynamic perturbation theory (TPT). Entire texts have been devoted to the development of these fundamental relations so we cannot reiterate the entire background here (cf. McQuarrie[15] and Hansen and McDonald[16]), but we can summarize a few of the key results in forms that aid comprehension of current practice. To illustrate the key points, we refer to the square well potential model as a simple characterization of the qualitative behaviors exhibited by detailed molecular models and real fluids. The advantage of assuming a specific potential model is that precise computations can shed light on experimental trends that may seem scattered or mysterious, such as the third virial coefficient, even if the accuracy of the computations is only qualitative.

6.3.1 The Virial Series and Multiparameter EOSs

The fundamental form of the virial EOS is expressed as a series in density.

$$Z = 1 + B_2\,\rho + B_3\,\rho^2 + B_4\,\rho^3 + \cdots \tag{6.3-1}$$

where B_i is the ith virial coefficient, ρ is the molar density. When truncated at second order in density, this series can be reverted to compute the density given pressure,

$$Z = 1 + B_2\left(\frac{P}{RT}\right) + \left(B_3 - B_2^2\right)\left(\frac{P}{RT}\right)^2 + \cdots \tag{6.3-2}$$

Although it is a simple Taylor series expansion about the ideal gas state, each virial coefficient can be related to a rigorous system of integrals over types of molecular interactions inherent in that virial coefficient. For instance, B_2 considers all ways that two molecules can interact, accurately accounting only for low density interactions; B_3 considers all ways that three molecules can interact and extend the accuracy to higher density.

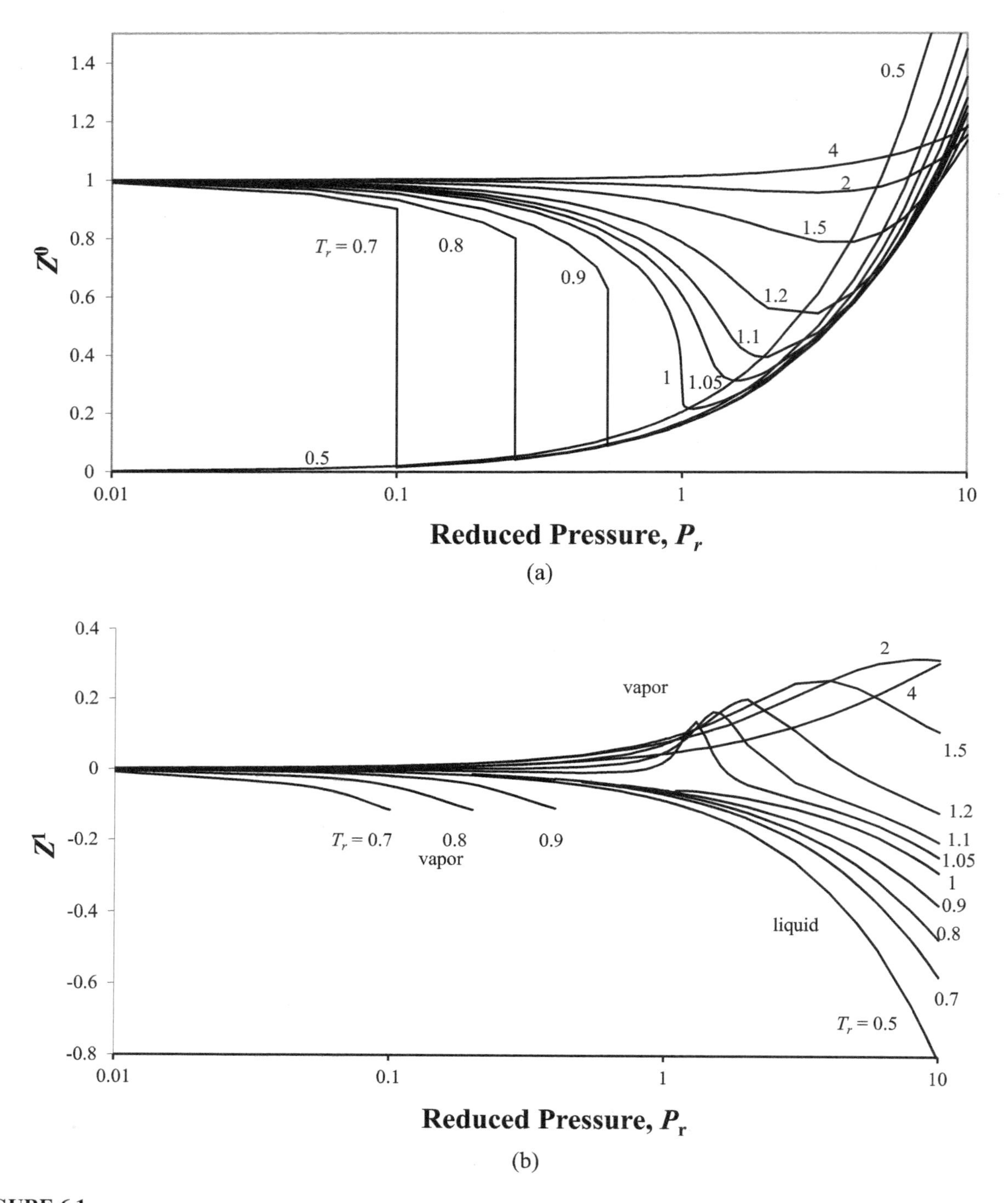

FIGURE 6.1
Contributions to the Lee-Kesler EOS. (a) The reference contribution, Z^0, for compounds with $\omega = 0$. (b) The correction term, Z^1, for $\omega > 0$.

As an example, the interactions for B_2 of square well spheres (SWS) are especially simple and the result is known exactly.

$$B_2^{SWS} = -0.5\int_0^\infty [\exp(-\beta u) - 1]4\pi r^2\, dr = 4b[1 - (\lambda^3 - 1)Y] \tag{6.3-3}$$

$$u(r) = \begin{cases} \infty & \text{if } r < \sigma \\ -\varepsilon^{SWS} & \text{if } \sigma \le r \le \lambda\sigma \\ 0 & \text{if } r > \sigma \end{cases}$$

where $b = N_A\pi\sigma^3/6$; σ is the spherical diameter, N_A is Avogadro's number, λ is the width of the square well relative to σ, u is the SWS potential model, $\beta \equiv 1/k_BT$, k_B is Boltzmann's constant, $Y = \exp(\beta\varepsilon^{SWS}) - 1$, ε^{SWS} is the depth of the SWS model. The expression for B_3 of SWS is a bit more complicated, involving a triple integral to address all the ways that three molecules can interact. Nevertheless, an exact solution for SWS is available when the potential function is assumed to follow the form $u_{tot}(r_1, r_2, r_3) = u(r_{12}) + u(r_{23}) + u(r_{31})$. Setting $f_{ij} = \exp[-\beta u(r_{ij})] - 1$,

$$B_3 = -\frac{1}{3V}\int_0^\infty\int_0^\infty\int_0^\infty f_{12}f_{23}f_{31}\, d\mathbf{r}_1 d\mathbf{r}_2 d\mathbf{r}_3 \tag{6.3-4}$$

$$B_3^{SWS} = b^2\,[10 + B_{31}Y + B_{32}Y^2 + B_{33}Y^3] \tag{6.3-5}$$

$B_{31} = 30 - 64\,\lambda^3 + 36\,\lambda^4 + 2\,\lambda^6$; $B_{32} = 32 - 36\,\lambda^2 - 64\,\lambda^3 + 72\,\lambda^4 - 4\,\lambda^6$; $B_{33} = 12 - 36\,\lambda^2 + 36\,\lambda^4 - 12\,\lambda^6$; $1 < \lambda < 2$. Expressions are also available for B_3^{SWS} when $\lambda > 2$. Elliott et al.[17] have computed the SWS virial coefficients through B_6, all of which can be expressed as polynomials in Y.

We compare experimental data to the results of the SWS model with $\lambda = 1.7$ in Fig. 6.2, after the manner of Sherwood and Prausnitz.[18] Figure 6.2 shows that B_2 is accurately characterized with this simple model over the entire range of experimental data. The characterization of B_3 requires more discussion. First, note that the curve for B_3 is a prediction rather than a correlation, since the values of $\lambda = 1.7$, $\sigma = 0.3067$ nm, and ε^{SWS}/k_B = 93.3 K were optimized for B_2. Another issue is that experimental data for B_3 exhibit the effects of nonadditivity in their potential of interaction. In other words, quantum mechanics leads to energies of interaction for three particles simultaneously that deviate from the assumption that the total energy is simply the sum of pairwise interaction energies. The SWS model applied here assumes pairwise additivity. Despite these caveats, we see that the prediction of B_3 is qualitatively correct. Without this result, one might suspect the maximum around the critical temperature results from high uncertainty in the experimental data, but this kind of maximum is observed consistently for many compounds. The divergence to negative values at low temperatures indicates that the molecules are exhibiting mutual attraction at long range. In this way, the theory helps us to better appreciate the experimental data. Figure 6.2 also shows the analogous results when the Lennard-Jones (LJ) potential model is applied. The results are much like the SWS results, except at high temperature. At very high temperatures, both B_2 and B_3 steeply approach zero, a reflection of the "softness" of the LJ model.[19] As the temperature approaches infinity, the LJ model effectively approaches the ideal gas model, indicating that two molecules can interpenetrate if the kinetic energy greatly exceeds the repulsive potential energy. The SWS model, on the other hand, approaches the hard sphere model in the high-temperature limit. The LJ model is more realistic and we should expect the experimental trends to resemble the LJ results if measurements at such temperatures ever become available.

The behavior of B_3 illustrated in Fig. 6.2 provides the first hint of how the utility of the virial series is limited. First, the availability of experimental data for B_3 is much more limited than for B_2, and availability gets progressively worse for B_4, B_5, etc. Every increase in virial coefficient requires greater care in interpretation, with greater sensitivity to uncertainty in the PvT measurements. Another issue is even more problematic. Theoretical computations of B_4, B_5, etc. show that the divergence to negative values is progressively steeper and initiates closer to the critical temperature. Altogether, these results suggest that if temperatures are subcritical, then the virial series should only be applied below the critical density.

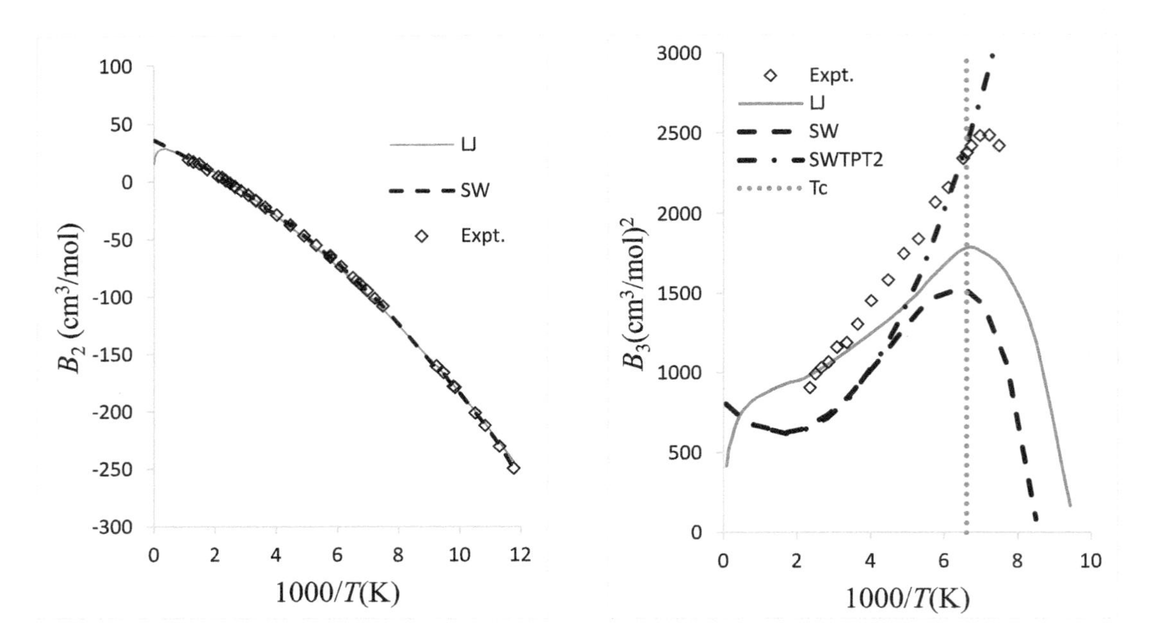

FIGURE 6.2
Second and third virial coefficients for argon according to pairwise additive treatments of the Lennard-Jones (LJ) and square well sphere (SWS) potential models. The dotted line shows the critical temperature (T_c) of argon. The dash-dotted line corresponds to treatment by the SWS model with second-order perturbation theory.[17] Experimental data were taken from the NIST-TDE database.[20]

One might expect to abandon the virial series perspective at that point, but there are a couple of twists to the story. In the early days of EOS development, several researchers assumed the virial series as a functional form for developing multiparameter correlations of PvT properties. With a few clever tricks, they could obtain good correlations, and even reproduce values of the (few) measured virial coefficients when available. One of the most common correlating equations was in the form of the Benedict-Webb-Rubin correlation.[21]

$$Z = 1 + f_1(T)\rho + f_2(T)\,\rho^2 + f_3(T)\,\rho^n + f_4(T)\,[(\alpha + \gamma\rho^2)\rho^m]\exp(-\gamma\rho^2) \tag{6.3-6}$$

where 30 coefficients might be used to characterize $f_1(T) - f_4(T)$ in addition to the parameters m, n, α, and γ. This form was commonly applied for multiparameter EOSs until the late 1990s. Since then, multiparameter EOSs have adopted the generalized form suggested by Wagner and coworkers. The following is adapted from Wagner and Pruss.[1]

$$F \equiv \frac{a^{\text{res}}}{RT} = \sum_i\sum_j n_{ij}\delta^{d_i}\tau^{t_j} + \sum_i\sum_j\sum_k n_{ijk}\delta^{d_i}\tau^{t_j}\exp(-\delta^{c_k}) + \sum_i n_i\delta^{d_i}\tau^{t_i}\exp\left[-\alpha_i(\delta - \varepsilon_i)^2 - \beta_i(\tau - \gamma_i)^2\right] \tag{6.3-7}$$

where $\delta = \rho/\rho_c$; $\tau = T_c/T$. A few equations added a nonanalytic term, F^{NA}, given by

$$F^{\text{NA}} = \sum_i\sum_j\sum_k\sum_l\sum_m n_{ijklm}\Delta^{b_j}\delta\psi \tag{6.3-8}$$

where $\Delta = \theta^2 + B_k[(1-\delta)^2]^{a_i}$; $\theta = (1-\tau) + A[(\delta-1)^2]^{1/2\beta}$; $\psi = \exp[-C_l(\delta-1)^2 - D_m(\tau-1)^2]$. Notably, the models developed for H_2O and CO_2 still retain the nonanalytic term. In other work, the nonanalytic term has been avoided, with fewer problems of implementation and no substantial sacrifice in accuracy.[22] The nonanalytic term is generally disfavored for future work. By expressing the EOS in terms of the Helmholtz departure function, properties like Z and U are easily inferred by differentiation.

One advantage of multiparameter EOSs might seem to be a disadvantage at first. Because there is no assumption about the equation of state being related to separate repulsive and attractive contributions, as in the perturbation perspective below, the improvement in accuracy of adding more terms (and parameters) is steadily progressive. In the perturbation perspective, one tends to get very close with just a few terms, then improve only marginally with higher-order terms. When correlating a large dataset to within the experimental uncertainty, the multiparameter approach is the most accurate (currently).

While the connection of Wagner's multiparameter EOS to the virial series is mostly historical, recent work by Kofke and coworkers may revitalize the virial series for future applications, comprising another twist to the story. Virial coefficients are evaluated by integrating the potential function over all possible molecular positions. For the second virial coefficient, the integration is performed for two molecules. Although the computation of B_2 is very simple, even for polymeric molecules, the resulting insights may be surprising.[23] For the third virial coefficient, the integration is for three molecules, and so forth. Schultz and Kofke[24] developed an algorithm that can automatically generate all the types of integrals that can occur and efficiently evaluate them. Note that virial coefficients are inherently more useful for engineering applications than a molecular simulation because the result is an EOS applicable over a range of state conditions whereas a simulation gives only a single state point. More recent work has shown how to compute TPT contributions from these same types of integrals, effectively building a bridge between TPT and the virial series.[17,19] Then the TPT/virial series can be applied for hydrocarbon densities less than 0.5 g/cm^3, for example, and simulation/TPT at higher densities, yielding a rigorous EOS for any molecule. This approach should substantially accelerate EOS development based on molecular modeling, but it has not yet reached a point of common application. The approach has not been extended to nonadditive potential models, which currently limits its implementation for *ab initio* potential models.

6.3.2 Thermodynamic Perturbation Theory (TPT) and PC-SAFT

The fundamental form of the EOS from a perturbation perspective is:

$$\frac{a^{\text{res}}}{RT} = A_0 + \frac{A_1}{T} + \frac{A_2}{T^2} + \frac{A_3}{T^3} + \frac{A_4}{T^4} + \cdots + A^{\text{polar}} + A^{\text{chem}} + \cdots \tag{6.3-9}$$

where A_i is the ith perturbation coefficient.

Ideally, the $\{A_i\}$ would be density-dependent only, as they are for the SWS model. More often, these coefficients are weak functions of temperature owing to the softness of potential models like the LJ fluid. The physical basis for perturbation theory is conveniently illustrated for the SWS model. In that case, A_0 simply corresponds to the hard sphere EOS, which is reasonably well approximated by the Carnahan-Starling EOS,[25] Eq. (6.3-10). The higher-order coefficients essentially form a Taylor series expansion around the hard sphere reference (for the SWS model) at infinite temperature. The A_1 contribution is then evaluated as the mean number "M" of hard spheres within the square well distance over an equilibrated ensemble of hard spheres. Similarly, the A_2 contribution is closely related to the variance of that mean. Just as the variance involves terms of $\langle M^2\rangle - \langle M\rangle^2$, where $\langle\,\rangle$ indicates an average, A_3 and A_4 involve terms like $\langle M^3\rangle$ and $\langle M^4\rangle$. Specifically, the terms in the expansion for SWS can be written as[26]

$$A_0^{\text{CS}} = \frac{\eta^{\text{P}}(4 - 3\eta^{\text{P}})}{(1 - \eta^{\text{P}})^2} \tag{6.3-10}$$

$$A_1 = -\frac{\langle M \rangle}{N} \tag{6.3-11}$$

$$A_2 = -\frac{\langle M \rangle^2 - \langle M \rangle^2}{2N} \tag{6.3-12}$$

$$A_3 = -\frac{\left\langle (M - \langle M \rangle)^3 \right\rangle}{6N} \tag{6.3-13}$$

$$A_4 = -\frac{\left\langle (M - \langle M \rangle)^4 \right\rangle - 3\left\langle \left\langle (M - \langle M \rangle)^2 \right\rangle^2 \right\rangle}{24N} \tag{6.3-14}$$

where $\eta^P = b\rho$ is the packing fraction and M represents the number of pairs of spheres located within the SW distance ($\sigma < r < \lambda\,\sigma$). These perturbation contributions have been determined by molecular simulation, so we know their qualitative features, as shown in Fig. 6.3. Note that each higher-order perturbation contribution is akin to a higher-order derivative property, like the virial coefficients. Accurate experimental measurements of these quantities would be challenging, so simulation is essential.

We observe several trends that can instruct us in evaluating the physical accuracy of the EOSs that we discuss later. First, the trends for the first-order term are nearly linear in density. van der Waals was right about that! Next, the second-order term is roughly an order of magnitude smaller than the first-order term, but care is required because the reference contribution (A_0) is similar in magnitude to the A_1 contribution at temperatures of interest, and opposite in sign. This means that the A_2 contribution may be quite significant relative to $A_0 + A_1$.

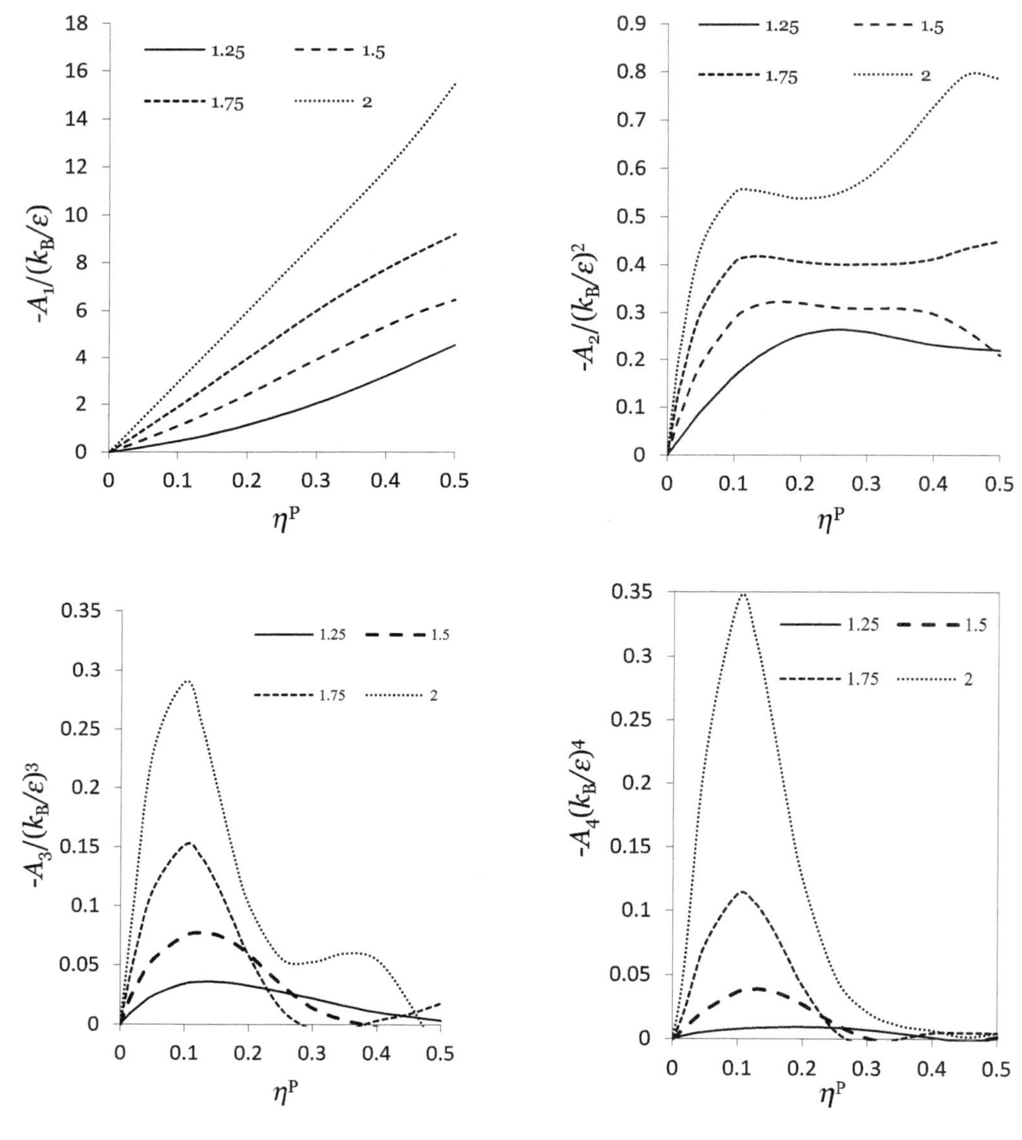

FIGURE 6.3
Perturbation contributions A_1–A_4 of SWS for $\lambda = 1.25$–2.0 from the simulations of Pavlyukhin.[27]

Last, we note that: (1) the A_3 and A_4 terms make contributions about 3 to 6 times smaller than the A_2 term. (2) A_3 and A_4 reach maximum magnitude near the critical density. (3) They decay steeply to zero at high density. The most interesting feature of the higher-order perturbations is their behavior near the critical density, informing our later discussion about the critical region. The A_2 contribution also has a peak near the critical density in most implementations, but it depends on details of the potential function and the choice of reference contribution. There is at least a shoulder in A_2 near the critical density in all cases, however. As we shall see in Sec. 6.3.3, these peaks near the critical shed light on one of the most challenging issues of EOS development. More broadly, the trends of the TPT contributions in the context of various theories and potential functions are important to understand when considering the resulting EOSs and their performance in computational efficiency and physical property accuracy. We elaborate to some degree on these trends in the discussion below.

Having established general trends of what the TPT contributions should look like, we can consider how their approximations have evolved historically. Keep in mind that our knowledge of how they should appear started to clarify in the 1970s for $A_0 - A_2$, and only became clear for higher-order contributions since about 2009. Until that time, approximations were largely based on speculation and regression of experimental data. This speculative basis is reflected in many of the cubic EOSs, including those in common use currently. Commonly applied cubic equations are effectively first-order perturbation theories, like the van der Waals equation. The Soave[2] and Peng-Robinson[3] equations simply modify the density dependence of the A_1 contribution, making it less linear. This tended to improve density correlations slightly, relative to the van der Waals equation, but we see little justification for it in the physical basis of A_1. Where Soave made his greatest contribution was in adding a compound-specific temperature dependence to the attractive term, multiplying by a factor that varies as $T^{1/2}$. There is no physical basis for this temperature dependence for nonassociating compounds, but it enabled him to match the vapor pressure curves for nonspherical molecules, while including a correlation dependent on acentric factor. Tailoring the vapor pressure of each compound to match the acentric factor and critical pressure was a substantial step forward because it meant the bubble pressure of the pure fluid was accurate. That is an essential prerequisite to accurate vapor-liquid equilibria (VLE) for mixtures. Peng and Robinson adopted the same temperature dependence. Many other variations in temperature and density dependence have been investigated, but improvements have been small. An important historical consideration is that application experts invest a lot of time into adapting a thermodynamic model for their process. If a new model cannot promise substantial improvement over an existing model (including customized parameters that may have been developed by the application experts), it is very difficult to justify the effort. This situation may change as new experts enter the field and standardized process simulators centralize the required effort to serve many applications,[28] but inertia will always be significant. Process simulation software developers have little motivation to implement new models if their customers do not demand them, and application experts will not try new models if they are not readily available.

Recent progress in perturbation theory for pure fluids has focused on a stronger physical basis, using the results of molecular simulations to suggest functional forms that can be adapted to engineering applications. Consistent with the philosophy advocated on Chap. 1 page 1, a stronger physical basis should eventually provide better models and deeper insights into peculiar molecular behaviors when they are observed. This outlook is well represented in the developments of the Statistical Associating Fluid Theory (SAFT) of EOSs, initially developed by Gubbins and coworkers[29] as an implementation of Wertheim's theory of association.[30] There have been many variations on the SAFT framework, but one example that has had broad application is the PC-SAFT implementation.[4] We examine this model in detail as a typical implementation of perturbation theory. PC-SAFT truncates the TPT expansion at second order and sets the reference term to the Carnahan-Starling equation for hard spheres.[25] The extension to tangent sphere chains is then implicit in Wertheim's theory by taking the mathematical limit as the association energy goes to infinity; this term is known as A^{chain}. A second association term (A^{chem}) is added to account for finite association energies like hydrogen bonding. We refer to this term as A^{chem} in anticipation of applications to mixtures, for which alternative theories address this generic chemical-physical interaction. To estimate the attractive contributions (A_1, A_2), PC-SAFT adapts an approximation for tangent sphere chains with a modified SWS potential, then fills in several adjustable parameters based on experimental data. The SWS model is modified by a repulsive shoulder step before the attractive well, leading to temperature dependence in the effective hard sphere diameter. The result is an EOS with the following form, where A^{chem} pertains to an alcohol for example,

$$\frac{a^{\text{res}}}{RT} = mA^{\text{HS}} + A^{\text{chain}} + \frac{A_1}{T} + \frac{A_2}{T^2} + A^{\text{chem}} + \cdots \tag{6.3-15}$$

where m is a shape parameter, theoretically equal to the number of spherical segments in a chain,

$$A_1 = -\frac{12\eta^{\text{P}} m I_1 \varepsilon}{k_{\text{B}}};\; A_2 = -6\eta^{\text{P}} m \left(\frac{\varepsilon}{k_{\text{B}}}\right)^2 \frac{I_2}{D_2};\; I_n = \sum_{j=0}^{6} a_{j,n} (\eta^{\text{P}})^j \tag{6.3-16}$$

$$a_{j,n} = a_{0j,n} + a_{1j,n}\frac{(m-1)}{m} + a_{2j,n}\frac{(m-1)^2}{m^2} \tag{6.3-17}$$

$$D_2 = \frac{1}{m} + \frac{8\eta g^{\text{HS}}}{(1-\eta)} - \frac{(m-1)}{m}\frac{\eta^{\text{P}}[20 + \eta^{\text{P}}(-27 + \eta^{\text{P}}(12 - 2\eta^{\text{P}}))]}{[(2-\eta^{\text{P}})(1-\eta^{\text{P}})]^2} \tag{6.3-18}$$

$$\eta^{\text{P}} = \frac{\pi}{6} m \rho \sigma^3;\; g^{\text{HS}} = \frac{\left(1 - \frac{\eta^{\text{P}}}{2}\right)}{(1-\eta^{\text{P}})^3} \tag{6.3-19}$$

$$A^{\text{chain}} = -(m-1)\ln[g^{\text{HS}}];\; A^{\text{chem}} = 2\ln X^A + (1 - X^A) \tag{6.3-20}$$

$$\frac{1}{X^A} = 1 + N_{\text{A}} \rho \Delta X^A;\; \Delta = g^{\text{HS}} K^{\text{AD}}[\exp(\beta\varepsilon^{\text{AD}}) - 1] \tag{6.3-21}$$

where $d^{\text{ehs}} = \sigma[1 - 0.12\exp(-3\varepsilon/k_{\text{B}}T)]$; X^A is the fraction of acceptor sites NOT bonded. For example, $X^A = 1$ implies zero association. g^{HS} is the radial distribution function (rdf) at contact of a hard sphere with effective diameter, d^{ehs}. The quantity g^{HS} plays a central role in Wertheim's theory.

Considering all the $a_{ij,n}$ coefficients in I_1 and I_2, 36 generalized parameters were optimized in relation to experimental data. Figure 6.4 illustrates the A_1 and A_2 terms for PC-SAFT. As one might hope, they do exhibit trends like those for SWS. The shoulders in A_2 at high density would be expected of a model conceived from the SWS potential. Figure 6.4 also shows how the TPT contributions evolve with increasing chain length, trends that are also supported (qualitatively) by molecular simulation. Note how the form of Eq. (6.3-17) naturally leads to asymptotic trends in the long-chain limit. The trend of the van der Waals estimate of A_0 is also shown in Fig. 6.4a. Recalling that the van der Waals EOS should be compared to the result for spheres, van der Waals was wrong about A_0. Unfortunately, this is the contribution that has been perpetuated in most cubic EOSs, compromising their physical basis. Another consideration pertaining to the virial series is that a second-order TPT like Eq. (6.3-15) cannot be expected to characterize the peculiar behavior of B_3 at subcritical temperatures, as illustrated by the TPT2 curve in Fig. 6.2b.

6.3.3 Challenges to EOS Models: The Critical, High Pressure, and Metastable Regions

Fluid properties in states near a pure compound's vapor-liquid critical point are the most difficult to obtain from both experiments and from models such as EOSs (see the collected articles in Kiran and Levelt Sengers[31]). The principal experimental difficulty is that the density of a near-critical fluid is so extremely sensitive to variations in P, T, that maintaining homogeneous and stable conditions takes extreme care.[32,33] Even gravity influences the measurements.

The principal model difficulty is that near-critical property variations do not follow the same mathematics as at conditions well-removed from the critical. For example, the difference of the saturation volumes, V_s from V_c near the critical point varies as

$$\lim_{T \to T_c} (V_s - V_c) \sim (T - T_c)^{\beta_c} \tag{6.3-22}$$

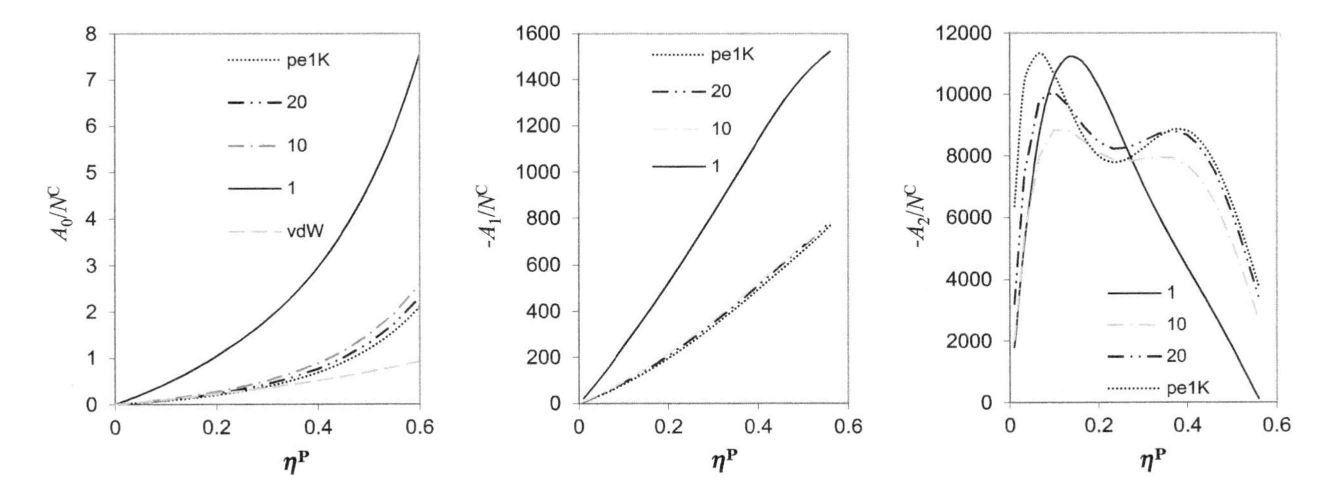

FIGURE 6.4
Illustration of the TPT contributions in PC-SAFT. (a) Z_0 shows a steady approach to its long-chain limit. (b) A_1 shows small variations in the approach to its long-chain limit. (c) The long-chain limit for A_2 shows a distinct difference from the shape of A_2 for monomer. C1, C10, and C20 correspond to methane, N-decane, and N-eicosane, respectively. pe1K corresponds to monodisperse polyethylene with molar mass of 996 g/mol.

Careful experiments have shown that $\beta_c = 0.325 \pm 0.01$. This is close to the results from theories that account for the molecular density fluctuations that cause critical opalescence. However, typical EOS models give a larger β_c value. Thus, for example, all cubic EOSs have $\beta_c = 0.5$. Also, the variation of P with V along the critical isotherm is found to be

$$\lim_{V \to V_c} (P - P_c) \sim (V - V_c)^{\delta_c} \quad for\ T = T_c \tag{6.3-23}$$

Careful experiments have shown that $\delta_c = 4.8 \pm 0.02$ Again, this is close to theoretical results, but EOS models give a smaller exponent. All cubic EOSs have $\delta_c = 3.0$. Differences also occur in the variation of C_V in the near-critical region where quite large values persist over a broad range of conditions, but cubic EOSs and other models do not show this.[34,35]

We can gain insight into the nature of the problem at the critical point by considering the impacts of higher-order TPT contributions. Figure 6.5a shows the progressive flattening of the pure-component binodal phase diagram for SWS as higher-order TPT contributions are included. The "infinite" order approximation incorporates the Gaussian extrapolation of Ghobadi and Elliott.[36] Briefly, Ghobadi and Elliott showed that the ratio $iA_i/A_{(i-1)}$ is roughly Gaussian and independent of i for SWS when $i > 3$. This observation facilitated a resummation of the TPT series to infinite order in closed form.

$$\frac{a^{res}}{RT} = A_0 + \frac{A_1}{T} + \frac{A_2}{T^2} + \frac{A_3}{T^3} A^+ \tag{6.3-24}$$

$$A^+ \equiv \left(1 + \frac{A_4}{TA_3}\left(1 + \frac{A_5}{TA_4} + \cdots\right)\right)$$

$$\approx \left\{\frac{6}{\Omega^3}\left[\exp\left(\frac{\Omega}{T}\right) - 1 - \Omega - \frac{\Omega^2}{2} - \frac{\Omega^3}{6}\right]\right\}$$

$$\Omega = \frac{\Omega_0 \exp\left[\Omega_1\left(1 - \left(1 - \frac{\rho}{\rho_0}\right)^2\right)\right]}{T}$$

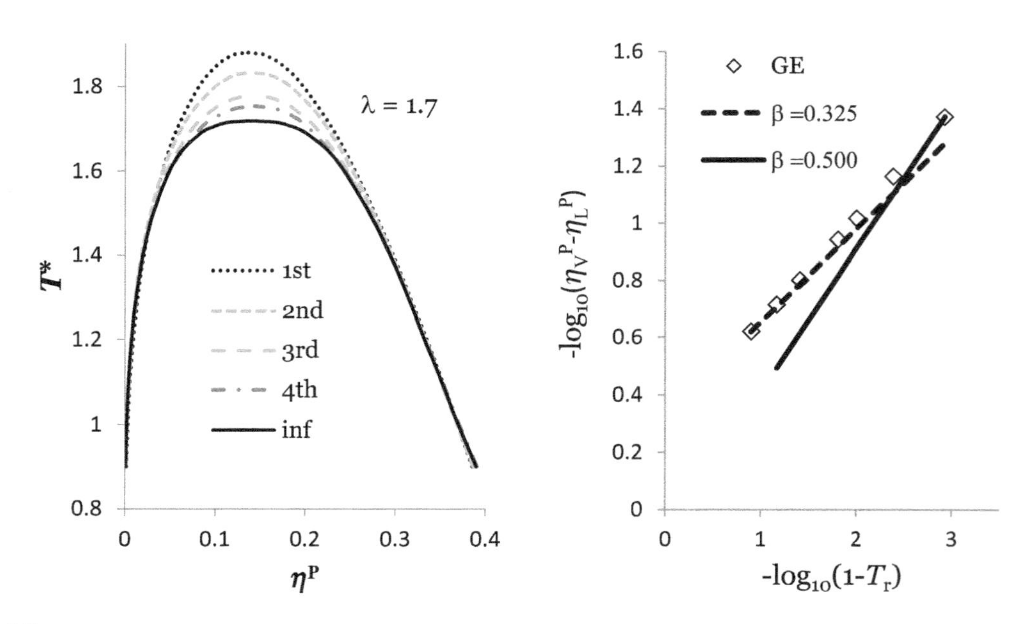

FIGURE 6.5
The flatness of the binodal curve for SW spheres with a well width of 1.7 near the critical point. (a) the impact of higher-order TPT contributions. (b) Inference of the apparent critical exponent β_c. The GE symbols refer to Gaussian extrapolation.

where Ω_0, Ω_1, and ρ_0 are adjustable parameters. Although this resummation is of infinite order in the sense of approximating the infinite order TPT contribution, it still fails to accurately characterize the β_c exponent when approaching to within 1% of the critical temperature, as shown in Fig. 6.5b. Nevertheless, the infinite order TPT outperforms the second-order TPT, which transitions to $\beta_c = 0.5$ near a reduced temperature of 0.9. Something quite peculiar is happening closer to the critical point. Note that any analytic expression, including the Gaussian extrapolation, can be expressed as a Taylor series and should be infinitely differentiable, but the exponent of $\beta_c = 0.325$ at T_c implies that derivatives would be raised to a negative exponent, leading to divergence in the derivative property when evaluated at T_c. These observations suggest that the behavior at the critical point is nonanalytic, requiring special functions to characterize the "cross-over" to nonanalytic behavior. Ideally, these crossover functions become significant in a narrow range near critical and decay to zero outside that region. Note that the Gaussian extrapolation and Eq. (6.3-7) both include Gaussian contributions to complement the lower-order EOS. This suggests something of a convergence between multiparameter EOSs and TPT, although Span and Wagner[33] tried including terms to represent reference contributions, then rejected them.

Only complex EOS expressions such as the form of Eq. (6.3-7) can capture these strong variations, but even they are not entirely rigorous very close to the critical point. To overcome this deficiency, a variety of EOS models that attempt to include both "classical" and "nonclassical" behavior have been developed. A couple of caveats must be considered, however, when contemplating EOS complexity. First is robustness. Currently common approaches to EOS development undervalue the impact of artificial roots in the metastable region. These roots can cause convergence to meaningless values that can cause downstream computations to fail. An approach to minimize inappropriate convergence is to identify and sort all roots, selecting just the meaningful roots, but that leads to the second caveat: computational speed. Wilhelmsen et al. have shown that the difference in computational speed for a complex EOS starts at a factor of 10 for pure component computations, then increases geometrically to a factor of 100 for a five-component system.[37] Recent work by Bell and Alpert has suggested ways that computational speed could be reduced for VLE calculations.[38]

TABLE 6.2 Estimates of the deviations in density for three EOSs applied to methane and *n*-butane[39].

Substance		Methane		*n*-Butane	
EOS	**Region**	**Critical**	**High P^+**	**Critical**	**High P^+**
Peng-Robinson[3]		8	15	10	8
Peng-Robinson with Translation*		10	1.5	12	4
Lee and Kesler[7]		3	2	1.5	1.5

*Tabulated values represent the maximum relative percent deviation.
+For $1000 < P < 2000$ bar, $400 < T < 500$ K.

The other region where EOSs are often inaccurate is at very high pressures, both above and below the critical temperature. The form of the *PV* isotherms of EOS functions often do not correspond to those which best correlate data, unless careful modifications are made.

To illustrate the difficulties of these two regions, Table 6.2 presents the maximum relative deviation in density for several EOSs applied to light hydrocarbons in the near-critical region and the high *P*, high *T* region.[39] Figure 6.6 illustrates results from a classical EOS, showing minimal deviations over all regions except the highest pressures. Similar plots of de Hemptinne and Ungerer[39] and de Sant'Ana et al.[40] suggest that the latter errors can increase as *T* decreases and that increasing errors are found for heavier hydrocarbons. Our evaluations in Sec. 6.13 indicate that the advantages of the Lee-Kesler model diminish for compounds other than light aliphatic hydrocarbons. Multiparameter models are preferable in these regions if available.

The effects of these errors in the *PvT* relation are carried through to all thermodynamic property variations because they involve derivatives. Major errors for the heat capacities, isothermal compressibility, and sound speed have been shown by Gregorowicz et al.[34]

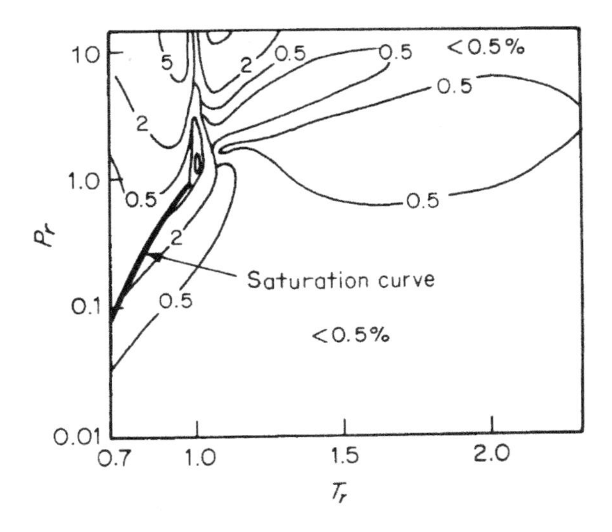

FIGURE 6.6
Contours of percent error in molar volume of CO_2 calculated from the Redlich-Kwong[41] EOS with parameters from Morris and Turek.[42]

The metastable region appears between the stable vapor and liquid densities below the critical temperature. Technically, metastable refers only to the conditions between the stable densities and the spinodal densities, where the pressure reaches its closest minimum or maximum to the stable densities. For a cubic EOS, there is only one spinodal vapor density and one spinodal liquid density, but multiparameter EOSs can exhibit multiple maxima and minima. Aursand et al.[43] have reviewed the implications of this behavior, including the implication of (unrealistic) stable fluid-fluid equilibria for pure fluids. For applications involving only bulk fluids, this can be considered as a technicality because unstable conditions are not of interest. On the other hand, integration of the Helmholtz energy through the unstable region is a key aspect of theories for characterizing surface tension. We return to this issue in Chap. 13. For now, we simply note that developing constraints to ensure reasonable behavior throughout the unstable region is an active area of research, especially for developers of multiparameter EOSs.

6.3.4 Summary

We have introduced several fundamental concepts in this section, some of which are quite sophisticated. Nevertheless, we have also seen how these fundamentals are simplified when reduced to practice, especially for the cubic EOSs. With this background, we hope to have established a certain degree of appreciation for how EOSs evolved to the present day and how they will continue to evolve. In the following sections, we illustrate specific EOSs with examples, leading to a grand evaluation of all EOSs against a database of thermodynamic properties.

This discussion is not comprehensive but does illustrate the immense amount of work that has been done in this area. Readers are referred to the papers of Deiters[44] for one description of how EOS models should be developed and communicated.

6.4 FUNDAMENTAL THERMODYNAMIC RELATIONSHIPS FOR PURE COMPOUNDS

Although most of our EOS discussion so far has focused on PvT expressions, we have occasionally referred to the Helmholtz energy, heat capacity, and other thermodynamic properties as if they were all one and the same. In fact, they are distinct properties, but they are all interrelated through Maxwell's relations and other classical thermodynamics. These relations include the PvT relationship of the EOS as also reflected in the compressibility and speed of sound (u_{ss}); the enthalpy (H), internal energy (U), and heat capacity (C_p) used in evaluating energy effects of processes through energy balances; the entropy (S) used in evaluating the properties of reversible processes and in evaluating the consequences of irreversibility in real processes; and the Gibbs energy (G) and fugacity (f) used for determining phase equilibrium conditions. Except for PvT, u_{ss}, and C_p, the above properties are not directly measurable; they may be called *conceptuals*.[45] Their changes can be inferred from experiment by using thermodynamic relations among *measurables* and they can be estimated from models for the EOS and for the ideal gas heat capacity, C_p°.

Section 6.4.1 gives general relations for these properties and shows how they are usually put into the most convenient form for calculations by using *departure functions* based on the EOS. The subsequent two sections describe general relationships among the departure functions and how to obtain departure function expressions with EOS models.

6.4.1 General Relations Between Thermodynamic Properties

Thermodynamic properties cannot always be defined in absolute terms. Enthalpy, internal energy, Helmholtz energy, and Gibbs energy must be defined in terms of a reference state. The absolute entropy of helium exemplifies an exception in this regard, as does the heat capacity, although these absolute properties pertain to esoteric conditions near 0 K. Fugacity is also defined in absolute terms because the pressure is absolute. Practically, only differences in these properties are useful. Their value is that they are *state properties.* This means that, unlike heat and work effects, changes in their values depend only on the initial and final states. For example, in evaluating the heat for steadily changing the temperature or phase of a pure component, the enthalpy change between the inlet and outlet states depends only on those state conditions, not on the details of the heating or cooling. The consequence of this also allows us to establish computational techniques for obtaining the changes from a minimum of information and with the use of the most readily accessible models.

The other advantage of these state properties is that many mathematical operations can be done to both interrelate them and to evaluate them. Partial derivatives and integrals are extensively used. Familiarity with such mathematics is useful, but not necessary, to fully understand the developments and applications. Since such procedures have been used for so long, the final formulae for the most interesting cases have been well established and can be directly used. However, subtle errors can arise, and reliable use of new models requires careful computer programming.

To illustrate path independence and the use of properties in establishing expressions for changes in conceptual properties, consider the molar enthalpy change of a pure component. The properties of H allow us to directly integrate the total differential of the enthalpy

$$H_2 - H_1 = \int_{H_1}^{H_2} dH \tag{6.4-1}$$

However, the way we characterize the two different states is by the variables T_1, P_1, and T_2, P_2. This implies that enthalpy is a function only of T and P, $H(T, P)$. This choice of variables is for convenience and essentially any two others such as V and C_p could be chosen. Then we use mathematics to obtain dH in terms of changes of T and P.

$$dH = \left(\frac{\partial H}{\partial T}\right)_P dT + \left(\frac{\partial H}{\partial P}\right)_T dP \tag{6.4-2}$$

So

$$H(T_2, P_2) - H(T_1, P_1) = \int_{T_1,P_1}^{T_2,P_2} \left[\left(\frac{\partial H}{\partial T}\right)_P dT + \left(\frac{\partial H}{\partial P}\right)_T dP\right] \tag{6.4-3}$$

The integration can be done in many ways but all must yield the same expression and, when calculated, the same numerical answer. To illustrate, we choose apparently convenient paths along isobars and isotherms. Figure 6.7 shows two possibilities, path ABC or path ADC.

$$\begin{aligned}[H(T_2, P_2) - H(T_1, P_1)]_{ABC} &= H(T_2, P_2) - H(T_2, P_1) + H(T_2, P_1) - H(T_1, P_1) \\ &= \int_{P_1}^{P_2} \left(\frac{\partial H}{\partial T}\right)_{T=T_2} dP + \int_{T_1}^{T_2} \left(\frac{\partial H}{\partial T}\right)_{P=P_1} dT \end{aligned} \tag{6.4-4}$$

$$\begin{aligned}[H(T_2, P_2) - H(T_1, P_1)]_{ADC} &= H(T_2, P_2) - H(T_1, P_2) + H(T_1, P_2) - H(T_1, P_1) \\ &= \int_{T_1}^{T_2} \left(\frac{\partial H}{\partial T}\right)_{P=P_2} dT + \int_{P_1}^{P_2} \left(\frac{\partial H}{\partial P}\right)_{T=T_1} dP \end{aligned} \tag{6.4-5}$$

Though the individual integrals in Eq. (6.4-5) are not the same, they must yield the same sum. Further, it is possible to use more apparently complicated paths such as $AEFGHC$ in Fig. 6.7 and still obtain the desired answer given the variations of the partial derivatives $(\partial H/\partial T)_P$ and $(\partial H/\partial P)_T$ over the states of the chosen path.

It turns out that the most convenient path is an artificial one, chosen because it only requires an EOS that relates P, V, and T, and the ideal gas heat capacity, C_p°, described in Chap. 4. This process transforms the fluid from interacting molecules at T_1, P_1 to an ideal gas (noninteracting molecules) at T_1, P_1. Then, the ideal gas is

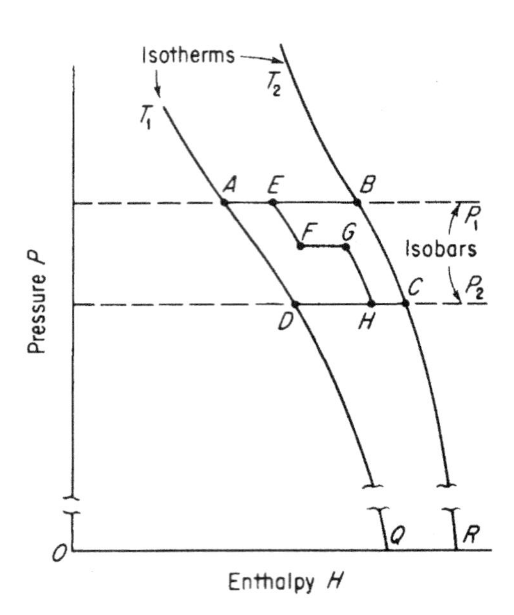

FIGURE 6.7
Schematic diagram of paths to evaluate changes in enthalpy, H, with changes in pressure, P, and temperature, T.

changed from T_1, P_1 to T_2, P_2. Finally, the ideal gas is returned to its real fluid state at T_2, P_2 (by restoring the intermolecular forces). We choose this path because we know how to evaluate the property changes of ideal gases between any two states, and this requires knowledge of only C_p° if $T_1 \neq T_2$. For H but no other conceptual properties, this path is equivalent to *ADQRC* in Fig. 6.7.

Thermodynamic manipulations yield the property changes for the molecular transformations in terms of *departure functions.* We define the departure function of a conceptual property, F, as $F^{\text{dep}} \equiv F(T, P) - F^{\text{ig}}(T, P)$. The distinction between departure functions as defined here and residual functions, F^{res}, is that $F^{\text{res}} = F(T, V) - F^{\text{ig}}(T, V)$. There are subtle differences between F^{d} and F^{res} that will be discussed further in the next section. Here, the enthalpy change of the process AC is found from the sum of (1) the departure function, $H^{\text{d}}(T_2, P_2)$, (2) the change in ideal gas enthalpy which, because the enthalpy of an ideal gas does not depend upon pressure, is an integral only over temperature, and (3) the departure function, $H^{\text{d}}(T_1, P_1)$

$$H(T_2, P_2) - H(T_1, P_1) = -[H^{\text{ig}}(T_2, P_2) - H(T_2, P_2)] + H^{\text{ig}}(T_2, P_2) - H^{\text{ig}}(T_1, P_1) + [H^{\text{ig}}(T_1, P_1) - H(T_1, P_1)]$$

$$= -\int_0^{P_2} \left(V - T\left(\frac{\partial V}{\partial T}\right)_P \right)_{T=T_2} dP + \int_{T_1}^{T_2} C_p^{\circ}\, dT + \int_0^{P_1} \left(V - T\left(\frac{\partial V}{\partial T}\right)_P \right)_{T=T_1} dP \tag{6.4-6}$$

For entropy, the relation is

$$S(T_2, P_2) - S(T_1, P_1) = -[S^{\text{ig}}(T_2, P_2) - S(T_2, P_2)] + S^{\text{ig}}(T_2, P_2) - S^{\text{ig}}(T_1, P_1) + [S^{\text{ig}}(T_1, P_1) - S(T_1, P_1)]$$

$$= -\int_o^{P_2} \left(\frac{R}{P} - \left(\frac{\partial V}{\partial T}\right)_P \right)_{T_2} dP + \int_{T_1}^{T_2} \frac{C_p^{\circ}}{T}\, dT - R\ln\left(\frac{P_2}{P_1}\right) + \int_o^{P_1} \left(\frac{R}{P} - \left(\frac{\partial V}{\partial T}\right)_P \right)_{T_1} dP \tag{6.4-7}$$

The departure functions need only the EOS for evaluation while the ideal gas change needs only C_p°. Chapter 4 gives methods for estimating C_p°; its temperature dependence may be significant. Section 6.4.2 describes departure functions for all the properties of interest, while Sec. 6.12 gives results for EOS models.

6.4.2 Residual and Departure Functions

This section gives general expressions for the departure functions for properties of interest in applications based on EOS models. Most textbooks in introductory chemical engineering thermodynamics derive the expressions given below (e.g., Elliott and Lira[5]). They can also be derived from molecular theory, from which model ideas for complex systems often arise.

The principal issue in expressing departure functions is the choice of the variable, P or v, at which the ideal gas property value is to be compared to the real fluid property value. Except for the reverted virial series (Eq. 6.3-2), EOS models are based on T and v as the independent variables because that is the natural way that multiple values of v at phase equilibrium are obtained from statistical thermodynamics. Thus, except for the reverted virial series, all the EOS model expressions are in a form where the compressibility factor can be expressed as

$$Z = PV/RT = f(T, V, \{y\}) \tag{6.4-8}$$

where $\{y\}$ is the set of composition variables such as mole fractions.

The Helmholtz energy ($A \equiv U - TS$) has become the key property in formulating modern multiparameter EOSs. Two important relations show why the Helmholtz energy is central to EOS derivations

$$Z - 1 = -V\left[\frac{\partial(a^{\text{res}}/RT)}{\partial V}\right]_{T,\{y\}} = \rho\left[\frac{\partial(a^{\text{res}}/RT)}{\partial \rho}\right]_{T,\{y\}} \tag{6.4-9}$$

$$\frac{u^{\text{res}}}{RT} = -T\left[\frac{\partial(a^{\text{res}}/RT)}{\partial T}\right]_{V,\{y\}} = \beta\left[\frac{\partial(a^{\text{res}}/RT)}{\partial \beta}\right]_{V,\{y\}} \tag{6.4-10}$$

Thus, the residual Helmholtz energy can be inferred from the internal energy or compressibility factor by integration.

$$\frac{a^{\text{res}}}{RT} = \int_V^\infty [Z-1]\frac{dV}{V} = \int_0^\rho [Z-1]\frac{d\rho}{\rho} \tag{6.4-11}$$

$$\frac{a^{\text{res}}}{RT} = \int_T^\infty \frac{u^{\text{res}}}{RT}\frac{dT}{T} + \frac{a_0^{\text{res}}(\rho)}{RT} = \int_0^\beta \frac{u^{\text{res}}}{RT}\frac{d\beta}{\beta} + \frac{a_0^{\text{res}}(\rho)}{RT} \tag{6.4-12}$$

where a_0^{res} is a temperature-independent integration "constant."

Another important formula derives from Maxwell's relations: $(\partial U/\partial V)_T = T(\partial P/\partial T)_V - P$.

$$\frac{u^{\text{res}}}{RT} = \int_V^\infty \left[-T\left(\frac{\partial Z}{\partial T}\right)_v\right]\frac{dV}{V} = \int_0^\rho \left[-T\left(\frac{\partial Z}{\partial T}\right)_v\right]\frac{d\rho}{\rho} \tag{6.4-13}$$

While Eq. (6.4-13) is often cited, it is usually more convenient to evaluate u^{res} from Eq. (6.4-10).

The major consequence is that changes in certain "natural" properties are determined from residual function integrals and then the others are found from algebraic relations. The most important residual and departure function relations are summarized below.

$$\frac{h^{\text{dep}}}{RT} = \frac{h^{\text{res}}}{RT} = \frac{u^{\text{res}}}{RT} + Z - 1 = \int_0^\rho -T\left(\frac{\partial Z}{\partial T}\right)_v \frac{d\rho}{\rho} + Z - 1 \tag{6.4-14}$$

$$\frac{s^{\text{res}}}{R} = \frac{u^{\text{res}}}{RT} - \frac{a^{\text{res}}}{RT} = \int_0^\rho \left[-T\left(\frac{\partial Z}{\partial T}\right)_v - (Z-1)\right]\frac{d\rho}{\rho} \tag{6.4-15}$$

$$\frac{s^{\text{dep}}}{R} = \frac{s^{\text{res}}}{R} + \ln(Z) = \int_0^{\rho}\left[-T\left(\frac{\partial Z}{\partial T}\right)_v - (Z-1)\right]\frac{d\rho}{\rho} + \ln(Z) \tag{6.4-16}$$

$$\frac{a^{\text{dep}}}{RT} = \frac{a^{\text{res}}}{RT} - \ln(Z) = \int_0^{\rho}[Z-1]\frac{d\rho}{\rho} - \ln(Z) \tag{6.4-17}$$

$$\ln\left(\frac{f}{P}\right) = \frac{g^{\text{dep}}}{RT} = \frac{a^{\text{dep}}}{RT} + Z - 1 = \int_0^{\rho}[Z-1]\frac{d\rho}{\rho} + Z - 1 - \ln(Z) \tag{6.4-18}$$

$$\frac{C_V^{\text{res}}}{R} = \int_0^{\rho} -\frac{T^2}{\rho RT}\left(\frac{\partial^2 P}{\partial T^2}\right)_v \frac{d\rho}{\rho} = \frac{u^{\text{res}}}{RT} + T\left[\frac{\partial\left(\frac{u^{\text{res}}}{RT}\right)}{\partial T}\right]_V \tag{6.4-19}$$

$$\frac{C_P}{R} = \frac{C_V}{R} - \frac{T}{R}\frac{\left(\frac{\partial P}{\partial T}\right)_V^2}{\left(\frac{\partial P}{\partial V}\right)_T} \tag{6.4-20}$$

$$\frac{C_P^{\text{ig}}}{R} = \frac{C_V^{\text{ig}}}{R} + 1 \tag{6.4-21}$$

$$\frac{C_P^{\text{res}}}{R} = \frac{C_V^{\text{res}}}{R} - 1 + \frac{\left[Z + T\left(\frac{\partial Z}{\partial T}\right)_\rho\right]^2}{\frac{1}{RT}\left(\frac{\partial P}{\partial \rho}\right)_T} = \frac{C_V^{\text{res}}}{R} - 1 + \frac{\left[Z + T\left(\frac{\partial Z}{\partial T}\right)_\rho\right]^2}{\left[Z + \rho\left(\frac{\partial Z}{\partial \rho}\right)_T\right]} \tag{6.4-22}$$

For the reverted virial series, departure functions can be written in terms of pressure.

$$\frac{H^{\text{dep}}}{RT} = \int_0^{P} - T\left(\frac{\partial Z}{\partial T}\right)_P \frac{dP}{P} + Z - 1 \tag{6.4-23}$$

$$\frac{S^{\text{dep}}}{R} = \int_0^{P}\left[-T\left(\frac{\partial Z}{\partial T}\right)_P - (Z-1)\right]\frac{dP}{P} \tag{6.4-24}$$

The $\ln(Z)$ terms in Eqs. (6.4-16)–(6.4-18) arise from differences between the departure and residual ideal gas reference states, as discussed in standard textbooks (e.g., Elliott and Lira,[5] §8.3).

6.5 VIRIAL EQUATIONS OF STATE

The virial EOS, Eq. (6.3-1), is a polynomial series in density that can be reverted to a series in pressure whose coefficients are functions only of T for a pure fluid. The reverted series to second order in pressure is more convenient, as given by Eq. (6.3-2). Except at high temperatures, B_2 is negative and, except at very low T, where they are of little utility, B_3 and higher coefficients are positive. This can be inferred from the behavior of the isotherms in Fig. 6.1. Formulae relating B_2 to molecular pair interactions, B_3 to molecular three-body interactions, etc., can be derived from statistical mechanics. Much has been written about the virial EOS; see especially Mason and Spurling[46] and Dymond and Smith.[47]

The virial equation is usually truncated at the second or third term and applied only to single-phase vapor or gas systems because (1) the virial expansion loses accuracy at higher pressures, (2) higher-order molecular force relations are intractable, and (3) alternative EOS forms are more accurate for dense fluids and liquids.

TABLE 6.3 Ranges of conditions for accurate Z values from virial equations using methane expressions from Setzmann and Wagner[48].

Equation	<1% Error*	<1% Error+	<5% Error*	<5% Error+
$Z = 1 + B_2\rho$	$\rho V_c < 0.18$	$T/T_c < 0.82$	$\rho V_c < 0.35$	$T/T_c < 0.9$
$Z = 1 + B_2 P/RT$	$\rho V_c < 0.1$	$T/T_c < 0.7$	$\rho V_c < 0.2$	$T/T_c < 0.8$
$Z = 1 + B_2/V + B_3/V^2$	$\rho V_c < 0.8$	$T/T_c < 0.95$	$\rho V_c < 1.5$	$T/T_c < 0.99$
$Z = 1 + B_2 P/RT + (B_3 - B_2^2)(P/RT)^2$	$\rho V_c < 0.15$	$T/T_c < 0.8$	$\rho V_c < 0.35$	$T/T_c < 0.9$

*Stated density conditions generally accurate when $T/T_c > 1.05$ or when $P/P_c < 5$.
+For saturated vapor.

The typical ranges of state for applying Eqs. (6.3-1) and(6.3-2) are given in Table 6.3; they were obtained by comparing very accurately correlated Z values of Setzmann and Wagner[48] with those computed by their highly accurate virial coefficients over the entire range of conditions that methane is described. When only B_2 is used, Eqs. (6.3-1) and (6.3-2) are equivalent at the lowest densities. Equation (6.3-2) is more accurate to somewhat higher densities but if it is used at higher pressures, it can yield negative Z values. Thus, it is common to use Eq. (6.3-2) if only the second virial, B_2, is known. If the term in B_3 is included, Eq. (6.3-1) is much more accurate than Eq. (6.3-2). Application ranges for virial equations have also been discussed elsewhere.[49,50] Another indication of the range covered by the second virial form of Eq. (6.3-2) is the initial relative linearity of isotherms.

Uncertainties in virial coefficients can affect user results. However, because errors affect $Z - 1$, which is often smaller than Z, tolerances in B_2 may be large. Absolute errors of $0.05V_c$ will generally cause the same level of error as truncating at B_2 rather than using B_3. Thus, for methane with $V_c \sim 100$ cm^3 mol^{-1}, an error of 5 cm^3 mol^{-1} in B_2 causes an error in Z of about 1% for both the saturated vapor at $T = 160$ K where $B_2 = -160$ cm^3 mol^{-1} and $Z = 0.76$ as well as at $T = 220$ K and $P = 30$ bar where $B_2 = -86$ cm^3 mol^{-1} and $Z = 0.85$. An error in B_2 of $0.25V_c$ is acceptable for estimating Z within 5% at these conditions. Since the ideal gas law would be wrong by 25% and 15% respectively, an approximate estimate of nonideality based on B_2 is likely to be better than assuming ideal gas behavior.

Extensive compilations of second virial coefficients are given by Frenkel and Marsh[51] as well as the DIPPR and TRC compilations. Values for alkanes, linear 1-alkanols, and alkyl ethers are given by Tsonopoulos and Dymond[52] and measurements using indirect thermodynamic methods have been reported by McElroy and coworkers[53] and Wormald and coworkers.[54] Methods for correlating and predicting second virial coefficients are presented in Chap. 5.

6.6 CUBIC EQUATIONS OF STATE

As pointed out above, an EOS used to describe both gases and liquids requires the form of Eq. (6.4-8) and it must be at least cubic in v. The term "cubic equation of state" implies that the function $f_V(T, v)$ has powers of v no higher than cubic. We refer readers to PGL 5ed for discussion of a quartic EOS that was feasible but offered no significant improvement over common cubic EOSs. Then, when T and P are specified, v can be found analytically rather than only numerically. We focus here on cubic EOSs because of their widespread use and simple form.

This section reviews two of the most popular cubic EOSs: the Soave-Redlich-Kwong (SRK) EOS and the Peng-Robinson (PR) EOS. Previous editions of PGL have explored the commonalities of all cubic EOSs, considering every possible permutation and combination. Briefly, the SRK and PR EOSs have been extensively developed over time and they provide correlation accuracy for physical properties that is at least comparable to other cubic equations.

6.6.1 Formulations of Cubic EOSs

With a slight adjustment, it is possible to formulate all possible cubic EOSs in a single general form with a total of nine parameters.[55] The general cubic form for Z is

$$Z = 1 + \frac{\zeta_6 b}{v - \zeta_5 b} - \frac{(a_c\alpha^S/RT)\ v(v - \zeta_4)}{(v - \zeta_3)(v^2 + \zeta_2 v + \zeta_1)} \tag{6.6-1}$$

where, depending upon the model, the parameters a_c, b, α^S, and $\{\zeta\}$ may be constants, including zero, or they may vary with T and/or composition. The parameters ζ_5 and ζ_6 were added to accommodate the ESD cubic form.[56] (We postpone further discussion of the ESD EOS to Sec. 6.8.) For example, the PR EOS obtains when $\zeta_2 = -2b$, $\zeta_1 = -b^2$ and $\zeta_i = 1$ for $i > 2$. The SRK EOS obtains when $\zeta_2 = b$, $\zeta_1 = 0$, and $\zeta_i = 1$ for $i > 2$. Thus, the distinctions among cubic EOS models for pure components depend on which of the parameters in Eq. (6.4-1) are nonzero and how they are made to vary with T. The most common variation is in the temperature dependence of the parameter α^S. This parameter is key to characterizing the vapor pressure.

Many variations of the cubic parameters are described in the 5th edition of PGL, but SRK and PR both use the same expression for α^S, developed initially by Soave,[2]

$$\alpha^S = \left[1 + \kappa^S\left(1 - T_r^{1/2}\right)\right]^2 \tag{6.6-2}$$

The characterization of the parameter κ requires careful attention if the SRK and PR EOSs are to achieve their broadest possible applications. The original correlations were given in terms of acentric factor, ω,

$$\kappa^S = 0.37464 + 1.54226\omega - 0.26992\omega^2;\quad \omega \le 0.491,\ \text{PR} \tag{6.6-3}$$

$$\kappa^S = 0.48 + 1.574\omega - 0.176\omega^2;\quad \text{SRK} \tag{6.6-4}$$

Note that ω is defined in terms of the vapor pressure at a reduced temperature, $T_r = 0.7$, and a_c is usually defined in terms of the vapor pressure at the critical temperature (i.e., critical pressure). Given the approximately linear Clausius-Clapeyron relation of vapor pressure to temperature and the two points specified by ω and a_c, the great improvement afforded by Soave's expression for α^S is not entirely surprising. Unfortunately, Eqs. (6.6-3) and (6.6-4) were developed for relatively small hydrocarbons and they should not be used when $\omega > 1$. In some applications, κ^S is treated as an adjustable parameter, specified for each component. When κ^S is approached in this manner, the SRK and PR EOSs can be applied to heavy molecules, and even to polymers.[57]

Le Guennec et al.[58] reformulated the PR model to include volume translation and an alternative expression for the temperature dependence originally developed by Twu et al.[59] Bell et al.[60] performed a similar development at about the same time.

$$\alpha^{\text{Twu}} = T_r^{N(M-1)} \exp\left[L\left(1 - T_r^{NM}\right)\right] \tag{6.6-5}$$

where L, M, N are compound-specific parameters. Twu's expression is designed to perform consistently from low temperatures to infinite temperatures. Hence Le Guennec et al.[58] referred to their model as the "translated consistent" PR (tcPR) model. With three compound-specific parameters, Pina-Martinez et al.[61] were able to correlate vapor pressures for 1700 compounds to a high degree of accuracy. Bell et al.[60] provide Twu parameters for an overlapping but broader set of 2500 compounds.

6.6.2 EOS Parameter Values from the Corresponding States Principle (CSP)

Most applications of cubic EOSs follow the original recommendations of determining a_c and b from van der Waals' critical criteria:

$$(\partial P/\partial \rho)_T = (\partial^2 P/\partial \rho^2)_T = 0 \tag{6.6-6}$$

whereby

$$a_c = \frac{\Omega_a R^2 T_c^2}{P_c};\ b = \frac{\Omega_b R T_c}{P_c}; \tag{6.6-7}$$

$$\Omega_b = \frac{2^{1/3}-1}{3};\ \Omega_a = \frac{1}{27\Omega_b};\ Z_c = 1/3;\ \text{SRK} \tag{6.6-8}$$

$$\Omega_b = \frac{\eta_c^P}{3+\eta_c^P};\ \Omega_a = \frac{8+40\eta_c^P}{49-37\eta_c^P};\ \eta_c^P = \left[1+(4-\sqrt{8})^{1/3}+(4+\sqrt{8})^{1/3}\right]^{-1} \tag{6.6-9}$$

$$Z_c = \frac{1}{1-\eta_c^P} - \frac{\Omega_a(3+\eta_c^P)}{1+\eta_c^P(2-\eta_c^P)};\ \text{PR}$$

$$\Omega_b = \frac{\eta_c^P}{3+\eta_c^P};\ \Omega_a = \frac{8+40\eta_c^P}{49-37\eta_c^P};\ \eta_c^P = \left[1+(4-\sqrt{8})^{1/3}+(4+\sqrt{8})^{1/3}\right]^{-1} \tag{6.6-10}$$

$$Z_c = \frac{1}{1-\eta_c^P} - \frac{\Omega_a(3+\eta_c^P)}{1+\eta_c^P(2-\eta_c^P)+2\left(\frac{c}{b}\right)\left(2+\eta_c^P\frac{c}{b}\right)\eta_c^P};\ \text{tcPR}$$

where the parameter "c" in the tcPR model is the volume translation. This approach recognizes the principle that all fluids have similar properties when represented in coordinates $T_r \equiv T/T_c$ and $P_r \equiv P/P_c$. We refer to this principle as the two-parameter corresponding states principle (CSP). Including the acentric factor in the characterization, as in Eqs. (6.6-3) and (6.6-4), or critical compressibility factor[56] (e.g., through volume translation) is referred to as three-parameter CSP or four-parameter CSP.

Other applications of CSP include the extended corresponding states (ECS) model[62] and Lee-Kesler model.[7] The Lee-Kesler and ECS models apply CSP by representing temperature and pressure as T_r and P_r while developing multiparameter EOSs that satisfy Eq. (6.6-6) when $T_r = P_r = 1$. Regardless of whether Eq. (6.6-6) is used directly for specific compounds or indirectly as with ECS, these approaches are considered as implementations of CSP.

6.6.3 Regression of Experimental Data for EOS Parameter Values

Many workers use a combination of critical criteria and fitting, as in the case of fitting κ^S as discussed above. Another variation is to regress the entire $\alpha^S(T)$ function; this may consist of regression of data over the entire liquid range or forced matching at a specific state such as the triple and boiling points and the critical temperature (if known). Zabaloy and Vera[63] present detailed discussion of such strategies. They also describe in depth the matching of saturation volumes to obtain EOS model parameters. The functional form of the SRK EOS causes systematic deviations in liquid density when applied from $T_r = 0.45$–1.0, overestimating V^L at high temperatures and underestimating it at low temperatures. Therefore, fitting to a narrow range of liquid density may lead to poor extrapolation. The problem is similar for the PR EOS, especially when applied to larger molecules.

6.6.4 Volume Translation

Fixing the EOS parameters to the critical point has advantages and disadvantages. The primary advantage is for applications to many compounds, like our EOS evaluations. When critical parameters are not available for a particular compound, correlations applicable to most compounds are given in Chap. 3. A disadvantage is that the characterization of liquid density is poor. Some inkling of this problem is perceived by noting that the critical compressibility factor, Z_c, is 1/3 for the SRK EOS and roughly 0.307 for the PR EOS, while typical values of

Z_c are 0.29 for spherical, nonpolar molecules, and progressively lower for longer and more polar compounds. Noting the significant role of Z_c in correlating liquid density by the original Rackett Equation (Chap. 5), the Z_c values of the SRK and PR EOSs foreshadow limitations in liquid density accuracy. One idea has been that of a "volume translation" where v computed from an original EOS is shifted so that the translated volume matches some experimental value(s) or values from an estimation method. It is common to express the shift by substituting v^{VT} for v in Eq. (6.6-1) where,

$$v - c^{VT} \equiv v^{VT} \tag{6.6-11}$$

Most forms of volume translation have been chosen to avoid changing the EOS vapor pressure (and mixture phase equilibria); Zabaloy and Brignole point out that care must be taken in the expression to ensure this and they give an example where the vapor pressure was affected by translation.[64] Pfohl warns of the dangers of making the translation parameter temperature-dependent.[65] It is recommended that a more sophisticated EOS be applied if the goal is accurate liquid densities, but volume translation may be useful for specific applications over narrow ranges of temperature. We evaluate volume translation as applied to the PR EOS with a general translation given by:

$$c^{VT} = v_{\text{ref}}^{\text{EOS}} - v_{\text{ref}}^{\text{expt}};\ v_{\text{ref}}^{\text{expt}} = v_{T_r=0.8}^{\text{SATL}} \tag{6.6-12}$$

An alternative formulation of generic cubic EOS is common when considering volume translation and tradeoffs between the van der Waals (vdW), SRK, and PR EOSs.

$$Z = 1 + \frac{\eta^P}{1-\eta^P} - \left(\frac{a_c\alpha^S}{bRT}\right)\frac{\eta^P}{(1-r_1\eta^P)(1-r_2\eta^P)} \tag{6.6-13}$$

where r_1 and r_2 are roots of the quadratic equation in the denominator. When $r_1 = r_2 = 0$, the vdW EOS is obtained and $Z_c = 0.375$. When $r_1 = -1$ and $r_2 = 0$, the SRK EOS is obtained and $Z_c = 1/3$. When $r_1 = -(2^{1/2}+1)$ and $r_2 = (2^{1/2} - 1)$ the PR EOS is obtained and Z_c is given by Eq. (6.6-9). Cismondi and Mollerup[66] noted this trend and that a common single relation could be obtained by defining,

$$r_1 r_2 = r_1 + r_2 + 1 \equiv c^{\text{RKPR}} \tag{6.6-14}$$

Along the axis of c^{RKPR}, values of Z_c are 0.375, 0.333, 0.307, and 0.29 at $c^{\text{RKPR}} = -1$, 0, 1, and 1.6, respectively. Since r_1 and r_2 multiply $\eta^P = b\rho$, the density responds reciprocally, providing the desired volume translation. We note that the volume translation is not additive, so this method does not maintain constant vapor pressure. Most work with this approach has focused on applications to heavy hydrocarbons. Tassin et al.[67] provide a recent update.

The form of Eq. (6.6-13) has also been favored by Le Guennec et al.[58] in their "translated consistent" PR (tcPR) EOS. Their methodology applies an additive constant like Eqs. (6.6-11) and (6.6-12). Their values for r_1 and r_2 are not constrained as in Eq. (6.6-14). The form of Eq. (6.6-13) also results in a convenient expression for the residual Helmholtz energy when adapting Huron-Vidal[68] mixing rules, as discussed in Chap. 7. Pina-Martinez et al.[61] provide supporting information detailing the mixing rule for b_m,

$$b_m = (x_1^2 b_{11} + 2x_1x_2b_{12} + x_2^2 b_{22});\ b_{ij}^{2/3} = \frac{\left(b_{ii}^{2/3} + b_{jj}^{2/3}\right)}{2}(1 - l_{ij}) \tag{6.6-15}$$

Example 6.1 Molar volumes of saturated propane

Find the molar volumes of saturated liquid and vapor propane at $T = 300$ K for the SRK and PR EOSs assuming $T_c = 369.8$ K, $P_c = 4.249$ MPa, $\omega = 0.152$, $Z_c = 0.281$.

Solution. The problem statement implies CSP to determine the EOS parameters. The solution requires iteration on pressure as well as the density to satisfy equality of the fugacity between the vapor and liquid phases while solving for the vapor and liquid densities at each pressure iteration. The values below correspond to the final iteration.

The solution is facilitated by a spreadsheet such as CH06SRKEOS.xlsx or CH06PREOS.xlsx on the PGL6ed website. The experimental values are: P = 0.99742 MPa, V^V = 2036.5 cm^3/mol; V^L = 88.334 cm^3/gmol. This temperature corresponds to, T_r = 0.8112.

SRK EOS: κ = 0.7152, α = 1.1471, a = 1091 J·L/mol^2; b = 62.70 cm^3/mol.

PR EOS: κ = 0.6028, α = 1.1233, a = 1143 J·L/mol^2; b = 56.30 cm^3/mol.

The %deviations correspond to 100(calc-expt)/expt:

EOS	P^{vp}	%dev	v^V	%dev	v^L	%dev
SRK	1.0106	1.32	2031.4	0.25	98.438	11.44
PR	0.9989	0.15	2035.1	−0.07	86.743	−1.80

It is evident from this example that the liquid volume is significantly overestimated for the SRK EOS, as expected from SRK's value of Z_c and $T_r > 0.7$. This performance relative to the PR EOS is expected when $Z_c > 0.26$, but the difference is less when $Z_c < 0.26$.

6.7 MULTIPARAMETER EQUATIONS OF STATE

The complexity of property behavior cannot be described with high accuracy by cubic EOSs. Though the search for better models began well before computers, the ability to rapidly calculate results or do parameter regression with complicated expressions has introduced increasing levels of complexity and numbers of fitted parameters. This section describes five approaches that are available for pure components. Two are strictly empirical: BWR/MBWR models and Wagner formulations. Two are semiempirical formulations based on theory: perturbation methods and chemical-physical association models. The last method attempts to account for the fundamentally different behavior of the near-critical region by using "crossover" expressions.

Until the late 1990s, the MBWR equation form was standard for IUPAC and NIST compilations of pure component fluid volumetric and thermodynamic properties. Kedge and Trebble[69] have investigated an MBWR expression with 16 parameters that provides high accuracy (within 0.3% of validated data for volumetric properties and P^{vp}). However, the Wagner models described below have become more prevalent in use, because the extension to added terms can be automated and they usually provide greater accuracy with fewer parameters.

6.7.1 MBWR: Lee-Kesler Models

The BWR expressions are based on the pioneering work of Benedict, Webb, and Rubin who combined polynomials in temperature with power series and exponentials of density into an eight-parameter form.[21.7C] Additional terms and parameters were later introduced by others to formulate modified Benedict-Webb-Rubin (MBWR) EOSs.

An MBWR correlation of particular interest is the Lee-Kesler generalized EOS:[7]

$$Z^{(i)} = 1 + \frac{B(T)}{V_r} + \frac{C(T)}{V_r^2} + \frac{D(T)}{V_r^n} + \frac{c_4}{T_r^3 V_r^2}\left(\beta + \frac{\gamma}{V_r^2}\right)\exp\left(-\frac{\gamma}{V_r^2}\right) \tag{6.7-1}$$

where $B = \Sigma_0^3 b_i T_r^i$; $C = \Sigma_0^3 c_i T_r^i$; $D = d_0 - d_1/T_r$; $V_r = P_c v/RT_c$; The parameters of the Lee-Kesler EOS are specified in Table 6.4.

TABLE 6.4 Parameters for the Lee-Kesler generalized MBWR EOS. $c_2^{(i)} \equiv 0$, so there are only 12 actual parameters per reference component.

Coeff	0	(r)	Coeff	0	(r)
b_0	0.1181	0.2027	c_3	0.0000	0.0169
b_1	–0.2657	–0.3315	c_4	0.0427	0.0416
b_2	–0.1548	–0.0277	$d_0(10^4)$	0.1555	0.4874
b_3	–0.0303	–0.2035	$d_1(10^4)$	0.6237	0.0740
c_0	0.0237	0.0313	β	0.6539	1.2260
c_1	–0.0187	–0.0504	γ	0.0602	0.0375
c_2	0	0			

The Lee-Kesler EOS further specifies that

$$Z = Z^{(0)} + \left(\frac{\omega}{0.3978}\right)(Z^{(r)} - Z^{(0)}) \tag{6.7-2}$$

Formulas for departure functions are given in the original reference. The designated procedure is to compute the values of $Z^{(0)}$ and $Z^{(r)}$ individually, then add them in accordance with Eq. (6.7-11).

6.7.2 MBWR: The TRAPP Model

The transport property prediction (TRAPP) model was conceived as a corresponding states method to calculate viscosities and thermal conductivities of pure fluids and mixtures. Accurate thermodynamic properties are an implicit part of the transport property predictions, however. In its original version,[71] it employed methane as a reference fluid. In the more recent version presented below for pure fluids, propane is the reference fluid. Other reference fluids could be chosen. For example, Huber and Ely use R134a as the reference fluid to describe the behavior of refrigerants.[72] The TRAPP method was originally developed only for nonpolar compounds, but there have been efforts to extend the method to polar compounds as well.[73]

The TRAPP model begins with an accurate EOS for propane of the form

$$P_0 = \rho RT + 1.1807\, T^{½}\rho^2 + \sum_{n=2}^{9} a_i(T_0)\rho_0^n + \exp(-\rho_r^2) \sum_{n=10}^{15} a_i(T_0)\,\rho_0^{2n-17} \tag{6.7-3}$$

where P_0 [=] MPa and, except for a_2, the expressions for $a_i(T)$ are given by

$$a_i = \sum_{j=-1}^{4} \frac{a_{ij}}{T_0^j} \tag{6.7-4}$$

The expression for a_2 effectively includes an extra term for $a_{25}T_0^{½} = 1.1807\ T_0^{½}$ Propane was chosen as the reference fluid because its reduced triple-point temperature is particularly low, ensuring coverage of the range of conditions necessary for all pure compounds. Coefficients are given in Table 6.5. Coefficients (in different units) with greater precision are available.[74]

To extend beyond the reference compound, the "shape factors" f and h were introduced.

$$f = \frac{T_c}{T_c^R}[1 + (\omega - \omega^R)(0.05203 - 0.7498 \ln T_r)] \tag{6.7-5}$$

$$h = \frac{\rho_c^R}{\rho_c}\frac{Z_c^R}{Z_c}[1 - (\omega - \omega^R)(0.1436 - 0.2822 \ln T_r)] \tag{6.7-6}$$

TABLE 6.5 Parameters for the TRAPP generalized MBWR EOS.[74] There are 32 parameters for propane including the coefficient of $T^{1/2}\rho^2$ in Eq. 6.7-3. T[K], P[bar], ρ[mol/L].

Coeff	−1	0	1	2	3	4
a_2	−0.0028043	−37.5633	5624.37	−935476		
a_3	−0.0004557	1.53004	−1078.11	221807		
a_4	6.629E-05	−0.06199	67.5421			
a_5		0.00647				
a_6			−0.68043	−97.2616		
a_7			0.050980			
a_8			−0.001005	0.43637		
a_9				−0.01249		
a_{10}				2644756	−7.944E+07	
a_{11}				−7299.92		5.38E+08
a_{12}				34.5022	9936.67	
a_{13}				−2.16670		−161210
a_{14}				−0.003633	11.0861	
a_{15}				−0.00013	−0.031577	1.42308

Given a temperature and molar density, the method computes T_0 and ρ_0 by

$$T_0 = T/f \tag{6.7-7}$$

$$\rho_0 = \rho h \tag{6.7-8}$$

Note that $\rho_r = \rho/\rho_c$ for all fluids because the h-factor cancels.
The method proceeds to solve Eq. (6.7-3) for P_0 then computes P for the real fluid from

$$P = \left(\frac{f}{h}\right)P_0 \tag{6.7-9}$$

Other thermodynamic properties are given by

$$\frac{a^{res}}{RT} = \frac{a_0^{res}(\rho_0, T_0)}{RT};\ \frac{h^{res}}{RT} = \frac{h_0^{res}(\rho_0, T_0)}{RT};\ \frac{s^{res}}{R} = \frac{s_0^{res}(\rho_0, T_0)}{R} \tag{6.7-10}$$

The TRAPP model was most extensively developed for 208 compounds ranging from argon and methane to n-tetracosane.[75] An alternative formulation was developed for 21 refrigerants using 1112 tetrafluoroethane (R134a) as the reference compound.[76] The model holds added interest because it forms the basis for correlations of transport properties as well as thermodynamic properties. In recent years, TRAPP has been largely superseded by REFPROP, which currently covers over 150 compounds, including transport properties, and is being actively extended to more compounds.

6.7.3 Wagner Models

Setzmann and Wagner[48] describe a computer-intensive optimization strategy for establishing highly accurate EOS models by a formulation for the residual Helmholtz energy,

$$a^{res}/RT = [A(T, v) - A^{ig}(T, v)]/nRT \tag{6.7-11}$$

where $A^{ig}(T, v)$ is the ideal gas Helmholtz energy at T and v. The model expressions contain large numbers of parameters whose values are obtained by regression on data for many properties over wide ranges of conditions. Recently, the trend is that empirical EOSs for pure components are based on this highly accurate methodology, provided sufficient data exist.

The technique first establishes a "bank of terms" that are functions of temperature and density in the forms $\Phi_{ij}(T/T_c, V/V_c) = n_{ij}\delta^{di}\tau^{tj}$ and $\Phi_{ijk}(T/T_c, V/V_c) = n_{ijk}\delta^{di}\tau^{tj}\exp(\delta^{ck})$, where n_{ijk} is a fitted coefficient, the reduced density is $\delta = V_c/V$, the reduced temperature is $\tau = T_c/T$, the d_i are integers, the t_j are integers and half-integers, and the c_k are integers. There can also be many additional terms designed to make significant contributions only near the critical point. Thus, up to 583 total terms and associated parameters may be considered for use.[33] The algorithm tests each term individually to determine if it makes a significant contribution to optimizing the objective function. If not, the term is discarded, along with any characteristic parameters. In the end, the number of terms and parameters is significantly reduced. Related algorithms continue to be advanced. For most applications, a generalized set of exponents is implemented for many components (e.g., nonpolar components), leaving roughly 10–12 parameters that are specific to a particular component.

Their optimization strategy is to regress all available and rigorously validated volumetric, calorimetric, and speed of sound data by finding optimal linear parameters, n_{ijk}, as different numbers of terms are included in the model. Ultimately only those terms which significantly improve the fit are included in the model. In the case of methane, the optimum was for 40 terms and parameters plus values of T_c and V_c. This number varies with different substances and ranges of data conditions. For methane they also fitted eight parameters to the ideal gas heat capacities to obtain the accurate temperature dependence of $A^{ig}(T, V)/RT$. The compressibility factor of Eq. (6.3-7) is found using a thermodynamic partial derivative

$$Z = 1 - V\frac{\partial\left(\frac{a^{res}}{RT}\right)}{\partial V}\Big|\tau$$

$$= 1 + \delta\left\{\sum_i\sum_j n_{ij}d_i\delta^{d_i - 1}\tau^{t_j} + \sum_i\sum_j\sum_k n_{ijk}\delta^{d_i - 1}\tau^{t_j}[\exp(-\delta^{c_k})][d_i - c_k\delta^{c_k}]\right\} \quad (6.7\text{-}12)$$

Additional terms, like u^{res}, could be differentiated from Eq. (6.3-8).

Equations in this form can describe all measured properties of a pure substance with an accuracy that may exceed that of the measurements. Measurements from multiple authors may display scatter that the model smooths, such that the model matches the "true" value better than the scattered data might indicate. It is straightforward to determine B_2 from Eq. (6.7-12), where B_2/v_c is all terms in the sums (inside the braces) when $\delta = 0$. It can also predict the properties of fluids at hyperpressures and hypertemperatures (accessible only at explosive conditions).[77] All other thermodynamic properties are straightforward derivatives of the terms in Eq. (6.7-11).

Thus, if the analysis and regression have been done for a substance, readers who wish benchmark descriptions of a common substance can use equations of this form with confidence. For example, roughly 75 are available in the NIST Webbook as of this writing and roughly 150 are available in the NIST REFPROP, including all 21 of the GERG-2008 compounds.[78] Additionally, most compounds can be fit to an abbreviated (12-term) Wagner form in the NIST TDE software,[20] but these should be used with caution because they may be based on fewer data and their extrapolation behavior is uncertain. As a guideline, Span and Wagner[33] achieved success with a (highly accurate) reduced dataset of four vapor pressures and saturated liquid densities from $T_r = [0.5–0.95]$, two saturated vapor densities, and 21 $P\rho T$ data points from $T_r = 0.5–1.6$ as a minimum for fitting a 12-term Wagner EOS. They also achieved success using only liquid data (and vapor pressures) to generate a complete equation of state. It is interesting to note that the linear form of Eq. (6.3-7) lends itself naturally to the Lee-Kesler interpolation approach, so initial guesses for fitting the parameters can be taken from that approach. After fitting, graphs of the data versus model should be generated and archived to validate the model.

6.8 PERTURBATION MODELS WITH CUSTOMIZED PARAMETERS

The concept of perturbation modeling is to use reference values for systems that are similar enough to the system of interest that good estimates of desired values can be made with small corrections to the reference values. For EOS models, this means that the residual Helmholtz energy of Eq. (6.7-12) is written as

$$a^{\mathrm{res}}/RT = [a^{\mathrm{res}}_{\mathrm{rep}}(T, V)/RT] + \sum [a^{\mathrm{res}}_{\mathrm{att}}(T, V)/RT]^{(i)} \tag{6.8-1}$$

where the form of the perturbation terms $[a^{\mathrm{res}}_{\mathrm{att}}(T, V)/RT]^{(i)}$ can be obtained from a rigorous or approximate theory. The result is that there are very many models obtained in this manner and expressed in this form. Although cubic EOSs are typically consistent with this form in a general sense, they have been discussed separately. Therefore, this section focuses primarily on the SAFT EOS family.

We would have liked to evaluate every SAFT model available in comparison to our standard database. Unfortunately, the process of characterizing many compounds for every model becomes a labor-intensive undertaking. Furthermore, a small doubt would always remain regarding whether the parameters we develop for a given method would be fully accepted by the methods' authors. Although many papers appear with compound characterizations included, the overall number of compounds for all papers usually totals to around 100 or less. As shown in our evaluation tables, highly accurate multiparameter EOSs are available for 150 compounds. To offer added value, alternative methods should be applicable to at least 200 compounds and cover a much greater diversity of chemical families. In future editions, we would hope to push that number over 1000. We reached out to the method developers in every case to ensure that our assessments were accurate, including offers of nondisclosure agreements. The descriptions and evaluations below represent our recommendations of methods available for current applications. Several reviews of SAFT EOSs and their evolutions have been published. The discussion below has benefitted especially from the reviews by Tan, Adidharma, and Radosz[79] and Papaioannou et al.[80]

6.8.1 PC-SAFT: A Couple of Caveats

Many of the sample calculations presented here focus on PC-SAFT. Our evaluations show that PC-SAFT is a viable implementation of SAFT for a range of applications that is at least as broad as other implementations. About 800 compounds have also been characterized by Gross and coworkers for ready applications, including polymers, ions, and ionic liquids. Furthermore, its authors have made a deliberate effort to make their implementation reproducible by independent researchers. For example, a FORTRAN code for nonpolar, nonassociating molecules has been downloadable for many years[81] and Joachim Gross has graciously granted permission to post an adaptation of his more general code on the PGL6ed website, complete with compound characterizations and many binary interaction parameters. Considering the importance of reproducibility to our recommendation philosophy, we favor PC-SAFT as an overall representative of this methodology. There are many other implementations of SAFT, each with their own advocates, and we outline the key features of these below.

Despite our recommendation of PC-SAFT, it should be noted that the original PC-SAFT formulation was found to exhibit unphysical behavior under certain conditions. Perhaps faults could be found in any model if it is applied sufficiently broadly. Two such faults in PC-SAFT include negative critical pressures for polymers and a pure-compound liquid-liquid phase separation at very high densities.[82] Figure 6.8a illustrates near-critical isotherms for several molar masses of polyethylene. The critical pressure at 8000 M_{m} is negative, and there is no viable vapor root. PC-SAFT can still work for polymer solutions because concentrations of polymer in the vapor phase are generally quite small, but the behavior is disturbing. Part of the problem is that the D_2 term in Eq. (6.3-18) approaches zero in the long chain, low density limit, as illustrated in Fig. 6.8b. Since D_2 appears in the denominator of A_2, the A_2 term diverges. In general, the A_2/T^2 term should be smaller than the A_1/T term to ensure convergence of the perturbation series, but this is not necessarily the case for PC-SAFT. A prospective solution to the problem of excess liquid roots at high density has been suggested by Polishuk.[83] Unfortunately, the vapor pressure of n-alkanes becomes inaccurate when his method is implemented. A related problem has been identified by Alsaifi et al.[84] Alsaifi's method points out similar problems with most common SAFT

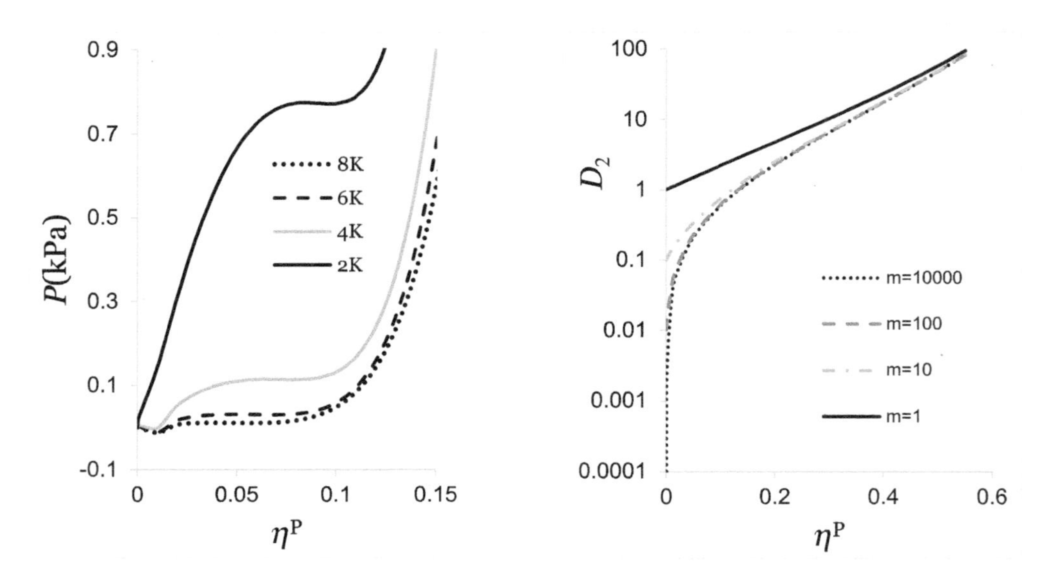

FIGURE 6.8 The peculiar behavior of PC-SAFT for long chains. (a) Near-critical isotherms where C2K corresponds to $M = 2000$ g/mol for polyethylene, C4K to $M = 4000$, etc. (b) The behavior of D_2 in Eq. (6.3-18).

implementations. Considering the broad applications of PC-SAFT in its given form, we choose to inform the reader of these caveats, but implement the method as described by its authors.

6.8.2 Wertheim's Theory of Association

Wertheim's original publications[26] treated relatively simple cases of hard spheres with one and two bonding sites. Much is known about the hard sphere fluid so it serves often as a useful reference. The extent of bonding is viewed as a perturbation on the hard sphere state. The bonding sites are carefully constructed such that only one bond can occur per bonding site (Fig. 6.9). Noting the perturbative nature of the theory and the limitation to one bond per bonding site, Wertheim referred to this approach as a first-order thermodynamic perturbation theory (TPT1). With two bonding sites, a polymer chain can be formed. In Wertheim's treatment, these polymer chains would be polydisperse, and he derived the chain EOS in terms of the average number of segments in the chain. Chapman and coworkers[85] showed that a slight alteration to the bonding scheme led to monodisperse chains, and this result coupled with reiteration of Wertheim's result with finite bonding energy to represent association was the basis of SAFT. Remarkably, Wertheim's formula for the chain EOS is identical to Chapman's if the average segment number is replaced by the exact number.

A key aspect of TPT1 is to recognize that strong interactions like hydrogen bonding or covalent bonding forces are short-ranged and orientationally specific. From this idea, Wertheim adopted a simple potential model that would exhibit these features. An illustration of Wertheim's potential model is given in Fig. 6.9. Note that the shape and position of the attractive "blister" is such that only two blisters can overlap; a third would not fit. This enabled Wertheim to ignore all attractive contacts that would correspond to multiple overlaps, a tremendous simplification. In the SWS potential model, for example, attractive contacts can come from every angle with a wide enough range to accommodate many simultaneous contacts. Not so for Wertheim's blister potential. Wertheim worked through the statistical mechanical simplifications that could be applied for a blister potential model, including the rigorous characterization of two distinct densities: the superficial density assuming zero association and the density of unbonded monomers after accounting for association. In the process, he invented a notation for exactly summing the perturbation contributions of this potential. From a chemical perspective, his results correspond to dimerization and oligomerization reactions. In this light, it is not entirely surprising that his results for linear association resemble the infinite equilibrium model of Heidemann and Prausnitz.[86]

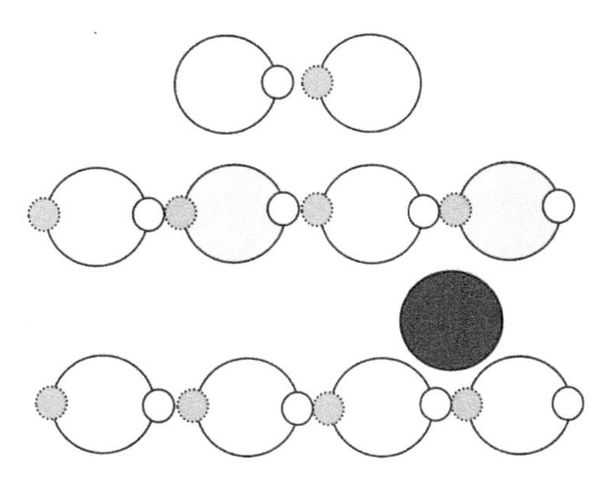

FIGURE 6.9
Illustrations of complexation interactions represented by Wertheim's blister potential model.

One benefit of Wertheim's approach, however, is that the accounting for bond formation does not require identification of specific oligomers, only a count of whether the bonding sites are unbonded. This greatly simplifies the extension to multicomponent mixtures, although accounting for every kind of bonding between multiple components with multiple bonding sites is still not entirely "simple."

A general implementation of Wertheim's theory requires a group contribution perspective regarding each molecule, because a single molecule might involve more than one bonding site. For example, monoethanolamine (MEA) involves both hydroxyl and amine functionality. Many applications may involve simpler molecules and streamlined implementations may be attractive, but we formulate the initial presentation in a general sense.

Michelsen and Hendriks[87] presented a general formalism that is especially efficient. They adopt a notation that can be applied to any number of bonding types, i.e., acceptors, donors, dimerizers, and possibly other types of bonding sites. As an alternative, we adopt notation that is explicit in terms of acceptors and donors, the most common situation. We also express the theory in multicomponent form to maintain consistency in future chapters.

$$\frac{A^{\text{chem}}}{RT} = \sum_{k=1}^{N_c} x_k \sum_{i}^{N_k^A} \ln\left(X_k^{A_i}\right) + \frac{1 - X_k^{A_i}}{2} + \sum_{i}^{N_k^D} \ln\left(X_k^{D_i}\right) + \frac{1 - X_k^{D_i}}{2} \tag{6.8-2}$$

where A^{chem} is the chemical-physical contribution to Helmholtz energy. In this form, we sum i over the number of acceptors and donors on the kth molecule. For example, MEA has distinct proton donors for the hydroxyl and amine functionalities. $X_k^{A_i}$ indicates the fraction of the ith acceptor on the kth molecule that is NOT bonded, x_k is the mole fraction of the kth component. For a pure alcohol modeled with one acceptor and one donor (such that $X^A = X^D$), Eq. (6.8-2) simplifies to Eq. (6.3-9). The values of $X_k^{A_i}$ are determined by an equation similar to a mass balance, noting that the sum of the unbonded fraction with all bonded fractions must be 1, and dividing through by the unbonded fraction of the ith acceptor on the mth molecule, X_{im}^A.

$$\frac{1}{X_m^{A_i}} = 1 + \sum_{k=1}^{N_c} x_k \sum_{j}^{N_k^D} X_k^{D_j} \rho \Delta_{m,k}^{A_i D_j} \tag{6.8-3}$$

By symmetry, a similar equation obtains for the unbonded donor fraction, X_{im}^D.

$$\frac{1}{X_m^{D_i}} = 1 + \sum_{k=1}^{N_c} x_k \sum_{j}^{N_k^A} X_k^{A_j} \rho \Delta_{m,k}^{D_i A_j} \tag{6.8-4}$$

Here, $\rho\Delta_{m,k}^{D_i A_j}$ plays a similar role to a reaction equilibrium constant for the transition from nonbonded to bonded. The summation in Eq. (6.8-3) accounts for every possible transition that the ith acceptor on the mth molecule can make. Wertheim showed how this "equilibrium constant" could be related to the probability of two bonding sites meeting times a probability of them staying bonded once they meet. The radial distribution function at contact, $g_{d,im,jk}^{\text{HS}}$, simply pertains to any unbonded spheres floating in solution, whether they have bonding sites or not. Hence, it is easily estimated by the Carnahan-Starling EOS extended to mixtures.[88] It is not necessary to account for the radial distribution function beyond contact because the bonding sites have such a short range. The presence of bonding sites is reflected by $K_{m,k}^{A_iD_j}$, which is related to the integral over the volume swept by the bonding sites, with dimensions of molar volume.

$$\rho\Delta_{m,k}^{A_iD_j} = g_{d,im,jk}^{\text{HS}}\,\rho K_{m,k}^{A_iD_j}\left[\exp\left(\frac{\varepsilon_{m,k}^{A_iD_j}}{k_{\text{B}}T}\right) - 1\right] \tag{6.8-5}$$

Here, $\varepsilon_{m,k}^{A_iD_j}$ is the bonding energy of the ith acceptor on the mth molecule with the jth donor on the kth molecule. We distinguish between $\varepsilon_{m,k}^{A_iD_j}$ and $\varepsilon_{m,k}^{D_iA_j}$ in Eqs. (6.8-3) and (6.8-4), recognizing that one functionality might be a strong proton acceptor and weak donor (e.g., 1 = methylamine) while another might be a weak acceptor and strong donor (e.g., 2 = phenol). Then $\varepsilon_{1,2}^{A_1D_1} >> \varepsilon_{1,2}^{D_1A_1}$. The analog of Eq. (6.8-5) is obtained by swapping the *AD* order.

The key to Michelsen's simplification is to recognize that Eqs. (6.8-2) to (6.8-4) can be simplified in terms of a central quantity, h.

$$h^{AD} = \sum_m^{N_c}\sum_k^{N_c} x_m x_k \sum_i^{N_m^A}\sum_j^{N_k^D} X_m^{A_i} X_k^{D_j}\rho\Delta_{m,k}^{A_iD_j} = \sum_m^{N_c}\sum_i^{N_m^A} x_m X_m^{A_i}\sum_k^{N_c} x_k \sum_j^{N_k^D} X_k^{D_j}\rho\Delta_{m,k}^{A_iD_j} = \sum_m^{N_c}\sum_i^{N_m^A} x_m\left(1 - X_m^{A_i}\right) \tag{6.8-6}$$

$$h^{DA} = \sum_m^{N_c}\sum_k^{N_c} x_m x_k \sum_i^{N_m^A}\sum_j^{N_m^D} X_m^{D_i} X_k^{A_j}\rho\Delta_{m,k}^{D_iA_j} = \sum_m^{N_c}\sum_i^{N_m^D} x_m\left(1 - X_m^{D_i}\right) \tag{6.8-7}$$

Then Eq. (6.8-2) can be rewritten as

$$\frac{a^{\text{chem}}}{RT} = \sum_{k=1}^{N_c} x_k\left[\sum_i^{N_i^A}\ln\left(X_k^{A_i}\right) + 1 - X_k^{A_i}\right] - \frac{h^{AD}}{2} + \sum_{k=1}^{N_c} x_k\left[\sum_i^{N_i^D}\ln\left(X_k^{D_i}\right) + 1 - X_k^{D_i}\right] - \frac{h^{DA}}{2} \tag{6.8-8}$$

Michelsen then proves that the derivatives of the bracketed terms evaluate to zero, so

$$Z^{\text{chem}} = -\frac{\rho}{2}\frac{\partial h^{AD}}{\partial\rho} - \frac{\rho}{2}\frac{\partial h^{DA}}{\partial\rho} \tag{6.8-9}$$

$$\frac{1}{RT}\frac{\partial P^{\text{chem}}}{\partial\rho} = -\frac{\rho^2}{2}\frac{\partial^2 h^{AD}}{\partial\rho^2} - \frac{\rho^2}{2}\frac{\partial^2 h^{DA}}{\partial\rho^2} \tag{6.8-10}$$

$$\mu_l^{\text{chem}} = \sum_{i=1}^{N_c}\ln\left(X_{il}^A\right) + \sum_{i=1}^{N_c}\ln\left(X_{il}^D\right) - \frac{1}{2}\frac{\partial(nh^{AD})}{\partial n_l} - \frac{1}{2}\frac{\partial(nh^{DA})}{\partial n_l} \tag{6.8-11}$$

By analogy to Eq. (6.8-9),

$$\frac{u^{\text{chem}}}{RT} = -\frac{\beta}{2}\frac{\partial h^{AD}}{\partial\beta} - \frac{\beta}{2}\frac{\partial h^{DA}}{\partial\beta} \tag{6.8-12}$$

$$\frac{C_V^{\text{chem}}}{R} = -\frac{\beta^2}{2}\frac{\partial^2 h^{AD}}{\partial\beta^2} - \frac{\beta^2}{2}\frac{\partial^2 h^{DA}}{\partial\beta^2} \tag{6.8-13}$$

These results might not look so simple at first, until you realize that the only dependence of Δ on density or mole number is through $g_{d,im,jk}^{\text{HS}}$. The results look much simpler if a formula is used for $g_{d,im,jk}^{\text{HS}}$ that is independent of subscript. Considering a case of specific assumptions may help to clarify the meanings of the terms. For example, the ESD and CPA EOSs assume,

$$g_{d,im,jk}^{\text{HS}} = \frac{1}{1 - 1.9\eta^P} => \frac{\rho\partial\left(\Delta_{im,jk}^{AD}\right)}{\Delta_{im,jk}^{AD}\,\partial\rho} = \frac{\eta^P\partial\left(\Delta_{im,jk}^{AD}\right)}{\Delta_{im,jk}^{AD}\,\partial\eta^P} = \frac{1.9\eta^P}{(1 - 1.9\eta^P)} \tag{6.8-14}$$

Then

$$Z^{\text{chem}} = -\frac{1.9\eta^P}{1-1.9}\frac{(h^{AD}+h^{DA})}{2} \tag{6.8-15}$$

$$\mu_k^{\text{chem}} = \sum_{i=1}^{N_c}\ln\left(X_{ik}^A\right) + \sum_{i=1}^{N_c}\ln\left(X_{ik}^D\right) - \frac{1.9b_k\rho}{1-1.9}\frac{(h^{AD}+h^{DA})}{2} \tag{6.8-16}$$

Additional efficiencies were derived by Elliott[89] for the special case of single matching site types per molecule when the following condition applies,

$$\Delta_{m,k}^{A_1D_1} = \left(\Delta_{m,m}^{A_1D_1} * \Delta_{k,k}^{A_1D_1}\right)^{1/2} \tag{6.8-17}$$

We refer to Eq. (6.8-17) as the geometric combining rule (GCR). In that case, $X_j^A = X_j^D$ and Eq. (6.8-3) can be written as

$$\left(\frac{1}{X_m^A}-1\right)\frac{1}{\left(\rho\Delta_m^{AD}\right)^{1/2}} = \sum_{k=1}^{N_c} x_k N_k^A X_k^A \left(\rho\Delta_k^{AD}\right)^{1/2} \equiv F^{AD} \tag{6.8-18}$$

where we have dropped the subscript and sum over site types when $N_k^A = N_k^D = 1$ for all k and m, but noted that the single site type can be repeated to the degree N_k^A as in polyvinyl alcohol. Note that the sum on the right-hand side of Eq. (6.8-18) is independent of component because it sums over all components. Replacing X_k^A with its definition in terms of F^{AD}, we have

$$X_k^A = \left[1 + F^{AD}\left(\rho\Delta_k^{AD}\right)^{1/2}\right]^{-1} \text{ and } F^{AD} = \sum_{k=1}^{N_c}\frac{N_k^A x_k \left(\rho\Delta_k^{AD}\right)^{1/2}}{\left[1 + F^{AD}\left(\rho\Delta_k^{AD}\right)^{1/2}\right]} \tag{6.8-19}$$

The GCR then results in a simple expression for Z^{chem}.

$$Z^{\text{chem}} = \frac{-(F^{AD})^2}{(1-1.9\eta)} \tag{6.8-20}$$

An exact solution of Eq. (6.8-3) for a single associating component applies $X^A = X^D$,

$$\frac{1}{X_k^A} = 1 + N_k^A X_k^A \rho\Delta_k^{AD} \Rightarrow X_k^A = X_k^D = \frac{2}{1+\left[1+4N_k^A\rho\Delta_k^{AD}\right]^{1/2}} \tag{6.8-21}$$

The equivalent result using Eq. (6.8-19) for a single associating component is

$$F^{AD} = \frac{N_k^A\left(\rho\Delta_k^{AD}\right)^{1/2}}{\left[1+F^{AD}\left(\rho\Delta_k^{AD}\right)^{1/2}\right]} \Rightarrow F^{AD} = \frac{2N_k^A\left(\rho\Delta_k^{AD}\right)^{1/2}}{1+\left[1+4N_k^A\rho\Delta_k^{AD}\right]^{1/2}} \tag{6.8-22}$$

For mixtures with more than one associating component, Eq. (6.8-19) comprises a single nonlinear equation with a single unknown, F^{AD}. It can be solved quickly by secant iteration taking initial guesses of $F^{AD} = 0$ (zero association) and $F^{AD} = 1$ (strong association). Note that this single result for F^{AD} implies estimates for all $\{X_k^A\}$ and $\{X_k^D\}$ (i.e., $\{X_k^D\}=\{X_k^A\}$) without directly solving the multidimensional nonlinear matrix system. We refer to the results following from the GCR as the "efficient" Wertheim theory, because they impose a minimal cost to computational speed (roughly 2x) relative to a nonassociative theory, regardless of the number of components. The GCR is reasonable for alcohols and water because $\varepsilon_a^{AD}/k_B \sim 2500$ K, so near 298 K: $[\exp(2500/298) - 1] = 4400 \approx \exp(2500/298) = 4401$, and $[\exp(\varepsilon_a^{AD}/k_BT)\exp(\varepsilon_b^{AD}/k_BT)]^{1/2} = \exp[(\varepsilon_a^{AD} + \varepsilon_b^{AD})/2k_BT]$. This means that the GCR corresponds (to four significant figures) to assuming the solvation energy between two alcohols is roughly an arithmetic average of the two individual association energies. It also works for hydrocarbons because Δ^{AD} for hydrocarbons is zero, so the solvation between water (or alcohols) and hydrocarbons is proportional to $[\Delta_w^{AD}\cdot 0]^{1/2} = 0$, indicating zero solvation between oil and water. The GCR is less accurate for systems that include ketones, esters, ethers, etc., but binary interaction parameters may suffice to achieve accurate results for those systems.

When the efficient assumptions are not applicable, Eqs. (6.8-3) and (6.8-4) must be solved in detail. The efficient solution can be taken as an initial guess, however, with subsequent iterations typically addressed by successive substitution. Sample calculations for chemical contributions are illustrated in Chap. 7, where mixtures are treated, and the implications of the assumptions are clearer.

6.8.3 Polar Contributions to Perturbation Models

Most current implementations of SAFT include an additional term to characterize the contributions of long-range multipolar forces. These contributions were developed initially by Jog et al.[90] The approach is to apply the "segment" concept of SAFT as a collection of independent spheres that are subsequently bound by Wertheim's chain term. In that context, the development of Twu and Gubbins[91] can be applied to the spherical segments. One specific implementation of interest is described in detail by Nguyen et al.[92]

$$a^{\text{polar}} = a_2^{\text{polar}}/(1 - a_3^{\text{polar}}/a_2^{\text{polar}}) \tag{6.8-23}$$

$$a_2^{\text{polar}} = A_2^{\text{polar}}(112) + A_2^{\text{polar}}(123) + A_2^{\text{polar}}(224) \tag{6.8-24}$$

$$A_2^{\text{polar}}(112) = -\frac{2}{3}\frac{\pi N\rho}{k_BT}\sum_{\alpha}^{n}\sum_{\beta}^{n} x_\alpha x_\beta x_{\text{p}\alpha}^{\mu} x_{\text{p}\beta}^{\mu} m_\alpha m_\beta \frac{\mu_\alpha^2\mu_\beta^2}{\sigma_{\alpha\beta}^3} J_{\alpha\beta}^{(6)} \tag{6.8-25}$$

$$A_2^{\text{polar}}(123) = -\frac{\pi N\rho}{k_{\text{B}}T}\sum_{\alpha}^{n}\sum_{\beta}^{n} x_\alpha x_\beta x_{\text{p}\alpha}^{\mu} x_{\text{p}\beta}^{Q} m_\alpha m_\beta \frac{\mu_\alpha^2 Q_\beta^2}{\sigma_{\alpha\beta}^5} J_{\alpha\beta}^{(8)} \tag{6.8-26}$$

$$A_2^{\text{polar}}(224) = -\frac{14}{5}\frac{\pi N\rho}{k_BT}\sum_{\alpha}^{n}\sum_{\beta}^{n} x_\alpha x_\beta x_{\text{p}\alpha}^{Q} x_{\text{p}\beta}^{Q} m_\alpha m_\beta \frac{Q_\alpha^2 Q_\beta^2}{\sigma_{\alpha\beta}^7} J_{\alpha\beta}^{(10)} \tag{6.8-27}$$

where x_α and x_β are the component mole fractions, $x_{\text{p}\alpha}^{\mu}$ is the dipolar segmental mole fraction (i.e., two or more segments on one molecule might be dipolar, and the total number of segments in solution includes polar and nonpolar segments from all components). We retain the mole fractions in these expressions because eliminating them does not significantly simplify the expressions; the segmental mole fractions still require multiple summations. In this way, we can simply refer to these expressions in Chap. 7. The terms for A_3^{polar} are

$$a_3^{\text{polar}} = A_{3A}^{\text{polar}} + A_{3B}^{\text{polar}} + 3A_{3C}^{\text{polar}} + 3A_{3D}^{\text{polar}} + A_{3E}^{\text{polar}} \tag{6.8-28}$$

$$A_{3A}^{\text{polar}} = \frac{144}{245}\frac{\pi N\rho}{(k_BT)^2}\sum_{\alpha}^{n}\sum_{\beta}^{n} x_\alpha x_\beta x_{\text{p}\alpha}^{Q} x_{\text{p}\beta}^{Q} m_\alpha m_\beta \frac{Q_\alpha^3 Q_\beta^3}{\sigma_{\alpha\beta}^{12}} J_{\alpha\beta}^{(16)} \tag{6.8-29}$$

$$A_{3B}^{\text{polar}} = \left(\frac{14336}{91125}\right)^{1/2}\frac{\pi N\rho^2}{(k_{\text{B}}T)^2}\sum_{\alpha}^{n}\sum_{\beta}^{n}\sum_{\gamma}^{n} x_\alpha x_\beta \chi^{\mu\mu\mu}\frac{\mu_\alpha^2\mu_\beta^2\mu_\gamma^2}{\sigma_{\alpha\beta}\sigma_{\beta\gamma}\sigma_{\gamma\alpha}} K_{\alpha\beta\gamma}^{(B)} \tag{6.8-30}$$

where $\chi^{abc} \equiv x_{\text{p}\alpha}^{a} x_{\text{p}\beta}^{b} x_{\text{p}\gamma}^{c} m_\alpha m_\beta m_\gamma$

$$A_{3C}^{\text{polar}} = \left(\frac{4096\pi^7}{33075}\right)^{1/2}\frac{\pi N\rho^2}{(k_{\text{B}}T)^2}\sum_{\alpha}^{n}\sum_{\beta}^{n}\sum_{\gamma}^{n} x_\alpha x_\beta \chi^{\mu\mu Q}\frac{\mu_\alpha^2\mu_\beta^2 Q_\gamma^2}{\sigma_{\alpha\beta}\sigma_{\beta\gamma}^2\sigma_{\beta\alpha}^2} K_{\alpha\beta\gamma}^{(C)} \tag{6.8-31}$$

$$A_{3D}^{\text{polar}} = -\left(\frac{22528\pi^7}{125278}\right)^{1/2}\frac{\pi N\rho^2}{(k_{\text{B}}T)^2}\sum_{\alpha}^{n}\sum_{\beta}^{n}\sum_{\gamma}^{n} x_\alpha x_\beta \chi^{\mu QQ}\frac{\mu_\alpha^2 Q_\beta^2 Q_\gamma^2}{\sigma_{\alpha\beta}\sigma_{\beta\gamma}^2\sigma_{\beta\alpha}^2} K_{\alpha\beta\gamma}^{(D)} \tag{6.8-32}$$

$$A_{3E}^{\text{polar}} = \left(\frac{4050\pi^7}{4.106}\right)^{1/2}\frac{\pi N\rho^2}{(k_{\text{B}}T)^2}\sum_{\alpha}^{n}\sum_{\beta}^{n}\sum_{\gamma}^{n} x_\alpha x_\beta \chi^{QQQ}\frac{Q_\alpha^2 Q_\beta^2 Q_\gamma^2}{\sigma_{\alpha\beta}^3\sigma_{\beta\gamma}^3\sigma_{\gamma\alpha}^3} K_{\alpha\beta\gamma}^{(E)} \tag{6.8-33}$$

The values of $J_{\alpha\beta}$ and $K_{\alpha\beta\gamma}$ are listed by Twu and Gubbins. A slightly different implementation has been described by Vrabec and Gross[93] (see also Tan et al.,[79]). Several researchers have noted that including the polar term improves the representation of mixture phase diagrams[94] and reduces the magnitudes of binary interaction parameters.[95]

6.8.3.1 HR-SAFT

At an early stage, Huang and Radosz[96] proposed an implementation of SAFT referred to here as HR-SAFT. This has the same form as Eq. (6.3-15), for the reference contribution, but applies the Chen and Kreglewski[97] term to estimate A^{dsp}. Chen and Kreglewski fitted the D_{ij} coefficients of Eq. (6.8-35) to argon. They assumed a square-well potential with a repulsive shoulder as the basis for developing their EOS, giving rise to a temperature-dependent effective hard sphere diameter, d^{ehs}, that is smaller than the diameter, σ^{ss}, where the potential crosses zero, like the theoretical basis of PC-SAFT, but A^{dsp} does not vary with m in HR-SAFT.

$$\frac{a^{res}}{RT} = mA^{HS} + A^{chain} + mA^{dsp} + \frac{a^{chem}}{RT} \tag{6.8-34}$$

where the covolume, $b = \pi N_A d^3/6$; $d^{ehs} = \sigma^{ss}[1 - 0.12\exp(-3u^{ss}/k_BT)]$; $u^{ss} = \varepsilon^{ss}(1 + e_1/k_BT)$; e_1 is a compound-specific parameter. The temperature dependence of the spherical repulsion term raises the question of what to use for the contact value of the radial distribution function, g^{HS}. Note that the definition of η^P in terms of d^{ehs} implies that the temperature-dependent diameter is used. Technically, this means that the bond length of the conceived tangent sphere chain is shrinking with temperature, as well as the segmental diameter. Similarly, the bond length to the associative acceptor and donor sites is also shrinking, conceptually. It would be difficult to apply Newton's laws to a potential model that followed this conceptual picture. Another distinction of PC-SAFT relative to HR-SAFT is that PC-SAFT sets $e_1 = 0$ for all compounds.

$$A^{dsp} = \sum_{i-1}^{m}\sum_{j}^{n} D_{ij}\left(\frac{\varepsilon^{ss}}{k_B}\right)^i\left(\frac{\eta^P}{\eta^P_{max}}\right)^j \tag{6.8-35}$$

where $\eta^P_{max} = 0.74048$. An advantage of HR-SAFT is that it fits the critical region of spherical molecules closely, owing to the multiparameter fit of Chen and Kreglewski.[97]

One way to conceive of the SAFT perspective is that any fluid is composed originally of spherical sites surrounded by their repulsive and attractive interactions, and Wertheim's chain term binds these segments into tangent sphere chains. We refer to this concept as "the segmental perspective." It is clear in the $m(A^{HS} + A^{dsp})$ contributions of Eq. (6.8-34). From this perspective, one might hope that the anomalous critical behavior characterized for spheres would be naturally included when the theory is extended to chains. Unfortunately, the accuracy of the HR-SAFT equation in the critical region deteriorates as the chain length increases until results are like those for PC-SAFT.

Overall, the HR-SAFT method has been broadly applied and is available in some process simulators. Its viability is like that of PC-SAFT and the choice between the two is largely a matter of whether parameters are available for the compounds of interest.

One legacy of the HR-SAFT development is a classification scheme for association models that is often applied today. This taxonomy provides a convenient shorthand for specifying the arrangement of bonding sites, as described (with slight modification) in Table 6.6.

6.8.4 The Elliott-Suresh-Donohue (ESD) Model

The ESD model of Elliott et al.[56] was initially articulated as an adaptation of chemical theory using the Heidemann and Prausnitz infinite equilibrium model.[86] Donohue and coworkers had developed the associated perturbed anisotropic chain theory (APACT) based on that approach.[99] The concept behind the ESD model was to simplify and make it cubic for nonassociating compounds, with the parameters evaluated using the critical

TABLE 6.6 Taxonomy of common bonding arrangements.

Type	Bonding	Typical assignments	Notes
1A	$\Delta^{AA} \neq 0$	Carboxylic acids	Solvation feasible with any other sites
1B	$\Delta^{AA} = 0$, $\Delta^{AD} \neq 0$	Ethers, esters, 3° amines...	Solvation without association
1C	$\Delta^{CC} \neq 0$	Carboxylic acids	C types can only solvate with other C types
2B	$\Delta^{ii} = 0$, all i; $\Delta^{AD} \neq 0$	Alcohols, 2° amines*	Acceptors bond with donors. Simple.
3B	$\Delta^{ii} = 0$, all i; $\Delta^{AD_1} = \Delta^{AD_2} \neq 0$	1° amines, "SPC" water	Two equivalent donor sites. SPC water merges the lone pairs into a single site.
4B	$\Delta^{ii} = 0$, all i; $\Delta^{AD_1} = \Delta^{AD_2} = \Delta^{AD_3} \neq 0$	ammonia	Three equivalent donor sites. A 3B model is also used for ammonia in some cases.
4C	$\Delta^{ii} = 0$, all i; $\Delta^{A_iD_j} \neq 0$; $i,j = 1,2$	glycols, "ST2" water[98]	Two each of donor and acceptor sites.

*abbreviation: 1° – primary; 2° – secondary; 3° – tertiary

criteria like other cubic EOSs. The ESD EOS was later shown to be consistent with Wertheim's theory subject to an assumption about the change in heat capacity due to association.[100] Elliott and Lira[5] have shown how to rearrange the ESD model into the SAFT format, clarifying that it is entirely consistent with that formalism subject to relatively simple assumptions about the segment and reference terms. To leverage the development in previous sections of this chapter, we present the method in the SAFT format here.

The method is not widely applied but it has been implemented in at least one process simulator. It has been developed for many binary systems.[101] The computer codes are relatively simple and are made available through the PGL6ed website.

The ESD model can be written as

$$\frac{a^{\text{res}}}{RT} = mA^{\text{HS}} + A^{\text{chain}} + mA^{\text{att}} + \frac{a^{\text{chem}}}{RT} \tag{6.8-36}$$

where

$$A^{\text{HS}} = -\frac{4}{1.9}\ln(1 - 1.9\eta^{\text{P}});\;\; A^{\text{att}} = -\frac{9.5}{k_1}\ln(1 + k_1 Y\eta^{\text{P}}); \tag{6.8-37}$$

$$A^{\text{chain}} = -(m-1)\ln\left[g_d^{\text{HS}}\right];\; g_d^{\text{HS}} = \frac{1}{(1 - 1.9\eta^{\text{P}})} \tag{6.8-38}$$

$$Y = \exp(\beta\varepsilon)\text{-}k_2 \tag{6.8-39}$$

$$\frac{a^{\text{chem}}}{RT} = 2\ln(X^A) + 1 - X^A \tag{6.8-40}$$

where $k_1 = 1.7745$ and $k_2 = 1.0617$. The equations for A^{chem} are analogous to those for PC-SAFT except that d^{ehs} is independent of temperature in the ESD treatment. Differentiating gives

$$Z = 1 + \frac{4m\eta^{\text{P}}}{1 - 1.9\eta^{\text{P}}} - (m-1)\frac{1.9\eta^{\text{P}}}{1 - 1.9\eta^{\text{P}}} - \frac{9.5mY}{1 + k_1 Y\eta^{\text{P}}} + Z^{\text{chem}} \tag{6.8-41}$$

The form of the equation was developed with several ideas in mind. First, the EOS should be cubic for nonassociating molecules, and the parameters should reflect the critical constraints. Second, the reference term should match hard sphere simulation data as well as possible for a cubic EOS. This is how the value of 1.9 was determined. Third, the attractive term should roughly represent the square well sphere attraction. The form of the attractive term was inspired by the work of Sandler and Lee.[102] Also, if $k_2 = 1.0$, the second virial coefficient of the square well fluid is closely matched for $\lambda = 1.5$. Finally, the EOS should provide a reasonable representation of the vapor pressure of spherical molecules. This is how the values of k_1 and k_2 were determined. Polarity contributions were not considered explicitly, but dipolar and induced-dipolar interactions are implicitly treated by regarding the σ and ε parameters as effective values.[103] Figure 6.10 shows qualitative agreement with simulation results and a substantial improvement relative to the van der Waals EOS. Results for PR and SRK EOSs would be like the vdW EOS, but their α^S function makes it difficult to infer their equivalent TPT terms.

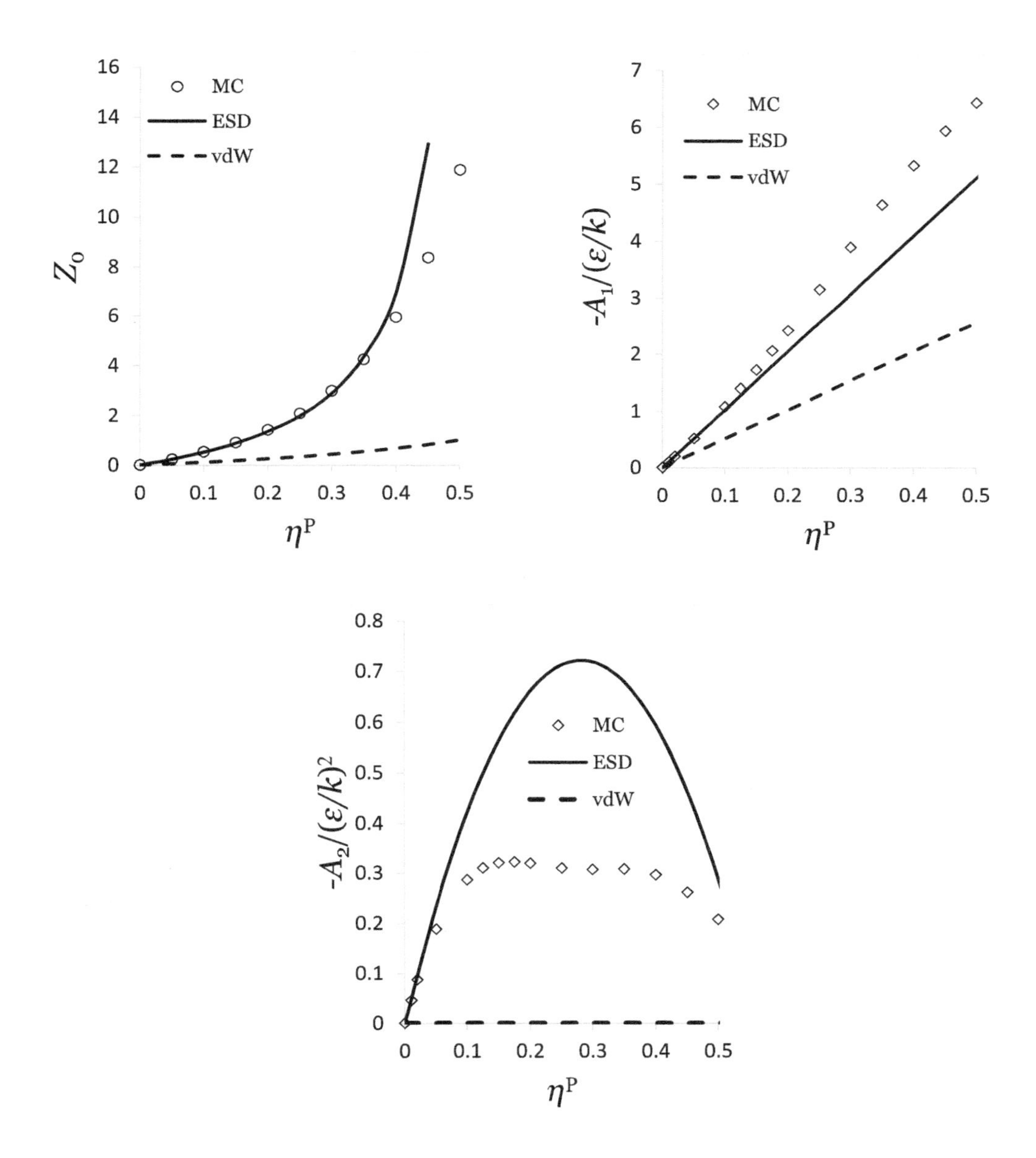

FIGURE 6.10
Perturbation contributions Z_0, A_1, A_2 of square well spheres for $\lambda = 1.5$. The solid curves show the ESD estimate and the dashed curves show the van der Waals estimate. Points correspond to molecular simulations of Pavlyukhin.[27]

The ESD model has been most fully developed with the efficient Wertheim assumptions of Elliott,[89] including the assumption of single matching donor and acceptor types per molecule, in which case,

$$Z^{\text{chem}} = -\sum_{k=1}^{N_c} x_k N_k^D \frac{[(1 - X_k^A)]}{1 - 1.9\eta^P} \tag{6.8-42}$$

For a single associating component this becomes

$$Z^{\text{chem}} = -\frac{N^D(1 - X^A)}{1 - 1.9\eta^P} \tag{6.8-43}$$

All evaluations in PGL6ed assume the efficient Wertheim implementation for the ESD EOS, such that Eq. (6.8-19) provides an exact solution for a pure compound. It may be of interest that the repulsive part of the ESD equation was originally written as $(mA^{\text{HS}} + A^{\text{chain}}) = 4c/1.9{*}\ln(1{-}1.9\eta^P)$ where $c = 1 + (m - 1)/1.90476$. Noting that $4/(4 - 1.9) = 1.90476$, it can be shown that Eq. (6.8-36) is consistent with the original article.

The pure component parameters of the ESD EOS are nearly as ubiquitous as for any cubic EOS. When a compound is nonassociating, the default is to estimate pure component parameters from T_c, P_c, ω using the equations given by Elliott and Lira.[5] A group contribution method is also available for estimating parameters when critical properties are not available, e.g., for polymers.[104] Customized parameters have been developed for about 155 compounds, including many associating compounds and 10 polymers. These are included with the code linked on the PGL6ed website.

6.8.5 The Cubic Plus Association (CPA) Model

Cubic EOSs have become such an ingrained part of property modeling that it is worthwhile to ask whether they can be retained in their traditional form for nonassociating compounds while simply adding the Wertheim contributions when necessary. This is the concept behind the CPA model.[105]

CPA incorporates the Wertheim contribution as an additive feature, without the chain term, adapting the Soave EOS to characterize the nonassociating contributions. It might be tempting to assume the value for g_d^{HS} implied by the van der Waals repulsive contribution, since that would be consistent with the Soave EOS, but the value of g_d^{HS} was first adapted from the Carnahan-Starling estimate, as for most SAFT implementations. The computation of the association contribution is also like other SAFT implementations,[105] but the composition dependence of g_d^{HS} is simplified such that the efficient Wertheim solution can be applied in most cases. This enhances the computational speed, especially for the multicomponent systems that might be encountered in reservoir or petrochemical applications. Later, Kontogeorgis et al. adopted the ESD estimate of g_d^{HS}, simplifying the model further.[106] The system of equations for pure fluids becomes

$$\frac{a^{\text{res}}}{RT} = -\ln(1 - b\rho) - \frac{a}{bRT}\ln(1 + b\rho) + \frac{a^{\text{chem}}}{RT} \tag{6.8-44}$$

$$Z = \frac{1}{1 - b\rho} - \frac{a}{bRT}\frac{b\rho}{(1 + b\rho)} + Z^{\text{chem}} \tag{6.8-45}$$

Here, A^{chem} and Z^{chem} are given by Eqs. (6.8-40) and (6.8-43) with $\eta^{\text{P}} \equiv b\rho/4$.

Although the pure component parameters of nonassociating compounds could be determined by the SRK EOS in principle, CPA's authors favor pure component parameters regressed from vapor pressure and density data, like other SAFT models. The CPA model applies a correlation for the attractive contribution that differs slightly from the SRK EOS.

$$a = a_0\left[1 + c_1\left(1 - \sqrt{T_r}\right)\right]^2 \tag{6.8-46}$$

where a_0 and c_1 are pure component parameters, along with the volume parameter, b. Pure component parameters are available for about 100 compounds.

6.8.6 Soft-SAFT

The concept behind Soft-SAFT[107] is to use a multiparameter (MBWR) EOS for Lennard-Jones (LJ) spheres[108] to represent a SAFT segment (instead of square-well spheres), then apply the Wertheim formalism to add the chain, association, and polar contributions. In this adaptation of the segmental perspective, both the spherical repulsion and attractive dispersion are accounted for by the LJ EOS and multiplied by the shape parameter, m. Accordingly, the LJ reference can be composed of various monomers, providing a basis for making the chains heteronuclear if so desired.

A key distinction of the Soft-SAFT approach is that the relevant contact rdf becomes g_σ^{LJ}, making the chain contribution temperature-dependent due to attractive forces as well as the temperature dependence of g_d^{HS}. Soft-SAFT applies a correlation by Johnson et al.[109] for g_σ^{LJ} as given by Eq. (6.8.47).

$$g_\sigma^{LJ} = 1 + \sum_{i=1}^{5}\sum_{j=1}^{5} a_{ij}(\rho^*)^i (T^*)^{1-j} \tag{6.8-47}$$

The coefficients $\{a_{ij}\}$ are given by Johnson et al.[109] The Soft-SAFT EOS can be written as,

$$\frac{a^{\text{res}}}{RT} = \frac{ma^{LJ}}{RT} + \frac{a^{\text{chain}}}{RT} + \frac{a^{\text{polar}}}{RT} + \frac{a^{\text{chem}}}{RT} \tag{6.8-48}$$

where $a^{\text{chain}}/RT = (m - 1)\ln(g_\sigma^{LJ})$. (Note: this expression is sometimes written as $(m - 1)\ln(y^{LJ})$, where $y^{LJ} = \exp(u^{LJ}/RT)g^{LJ}$, but $u^{LJ} = 0$ at $r = \sigma$, so there is no difference.)

The Soft-SAFT model has been applied most extensively to ionic liquids.[110,111] A challenge with ionic liquids is that experimental data are mostly limited to liquid densities since the vapor pressures are almost immeasurably low. Therefore, it is valuable to "transfer" characterizations of certain atoms or groups of atoms from other molecular characterizations, like a transferable force field in molecular simulation. The Soft-SAFT model lends itself well to this kind of transferability. Vega and coworkers provide many illustrations of this approach applied to Soft-SAFT.[110] It is an active subject of research and future editions of PGL may be able to consider the model in more detail.

6.8.7 SAFT-VR

The original concept behind SAFT-VR is a segmental approach like that of Soft-SAFT, except that a generalized perspective was brought to the potential model instead of LJ spheres. The key attribute of this perspective was characterizing the range of attraction as being of variable range (VR).[112] The simplest articulation of this concept would be the well width of a SW potential, but another common articulation is in the exponents of the Mie potential expressed as

$$u(r) = -C_{mn}\varepsilon\left[\left(\frac{\sigma}{r}\right)^n - \left(\frac{\sigma}{r}\right)^m\right] \tag{6.8-49}$$

$$C_{mn} = \frac{n}{n-m}\left(\frac{n}{m}\right)^{\frac{m}{n-m}} \tag{6.8-50}$$

where C_{mn} is generally specified such that $u(r) = -\varepsilon$ at the potential's minimum. When $n = 12$ and $m = 6$, $C_{mn} = 4$ and the LJ model is recovered. Most implementations use $n > 12$, making the repulsion steeper and the minimum sharper and deeper. Setting $m < 6$ extends the range of the potential, which can be useful if describing a molecule like CO_2 with a single spherical site.

The SAFT-VR model has gone through several evolutions over the years. Early implementations favored the SWS perspective, while more recent implementations favor the Mie perspective. We focus here on the SAFT-VR-Mie model, as described by Lafitte et al.[113] A feature of Lafitte's implementation includes the observation that thermodynamic derivative properties are especially sensitive to the repulsive exponent, n, of the Mie potential. Hence, careful attention is paid to both exponents. Typically, even integers are favored for these exponents such that all exponents are integers when mixtures of molecules with different exponents are modeled.

Comparisons of SAFT-VR-Mie to CPA and PC-SAFT suggest that SAFT-VR-Mie offers improvements over PC-SAFT and CPA for derivative properties, especially for n-alkanes.[114,115] Lafitte's implementation applies the Barker-Henderson (BH) split of the repulsive and attractive portions of the potential.[116]

$$u_0 = u^{\mathrm{Mie}} \text{ at } r \leq \sigma, 0 \text{ otherwise}; \; u_1 = u^{\mathrm{Mie}} \text{at } r > \sigma, 0 \text{ otherwise.} \tag{6.8-51}$$

Another alternative is to split the potential at its minimum, which we discuss later. A distinct improvement of SAFT-VR-Mie[113] is that they include the third-order TPT contribution.

$$\frac{a^{\mathrm{res}}}{RT} = \frac{ma^{\mathrm{MS}}}{RT} + \frac{a^{\mathrm{chain}}}{RT} + \frac{a^{\mathrm{polar}}}{RT} + \frac{a^{\mathrm{chem}}}{RT} \tag{6.8-52}$$

$$\frac{A^{\mathrm{MS}}}{RT} = A^{\mathrm{HS}} + \frac{A_1}{T} + \frac{A_2}{T^2} + \frac{A_3}{T^3} \tag{6.8-53}$$

where A^{MS} corresponds to a "monomer segment," and A^{HS} is evaluated at the equivalent hard sphere diameter, d, given by

$$d = \int_0^\infty [1 - \exp(-\beta u_0)]\, dr \tag{6.8-54}$$

Then A^{HS} is given by the Carnahan-Starling formula,

$$A^{\mathrm{HS}} = \frac{\eta^{\mathrm{P}}(4 - 3\eta^{\mathrm{P}})}{(1 - \eta^{\mathrm{P}})^2} \tag{6.8-55}$$

where η is evaluated using d, as given in Eq. (6.3-19). Continuing with Barker-Henderson TPT,[116]

$$A_1 = 12\eta^P \int_0^\infty g^{\mathrm{HS}}(r) u_1 x^2\, dx \approx I_1^A + I_2^B;\; x \equiv \frac{r}{d} \tag{6.8-56}$$

$$I_1^A = C_{mn}\left[x_0^m a_1^{\mathrm{S}}(m) - x_0^n a_1^{\mathrm{S}}(n)\right];\; a_1^{\mathrm{S}}(j) = -12\eta^P\left(\frac{\varepsilon}{k_B}\right) g_d^{\mathrm{HS}}(\eta_{\mathrm{eff}})$$

$$I_1^B = C_{mn}\left[x_0^m B(m) - x_0^n B(n)\right];$$

$$B(j) = 12\eta^{\mathrm{P}}\left(\frac{\varepsilon}{k_B}\right) g_d^{HS}\left[a_1^{B_1}(j) + \frac{9\eta^{\mathrm{P}}(1+\eta^{\mathrm{P}})}{2-\eta^{\mathrm{P}}} a_1^{B_2}(j)\right]$$

$$a_1^{B_1}(j) = -\frac{x_0^{3-j} - 1}{j-3};\; a_1^{B_2}(j) = -\frac{x_0^{4-j}(j-3) - x_0^{3-j}(j-4) - 1}{(j-3)(j-4)}$$

Key terms in Eq. (6.8-56) are a_1^{S} and the definition of η_{eff}. Evaluation of a_1^{S} follows the method of Gil-Villegas et al.,[112] identifying η_{eff} as the density at which g_d^{HS} takes on a mean value that satisfies equivalence of the integral. It is effectively a correlation such that,

$$\int_1^\infty \left(-\frac{1}{x^j}\right) g^{\mathrm{HS}}(r, \eta^{\mathrm{P}}) x^2 dx = \frac{g_d^{\mathrm{HS}}(\eta_{\mathrm{eff}}(j))}{j-3};\; \eta_{\mathrm{eff}}(j) = \sum_{k=0}^{3}\sum_{i=1}^{4} c_{ik}(\eta^{\mathrm{P}})^i j^{-k}; \tag{6.8-57}$$

Hence, there are 16 coefficients in the correlation of a_1^{S}, but these are all derived from fitting the numerical integral to enhance computational speed, not by fitting experimental data. This approach puts the characterization of A_1^{MS} on a physical basis that is traceable in statistical mechanical terms.

The estimate of A_2 adapts a corrected version of the macroscopic compressibility approximation introduced by Barker and Henderson.[116]

$$A_2 = 6\eta^P K^{HS}(1+\chi)\int_0^\infty g^{HS}(r)u_1^2 x^2\,dx; \tag{6.8-58}$$

where χ is the correction term and K^{HS} is given by

$$K^{HS} = \frac{(1-\eta^P)^4}{1+4\eta^P+4(\eta^P)^2-4(\eta^P)^3+(\eta^P)^4} \tag{6.8-59}$$

And the empirical correlation for χ is fit to simulation data and given by

$$\chi = \eta^P x_0^3 f_1 + \left(\eta^P x_0^3\right)^5 f_2 + \left(\eta^P x_0^3\right)^8 f_3 \tag{6.8-60}$$

where $x_0 = \sigma/d$ and the functions f_i are defined in terms of a polynomial ratio

$$f_i = \frac{\sum_{j=1}^{3} c_{ij}\alpha^j}{\left[\sum_{j=4}^{6} c_{ij}\alpha^{j-3}\right]} \tag{6.8-61}$$

$$\alpha = \frac{1}{\varepsilon\sigma^3}\int_\sigma^\infty u_1 r^2\,dr; \tag{6.8-62}$$

The integral of Eq. (6.8-58) is like that of Eq. (6.8-56), so the algebraic characterization of A_2 borrows terms from the characterization of A_1,

$$\begin{aligned} A_2 = \tfrac{1}{2} K^{HS}(1+\chi)(C\varepsilon/k_B)^2\{x_0^{2m}[a_1^S(2m)+B(2m)] \\ -2x_0^{n+m}[a_1^S(n+m)+B(n+m)]+x_0^{2n}[a_1^S(2n)+B(2n)]\} \end{aligned} \tag{6.8-63}$$

The third-order term is fit to simulation data in combination with scaling law critical point estimates (Eq. 6.3-22) using various Mie exponents,

$$A_3 = -\left(\frac{\varepsilon}{k_B}\right)^3 f_4\eta^P x_0^3 \exp\left[\eta^P\left(f_5 x_0^3+\eta^P f_6 x_0^6\right)\right] \tag{6.8-64}$$

The 41 coefficients of $\{f_i\}$ are tabulated by Lafitte et al.[113] Once again, these coefficients are fit to simulation data for computational speed. They are not fit to experimental data.

The remaining terms for A^{chain} and A^{chem} follow the usual Wertheim formulation, but it remains to characterize the contact value of $g^{Mie}(\sigma)$. Lafitte et al. recognized that the bond length should be constant with respect to temperature. Therefore, it is necessary to evaluate $g^{Mie}(\sigma)$ at a slightly different distance from g_d^{HS}. They account for this discrepancy with a temperature and density-dependent correlation

$$\begin{aligned} g^{Mie}(\sigma) &= g_0(\sigma)\exp\left(\frac{\beta\varepsilon g_1}{g_0}+\frac{(\beta\varepsilon)^2 g_2}{g_0}\right) \\ g_0(\sigma) &= \exp\left(\sum_{i=0}^{3} k_i x_0^i\right) \end{aligned} \tag{6.8-65}$$

Formulas for g_1 and g_2 are quite complex and are given by Eqs. (64)–(66) of Lafitte et al. along with coefficients for $\{k_i\}$. Like Soft-SAFT, the temperature dependence of $g^{Mie}(\sigma)$ is reflected in the chain term owing to the assumption of a segmental perspective at the outset.

Lafitte et al. go on to demonstrate the accuracy of their theory for LJ and Mie fluids. The comparisons to molecular simulations of tangent sphere chains show good agreement in most properties. Jackson and coworkers have adopted the SAFT-VR Mie framework as the EOS basis for their generalized Mie chain models of thermodynamic properties. Their strategy is to characterize transferable force fields for many molecules in terms of tangent sphere chains. The basis of the model is molecular simulation with the SAFT-VR-Mie model as the EOS interface for summarizing their results. They have adapted this approach for interfacial properties and transport properties. It is inherently a coarse-grained approach because the tangent sphere segments comprise several united heavy atoms, not just uniting hydrogens on a single heavy atom. It is an active subject of research and future editions of PGL may be able to consider the model in more detail. A recent open-source code makes many aspects of this model more accessible, but its publication was too late for detailed consideration.[117]

6.9 PERTURBATION MODELS WITH TRANSFERABLE PARAMETERS

As discussed in relation to Fig. 6.5, it is not expected for second-order TPT to match the critical point. Therefore, it is logical to argue that the parameters $\{\varepsilon/k_B, \sigma^3, m\}$ should not be fit to the critical point (as they would be analogous to a, b, and κ of the PR-EOS). Therefore, regression of the liquid density and vapor pressure for $T_r < 0.9$ is the most common manner of estimating SAFT parameters. The DIPPR, DECHEMA, and NIST/TDE databases provide good resources of these data for roughly 3000 compounds in DIPPR's case and 35,000 compounds for NIST/TDE, but one must still be careful. For PC-SAFT, NIST/TDE provides a utility to fit parameters while interactively selecting the data to include in the fit. The parameters can be accessed by right clicking a plot of vapor pressure after regression. In DIPPR's temperature-dependent equations, the stated application limits often exceed the range of experimental data. The equations can be used to smooth multiple datasets, but care is required to limit the application range to temperatures where measurements have been performed. One must examine all sets of accepted experimental data to determine the maximum and minimum temperatures of available vapor pressure and liquid density data individually. Also beware of data cited as "Othmer-Yu" or "Riedel"; these are not usually traceable to experimental measurements.

Predicting the parameters of cubic EOSs simply relies on predicting critical properties, but each SAFT EOS requires its own method for parameter prediction. Like methods of predicting critical properties, group contribution (GC) methods play the primary role in these predictions. We discuss three GC methods for predicting SAFT parameters: Tihic et al.,[118] Emami et al.,[104] and DeHemptinne and coworkers.[119]

6.9.1 Tihic and Emami Methods

Tihic et al.[118] reported 45 first-order UNIFAC groups applicable to PC-SAFT for nonassociating compounds. These groups included increments for m, $m\sigma^3$, and $m\varepsilon/k_B$. Noting that m is linear in molecular weight, these correlations effectively result in first-order Pade approximants for σ^3 and ε/k_B; they each approach asymptotic values for polymers. Hydrogen bonding parameters were not considered, so the correlation has limited use.

Emami et al.[104] reported 86 first-order UNIFAC groups for PC-SAFT, including groups for associating components. The hydrogen bonding parameters for associating groups were somewhat simple, but the GC correlation customized each group's m, σ^3, and ε/k_B parameters. The hydrogen bonding energy ($N_A\varepsilon^{AD}$) was taken as 4.0 kcal/mol for hydroxyl groups, and 1.25 for amine (1° and 2°), amide, nitrile, and aldehyde groups. The bonding volume was taken as K^{AD} (cm^3/mol) = 0.035 b/m, where $b = N_A\pi m\sigma^3/6$ is the PC-SAFT molecular volume. Emami et al.[104] do not correlate m, σ^3, and ε/k_B directly, as in the manner of Tihic et al. Instead, they adapt existing GC correlations for liquid volume at 25 °C, V_L^{298}, and solubility parameter, δ. This narrows the requirements for SAFT GC correlation to just the shape parameter, m. These three target values suffice to determine the values of m, σ^3, and ε/k_B through simple nonlinear equations. An advantage of the Emami approach is that experimental values can be substituted for the GC values if available. For example, if V_L is known for a polymer at 400 K, it is a simple matter to substitute these values of volume and temperature. If a vapor pressure point is known, the recommended approach is to iterate on the value of m to match the target vapor pressure, liquid density, and solubility parameter. Solubility parameter is considered as a target because it often plays a significant role in predictions of polymer thermodynamics. An equivalent GC correlation was given by Emami et al. for the shape parameter "c" of the ESD EOS. It is useful to evaluate the Emami correlations when no experimental data are used, but it would be valuable to know how much improvement is achievable when a single measurement is available. In Sec. 6.13, EGC-PC-SAFT(T_b) and EGC-ESD(T_b) refer to the Emami GC correlations when the EOS parameters have been adjusted to fit the experimental value of boiling temperature along with the GC correlations for V_L^{298} and δ.

6.9.2 GC-PPC-SAFT

DeHemptinne and coworkers[120] have proposed a more sophisticated GC correlation when considering polar PC-SAFT applications. Their method is designated GC-PPC-SAFT. When incorporating the polar contribution into PC-SAFT, it was observed that the pure component ε/k_B was strongly correlated with the polarity parameters, μ and Q. Consequently, multiple sets of parameters fit the pure component data equally well, but some parameters performed much better in predictions of binary VLE. Developing GC correlations identified trends favoring the parameters that were more successful in VLE predictions. They included a few representative binary mixtures when developing their GC correlations to guide the GC optimization for pure components. Their method is applicable to $T_r > 0.4$, which is a greater range than most Peng-Robinson implementations.

The key equations for their GC method applicable to n-alkanes are listed below:

$$\ln(\varepsilon) = \frac{\sum_{i=1}^{n_G} n_i \ln(\varepsilon_i)}{\sum_{i=1}^{n_G} n_i} \tag{6.9-1}$$

$$m = \frac{\sum_{i=1}^{n_G} n_i R_i}{\sum_{i=1}^{n_G} n_i} \tag{6.9-2}$$

$$\sigma = \frac{\sum_{i=1}^{n_G} n_i \sigma_i}{\sum_{i=1}^{n_G} n_i} \tag{6.9-3}$$

For polar molecules, they apply Eqs. (6.8-23)–(6.8-33). The values of μ and Q in Eqs. (6.8-23)–(6.8-33) are assumed from values determined in vacuum, while values of x_p^{μ} and x_p^{Q} are treated as pure component parameters. An extensive list of GC parameters is given by Nguyen et al.[120]

6.9.3 The SPEADMD Model

The concept behind step potentials for equilibria and discontinuous molecular dynamics (SPEADMD, pronounced *speed-em-dee*) is that assuming tangent sphere chains as the model of every molecule may be inaccurate, and unnecessary. Performing molecular simulations of a compound like squalane takes about 24 hours to cover all densities of interest. Putting together a model that can extrapolate a polymer to its long chain limit requires multiple simulations over the course of a week. Through the formalism of SPEADMD, the simulated A_0, A_1, and A_2 automatically form the EOS with no further simulation required. The SPEADMD EOS operates like any other EOS in practice. The simulation time is therefore a trivial investment.

The benefit of SPEADMD is that the simulated molecules include all the details of bond angles, bond lengths, rings, and branches. The dihedral angle distribution of *n*-alkanes is simplified relative to more rigorous models, however. Having access to this level of molecular detail opens the door to nanoscale analysis of EOS contributions, all while performing computations like any other EOS. The methodology has been reviewed extensively in two papers. The first paper focuses on the simulation data and types of molecules covered for over 500 compounds that have been characterized, and the characterization of 197 transferable site types including 17 associating and 19 solvating site types.[121] The second paper reviews the methodology and the implications of several details in molecular architecture.[122]

SPEADMD is founded on step potential models. Unlike the shoulder well potential that forms the basis of HR-SAFT and PC-SAFT, SPEADMD assumes four attractive wells, as shown in Fig. 6.11. The TPT terms are computed with formulas analogous to Eqs. (6.3-11) and (6.3-12). Note that the depths of the wells are specified separately from the ensemble averages of the pair distributions, so the depths can be assigned post-simulation and optimized by fitting to a large database in the manner of group contribution characterization. Wertheim's theory is applied for A^{chem} so that part of the model is comparable to any SAFT model. No polar contributions have been tested for SPEADMD. Hence the SPEADMD EOS looks the same as Eq. (6.3-9) terminated after A_2.

One significant distinction between SPEADMD and PC-SAFT is that the well depths are based on individual sites, not the entire molecule. SPEADMD is constructed as a "group contribution" model at the outset, where group contributions at the atomistic level are referred to as transferable "force fields." Thus, the CH3 and CH2 potentials in Fig. 6.11 apply equally to *n*-octane and *n*-octacontane.

The EOS for characterizing the reference fluid is adapted from the Carnahan-Starling equation

$$Z_0 = \frac{1 + \eta^{\text{P}}[z_1 + \eta^{\text{P}}(z_2 + \eta^{\text{P}} z_3)]}{(1 - \eta^{\text{P}})^3} \tag{6.9-4}$$

where z_1_z_3 are regression coefficients to match the simulated pressures. The Helmholtz energy is obtained by analytical integration.

$$A_0 = \frac{c_1 + c_2 + c_3}{2(1 - \eta^{\text{P}})^2} - \frac{c_2 + 2c_3}{(1 - \eta^{\text{P}})} - c_3 \ln(1 - \eta^{\text{P}}) - \frac{c_1 - c_2 - 3c_3}{2} \tag{6.9-5}$$

where $c_1 = z_1 + 3$, $c_2 = z_2 - 3$, $c_3 = z_3 + 1$.

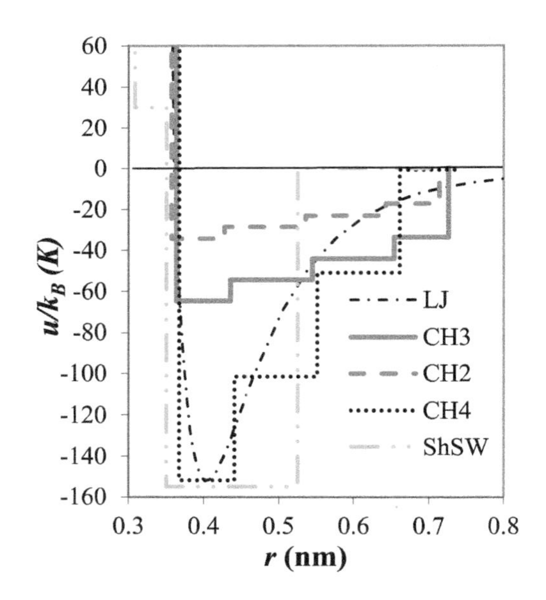

FIGURE 6.11
CHx step potential models compared to LJ and shoulder SW (ShSW) models.

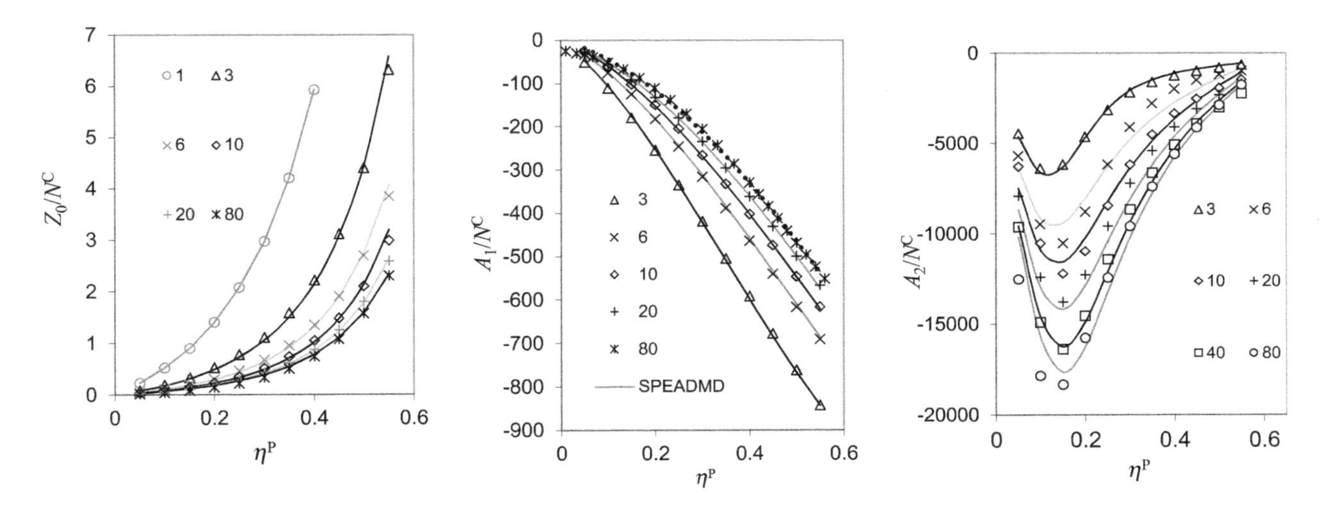

FIGURE 6.12
Illustration of the TPT contributions in SPEADMD. Points are simulation results with the given carbon number. Lines show the trend of Eqs. (6.9-6) and (6.9-7). (a) Z_0 shows a steady approach to its long- chain limit, like PC-SAFT. (b) A_1 trends are like PC-SAFT but with greater sensitivity to carbon number. (c) The trend for A_2 shows a qualitative difference from PC-SAFT.

Ghobadi and Elliott[123] studied functional forms for A_1 and A_2 that could provide reasonable derivative behavior. Their suggestion was

$$A_1 = a_{11}\eta^P + a_{12}\eta^P \exp[-\delta_1(\eta^P)^3] + \frac{a_{13}(\eta^P)^3}{\delta_2 + \eta^P} \tag{6.9-6}$$

$$A_2 = \frac{\eta^P\left[a_{21} + \eta^P\left(a_{22} + \eta^P\left(a_{23} + \eta^P a_{24}\right)\right)\right]}{1 + 50\left(\eta^P\right)^3} \tag{6.9-7}$$

where $\{a_{ij}\}$, δ_1, and δ_2 are constants fit to simulation data. Default values are $\delta_1 = 2$ and $\delta_2 = 0.2$. Additional constraints were that $A_2 < 0$ for $\eta^P < 1$ and $A_2 = 0$ for $\eta^P = 1$. Trends for the TPT contributions of polyethylene according to SPEADMD are illustrated in Fig. 6.12. The trends for A_0 and A_1 are like Fig. 6.4, but not for A_2, in which case the high-density shoulder is missing. Isotherms for polyethylene exhibit single maximum/minimum behavior, and the SPEADMD EOS displays no bifurcations when the number of carbons is greater than 9 at temperatures greater than 40 K.[84] The maximum bifurcation temperature is lower for smaller carbon number. This suggests that eliminating artificial roots from TPT EOSs may not be so difficult if given sufficient emphasis.

SPEADMD parameters are initially determined by application of a transferable force field (TFF). For the customized parameters, the molecular volume is adjusted by a factor (1-l_{ii}) and disperse attractive contributions (A_1, A_2) are multiplied by a factor ($1 - k_{ii}$), where l_{ii} and k_{ii} are molecular parameters as opposed to site parameters. The customized form is fit to vapor pressure and liquid density data like PC-SAFT, but only two parameters are adjusted compared to five for PC-SAFT. For a consistent basis of comparison, TFF-SPEADMD should be compared to EGC-PC-SAFT or EGC-ESD whereas SPEADMD should be compared to PC-SAFT.

6.9.4 SAFTγ

The origins of the SAFTγ formalism can be traced to Lymperiadis et al.[124] It was conceived as a heterosegmented group contribution method for fused sphere segments. In principle, intramolecular energies like bond bending and torsion can also be included, so it is a very general formalism. A key distinction of SAFTγ is the inclusion of shape factors to adjust the impact of a particular functional group on the overall molecular properties. For example, the molecular volume, b, can be written for a single component as

$$b = \frac{N_A \pi}{6} \sum_{k=1}^{N_G} \nu_k S_k d_{kk}^3 \tag{6.9-8}$$

where S_k is the shape factor, ν_k is the number of occurrences of the kth group in the molecule, and d_{kk} is the (temperature-dependent) effective hard sphere diameter for the kth group. For example, the molecular volume of n-pentane would be

$$b = \frac{N_A \pi}{6}(2(0.748)0.367^3 + 3(0.242)0.354^3) = 33.5 \text{ cm}^3/\text{mol} \tag{6.9-9}$$

given $S_{CH3} = 0.748$, $d_{CH3} = 0.367$ nm, $S_{CH2} = 0.242$, $d_{CH2} = 0.354$. The smaller value of S_{CH2} recognizes that CH_2 contributes less to the volume owing to the high degree of overlaps for CH_2. Interestingly, the same value of S_k is applied when computing multiple properties in the SAFTγformalism.

The original formalism was developed by Jackson and coworkers and focused on fused sphere segments, even though it still applied the segmental perspective when formulating the chain EOS.[124] More recently, Jackson and coworkers have directed their emphasis toward SAFTγ-CG-Mie, where the CG stands for "coarse grained." In the CG perspective, molecules are once again represented as tangent sphere segments, and the GC aspects of SAFT$\gamma$are the primary interest. In the presentation below, we refer particularly to Papaioannou et al.,[80] which retains the fused sphere perspective in principle but adopts the tangent sphere convention in CG practice. Apparently, $S_k = 1$ for all k in the tangent sphere convention. An alternative implementation of SAFTγ by Ghobadi and Elliott[125–127] retains the fused sphere basis but derives its reference thermodynamics from molecular simulation rather than the segmental perspective. Both methods are outlined below. Neither method is quite ready for large-scale evaluation, but they illustrate two ways of interfacing with molecular simulation through use of an engineering EOS.

6.9.4.1 SAFTγ-CG-Mie

The SAFTγ-CG-Mie model is based largely on the SAFT-VR-Mie model. The expressions for A_1–A_3 and g_σ^{ref} are nearly identical, but the expression for A^{HS} is different,

$$\frac{A^{\text{res}}}{RT} = A^{\text{mono}} + A^{\text{chain}} + A^{\text{chem}} \tag{6.9-10}$$

$$A^{\text{mono}} = A^{\text{HS}} + \frac{A_1}{T} + \frac{A_2}{T^2} + \frac{A_3}{T^3}$$

$$A^{\text{HS}} = \frac{6\pi}{\rho}\left[\left(\frac{\zeta_2^3}{\zeta_3^2} - \zeta_0\right)\ln(1-\zeta_3) + \frac{3\zeta_2\zeta_3}{1-\zeta_3} + \frac{\zeta_2^3}{(1-\zeta_3)^2}\right] \tag{6.9-11}$$

where

$$\zeta_m = \frac{\pi \rho_s}{6}\sum_{k=1}^{N_G} x_{s,k} d_{kk}^m \tag{6.9-12}$$

$$x_{s,k} = \frac{\nu_k S_k}{\sum_{k=1}^{N_G} \nu_k S_k}$$

$$A^{\text{chain}} = -\ln\left(g_\sigma^{\text{Mie}}\right)\sum_{k=1}^{N_G} x_{s,k}(\nu_k S_k - 1) \tag{6.9-13}$$

For the higher-order TPT contributions,

$$A_1 = a_1 \sum_{k=1}^{N_G} x_{s,k} S_k \tag{6.9-14}$$

$$a_1 = \sum_{k=1}^{N_G} \sum_{l=1}^{N_G} x_{s,k} x_{s,l} a_{1;kl}$$

where $a_{1:kl}$ is given by Eq. (6.8-56) evaluated with $\eta^{P} = \zeta_x$, where

$$\zeta_x = \frac{\pi \rho_s}{6} \sum_{k=1}^{N_G} \sum_{l=1}^{N_G} x_{s,k} x_{s,l} d_{kl}^3 \tag{6.9-15}$$

and $\eta^{\text{eff}}(j) = \zeta_{kl}^{\text{eff}}$ in Eq. 6.8-57 where $j = m_{kl}$ or n_{kl} as appropriate. The TPT contributions for A_2 and A_3 are adapted in a manner like for A_1. Altogether, the SAFTγ-CG-Mie EOS can be viewed as group contribution version of the SAFT-VR-Mie EOS.

The close relationship of the SAFT-γ-CG-Mie EOS to its molecular basis suggests using the EOS as a surrogate for molecular simulations with the same Mie potential functions. For example, one might use the EOS to regress $\{m, n, \sigma, \varepsilon\}$ parameters for all the site types required to describe n-alkanes,[113] alkylbenzenes,[128] and perfluoroalkylalkanes (PFAAs),[129] then perform molecular simulations with those parameters to study fundamental phenomena like interfacial behavior of PFAAs.[130] Jackson and coworkers refer to this regression practice as a "top-down" approach because it uses macroscopic properties to infer EOS parameters with a somewhat more direct connection to an intermolecular potential function than is the case with PC-SAFT. Preliminary demonstrations of the methodology have shown promise as a bridge between molecular simulations and traditional engineering EOSs.

6.9.4.2 SAFTγ-WCA

The primary distinction between the WCA implementation of SAFTγ and the adaptations by Jackson and coworkers is the split of the potential function defined by Eq. (6.8-51). Whereas the BH split occurs at $r = \sigma$, the Weeks-Chandler-Andersen (WCA)[131] split occurs at $r = r_{\text{min}}$, the minimum of the potential function, as shown in Eq. (6.9-16). The WCA split makes A_0 more sensitive to temperature, but the higher-order TPT contributions tend to be simpler and less temperature-sensitive. The BH and WCA splits are illustrated in Fig. 6.13.

$$\begin{aligned} u_0 &= u^{\text{Mie}} + \varepsilon \text{ at } r \leq \sigma, \text{ 0 otherwise;} \\ u_1 &= u^{\text{Mie}} \text{ at } r > \sigma, -\varepsilon \text{ otherwise} \end{aligned} \tag{6.9-16}$$

To construct a chain EOS in the SAFT perspective, one must accurately characterize the monomer contribution. The HS perspective is a useful starting point, which returns us to defining an effective HS diameter (d). Although Weeks et al. suggested making d a function of density as well as temperature, most engineering applications favor the temperature-dependent definition of Eq. (6.8-54), which is also consistent with the HR-SAFT and PC-SAFT definitions in the context of a square shoulder potential model. Heyes and Okumura[132] showed that a slight modification in the Carnahan-Starling (CS) EOS was accurate for LJ spheres with a WCA split and $d(T)$.

$$a^{\text{WCA}} = \frac{1}{\rho}\left(-\zeta_0 \ln(1-\zeta_3) + \frac{\zeta_1 \zeta_2 k_1}{(1-\zeta_3)} + f_3 \zeta_2^3\right) \tag{6.9-17}$$

$$f_3 = \frac{(3k_2 - 2k_3)\zeta_3 - 1.5(k_2 - k_3)\left[2\zeta_3 - \zeta_3^2\right] + k_3(1-\zeta_3)^2 \ln(1-\zeta_3)}{36\pi \zeta_3^2 (1-\zeta_3)^2}$$

$$Z^{\text{WCA}} = \frac{\zeta_3}{1-\zeta_3} + \frac{\zeta_1 \zeta_2 k_1}{\zeta_0 (1-\zeta_3)^2} + \frac{\zeta_2^3 k_2}{12\pi (1-\zeta_3)^3} - \frac{\zeta_3 \zeta_2^3 k_3}{36\pi (1-\zeta_3)^3} \tag{6.9-18}$$

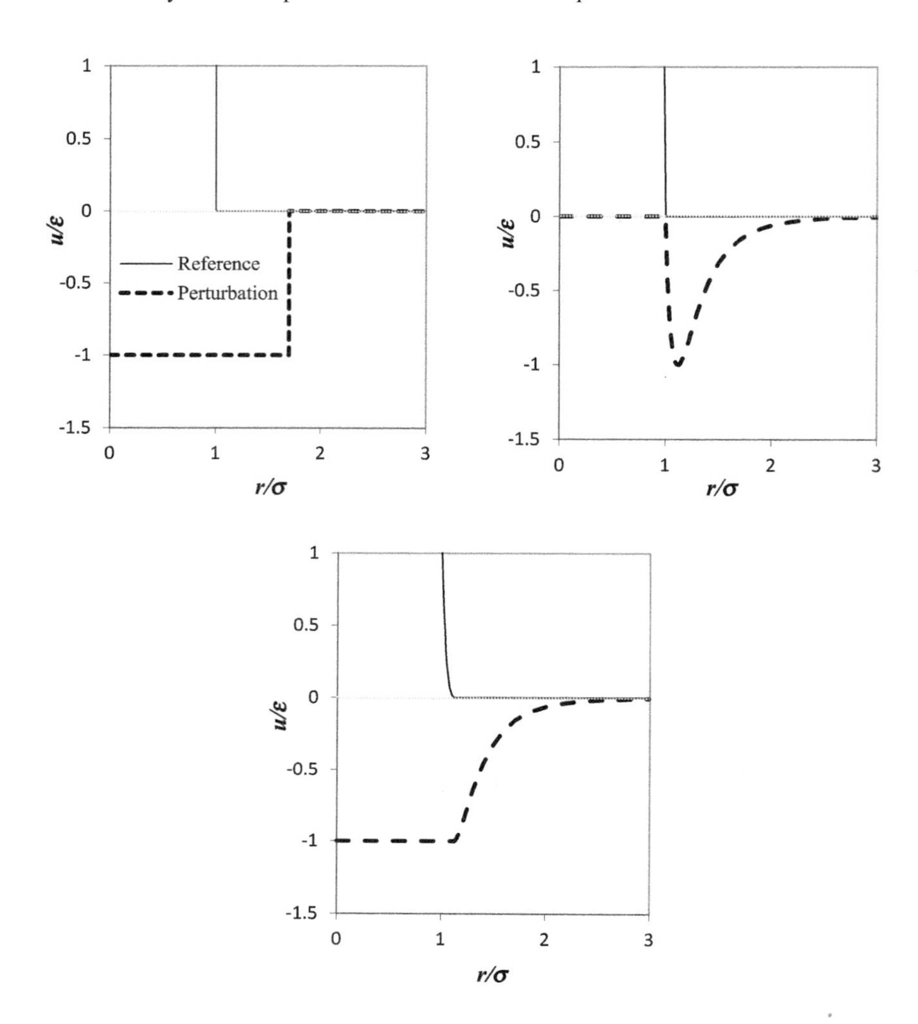

FIGURE 6.13
The definition of reference (solid line) and perturbation potential (dash line) for SW (a), BH (b), and WCA (c) perturbation theories.

$$\zeta_m = \rho M^{(m)}, \quad m = 0 - 3 \tag{6.9-19}$$

$$\mathrm{M}^{(0)} = \sum_k^{NS} \nu_{k,i} S_k, \; M^{(1)} = \frac{1}{2}\sum_k^{NS} \nu_k S_k d_{kk}, \; M^{(2)} = \pi \sum_k^{NS} \nu_k S_k d_{kk}^2, \; M^{(3)} = \frac{\pi}{6}\sum_k^{NS} \nu_k S_k d_{kk}^3$$

where Z^{WCA} is the compressibility factor of WCA spheres and k_i ($i = 1$–3) are adjustable constants to fit simulation data of WCA spheres ($k_1 = 0.94$; $k_2 = 1.45$; $k_3 = 4.50$). Equation (6.9-17) reduces to the Carhahan-Starling equation when $k_i = 1$ for all i, and is similar to Eq. (6.9-11). The $M^{(m)}$ variables are selected to be consistent with fundamental measure theory for interfacial thermodynamics.[133] To extend Eq. (6.9-18) to $n - 6$ Mie potentials, SAFTγ-WCA applies Eq. (6.8-54) using the analytical formula of Elliott and Daubert,[134]

$$\frac{d}{r_{\min}} = [n(0.0093n - 0.0592)(T^*)^2 - (n-1)T^* + 1]^{\frac{-1}{2n+1}} \tag{6.9-20}$$

$$\frac{r_{\min}}{\sigma} = \left(\frac{n}{6}\right)^{\frac{1}{(n-6)}} \tag{6.9-21}$$

SAFTγ-WCA treats n as an adjustable parameter for each united-atom site type, even if the value of n in the molecular potential model is fixed. The authors rationalize this choice by observing that different site types present different degrees of softness owing to repulsive overlaps of fused spheres.

Noting that the term in parentheses of Eq. (6.9-13) can be zero for some values of shape factor, SAFTγ-WCA adopts a different expression for the chain term. For example, if for a two-site molecule (nitrogen, ethane …) shape factor happens to be 0.5, a^{ch} would be zero. When each site type has a different contribution to the chain formation, Wertheim's TPT1 for a heteronuclear chain can be rewritten as:

$$a^{\text{ch}} = -\left(\frac{1}{2}\sum_{k=1}^{NS} \nu_{k,i} S_k N_k^B\right) \ln g_d^{\text{WCA}} \tag{6.9-22}$$

where N_k^B stands for number of bonding sites for site-type k and g_d^{WCA} is the rdf at contact for WCA fluid (attractive forces not included). This chain contribution is valid for any arbitrary set of site parameters and is only zero when N_k^B is zero for all k. The expression of g_d^{WCA} is

$$g_d^{\text{WCA}} = \frac{1}{1-\zeta_3} + \frac{(k_1+k_2)\zeta_2}{4(1-\zeta_3)^2}\left(\frac{\bar{d}_{ii}\bar{d}_{jj}}{\bar{d}_{ii}+\bar{d}_{jj}}\right) + \frac{(3k_2-k_3)\zeta_2^2}{36(1-\zeta_3)^3}\left(\frac{\bar{d}_{ii}\bar{d}_{jj}}{\bar{d}_{ii}+\bar{d}_{jj}}\right)^2 \tag{6.9-23}$$

With the effective size parameter of the ith site ($\bar{d}_{ii}$) defined as

$$\bar{d}_{ii}^3 = \frac{\sum_{k=1}^{NS} \nu_{k,i} S_k d_{kk}^3}{\sum_{k=1}^{NS} \nu_{k,i} S_k} \tag{6.9-24}$$

In summary, the SAFT-γ WCA EOS has three molecular constants (k_i, i = 1–3) and three adjustable parameters per site type (n, σ, S). The entire reference term is fit to simulation data of WCA molecules, so the segmental perspective is not involved.

Like the SAFTγ-CG-Mie and SAFT-VR methodologies, the mean value theorem (MVT) is used to estimate the first-order perturbation contribution.

$$A_1 = -\rho \sum_{k=1}^{\text{NS}} \sum_{l=1}^{\text{NS}} \nu_k \nu_l S_k S_l \alpha_{kl}^{\text{WCA1}} g_{0,kl}^{\text{WCA}}\left(\eta_{1,kl}^{\text{eff}}\right) \tag{6.9-25}$$

However, one should note that in SAFT-VR and SAFTγ-CG-Mie equations, perturbation terms are evaluated for monomers and then are scaled by the number of segments in the chain. In SAFTγ-WCA, like PC-SAFT and SPEADMD, perturbation terms are directly evaluated at the molecular level to be consistent with the definition of the reference term. In Eq. (6.9-25), $g_{0,kl}^{\text{WCA}}\left(\eta_{1,kl}^{\text{eff}}\right)$ is evaluated at an effective density:

$$g_{0,kl}^{\text{WCA}}\left(\eta_{1,kl}^{\text{eff}}\right) = \frac{4\left(1-\eta_{1,kl}^{\text{eff}}\right)^2 + 3(k_1+k_2)\left[\eta_{1,kl}^{\text{eff}} - \left(\eta_{1,kl}^{\text{eff}}\right)^2\right] + (3k_2-k_3)\left(\eta_{1,kl}^{\text{eff}}\right)^2}{4\left(1-\eta_{1,kl}^{\text{eff}}\right)^3} \tag{6.9-26}$$

$$\eta_{1,kl}^{\text{eff}} = \frac{\frac{\eta_{1,kk}^{\text{eff}}}{(\sigma_k\varepsilon_k)} + \frac{\eta_{1,ll}^{\text{eff}}}{(\sigma_l\varepsilon_l)}}{\frac{1}{(\sigma_k\varepsilon_k)} + \frac{1}{(\sigma_l\varepsilon_l)}} \tag{6.9-27}$$

$$\eta_{1,ii}^{\text{eff}} = a_i^{11} + a_i^{12}\zeta_3 + a_i^{13}\zeta_3^2 \tag{6.9-28}$$

In Eq. (6.9-28), (a_i^{1w}, w = 1–3) are adjustable parameters of the ith site type for A_1 term. In Eq. 6.9-25, $\alpha_{kl}^{\text{WCA1}}$ is given by:

$$\alpha_{kl}^{\text{WCA1}} = \pi\left(\frac{16}{9}\sqrt{2} - \frac{2}{3}\right)\frac{\varepsilon_{kl}}{k_B}\sigma_{kl}^3 \tag{6.9-29}$$

Finally, the combining rules for ε and σ complete the functional form of the first-order term:

$$\varepsilon_{kl} = \sqrt{\varepsilon_k \varepsilon_l}, \quad \sigma_{kl} = \sqrt{\sigma_k \sigma_l} \tag{6.9-30}$$

This set of combining rules eliminates the necessity of cross group energy parameters, ensures the transferability of adjustable parameters defined for each united atom site type, and leads to superior results for mixtures without introducing any binary interaction parameter.

The A_2 term is implemented like A_1, to obtain:

$$A_2 = -\rho K^{\mathrm{WCA}} \sum_{k=1}^{\mathrm{NS}} \sum_{l=1}^{\mathrm{NS}} v_k v_l S_k S_l \alpha_{kl}^{\mathrm{WCA2}} g_{0,kl}^{\mathrm{WCA}}\left(\eta_{2,kl}^{\mathrm{eff}}\right) \tag{6.9-31}$$

where $g_{0,kl}^{\mathrm{WCA}}(\eta_{2,kl}^{\mathrm{eff}})$ is given by Eq. (6.9-26) with an effective packing fraction $\eta_{2,kl}^{\mathrm{eff}}$ that is optimized to reproduce the A_2 computed by simulation. The combining rules for $\eta_{2,kl}^{\mathrm{eff}}$ resemble those of A_1 and are not repeated here. The effective density that is used in definition of $g_{0,kl}^{WCA}(\eta_{2,kl}^{\mathrm{eff}})$ introduces three more adjustable parameters per site type (a_{w}^{2w}, $w = 1$–3) for the A_2 term. The isothermal compressibility (K^{WCA}) corresponding to Eq. (6.9-18) reads as

$$K^{\mathrm{WCA}} = \frac{(1-\zeta_3)^4}{1+(6k_1-2)\zeta_3+(1+9k_2-6k_1)\zeta_3^2+k_3\left(\zeta_3^4-4\zeta_3^3\right)} \tag{6.9-32}$$

The interaction energy for the second-order term is computed by

$$\alpha_{kl}^{\mathrm{WCA2}} = \pi\left(\frac{176\sqrt{2}}{315}-\frac{1}{3}\right)\left(\frac{\varepsilon_{kl}}{k_B}\right)^2 \sigma_{kl}^3 \tag{6.9-33}$$

Note that both $\alpha_{kl}^{\mathrm{WCA1}}$ and $\alpha_{kl}^{\mathrm{WCA2}}$ are positive numbers.

For the A_3 term, a Gaussian factor, Γ_{kl}, multiplies A_2 such that

$$A_3 = -\rho K^{\mathrm{WCA}} \sum_{k=1}^{\mathrm{NS}} \sum_{l=1}^{\mathrm{NS}} v_k^2 v_l s_k^2 s_l \alpha_{kl}^{\mathrm{WCA2}} g_{0,kl}^{\mathrm{WCA}}\left(\eta_{2,kl}^{\mathrm{eff}}\right)\Gamma_{kl} \tag{6.9-34}$$

$$\Gamma_{kl} = \frac{\frac{\Gamma_k}{(\sigma_k \varepsilon_k)}+\frac{\Gamma_l}{(\sigma_l \varepsilon_l)}}{\frac{1}{(\sigma_k \varepsilon_k)}+\frac{1}{(\sigma_l \varepsilon_l)}} \tag{6.9-35}$$

$$\Gamma_{ii} = \frac{\varepsilon_{kl}}{k_B} a_{ii}^{31} \exp\left(-a_{ii}^{32}\left(n_3-a_{ii}^{33}\right)^2\right) \tag{6.9-36}$$

where (a_{ii}^{3w}, $w = 1$–3) are adjustable parameters of the ith site type for the A_3 term.

With these equations, the SAFTγ-WCA EOS is completely defined. At first glance, it may seem like the number of parameters is quite large in this EOS. There are three parameters per site type at each order, plus the parameters for n, σ_k, ε_k, and S_k. For n-pentane, as an example, that would come to 26 parameters. On the other hand, all of those parameters should remain the same for n-hexane through polyethylene. In comparisons to simulations of the TraPPE-UA force field, Ghobadi and Elliott[118] achieved near-quantitative agreement with simulation results for methane-dodecane, with two caveats. To match TraPPE-UA force field, $\sigma_{\mathrm{CH3}}^{\mathrm{WCA}} = 1.01\ \sigma_{\mathrm{CH3}}^{\mathrm{TraPPE}}$ and $\sigma_{\mathrm{CH3}}^{\mathrm{WCA}} = 1.37\ \sigma_{\mathrm{CH3}}^{\mathrm{TraPPE}}$. Perhaps a better characterization of the number of parameters in SAFTγ-WCA comes from examining the variation in the EOS parameters relative to the force field parameters. By characterizing Jacobian matrices for $\partial a_k^{ij}/\partial \sigma_k$ and $\partial a_k^{ij}/\partial \varepsilon_k$, Ghobadi and Elliott showed that the NERD and OPLS-UA force fields could be reproduced simply by substituting their values for $\{\varepsilon_k, \sigma_k\}$. In that sense, only the four parameters $\{\varepsilon_k, \sigma_k\}$ of the force field itself are adjustable, and the "top-down" approach of Jackson and coworkers would be applicable at the fused sphere united atom (UA) level, as well as at the tangent sphere CG level. Since the UA site types are more fundamental, transferability should be applicable to a broader class of molecules than at the CG level.

The distinction between the BH and WCA splits of the potential function gives rise to distinct trends in the TPT contributions. While the A_0 and A_1 contributions are similar regardless of the assumed split, the A_2 and A_3 contributions exhibit significant qualitative differences. We are now able to illustrate those differences and understand their physical basis. Figure 6.14 displays the trends in A_1–A_3 for LJ spheres with the two different split approaches. With the BH split, the A_2 contribution exhibits a shoulder region at high density whereas the A_3 contribution changes sign at high density. [Note that Eq. (6.8-64) precludes characterization of the sign change in A_3.] van Westen and Gross[135] have considered these differences and concluded that both approaches

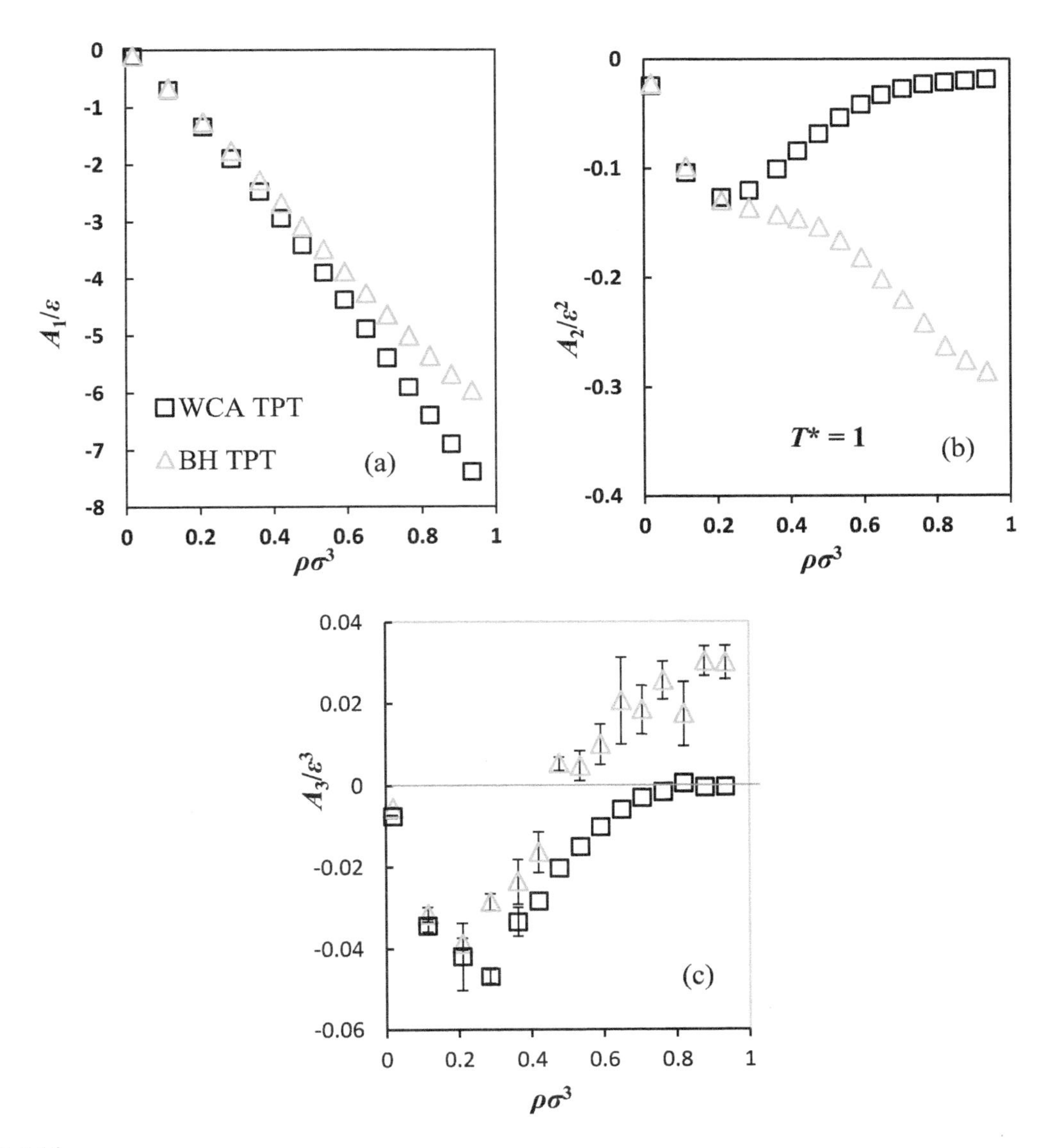

FIGURE 6.14
First- (a), second- (b), and third- (c) order perturbation contributions of monomer from BH and WCA perturbation theories obtained by MC simulation at NVT ensemble with $N = 500$ and $T^* = 1$. If not shown, error bars are smaller than the symbol size.

are viable. They suggest that the BH split appears to lead to faster convergence, such that truncation at A_3 may suffice, but the trends for A_4, A_5, etc. are more predictable with the WCA split. In this context, it is useful to simply understand both approaches and what gives rise to the differences.

We have seen shoulders in the density trends of A_2 for SW spheres and LJ spheres with the BH split. The shoulders disappear for step potentials and LJ spheres with the WCA split. Reflecting on Eq. (6.3-12), we note that it takes the form of a variance, which is known in thermodynamics as a fluctuation. It makes sense that fluctuations in energy should be large at the critical density, but what gives rise to the large fluctuations at high density for SW spheres and the BH split? In the SWS case, the sharp change in the potential at $r = \lambda\sigma^{SWS}$ gives rise to large changes in energy as molecules transition back and forth across that large energy change. Those changes are larger for large λ because there is more volume in which to make the transition. At extremely high density, the molecules get locked into place and A_2 decays to zero as expected. In the case of the BH split, a similar sharp transition in u_1 appears, this time near $r = \sigma$ (cf. Fig. 6.13). The WCA split has only smooth transitions in u_1. For step potentials, it can be shown that two steps suffice to eliminate the high-density shoulder.

From a physical perspective, both WCA and BH splits are viable and understandable. From an engineering perspective, it remains to be seen which perspective leads to the greatest accuracy, efficiency, and freedom from artifacts.

6.9.5 Coming Attractions: SAFT for Branched Interactions

For the most part, Chapman and coworkers have focused their attention on extending the SAFT perspective to broader ranging applications and the underlying fundamentals of the SAFT perspective. Recently, Zhang and Chapman[136] have extended to methodology to include branched molecules and branched chemical-physical interactions. The basic idea involves relaxing the assumption that only one bond can occur at any single bonding site. For example, the protons in water operate clearly as separate sites, but the proton acceptor(s) are less distinct.[137] Several theoretical and experimental studies suggest that three bonds are most common at temperatures around 50 °C and four bonds are more common around 0 °C. Marshall has developed a model that characterizes these trends, resulting in qualitative accuracy for water's density, heat capacity, and compressibility anomalies.[138] It remains to be seen whether correct characterization of pure water's anomalies can improve accuracy for mixed phase anomalies like the solubility of *n*-alkanes in water.

6.9.6 Crossover EOS Models for the Near-Critical Region

Conditions near the vapor-liquid critical point of a substance show significantly different behavior from simple extrapolations of the EOS models described so far in this chapter. The shape of the critical isotherm, the variations of C_v, the isothermal compressibility, and the vapor-liquid coexistence curve are all different than that given by most EOS models. This is because the molecular correlations are much longer ranged and fluctuate differently in this region. Unlike the "classical" region where Taylor series expansions can be taken of properties about a given state, such a mathematical treatment appears to break down near the "nonclassical" critical point. Research into this effect shows that certain universalities appear in the properties, but substance-specific quantities also are involved.

There are a variety of ways to define the "critical region." Anisimov et al.[139] define a criterion of $0.96 < T/T_c < 1.04$ along the critical isochore with effects on derivative properties felt at densities as far as 50 to 200% from ρ_c.

Considerable work has been done to develop EOS models that will suitably bridge the two regimes. Several approaches have been taken. An early approach was to use a "switching function" that decreases the contribution to the pressure of the classical EOS and increases that from a nonclassical term (e.g., Chapela and Rowlinson[140]). The advantage of this method is that no iterative calculations are needed. A similar approach is illustrated by Wagner and Pruss[1] in the IAPWS EOS for steam. Equation (6.3-8) includes contributions with noninteger exponents and a Gaussian factor that rapidly approaches zero outside the critical region. Another approach is to "renormalize" T_c and ρ_c from the erroneous values that a suitable EOS for the classical region gives to the correct ones. Examples of this method include Kiselev and coworkers,[141,142] Fox,[143] Pitzer and Schreiber,[144] Chou and Prausnitz,[145] Vine and Wormald,[146] Solimando et al.,[147] Lue and Prausnitz,[148] and Fornasiero et al.[149] These have different levels of rigor, but all involve approximations and iterative calculations. The technique of Fornasiero et al. was applied to the corresponding states forms of the van der Waals, Soave,[2] and Peng-Robinson[3] cubic EOS models described in Sec. 4.6 and used Z_c as an additional piece of data. Comparisons of saturated liquid densities with data for 17 normal fluids and 16 polar and associating substances showed RMS deviations of 1 to 5%, which appears to be comparable with the direct methods described above and the liquid density correlations described below.

The final approach to including nonclassical behavior has been the more rigorous approach via crossover functions of Sengers and coworkers.[139,150,151] The original method was to develop an EOS model that was accurate from the critical point to well into the classical region but did not cover all conditions. Anisimov et al.[139] and Tang et al.[152] show results for several substances. Recent efforts with this method have led to EOS models applicable to all ranges. Though not applied extensively yet, indications are that it should be broadly applicable with accuracies comparable to the scaling methods. In addition, theoretical analyses of this group[150] have considered the differences among approaches to the critical point of different kinds of systems such as electrolytes, micelles and other aggregating substances, and polymers where the range of the nonclassical region is smaller than molecular fluids and the transition from classical to nonclassical can be sharper and even non-universal.

All the above benefit if the region where nonanalytic corrections are required can be minimized. As more is known about higher-order contributions, characterization of EOS behavior in the critical region should systematically improve, as illustrated by Fig. 6.5 and Eq. (6.3-24).

6.10 CHEMICAL THEORY EOSs

It is technically feasible in some situations to solve the system of chemical reaction and phase equilibrium equations with explicit reactions for all species. This approach is exemplified by the work of Grensheuzer and Gmehling,[153] in which dimerization was assumed for all associating species. Such an approach fails to recognize the trimers and higher oligomers of alcohols etc., so it has received little attention in recent years, while Wertheim's theory has taken precedence.

We have already discussed solvation and association in the context of Wertheim's theory, but there are situations in which Wertheim's theory may not suffice. For example, spectroscopic evidence suggests that acetic acid forms associating chains in the liquid phase while dimers strongly dominate in the vapor phase. Alcohols appear to form more weakly bonded dimers than the higher oligomers. Hydrofluoric acid (HF) forms dimers, hexamers, and octamers, but other oligomers are less populated than would be expected from Wertheim's TPT1 formulation, although TPT2 may account for some of HF's peculiarities.[154] These observations suggest a possible need for explicit chemical reaction models, possibly superposed on the TPT1 or TPT2 model. Although it may seem redundant at first glance, valid fundamental reasons exist for doing so, and there is no computational reason why it cannot be done.

It is true that redundancy should be avoided, so it is important to appreciate the relationship between chemical theory and Wertheim's theory. Heidemann and Prausnitz[86] showed how chemical reactions can provide a theory of chain thermodynamics very similar to Wertheim's 2B model. The basis of their model was to assume that K_i^a is a constant for each monomer addition.

$$2A_1 = A_2;\ K_2^a = \frac{\hat{a}_2}{\hat{a}_1^2};\ A_1 + A_2 = A_3;\ K_3^a = \frac{\hat{a}_3}{\hat{a}_1\hat{a}_3};\cdots \tag{6.10-1}$$

When $K_2^a = K_3^a = K_4^a =$ constant, Heidemann and Prausnitz called this the infinite equilibrium model (IEM). When characterizing the equilibrium constants, it is necessary to specify their temperature dependence. A common assumption in chemical theory was

$$K_i^a \sim K_*^a \exp\left[H\left(1-\frac{1}{T_r}\right)\right] = \frac{K_c^* RT_c}{b} T_r^{\frac{\Delta C_p}{R}+1} \exp\left[\left(\frac{\Delta H}{RT_c} - \frac{\Delta C_p}{R}\right)\left(1-\frac{1}{T_r}\right)\right] \tag{6.10-2}$$

When $\Delta C_p/R = -1$, Eq. (6.10-2) is especially simple and this was the most common assumption. Close inspection shows that Wertheim's theory corresponds to a slightly different temperature dependence.

$$K_i^a \sim K_*^a\left[\exp\left(\frac{\varepsilon_{AD}}{T}\right) - 1\right]; \qquad \text{Wertheim's theory} \tag{6.10-3}$$

Suresh and Elliott[100] showed that the IEM is equivalent to a 2B Wertheim model subject to a somewhat arcane assumption about $\Delta C_p/R$:

$$\frac{\Delta C_p}{R} = \frac{\ln\left[\exp\left(\frac{\varepsilon^{AD}}{k_B T}\right) - 1\right] + \frac{\varepsilon^{AD}}{k_B T_c}\left(1-\frac{1}{T_r}\right) - \ln(T_r)}{\ln(T_r) - \left(1-\frac{1}{T_r}\right)} \tag{6.10-4}$$

Computationally, Eq. (6.10-2) would perform about the same as Eq. (6.10-3), so the choice is arbitrary from that perspective. From a theoretical perspective, there is no fundamental basis for Eq. (6.10-2), but Wertheim's theory has a strong fundamental footing. It would seem logical to use Wertheim's expression for characterizing the temperature dependence of K_i^a, even if applying an explicit chemical theory.

Given that explicit chemical theory can be equivalent to Wertheim's theory, the next question is how to manage the two approaches to the best effect. Detailed investigations of carboxylic acids have been performed in this regard, meriting a separate subsection. Other systems have been investigated less thoroughly and we simply outline a few of the results.

6.10.1 Carboxylic Acids

Considering acetic acid as a sample case, the strong vapor dimerization suggests there are two ways in which carboxylic acids form dimers, the first being singly bonded and the second being doubly bonded through closure of two carbonyl-hydroxyl hydrogen bonds to form a hexagonal ring. At high density, the singly bonded form offers the prospect of long chains with more total bonds. At low density, the closure of the ring doubles the association energy while finding a bonding partner in proximity. From this perspective, it is reasonable to write,

$$K_2^a = K_2^{AD} + K_2^C \tag{6.10-5}$$

where the contribution of K_2^{AD} addresses chain formation in the usual manner of the 2B Wertheim model, and K_2^C addresses the doubly bonded dimer specific to carboxylic acids. It would not be redundant in that case to apply Wertheim's theory to chain formation and explicit chemical theory as a supplementary contribution. The next question would be whether the doubly bonded dimer can participate in any further reactions. Doubly bonded dimers would not bond strongly with other species like alcohols or esters, so those kinds of reactions could be ignored. On the other hand, if a doubly bonded dimer of one acid could react with another carboxylic acid to solvate as a doubly bonded dimer, that would need to be represented explicitly in the context of chemical theory. Note that the compositions treated by Wertheim's theory would need to be adjusted relative to the superficial compositions to account for the presence of doubly bonded dimers, and the compositions treated by chemical theory would likewise need to be adjusted to reflect the competition with chain formation. To our knowledge, no evaluations of this approach are available in the literature.

In this context, the chemical+Wertheim theory needs to be considered in relation to various implementations of Wertheim's theory. Most implementations for carboxylic acids in Wertheim's theory use the 1A model, which can solvate with other species, including alcohols for example, but cannot form chains because they have only one bonding site per molecule. The CPA model uses the 1A approach,[155,156] as does the PC-SAFT model.[157] When the CPA model is combined with Huron-Vidal mixing rules, highly accurate correlations are achieved, but caution is advised if extrapolating this model to multicomponent mixtures.

Another approach is to introduce a 1C site type, which can only bond with other 1C site types, then include the usual A and D sites to account for chain oligomerization. Vahid and Elliott[158] applied this approach in the SPEADMD model. In this way, carboxylic acids can solvate with each other or form chains, all in the context of Wertheim's theory such that solvation with other acids is naturally included. This method does not account for the competition between dimerization and chain formation but allows both to occur simultaneously. They assumed characterizations of the A, D, and C types that were transferable to all carboxylic acids and were able to explain the evolution of saturated vapor dimerization with increasing aliphatic chain length. Their study included 42 carboxylic acids and 33 binary mixtures.

As a final example, Janecek and Paricaud[159] proposed a very sophisticated model that accounts directly for formation of doubly bonded dimers within the Wertheim formalism. The method requires solution of an additional association equation for pure acids, accounting for the fraction of molecules not bonded at either site, which can be related to cycle formation.[160] The additional association equation depends on doubly bonded dimer formation such that it competes naturally with other bonding in the Wertheim model, like the composition adjustment that would be required for a chemical+Wertheim theory. Additional explicit equations are required for each added acid in the mixture as well. And a special association equation is required for treating water. Altogether, this model might be very similar to a chemical+Wertheim theory in both computational speed and correlation performance, while naturally retaining compositional consistency. Results of their comparisons to the 2B model were mixed.

6.10.2 Other Adaptations of Chemical Theory

Other applications of chemical theory are much more specialized than the work available for carboxylic acids. For hydrogen fluoride, the work of Lee and Kim[161] summarizes work accounting directly for dimer, hexamer, and octamer. Their model can also be applied to mixtures by the Gibbs excess mixing rules.

It might be interesting to consider a chemical+Wertheim theory where the chemical component is applied to hexamers, which dominate the vapor phase ideality deviations. A compound of commercial significance with complex chemical and phase behavior relations is formaldehyde. If dehydrated carefully, formaldehyde can take the form of the hexagonal ring known as trioxane. Like the confusing coexistence of $2NO_2=N_2O_4$, and $6HF=(HF)_6$, the $3CH_2O$=trioxane relation makes it difficult to know whether one is considering a pure compound or a mixture.

6.10.3 Discussion

We include this section on chemical theory to clarify its relation to Wertheim's theory. In the process, we identify a few systems where Wertheim's theory may fall short, providing opportunities for chemical theory to provide the best available model. We consider these cases to be quite specialized and beyond the broad evaluation scope of this edition of PGL, but we welcome suggestions for future editions.

6.11 MOLECULAR SIMULATION MODELS

PvT and thermodynamic properties are some of the most basic properties that can be obtained from molecular simulation models. In the EOS context, molecular simulation with a "transferable" force field is pedagogically equivalent to a group contribution EOS. The primary drawback of molecular simulation is that it is extremely slow. Iterative evaluation of the density given temperature and pressure is not an option when each iteration requires several hours. The advantage of molecular simulation is that equilibrium and transport properties can be computed self-consistently from a single simulation and the physical basis is quite rigorous. The challenge is to make the greatest possible use from molecular simulation at any given time. Three basic concepts are available: (1) Use molecular simulation as "suggestive" for functional forms in a semiempirical EOS, then fit the parameters of the EOS to experimental data. The SAFT family is generally a good example of this approach, particularly the PC-SAFT model. The Carnahan-Starling term is very accurate for the hard sphere contribution and the chain term for tangent hard spheres is reasonable for liquids. The A_1 and A_2 contributions of PC-SAFT are fit to experimental data, but reflect the qualitative behavior exhibited in molecular simulations. (2) "Direct" molecular simulation (aka brute force) to get the desired properties from simulation at specified conditions. This approach is often demonstrated in the Industrial Fluid Properties Simulation Challenge (IFPSC) events. (3) Use a "surrogate" EOS to characterize the molecular simulations and accelerate application. This approach is exemplified by Thol et al.[162] and the SPEADMD method.[122] We have discussed SAFT approaches in previous sections, so we focus here on the latter approaches.

Although molecular simulation can be quite rigorous in scaling from a potential model to macroscopic properties, it is only as good as the potential model. Several aspects of current practice in potential modeling can be questioned. We describe the caveats of transferable potential models below. As another example, hydrogen bonding molecules are often represented with point charge models, but point charges are long-ranged while hydrogen bonds are short-ranged. Multiple manipulations are possible to mimic key features of hydrogen bonding, but there is always some sense of fitting a square peg in a round hole. Even for representing the polar contributions, the effective dipole moment may vary from vacuum to dense liquid, while most point charge models represent it statically. Finally, molecular simulations almost always assume a pairwise additive potential model, while real intermolecular potentials are not, as evidenced in Fig. 6.2b. Incorporating nonadditive effects into virial coefficients is much more feasible,[163] so both of these methodologies must work together to make systematic progress in molecular thermodynamics.

6.11.1 Direct Simulation

The direct approach is steadily gaining ground as computing power grows and methods become more advanced. The IFPSC results are published on roughly a biennial basis and cover topics like viscosity, interfacial tension, adsorption, LLE, VLE, vapor pressure, and density. (See the Publications links on the Challenge tab at fluidproperties.org.) We focus in this chapter on density and thermodynamic properties.

The fundamental basis of any molecular simulation is the potential model, also known as the force field. Potential models can be of the explicit atom, united atom, or coarse-grained varieties. As one might expect, explicit atom (EA) models are more detailed but slower to simulate and more difficult to characterize. United atom (UA) models group the hydrogen atoms of a functionality like CH2 into a single spherical site. This approach can be 2–3 times faster than EA approaches and can be relatively easy to characterize. Coarse-grained (CG) potentials can be an order of magnitude faster than UA models but requires inferring specific potential models for arbitrary groupings of atoms.

A key feature of any potential model is transferability. Like group contributions, transferability assumes that the potential energy around an interaction site like a CH2 in *n*-pentane is equivalent to the CH2 sites in *n*-pentadecane. The difference is that more than just the potential energy is implicitly transferred. Through Newton's laws, molecular dynamics can compute all the *PvT*, thermodynamic, interfacial, and transport properties simultaneously, not just for pure fluids but for mixtures as well. The breadth of coverage and rigorous physics intrinsic in the molecular simulation approach are big parts of the attraction for molecular simulation.

Returning to the choice amongst EA, UA, and CG, UA hits a sweet spot between molecular detail and transferability. EA is slower to simulate but offers increased accuracy. For example, an application that is sensitive to EA details is SLE of cyclohexane. The energy of chair and boat conformations and their packing are sensitive to the EA details. Among the EA models, the TraPPE-EA model[164] would generally have the best accuracy for engineering applications. It is more accurate than the TraPPE-UA model[165] for vapor pressure, but there are a couple of UA models that are more accurate than TraPPE-EA. One limitation with CG models is that every site is distinct. For example, the sites in *n*-heptane and *n*-octane would require distinct values for the potential parameters. Another limitation is that there is no widely accepted procedure for mapping from an atomistic model to the CG parameters. It is unclear how to evaluate such a model currently. Therefore, we choose to focus our evaluations on transferable UA models at present.

It is challenging to demand accurate values for all thermodynamic and transport properties from a single set of "group contributions." Common group contribution methods have different parameters for every property. Beyond the challenge of multiproperty optimization, optimizing the parameters of the intermolecular potential is hindered by the speed of computation for each prospective parameter iteration. A recent advance in this regard has been the MBAR methodology of Shirts and coworkers.[166] Briefly, MBAR provides a procedure for reweighting the configurations of a previous simulation to determine the thermodynamic properties applicable to a simulation with a different potential model. For example, the vapor pressure and liquid density of the OPLS-UA model could be predicted from simulation of the TraPPE-UA model and application of MBAR. This advance will accelerate the development of transferable potential models significantly.

There are several UA potential models to choose from. TraPPE-UA has the broadest coverage of molecular functionality, including sulfides and phosphonates as well as branched, ring, and chain hydrocarbons, alcohols, ethers, esters, etc. The anisotropic UA model (AUA) of Ungerer and coworkers[167,168] provides slightly better accuracy than TraPPE-UA, especially for heavy hydrocarbons,[169] but less coverage of functionalities like sulfides and phosphonates. A couple of recently developed models that provide greater accuracy are the transferable anisotropic Mie (TAMie)[170] and Potoff-Mie[171] models. We illustrate some typical results for the Potoff-Mie model in the evaluations below.

6.11.2 Simulation Surrogates

An alternative to direct simulation is to simulate a potential model in a generalized form that can be refined to fit a specific molecule at a later stage. This is a broad definition that requires further clarification to distinguish it from methods like SAFT. It could be argued that SAFT applies the hard sphere potential model in a broad way with refinement to fit specific molecules, but there is not usually a potential model for the sites in the molecule at the end of the SAFT development. A better argument could be made that the MBAR method satisfies this definition, but MBAR requires several simulations in the proximity of the optimal parameter space to work reliably. This could be called a gray area. What we intend by this definition is better exemplified by corresponding states analysis like that of Thol et al., the SAFT-VR-Mie tangent sphere model, and the TPT formulation of SPEADMD.

The analysis by Thol et al. results in a multiparameter EOS for the LJ fluid with analytic tail corrections. By scaling the temperature, pressure, and density with potential parameters $\{\varepsilon/k_B, \sigma\}$ the optimal potential parameters for a given molecule can be determined by fitting them to the entirety of data for that molecule. This concept has been implemented in many forms over the years.[108,172,173] Three things distinguish the Thol et al. EOS: (1) It is expressed in the form of a Wagner-type EOS. (2) It includes accurate simulations of derivative properties like the speed of sound. (3) They do not simulate the saturation properties directly but infer them from accurate treatment of the bulk phases. Nevertheless, their comparisons to literature data for the LJ EOS show good agreement with the most reliable simulations. The principal advantage is the availability of speed of sound, which has been shown to narrow the uncertainty in the best fitting potential parameters. If similar

EOSs could be developed for spherical Mie potential models, good accuracy could be achieved for spherical molecules generally.

Another surrogate model that we would like to evaluate is the SPEADMD model. When simulations with the full potential were compared to results from TPT for this model, agreement was quantitative outside the critical region. This means that values for the depths of the steps in the attractive potential can be inferred from TPT without explicitly simulating those step depths. Direct simulation of the hard reference fluid is still required with all the molecular details of bond angles, bond lengths, branches, and rings. Only the attractive potential is treated with TPT. A review of the methodology is available and over 800 compounds have been characterized with a wide range of molecular functionalities. In this evaluation, we include only the second-order version of the theory, ignoring the Gaussian extrapolation correction.

Finally, it is feasible to use molecular simulations to supplement experimental data in regions that are either unavailable or inaccessible. This can be especially valuable in the development of multiparameter EOSs, which might extrapolate poorly if not informed in this way. An example of this hybrid approach is provided by Thol et al.[174]

6.12 RESIDUAL FUNCTIONS FOR EVALUATED MODELS

We have evaluated implementations of several models for their accuracy in reproducing thermodynamic properties from a standard database. In most cases, expressions for the compressibility factor and the residual internal energy suffice for our evaluations. When additional derivative properties are required, numerical differentiation is usually applied. Expressions for residual internal energy of several models are listed in Table 6.7. For the REFPROP, CPA, and GC-PC-SAFT models, the software developers should be consulted.

TABLE 6.7 Useful residual functions for attractive dispersion interactions.

Model	Expression	Eq.
PR	$\dfrac{u^{\text{res}}}{RT} = \dfrac{-A}{B\sqrt{8}}\left(1 + \dfrac{\kappa^{\text{PR}}\sqrt{T_r}}{\sqrt{\alpha^S}}\right)\ln\left(\dfrac{Z + (1+\sqrt{2})\,B}{Z + (1-\sqrt{2})\,B}\right)$	(6.12-1)
SRK	$\dfrac{u^{\text{res}}}{RT} = \dfrac{-A}{B}\left(1 + \dfrac{\kappa^{S}\sqrt{T_r}}{\sqrt{\alpha^S}}\right)\ln\left(\dfrac{Z + B}{Z}\right)$	(6.12-2)
tcPR	$\dfrac{u^{\text{res}}}{RT} = \dfrac{-A}{B\sqrt{8}}\left(1 + \dfrac{\kappa^{\text{PR}}\sqrt{T_r}}{\sqrt{\alpha^S}}\right)\ln\left(\dfrac{Z + (1+\sqrt{2})\,B}{Z + (1-\sqrt{2})\,B}\right)$	(6.12-3)
ESD	$\dfrac{u^{\text{res}}}{RT} = \dfrac{9.5\beta\varepsilon q\eta^P(Y + 1.0617)}{1 + 1.7745Y\eta^P}$	(6.12-4)
SPEADMD	$\dfrac{u^{\text{res}}}{RT} = \dfrac{A_1}{T} + \dfrac{2A_2}{T^2}$ $\dfrac{C_V - C_V{}^{\text{ig}}}{RT} = \dfrac{2A_2}{T^2}$	(6.12-5)
PC-SAFT	$\dfrac{u^{\text{res}}}{RT} = \dfrac{A_0'}{T} + \dfrac{A_1}{T} + \dfrac{A_1'}{T^2} + \dfrac{2A_2}{T^2} + \dfrac{2A_2'}{T^3}$	(6.12-6)

where the A_i' represents derivatives affected by the temperature dependence of hard sphere diameter. These derivatives are quite tedious and readers are referred to the source code for further information.

The residual energy of the chemical contribution is generally more involved and not as easy to tabulate as in the case of the dispersion contributions. An exception to the general rule is the efficient ESD formula, which is simplified by noting for a pure fluid

$$\left(\frac{1}{X^{AD}} - 1\right)(\alpha^{AD})^{-½} \equiv F^{AD} = N^{AD}X^{AD}(\alpha^{AD})^{½} \tag{6.12-7}$$

where the superscript AD indicates that $X^A = X^D$ in the ESD formulation. Multiplying the left-hand side by the right-hand side gives $(F^{AD})^2 = N^{AD}(1\text{-}X^{AD})$. Elliott and Natarajan[175] then show

$$\frac{u^{\text{chem}}}{RT} = [N^{AD} + (F^{AD})^2]\frac{\beta\varepsilon^{AD}F^{AD}(\alpha^{AD})^{½}}{1 + 2F^{AD}(\alpha^{AD})^{½}}\frac{1 + Y^{AD}}{Y^{AD}} \tag{6.12-8}$$

More generally, the derivative of Eq. (6.8-2) is best expressed through the formalism of Michelsen and Hendriks.[87] The expression for a pure fluid is

$$\frac{u^{\text{chem}}}{RT} = -\frac{\rho}{2}\sum_{k}^{N_1^{\text{type}}}\sum_{l}^{N_1^{\text{type}}} N_{1,k}^{d}X_{1,k}N_{1,l}^{d}X_{1,l}\frac{\beta d\Delta_{11,kl}}{d\beta} \tag{6.12-9}$$

$$\frac{\beta d\Delta_{11,kl}}{d\beta} = \beta\varepsilon_{11,kl}^{AD}g_{11,kl}\frac{Y_{11,kl}^{AD}+1}{Y_{11,kl}^{AD}} + \beta g'_{11,kl}\Delta_{11,kl}$$

where there are typically two values for N_1^{type}: A and D. For SPEADMD, $g'_{11,kl} = 0$. For PC-SAFT the temperature dependence of d^{hs} affects $g'_{11,kl}$ in a tedious manner once again.

6.13 EVALUATIONS OF EQUATIONS OF STATE

In this section, we discuss the evaluations of the EOSs described above for both PvT and thermodynamic properties. In the low-density regions, distinctions are small between most EOSs. In the critical region, REFPROP and Lee-Kesler provide clearly superior accuracy. At high densities, distinctions are apparent, and it is helpful to interpret the meanings of those distinctions. When considering how to characterize the accuracy of thermodynamic properties, one must consider how to select representative properties. Referring to Fig. 6.5, we can define the critical region with one line running from a $T_r = 0.9$ and the saturated vapor density to $T_r = 1.2$ and the critical density, and another line running from $T_r = 0.9$ and the saturated liquid density to $T_r = 1.2$ and the critical density. The liquid region is at densities above critical and outside the critical region. The vapor/gas region is at densities below critical and outside the critical region.

The database for this evaluation is composed of fluid classes as defined in Chap. 1. We have separate tables based on whether properties are in the vapor/gas, critical, or liquid regions. Each table includes deviations for each class of fluid. Estimations of the density, heat capacity, and compressibility are based on the given T and P. We choose these variables as the basis for evaluation because they are measurable. Although our interest is typically in enthalpy and entropy for pure fluids (e.g., for refrigeration or power cycle design), these properties cannot be measured directly. Rather, they can be inferred by integrating properties like heat capacity and compressibility. Any EOS naturally provides this integration implicitly. For evaluation purposes, it is most reasonable to compare directly to the available experimental data. The expected uncertainty in the properties of enthalpy and entropy can then be estimated through Monte Carlo methods applied to the observed deviations in the measurable properties.

The measures of heat capacity and compressibility require further clarification because the experimental measurements may be performed in various ways. For thermal properties, the measurements may be in the form of C_p, C_v, C_σ, finite changes in enthalpy with respect to temperature, and the isobaric coefficient of expansion. The measures of C_p and C_v have their familiar definitions, but C_σ may be unfamiliar so it is defined by Eq. (6.13-1).

All deviations for these measures are grouped as "thermal properties." Furthermore, we seek to distinguish between the quality of the EOS versus the ideal gas property. Therefore, all comparisons are based on the departures of the measured property from its ideal gas value, e.g., $(C_p - C_p^\circ)$, not C_p itself.

$$C_\sigma = \left(\frac{\partial h}{\partial T}\right)_{\text{sat}} \tag{6.13-1}$$

Compressibility measures include the Joule-Thompson coefficient and the speed of sound. The Joule-Thomson coefficient can be written as

$$\mu_{JT} = \left(\frac{\partial T}{\partial P}\right)_H = \frac{\left(T\left(\frac{\partial v}{\partial T}\right)_H - v\right)}{C_P} \tag{6.13-2}$$

and the speed of sound is

$$u_{ss} = \sqrt{\left(\frac{\partial P}{\partial \rho}\right)_S} = \sqrt{\frac{C_P}{C_v}\left(\frac{\partial P}{\partial \rho}\right)_T} \tag{6.13-3}$$

where ρ is the molar density.

The primary differences among the myriad of forms are computational complexity and quality of the results at high pressures, for liquids, for polar, and for associating substances. While EOSs were previously limited to the vapor phase properties, they now are commonly applied to the liquid phase as well. Thus, the most desirable expressions give the PvT behavior of both vapor and liquid phases and all other pure-component properties with extensions to mixtures while remaining as simple as possible for computation. Of course, since not all these constraints can be satisfied simultaneously, the reader's judgment and optimization for their application are required to determine which model to use.

A shorthand is used to represent the various methods. In most cases the shorthand is obvious, but a few methods may need clarification. Some methods can be applied as correlations or predictions. Predictive methods are indicated with a "*." P_C-SAFT refers to the F90 implementation shared by Joachim Gross with code. tcPR refers to the volume-translated F90 implementation shared by Jean-Noel Jaubert, which also includes numerous alternative options like PR and SRK that were not evaluated. The source codes for these and the ESDREFPROPMD, PR, SRK, and vtPR methods from the University of Akron are linked on the PGL6ed website. Group contribution methods are indicated with the GC prefix. The EGC methods refer to Emami et al.[113] The suffix "(T_b)" indicates that the boiling temperature was used to adjust the ε^{ss} parameter in the designated EOS and all other parameters were taken from the GC estimation. $P_C$he EGC-PC-SAFT versions apply Gross's F90 code after estimating the parameters with EGC methods. The CPA method is used with the volume translation option. Although CPA covers a relatively small number of compounds, it should be recalled that it defaults to the SRK method when customized parameters are not available. The SW12-TDE method is an abbreviated 12-parameter version of the Span-Wagner[29] multiparameter EOS. Aspen-TDE-SAFT refers to TDE's implementation of SAFT in the ASPEN process simulator.[176] GC-PPC-SAFT(IFP) refers to the method of Nguyen-Huynh et al.[115] and includes a substantial number of compounds with customized parameters.

6.13.1 Discussion

Table 6.8 is sorted first according to the degree of predictivity of the method, with more predictive methods at the bottom. Then the methods are sorted according to accuracy for correlation of vapor pressure. The same sorting is maintained in the remaining tables in the hope that it may facilitate comparisons from one property to another. The number of compounds is indicated by N. This number drops substantially for some methods because the intersection of the compounds characterized by method developers with the compounds listed in Appendix A was reduced. When N drops substantially, it becomes difficult to compare methods objectively. For this reason, Aspen-TDE-SAFT, SW12-TDE, CPA, and GC-PPC-SAFT are omitted from discussion below.

TABLE 6.8 Comparisons of EOS models for vapor pressure deviations.

	Normal		Heavy		Polar		Assoc		Overall		
	N	%	*N*	%	*N*	%	*N*	%	*N*	*NPTS*	%
REFPROP	92	0.56	—	—	9	0.56	9	0.92	110	49998	0.62
Aspen-TDE-SAFT	67	0.84	2	1.72	6	1.04	9	1.36	84	16077	0.95
tcPR	249	0.87	18	3.48	51	1.25	105	2.40	423	85999	1.35
SW12-TDE general	47	1.47	1	1.63	13	1.29	5	3.34	66	13494	1.47
PC-SAFT	220	1.37	10	11.31	46	1.64	94	3.40	370	93816	2.13
SPEADMD	204	1.30	14	6.13	44	1.76	110	3.62	372	89579	2.12
CPA	50	1.50	8	3.94	6	2.58	34	3.76	98	48171	2.50
ESD	248	2.04	20	7.27	54	3.05	106	5.25	428	92737	3.10
SRK	252	1.79	20	11.25	54	2.46	122	12.84	448	101222	5.31
vtPR⁺	252	1.97	20	30.74	54	1.80	122	17.76	448	100874	7.25
PR	252	1.97	20	30.74	54	1.80	122	17.76	448	100874	7.25
Lee-Kesler	251	7.78	11	23.49	53	6.46	99	29.75	414	97081	13.87
GC-PPC-SAFT(IFP)	98	11.18	9	19.88	22	7.44	69	17.15	198	59318	13.23
EGC-ESD(T_b)	152	6.47	20	13.67	32	4.66	70	6.27	274	57751	6.38
EGC-PC-SAFT(T_b)	151	6.55	19	17.45	32	5.10	70	14.61	272	60570	9.09
TFF-SPEADMD*	110	10.87	14	14.17	21	8.64	62	13.27	207	47874	11.41
EGC-ESD*	152	26.22	20	31.08	32	45.87	70	40.17	274	57857	32.62
EGC-PC-SAFT*	151	25.14	19	41.38	32	46.14	70	49.69	272	61387	35.14

*indicates predictive methods. ⁺volume-translated-PR (vt-PR) has the same vapor pressure as PR.

Note that *N* drops substantially for all the predictive methods, but this is because Appendix A includes many compounds that cannot be characterized transferably. For example, water, methanol, CO_2, and methane are small molecules for which GC and TFF methods cannot be applied. The reduced *N* for these methods reflects the omission of nontransferable compounds.

The "best" method depends on the property and class of interest of course, but the REFPROP package stands out when parameters are available, with typical deviations less than 1%. A recent publication by Huber et al. provides extended background on the REFPROP package, its history, and its future plans. Huber et al. also describe how to access the source code and many interfacing strategies.[177] A notable exception is the vapor pressure of heavy compounds, which has been omitted for REFPROP because results for *n*-hexadecane and *n*-docosane appeared to be outliers that might be affected by uncertainty in the experimental data. REFPROP's success should not be surprising owing to the large number of parameters in the REFPROP EOSs and the high degree of overlap between the evaluation database and the training set for REFPROP models. Similarly, the tcPR method stands out for vapor pressure because its training set for 1700 compounds overlaps substantially with the evaluation database and because it restricts coverage to $P^{vp} > 10$ kPa. Among the other methods, PC-SAFT and SPEADMD perform similarly with deviations around 2% compared to 1.5% for tcPR, while extending the range to $P^{vp} > 0.1$ kPa. Lower temperatures and pressures are often associated with higher uncertainty in vapor pressure. The ESD deviations increase to 3% and other cubic EOSs increase to 6%. Deviations increase substantially for the predictive methods, of course, but provide a baseline for comparison. Deviations approaching 30% for classical GC methods may be a reasonable expectation for some time to come. Deviations of 10 to 15% for molecular simulation with transferable force fields may represent a limitation of the transferability assumption when applied to a large database. Alternatively, Mick et al.[178] report an average deviation of about 3% for 31 branched hydrocarbons treated with a transferable Mie potential model with $T > 0.64\ T_c$. Deviations are closer to 30% for the transferable TraPPE-UA potential model, which applies to broader classes of compounds. A notable disappointment over all properties is the Lee-Kesler method. It seems this method works reasonably well for densities of normal compounds but alternative methods should be preferred generally.

The story is similar for saturated liquid density, except that REFPROP sets the standard for all classes at around 0.1%. Deviations jump to 1% for PC-SAFT and SPEADMD, 2% for tcPR, and 6% for other cubic EOSs. Once again, these differences track with the degree of overlap between the training and evaluation datasets, and the adaptability of the models. For example, tcPR and SPEADMD have a comparable number of customizable parameters but SPEADMD is not cubic and has a form that naturally adapts better to liquid density behavior. Among the predictive methods, SPEADMD stands out at 3% deviation with 6% more common for the other methods. Results for compressed liquid density are similar.

Vapor densities show deviations around 2 to 3% for all methods except SPEADMD (6%) and REFPROP (0.4%). Results for associating compounds are exceptionally poor for SPEADMD and should benefit from more in-depth analysis. Trends for density in the critical region are like those for vapor density, except the deviations are about an order of magnitude larger for all methods. Density in the critical region is one of few cases where the Lee-Kesler method performs relatively well, but primarily for normal compounds.

Deviations for compressibility properties are exceptionally large for tcPR and PC-SAFT at 30%, whereas cubic EOSs are close to 20% and SPEADMD is around 15%. None of these could be considered "good." REFPROP, with deviations near 0.5%, is the only commendable method for compressibility properties.

For thermal properties, even REFPROP deviations are relatively large at 1.5%, which may reflect uncertainty in the experimental data again. The TPT1 implementations (PC-SAFT, SPEADMD, and ESD) seem to have a slight edge over cubic equations, comparing 6–7% deviations to 11–14% deviations. One possible explanation is the behavior of vapor heat capacity at atmospheric pressure. In most cases, one might assume that vapor deviations from ideal gas behavior are negligible at ambient conditions, and that is true in some sense. On the other hand, if association is significant, the apparent deviation can be significant even if each individual species behaves ideally. Figure 6.15 illustrates the behavior of methanol. Similar behavior is exhibited by ethanol and acetic acid, among others. The experimental data deviate from ideality by almost 100% at 345 K. The association models are not quantitative, but do provide a qualitative prediction of the effect, noting that the association models did not include these data in their training sets. Even the best cubic models are qualitatively incorrect. Alternatively, REFPROP includes extra parameters and extensive training to account for this effect.

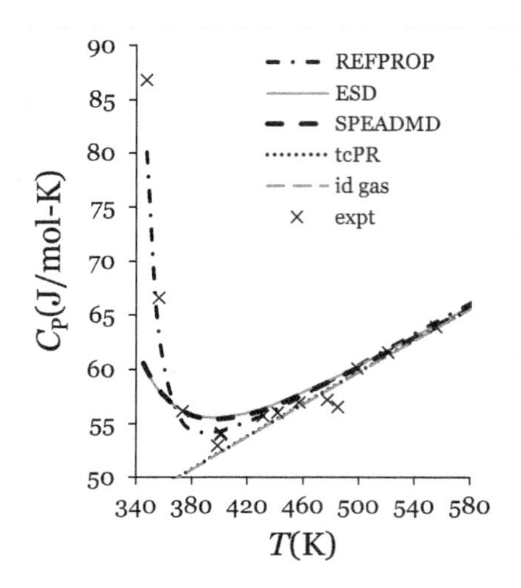

FIGURE 6.15
Vapor phase heat capacity of methanol at 101 kPa. Individual species may behave ideally, but association predicts speciation. Experimental data of Ref 179.

Nuances like those for vapor heat capacity suggest that the evaluations exemplified by Table 6.8 through Table 6.14 cannot tell the complete story. Every reader must be alert to effects that may impact their applications. Then they must seek experimental data to support their analysis and theories that provide the best predictions when experimental data are limited.

Overall, REFPROP is clearly the best choice for pure compound properties when parameters are available. The only exception is the vapor pressure of heavy compounds, which may require an update. Otherwise, tcPR, PC-SAFT, and SPEADMD provide similar accuracy for multiple properties and broad compound coverage, with slight preferences depending on the property of interest and class of compounds. Note that the tcPR method is limited to $P > 10$ kPa while PC-SAFT, SPEADMD, and ESD extend to 0.1 kPa. The large vapor pressure deviations for heavy compounds with these latter methods may be due to uncertainty in the experimental data as well as shortcomings in the methods. In most cases, SPEADMD is comparable to PC-SAFT, except for critical density and vapor density. The EOS critical temperature is substantially larger than the experimental value for the SPEADMD TPT2 method owing to inferior critical scaling. This might be improved if the Gaussian extrapolation method was implemented, but Gaussian extrapolation could also improve the results for PC-SAFT.

Cubic EOSs have often been chosen as the optimal forms because the accuracy is adequate and the analytic solution for the phase densities is reliable and computationally efficient. When selecting a cubic EOS for PvT properties, users should first evaluate what errors they can accept for the substances and conditions of interest, as well as the effort it would take to obtain parameter values if they are not available in the literature. Sometimes this takes as much effort as implementing a more complex, but accurate model such as a multiparameter EOS. If committed to a cubic EOS, the tcPR implementation stands out.

TABLE 6.9 Comparisons of EOS models for saturated liquid density deviations.

	Normal		Heavy		Polar		Assoc		Overall		
	N	%	*N*	%	*N*	%	*N*	%	*N*	*NPTS*	%
REFPROP	90	0.09	4	0.12	8	0.17	9	0.05	111	32454	0.08
Aspen-TDE-SAFT	66	0.44	2	0.75	6	0.56	9	0.37	83	8806	0.44
tcPR	229	1.43	13	1.11	46	2.09	95	4.66	383	33999	2.46
SW12-TDE general	47	0.13	1	0.40	13	0.09	5	0.39	66	10620	0.12
PC-SAFT	218	0.86	10	0.65	43	1.26	92	1.34	363	64476	1.06
SPEADMD	197	1.02	14	1.05	41	1.23	103	1.08	355	51304	1.07
CPA	50	1.34	8	5.28	6	0.70	34	0.99	98	41297	1.24
ESD	243	2.26	17	3.95	51	7.04	93	11.36	404	57659	5.82
SRK	250	10.94	19	23.51	51	17.22	116	18.37	436	70897	14.62
vtPR[+]	250	2.29	19	5.22	51	6.44	116	9.66	436	70897	5.52
PR	250	4.16	19	16.08	51	7.03	116	8.97	436	70897	6.50
Lee-Kesler	249	3.44	10	13.87	51	8.29	94	16.13	404	67393	8.58
GC-PPC-SAFT(IFP)	98	2.77	9	1.39	22	2.47	66	3.12	195	48199	2.86
EGC-ESD(T_b)	149	7.15	17	4.01	32	5.23	58	3.12	256	41812	5.69
EGC-PC-SAFT(T_b)	150	7.09	18	6.78	32	5.62	65	3.12	265	49710	5.67
TFF-SPEADMD*	103	3.04	14	6.05	21	1.76	55	2.70	193	32694	2.88
EGC-ESD*	149	17.47	17	3.71	32	5.23	58	3.24	256	41405	11.87
EGC-PC-SAFT*	150	7.11	18	6.94	32	5.61	65	3.20	265	49699	5.71

*indicates predictive methods. [+]volume-translated-PR (vt-PR) has the same vapor pressure as PR.

TABLE 6.10 Comparisons of EOS models for vapor density deviations.

	Normal		Heavy		Polar		Assoc		Overall		
	N	%	*N*	%	*N*	%	*N*	%	*N*	*NPTS*	%
REFPROP	71	0.33	—	—	6	0.44	5	0.88	82	32961	0.36
Aspen-TDE-SAFT	19	1.20	—	—	2	0.59	1	1.36	22	9004	1.10
tcPR	79	1.03	—	—	8	1.29	16	4.10	103	34453	1.19
SW12-TDE general	9	1.15	—	—	2	0.16	1	0.24	12	1274	0.99
PC-SAFT	71	1.04	—	—	8	1.76	17	9.29	96	31484	1.64
SPEADMD	74	4.32	—	—	8	5.65	19	14.52	101	34219	5.10
CPA	23	1.13	—	—	2	1.72	11	6.68	36	11143	1.87
ESD	78	1.58	—	—	8	2.54	15	4.49	101	34078	1.83
SRK	80	1.22	—	—	8	2.16	19	9.53	107	35350	1.82
vtPR+	80	0.97	—	—	8	1.38	19	9.06	9.06	9.06	9.06
PR	80	1.41	—	—	8	1.41	19	9.07	107	35389	1.86
Lee-Kesler	80	0.65	—	—	8	1.21	16	10.28	104	35003	1.19
GC-PPC-SAFT(IFP)	—	—	—	—	—	—	—	—	0	0	0.00
EGC-ESD(T_b)	23	2.77	—	—	1	0.83	5	2.11	29	3390	2.67
EGC-PC-SAFT(T_b)	23	2.55	—	—	1	0.79	5	2.78	29	3408	2.57
TFF-SPEADMD*	22	2.04	—	—	1	0.94	4	6.02	27	3115	2.53
EGC-ESD*	23	3.16	—	—	1	0.95	5	2.13	29	3380	3.00
EGC-PC-SAFT*	23	2.62	—	—	1	0.89	5	2.54	29	3412	2.60

*indicates predictive methods. +volume-translated-PR (vt-PR) has the same vapor pressure as PR.

TABLE 6.11 Comparisons of EOS models for compressed liquid density deviations.

	Normal		Heavy		Polar		Assoc		Overall		
	N	%	*N*	%	*N*	%	*N*	%	*N*	*NPTS*	%
REFPROP	86	0.16	2	0.19	7	0.11	8	0.13	103	112246	0.15
Aspen-TDE-SAFT	44	0.82	2	1.32	5	0.80	5	0.29	56	32866	0.79
tcPR	146	2.40	8	2.48	25	2.82	40	3.24	219	114182	2.55
SW12-TDE general	39	0.29	—	—	13	0.31	2	0.70	54	10900	0.30
PC-SAFT	143	1.03	6	1.20	27	1.42	53	1.55	229	143073	1.15
SPEADMD	143	2.25	7	1.17	28	2.92	50	1.78	228	131622	2.21
CPA	45	2.41	6	5.85	4	1.79	24	1.38	79	80622	2.21
ESD	156	5.74	9	5.39	29	10.68	50	11.42	244	139006	7.00
SRK	159	9.48	10	19.37	29	17.60	64	18.90	262	154418	11.81
vtPR+	159	3.43	10	5.09	29	8.34	64	9.19	262	152706	4.79
PR	159	5.22	10	13.80	29	8.11	64	9.23	262	152211	6.24
Lee-Kesler	158	2.80	5	15.47	29	9.90	51	14.46	243	147689	5.25
GC-PPC-SAFT(IFP)	—	—	—	—	—	—	—	—	0	0	0.00
EGC-ESD(T_b)	86	12.60	9	8.69	16	7.79	33	3.99	144	70703	10.42
EGC-PC-SAFT(T_b)	86	9.96	9	7.66	16	6.38	34	2.53	145	78387	8.12
TFF-SPEADMD*	73	3.45	8	6.68	11	2.41	30	2.82	122	66772	3.38
EGC-ESD*	86	23.58	9	8.35	16	8.15	33	3.97	144	70617	18.36
EGC-PC-SAFT*	86	10.05	9	7.61	16	6.52	34	2.72	145	78346	8.24

*indicates predictive methods. +volume-translated-PR (vt-PR) has the same vapor pressure as PR.

TABLE 6.12 Comparisons of EOS models for density deviations in the critical region.

	Normal		Heavy		Polar		Assoc		Overall		
	N	%	*N*	%	*N*	%	*N*	%	*N*	*NPTS*	%
REFPROP	82	1.88	2	3.59	8	1.63	6	1.58	98	21304	1.84
Aspen-TDE-SAFT	43	9.31	1	4.93	3	19.01	4	19.66	51	4772	10.65
tcPR	151	8.86	6	11.80	22	10.82	33	14.38	212	22179	9.59
SW12-TDE general	28	2.00	1	1.34	4	2.85	2	4.61	35	1433	2.05
PC-SAFT	143	8.91	1	1.86	21	16.70	33	14.00	198	21789	10.25
SPEADMD	143	32.80	5	17.75	18	37.83	38	37.47	204	22404	33.77
CPA	41	15.11	4	13.34	5	10.37	21	22.30	71	10945	16.41
ESD	150	14.73	7	21.49	22	20.04	35	24.47	214	22264	16.23
SRK	158	13.14	7	11.08	22	17.16	38	21.25	225	20974	14.26
vtPR⁺	158	8.84	7	5.40	23	12.36	39	15.10	227	23102	9.76
PR	159	7.63	7	5.44	22	11.83	39	16.48	227	23902	8.80
Lee-Kesler	160	4.12	7	12.10	24	7.06	36	14.59	227	24419	5.37
GC-PPC-SAFT(IFP)	—	—	—	—	—	—	—	—	0	0	0.00
EGC-ESD(T_b)	77	31.35	7	18.26	12	16.46	19	18.76	115	4994	29.42
EGC-PC-SAFT(T_b)	84	34.22	7	18.35	13	18.79	21	24.24	125	5578	32.38
TFF-SPEADMD*	75	31.07	5	28.24	6	24.46	22	37.18	108	5484	31.94
EGC-ESD*	77	49.95	7	22.96	12	16.91	20	19.65	116	4702	45.62
EGC-PC-SAFT*	79	33.89	7	37.88	13	22.58	22	37.56	121	5567	34.37

*indicates predictive methods. ⁺volume-translated-PR (vt-PR) has the same vapor pressure as PR.

TABLE 6.13 Comparisons of EOS models for deviations in measures of compressibility#.

	Normal		Heavy		Polar		Assoc		Overall		
	N	%	*N*	%	*N*	%	*N*	%	*N*	*NPTS*	%
REFPROP	74	0.51	4	0.68	7	0.63	8	0.40	93	43937	0.51
Aspen-TDE-SAFT	45	7.26	2	16.00	5	4.93	8	7.03	60	9701	7.14
tcPR	28	24.69	1	63.85	10	28.21	13	39.48	52	633	28.70
SW12-TDE general	39	1.89	1	12.22	13	1.31	4	1.35	57	5860	1.73
PC-SAFREFPROP		5.71	1	39.12	17	15.32	32	29.27	95	1692	26.20
SPEADMD	38	7.19	1	5.26	13	12.60	27	28.12	79	1407	14.57
CPA	21	27.88	1	9.52	5	27.67	23	12.83	50	862	19.59
ESD	43	137.23	1	176.79	17	141.65	29	60.86	90	1462	118.43
SRK	158	12.81	15	38.09	37	10.23	85	44.75	295	62347	19.02
vtPR⁺	158	11.35	15	23.68	37	9.57	85	42.99	295	62289	17.20
PR	158	12.36	15	36.80	37	9.72	85	49.29	295	62424	19.43
Lee-Kesler	157	5.93	6	3.88	37	6.69	72	87.19	272	60347	20.04
GC-PPC-SAFT(IFP)	—	—	—	—	—	—	—	—	0	0	0.00
EGC-ESD(T_b)	36	143.59	1	151.74	13	139.36	23	94.10	73	1002	137.41
EGC-PC-SAFREFPROP	9	27.43	1	25.34	14	10.89	28	24.25	82	1143	25.90
TFF-SPEADMD*	32	8.07	1	12.89	7	7.54	15	11.22	55	929	8.47
EGP_c-ESD*	36	156.47	1	139.67	13	144.77	23	90.52	73	1002	147.67
EGC-PC-SAFT*	39	34.06	1	21.67	14	16.08	28	21.44	82	1144	30.81

*indicates predictive methods. ⁺volume-translated-PR (vt-PR) has the same vapor pressure as PR.
#these properties are composed of Joule-Thomson coefficient and speed of sound.

TABLE 6.14 Comparisons of EOS models for deviations in measures of thermal properties.

	Normal		Heavy		Polar		Assoc		Overall		
	N	%	*N*	%	*N*	%	*N*	%	*N*	*NPTS*	%
REFPROP	84	1.50	3	0.79	9	1.25	8	1.59	104	30395	1.49
Aspen-TDE-SAFT	63	10.36	2	3.55	5	4.04	9	18.43	79	7029	11.67
tcPR	191	10.29	6	2.98	36	6.25	68	17.26	301	24069	11.40
SW12-TDE general	43	10.27	—	—	13	1.33	5	7.60	61	5179	8.68
PC-SAFT	192	4.09	10	2.26	39	5.47	77	10.23	318	37373	5.70
SPEADMD	180	5.57	9	3.82	35	8.78	78	10.13	302	29686	6.99
CPA	48	4.92	7	5.90	5	6.34	33	7.52	93	20426	5.83
ESD	212	5.92	11	5.75	44	7.45	81	10.38	348	30818	7.08
SRK	218	13.55	17	3.64	44	9.58	106	18.69	385	48711	14.60
vtPR+	218	12.73	17	5.30	44	8.24	106	16.19	385	48297	13.28
PR	218	12.73	17	5.30	44	8.24	106	16.41	385	48378	13.35
Lee-Kesler	217	14.56	8	10.89	44	10.50	84	24.38	353	46226	16.77
GC-PPC-SAFT(IFP)	—	—	—	—	—	—	—	—	0	0	0.00
REFPROPGC-PC-SAFT(T_b)	134	5.38	16	3.20	25	3.29	56	8.72	231	23397	6.10
TFF-SPEADMD*	96	12.19	13	1.91	17	2.13	43	14.72	169	15438	12.04
EGC-ESD*	132	9.47	11	8.24	25	6.44	49	9.26	217	17427	9.27
EGC-PC-SAFT*	134	5.39	16	3.52	25	3.07	56	8.80	231	23410	6.12

*indicates predictive methods. +volume-translated-PR (vt-PR) has the same vapor pressure as PR.

#C_p, C_v, C_σ, finite changes in enthalpy with respect to temperature, and the isobaric coefficient of expansion comprise thermal properties in this evaluation.

There are now both cubic and other EOS models that can be used to correlate the *PvT* behavior of polar molecules with accuracy approaching that of models in Chap. 5. Complex substances require more than three parameters, but when these are obtained from critical properties and measured liquid volumes and vapor pressures, good agreement can be obtained.

For predictions, the EGC-ESD method of Emami et al.[104] provides a fair representation of what can be expected from group contribution methods. Accuracy for predicted vapor pressure is comparable to the TraPPE-UA method, although the GC-ESD method is much faster and covers more compounds. The TFF methods of SPEADMD and Mick et al. are more promising, with vapor pressure deviations around 10 to 15%. Vapor pressure deviations drop from 30% to 6% if a single measurement like T_b is available. If one wishes to calculate only saturated or compressed liquid volumes using T_c, P_c, and T_b as inputs, one of the correlations from Chap. 5 may be the best choice. It is difficult to compare vapor pressure deviations directly with Chap. 5 because a larger database was applied and the temperature range typically extended to the triple point, but it should be emphasized that the "predictions" in Chap. 5 require T_c, P_c, and T_b as inputs. That requirement is like the ESD EOS which takes T_c, P_c, and ω as inputs, but ESD accuracy in Table 6.8 is roughly 3%, compared to roughly 10% for the predictions in Chap. 5. To clarify this comparison, we repeated the evaluation of the Riedel method in Chap. 5 using the compounds from Appendix A and constraining $T > 0.4\ T_c$ and $P > 0.1$ kPa, as stipulated for the ESD EOS. This resulted in an average deviation of 2% for the Riedel method of Chap. 5. This finding highlights the importance of accurate critical properties and low-temperature vapor pressures when assessing these methods.

6.14 NOTATION

In many equations in this chapter, special constants or parameters are defined and usually denoted as $a, b, \ldots,$ $A, B, \ldots$. They are not defined in this section because they apply only to the specific equation and are not used anywhere else in the chapter.

A	molar Helmholtz energy, J mol^{-1}
b	cubic EOS molecular volume parameter
B_2	second virial coefficient, cm^3 mol^{-1}
C_v	heat capacity at constant volume, J mol^{-1} K^{-1}
C_p	heat capacity at constant pressure, J mol^{-1} K^{-1}
EOS	Equation of state
ΔH_b	enthalpy change of boiling, kJ mol^{-1}
k_B	Boltzmann's constant
M	molecular weight, g mol^{-1}
n^G	number of groups in a molecule
N	number of atoms in molecule
N_A	Avogadro's number, $= 6.022142 \times 10^{23}$ mol^{-1}
P	pressure, MPa
P^{vp}	vapor pressure, MPa
r_i	number of segments in a chain molecule
R	universal gas constant
T	temperature, K
V	total volume, cm^3
v	molar volume, cm^3 mol^{-1}
y_i	mole fraction
Z	compressibility factor, $= Pv/RT$

Greek

α	cubic EOS attractive energy parameter
β_c	critical exponent for liquid-vapor density difference
δ_c	critical isotherm exponent
ε/k_B	molecular scale attractive dispersion energy, K
μ	dipole moment
Δv	molar volume group contribution
ρ	molar density, mol·cm^{-3}
σ	spherical segment diameter, nm
ω	acentric factor

Superscripts

$\circ$	ideal gas
chem	chemical-physical (aka Association) contribution
L	liquid phase
V	vapor phase
res	residual property

Subscripts

att attractive perturbation
c critical property
liq liquid property
b boiling state
calc calculated value
D dimer in chemical theory
expt experimental value
M monomer in chemical theory
r reduced property
rep repulsive reference state for perturbation expansions

6.15 REFERENCES

1. W. Wagner, A. Pruss, *J. Phys. Chem. Ref. Data* **2002**, *31*, 387–535.
2. G. Soave, *Chem. Eng. Sci.* **1972**, *27*, 1197–1203.
3. D. Peng, D. B. Robinson, *Ind. Eng. Chem. Fund.* **1976**, *15*, 59–64.
4. J. Gross, G. Sadowski, *Ind. Eng. Chem. Res.* **2001**, *40*, 1244–1260.
5. J. R. Elliott, C. T. Lira, *Introductory Chemical Engineering Thermodynamics*, Prentice-Hall, Englewood Cliffs, NJ, **2012**.
6. A. Thompson, B. N. Taylor, "Guide for the Use of the International System of Units (SI), NIST Special Publication 811, 2008 Edition, U.S. Government Printing Office, Washington," can be found under https://www.nist.gov/pml/special-publication-811), **2008**.
7. B. I. Lee, M. G. Kesler, *AIChE J.* **1975**, *21*, 510–527.
8. J. V. Sengers, R. F. Kayser, C. J. Peters, H. J. White, *Equations of State for Fluids and Fluid Mixtures*, Elsevier, Amsterdam, **2000**.
9. G. M. Kontogeorgis, G. K. Folas, *Thermodynamic Models for Industrial Applications: From Classical and Advanced Mixing Rules to Association Theories*, Wiley, West Sussex, UK, **2010**.
10. J. D. van der Waals, *On the Continuity of the Gas and Liquid State*, Leiden, **1873**.
11. H. Kammerlingh-Onnes, *Konink. Akad. Weten. Amsterdam* **1902**, *5*, 125–127.
12. L. Boltzmann, *Nederlandse Akad. Wtensch.* **1899**, *7*, 484.
13. J. E. Mayer, M. G. Mayer, *Statistical Mechanics*, John Wiley & Sons, Ltd, New York, **1940**.
14. R. W. Zwanzig, *J. Chem. Phys.* **1954**, *22*, 1420.
15. D. A. McQuarrie, *Statistical Mechanics*, Harper and Row, New York, **1976**.
16. J.-P. Hansen, I. D. McDonald, *Theory of Simple Liquids, 2ed*, Elsevier, London, **1986**.
17. J. R. Elliott, A. J. Schultz, D. A. Kofke, *J. Chem. Phys.* **2015**, *143*, 114110.
18. A. E. Sherwood, J. M. Prausnitz, *J. Chem. Phys.* **1964**, *41*, 429–437.
19. J. R. Elliott, A. J. Schultz, D. A. Kofke, *J. Chem. Phys.* **2019**, *151*, 204501.
20. V. Diky, R. D. Chirico, C. D. Muzny, A. F. Kazakov, K. Kroenlein, J. W. Magee, I. Abdulagatov, J. W. Kang, R. Gani, M. Frenkel, *J. Chem. Inf. Model.* **2013**, *53*, 249–266.
21. M. Benedict, G. B. Webb, L. C. Rubin, *J. Chem. Phys.* **1942**, *10*, 747.
22. R. Span, E. Lemmon, W. Wagner, *J. Chem. Eng. Data* **2009**, *54*, 3141–3180.
23. J. R. Elliott, *J. Phys. Chem. B* **2021**, *125*, 4494–4500.
24. A. J. Schultz, D. A. Kofke, *J. Chem. Phys.* **2010**, *133*, 104101.
25. N. F. Carnahan, K. E. Starling, *J. Chem. Phys.* **1969**, *51*, 635–636.
26. R. Espíndola-Heredia, F. del Río, A. Malijevsky, J. Chem. Phys. 2009, 130, DOI 10.1063/1.3054361.
27. Yu. T. Pavlyukhin, *J. Struct. Chem.* **2012**, *53*, 476–486.
28. E. Hendriks, G. M. Kontogeorgis, R. Dohrn, J. C. de Hemptinne, I. G. Economou, L. F. Ẑilnik, V. Vesovic, *Ind. Eng. Chem. Res.* **2010**, *49*, 11131–11141.
29. W. G. Chapman, K. E. Gubbins, G. Jackson, M. Radosz, *Ind. Eng. Chem. Res.* **1990**, *29*, 1709–1721.
30. M. S. Wertheim, *J. Stat. Phys.* **1984**, *35*, 19–34.

31. E. Kiran, J. M. H. Levelt-Sengers, *Supercritical Fluids-Fundamentals for Application*, Kluwer Academic Publishers, Dordrecht, **1994**.
32. K. Kurzeja, T. Thielkes, W. Wagner, *Int. J. Thermophys.* **1999**, *20*, 531–561.
33. R. Span, W. Wagner, *Int. J. Thermophys.* **2003**, *24*, 1–39.
34. J. Gregorowicz, J. P. O'Connell, C. J. Peters, *Fluid Phase Equilib.* **1999**, *116*, 94–101.
35. R. Haghbakhsh, M. Konttorp, S. Raeissi, C. J. Peters, J. P. O'Connell, *J. Phys. Chem. B* **2014**, *118*, 14397–14409.
36. A. F. Ghobadi, J. R. Elliott, *J. Chem. Phys.* **2015**, *143*, 114107.
37. Ø. Wilhelmsen, A. Aasen, G. Skaugen, P. Aursand, A. Austegard, E. Aursand, M. A. Gjennestad, H. Lund, G. Linga, M. Hammer, *Ind. Eng. Chem. Res.* **2017**, *56*, 3503–3515.
38. I. H. Bell, B. K. Alpert, *Int. J. Thermophys.* **2021**, *42*, 75.
39. J.-C. de Hemptinne, P. Ungerer, *Fluid Phase Equilib.* **1995**, *106*, 81–109.
40. H. B. de Sant'Ana, P. Ungerer, J. C. de Hemptinne, *Fluid Phase Equilib.* **1999**, *154*, 193–204.
41. O. Redlich, J. N. S. Kwong, *Chem. Rev.* **1949**, *44*, 233–244.
42. R. W. Morris, E. A. Turek, *ACS Symp. Series* **1986**, *300*, 389.
43. P. Aursand, M. Aa, E. Aursand, M. Hammer, Ø. Wilhelmsen, *Fluid Phase Equilib.* **2017**, *436*, 98–112.
44. U. K. Deiters, *Fluid Phase Equilib.* **1999**, *161*, 205–219.
45. J. P. O'Connell, J. M. Haile, *Thermodynamics: Fundamentals for Applications*, Cambridge University Press, Cambridge, **2005**.
46. E. A. Mason, T. H. Spurling, *The Virial Equation of State*, Pergamon, New York, **1968**.
47. J. H. Dymond, E. B. Smith, *The Virial Coefficients of Pure Gases and Mixtures; A Critical Compilation*, Clarendon Press, Oxford, **1980**.
48. U. Setzmann, W. Wagner, *J. Phys. Chem. Ref. Data* **1991**, *20*, 1061–1155.
49. P. L. Chueh, J. M. Prausnitz, *AIChE J.* **1967**, *13*, 896–902.
50. H. C. van Ness, M. M. Abbott, *Classical Thermodynamics of Non-Electrolyte Solutions*, McGraw-Hill, New York, **1982**.
51. M. Frenkel, K. N. Marsh, *Virial Coefficients of Pure Gases and Mixtures*, Springer, **2002**.
52. C. Tsonopoulos, J. H. Dymond, *Fluid Phase Equilib.* **1997**, *133*, 11–34.
53. P. J. McElroy, J. Moser, *J. Chem. Thermo.* **1995**, *27*, 267–271.
54. M. Massucci, C. J. Wormald, *J. Chem. Thermo.* **1998**, *30*, 919–927.
55. M. M. Abbott, *Adv. Chem. Ser.* **1979**, *182*, 47.
56. J. R. Elliott, S. J. Suresh, M. D. Donohue, *Ind. Eng. Chem. Res.* **1990**, *29*, 1476–1485.
57. G. M. Kontogeorgis, R. Privat, J.-N. Jaubert, *J. Chem. Eng. Data* **2019**, *64*, 4619–4637.
58. Y. le Guennec, R. Privat, J.-N. Jaubert, *Fluid Phase Equilib.* **2016**, *429*, 301–312.
59. C. H. Twu, D. Bluck, J. R. Cunningham, J. E. Coon, *Fluid Phase Equilib.* **1991**, *69*, 33–50.
60. I. H. Bell, M. A. Satyro, E. W. Lemmon, *J. Chem. Eng. Data* **2018**, *63*, 2402–2409.
61. A. Pina-Martinez, Y. le Guennec, R. Privat, J.-N. Jaubert, P. M. Mathias, *J. Chem. Eng. Data* **2018**, *63*, 3980–3988.
62. J. Ely, M. Huber, *Int. Ref. Air Cond. Conf.* **1990**.
63. M. S. Zabaloy, J. H. Vera, *Ind. Eng. Chem. Res.* **1998**, *37*, 1591–1597.
64. M. S. Zabaloy, E. A. Brignole, *Fluid Phase Equilib.* **1997**, *140*, 87.
65. O. Pfohl, *Fluid Phase Equilib.* **1999**, *163*, 157.
66. M. Cismondi, J. Mollerup, *Fluid Phase Equilib.* **2005**, *232*, 74–89.
67. N. G. Tassin, S. B. R. Reartes, M. Cismondi, *J. Chem. Eng. Data* **2019**, *64*, 2093–2109.
68. M. J. Huron, J. Vidal, *Fluid Phase Equilib.* **1979**, *3*, 255.
69. C. J. Kedge, M. A. Trebble, *Fluid Phase Equilib.* **1999**, *158–60*, 219.
70. M. Benedict, G. B. Webb, L. C. Rubin, *J. Chem. Phys.* **1940**, *8*, 334.
71. J. F. Ely, H. J. M. Hanley, *Ind. Eng. Chem. Fundam.* **1981**, *20*, 323–332.
72. M. L. Huber, D. G. Friend, J. F. Ely, *Fluid Phase Equilib.* **1992**, *80*, 249–261.
73. M. J. Hwang, W. B. Whiting, *Ind. Eng. Chem. Res.* **1987**, *26*, 1758–1766.
74. B. A. Younglove, J. F. Ely, *J. Phys. Chem. Ref. Data* **1987**, *16*, 577–798.
75. M. E. Baltatu, R. A. Chong, M. L. Huber, A. Laesecke, *Int. J. Thermophys.* **1999**, *20*, 85–95.
76. M. L. Huber, J. F. Ely, *Int. J. Refrig.* **1994**, *17*, 18–31.
77. R. Span, W. Wagner, *Int. J. Thermophys.* **1997**, *18*, 1415–1443.
78. O. Kunz, W. Wagner, *J. Chem. Eng. Data* **2012**, *57*, 3032–3091.
79. S. P. Tan, H. Adidharma, M. Radosz, *Ind. Eng. Chem. Res.* **2008**, *47*, 8063–8082.
80. V. Papaioannou, T. Lafitte, C. Avendaño, C. S. Adjiman, G. Jackson, E. A. Müller, A. Galindo, *J. Chem. Phys.* **2014**, *140*, 1–29.

81. G. Sadowski, "TU Dortmund > Fakultät Bio- und Chemieingenieurwesen > Thermodynamik > Aktuelles > PC-SAFT > Download," can be found under https://www.th.bci.tu-dortmund.de/cms/de/Aktuelles/PC-SAFT/Download/index.html, **2020**.
82. R. Privat, R. Gani, J. Jaubert, *Fluid Phase Equilib.* **2010**, *295*, 76–92.
83. I. Polishuk, *Ind. Eng. Chem. Res.* **2014**, *53*, 14127–14141.
84. N. M. Alsaifi, M. Alkhater, H. Binous, I. al Aslani, Y. Alsunni, Z. G. Wang, *Ind. Eng. Chem. Res.* **2019**, *58*, 1382–1395.
85. W. G. Chapman, G. Jackson, K. E. Gubbins, *Mol. Phys.* **1988**, *65*, 1057.
86. R. A. Heidemann, J. M. Prausnitz, *Proc. Nat. Acad. Sci.* **1976**, *73*, 1773–1776.
87. M. L. Michelsen, E. M. Hendriks, *Fluid Phase Equilib.* **2001**, *180*, 165–174.
88. G. A. Mansoori, N. F. Carnahan, K. E. Starling, T. W. Leland, *J. Chem. Phys.* **1971**, *54*, 1523–1525.
89. J. R. Elliott, *Ind. Eng. Chem. Res.* **1996**, *35*, 1624–1629.
90. P. K. Jog, S. G. Sauer, J. Blaesing, W. G. Chapman, *Ind. Eng. Chem. Res.* **2001**, *40*, 4641–4648.
91. C. H. Twu, K. E. Gubbins, *Chem. Eng. Sci.* **1978**, *33*, 879.
92. D. Nguyen-Huynh, J. P. Passarello, P. Tobaly, J. C. de Hemptinne, *Fluid Phase Equilib.* **2008**, *264*, 62–75.
93. J. Vrabec, J. Gross, *J. Phys. Chem. B* **2008**, *112*, 51–60.
94. J. Gross, J. Vrabec, *AIChE J.* **2006**, *52*, 1194–1204.
95. A. Dominik, W. G. Chapman, M. Kleiner, G. Sadowski, *Ind. Eng. Chem. Res.* **2005**, *44*, 6928–6938.
96. S. H. Huang, M. Radosz, *Ind. Eng. Chem. Res.* **1990**, *29*, 2284–2294.
97. S. S. Chen, A. Kreglewski, *Ber. Buns. Phys. Chem.* **1977**, *81*, 1048–1052.
98. F. H. Stillinger, A. Rahman, *J. Chem. Phys.* **1974**, *60*, 1545–1557.
99. P. Vimalchand, G. D. Ikonomou, M. D. Donohue, *Fluid Phase Equilib.* **1988**, *43*, 121–135.
100. S. J. Suresh, J. R. Elliott, *Ind. Eng. Chem. Res.* **1992**, *31*, 2783–2794.
101. A. S. Puhala, J. R. Elliott, *Ind. Eng. Chem. Res.* **1993**, *32*, 3174–3179.
102. S. I. Sandler, K.-H. Lee, *Fluid Phase Equilib.* **1986**, *30*, 135–142.
103. K. E. Gubbins, *Fluid Phase Equilib.* **2016**, *416*, 3–17.
104. F. S. Emami, A. Vahid, J. R. Elliott, F. Feyzi, *Ind. Eng. Chem. Res.* **2008**, *47*, 8401–8411.
105. G. M. Kontogeorgis, E. C. Voutsas, I. v Yakoumis, D. P. Tassios, *Ind. Eng. Chem. Res.* **1996**, *35*, 4310.
106. G. M. Kontogeorgis, I. v. Yakoumis, H. Meijer, E. Hendriks, T. Moorwood, *Fluid Phase Equilib.* **1999**, *158–160*, 201–209.
107. F. J. Blas, L. F. Vega, *Mol. Phys.* **1997**, *92*, 135–150.
108. J. K. Johnson, J. A. Zollweg, K. E. Gubbins, *Mol. Phys.* **1993**, *78*, 591–618.
109. J. K. Johnson, E. A. Muller, K. E. Gubbins, *J. Phys. Chem.* **1994**, *98*, 6413–6419.
110. J. S. Andreu, L. F. Vega, *J Phys. Chem. C.* **2007**, *111*, 16028–16034.
111. M. Stuckenholz, E. A. Crespo, L. F. Vega, P. J. Carvalho, J. A. P. Coutinho, W. Schröer, J. Kiefer, B. Rathke, *J. Phys. Chem. B* **2018**, *122*, 6017–6032.
112. A. Gil-Villegas, A. Galindo, P. J. Whitehead, S. J. Mills, G. Jackson, A. N. Burgess, *J. Chem. Phys.* **1997**, *106*, 4168–4186.
113. T. Lafitte, A. Apostolakou, C. Avendano, A. Galindo, C. S. Adjiman, E. A. Müller, G. Jackson, *J. Chem. Phys.* **2013**, *139*, 154504.
114. T. Lafitte, M. M. Piñeiro, J. L. Daridon, D. Bessières, *J. Phys. Chem. B* **2007**, *111*, 3447–3461.
115. A. J. de Villiers, C. E. Schwarz, A. J. Burger, G. M. Kontogeorgis, *Fluid Phase Equilib.* **2013**, *338*, 1–15.
116. J. A. Barker, D. Henderson, *J. Chem. Phys.* **1967**, *47*, 4714–4721.
117. A. Mejía, E. A. Müller, G. Chaparro-Maldonado, *J. Chem. Inf. Model.* **2021**, *61*, 1244–1250.
118. A. Tihic, G. M. Kontogeorgis, N. von Solms, M. L. Michelsen, L. Constantinou, *Ind. Eng. Chem. Res.* **2008**, *47*, 5092–5101.
119. T. B. Nguyen, J. C. de Hemptinne, B. Creton, G. M. Kontogeorgis, *Ind. Eng. Chem. Res.* **2013**, *52*, 7014–7029.
120. D. NguyenHuynh, J. P. Passarello, J. C. de Hemptinne, P. Tobaly, *Fluid Phase Equilib.* **2011**, *307*, 142–159.
121. A. D. Sans, A. Vahid, J. R. Elliott, *J. Chem. Eng. Data* **2014**, *59*, 3069–3079.
122. J. R. Elliott, *Fluid Phase Equilib.* **2016**, *416*, 27–41.
123. A. F. Ghobadi, J. R. Elliott, *Fluid Phase Equilib.* **2011**, *306*, 57–66.
124. A. Lymperiadis, C. S. Adjiman, A. Galindo, G. Jackson, *J. Chem. Phys.* **2007**, *127*, 234903.
125. A. F. Ghobadi, J. R. Elliott, *J. Chem. Phys.* **2013**, *139*, 234104.
126. A. F. Ghobadi, J. R. Elliott, *J. Chem. Phys.* **2014**, *141*, 024708.
127. A. F. Ghobadi, J. R. Elliott, *.J Chem. Phys.* **2014**, *141*, 094708.
128. T. Lafitte, C. Avendano, V. Papaioannou, A. Galindo, C. S. Adjiman, G. Jackson, E. A. Müller, *Mol. Phys.* **2012**, *110*, 1189–1203.

129. P. Morgado, J. Barras, A. Galindo, G. Jackson, E. J. M. Filipe, *J. Chem. Eng. Data* **2020**, *65*, 5909–5919.
130. P. Morgado, O. Lobanova, E. A. Müller, G. Jackson, M. Almeida, E. J. M. Filipe, *Mol. Phys.* **2016**, *114*, 2597–2614.
131. J. D. Weeks, D. Chandler, H. C. Andersen, *J. Chem. Phys.* **1971**, *54*, 5237.
132. D. M. Heyes, H. Okumura, *J. Chem. Phys.* **2006**, *124*, 164507.
133. R. Roth, R. Evans, A. Lang, G. Kahl, *J. Phys.: Condens. Matter* **2002**, *14*, 12063–12078.
134. J. R. Elliott, T. E. Daubert, *Fluid Phase Equilib.* **1986**, *31*, 153–160.
135. T. van Westen, J. Gross, T. van Westen, J. Gross, *J. Chem. Phys.* **2017**, *147*, 014503.
136. Y. Zhang, W. G. Chapman, *Ind. Eng. Chem. Res.* **2018**, *57*, 1679–1688.
137. A. Haghmoradi, D. Ballal, W. A. Fouad, L. Wang, W. G. Chapman, *AIChE J.* **2021**, *67*, 17146.
138. B. D. Marshall, *AIChE J.* **2021**, *67*, e17342.
139. M. A. Anisimov, S. B. Kiselev, J. v. Sengers, S. Tang, *Physica A* **1992**, *188*, 487–525.
140. G. A. Chapela, J. S. Rowlinson, *J. Chem. Soc., Faraday Trans.* **1974**, *70*, 584–593.
141. S. B. Kiselev, J. F. Ely, *Ind. Eng. Chem. Res.* **1999**, *38*, 4993–5004.
142. S. B. B. Kiselev, J. F. F. Ely, J. R. Elliott, *Mol. Phys.* **2006**, *104*, 2545–2559.
143. J. R. Fox, *Fluid Phase Equilib.* **1983**, *14*, 45–53.
144. K. S. Pitzer, D. R. Schreiber, *Fluid Phase Equilib.* **1988**, *41*, 1–17.
145. G. F. Chou, J. M. Prausnitz, *AIChE J.* **1989**, *35*, 1487–1496.
146. M. D. Vine, C. J. Wormald, *J. Chem. Soc., Faraday Trans.* **1993**, *89*, 69–75.
147. R. Solimando, M. Rogalski, E. Neau, A. Péneloux, *Fluid Phase Equilib.* **1995**, *106*, 59–80.
148. L. Lue, J. M. Prausnitz, *AIChE J.* **1998**, *44*, 1455–1466.
149. F. Fornasiero, L. Lue, A. Bertucco, *AIChE J.* **1999**, *45*, 906–915.
150. M. A. Anisimov, A. A. Povodyrev, J. v Sengers, *Fluid Phase Equilib.* **1999**, *158*, 537–547.
151. A. K. Wyczalkowska, M. A. Anisimov, J. v Sengers, *Fluid Phase Equilib.* **1999**, *158*, 523–535.
152. S. Tang, J. v. Sengers, Z. Y. Chen, *Physica A* **1991**, *179*, 344–377.
153. P. Grenzheuser, J. Gmehling, *Fluid Phase Equilib.* **1986**, *25*, 1–29.
154. A. Haghmoradi, W. G. Chapman, *J. Chem. Phys.* **2019**, *150*, 174503.
155. M. P. Breil, G. M. Kontogeorgis, P. K. Behrens, M. L. Michelsen, *Ind. Eng. Chem. Res.* **2011**, *50*, 5795–5805.
156. S. O. Derawi, J. Zeuthen, M. L. Michelsen, E. H. Stenby, G. M. Kontogeorgis, *Fluid Phase Equilib.* **2004**, *225*, 107–113.
157. M. Kleiner, F. Tumakaka, G. Sadowski, H. Latz, M. Buback, *Fluid Phase Equilib.* **2006**, *241*, 113–123.
158. A. Vahid, J. R. Elliott, *AIChE J.* **2010**, *56*, 485.
159. J. Janeček, P. Paricaud, *J. Phys. Chem. B* **2012**, *116*, 7874–7882.
160. R. P. Sear, G. Jackson, *Phys. Rev. E* **1994**, *50*, 386.
161. J. Lee, H. Kim, *Fluid Phase Equilib.* **2001**, *190*, 47–59.
162. M. Thol, G. Rutkai, A. Köster, R. Lustig, R. Span, J. Vrabec, *J. Phys. Chem. Ref. Data* **2016**, *45*, 023101.
163. A. Aasen, M. Hammer, Å. Ervik, E. A. Müller, Ø. Wilhelmsen, *J. Chem. Phys.* **2019**, *151*, 064508.
164. B. Chen, J. I. Siepmann, *J. Phys. Chem. B* **1999**, *103*, 5370–5379.
165. M. G. Martin, J. I. Siepmann, *J. Phys. Chem. B* **1998**, *102*, 2569–2577.
166. R. A. Messerly, S. M. Razavi, M. R. Shirts, *J. Chem. Theory Comp.* **2018**, *14*, 3144–3162.
167. P. Ungerer, C. Nieto-Draghi, B. Rousseau, G. Ahunbay, V. Lachet, *J. Mol. Liq.* **2007**, *134*, 71–89.
168. C. Nieto-Draghi, G. Fayet, B. Creton, X. Rozanska, P. Rotureau, J. C. de Hemptinne, P. Ungerer, B. Rousseau, C. Adamo, *Chem. Rev.* **2015**, *115*, 13093–13164.
169. M. G. Ahunbay, S. Kranias, P. Ungerer, *Fluid Phase Equilib.* **2005**, *228–229*, 311–319.
170. A. Hemmen, J. Gross, *J. Phys. Chem. B* **2015**, *119*, 11695–11733.
171. J. J. Potoff, D. A. Bernard-Brunel, **2009**, *J. Phys. Chem. B.*, 14725–14731.
172. J. J. Nicolas, K. E. Gubbins, W. B. Streett, D. J. Tildesley, *Mol. Phys.* **1979**, *37*, 1429–1454.
173. J. Kolafa, I. Nezbeda, *Fluid Phase Equilib.* **1994**, *100*, 1–34.
174. M. Thol, G. Rutkai, A. Köster, S. Miroshnichenko, W. Wagner, J. Vrabec, R. Span, *Mol. Phys.* **2017**, *115*, 1166–1185.
175. J. R. Elliott, R. N. Natarajan, *Ind. Eng. Chem. Res.* **2002**, *41*, 1043–1050.
176. "Aspen Physical Property System V12 Physical Properties Methods Reference Manual," can be found under https://esupport.aspentech.com/S_Article?id=000097682, **2022**.
177. M. L. Huber, E. W. Lemmon, I. H. Bell, M. O. McLinden, *Ind. Eng. Chem. Res.* **2022**, DOI 10.1021/acs.iecr.2c01427.
178. J. R. Mick, M. S. Barhaghi, B. Jackman, L. Schwiebert, J. J. Potoff, *J. Chem. Eng. Data* **2017**, *62*, 1806–1818.
179. E. Stromsoe, H. G. Ronne, A. L. Lydersen, *J. Chem. Eng. Data* **1970**, *15*, 286–2990.

7

Thermodynamic Properties of Mixtures

7.1 SCOPE*

Methods are presented in this chapter to estimate the volumetric, enthalpic, and entropic properties and their derivatives for mixtures of gases and liquids as a function of temperature, pressure, and composition as expressed in mole, mass, or volume fractions. Formulas are also presented for computing the fugacity coefficients of several commonly used equations of state (EOSs). We organize the presentation in a slightly different manner than in the 5th edition. We begin with theoretical foundations in Sec. 7.3, including the segmental SAFT perspective that was initiated in Chap. 6 and an introduction to the activity models that underpin Gibbs excess (g^E) mixing rules. The familiar quadratic mixing rules and other results are inferred as simplifications of the general formalism. Perturbation theories are presented in Sec. 7.4 and exemplified by the PC-SAFT EOS (Gross and Sadowski, 2002), the ESD EOS (Elliott et al., 1990), and the SPEADMD model (Unlu et al., 2004). Section 7.5 describes the g^E mixing rules pioneered by Huron and Vidal (1979), exemplified by the tcPR-GE EOS. Mixing rules for multiparameter EOSs are presented in Sec. 7.6 as applied in REFPROP. Mixing rules for the virial equation are presented in Sec. 7.7. Section 7.8 summarizes our evaluation of mixing rules for thermodynamic properties. In general, our objective is to enable the reader to understand the conceptual foundations of the various models, and to implement models that perform well. We omit empirical methods that are focused on specific properties like densities at the bubble point pressure because little has changed since the 5th edition regarding these methods.

As in Chap. 6 for pure components, the discussion here is not comprehensive, and focuses on commonly applied formulations of models for mixtures. Chapter 6 includes application of EOS methods to calorimetric, free energy, and derivative properties, which are required for the phase equilibria modeling of pure fluids. Mixing rules are also used in the estimation of transport properties and the surface tension of mixtures as in Chaps. 10 through 13. Thus, the mixing rules developed here appear in much of the rest of the book.

Several reviews of this literature have been published over the years. A nice historical perspective on the evolution of empirical EOS models is given by Soave (1993). Kontogeorgis and Folas (2010) have reviewed the empirical approach more recently, although their primary emphasis pertains more to phase behavior, as we discuss in Chaps. 8 and 9. Reviews of SAFT implementations have been reported by Muller and Gubbins (2001), Tan et al. (2008), and Kontogeorgis et al. (2020).

*With special contributions from V. Diky.

7.2 MIXTURE PROPERTIES—GENERAL DISCUSSION

Typically, a model for a pure compound physical property contains parameters that are constant or temperature-dependent and found either by fitting to data or by the corresponding states principle (CSP). Thus, the EOS models of Chap. 6 express the relationship among the variables P, V, and T (or equivalently A, V, and T). To describe mixture properties, it is necessary to include composition dependence which adds considerable richness to the behavior, and thus complicates modeling. Therefore, a mixture EOS is an algebraic relation between P, V, T, and $\boldsymbol{x}$, where $\boldsymbol{x}$ is the set of $N_c - 1$ independent mole fractions of the mixture's N_c components.

The fundamentals of mixture modeling have not changed much since PGL5ed. Despite a proliferation of various SAFT models, the trade-offs remain between Gibbs excess mixing rules like Huron-Vidal (1979) and Wong and Sandler (1992) vs perturbation models like PC-SAFT (Gross and Sadowski, 2002). Models like the cubic plus association (CPA) model attempt to blend the benefits of both modeling approaches. Pertaining to Gibbs excess mixing rules, preferences have moved from Huron-Vidal to Wong-Sandler and back to Huron-Vidal. The Gibbs excess model evaluated here represents the Huron-Vidal mixing rules (tcPR-GE). Within the SAFT world, there has been some progress since 2001 integrating characterizations of polarity into the formalism. This occasionally helps to reduce the magnitudes of binary interaction parameters (BIPs) between components like CO_2 and ethane, since the unlike interaction between these types of species is reflected in the polarity mixing rules specified between these unlike components. Reducing the magnitude of BIPs should help to improve predictions when EOSs evolve to be competitive with group contribution (GC) models of the UNIFAC variety, but the SAFT models have not yet advanced to that level of competitiveness generally.

Predictive EOS models for mixtures still rely primarily on some form of the "solution of groups" concept, like UNIFAC or ASOG, as the basis for their mixture predictions. An exception is the FSAC-Phi EOS, which combines aspects of COSMO-RS/SAC and group contributions (Soares et al., 2019). Then the BIPs of the EOS model are tuned to match the preferred GC activity model, and the applications proceed. This is roughly the basis for the PSRK and USRK models of Gmehling and coworkers and for the predictive implementation.

This brings us back to the prevailing perspective on mixture properties: activity modeling. Activity models are primarily focused on modeling phase equilibria in mixtures, especially vapor-liquid equilibria (VLE) with more recent emphasis on liquid-liquid equilibria (LLE). We reserve detailed discussion of activity models and phase equilibria for Chaps. 8 and 9. Nevertheless, the characterizations of mixtures designed for activity models have implications for predictions of temperature and density derivatives through Maxwell's relations, which we consider in this chapter. Therefore, the conceptual basis of the UNIQUAC model is introduced here in Chap. 7 to facilitate understanding the bigger picture of mixing rules versus activity models, noting that UNIFAC is closely related to the UNIQUAC model.

Historically, derivations of activity models begin with analyses of the theoretical foundations of the energy and entropy of mixing. The resulting expressions are then combined to obtain the Gibbs energy and composition derivatives, yielding the desired activity coefficients. Thus, it is a little ironic that the accuracy of the models with respect to energy and entropy has been treated as something of an afterthought, while phase equilibrium accuracy dominates the discussion. The prevailing approach is that two extra parameters are added to characterize the temperature dependence of each binary group interaction and these parameters are tuned to simultaneously fit the activities and heats of mixing for a large database. This is the basis for the NIST (Kang et al., 2015) and Gmehling (cf. Constantinescu and Gmehling, 2016) "modified-UNIFAC" models and the NIST-KT (Kang et al., 2011) and Lyngby UNIFAC (Kang et al., 2002) models. We report on the predictions of these models for heats of mixing in this chapter while reporting on phase equilibrium predictions in Chaps. 8 and 9.

Progress of molecular simulations for mixtures has not generally shed light on the kinds of correlations relevant to *The Properties of Gases and Liquids*. Molecular simulations of phase equilibrium predictions have generally constituted proofs of principle that the force fields developed for pure fluids result in qualitatively reasonable predictions for mixtures. There has been little suggestion that direct molecular simulation methods can compete with GC methods on the scale of chemical and engineering applications. This is not to say that molecular simulation has not been applied to mixtures. In fact, most applications of molecular modeling pertain to mixture applications, but these applications are often specific and not germane to chemical and engineering processes. We do not consider molecular simulation models *per se* in the evaluation of mixture models of this edition, except to the extent that the SPEADMD model draws its basis from molecular simulation of the reference fluid with a transferable potential model. The evaluations here consider only the customized version of SPEADMD, where the transferable parameters have been optimized for vapor pressure and liquid density.

7.2.1 The Challenges of Multiphase Equilibria, the Critical Region, and High Pressures

The composition dependence of the properties of liquid mixtures is fundamentally different from that of a vapor or gas. The strongest effect on gaseous fluids is caused by changes in system density from changes in pressure; composition effects are usually of secondary importance, especially when mixing is at constant volume. Except at high pressures, vapors are not dissimilar to ideal gases and deviations from ideal mixing (Van Ness and Abbott, 1982) are small. However, changes in pressure on liquids make little difference to the properties, and volumetric, calorimetric, and phase variations at constant T and P are composition-dominated. The extreme example is at a composition near infinite dilution where the solute environment is both highly dense and far from the pure compound. These phenomena mean that comprehensive property models such as EOSs must show different composition/pressure connections at low and high densities.

This distinction between low- and high-density phases is most obviously seen in the liquid-liquid and vapor-liquid-liquid systems discussed in Chap. 9, especially at low concentrations of one or more species. The standard state for EOS models is the ideal gas where no phase separation can occur. As a result, when a model must quantitatively predict deviations from ideal liquid solution behavior from subtle differences between like and unlike interactions, complex relationships among the parameters are usually required. Several issues in these formulations, such as inconsistencies and invariance in multicomponent systems, are discussed in Sec. 7.3. However, the advent of Gibbs excess mixing rules has established useful expressions and computational tools for EOSs to yield reliable results for many complex systems.

Fluid properties in states near a mixture vapor-liquid critical point are less difficult to obtain from experiment than near-pure component critical points, since the fluid compressibility is no longer divergent. However, there are composition fluctuations that lead to both universalities and complex near-critical phase behavior (see the collected articles in Kiran and Levelt Sengers, 1994, reviews such as Rainwater, 1991, and work by Sengers and coworkers such as Jin et al., 1993 and Anisimov et al., 1999). Describing the crossover from nonclassical to classical behavior is even more difficult than for pure components because of the additional degrees of freedom from composition variations (see, for example, Kiselev and Friend, 1999). A brief discussion of such treatments was made in the 5th edition.

As for pure compounds, mixture EOS expressions are often inaccurate at very high pressures, both above and below the critical temperature. The forms of EOS PV isotherms at constant composition often do not correspond to those which best correlate data more generally. The effects of errors in $PVTy$ relations are carried through to all thermodynamic property variations because they involve derivatives, including those with respect to composition.

7.2.2 Composition Variations

Typically, composition is specified by some fractional weighting property such as the mole fraction, y_i, the mass fraction, w_i, or the superficial volume fraction, Φ_i

$$y_i = \frac{N_i}{N_{\mathrm{m}}} \tag{7.2-1}$$

$$w_i = \frac{y_i M_i}{M_{\mathrm{m}}} \tag{7.2-2}$$

$$\Phi_i = \frac{y_i V_i^\circ}{V_{\mathrm{m}}^\circ} \tag{7.2-3}$$

where N_i is the number of moles of component i, M_i is the molecular weight of component i, and V_i° is the pure-component molar volume of component i, typically taken at 25 °C, but it could be taken at the system temperature in some cases. The denominators in Eqs. (7.2-1)–(7.2-3) perform the normalization function by summing the numerators over all components. Thus, $N_m = \sum_{i=1}^{N_c} N_i$; $M_m = \sum_{i=1}^{N_c} y_i M_i$, $V_m^{\circ} = \sum_{i=1}^{N_c} y_i V_i^{\circ}$. Often, representation of the properties of mixtures are via plots versus the mole fraction of one of the components as expressed by theories. However, experimental data are often reported in mass fractions. Sometimes, asymmetries in these plots can be removed if the composition variable is the volume fraction, which allows simpler correlations.

There are two principal ways to extend the methods of Chap. 6 to include composition variations. One is based on molecular theory which adds contributions from terms that are associated with interactions or correlations of properties among pairs, trios, etc., of the components. The virial equation of state described in Chap. 6 and Sec. 7.7 is an example of this approach; the mixture expression contains pure-component and "cross" virial coefficients in a quadratic, cubic, or higher-order summation of mole fractions. The other approach to mixtures, which is more convenient, uses the same equation formulation for a mixture as for pure components, and composition dependence is included by making the parameters vary with composition. This leads to *mixing rules* as discussed in Secs. 7.4 to 7.6. Essentially all the models for pure components discussed in Chap. 6 have been extended to mixtures, often within the original articles; we will not cite them again in this chapter. Generally, a loose physical basis pertains to the mixing rule composition dependence and the parameter *combining rules* that bring together pairwise or higher-order contributions from the interactions of different components. The empiricism of this situation yields many possibilities, which must be evaluated individually for accuracy and reliability.

7.2.3 Mixing and Combining Rules

The concept of a *one-fluid* mixture is that, for fixed composition, the mixture properties and their variations with T and P are the same as some pure component with appropriate parameter values. To describe all pure components as well as mixtures, the mixture parameters must vary with composition so that if the composition is actually for a pure component, the model describes that substance. Though other variations are possible, a common *mixing rule* for a parameter Q is to have a quadratic dependence on mole fractions of the components in the phase, y_i

$$Q_m = \sum_{i=1}^{N_c} \sum_{j=1}^{N_c} y_i y_j Q_{ij} \tag{7.2-4}$$

In Eq. (7.2-7), the parameter value of pure component i would be Q_{ii}. The one-fluid concept traces its roots to van der Waals and is associated with quadratic mixing rules. We refer to this perspective henceforth as the "vdw 1-fluid" mixing rules.

Depending upon how the "interaction" parameter, Q_{ij} for $i \neq j$ is obtained from a *combining rule*, the resulting expression can be simple or complicated. For example, linear mixing rules arise from arithmetic and geometric combining rules.

$$\text{For } Q_{ij}^{(a)} = \frac{Q_{ii} + Q_{jj}}{2} \qquad Q_m = \sum_{i=1}^{N_c} y_i Q_{ii} \tag{7.2-5}$$

$$\text{For } Q_{ij}^{(g)} = (Q_{ii} Q_{jj})^{1/2} \qquad Q_m = \left(\sum_{i=1}^{N_c} y_i Q_{ii}^{1/2} \right)^2 \tag{7.2-6}$$

There is also the harmonic mean combining rule $Q_{ij}^{(h)} = 2/[1/Q_{ii}) + (1/Q_{jj})]$, but no linear relationship arises with it. The order of values for positive Q_{ii} and Q_{jj} is $Q_{ij}^{(h)} < Q_{ij}^{(g)} < Q_{ij}^{(a)}$.

However, these relationships are not adequate to describe most composition variations, especially those in liquids. Thus, it is common to use parameters that only apply to mixtures and whose values are obtained by fitting mixture data or from some correlation that involves several properties of the components involved. Examples include BIPs, which modify the combining rules. These parameters can appear in many different forms. They may be called simply binary parameters or interaction parameters, and they are often given symbols such as k_{ij} and l_{ij}.

The reader is cautioned to know precisely the definition of BIPs in a model of interest, since the same symbols may be used in other models, but with different definitions. Further, values may be listed for a specific formulation, but are likely to be inappropriate for another model even though the expressions are superficially the same. For instance, consider Eqs. (7.2-7) and (7.2-8). It is expected that in Eqs. (7.2-7) and (7.2-8) the values would be close to zero. Significant errors would be encountered if a value for the wrong parameter were used.

$$Q_{ij} = (Q_{ii}Q_{jj})^{1/2}(1 - k_{ij}) \tag{7.2-7}$$

$$Q_{ij} = \frac{Q_{ii} + Q_{jj}}{2}(l - l_{ij}) \tag{7.2-8}$$

These BIPs may be constants, functions of T, or even functions of the mixture density, $\rho = 1/V$; model formulations have been made with many different types.

The sensitivity of solution properties to BIPs can be very high or be negligible, depending upon the substances in the system and the property of interest. For example, mixture volumes from EOSs change very little with k_{ij} of Eq. (7.2-7) if it is used for the "a" parameter of a cubic EOS, but l_{ij} of Eq. (7.2-8) can be quite important for "b" parameter of a cubic EOS when the substances are very different in size or at high pressures (Arnaud et al., 1996). On the other hand, partial properties, such as fugacities, are very sensitive to k_{ij} in the "a" parameter and change little with l_{ij} in the "b" parameter.

In addition to one-fluid mixing rules for EOSs, recent research has generated many different ways to connect EOS mixture parameters to liquid properties such as excess Gibbs energies. These are described in detail in Sec. 7.5.

7.3 THEORY OF MIXTURE MODELING

The energy equation for mixtures provides a convenient starting point to discuss mixture modeling. This is a simple intuitive equation that connects the intermolecular potential, $u(r)$, to the residual internal energy, u^{res}. For pure fluids, it is given by (cf. Elliott and Lira, 2012, Chap. 7)

$$\frac{u^{\text{res}}}{RT} = N_{\text{A}} \int_0^{\infty} \frac{u(r)\rho(r)}{2k_{\text{B}}T} 4\pi r^2 dr \tag{7.3-1}$$

where $\rho(r)$ is the local density around the particle of interest. The factor of 2 in the denominator is necessary to avoid double counting the pairwise energy when counting particle j around i and particle i around j. The local density is easily understood in terms of the coordination number or number of neighboring particles, or, alternatively, the radial distribution function (Elliott, 1993)

$$N_{\lambda} = \int_0^{\lambda^R} \rho(r) 4\pi r^2 dr = \rho \int_0^{\lambda^R} g(r) 4\pi r^2 dr \tag{7.3-2}$$

where $g(r)$ is the radial distribution function. For example, $N_{\lambda} = 8$ for a body-centered-cubic (bcc) crystal when $\lambda^R = \sigma^{\text{ss}}$ and $N_{\lambda} = 14$ for a bcc crystal when $\lambda^R = \sigma^{\text{ss}}\sqrt{3}$, where σ^{ss} is the diameter of a spherical site. For a crystal, $g(r)$ resembles a sequence of spikes, but it is more amorphous for a fluid, oscillating around then decaying to 1 at large r. Elliott and Lira (2012, Chapter 12) provide additional introductory background on the radial distribution function and energy equation.

When extended to mixtures, the energy equation becomes

$$\frac{u^{\text{res}}}{RT} = \sum_i \sum_j x_i x_j \frac{N_A \rho}{2k_B T} \int_0^\infty u_{ij}(r) g_{ij}(r) 4\pi r^2 dr \tag{7.3-3}$$

Noting that $u_{ij}(r)$ is proportional to its minimum value, $-\varepsilon_{ij}^{\text{ss}}$, and $r/\sigma_{ij}^{\text{ss}}$ defines a dimensionless scale for distance, the energy equation can be rewritten as

$$\frac{u^{\text{res}}}{RT} = -\frac{N_A \rho}{2} \sum_i \sum_j x_i x_j \frac{\varepsilon_{ij}^{\text{ss}} \left(\sigma_{ij}^{\text{ss}}\right)^3}{k_B T} \int_0^\infty \frac{u_{ij}(r)}{\varepsilon_{ij}^{\text{ss}}} g_{ij}(r) 4\pi \frac{r^2 dr}{\left(\sigma_{ij}^{\text{ss}}\right)^3} \tag{7.3-4}$$

Various approximations of the dimensionless integral can be defined by $a_{ij}(T,\rho,\boldsymbol{x})$,

$$\frac{u^{\text{res}}}{RT} = \frac{-\rho}{RT} \sum_i \sum_j x_i x_j a_{ij} \equiv -\frac{a_m \rho}{RT} \tag{7.3-5}$$

where

$$a_{ij} \equiv \frac{N_A^2 \varepsilon_{ij}^{\text{ss}} \left(\sigma_{ij}^{\text{ss}}\right)^3}{2} \int_0^\infty \frac{u_{ij}(r)}{\varepsilon_{ij}^{\text{ss}}} g_{ij}(r) 4\pi \frac{r^2 dr}{\left(\sigma_{ij}^{\text{ss}}\right)^3} \tag{7.3-6}$$

If $a_{ij}(T,\rho,\boldsymbol{x})$ is independent of $\boldsymbol{x}$, the well-known quadratic mixing rule is immediately recognized.

7.3.1 Mixing Rules for Cubic EOSs and Inference of van der Waals Activity Models

The van der Waals EOS is obtained from Eq. (7.3-5) by integrating the internal energy to obtain the Helmholtz energy and adding $-\ln(1-b\rho)$ as the athermal integration coefficient. When $a_{ij}(T,\rho,\boldsymbol{x}) = a_{ij}(T,\rho)$, i.e., a_{ij} is independent of $\boldsymbol{x}$, the most common adaptation of Eq. (7.3-5) is:

$$a_m = \sum_i \sum_j x_i x_j a_{ij} \tag{7.3-7}$$

The strict consistency of this assertion with Eq. (7.3-5) is not generally considered, however, noting how various temperature dependencies are implemented. The SRK (Soave, 1972), PR (Peng and Robinson, 1976), and PPR78 (Jaubert and Mutelet, 2004) models all apply Eq. (7.3-7). Other EOSs, and activity models, simply comprise different assumptions about $a_{ij}(T,\rho,\boldsymbol{x})$, all of which would have Eq. (7.3-5) as a logical starting point for describing their mixture behavior.

7.3.2 Introduction to UNIQUAC as a Basis for Gibbs Excess Mixing Rules

Many details pertaining to activity models are discussed in Sec. 8.7, but Eq. (7.3-5) suffices as a basis for most familiar models. It is necessary to briefly introduce activity models here because some EOSs formulate their mixing rules based on popular activity models, as detailed in Sec. 7.5. In any activity model, the expression for g^E is well known. Given g^E, it is feasible to equate $g^E \approx a^E$ for an EOS in a liquid state and apply the inferred mixing rule to the EOS at any state. Further details are given in Sec. 7.5.

Maurer and Prausnitz (1978) show how the UNIQUAC model can be derived from Eq. (7.3-5). We adapt that presentation here. A more introductory adaptation is given in Chap. 13 of Elliott and Lira (2012). In addition to the excess internal energy, u^E, UNIQUAC incorporates an athermal term to represent the excess entropy, s_{GSA}^E, described by Guggenheim (1952) and Staverman (1950) to account for branching as well as simple size differences. (We discuss the accuracy of that term relative to molecular simulation in Sec. 7.4.)

$$\left(\frac{a^E_{\text{UNI}}}{RT}\right)\bigg|_{T=\infty} = -\frac{s^E_{\text{GSA}}}{R} = -\frac{s^E_{\text{FH}}}{R} + \frac{N^C_{\text{ref}}}{2}\sum x_i Q_i \ln\left(\frac{\Theta_i}{\Phi_i}\right) \tag{7.3-8}$$

$$-\frac{s^E_{\text{FH}}}{R} = \sum x_i \ln\left(\frac{\Phi_i}{x_i}\right)$$

where s^E_{FH} is the Flory-Huggins (1941) estimate of the excess entropy, Θ_i is the surface area fraction, and Φ_i is the volume fraction, N^C_{ref} is the coordination number of a reference site, and Q_i represents the molecular surface area relative to the reference site. Specifically, UNIQUAC takes the molecular fragment "CH3—" as its "reference site." Molecular volumes and surface areas are computed using group temperature-independent contributions in the UNIQUAC method.

Maurer and Prausnitz (1978) also define a specific composition dependence to $a_{ij}(T,\rho,\mathbf{x})$ in the form of "local composition theory." A moment's thought about the meaning of g_{ij} in terms of neighboring atoms reveals that "local compositions" are implicit in that meaning, but "local composition theory" has taken on the connotative meaning of a particular conjecture by Wilson (1964). The approach begins with defining local compositions as,

$$N_{ij} \equiv \int_0^{\lambda_{ij}^R} g_{ij}(r) 4\pi r^2 dr \tag{7.3-9}$$

$$x_{ij} \equiv \frac{N_{ij}}{\sum_i N_{ij}} \equiv \frac{N_{ij}}{N_j^C} \tag{7.3-10}$$

where the N_j^C defines a coordination number for the jth particle. In local composition theory, the second subscript of N_{ij} corresponds conceptually to the central particle, so the meaning of N_{ij} is the number of "i" particles around j particles. Obviously, as particle j grows larger relative to i particles, more particles can fit around it. These local compositions can be conveniently related to the bulk compositions in terms of weighting factors that we call Ω_{ij}, such that

$$\frac{x_{ij}}{x_{jj}} \equiv \frac{x_i}{x_j}\Omega_{ij} \tag{7.3-11}$$

where Ω_{ij} is rigorously defined in terms of Eq. (7.3-11). The energy equation can then be rearranged in terms of these local compositions (e.g., Elliott and Lira, 2012, Eq. [13.16]),

$$\frac{u^E}{RT} = \frac{N_A}{2}\left[\frac{x_1 x_2 \Omega_{21} N_1^C(\varepsilon_{21} - \varepsilon_{11})}{x_1 + x_2\Omega_{21}} + \frac{x_2 x_1 \Omega_{12} N_2^C(\varepsilon_{12} - \varepsilon_{22})}{x_2 + x_1\Omega_{12}}\right] \tag{7.3-12}$$

where ε_{ij} is the well depth if a square well potential model is assumed, or an average potential energy defined by the integral mean value for any other potential. So far, this derivation comprises definitions and rearrangements that do not involve empiricism in U^E. The defining characteristic of "local composition theory" is the bold assertion first articulated by Wilson (1964) that

$$\Omega_{ij} = \frac{q_i}{q_j}\exp\left(-\frac{A_{ij}}{RT}\right) \tag{7.3-13}$$

where q_i, q_j, N_j^C, and A_{ij} are constants related to the $\{\sigma^{ss}\}$ and $\{\varepsilon^{ss}\}$ of the components. Then,

$$\frac{a^E}{RT} = \int_\infty^T \frac{-u^E}{RT}\frac{dT}{T} + \left(\frac{a^E}{RT}\right)_{|\infty} = -\sum_i x_i \ln\left(\sum_j \Theta_j \Omega_{ij}\right) + \left(\frac{a^E}{RT}\right)\bigg|_{T=\infty} \tag{7.3-14}$$

where $\Theta_j = x_j q_j / \Sigma x_i q_i$ and $\Omega_{ii} = 1$. Specific assumptions for $a^E_{|\infty}$, q_i, q_j, N_j^C, and A_{ij} yield specific activity models. For example, Wilson's activity model is obtained from $q_i = V_i^\circ$, $A_{ji} = N_j^C(\varepsilon_{ij} - \varepsilon_{jj})/2$, and $a^E_{|\infty} = -Ts^E_{\text{FH}}$.

Substituting Eq. (7.3-13) into Eq. (7.3-14) and taking the derivative verifies that Eq. (7.3-12) is recovered. The UNIQUAC model is obtained from $q_i = Q_i$, $A_{ji} = N_j^{\mathrm{C}}(\varepsilon_{ij} - \varepsilon_{jj})/2$, and $a^E|_\infty = -Ts^E_{\mathrm{GSA}}$. Abrams and Prausnitz (1975) show how NRTL and other familiar activity models can be derived from the UNIQUAC model subject to specific assumptions.

In summary, the theoretical foundations of most mixture modeling can be traced to the energy equation for mixtures, Eq. (7.3-3). The current state of the art regarding mixture modeling may occasionally seem perplexing, especially when empiricism comes into play, but the theoretical foundations are not really all that complicated.

7.3.3 Thermodynamic Consistency and Invariance

A subject of importance in mixing rules is that of *invariance.* Thermodynamics requires that some properties should not change when another property or parameters vary. An example is the volume translation of a cubic equation of state that should not change the vapor pressure (Huron and Vidal, 1979). Another is that if one of the components of a mixture is divided into two identical but distinct subcomponents with the same characteristic properties, the mixture properties should not change (Michelsen and Kistenmacher, 1990). Finally, if a mixing rule has multiples of more than two mole fractions in a double summation that are not normalized, as the number of components increases, the importance of the term decreases—the so-called "dilution" effect. Michelsen and Kistenmacher (1990) first noted the mixture issues by pointing out that the mixing rule of Schwarzentruber and Renon (1989) does not meet these requirements. Neither does the rule of Panagiotopoulos and Reid (1986). In current practice, the GERG-2008 and hence the REFPROP mixing rules do not comply (Kunz and Wagner, 2012). There are several articles that discuss the subject of invariance in detail (Mathias et al., 1991; Leibovici, 1993; Brandani and Brandani, 1996; Zabaloy and Vera, 1996) and list mixing rules that do not meet thermodynamic constraints. In addition to mathematical errors, programming errors can also be made. Mollerup and Michelsen (1992) describe useful relationships to numerically check code for errors. Brandani and Brandani (1996) also point out an inconsistency that arises for infinite pressure matching.

7.4 PERTURBATION MODELS

Thermodynamic perturbation theory (TPT) was introduced in Chap. 6, including site-based contributions for polarity and chemical-physical interactions that are straightforwardly extended to mixtures at the molecular level. The multiple summations in those expressions are consistent with the energy equation for mixtures (Eq. 7.3-3). Briefly, perturbation models take the form

$$\frac{a^{\mathrm{res}}}{RT} = \frac{a^{\mathrm{rep}}}{RT} + \frac{a^{\mathrm{att}}}{RT} + \frac{a^{\mathrm{chain}}}{RT} + \frac{a^{\mathrm{polar}}}{RT} + \frac{a^{\mathrm{chem}}}{RT} \tag{7.4-1}$$

$$\frac{a^{\mathrm{att}}}{RT} = \frac{A_1}{T} + \frac{A_2}{T^2} + \frac{A_3}{T^3} + \cdots \tag{7.4-2}$$

In this chapter, we focus on expressions for the reference contribution and the dispersion contributions. The chemical-physical contributions are briefly summarized in their extended form. The dipole-quadrupole interaction for PC-SAFT is provided here to supplement the presentation in Chap. 6, but derivatives of the polar contributions are relegated to the original literature on these contributions.

7.4.1 Repulsive Contribution for SAFT Models

The most common mixture reference is that for hard spheres by Boublik (1970) and Mansoori et al. (1971) (BMCSL), which is typically written in terms of the compressibility factor. As with pure components, this idea has been expanded to deal with nonspherical and flexible molecules. Alternatives are presented in PGL5ed, but the prevailing approach is that of Chapman et al. (1990), based on the SAFT perspective (see also Gross and Sadowski, 2002).

$$Z^{rep} = Z^{HC} = m_m Z^{HS} - \sum_i x_i(m_i - 1)\rho \frac{\partial \ln\left(g_{ii}^{HS}\right)}{\partial \rho} \tag{7.4-3}$$

where $m_m = \Sigma x_i m_i$. This factor of m_m multiplying the hard-sphere value exemplifies the segmental SAFT perspective, with

$$Z^{HS} = \frac{\xi_3}{(1-\xi_3)} + \frac{3\xi_1\xi_2}{\xi_0(1-\xi_3)^2} + \frac{\xi_2^3(3-\xi_3)}{\xi_0(1-\xi_3)^3} \tag{7.4-4}$$

where the covolumes are

$$\xi_j = \frac{\pi\rho}{6}\sum_{i=1}^{n} y_i m_i (d_i^{ehs})^j \qquad j = 0, 1, 2, 3 \tag{7.4-5}$$

with d_i being the equivalent hard sphere diameter of species i. The general expression for g_{ij}^{HS} is

$$g_{ij}^{HS} = \frac{1}{(1-\xi_3)} + \left(\frac{d_i^{ehs}d_j^{ehs}}{d_i^{ehs}+d_j^{ehs}}\right)\frac{3\xi_2}{\xi_0(1-\xi_3)^2} + \left(\frac{d_i^{ehs}d_j^{ehs}}{d_i^{ehs}+d_j^{ehs}}\right)^2\frac{2\xi_2^2}{\xi_0(1-\xi_3)^3} \tag{7.4-6}$$

where g_{ii}^{HS} is obtained by setting $d_j^{ehs} = d_i^{ehs}$. The BMCSL expression is recovered when $m_i = 1$.

Note that these mixing rules for the reference fluid have important implications about the excess entropy of mixing, S^E. Since $A = U - TS$, and $U = 0$ for hard chain (aka athermal) systems, $A^E/RT = -S^E/R$. Therefore, the integral of Z^{HC} to obtain A^{res} is directly related to S^E. At constant packing fraction, the SAFT mixing rules indicate negative deviations from ideality, especially for asymmetric mixtures, as shown in Fig. 7.1. The figure is plotted as $-S^E$ to clarify that negative values in this plot correspond to negative deviations of A^E from ideality. Mixing trends for Z^{HC} are like those for A^E.

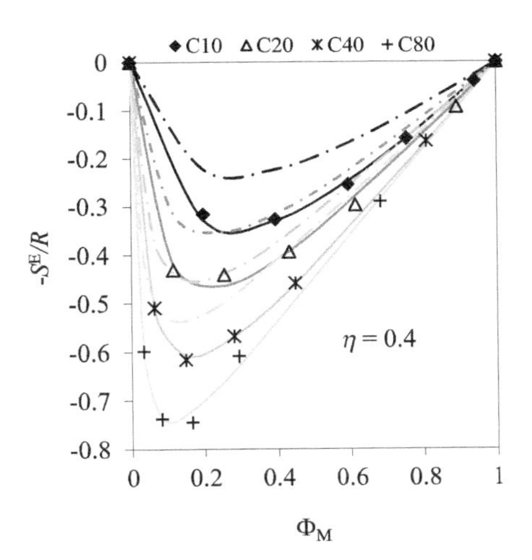

FIGURE 7.1
Athermal excess entropy of mixing for methane (M) + "polyethylene" (P) at a constant packing fraction of 0.4. The legend indicates the number of carbons in each polyethylene chain. Points refer to molecular simulations. The dash-dot lines refer to PC-SAFT and the solid lines represent the SPEADMD mixing rule (Vahid et al., 2014).

Simplified versions of these mixing rules have been explored. For example, the ESD EOS assumes

$$Z^{HC} = \frac{4c_m\xi_3}{(1-1.9\xi_3)} \tag{7.4-7}$$

where $c_m = \Sigma x_i c_i$ and the simplified PC-SAFT EOS (von Solms et al., 2006) applies,

$$Z^{HC} = \frac{4m_m\xi_3(1-\xi_3/2)}{(1-\xi_3)^3} - \frac{(m_m-1)\xi_3(5-2\xi_3)}{(2-\xi_3)(1-\xi_3)} \tag{7.4-8}$$

Noting that ξ_3 is the packing fraction, commonly denoted by η^P, both models give zero deviation from ideality at constant packing fraction, making them equivalent to the van der Waals EOS in this regard, as is the EPPR78 model (Xu et al., 2017). Positive values for l_{ij} in Eq. (7.2-8) would also give negative deviations from ideality, providing phenomenological justification for empirical analyses that include l_{ij}, as the CPA model does in some instances.

The discussion so far has omitted details of temperature dependence in the effective hard sphere diameter, d^{ehs}. For the ESD, SPEADMD, tcPR, and CPA models, d^{ehs} is treated as constant for most compounds. Hydrogen and helium are occasional exceptions. On the other hand, most SAFT implementations treat d^{ehs} as temperature-dependent, usually in the form,

$$\frac{d^{ehs}}{\sigma^{ss}} = \left[1 - 0.12\exp(-\beta\varepsilon^{ss})\right] \tag{7.4-9}$$

For PC-SAFT, $\varepsilon^{ss}/k \approx 250$ K for alkanes with 10 or more carbons, in which case this function is close to 0.99 at room temperature, 0.97 at 500 K, and 0.88 at infinite temperature. At most temperatures of practical interest, this temperature variation has a small impact on the excess entropy.

7.4.2 Attractive Contributions for PC-SAFT and ESD

The perturbation terms, or those which consider the attractive forces between the molecules, have ranged from very simple to extremely complex. For example, the simplest form is that of van der Waals (1890) which in terms of the Helmholtz energy is

$$\left[a^{res}(T, V, \boldsymbol{x})/RT\right]_{att}^{(vdW)} = -a_m\rho/RT \tag{7.4-10}$$

and which leads to an attractive contribution to the compressibility factor of

$$Z_{att}^{(vdW)} = a_m\rho/RT \tag{7.4-11}$$

Here, the parameter a_m is usually a function of T and $\boldsymbol{x}$ and obtained with mixing and combining rules such as the quadratic mixing rules discussed in relation to Eq. (7.3-5). This form would be appropriate for simple fluids, though it has also been used with a variety of expressions for the repulsive contribution. Several alternative forms have been discussed in the 5th edition, but they generally follow a similar form to the mixing rules exemplified by PC-SAFT.

$$\frac{A_1}{RT} = -2\pi\rho I_1(\eta^P, m_m)\sum_i\sum_j x_i x_j m_i m_j \frac{\varepsilon_{ij}^{ss}}{k_B T}\left(\sigma_{ij}^{ss}\right)^3 \tag{7.4-12}$$

$$\frac{A_2}{RT} = \frac{-\pi\rho m_m I_2(\eta^P, m_m)\sum_i\sum_j x_i x_j m_i m_j\left(\dfrac{\varepsilon_{ij}^{ss}}{k_B T}\right)^2\left(\sigma_{ij}^{ss}\right)^3}{\left(1 + Z^{HC} + \rho\dfrac{\partial Z^{HC}}{\partial\rho}\right)} \tag{7.4-13}$$

where $\sigma_{ij}^{ss} = (\sigma_{ii}^{ss} + \sigma_{jj}^{ss})/2$ and $\varepsilon_{ij}^{ss} = (\varepsilon_{ii}^{ss}\varepsilon_{jj}^{ss})^{1/2}(1 - k_{ij})$. The denominator of Eq. (7.4-13) is evaluated at η^P and m_m like I_1 and I_2. The GC-PPC-SAFT model applies the same mixing and combining rules.

The mixing rules for the ESD EOS are similarly based on quadratic mixing.

$$(qY\eta^{P})_{m} = \rho\sum_{i}\sum_{j} x_i x_j Y_{ij}(q_i b_j + q_j b_i)/2 \tag{7.4-14}$$

$$(Y\eta^{P})_{m} = \rho\sum_{i} x_i Y_i b_i \tag{7.4-15}$$

$$Y_{ij} = \exp(\varepsilon_{ij}/kT) - 1.0617;\ \varepsilon_{ij} = (\varepsilon_i\,\varepsilon_j)^{1/2}(1 - k_{ij});\ q_m = \Sigma x_i q_i \tag{7.4-16}$$

7.4.3 Athermal Mixing Rules Determined by Molecular Simulation

The mixing rules for SPEADMD basically follow the usual vdW 1-fluid approach, but they incorporate results from molecular simulations of mixtures. The SPEADMD eos is given by

$$\frac{a^{\text{res}}}{RT} = \frac{a_0}{RT} + \frac{A_1}{T} + \frac{A_2}{T^2} + \frac{a^{\text{chem}}}{RT} \tag{7.4-17}$$

Remarkably, the vdW 1-fluid perspective appears to characterize the available simulation results surprisingly well. Possibly, this is an indication that we do not yet have sufficiently precise simulation results to distinguish the deviations from 1-fluid behavior. The mixture simulations for SPEADMD are limited to hard reference potentials, so they only address the a_0, A_1 and A_2 terms, and say nothing about polarity contributions.

We alluded to simulation results for S^E in the discussion of Fig. 7.1. These results are derived from analysis of A_0 which is computed by

$$\frac{a_0}{RT} = \int_0^{\eta^P} Z_0 - 1\frac{d\eta^P}{\eta^P} \tag{7.4-18}$$

The process is straightforward but tedious, requiring simulations at many compositions and packing fractions, as represented in Fig. 7.2a. After integrating with respect to η^P at each composition, the a_0 values are tabulated and $a^E/RT = -s^E/R$ values are plotted in Fig. 7.2b.

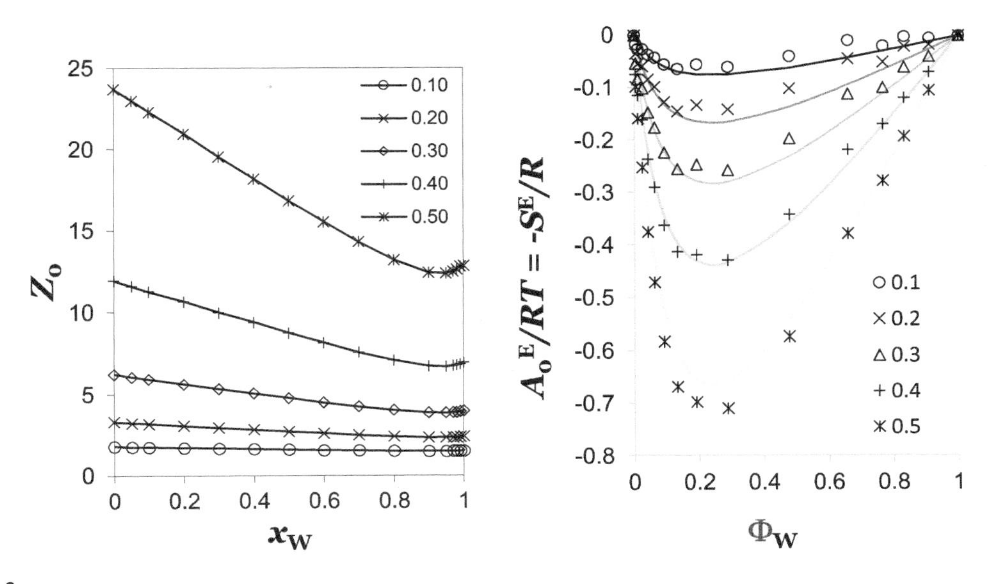

FIGURE 7.2
Reference contributions for the n-octanol + water system. x_w represents the mole fraction of water. Five sample packing fractions are listed in the legend. (a) Compressibility factor. (b) Excess entropy. Symbols are simulation results. Curves represent the correlation (Vahid et al., 2014).

The curves in Fig. 7.2b are generated by application of the vdw 1-fluid approach.

$$\frac{a_0^E}{RT} = \frac{\sum_i \sum_j x_i x_j a_{0,ij} (b_i b_j)^{1/2}}{RT \sum_i x_i b_i} - \sum_i x_i \frac{a_{0,i}}{RT} = \frac{-s_0^{\mathrm{E}}}{R} \tag{7.4-19}$$

where $a_{0,ij} = (a_{0,i}\, a_{0,j})^{1/2}(1 - k_{ij}^{S})$ and $\Phi_i^b = x_i b_i / \Sigma x_i b_i$; and

$$k_{ij}^{\mathrm{S}} = k_{ij}^{\mathrm{S}^\circ} \left(\delta_i^{\mathrm{S}^\circ} - \delta_j^{\mathrm{S}^\circ}\right)^2 \frac{(b_i + b_j)}{2 \cdot 298.15 \cdot R} \tag{7.4-20}$$

where $k_{ij}^{\mathrm{S}^\circ} = -0.016 \cdot 18/b_{\mathrm{avg}}$; $b_{\mathrm{avg}} = \Sigma b_i / N_c$; $\delta_i^{\mathrm{S}^\circ} = (298.15 \cdot R \cdot A_{0,i}/b_i)^{1/2}$ evaluated at $\eta^{\mathrm{P}} = 0.4$. The entropic solubility parameter, $\delta_i^{\mathrm{S}^\circ}$, characterizes the entropy density for a given molecule, defined to make it comparable in scale to the usual (energetic) solubility parameter, δ_i. Large molecules tend to have lower entropy density. The quantity $k_{ij}^{\mathrm{S}^\circ}$ is usually quite small. Noting that $k_{ij}^{\mathrm{S}^\circ} \to 0$ if polymers are present, $k_{ij}^{S} \to 0$ in that case. Equations (7.4-19) and (7.4-20) comprise the correlation represented in Fig. 7.2b. If simulation results indicate significant deviations from Eq. (7.4-20), then k_{ij}^{S} can be treated as an adjustable parameter. For completeness, we note that

$$Z_0 = \frac{\sum_i \sum_j x_i x_j Z_{0,ij} (b_i b_j)^{1/2}}{\sum_i x_i b_i}; \; Z_{0,ij} = \frac{(Z_{0,i}\, a_{0,j} + Z_{0,j}\, a_{0,i})}{2(a_{0,i}\, a_{0,j})^{1/2}} \left(1 - k_{ij}^{\mathrm{S}}\right) \tag{7.4-21}$$

While Fig. 7.2 refers to mixing with a linear chain, Vahid et al. (2014) also considered branched chains and rings as well as a range of degrees of polymerization. The correlation of Eqs. (7.4-19) and (7.4-20) was reasonably accurate in all cases. Figure 7.3 illustrates the trends for ring and branched molecules, where k_{ij}° was treated as an adjustable parameter. Note that the correlation of Eq. (7.4-20) leads to significant percentage deviations when the magnitude of S_0^E is small, but this has little effect on property predictions because the entire effect is small. The systems in Fig. 7.3 exhibit moderate nonideality, and deviations from Eq. (7.4-20) probably reflect influences of rings and branching in these systems.

It would be expected that nonideality should increase as the size ratio of polymer to solvent increases. Mixtures with large size ratios are often referred to as "asymmetric." Figure 7.4 shows that size matters, but

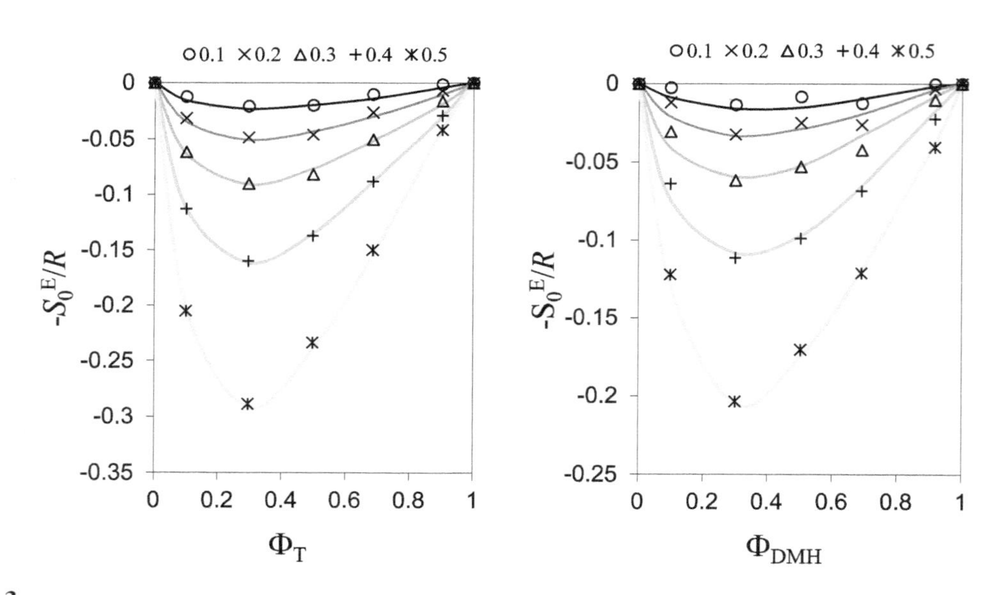

FIGURE 7.3
Excess entropy for ringed and branched molecules. (a) toluene+polystyrene(C40). (b) 2,4-dimethylhexane(DMH)+polypro pylene(C39). Points refer to molecular simulations. Curves refer to Eq. (7.4-20).

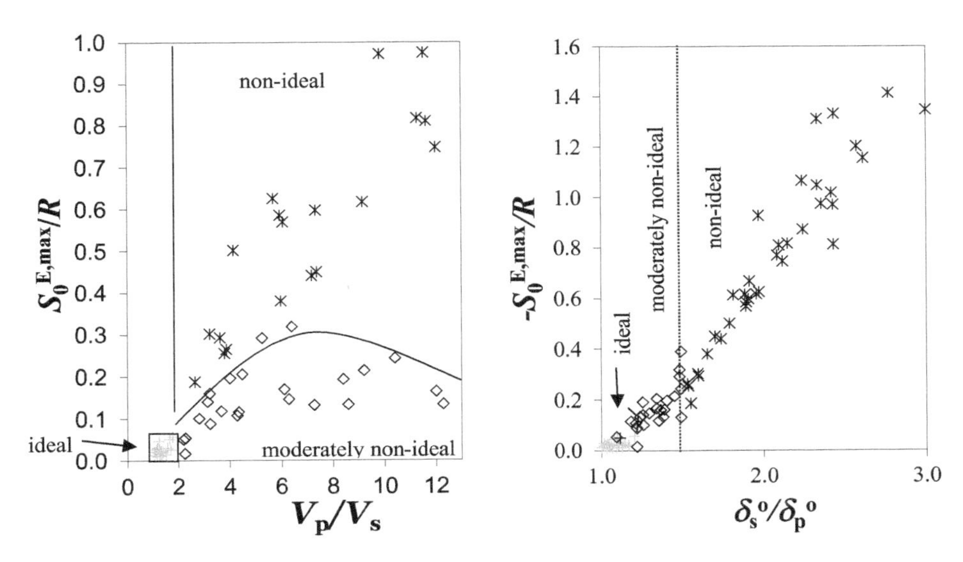

FIGURE 7.4
Trends in the magnitude of the athermal excess entropy from molecular simulation. (a) Compared to polymer/solvent volume ratio, V_p/V_s. (b) Compared to solubility parameter ratio, $\delta_s^\circ/\delta_p^\circ$. Symbols were inferred after smoothing with Eq. (7.4-19).

further consideration is required. A more consistent trend is observed when plotting the maximum magnitude of athermal excess entropy versus entropic solubility parameter.

Figure 7.5 shows comparisons to alternative theories, like the Guggenheim-Stavermann athermal (GSA) contribution in UNIQUAC and UNIFAC. None of the alternative theories scale with density, so the choice of $\eta^P = 0.5$ for the packing fraction in this analysis is arbitrary. A method of scaling with density would be

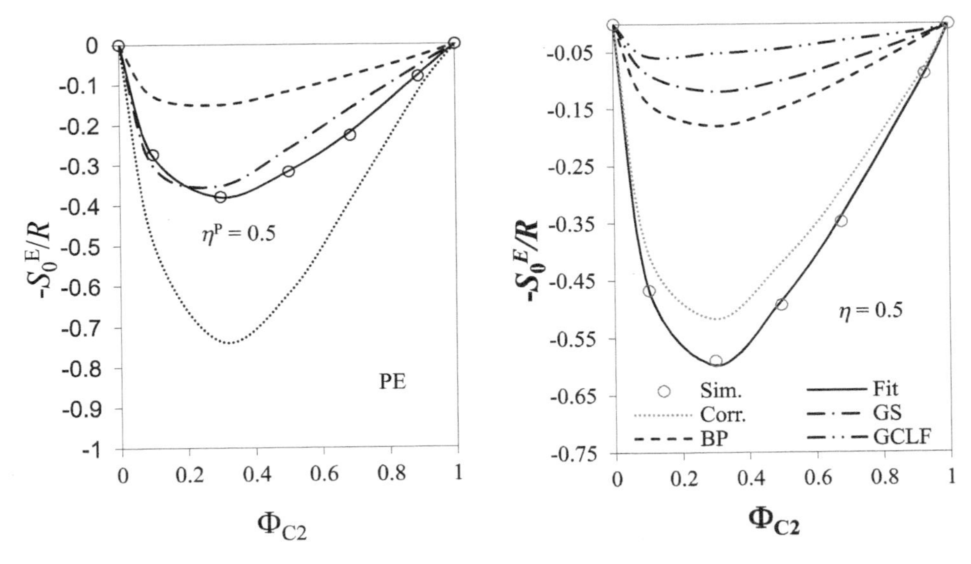

FIGURE 7.5
Alternative theories of the athermal excess entropy. (a) Ethane+*n*-hexadecane, V_p/V_s = 5.98, $\delta_s^\circ/\delta_p^\circ$ = 1.65 (b) Ethane+polystyrene(C24), V_p/V_s = 7.34, $\delta_s^\circ/\delta_p^\circ$ = 1.91 Symbols from molecular simulations. Dotted red curve is the correlation of Eq. (7.4-20). Solid blue curve is the fit of Eq. (7.4-19). BP applies the universal value of Blanks and Prausnitz. GS corresponds to GSA term. GCLF is the theory of Dudowicz and Freed (1991).

necessary for all these alternative theories. The theory of Blanks and Prausnitz (BP) (1964) assumes a universal value of $\chi_{ps}{}^0 = -0.34/b_s$, where b_s is the molar volume of the solvent. A single universal value of $\chi_{ps}{}^0$ is incapable of describing the trends of Fig. 7.4, but this approach does reflect the vdw 1-fluid composition dependence that was observed by Vahid et al. (2014). The correlation of Eq. (7.4-20) deviates more in the presence of branching and rings. The trend of the GS contribution correctly predicts the qualitative effect of branching, but simply adding this contribution to Eq. (7.4-20) would overestimate the effect of branching. The group contribution lattice fluid (GCLF) theory of Dudowicz and Freed (1991) applies a very detailed accounting of polymer chain structure. Apparently, the GCLF theory is qualitatively inconsistent with the branching effects observed by Vahid et al. (2014).

Like BP, many applications of Flory-Huggins theory have applied constant values of χ^0 to characterize experimental observations, but we are unaware of a rigorous theory that accounts for the vdw 1-fluid composition dependence observed by Vahid et al. (2014). In that context, it is quite remarkable that the vdw 1-fluid perspective describes the molecular simulations so well. Therefore, the SPEADMD model effectively applies an empirical correlation of molecular simulation data as the basis for further empirical correlations of experimental data. Readers are forgiven if they find this approach perplexing, but it is a convenient expedient to manage the distinction between molecular simulation data and experimental data.

Note that the contribution of excess entropy competes with the contribution of excess enthalpy in the Gibbs excess energy. When both contributions adhere to vdw 1-fluid composition dependence, the net effect is to impose a temperature variation in the Gibbs excess function that could be mimicked by a temperature-dependent k_{ij} parameter. Therefore, many readers may prefer to implement a temperature-dependent k_{ij} parameter and ignore the entire discussion of athermal excess entropy. The approach of SPEADMD, and molecular modeling in general, is to account directly for as many known effects as possible with the hope of minimizing BIPs like k_{ij} and maximizing predictive capability in the long run. On the other hand, the close interaction between athermal excess entropy and a temperature-dependent k_{ij} does counsel against investing heavily in refining the correlation of Eq. (7.4-20) until predictions from perturbation theory improve beyond their current capabilities.

7.4.4 Attractive Mixing Rules Determined by Molecular Simulation

Molecular simulations of the reference fluid simultaneously provide the attractive perturbations as well as the reference contributions. These can be plotted and correlated like the reference contributions. For the first-order attraction, it should not be surprising that vdw 1-fluid behavior is observed. The energy equation for the first-order attraction is subtly altered from the general form.

$$\frac{U - U^{ig}}{RT} = \frac{A_1}{T} + \frac{2A_2}{T^2} + \cdots = \sum_i \sum_j x_i x_j \frac{N_A \rho}{2k_B T} \int_0^\infty u_{ij}(r) g_{ij}^0(r) 4\pi r^2 dr + \cdots \tag{7.4-22}$$

The only change is the superscript "0" on g_{ij}, denoting the radial distribution function of the reference fluid instead of the full potential. According to the perspective of perturbation theory, repulsive influences dominate the structure of dense liquids because they control the packing, and packed molecules fluctuate over a narrow range of locations. Therefore, it is very important to get A_1 right when modeling liquids.

The good news is that A_1 follows vdw 1-fluid behavior and the excess A_1 can be written as

$$\frac{A_1^E}{T} = \frac{\sum_i \sum_j x_i x_j A_{1,ij} (b_i b_j)^{1/2}}{T \sum_i x_i b_i} - \sum_i x_i \frac{A_{1,i}}{T}; \; A_{1,ij} = (A_{1,i} \cdot A_{1,j})^{1/2} (1 - k_{ij}) \tag{7.4-23}$$

This equation is analogous to Eq. (7.4-19) except that T takes the place of R when scaling A_1 versus S_0. To test this hypothesis, $A_1{}^E$ can be plotted from simulation results. Figure 7.6a shows the typical behavior exemplified by "water+n-octanol." Note that only the sizes and shapes of the molecules matter for these simulations of the reference potentials; "water" is simply a hard sphere in this system. The familiar behavior is observed, but

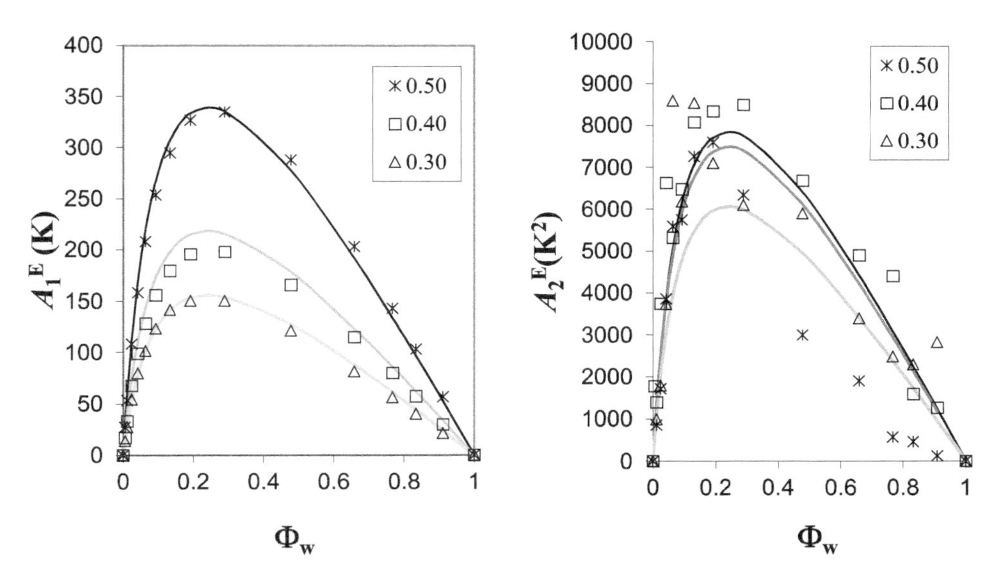

FIGURE 7.6
Mixture behavior of the attractive perturbations in the SPEADMD model for water+*n*-octanol. The legend indicates packing fractions. Points are simulation results and curves correspond to vdw 1-fluid mixing.

the deviations from ideality are opposite in sign from the repulsive contribution. Not only are the a_0 and A_1 terms of equal magnitude and opposite in sign for the pure fluids, but also for the excess contributions. On the other hand, A_1^E/T is a function of density and scales with reciprocal temperature, such that the temperature, density, and composition where a_0 and A_1 cancel is still complicated. The near cancellation of a_0 and A_1 enhances the importance of A_2 and higher perturbations. Pavlyukhin (2012) has argued that a_0 and A_1 should be considered in combination as the reference of the perturbation series. After all, Eq. (7.4-22) is completely specified by g_{ij}^0, and g_{ij}^0 is a property of the reference fluid. Unfortunately, Fig. 7.6b shows that uncertainty is greater for A_2^E. Furthermore, the uncertainty is roughly constant with increasing density, while the magnitude of A_2 is decreasing, such that relative uncertainty is quite significant at liquid densities. This makes it difficult to justify any deviation from vdw 1-fluid mixing with zero binary interaction parameter. The higher-order TPT contributions are likely to cause the kind of clustering that would lead to deviations from vdw mixing because of cooperative attraction effects (Vasudevan and Elliott, 1993). Cooperative attraction happens when the attractive influence on fluid structure due to one pair of sites is enhanced by dragging other similar sites into the attracted region through intramolecular bonds. These influences are especially apparent in treatments by integral equation theory, where precision is high, but the accuracy of the theory is not assured. Progress in making perturbation methods as accurate as the best empirical correlations for very complex mixtures will probably rely on improved analysis of composition dependence for higher-order TPT terms, as well as improved potential models.

7.4.5 Chemical-Physical Contributions

In many practical systems, strong interactions occur due to charge-transfer and hydrogen bonding. This occurs in mixtures of alcohols, amines, amides, ethers, carboxylic acids, water, HF, etc. The chemical-physical approach has been to consider that these interactions are so strong that new "chemical species" are formed. Then the thermodynamic treatment assumes that the properties deviate from an ideal gas mainly due to the "speciation" (the actual molecular weight is other than that of pure monomer) plus some physical effects. The EOS then becomes

$$Z = 1 + Z^{\text{rep}} + Z^{\text{att}} + Z^{\text{chem}} \tag{7.4-24}$$

where Z^{chem} accounts for the chemical-physical contribution.

It is assumed that all the species are in reaction equilibrium. Thus, their concentrations can be determined from equilibrium constants having parameters such as enthalpies and entropies of reaction as well as parameters for species physical interactions. In mixtures, an example is the vapor formation of dimers (D) from two different monomers (M_1 and M_2)

$$(M_1 + M_2) = D \tag{7.4-25}$$

The equilibrium constant for this reaction can be approximately related to the second cross virial coefficient

$$K_D = \frac{y_D}{y_{M_1} y_{M_2} P} = -2\frac{B_{12}}{RT} \tag{7.4-26}$$

where the factor of 2 arises here because of the distinct monomers of Eq. (7.4-25). The model of Hayden and O'Connell (1975) described in Chap. 5 explicitly includes such contributions so that it can also predict the properties of strongly interacting unlike substances.

While models based on TPT1 theory like SAFT treat chemical-physical interactions implicitly, it is occasionally advantageous to treat the effects explicitly. For example, Anderko's (1991) explicit treatment of systems with speciation has been extended to mixtures, as has that of Gmehling et al. (1979). Economou and Donohue (1992) and Anderko et al. (1993) show that care must be exercised in treating the chemical and physical contributions to equations of state since some methods introduce thermodynamic inconsistencies. Maurer and coworkers have demonstrated such an approach for formaldehyde-containing systems, limiting the treatment to a few specific oligomers. Coto et al. (2003) provide an edifying review of the sophisticated approach applied to this system and its process implications, with supporting spectroscopic data for the speciation and reaction kinetics. If a high degree of accuracy is required for such a system, it may be worthwhile to invest in such a detailed "user added" model for process simulation. There is ample opportunity for the parameters in such a model to be tuned if extensive experimental data are available but expect slow computational speeds and be very careful about extrapolating such a model beyond the range of the experimental data.

The method of Heidemann and Prausnitz (1976) includes explicit speciation but extends the reaction to infinitely long oligomers (in principle) by envisioning reactions like:

$$K_{i+1}^{j,k} = \frac{x_{i+1}^{j,k}}{x_{M_1}^{j} x_i^{k}} \frac{\varphi_{i+1}^{j,k}}{\varphi_{M_1}^{j} \varphi_i^{k} P} \tag{7.4-27}$$

where this reaction envisions adding a monomer of type 1 from a molecule of type j to an oligomer of the ith degree from a molecule of type k. Through the summation over infinitely long oligomers, exact solutions of the reaction equilibria can be obtained that match TPT1 in many regards. The ESD EOS was originally formulated in this manner, as was the "associating perturbed anisotropic chain theory" (APACT) of Ikonomou and Donohue (1988). This approach is feasible in principle but developing combining rules and reaction expressions to cover all combinations of mixed solvating species quickly becomes arbitrary and very complicated (Suresh and Elliott, 1991).

Finally, we turn to EOSs based on Wertheim's TPT1 approach, which treat speciation implicitly. While the moles of species are not computed explicitly, they are implied by the fraction of donor (X_i^D) and acceptor (X_i^A) sites NOT bonded. It is feasible to compute the true moles of the species at any point in the computation, however. The mixing formulas vary in complexity according to assumptions about the contact probability for two unbonded sites, g_{ij}. We begin with the ESD and CPA models, where g_{ij} is simply a function of packing fraction, η^P, independent of i or j, and the formulas are relatively simple. The ESD model, for example, assumes

$$g_{ij} = \frac{1}{1 - 1.9\eta^P} \quad \text{for all } i, j \tag{7.4-28}$$

The interactions between donors and acceptors on the same and different molecules are much more tractable in Wertheim's approach than for explicit speciation, but still non-trivial,

$$1 - X_i^{A_k} = X_i^{A_k} \sum \sum x_j N_j^{D_l} X_j^{D_l} \rho \Delta_{ij}^{k,l} \tag{7.4-29}$$

$$1 - X_i^{D_k} = X_i^{D_k} \sum \sum x_j N_j^{A_l} X_j^{A_l} \rho \Delta_{ij}^{l,k} \tag{7.4-30}$$

Equation (7.4-29) indicates that the fraction of *bonded* acceptors (i.e., 1 − *unbonded*) of type k on the ith molecule is a sum of bonding probabilities over all donors on all molecules, including the ith molecule. Equation (7.4-30) applies to all donors. Together they form a nonlinear system of equations that can be solved for all $\{X_i^{A_k}, X_j^{D_l}\}$. Solution generally follows an approach like successive substitution, possibly including a Picard factor to improve convergence stability. It may sound complicated, but it is straightforward to program in a general manner. The PC-SAFT and SPEADMD codes available through the PGL6ed website demonstrate this approach. Solving for A^{chem} and Z^{chem} is straightforward once $\{X_i^{A_k}, X_j^{D_l}\}$ is determined.

The general formulas for A^{chem} and Z^{chem} are given as Eqs. (7.4-36) to (7.4-48), but they are complicated. Simple insights and efficient computations can be facilitated by recognizing a factorization that occurs for many types of strong interactions. If the interactions are strong,

$$\Delta_{ij}^{k,l} = \exp\left(\beta \varepsilon_{ij}^{k,l}\right) - 1 \approx \exp\left(\beta \varepsilon_{ij}^{k,l}\right) \tag{7.4-31}$$

For example, a typical bonding energy for alcohols is $\varepsilon^{AD}/k = 2100$ K, so $\beta\varepsilon^{AD} = 7$ at room temperature and $\exp(\beta\varepsilon^{AD}) = 1097$. Equation (7.4-31) results in less than 0.1% error then. On the other hand, many mixtures of interest involve strongly interacting species like alcohols with "inert" species like hydrocarbons. For inert compounds, there is no bonding between molecules and

$$\Delta_{ij}^{k,l} = \exp(0\beta) - 1 = 0 \tag{7.4-32}$$

Elliott (1996) recognized this factorization as a way of greatly simplifying Wertheim's approach when molecular interactions are either strong or nonexistent.

$$\rho \Delta_{ij}^{k,l} = (\alpha_i \alpha_j)^{½} \tag{7.4-33}$$

where $\alpha_i = \rho \Delta_{ii}^{AD}$ and the number of acceptors and donors is assumed to be equal on each molecule and only one type of acceptor or donor is allowed per molecule. We refer to this approximation as the "geometric combining rule (GCR)." Noting $X_i^A = X_i^D$ in this context, substitution yields

$$\left(\frac{1}{X_i^A} - 1\right)\frac{1}{\alpha_i^{½}} = \sum x_j X_j^A N_j^{AD} \alpha_j^{½} \equiv F^{AD} = \sum \frac{x_j N_j^{AD} \alpha_j^{½}}{1 + \alpha_j^{½} F^{AD}} \tag{7.4-34}$$

where N_j^{AD} is the number of acceptors (equal to the number of donors) on the jth molecule. For example, $N^{AD} = 1$ for alcohols like methanol and ethanol, N^{AD} = degree of polymerization for polyvinylphenol.

Note that F^{AD} is independent of subscript regardless of how many components are present, so solving a single nonlinear "master" equation provides all values of $X_i^A = X_i^D$. Also, $F^{AD} = 0$ at low density or when interactions are weak and $F^{AD} \approx (N^{AD})^{½}$ when interactions are strong, so taking these limits as initial guesses leads to rapid and robust Newtonian convergence. For example, if there is only one associating component, Eq. (7.4-34) is quadratic in F^{AD} and the solution is exact. When the GCR is combined with Eq. (7.4-28), the EOS contribution is especially simple:

$$Z^{\text{chem}} = \frac{-(F^{AD})^2}{1 - 1.9\eta^P} \tag{7.4-35}$$

Having such a simple (but approximate) result may help to appreciate the physical basis of models like SAFT that are based on Wertheim's TPT1 approach. Although not a strict requirement of the ESD model, all evaluations of the ESD model in PGL6ed conform to Eq. (7.4-35). Note that hydrogen bonding between alcohols and ethers, for example, is ignored by this approach, but those interactions can be roughly characterized by a negative value for k_{ij} of the dispersion term.

When chemical-physical interactions violate the assumptions of the GCR, it is necessary to solve the TPT1 relations iteratively. This is often the case for the SPEADMD and PC-SAFT models. Given the $\{X_i^{A_k}, X_j^{D_l}\}$ from solving Eq. (7.4-34), the thermodynamic properties can be computed by extending the relations in Chap. 6 to multiple components

$$\frac{a^{\text{chem}}}{RT} = \sum_{k=1}^{n} x_k \sum_{i}^{N_k^A} \ln\left(X_{ik}^A\right) + \frac{1 - X_{ik}^A}{2} + \sum_{i}^{N_k^D} \ln\left(X_{ik}^D\right) + \frac{1 - X_{ik}^D}{2} \tag{7.4-36}$$

$$\frac{1}{X_{im}^A} = 1 + \sum_{k=1}^{n} x_k \sum_{j}^{N_s} X_{jk}^D \rho \Delta_{im,jk}^{AD} \tag{7.4-37}$$

$$\frac{1}{X_{im}^D} = 1 + \sum_{k=1}^{n} x_k \sum_{j}^{N_s} X_{jk}^A \rho \Delta_{im,jk}^{DA} \tag{7.4-38}$$

$$\rho \Delta_{im,jk}^{AD} = g_{d,im,jk}^{HS}\, \rho K_{im,jk}^{AD} \left[\exp\left(\frac{\varepsilon_{im,jk}^{AD}}{k_B T} \right) - 1 \right] \tag{7.4-39}$$

From Michelsen and Hendriks (2001), the quantity, h, becomes

$$h^{AD} = \sum_{m}^{N_c} \sum_{i}^{N_s} x_m \left(1 - X_{im}^A\right) \tag{7.4-40}$$

$$h^{DA} = \sum_{m}^{N_c} \sum_{i}^{N_s} x_m \left(1 - X_{im}^D\right) \tag{7.4-41}$$

Then Eq. (7.4-36) can be rewritten as

$$\frac{a^{\text{chem}}}{RT} = \sum_{k=1}^{N_c} x_k \left[\sum_{i}^{N_i^A} \ln\left(X_{ik}^A\right) + 1 - X_{ik}^A \right] - \frac{h^{AD}}{2} + \sum_{k=1}^{N_c} x_k \left[\sum_{i}^{N_i^D} \ln\left(X_{ik}^D\right) + 1 - X_{ik}^D \right] - \frac{h^{DA}}{2} \tag{7.4-42}$$

Michelsen and Hendriks (2001) then prove that the derivatives of the bracketed terms evaluate to zero, so

$$Z^{\text{chem}} = -\frac{1}{2} \left(\frac{\rho}{2} \frac{\partial h^{AD}}{\partial \rho} + \frac{\rho}{2} \frac{\partial h^{DA}}{\partial \rho} \right) \tag{7.4-43}$$

$$\frac{1}{RT} \frac{\partial P^{\text{chem}}}{\partial \rho} = -\frac{\rho^2}{2} \frac{\partial^2 h^{AD}}{\partial \rho^2} - \frac{\rho^2}{2} \frac{\partial^2 h^{DA}}{\partial \rho^2} \tag{7.4-44}$$

$$\mu_l^{\text{chem}} = \sum_{i=1}^{N_c} \ln\left(X_{il}^A\right) + \sum_{i=1}^{N_c} \ln\left(X_{il}^D\right) - \frac{1}{2} \frac{\partial (n h^{AD})}{\partial n_l} - \frac{1}{2} \frac{\partial (n h^{DA})}{\partial n_l} \tag{7.4-45}$$

$$\rho \frac{\partial h^{AD}}{\partial \rho} = h^{AD} + \left[\sum_m^{N_c}\sum_k^{N_c} x_m x_k \sum_i^{N_s}\sum_j^{N_s} X_{im}^A X_{jk}^D \rho \Delta_{im,jk}^{AD} \frac{\rho \partial\left(\Delta_{im,jk}^{AD}\right)}{\Delta_{im,jk}^{AD} \partial \rho} \right] \tag{7.4-46}$$

$$\frac{\partial(nh^{AD})}{\partial n_l} = \left[\sum_m^{N_c}\sum_k^{N_c} x_m x_k \sum_i^{N_s}\sum_j^{N_s} X_{im}^A X_{jk}^D \rho \Delta_{im,jk}^{AD} \frac{n\partial\left(\Delta_{im,jk}^{AD}\right)}{\Delta_{im,jk}^{AD} \partial n_l} \right] \tag{7.4-47}$$

$$\frac{\rho\, \partial\left(\Delta_{im,jk}^{AD}\right)}{\Delta_{im,jk}^{AD}\, \partial \rho} = 1 + \frac{\rho\, \partial g(\sigma^{ss})}{g(\sigma^{ss})\, \partial \rho} \tag{7.4-48}$$

Once again, we reiterate that it looks complicated, but it is straightforward to program in a general manner and codes are available.

Example 7.1 Mixing rules with the ESD EOS

Use the ESD mixing rules and efficient Wertheim implementation to calculate the excess volume for the binary methanol (1) + benzene (2) at 60 °C, $P = 0.1$ MPa, and $x_1 = 0.4$. Assume $k_{ij} = 0.0196$ for the BIP.

Solution. We demonstrate calculation for the final density iteration.

The ESD EOS can be written as

$$Z = 1 + \frac{4q_m\eta^P}{1-1.9\eta^P} - \frac{(q_m - 1)1.9\eta^P}{1-1.9\eta^P} - \frac{9.5(qY\eta^P)_m}{1+1.7745(Y\eta^P)_m} - \frac{(F^{AD})^2}{1-1.9\eta^P} \tag{7.4-49}$$

ID	NAME	q	b(cm³/mol)	ε/k(K)	Y(Eq. [7.4-16])	N^D	K^{AD}(nm³)	ε^{AD}/k(K)
1	METHANOL	1.512	19.96	312.32	1.4918	1	0.0007	2516
2	BENZENE	2.461	29.61	336.72	1.6859	0	0	0

Applying Eqs. (7.4-14) and (7.4-16), $Y_{12} = \exp[(312.32 \cdot 336.72)^{1/2}(1 - 0.0196)] - 1.0617 = 1.5351$ and,

$$\frac{(qY\eta^P)_m}{\rho} = 0.4^2 \cdot 1.512 \cdot 1.4918 \cdot 19.96 + 2 \cdot 0.4 \cdot 0.6 \cdot \frac{1.5351(1.512 \cdot 29.61 + 2.461 \cdot 19.96)}{2} + 0.6^2 \cdot 2.461 \cdot 1.6859 \cdot 29.61 = 86.029 \tag{7.4-50}$$

$q_m = 0.4 \cdot 1.512 + 0.6 \cdot 2.461 = 2.0815$; $b_m = 25.7479$. From Eq. (7.4-15),

$$\frac{(Y\eta^P)_m}{\rho} = 0.4 \cdot 1.4918 \cdot 19.96 + 0.6 \cdot 1.6859 \cdot 29.61 = 41.862 \tag{7.4-51}$$

At a density of 0.012302 mol/cm³, $\eta^P = 0.31676$, and

$$Z^{rep} = q_m Z^{HS} + Z^{chain} = \frac{(4 \cdot 2.0815 - 1.0815 \cdot 1.9) \cdot 0.31676}{1 - 1.9 \cdot 0.31676} = 4.9891 \tag{7.4-52}$$

$$Z^{att} = -\frac{9.5 \cdot 86.029 \cdot 0.012302}{1 + 1.7745 \cdot 41.862 \cdot 0.012302} = -5.2533 \tag{7.4-53}$$

Applying Eq. (7.4-34), noting that $\alpha_{12}^{AD} = \alpha_{22}^{AD} = 0$,

$$\frac{x_1 N_1^D \left(\alpha_{11}^{AD}\right)^{1/2}}{1 + \left(\alpha_{11}^{AD}\right)^{1/2} F^{AD}} = F^{AD} \tag{7.4-54}$$

where α_{11}^{AD} = 0.012302*0.0007*602.22*[exp(2516/333.15) – 1]/(1 – 1.9*0.012302) = 24.797.

$$\left(\alpha_{11}^{AD}\right)^{1/2} = 4.980 \text{ and } N_1^D = 1.$$

Solving the quadratic equation and completing the square,

$$F^{AD} = \frac{2x_1 N_1^D \left(\alpha_{11}^{AD}\right)^{1/2}}{1 + \left(1 + 4x_1 N_1^D \alpha_{11}^{AD}\right)^{1/2}} = \frac{0.8 * 4.980}{1 + (1 + 1.6 * 24.797)^{1/2}} = 0.5400 \tag{7.4-55}$$

Collecting terms,

$$Z = 1 + 4.9891 - 5.2533 - \frac{0.54^2}{1 - 1.9 \cdot 0.31676} = 0.003 \tag{7.4-56}$$

Checking, P = 0.003*0.012302*8.31447*333.15 = 0.102; V_m = 1/0.012302 = 81.29 cm³/mol. Note that values of Z and P are very sensitive to roundoff error, but the value of ρ is not. Repeating for the pure fluids, $V_1 = 56.03$, $V_2 = 95.39$ cm³/mol. Finally, the excess volume is,

$$V^E = 81.29 - (0.4 \cdot 56.03 + 0.6 \cdot 95.39) = 1.66 \text{ cm}^3/\text{mol}.$$

Example 7.2 Mixing rules for the SPEADMD EOS

Use the SPEADMD mixing rules to calculate the excess volume for the binary methanol (1) + benzene (2) at 60 °C, $P = 0.1$ MPa, and $x_1 = 0.4$. Assume $k_{ij} = 0.0240$ for the BIP.

Solution. We demonstrate calculation for the final density iteration.

The SPEADMD EOS can be written as

$$Z = Z_0 + \frac{Z_1}{T} + \frac{Z_2}{T^2} + Z^{\text{chem}} \tag{7.4-57}$$

The parameters are given by

ID	NAME	z_1	z_2	z_3	a_{11}	a_{12}	a_{13}	$10^{-3}a_{21}$	$10^{-3}a_{22}$	$10^{-3}a_{23}$	$10^{-3}a_{24}$
1	MeOH	0.9787	2.4338	−3.5547	−44.71	−3305	−4698	−288.2	1541	−3094	1838
2	Benzene	1.3628	2.8002	−0.8613	5406	−9630	−15898	−1153	5442	−9309	5011

SPEADMD requires additional parameters to characterize the chemical-physical terms

ID	NAME	b(cm³/mol)	K^{AD}(nm³)	ε^{AD}/k(K)
1	MeOH	18.73	0.001	2516
2	Benzene	43.08	0	0

At a density of 0.013434 mol/cm^3, the mixing rules require the TPT terms for the pure fluids at $\eta^P = 0.4479$.

ID	NAME	A_0	A_1	$A_2 10^{-3}$	Z_0	Z_1	$Z_2 10^{-3}$	δ^S(J/cm^3)$^{1/2}$
1	MeOH	4.065	−1908	−4364	8.551	−2094	1795	20.71
2	Benzene	4.925	−3387	−10846	11.448	−4330	20127	14.87

The computation of k_{ij}^S requires the entropic solubility parameters, $\delta_i^{S^\circ}$, defined below Equation (7.4-20) and tabulated above. For a binary mixture, $b_{avg} = \Sigma b_i / N_c = (b_1 + b_2)/2$. Then,

$$k_{ij}^S = -0.016 \cdot 18 \cdot \frac{(20.71 - 14.87)^2}{8.31446 \cdot 298.15} = -0.00397 \tag{7.4-58}$$

Then $a_{0.ij} = (4.065 \cdot 4.925)^{1/2}(1 + 0.00397) = 4.492$ and from Eq. (7.4-19)

$$a_0 = \frac{0.4^2 \cdot 4.065 \cdot 18.73 + 2 \cdot 0.4 \cdot 0.6 \cdot 4.492(18.73 \cdot 43.08)^{1/2} + 0.6^2 \cdot 4.925 \cdot 43.08}{0.4 \cdot 18.73 + 0.6 \cdot 43.08} \tag{7.4-59}$$

Then $a_0 = 4.493$. By Eq. (7.4-21)

$$Z_{0,ij} = \frac{8.551 \cdot 4.065 + 11.448 \cdot 4.925}{2(4.065 \cdot 4.925)^{1/2}}(1.00397) = 10.225 \tag{7.4-60}$$

$$Z_0 = \frac{0.4^2 \cdot 8.551 \cdot 18.73 + 2 \cdot 0.4 \cdot 0.6 \cdot 10.225(18.73 \cdot 43.08)^{1/2} + 0.6^2 \cdot 11.448 \cdot 43.08}{0.4 \cdot 18.73 + 0.6 \cdot 43.08} \tag{7.4-61}$$

Then $Z_0 = 10.165$. Turning to A_1, $A_{1.ij} = -(1908 \cdot 3387)^{1/2}(1 - 0.0240) = -2481$ and Eq. (7.4-23),

$$A_1 = -\frac{0.4^2 \cdot 1908 \cdot 18.73 + 2 \cdot 0.4 \cdot 0.6 \cdot 2481(18.73 \cdot 43.08)^{1/2} + 0.6^2 \cdot 3387 \cdot 43.08}{0.4 \cdot 18.73 + 0.6 \cdot 43.08} \tag{7.4-62}$$

Then $A_1 = -2762$. By Eq. (7.4-21),

$$Z_{1,ij} = -\frac{2094 \cdot 3387 + 4334 \cdot 1908}{2(1908 \cdot 3387)^{1/2}}(1 - 0.0240) = -2949 \tag{7.4-63}$$

$$Z_1 = -\frac{0.4^2 \cdot 2094 \cdot 18.73 + 2 \cdot 0.4 \cdot 0.6 \cdot 2949(18.73 \cdot 43.08)^{1/2} + 0.6^2 \cdot 4334 \cdot 43.08}{0.4 \cdot 18.73 + 0.6 \cdot 43.08} \tag{7.4-64}$$

Then $Z_1 = -3409$. Turning to A_2, $A_{2,ij}10^{-3} = -(4364 \cdot 10846)^{1/2}(1 - 0.0240) = -6715$ and Eq. (7.4-23),

$$10^{-3}A_2 = -\frac{0.4^2 \cdot 4364 \cdot 18.73 + 2 \cdot 0.4 \cdot 0.6 \cdot 6715(18.73 \cdot 43.08)^{1/2} + 0.6^2 \cdot 10486 \cdot 43.08}{0.4 \cdot 18.73 + 0.6 \cdot 43.08} \tag{7.4-65}$$

Then $A_2 = -8184$. By Eq. (7.4-21),

$$Z_{2,ij} = -\frac{1795 \cdot 10846 + 20127 \cdot 4364}{2(4364 \cdot 10846)^{1/2}}(1 - 0.0240) = 7611 \tag{7.4-66}$$

$$10^{-3}Z_2 = \frac{0.4^2 \cdot 1795 \cdot 18.73 + 2 \cdot 0.4 \cdot 0.6 \cdot 7611(18.73 \cdot 43.08)^{1/2} + 0.6^2 \cdot 20127 \cdot 43.08}{0.4 \cdot 18.73 + 0.6 \cdot 43.08} \tag{7.4-67}$$

$$10^{-3}Z_2 = 12636.$$

Like ESD, the computation of $X_1{}^A = X_1{}^D$ applies Eq. (7.4-34), noting that $\alpha_{12}^{AD} = \alpha_{22}^{AD} = 0$,

$$\alpha_{11}^{AD} = \rho K_1^{AD} N_A Y^{AD} g(\sigma^{ss});\ g(\sigma^{ss}) = \frac{(1 - \eta^P/2)}{(1 - \eta^P)^3} \tag{7.4-68}$$

so $g(\sigma) = 4.612$. Hence $\alpha_{11}^{AD} = 0.013434*0.001*602.22*[\exp(2516/333.15) - 1]4.612 = 71.036$. $\left(\alpha_{11}^{AD}\right)^{1/2} = 8.428$ and $N_1^D = 1$.

Solving the quadratic equation using from α_{11}^{AD} SPEADMD,

$$F^{AD} = \frac{2x_1N_1^D\left(\alpha_{11}^{AD}\right)^{1/2}}{1 + \left(1 + 4x_1N_1^D\alpha_{11}^{AD}\right)^{1/2}} = \frac{0.8 \cdot 8.428}{1 + (1 + 1.6 \cdot 71.036)^{1/2}} = 0.5759 \tag{7.4-69}$$

So, $X_1{}^A = X_1{}^D = \left(1 + \left(\alpha_{11}^{AD}\right)^{1/2} F^{AD}\right)^{-1} = 0.1708$ and from Eqs. (7.4-40) and (7.4-41),

$$h^{AD} = h^{DA} = 0.4\,(1 - 0.1708) = 0.3317 \tag{7.4-70}$$

Taking the derivative,

$$Z^{\text{chem}} = -\frac{1}{2}(h^{DA} + h^{AD})\left(1 + \frac{\rho}{g}\frac{dg}{d\rho}\right) = -1.0433 \tag{7.4-71}$$

Collecting terms,

$$Z = 1 + 10.165 - \frac{3409}{333.15} + \frac{12636}{333.15^2} - 1.0433 = 0.003 \tag{7.4-72}$$

Checking, $P = 0.003*0.013434*8.31447*333.15 = 0.112$; $V_m = 1/0.013434 = 74.44$ cm^3/mol. We reiterate that values of Z and P are very sensitive to roundoff error. Repeating for the pure fluids, $V_1 = 42.90$, $V_2 = 94.94$ cm^3/mol. Finally, the excess volume is,

$$V^E = 74.44 - (0.4 \cdot 42.90 + 0.6 \cdot 94.94) = 0.31 \text{ cm}^3/\text{mol}.$$

7.4.6 Polarity Contributions

Polarity contributions represent interactions that are intermediate in strength between dispersion interactions and strong interactions like hydrogen bonding. These are generally represented by dipole-dipole, quadrupole-quadrupole, and dipole-quadrupole contributions. In most molecules, these interactions are coupled with chemical-physical interactions and it is not obvious how to treat them separately. Vrabec and Gross (2008) provide one approach; it is the method implemented in the polar PC-SAFT code available through the PGL6ed website. The key step is to write

$$Z = 1 + Z^{\text{rep}} + Z^{\text{att}} + Z^{\text{chem}} + Z^{\text{DD}} + Z^{\text{QQ}} + Z^{\text{DQ}} \tag{7.4-73}$$

Vrabec and Gross (2008) then performed molecular simulations of two-center Lennard-Jones molecules (2CLJ) with several combinations of bond length, dipole strength, and quadrupole strength. Their simulations were combined with previous simulations in the literature that treated dipoles or quadrupoles individually or treated spheres without a second center. The dipoles and quadrupoles were placed at the middle of the bond between the two LJ centers. Their correlating equations were expressed as contributions to the Helmholtz energy with compressibility factors implied by differentiation. For example, the dipole-quadrupole contribution is

$$\frac{a^{\text{DQ}}}{RT} = \frac{a_2^{\text{DQ}}/RT}{1 - a_3^{\text{DQ}}/a_2^{\text{DQ}}} \tag{7.4-74}$$

$$\frac{a_2^{\text{DQ}}}{RT} = -\frac{9\pi}{4}\rho\sum\sum x_i x_j \beta\varepsilon_{ii}^{TS}\beta\varepsilon_{jj}^{TS}\mu_i^{*,TS2}Q_j^{*,TS2}J_{2,ij}^{\text{DQ}}\sigma_{ii}^3\sigma_{jj}^5/\sigma_{ij}^5 \tag{7.4-75}$$

$$\frac{a_3^{\text{DQ}}}{RT} = -\rho^2\sum\sum x_i x_j x_k \beta\varepsilon_{ii}^{TS}\beta\varepsilon_{jj}^{TS}\beta\varepsilon_{kk}^{TS}\frac{\sigma_{ii}^4\sigma_{jj}^4\sigma_{kk}^4}{\sigma_{ij}^2\sigma_{jk}^5\sigma_{ki}^5}$$
$$\times\left(\mu_i^{*,TS2}\mu_j^{*,TS2}Q_k^{*,TS2} + \alpha^{DQ}\mu_i^{*,TS2}Q_j^{*,TS2}Q_k^{*,TS2}\right)J_{3,ij}^{\text{DQ}} \tag{7.4-76}$$

where $\mu_i^{*,TS2} = \mu_i^2/\left(m_i\varepsilon_{ii}^{TS}\sigma_{ii}^3\right)$ represents the dimensionless squared dipole moment in the tangent-sphere (TS) framework, and $Q_i^{*,TS2} = Q_i^2/\left(m_i\varepsilon_{ii}^{TS}\sigma_{ii}^5\right)$ is the analogous quadrupole moment. Lorentz-Berthelot combining rules were used for $\varepsilon_{ij}^{TS} = \left(\varepsilon_{ii}^{TS}\varepsilon_{jj}^{TS}\right)^{1/2}$ and $\sigma_{ij}^{TS} = ½\left(\sigma_{ii}^{TS} + \sigma_{jj}^{TS}\right)$, and σ_{ii}^{TS} refers to the diameter of the ith LJ center in the TS framework. The J^{DQ} symbols represent integrals that have been numerically integrated and correlated as

$$J_{2,ij}^{\text{DQ}} = \sum_{n=0}^{4}(a_{n,ij} + b_{n,ij}\beta\varepsilon_{ij})(\eta^{\text{P}})^n \tag{7.4-77}$$

$$J_{3,ijk}^{\text{DQ}} = \sum_{n=0}^{4}c_{n,ijk}(\eta^{\text{P}})^n \tag{7.4-78}$$

The coefficients $a_{n,ij}$, $b_{n,ij}$, and $c_{n,ijk}$ have been correlated with the effective tangent sphere chain length, m^{Eff}, of each molecule

$$J_{3,ijk}^{\text{DQ}} = \sum_{n=0}^{4}c_{n,ijk}(\eta^{\text{P}})^n \tag{7.4-79}$$

$$J_{3,ijk}^{\text{DQ}} = \sum_{n=0}^{4}c_{n,ijk}(\eta^{\text{P}})^n \tag{7.4-80}$$

Contributions for dipolar and quadrupolar are like those given in Chap. 6. The exact forms for Gross's PC-SAFT implementation are given by Gross (2005) and Vrabec and Gross (2008). Once again, these contributions look complicated but they are straightforward to code, and the code is available. The key consideration for considering the physical basis of these contributions is that they have been validated for relatively small molecules like refrigerants. Gross (2005) shows that assigning a significant component of the attractive energy to these polarity contributions provides better predictive capability when k_{ij} is set to zero because contributions like Eq. (7.4-75) naturally lead to positive deviations in the excess energy when one of the components is nonpolar.

Example 7.3 Mixing rules for the PC-SAFT EOS

Use the polar PC-SAFT (PPC-SAFT) model for the methanol (1)+benzene (2) binary at $x_1 = 0.1031$. For the binary interaction parameter, assume $k_{12} = -0.01075$ obtained by minimizing deviations for this system for all data in the Jaubert et al. (2020) VLE database. Compute the excess volume at 60 °C.

Solution. The PC-SAFT model includes accounting for quadrupole and dipole moments, as well as for the chemical contribution of all TPT1 models. In this case, only benzene's quadrupole matters.

ID	NAME	m	σ(nm)	ε/k(K)	μ	q	N^d	K^{AD}	ε^{AD}/k(K)
1	METHANOL	1.507	0.3325	211.60	1.7	0	1	0.03	2519.7
2	BENZENE	2.291	0.3756	294.06	0	5.5907	0	0	0

Once again, we detail the computation for the last iteration of density where $\rho = 0.013497$ cm^3/mol. Note that methanol has a dipole and benzene has a quadrupole. The terms Z^{DD}, Z^{QQ}, and Z^{DQ} correspond to the dipole-dipole, quadrupole-quadrupole, and dipole-quadrupole contributions, respectively, and all are non-zero.

x_M	η^P	mZ^{HS}	Z^{HC}	$Z_1/T + Z_2/T^2$	Z^{chem}	Z^{DD}	Z^{QQ}	Z^{DQ}	Z^{tot}
0.0	0.39455	13.0771	−2.2065	−11.4998	0.0000	0.0000	−0.3674	0.0000	0.0034
0.4	0.39158	10.9820	−1.6524	−9.0504	−0.8515	−0.0830	−0.1768	−0.1653	0.0027
1.0	0.39088	8.3880	−0.8529	−5.5365	−2.4356	−0.5614	0.0000	0.0000	0.0015

The volumes for the pure fluids correspond to $V_1 = 42.339$ and $V_2 = 94.590$. The mixed density of 0.013497 corresponds to $V_m = 74.091$, so $V^E = 0.40$ cm^3/mol. Once again, it may seem that the polar contributions are quite small, but the contributions of $Z^{HS} + Z^{HC} + Z_1/T + Z_2/T^2$ nearly cancel, so these contributions can be significant. A similar argument holds for the contributions to activity coefficient, as discussed in Chapter 8.

7.4.7 EOS Models for the Near-Critical Region

Conditions near a mixture vapor-liquid critical point show significantly different behavior from simple extrapolations of the EOS models described so far in this chapter. The molecular correlations mentioned in Chap. 6 are long-ranged and concentration fluctuations dominate. The formulation must be in terms of chemical potentials as the independent variables, not the composition variables. Research into this effect shows that certain universalities appear in the properties, but substance-specific quantities also are involved.

Kiselev (1998) has published a general procedure for adapting equations of state to describe both classical and near-critical regions of mixtures. This has been applied to cubic EOSs by Kiselev and Friend (1999). Their model predicts two-phase behavior and excess properties (see Chap. 6) using parameters that are fitted only to volumetric behavior in the one-phase region.

It has also been found that complex solutions such as ionic solutions and polymers do not have the same universalities as simpler fluids (Anisimov et al., 1999). This is because the long-range forces among such species also affect long-range correlations. Thus, crossovers in these systems commonly appear at conditions closer to the critical than for small and nonionic systems. As a result, classical models should apply over larger ranges of conditions, but the crossovers can cause very sharp property changes.

7.5 EXCESS GIBBS ENERGY MIXING RULES

As mentioned above, to treat more complex solutions, mixing rules based on liquid activity models have been developed, especially for phase equilibrium calculations. The first widely recognized analysis of this approach was by Huron and Vidal (1979). Since that time, a very large literature has arisen with many different expressions; a relatively recent review has been given, for example, by Kontogeorgis and Folas (2010).

The concept is that g^E is typically correlated by activity models like the UNIQUAC model. However, thermodynamics also allows it to be computed from EOS expressions. Thus, an EOS mixing rule for the liquid phase can reflect the composition variations of a desirable activity model (AM) by setting $a^E_{\text{EOS}} = g^E_{AM} + \Delta g(P_{\text{match}})$. To accomplish this, the P, V_{m}, and V_i at the matching condition must be specified.

Precise matching cannot be made over all conditions because activity models do not include density dependence as EOS models do. However, many strategies have developed for making the connection. We will briefly outline the procedure; full details must be obtained from the literature. Since there are many subtle consequences of these analyses (see, e.g., Michelsen, 1996; Michelsen and Heidemann, 1996), care in implementation should be exercised.

There have been many strategies developed to select the optimal conditions for matching. Fischer and Gmehling (1996) show that the general process is to select a pressure and the g^E function (which we denote g^{E_0} or g^{E_∞}, depending upon the reference pressure). Then values of the inverse packing fraction, $1/\eta^{\text{P}} = V_{\text{m}}/b_{\text{m}}$ and $1/\eta_i^P = V_i/b_i$, are selected so that the parameter mixing rule that gives Z_{m} and V_{m} can be found. Twu and Coon (1996) also discuss matching with constraints.

Here we show some of the more popular expressions. We focus on cubic EOS models (such as those in Chap. 6), since most g^E mixing rules apply to those models. Twu et al. (1999) offer an interesting perspective on the issue of matching.

For cubic equations, the EOS expressions for A^E_{EOS} can be written as

$$\frac{A^E_{EOS}}{RT} = \sum_{i=1}^{n} y_i \left\{ -\ln\frac{Z_m}{Z_i} - \ln\left(\frac{1-\eta^{\text{P}}}{1-\eta_i^{\text{P}}}\right) + \left[\frac{a_m}{b_m RT}C(\eta^P) - \frac{a_i}{b_i RT}C(\eta_i^P)\right]\right\} \tag{7.5-1}$$

Here an EOS-dependent dimensionless function appears for both the mixture, $C(\eta^{\text{P}})$, and the pure component, $C(\eta_i^P)$,

$$C(\eta^{\text{P}}) = \frac{b}{\sqrt{\zeta_2^2 - \zeta_1}} \ln\left(\left[\frac{2b + \eta^{\text{P}}(\zeta_2 - \sqrt{\zeta_2^2 - \zeta_1})}{2b + \eta^{\text{P}}(\zeta_2 + \sqrt{\zeta_2^2 - \zeta_1})}\right]\right) \tag{7.5-2}$$

where ζ_1 and ζ_2 are generalized cubic EOS constants (Chap. 6). Table 7.1 shows the various matching conditions and relations that were tabulated by Fischer and Gmehling (1996). For illustration we also tabulate the results of Eq. (7.5-2) for the Soave (1972) (SRK) model [$\zeta_2^{\text{SRK}} = b$ and $\zeta_1^{\text{SRK}} = 0$ gives $C_{\text{SRK}}(\eta^{\text{P}}) = -\ln(1 + \eta^{\text{P}})$.

Some specific results from Table 7.1 are:

7.5.1 Huron-Vidal

$$b_{\text{m}} = \sum_{i=l}^{n} y_i b_i \tag{7.5-3}$$

with $a^{E_\infty}_{\text{EOS}} = g^{E_\infty}$, and

$$\Theta^\infty_{\text{m}} = b_{\text{m}}\left[\sum_{i=1}^{n}\frac{y_i a_i}{b_i} + \frac{a^{E_\infty}_{EOS}}{C(\eta^{\text{P}})}\right] = b_{\text{m}}\left[\sum_{i=1}^{n}\frac{y_i a_i}{b_i} + \frac{g^{E_\infty}}{C(\eta^{\text{P}})}\right] \tag{7.5-4}$$

Equations (7.5-3) and (7.5-4) do not satisfy quadratic composition dependence for the second virial, and parameters obtained from data are not necessarily appropriate for high pressures.

TABLE 7.1 EOS-g^E matching conditions (after Fischer and Gmehling, 1996).

Mixing Rule	*P* Match	η^P Match	g^E Match*	b_m	$C_{SRK}(\eta^P)$
Huron/Vidal (1979)	∞	1.0000	$g^{E\infty}/RT$	$\Sigma y_i b_{ii}$	−0.693
MHV1 (Michelsen, 1990)	0	0.8097	$g^{E0}/RT + \sum_{i=1}^{n} y_i \ln \frac{b_m}{b_i}$	$\Sigma y_i b_{ii}$	−0.593
SRK/PSRK (Horstmann et al., 2005)	1 atm	0.9091	$g^{E0}/RT + \sum_{i=1}^{n} y_i \ln \frac{b_m}{b_i}$	$\Sigma y_i b_{ii}$	−0.647
MHV2 (Dahl and Michelsen, 1990)	0	0.6127	$\left(g^{E0}/RT + \sum_{i=1}^{n} y_i \ln \frac{b_m}{b_i}\right)^a$	$\Sigma y_i b_{ii}$	−0.478
LCVM (Boukouvalas, et al., 1994)	0		$\left(\frac{g^{E0}}{RT} + \frac{1-\lambda}{C(0.8097)} \sum_{i=1}^{n} y_i \ln\left(\frac{b_m}{b_i}\right)\right)^b$	$\Sigma y_i b_{ii}$	−0.553
Wong and Sandler (1992)	∞	1.0000	$(g^{E0}/RT)^c$	+	−0.693
Tochigi, et al. (1994)	0	0.8097	$g^{E0}/RT + \sum_{i=1}^{n} y_i \ln \frac{b_m}{b_i}$	+	−0.593
Orbey and Sandler (1995)	∞	1.0000	$g/RT + \sum_{i=1}^{n} y_i \ln \frac{b_m}{b_i}$	+	−0.693

*Expression to set equal to $\frac{g^E_{EOS}}{RT}$. For those not indicated ^{abc}, $\frac{g^E_{EOS}}{RT} = C(\eta^P)\left[\frac{a_m}{b_m RT} - \sum_{i=1}^{n} y_i \frac{a_i}{b_i RT}\right]$

a. $\frac{g^E_{EOS}}{RT} = C(u)\left[\frac{a_m}{b_m RT} - \sum_{i=1}^{n} y_i \frac{a_i}{b_i RT}\right] + A_2\left[\left(\frac{a_m}{b_m RT}\right)^2 - \sum_{i=1}^{n} y_i \left(\frac{a_i}{b_i RT}\right)^2\right]$; $(A_2)_{SRK} = -0.0047$

b. $\frac{g^E_{EOS}}{RT} = F(\lambda)\left[\frac{\Theta_m}{b_m RT} - \sum_{i=1}^{n} y_i \frac{\Theta_i}{b_i RT}\right]$; $F(\lambda) = 1\Big/\left[\frac{\lambda}{C(1)} + \frac{1-\lambda}{C(1.235)}\right]$; $\lambda_{SRK} = 0.36$

c. $\frac{g^E_{EOS}}{RT} = C(\eta^P)\left\{\frac{\Theta_m}{b_m RT} - \frac{1}{2b_m}\sum_{i=1}^{n}\sum_{j=1}^{n} y_i y_j\left[\left(b - \frac{\Theta}{RT}\right)_{ii} + \left(b - \frac{\Theta}{RT}\right)_{jj}\right]\right\} + b_m$

$= \left[\sum_{i=1}^{n}\sum_{j=1}^{n} y_i y_j \left(b - \frac{\Theta}{RT}\right)_{ij}\right]\Big/\left[1 - \left(\frac{G^{E_0}}{C(1.0)} + \sum_{i=1}^{n} y_i \frac{\Theta_i}{b_i RT}\right)\right]$

7.5.2 Wong-Sandler

$$b_m - \frac{a_m}{RT} = \sum_{i=1}^{n}\sum_{j=1}^{n} y_i y_j \left(b - \frac{a}{RT}\right)_{ij} \tag{7.5-5}$$

With $A^{E\infty}_{EOS} = G^{E_0}$, and

$$a_m = b_m\left[\sum_{i=1}^{n} \frac{y_i a_{ii}}{b_{ii}} + \frac{A^{E\infty}_{EOS}}{C(\eta^P)}\right] = b_m\left[\sum_{i=1}^{n} \frac{y_i a_{ii}}{b_{ii}} + \frac{G^{E_0}}{C(\eta^P)}\right] \tag{7.5-6}$$

After selecting a combining rule for $(b - a/RT)_{ij}$ and substituting in Eq. (7.5-6) to eliminate a_m from Eq. (7.5-5), b_m is found. Then a_m is found from Eq. (7.5-6). Unlike for Eqs. (7.5-3) and (7.5-4), the combining rules for b_{ij} and a_{ij} can be chosen independently so the second virial relation is preserved. Note that all $(b - a/RT)$ values must be positive to avoid b_m becoming zero or negative (Orbey and Sandler, 1995, 1998).

Recognizing that there will not be an exact match of computed and the input experimental or correlated G^{E0}, Wong and Sandler (1992) and Wong et al. (1992) suggest that the combining rule be of the form

$$1 - k_{ij} = \frac{2(b - a/RT)_{ij}}{(b_{ii} - a_{ii}/RT) + (b_{jj} - a_{jj}/RT)} \tag{7.5-7}$$

where the optimal value of k_{ij} is obtained by minimizing the difference between calculated and input G^E over the whole composition range. Another choice is to match EOS second cross virial coefficients and pure virial coefficients to experiment with the relation of Eubank et al. (1995),

$$\frac{2(b - a/RT)_{ij}}{(b_{ii} - a_{ii}/RT) + (b_{jj} - a_{jj}/RT)} = \frac{2B_{ij}}{B_{ii} + B_{jj}} \tag{7.5-8}$$

Kolar and Kojima (1994) match infinite dilution activity coefficients from the input experiment or correlation to those computed from the EOS. There have been revisions of the original Wong-Sandler rule (Orbey and Sandler, 1995; Satyro and Trebble, 1996, 1998).

7.5.3 MHV1 and MHV2

A linear or quadratic function q is used to cover the variation of $\alpha^{MHV} = a/(bRT)$ over all possible values

$$q = q_o + q_1\alpha^{MHV} + q_2(\alpha^{MHV})^2 \tag{7.5-9}$$

for the ideal solution, $\alpha_m^{MHV,id} = \sum_{i=1}^{n} y_i a_i/(b_i RT)$ and the real solution $\alpha_m^{MHV} = a_m/(b_m RT)$. The value of q_1 is $C(0.8097)$. The "MHV1" ($q_2 = 0$) and "MHV2" (q_2 optimized) mixing rules that result are (Michelsen, 1990; Dahl and Michelsen, 1990)

$$q_m = q_m^I + \frac{G^{E_0}}{RT} + \sum_{i=1}^{n} y_i \ln\left(\frac{b_m}{b_i}\right) \tag{7.5-10}$$

The LCVM model is a linear combination of the Huron-Vidal and MHV1 mixing rules with the coefficients optimized for application. Other models with a G^E basis are those of Heidemann and Kokal (1990) which involve an iterative calculation to obtain the parameters and Kolar and Kojima (1993) where the η^P matching involves parameters.

Example 7.4 Wong-Sandler mixing rules with PRSV EOS

Use the Wong-Sandler mixing rules and the PRSV EOS to calculate the optimal value of k_{12} for the binary 2-propanol (1)-water (2) at 80 °C. To calculate g^E, use the NRTL equation with parameters fitted to data at 30 °C.

Solution. The pure-component parameter κ_i for the PRSV EOS is found from

$$\kappa_i = 0.378893 + 1.4897\omega_i - 0.17138\omega_i^2 + 0.0196554\omega_i^3 + \kappa_i^{(1)}\left[1 + \left(\frac{T}{T_{c,i}}\right)^{1/2}\right]\left[0.7 - \frac{T}{T_{c,i}}\right] \tag{7.5-11}$$

where ω_i is the component's acentric factor. The parameter $\kappa_i^{(1)}$ is found by fitting experimental P^{vp} data over some temperature range. In this case, $\kappa_1^{(1)} = 0.2326$ and $\kappa_2^{(1)} = -0.0664$. The mixing rules of Eqs. (7.5-5) and (7.5-6) give the mixture parameters a^V, a^L, b^V, and b^L as

$$a^V = b^V\left(\frac{g^{E,V}}{C} + y_1\frac{a_{11}}{b_1} + y_2\frac{a_{22}}{b_2}\right); \quad a^L = b^L\left(\frac{g^{E,L}}{C} + x_1\frac{a_{11}}{b_1} + x_2\frac{a_{22}}{b_2}\right) \tag{7.5-12}$$

$$b^V = \frac{\sum_{i=1}^{2}\sum_{j=1}^{2} y_i y_j (b - a/RT)_{ij}}{1 - \frac{1}{RT}\left(\sum_{i=1}^{2} y_i \frac{a_{ii}}{b_i} + \frac{g^{E,V}}{C}\right)}; \quad b^L = \frac{\sum_{i=1}^{2}\sum_{j=1}^{2} x_i x_j \left(b - \frac{a}{RT}\right)_{ij}}{1 - \frac{1}{RT}\left(\sum_{i=1}^{2} x_i \frac{a_{ii}}{b_i} + \frac{g^{E,L}}{C}\right)} \tag{7.5-13}$$

where g^{EV} is the excess Gibbs energy at the vapor composition and g^{EL} is that at the liquid composition. The constant C for the PRSV EOS is −0.623. It is evident that these are more complex than the vdW1 rule of Eq. (8.12-23).

With the Wong-Sandler mixing rules, we can use any convenient model for g^E; here we use the NRTL expression which for a binary is

$$\frac{g^E}{RT} = x_1x_2\left(\frac{\tau_{21}G_{21}}{x_1 + G_{21}x_2} + \frac{\tau_{12}G_{12}}{x_2 + G_{12}x_1}\right) \tag{7.5-14}$$

with $G_{ij} = \exp(-\alpha_N\tau_{ij})$. There are three parameters: α_N, τ_{12}, and τ_{21}. Typically, α_N is fixed independently. Here the parameter values are those obtained by Gmehling and Onken (1977) by fitting the data of Udovenko and Mazanko (1967). The results are $\alpha_N = 0.2893$, $\tau_{12} = 0.1759$, and $\tau_{21} = 2.1028$.

The cross parameter of Eq. (8.12-32*b*) is

$$(b - a/RT)_{12} = \frac{1}{2}(b_1 + b_2) - \frac{1}{RT}(a_{11}a_{22})^{1/2}(1 - k_{12}) \tag{7.5-15}$$

Among the two approaches for obtaining a value of the parameter k_{12} described in Sec. 5.5, we choose to match the g^E model by minimizing the objective function F, rather than match second cross virial coefficients as suggested by Kolar and Kojima (1994) and others.

$$F = \sum_{\text{data}}\left|\frac{a^E_{\text{EOS}}}{RT} - \frac{g^{E_L}}{RT}\right| \tag{7.5-16}$$

The summation of Eq. (7.5-16) is for all data points at the specified temperature of 30 °C and evaluated at the liquid compositions. The molar excess Helmholtz energy from the EOS a^E_{EOS}, is

$$a^E_{\text{EOS}} = \frac{a^L}{b^L} - x_1\frac{a_{11}}{b_1} - x_2\frac{a_{22}}{b_2} \tag{7.5-17}$$

Optimizing Eq. (7.5-16), we obtain $k_{12} = 0.3644$ for use in Eq. (7.5-15).

Example 7.5 Huron-Vidal mixing rules with tcPR EOS

Use the Huron-Vidal mixing rules and the tcPR EOS to calculate the excess volume for the binary methanol (1)-benzene (2) at 60 °C and $x_1 = 0.4$. To calculate g^E, use the Wilson equation with parameters $a_{12} = 1088.3$ and $a_{21} = 47.8$.

Solution. The tcPR EOS is from Chap. 6,

$$Z = 1 + \frac{\eta^{\text{P}}}{1 - \eta^{\text{P}}} - \left(\frac{a_c\alpha^S}{\text{bRT}}\right)\frac{\eta^{\text{P}}}{(1 - r_1\eta^{\text{P}})(1 - r_2\eta^{\text{P}})} \tag{7.5-18}$$

where $r_1 = -(2^{½} + 1)$ and $r_2 = (2^{½} - 1)$. The pure-component parameters for the tcPR EOS from Pina-Martinez et al. (2018) are

$$a_i = \frac{0.45723553\alpha_i^{tc} R^2 T_{c,i}^2}{P_c} \tag{7.5-19}$$

$$b_i = \frac{0.07779607 R T_{c,i}}{P_{c,i}} - c_i^{VT} \tag{7.5-20}$$

with

$$c_i^{VT} = V_{\text{liq},i}^{PR}\left(T = 0.8T_{c,i}\right) - V_{\text{liq},i}^{\text{expt}}\left(T = 0.8T_{c,i}\right) \tag{7.5-21}$$

$$\alpha_i^{tc} = T_{r,i}^{N_i^{tc}(M_i^{tc}-1)} \exp\left[L_i^{tc}\left(1 - T_{r,i}^{M_i^{tc}N_i^{tc}}\right)\right]; T_{r,i} \equiv T/T_{c,i} \tag{7.5-22}$$

The pure-component parameters are:

Component	T_c (K)	P_c (bar)	L^{tc}	M^{tc}	N^{tc}	a^{PR}	b^{PR}	c^{VT}(cm^3/mol)
METHANOL	512.50	80.84	0.7682	0.9348	1.5716	1604381	41.01	9.177
BENZENE	562.05	48.95	0.1349	0.8482	2.5790	2827052	74.27	−1.439

where a_i^{PR} [=] J-cm^3/mol^2 and b_i^{PR} [=] c_i^{VT} [=] cm^3/mol.

The mixing rules of Sec. 7.3 give the mixture parameters a_m and b_m as

$$b_m = \left(x_1^2 b_{11} + 2x_1x_2b_{12} + x_2^2 b_{22}\right); b_{ij}^{⅔} = \frac{\left(b_{ii}^{⅔} + b_{jj}^{⅔}\right)}{2} \tag{7.5-23}$$

$$a_m = b_m\left(\frac{g^E}{C^{GE}(\eta^P)} + x_1\frac{a_{11}}{b_1} + x_2\frac{a_{22}}{b_2}\right) \tag{7.5-24}$$

where g^E is the excess Gibbs energy at the given composition. At the condition $\eta^P = 1$,

$$C^{GE}(\eta^P) = \frac{1}{r_2 - r_1}\ln\left(\frac{1 - r_1}{1 - r_2}\right) = \frac{1}{8^{½}}\ln\left(\frac{2 - 2^{½}}{2 + 2^{½}}\right) = -0.623 \tag{7.5-25}$$

With the Huron-Vidal mixing rules, we can use any convenient model for g^E; here we use the Wilson expression, which for a volume translated binary mixture is

$$\frac{g^E}{RT} = -x_1\ln\left(\Phi_1^{VT} + \Phi_2^{VT}\Lambda_{21}\right) - x_2\ln\left(\Phi_2^{VT} + \Phi_1^{VT}\Lambda_{12}\right) \tag{7.5-26}$$

with $\Lambda_{ij} = \exp(-a_{ij}/T)$, $a_{12} = 1088.3$, $a_{21} = 47.8$, $\Phi_j^{VT} = x_j b_j^{VT}/b_m^{VT}$, $b_j^{VT} = b_j^{PR} - c_j^{VT}$, $b_m^{VT} = \Sigma x_i b_i^{VT}$

Substitution shows that $b_m = 60.57$ cm^3/mol, $b_m^{VT} = 57.77$ cm^3/mol, $\Sigma x_i a_i/b_i RT = 14.78$, $g^E/RT = -0.9381$, and $a_m = 2183583$ J-cm^3/mol^2. It is evident that these mixing rules are more complex than the typical quadratic mixing rule. Solving gives $Z = 0.0028089$, so $V_m = 73.13$ cm^3/mol. Computing the saturated liquid volume at 313.15 K for each pure fluid gives $V_1 = 39.39$ and $V_2 = 89.94$, so $V^E = 3.41$ cm^3/mol.

7.6 MIXING RULES FOR MULTIPARAMETER EOS

As described in Chap. 6, the complexity of property behavior cannot generally be described with high accuracy by cubic or quartic EOSs that can be solved analytically for the volume when given T, P, and y. Though the search for better models began well before computers, the ability to rapidly calculate results or do parameter regression with complicated expressions has introduced increasing levels of complexity and numbers of fitted parameters.

This section briefly covers the mixture forms of the multiparameter EOS models of Chap. 6. The MBWR form is omitted from this section because its primary use is for the Lee-Kesler model, which now appears to be of limited use relative to alternative methods. We retain coverage of the CSP mixing rules as examples comparable to the TRAPP ECS model of Ely and Hanley (1981) that is still useful for estimating transport properties. We also include mixing rules for the REFPROP models, which derive largely from the GERG-2008 description by Kunz and Wagner (2012).

7.6.1 Two-Parameter and Three-Parameter CSP

The assumptions about intermolecular forces that allow CSP use for mixtures are the same as for pure components (Chap. 6). However, here it is necessary to deal with the effects of interactions between unlike species as well as between like species. As described above, this is commonly done with mixing and combining rules. The primary model of interest in this context is the ECS model of Ely and Hanley (1981).

For the pseudocritical temperature, T_{cm}, the simplest mixing rule is a mole-fraction average method. This rule, often called one of Kay's rules (Kay, 1936), can be satisfactory.

$$T_{c,m} = \sum_{i=1}^{n} y_i T_{c,i} \tag{7.6-1}$$

Comparison of $T_{c,m}$ from Eq. (7.6-1) with values determined from other, more complicated rules considered below shows that the differences in $T_{c,m}$ are usually less than 2% if, for all components the pure-component critical properties are not extremely different. Thus, Kay's rule for $T_{c,m}$ is probably adequate for $0.5 < T_{c,i}/T_{ij} < 2$ and $0.5 < P_{c,i}/P_{c,j} < 2$ (Reid and Leland, 1965).

For the pseudocritical pressure, $P_{c,m}$, a mole-fraction average of pure-component critical pressures is normally unsatisfactory. This is because the critical pressure for most systems goes through a maximum or minimum with composition. The simplest rule which can give acceptable P_{cm} values for two-parameter or three-parameter CSP is the modified rule of Prausnitz and Gunn (1958)

$$P_{c,m} = \frac{Z_{c,m} R T_{c,m}}{V_{c,m}} = \frac{\left(\sum_{i=1}^{n} y_i Z_{c,i}\right) R \left(\sum_{i=1}^{n} y_i Z_{c,i}\right)}{\left(\sum_{i=1}^{n} y_i V_{c,i}\right)} \tag{7.6-2}$$

where all the mixture pseudocriticals $Z_{c,m}$, $T_{c,m}$, and $V_{c,m}$ are given by mole-fraction averages (Kay's rule) and R is the universal gas constant.

For three-parameter CSP, the mixture pseudo acentric factor is commonly given by a mole fraction average (Joffe, 1971)

$$\omega_m = \sum_{i=1}^{n} y_i \omega_i \tag{7.6-3}$$

though others have been used (see, e.g., Brule et al., 1982). While no empirical binary (or higher-order) interaction parameters are included in Eqs. (7.6-1) to (7.6-3), good results may be obtained when these simple pseudomixture parameters are used in corresponding-states calculations for determining mixture properties.

Example 7.6 Molar volume estimation

Estimate the molar volume of an equimolar mixture of methane (1) and propane (2) at $T = 310.92$ K, $P = 206.84$ bar and mixtures of 22.1 and 75.3 mole percent methane at $T = 153.15$ K, $P = 34.37$ bar using CSP. Literature values are 79.34, 60.35, and 48.06 cm^3 mol^{-1}, respectively (Huang et al., 1967).

Solution. The characteristic properties of methane and propane from Appendix A are listed in the table below. Also, the computed pseudoproperties from Eqs. (7.6-1) to (7.6-3) for the three cases are given.

Pure Component/Property	T_c, K	P_c, bar	V_c, cm^3 mol^{-1}	Z_c	ω
Methane	190.56	45.99	98.60	0.286	0.011
Propane	369.83	42.48	200.00	0.276	0.152
Mixture Pseudoproperty	$T^+_{c,m}$, K	$P^+_{c,m}$, bar	$V^*_{c,m}$, cm^3 mol^{-1}	$Z^*_{c,m}$	$\omega^\#_m$
$y_1 = 0.5$	280.20	44.24	149.30	0.281	0.082
$y_1 = 0.221$	330.21	43.26	177.59	0.278	0.121
$y_1 = 0.753$	234.84	45.12	123.65	0.284	0.046

*Mole fraction average as in Eq. (7.6-1)
+Eq. (7.6-2)
Eq. (7.6-3)

The value of Z can be found from Fig. 6.1 only for the first case, but Tables 3.2 and 3.3 of the 4th edition give values of $Z^{(0)}$ and $Z^{(1)}$ to use in $Z = Z^{(0)} + \omega Z^{(1)}$.

The errors for these compressed fluid mixtures with components having significantly different T_c's and V_c's are typical for CSP. In general, accuracy for normal fluid mixtures is slightly less than for pure components unless one or more binary interaction parameters are used.

T K	P bar	y_1	T_r	P_r	Fig. 6.1 Z	Fig. 6.1 v, cm^3 mol^{-1}	Error %	$Z^{(0)}$	$Z^{(1)}$	Z	v, cm^3 mol^{-1}	Err. %
310.9	206.8	0.500	1.11	4.68	0.64	79.99	0.8	0.66	−0.092	0.65	80.92	2.0
153.1	34.37	0.753	0.65	0.76	—	—	—	0.14	0.005	0.14	49.62	3.2
153.1	34.37	0.221	0.46	0.80	—	—	—	0.17	0.007	0.16	60.80	0.7

As discussed in Chap. 6, CSP descriptions are less reliable for substances with strong dipoles or showing molecular complexation (association). The same limitations apply to mixtures of such compounds. Mixtures can also bring in one additional dimension; there can be mixtures involving normal substances with complex substances. Though the interactions between a nonpolar species and a polar or associating species involve only nonpolar forces (Prausnitz et al., 1999), because the critical or other characteristic properties of the polar species involve more than just the nonpolar forces, combining rules such as Eqs. (7.6-1) to (7.6-3) are usually in error. The common approach to treating polar/nonpolar systems, and mixtures of normal compounds where the sizes are significantly different, is to use binary interaction parameters as described in Sec. 7.3.

Regardless of which combining rule is used, it is common for the fitted binary interaction parameter of polar/nonpolar systems to reduce the value of T_{cij} to less than that of the geometric or arithmetic mean of the pure component values. The need for a binary interaction parameter may be overcome by a combination of the geometric and harmonic means as in the second virial coefficient correlation of Hayden and O'Connell (1975).

7.6.2 REFPROP and Related Models

The most used multiparameter models can be exemplified in the form of the REFPROP model (Lemmon et al., 2018). REFPROP includes the GERG-2008 (Kunz and Wagner, 2012) compounds and characterizations as well as CoolProp (Bell et al., 2014) compounds. At the outset, the mixture model applies quadratic mixing to the Helmholtz energy of the individual components:

$$a_m^{res}(\delta_r, \tau_r, \boldsymbol{x}) = \sum_{i=1}^{N_c} x_i a_i^{res} + \sum_{i=1}^{N_c-1} \sum_{j=i+1}^{N_c} x_i x_j a_{ij}^{res}(\delta_r, \tau_r, \boldsymbol{x}) \tag{7.6-4}$$

where $\delta_r = \rho/\rho_{ref}(\mathbf{x})$; $\tau_r = T_{ref}(\mathbf{x})/T(\mathrm{K})$,

$$\frac{1}{\rho_{ref}(\boldsymbol{x})} = \sum_{i=1}^{N_c} \frac{x_i^2}{\rho_{c,i}} + \sum_{i=1}^{N_c-1} \sum_{j=i+1}^{N_c} \frac{x_i x_j \beta_{v,ij} \gamma_{v,ij}}{4} \frac{x_i + x_j}{\beta_{v,ij}^2 x_i + x_j} \left(\frac{1}{\rho_{c,i}^{1/3}} + \frac{1}{\rho_{c,j}^{1/3}} \right)^3 \tag{7.6-5}$$

$$T_{ref}(\boldsymbol{x}) = \sum_{i=1}^{N_c} x_i^2 T_{c,i} + \sum_{i=1}^{N_c-1} \sum_{j=i+1}^{N_c} 2 x_i x_j \beta_{T,ij} \gamma_{T,ij} \frac{x_i + x_j}{\beta_{T,ij}^2 x_i + x_j} (T_{c,i} T_{c,j})^{1/2} \tag{7.6-6}$$

$$a_{ij}^{res}(\delta_r, \tau_r, \boldsymbol{x}) = \sum_{k=1}^{K_{ij}^{pol}} n_{ij,k} \delta_r^{d_{ij,k}} \tau_r^{t_{ij,k}} + \sum_{k=K_{ij}^{pol}+1}^{K_{ij}^{pol}+K_{ij}^{exp}} n_{ij,k} \delta_r^{d_{ij,k}} \tau_r^{t_{ij,k}} \exp\left[\eta_{ij,k}(\delta_r - \varepsilon_{ij,k})^2 + \beta_{ij,k}(\delta_r - \gamma_{ij,k}) \right] \tag{7.6-7}$$

The most significant BIPs are $\beta_{v,ij}$, $\gamma_{T,ij}$, $\beta_{T,ij}$, $\gamma_{T,ij}$, used in correlating ρ_{ref} and T_{ref}. The doubly summed contribution in Eq. (7.6-4) is usually small and sometimes neglected entirely. This contribution can add 76 more BIPs in the case of methane+ethane, for example, a case where experimental data are plentiful. The multicomponent behavior is predicted from the binary mixture data. For the GERG-2008 model, the values of the BIPS were characterized by minimizing deviations for roughly 125,000 values of experimental data from roughly 650 sources. The experimental data covered thermal and caloric properties as well as VLE at temperatures ranging from 16 to 2500 K and to pressures as high as 2000 MPa. These mixing rules do not comply with the invariance condition of Michelsen and Kistenmacher (1990). Kunz and Wagner (2012) suggest that noncompliance does not affect the accuracy of their regression and is of minor consequence, possibly because all the compounds in their analysis were sufficiently similar in size. Readers should be aware of possibly artificial behavior if adapting this approach to broader classes of mixtures.

7.7 VIRIAL EQUATIONS OF STATE FOR MIXTURES

As described in Chap. 6, the virial equation of state is a polynomial series in pressure or in inverse volume, but for mixtures the coefficients are functions of both T and $\mathbf{y}$. The consistent forms for the initial terms are

$$Z = 1 + B_2\left(\frac{P}{RT}\right) + \left(B_3 - B_2^2\right)\left(\frac{P}{RT}\right)^2 + \cdots \tag{7.7-1}$$

$$= 1 + \frac{B_2}{V} + \frac{B_3}{V^2} + \cdots \tag{7.7-2}$$

where the coefficients B_2, B_2, ... are called the second, third, ... virial coefficients. Except at high temperatures, B_2 is negative and, except at very low T where they are of little importance, B_3 and higher coefficients are positive. Mixture isotherms at constant composition are like those of Fig. 6.1. Formulae relating B_2 to molecular pair interactions, B_3 to molecular trio interactions, etc., can be derived from statistical mechanics. Their composition dependence is rigorous.

$$B_2(T, \boldsymbol{y}) = \sum_{i=1}^{n} \sum_{j=1}^{n} y_i y_j B_{2,ij}(T) \tag{7.7-3}$$

$$B_3(T, \boldsymbol{y}) = \sum_{i=1}^{n} \sum_{j=1}^{n} \sum_{k=1}^{n} y_i y_j y_k B_{3,ijk}(T) \tag{7.7-4}$$

where the virial coefficients for pure component i would be $B_{2,ii}$ and $B_{3,iii}$ with the pairs and trios being the same substance. When $i \neq j, k$, the pairs and trios are unlike and the coefficients are called *cross coefficients.*

There is symmetry with the subscripts so that $B_{2,ij} = B_{2,ji}$ and $B_{3,iij} = B_{3,jii} = B_{3,iji}$. In the case of a three-component system

$$B_2(T, \mathbf{y}) = y_1^2 B_{2,11}(T) + 2y_1y_2B_{2,12}(T) + y_2^2B_{2,22}(T) + 2y_1y_3B_{2,13}(T) + 2y_2y_3B_{2,23}(T) + y_3^2B_{2,33}(T) \quad (7.7\text{-}5)$$

while in the case of a two-component system

$$B_3(T, \mathbf{y}) = y_1^3B_{3,111}(T) + 3y_1^2y_2B_{3,112}(T) + 3y_1y_2^2B_{3,122}(T) + y_2^3B_{3,222}(T) \quad (7.7\text{-}6)$$

Much has been written about the virial EOS; see especially Mason and Spurling (1968) and Dymond and Smith (1980).

The general ranges of state for applying Eqs. (7.7-1) and (7.7-2) to mixtures are the same as described in Chap. 6 for pure fluids; the virial equation should be truncated at the second or third term and applied only to single-phase gas systems.

The most extensive compilations of second cross virial coefficients are those of Frenkel and Marsh (2002). Some third cross virial coefficient values are also given by Dymond and Smith (1980). Iglesias-Silva et al. (1999) discuss methods to obtain cross coefficients from density measurements.

7.7.1 Estimation of Second Cross Virial Coefficients

Our treatment of cross virial coefficients is the same as for pure coefficients in Chap. 6. All the methods there can be used here if the parameters are suitably adjusted. As before, the formulation is in CSP for all pairs of components in the mixture, i and j.

$$\frac{B_{2,ij}(T)}{V_{ij}^*} = \sum_m a_{mij} f^{(m)}\left(\frac{T}{T_{ij}^*}\right) \quad (7.7\text{-}7)$$

where V_{ij}^* is a characteristic volume for the pair, the a_{mij} are strength parameters for various pair intermolecular forces described in Chap. 6, and the $f^{(m)}$ are sets of universal functions of reduced temperature, and T/T_{ij}^* with T_{ij}^* a characteristic temperature for the pair. Then, $f^{(0)}$ is for simple substances with a_0 being unity, $f^{(1)}$ corrects for nonspherical shape and globularity of normal substances with a_1 commonly being ω_{ij}. If one or both of the components are dipolar, $f^{(2)}$ takes account of polarity with a_2 being a function of the dipole moments (see Chap. 3), μ_i and μ_j, when both are dipolar, or, if only one species is dipolar, another function of the dipole of the polar species and the polarizability of the other component. Finally, $f^{(3)}$ takes account of association among like molecules or solvation among unlike molecules with a_3 an empirical parameter. The value of a_3 may be the same for cross coefficients as for pure coefficients among substances of the same class such as alcohols. On the other hand, an a_3 value can be required even if none exists for pure interactions, should strong interactions exist as in $CHCl_3$ and $(CH_3)_2C{=}O$.

When treating cross coefficients, most of the methods of Chap. 6 use combining rules for T_{ij}^* of the form of Eq. (7-2.4b) with a constant BIP, k_{ij}. Often, there are methods for estimating k_{ij} such as the same value for all pairs of compound classes. These methods commonly omit the polar/associating contribution for polar/nonpolar pairs and may use an empirical parameter for solvation. There is normally considerable sensitivity to the values of the parameters; Stein and Miller (1980) discuss this issue with the Hayden-O'Connell (1975) model and provide useful guidance about obtaining solvation parameters.

Detailed discussion of second virial coefficient correlations is given in Chap. 5. Just as for pure components, no single technique is significantly better than the others for cross coefficients, except for systems involving very strongly associating/solvating species such as carboxylic acids where the correlation of Hayden and O'Connell (1975) is the only one that applies. We illustrate the expressions and use in detail only the Tsonopoulos (1974, 1975) correlation since it is one of the most popular and reliable.

For second cross coefficients, the Tsonopoulos correlation uses $V_{ij}^* = RT_{cij}/P_{cij}$ and $T_{ij}^* = T_{cij}$. The substance-dependent strength coefficients are $a_{1ij} = \omega_{ij}$, $a_2 = a_{ij}$, and $a_3 = b_{ij}$. Values of the pure compound parameters are given in Chap. 6. The following combining rules were established (these expressions are rearrangements of the original expressions):

$$T_{ij}^* = T_{c,ij} = (T_{c,ii}T_{c,jj})^{1/2}(1 - k_{ij}) \tag{7.7-8}$$

$$V_{ij}^* = \frac{(V_{c,ii}^{1/3} + V_{c,jj}^{1/3})^3}{4(Z_{c,ii} + Z_{c,jj})} = \frac{RT_{c,ij}}{P_{c,ij}} = \frac{R(V_{c,ii}^{1/3} + V_{c,jj}^{1/3})^3}{4(P_{c,ii}V_{c,ii}/T_{c,ii} + P_{c,jj}V_{c,jj}/T_{c,jj})} \tag{7.7-9}$$

$$a_{1ij} = \omega_{ij} = (\omega_{ii} + \omega_{jj})/2 \tag{7.7-10}$$

where a binary interaction parameter, k_{ij}, has been included. For either i or j or both without a significant dipole moment

$$a_{ij} = 0 = b_{ij} \qquad \mu \sim 0 \text{ and/or } \mu_j \sim 0 \tag{7.7-11}$$

For both i and j having a significant dipole moment

$$a_{ij} = (a_{ii} + a_{jj})/2 \qquad \mu_i \neq 0 \neq \mu_j \tag{7.7-12}$$

$$b_{ij} = (b_{ii} + b_{jj})/2 \tag{7.7-13}$$

Values of the BIP, k_{ij}, are given in the references cited in Chap. 6. Estimations of k_{ij} for nonpolar pairs usually involve critical volumes. For example, Tsonopoulos et al. (1989) reconfirm the relationship of Chueh and Prausnitz (1967b) for nonpolar pairs which is apparently reliable to within ±0.02:

$$k_{ij} = 1 - \left[\frac{2(V_{c,ii}V_{c,jj})^{1/6}}{\left(V_{c,ii}^{1/3} + V_{c,jj}^{1/3}\right)}\right]^3 \qquad \mu_i \approx \mu_j \approx 0 \tag{7.7-14}$$

For polar/nonpolar pairs, constant values of k_{ij} are used. Thus, for 1-alkanols with ethers, Tsonopoulos and Dymond (1997) recommend $k_{ij} = 0.10$.

In many practical systems, strong interactions occur due to charge-transfer and hydrogen bonding. This occurs in mixtures if alcohols, amines, amides, ethers, carboxylic acids, water, HF, etc. are present. The chemical approach has been to consider the interactions so strong that new "chemical species" are formed. Then the thermodynamic treatment assumes that the properties deviate from an ideal gas mainly due to the "speciation" (the actual number of molecules in the system is not the number put in) plus some physical effects. It is assumed that all the species are in reaction equilibrium. Thus, their concentrations can be determined from equilibrium constants having parameters such as enthalpies and entropies of reaction as well as parameters for species physical interactions. In mixtures, an example is the formation of unlike dimers (D) from two different monomers (M_1 and M_2):

$$M_1 + M_2 = D \tag{7.7-15}$$

The equilibrium constant for this reaction can be related to the second cross virial coefficient

$$K_D = \frac{y_D}{y_{M_1}y_{M_2}P} = -2\frac{B_{12}}{RT} \tag{7.7-16}$$

where the factor of 2 arises here because of the distinct monomers of Eq. (7.4-25). The model of Hayden and O'Connell (1975) described in Secs. 6.5 and 7.4 explicitly includes such contributions so that it can also predict the properties of strongly interacting unlike substances.

Computations of second cross virial coefficients from molecular theory can add more value than similar calculations for pure compounds. Measurement of cross coefficients is more difficult than for pure fluids and estimation methods are less reliable than for pure fluids. Computation of cross coefficients using the method of Elliott (2021) is no more difficult than for pure fluids. Hellman (2020) provides an example of computation from first principles.

Example 7.7 Quadratic mixing of the second virial coefficient

Estimate the second virial coefficient of a 40 mole percent mixture of ethanol (1) with benzene (2) at 403.2 K and 523.2 K using the Tsonopoulos method and compare the results with the data of Wormald and Snowden (1997).

Solution. From Appendix A, the critical properties and acentric factors of the substances are given in the table below. In addition, from Table 6.5, $a_{11} = 0.0878$ and with $\mu_r = 66.4$, $b_{11} = 0.0553$, while these quantities are zero for both B_{12} and B_{22}. The recommendation (Tsonopoulos, 1974) for the binary interaction constant is $k_{12} = 0.20$; however, this seems too large and an estimate of $k_{12} = 0.10$ is more consistent with later analyses such as by Tsonopoulos et al. (1989). The values of T^*_{ij}, V^*_{ij}, and ω_{ij}, are given for each pair followed by all computed quantities for the B_{ij} values.

Quantity	For 1-1 pair	For 1-2 pair	For 2-2 pair
$T_{c,ij}$, K	513.92	483.7	562.05
$P_{c,ij}$, bar	61.48	—	48.95
$v_{c,ij}$, cm^3 mol^{-1}	167.00	208.34	256.00
$Z_{c,ij}$	0.240	0.254	0.268
v^*_{ij}, cm^3 mol^{-1}	695.8	820.2	955.2
ω_{ij}	0.649	0.429	0.209
$T/T_{c,ij}$	0.785	0.834	0.717
$f^{(0)}_{ij}$	−0.530	−0.474	−0.626
$f^{(1)}_{ij}$	−0.330	−0.225	−0.553
$f^{(2)}_{ij}$	4.288	2.981	7.337
$f^{(3)}_{ij}$	−6.966	−4.290	−14.257
$B_{2,ij}/v^*_{ij}$	−0.753	−0.571	−0.742
$B_{2,ij}$, cm^3 mol^{-1}	−524	−468	−708
$B_{2,ij}$ (exp), cm^3 mol^{-1}	−529	−428	−717
Error, cm^3 mol^{-1}	-5	40	-9

The computed mixture value is B_{2m} (calc) = −591 cm^3 mol^{-1} while the experimental value, B_{2m} (exp) = −548 cm^3 mol^{-1}. The difference of 43 cm^3 mol^{-1} is almost within the experimental uncertainty of ±40 cm^3 mol^{-1}.

If the same procedure is used at $T = 523.2$ K, the results are:

Quantity	For 1-1 pair	For 1-2 pair	For 2-2 pair
$T/T_{c,ij}$	1.018	1.082	0.931
$f_{ij}^{(0)}$	−0.325	−0.289	−0.386
$f_{ij}^{(1)}$	−0.025	0.008	−0.093
$f_{ij}^{(2)}$	−0.898	0.624	1.537
$f_{ij}^{(3)}$	−0.867	−0.534	−1.774
$B_{2,ij}/v_{ij}^*$	−0.310	−0.285	−0.405
$B_{2,ij}$, cm^3 mol^{-1}	−216	−234	−387
$B_{2,ij}$ (exp), cm^3 mol^{-1}	−204	−180	−447
Error, cm^3 mol^{-1}	12	54	−60

The agreement for the individual coefficients is not as good as at $T = 403.2$ K, but the errors compensate and the calculated mixture value, B_{2m} (calc) = −278 cm^3 mol^{-1} is very close to the experimental value, B_{2m} (exp) = −280 cm^3 mol^{-1}, though this would not occur when y_2 is near 1. To reproduce $B_{2,12}$ (exp) precisely, $k_{12} = 0.18$. This example illustrates the sensitivity of the calculations to the value of the BIP and how results may appear accurate at certain conditions but not at others.

See the discussion in Chap. 5 for updates to the methods for second virial coefficients. Lee and Chen (1998) have revised the Tsonopoulos expression. The model of Hayden and O'Connell (1975) has been discussed by several authors as noted in Chap. 5. Stein and Miller (1980) made improvements in this model to treat solvating systems such as amines with methanol.

Literature discussion and our own comparisons show that none of the correlations referenced above is significantly more accurate or reliable than the others except for systems with carboxylic acids where the Hayden-O'Connell method is best. Thus, for the range of conditions that the second virial coefficient should be applied to obtain fluid properties, all models are likely to be adequate.

7.7.2 Estimation of Third Cross Virial Coefficients

The limitations on predicting third cross virial coefficients are the same as described in Sec. 6.5 for pure third virial coefficients. In particular, no comprehensive models have been developed for systems with polar or associating substances. Further, there are very few data available for the $B_{3,ijk}$ or for B_{3m}.

For third cross coefficients of nonpolar substances, the CSP models of Chueh and Prausnitz (1967a), De Santis and Grande (1979), and Orbey and Vera (1983) can be used. In all cases, the approach is

$$B_{3,ijk} = (B_{3,ij} B_{3,jk} B_{3,ik})^{1/3} \tag{7.7-17}$$

where the pairwise $B_{3,ij}$ are computed from the pure component formula with characteristic parameters obtained from pairwise combining rules including BIPs. The importance of accurate values of $B_{3,112}$ for describing solid-fluid equilibria of a dilute solute (2) in a supercritical solvent (1) is nicely illustrated in Chueh and Prausnitz (1967a).

7.8 RESIDUAL FUNCTIONS FOR EVALUATED MODELS

The enthalpy of a mixture can be computed directly from the EOS through the mixing rules and temperature derivative of the Helmholtz energy. The heat of mixing can then be computed by subtracting the molar average of the pure component enthalpies at the same temperature and pressure. Therefore, the only partial molar property of significant interest is the chemical potential or, alternatively, the fugacity coefficient for each component in the mixture. Table 7.2 summarizes useful residual functions for the present evaluations.

TABLE 7.2 Useful residual functions for dispersion interactions.

a. Contributions from dispersion interactions		
PR	$\frac{u^{\text{res}}}{RT} = \frac{-A}{B\sqrt{8}}\left(1+\frac{T}{a}\frac{da}{dT}\right)\ln\left(\frac{Z+\left(1+\sqrt{2}\right)B}{Z+\left(1-\sqrt{2}\right)B}\right)$ $\frac{T}{a}\frac{da}{dT} = \sum_i\sum_j \frac{x_i x_j a_{ij}}{2a}\left(\frac{\kappa_i^{PR}\sqrt{T}}{\sqrt{T_{c,i}\alpha_i^S}} + \frac{\kappa_j^{PR}\sqrt{T}}{\sqrt{T_{c,j}\alpha_j^S}}\right)$	(7.8-1)
SRK	$\frac{u^{\text{res}}}{RT} = \frac{-A}{B}\left(1+\frac{T}{a}\frac{da}{dT}\right)\ln\left(\frac{Z+B}{Z}\right)$	(7.8-2)
tcPRq	$\frac{u^{\text{res}}}{RT} = \frac{-A}{B\sqrt{8}}\left(1+\frac{T}{a}\frac{da}{dT}\right)\ln\left(\frac{Z+\left(1+\sqrt{2}\right)B}{Z+\left(1-\sqrt{2}\right)B}\right)$ $\frac{T}{a}\frac{da}{dT} = \sum_i\sum_j \frac{x_i x_j a_{ij}}{2a}\left[\left(\frac{T}{\alpha_i}\right)*\left(\frac{d\alpha_i}{dT}\right)+\left(\frac{T}{\alpha_j}\right)*\left(\frac{d\alpha_j}{dT}\right)\right]$	(7.8-3)
ESD	$\mu_{JT} = \left(\frac{\partial T}{\partial P}\right)_H = \frac{\left(T\left(\frac{\partial V}{\partial T}\right)_H - V\right)}{C_P}$	(7.8-4)
PC-SAFT	$\mu_{JT} = \left(\frac{\partial T}{\partial P}\right)_H = \frac{\left(T\left(\frac{\partial V}{\partial T}\right)_H - V\right)}{C_P}$	(7.8-5)
SPEADMD	$\frac{u^{\text{res}}}{RT} = \frac{A_1}{T} + \frac{2A_2}{T^2}$ $\quad \frac{C_V^{\text{res}}}{RT} = \frac{2A_2}{T^2}$	(7.8-6)

For models that include chemical contributions, additional terms are required to fully express the residual energy and related functions. Continuing Eqs. (7.4-36) to (7.4-48) for the residual internal energy,

$$\frac{u^{\text{chem}}}{RT} = -\frac{\beta}{2}\frac{\partial h^{AD}}{\partial \beta} - \frac{\beta}{2}\frac{\partial h^{DA}}{\partial \beta} \tag{7.8-7}$$

$$\beta\frac{\partial h^{AD}}{\partial \beta} = \left[\sum_m^{N_c}\sum_k^{N_c} x_m x_k \sum_i^{N_s}\sum_j^{N_s} X_{im}^A X_{jk}^D \rho g_{im,jk}^{AD} \beta\varepsilon_{im,jk}^{AD} \exp(\beta\varepsilon_{im,jk}^{AD})\right] \tag{7.8-8}$$

$$\frac{C_V^{\text{chem}}}{R} = -\frac{\beta^2}{2}\frac{\partial^2 h^{AD}}{\partial \beta^2} - \frac{\beta^2}{2}\frac{\partial^2 h^{DA}}{\partial \beta^2} \tag{7.8-9}$$

$$\frac{\beta^2}{2}\frac{\partial^2 h^{AD}}{\partial \beta^2} = \left[\sum_m^{N_c}\sum_k^{N_c} x_m x_k \sum_i^{N_s}\sum_j^{N_s} X_{im}^A X_{jk}^D \rho g_{im,jk}^{AD} \left(\beta\varepsilon_{im,jk}^{AD}\right)^2 \exp\left(\beta\varepsilon_{im,jk}^{AD}\right)\right] \tag{7.8-10}$$

Similar results apply for h^{DA}. These equations are numerous, but they cover many quantities of interest and they are straightforward to apply when $\{X_{jk}^B\}$ has been determined.

Derivative properties for the polar contributions are even more tedious than the ones for the chemical contributions. Readers are referred to the paper by Vrabec and Gross (2008) and citations therein for those formulas.

7.9 EMPIRICAL CORRELATIONS FOR MIXTURE PROPERTIES

To extend empirical methods like those in Chap. 5 to mixtures, mixing rules are required. Such extensions should be used with care because they may require specific adaptations according to specific situations. Constraints on such methods may lead to confounding results when applied as a small part of a generalized procedure. These kinds of methods were reviewed in PGL5ed, but they have not been evaluated here. Our expectation is that the best methods from our current evaluations should provide better accuracy over broader ranges of application.

7.10 EVALUATIONS AND RECOMMENDATIONS

Mixture EOS models must reflect the complexity of real mixtures. As a result, the expressions can be very complicated. Fortunately, the coding is straightforward and codes for several representative models are available through the PGL6ed website.

The PRLorraine module developed by Jaubert and coworkers includes coverage of several mixing rules applied to the tcPR model as the cubic EOS basis. As seen in Chap. 6, the tcPR model provides accuracy for vapor pressure of about 1%, approaching the uncertainty in the experimental data for many compounds. Given the flexibility of the g^E mixing rules, it should be feasible to find a specific formulation that works well for most applications. For example, Xu et al. (2017) describe a predictive implementation that describes a broad range of complex phase behaviors related to carbon capture and storage. Bell et al. (2019) describe refrigerant mixtures with a similar model. Our evaluations are limited to the Wilson g^E model applied at the infinite-pressure reference state, but many alternatives are available, including UNIQUAC and NRTL as the basis for the g^E mixing rules. We chose the Wilson g^E model because it comes closest to a consistent representation of the athermal contribution within the EOS framework, as discussed in Chap. 8.

The PC-SAFT module comprises a sophisticated FORTRAN90 implementation by Gross and coworkers. It includes code for multicomponent TPT1 computations and for the *DD*, *QQ*, and *DQ* polarity contributions as well as simpler mixing rules applied to the polar PC-SAFT model. It also includes code for checking phase stability in accordance with the guidelines of Michelsen and Mollerup (2007).

The ESD96 module is a self-contained implementation of the ESD model with the efficient implementation of Wertheim's theory (Elliott, 1996). It includes code for multicomponent TPT1 computations as well as quadratic mixing rules for the cubic component of the ESD model.

The SPEADMD module includes code for the repulsive and attractive perturbation contributions, with calls to a Wertheim module for TPT1 computations. In principle, it should be feasible to adapt the Wertheim module to TPT1 computations with other models. The repulsive contribution incorporates the behavior illustrated in Fig. 7.5.

These modules are accompanied by a few wrapper and utility codes to define global constants and uniform calling arguments from the wrapper. Databases for each module include the pure compound and BIPs required in Chaps. 6 to 9. The BIP files are organized to favor the LLE values by default, followed by VLE, and finally solid-liquid equilibrium (SLE) values. Readers can rearrange these open-source files, of course. These codes are provided to illustrate the manner of implementation and no guarantees are implied. We welcome comments about errors or bugs.

The NIST-mod- and NIST-KT-UNIFAC models are described by Kang et al. (2011) and Kang et al. (2015). All UNIFAC activity models are also described in more detail in Chap. 8. The REFPROP package is available from NIST as Database 23. Version 10 provides a recent example (Lemmon et al., 2018). Portions of related code are available through the CoolProp python project (Bell et al., 2014).

Several open-source models have been contributed by R.D.P. Soares. These models comprise the University Federal Rio Grande Sul (UFRGS) module, available through the GitHub mentioned in several of their manuscripts. The UFRGS models include the COSMO-RS/SAC-Phi model of Soares et al. (2019), the COSMO-RS/SAC-GAMESS model of Ferrarini et al. (2018), and the COSMO-RS/FSAC model of Soares et al. (2013). Professor Soares also provided results from his implementation of the PSRK model (Horstmann et al., 2005).

Characterizing thermodynamic properties for mixtures is a lot like the characterization for pure compounds. The favored methods are also similar. The data sources tend to be less systematic, however, and more difficult to characterize. Data sources tend to favor many measurements for similar compounds. For example, *n*-alkanes and gases dominate mixtures of normal fluids and alcohols dominate mixtures involving associating compounds. Characterization of the data classes is made instantly more difficult because every binary combination must be considered. Therefore, the tables in this chapter are transposed relative to those in Chap. 6 and only the best representative methods from Chap. 6 are included.

The relative deviations for excess properties can be positive or negative, or zero in principle. To avoid artificial emphasis on excess properties near zero, we apply offsets of 2 cm^3/mol for excess volume and 1 kJ/mol for excess enthalpy. These offset values were determined by raising their values until the %off deviations were on the order of 10% for the best models. Therefore, the tabulated relative deviations are given below. For excess volume,

$$\%\text{off}(V^E) = \sum \frac{(\text{calc} - \text{expt})}{N^{\text{pts}}\left(\text{expt} + 2.0\,\dfrac{\text{cm}^3}{\text{mol}}\right)} * 100\% \tag{7.10-1}$$

For excess enthalpy,

$$\%\text{off}(H^E) = \sum \frac{(\text{calc} - \text{expt})}{N^{\text{pts}}\left(\text{expt} + 1000\,\dfrac{\text{J}}{\text{mol}}\right)} * 100\% \tag{7.10-2}$$

7.10.1 Discussion and Recommendations

None of the models stood out for predicting excess volume (Table 7.3). The COSMO-RS/SAC-Phi and REFPROP models covered too few systems to be considered comparable to other models. The deviations for other models were all in the range of 13 to 15% off, except for the EGC-PC-SAFT model at 18%. One might have hoped that detailed accounting for the effects of strong interactions would improve predictions of excess volume, but no clear indication was apparent. Further research is required to develop a clear improvement in modeling excess volumes. Without a clearly superior overall model, readers should choose their model based on the detailed results. For example, if readers are interested in systems that include heavy molecules, they should choose the PC-SAFT model, but the tcPR-GE model should be acceptable for other types of mixtures. In no case should accuracy better than 2 cm^3/mol be expected.

Gradations in quality for predictions of excess enthalpy were more apparent (Table 7.4). Once again, the COSMO-RS/SAC-Phi and REFPROP models covered too few systems to be considered comparable to other models. Both NIST implementations of UNIFAC showed the smallest deviations at around 15%, but they include extra temperature-dependent parameters and their training dataset was very similar to our evaluation database. The two COSMO-RS/SAC implementations were next at about 20% deviation. This suggests that the physical basis of their activity models is at least self-consistent with the excess enthalpy, noting that neither model was trained with excess enthalpy in mind. The ESD and SPEADMD models showed about 25% deviation, both for customized and transferable forms, which was in the mid-range of all models tested. The tcPR-GE model showed larger deviations, around 40%. This may suggest that successes of this model in correlating low pressure VLE data (Chap. 8) come at the expense of self-consistent predictions of excess enthalpy. The most surprising result was for the PC-SAFT implementation, which exhibited deviations around 45%. This is a model that accounts for polarity as well as dispersion and strong interactions, so it should have a strong physical basis. It may be that the values of parameters for polarity and association energy need to be considered more systematically for the PC-SAFT model to provide reliable predictions of excess enthalpy. Polar-polar mixtures exhibited unusually large deviations for the PC-SAFT model. Overall, any of the models with deviations of 25% or less are recommended for excess enthalpy predictions. As a relatively simple EOS, the ESD model provided steady performance for all types of mixtures. Note that COSMO-RS/SAC models include detailed accounting for strong interactions like hydrogen bonding, as discussed in Chap. 8, as do the ESD and SPEADMD models. Altogether, it seems justified to recommend that readers choose such a model if heats of mixing are a concern. As of this writing,

TABLE 7.3 Comparisons of EOS models for excess volume deviations. The classes (Normal, Heavy, Polar, and Associating) are abbreviated by their first letter.

 # points

Type	AA	AH	AN	AP	HH	HN	HP	NN	NP	PP	total
REFPROP	7466	—	5319	2806	—	17	—	16394	3428	30	35460
SPEADMD	44386	429	64683	21714	30	2692	415	47466	26752	2061	210628
ESD	42888	429	65695	22742	39	3169	447	51224	28865	2191	217689
TFF-SPEADMD	7793	358	35891	7044	30	2264	232	30470	10569	162	94813
tcPR-GE	43260	429	66422	21199	39	3125	447	51218	27404	1930	215473
EGC-ESD	11051	429	43463	11235	39	2921	388	36333	16145	477	122481
EGC-ESD(T_b)	11051	429	43463	11235	39	2921	388	36333	16145	477	122481
PC-SAFT	38400	429	59740	19295	24	2440	431	47700	23678	1783	193920
EGC-PC-SAFT	11051	429	43463	11235	30	2505	376	36276	16145	477	121987
COSMO-RS/SAC-Phi)	18783	—	19068	5024	6	179	—	12241	5598	281	61180
PSRK	43023	429	64796	19674	30	2679	431	46510	24217	1621	203410

 # systems

Type	AA	AH	AN	AP	HH	HN	HP	NN	NP	PP	total
REFPROP	8	—	31	6	—	1	—	134	24	1	205
SPEADMD	528	10	1109	433	3	83	16	983	600	80	3845
ESD	503	10	1131	471	4	99	19	1111	686	87	4121
TFF-SPEADMD	193	9	660	159	3	70	9	567	253	9	1932
tcPR-GE	525	10	1170	434	4	96	19	1110	648	75	4091
EGC-ESD	243	10	767	250	4	89	16	728	384	24	2515
EGC-ESD(T_b)	243	10	767	250	4	89	16	728	384	24	2515
PC-SAFT	416	10	986	387	2	71	17	1010	585	69	3553
EGC-PC-SAFT	243	10	767	250	3	78	15	725	384	24	2499
COSMO-RS(SAC-Phi)	52	—	145	45	1	6	—	123	63	8	443
PSRK	516	10	1119	435	3	86	17	994	617	65	3862

% off 2 cm^3/mol

Type	AA	AH	AN	AP	HH	HN	HP	NN	NP	PP	total
REFPROP	6	—	18	55	—	164	—	10	8	3	14
SPEADMD	9	8	12	17	10	16	16	12	15	21	13
ESD	13	19	18	23	10	13	5	12	14	8	15
TFF-SPEADMD	11	6	12	21	10	15	22	9	16	33	12
tcPR-GE	17	51	15	15	9	44	18	11	11	6	14
EGC-ESD	12	23	16	22	9	17	9	9	17	25	14
EGC-ESD(T_b)	11	23	15	23	10	15	9	9	17	25	14
PC-SAFT	15	9	15	13	4	9	18	10	15	7	13
EGC-PC-SAFT	11	10	15	11	10	30	160	23	26	25	19
COSMO-RS(SAC-Phi)	15	—	11	13	37	60	—	9	13	10	12
PSRK	14	81	19	13	15	54	38	11	11	10	15

TABLE 7.4 Comparisons of EOS models for excess enthalpy deviations.

Points

Type	AA	AH	AN	AP	HH	HN	HP	NN	NP	PP	total
REFPROP	3055	—	6294	1677	—	7	—	13187	935	17	25172
SPEADMD	16612	564	40837	10167	—	2721	417	37882	14079	715	123994
ESD	15408	564	40603	10930	—	3035	445	40583	15630	789	127987
TFF-SPEADMD	4020	413	22482	2991	—	2217	263	22149	5819	72	60426
tcPR-GE	16138	564	41433	10763	—	2994	445	40574	15375	775	129061
EGC-ESD	4721	564	26865	4523	—	2614	388	26039	8295	102	74111
EGC-ESD(T_b)	4721	564	26865	4523	—	2614	388	26039	8295	102	74111
PC-SAFT	14895	555	37812	9793	—	2410	436	36667	13810	580	116958
EGC-PC–SAFT	4721	564	26865	4523	—	2380	379	26032	8295	102	73861
NIST-KT-UNIFAC	16271	553	36874	10430	—	3092	425	41438	13366	825	123274
NIST-mod-UNIFAC	16365	642	38495	12075	—	3187	388	41896	15301	1104	129453
COSMO-RS(SAC-Phi)	6733	—	17424	3498	—	133	—	11043	3103	129	42063
PSRK	16125	564	40796	9613	—	2761	436	37338	13917	549	122099
COSMO-RS/SAC/GAMESS	17006	564	41479	9698	—	2670	424	36544	13514	465	122364
COSMO-RS/FSAC	7509	400	25542	4697	—	1565	118	24211	8781	313	73136
COSMOtherm	14227	528	33461	8562	—	2947	391	36513	13266	695	110590

systems

Type	AA	AH	AN	AP	HH	HN	HP	NN	NP	PP	total
REFPROP	7	—	43	6	—	1	—	126	25	1	209
SPEADMD	383	18	965	297	—	87	19	969	509	34	3281
ESD	337	18	958	314	—	107	21	1104	577	39	3475
TFF-SPEADMD	124	17	529	106	—	70	10	507	207	4	1574
tcPR-GE	376	18	1001	299	—	105	21	1103	562	37	3522
EGC-ESD	151	18	610	165	—	93	17	667	302	6	2029
EGC-ESD(T_b)	151	18	610	165	—	93	17	667	302	6	2029
PC-SAFT	319	17	856	285	—	76	20	955	502	26	3056
EGC-PC-SAFT	151	18	610	165	—	77	16	666	302	6	2011
NIST-KT-UNIFAC	360	25	977	280	—	109	19	992	471	20	3253
NIST-mod-UNIFAC	372	25	1001	317	—	114	21	1015	540	37	3442
COSMO-RS(SAC-Phi)	49	—	161	46	—	6	—	134	69	5	470
PSRK	371	18	979	290	—	89	20	987	511	25	3290
COSMO-RS/SAC/GAMESS	405	18	996	295	—	88	19	923	486	24	3254
COSMO-RS/FSAC	100	10	427	112	—	47	7	531	273	11	1518
COSMOtherm	375	18	955	306	—	107	20	1078	537	35	3431

(*Continued*)

TABLE 7.4 Comparisons of EOS models for excess enthalpy deviations. (*Continued*)

deviations (%off, 1 kJ/mol)											
Type	**AA**	**AH**	**AN**	**AP**	**HH**	**HN**	**HP**	**NN**	**NP**	**PP**	**total**
REFPROP	30	—	49	60	54	—	6	9	3	48	27
tcPR-GE	80	33	37	51	16	27	16	27	11	151	42
SPEADMD	48	12	18	59	19	38	18	31	37	57	29
PC-SAFT	96	24	29	36	13	44	17	28	13	198	44
ESD	37	21	21	26	33	57	22	34	12	34	26
TFF-SPEADMD	49	13	15	79	19	45	16	31	120	0	23
EGC-ESD	40	27	23	20	31	58	18	35	18	0	24
EGC-ESD(T_b)	40	27	23	19	30	56	18	35	17	0	24
EGC-PC-SAFT	40	22	32	28	53	740	95	57	23	0	61
NIST-KT-UNIFAC	13	8	12	8	8	6	16	25	3	24	15
NIST-mod-UNIFAC	13	8	13	11	11	7	14	11	12	12	13
COSMO-RS(SAC-Phi)	38	—	29	16	53	—	15	17	15	52	28
PSRK	34	17	19	24	13	8	15	17	11	43	21
COSMO-RS/SAC/GAMESS	28	11	24	23	16	7	11	12	10	36	20
COSMO-RS/FSAC	15	12	14	11	14	9	8	10	8	17	12
COSMOtherm(3ds)	21	12	19	24	22	21	13	12	6	25	17

COSMO-RS/SAC models are only available as predictive models, with roughly double the deviations in modeling VLE when compared to ESD or SPEADMD. Therefore, we conclude that either the ESD or SPEADMD models should be selected if readers are interested in accurately correlating VLE while obtaining the best possible estimates of mixing enthalpies. The PC-SAFT model could be included in this recommendation if not for the unexpected deviations observed for polar-polar mixtures. Detailed investigation of that issue may lead to a resolution that makes the PC-SAFT model equally viable.

It may seem artificial to contemplate mixture volumes and enthalpies without simultaneously considering phase behavior. On the other hand, the interplay of VLE, LLE, SLE, and activity coefficients can be complicated, especially at pressures above 1 MPa. Thus, we must defer further comments about the impacts of mixing rules and mixture models and phase behavior to Chaps. 8 and 9.

7.11 NOTATION

In many equations in this chapter, special constants or parameters are defined and usually denoted $a, b, \ldots, A, B, \ldots$. They are not defined in this section because they apply only to the specific equation and are not used anywhere else in the chapter.

a^E	excess Helmholtz energy, J mol^{-1}
a_{mij}	coefficient for terms in second virial coefficient correlations
b	cubic EOS covolume parameter (cm^3/mole)
$B_{2,ij}$	second virial coefficient, cm^3 mol^{-1}
$B_{3,ijk}$	third virial coefficient, cm^6 mol^{-2}

$C(u)$ EOS matching variable for G^E mixing rules
g^E excess Gibbs energy, J mol^{-1}
k_{ij}, l_{ij} binary interaction parameters
n number of components in a mixture
P pressure, MPa
P^{vp} vapor pressure, MPa
Q generalized property
q quantity in MHV1 and MHV2 mixing rules
R gas constant, Table 6.1
r number of segments in a chain
T temperature, K
u reciprocal packing fraction = v/b
v molar volume, cm^3 mol^{-1}
V volume, cm^3
w weight fraction
x, y mole fraction
Z compressibility factor = Pv/RT

Greek

α quantity in MHV1 and MHV2 mixing rules = Θ/bRT
δ, ε EOS variables,
η^{P} packing fraction
Φ volume fraction
Θ EOS variable
μ dipole moment
ρ molar density
σ^{HS} hard sphere diameter
σ^{LJ} Lennard-Jones sphere diameter
ξ mixture hard-sphere packing fraction
ω acentric factor

Superscript

0 zero pressure
∞ infinite pressure
id ideal solution
liq liquid
MCSL Mansoori, et al. (1971) hard-sphere EOS,
$\circ$ pure component property
res residual property
vap vapor

Subscripts

att	attractive forces
c	critical
cm	mixture pseudocritical
EOS	equation of state
i	component *i*
ij	component pair *i* and *j*
m	mixture
rep	repulsive

7.12 REFERENCES

Abrams, D. S., and J. M. Prausnitz: *AIChE J.,* **21:** 116 (1975).
Anderko, A: *Fluid Phase Equil.,* **65:** 89 (1991).
Anderko, A., I. G. Economou, and M. D. Donohue: *Ind. Eng. Chem. Res.,* **32:** 245 (1993).
Anisimov, M. A., A. A. Povodyrev, and J. V. Sengers: *Fluid Phase Equil.,* **158–160:** 537 (1999).
Arnaud, J. F., P. Ungerer, E. Behar, B. Moracchini, and J. Sanchez: *Fluid Phase Equil.,* **124:** 177 (1996).
Bell, I. H., J. Wronski, S. Quoilin, and V. Lemort: *Ind. Eng. Chem. Res.,* **53:** 2498–2508 (2014).
Bell, I. H., J. Welliquet, M. E. Mondejar, A. Bazyleva, S. Quoilin, and F. Haglind: *Int. J. Refrig.,* **103:** 316–328 (2019).
Blanks, R. F., and J. M. Prausnitz: *Ind. Eng. Chem.,* **3:** 1 (1964).
Boublik, T.: *J. Chem. Phys.,* **53:** 471 (1970).
Boukouvalas, C., N. Spiliotis, P. Coutsikos, and N. Tzouvaras: *Fluid Phase Equil.,* **92:** 75 (1994).
Brandani, S. and V. Brandani: *Fluid Phase Equil.,* **121:** 179 (1996).
Brule, M. R., C. T. Lin, L. L. Lee, and K. E. Starling: *AIChE J.,* **28:** 616 (1982).
Chapman, W. G., K. E. Gubbins, G. Jackson, and M. Radosz: *Ind. Eng. Chem. Res.,* **29:** 1709 (1990).
Chueh, P. L., and J. M. Prausnitz, *AIChE J.,* **13:** 896 (1967a).
Chueh, P. L., and J. M. Prausnitz: *Ind. Eng. Chem. Fundam.,* **6:** 492 (1967b).
Constantinescu, D., and J. Gmehling: *J. Chem. Eng. Data,* **61:** 2738–2748 (2016).
Coto, B., R. Peschla, C. Kreiter, and G. Maurer: *Ind. Eng. Chem. Res.,* **42:** 2934–2939 (2003).
Dahl, S., and M. L. Michelsen: *AIChE J.,* **36:** 1829 (1990).
De Santis, R., and B. Grande: *AIChE J.,* **25:** 931 (1979).
Dudowicz, J., and K. F. Freed: *Macromolecules,* **24:** 5076–5095 (1991).
Dymond, J. H., and E. B. Smith: *The Virial Coefficients of Pure Gases and Mixtures: A Critical Compilation*, Clarendon Press, Oxford, 1980.
Economou, I. G., and M. D. Donohue: *Ind. Eng. Chem. Res.,* **31:** 1203 (1992).
Elliott, J. R., Suresh, S.J., Donohue: *Ind. Eng. Chem. Res.,* **29:** 1624–1629 (1990).
Elliott, J. R.: *Chem. Eng. Ed.,* **27:** 44–51 (1993).
Elliott, J. R., *Ind. Eng. Chem. Res.,* **35:** 1624–1629 (1996).
Elliott, J. R., and C. T. Lira: *Introductory Chemical Engineering Thermodynamics*, Prentice-Hall, Englewood Cliffs, NJ, 2012.
Elliott, J. R.: *J. Phys. Chem. B,* **125:** 4494–4500 (2021).
Ely, J. F., and H. J. M. Hanley: *Ind. Eng. Chem. Fund.,* **20:** 323–332 (1981).
Eubank, P. T., G-S. Shyu, and N. S. M. Hanif: *Ind. Eng. Chem. Res.,* **34:** 314 (1995).
Ferrarini, F., G. B. Flôres, A. R. Muniz, and R. P. D. Soares: *AIChE J.,* **64:** 3443–3455 (2018).
Fischer, K., and J. Gmehling: *Fluid Phase Equil.,* **121:** 185 (1996).
Flory, P. J.: *J. Chem. Phys.,* **9:** 660 (1941).
Frenkel, M., and K. N. Marsh (eds): "Virial Coefficients of Pure Gases and Mixtures," Landolt-Börnstein - Group IV Physical Chemistry Volume 21A, Springer-Materials, Springer-Nature, Switzerland AG (2002).
Gmehling, J., and U. Onken: *Dechema Chemistry Data Series,* **1:** Part 1: 49 and Part 2a: 77 (1977).
Gmehling, J.: D. D. Liu, and J. M. Prausnitz, *Chem. Eng. Sci.,* **34:** 951 (1979).
Gross, J., and G. Sadowski: *Ind. Eng. Chem. Res.,* **41:** 5510–5515 (2002).

Gross, J.: *AIChE J.,* **51:** 2556 (2005).
Guggenheim, E. A.: *Mixtures*, Clarendon, Oxford, 1952.
Hayden, J. G., and J. P. O'Connell: *Ind. Eng. Chem. Process Des. Dev.,* **14:** 209 (1975).
Heidemann, R. A., J. M. Prausnitz, *Proc. Nat. Acad. Sci.,* **73:** 1773–1776 (1976).
Heidemann, R. A., and S. L. Kokal: *Fluid Phase Equil.,* **56:** 17 (1990).
Helmann, R., *J. Chem. Eng. Data,* **65:** 4130–4141 (2020).
Horstmann, S., A. Jabłoniec, J. Krafczyk, K. Fischer, and J. Gmehling: *Fluid Phase Equilib.,* **227:** 157–164 (2005).
Huang, E. T. S., G. W. Swift, and F. Kurata: *AIChE J.,* **13:** 846 (1967).
Huggins, M.L., *J. Chem. Phys.,* **9:** 440 (1941).
Huron, M.-J., and J. Vidal: *Fluid Phase Equil.,* **3:** 255 (1979).
Iglesias-Silva, G. A., M. S. Mannan, F. Y. Shaikh, and K. R. Hall: *Fluid Phase Equil.,* **161:** 33 (1999).
Ikonomou, G. D., and M. D. Donohue: *Fluid Phase Equilib.,* **39:** 129 (1988).
Jaubert, J.-N., F. Mutelet, *Fluid Phase Equilib.,* **224:** 285–304 (2004).
Jaubert, J.-N., Y. Le Guennec, A. Pina-Martinez, N. Ramirez-Velez, S. Lasala, B. Schmid, I. K. Nikolaidis, I. G. Economou, and R. Privat: *Ind. Eng. Chem. Res.,* **59:**14981–15027 (2020).
Jin, G. X., S. Tang, and J. V. Sengers: *Phys. Rev. A,* **47:** 388 (1993).
Joffe, J.: *Ind. Eng. Chem. Fundam.,* **10:** 532 (1971).
Kang, J. W., J. Abildskov, R. Gani, and J. Cobas: *Ind. Eng. Chem. Res.,* **41:** 3260–3273 (2002).
Kang, J. W., V. Diky, R. D. Chirico, J. W. Magee, C. Muzny, I. Abdulagatov, A. F. Kazakov, and M. Frenkel: *Fluid Phase Equilib.,* **309:** 68–75 (2011).
Kang, J. W., V. Diky, M. Frenkel: *Fluid Phase Equilib.,* **388:** 128–141 (2015).
Kay, W. B.: *Ind. Eng. Chem.,* **28:** 1014 (1936).
Kiran, E., and J. M. H. Levelt Sengers (eds): *Supercritical Fluids: Fundamentals for Applications,* NATO ASI Series E: **273:** Kluwer Academic Publishers, Dordrecht, Holland, 1994.
Kiselev, S. B., *Fluid Phase Equil.,* **147:** 7 (1998).
Kiselev, S. B., and D. G. Friend: *Fluid Phase Equil.,* **162:** 51 (1999).
Kolar, P., and K. Kojima: *J. Chem. Eng. Japan,* **26:** 166 (1993).
Kolar, P., and K. Kojima: *J. Chem. Eng. Japan,* **27:** 460 (1994).
Kontogeorgis, G. M., and G. Folas: *Thermodynamic Models for Industrial Applications*, John Wiley & Sons, West Sussex, UK, 2010.
Kontogeorgis, G. M., X. Liang, A. Arya, and I. Tsivintzelis: *Chem. Eng. Sci: X,* **7:** 100060 (2020).
Kunz, O., and W. Wagner: *J. Chem. Eng. Data,* **57:** 3032 (2012).
Lee, M.-J., and J.-T. Chen: *J. Chem. Eng. Japan,* **31:** 518 (1998).
Leibovici, C. F.: *Fluid Phase Equil.,* **84:** 1 (1993).
Lemmon, E. W., I. H. Bell, M. L. Huber, and M. O. McLinden: NIST Standard Reference Database 23: Reference Fluid Thermodynamic and Transport Properties-REFPROP, Version 10.0, National Institute of Standards and Technology, 2018. http://www.nist.gov/srd/nist23. cfm
Mansoori, G. A., N. F. Carnahan, K. E. Starling, and T. W. Leland, Jr.: *J. Chem. Phys.,* **54:** 1523 (1971).
Mason, E. A., and T. H. Spurling: *The Virial Equation of State,* Pergamon, New York, 1968.
Mathias, P. M., H. C. Klotz, and J. M. Prausnitz: *Fluid Phase Equil.,* **67:** 31 (1991).
Maurer, G., Prausnitz, J.M. *Fluid Phase Equil.* **2:** 91 (1978).
Michelsen, M. L.: *Fluid Phase Equil.,* **60:** 47, 213 (1990).
Michelsen, M. L., and H. Kistenmacher: *Fluid Phase Equil.,* **58:** 229 (1990).
Michelsen, M. L.: *Fluid Phase Equil.,* **121:** 15 (1996).
Michelsen, M. L., R. A. Heidemann: *Ind. Eng. Chem. Res.,* **35:** 278–287 (1996).
Michelsen, M. L., E. M. Hendriks: *Fluid Phase Equilibria,* **180:** 165–174 (2001).
Michelsen M. L., Mollerup J.: *Thermodynamic Modelling: Fundamentals and Computational Aspects*, 2nd ed., Tie-Line Publications, 2007.
Mollerup, J., and M. L. Michelsen: *Fluid Phase Equil.,* **74:** 1 (1992).
Muller, E. A., and K. E. Gubbins: *Ind. Eng. Chem. Res.,* **40:** 2193–2211 (2001).
Orbey, H., and S. I. Sandler: *Fluid Phase Equil.,* **111:** 53 (1995).
Orbey, H., and S. I. Sandler: *Modeling Vapor-Liquid Equilibria,* Cambridge University Press, Cambridge, 1998.
Orbey, H., and J. H. Vera: *AIChE J.,* **29:** 107 (1983).
Panagiotopoulos, A. Z., and R. C. Reid: *Adv. in Chem. Ser.,* **300:** 571 (1986).
Pavlyukhin, Y. T., *J. of Struct. Chem.,* **53**:476–486 (2012).
Peng, D. Y., and D. B. Robinson: *Ind. Eng. Chem. Fundam.,* **15:** 59 (1976).

Pina-Martinez, A., Y. le Guennec, R. Privat, J.-N. Jaubert, and P. M. Mathias: *J. Chem. Eng. Data,* **63:** 3980–3988 (2018).
Prausnitz, J. M., and R. D. Gunn: *AIChE J.,* **4:** 430, 494 (1958).
Prausnitz, J. M., R. N. Lichtenthaler, and E. G. de Azevedo: *Molecular Thermodynamics of Fluid-Phase Equilibria,* 3rd ed., Prentice-Hall, Upper Saddle River, NJ, 1999.
Rainwater, J. C.: in Ely, J. F., and T. J. Bruno (eds): *Supercritical Fluid Technology,* CRC Press, Boca Raton, FL, 1991, p. 57.
Reid, R. C., and T. W. Leland, Jr.: *AIChE J.,* **11:** 228 (1965), **12:** 1227 (1966).
Satyro, M. A., and M. A. Trebble: *Fluid Phase Equil.,* **115:** 135 (1996).
Satyro, M. A., and M. A. Trebble: *Fluid Phase Equil.,* **143:** 89 (1998).
Schwartzentruber, J., and H. Renon: *Ind. Eng. Chem. Res.,* **28:** 1049 (1989).
Soares, R. D. P., L.F. Baladao, P.B. Staudt, *Fluid Phase Equilib.,* **488:** 13–26 (2019).
Soares, R. D. P., R. P. Gerber, L. F. K. Possani, and P. B. Staudt: *Ind. Eng. Chem. Res.,* **52:** 11172–11181 (2013).
Soave, G.: *Fluid Phase Equil.,* **82:** 345 (1993).
Staverman, A. J.: *Rec. Trav. Chem. Pays-Bas.,* **69:** 163 (1950).
Stein, F. P., and E. J. Miller: *Ind. Eng. Chem. Process Des. Dev.,* **19:** 123 (1980).
Suresh, S. J., and J. R. Elliott: *Ind. Eng. Chem. Res.,* **30:** 524–532 (1991).
Tan, Sugata P., H. Adidharma, and M. Radosz: *Ind. Eng. Chem. Res.,* **47:** 8063–8082 (2008).
Tochigi, K., P. Kolar, T. Izumi, and K. Kojima: *Fluid Phase Equil.,* **96:** 53 (1994).
Tsonopoulos, C.: *AIChE J.,* **20:** 263 (1974).
Tsonopoulos, C.: *AIChE J.,* **21:** 827 (1975).
Tsonopoulos, C., J. H. Dymond, and A. M. Szafranski: *Pure. Appl. Chem.,* **61:** 1387 (1989).
Tsonopoulos, C., and J. H. Dymond: *Fluid Phase Equil.,* **133:** 11 (1997).
Twu, C. H., and J. E. Coon: *AIChE J.,* **42:** 2312 (1996).
Twu, C. H., J. E. Coon, D. Bluck, and B. Tilton: *Fluid Phase Equil.,* **158–160:** 271 (1999).
Udovenko, V. V., and T. F. Mazanko: *Zh. Fiz. Khim.,* **41:** 1615 (1967).
Unlu, O., N. H. Gray, Z. N. Gerek, and J. R. Elliott: *Ind. Eng. Chem. Res.,* **43:** 1788–1793 (2004).
Vahid, A. V., N. H. Gray, and J. R. Elliott: *Macromolecules,* **47:** 1514–1531 (2014).
van der Waals, J. H.: *Zh. Phys. Chem.,* **5:** 133 (1890).
Van Ness, H. C., and M. M. Abbott: *Classical Thermodynamics of Non-Electrolyte Solutions,* McGraw-Hill, New York, 1982.
Vasudevan, V. J., and J. R. Elliott: *Fluid Phase Equil.,* **83:** 33–41 (1993).
von Solms, N., I. A. Kouskoumvekaki, M. L. Michelsen, and G. M. Kontogeorgis: *Fluid Phase Equil.,* **241:** 344–353 (2006).
Vrabec, J., and J. Gross: *J. Phys. Chem. B,* **112:** 51–60 (2008).
Wilson, G. M., *J. Am. Chem. Soc.,* **86:** 127, 133 (1964).
Wong, D. S. H., and S. I. Sandler: *AIChE J.,* **38:** 671 (1992).
Wong, D. S. H., H. Orbey, and S. I. Sandler: *Ind. Eng. Chem. Res.,* **31:** 2033 (1992).
Wormald, C. J., and C. J. Snowden: *J. Chem. Thermo.,* **29:** 1223 (1997).
Xu, X., S. Lasala, R. Privat, and J.-N. Jaubert, *Int. J. Greenhouse Gas Control,* **56:**126–154 (2017).
Zabaloy, M. S., and J. H. Vera: *Fluid Phase Equil.,* **119:** 27 (1996).

8

Vapor-Liquid Equilibria in Mixtures

8.1 SCOPE

In the chemical process industries, fluid mixtures are often separated into their components by unit operations such as distillation, absorption, and stripping. Design of such separation operations requires quantitative estimates of the vapor-liquid equilibrium (VLE) properties of fluid mixtures. Whenever possible, such estimates should be based on reliable experimental data for the mixture at conditions of temperature, pressure, and composition corresponding to those of interest. Since most designs are supported by chemical process simulators, which rarely use experimental data directly, the available models must be correlated to reproduce the available experimental data. Unfortunately, such data are rarely plentiful. In typical cases, only fragmentary data are at hand, and it is necessary to reduce and correlate the limited data to make the best possible interpolations and extrapolations. This chapter discusses some techniques that are useful toward that end. In this edition, the focus is on nonelectrolytes. Volumes have been written about electrolyte solutions (e.g., Zemaitis et al., 2010) and including their evaluations in PGL6ed would be overwhelming. Emphasis is given to the calculation of fugacities in liquid solutions. Vapor and supercritical fugacities are typically treated through equations of state. By putting evaluations of equations of state on the same basis as activity models, our expectation is that equations of state that perform well are accurately representing vapor fugacities.

The scientific literature on fluid phase equilibria goes back well over 170 years and has reached monumental proportions, including thousands of articles and hundreds of books and monographs. The fifth edition of *The Properties of Gases and Liquids* (PGL5ed) tabulates lists of authors and titles of some books useful for obtaining data and for more detailed discussions. The lists are not exhaustive; they are restricted to publications likely to be useful to the practicing engineer in the chemical process industries. Since 2012, the National Institute of Standards and Technology (NIST) has maintained an online archive of literature references with experimental data for many mixtures, as well as for pure fluid properties (Kazakov et al., 2012). This "NIST ThermoLit" website can be very useful in quickly identifying whether data are available in the literature. Other resources like the NIST/TDE interface (also available in the Aspen process simulator) or the DECHEMA database can provide the experimental data as well as the literature sources.

The important step is to compile the available experimental data for as many relevant mixtures as possible and to compare various models to the data and arrive at recommended models. The level of relevance may be different for various stages of a process. For example, a bioethanol process might consist of multiple distillation columns to first approach and break the azeotrope, then recover any solvents or entrainers. Each distillation column would operate with different key components and at different conditions. Ideally, experimental data could be acquired that reflected those key components at their process conditions and graphs could be prepared for each case. The more likely case is that each binary combination of compounds in the mixture has been studied at conditions that might differ considerably from the actual process. Even then, some important binary combinations may lack experimental data entirely.

In principle, it should not be a problem to infer binary interaction parameters (BIPs) at one temperature, pressure, and composition and then apply them at another. For example, the temperature dependence of the BIPs is an inherent part of each model's development by its original authors. On the other hand, this raises questions about the underlying theory. Activity models like the Redlich-Kister and Margules models are like polynomial regressions; there is no theory *per se*. Other models like PC-SAFT include sophisticated theories, with significant contributions that have been validated with molecular simulation. Other familiar models like UNIQUAC, NRTL, and Wilson's equation lie somewhere between pure empiricism and validated statistical mechanics. Therefore, the engineer must know enough about the models to assess their reliable extrapolation and include that assessment when arriving at a recommendation. Consequently, we devote section 8.7 of this chapter to understanding activity models.

We illustrate a typical analysis in Fig. 8.1. The components are methanol and benzene at 55 °C. More relevant conditions for distillation might be at a constant pressure near atmospheric, implying slightly higher temperatures, but isothermal data are preferred when focusing on composition effects. The Margules 2-parameter (Margules2), UNIQUAC (Abrams and Prausnitz, 1975), and Flory-Huggins (1941) (FH) models should be familiar from a course in introductory chemical engineering thermodynamics (e.g., Chaps. 11–13 of Elliott and Lira, 2012). The PC-SAFT (Gross and Sadowski, 2002) equation of state (EOS) is relatively new and may be less familiar. Quasichemical theory (QCT, Guggenheim, 1952) is quite old, but rarely discussed in introductory coursework, although it provides the fundamental basis for the COSMO-RS model (Klamt et al., 1995, 1998, 2002) as well as UNIQUAC. All the models shown except UNIQUAC and Margules2 include a single BIP. The Wilson (1964) and NRTL (Renon and Prausnitz, 1969) models, with two BIPs each, are not shown but they perform much like the UNIQUAC model. The SPEADMD EOS (Unlu et al., 2004) and PC-SAFT EOS perform much like the ESD EOS (Elliott et al., 1990) for this system.

Table 8.1 provides quantitative comparisons for all these models as well as the "translated consistent" Peng-Robinson EOS (Le Guennec et al., 2016) with a quadratic mixing rule (tcPRq) and with Gibbs excess mixing rules (tcPR-GE). All EOS models have been introduced in Chap. 6 with mixing rules presented in Chap. 7. Summary expressions for the above activity models are included in Table 8.2.

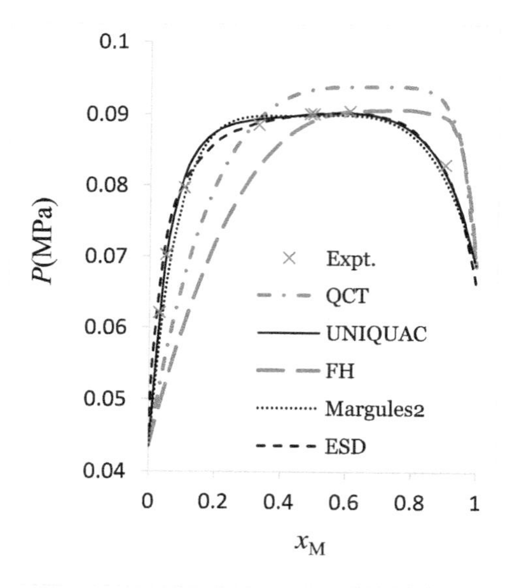

FIGURE 8.1
Illustration of model comparisons for the system methanol+benzene at 55 °C. Model details are given in Table 8.1.

TABLE 8.1 Brief comparison of several models for the methanol+benzene system at 55 °C.

One-parameter models:

Model	QCT	FH	PC-SAFT	SPEADMD	ESD	tcPRq
%AADP	3.9	11.4	3.2	3.2	1.6	8.9
BIP	565	0.1318	−0.0031	0.033	0.022	0.085

Two-parameter models:

Model	Wilson	UNIQUAC	NRTL	Margules2	tcPR-GE
%AADP	0.33	0.73	0.34	2.54	0.70
BIP12	643.5	−47.76	2.204	2.193	1056
BIP21	357.1	620.0	1.322	1.665	78.8

TABLE 8.2 Common expressions for g^E.

Name	g^E	BIPs	ln γ_1 and ln γ_2
One-parameter Margules	$g^E = x_1 x_2 A$	A	$RT\ln(\gamma_1) = A(1-x_1)^2$ $RT\ln(\gamma_2) = A(1-x_2)^2$
Two-parameter Margules	$g^E = x_1 x_2[A + B(x_1 - x_2)]$	A, B	$RT\ln(\gamma_1) = (A+3B)x_2^2 - 4Bx_2^3$ $RT\ln(\gamma_2) = (A-3B)x_1^2 + 4Bx_1^3$
Three-parameter Margules	$g^E = x_1 x_2[A + B(x_1 - x_2) + C(x_1 - x_2)^2]$	A, B, C	$RT\ln(\gamma_1) = (A+3B+5C)x_2^2 - 4(B+4C)x_2^3 + 12Cx_2^4$ $RT\ln(\gamma_2) = (A-3B+5C)x_1^2 - 4(B-4C)x_1^3 + 12Cx_1^4$
van Laar	$g^E = \dfrac{Ax_1x_2}{x_1(A/B) + x_2}$	A, B	$RT\ln(\gamma_1) = A\left(1 + \dfrac{A}{B}\dfrac{x_1}{x_2}\right)^{-2}$ $RT\ln(\gamma_1) = B\left(1 + \dfrac{B}{A}\dfrac{x_2}{x_1}\right)^{-2}$
Wilson	$\dfrac{g^E}{RT} = -x_1\ln(x_1 + \Lambda_{12}x_2) - x_2\ln(x_2 + \Lambda_{21}x_1)$	A_{12}, A_{21}	$\ln(\gamma_1) = -\ln(x_1 + \Lambda_{12}x_2) + x_2\left(\dfrac{\Lambda_{12}}{x_1 + \Lambda_{12}x_2} - \dfrac{\Lambda_{21}}{\Lambda_{21}x_1 + x_2}\right)$ $\ln(\gamma_2) = -\ln(x_2 + \Lambda_{21}x_1) - x_1\left(\dfrac{\Lambda_{12}}{x_1 + \Lambda_{12}x_2} - \dfrac{\Lambda_{21}}{\Lambda_{21}x_1 + x_2}\right)$

(Continued)

TABLE 8.2 Common expressions for g^E. (*Continued*)

Name	g^E	BIPs	$\ln\gamma_1$ and $\ln\gamma_2$
NRTL	$\frac{g^E}{RT} = x_1x_2\left(\frac{\tau_{21}G_{21}}{x_1 + x_2G_{21}} + \frac{\tau_{12}G_{12}}{x_2 + x_1G_{12}}\right)$	A_{12}, A_{21}	$\ln(\gamma_1) = x_2^2\left[\frac{\tau_{12}G_{12}}{(x_2 + G_{12}x_1)^2} + \tau_{21}\left(\frac{G_{21}}{G_{21}x_2 + x_1}\right)^2\right]$ $\ln(\gamma_2) = x_1^2\left[\frac{\tau_{21}G_{21}}{(x_1 + G_{21}x_2)^2} + \tau_{12}\left(\frac{G_{12}}{G_{12}x_1 + x_2}\right)^2\right]$
UNIQUAC[e]	$g^E = g^E(\text{combinatorial}) + g^E(\text{res})$ $\frac{g^E(\text{res})}{RT} = -q_1x_1\ln[\theta_1 + \theta_2\tau_{21}] - q_2x_2\ln[\theta_2 + \theta_1\tau_{12}]$ $\Phi_1 = \frac{x_1r_1}{x_1r_1 + x_2r_2};\ \theta_1 = \frac{x_1q_1}{x_1q_1 + x_2q_2}$	$\Delta u_{12}, \Delta u_{21}$	$\ln(\gamma_i) = \ln(\gamma_i^{\text{comb}}) + \ln(\gamma_i^{\text{res}})$ $\ln(\gamma_i^{\text{res}}) = -q_i\ln(\theta_i + \theta_j\tau_{ji}) + \theta_jq_i\left(\frac{\tau_{ji}}{\theta_i + \theta_j\tau_{ji}} - \frac{\tau_{ij}}{\theta_j + \theta_i\tau_{ij}}\right)$ where $i = 1,\ j = 2$ or $i = 2,\ j = 1$
GSA	$\frac{g^E(\text{combinatorial})}{RT} = x_1\ln\left(\frac{\Phi_1}{x_1}\right) + x_2\ln\left(\frac{\Phi_2}{x_2}\right) + \frac{z}{2}\left(q_1x_1\ln\left(\frac{\theta_1}{\Phi_1}\right) + q_2x_2\ln\left(\frac{\theta_2}{\Phi_2}\right)\right)$		$\ln(\gamma_i^{\text{comb}}) = \ln\left(\frac{\Phi_i}{x_i}\right) + \frac{z}{2}q_i\ln\left(\frac{\theta_i}{\Phi_i}\right) + \Phi_j\left(l_i - \frac{r_i}{r_j}l_j\right)$ $l_i = \frac{z}{2}(r_i - q_i) - (r_i - 1)$ $l_j = \frac{z}{2}(r_j - q_j) - (r_j - 1)$
mod-GSA	$\frac{g^E(\text{combinatorial})}{RT} = x_1\ln\left(\frac{\Phi'_1}{x_1}\right) + x_2\ln\left(\frac{\Phi'_2}{x_2}\right) + \frac{z}{2}\left(q_1x_1\ln\left(\frac{\theta_1}{\Phi'_1}\right) + q_2x_2\ln\left(\frac{\theta_2}{\Phi'_2}\right)\right)$		$\ln(\gamma_i^{\text{comb}}) = \ln\left(\frac{\Phi'_i}{x_i}\right) + \frac{z}{2}q_i\ln\left(\frac{\theta_i}{\Phi'_i}\right) + \Phi'_j\left(l_i - \frac{r_i}{r_j}l_j\right)$ $\Phi'_1 = \frac{x_1r_1^{3/4}}{x_1r_1^{3/4} + x_2r_2^{3/4}}$
LSG	Same as UNIQUAC except: $\frac{g^E(\text{res})}{RT} = -\frac{q_1x_1z}{2}\ln[\theta_1 + \theta_2\tau_{21}] - \frac{q_2x_2z}{2}\ln[\theta_2 + \theta_1\tau_{12}]$		$\ln(\gamma_i^{\text{res}}) = -\frac{q_iz}{2}\ln(\theta_i + \theta_j\tau_{ji}) + \theta_j\frac{q_iz}{2}\left(\frac{\tau_{ji}}{\theta_i + \theta_j\tau_{ji}} - \frac{\tau_{ij}}{\theta_j + \theta_i\tau_{ij}}\right)$

In Table 8.1, the average absolute percent deviation in pressure is:

$$\%\text{AADP} \equiv 100 \cdot \text{NPTS}^{-1}\sum |\text{calc-expt}|/\text{expt} \tag{8.1-1}$$

A value of $\alpha_N = 0.43$ was used for the NRTL model, as illustrated in Example 8.2. All models were constrained to avoid liquid-phase splitting. The numerical results indicate that the Wilson model is much more accurate than the FH or tcPRq models and other 1-parameter models are of intermediate accuracy. That interpretation is oversimplified, however. Relatively few experimental data are considered in this example, and only a single temperature. It is safer to say that the models with deviations smaller than 1% are quite accurate, primarily because their two BIPs enable free adjustment of skewness and magnitude of the Gibbs excess energy.

The ESD model is listed in Table 8.1 with deviations of 1.6%, which is close to the 1% threshold despite being a one-parameter model. The ESD, SPEADMD, and PC-SAFT equations of state (EOSs) all adapt Wertheim's (1984) first-order thermodynamic perturbation theory (TPT1) to represent chemical-physical interactions like hydrogen bonding. The initial adaptation of TPT1 was called the statistical associating fluid theory (SAFT) by Chapman et al. (1990). As described in Chap. 6, the SAFT acronym has been applied to many EOSs since then, but not all implementations of TPT1 include SAFT in their names. Attempting to minimize confusion, we refer to adaptations of Wertheim's first-order theory generically as TPT1. For the evaluations in this chapter, efforts were made to acquire working codes for several implementations of TPT1, including the CPA model (Kontogeorgis et al., 1996), the GC-PPC-SAFT model (Nguyen-Huynh et al., 2011), SAFT-VR (Gil Villegas et al., 1997), and the Soft-SAFT model (Blas et al., 1997). Unfortunately, only the ESD, SPEADMD, and PC-SAFT models could be implemented as representative of TPT1 in time for publication. The PC-SAFT model is representative of models that include explicit treatment of polar and quadrupolar contributions. The ESD EOS can be likened to simplified SAFT models like CPA; it is cubic in the absence of the TPT1 term and still artifact-free when TPT1 is included. The SPEADMD model can be likened to a group-contribution SAFT model, although adding a new compound requires molecular simulation. FORTRAN codes for these models are available through the PGL6ed website as discussed in Chap. 7. We hope to include a broader representation of TPT1 implementations in PGL7ed. The smaller deviations of ESD for the system in Table 8.1 are fortuitous, probably owing to a cancellation of errors between vapor pressure and TPT1. Graphing the TPT1 models would show they are similar.

Briefly, the TPT1 models vary from 2% to 3% deviation with QCT close to 4% and tcPRq and FH around 10%. Overall, this example suggests that TPT1 has a slight advantage over QCT in the presence of strong interactions like hydrogen bonding. For two-parameter models, doubling the number of BIPs offers a significant advantage for fitting experimental data. Questions remain about whether this advantage conveys to extrapolations beyond the fitted data, however.

We shall see that these broad observations are consistent with observations based on deeper analysis of the theories and evaluations with large databases. While Fig. 8.1 pertains only to vapor-liquid equilibrium (VLE) for a single binary system, we shall see that the broader conclusions for large databases of VLE, liquid-liquid equilibrium (LLE), and solid-liquid equilibrium (SLE) are similar. Hence this single figure is very instructive about all kinds of phase behavior modeling and sets the tone for Chap. 9 as well as Chap. 8.

In one or two chapters it is not possible to present a complete review of such a large subject. Since the variety of mixtures is extensive, and since mixture conditions (temperature, pressure, and composition) cover many possibilities, and finally, since there are large variations in the availability, quantity, and quality of experimental data, it is not possible to recommend to the reader simple, unambiguous rules for obtaining quantitative answers to a particular phase equilibrium problem. The reader cannot escape responsibility for using judgment, which, ultimately, is obtained only by experience. This chapter and the next, therefore, are qualitatively different from the others in this book. They do not give specific advice on how to calculate specific quantities. They provide only an introduction to some (not all) of the tools and techniques which may be useful for an efficient strategy toward calculating phase equilibria for a particular process design.

8.2 A NOTE ABOUT THE MODELING OF TEMPERATURE EFFECTS

Note that we restrict the treatment of temperature effects in all models according to the specifications of the original authors. It is often the practice to express the parameters of various models as functions of temperature. In our opinion, this practice can lead to unreliable extrapolations, and it obfuscates the distinctions between the models according to the authors' intentions and the theory on which the models are based. Furthermore, it is impractical for our evaluations to consider every conceivable variation of temperature for every parameter in every model. The most promising models should be logical starting points for more broadly empirical variations, but we leave it to readers to (cautiously) introduce those variations. We also hope that theories may soon improve to eliminate the need for such measures.

8.3 THERMODYNAMICS OF VAPOR-LIQUID EQUILIBRIA

We are concerned with a liquid mixture that, at temperature T and pressure P, is in equilibrium with a vapor mixture at the same temperature and pressure. The quantities of interest are the temperature, the pressure, and the compositions of both phases. Given some of these quantities, our task is to calculate the others.

For every component i in the mixture, the condition of thermodynamic equilibrium is given by

$$f_i^{\text{vap}} = f_i^{\text{liq}} \tag{8.3-1}$$

where $f \equiv$ fugacity, vap = vapor, liq = liquid.

The fundamental problem is to relate these fugacities to mixture composition. In the subsequent discussion, we neglect effects due to surface forces, gravitation, electric or magnetic fields, semipermeable membranes, or any other special conditions.

The fugacity of a component in a mixture depends on the temperature, pressure, and composition of that mixture. In principle, any measure of composition can be used. For the vapor-phase, the composition is nearly always expressed by the mole fraction $\boldsymbol{y}$. To relate f_i^{vap} to temperature, pressure, and mole fraction, it is useful to introduce the vapor-phase fugacity coefficient φ_i^{vap}

$$\varphi_i^{\text{vap}} = \frac{f_i^{\text{vap}}}{y_i P} \tag{8.3-2}$$

which can be calculated from vapor-phase $PVTy$ data, usually given by an equation of state as discussed in Sec. 6.7. For ideal gases $\varphi_i^{\text{vap}} = 1$.

The fugacity coefficient φ_i^{vap} depends on temperature and pressure and, in a multicomponent mixture, on *all* mole fractions in the vapor-phase, not just y_i. The fugacity coefficient is, by definition, normalized such that as $P \to 0$, $\varphi_i^{\text{vap}} \to 1$ for all i. At low pressures, therefore, it is usually a good assumption to set $\varphi_i^{\text{vap}} = 1$. But just what "low" means depends on the composition and temperature of the mixture. For typical mixtures of non-associating fluids at a temperature near or above the normal boiling point of the least volatile component, "low" pressure means a pressure less than 5 bars. However, for mixtures containing a strongly associating carboxylic acid, e.g., acetic acid+water at 25 °C, fugacity coefficients may differ appreciably from unity at pressures much less than 1 bar.[1] For mixtures containing one component of very low volatility and another of high volatility, e.g., n-decane+methane at 25 °C, the fugacity coefficient of the light component may be close to unity for pressures up to 10 or 20 bar while at the same pressure the fugacity coefficient of the heavy component is typically much less than unity. A detailed discussion is given in Chap. 5 of Prausnitz et al. (1999) and O'Connell (2019).

[1]For moderate pressures, fugacity coefficients can often be estimated with good accuracy using virial coefficients as discussed for example, by Prausnitz, et al., (1980).

The fugacity of component i in the liquid-phase is generally calculated by one of two approaches: the equation of state approach or the activity coefficient approach. In the former, the liquid-phase fugacity coefficient, φ_i^{liq}, is introduced

$$\varphi_i^{\text{liq}} = \frac{f_i^{\text{liq}}}{x_i P} \tag{8.3-3}$$

where x_i is the liquid-phase mole fraction. Certain equations of state can accurately represent liquid-phase, as well as vapor-phase behavior.

In the activity coefficient approach, the fugacity of component i in the liquid-phase is related to the composition of that phase through the activity coefficient γ_i. In principle, any composition scale may be used; the choice is strictly a matter of convenience. For some aqueous solutions, frequently used scales are molality (moles of solute per 1000 g of solvent) and molarity (moles of solute per liter of solution); for polymer solutions, useful scales are the volume fraction or weight fraction. However, for typical solutions containing nonelectrolytes of normal molecular weight (including water), the most useful measure of concentration is mole fraction $\boldsymbol{x}$. Activity coefficient γ_i is related to x_i and to standard-state fugacity f_i^o by

$$\gamma_i \equiv \frac{a_i}{x_i} = \frac{f_i^{\text{liq}}}{x_i f_i^o} \tag{8.3-4}$$

where a_i is the activity of component i. The standard-state fugacity f_i^o is the fugacity of component i at the temperature of the system, i.e., the mixture, and at some arbitrarily chosen pressure and composition. The choice of standard-state pressure and composition is dictated only by convenience, but it is important to bear in mind that the numerical values of γ_i and a_i have no meaning unless f_i^o is clearly specified.

While there are some important exceptions, activity coefficients for most typical solutions of nonelectrolytes are based on a standard state where, for every component i, f_i^o is the fugacity of *pure* liquid i at system temperature and pressure, i.e., the arbitrarily chosen pressure is the total pressure P, and the arbitrarily chosen composition is $x_i = 1$. Frequently, this standard-state fugacity refers to a hypothetical state since it may happen that component i cannot physically exist as a pure liquid at system temperature and pressure. Fortunately, for many common mixtures it is possible to calculate this standard-state fugacity by modest extrapolations with respect to pressure; and since liquid-phase properties remote from the critical region are not sensitive to pressure (except at high pressures), such extrapolation introduces little uncertainty. For example, the extrapolation for liquid water at 270 K in a liquid methanol+water mixture would be modest. In some mixtures, however, namely, those that contain supercritical components, extrapolations with respect to temperature are required, and these, when carried out over an appreciable temperature range, may lead to large uncertainties. Whenever the standard-state fugacity is that of the pure liquid at system temperature and pressure, we obtain the limiting relation that $\gamma_i \to 1$ as $x_i \to 1$. The various options for reference fugacities and Poynting corrections are thoroughly described as the "Five Famous Fugacity Formulae" in Sec. 6.4 of O'Connell and J. M. Haile (2005).

8.4 FUGACITY OF A PURE LIQUID

To calculate the fugacity of a pure liquid at a specified temperature and pressure, we may use an equation of state capable of representing the liquid-phase and first calculate φ_i^{liq} (see Chap. 6) and then use Eq. (8.3-3). Alternatively, we may use the two primary thermodynamic properties: the saturation (vapor) pressure, which depends only on temperature, and the liquid density, which depends primarily on temperature and to a lesser extent on pressure. Unless the pressure is very large, it is the vapor pressure which is by far the more important of these two quantities. In addition, we require volumetric data (equation of state) for pure vapor i at system temperature, but unless the vapor pressure is high or unless there is strong dimerization in the vapor-phase, this requirement is of minor, often negligible, importance.

The fugacity of pure liquid i at temperature T and pressure P is given by

$$f_i^{\text{liq}}(T, P, x_i = 1) = P_i^{\text{vp}}(T)\varphi_i^{\text{sat}}(T)\exp\left(\int_{P_i^{\text{vp}}}^{P} \frac{V_i^{\text{liq}}(T, P)}{RT} dP\right) \tag{8.4-1}$$

where P_i^{vp} is the vapor pressure and superscript "vp" stands for saturation vapor pressure. The fugacity coefficient φ_i^{sat} is typically calculated from a vapor-phase EOS; for nonassociated fluids at temperatures well below the critical point, φ_i^{sat} is close to unity. Further, for nonpolar (or weakly polar) liquids, the ratio of fugacity to pressure can be estimated from generalized (corresponding states) methods, including cubic EOSs and related methods from Chap. 6.

The exponential term in Eq. (8.4-1) is called the *Poynting factor.* The liquid molar volume V_i^{liq} is the ratio of the molar mass to the density, where the latter is expressed in units of mass per unit volume. At temperatures well below critical, liquids are nearly incompressible so $V_i^{\text{liq}} \approx$ constant, simplifying the integral to $(V_i^{\text{liq}}\,\Delta P)/RT$. Calculation shows that the effect of pressure on the Poynting factor is not large unless the pressure is very high, or the temperature is very low. To illustrate Eq. (8.4-1), consider the ratio of the fugacity of pure liquid water to the vapor pressure (equal to the product of φ^{sat} and the Poynting factor) at 250 °C according to the IAPWS steam tables (Wagner and Pruss, 2002). For the pure, saturated liquid, $\varphi^{\text{sat}} = 0.859$ and the ratio is less than 1. However, at 35 MPa, the product of φ^{sat} and the Poynting factor is 1.009, and then the fugacity is larger than the vapor pressure. Sometimes it is necessary to calculate a liquid fugacity for conditions when the substance does not exist as a liquid. At 250 °C, the vapor pressure of water is 3.9762 MPa, and therefore pure liquid water cannot exist at this temperature and 2 MPa. Nevertheless, the ratio of fugacity to pressure at these conditions can be calculated by Eq. (8.4-1) as 0.851 if we neglect the effect of pressure on molar liquid volume in the Poynting factor.

The vapor pressure is the primary quantity in Eq. (8.4-1). When data are not available, the vapor pressure can be estimated, as discussed in Chap. 5. This point merits emphasis. VLE is often the primary concern for many engineers, owing to its importance in distillation, for instance. Nevertheless, accurate VLE for mixtures requires accurate VLE for the pure fluids. The VLE of a pure fluid is simply the vapor pressure at a given temperature.

8.5 SIMPLIFICATIONS IN THE VAPOR-LIQUID EQUILIBRIUM RELATION

Equation (8.3-1) gives the rigorous, fundamental relation for VLE. Equations (8.3-2), (8.3-3), and (8.4-1) are also rigorous, without any simplifications beyond those indicated in the paragraph following Eq. (8.3-1). Substitution of Eqs. (8.3-2), (8.3-3), and (8.4-1) into Eq. (8.3-1) gives

$$y_i\, P = x_i\, \gamma_i\, P_i^{\text{vp}}\, F_i \tag{8.5-1}$$

where

$$F_i = \frac{\varphi_i^{\text{sat}}}{\varphi_i^{\text{vap}}}\exp\left(\int_{P_i^{\text{vp}}}^{P} \frac{V_i^{\text{liq}}\,dP}{RT}\right) \tag{8.5-2}$$

In Eq. (8.5-1), both φ_i^{vap} and γ_i depend on temperature, composition, and pressure. However, remote from critical conditions, and unless the pressure is large, the effect of pressure on γ_i is usually small. For subcritical components, the correction factor F_i is often near unity when the total pressure P is sufficiently low. However, even at moderate pressures, we are nevertheless justified in setting $F_i = 1$ if only approximate results are required and, as happens so often, if experimental information is sketchy, giving large uncertainties in γ. Even at pressures of 1 MPa, where the fugacity coefficients deviate from unity, the ratio often does not, unless small carboxylic acids are involved. If, in addition to setting $F_i = 1$, we assume that $\gamma_i = 1$, Eq. (8.5-1) reduces to the familiar relation known as *Raoult's law.* If, we only set $F_i = 1$, and assume that $\gamma_i \neq 1$, Eq. (8.5-1) is referred to as *modified Raoult's law.* Summing Eq. (8.5-1) for all components leads to the equation used for computing bubble pressure for low-pressure VLE,

$$P = \sum x_i \gamma_i P_i^{\text{vp}} \tag{8.5-3}$$

The arguments leading to $F_i = 1$ are equally applicable to EOSs. At low pressure, it may be useful to compare models based on their activity coefficients. This requires an extra step for EOSs, but it is straightforward.

$$\ln(\gamma_i) = \ln\left[\varphi_i(T, P, \boldsymbol{x})\right] - \ln\left[\varphi_i^{\circ}(T, P)\right] \tag{8.5-4}$$

Where φ_i° is computed for each pure fluid in advance of the φ_i computation. For example, our evaluations for predictions of ternary low-pressure VLE apply Eqs. (8.5-3) and (8.5-4) for EOSs as a means of expediting the bubble pressure computation.

8.6 ACTIVITY COEFFICIENTS; GIBBS-DUHEM EQUATION, AND EXCESS GIBBS ENERGY

In typical mixtures, Raoult's law provides no more than a rough approximation; only when the components in the liquid mixture are similar, e.g., a mixture of *n*-butane and isobutane, can we assume that γ_i is essentially unity for all components at all compositions. The activity coefficient, therefore, plays a key role in the calculation of VLE.

Classical thermodynamics has little to tell us about the activity coefficient; as always, thermodynamics does not give us the experimental quantity we desire but only relates it to other experimental quantities. Thus, thermodynamics relates the effect of pressure on the activity coefficient to the partial molar volume, and it relates the effect of temperature on the activity coefficient to the partial molar enthalpy, as discussed in many thermodynamics texts. These relations are of limited use because precise data for the partial molar volume and for the partial molar enthalpy are rare.

However, there is one thermodynamic relation that provides a useful tool for correlating and extending limited experimental data: the Gibbs-Duhem equation. This equation is not a panacea, but, given some experimental results, it enables us to use these results efficiently. In essence, the Gibbs-Duhem equation says that, in a mixture, the activity coefficients of the individual components are not independent of one another but are related by a differential equation. In a binary mixture the Gibbs-Duhem relation is

$$x_1\left(\frac{\partial \ln \gamma_1}{\partial x_1}\right)_{T,P} = x_2\left(\frac{\partial \ln \gamma_2}{\partial x_2}\right)_{T,P} \tag{8.6-1}$$

Equation (8.6-1) has several important applications.

1. If we have γ_1 as a function of x_1 from experimental data, we can integrate Eq. (8.6-1) and calculate γ_2 as a function of x_2. That is, in a binary mixture, activity coefficient data for one component can be used to predict the activity coefficient of the other component.
2. If we have extensive experimental data for *both* γ_1 and γ_2 as a function of composition, we can test the data for thermodynamic consistency by determining whether the data obey Eq. (8.6-1). If the data show serious inconsistencies with Eq. (8.6-1), we may conclude that they are unreliable.
3. If we have limited data for γ_1 and γ_2, we can use an integral form of the Gibbs-Duhem equation; the integrated form provides us with thermodynamically consistent equations that relate γ_1 and γ_2 to $\boldsymbol{x}$. These equations contain a few adjustable parameters that can be determined from the limited data. It is this application of the Gibbs-Duhem equation which is of particular use to chemical engineers. However, there is no *unique* integrated form of the Gibbs-Duhem equation; many forms are possible. To obtain a particular relation between γ and $\boldsymbol{x}$, we must assume some model consistent with the Gibbs-Duhem equation.

For practical work, the utility of the Gibbs-Duhem equation is best realized through the concept of excess Gibbs energy, i.e., the inferred Gibbs energy of a mixture relative to what it would be for an ideal solution at the same temperature, pressure, and composition. By definition, an ideal solution is one where all $\gamma_i = 1$. The *total* excess Gibbs energy G^E for a binary solution, containing n_1 moles of component 1 and n_2 moles of component 2, is defined by

$$G^E = RT\,(n_1 \ln \gamma_1 + n_2 \ln \gamma_2) \tag{8.6-2}$$

Equation (8.6-2) gives G^E as a function of *both* γ_1 and γ_2. Upon applying the Gibbs-Duhem equation, we can relate the *individual* activity coefficients γ_1 or γ_2 to G^E by differentiation

$$RT \ln \gamma_1 = \left(\frac{\partial G^E}{\partial n_1}\right)_{T,P,n_2} \tag{8.6-3}$$

$$RT \ln \gamma_2 = \left(\frac{\partial G^E}{\partial n_2}\right)_{T,P,n_1} \tag{8.6-4}$$

Note that these are derivatives with respect to mole number, not mole fraction, because it is impossible to hold the mole fraction of all other components constant while varying one of the mole fractions. Equations (8.6-2) to (8.6-4) are useful because they enable us to interpolate and extrapolate limited data with respect to composition. To do so, we must first adopt some mathematical expression for G^E as a function of composition. Second, we fix the numerical values of the constants in that expression from the limited data; these constants are independent of $\boldsymbol{x}$, but they usually depend on temperature. Third, we calculate activity coefficients at any desired composition by differentiation, as indicated by Eqs. (8.6-3) and (8.6-4).

To illustrate, consider a simple binary mixture. Suppose that we need activity coefficients for a binary mixture over the entire composition range at a fixed temperature T. However, we have experimental data for only one composition, say x_1', and $x_2' = 1 - x_1'$. From that one datum, modified Raoult's law gives $\gamma_1 = y_1 P / x_1' P_1^{vp}$, and $\gamma_2 = y_2 P / x_2' P_2^{vp}$.

We must adopt an expression relating G^E to the composition subject to the conditions that at fixed composition G^E is proportional to $n_1 + n_2$ and that $G^E = 0$ when $x_1 = 0$ or $x_2 = 0$. The simplest expression we can construct is known as the Margules 1-parameter (M1) model,

$$G^E = (n_1 + n_2) g^E = (n_1 + n_2) A x_1 x_2 \tag{8.6-5}$$

where g^E is the excess Gibbs energy per mole of mixture and A is a constant depending on temperature. The mole fraction $\boldsymbol{x}$ is simply related to mole number n by

$$x_1 = \frac{n_1}{n_1 + n_2} \tag{8.6-6}$$

$$x_2 = \frac{n_2}{n_1 + n_2} \tag{8.6-7}$$

The constant A can be found from substituting Eq. (8.6-5) into Eq. (8.6-2) and using the experimentally determined γ_1^{expt} and γ_2^{expt} at the composition midpoint:

$$A = \frac{RT}{x_1' x_2'} \left[x_1' \ln\left(\gamma_1^{expt}\right) + x_2' \ln\left(\gamma_2^{expt}\right) \right] \tag{8.6-8}$$

Upon differentiating Eq. (8.6-5) as indicated by Eqs. (8.6-3) and (8.6-4), we find

$$RT \ln \gamma_1^{calc} = A x_2^2 \tag{8.6-9}$$

$$RT \ln \gamma_2^{calc} = A x_1^2 \tag{8.6-10}$$

TABLE 8.3 Multicomponent g^E expressions for the Wilson, NRTL, and UNIQUAC models.

Name	g^E	$\ln \gamma_i$
Wilson	$\frac{g^E}{RT} = -\sum_i^{N_c} x_i \ln\left(\sum_j^{N_c} x_j \Lambda_{ij}\right)$	$\ln(\gamma_i) = -\ln\left(\sum_1^{N_c} x_j \Lambda_{ij}\right) + 1 - \sum_k^{N_c} \frac{x_j \Lambda_{ki}}{\sum_j^{N_c} x_j \Lambda_{kj}}$
NRTL	$\frac{g^E}{RT} = \sum_i^{N_c} x_i \frac{\sum_j^{N_c} \tau_{ji} G_{ji} x_j}{\sum_k^{N_c} G_{ki} x_k}$	$\ln(\gamma_i) = \frac{\sum_j^{N_c} \tau_{ji} G_{ji} x_j}{\sum_k^{N_c} G_{ki} x_k} + \sum_j^{N_c} \frac{x_j G_{ij}}{\sum_k^{N_c} G_{kj} x_k}\left(\tau_{ij} - \frac{\sum_k^{N_c} x_k \tau_{kj} G_{kj}}{\sum_k^{N_c} G_{kj} x_k}\right)$
UNIQUAC[e]	$g^E = g^E(\text{combinatorial}) + g^E(\text{res})$ $\frac{g^E(\text{res})}{RT} = -\sum_i^{N_c} q_i x_i \ln\left(\sum_j^{N_c} \theta_i \tau_{ji}\right)$	$\ln(\gamma_i) = \ln\left(\gamma_i^{\text{comb}}\right) + \ln\left(\gamma_i^{\text{res}}\right)$ $\ln\left(\gamma_i^{\text{res}}\right) = -q_i \ln\left(\sum_j^{N_c} \theta_j \tau_{ji}\right) + q_i - q_i \sum_j^{N_c} \frac{\theta_j \tau_{ij}}{\sum_k^{N_c} \theta_k \tau_{kj}}$
GSA	$\frac{g^E(\text{combinatorial})}{RT} = \sum_i^{N_c} x_i \ln\left(\frac{\Phi_i}{x_i}\right) + \frac{z}{2}\sum_i^{N_c} q_i x_i \ln\left(\frac{\theta_i}{\Phi_i}\right)$	$\ln\left(\gamma_i^{\text{comb}}\right) = \ln\left(\frac{\Phi_i}{x_i}\right) + \frac{z}{2} q_i \ln\left(\frac{\theta_i}{\Phi_i}\right) + \Phi_j\left(l_i - \frac{r_i}{r_j} l_j\right)$

Note that $\gamma_1^{\text{calc}} \neq \gamma_1^{\text{expt}}$ even at x' owing to imprecision in this very simple G^E model. With these relations, we can now calculate activity coefficients γ_1 and γ_2 at any desired x_1 even though experimental data were obtained only at one point, namely, x_1'.

This simple example illustrates how the concept of excess function, coupled with the Gibbs-Duhem equation, can be used to interpolate, or extrapolate experimental data with respect to composition. Unfortunately, the Gibbs-Duhem equation tells nothing about interpolating or extrapolating such data with respect to temperature or pressure.

8.6.1 Common Activity Models

Equations (8.6-2) to (8.6-4) indicate the intimate relation between activity coefficients and excess Gibbs energy, G^E. Many expressions relating g^E (per mole of mixture) to composition have been proposed, and a few are given in Table 8.3. All these expressions contain adjustable parameters which, at least in principle, depend on temperature, which often varies, such as in an isobaric distillation column. That temperature dependence may in many cases be neglected, especially if the temperature interval is not large. In practice, the number of adjustable constants per binary mixture is typically two or three; the larger the number of constants, the better the representation of the data but, at the same time, the larger the number of reliable experimental data points required to determine the constants. One parameter, like the parameter A in Eqs. (8.6-5)–(8.6-10), suffices to control the magnitude of g^E. Two parameters suffice to adjust both the magnitude and skewness. Extensive and highly accurate experimental data are required to justify more than two empirical constants for a binary mixture at a fixed temperature.[2]

[2]The models shown in Table 8.2 are not applicable to solutions of electrolytes.

For many moderately nonideal binary mixtures, all equations for g^E containing two (or more) binary parameters give good results; there is little empirical reason to choose one over another. The older ones (e.g., Margules and van Laar) are mathematically easier to handle than the newer ones (Wilson, NRTL, UNIQUAC), but the newer ones extend to multicomponent mixtures more reliably with fewer parameters. The one-parameter Margules equation in Eqs. (8.6-5)–(8.6-10) is applicable only to simple mixtures where the components are similar in chemical nature and in molecular size.

For VLE of strongly nonideal binary mixtures, e.g., solutions of alcohols with hydrocarbons, the equation of Wilson is probably the most useful because, unlike the NRTL equation, it contains only two adjustable parameters, it is mathematically simpler than the UNIQUAC equation, and it inherently avoids artificial indications of LLE if it is known that none exists. Alternatively, when $\alpha_N = 0.43$ in the NRTL model, it behaves somewhat like the Wilson model and admits the possibility of LLE if other parts of the process operate at colder temperatures. The value of the α_N parameter is generally varied between 0.06 and 0.43 with a default value of 0.3. Guidelines for selecting α_N are described in more detail in Chap. 9, where the strongly nonideal solutions associated with LLE can cause anomalies if α_N is not chosen carefully. For strongly nonideal mixtures like small alcohols with hydrocarbons, the two-parameter Margules equation, and the van Laar equation are likely to represent the data with less success, especially in the region dilute with respect to alcohol, where the Wilson equation is particularly suitable. With rare exceptions, the three-parameter Margules equation has no significant advantages over the three-parameter NRTL equation. Numerous articles in the literature use the Redlich-Kister expansion for g^E. This expansion is mathematically identical to the Margules model. The Redlich-Kister, Margules, and van Laar models have been disfavored in recent years as the local composition models provide a more straightforward basis for extension to multicomponent mixtures.

The Wilson equation is not applicable to a mixture which exhibits a miscibility gap; it is inherently unable, even qualitatively, to account for phase splitting. Nevertheless, Wilson's equation may be useful even for those mixtures where miscibility is incomplete provided attention is confined to regions with a single liquid-phase.

Unlike Wilson's equation, the NRTL and UNIQUAC equations are applicable to *both* VLE and LLE.[3] Therefore, mutual solubility data can be used to determine NRTL or UNIQUAC parameters but not Wilson parameters. While UNIQUAC is mathematically more complex than NRTL, it has three advantages: (1) it has only two (rather than three) adjustable parameters, (2) UNIQUAC's parameters often have a smaller dependence on temperature, and (3) because the primary concentration variable is a surface fraction (rather than mole fraction), UNIQUAC is applicable to solutions containing small or large molecules, including polymers.

8.7 THEORY OF ACTIVITY MODELS

The theory of activity models is more detailed than most theory sections in PGL6ed. We introduced this theory in Chap. 7 with the energy equation for mixtures, leading to the outline of the local composition activity models that form the basis for G^E mixing rules. That analysis also informs the presentation in this chapter, but it omits treatment of strong interactions like hydrogen bonding. One of the significant advances since PGL5ed has been the integration of strong interaction treatment into mainstream thermodynamics applications like chemical process simulators. Examples of those treatments are the SAFT and COSMO-RS/SAC models, each with multiple implementations by highly respected authors. Wertheim's first-order theory (TPT1) was introduced in Chap. 6, but the activity model inherent in COSMO-RS/SAC is new to this chapter. The activity model in COSMO-RS/SAC is based on quasichemical theory (QCT). Based on that name, one might expect a relation with "universal" quasichemical theory (aka UNIQUAC), but UNIQUAC and QCT are quite contradictory in some situations and hydrogen bonding is not recognized explicitly in the UNIQUAC model. In that sense, QCT is more like TPT1 than like UNIQUAC.

[3]Wilson (1964) has given a three-parameter form of his equation that is applicable also to LLE. The three-parameter Wilson equation has not received much attention, primarily because it is not readily extended to multicomponent systems.

Understanding these theories is nontrivial, but acronyms and misconceptions contribute to the confusion. If the reader can focus foremost on the accounting methods for various interaction energies, the acronyms can come later. Four types of molecular interaction energies play significant roles in solution thermodynamics: repulsive dispersion, attractive dispersion, polar, and strong interactions. When a model omits explicit accounting for any of these specific interactions, the hope is that those interactions can be accounted for "effectively" through characterizations of the interactions explicitly treated in the model. The second step is to assess how accurately the theory accounts for the interactions that it does treat explicitly. This assessment is best achieved through molecular simulation.

A proper understanding of the similarities and differences between the various activity models is not widely appreciated as of this writing, but it is important. When we refer to the "physical basis" of various models, we are referring to the descriptions contained in this section. Theories with a strong physical basis have few caveats and multiple validations through molecular simulations and nanoscopic measurements (e.g., spectroscopy), as well as accurate comparisons to macroscale experimental measurements like ebulliometry. For many readers, it may seem satisfactory to simply follow the procedures to obtain a workable model of the phase equilibria for their mixture(s) of interest. Other sections of this chapter address that approach in detail. On the other hand, the PGL mission includes characterizing the quality of the available methods "such that the future progress of the field is advanced." One aspect of the quality of a method is how well its assumptions can be validated with fundamental analysis and comparison to experiments on the nanoscale. We presume that future progress will be advanced when inaccuracies in those assumptions are minimized. Therefore, this section is devoted to a brief analysis of the most popular activity models.

8.7.1 Preliminary Fundamentals

To begin, numerous sources have described how the Flory-Huggins model derives from the van der Waals equation of state applied at constant packing fraction. (e.g., Hildebrand, 1947; Mollerup, 1981; Elliott, 1993). The perspective offered by constant packing fraction is useful, so a brief reprise of that perspective is presented here. Two preliminary fundamentals are necessary: the role of ln(Z) and the relation of constant pressure derivatives to constant volume derivatives.

The ln(Z) term appears first in the Gibbs energy departure function, as a correction from a constant volume ideal gas reference to a constant pressure reference (Sec. 3.5 of O'Connell and Haile, 2005; Sec. 9.5 of Elliott and Lira, 2012)

$$\frac{G(T,P) - G^{\text{ig}}(T,P)}{RT} = \frac{G^{\text{dep}}}{RT} = \frac{A(T,V) - A(T,V)^{\text{ig}}}{RT} + Z - 1 - \ln(Z) \tag{8.7-1}$$

Elliott and Lira (Chap. 12) then show how the ln(Z) term leads to the Flory-Huggins athermal (FHA) contribution,

$$\ln\left(\gamma_k^{\text{FHA}}\right) = \ln\left(\frac{\Phi_k}{x_k}\right) + 1 - \frac{\Phi_k}{x_k} \tag{8.7-2}$$

where Φ_k is defined with the UNIQUAC model in Table 8.2. This result was first reported by Hildebrand (1947). Remarkably, this contribution for dense liquids derives from reference state considerations of the ideal gas. The term is often attributed to entropic considerations, and it is, but in an almost trivial manner. A common misconception is to think that this contribution arises from repulsive effects. To isolate the repulsive part, attractive (dispersion) part, and chemical-physical (attractive) part, one must apply the expansion rule as given by Bala and Lira (2016).

$$\ln\left(\gamma_k^{\text{part}}\right) = \left(\frac{\partial \dfrac{n(A^{\text{part}} - A^{\text{ig}})}{RT}}{\partial n_k}\right)_{T,nV,n_{j\neq k}} - Z^{\text{part}}\frac{V_k}{V} - \left[\left(\frac{(A^{\text{part}} - A^{\text{ig}})^{\circ}}{RT} - Z^{\text{part}\circ}\frac{V_k}{V_k}\right)\right] \tag{8.7-3}$$

Equation (8.7-3) provides the capability to analyze contributions to the activity coefficient at any conditions for any model including EOSs, and not just at constant pressure. For example, activity coefficients at constant temperature are often of interest in VLE measurements. Furthermore, the equations are applicable to equations of state as well as more limited activity models. To demonstrate the utility of this perspective, consider analysis of the system benzene(1)+methanol(2) with the ESD EOS. Figure 8.2a shows the component parts of the activity coefficient at a constant pressure of 0.79 MPa. Figure 8.2b shows the same system at a constant packing fraction of 0.35. Note that the packing fraction decreases with increasing x_1 in Fig. 8.2a while the pressure increases in Fig. 8.2b. The behavior of the repulsive contribution is particularly interesting. At constant pressure, this contribution varies in a complex manner, but it is identically zero at constant packing fraction. To see how this is true, note that the repulsive contribution to the ESD EOS can be written as

$$Z^{\text{rep}} = \frac{4\eta^{\text{P}}\sum x_i c_i^{\text{ESD}}}{1 - 1.9\eta^{\text{P}}} \tag{8.7-4}$$

When η^{P} = constant, and a simple molar average applies to the shape factor, c^{ESD}, the mixture and the molar average of the ideal solution clearly cancel, no matter how you integrate and differentiate. Similar observations apply to any EOS that applies the van der Waals repulsive contribution with molar average mixing rules on the volume parameter, b^{vdW}. As discussed in Chap. 7, the mixing rules for PC-SAFT and SAFT-VR apply a segmental perspective with the Boublik (1970)-Mansoori-Carnahan-Starling-Leland (1971) (BMCSL) rules for hard sphere mixtures to describe the repulsive terms. Also, the SPEADMD model applies a repulsive mixing rule based on the excess entropy of chain, branched, and ring compounds (Vahid et al., 2014). So, the repulsive contribution is more complicated for those cases. Nevertheless, the perspective of constant packing fraction helps to focus attention on the parts of the various models that distinguish them, and their relations to activity models. The capability to analyze individual components of each model facilitates an understanding of the assumptions behind the models in ways that can be systematically improved.

It should be kept in mind, however, that many of the models in common use have developed through trial and error, and their success may depend on a cancellation of errors. The implication is that fixing errors in one aspect of the model, without fixing errors in other aspect(s), could lead to a less successful model. Analysis of a model's individual aspects is valuable, but the holistic perspective is essential.

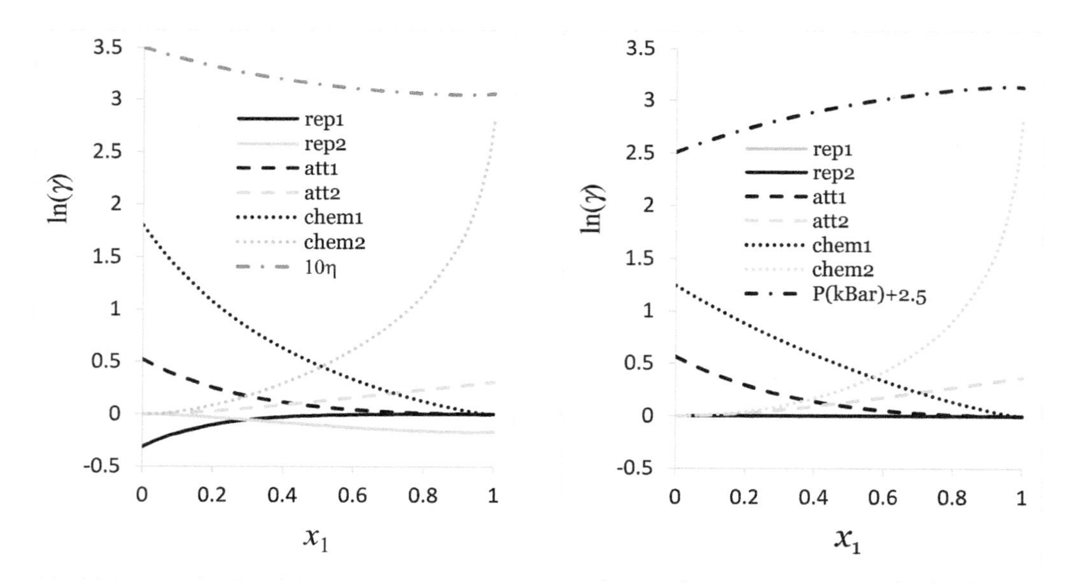

FIGURE 8.2
Components of the activity coefficients for benzene(1)+methanol(2) at 345 K according to the ESD EOS. (a) Constant pressure of 0.79 MPa. Note η^{P} = [0.31,0.35], (b) Constant packing fraction of η^{P} = 0.35. Note P(kbar) = [0.001,0.60].

8.7.2 Effects of Attraction Through Quadratic Mixing Rules

Quadratic mixing rules arise from the assumption that molecules are distributed randomly. Elliott and Lira (Chap. 12) show how this leads to

$$RT\ln\left(\gamma_1^{att}\right) = \frac{-2(x_1a_{11} + x_2a_{12})}{V} + \frac{aV_1}{V^2} + \left(\frac{a_{11}}{V_1}\right) \tag{8.7-5}$$

Defining $\delta_i^2 \equiv a_{ii}/V_i^2$, $\Phi_i \equiv x_iV_i/\Sigma x_jV_j$, $a_{12} = a_{11}a_{22}(1-k_{12})$, this expression can be rearranged as

$$RT\ln\left(\gamma_1^{att}\right) = V_1(1-\Phi_1)^2\left[(\delta_2-\delta_1)^2 + 2k_{12}\delta_1\delta_2\right] \tag{8.7-6}$$

This contribution is immediately recognized as the Scatchard (1931)-Hildebrand (1929) contribution. At constant packing fraction, the PR and SRK EOSs give the same result, which can be generalized to

$$RT\ln\left(\gamma_k^{att}\right) = V_k(1-\Phi_k)^2\left[(\delta_k-\langle\delta\rangle)^2 + 2\langle k_{mk}\rangle\delta_k - \langle\langle k_{mm}\rangle\rangle\right] \tag{8.7-7}$$

where $\langle k_{mk}\rangle = \Sigma x_j\Phi_j\, k_{ij}$ and $\langle\langle k_{mm}\rangle\rangle = \Sigma x_j\Phi_j\, k_{mj}$. In this way, we recognize that most EOSs with quadratic mixing rules are essentially equivalent to a Scatchard-Hildebrand (SH) activity model combined with the FHA term. In polymer literature, this model is often called "the Flory-Huggins model," with the solvent volume and solubility parameters lumped together,

$$RT\ln\left(\gamma_P^{att}\right) = (1-\Phi_P)^2\,\chi^{FH}\times r_P \tag{8.7-8}$$

where χ^{FH} is called the Flory-Huggins parameter and r_P is the ratio of polymer volume to solvent volume. In future discussions, we refer to activity models that combine the FH athermal term with SH attraction as simply "FH."

One distinction when using an EOS instead of an activity model is that the liquid volumes in an EOS are determined at the conditions of interest, rather than at 25 °C like the SH model. Similarly, the solubility parameters are defined in terms of the EOS parameters. This gives an advantage to EOSs for gaseous compounds because their EOS parameters are well defined even if their volume and solubility parameters at 25 °C are not.

There was a time when vapor pressures computed from EOSs were significantly inferior to correlations devoted to vapor pressure alone. This circumstance greatly favored activity models over EOSs. On the other hand, our evaluations in Chap. 6 have shown that vapor pressure deviations of 1–3% are typical with newer EOSs like tcPR, PC-SAFT, and SPEADMD. In that case, vapor pressure is rarely the limiting factor in characterizing accurate VLE. If uncertainty in the available experimental data cannot support more than one binary parameter, an EOS with accurate vapor pressures and a single adjustable parameter might be the best choice. Such a situation might occur if only the Henry's volatility was known.

8.7.3 Chemical-Physical Interactions: Association and Strong Solvation

Technically, association refers to strong stereospecific complexation between molecules of the same type and strong solvation refers to complexation between unlike molecules. For example, methanol and ethanol could be said to associate, but acetone and chloroform would be said to solvate. Nevertheless, it is common in the literature to refer to strong solvation as cross-association and we may do so here occasionally. In general, we refer to these types of contributions as "chemical-physical" contributions, to distinguish them from attractive dispersion forces. The most applied theory for these types of interactions was first articulated by Wertheim, extended to mixtures, and adapted for chemical engineering applications by Chapman et al. (1990) in the form of statistical associating fluid theory (SAFT). For pure fluids, it has been established that Wertheim's theory is equivalent to the chemical-physical theory of solving explicit systems of simultaneous reaction and phase equilibria, subject to certain assumptions like the change of heat capacity due to reaction. For mixtures, Chapman's adaptation greatly facilitates the analysis. Before 1990, extensions of chemical-physical theory to mixtures involved many conjectures about speciation and complicated equations. The adaptation of Wertheim's perspective in the form of SAFT transforms the extension to mixtures into a straightforward application of consistent accounting principles.

Panayiotou et al. (2007) and Panayiotou and Sanchez (1991) have worked extensively on a lattice theory related to work by Barker (1952), Wheeler and Andersen (1980), and Veytsmann (1990). The resulting activity model is qualitatively like the TPT1 result, and we would expect similar correlations and predictions when similar supporting equations are used. When the same physical principles are extended to an EOS, however, differences arise. For example, the SAFT methodology implements the BMCSL model for hard sphere interactions whereas lattice implementations favor equations like that of Sanchez and Lacombe (1978). In off-lattice applications, the BMCSL model is demonstrably superior when compared to molecular simulations. The discussion below adapts the SAFT perspective and notation.

For this section, we seek to understand the effects of chemical-physical interactions on activity coefficients in a succinct manner, like our previous observations about the Flory-Huggins athermal term, repulsive contributions, and attractive contributions. Brief inspection of Eq. (8.7-7) can inform the reader quite quickly about the effects of changes in the molecular volume, differences in cohesive energy density, and the impact of variation in the BIPs. Similar insights about the impacts of chemical interactions would be desirable.

The SAFT development begins with an expression for Helmholtz energy inspired by Wertheim,

$$\frac{A^{\text{chem}}}{RT} = \sum_{i}^{N_c}\sum_{j}^{N_c} x_i N_i^{A_j}\left[\ln\left(1 - X_i^{A_j}\right) + \frac{\left(1 - X_i^{A_j}\right)}{2}\right] + \sum_{j}^{N_c} x_i N_i^{D_j}\left[\ln\left(1 - X_i^{D_j}\right) + \frac{\left(1 - X_i^{D_j}\right)}{2}\right] \tag{8.7-9}$$

where $X_i^{A_j}$ and $X_i^{D_j}$ are the mole fractions of electron acceptors and donors NOT bonded, $N_i^{A_j}$ and $N_i^{D_j}$ are the numbers of acceptors and donors of type j on the ith molecule. The $\{X_i^{A_j}, X_i^{D_j}\}$ is determined by

$$1 - X_i^{A_k} = X_i^{A_k}\sum\sum x_j N_j^{D_l} X_j^{D_l}\rho\Delta_{ij}^{k,l} \tag{8.7-10}$$

$$1 - X_i^{D_k} = X_i^{D_k}\sum\sum x_j N_j^{A_l} X_j^{A_l}\rho\Delta_{ij}^{l,k} \tag{8.7-11}$$

To simplify matters, we adopt the ESD EOS, but similar results are obtained with any EOS that satisfies the conditions that make the repulsive contribution to activity coefficient negligible at constant packing fraction. We also adopt a geometric mean to describe strong solvation interactions. In that case, a nonassociating compound has zero solvation as well as zero association. Applying these assumptions to the results of Elliott (1996),

$$\ln\left(\gamma_k^{\text{chem}}\right) = 2N_k^{AD}\ln\left(\frac{1 + F_{k^\circ}^{AD}\left(\alpha_{kk}^{AD}\right)^{1/2}}{1 + F^{AD}\left(\alpha_{kk}^{AD}\right)^{1/2}}\right) + \left(F^{AD}\right)^2\frac{V_k}{V} - \left(F_{k^\circ}^{AD}\right)^2 \tag{8.7-12}$$

$$F^{AD} \equiv \left(-1 + \frac{1}{X_k^A}\right)\frac{1}{\left(\alpha_{kk}^{AD}\right)^{1/2}} = \sum x_j N_j^{AD} X_j^D\left(\alpha_{jj}^{AD}\right)^{1/2} \tag{8.7-13}$$

where X_k^A is the fraction of acceptor sites NOT bonded, N_j^{AD} is the number of acceptors (or donors) on the ith component, and α_{ij}^{AD} is defined by Eq. (8.7-14)

$$\alpha_{ij}^{AD} = \rho g_{ij}(\sigma)K_{ij}^{AD}\left[\exp\left(\frac{\varepsilon_{ij}^{AD}}{kT}\right) - 1\right] = \frac{\rho K_{ij}^{AD}}{1 - 1.9\eta}\left[\exp\left(\frac{\varepsilon_{ij}^{AD}}{kT}\right) - 1\right] \tag{8.7-14}$$

where K_{ij}^{AD} is an adjustable parameter called the "bonding volume," ε_{ij}^{AD} is the bonding energy, and the ESD value for $g_{ij}(\sigma)$ has been applied. Note that the factorization is such that F^{AD} is independent of component, although the quantities in its definition are component-dependent. This means that solution for F^{AD} can be obtained by Newton iteration on a single variable and all the component quantities are obtained in a simple loop after convergence. Normally, Eqs. (8.7-10) and (8.7-11) comprise a coupled nonlinear system of equations would need to be solved for each X_k^A and X_k^D. To illustrate the meaning of these equations, consider the infinite-dilution activity coefficients (IDACs) of benzene(1) + methanol(2) at 345 K and $\eta^P = 0.35$. In that case, $N_1^{AD} = N_2^{AD} = 1$, and $(\alpha_{22}^{AD})^{1/2} = 6.2$. When methanol is pure, $F^{AD} = F_{2^\circ}^{AD} \approx (\alpha_{22}^{AD})^{1/2}/[1 + (\alpha_{22}^{AD})^{1/2}] \approx 1$. This value of F^{AD} applies to the pure methanol and the infinitely dilute benzene, so we introduce a new notation for these cases,

$$F_{2^\circ}^{AD} \approx 1 \approx F^{AD}(1^\infty);\ F_{1^\circ}^{AD} \approx 0 \approx F^{AD}(2^\infty); \tag{8.7-15}$$

Substituting, and noting that $\alpha_{11}^{AD} = 0 = F_{1^\circ}^{AD}$, gives

$$\ln\left(\gamma_1^{\text{chem}}\right)^\infty = 2Nd_1 \ln\left(\frac{1 + F_{1^\circ}^{AD}\left(\alpha_{11}^{AD}\right)^{1/2}}{1 + F_{AD}(1^\infty)\left(\alpha_{11}^{AD}\right)^{1/2}}\right) + \left(F_{AD}(1^\infty)\right)^2\frac{V_1}{V_2} - \left(F_{1^\circ}^{AD}\right)^2 \approx \frac{V_1}{V_2} \tag{8.7-16}$$

and by similar substitutions

$$\ln\left(\gamma_2^{\text{chem}}\right)^\infty \approx 2\ln\left(1 + \left(\alpha_{22}^{AD}\right)^{1/2}\right) - 1 \tag{8.7-17}$$

This result shows that the primary impact of the inert component is to create a hole in the association network at infinite-dilution equivalent to the size ratio of the inert to associating compound. If the inert compound is very small, then it can fit into the free volume without disrupting the association network and its activity is unaffected by strong interactions. The activity of the associating compound, on the other hand, is proportional to ε^{AD}/kT for large α_2, where $\varepsilon^{AD}/k \approx 2000$ K for strongly associating compounds.

As a second case, consider a mixture of ethanol(1)+methanol(2). In this case, $F^{AD} \approx 1$ for the entire range of compositions, and

$$\ln\left(\gamma_1^{\text{chem}}\right)^\infty = 2Nd_1 \ln\left(\frac{1 + F_{1^\circ}^{AD}\left(\alpha_{11}^{AD}\right)^{1/2}}{1 + F_{AD}(1^\infty)\left(\alpha_{11}^{AD}\right)^{1/2}}\right) + \left(F_{AD}(1^\infty)\right)^2\frac{V_1}{V_2} - \left(F_{1^\circ}^{AD}\right)^2 \approx \frac{V_1}{V_2} - 1 \tag{8.7-18}$$

For two strongly associating molecules of roughly equal size, the chemical contribution is near zero. If the attractive contribution is also near zero, it is easy to see how a mixture of two associating species can result in an ideal solution. Thus, statements like "nonideality is caused by hydrogen bonding or polarity" are oversimplified; asymmetry of the chemical contribution causes nonideality.

8.7.4 Quasichemical Theory

Quasichemical theory forms the basis for many of the most successful activity models. Not only did it provide inspiration for Wilson's model, UNIFAC, and UNIQUAC, but it also forms the theoretical basis for the various implementations of COSMO-RS/SAC theory. Understanding the provenance of this theory and how its conjectures can be improved is important for the advancement of phase equilibrium research.

Quasichemical theory (QCT) has its roots in theories of metal lattices from the 1930s. Guggenheim (1952) provided an excellent description of the theory in his book on "Mixtures." His reasoning was that liquids are dense like solids, so their attractive dispersion interactions could be approximated with a similar lattice treatment, accounting only for nearest neighbor interactions. His concept started with two pure fluids and swapped molecules to form a mixture. Guggenheim proposed modeling the change in energy due to mixing as a chemical reaction:

$$AA + BB = 2AB \tag{8.7-19}$$

The change in internal energy can then be computed by adding the interactions of nearest neighbors. For the reaction equilibrium constant, the reference temperature is chosen as infinity, such that the particles are indistinguishable, and the reference value is 1. The change in energy due to "reaction" is

$$zw \equiv z\left(\varepsilon_{AA}^{cc} + \varepsilon_{BB}^{cc} - 2\varepsilon_{AB}^{cc}\right)/2 \tag{8.7-20}$$

where z is the coordination number and ε_{ij}^{cc} represents the contact energy between two elements of surface area, which characterizes a negative value for potential energy. In a lattice, these contacts occur on faces of the unit cell.

If all faces are the same, as above, multiplication by z is required. We can neglect the "PV" contribution to the reaction enthalpy because V is small for a liquid or solid. The sign convention here and "w" definition are different from some presentations, to facilitate comparisons with off-lattice molecular simulation and other theories. The square well model is well suited to the QCT perspective, and this presentation refers to it occasionally. Guggenheim's conjecture leads to the equilibrium equation

$$y_{AB}^2 = y_{AA}y_{BB}\exp\frac{(-2\beta w)}{4} \equiv y_{AA}y_{BB}\psi_{AB}^2 \tag{8.7-21}$$

where $\psi_{AB} \equiv \exp[-\beta w]/2$, $\beta = 1/kT$, $y_i = (n_i^\circ - X)/(n_A^\circ + n_B^\circ)$, $i \in \{A,B\}$ is the hypothetical mole fraction of the ith "compound," X is the extent of reaction, and n_i° is the initial moles of the ith compound. This expression characterizes deviations from the random assumption of quadratic mixing rules. If $\varepsilon_{AB}^{cc} < (\varepsilon_{AA}^{cc} + \varepsilon_{BB}^{cc})/2$, then $w > 0$ and AB interactions are disfavored.

Later discussion in this section suggests that we should clarify here the meaning of coordination number and the generalization to mixtures of different sized molecules. Guggenheim divides by z in the exponent to put the energy of reaction on a per-pair basis. For the general case, it is more convenient to consider each molecular surface as being composed of what Guggenheim called surface "elements." These "elements" have been called "segments" by Klamt (1995). The choice is arbitrary, so we favor the term "segments" henceforth. The coordination number can be considered as the number of segments for a reference monomer, N_{ref}^{seg}. In the generalization, it is conventional to express the number of segments per molecule relative to N_{ref}^{seg}, represented by the symbol q_i. Therefore, we recognize,

$$q_i z \equiv \frac{N_i^{seg}}{N_{ref}^{seg}} N_{ref}^{seg} = N_i^{seg} \tag{8.7-22}$$

The conventional notation has led to confusion over the years, leading to "rediscovery" of QCT in various guises (e.g., Prange et al., 1989), including by Klamt himself. We can only hope that our use of the segmental perspective, and its relation to coordination number, adds more clarification than confusion. In particular, the segmental perspective should clarify that the contact interaction energy of the quasichemical "reaction" should be expressed on a per segment basis because only one segment per molecule is participating in any specific "reaction." If another segment in the molecule is also "bound," that is a different "reaction."

Guggenheim then solves for the extent of "reaction," which he expresses as

$$\frac{X}{n_A^\circ + n_B^\circ} = x_A x_B \frac{2}{1 + B} \tag{8.7-23}$$

where $x_i = n_i^\circ/(n_A^\circ + n_B^\circ)$ is the typical definition of mole fraction, and

$$B = \left(1 + 4x_A x_B \Delta^{AB}\right)^{1/2} \tag{8.7-24}$$

where $\Delta^{AB} = \exp(-2\beta w) - 1$. Note that $\Delta^{AB} \to -1$ when $\beta w \to \infty$.

The heat of mixing is equal to the extent of reaction times the energy of reaction.

$$\frac{U^E}{RT} = z x_A x_B \frac{2\beta w}{1 + B} \tag{8.7-25}$$

The factor of z recognizes that all segments of the molecule can participate in this "reaction" of dispersion interactions. The exact integration to obtain free energy is complicated, but $U^E/RT = \beta\partial(A_{att}^E/RT)/\partial\beta$ so we can (paradoxically) approximate the integral by taking the derivative

$$\frac{A_{att}^E}{RT} \approx 0 + \frac{U^E}{RT} + \frac{\beta^2}{2}\left\{\frac{d^2\left(\frac{A_{att}^E}{RT}\right)_\rho}{d\beta^2}\right\}_\rho + \cdots = \frac{U^E}{RT}(1 - \beta w x_A x_B + \cdots) \tag{8.7-26}$$

We refer to this result as "first-order QCT," although it involves the exact QCT result for U^E. To clarify matters, consider a mixture of argon(1)+xenon(2) at 167 K modeled as square-well spheres with a well width of $\lambda^{SW} = 1.5\sigma$. The critical properties for this fluid are $kT_c/\varepsilon = 1.22$ and $P_c\sigma^3/\varepsilon = 0.097$ (Orkoulas and Panagiotopoulos, 2002; Kiselev et al., 2006). These imply $\varepsilon_{11}/k = 123$ K, $\sigma_{11} = 0.323$ nm for argon and $\varepsilon_{22}/k = 237$ K, $\sigma_{22} = 0.379$ nm for xenon. We assume $k_{ij} = 0.2$ for the mixture to magnify the local composition effects, so $\varepsilon_{12}/k = 137$ K. Then Eq. (8.7-26) deviates 1% from the exact result at $\beta w = 1$, $x_A = 0.5$ and the correction to U^E is 25% of U^E. The correction term is smaller for unequal mole fractions. Hence, we see that the impact of QCT is to decrease the excess Helmholtz energy relative to regular solution theory, because atoms redistribute from random positions to avoid unfavorable interactions. Chemical-physical theories like TPT1 achieve a similar effect by separating complexation interactions from dispersion interactions and accounting for them separately. In QCT, dispersion interactions and strong interactions like hydrogen bonding are treated within the same framework.

For nonspherical molecules of unequal size, Guggenheim generalizes contacts between spherical molecules to contacts between segments in his Chap. 11. This perspective is especially important for the COSMO-RS/SAC conception of molecular interactions. Thus, Guggenheim's generalization for two segment types is

$$\frac{U^E}{RT} = z\beta w\Theta_A\Theta_B\frac{2}{1+B} \tag{8.7-27}$$

$$\frac{A^E}{RT} \approx \frac{U^E}{RT}\left(1 - \beta w\Theta_A\Theta_B\right) \tag{8.7-28}$$

where $B^2 = 1 + 4\,\Theta_A\Theta_B\,\Delta^{AB}$, $\Theta_i = x_iq_i/\Sigma x_jq_j$, and $\Delta^{AB} = \exp(-2\beta w_{AB}) - 1$.

Note that w is the energy per segment. In principle, the number of surface segment types is not limited to two for two molecules. For example, benzene + dimethyl ether could be represented with three segment types: ACH, CH_3, and O. The surface fraction of each segment type would then need to be estimated. When each single molecule is composed of a single distinct segment type, as in the case of chlorine + xenon for example, Guggenheim refers to the mixture as "homogeneous." Formulas for activity coefficients of multicomponent mixtures with polysegmented molecules are not readily available from Guggenheim's approach, but he does provide the result for a two-component homogeneous mixture (cf. Eq. 11.08.9 of Guggenheim, 1952):

$$\ln(\gamma_1) = \frac{z}{2}q_1\ln\left\{\frac{(B+1-2\Theta_B)}{[\Theta_A(B+1)]}\right\} \equiv N_1^{seg}\ln(\Gamma_{AA})/2 \tag{8.7-29}$$

$$\Gamma_{AA} = \left\{\frac{(B+1-2\Theta_B)}{[\Theta_A(B+1)]}\right\} \tag{8.7-30}$$

In this form, we can recognize γ_1 as the overall activity coefficient and $\Gamma_{AA}^{1/2}$ as the "pseudo-activity coefficient" of a segment of type A. Since this mixture is homogeneous, $\Gamma_{AA}{}^\circ = 1$, so the relation between γ_1 and Γ_{AA} is simpler than would be expected in a truly general case. Recognition of the molecular chemical potential as a sum over segmental pseudo-chemical potentials is equivalent to the conjecture of Wilson and Deal (1962), later proved by Panayiotou and Vera (1980).

$$\ln(\gamma_k) = \ln\left(\gamma_k^{GSA}\right) + \frac{1}{2}\sum v_j^{(k)}zq_j\left[\ln\left(\Gamma_{jj}\right) - \ln\left(\Gamma_{jj}^{(k)}\right)\right] \tag{8.7-31}$$

where γ_k^{GSA} designates the Guggenheim (1952)-Staverman (1950) athermal (GSA) term from Table 8.2, the superscript (k) indicates the solution for $\{\Gamma_{jj}\}$ when the kth component is pure, and $v_j^{(k)}$ is the frequency of the jth segment type in the kth component. Note that $v_j = \sum_k v_j^{(k)}$ and zq_j corresponds to the surface area of the jth segment type. In their off-lattice formulation, Vera and Wilczek-Vera (1990) combined this quantity into a single variable.

Equation (8.7-31) also reflects Panayiotou and Vera's (1980) extension of QCT to multicomponent mixtures with polysegmented molecules, where the Γ_{jj} values are determined by

$$\frac{1}{\Gamma_{jj}^{1/2}} = \sum \Theta_i \Gamma_{ii}^{1/2} \psi_{ij} \tag{8.7-32}$$

This is equivalent to Eq. (18) of Larsen and Rasmussen (1986) who discuss its convergence in detail. Furthermore, these formulas are equivalent to those of "SS-QCT" as discussed below, but they retain the conception of a lattice and segments are conceived as functional groups. The distinction is somewhat arbitrary, but we prefer to discuss the broader generalization in which segments are conceived as smaller patches of surface area under the heading "SS-QCT."

8.7.5 "Small Segment" Quasichemical Theory (SS-QCT)

During the 1990s, Klamt and coworkers sought to compute liquid activity coefficients from quantum density functional theory (QDFT). They named their methodology "the conductor-like screening model for real solvents" (COSMO-RS). Although often considered as a single methodology, the COSMO application of QDFT is quite distinct from the "real solvent" computation of the activity coefficients. We discuss COSMO-RS in more detail later, but we focus in this subsection on the activity model, which was initially developed by Klamt and coworkers without knowledge of Vera's or Guggenheim's work. We refer to this activity model as SS-QCT owing to its connection with Guggenheim's (1935) QCT (GQCT) and his recognition that each face of his unit cell could be recognized as a different "element." Indeed, this recognition is essential to Barker's (1952) hydrogen bonding adaptation of the theory using Bethe's (1935) equivalent formalism. Klamt et al. (2002) and Lin et al. (2009) noted similarities between their activity model and Eqs. (8.7-31) and (8.7-32) but confined their comparisons to $z = 1$. Panayiotou (2003) notes that the statistical mechanics of Klamt et al. (2002) and Lin and Sandler (2002) is "essentially equivalent," and consistent with QCT generally.

Several formulations of the QCT relations are available. Vera and Wilczek-Vera (1990) classify formulations as (1) neighbor swapping (e.g., Barker, 1952), (2) contact counting (e.g., Guggenheim, 1952), and (3) weighting nonrandom factors (e.g., Payayiotou and Vera, 1980). Lin and Sandler (2002) provide a succinct example of the third approach as part of their segmental activity coefficient (SAC) implementation. Note that all recent formulations are slightly different from Guggenheim's conception in the sense that it makes no reference to a lattice. Vera and Wilczek-Vera (1990) noted that the lattice restriction was superfluous and the presentations of Klamt (1995) and Lin and Sandler (2002) never invoke a lattice from the outset. Soares and Staudt (2022) provide a derivation that simplifies the notation by distinguishing "*mn*" pairs from "*nm*" pairs in their nonrandom factors. They also show how activity models like Wilson, NRTL, and UNIQUAC correspond to assumptions about contact probabilities. Another feature of SS-QCT as treated in this chapter is that contact energies are classified in accordance with their surface polarization values, as discussed in Sec. (8-11). Synergy with this connection to the COSMO/QDFT output was a prime motivation for the "small segment" aspect of SS-QCT. The discussion below refers only to the attractive or "residual" contribution to the free energy, not the athermal contribution.

The key step in SS-QCT is to rewrite the probability of pairwise surface contacts as

$$y_{AB} \equiv \frac{N_{AB}}{N_{\text{tot}}^{\text{seg}}} = \Theta_A \Gamma_A \Theta_B \Gamma_B \psi_{AB} \tag{8.7-33}$$

where $\Theta_A = N_A^{\text{seg}}/N_{\text{tot}}^{\text{seg}}$ and Γ_i can be considered for the moment as a segmental scaling factor to be determined by a mass balance. Equation (8.7-33) uses an amalgam of the notation used by Guggenheim with that of Klamt and Lin and Sandler. The definition of ψ_{AB} is where assumptions are most subtle.

$$\psi_{AB} = \exp(-\beta w)/2^{\delta_{AB}} \tag{8.7-34}$$

where δ_{AB} is the Kronecker delta. The factor of $2^{\delta_{AB}}$ relates to the factor of 2^t in Eq. (A1.4) of Lin and Sandler (2002). This factor is eliminated by the formulation of Soares and Staudt (2022). An implicit assumption in Eq. (8.7-34) is that only the contact energies influence the contact probability, and the full contact energy is expressed. If, for example, the influence of contact energy was altered by packing effects, this assumption might require refinement. There is no issue of coordination number in SS-QCT because all interactions are segmental and there is no limit to the number of segments per molecule. Division confirms a QCT relation, noting $\psi_{ii} = 1$

$$\frac{y_{AB}^2}{y_{AA}y_{BB}} = \frac{\psi_{AB}^2}{4} \tag{8.7-35}$$

At zero interaction energy, $\psi_{AB} = 1$ and $y_{AB} = \Theta_A\Theta_B$, implying that $\Gamma_A = \Gamma_B = 1$ in this limit. The mass balance for all "ν" segments in the mixture is

$$\Theta_\nu = \sum y_i = \sum \Theta_i \Gamma_i \Theta_\nu \Gamma_\nu \psi_{i\nu} \tag{8.7-36}$$

Dividing by $\Theta_\nu\Gamma_\nu$ gives Klamt's "self-consistency" relation.

$$\frac{1}{\Gamma_\nu} = \sum \Theta_i \Gamma_i \psi_{i\nu} \tag{8.7-37}$$

Equation (8.7-37) is equivalent to (8.7-32) where $\Gamma_{jj}^{1/2} \equiv \Gamma_j$.

A key feature of SS-QCT is that it conceives of (small) segment types classified by spanning a spectrum of $\nu^{\max}$ segments. The COSMO-RS/SAC formalism relates the contact energies of these segments to bins of surface polarization values, as discussed in Sec. (8-11). Nonpolar molecules like hydrocarbons may be composed of many segments but spanning a relatively small number of (nearly neutral) bins. Polar molecules span more bins. Recalling Guggenheim's elemental conception originally, $\nu^{\max}$ "elements" comprise the "periodic table" of SS-QCT.

Equation (9) of Lin and Sandler (2002) is helpful in establishing the relation between Γ_ν and molecular activity coefficient, resulting in

$$\ln(\gamma_k) = \ln(\gamma_k^{A}) + \sum_{1}^{\nu^{\max}} p_\nu^{(k)} \left(\ln(\Gamma_\nu) - \ln(\Gamma_\nu^{(k)})\right) / A_{\text{eff}} \tag{8.7-38}$$

where γ_k^A is the athermal term and $p_\nu^{(k)}/A_{\text{eff}}$ is the normalized area of segments of type ν on the kth molecule. Defining $zq_k\nu_j^{(k)}$ as in Eq. (8.7-39) shows that Eq. (8.7-38) is identical to Eq. (8.7-31) except that it is written in factorized form so the leading factor of ½ is omitted.

$$N_j^{\text{seg}}(k) = zq_k\nu_j^{(k)} \equiv p_j^{(k)}/A_{\text{eff}} \tag{8.7-39}$$

The quantity $\nu_j^{(k)}$ is necessary in Eq. (8.7-39) because Panayiotou and Vera (1980) thought in terms of functional groups. Computing the internal energy turns out to be remarkably simple. Multiplying the contact probability by the contact energy and summing

$$\frac{U}{RT} = \frac{1}{2}\sum\sum \Theta_i \Gamma_i \Theta_j \Gamma_j \psi_{ij} \beta \varepsilon_{ij} \tag{8.7-40}$$

Klamt et al. (2002) report this result as their Eq. (36), but their notation is difficult to interpret. The comparable equation for GQCT is given by Eq. (33) of Abusleme and Vera (1985). Note that the derivations here treat ε_{ij} as temperature-independent, but the formulas given by Klamt et al. (2002) and Abusleme and Vera (1985) are more general. Applying Eq. (8.7-40) to the ith pure fluid results in the energy $U^{(i)}$ and the excess internal energy is

$$\frac{U^{\text{E}}}{RT} = \frac{U}{RT} - \sum x_i \frac{U^{(i)}}{RT} \tag{8.7-41}$$

8.7.6 Lattice Fluid Hydrogen Bonding (LFHB) Theory

Tracing back (at least) to early work by Barker (1952), Veytsmann (1990), Panayiotou and Sanchez (1991) and others have developed an alternative formulation of QCT that shares elements of SS-QCT and TPT1. For example, Veytsmann's formula for the free energy of strong solvation can be written as

$$\frac{A^{\text{chem}}}{RT} = x_2 \ln\left(X_2^D\right) \tag{8.7-42}$$

$$\left(-1 + \frac{1}{X_2^D}\right) = x_1 X_1^A \alpha_{12}^{AD} \tag{8.7-43}$$

where X_k^A is the fraction of acceptor sites NOT bonded and

$$\alpha_{ij}^{AD} = \kappa_{ij}^{AD}\left[\exp\left(\frac{\varepsilon_{ij}^{AD}}{kT}\right)\right] \tag{8.7-44}$$

where κ_{ij}^{AD} is a dimensionless adjustable parameter. This result is comparable to TPT1 as given by Eq. (8.7-14). Like SS-QCT, the key step in LFHB is to recognize that strong interactions occur stereospecifically on a single face of the unit cell (cf. Barker, 1952). There are subtle differences between TPT1 and LFHB, however (Gupta et al., 1992). The explicit QCT treatment of hydrogen bonding in LFHB can also be computed within the SS-QCT formalism. Nevertheless, LFHB developers prefer to treat strong interactions separately so they can relate the parameters to spectroscopic measurements like those in the Kamlet-Taft model. For purposes of this discussion, it is convenient to subsume LFHB consideration within SS-QCT analysis.

Example 8.1 Strong solvation

Interpreting the most general forms of GQCT, TPT1, and SS-QCT is nontrivial. To gain some understanding of the meanings of the terms, the nature of the procedures, and the relation of SS-QCT to GQCT and TPT1, we focus here on the chemical aspect of strong solvation. This has the advantage of simplifying the number of interactions and clarifying exactly how the chemical contribution is computed. We consider an equimolar mixture of equal-sized molecules at $T = 333$ K with one acceptor "A" on molecule 1 and one donor "D" on molecule 2. The A and D share an energy of $E^{\text{chem}}/k = 2000$ K when "bonded," but all other molecular interaction energies are set to zero. For QCT, we assume $z = 10$, but only one cell "face" possesses the A or D. For SS-QCT, we assume 10 equal surface segments per molecule with the A or D occupying a single segment.

For TPT1, we assume $K^{AD} = 0.0015$ nm^3, a packing fraction of 0.35, and $\sigma = 0.36$ nm for spherical molecules, and apply ESD for $g(\sigma)$. Equation (8.7-14) results in $\alpha^{AD} = \rho K^{AD} g(\sigma) Y = 26$ at 333 K, where $Y = \exp(\beta\varepsilon^{AD}) - 1$. TPT1 gives the same mass conservation relation as QCT.

$$x_1 x_2 X^A X^D \alpha^{AD} = x_1(1 - X^D) = x_2(1 - X^A) \tag{8.7-45}$$

Solving Eq. (8.7-45) when $x_1 = x_2 = ½$ leads to the quadratic equation,

$$(X^A)^2 \alpha^{AD} + 2X^A - 2 = 0 \tag{8.7-46}$$

and

$$X^A = \frac{2}{\left\{1 + [1 + 2\alpha^{AD}]^{½}\right\}} \tag{8.7-47}$$

where $Y = \exp(\beta E^{\text{chem}}) - 1$, The Helmholtz energy is easily written in Wertheim's formalism,

$$\frac{A^{\text{E}}}{RT} = x_1 \ln(X^A) + x_2 \ln(X^D) + \frac{x_1}{2}(1 - X^A) + \frac{x_2}{2}(1 - X^D) \tag{8.7-48}$$

Substituting the equimolar expression for $X^A = X^D$ and taking the derivative $\beta[(\partial(\beta A^E)/\partial\beta]_\rho$,

$$\frac{U^E(x_1 = x_2)}{RT} = \frac{\beta w}{2}\left(1 - \frac{X^A}{2}\right)\frac{Y+1}{Y}\left[\frac{\alpha^{AD}X^A}{(1+2\alpha^{AD})^{1/2}}\right] \tag{8.7-49}$$

Simplification is easier if we define $D \equiv 1 + (1 + 2\alpha^{AD})^{1/2}$. Then by Eq. (8.7-46), $\alpha^{AD}X^A = D - 2$ and

$$\left(1 - \frac{X^A}{2}\right)\frac{\alpha X^A}{(1+2\alpha)^{1/2}} = \left(1 - \frac{1}{D}\right)\frac{D-2}{D-1} = \left(\frac{D-1}{D}\right)\frac{D-2}{D-1} = 1 - X^A \tag{8.7-50}$$

Substitution gives the remarkably simple result:

$$\frac{U^E(x_1 = x_2)}{RT} = \frac{\beta w}{2}(1 - X^A)\frac{Y+1}{Y} \tag{8.7-51}$$

There is no problem in principle with applying GQCT to the case when two surface elements interact strongly while their self-interactions are negligible. The surface fractions of two specific segments in Eq. (8.7-27) need not sum to 1. For this analysis, $q_A/\Sigma\nu_i q_i = q_D/\Sigma\nu_i q_i = \text{constant} \equiv q$. Noting that $(n_A^\circ - X)/n_A^\circ = X_Q^A$ and $(n_D^\circ - X)/n_D^\circ = X_Q^D$ in Wertheim's notation, Eqs. (8.7-25) and (8.7-21) can be rewritten as

$$\frac{U^E(x_1 = x_2)}{RT} = \beta w \frac{X}{\overset{\circ}{n}_A + \overset{\circ}{n}_D} = \frac{\beta w}{2}\left(1 - X_Q^A\right) \tag{8.7-52}$$

$$X_Q^A X_Q^D = \left(1 - X_Q^A\right)\left(1 - X_Q^D\right)\alpha_Q^{-1} \tag{8.7-53}$$

where $\alpha_Q^{-1} \equiv 4q^2\exp(-\beta w)$. Note that the factor of z is omitted from Eq. (8.7-52) because only one segment of the 10 can participate in this strong solvation reaction. Nevertheless, βw has a large magnitude owing simply to the strength of strong solvation. This observation is related to Wertheim's steric hindrance condition. For an equimolar mixture in the limit where $X \to n_A^\circ$, $U^E/RT = \beta w/2$. The solution for X_Q^A can be obtained by multiplying both sides of Eq. (8.7-53) by $x_A x_D \alpha_Q$ and applying the mass balance relation of Eq. (8.7-45). For the equimolar case,

$$X_Q^A = \frac{1}{1 + \alpha_Q^{1/2}} \tag{8.7-54}$$

Comparison of Eq. (8.7-49) to Eq. (8.7-52) shows how similar the results are in functional form and Fig. 8.3 shows that the numerical results are also similar.

For SS-QCT, $\Theta_A = \Theta_D = 0.05$, $\beta w = 2000/333 = 6.0$, and $\tau_{AD} = \tau_{DA} = \exp(6) = 403$. Note that we set $\tau_{AA} = \tau_{DD} = 1$ in this example because the interaction is defined as zero unless donor and acceptor overlap. It would be straightforward to include nonzero interactions within the SS-QCT formalism, but then $\tau_{AA} = \tau_{DD} = \exp(-6) \approx 0$, so there is little impact. We mention this because it is more common to treat the interactions as nonzero in the normal practice of SS-QCT. The neutral segments on each molecule are indistinguishable from the perspective of segmental mixing, so their surface areas combine to $\Theta_0 = 0.90$. Then the segmental activities are given by

$$\Gamma_A = \left\{\Theta_A\Gamma_A + \Theta_D\Gamma_D\tau + \Theta_0\Gamma_0\right\}^{-1} \tag{8.7-55}$$

$$\Gamma_D = \left\{\Theta_A\Gamma_A + \Theta_D\Gamma_D\tau + \Theta_0\Gamma_0\right\}^{-1} \tag{8.7-56}$$

$$\Gamma_0 = \left\{\Theta_A\Gamma_A + \Theta_D\Gamma_D\tau + \Theta_0\Gamma_0\right\}^{-1} \tag{8.7-57}$$

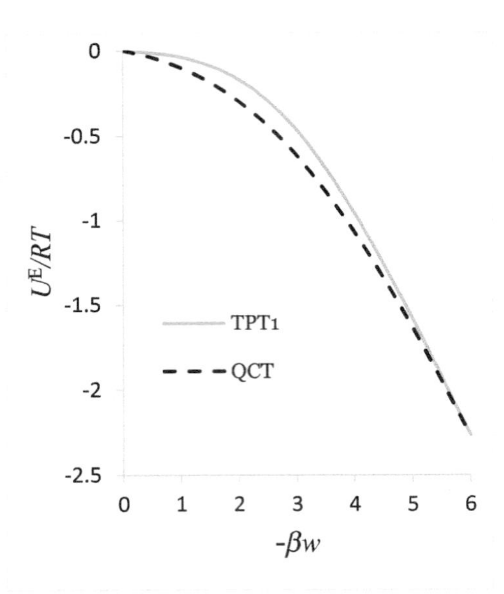

FIGURE 8.3
Comparison of QCT to TPT1 for excess energy of strong solvation.

In general, this would give a quartic equation, but for this contrived system, $\Gamma_A = \Gamma_D$ and

$$\Gamma_A = \Gamma_D = \frac{\left(\frac{1}{\Gamma_0} - \Theta_0\Gamma_0\right)}{1 - \Theta_0} \tag{8.7-58}$$

Substitution results in a quadratic equation for Γ_0^2

$$\Gamma_0^2 = \frac{2t}{\left\{\left(2\Theta_0 t + 1 - 2\Theta_0\right) - \left[\left(2\Theta_0 t + 1 - 2\Theta_0\right)^2 - 4\Theta_0^2(t-1)t\right]^{1/2}\right\}} \tag{8.7-59}$$
$$= 1.0880 \Rightarrow \Gamma_A = \Gamma_D = 0.1998$$

where $t \equiv 0.5(\psi_{AD} + 1)$ to make the equation more compact. For pure compound 1,

$$\overset{\circ}{\Gamma}_A = \left\{\Theta_A\overset{\circ}{\Gamma}_A + \Theta_0\overset{\circ}{\Gamma}_0\right\}^{-1} = \overset{\circ}{\Gamma}_0 = 1 \tag{8.7-60}$$

An equivalent result applies to the pure donor compound. Returning to the equimolar case and summing up,

$$\ln(\gamma_1) = 10\left[0.1\left(\ln(0.1998) - \ln(1)\right) + 0.9(\ln(1.0430) - \ln(1))\right] = -1.231 = \ln(\gamma_2) \tag{8.7-61}$$

The free energy is simply

$$\frac{A^E}{RT} \approx \frac{G^E}{RT} = x_1 \ln(\gamma_1) + x_2 \ln(\gamma_2) = -1.231 \tag{8.7-62}$$

For this example, only A and D in the mixture contribute to the energy

$$\frac{U^E}{RT} = z\left(\beta w_{AD}\Theta_A\Gamma_A\Theta_D\Gamma_D\tau_{AD}\right) = -(10)6(0.05)^2(0.1998)^2\exp(6) = -2.413 \tag{8.7-63}$$

Comparing Eq. (8.7-27) to Eq. (8.7-63) shows SS-QCT is equivalent to GQCT for this system if we recognize $y_{AD} = \Theta_A\Gamma_A\Theta_D\Gamma_D\tau_{AD}$.

In Fig. 8.3, GQCT and SS-QCT exhibit equivalent numerical results and are shown as a single curve designated "QCT." More notably, GQCT and TPT1 show remarkable similarities in both functional form and numerical results. This suggests a remarkable advantage for QCT. Whereas TPT1 requires a separate solution for the monomer interactions and bonding interactions, QCT simply includes the bonding interactions into the same (robust) solution methodology as the rest of the interactions. Like TPT1, the bonding segment fraction of the reactive segment in SS-QCT must be small relative to the reference monomer, like 0.1 in this example. In this regard, the SS-QCT perspective stands out over functional group implementations of QCT. Consequently, SS-QCT and TPT1 also provide explicit recognition of the difference in temperature scaling that occurs between strong interactions and dispersion interactions.

Perhaps this result should not be surprising. Both theories are based on a chemical reaction at their foundation. Both theories also express that temperature dependence as an exponential involving the energy of "reaction." QCT explicitly accounts for only nearest neighbor interactions, but short-range interactions like hydrogen bonding should qualify. So, the resulting functional forms are similar, as are the numerical results.

On the other hand, the reaction stoichiometry of QCT (Eq. 8.7-53) is different from that of TPT1 (Eq. 8.7-45). From Wertheim's perspective, the reaction is from two unbonded molecules to a single-bonded molecule. From Guggenheim's perspective, an AA dimer reacts with a BB dimer to form two AB dimers. This difference in perspectives might be substantial at low density, especially for entropic implications. Perhaps this explains why adaptations of QCT to EOS development led to expressions for the repulsive (~athermal) part of the EOS that are inconsistent with molecular simulation (Panayiotou and Sanchez, 1991). Gubbins (2016) has reviewed numerous studies showing that lattice theories are deficient in the treatment of density effects for mixtures, especially for repulsive contributions like hard sphere mixtures. A modernized QCT would need to be coupled with an accurate theory for repulsive contributions at low densities and it would need to recognize the screening of attractive influences on the fluid structure at high density. The approximate equations of Lee, Sandler, and Monson (1986) could serve that purpose on a semiempirical basis.

Further insight into relationships between these theories can be obtained by considering developments by Pratt and coworkers. They extended QCT in an off-lattice framework, including higher-order terms beyond the nearest neighbor. They refer to their methodology as "molecular" QCT (MQCT). Their work has been reviewed by Beck et al. (2006). Although they do not mention TPT1 in discussing their Eq. (7.8), adapting MQCT for strong solvation shows that

$$\frac{A^{\mathrm{E}}}{RT} = x_A \ln(X^A) + x_D \ln(X^D) + \text{higher-order terms} \tag{8.7-64}$$

where

$$X^A = \left[1 + x_D\text{“}K\text{”}\right]^{-1} \tag{8.7-65}$$

and "K" is related to the solvation equilibrium constant. The first two terms in this expansion involve only nearest neighbor interactions while the MQCT formalism extends to all interactions. MQCT can also be considered as a generalized form of TPT1 that includes TPT2 and permits analysis of general cases when Wertheim's steric hindrance condition breaks down.

In summary, QCT shows remarkable similarities to a combination of TPT1 with Flory-Huggins theory, especially in the SS-QCT formulation. For condensed phase mixtures, QCT's treatment of dispersion forces incorporates local composition effects in a way that is simple and arguably superior to the Flory-Huggins model. In the presence of strong interactions like hydrogen bonding, SS-QCT provides similar results to TPT1 while integrating the effects into a single computational framework. Although QCT and SS-QCT are limited to nearest neighbor interactions, as discussed by Pratt and coworkers, these theories may have more in common with mainstream engineering models than is widely appreciated, especially for short-range interactions like hydrogen bonding.

8.7.7 Local Composition Theories and QCT Caveats

Local composition theories account for nonrandom mixing in a slightly different manner than QCT. Although UNIQUAC is an acronym for "universal quasichemical theory," its results can deviate quite dramatically from Guggenheim's QCT. We can illustrate the distinctions by considering a mixture of argon + xenon in the context of the UNIQUAC model. Wilson's model, the van Laar model, Margules' models, and the NRTL model can all be considered as special cases of the UNIQUAC model, as described by Abrams and Prausnitz (1975). Shortly after Abrams and Prausnitz published the original development of the UNIQUAC model as inspired by QCT, Maurer and Prausnitz (1978) published an alternative derivation based on local compositions like the presentation below. We refer to both but adopt notation consistent with the above presentation for QCT. An article by Fischer (1983) is instructive and we begin by adapting his analysis.

The UNIQUAC model can be written as

$$\frac{U^{E}}{RT} = \frac{-z}{2}\{q_1 x_1(\varepsilon_{21} - \varepsilon_{11})x_{21} + q_2 x_2(\varepsilon_{12} - \varepsilon_{22})x_{12}\} \tag{8.7-66}$$

where $\varepsilon_{ij} = \varepsilon_{ji}$, q_i is a reduced surface area relative to a universal coordination number, z. Here, we maintain explicit representation of the coordination number, z. The local mole fraction of an ith molecule around a jth molecule, x_{ij}, is defined by

$$\frac{x_{ij}}{x_{jj}} \equiv \frac{x_i}{x_j}\Omega_{ij} \tag{8.7-67}$$

where Ω_{ij} is a factor characterizing deviations from random mixing. The adaptation of local mole fractions was introduced by Wilson (1964). The integration to obtain free energy is made simpler by using local compositions. The characterization of the local compositions is inspired by QCT. In the UNIQUAC model, the assumption is

$$\Omega_{ij} = \frac{q_i}{q_j}\exp\left(\frac{z(\varepsilon_{ij} - \varepsilon_{jj})}{2kT}\right) \equiv \frac{q_i}{q_j}\exp\left(\frac{-a_{ij}}{T}\right) \equiv \frac{q_i}{q_j}\tau_{ij} \tag{8.7-68}$$

Noting that, for example, $x_{12} + x_{22} = 1$, and $\Omega_{ii} = 1$ for all i,

$$x_{ij} = \frac{x_i\Omega_{ij}}{\sum x_k\Omega_{kj}} = \frac{x_i q_i / q_j \tau_{ij}}{(x_j + x_i\tau_{ij}q_i/q_j)} = \frac{\Theta_i\tau_{ij}}{(\Theta_j + \Theta_i\tau_{ij})} \tag{8.7-69}$$

The internal energy can be rewritten as

$$\frac{U^{E}}{RT} = \left\{-\frac{q_1x_1\Theta_2\frac{z\beta}{2}(\varepsilon_{21} - \varepsilon_{11})\tau_{21}}{(\Theta_1 + \Theta_2\tau_{21})} - \frac{q_2x_2\Theta_1\frac{z\beta}{2}(\varepsilon_{12} - \varepsilon_{22})\tau_{12}}{(\Theta_1\tau_{12} + \Theta_2)}\right\} \tag{8.7-70}$$

Abrams and Prausnitz then integrate to obtain the free energy

$$\frac{A^{E}_{\text{att}}}{RT} = \left\{-q_1x_1\ln(\theta_1 + \theta_2\tau_{21}) - q_2x_2\ln(\Theta_1\tau_{12} + x_2)\right\} \tag{8.7-71}$$

The Wilson model is recovered when $q_i = 1$ for all i and $z = 2$. Although the integration is simple, the second-order expansion is still informative,

$$\frac{A^{E}_{\text{att}}}{RT} \approx \frac{z\Theta_1\Theta_2}{2}\left[2\beta w - \Theta_1(\beta\varepsilon_{21} - \beta\varepsilon_{11})^2 - \Theta_2(\beta\varepsilon_{12} - \beta\varepsilon_{22})^2\right] \tag{8.7-72}$$

Returning to the example of hypothetical argon+xenon introduced below Eq. (8.7-26), the first-order term looks very similar to QCT, but the exact integration results in an excess free energy that is *opposite in sign*, as illustrated in Fig. 8.4a. The problem is that the coordination number has been subsumed in the exponential, exaggerating the temperature sensitivity of the τ_{ij} parameters. This assignment is not based on comparison to QCT

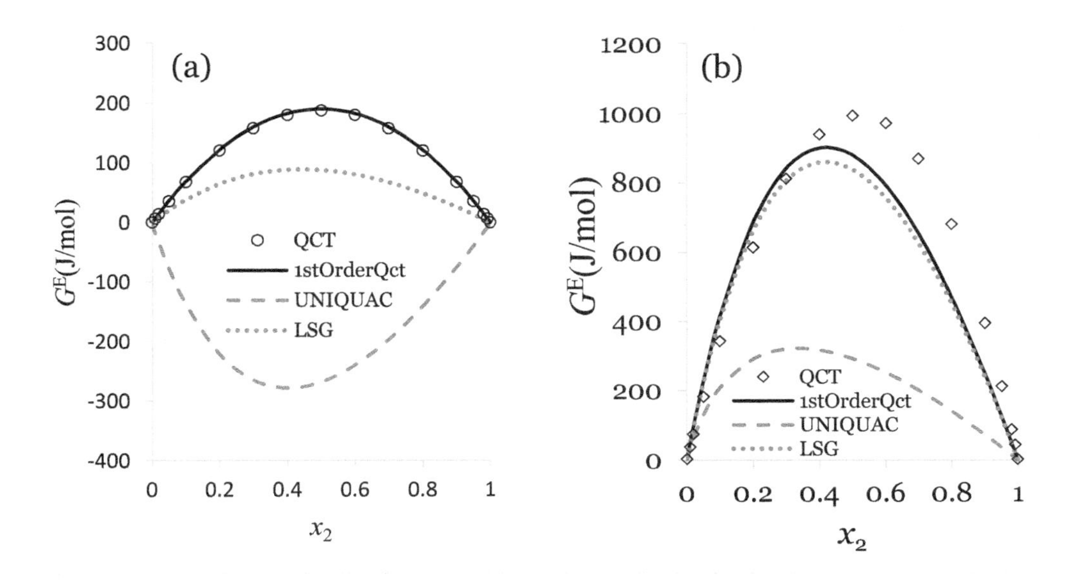

FIGURE 8.4
Gibbs excess energy for a hypothetical mixture of argon(1)+xenon(2) at 167 K. (a) $k_{ij} = 0$. (b) $k_{ij} = 0.2$.

or molecular simulation. With a slightly different assumption, agreement with QCT and molecular simulation is improved. For example,

$$\exp\left(\frac{(\varepsilon_{ij} - \varepsilon_{jj})}{kT}\right) \equiv \tau_{ij}^{\text{mod}} \tag{8.7-73}$$

leads to the expression

$$\frac{A_{\text{att}}^{\text{E}}}{RT} = \frac{z}{2}\left\{-q_1 x_1 \ln\left(\Theta_1 + \Theta_2 \tau_{21}^{\text{mod}}\right) - q_2 x_2 \ln\left(\Theta_1 \tau_{12}^{\text{mod}} + x_2\right)\right\} \tag{8.7-74}$$

This is the "local surface Guggenheim" (LSG) model of Vera et al. (1977). As illustrated in Fig. 8.4b for argon+xenon with $k_{ij} = 0.2$, LSG matches QCT to within about 26% at $\beta w = 0.15$, while UNIQUAC gives a deviation of 59%. Fischer includes comparisons of QCT with molecular simulation that show roughly 20% deviations for $k_{ij} = 0$, whereas the deviation of UNIQUAC cannot be quantified as a percentage. Klamt et al. (2002) included an analysis of UNIQUAC from the perspective of QCT and noted several other inconsistencies. Of course, these comparisons treat $\{\varepsilon_{ij}\}$ as fixed values rather than adjustable parameters, so it would be feasible to achieve a similar result for $A_{\text{att}}^{\text{E}}$ with a different set of parameters. The a_{ij} parameters for the LSG model would simply be smaller than those of the UNIQUAC model. The value of U^{E} would be very different in that case, however, requiring parameters like temperature-dependent εs for UNIQUAC to match $A_{\text{att}}^{\text{E}}$ and U^{E}. Also, the ability to validate the assumptions of the theory by comparisons with simulation would be improved with the LSG model instead of UNIQUAC. Larsen and Rasmussen (1986) go further, showing that all local composition models, including LSG, NRTL, and Wilson as well as UNIQUAC, deviate fundamentally from the proper trends in the presence of strong solvation interactions like hydrogen bonding. They trace this error to approximating Eq. (8.7-32) as

$$\ln\left(\Gamma_{jj}^{1/2}\right) = -\ln\left(\sum \Theta_i \Gamma_{ii}^{1/2} \psi_{ij}\right) \approx 1 - \sum \Theta_i \Gamma_{ii}^{1/2} \psi_{ij} \tag{8.7-75}$$

which is inherent in all these models. This approximation is off by about 50% when $\psi_{ij} = 2$, and the error grows exponentially for higher values of ψ_{ij}. Recall that $\psi_{AD} \approx 600$ for the strong solvation example. Therefore, local composition models are fundamentally flawed when it comes to modeling strong interactions, whereas QCT is immune to this flaw. It would be more appropriate to apply TPT1 or QCT for the strong interactions and a local composition model to the dispersion interactions. Either approach should provide the quantitative accuracy that engineers require without introducing artificial temperature-dependent parameters. Hao and Chen (2021) have recently demonstrated the benefits of combining NRTL with TPT1 in this manner, including demonstrations of lower critical solution behavior (aka "Type VI") and correlations of ternary LLE that are much more accurate than with NRTL alone.

Based on the presumption that progress in the field is enhanced when consistency with molecular physics is improved, we should devote a few words to what is known about local compositions from molecular simulation. Lee et al. (1986) performed extensive simulations of square well sphere mixtures and concluded that the Ornstein-Zernike equation with closure by the mean spherical approximation (MSA) was quite accurate, especially at liquid densities. This is convenient because the MSA model can be computed at any conditions of temperature, pressure, composition, and packing fraction in just a few seconds. The internal energy and Ω_{ij}s can then be computed from the determined radial distribution functions (rdf's) and $\{g_{ij}(r)\}$.

$$\frac{U^E}{RT} = -\tfrac{1}{2}\{z_1x_1(\varepsilon_{21}-\varepsilon_{11})x_{21} + z_2x_2(\varepsilon_{12}-\varepsilon_{22})x_{12}\} \tag{8.7-76}$$

where

$$z_j = \sum N_{kj} \tag{8.7-77}$$

$$N_{ij} = \rho\sigma_{ij}^3 x_i \int_1^{\lambda} g_{ij} 4\pi x^2 dx \tag{8.7-78}$$

$$\frac{x_{ij}}{x_{jj}} = \frac{N_{ij}}{N_{jj}} = \frac{x_i}{x_j}\frac{\sigma_{ij}^3\int_1^{\lambda} g_{ij}4\pi x^2 dx}{\sigma_{ij}^3\int_1^{\lambda} g_{ij}4\pi x^2 dx} = \frac{x_i}{x_j}\Omega_{ij} \tag{8.7-79}$$

We examine trends with temperature in Fig. 8.5 for the argon+xenon mixture with $k_{ij} = 0.2$. Figure 8.5a shows that the temperature dependence of the residual energy is weak and linear in β, supporting TPT expansion in β. Figure 8.5b shows how the Ω_{ij}s vary with temperature in relation to the diameter ratio: $S_{ij}^3 \equiv (\sigma_{ij}/\sigma_{jj})^3$. The Ω_{ij}s

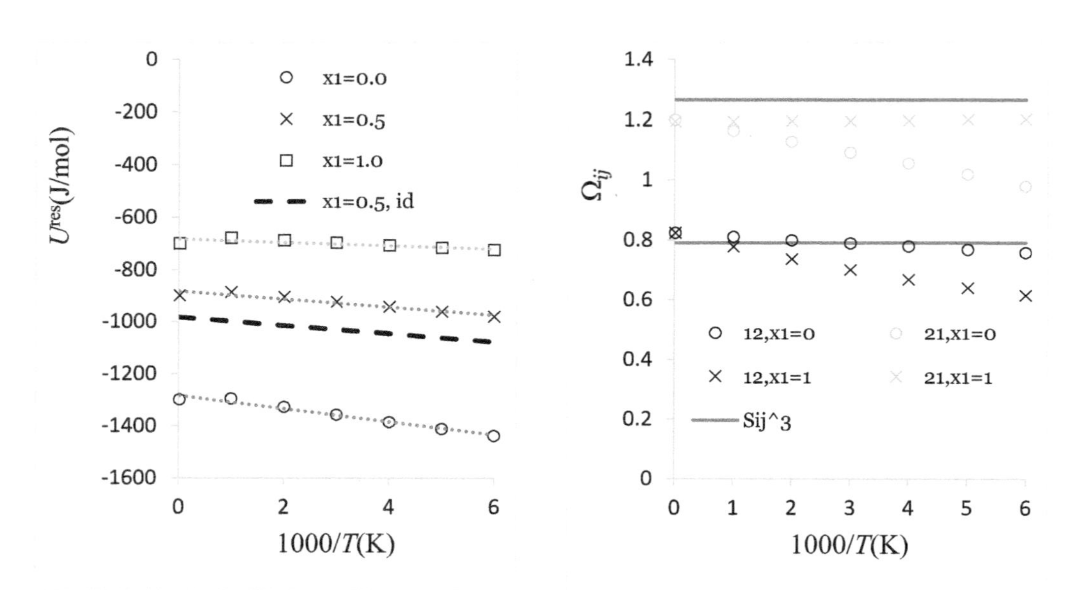

FIGURE 8.5
Mixing inferences from the MSA model of square well spheres.

at infinite temperature are determined by the hard sphere rdfs for a mixture. This analysis shows that the Ω_{ij}s are practically independent of temperature when the ith component is dilute. A single molecule in an ocean of its opposite can only exert its hard sphere packing diameter. The concentrated compound, on the other hand, can choose its partners. When the interactions are unfavorable, this choice effectively generates a larger "exclusion zone" around the unlike molecule. When the small molecule is dilute, it appears larger as the attractive interactions strengthen, causing those local compositions to approach the bulk compositions. When the large molecule is dilute, it appears even larger and more asymmetric with stronger attractive interactions.

These observations have several implications for the assumed temperature dependencies of the Ω_{ij}s. First, their slopes have the same sign, but ε_{12}-ε_{11} and ε_{21}-ε_{22} have opposite signs. Second, the variation with temperature is much weaker than $z/2$. Taking the average of the Ω_{ij}s for all temperatures, $\ln\Omega_{ij} \sim \beta w/10$, that's a factor of ~50 discrepancy with the typical assumption. Whereas QCT and UNIQUAC express the entire magnitude of $\{\varepsilon_{ij}\}$ in their Boltzmann factors, repulsive forces dominate packing (and local compositions) at liquid densities. The quasichemical assumption about the Boltzmann factors is appropriate at low density where pairwise contacts predominate, but a multitude of (mostly repulsive) neighboring interactions screens the effects of attractive forces at high density. Third, there is a significant variation of Ω_{ij} with composition. All these caveats apply to QCT in general as well as all current local composition models.

On the other hand, the QCT and UNIQUAC assumptions about coordination number are remarkably accurate for most liquids. For the argon+xenon mixture at a packing fraction of 0.4, $z_1 = 10.4$ and $z_2 = 13.4$. These values could be accommodated easily in the UNIQUAC formalism with a "reference coordination number" of $z = 10$, $q_1 = 1.04$, and $q_2 = 1.34$.

Altogether, it should be feasible to improve local composition activity models in a systematic manner by adapting what can be learned from molecular simulation. Inferring the energy of mixing from molecular simulation is simple. Integration to obtain the free energy would be more complicated but estimating the free energy through Eq. (8.7-26) should be quite accurate for dispersion interactions. Nevertheless, we should be prepared for the possibility that pairwise potential models based on interaction sites might not characterize molecular interactions to the level of precision required to match experimental data like current semiempirical models. It should be possible to steadily shrink the magnitude of the contribution but including a term for these "unaccounted" interactions may be necessary for many years to come. The functional form of local composition theories has proved to be useful for this kind of empiricism and should be reliable if the magnitude of the contribution is indeed small. Briefly, account for all you know, but be prepared for what you do not.

8.7.8 Separation of Cohesive Energy Density and Linear Free Energy Relations

The Scatchard-Hildebrand perspective is simple and instructive: barring any information to the contrary, deviations from ideal solution behavior are characterized by $(\Delta\delta)^2$. We have seen the mathematics, but what does it mean intuitively? The key quantity is $(\Delta\delta)^2$, which has units of energy density. To distinguish from the explosive energy density of, say nitroglycerin, this quantity is called the *cohesive* energy density. Intuitively, it means that a small molecule with a high amount of attractive energy, like water, can achieve lower Gibbs energy relative to a weakly attracting but large molecule like perfluorobenzene, by clustering together as much as possible, perhaps even forming a separate liquid-phase. In other words, the large negative internal energy of forming an aqueous phase with high cohesive energy density overwhelms the entropy of mixing.

On the other hand, hydrogen bonding in water is a well-known cause of its peculiar properties. Although hydrogen bonding is an attractive energy and contributes to water's cohesive energy density, it is a bit different from the typical dispersion energy that forms the basis of Scatchard-Hildebrand theory. This is the motivation for several activity models that apply separation of cohesive energy density (SCED) in various forms. For example, the modified SCED (MOSCED) model of Lazzaroni et al. (2005) correlates IDACs in terms of acidity, basicity, polarity, and dispersion contributions to the solubility parameter. By recognizing separate acidity and basicity contributions, the MOSCED model can predict negative deviations from ideality, whereas the Scatchard-Hildebrand model predicts only positive deviations. Another approach is to simply recognize hydrogen bonding, polarity, and dispersion interactions, as in the Hansen (1967) solubility parameter method. Then deviations from ideality can be expressed in terms of a Cartesian distance: $d = [(\delta_2^{HB} - \delta_1^{HB})^2 + (\delta_2^{P} - \delta_1^{P})^2 + (\delta_2^{d} - \delta_1^{d})^2]^{1/2}$. Hansen predicts a remarkable variety of properties using this approach. Like the Scatchard-Hildebrand model, however, Hansen's method can only represent positive deviations from ideality.

Hait et al. (1993) revised an early version of the MOSCED approach to reduce the number of adjustable parameters, referring to their newer method as SPACE. An alternative is to predict the MOSCED parameters from quantum mechanics (cf. Phifer et al., 2017) or alternatively to correlate them using COSMO sigma-profiles (e.g., Gnap and Elliott, 2018). To date, all SCED methods have retained the FH activity model as their fundamental basis, but it would be straightforward to implement TPT1 for the strong interaction contributions as in Eq. (8.7-12). Another alternative would be to use LSG for the attractive dispersion forces instead of Flory-Huggins, then add TPT1. When SS-QCT is used with COSMO sigma profiles as described by Klamt and coworkers, this is the COSMO-RS/SAC method, of course. We provide more details of the COSMO-RS/SAC method in the section on predicting activity coefficients.

A related set of methods has been developed to characterize acidity, basicity, and polarity based on spectroscopic measurements. Kamlet and Taft (1976) were early advocates of this approach. Extensive tables of Kamlet-Taft parameters have been developed (cf. Abraham, 1993), and that development continues. A common application is to develop models of the free energy of solvation in terms of linear combinations of the Kamlet-Taft parameters. The free energy of solvation is closely related to the IDAC. These methods are known as linear free energy relations (LFER) and they often form the basis for predictions regarding environmental contaminants. Panayiotou et al. (2017) have described several related methods and their attempts to unify several measures of acidity and basicity.

8.7.9 Conclusions

In summary, there is still considerable room for improvement in activity models, despite decades of research. Local composition models remain the method of choice, primarily because they provide two adjustable parameters per binary mixture and that suffices to achieve 1% accuracy or better. Nevertheless, the advent of molecular simulation has demonstrated deficiencies in local composition models. Furthermore, analyses of the theoretical foundations of local composition models by Fischer (1983) and Klamt et al. (2002) raise several important questions. It may be possible to address some of the foundational issues with simple adjustments like the LSG model but treating strong interactions like hydrogen bonding requires either a rigorous implementation of QCT as exemplified by the COSMO-RS/SAC method, or combination of a local composition model with TPT1 theory as recently demonstrated by Hao and Chen (2021). Long-term progress would be well served by moving in the direction of application models that are at least qualitatively consistent with fundamental analysis and molecular simulations.

The presentation here hopefully clarifies the relationships between the various theories. EOSs with quadratic mixing rules are effectively like Flory-Huggins theory. Quasichemical theories are like Flory-Huggins theory in that they only have one BIP per binary interaction, but they shift the Gibbs excess curve to reflect redistribution of the molecules to find more favorable interactions, with substantial reduction in deviations from experimental data as shown in Fig. 8.1. In the presence of strong interactions like hydrogen bonding, SS-QCT and TPT1 provide similar results, but TPT1 requires a separate accounting for the dispersion interactions while SS-QCT accounts for all interactions holistically. Combination of SS-QCT or TPT1 with a model like LSG for the dispersion interactions could provide a meaningful semiempirical methodology that matches the correlation accuracy of current empirical models, especially when these systems exhibit peculiar phase behavior like lower critical solution temperatures. Theoretical developments by Barker (1952), Veytsman (1990), and Panayiotou and Sanchez (1991) characterize strong interactions with QCT terms that resemble TPT1. Although we have not investigated those models in detail, our expectation is that results would resemble implementations of SS-QCT or TPT1 that are represented in our evaluations. Observations about the similarities between SS-QCT and TPT1 for liquid mixtures should not be extended to equations of state at lower than liquid densities without further research, because the fundamentals of the two methods differ in ways that might be sensitive to density.

Meanwhile, the improvements in EOS models, particularly TPT1 models, suggest a more prominent role for these methods than was common 20 years ago. EOS models that offer two BIPs can provide similar correlation accuracy to activity models, while offering straightforward extensions to gaseous compounds and high temperature and pressure. The various implementations of TPT1, including adaptations for density-independent activity models, provide capabilities for phase diagrams that were previously inaccessible without crass empiricism, like systems exhibiting lower critical solution temperatures, for example.

8.8 CORRELATING LOW-PRESSURE BINARY VAPOR-LIQUID EQUILIBRIA

Modeling the behavior of binary mixtures requires minimizing the deviations between experimental data and model computations. Experimental data are generally either isothermal or isobaric and may include vapor-phase compositions as well as liquid compositions. If experimental data are only available at a single azeotropic point, a special procedure is necessary. We begin by describing the most common case, where temperature, pressure, and composition are measured for a series of points.

8.8.1 Correlating Multiple Measurements Using Activity Models

We begin by considering the case in which liquid composition, temperature, and pressure have been reported. The following procedure should suffice whether the data are isothermal or isobaric. We wish to construct two diagrams: y_1 versus x_1 and P or T versus x_1. We assume that, since the pressure is low, we can use Eq. (8.5-3). As indicated by numerous authors, notably Abbott and Van Ness (1975), experimental errors in vapor composition y are usually larger than those in experimental pressure P, temperature T, and liquid-phase composition x. Therefore, a relatively simple fitting procedure is provided by reducing only P-x-T data; y data, even if available, are not used.[4] The essential point is to minimize the deviation between calculated and observed pressures. The steps in the procedure are:

1. Find the pure liquid vapor pressures P_1^{vp} and P_2^{vp} at every composition given T.
2. For each point, calculate γ_1 and γ_2 according to the chosen activity model.
3. For each point, calculate the bubble pressure according to Eq. (8.5-3).

$$P = \sum x_i \gamma_i P_i^{vp} \tag{8.5-3}$$

4. Compute the root mean square deviation (rmsd) between calculated and experimental pressures. It may be preferable to express these deviations as percentage relative deviations if the vapor pressures of the compounds differ substantially.
5. It is advisable to estimate the second derivative of the Gibbs excess energy for all compositions to check whether the model represents a stable liquid phase. If it is known that the liquid phase is stable, then constraining $\min(d^2g/dx_1^2) \geq 0$ during the optimization should be applied. Numerical estimation of the second derivative should suffice for this determination. A sample spreadsheet called CH08ActCoeff.xlsx is provided at the PGL6ed website to demonstrate.
6. Minimize the rmsd by varying the parameters of the activity model, constrained to a stable liquid phase if applicable.

The simple steps outlined above provide a rational, thermodynamically consistent procedure for interpolation and extrapolation with respect to composition. The crucial step is 4. Judgment is required to obtain the best, i.e., the most representative, constants in the expression chosen for g^E. To do so, it is necessary to decide on how to weight the individual experimental data; some may be more reliable than others. For determining the constants, the experimental points that give the most information are those at the ends of the composition scale, that is, y_1 when x_1 is small and y_2 when x_2 is small. Unfortunately, however, these experimental data are often the most difficult to measure. Thus, it frequently happens that the data that are potentially most valuable are least likely to be accurate.

[4]This technique is commonly referred to as *Barker's method*, which may include corrections for Poynting factors and second virial coefficients.

Calculation of isothermal or isobaric VLE can be efficiently performed with a computer. Spreadsheets have become ubiquitous tools for such computations. The spreadsheet called CH08ActCoeff.xlsx provides an example of such a tool. It includes examples for the activity models presented in Fig. 8.1, as well as the examples presented below.

When the procedures outlined above are followed, the accuracy of any VLE calculation depends primarily on the extent to which the activity model accurately represents the behavior of the mixture at the conditions (temperature, pressure, composition) for which the calculation is made. This accuracy of representation often depends not so much on the algebraic form the activity model as on the reliability of the constants appearing in that expression. This reliability, in turn, depends on the quality and quantity of the experimental data used to determine the constants.

Some activity models have a better theoretical foundation than others, but all have a strong empirical flavor. Experience has indicated that the Wilson, NRTL, and UNIQUAC models (see Table 8.1) are consistently reliable in the sense that they can usually characterize nonideal behavior by using only two or three adjustable parameters. An exception is the description of systems exhibiting lower consolute temperature (LCT) behavior, also known as "Type VI" behavior, in which case some explicit treatment of hydrogen bonding should be incorporated. For example, Hao and Chen (2021) have described adaptation of the NRTL model with TPT1 to characterize the behavior of the system water+EGBE (ethylene glycol, *n*-butyl ether). An advanced adaptation of the procedure described here would be required to characterize systems with LCT phase diagrams.

It is desirable to use an activity model that is relatively simple, and which contains only two (or at most three) adjustable binary parameters. Experimental data are then used to find the "best" binary parameters. Since experimental data are always of limited accuracy, it often happens that several sets of binary parameters may equally well represent the data within experimental uncertainty. Only in rare cases, when experimental data are both plentiful and highly accurate, is there any justification for using more than three adjustable binary parameters. As discussed in Sec. 8.2, we restrict parameter characterizations to the constants defined by the original authors without extended temperature dependencies.

Example 8.2 Estimating isothermal VLE data with the NRTL model

To illustrate, the system methanol (1) + benzene (2) has been studied at 55 °C by Scatchard et al. (1946). This dataset is included in the Jaubert et al. (2020) VLE database linked at the PGL6ed website. Vapor pressures are calculated as 0.06875 MPa for methanol and 0.04362 for benzene. A value of $\alpha_N = 0.43$ is applied here, which is near the optimal value. We illustrate the computations with $g_{12} = 330.3$ K and $g_{21} = 605.6$ K, which were found to minimize the rmsd for this example. Readers can validate these parameters by typing the data and equations into a spreadsheet. This leads to $\tau_{12} = 1.0064$ and $\tau_{21} = 1.8454$ for this isothermal example, and $G_{12} = 0.6487$ and $G_{21} = 0.4522$. For example, referring to the NRTL model in Table 8.2 with the composition $x_1 = 0.3297$, $x_2 = 0.6703$ gives

$$\gamma_1 = \exp\left\{x_2^2\left[\tau_{21}\left(\frac{G_{21}}{x_1 + x_2G_{21}}\right)^2 + \tau_{12}\frac{G_{12}}{\left(x_2 + x_1G_{12}\right)^2}\right]\right\} = 2.223 \tag{8.8-1}$$

$$\gamma_2 = \exp\left\{x_1^2\left[\tau_{12}\left(\frac{G_{12}}{x_2 + x_1G_{12}}\right)^2 + \tau_{21}\frac{G_{21}}{\left(x_1 + x_2G_{21}\right)^2}\right]\right\} = 1.330 \tag{8.8-2}$$

The total pressure is

$$P = 0.3297(0.06875)2.223 + 0.6703(0.04362)1.330 = 0.08927 \tag{8.8-3}$$

This compares to the experimentally reported value of 0.0886 MPa with a deviation of 0.81%. The overall average deviation for all eight points in this dataset is 0.58%. The estimated value of $\min(d^2g/dx_1^2)$ is 0.32 for this set of parameters, so liquid stability is not a problem.

Application of the procedure to correlate isobaric VLE data is straightforward, though perhaps more tedious. The temperature dependence of vapor pressures and model parameters must be articulated at each composition since the temperature varies with composition for an isobaric system. The computed pressures relative to the experimental data deviate from the specified pressure, of course, but minimizing those deviations suffices to provide optimal model parameters. The model can then be applied in constant pressure mode to compute the *T-x-y* diagram. We illustrate the procedure with Example 8.3.

Example 8.3 Correlating isobaric VLE data with the NRTL model

Given five experimental vapor-liquid equilibrium data for the binary system propanol (1)-water (2) at 101 kPa, predict the *T-y-x* diagram for the same system at 133 kPa.

Experimental data at 101 kPa (Murti and Van Winkle, 1978)

$100\,x_1$	$100\,y_1$	T, °C
7.5	37.5	89.05
17.9	38.8	87.95
48.2	43.8	87.80
71.2	56.0	89.20
85.0	68.5	91.70

Solution. To represent the experimental data, we choose the NRTL model once more. We assume constant values for $g_{12} = 202.1$ K, $g_{21} = 896.1$ K, and $\alpha_N = 0.4$, which were found to minimize the rmsd (0.05%) for this example. The estimated value of $\min(d^2g/dx_1^2)$ is 0.18 for this set of parameters, confirming the absence of LLE. The temperatures are computed by iterating all values to minimize the pressure deviations while holding the parameters constant. Fig. 8.6 illustrates the model validation relative to experimental data at 101 kPa and the prediction at 133 kPa.

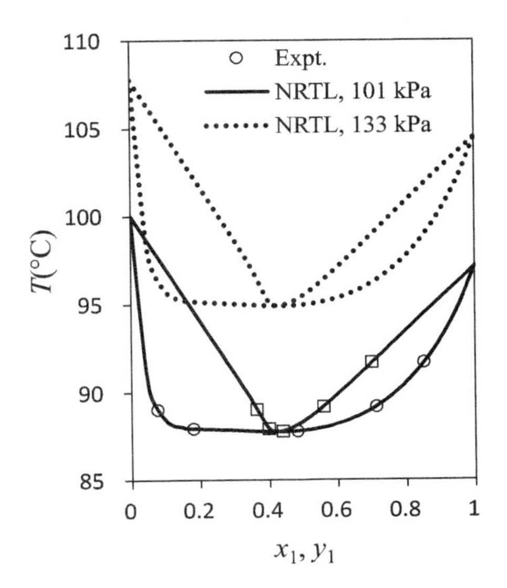

FIGURE 8.6
T-x-y diagram for 1-propanol+water at 101 kPa and 133 kPa. Experimental data of Murti and van Winkle (1978).

This is a convenient way to compute all the temperatures simultaneously; the "minimum" is practically zero, of course. For example, at the composition $x_1 = 0.65$, the computed temperature at 133 kPa is 368.87. The remaining quantities are: $\gamma_1 = 1.109$, $\gamma_2 = 2.102$, $P_1^{vp} = 0.960$; $P_2^{vp} = 0.868$, the total pressure 133 kPa.

The two simple examples above illustrate the essential steps for correlating VLE from limited experimental data. Because of their illustrative nature, these examples are intentionally simplified. For more general cases, it may be desirable to replace some of the details by more sophisticated techniques. For example, it may be worthwhile to include corrections for vapor-phase nonideality and perhaps the Poynting factor, i.e., to relax the simplifying assumption $F_i = 1$ in Eq. (8.5-1). At the modest pressures encountered here, however, such modifications are likely to have little effect. A more important change could be an alternative equation for the activity coefficients, e.g., the Wilson equation or the UNIQUAC equation. If this is done, the calculational procedure is the same but the details of computation are more complex. The deviations with the NRTL equation are quite small however, so little improvement is likely.

In Examples 8.2 and 8.3, we have not only made simplifications in the thermodynamic relations but have also neglected to take into quantitative consideration the effect of experimental error. It is beyond the scope of this chapter to discuss in detail the highly sophisticated statistical methods now available for optimum reduction of VLE data. Nevertheless, a very short discussion may be useful as an introduction for readers who want to obtain the highest possible accuracy from the available data.

A particularly effective data reduction method is described by Anderson, Abrams, and Grens (1978) who base their analysis on the principle of maximum likelihood while considering probable experimental errors in all experimentally determined quantities.

To illustrate the general ideas, we define a calculated pressure (constraining function) by

$$P^{\text{calc}} = \exp\left[x_1 \ln\left(\frac{x_1\gamma_1 P_1^{vp}}{y_1} F_1\right) + x_2 \ln\left(\frac{x_2\gamma_2 P_2^{vp}}{y_2} F_2\right)\right] \tag{8.8-4}$$

where F_i is given in Eq. (8.5-2). The most probable values of the parameters (for the function chosen for g^E) are those which minimize the function I:

$$I = \sum_{\text{data}}\left[\frac{(x_i^o - x_i^M)^2}{\sigma_{xi}^2} + \frac{(y_i^o - y_i^M)^2}{\sigma_{yi}^2} + \frac{(P_i^o - P_i^M)^2}{\sigma_{Pi}^2} + \frac{(T_i^o - T_i^M)^2}{\sigma_{Ti}^2}\right] \tag{8.8-5}$$

In Eq. (8.8-5), the superscript ^{M} means a measured value of the variable and $^\circ$ means a statistical estimate of the true value of the variable which is used to calculate all the properties in Eq. (8.8-4). The σs provide estimates of the variances of the variable values, i.e., an indication of the experimental uncertainty. These may or may not be varied from data point to data point.

By using experimental P-T-x-y data and the UNIQUAC equation with estimated parameters $u_{12} - u_{22}$ and $u_{21} - u_{11}$, we obtain estimates of x_i°, y_i°, T_i°, and P_i°. The last of these is found from Eq. (8.8-4) with true values, x_i°, y_i°, and T_i°. We then evaluate I, having previously set variances σ_x^2, σ_y^2, σ_P^2, and σ_T^2 from a critical inspection of the data's quality. Upon changing the estimate of UNIQUAC parameters, we calculate a new I; with a suitable computer program, we search for the parameters that minimize I. Convergence is achieved when, from one iteration to the next, the relative change in I is less than 10^{-5}. After the last iteration, the variance of fit σ_F^2 is given by

$$\sigma_F^2 = \frac{I}{D - L} \tag{8.8-6}$$

where D is the number of data points and L is the number of adjustable parameters.

Since all experimental data have some experimental uncertainty, and since any equation for g^E can provide only an approximation to the experimental results, it follows that the parameters obtained from data reduction are not unique; there are many sets of parameters which can equally well represent the experimental data within experimental uncertainty. To illustrate this lack of uniqueness, Fig. 8.7 shows results of data reduction for the binary mixture ethanol(1)–water(2) at 70 °C. Experimental data reported by Mertl (1972) were reduced using the UNIQUAC equation with the variances

$$\sigma_x = 10^{-3} \qquad \sigma_y = 10^{-2} \qquad \sigma_P = 6.7 \times 10^{-2}\ \text{kPa} \qquad \sigma_T = 0.1\ \text{K}$$

For this binary system, the fit is very good; $\sigma_F^2 = 5 \times 10^{-4}$.

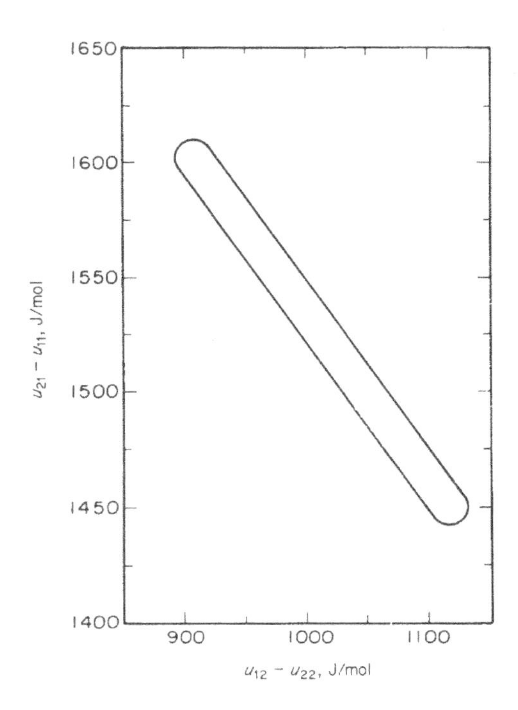

FIGURE 8.7
The 99% confidence ellipse for UNIQUAC parameters in the system ethanol(1)+water(2) at 70 °C.

The ellipse in Fig. 8.7 clearly shows that, although parameter $u_{21} - u_{11}$ is strongly correlated with parameter $u_{12} - u_{22}$, there are many sets of these parameters that can equally well represent the data. The experimental data used in data reduction are not sufficient to fix a unique set of "best" parameters. Realistic data reduction can determine only a region of parameters.[5]

While Fig. 8.7 pertains to the UNIQUAC equation, similar results are obtained when other equations for g^E are used; only a region of acceptable parameters can be obtained from a given set of P-T-y-x data. For a two-parameter equation, this region is represented by an area; for a three-parameter equation, it is represented by a volume. If the equation for g^E is suitable for the mixture, the region of acceptable parameters shrinks as the quality and quantity of the experimental data increase. However, considering the limits of both theory and experiment, it is unreasonable to expect this region to shrink to a single point.

8.8.2 Correlating Azeotropic Data Using Activity Models

Many binary systems exhibit azeotropy, i.e., a condition in which the composition of a liquid mixture is equal to that of its equilibrium vapor. When the azeotropic conditions (temperature, pressure, composition) are known, activity coefficients γ_1 and γ_2 at that condition are readily found. These activity coefficients can then be used to calculate two parameters in an activity model (Table 8.3). Extensive compilations of azeotropic data are available (Horsley, 1952, 1962, 1973; Gmehling et al., 1994).

[5]Instead of the constraint given by Eq. (8.8-4), it is sometimes preferable to use instead two constraints: first, Eq. (8.5-3), and second,

$$y_1 = \frac{x_1\gamma_1 P_1^{vp} F_1}{x_1\gamma_1 P_1^{vp} F_1 + x_2\gamma_2 P_2^{vp} F_2}$$

or the corresponding equation for y_2.

For a binary azeotrope, $x_1 = y_1$ and $x_2 = y_2$; therefore, Eq. (8.5-1), with $F_i = 1$, becomes

$$\gamma_1 = \frac{P}{P_1^{vp}} \quad \text{and} \quad \gamma_2 = \frac{P}{P_2^{vp}} \tag{8.8-7}$$

Knowing total pressure P and pure-component vapor pressures P_1^{vp} and P_2^{vp}, we determine γ_1 and γ_2. With these activity coefficients and the azeotropic composition x_1 and x_2, it is possible to find two parameters by simultaneous solution of two equations. Some activity models are sufficiently simple that analytical solutions for the parameters can be derived. More generally, the solution must be obtained numerically. One simple method is to set the calculated pressure equal to the azeotropic pressure while minimizing the difference between the vapor and liquid compositions. This method is demonstrated on the NRTL tab of the CH08ActCoeff spreadsheet.

Example 8.4 Fitting azeotropic data with the NRTL model

For the system methanol+benzene, Scatchard et al. (1946) report the azeotropic point as $x_1 = y_1 = 0.6079$ at 55 °C and 0.09045 MPa. This point is included in the Jaubert et al. (2020) VLE database listed on the PGL6ed website. Vapor pressures can be estimated as 0.06875 MPa for methanol and 0.04362 for benzene. Assuming $\alpha_N = 0.43$, determine the NRTL parameters based on the azeotropic point and compare them to the values using the entire dataset from Scatchard et al (1946). Also, compare the overall deviations for both methods.

Solution. Set a cell to the squared deviation of the calculated vapor composition from the given liquid azeotropic composition of 0.6079. Minimize this cell subject to the constraint that the calculated pressure, $P = 0.09045$ MPa, while changing g_{12} and g_{21}. The result is: $g_{12} = 363.3$ K and $g_{21} = 532.7$ K. This leads to $\tau_{12} = 1.1071$ and $\tau_{21} = 1.6234$, and $G_{12} = 0.6212$ and $G_{21} = 0.4976$. At the azeotropic composition $x_1 = 0.6079$, $x_2 = 0.3921$:

$$\gamma_1 = \exp\left\{x_2^2\left[\tau_{21}\left(\frac{G_{21}}{x_1 + x_2G_{21}}\right)^2 + \tau_{12}\frac{G_{12}}{(x_2 + x_1G_{12})^2}\right]\right\} = 1.3156 \tag{8.8-8}$$

$$\gamma_2 = \exp\left\{x_1^2\left[\tau_{12}\left(\frac{G_{12}}{x_2 + x_1G_{12}}\right)^2 + \tau_{21}\frac{G_{21}}{(x_1 + x_2G_{21})^2}\right]\right\} = 2.0738 \tag{8.8-9}$$

The total pressure is

$$P = 0.6079(0.06875)1.3156 + 0.3921(0.04362)2.0738 = 0.09045 \tag{8.8-10}$$

The vapor composition is

$$y_1 = \frac{0.6079(0.06875)1.3156}{0.09045} = 0.6079 \tag{8.8-11}$$

The values of the optimal g_{ij} parameters were given in Example 8.2. The parameters determined from the azeotropic point deviate about 20% from the optimal values. The overall deviation for the eight points in this dataset is 1.41% compared to 0.58% for the optimal values of the g_{ij} parameters. It should be noted that this dataset and these two examples in general provide unusually small uncertainties and deviations. If the only data available were the azeotropic point, then the default value of $\alpha_N = 0.3$ would be a more likely assumption, in which case the parameters determined from the azeotropic condition are: $g_{12} = 306.6$ K and $g_{21} = 463.9$ K, and the deviation for the overall dataset is 2.93%. That level of deviation is still quite reasonable from a single experimental measurement.

Figure 8.8 shows good overall agreement between experimental and calculated pressures. Generally, fair agreement is found if the azeotropic data are accurate, if the binary system is not highly complex, and, most important, if

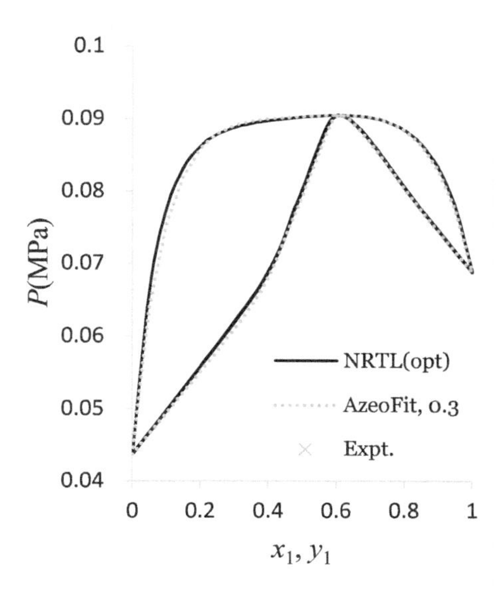

FIGURE 8.8
Alternative fits to the system methanol+benzene using the NRTL model.

the azeotropic composition is in the midrange $0.25 < x_1$ (or x_2) < 0.75. If the azeotropic composition is at either dilute end, azeotropic data are of much less value for estimating activity coefficients over the entire composition range. This negative conclusion follows from the limiting relation $\gamma_1 \rightarrow 1$ as $x_1 \rightarrow 1$. Thus, if we have an azeotropic mixture where $x_2 << 1$, the experimental value of γ_1 gives us very little information, since γ_1 is necessarily close to unity. For such a mixture, only γ_2 supplies significant information, and therefore we cannot expect to calculate two meaningful adjustable parameters when we have only one significant datum. However, if the azeotropic composition is close to unity, we may, nevertheless, use the azeotropic data to find one activity coefficient, namely, γ_2 (where $x_2 << 1$), and then use that γ_2 to determine the single adjustable parameter of a one-parameter activity model or EOS.

8.9 EFFECT OF TEMPERATURE ON LOW-PRESSURE VAPOR-LIQUID EQUILIBRIA

A particularly troublesome question is the effect of temperature on the molar excess Gibbs energy g^E. This question is directly related to s^E, the molar excess entropy of mixing about which little is known.[6] In practice, either one of two approximations is frequently used.

(a) ***Athermal solution.*** This approximation sets $g^E = -Ts^E$, which assumes that the components mix at constant temperature without change of enthalpy ($h^E = 0$). This assumption leads to the conclusion that, at constant composition, $\ln(\gamma_i)$ is independent of T or, its equivalent, that g^E/RT is independent of temperature. Such an approximation may be applicable to mixtures of small and large n-alkanes like propane and n-octadecane.

(b) ***Regular solution.*** This approximation sets $g^E = h^E$, which is the same as assuming that $s^E = 0$. This assumption leads to the conclusion that, at constant composition, $\ln(\gamma_i)$ varies as $1/T$ or, its equivalent, that g^E is independent of temperature.

[6]From thermodynamics, $s^E = -(\partial g^E/\partial T)_{P,x}$ and $g^E = h^E - Ts^E$.

Neither one of these extreme approximations is reliable, although the second one is often better than the first. Good experimental data for the effect of temperature on activity coefficients are rare, but when such data are available, they suggest that, for a moderate temperature range, they can be expressed by an empirical equation of the form

$$\ln(\gamma_i) = c(\boldsymbol{x}) + d(\boldsymbol{x})/T \tag{8.9-1}$$

where c and d are empirical constants that depend on the composition vector ($\boldsymbol{x}$). In most cases, constant d is positive as suggested by the findings of Flory-Huggins, Guggenheim-Staverman, and Vahid et al. (2014). It is evident that, when $d = 0$, Eq. (8.9-1) reduces to an athermal solution and, when $c = 0$, it reduces to the regular solution assumption. Unfortunately, in typical cases c and d/T are of comparable magnitude. On a brighter note, life on earth might not be possible without this near balance in thermal driving forces. This is literally a situation that we must learn to live with!

Thermodynamics relates the effect of temperature on γ_i to the partial molar enthalpy $\bar{h}_i$

$$\left[\frac{\partial \ln(\gamma_i)}{\partial(1/T)}\right]_{x,p} = \frac{\bar{h}_i - h_i^o}{R} = \frac{\bar{h}_i^{\mathrm{E}}}{R} \tag{8.9-2}$$

where h_i^o is the enthalpy of liquid i in the standard state, usually taken as pure liquid i at the system temperature and pressure. Experimental data for h^{E} may be available, from which $\bar{h}_i^{\mathrm{E}}$ can be inferred; if so, they can be used to provide information on how the activity coefficient changes with temperature. However, even if such data are at hand, Eq. (8.9-2) must be used with caution because $\bar{h}_i^{\mathrm{E}}$ depends on temperature and often strongly so.

Some of the expressions for g^{E} shown in Table 8.2 contain T as an explicit variable. However, one should not therefore conclude that the parameters appearing in those expressions are independent of temperature. The explicit temperature dependence indicated provides only an approximation. This approximation is usually, but not always, better than approximation (*a*) or (*b*), but, in any case, it is not exact.

Fortunately, the primary effect of temperature on VLE is contained in the pure-component vapor pressures or, more precisely, in the pure-component liquid fugacities. While activity coefficients depend on temperature as well as composition, the temperature dependence of the activity coefficient is usually small when compared with the temperature dependence of the pure-liquid vapor pressures. In a typical mixture, a rise of 10 °C increases the vapor pressures of the pure liquids by a factor of 1.5 or 2, but the change in activity coefficient is likely to be only a few percent, often less than the experimental uncertainty. Therefore, unless there is a large change in temperature, it is frequently satisfactory to neglect the effect of temperature on g^{E} when calculating VLE. However, in calculating LLE, vapor pressures play no role at all, and therefore the effect of temperature on g^{E}, although small, may seriously affect LLE. Even small changes in activity coefficients can have a large effect on multicomponent LLE.

These relationships between VLE, LLE, and h^{E} motivate including temperature dependence in the activity model parameters, but large amounts of reliable experimental data are required to infer meaningful values. Recent implementations of the UNIFAC predictive model have adopted the approach of simultaneously correlating VLE, LLE, and h^{E} with a single consistent set of model parameters (Kang et al., 2011, 2015; Constantinescu and Gmehling, 2016).

8.10 MULTICOMPONENT VAPOR-LIQUID EQUILIBRIA AT LOW PRESSURE

The equations required to calculate VLE in multicomponent systems are, in principle, the same as those required for binary systems. In a system containing N_{C} components, we must solve N_{C} equations simultaneously: Eq. (8.5-1) for each of the N_{C} components. We require the saturation (vapor) pressure of each component, as a pure liquid, at the temperature of interest. If all pure-component vapor pressures are low, the total pressure also is low. In that event, the factor F_i [Eq. (8.4-2)] can often be set equal to unity.

Activity coefficients γ_i are found from an expression for the excess Gibbs energy, as discussed in Sec. 8.8. For a mixture of N_{C} components, the total excess Gibbs energy G^{E} is defined by

$$G^{\mathrm{E}} = RT\sum_{i=1}^{N_{\mathrm{C}}} n_i \ln \gamma_i \tag{8.10-1}$$

where n_i is the number of moles of component i. The molar excess Gibbs energy g^E is simply related to G^E by

$$g^{E} = \frac{G^{E}}{n_{T}} \tag{8.10-2}$$

where n_T, the total number of moles, is equal to $\sum_{i=1}^{N_C} n_i$.

Individual activity coefficients can be obtained from G^E upon introducing the Gibbs-Duhem equation for a multicomponent system at constant temperature and pressure. That equation is

$$\sum_{i=1}^{N_C} n_i \, \mathrm{d} \ln \gamma_i = 0 \tag{8.10-3}$$

The activity coefficient γ_i is found by a generalization of Eq. (8.6-3):

$$RT \ln \gamma_i = \left(\frac{\partial G^{E}}{\partial n_i} \right)_{T,P,n_{j \neq i}} \tag{8.10-4}$$

where $n_{j \neq i}$ indicates that all mole numbers (except n_i) are held constant in the differentiation.

The key problem in calculating multicomponent vapor-liquid equilibria is to find an expression for g^E that provides a good approximation for the properties of the mixture. Toward that end, the expressions for g^E for binary systems, shown Table 8.2, can be extended to multicomponent systems. A few of these are shown in Table 8.3. All these local composition models involve only binary parameters so the extension is straightforward.

The excess Gibbs energy concept is particularly useful for multicomponent mixtures because in many cases, to a good approximation, extension from binary to multicomponent systems can be made in such a way that only binary parameters appear in the final expression for g^E. When that is the case, a large saving in experimental effort is achieved, since experimental data are then required only for the mixture's constituent binaries, not for the multicomponent mixture itself. For example, activity coefficients in a ternary mixture (components 1, 2, and 3) can often be calculated with good accuracy by using only experimental data for the three binary mixtures: components 1 and 2, components 1 and 3, and components 2 and 3. The accuracy of this extrapolation from binary to ternary systems is not well established, but it typically deteriorates relative to the correlation for the binary mixture.

Most activity models for a binary system consider only two-body intermolecular interactions, i.e., interactions between two (but not more) molecules. Because of the short range of molecular interaction between non-electrolytes, it is often permissible to consider only interactions between molecules that are first neighbors and then to sum all the two-body, first-neighbor interactions. A useful consequence of these simplifying assumptions is that extension to ternary (and higher) systems requires only binary, i.e., two-body, information; no ternary (or higher) constants appear. However, not all physical models use this simplifying assumption, and those which do not may require additional simplifying assumptions if the final expression for g^E is to contain only constants derived from binary data. Conversely, if the accuracy of extrapolation to the multicomponent mixture is insufficient, models based on binary parameters have limited capability to adapt. The only recourse is to optimize binary and multicomponent measurements simultaneously. Such deviations, however, are significant only if they exceed experimental uncertainty. To detect significant deviations, data of high accuracy are required, and such data are rare for ternary systems; they are extremely rare for quaternary (and higher) systems.

Typical results for extension from binary to ternary mixtures have been reported by Dadmohammadi et al. (2016). They studied 57 ternary mixtures, correlating the binary parameters for NRTL and UNIQUAC then applying them to predict ternary VLE. They obtained 1.7%AAD deviation in pressure for the binary systems but the multicomponent extension yielded 3%AAD for the NRTL model and 2.9% for UNIQUAC. Remarkably, they observed 2.5%AAD when predicting the ternary VLE using the Dortmund modified UNIFAC model. Graczova et al. (2004) have reported similar findings in a study of 103 ternary systems. Focke et al. (2021) showed how classical models like Margules and van Laar could be recast as special cases of a "mean mixture model" that also included the Wilson and NRTL models as special cases. Nevertheless, their analysis of nine ternary systems showed that the Wilson model was most reliable. Our evaluations for ternary systems are reported in Sec. 8.14, but briefly our results support these previous studies. It is noteworthy that relative deviations for single-parameter models are also near 3%. Predictions for ternary systems with these models have not been previously reported, but we include a few such models in Sec. 8.14.

To further improve predictions for multicomponent mixtures, it is necessary to incorporate ternary parameters. An additive approach to making this extension is often applied. For the ternary, Abbott et al. (1975) expressed the excess Gibbs energy by

$$\frac{g^E_{123}}{RT} = \frac{g^E_{12}}{RT} + \frac{g^E_{13}}{RT} + \frac{g^E_{23}}{RT} + (C_0 - C_1x_1 - C_2x_2 - C_3x_3)x_1x_2x_3 \tag{8.10-5}$$

where C_0, C_1, C_2, and C_3 are ternary constants and g^E_{ij} is given by the optimized ij binary model. Equation (8.10-5) successfully reproduced the ternary data within experimental error (rms $\Delta P = 0.0012$ bar).

Abbott et al. considered two simplifications:

Simplification a: $C_0 = C_1 = C_2 = C_3 = 0$

Simplification b: $C_1 = C_2 = C_3 = 0 \qquad C_0 = \frac{1}{2}\sum_{i \neq j}\sum A'_{ij}$

where the A'_{ij}s are the binary parameters.

Simplification b was first proposed by Wohl (1953) on semitheoretical grounds. When calculated total pressures for the ternary system were compared with experimental results, the deviations exceeded the experimental uncertainty.

Simplification	rms ΔP, bar
a	0.0517
b	0.0044

These results suggest that Wohl's approximation (simplification b) provides significant improvement over the additivity assumption for g^E (simplification a). However, one cannot generalize from results for one system (cf. Adler et al., 1966). Abbott et al. (1975) made similar studies for another ternary (acetone-chloroform-methanol) and found that for this system simplification a gave significantly better results than simplification b, although both simplifications produced errors in total pressure beyond the experimental uncertainty.

Although the results of Abbott and coworkers illustrate the limits of predicting ternary (or higher) VLE for nonelectrolyte mixtures from binary data only, these limitations are rarely serious for engineering work unless the system contains an azeotrope. As a practical matter, it is common that experimental uncertainties in binary data are as large as the errors that result when multicomponent equilibria are calculated with some model for g^E using only parameters obtained from binary data. Graczova et al. provide a tabulation of ternary parameters.

Example 8.5 Multicomponent isobaric VLE using the UNIQUAC model

A liquid mixture at 1.013 bar contains 4.7 mole% ethanol (1), 10.7 mole% benzene (2), and 84.5 mole% methylcyclopentane (3). Find the bubble-point temperature and the composition of the equilibrium vapor using the UNIQUAC model combined with the virial model for the vapor-phase.

Solution. There are three unknowns: the bubble-point temperature and two vapor-phase mole fractions. To find them we use three equations of equilibrium:

$$y_i\varphi_iP = x_i\gamma_if_i^{o,\text{liq}} \qquad i = 1, 2, 3 \tag{8.10-6}$$

where $\boldsymbol{y}$ is the vapor-phase mole fraction and $\boldsymbol{x}$ is the liquid-phase mole fraction. Fugacity coefficient ϕ_i is given by the truncated virial equation of state

$$\ln \varphi_i = \left(2\sum_{j=1}^{3} y_j B_{ij} - B_{\mathrm{M}}\right)\frac{P}{RT} \tag{8.10-7}$$

where subscript M stands for mixture:

$$B_{\mathrm{M}} = y_1^2 B_{11} + y_2^2 B_{22} + y_3^2 B_{33} + 2y_1 y_2 B_{12} + 2y_1 y_3 B_{13} + 2y_2 y_3 B_{33} \tag{8.10-8}$$

All second virial coefficients B_{ij} are found from the correlation of Hayden and O'Connell (1975).

The standard-state fugacity $f_i^{o,\mathrm{liq}}$ is the fugacity of pure liquid i at system temperature T and system pressure P.

$$f_i^{o,\mathrm{liq}} = P_i^{\mathrm{vp}} \varphi_i^{\mathrm{sat}} \exp \frac{V_i^{\mathrm{liq}}(P - P_i^{\mathrm{vp}})}{RT} \tag{8.10-9}$$

where P_i^{vp} is the saturation pressure (i.e., the vapor pressure) of pure liquid i, φ_i^{sat} is the fugacity coefficient of pure saturated vapor i, and V_i^{liq} is the liquid molar volume of pure i, all at system temperature T.

Activity coefficients are given by the UNIQUAC equation with the following parameters:

Pure-Component Parameters

Component	r	q	q'
1	2.11	1.97	0.92
2	3.19	2.40	2.40
3	3.97	3.01	3.01

Binary Parameters

$\tau_{ij} = \exp\left(-\frac{a_{ij}}{T}\right)$ and $\tau_{ji} = \exp\left(-\frac{a_{ji}}{T}\right)$

i	j	a_{ij}, K	a_{ji}, K
1	2	−128.9	997.4
1	3	−118.3	1384
2	3	−6.47	56.47

For a bubble-point calculation, a useful objective function $F(1/T)$ is

$$F\left(\frac{1}{T}\right) = \ln\left[\sum_{i=1}^{3} K_i x_i\right] \rightarrow \text{zero}$$

where $K_i = y_i/x_i$. In this calculation, the important unknown is T (rather than y) because P_i^{vp} is a strong function of temperature, whereas φ_i is only a weak function of y.

A suitable program for these iterative calculations uses the Newton-Raphson method, as discussed, for example, by Prausnitz et al. (1980). This program requires initial estimates of T and y.

The calculated bubble-point temperature is 335.99 K. At this temperature, the second virial coefficients (cm³/mole) and liquid molar-volumes (cm³/mole) are:

$$\begin{aligned} B_{11} &= -1155 \\ B_{12} = B_{21} &= -587 \\ B_{22} &= -1086 \\ B_{23} = B_{32} &= -1134 \\ B_{33} &= -1186 \\ B_{31} = B_{13} &= -618 \\ B_M &= -957.3 \end{aligned}$$

$$V_1^{\text{liq}} = 61.1;\ V_2^{\text{liq}} = 93.7;\ V_3^{\text{liq}} = 118$$

The detailed results at 335.99 K are:

				100 y_i	
Component	γ_i	$f_i^{\text{o,liq}}$**(kPa)**	φ_i	**Calculated**	**Observed**
1	10.58	52.1	0.980	26.1	25.8
2	1.28	56.4	0.964	7.9	8.4
3	1.03	73.9	0.961	66.0	65.7

The experimental bubble-point temperature is 336.15 K. Experimental results are from Sinor and Weber (1960). In this case, there is very good agreement between calculated and experimental results. Such agreement is not unusual, but it is, unfortunately, not guaranteed.

8.11 PREDICTING ACTIVITY COEFFICIENTS

As discussed in the preceding sections, activity coefficients in binary liquid mixtures can often be estimated from a few experimental VLE data for the mixtures by using some empirical (or semiempirical) activity model. The excess functions provide a thermodynamically consistent method for interpolating and extrapolating limited binary experimental mixture data and for extending binary data to multicomponent mixtures. Frequently, however, few or no mixture data are at hand, and it is necessary to estimate activity coefficients from some suitable prediction method. Unfortunately, entirely reliable prediction methods have not been established yet. Theoretical understanding of liquid mixtures continues to evolve. Therefore, the few available prediction methods are essentially empirical. This means that estimates of activity coefficients can be made only for systems like those used to establish the empirical prediction method. Even with this restriction, with few exceptions, the accuracy of prediction is not likely to be high whenever predictions for a binary system do not utilize at least some reliable binary data for that system or for another that is closely related. In the following sections we summarize a few of the activity-coefficient prediction methods useful for chemical engineering applications.

Previous editions of PGL have discussed the use of regular solution theory and the ASOG model for VLE predictions. As discussed in Sec. 8.7, the primary role for Scatchard-Hildebrand theory currently is the basis for FH theory and theories that separate the cohesive energy density (SCED) into dispersive, polar, and strong solvation/association contributions. The form of FH theory, combined with the SCED perspective, has been adapted somewhat successfully as the basis for predicting IDACs, however. IDACs are independent of any composition dependence and using a solvent-dependent polarity parameter enables adaptations that would be difficult if they had to relate at all compositions in a consistent manner. Kamlet-Taft theory, and the MOSCED and SPACE models are examples of this approach. We evaluate correlations of IDACs in Chap. 9, with models of LLE and SLE, because they relate more strongly to IDACs than VLE does.

PGL5ed discusses the use of IDACs with the Wilson model to predict VLE. If highly accurate IDACs are available, this approach provides accuracy like that from fitting the azeotrope. Unfortunately, the IDACs determined from predictive correlations are not highly accurate, especially for aqueous systems. Therefore, we omit Scatchard-Hildebrand, MOSCED, and IDACs from consideration for VLE predictions in favor of UNIFAC and COSMO-RS/SAC methods.

Multiple implementations of the UNIFAC and COSMO-RS/SAC methods have been advanced since PGL5ed. The Dortmund-modified UNIFAC (UNIFAC-Do, cf. Constantinescu and Gmehling, 2016), NIST-KT-UNIFAC (Kang et al., 2011), and NIST-mod-UNIFAC (Kang et al., 2015) have been regressed against large databases, including temperature-dependent parameters that simultaneously correlate VLE, LLE, and excess enthalpy data. We evaluate all three models as well as the original UNIFAC (Fredenslund et al., 1975). We also consider three implementations of the COSMO-RS/SAC model: COSMOtherm (Klamt et al., 1998), COSMO-RS/SAC/VT (Mullins et al., 2005), and COSMO-RS/SAC/GAMESS (Ferrarini et al., 2018).

8.11.1 The UNIFAC Method

For correlating thermodynamic properties, it is often convenient to regard a molecule as an aggregate of functional groups; as a result, some thermodynamic properties of pure fluids, e.g., heat capacity (Chaps. 4–6) and critical volume (Chap. 3), can be calculated by summing group-contributions. Extension of this concept to mixtures was suggested long ago by Langmuir (1925), and several attempts have been made to establish group-contribution methods for heats of mixing and for activity coefficients.

In any group-contribution method, we assume that a physical property of interest is the sum of contributions made by the molecule's functional groups. Any group-contribution method is necessarily approximate because the contribution of a given group in one molecule is not necessarily the same as that in another molecule. The fundamental assumption of a group-contribution method is additivity: the contribution made by one group within a molecule is assumed to be linearly independent of that made by any other group in that molecule. This assumption is valid only when the influence of any one group in a molecule is not affected by the nature of other groups within that molecule (Wilson and Deal, 1962; Perry et al., 1981). For example, we would not expect the contribution of a carbonyl group in a ketone (say, acetone) to be the same as that of a carbonyl group in an organic acid (say, acetic acid). On the other hand, experience suggests that the contribution of a carbonyl group in, for example, 3-pentanone, is close to (although not identical with) the contribution of a carbonyl group in another ketone, say 4-heptanone.

Accuracy of correlation improves with increasing distinction of groups; in considering, for example, aliphatic alcohols, in a first approximation no distinction is made between the position (primary or secondary) of a hydroxyl group, but in a second approximation such a distinction is desirable. In the limit as more and more distinctions are made, we recover the ultimate group, namely, the molecule itself. In that event, the advantage of the group-contribution method is lost. For practical utility, a compromise must be attained. The number of distinct groups must remain small but not so small as to neglect significant effects of molecular structure on physical properties.

Extension of the group-contribution idea to mixtures is attractive because, although the number of pure fluids in chemical technology is already very large, the number of different mixtures is larger by many orders of magnitude. Thousands, perhaps millions, of multicomponent liquid mixtures of interest in the chemical industry can be constituted from 100 functional groups.

The fundamental idea of a solution-of-groups model is to utilize existing phase equilibrium data for predicting phase equilibria of systems for which no experimental data are available. In concept, activity coefficients in mixtures are related to interactions between structural groups. The essential features are:

1. Suitable reduction of experimentally obtained activity-coefficient data to yield parameters characterizing interactions between pairs of structural groups in nonelectrolyte systems.
2. Use of those parameters to predict activity coefficients for other systems that have not been studied experimentally but that contain the same functional groups.

The molecular activity coefficient is separated into two parts: the combinatorial part provides the contribution due to differences in molecular size and shape, and the residual part provides the contribution due to molecular interactions. The GSA part of the UNIQUAC model is inherently a group-contribution model and can be carried over to UNIFAC (Fredenslund et al., 1975) without modification. The residual contribution in UNIFAC applies the UNIQUAC residual term to a solution of groups, then corrects the pseudoactivity coefficients for the groups such that the activity coefficient approaches unity for each pure molecule.

The residual part of the activity coefficient is replaced by the solution-of-groups concept in the form of Eq. (8.7-31), which we reiterate here as Eq. (8.11-1).

$$\ln \gamma_i^{\text{res}} = \sum_{\substack{k \\ \text{all groups}}} \nu_k^{(i)}\left[\ln(\Gamma_k) - \ln\left(\Gamma_k^{(i)}\right)\right] \tag{8.11-1}$$

where Γ_k is the group residual activity coefficient and $\Gamma_k^{(i)}$ is the residual activity coefficient of group k in a reference solution containing only molecules of type i. In Eq. (8.11-1), the term $\ln(\Gamma_k^{(i)})$ is necessary to attain the normalization that activity coefficient γ_i becomes unity as $x_i \to 1$. The activity coefficient for group k in molecule i depends on the molecule i in which k is situated. For example, $\Gamma_k^{(i)}$ for the COH group[7] in pure ethanol refers to a "solution" containing 50 group percent COH and 50 group percent CH_3 at the temperature of the mixture, whereas $\Gamma_k^{(i)}$ for the COH group in n-butanol refers to a "solution" containing 25 group percent COH, 50 group percent CH_2, and 25 group percent CH_3.

The group pseudoactivity coefficient Γ_k is found from the UNIQUAC expression:

$$\ln(\Gamma_k) = Q_k\left[1 - \ln\left(\sum_m \Theta_m \Psi_{mk}\right) - \sum_m \frac{\Theta_m \Psi_{km}}{\sum_n \Theta_n \Psi_{nm}}\right] \tag{8.11-2}$$

Equation (8.11-2) also holds for $\ln \Gamma_k^{(i)}$. In Eq. (8.11-2), Θ_m is the area fraction of group m, and the sums are over all different groups. Θ_m is calculated in a manner like that for Θ_i:

$$\Theta_m = \frac{Q_m X_m}{\sum_n Q_n X_n} \tag{8.11-3}$$

where X_m is the mole fraction of group m in the mixture. The group-interaction parameter Ψ_{mn} is given by

$$\Psi_{mn} = \exp\left(-\frac{U_{mn} - U_{nn}}{RT}\right) = \exp\left(-\frac{a_{mn}}{T}\right) \tag{8.11-4}$$

where U_{mn} is a measure of the energy of interaction between groups m and n. The group interaction parameters a_{mn} must be evaluated from experimental phase equilibrium data. Note that a_{mn} has units of Kelvins and $a_{mn} \neq a_{nm}$. Parameters a_{mn} and a_{nm} are obtained from a database using a wide range of experimental results. Sample values are shown in Table 8.4 for VLE predictions with the original UNIFAC method (Fredenslund et al., 1975). Complete tables of parameters are given in the original article and made available through the PGL6ed website.

Efforts toward updating and extending UNIFAC group-contributions have continued, often with modifications to the original equations or to the manner of defining the functional groups. We refer to Gmehling's adaptations as Dortmund modified (Do-mod-) UNIFAC (cf. Constantinescu and Gmehling, 2016). We refer to the version developed by the Kemi Teknik Dept of Danish Technical University as KT-UNIFAC (Kang et al., 2002).

[7]COH is shorthand notation for CH_2OH.

TABLE 8.4 Sample values of original UNIFAC VLE residual group-contribution parameters.

		CH_3	ACH	OH	CH_3OH	H_2O	CH_3CO	CH_3COO	$CHCL_3$
Main	**Type**	**1**	**3**	**5**	**6**	**7**	**9**	**11**	**23**
CH_3	1	0.0	61.1	986.5	697.2	1318.0	476.4	232.1	24.9
ACH	3	−11.1	0.0	636.1	637.3	903.8	25.8	6.0	−231.9
OH	5	156.4	89.6	0.0	−137.1	353.5	84.0	101.1	−98.1
CH_3OH	6	16.5	−50.0	249.1	0.0	−181.0	23.4	−10.7	−139.4
H_2O	7	300.0	362.3	−229.1	289.6	0.0	−195.4	72.9	353.7
CH_3CO	9	26.8	140.1	164.5	108.7	472.5	0.0	−213.7	−354.6
CH_3COO	11	114.8	85.8	245.4	249.6	200.8	372.2	0.0	−209.7
$CHCL_3$	23	36.7	228.5	742.1	649.1	826.8	552.1	176.5	0.0

More recently, Kang et al. (2011, 2015) developed two new correlations applying the formats of Do-mod- and KT-UNIFAC but implementing the extensive TRC database of mixture data for excess heats of mixing, LLE, and VLE. We refer to those correlations as NIST-KT-UNIFAC and NIST-mod-UNIFAC. The parameters are expressed as temperature-dependent during their optimization ($a_{mn} = a^0_{nm} + a^1_{mn}/T$). Another feature of the NIST implementations is that an effort has been made to unambiguously define UNIFAC groups for any molecule, with rare exceptions. For both NIST versions of UNIFAC, all parameters are always made public as new characterizations are added. Similarly, the latest version of Do-mod-UNIFAC continues to be developed based on the extensive Dortmund Data Base, and it applies a single set of (temperature-dependent) parameters to all three properties. For the reader's convenience, the PGL6ed website provides defined descriptors for roughly 30,000 molecules each for both the NIST-KT-UNIFAC and NIST-mod-UNIFAC versions. We present examples that illustrate (1) the nomenclature and use of Table 8.4 and (2) the "original UNIFAC" method for calculating activity coefficients.

Example 8.6 Computing activity coefficients using the UNIFAC model

Estimate activity coefficients for the acetone (1) + *n*-pentane (2) system at 307 K and $x_1 = 0.047$.

Solution. Acetone has one ($\nu_1 = 1$) CH_3 group (main group 1, secondary group 1) and one ($\nu_9 = 1$) CH_3CO (main group 9, secondary group 18). *n*-Pentane has two ($\nu_1 = 2$) CH_3 groups (main group 1, secondary group 1), and three ($\nu_1 = 3$) CH_2 groups (main group 1, secondary group 2).

Based on the information in Table 8.4, we can construct the following table:

	Group Identification					
Molecule	**Name**	**Main No.**	**Sec. No.**	$\nu_j^{(i)}$	R_j	Q_j
Acetone (1)	CH_3	1	1	1	0.9011	0.848
	CH_3CO	9	9	1	1.6724	1.488
n-Pentane (2)	CH_3	1	1	2	0.9011	0.848
	CH_2	1	2	3	0.6744	0.540

We can now write:

$$r_1 = (1)(0.9011) + (1)(1.6724) = 2.5735$$

$$q_1 = (1)(0.848) + (1)(1.488) = 2.336$$

$$\Phi_1 = \frac{(2.5735)(0.047)}{(2.5735)(0.047) + (3.8254)(0.935)} = 0.0321$$

$$\Theta_1 = \frac{(2.336)(0.047)}{(2.336)(0.047) + (3.316)(0.953)} = 0.0336$$

$$l_1 = (5)(2.5735 - 2.336) - 1.5735 = -0.3860$$

or in tabular form:

Molecule (i)	r_i	q_i	**100 Φ_i**	**100 Θ_i**	l_i
Acetone (1)	2.5735	2.336	3.21	3.36	−0.3860
n-Pentane (2)	3.8254	3.316	96.79	96.64	−0.2784

We can now calculate the combinatorial contribution to the activity coefficients:

$$\ln \gamma_1^{\text{GSA}} = \ln\left(\frac{0.0321}{0.047}\right) + (5)(2.336)\ \ln\left(\frac{0.0336}{0.0321}\right) - 0.3860$$

$$+ \frac{0.0321}{0.047}[(0.047)(0.3860) + (0.953)(0.2784)] = -0.0403$$

$$\ln \gamma_2^{\text{GSA}} = -0.0007$$

Next, we calculate the residual contributions to the activity coefficients. Since only two main groups are represented in this mixture, the calculation is relatively simple. The group interaction parameters, a_{mn}, are obtained from Table 8.4.

$$a_{1,9} = 476.40$$

$$\psi_{1,9} = \exp\left(\frac{-476.40}{307}\right) = 0.2119$$

$$a_{9,1} = 26.760$$

$$\psi_{9,1} = \exp\left(\frac{-26.760}{307}\right) = 0.9165$$

Note that $\psi_{1,1} = \psi_{9,9} = 1.0$, since $a_{1,1} = a_{9,9} = 0$. Let 1 = CH_3, 2 = CH_2, and 18 = CH_3CO.

Next, we compute $\Gamma_k^{(i)}$, the residual activity coefficient of group k in a reference solution containing only molecules of type i. For pure acetone (1), the mole fraction of group m, X_m, is

$$X_1^{(1)} = \frac{\nu_1^{(1)}}{\nu_1^{(1)} + \nu_{18}^{(1)}} = \frac{1}{1+1} = \frac{1}{2} \qquad X_{18}^{(1)} = \frac{1}{2}$$

Hence,

$$\Theta_1^{(1)} = \frac{\frac{1}{2}(0.848)}{\frac{1}{2}(0.848) + \frac{1}{2}(1.488)} = 0.363 \qquad \Theta_{18}^{(1)} = 0.637$$

$$\ln\left(\Gamma_1^{(1)}\right) = 0.848\left\{1 - \ln[0.363 + (0.637)(0.9165)] - \left[\frac{0.363}{0.363 + (0.637)(0.9165)} + \frac{(0.637)(0.2119)}{(0.363)(0.2119) + 0.637}\right]\right\} = 0.409$$

$$\ln\left(\Gamma_{18}^{(1)}\right) = 1.488\left\{\begin{matrix} 1 - \ln[(0.363)(0.2119) + 0.637] - \\ \left[\frac{(0.363)(0.9165)}{0.363 + (0.637)(0.9165)} + \frac{0.637}{(0.363)(0.2119)} + 0.637\right] \end{matrix}\right\} = 0.139$$

For pure n-pentane (2), the mole fraction of group m, X_m, is

$$X_1^{(2)} = \frac{\nu_1^{(2)}}{\nu_1^{(2)} + \nu_2^{(2)}} = \frac{2}{2+3} = \frac{2}{5} \qquad X_2^{(2)} = \frac{3}{5}$$

Since only one main group is in n-pentane (2),

$$\ln\left(\Gamma_1^{(2)}\right) = \ln\left(\Gamma_2^{(2)}\right) = 0.0$$

The group residual activity coefficients can now be calculated for $x_1 = 0.047$:

$$X_1 = \frac{(0.047)(1) + (0.953)(2)}{(0.047)(2) + (0.953)(5)} = 0.4019$$

$$X_2 = 0.5884 \qquad X_{18} = 0.0097$$

$$\Theta_1 = \frac{(0.848)(0.4019)}{(0.848)(0.4019) + (0.540)(0.5884) + (1.488)(0.0097)} = 0.5064$$

$$\Theta_2 = 0.4721 \qquad \theta_{18} = 0.0214$$

$$\ln(\Gamma_1) = 0.848\left\{1 - \ln[0.5064 + 0.4721 + (0.0214)(0.9165)] - \left[\frac{0.5064 + 0.4721}{0.5064 + 0.4721 + (0.0214)(0.9165)} + \frac{(0.0214)(0.2119)}{(0.5064 + 0.4721)(0.2119) + 0.0214}\right]\right\}$$
$$= 1.45 \times 10^{-3}$$

$$\ln(\Gamma_2) = 0.540\left\{1 - \ln[0.5064 + 0.4721 + (0.0214)(0.9165)] - \left[\frac{0.5064 + 0.4721}{0.5064 + 0.4721 + (0.0214)(0.9165)} + \frac{(0.0214)(0.2119)}{0.5064 + 0.4721 + 0.2119) + 0.0214}\right]\right\} = 9.26 \times 10^{-4}$$

$$\ln(\Gamma_{18}) = 1.488\left\{1 - \ln[(0.5064 + 0.4721)(0.2119) + 0.0214] - \left[\frac{(0.5064 + 0.4721)(0.9165)}{0.5064 + 0.4721 + (0.0214)(0.9165)} + \frac{0.0214}{(0.5064 + 0.4721)(0.2119) + 0.0214}\right]\right\} = 2.21$$

The residual contributions to the activity coefficients follow:

$$\ln\left(\gamma_1^R\right) = (1)(1.45 \times 10^{-3} - 0.409) + (1)(2.21) - (0.139) = 1.66$$

$$\ln\left(\gamma_2^R\right) = (2)(1.45 \times 10^{-3} - 0.0) + (3)(9.26 \times 10^{-4} - 0.0) = 5.68 \times 10^{-3}$$

Finally, we calculate the activity coefficients:

$$\ln\left(\gamma_1\right) = \ln\left(\gamma_1^{\text{GSA}}\right) + \ln \gamma_1^R = -0.0403 + 1.66 = 1.62$$

$$\ln\left(\gamma_2\right) = \ln\left(\gamma_2^{\text{GSA}}\right) + \ln \gamma_2^R = -0.0007 + 5.68 \times 10^{-3} = 4.98 \times 10^{-3}$$

Hence,

$$\gamma_1 = 5.07$$
$$\gamma_2 = 1.01$$

Retaining more significant figures in the values for ϕ and θ, as would be the case if calculations were done on a computer, leads to slightly different answers. In this case, $\ln\left(\gamma_1^{\text{GSA}}\right) = -0.0527$, $\ln\left(\gamma_2^{\text{GSA}}\right) = -0.0001$, $\gamma_1 = 4.99$, and $\gamma_2 = 1.005$.

The corresponding experimental values of Lo et al. (1962) are:

$$\gamma_1 = 4.41$$
$$\gamma_2 = 1.11$$

Although agreement with experiment is not as good as we might wish, it is not bad, and it is representative of what UNIFAC can do. The main advantage of UNIFAC is its wide range of application for VLE of nonelectrolyte mixtures.

Because UNIFAC is so popular in the chemical (and related) industries, and because it is used so widely to estimate VLE for a large variety of mixtures, three illustrative examples are presented to indicate both the power and the limitations of the UNIFAC method.

Example 8.7 Ternary VLE using UNIFAC at 45 °C

Using original UNIFAC, calculate VLE for the ternary acetone (1)–methanol (2)–cyclohexane (3) at 45 °C.

Solution. The three governing equations at equilibrium are

$$y_i P = x_i \gamma_i P_i^{\text{vp}} \qquad (i = 1, 2, 3)$$

Because the total pressure is low, the Poynting factors as well as corrections for vapor-phase nonideality are neglected.

To obtain phase equilibria at 45 °C:

Step 1. Assign values of x_1 and x_2 from 0 to 1. To facilitate comparison with experiment, we choose sets (x_1, x_2) identical to those used by Marinichev and Susarev (1965). At each set (x_1, x_2), we have three equations with three unknowns: P, y_1, and y_2. Mole fractions x_3 and y_3 are not independent variables because they are constrained by material balances $x_1 + x_2 + x_3 = 1$ and $y_1 + y_2 + y_3 = 1$.

Step 2. At each set (x_1, x_2), use UNIFAC to calculate the activity coefficients for the three components at 45 °C. The three molecules are broken into groups as follows:

Component	Constitutive Groups
1	$CH_3 + CH_3CO$
2	CH_3OH
3	$6CH_2$

Group-volume (R_k) and surface-area (Q_k) parameters are

Group	R_k	Q_k
CH_3	0.9011	0.848
CH_2	0.6744	0.540
CH_3CO	1.6724	1.448
CH_3OH	1.4311	1.432

Group-group interaction parameters (in K) are

Group	CH_3	CH_2	CH_3CO	CH_3OH
CH_3	0	0	476.4	697.2
CH_2	0	0	476.4	679.2
CH_3CO	26.76	26.76	0	108.7
CH_3OH	16.51	16.51	23.39	0

Step 3. Calculate the equilibrium total pressure from

$$P = (y_1 + y_2 + y_3)P = x_1\gamma_1 P_1^{vp} + x_2\gamma_2 P_2^{vp} + x_3\gamma_3 P_3^{vp}$$

with γ_i from Step 2.

Step 4. Evaluate mole fractions y_1 and y_2 from

$$y_i = \frac{x_i\gamma_i P_i^{vp}}{P}$$

with γ_i from Step 2 and P from Step 3.

Step 5. Return to Step 2 with the next set (x_1, x_2).

Following these steps, Table 8.5 gives calculated results. Also shown are experimental data at the same temperature from Marinichev and Susarev (1965). Figure 8.9 compares calculated and observed results.

TABLE 8.5 VLE of acetone(1) + methanol(2) + cyclohexane at 45 °C.

		Calculated			Experimental			(Calculated-Experimental)		
x_1	x_2	y_1	y_2	P (mmHg)	y_1	y_2	P (mmHg)	Δy_1	Δy_2	ΔP (mmHg)
0.117	0.127	0.254	0.395	558	0.276	0.367	560	−0.022	0.028	−1.6
0.118	0.379	0.169	0.470	559	0.191	0.452	568	−0.022	0.018	−8.9
0.123	0.631	0.157	0.474	554	0.176	0.471	561	−0.019	0.003	−6.8
0.249	0.120	0.400	0.272	567	0.415	0.252	594	−0.016	0.020	−27.0
0.255	0.369	0.296	0.375	574	0.312	0.367	585	−0.016	0.008	−10.8
0.250	0.626	0.299	0.435	551	0.325	0.412	566	−0.026	0.023	−14.6
0.382	0.239	0.418	0.281	580	0.433	0.267	593	−0.015	0.014	−13.0
0.379	0.497	0.405	0.367	563	0.414	0.343	579	−0.009	0.024	−16.3
0.537	0.214	0.521	0.222	581	0.526	0.202	597	−0.005	0.020	−16.5
0.669	0.076	0.656	0.098	569	0.654	0.088	584	0.002	0.010	−14.7
0.822	0.054	0.772	0.064	557	0.743	0.076	574	0.029	−0.012	−17.2
							Stdv.	**0.015**	**0.011**	**6.5**

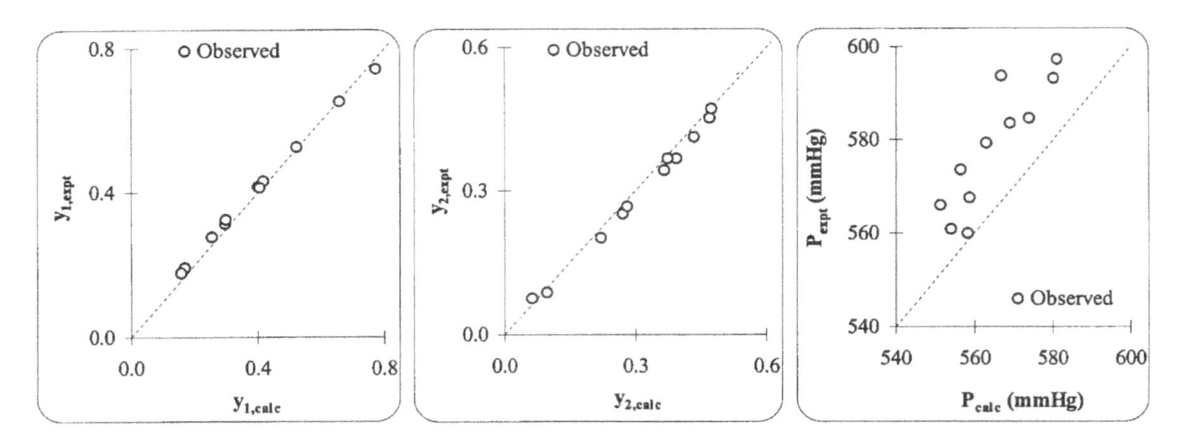

FIGURE 8.9
Vapor-liquid-equilibrium data for acetone (1)–methanol (2)–cyclohexane (3).

While UNIFAC gives a good representation of vapor-phase mole fractions, there is appreciable error in the total pressure.

Example 8.8 Ternary VLE using UNIFAC at 760 mmHg

Using the original UNIFAC, calculate vapor-liquid equilibria *T-x-y* for the ternary acetone (1)+2-butanone (2)+ethyl acetate (3) at 760 mmHg.

Solution. The three equations of equilibrium are

$$y_i P = x_i \gamma_i P_i^{vp} \qquad (i = 1, 2, 3)$$

From Gmehling et al. (1979), coefficients in Antoine's equation for P_i^{vp} in mmHg are

Component	*A*	*B*	*C*
1	7.117	1210.6	229.7
2	7.064	1261.3	222.0
3	7.102	1245.0	217.9

where

$$\log_{10}\left(P_i^{vp}\right) = A_i - \frac{B_i}{t + C_i} \tag{8.11-5}$$

To obtain phase equilibria at 760 mmHg, we use an iteration procedure (Step 1 to Step 9) like that in Example 8.3.

Step 1. Assign values of x_1 and x_2 from 0 to 1. We choose sets (x_1, x_2) identical to those used by Babich et al. (1969). At each set (x_1, x_2), we have three equations with three unknowns: t, y_1, and y_2. At each set (x_1, x_2), the iteration is as follows:

Step 2. Make an initial guess of the temperature t (°C), e.g.:

$$t = x_1 t_1 + x_2 t_2 + x_3 t_3$$

where t_1, t_2, and t_3 are calculated from Antoine's equation using $P_i^{vp} = 760$ mmHg.

$$t_i = \frac{B_i}{A_i - \log_{10} P_i^{vp}} - C_i \qquad (i = 1, 2, 3)$$

Step 3. Calculate the three pure-component vapor pressures from Antoine's equation.

Step 4. Compute the liquid-phase activity coefficients at $T = t + 273.15$.

Step 5. Solve for $P^{calc} = 760$ while changing t, where $P^{calc} = \sum x_i \gamma_i P_i^{vp}$.

Molecules of the three components are broken into groups as follows:

Component	Constitutive Groups
1	$CH_3 + CH_3CO$
2	$CH_3 + CH_2 + CH_3CO$
3	$CH_3 + CH_2 + CH_3COO$

Group-volume (R_k) and surface-area (Q_k) parameters are

Group	R_k	Q_k
CH_3	0.9011	0.848
CH_2	0.6744	0.540
CH_3CO	1.6724	1.448
CH_3COO	1.9031	1.728

Group-group interaction parameters (in K) are

Group	CH_3	CH_2	CH_3CO	CH_3COO
CH_3	0	0	476.4	232.1
CH_2	0	0	476.4	232.1
CH_3CO	26.76	26.76	0	–213.7
CH_3COO	114.8	114.8	372.2	0

Table 8.6 gives calculated and experimental results at 760 mmHg. Experimental data are from Babich et al. (1969).

TABLE 8.6 Comparison of UNIFAC estimates to the experimental data of Babich et al. (1969).

		Calculated			Experimental			(Calculated-Experimental)		
x_1	x_2	y_1	y_2	t (°C)	y_1	y_2	t (°C)	Δy_1	Δy_2	Δt (°C)
0.200	0.640	0.341	0.508	72.5	0.290	0.556	72.6	0.051	–0.048	–0.1
0.400	0.480	0.583	0.322	67.5	0.525	0.370	67.6	0.058	–0.048	–0.1
0.600	0.320	0.761	0.185	63.2	0.720	0.215	63.5	0.041	–0.031	–0.3
0.800	0.160	0.896	0.081	59.4	0.873	0.095	59.8	0.023	–0.014	–0.4
0.200	0.480	0.336	0.374	71.8	0.295	0.420	71.8	0.041	–0.046	0.0
0.400	0.360	0.576	0.238	67.1	0.535	0.285	67.3	0.041	–0.047	–0.3
0.600	0.240	0.755	0.137	63.0	0.725	0.170	63.3	0.030	–0.033	–0.3
0.800	0.120	0.893	0.060	59.3	0.880	0.075	59.6	0.013	–0.015	–0.3
0.200	0.320	0.334	0.248	71.2	0.302	0.276	71.3	0.032	–0.029	–0.1
0.400	0.240	0.571	0.157	66.6	0.540	0.180	67.0	0.031	–0.023	–0.4
0.600	0.160	0.750	0.091	62.7	0.729	0.105	62.9	0.021	–0.014	–0.2
0.200	0.160	0.339	0.125	70.7	0.295	0.145	71.1	0.044	–0.020	–0.4
0.400	0.120	0.570	0.078	66.2	0.530	0.095	66.5	0.040	–0.017	–0.3
0.600	0.080	0.746	0.045	62.5	0.720	0.095	62.7	0.026	–0.050	–0.2
0.800	0.040	0.887	0.020	59.1	0.873	0.024	59.6	0.014	–0.004	–0.4
							Stdv.	**0.013**	**0.015**	**0.1**

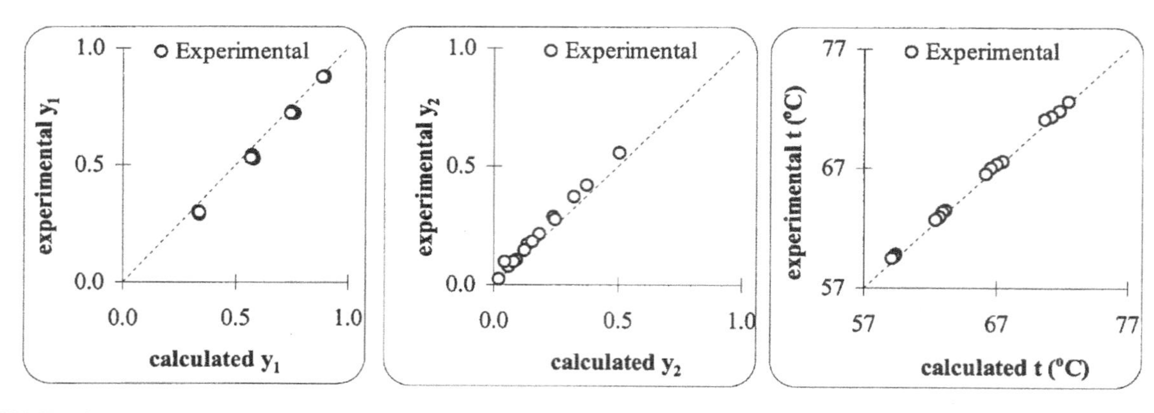

FIGURE 8.10
VLE for acetone(1)+2-butanone(2)+ethylacetate(3) at 760 mmHg.

Figure 8.10 compares calculated and observed results. Agreement is good. However, we must not conclude from this example that agreement will necessarily be equally good for other systems.

Example 8.9 VLE of sugar solutions

Peres and Macedo (1997) studied mixtures containing D-glucose, D-fructose, sucrose, and water. From these studies, they proposed new groups for the KT-modified-UNIFAC method (Larsen et al., 1987) to calculate VLE and SLE for these mixtures.

In this example, the modified UNIFAC method is used to calculate VLE for the ternary D-glucose (1)–sucrose (2)–water (3) at 760 mmHg.

Solution. Because the vapor phase contains only water, and because the pressure is low, the vapor-phase can be assumed to be ideal. At equilibrium,

$$P = x_3 \gamma_3 P_3^{\text{vp}}$$

where P and P_3^{vp} are total pressure and water vapor pressure, respectively; x is liquid-phase mole fraction; γ is the liquid-phase activity coefficient. The temperature dependence of P^{vp} is expressed by Antoine's equation, Eq. (8.11-5).

$$\log_{10}\left(P_3^{\text{vp}}\right) = A_3 - \frac{B_3}{t + C_3}$$

where P_3^{vp} is in mmHg and t is in °C.

From Gmehling et al. (1981), Antoine parameters are

	A_3	B_3	C_3
Water	8.071	1730.6	233.4

In the modified UNIFAC, the GSA part of the activity coefficient is calculated from Larsen et al. (1987).

$$\ln\left(\gamma_i^{\text{GSA}}\right) = \ln\left(\frac{\Phi_i}{x_i}\right) + 1 - \frac{\Phi_i}{x_i}$$

where x_i is mole fraction of component i.

Volume fraction Φ_i of component i is defined as

$$\Phi_i = \frac{x_i r_i^{2/3}}{\sum_{j=1}^{3} x_j r_j^{2/3}}$$

The volume parameter r_i is calculated from

$$r_i = \sum_k R_k \nu_k^{(i)}$$

where R_k is the volume parameter of group k; $\nu_k^{(i)}$ is the number of occurrences of group k in molecule i. The residual activity coefficient is calculated as in the original UNIFAC method.

To break the two sugars into groups, Peres and Macedo (1997) used the three groups proposed by Catte et al. (1995): pyranose ring (PYR), furanose ring (FUR),[8] and osidic bond (—O—), that is, the ether bond connecting the PYR and FUR rings in sucrose. Further, because there are many OH groups in D-glucose and sucrose and because these OH groups are close to one another, their interactions with other groups are different from those for the usual alcohol group. Peres and Macedo (1997) proposed a new OH_{ring} group. The two sugars and water are broken into groups as

Component	Constitutive Groups
D-glucose	CH_2 + PYR + 5OH_{ring}
Sucrose	3CH_2 + PYR + FUR + (—O—) + 8OH_{ring}
water	H_2O

Group-volume and surface area parameters are

Group	R_k	Q_k
CH_2	0.6744	0.5400
PYR	2.4784	1.5620
FUR	2.0315	1.334
(—O—)	1.0000	1.200
OH_{ring}	0.2439	0.442
H_2O	0.9200	1.400

Group-group interaction parameters (in K) suggested by Peres and Macedo (1997)

	CH_2	PYR	FUR	(—O—)	OH_{ring}	H_2O
CH_2	0	0	0	0	0	0
PYR	0	0	0	0	0	−43.27
FUR	0	0	0	0	0	−169.23
(—O—)	0	0	0	0	0	0
OH_{ring}	0	0	0	0	0	591.93
H_2O	0	−599.04	−866.91	0	−102.54	0

[8] PYR = OH, OH — OH, OH, OH — O (six-membered ring); FUR = OH, OH — OH, O, O (five-membered ring)

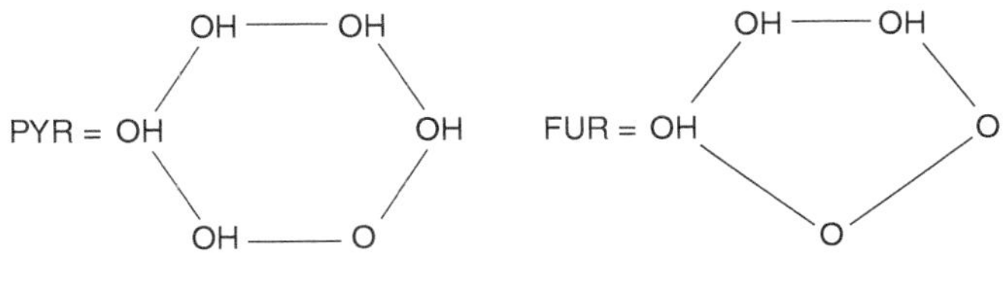

TABLE 8.7 VLE for D-glucose(1)+sucrose(2)+water(3) at 760 mmHg.

x_1	x_2	Expt. T (°C)	Calc. T (°C)
0.0014	0.0147	101.0	100.2
0.0023	0.0242	102.0	100.6
0.0036	0.0390	103.0	101.2
0.0054	0.0579	104.0	102.2
0.0068	0.0714	105.0	103.0
0.0075	0.0803	105.5	103.5
0.0098	0.1051	107.0	105.2
0.0150	0.1576	108.5	109.4
0.0167	0.1756	107.0	111.0

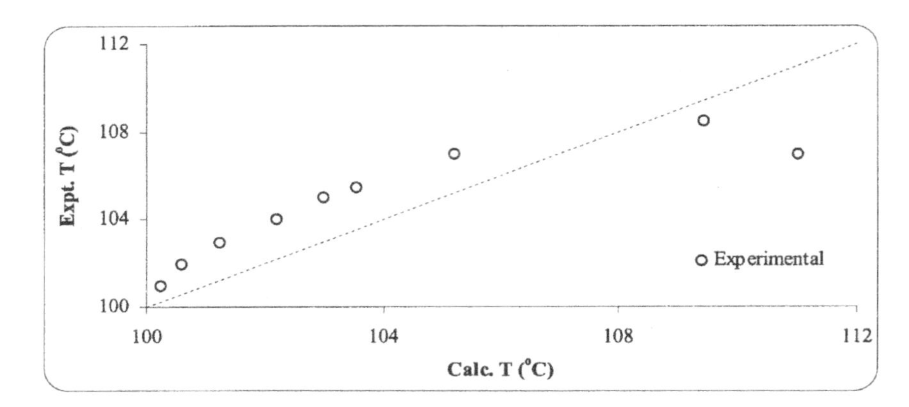

FIGURE 8.11
VLE for D-glucose(1)+sucrose(2)+water(3) at 760 mmHg.

For any given (x_1, x_2), we first substitute γ_3 into the equilibrium equation and solve for P_3^{vp}. We then use this P_3^{vp} in the Antoine equation and solve for T. Mole fraction $x_3 = 1 - x_1 - x_2$. Table 8.7 shows calculated boiling temperatures. Experimental data are from Abderafi and Bounahmidi (1994). Figure 8.11 compares calculated and experimental results.

8.11.2 COSMO-RS/SAC THEORY

In principle, quantum mechanics can explain all molecular interactions. So why have we not discussed activity models based on quantum mechanics? Before 1993, there was not much to talk about, because the rare activity predictions based on quantum mechanics were not sufficiently accurate for engineering applications. The COSMO-RS (Klamt, 1995) development changed that.

COSMO-RS stands for conductor-like screening model for real solvents. COSMO-RS represents two distinct developments. The COSMO component applies a boundary condition to solution of Schrodinger's equation that corresponds to an infinitely conductive medium. This turns out to be a relatively easy problem to solve with QDFT. The infinitely conductive medium may sound surprising at first, but it better approximates a liquid than the alternative vacuum boundary condition does. A COSMO calculation yields a molecular surface with screening charge density, σ^c, on each segment, which in a second step is converted into a histogram $p(\sigma^c)$. Negative values of σ^c indicate positively charged regions on the molecular surface to balance the negativity of the medium. The integral of $p(\sigma^c)$ over all values of σ^c is equal to the total (normalized) surface area of the molecule. For practical purposes, $p(\sigma^c)$ is discretized over the typical range of polarization charge densities such that $p_i = p(\sigma^c)$ at the ith element in the range. To be specific, we assume here that $\sigma^c \in [-0.025, 0.025]$ with $\nu^{max} = 51$ values and a step size of 0.001, as suggested by the segmental activity coefficient (SAC) implementation of Mullins et al. (2006). In general, alternative ranges are possible and may be relevant in treating charged or inorganic molecules.

The activity coefficient is determined from SS-QCT, as discussed in Sec. 8.7. Like UNIFAC, the logarithmic activity of the pure fluid is subtracted from the computed activity of the mixed compound. In the case of COSMO-RS/SAC, however, this difference has the interpretation of exchanging the infinite conductor for an environment of the actual solution in one case compared to exchanging for an environment of the pure compound. Rather than considering the molecules as the sources of dispersion energy, the 51 surface segments comprise the mixture. All mixtures are composed of these same 51 segments, and the number of ith segments per molecule is determined from the sigma profile of the kth molecule. The key equations are:

$$N_{i,k}^{seg} = \frac{p_i(k)}{A^{eff}} \tag{8.11-6}$$

$$\ln(\gamma_k) = \ln(\gamma_k^A) + \sum_1^{51} p_i^{(k)} \left(\ln(\Gamma_i) - \ln(\Gamma_i^{(k)}) \right) / A^{eff} \tag{8.11-7}$$

$$\Gamma_i = \left\{ \sum_1^{51} \Theta(j) \Gamma_j \exp[-\beta w(i,j)] \right\}^{-1} \equiv \left\{ \sum_1^{51} \Theta(j) \Gamma_j \psi_{ji} \right\}^{-1} \tag{8.11-8}$$

$$w(i,j) = [8233\left(\sigma^c(i) + \sigma^c(j)\right)^2 + 85580\alpha^c(i)\beta^c(j)]/0.001987 \tag{8.11-9}$$

where $p_i(k)$ is the ith discrete element of the sigma profile for the kth compound, A^{eff} is a universal constant to normalize all surface areas to a single reference value, and $\ln(\gamma_k^A)$ is the athermal contribution, which is not necessarily the GSA term. The surface fraction, Θ, is:

$$\Theta(j) = \frac{x_1 p(j) + x_2 p(j)}{x_1 q_1 + x_2 q_2} \tag{8.11-10}$$

$$q_k \equiv \sum_1^{51} p_i^{(k)} \tag{8.11-11}$$

The α^c and β^c quantities reflect the acidity and basicity contributions to strong interactions. In the simplest implementation of COSMO-RS/SAC, they are given by σ^c,

$$\alpha_i^c \equiv \max\left\{\left[\min\left(\sigma_i^c, \sigma_j^c\right) + 0.0084\right], 0\right\} \tag{8.11-12}$$

$$\beta_j^c \equiv \max\left\{\left[\max\left(\sigma_i^c, \sigma_j^c\right) - 0.0084\right], 0\right\}$$

Although Eq. (8.11-12) looks complicated, it simply indicates that strong solvation occurs above a threshold of 0.0084 $|\sigma^c|$ units, and a segment's complexation strength is proportional to the magnitude of σ^c beyond the threshold. Keep in mind that α^c pertains to the negative domain of σ^c values because this range corresponds to positive surface polarization to cancel the negative dielectric medium. The proportionality of $w(i, j)$ in Eq. (8.11-9) is a relatively simple conjecture, and the coefficients (8233 and 85580) are adjustable parameters that are optimized for a large database. More elaborate mapping of segmental properties to energy could incorporate additional descriptors. For example, proportionality constants might change if sulfur replaced oxygen as in a thiol versus alcohol. Another descriptor could be the polarization or screening environment generated by nearby functionalities (cf. Klamt, 1998). Multiple authors have proposed refinements of the hydrogen bonding description. For example, Hsieh et al. (2010) propose a simple scheme based on alcohols and "other" hydrogen bonds. Soares et al. (2013) propose a more elaborate scheme explicitly recognizing multiple types of donors and

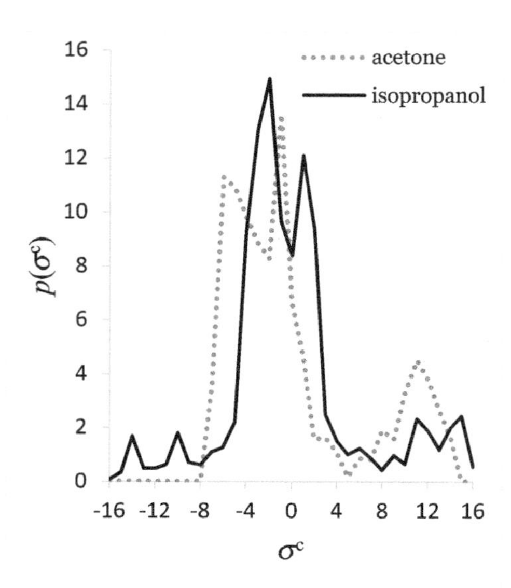

FIGURE 8.12
Sigma profiles for 2-propanol and acetone. Solid vertical bars designate the regions where hydrogen bonding interactions are indicated. Dashed vertical bars designate the effective acceptor and donor segments assigned to strong solvation.

acceptors, among which specifications for water stand out. On a related note, Possani et al. (2014) report notable characterizations of hydrocarbon+water systems. Paulechka et al. (2015) have developed another parameterization based on a small but critically evaluated database. The parameter A^{eff} is also optimized, but a range of values from 0.040 to 0.085 nm^2 would probably give similar results if the other coefficients were optimized with any set value of A^{eff}. The accuracy of COSMO-RS/SAC depends on the accuracy of two conjectures, particularly as they pertain to strong interactions: SS-QCT and Eq. (8.11-9).

To stabilize convergence, new and old values of Γ are averaged in the manner of a Picard iteration where α^{P} controls the weighting on the new estimate compared to the old. For example,

$$\Gamma_k^{\text{new}} = \left(1 - \alpha^{\text{P}}\right)\Gamma_i^{\text{old}} + \alpha^{P}\left(\sum \Theta(j)\Gamma_j \psi_{ji}\right)^{-1} \tag{8.11-13}$$

where $\alpha^{\text{P}} \sim 0.5$–$0.7$. Alternatively, a geometric mean could be applied, such as

$$\ln\left(\Gamma_k^{\text{new}}\right) = \left(1 - \alpha^{\text{P}}\right)\ln\left(\Gamma_i^{\text{old}}\right) - \alpha^{\text{P}}\ln\left(\sum \Theta(j)\Gamma_j \psi_{ji}\right) \tag{8.11-14}$$

Typical sigma profiles are illustrated in Fig. 8.12. The profile for 2-propanol illustrates peaks at both ends of the hydrogen bonding thresholds, indicating association. Acetone only displays positive polarization above the threshold, indicating proton acceptor tendency.

Example 8.10 COSMO-RS/SAC prediction of activity coefficients

Estimate the activity coefficients of acetone(1)+isopropanol(2) at 55 °C and $x_1 = 0.25$ using the COSMO-RS/SAC implementation given by Mullins et al. (2006).

Solution. Sigma profiles are tabulated below to one decimal to conserve space. Mullins et al. tabulate values to three decimals and those values are used in the computations. The values of σ^c are multiplied by 1000 and $p_i(k)$ values outside the range of $\sigma^c \in [-0.016, 0.016]$ happen to be zero for these two molecules.

σ^c	−16	−15	−14	−13	−12	−11	−10	−9	−8	−7	−6	−5	−4	−3	−2	−1	0
1	0	0	0	0	0	0	0	0	0	3.5	11.3	10.9	9.8	8.8	8.3	13.6	6.6
2	0.1	0.3	1.7	0.5	0.5	0.6	1.8	0.7	0.6	1.1	1.2	2.2	9.3	13.1	14.9	9.7	8.4

σ^c	0	1	2	3	4	5	6	7	8	9	10	11	12	13	14	15	16
1	6.6	4.7	1.6	1.6	1.1	0.2	0.8	0.8	1.9	1.5	3.3	4.5	3.8	2.6	1.6	0	0
2	8.4	12.1	9.3	2.5	1.5	1.0	1.2	0.9	0.4	1.0	0.6	2.3	1.9	1.2	2.0	2.4	0.5

We detail calculations for $i = \{-14, 5, 11\}$. The matrix of $w(i, j)$ can be tabulated in advance. We illustrate three exemplary cases.

Case 1, $i = -14, j = -14$: $w(i, j) = [8233(-0.014 - 0.014)^2 + 85580(0)]/0.001987 = 3248$ K

Case 2, $i = -14, j = 5$: $w(i, j) = [8233(-0.014 + 0.005)^2 + 85580(0)]/0.001987 = 335.6$ K

Case 3, $i = -14, j = 11$: $w(i, j) = [(8233(-0.014 + 0.011)^2 + 85580(0.011 - 0.0084)*(-0.014 + 0.0084)]/0.001987$
$= -589.8$ K

Values of $\psi_{ij} = \exp(-w\,(i, j)/T)$ can also be tabulated in advance, where $T = 55 + 273.15$.

	ψ_{ij}		
$\sigma_i^c \backslash \sigma_j^c$	−0.014	0.005	0.011
−0.014	0.00005	0.360	6.034
0.005	0.360	0.283	0.039
0.011	6.034	0.039	0.002

Note that values of $\psi_{ij} >> 1$ occur when strong solvation is favorable. Values of $\psi_{ij} < 1$ indicate unfavorable dispersion interactions. With ψ_{ij} tabulated for all i, j and setting $\Gamma_i = 1$ for all i, Eq. (8.11-8) yields the values of Γ_i° tabulated below. Applying Eq. (8.11-13) and iterating until convergence yields the values of Γ_i. (These calculations are illustrated on the CosmoRsSac tab of CH08ActCoeff.xlsx.) Setting $\Theta_i = p_i(k)/q_k$, where $q_k = \Sigma p_i(k)$, and iterating for each pure compound, Eq. (8.11-8) gives the values of $\Gamma_i^{(k)}$ tabulated below. Applying Eq. (8.11-7) yields the residual values of the activity coefficients. Estimates of the volumetric parameters are given by Mullins et al. (2006) for application of the GSA term.

σ_j^c	−0.014	0.005	0.011
Γ_i°	0.1925	1.4585	2.2611
Γ_i	0.0769	1.2283	2.4822
$\Gamma_i^{(1)}$	0.1222	1.1045	2.1232
$\Gamma_i^{(2)}$	0.0781	1.2659	2.5786

Combining the various contributions leads to the following estimates:

Name	x_k	V_k	q_k	Φ_k	Θ_k	$\ln(\gamma_k)^{GSA}$	$\ln(\gamma_k)^{res}$	γ_k
Acetone	0.25	86.42	102.65	0.2376	0.2419	0.000227	0.1596	1.1728
Isopropanol	0.75	92.41	107.20	0.7624	0.7581	0.000026	0.0116	1.0117

For comparison, UNIFAC gives values of $\gamma_1 = 1.474$ and $\gamma_2 = 1.045$. Correlating the data of Freshwater and Pike (1967) with NRTL and $\alpha_N = 0.43$ gives $\gamma_1 = 1.50$ and $\gamma_2 = 1.06$.

8.12 PHASE EQUILIBRIUM WITH HENRY'S LAW

Although the compositions of liquid mixtures may span the entire composition range from dilute up to the pure component, many multiphase systems contain compounds only in the dilute range ($x_i < 0.1$). This is especially true for components where the system T is above their critical T_c (gases) or where their pure-component vapor pressure, P^{vp}, is well above the system pressure. Liquid-liquid and solid-liquid systems also often do not span the entire composition range. In such cases, the thermodynamic description using the pure-component standard state may not be most convenient. This section describes methods based on the Henry's law standard state. Details are given by Prausnitz et al. (1999) and Elliott and Lira (2012, Sec. 11.12).

At modest pressures, most gases are only sparingly soluble in typical liquids. For example, at 25 °C and a partial pressure of 101 kPa, the (mole fraction) solubility of nitrogen in cyclohexane is $x = 7.6 \times 10^{-4}$ and that in water is $x = 0.18 \times 10^{-4}$. Although there are some exceptions (notably, hydrogen), the solubility of a gas in typical solvents usually falls with rising temperature. However, at higher temperatures, approaching the critical temperature of the solvent, the solubility of a gas usually rises with temperature, as illustrated in Fig. 8.13.

Experimentally determined solubilities have been reported in the chemical literature for over 100 years, but many of the data are of poor quality. Although no truly comprehensive and critical compilation of the available data exists, Table 8.8 gives some useful data sources.

Unfortunately, a variety of units have been employed in reporting gas solubilities. Two dimensionless coefficients were common in older literature: *Bunsen coefficient*, defined as the volume (corrected to 0 °C and 1 atm) of gas dissolved per unit volume of solvent at system temperature T when the partial pressure (mole fraction times total pressure, yP) of the solute is 1 atm; *Ostwald coefficient*, defined as the volume of gas at system temperature T and partial pressure p dissolved per unit volume of solvent. If the solubility is small and the gas phase is ideal, the Ostwald and Bunsen coefficients are simply related by

$$\text{Ostwald coefficient} = \frac{T}{273.15} \text{ (Bunsen coefficient)}$$

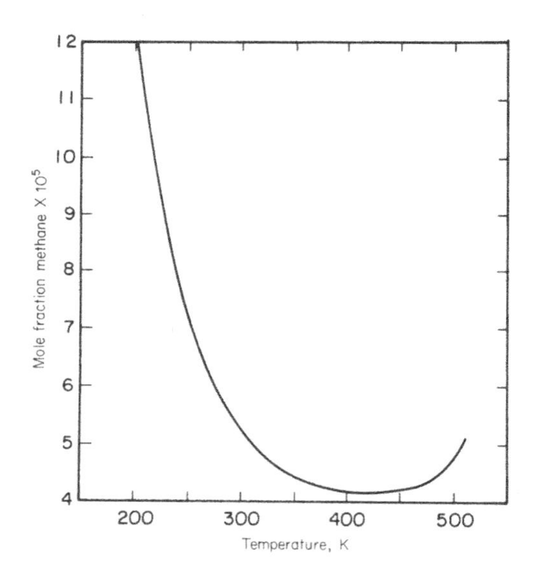

FIGURE 8.13
Solubility of methane in n-heptane when vapor-phase fugacity of methane is 0.0101 bar (Prausnitz et al. 1999).

TABLE 8.8 Data sources for solubilities of gases in liquids.

Kazakov, A., C. D. Muzny, K. Kroenlein, V. Diky, R. D. Chirico, J.W. Magee, I. M. Abdulagatov, M. Frenkel. NIST/TRC SOURCE Data Archival System: The Next-Generation Data Model for Storage of Thermophysical Properties. *Int. J. Thermophys.*, **33:** 22–33 (2012).

IUPAC: *Solubility Data Series,* https://srdata.nist.gov/solubility/IUPAC/iupac.aspx (1979–present).

Washburn, E. W. (ed.): *International Critical Tables,* McGraw-Hill, New York, 1926.

Markam, A. E., and K. A. Kobe: *Chem. Rev.,* **28:** 519 (1941).

Seidell, A.: *Solubilities of Inorganic and Metal-Organic Compounds,* Van Nostrand, New York, 1958, and *Solubilities of Inorganic and Organic Compounds, ibid.,* 1952.

Linke, W. L.: *Solubilities of Inorganic and Metal-Organic Compounds,* 4th ed., Van Nostrand, Princeton, N.J., 1958 and 1965, Vols. 1 and 2. (A revision and continuation of the compilation originated by A. Seidell.)

Stephen, H., and T. Stephen: *Solubilities of Inorganic and Organic Compounds,* Vols. 1 and 2, Pergamon Press, Oxford, and Macmillan, New York, 1963 and 1964.

Battino, R., and H. L. Clever: *Chem. Rev.,* **66:** 395 (1966).

Wilhelm, E., and R. Battino: *Chem. Rev.,* **73:** 1 (1973).

Clever, H. L., and R. Battino: "The Solubility of Gases in Liquids," in M. R. J. Dack (ed.), *Solutions and Solubilities,* Vol. 8, Part 1, Wiley, New York, 1975, pp. 379–441.

Kertes, A. S., O. Levy, and G. Y. Markovits: "Solubility," in B. Vodar (ed.), *Experimental Thermodynamics of Nonpolar Fluids,* Vol. II, Butterworth, London, 1975, pp. 725–748.

Gerrard, W.: *Solubility of Gases and Liquids,* Plenum, New York, 1976.

Landolt-Börnstein: 2. Teil, Bandteil b, *Lösungsgleichgewichte I,* Springer, Berlin, 1962; IV. Band, Technik, 4. Teil, Wärmetechnik; Bandteil c, *Gleichgewicht der Absorption von Gasen in Flüssigkeiten, ibid.,* 1976.

Gerrard, W.: *Gas Solubilities, Widespread Applications,* Pergamon, Oxford, 1980.

Chang, A. Y., K. Fitzner, and M. Zhang: "The Solubility of Gases in Liquid Metals and Alloys," *Progress in Materials Science,* Vol. 32. No. 2–3, Oxford, New York, Pergamon Press, 1988.

Fogg, P. G. T., and W. Gerrard: *Solubility of Gases in Liquids: A Critical Evaluation of Gas/Liquid Systems in Theory and Practice,* Chichester, New York, J. Wiley, 1991.

where T is in Kelvins. Adler (1983), Battino (1971, 1974, and 1984), Carroll (1999), and Friend and Adler (1957) have discussed these and other coefficients for expressing solubilities as well as some of their applications for engineering calculations.

These coefficients are often found in older articles. In recent years it has become more common to report solubilities in units of mole fraction given the solute partial pressure or as Henry's constants. The NIST Chemistry Webbook lists Henry's constants for many common fluids (Linstrom and Mallard, 2022). Gas solubility is a case of phase equilibrium where Eq. (8.3-1) holds. We use Eq. (8.3-2) for the gas phase, mostly dominated by the normally supercritical solute (2), but for the liquid dominated by one or more subcritical solvents (1, 3, . . .) since x_2 is small, Eq. (8.3-4) is not convenient. As a result, instead of using an ideal solution model based on Raoult's law with standard-state fugacity at the pure-component saturation condition, we use the Henry's law ideal solution with a standard-state fugacity based on the infinitely dilute solution. Henry's law for a binary system need not assume an ideal gas phase; we write it as

$$y_g \varphi_g^{vap} P = x_g \gamma_g^* H_v^{px} \tag{8.12-1}$$

Subscript "g" indicates that Henry's law volatility constant H is for solute "g" which is often a gas. Then, Henry's volatility in a single solvent is rigorously defined by a limiting process

$$H_{\mathrm{v}}^{\mathrm{px}} = \lim_{x_g \to 0} \left(\frac{y_g \varphi_g^{\mathrm{vap}} P}{x_g} \right) \tag{8.12-2}$$

where in the limit, $P = P_{\mathrm{solv}}^{\mathrm{vp}}$, "solv" indicating solvent. When $y_g \approx 1$, as in Fig. 8.14, Henry's volatility is proportional to the inverse of the solubility.

Note that we have applied the recently adopted IUPAC standard notation for $H_{\mathrm{v}}^{\mathrm{px}}$ (Sander et al., 2022). Owing to the multitude of notations for Henry's law, it is necessary to clearly specify what quantity is being referenced. Some definitions refer to the Henry's law solubility constant, which is the reciprocal of the Henry's law volatility constant favored here, as indicated by the subscript "v." Variability can also occur depending on the characterizations of concentration in each phase. The first superscript, "p," indicates that partial pressure is the concentration variable of the solute in the vapor, as indicated by Eq. (8.12-2). The second superscript indicates that mole fraction is the concentration variable of the solvent. Note that mole fraction would be inappropriate for expressing a Henry's volatility when the solvent is a polydisperse polymer. Weight fraction, "w," should be favored in that case because the mole fraction is not precisely known. We refer to "Henry's volatility" as a concise form of "Henry's law volatility constant." Since usage of Henry's law solubility constant is relatively rare, historical mention of "Henry's constant" in the thermochemical literature most probably refers to Henry's volatility.

Like vapor pressure, $H_{\mathrm{v}}^{\mathrm{px}}$ depends only on T, but often strongly, as Fig. 8.14 shows (see also Prausnitz et al., 1999).

If total pressure is not low, Eq. (8.12-1) must include consideration of φ_i^{v} and γ_g^*. High pressure is common in gas-liquid systems. The effects of pressure in the vapor are accounted for by φ_i^{v}, while for the liquid the effect of pressure is in the Poynting factor which contains the partial molar volume. For typical dilute solutions, this is close to the infinite-dilution value, $\bar{V}_{g,\mathrm{solv}}^{\infty}$. However, in addition, an effect of liquid nonideality can often

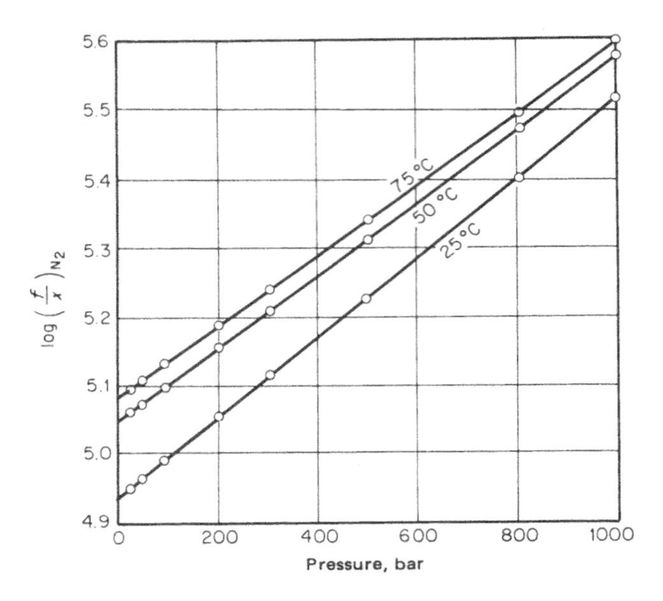

FIGURE 8.14
Solubility of nitrogen in water at high pressures. (Prausnitz, et al., 1999)

occur because as P increases at constant T, so must x_g. This nonideality is considered by an activity coefficient γ_g^* which is usually *less* than unity and has the limit (Eq. 8.12-2)

$$\lim_{x_g \to 0} \gamma_g^* = 1 \tag{8.12-3}$$

At high pressures, Eq. (8.12-1) becomes

$$y_g \varphi_g^v P = x_g \gamma_g^* H_v^{px} \exp\left[\int_{P_{solv}^{vp}}^{P} \frac{\bar{V}_{g,solv}^{\infty}}{RT} dP\right] \tag{8.12-4}$$

A convenient form of Eq. (8.12-4) is

$$\ln\left(\frac{y_g \varphi_g^v P}{x_g}\right) = \ln\left(\gamma_g^* H_v^{px}\right) + \frac{\bar{V}_{g,solv}^{\infty}(P - P_{solv}^{vp})}{RT} \tag{8.12-5}$$

where it has been assumed that $\bar{V}_{g,solv}^{\infty}$ is independent of pressure. The issues associated with using Eq. (8.12-5) are described in detail by Prausnitz et al. (1999) and Mathias and O'Connell (1979). Briefly, for $T_{c,g} << T << T_{c,solv}$, the last term in Eq. (8.12-5) makes the solubility less than expected from Eq. (8.12-1). Figure 8.14 shows this effect for nitrogen in water near ambient temperatures. It is possible to obtain useful values of $H_{2,1}$ from such plots and under the right conditions, infer values of $\bar{V}_{g,solv}^{\infty}$. Alternatively, the volume change of a solution upon dissolution of a gas, by dilatometry, also gives experimental data for $\bar{V}_{g,solv}^{\infty}$.

Table 8.9 shows typical values of $\bar{V}_{g,solv}^{\infty}$ for gases in liquids at 25 °C; they compare approximately to the pure liquid volume of the solute at its normal boiling point, $T_{b,2}$. This means that the noncondensable molecule has liquid-like properties in solution. Correlations for $\bar{V}_{g,solv}^{\infty}$ have been developed by Brelvi and O'Connell (1972, 1975), Campanella et al. (1987), Lyckman et al. (1965), and Tiepel and Gubbins (1972, 1973).

As T approaches $T_{c,solv}$, $\bar{V}_{g,solv}^{\infty}$ diverges to infinity as the isothermal compressibility of the solvent, $\kappa_g^{ic} = -1/V(\partial V/\partial P)_T$ also diverges (Levelt Sengers, 1994; O'Connell, 1994). Under these conditions, the integral of Eq. (8.12-4) is not a simple function of pressure such as in Fig. 8.14. Sharygin et al. (1996) show that an excellent correlation of $\bar{V}_{g,solv}^{\infty}$ for nonelectrolyte gases in aqueous solution from ambient conditions to well above the critical point of pure water can be obtained from correlating $\bar{V}_{g,solv}^{\infty}/\kappa_g^{ic}RT$ with the density of water. This has been extended to other properties such as fugacity coefficients, enthalpies, and heat capacities by Sedlbauer et al. (2000), Plyasunov et al. (2000), and Plyasunov et al. (2019).

TABLE 8.9 Partial molal volumes, $\bar{V}_{g,solv}^{\infty}$, of gases in liquid solution at 25 °C.

	H_2	N_2	CO	O_2	CH_4	C_2H_2	C_2H_4	C_2H_4	CO_2	SO_2
Ethyl ether	50	66	62	56	58					
Acetone	38	55	53	48	55	49	58	64	...	68
Methyl acetate	38	54	53	48	53	49	62	69	...	47
Carbon tetrachloride	38	53	53	45	52	54	61	67	...	54
Benzene	36	53	52	46	52	51	61	67	...	48
Methanol	35	52	51	45	52	...	...	...	43	
Chlorobenzene	34	50	46	43	49	50	58	64	...	48
Water	26	40	36	31	37	...	...	...	33	
$\bar{V}_g^{\circ}$ (T_b)	28	35	35	28	39	42	50	55	40	45

*J. H. Hildebrand and R. L. Scott (1950).

As discussed by Orentlicher and Prausnitz (1964), Campanella et al. (1987), Mathias and O'Connell (1979), and Van Ness and Abbott (1982), when x_g increases because y_iP is large, or because the system nears $T_{c,solv}$ or due to solvation effects as with CO_2 in water, the middle term of Eq. (8.12-5) can become important and partially cancel the last term. Then, though the variation of $\ln(f_g/x_g)$ may be linear with P, the slope will not be $\bar{V}^{\infty}_{g,solv}/RT$. Using only a volumetrically determined $\bar{V}^{\infty}_{g,solv}$ will underestimate the solubility at elevated pressures and solute compositions.

A crude estimate of Henry's constant can be obtained by extrapolating the vapor pressure of the gaseous solute on a plot of log P^{vp} versus $1/T$ and estimating the IDAC.

$$H_v^{px} = \frac{y_g P}{\gamma^{\infty}_{g,solv} P_g^{vp,eff}} \tag{8.12-6}$$

where $P_2{}^{vp,eff}$ is the (effective) vapor pressure of the solute at temperature T and $\gamma^{\infty}_{g,solv}$ is the IDAC. The so-called *ideal solubility* is obtained when $\gamma^{\infty}_{g,solv} = 1$ and $y_g = 1$. Prausnitz and Shair (1961) provided a graphical correlation for $P_g^{vp,eff}$ that has been recast in equation form by Elliott and Lira (2012, Eq. 11.68). Equation (8.12-7) matches Prausnitz and Shair's graph when $\omega_g = 0.21$.

$$\log_{10}\left(\frac{P_g^{vp,eff}}{P_{c,g}}\right) = \frac{7}{3}(1+\omega_g)\left(1-\frac{T_{c,g}}{T}\right) - 3\exp\left(-E^H \frac{T_{c,g}}{T}\right) \tag{8.12-7}$$

where the default value is $E^H = 3.0$, but the value for hydrogen is $E^H = 14.1$ and the pseudocritical constants of Prausnitz et al. (1999) should be used ($T_c = 42$ K, $P_c = 1.9$ MPa, $\omega = 0$). The ideal solubility is a function of temperature, but it is independent of the solvent. Table 8.10 shows that for many typical cases, the ideal solubility provides an order-of-magnitude estimate. The experimentally measured solubilities are lower because $\gamma^{\infty}_{g,solv} > 1$. This method breaks down as the solvent's critical temperature is approached.

A slightly more advanced guideline takes the form of Just's (1901) rules:

(a) The ratio of Henry's constants for a given gas in two different solvents varies little for many gases. For example,

$$\frac{H_v^{px}(O_2, \text{methanol})}{H_v^{px}(O_2, \text{ethanol})} \approx \frac{H_v^{px}(N_2, \text{methanol})}{H_v^{px}(N_2, \text{ethanol})} \tag{8.12-8}$$

(b) The ratio of Henry's constants for two gases in a single solvent varies little for many solvents. For example,

$$\frac{H_v^{px}(N_2, \text{ethanol})}{H_v^{px}(CO_2, \text{ethanol})} \approx \frac{H_v^{px}(N_2, \text{methanol})}{H_v^{px}(CO_2, \text{methanol})} \tag{8.12-9}$$

These two rules can often be applied to estimate solubility of gases for new solvents. However, they work best when the two solvents are similar in polarity and the two gases are similar in polarity. Substitution of Eq. (8.12-6) shows that the $P_i^{vp,eff}$ values cancel in Eq. (8.12-8) so the approximation comes down to estimating ratios

TABLE 8.10 Solubilities of gases in several liquid solvents at 25 °C and 101 kPa partial pressure. Mole fraction × 10⁴.

	Ideal*	***n*-C7F16**	***n*-C7H16**	**CCl4**	**CS_2**	**$(CH_3)_2CO$**
H_2	13	14.01	6.88	3.19	1.49	2.31
N_2	82	38.7	. . .	6.29	2.22	5.92
CH_4	84	82.6	. . .	28.4	13.12	22.3
CO_2	214	208.8	121	107	32.8	

*Equation (8.12-6).

of $\gamma^{\infty}_{g,solv}$. Equation (8.12-9) is simply a cross-multiplied form of Eq. (8.12-8), but $P_i^{vp,eff}$ values come into play. For example, Just (1901) lists the ratio for N_2/CO_2 at room temperature as varying from 15 (*n*-pentanol) to 51 (water) with a mean of 30 ± 11 where the uncertainty represents one standard deviation. Other ratios covered are 1.70 ± 0.6 for H_2/N_2, 2.5 ± 0.7 for H_2/CO, 47 ± 19 for H_2/CO_2, 1.46 ± 0.09 for N_2/CO, and 20 ± 7 for CO/CO_2.

Many attempts have been made to correlate gas solubilities more accurately, but success has been severely limited because, on the one hand, a satisfactory theory for gas-liquid solutions has not been established and, on the other, reliable experimental data are not plentiful, especially at temperatures remote from 25 °C. Among others, Battino and Wilhelm (1971, 1974) have obtained some success in correlating solubilities in nonpolar systems near 25 °C by using concepts from perturbed-hard-sphere theory, but these are of limited use for engineering work yet. A more useful graphical correlation, including polar systems, was prepared by Hayduk et al. (1970, 1971), and a correlation based on regular solution theory for nonpolar systems was established by Prausnitz and Shair (1961) and, in similar form, by Yen and McKetta (1962). The regular solution correlation is limited to nonpolar (or weakly polar) systems, and although its accuracy is not high, it has two advantages: it applies over a wide temperature range, and it requires no mixture data. Correlations for nonpolar systems, near 25 °C, are given by Hildebrand and Scott (1962). The correlations for nonpolar systems can be adapted to polar systems by treating the nonpolar prediction as a reference state and applying activity coefficients. A more sophisticated method for predicting Henry's constant in a different solvent from H_v^{px} at the same T is given by Campanella et al. (1987).

Gas solubility in mixed solvents and, therefore, Henry's volatility, varies with solvent composition. The simplest approximation for this is (Prausnitz et al., 1999)

$$\ln\left(H^{px}_{v,mix}\right) = \sum_{i=1}^{\#\,solvents} x_i \ln\left(H^{px}_{v,i}\right) \tag{8.12-10}$$

The next-order estimate (Prausnitz et al., 1999), which usually gives the correct sign to deviations from Eq. (8.12-10) except for aqueous alcohols is

$$\ln\left(H^{px}_{v,mix}\right) = \sum_{i=1}^{\#\,solvents} x_i \ln\left(H^{px}_{v,i}\right) - \frac{g^E_{solvents}}{RT} \tag{8.12-11}$$

where $g^E_{solvents}/RT$ is the excess Gibbs energy for the solvent mixture found from an activity model. The method of Campanella et al. (1987) can also be applied to mixed-solvent systems including aqueous alcohols and ternary solvents with much greater reliability.

Henry's law can give reliable results for many systems, and it is convenient for hand calculations. When nonidealities do arise, or when using a process simulator, it is common to use EOS methods. It is important to validate the output of EOS methods, and to tune the BIPs if necessary, especially when using a process simulator. Ideally, literature data (e.g., ThermoLit) or sources like those in Table 8.8 can provide the needed validation. In the absence of experimental data, the correlations presented here can provide guidance, but are no substitute for experimental measurement.

8.12.1 Other Dilute-Solution Methods

In some practical situations, especially those concerned with water-pollution abatement, it is necessary to estimate the solubilities of volatile organic solutes in water. While such estimates could be obtained with the methods of Secs. 8.3 to 8.9, the Henry's law approach is preferable. The basis is that Henry's constant is related to the fugacity of the pure compound by the IDAC. Methods for correlating and predicting IDACs are presented in Chap. 9. Briefly, ambient conditions prevail for pollutants and

$$H^{px}_v = \gamma^{\infty}_{p,solv} P^{vp}_p \tag{8.12-12}$$

where subscript "p" indicates the solute, which is typically a pollutant well below its critical temperature.

A common method for obtaining $H_{\mathrm{v}}^{\mathrm{px}}$ is to estimate $\gamma_{\mathrm{p,solv}}^{\infty}$ from experimental measurements and then use Eq. (8.12-12). Alternatively, it is possible to use a direct estimation method, such as machine learning. Since this approach involves estimated IDACs and water pollutants generally exhibit LLE, we refer the reader to Chap. 9 for further discussion of those details.

8.13 VAPOR-LIQUID EQUILIBRIA WITH EQUATIONS OF STATE

Thermodynamics provides the basis for using EOS not only for the calculation of *PVT* relations and caloric property relations but also for computing phase equilibria among fluid phases. We consider this alternative in detail because there are several advantages, especially for high-pressure systems.

1. While liquid properties are generally insensitive to pressure changes, maintaining VLE often implies higher temperatures as well, making the liquid more compressible. This is especially true near the solution critical point.
2. Common low-pressure models for the vapor fugacity, such as the ideal gas ($\varphi_i^{\mathrm{vap}} = 1$) and second virial model, become inaccurate and must be replaced by models valid at higher pressures.
3. The presence of components at $T > T_{\mathrm{c}}$ prevents us from directly obtaining the commonly used pure-component standard-state fugacity that is determined primarily by the vapor pressure. The supercritical standard state can be defined by a Henry's constant with the unsymmetric convention for activity coefficients and some correlations for engineering use have been established on that basis. However, because conceptual complexities arise in ternary and higher-order systems and because computational disadvantages can occur, this approach has not been popular.
4. The use of different formulae for computing fugacities in the vapor and liquid phases leads to a discontinuity as the mixture critical point is approached. This can cause considerable difficulty in computational convergence as well as large inaccuracies, especially for liquid-phase properties.
5. Complex phase behavior such as retrograde condensation and lower critical solution temperatures can make it necessary to maintain a consistent single model for all phases to describe the behavior even qualitatively.
6. Some EOSs with advanced mixing rules can describe low-pressure phase behavior as accurately as activity models. This suggests that a simpler "one size fits all" approach may be practical for many applications. Understanding activity models suggests their close relationship with EOSs. If EOSs can do anything activity models can do, while maintaining the capability to do things that activity models cannot do, the role of activity models becomes more narrowly confined.

As a result of these advantages, VLE calculations using the same EOS model for both phases have become popular with an enormous number of articles describing models, methods, and results. Detailed examples of bubble, dew, and flash computations with EOSs like the PR76 model are commonly available in introductory chemical engineering thermodynamics textbooks (e.g., Elliott and Lira, 2012).

The basis of methods for the calculation of phase equilibria is Eq. (8.2-1) with vapor and liquid fugacity coefficients, φ_i^{vap} and φ_i^{liq}:

$$f_i^{\mathrm{vap}} = y_i \varphi_i^{\mathrm{vap}} P = x_i \varphi_i^{\mathrm{liq}} P = f_i^{\mathrm{liq}} \tag{8.13-1}$$

The *K*-factor commonly used in calculations for process simulators is then simply related to the fugacity coefficients of Eq. (8.13-1).

$$K_i = \frac{y_i}{x_i} = \frac{\varphi_i^{\mathrm{liq}}}{\varphi_i^{\mathrm{vap}}} \tag{8.13-2}$$

To obtain φ_i^{vap}, we need the vapor composition, $\boldsymbol{y}$, and volume, V^{vap}, while for the liquid-phase, φ_i^{liq} is found using the liquid composition, $\boldsymbol{x}$, and volume, V^{liq}. Since state conditions are usually specified by T and P, the volumes must be found by solving the PVT relationship of the EOS.

$$P = P(T, V^{\mathrm{vap}}, \boldsymbol{y}) = P(T, V^{\mathrm{liq}}, \boldsymbol{x}) \tag{8.13-3}$$

In principle, Eqs. (8.13-1)–(8.13-3) with well-defined mixing rules are sufficient to find all K-factors in a multicomponent system of two or more phases. This kind of calculation is not restricted to high-pressure systems. A great attraction of the EOS method is that descriptions developed from low-pressure data can often be used for high-temperature, high-pressure situations with little adjustment. One difficulty is that EOS relations can be highly nonlinear and thus can require sophisticated numerical initialization and convergence methods to obtain final solutions.

To fix ideas, consider a two-phase (vapor-liquid) system containing m components at a fixed total pressure P. The variable mole fractions in the liquid-phase are $x_1, x_2, \ldots, x_{m-1}$. We want to find the bubble-point temperature T and the vapor-phase mole fractions $y_1, y_2, \ldots, y_{m-1}$. The total number of unknowns, therefore, is m. However, to obtain φ_i^{vap} and φ_i^{liq}, we also must know the molar volumes V^{liq} and V^{vap}. Therefore, the total number of unknowns is $m + 2$.

To find $m + 2$ unknowns, we require $m + 2$ independent equations. These are:

Equation (8.12-2) for each component i:	m equations
Equation (8.12-3), once for the vapor-phase and once for the liquid-phase:	2 equations
Total number of independent equations:	$m + 2$

This case, in which P and $\boldsymbol{x}$ are given and T and y are to be found, is called a bubble-point T problem. Other common cases are:

Given Variables	Variables to be Found	Name
$P, \boldsymbol{y}$	$T, \boldsymbol{x}$	Dew-point T
$T, \boldsymbol{x}$	$P, \boldsymbol{y}$	Bubble-point P
$T, \boldsymbol{y}$	$P, \boldsymbol{x}$	Dew-point P

However, a common way to calculate phase equilibria in process design and simulation is to solve the "isothermal flash" problem. In this case, we are given P, T, and the mole fractions, z, of a feed to be split into fractions L/F of liquid and $(1 - L/F)$ of vapor. The "adiabatic flash" problem is very similar. We defer to introductory textbooks for details about the procedure.

A particularly useful text about complex phase equilibrium computations is the monograph by Michelsen and Mollerup (2007). They describe systematic procedures that carefully identify the relevant stable phases, their densities, and compositions. Procedures are also available for mapping phase diagrams over broad ranges of conditions, including the identification of multiphase equilibria, critical end points, and critical lines for the various types of phase diagrams classified by the scheme of van Konynenberg and Scott (1980). The methodology is quite sophisticated, but Cismondi-Duarte and Michelsen (2006) have developed a software package called GPEC (ipqa.unc.edu.ar/en/downloads/). GPEC demonstrates what is possible for the PC-SAFT, PR76, and "RKPR" model. The RKPR model is like the tcPRq model. GPEC provides 2D and 3D graphics. The 3D graphics can be very helpful when rotated to visualize various cross-sections of the phase diagram. Tang et al. (2012) have described an alternative approach based on the equal area method, but it has not been as fully developed as Michelsen's methodology. Interested readers should consult these resources to obtain a detailed understanding of this specialized field of expertise.

For purposes of the evaluations in PGL6ed, we favor computing the bubble pressure. It is the simplest of the "flash" computations, providing convergence for most of our evaluations, and it generally works well with the uncertainties in the experimental data. In other words, taking the temperature as given is a good reflection of its relatively small uncertainty. Between vapor and liquid composition, uncertainty is typically smaller in the liquid composition. Our evaluations limit the pressure to 5 MPa or less. This avoids convergence issues in most cases, while still providing a reasonable sense of model performance at elevated pressures. Higher pressures can lead to phase behavior that is qualitatively incorrect requiring a high degree of adaptation to specific systems. This kind of fine tuning is often required for specialized applications, but that degree of specialization is beyond the scope of our evaluations.

Example 8.11 VLE *K*-factor using the tcPR EOS

Estimate the K-factor for a 10.31 mol% methanol (1) + benzene (2) mixture at 55 °C and 0.07966 MPa as the first step in a bubble-point P calculation. (a) Use the tcPRq model with $k_{ij} = 0.0980$. (b) Use the tcPR-GE model with $k_{12} = 1088.29$ and $k_{21} = 47.80$.

Solution. The vapor composition must also be guessed to obtain the vapor fugacity coefficient. For this example, we assume the experimental value of $y_1 = 0.4841$ as reported by Scatchard et al. (1946). The pure compound parameters are the same for the two models, as tabulated below. The fugacity coefficients are then computed for each model using the formula

$$\ln(\varphi_k) = -\ln(Z - B) + (Z-1)\frac{A^{\text{att}}}{RT} \times \left[\frac{1}{na}\left(\frac{\partial n^2 a}{\partial n_k}\right)_{T,n_{j\neq k}} - \frac{1}{b}\left(\frac{\partial nb}{\partial n_k}\right)_{T,n_{j\neq k}}\right] \quad (8.13\text{-}4)$$

$$\frac{A^{\text{att}}}{RT} = \frac{a}{2\sqrt{2}bRT}\ln\left(\frac{1+\left(1-\sqrt{2}\right)\eta^P}{1+\left(1+\sqrt{2}\right)\eta^P}\right) \quad (8.13\text{-}5)$$

	T_c (K)	P_c(MPa)	α_L	α_M	α_N	$\alpha(T)$	b(cm^3/mol)	c(cm^3/mol)
Methanol	512.50	8.084	0.7682	0.9348	1.5716	1.5141	41.01	9.18
Benzene	562.05	4.895	0.1348	0.8481	2.5791	1.3554	74.27	−1.44

(a) The phase-dependent contributions are listed below, where $\eta^P = \Sigma b_i/V$ is the converged value to match the specified pressure. In this formula, $A = aP/(R^2T^2)$ and $B = bp/RT$.

	b	c	η^P	Z^{rep}	Z^{att}	A	B	A^{att}/RT
V	58.17	3.70	0.00162	0.00163	−0.02213	0.02176	0.00159	−0.02217
L	70.84	−0.34	0.82161	4.60584	−5.603306	0.02767	0.00208	−7.13375

The component-specific properties are:

	Vapor			Liquid			
	$\Sigma x_j a_{ij}/a$	$n\partial(c/b)/\partial n_i$	$\ln(\varphi^{\text{vap}})$	$\Sigma x_j a_{ij}/a$	$n\partial(c/b)/\partial n_i$	$\ln(\varphi^{\text{liq}})$	K
(1)	0.844142	0.128834	−0.0141	0.709663	0.131124	1.130178	3.1402
(2)	1.14625	−0.12089	−0.02618	1.03337	−0.01507	−0.6068	0.5595

(b) Not all the intermediate computations are comparable with the g^E model. The most important contributions are listed below, where $\lambda^{HV} = -0.6232$, $E = -\sum x_i \ln[\sum \psi_{ij} \exp(-\beta a_{ij})]/\lambda$,

	b	*c*	η^P	Z^{rep}	Z^{att}	*A*	*B*	*a/bRT*	*E*
V	57.76	3.70	0.00172	0.00172	−0.02200	0.02163	0.001687	12.822	−0.94486
L	70.54	−0.34	0.81843	4.50750	−5.50519	0.02734	0.002060	13.244	−0.42540

The component-specific properties are

	Vapor			**Liquid**			
	$n\partial E/\partial n_k$	$n\partial(a/bRT)/\partial n_k$	$\ln(\varphi^{vap})$	$n\partial E/\partial n_k$	$n\partial(a/bRT)/\partial n_k$	$\ln(\varphi^{liq})$	*K*
(1)	−0.91828	12.9804	−0.01461	−3.2785	10.620	1.4797	4.4563
(2)	−0.96979	12.6737	−0.02546	−0.0974	13.546	−0.5645	0.5833

This result shows how the *K*-value for methanol is much higher with the g^E model at this composition, a direct impact of the extra fitting parameter. Comparing to Fig. 8.1 shows how this makes sense.

Example 8.12 Bubble pressure calculation using the ESD model

Use the 1996 efficient implementation of the ESD model to calculate the bubble-point pressure and vapor-phase composition for 10.31 mole% methanol (1) + benzene (2) at 55 °C. Assume a value of $k_{ij} = 0.0196$ for this mixture.

Solution. In TPT1 models, a water-methanol mixture contains an infinite number of species: methanol monomers, dimers, trimers, etc. The mole fractions of these oligomers are implied by the equilibrium constant, characterized by the α contribution. Because the fugacities of the oligomers are tied to the fugacity of the monomer through the reaction equilibria, it suffices to determine the mole fraction of monomer, or from Wertheim's perspective, the mole fraction of unbonded sites, X^A and X^D. Therefore, the determination of density requires two steps at each density iteration: (1) iteratively compute X^A and X^D using reaction constraints, (2) sum Z^{rep}, Z^{att}, and Z^{chem} then compare the pressure to that specified. The specification of pressure requires an outer loop of iteration where the pressure, P, and vapor composition, $\mathbf{y}$, are guessed, and K-values computed. New guesses for $\mathbf{y}$ are computed from $y_i = x_i K_i$ and the objective function is computed as $f = 1 - \sum y_i$. Iteration is declared as converged when $|f| < 10^{-7}$. The computations below illustrate the final iteration on pressure, 0.0791 MPa. The EOS parameters for methanol have been characterized by optimizing vapor pressure behavior. The EOS parameters for benzene are computed from the experimental critical properties and the critical point constraints of the ESD EOS. The converged value of the vapor mole fraction is $y_1 = 0.4685$.

ID	NAME	*c*	ε/k(K)	*b*	N^{AD}	K^{AD}	ε^{AD}/k(K)
1	METHANOL	1.269	312.32	19.96	1	0.0007	2516
2	BENZENE	1.767	336.72	29.61	0	0	0

The key equation in the law of mass action is

$$1 - X_i^A = \sum x_j \alpha_{ij}^{AD} X_i^A X_j^D \tag{8.13-6}$$

where $\alpha_{ij}^{AD} = \rho\, g_{ij}(\sigma)\, K_{ij}^{AD}\, [\exp(\beta\varepsilon_{ij}^{AD}) - 1]$ characterizes the equilibrium constant. Equation (8.13-3) essentially states that the fraction of bonded sites is proportional to the equilibrium constant times the product of unbonded fractions, summed over all reactive pairs. Note that only methanol associates in this mixture. Since methanol has only one donor and one acceptor, symmetry leads to $X_1^A = X_1^D$. Therefore, Eq. (8.13-3) suffices to characterize all chemical interactions for this mixture. Nevertheless, further simplification is possible. Elliott (1996) showed that the assumption of a geometric combining rule (GCR) for α^{AD} leads to a factorization of Eq. (8.13-3) that is component-independent. That is, when $\alpha_{ij}^{AD} = (\alpha_{ii}^{AD}\alpha_{jj}^{AD})^{1/2}$, Eq. (8.13-3) can be rewritten as

$$\left(\frac{1}{X_i^A} - 1\right)\frac{1}{\left(\alpha_{ij}^{AD}\right)^{1/2}} = \sum x_j N d_j \left(\alpha_{jj}^{AD}\right)^{1/2} X_j^D \equiv F^{AD} \tag{8.13-7}$$

Note that F^{AD} defined by Eq. (8.13-7) is independent of subscript, despite the subscripts on the left-hand side. This is no problem for nonreactive compounds because $\alpha_{ii}^{AD} = 0$ in that case, leading to $X_i^A = 1$ with the implication that all (imaginary) acceptors on a nonreactive compound are unbonded. Noting that $\alpha_{22}^{AD} = 0$ for benzene and substituting $X_1^D = X_1^A = [1 + \left(\alpha_{ii}^{AD}\right)^{1/2} F^{AD}]^{-1}$ gives

$$\frac{x_1 N_1^{AD}\left(\alpha_{11}^{AD}\right)^{1/2}}{1 + \left(\alpha_{11}^{AD}\right)^{1/2} F^{AD}} = F^{AD} \tag{8.13-8}$$

This is a quadratic equation that can be solved exactly for F^{AD} in this simple example. In general, taking the limit for large and small α^{AD} leads to $F^{AD} \approx 0$ when α^{AD} is small (e.g., for vapors) and $(F^{AD})^2 \approx \sum x_i N_i^{AD}$ for large α^{AD} (e.g., for pure associating liquid compounds) and secant method converges rapidly with these two initial guesses. Key representative contributions are tabulated below, leading to the fugacity coefficients and K-values.

	b	η^P	α_1^{AD}	F^{AD}	X_1^A	$qYb\rho$	Z^{rep}	Z^{att}	Z^{chem}
V	25.09	0.00075	0.0268	0.0758	0.9877	0.00245	0.00458	−0.02324	−0.00575
L	28.62	0.31261	24.249	0.2352	0.46335	1.2613	5.281492	−6.1426	−0.1362

The component-specific properties are

	Vapor				**Liquid**				
	ln(φ^{rep})	**ln(φ^{att})**	**ln(φ^{chem})**	**ln(φ^{vap})**	**ln(φ^{rep})**	**ln(φ^{att})**	**ln(φ^{chem})**	**ln(φ^{liq})**	**K**
(1)	0.0074	−0.0343	−0.0247	−0.0269	6.092	−8.942	−1.595	1.487	4.542
(2)	0.0107	−0.0572	0.0000	−0.0219	8.817	−15.211	−0.084	−0.546	0.592

Comparing to the tcPRq and tcPR-GE models, noting the slightly lower value of pressure, these K-values are much closer to the values from the tcPR-GE model, despite using a single BIP and that value is much closer to zero when compared to the tcPRq model which used $k_{ij} = 0.0994$. These observations are typical of TPT1 models. TPT1 correlations use fewer parameters but achieve accuracies approaching models that have been highly parameterized.

Example 8.13 Bubble pressure calculation using the PPC-SAFT model

Use the polar PC-SAFT (PPC-SAFT) model for the methanol (1)+benzene (2) binary at $x_1 = 0.1031$. For the BIP, assume $k_{12} = -0.01075$ obtained by minimizing deviations for this system for all data in the Jaubert et al. (2020) database. Compute the bubble pressure at 55 °C.

Solution. The PC-SAFT model includes accounting for quadrupole and dipole moments, as well as for the chemical contribution of all TPT1 models.

ID	NAME	m	σ(nm)	ε/k(K)	μ	q	N^{AD}	K^{AD}	ε^{AD}/k(K)
1	METHANOL	1.507	0.3325	211.60	1.7	0	1	0.03	2519.7
2	BENZENE	2.291	0.3756	294.06	0	5.5907	0	0	0

Once again, we demonstrate the computation for the last iteration of part (a), where P = 0.0836 MPa and y_1 = 0.4987. Key representative contributions are tabulated below, leading to the fugacity coefficients and K-values.

	η^P	Z^{HS}	Z^{HC}	Z_1/T	$Z_2/T2$	Z^{chem}	Z^{DD}	Z^{QQ}	Z^{DQ}
V	0.00085	0.0064	−0.0019	−0.0123	−0.0055	−0.0108	−0.00037	−0.00014	−0.00019
L	0.39596	12.7135	−2.0821	−11.4335	0.3822	−0.1786	−0.00553	−0.32830	−0.06502

The terms Z^{DD}, Z^{QQ}, and Z^{DQ} correspond to the dipole-dipole, quadrupole-quadrupole, and dipole-quadrupole contributions, respectively. It is interesting to note that these terms are similar in magnitude and opposite in sign to the Z_2 contribution at these conditions. The quadrupole contribution from benzene appears to be the most significant of these contributions. The component-specific properties are

	Vapor				Liquid				
	ln(φ^{rep})	**ln(φ^{att})**	**ln(φ^{chem})**	**ln(φ^{vap})**	**ln(φ^{rep})**	**ln(φ^{att})**	**ln(φ^{chem})**	**ln(φ^{liq})**	**K**
(1)	0.0067	−0.0238	−0.0436	−0.03568	8.829	−10.996	−2.198	1.5406	4.837
(2)	0.0112	−0.0499	0.0000	−0.01364	16.602	−22.985	−0.117	−0.5954	0.559

For brevity, the hard sphere, chain, dispersion, and polar contributions have been combined into φ^{att} in this tabulation. Comparing to the tcPRq and tcPR-GE models, noting the slightly higher value of pressure, these K-values are closer to the values from the tcPR-GE and ESD models. All the φ contributions are larger for PPC-SAFT than for ESD, but the sign differences lead to cancellations, so the K-values are quite similar.

Example 8.14 Bubble pressure calculation using the SPEADMD model

Use the SPEADMD model for the methanol (1)+benzene (2) binary at x_1 = 0.1031. For the BIP, assume k_{12} = 0.0240 obtained by minimizing deviations for this system for all data in the Jaubert et al. (2020) database. Compute the bubble pressure at 55 °C.

Solution. The SPEADMD model shares elements of the PC-SAFT and ESD models. Like the ESD model, it neglects the polar contributions, effectively lumping those effects into the dispersion and chemical contributions. Like the PC-SAFT model, it implements second-order TPT for the dispersion contributions. Distinctive to the SPEADMD model, it does not apply a universal correlation for the A_0, A_1, and A_2 terms. These terms are inferred from the transferable force field (TFF), including effects of bond length, chain stiffness, branching, and ring structures. The SPEADMD model scales the molecular volume (b) and A_1 contributions relative to the SPEADMD-TFF values to match experimental liquid density and vapor pressure data when available. Multiple parameters are applied to match the simulation data, but only the two scaling parameters are adjustable during customization.

ID	NAME	z_1	z_2	z_3	a_{11}	a_{12}	a_{13}	$10^{-3}a_{21}$	$10^{-3}a_{22}$	$10^{-3}a_{23}$	$10^{-3}a_{24}$
1	MeOH	0.9787	2.4338	−3.5547	−44.71	−3305	−4698	−288.2	1541	−3094	1838
2	Benzene	1.3628	2.8002	−0.8613	5406	−9630	−15898	−1153	5442	−9309	5011

SPEADMD requires additional parameters to characterize the chemical-physical terms

ID	NAME	b(cm³/mol)	K^{AD}(nm³)	ε^{AD}/k(K)
1	MeOH	18.73	0.001	2516
2	Benzene	43.08	0	0

Once again, we demonstrate the computation for the last iteration of part (a), where P = 0.0788 MPa and y_1 = 0.4708. Key representative contributions are tabulated below, leading to the fugacity coefficients and K-values.

	b	η^P	Z^{rep}	Z_1/T	Z_2/T^2	X_1^A	Z^{chem}	A^{att}/RT
V	31.62	0.00093	0.00380	–0.0105	–2.05614	0.982684	–0.00817	–0.0168
L	40.57	0.45475	11.703282	–12.6293	51.28752	0.31154	–0.22769	–10.1076

The component specific properties are:

	Vapor				Liquid				
	$\ln(\varphi^{rep})$	$\ln(\varphi^{att})$	$\ln(\varphi^{chem})$	$\ln(\varphi^{vap})$	$\ln(\varphi^{rep})$	$\ln(\varphi^{att})$	$\ln(\varphi^{chem})$	$\ln(\varphi^{liq})$	K
(1)	3.1E-06	–0.01543	–0.0349	–0.029	3.102	–5.169	–2.405	1.4901	4.568
(2)	7.0E-06	–0.01403	0.0000	–0.0219	7.138	–13.474	–0.166	–0.5411	0.595

The chemical contributions in SPEADMD are stronger than in the other two TPT1 models as evidenced by the values of $\ln(\varphi^{chem})$, while the physical contributions resemble those in ESD. The repulsive contributions to the vapor fugacity are quite small in the SPEADMD model. Comparing to the other models, noting the slightly lower value of pressure due to the characterized BIP for this overall system, these K-values are closer to the values from the tcPR-GE and the TPT1 models than to the tcPRq model.

Example 8.15 Bubble pressure calculation including gaseous compounds

Use the tcPR-GE model to calculate VLE for the binary CO_2 (1)-propane (2) at 37.8 °C. For the molar excess Gibbs energy, g^E, use the Wilson model with a_{12} = 232.4 and a_{21} = 132.1.

Solution. The procedure to compute the bubble pressure is the same as before. Therefore, we simply summarize the results of the computations in the table below. The experimental data were taken from Reamer et al. (1951).

	calc		expt	
x_1	y_1	P(MPa)	y_1	P(MPa)
0	0.000	1.3139	0	1.301
0.093	0.341	2.1037	0.351	2.068
0.178	0.485	2.7732	0.499	2.758
0.271	0.578	3.4519	0.588	3.447
0.369	0.641	4.1149	0.651	4.137
0.474	0.690	4.7815	0.701	4.826
0.581	0.730	5.4375	0.75	5.516
0.686	0.764	5.6188	0.78	6.205
0.736	–	–	0.79	6.550

The tcPR-GE model is very accurate at $x_1 = 0.581$ or less, but the last two points reflect that the model closes the phase envelope at a lower composition than the experimental data. This is typical for these kinds of models at $P > 5$ MPa. Note that the BIPs in this case were characterized with data over a wide range of conditions. If greater accuracy is required at this specific temperature, readers may wish to adjust the BIPs accordingly.

Example 8.16 Bubble pressure calculation for ternary systems

Use the Wong-Sandler mixing rules (with the NRTL expression for g^E) and the PRSV EOS to calculate vapor-liquid equilibria for the ternary acetone (1)+methanol (2)+water (3) at 100 °C. Use the NRTL parameters from fitting the three binaries and the ternary as reported by Gmehling and Onken (1977).

Solution. The procedure is like that in previous Examples 8.13 to 8.15 for binary systems, except that now there are three phase-equilibrium relations, resulting in three K-values (Eq. 8.13-2) to compute three vapor mole fractions ($\boldsymbol{y}$), and three instances of Eq. (8.13-1) to be checked for convergence. Finite difference Newton iteration usually suffices for pressure and successive substitution is applied to iteration on $\boldsymbol{y}$ below 5 MPa. Since these components are near their boiling temperatures, we anticipate the computed pressure to be less than 0.2 MPa.

Here the PRSV fitted parameters are $\kappa_1^{(1)} = -0.0089$, $\kappa_2^{(1)} = -0.1682$, and $\kappa_3^{(1)} = -0.0664$. The mixing rules of Eq. (7.5-10) give the mixture parameters a^V, a^L, b^V, and b^L as

$$a^V = b^V\left(\frac{g^{EV}}{C} + \sum_{i=1}^{3} y_i \frac{a_{ii}}{b_i}\right) \qquad a^L = b^L\left(\frac{g^{EL}}{C} + \sum_{i=1}^{3} x_i \frac{a_{ii}}{b_i}\right) \tag{8.13-9}$$

$$b^V = \frac{\sum_{i=1}^{3}\sum_{j=1}^{3} y_i y_j\left(b - \frac{a}{RT}\right)_{ij}}{1 - \frac{1}{RT}\left(\sum_{i=1}^{3} y_i \frac{a_{ii}}{b_i} + \frac{g^{EV}}{C}\right)} \qquad b^L = \frac{\sum_{i=1}^{3}\sum_{j=1}^{3} x_i x_j\left(b - \frac{a}{RT}\right)_{ij}}{1 - \frac{1}{RT}\left(\sum_{i=1}^{3} x_i \frac{a_{ii}}{b_i} + \frac{g^{EL}}{C}\right)} \tag{8.13-10}$$

where g^{EV} is the excess Gibbs energy at the vapor composition and g^{EL} is that at the liquid composition. The constant C for the PRSV EoS is −0.623.

The Wong-Sandler mixing rules adapted to the NRTL expression give

$$\frac{g^E}{RT} = \sum_{i=1}^{3}\sum_{j>i}^{3} x_i x_j\left(\frac{\tau_{ji}G_{ji}}{\sum x_k G_{ki}} + \frac{\tau_{ij}G_{ij}}{\sum x_k G_{kj}}\right) \tag{8.13-11}$$

with $G_{ij} = \exp(-\alpha_N \tau_{ij})$. The covolume cross parameter is given by

$$\left(b - \frac{a}{RT}\right)_{ij} = \frac{(b_i + b_j)}{2} - \frac{(a_{ii}a_{jj})^{½}}{RT}(1 - k_{ij}) \tag{8.13-12}$$

We match the g^E model by minimizing the objective function F_{WS}.

$$F_{WS} = \sum_{\text{data}}\left|\frac{a^E_{EOS}}{RT} - \frac{g^{EL}}{RT}\right| \tag{8.13-13}$$

The summation is for all data points at the specified temperature of 30 °C and evaluated at the liquid compositions. The molar excess Helmholtz energy from the EOS is:

$$a^E_{EOS} = b^{vap}\left(\frac{a^{liq}}{b^{liq}} - \sum_{i=1}^{3} x_i \frac{a_{ii}}{b_i}\right) \tag{8.13-14}$$

TABLE 8.11 Vapor-liquid equilibria for acetone (1), methanol (2), and water (3) at 100 °C.

Liquid		Experiment			From NRTL Binary Fitting			Predicted by EOS		
x_1	x_2	y_1	y_2	P	Δy_1	Δy_2	ΔP	Δy_1	Δy_2	ΔP
0.001	0.019	0.11	0.045	1.234	−0.068	−0.007	0	0.001	0.006	0.047
0.019	0.029	0.238	0.132	1.545	−0.045	−0.051	0.009	0.005	0.017	−0.058
0.066	0.119	0.341	0.271	2.267	−0.026	−0.049	0.024	0.001	0.006	0.09
0.158	0.088	0.525	0.151	2.668	−0.001	−0.05	0.021	−0.041	0.055	−0.132
0.252	0.243	0.47	0.3	3.103	0.036	−0.038	0.03	0.011	−0.004	0.053
0.385	0.479	0.466	0.47	3.818	0.077	−0.02	0.021	−0.053	0.071	−0.103
0.46	0.171	0.62	0.2	3.398	0.026	−0.024	0.023	−0.064	0.059	−0.256
0.607	0.33	0.622	0.345	4.013	0.128	−0.013	0.015	−0.071	0.074	−0.174
0.77	0.059	0.813	0.081	3.638	0.064	0.01	0.003	−0.076	0.034	−0.151
0.916	0.05	0.902	0.075	3.811	0.1	0.011	−0.006	−0.091	0.076	−0.076
RMSE(based on binary)					0.069	0.084	0.043	0.039	0.031	0.111
RMSE(fit to ternary)					0.034	0.016	0.062			

The NRTL binary parameters obtained by fitting available binary data are

i	j	α_N	τ_{ij}	τ_{ji}
1	2	0.3014	1.4400	−0.5783
1	3	0.2862	0.1835	2.0009
2	3	0.3004	−0.5442	1.5011

Minimizing the objective function in Eq. (8.13-13), we obtain the EOS binary parameters, $k_{12} = 0.127$, $k_{13} = 0.189$, and $k_{23} = 0.100$. Then we use Eq. (8.13-14).

Table 8.11 shows results calculated from the fitting and from the EOS predictions and experimental results at the x_1 and x_2 values of Griswold and Wong (1952). The EOS with the binary parameters is better than the prediction using the NRTL g^E model. For comparison, we note that when ternary data are included in the fitting of NRTL parameters, the standard deviations are somewhat better than those from the EOS results.

While the agreement with P is good, the vapor mole fraction comparisons are only fair.

A variety of other methods based on g^E-mixing rules have been proposed to calculate VLE for mixtures containing one or more polar or hydrogen-bonding components. An issue that has been prominent is the appropriate standard-state pressure to match the EOS to the g^E expression. Several researchers have discussed this issue at great length, describing the options that various workers have chosen (e.g., Kontogeorgis and Folas, 2010).

8.14 EVALUATIONS

Knapp et al. (1982) have presented a comprehensive monograph on EOS calculations of binary VLE for relatively ideal mixtures using quadratic mixing rules. It contains an exhaustive literature survey (1900–1980) for binary mixtures encountered in natural-gas and petroleum technology: hydrocarbons, common gases, chlorofluorocarbons, and a few oxygenated hydrocarbons. The survey has been extended in a series of articles by Dohrn and coworkers (Dohrn et al.,1995; Peper et al., 2019). Broadly, their finding was that the Peng-Robinson (1976) (PR76) EOS was representative of the evaluated models, with deviations comparable to the other models or better. Based on our evaluations, we would expect improvement using the tcPRq model for these types of mixtures.

Experimental data for VLE have been reported for many systems. In many of these cases, the same systems have been studied by multiple researchers. This redundancy is a disadvantage in the efficient collection of VLE data, but it is helpful in illuminating levels of uncertainty. This illumination shows that uncertainty varies in a complicated manner. Correlation of VLE data can be quite precise, with less than 1% deviation for some models. Therefore, it is important to construct a database for comparison with less than 0.5% uncertainty if the distinctions between models are to be properly identified.

Another consideration is that the types of mixture combinations studied experimentally tend to be redundant. Industrial interest has motivated much of this research, which focuses interest on a relatively small number of molecular types. Petroleum production and refining has contributed actively to VLE data. Without deliberation, a database might put excessive emphasis on hydrocarbons, gases, and water, for example.

Fortunately, these considerations have been recognized in the construction of two databases that we find useful for the evaluations in PGL6ed. Danner and Gess (1990) compiled a database for low-pressure VLE of 103 binary mixtures classified according to nonpolar, polar, and associating combinations. Recently, Jaubert et al. (2020) extended the concept to include gases and high-pressure data for 203 binary mixtures, several of which overlap with the Danner-Gess database. Both databases are publicly available, with links to the data on the PGL6ed website. We hope that the open-source nature of these databases facilitates the reproducibility of our analysis and of future analyses.

8.14.1 Binary Mixtures

The results of our evaluations for binary mixtures are presented in Table 8.12 and Table 8.13. Table 8.12 shows that the quadratic mixing model of the tcPRq model provides a substantial improvement over the quadratic mixing with the PR76 model. This confirms that improving the vapor pressure correlation improves VLE correlation

TABLE 8.12 Evaluations of binary VLE models at pressures below 0.5 MPa.

Danner-Gess DB	%AAD(All)	N^{systs}	NPTS
REFPROP	2.29	13	162
LSG	0.63	103	1363
NRTL	0.67	103	1363
UNIQUAC	0.67	103	1364
Wilson	0.73	103	1364
tcPR-GE	1.01	103	1548
tcPRq	2.77	95	1415
PR76	4.92	103	1537
PC-SAFT	2.85	103	1548
ESD96	2.83	103	1548
SPEADMD	2.17	103	1548
UNIFAC-NIST-Mod	5.79	101	1512
UNIFAC-NIST-KT	4.53	103	1531
UNIFAC-Original	5.02	103	1536
PSRK	5.17	103	1546
COSMO-RS/Therm(3ds)	6.73	103	1548
COSMO-RS/SAC-GAMESS	9.42	100	1506
COSMO-RS/SAC-Phi	9.04	47	723
Ideal Solution Model	12.58	103	1517

TABLE 8.13 Evaluations of binary VLE models at pressures up to 5 MPa.

	%dev(NoAq)	N^{systs}	NPTS	%dev(Aq)	N^{systs}	NPTS	N^{systs} (tot)
REFPROP	4.68	86	9605	7.18	10	1514	96
tcPR-GE	2.59	181	12195	8.97	19	2127	200
SPEADMD	3.00	182	12252	10.38	19	2139	201
ESD96	3.25	181	12210	10.62	19	2131	200
tcPRq	3.51	181	12275	15.45	19	2023	200
PC-SAFT	3.72	181	12339	18.29	19	1965	200
PR76	3.88	181	12331	18.57	19	1969	200

for mixtures and it suggests that the tcPR model does provide a meaningful update regardless of mixing rule. When combined with the two-parameter mixing rule, the tcPR-GE model provides modeling accuracy that rivals the two-parameter activity models at low pressure. Among the models with a single BIP, SPEADMD appears to have a slight advantage, but all the models are similar. The tcPRq model fails to converge for several of the aqueous systems in this database, however, and that causes its deviations to appear to be artificially low. The UNIFAC models exhibit deviations around 5%, with a slight advantage to the NIST/KT UNIFAC model. The PSRK model exhibits deviations like the UNIFAC models while providing an EOS with applicability to gases and higher pressures. The COSMO-RS/SAC models generally exhibit larger deviations than the UNIFAC models, but the number of adjustable parameters is much smaller for COSMO-RS/SAC. The COSMO-RS/Therm version from 3ds.com derives from the original development by Klamt and coworkers. It is proprietary now so details of its refinement are unavailable, but its performance is significantly better than other COSMO-RS/SAC implementations, and very close to the performance of UNIFAC models. We compare to two COSMO-RS/SAC implementations by Soares and coworkers. The COSMO-RS/SAC-GAMESS model is based on the work of Ferrarini et al. (2018). This model has the advantage of applying the open source GAMESS software for the QDFT computation of the COSMO sigma profiles. The COSMO-RS/SAC-Phi implementation is based on the work of Soares et al. (2019), which comprises an EOS extension of the COSMO-RS/SAC formalism.

Table 8.13 shows similar trends to the low-pressure analysis, but the advantage of the tcPR-GE model is much diminished. All the models exhibit around 3% deviation for nonaqueous systems. For aqueous systems, the average deviations are around 10% for the best models, but it needs to be kept in mind that some of these aqueous systems include two associating compounds, for which deviations are small. Deviations with a gaseous or hydrocarbon compound in an aqueous system are typically around 20% for all models. These systems provide opportunities for substantial improvement through further research.

8.14.2 Ternary Mixtures

In addition to the evaluation of models for binary VLE, we have also considered extrapolations to ternary VLE. Specifically, we evaluated the ability to predict ternary low-pressure VLE based on binary systems that we correlated. Similar studies have been conducted by Gasem and coworkers for a database of about 57 systems and Graczova and coworkers for a database of about 103 systems. For our evaluation, we constructed a database of 100 ternary systems based on compounds included in Appendix A. Binary parameters were correlated for each binary system in any ternary combination based on minimizing bubble pressure deviations. For EOSs, predictions of ternary low-pressure VLE apply Eqs. (8.5-3) and (8.5-4) as a means of expediting the bubble pressure computation. Table 8.14 shows that the Wilson model generally provides the smallest deviations. For many types of mixtures, the deviations are close to 2%. Deviations exceed 3% when all compounds are associating, or when two polar compounds are mixed with an associating or normal compound. None of the mixtures in this database include heavy compounds.

TABLE 8.14 Predictions of ternary VLE at pressures below 0.5 MPa.

Class*	Wilson	UNIQUAC	NRTL
AAA	3.1%	3.4%	3.7%
AAN	1.3%	1.9%	2.2%
AAP	1.9%	2.0%	1.5%
ANN	1.9%	1.4%	3.8%
ANP	1.8%	2.0%	4.5%
APP	3.3%	3.4%	3.4%
NNN	1.7%	1.6%	0.8%
NNP	2.3%	1.9%	1.0%
NPP	3.0%	4.0%	2.8%
PPP	1.0%	0.9%	0.9%
Overall	2.1%	2.1%	1.9%
N^{systs}	97	87	60

*The classes (Normal, Heavy, Polar, and Associating) are abbreviated here. No heavy compounds were represented in the ternary VLE database.

8.14.3 Discussion

Altogether, the tcPR-GE model appears to provide the best capability for correlating VLE data. Its accuracy for correlating low-pressure VLE is very close to that of conventional activity models. As an EOS, it provides seamless treatment of systems containing gases. As a cubic EOS, its implementation should be familiar to those who have used cubic EOSs in the past. If predictions are required for mixtures including compounds that have not been experimentally measured, one of the UNIFAC models can be used to generate binary pseudo-data at the conditions of interest and BIPs can be characterized for the binary combinations that have not been measured. Keep in mind that activity models assume that liquids exist at a single value of density, whereas EOSs inherently include variations in properties like density as the liquid-phase varies.

One might conclude that this is not much of an advancement from the models that were available 20 years ago. Cubic EOSs with Gibbs excess mixing rules were available at that time as well. That is fair, but it overlooks the competitiveness provided by models that incorporate more physical insight. For systems that include elevated pressures, the TPT1 models provided comparable accuracy to the tcPR-GE model, while implementing a single BIP. We have not explored the phase diagrams extensively, but high-pressure systems can be very sensitive to the mixing rules. Artificial multiphase regions might be more likely for a model that includes multiple BIPs tuned to uncertain experimental data and a weak physical basis. The TPT1 models also provide certain capabilities that are beyond the scope of Gibbs excess models like Wilson, NRTL, and UNIQUAC. For example, peculiar phase diagrams like in isobutanol+water are more amenable to treatment with TPT1 models.

Going forward, an obvious approach would be to combine TPT1 with two-parameter (mixing rules) mixing rules for the disperse attractive contribution. This is the concept behind the CPA model. Unfortunately, that model is sufficiently complex that we could not reproduce its implementation ourselves and attempts to implement the package from its developers met with limited success. Recent work by Hao and Chen (2021) suggests that combination of the NRTL model with TPT1 provides interesting capability however: (1) quantitative treatment of VLE and LLE with consistent parameters, (2) closed-loop phase diagrams, (3) accurate analysis of ternary phase diagrams over a wide temperature range. We hope to provide a more rigorous evaluation of these kinds of models in PGL7ed. Our evaluation databases may require considerable extension to cover the entire scope of phase behavior in a properly balanced manner, however.

Another advance that might be overlooked is the advancement of COSMO-RS/SAC models. The COSMO-RS/therm package provides comparable accuracy to most of the UNIFAC models while using substantially fewer parameters. For readers exploring phase behavior of systems that include interactions of functional groups not characterized by UNIFAC, a COSMO-RS/SAC model would be recommended. Currently available COSMO-RS/SAC models lack a feature for customizing predictive results to correlate experimental data when available, but there should be no problem in principle to add such a feature. For example, a local composition activity model like LSG for molecular-scale customized dispersion interactions could be added to the COSMO-RS/SAC prediction, retaining the COSMO-RS/SAC predictive treatment of the strong interactions. It is physically reasonable that segments may sense a (small) distinct correction in dispersion energy when assembled into one complete molecule versus another, and the LSG model should be applicable to such small corrections, positive or negative. Such a model would be comparable to the way Hao and Chen (2021) combined the NRTL model with TPT1 and a predictive correlation of strong interactions, but it would retain the SS-QCT framework for the activity model. Although the COSMO-RS/therm package is proprietary and only available through purchase, the implementation of COSMO-RS/SAC by Ferrarini et al. (2018) applies open-source codes and sigma-profiles to the open-source GAMESS package for QDFT calculations, facilitating adaptations and improvements globally.

8.15 CONCLUDING REMARKS

This chapter on phase equilibria has presented no more than a brief introduction to a very broad subject. The variety of mixtures encountered in the chemical industry is extremely large, and, except for general thermodynamic equations, there are no quantitative relations that apply rigorously to all, or even to a large fraction, of these mixtures. Thermodynamics provides only a coarse but reliable framework; the details must be supplied by physics and chemistry, which ultimately rest on experimental data.

For each mixture, it is necessary to construct an appropriate mathematical model for representing the properties of that mixture. Whenever possible, such a model should be based on physical concepts, but since our fundamental understanding of fluids is limited, any useful model is inevitably influenced by empiricism. While at least some empiricism cannot be avoided, the strategy of the process engineer must be to use enlightened rather than blind empiricism. This means foremost that critical and informed judgment must always be exercised. While such judgment is attained only by experience, we conclude this chapter with a few guidelines.

1. Face the facts: You cannot get something from nothing. Do not expect magic from thermodynamics. If you want reliable results, you will need reliable experimental data. You may not need much, but you do need some. The required data need not necessarily be for the system of interest; sometimes they may come from experimental studies on closely related systems, perhaps represented by a suitable correlation. It is the exception rather than the rule that accurate partial thermodynamic properties in a mixture, e.g., activity coefficients, can be found from pure-compound data alone.
2. Correlations provide the easy route, but they should be used last, not first. The preferred first step should always be to obtain *reliable* experimental data, either from the literature or from the laboratory. Do not at once reject the possibility of obtaining a few crucial data yourself. Laboratory work is more tedious than pushing a computer button, but a few simple measurements may provide more systematic progress when compared to a multitude of hypothetical calculations. A small laboratory with a few analytical instruments (especially a balance, a chromatograph, and a simple boiling-point apparatus) can often save both time and money. If you cannot do the experiment yourself, consider the possibility of having someone else do it for you.
3. It is always better to obtain a few well-chosen and reliable experimental data than to obtain many data with high uncertainty. Beware of statistics, which may be the last refuge of a poor experimentalist.
4. Always regard published experimental data with skepticism. Many experimental results are of high quality, but many are not. Just because a number is reported by someone and printed by another, do not automatically

assume that it must therefore be correct. Curated databases that facilitate comparisons between multiple independent authors can reveal realistic uncertainties.

5. When choosing a mathematical model for representing mixture properties, give preference, if possible, to those which have some physical basis.
6. Seek simplicity; beware of models with many adjustable parameters. When such models are extrapolated even mildly into regions other than those for which the constants were determined, highly erroneous results may be obtained.
7. In reducing experimental data, keep in mind the probable experimental uncertainty of the data. Whenever possible, give more weight to those data which you have reason to believe are more reliable.
8. If you do use a correlation, be sure to note its limitations. Extrapolation outside its domain of validity can lead to large error.
9. Never be impressed by calculated results merely because they come from a computer. The virtue of a computer is speed, not intelligence. Always validate the VLE correlations coming from the computer by comparing them to experimental data. Prepare comparisons for each binary combination of components in the mixture and for the multicomponent case if data are available. The NIST/TRC or DECHEMA databases provide convenient resources if available. Otherwise, consult the NIST/ThermoLit resource and track down the most relevant experimental data in the literature. If you are running a distillation column, compare experimentally determined tray compositions to the simulated values. A Hengstebeck diagram can be a convenient basis for such comparisons (Kister et al., 1992). Observations from the actual process under consideration comprise the most important experimental data of all.
10. Maintain perspective. Always ask yourself: Is this result reasonable? Do other similar systems behave this way? If you have limited experience, get help from someone who has more experience. Phase equilibria in fluid mixtures is not a simple subject. Do not hesitate to ask for advice.

8.16 ACRONYMS

AAD__	Absolute average deviation defined in Chap. 1 and reiterated in Eq. (8.1-1). The suffix indicates the quantity being averaged. For example, AADP represents the AAD in bubble point pressure.
BIP	Binary interaction parameter.
BMCSL	Boublik (1970)-Mansoori, Carnahan, Starling, Leland (1971) EOS for hard sphere mixtures.
COSMO-RS	Conductor-like screening model for real solvents. A combination of quantum density functional theory to compute surface polarizations with SS-QCT to predict activity coefficients from first principles.
EOS	Equation of state.
ESD	Elliott, Suresh, and Donohue EOS, Elliott et al. (1990). See also Elliott (1996).
FH	Flory-Huggins theory. See Eq. (8.6-23). See also Chap. 12 of Elliott and Lira for detailed development.
FHA	Flory-Huggins athermal term. See Eq. (8.6-23).
GSA	Guggenheim-Staverman athermal term. See Table 8.2.
LLE	Liquid-liquid equilibrium
LSG	Lattice surface Guggenheim model of Vera et al. (1977). It is functionally like UNIQUAC but matches QCT more closely.
MOSCED	Modified separation of cohesive energy density model. Cf. Lazzaroni et al. (2005).
NPTS	Number of points in a data set
NRTL	Non-random two-liquid theory of Renon and Prausnitz (1969)
PC-SAFT	EOS developed by Gross and Sadowski (2002) as an implementation of SAFT based on a conception of tangent sphere chains as the reference for the perturbation theory

PGL__ed	*The Properties of Gases and Liquids* book series. The suffix indicates the edition.
PR	Peng-Robinson (1976) EOS
PVT__	Pressure-volume-temperature. The suffix "x" indicates inclusion of mole fraction. The suffix "y" specifies vapor mole fraction.
QCT	Generic for quasichemical theory initially developed by Guggenheim (1952)
SAFT	Statistical associating fluid theory dubbed by Chapman et al. (1990) as an adaptation of Wertheim's (1984) TPT1 theory
SH	Scatchard (1931)-Hildebrand (1929) contribution. See Eqs. (8.6-21) and (8.6-22).
SLE	Solid-liquid equilibrium
SPEADMD	Step potentials for equilibrium and discontinuous molecular dynamics model of Unlu et al. (2004). See Elliott (2016) for a review.
SRK	Soave (1972) EOS
SS-QCT	"Real solvent" QCT developed initially by Klamt et al. (1998) for representing the thermodynamic component of his COSMO-RS model
tcPR__	"Translated consistent" adaptation of the Peng-Robinson (1976) EOS by Le Guennec et al. (2016). The suffix "q" indicates the use of quadratic mixing rules. The suffix "-GE" indicates the use of a Gibbs excess mixing rule with Wilson's (1964) activity model as its basis.
TPT1	Wertheim's (1984) first-order thermodynamic perturbation theory for chemical-physical treatment of strong interactions like hydrogen bonding.
UNIQUAC	"Universal quasichemical theory" of Abrams and Prausnitz (1975)
VLE	Vapor-liquid equilibrium

8.17 NOTATION

a^E	excess molar Helmoltz energy
a, b, c	empirical coefficients
a_i	activity of component i
a_{mn}	group interaction parameter
A, B, C	empirical constants
B_{ij}	second virial coefficient for the ij interaction
c, d	empirical constants
c_{ij}	cohesive energy density for the ij interaction
c_{12}	empirical constant
C	empirical constant
C_P	molar-specific heat at constant pressure
D	empirical constant
f_i	fugacity of component i
f	a function
F_i	nonideality factor
g_{ij}	empirical constant
g^E	molar excess Gibbs energy
G^E	total excess Gibbs energy
G_{ij}	empirical constant (Table 8.2)
h^E	molar excess enthalpy

h_i	partial molar enthalpy of component i
H_v^{px}	Henry's volatility constant
ΔH_v	enthalpy of vaporization
k_{12}	binary parameter in Sec. 8.12
K	y/x; distribution coefficient
L/F	fraction liquid
l_i	constant defined in Table 8.2
l_{12}	empirical constant
m	defined after
m_i	molality
n_i	number of moles of component i
n_T	total number of moles
N_C	number of components
p_i	partial pressure of the ith component
P	total pressure
P^{vp}	vapor pressure
q	molecular surface parameter, an empirical constant (Table 8.2)
Q_k	group surface parameter
r	molecular-size parameter, an empirical constant (Table 8.2)
R_{gas}	gas constant
R_k	group size parameter
s^E	molar excess entropy
s_j	number of size groups in molecule j
t	temperature, °C
T	absolute temperature, K
T_{fp}	melting point temperature
T_{tpt}	triple-point temperature
u_{ij}	empirical constant (Table 8.2)
ΔU	change in internal energy
V	molar volume
w_k	weight fraction of component k
x_i	liquid-phase mole fraction of component i
x_M	mole fraction of monomer in a chemical-physical treatment of strong interactions
X_k	group mole fraction for group k
y_i	vapor-phase mole fraction of component i
z	coordination number (Table 8.2)
z_i	overall mole fraction
Z	compressibility factor

Greek

α^{AD}	TPT1 parameter characterizing interaction strength between an acceptor and donor
α_N	NRTL nonrandomness parameter
γ_i	activity coefficient of component i

Γ_k activity coefficient of group k
δ solubility parameter
$<\delta>$ volume-averaged solubility parameter
ε Molecular interaction energy
η^{P} packing fraction
θ parameter
Θ_i surface fraction of component i
$\lambda_{\iota\varphi}$ empirical constant
$\Lambda_{\iota\varphi}$ empirical constant in Table 8.3
$\nu_k^{(i)}$ number of groups of type k in molecule i
ν_{kj} number of interaction groups k in molecule j
ρ density, g/cm^3
τ_{ij} empirical constant in Table 8.3
φ_i fugacity coefficient of component i
Φ_i site fraction (or volume fraction) of component i
χ Flory interaction parameter
$\Psi_{\mu\nu}$ group interaction parameter
ψ_{12} binary (induction) parameter

Superscripts

AD acceptor-donor strong interaction
calc calculated
chem chemical-physical contributions for strong interactions like hydrogen bonding
expt experimental, Eq. (8.12-4)
E excess
FHA Flory-Huggins athermal contribution, Eq. (8.6-1)
GSA Guggenheim-Staverman athermal contribution, Table 8.2
G group
H solute parameter
KT Kamlet-Taft solvent parameter
liq liquid-phase
M measured value, Eq. (8.8-16)
$\circ$ standard state as in f_i°, estimated true value
res residual
sat saturation, susceptibility parameter
sub sublimation
vap vapor-phase
∞ infinite dilution

8.18 REFERENCES

Abbott, M. M., J. K. Floess, G. E. Walsh, H. C. Van Ness, *AIChE J.*, **21:** 72 (1975).
Abbott, M. M., H. C. Van Ness, *AIChE J.*, **21:** 62 (1975).
Abderafi, S., T. Bounahmidi, *Fluid Phase Equil.*, **93:** 337 (1994).

Abraham, M. H., *Chem. Soc. Rev.,* **22:** 73 (1993).
Abrams, D. S., J. M. Prausnitz, *AIChE J.,* **21:** 116 (1975).
Abusleme, J. A., J. H. Vera, *Fluid Phase Equil.,* **22:** 123–138 (1985).
Adler, S. B., L. Friend, R. L. Pigford, *AIChE J.,* **12:** 629–637 (1966).
Adler, S. B., *Hydrocarbon Process. Intern. Ed.,* **62**(5): 109, **62**(6): 93 (1983).
Anderson, T. F., D. S. Abrams, E. A. Grens, *AIChE J.,* **24:** 20 (1978).
Babich, S. V., R. A. Ivanchikova, L. A. Serafimov, *Zh. Prikl. Khim.,* **42:** 1354 (1969).
Bala, A., C. T. Lira, *Fluid Phase Equil.,* **430:** 47–56 (2016).
Barker, J. A., *J. Chem. Phys.,* **20:** 1526 (1952).
Battino, R., *Fluid Phase Equil.,* **15:** 231 (1984).
Battino, R., E. Wilhelm, *J. Chem. Thermodyn.,* **3:** 379 (1971).
Beck, T. L., M.E. Paulaitis, L. R. Pratt, *The Potential Distribution Theorem and Models of Molecular Solutions*, Cambridge University Press, New York, 2006.
Bell, I. H., E. Mickoleit, C. M. Hsieh, S. T. Lin, J. Vrabec, C. Breitkopf, A. Jäger, *J Chem. Theory Comp.,* **16:** 2635–2646 (2020).
Bethe, H., *Proc. Roy. Soc. London A,* **150:** 552 (1935).
Blas, F. J., L. F. Vega, *Mol. Phys.,* **92:**135 (1997).
Boublik, T., *J. Chem. Phys.,* **53:** 471 (1970).
Brelvi, S. W., J. P. O'Connell, *AIChE J.,* **18:** 1239 (1972).
Brelvi, S. W., J. P. O'Connell, *AIChE J.,* **21:** 157 (1975).
Campanella, E. A., P. M. Mathias, J. P. O'Connell, *AIChE J.,* **33:** 2057 (1987).
Carroll, J. J., *Chem. Eng. Prog.,* **95:** 49 (1999).
Catte, M., C. G. Dussap, C. Archard, J. B. Gros, *Fluid Phase Equil.,* **105:** 1 (1995).
Chapman, W. G., K. E. Gubbins, G. Jackson, M. Radosz, *Ind. Eng. Chem. Res.,* **29:** 1709 (1990).
Cismondi Duarte, M., M. L. Michelsen, *J. Supercritical Fluids,* **39:** 287–295 (2006).
Constantinescu, D., J. Gmehling, *J. Chem. Eng. Data,* **61:** 2738–2748 (2016).
Dadmohammadi, Y., S. Gebreyohannes, B. J. Neely, K. A. M. Gasem, *Fluid Phase Equil.,* **409:** 318 (2016).
Danner, R. P., M. A. Gess, *Fluid Phase Equil.,* **56:** 285–301 (1990).
Dohrn, R., G. Brunner, *Fluid Phase Equil.,* **106:** 213 (1995).
Elliott, J. R., *Chem. Eng. Ed.,* **27:** 44–51 (1993).
Elliott, J. R., *Ind. Eng. Chem. Res.,* **35:** 1624–1629 (1996).
Elliott, J. R., *Fluid Phase Equilib.,* **416:** 27–41 (2016).
Elliott, J. R., C. T. Lira, *Introductory Chemical Engineering Thermodynamics,* 2nd ed., Prentice-Hall, Englewood Cliffs, NJ, 2012.
Elliott, J. R., C. T. Lira, T. C. Frank, P. M. Mathias, Section 4 in: "Perry's Chemical Engineers' Handbook 9th ed.," D. Green and M. Z. Southard (eds.), McGraw-Hill Education, New York, 2019.
Elliott, J. R., Suresh, S. J., M. D. Donohue, *Ind. Eng. Chem. Res.,* **29:** 1624–1629 (1990).
Ferrarini, F., G. B. Flôres, A. R. Muniz, R. P. D. Soares, *AIChE J.,* **64:** 3443–3455 (2018).
Field, L. R., E. Wilhelm, R. Battino, *J. Chem. Thermodyn.,* **6:** 237 (1974).
Fischer, J. C., *Fluid Phase Equilib.,* **10:** 1–7 (1983).
Flory, P. J., *J. Chem. Phys.,* **9:** 660 (1941).
Fredenslund, A., R. L. Jones, J. M. Prausnitz, *AIChE J.,* **21:** 1086 (1975).
Freshwater, D. C.; Pike, K. A., *J. Chem. Eng. Data,* **12:** 179 (1967).
Friend, L., S. B. Adler: *Chem. Eng. Progr.,* **53:** 452 (1957).
Gil-Villegas, A., A. Galindo, P. J. Whitehead, S. J. Mills, G. Jackson, G., A. N. Burgess, *J. Chem. Phys.,* **106:** 4168 (1997).
Gmehling, J., U. Onken: *Dechema Chemistry Data Series,* **1:** Part 1: 49 and Part 2a: 77 (1977).
Gmehling, J., U. Onken, W. Arlt: *Dechema Chemistry Data Series,* **1:** Parts 3 & 4: 389 (1979).
Gmehling, J., U. Onken, W. Arlt: *Dechema Chemistry Data Series,* **1:** Part 1a: 344 (1981).
Gmehling, J., et al.: *Azeotropic Data*, Wiley-VCH, Weinheim and New York, 1994.
Gnap, M., Elliott, J. R., *Fluid Phase Equil.,* **470:** 241–248 (2018).
Graczová, E., P. Steltenpohl, L. Bálintová, E. Kršačková, *Chem. Pap.,* **58:** 442–446 (2004).
Griswold, J., S. Y. Wong: *Chem. Eng. Progr., Symp. Ser.* **48**(3): 18 (1952).
Gross, J., G. Sadowski, *Ind. Eng. Chem. Res.,* **41:** 5510–5515 (2002).
Gubbins, K. E., *Fluid Phase Equilib.,* **416:** 3–17 (2016).
Guggenheim, E. A., *Proc. Royal Soc. London A*, **148:** 304 (1935).
Guggenheim, E. A., *Mixtures*, Clarendon, Oxford, 1952.
Gupta, R. B., C. Panayiotou, I. C. Sanchez, K. P. Johnston, *AIChE J.,* **38:** 1243–1253 (1992).

Hait, M. J., C. L. Liotta, C. A. Eckert, D. L. Bergmann, A. M. Karachewski, A. J. Dallas, D. I. Eikens, J. J. Li, P. W. Carr, R. B. Poe, and S. C. Rutan, *Ind. Eng. Chem. Res.,* **32:** 2905 (1993).

Hansen, C. M., *J. Paint Technol.,* **39:** 104, 505 (1967).

Hao, Y., C.-C. Chen, *AIChE J.,* **67:** e17061 (2021).

Hayden, J. G., J. P. O'Connell, *Ind. Eng. Chem. Proc. Des. Dev.,* **14:** 3 (1975).

Hayduk, W., S. C. Cheng, *Can. J. Chem. Eng.,* **48:** 93 (1970).

Hayduk, W., W. D. Buckley, *Can. J. Chem. Eng.,* **49:** 667 (1971).

Hayduk, W., H. Laudie, *AIChE J.,* **19:** 1233 (1973).

Heidemann, R. A., *Fluid Phase Equil.,* **14:** 55 (1983).

Hildebrand, J. H., *J. Am. Chem. Soc.,* **51:** 66 (1929).

Hildebrand, J. H., *J. Chem. Phys.,* **15:** 225–228 (1947).

Hildebrand, J. H., R. L. Scott, *Regular Solutions,* Prentice-Hall, Englewood Cliffs, N.J., 1962.

Horsley, L. H., "Azeotropic Data," *Advan. Chem. Ser.,* **6:** (1952), **35:** (1962), **116:** (1973).

Huggins, M. L., *J. Chem. Phys.,* **9:** 440 (1941).

Hwang, Y.-L., G. E. Keller, J. D. Olson, *Ind. Eng. Chem. Res.,* **31:** 1753, 1759 (1992).

Jaubert, J.-N., Y. Le Guennec, A. Pina-Martinez, N. Ramirez-Velez, S. Lasala, B. Schmid, I. K. Nikolaidis, I. G. Economou, R. Privat, *Ind. Eng. Chem. Res.,* **59:** 14981–15027 (2020).

Just, G., *Z. Phys. Chem.,* **37U:** 342–367 (1901).

Kamlet, M. J., R.W. Taft, *J. Am. Chem. Soc.,* **98:** 377 (1976).

Kang, J. W., J. Abildskov, R. Gani, J. Cobas, *Ind. Eng. Chem. Res.,* **41:** 3260–3273 (2002).

Kang, J. W., V. Diky, R. D. Chirico, J. W. Magee, C. Muzny, I. Abdulagatov, A. F. Kazakov, M. Frenkel, *Fluid Phase Equilib.,* **309:** 68–75 (2011).

Kang, J. W., V. Diky, M. Frenkel, *Fluid Phase Equilib.,* **388:** 128–141 (2015).

Kazakov, A., C. D. Muzny, K. Kroenlein, V. Diky, R. D. Chirico, J. W. Magee, I. M. Abdulagatov, M. Frenkel. NIST/TRC SOURCE Data Archival System: The Next-Generation Data Model for Storage of Thermophysical Properties. *Int. J. Thermophys.,* **33:** 22–33 (2012).

Kiselev, S. B., J. F. Ely, J. R. Elliott, *Mol. Phys.,* **104:** 2545–2559 (2006).

Kister H. Z., J. R. Haas, D. R. Hart, D. R. Gill., *Distillation Design*, McGraw-Hill, New York, 1992.

Klamt, A., *J. Phys. Chem.,* **99:** 2224–2235 (1995).

Klamt, A., V. Jonas, T. Burger, J. C. W. Lohrenz, *J. Phys. Chem.,* **102:** 5074–5085 (1998).

Klamt, A., G. J. P. Krooshof, R. Taylor, *AIChE J.,* **48:** 2332–2349 (2002).

Knapp, H., R. Döring, L. Oellrich, U. Plöcker, J. M. Prausnitz, *Chemistry Data Series,* Vol. VI: *VLE for Mixtures of Low Boiling Substances,* D. Behrens and R. Eckerman (eds.), DECHEMA, Frankfurt a. M., 1982.

Kontogeorgis, G. M., E. C. Voutsas, I. V. Yakoumis, D. P. Tassios, *Ing. Eng. Chem. Res.,* **35:** 4310–4318 (1996).

Kontogeorgis, G. M., G. Folas, *Thermodynamic Models for Industrial Applications,* John Wiley & Sons, West Sussex, U.K., 2010.

Langmuir, I., *Colloid Symposium Monograph,* **3:** 48 (1925).

Larsen, B. L., P. Rasmussen, *Fluid Phase Equil.,* **28:** 1–11 (1986).

Larsen, B. L., P. Rasmussen, Aa. Fredenslund, *Ind. Eng. Chem. Res.,* **26:** 2274 (1987).

Lazzaroni, M. J., D. Bush, C. A. Eckert, T. C. Frank, S. Gupta, J. D. Olson, *Ind. Eng. Chem. Res.,* **44:** 4075 (2005).

Le Guennec, Y., R. Privat, J.-N. Jaubert, *Fluid Phase Equil.,* **429:** 301–312 (2016).

Lee, K.-H., S. I. Sandier, P. A. Monson, *Int. J. Thermophys.,* **7:** 367–379 (1986).

Levelt Sengers, J. M. H., in Kiran, E., J. M. H. Levelt Sengers (eds.), *Supercritical Fluids: Fundamentals for Applications,* NATO ASI Series E: **273:** Kluwer Academic Publishers, Dordrecht, Holland, 1994, p. 1.

Lin, S.-T., S. I. Sandler, *Ind. Eng. Chem. Res.,* **41:** 899–913 (2002).

Linstrom, P. J., W. G. Mallard, *NIST Chemistry WebBook*, NIST Standard Reference Database Number 69, Eds. National Institute of Standards and Technology, Gaithersburg MD, 20899, (retrieved April 13, 2022).

Lyckman, E. W., C. A. Eckert, J. M. Prausnitz, *Chem. Eng. Sci.,* **20:** 685 (1965).

Mansoori, G. A., N. F. Carnahan, K. E. Starling, T. W. Leland, Jr., *J. Chem. Phys.,* **54:** 1523 (1971).

Marinichev, A. N., M. P. Susarev, *Zh. Prikl. Khim.,* **38:** 1054 (1965).

Mathias, P. M., J. P. O'Connell: Chapter 5 in: "Equations of State in Engineering and Research," K. C. Chao and R. L. Robinson (eds.), *Adv. Chem. Ser.,* **182:** 97 (1979).

Maurer, G., J. M. Prausnitz, *Fluid Phase Equil.* **2:** 91 (1978).

Mertl, I., *Coll. Czech. Chem. Commun.,* **37:** 366 (1972).

Michelsen, M. L., *Fluid Phase Equil.,* **9:** 21 (1982).

Michelsen M. L., J. Mollerup, *Thermodynamic Modelling: Fundamentals and Computational Aspects*, 2nd ed., Tie-Line Publications; 2007.

Mollerup J., *Fluid Phase Equil.,* **7:** 121–138 (1981).

Mullins, E., R. Oldland, Y. A. Liu, S. Wang, S. I. Sandler, C.-C. Chen, M. Zwolak, K. C. Seavey, *Ind. Eng. Chem. Res.,* **45:** 4389–4415 (2006).

Murti, P. S., M. van Winkle, *Chem. Eng. Data Ser.,* **3:** 72 (1978).

Nguyen-Huynh, D., J.-C. de Hemptinne, R. Lugo, J.-P. Passarello, P. Tobaly, *Ind. Eng. Chem. Res.,* **50:** 7467–7483 (2011).

O'Connell, J. P., in Kiran, E., J. M. H. Levelt Sengers (eds.), '*Supercritical Fluids: Fundamentals for Applications*,' NATO ASI Series E: **273:** Kluwer Academic Publishers, Dordrecht, Holland, 1994, p. 191.

O'Connell, J. P., J. M. Haile, *Thermodynamics: Fundamentals for Applications,* Cambridge University Press, New York, 2005.

Orentlicher, M., J. M. Prausnitz, *Chem. Eng. Sci.,* **19:** 775 (1964).

Orkoulas, G. E., A. Z. Panagiotopoulos, *Phys. Rev. E,* **63:** 051507 (2002).

Panayiotou, C., I. Tsivintzelis, I. G. Economou, *Ind. Eng. Chem. Res.,* **46:** 2628 (2007).

Panayiotou, C., I. C. Sanchez, *J. Phys. Chem.,* **95:**10090 (1991).

Panayiotou, C., J. H. Vera, *Fluid Phase Equil.,* **5:** 55–80 (1980).

Panayiotou, C, S. Mastrogeorgopoulos, V. Hatzimanikatis, *J. Chem. Thermo.,* **110:** 3 (2017).

Paricaud, P., A. Galindo, G. Jackson, *Mol. Phys.,* **101:** 2575–2600 (2003).

Paulechka, E., V. Diky, A. Kazakov, K. Kroenlein, M. J. Frenkel, *J. Chem. Eng. Data,* **60:** 3554–3561 (2015).

Peng, D. Y., D. B. Robinson: *Ind. Eng. Chem. Fundam.,* **15:** 59 (1976).

Peper, S., J. M. S Fonseca, R. Dohrn, *Fluid Phase Equil.,* **484:** 126–224 (2019).

Peres, A. M., E. A. Macedo, *Fluid Phase Equil.,* **139:** 47 (1997).

Perry, R. L., J. C. Telotte, J. P. O'Connell, *Fluid Phase Equil.,* **5:** 245–277 (1981).

Phifer, J. R., C. E. Cox, L. F. da Silva, G. C. Nogueira, A. Karolyne, P. Barbosa, R. T. Ley, Samantha M. Bozada, Elizabeth J. O'Loughlin, A. S. Paluch, *Mol. Phys.,* **115:** 1286–1300 (2017).

Plyasunov, A., J. P. O'Connell, R. H. Wood, *Geochimica et Cosmochimica Acta,* **64:** 495 (2000).

Plyasunov, A., V. S. Korzhinskaya, J. P. O'Connell, *Fluid Phase Equilib.,* **498:** 9–22 (2019).

Prange, M. M., H. H. Hooper, J. M. Prausnitz, *AIChE J.,* **35:** 803–813 (1989).

Prausnitz, J. M., R. N. Lichtenthaler, E. G. Azevedo, *Molecular Thermodynamics of Fluid-Phase Equilibria,* 3rd ed., Prentice Hall, Englewood Cliffs, N.J., 1999.

Prausnitz, J. M., T. A. Anderson, E. A. Grens, C. A. Eckert, R. Hsieh, J. P. O'Connell: *Computer Calculations for Multicomponent Vapor-Liquid and Liquid-Liquid Equilibria,* Prentice-Hall, Englewood Cliffs, N.J., 1980.

Prausnitz, J. M., F. H. Shair, *AIChE J.,* **7:** 682 (1961).

Reamer, H. H., B. H. Sage, W. N. Lacey, *Ind. Eng. Chem.,* **43:** 2515 (1951).

Renon, H., J. M. Prausnitz, *Ind. Eng. Chem. Process Design Develop.,* **8:** 413 (1969).

Sanchez, I. C., R. Lacombe, *Macromolecules,* **11:** 1145 (1978).

Scatchard, G., *Chem. Rev.,* **8:** 321–333 (1931).

Scatchard, G., S. E. Wood, J. M. Mochel, *J. Am. Chem. Soc.,* **68:** 1957–1960 (1946).

Sedlbauer, J., J. P. O'Connell, R. H. Wood, *Chem. Geology,* **163:** 43 (2000).

Sharygin, A. V., J. P. O'Connell, R. H. Wood, *Ind. Eng. Chem. Res.,* **35:** 2808 (1996).

Sinor, J. E., J. H. Weber, *J. Chem. Eng. Data,* **5:** 243 (1960).

Soares, R. D. P., R. P. Gerber, L. F. K. Possani, P. B. Staudt, *Ind. Eng. Chem. Res.,* **52:** 11172–11181 (2013).

Soares, R. D. P., L. F. Baladao, P. B. Staudt, *Fluid Phase Equilib.,* **488:** 13–26 (2019).

Soares, R.D. P., P. B. Staudt, *Fluid Phase Equilib.,* 113611 (2022)

Soave, G., *Chem. Eng. Sci.,* **27:** 1197 (1972).

Staverman, A. J., *Rec. Trav. Chem. Pays-Bas.,* **69:** 163 (1950).

Tang, Y-P., G. Stephenson, E. Zhao, M. Agrawal, S. Saha, *AIChE J.,* **58:** 591 (2012).

Tiepel, E. W., K. E. Gubbins, *Ind. Eng. Chem. Fundam.,* **12:** 18 (1973); *Can J. Chem. Eng.,* **50:** 361 (1972).

Unlu, O., N. H. Gray, Z. N. Gerek, J. R. Elliott, *Ind. Eng. Chem. Res.,* **43:** 1788 (2004).

Vahid, A. V., N. H. Gray, J. R. Elliott, *Macromolecules,* **47:** 1514 (2014).

van Konynenburg, P. H., R. L. Scott, *Phil. Trans. Roy. Soc. London, Ser. A,* **298** (1442): 495 (1980).

Van Ness, H. C., M. M. Abbott,. *Classical Thermodynamics of Nonelectrolyte Solutions,* McGraw-Hill Book Co., New York, 1982.

Vera, J. H., S.G. Sayegh, G. A. Ratcliff, *Fluid Phase Equil.,* **1:** 113 (1977).

Vera, J. H., G. Wilczek-Vera, *Fluid Phase Equil.,* **59:** 15 (1990).

Veytsmann, B. A., *J. Phys. Chem.,* **94:** 8499 (1990).
Wertheim, M. S., *J. Stat. Phys.,* **35:** 19 (1984), *ibid.* **35:** 35 (1984).
Wheeler, J. C., G. R. Andersen, *J. Chem. Phys.,* **73:** 5778 (1980).
Wilson, G. M., C. H. Deal, *Ind. Eng. Chem. Fundam.,* **1:** 20 (1962).
Wilson, G. M., *J. Am. Chem. Soc.,* **86:** 127, 133 (1964).
Wohl, K., *Chem. Eng. Progr.,* **49:** 218 (1953).
Yen, L., J. J. McKetta, *AIChE J.,* **8:** 501 (1962).
Zemaitis, J.F., D.M. Clark, M. Rafal, N.C. Scrivner, Handbook of aqueous electrolyte thermodynamics: Theory & application, Wiley, New York (2010).

9

Specialized Phase Behavior in Mixtures

9.1 SCOPE*

Vapor-liquid equilibrium (VLE) in mixtures is strongly influenced by the vapor pressures of the pure components, especially when one compound is highly concentrated. This means that the bubble pressure is relatively insensitive to infinite dilution activity coefficients (IDACs). For liquid-liquid equilibria (LLE) and solid-liquid equilibria (SLE), however, the vapor pressures play an insignificant role, and the mixtures are dilute in most cases. Since activity models and equations of state are imperfect in their characterizations of fugacity, accuracy in the midrange of compositions is no guarantee of accuracy for dilute components. Therefore, we focus this chapter on IDACs, LLE, and SLE prepared for the possibility that findings in these applications may not entirely coincide with findings for VLE, although the thermodynamic constraints and model equations are the same.

As was the case for VLE, equality of the fugacities of all components in all phases provides the thermodynamic constraints for LLE and SLE. On the other hand, LLE and SLE are relatively insensitive to pressure, so bubble pressure computations are not helpful. For LLE, a binary flash computation is required, iterating both phase compositions until the equilibrium constraints are satisfied. These tend to be more difficult to converge than bubble pressure computations. For SLE, the solid phase is typically assumed to be pure, leading to iteration on the liquid composition at a given temperature and convergence is usually robust.

IDACs are useful in characterizing deviations from ideality with a single value per solute-solvent combination. Contrary to the Margules one-parameter model, these coefficients are not equivalent when solute and solvent are swapped. As discussed in Sec. 8.7, many of the details of activity models and EOSs can be set aside at infinite dilution, making the key sources of nonideality more readily apparent. This can be very useful if an engineering objective is to manipulate the nonideality to improve process outcomes. For example, adding a cosolvent or antisolvent would be one means of altering the solution nonideality. Moreover, IDACs can be closely related to LLE when mutual solubilities are low. In those cases, $x_i^\infty \approx 1/\gamma_i^\infty$, where x_i^∞ is the mole fraction of the dilute component and γ_i^∞ is the IDAC. This observation is a major motivation for treating IDACs, LLE, and SLE in one cohesive chapter.

Altogether, both the thermodynamic computations and the correlation capability are significantly more challenging in this chapter than for VLE. While the topics are more challenging, however, the opportunities for improved models are greater.

*With special contributions from V. Diky.

9.2 INFINITE DILUTION ACTIVITY COEFFICIENTS

Experimental activity coefficients at infinite dilution can be useful for calculating the parameters needed in an expression for the excess Gibbs energy (Table 8.2). In a binary mixture, suppose experimental data are available for IDACs γ_1^∞ and γ_2^∞. These can be used to evaluate two adjustable constants in any desired expression for g^E. For example, consider the van Laar equation

$$g^E = Ax_1x_2\left(x_1\frac{A}{B} + x_2\right)^{-1} \tag{9.2-1}$$

Taking derivatives as indicated in Sec. 8.6 gives

$$RT \ln \gamma_1 = A\left(1 + \frac{A}{B}\frac{x_1}{x_2}\right)^{-2} \tag{9.2-2}$$

and

$$RT \ln \gamma_2 = B\left(1 + \frac{B}{A}\frac{x_2}{x_1}\right)^{-2} \tag{9.2-3}$$

In the limit as $x_1 \to 0$ or as $x_2 \to 0$, Eqs. (9.2-2) and (9.2-3) become

$$RT \ln \gamma_1^\infty = A \tag{9.2-4}$$

$$RT \ln \gamma_2^\infty = B \tag{9.2-5}$$

The van Laar equation is closely related to the Scatchard (1931)-Hildebrand (1929) (SH) model where the volume ratio has been treated as an adjustable parameter (cf. Elliott and Lira, 2012, Chap. 12). Alternatively, if one solubility parameter is unknown (or both), the terms involving solubility parameter can be lumped together as an adjustable parameter to obtain:

$$\ln\left(\gamma_k^\infty\right) = (1 - \Phi_k)^2 A \tag{9.2-6}$$

The advantage of Eq. (9.2-6) over the van Laar equation is that it has only one parameter. The advantage over the M1 model is that it represents the effect of volume ratio by applying volume fraction instead of mole fraction. Molecular volume can be estimated using the methods of Chap. 3. Calculation of parameters from γ^∞ data is particularly simple for the van Laar equation, or similarly for the Margules two-parameter model, but in principle, similar calculations can be made by using any two-parameter equation for the excess Gibbs energy. If a three-parameter equation, e.g., NRTL (Renon and Prausnitz, 1969), is used, an independent method must be chosen to determine the third parameter α_N. A convenient correlation for α_N is described in Sec. 9.3.

Relatively simple experimental methods have been developed for rapid determination of IDACs, and access to these values can be quite useful. Experimental methods are based on gas-liquid chromatography and on ebulliometry. Schreiber and Eckert (1971) have shown that if reliable values of γ_1^∞ and γ_2^∞ are available, either from direct experiment or from a correlation, it is possible to predict VLE over the entire range of composition with fair accuracy. They reported roughly 2% deviations in vapor composition whereas optimal parameters provided roughly 1% deviations. Although LLE and SLE tend to be more sensitive than VLE to the parameter values, their applications often operate in dilute composition ranges, in which cases IDACs may also suffice.

IDACs have been reported and compiled by numerous sources. One readily available source was reported by Lazzaroni et al. (2005) as supplemental material for their MOSCED update. That resource includes roughly 5000 IDAC values, but it is relatively sparse for aqueous systems. Yaws (2012) provides a database with aqueous solubilities of over 17,000 organic compounds and water solubility in over 7600 organic compounds at temperatures generally near ambient. When these solubilities are less than 2 wt% for liquid compounds at ambient temperature, the M1 model can be used to estimate the IDAC as

$$\ln\left(\gamma_k^{\infty}\right) = \frac{1}{1-2x_k^{\delta}} \ln\left[\frac{\left(1-x_k^{\delta}\right)}{x_k^{\delta}}\right] \tag{9.2-7}$$

Where x_k^{δ} is the mole fraction of the dilute solute k, and γ_k^{∞} is the IDAC. This formula was tested by comparing to IDAC values computed using volume fractions and the Flory-Huggins parameter. Both methods were consistent to 0.1% in $\ln(\gamma_k^{\infty})$ when $b_k/b_s > 0.1$, where b_k and b_s are the solute and solvent molecular volumes, respectively. This accuracy holds despite the implicit assumption that $b_k/b_s = 1$ in the derivation of the M1 model. For common organic compounds, $b_k/b_s \approx M_k/M_s$ suffices to assess whether $b_k/b_s > 0.1$, where M_k and M_s are the solute and solvent molar masses. Roughly 60% of organic compounds in the Yaws database have aqueous solubilities less than 2 wt%. Additionally, Moine et al. (2017) have published a database of over 69,000 values of solvation free energies. The Moine database includes original source citations over a range of temperatures with multiple values when reported by multiple sources. Gibbs energies of solvation can be converted to IDACs as shown in Table 9.1. The key equation for the conversion is

$$\ln\left(\gamma_i^{\infty}\right) = \frac{RT\rho_{\text{pure},j,\text{liq}}}{f_{\text{pure},i,\text{liq}}} \exp\left(\frac{\Delta_{\text{solv}} g_i^{\infty}}{RT}\right) \tag{9.2-8}$$

where $f_{\text{pure},i,\text{liq}} = P^{vp} \cdot \exp(2B_2P^{vp}/RT)/(1 + B_2P^{vp}/RT)$ if $P < 1000$ kPa, for example.

We compiled data from these three sources as they pertain to the organic compounds in Appendix A, then we performed an informal analysis to eliminate obvious outliers. For non-aqueous systems, the Lazzaroni database was used except for a few values that appeared to be outliers when compared to the Thermodynamics Data Engine (TDE) database. For aqueous systems, the analysis comprised comparing the reported IDACs to each other and to the TDE database. If any two of the sources were consistent to within 40%LD, where %LD ≡ 100%·$\ln(\gamma_{\text{base1}}^{\infty}/\gamma_{\text{base2}}^{\infty})$, a value was included in the database. The priority was to include the value of Moine et al. first. If no consistent value was available from Moine et al., the value from Yaws was included, but the source citation from the TDE database was added. The resulting database included 3272 values for organic compounds in water and 519 values for water in organic solvents. The average inconsistency for all aqueous data was 10%AALD, where %AALD is defined in Chapter 1.

TABLE 9.1 Sample conversion of solvation free energy to IDAC for water (solute) in benzene (solvent). Experimental data of Cooling et al. (1992).

T(K)	293.15	308.15	323.15
ΔG, kcal/mol	−2.25	−2.07	−1.83
$\exp(\Delta G/RT)$	0.02102	0.03403	0.05786
Solute P^{vp}/kPa	2.339	5.629	12.352
Solute B_2(L/mol)	−1.2507	−0.98973	−0.80171
Solute φ^{sat}	0.9988	0.9978	0.9963
Solute f^{sat}	2.336	5.617	12.307
Solvent MW, g/mol	78.114	78.114	78.114
Solvent density, kg/m^3	878.80	863.01	846.96
IDAC (Eq. 9.2-8)	246.7	171.5	137.0
IDAC (Cooling et al.)	245.4	170.0	135.5

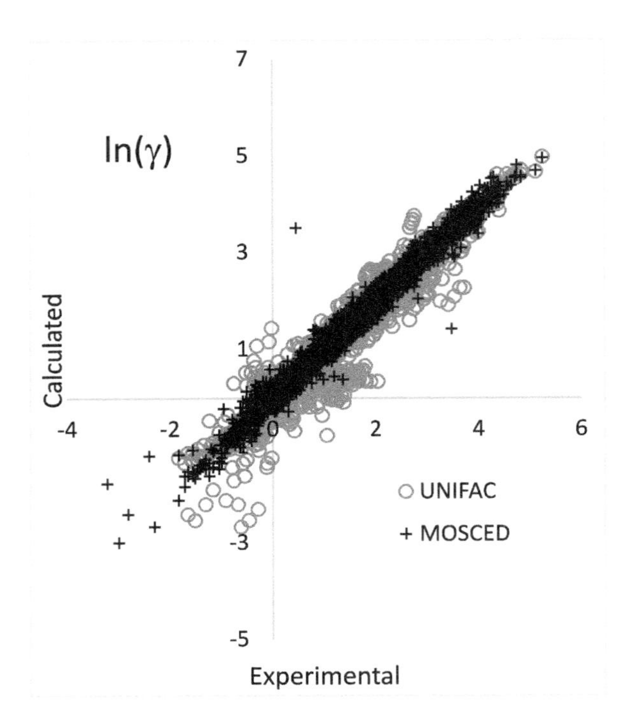

FIGURE 9.1
Parity plot comparing accuracy of MOSCED to original UNIFAC for IDACs in the database of Lazzaroni et al. (2005).

One correlation for IDACs in binary systems has been presented by Pierotti et al. (1959). This correlation can be used to predict IDACs for water, hydrocarbons, and typical organic compounds, e.g., esters, aldehydes, alcohols, ketones, nitriles, at temperatures of 25, 60, and 100 °C. The pertinent equations and tables are summarized by Treybal (1963) and, with slight changes, are reproduced in PGL5ed. The method is based on broad classifications such as aldehyde solutes in *n*-alcohol solvents. As such, many combinations of classes are uncovered, and many temperatures require interpolation. The accuracy of the correlation varies considerably from one system to another and is unreliable when IDAC > 10. Noting that ln(10) = 2.3 and referring to Fig. 9.1, IDAC < 10 covers a relatively small range. For aqueous systems, such a restriction is even more serious. We have not performed a detailed analysis of that method here.

9.2.1 Kamlet-Taft and Solvatochromic Methods

One of the most serious limitations of the majority of the g^E expressions such as Wilson, NRTL, and UNIQUAC, and thus of the various versions of UNIFAC for γ^∞ estimation is the absence of any explicit accounting for strong interactions, such as hydrogen bonds. Since many separation processes seek specifically to take advantage of these interactions, such as extraction or extractive distillation, it is useful to have available methods for property estimation that account for strong, specific interactions. Perhaps the most useful of these is the method of solvatochromism (Kamlet and Taft, 1976; Kamlet et al., 1983, Abraham, 1993).

The basic Kamlet-Taft (1976) multiparameter approach gives any configurational property *XYZ* in terms of the sum of an intercept XYZ_o, a cavity-formation term related to the energy required to make a cavity in the solvent large enough to accommodate a solute molecule, and a term summing the solvent-solute intermolecular interactions.

$$XYZ = XYZ_o + \text{cavity formation term} + \sum(\text{solvent-solute interactions}) \qquad (9.2\text{-}9)$$

Most often this is expressed in terms of parameters π^* (polarity/polarizability), α^{KT} (acidity or hydrogen-bond donor strength), and β^{KT} (basicity or hydrogen-bond acceptor strength). In its simplest form, it is

$$XYZ = (XYZ)_0 + s\pi^* + a\alpha^{KT} + b\beta^{KT} \tag{9.2-10}$$

XYZ	a solvent-dependent physicochemical property such as $\ln \gamma^\infty$
$(XYZ)_0$	the property in the gas phase or in an inert solvent
s, a, b	relative susceptibilities of the property XYZ to the solvent parameters
π^*	dipolarity/polarizability scale (dispersive, inductive, and electrostatic forces)
α^{KT}	hydrogen-bond donor (HBD)/electron-pair acceptor (EPA) scale
β^{KT}	hydrogen-bond acceptor (HBA)/electron-pair donor (EPD) scale

Equation (9.2-10) is an example of a linear solvation-free energy relationship (LSER) and is successful for describing a wide variety of medium-related processes, including γ^∞ in ambient water over more than six orders of magnitude variation (Sherman et al., 1996) and extending to such diverse applications as predictions of dipole moments, fluorescence lifetimes, reaction rates, NMR shifts, solubilities in blood, and biological toxicities (Kamlet et al., 1988; Carr, 1993; Taft et al., 1985). Tables are available in the literature for the parameters of many solvents. For example, Abraham (1993) provides parameters for over 400 solvents. Solute parameters sometimes vary somewhat from those for the substance as a solvent, as the solute parameter represents the forces for a single molecule, and the solvent parameters for the aggregate. For example, the acidity of a hydrogen-bonded alcohol is different from that for an unbonded alcohol. Table 9.2 gives the best current values of these parameters for samples of 10 solvents. In Table 9.2, π^{*KT}, α^{KT}, and β^{KT} represent values when a substance is in the solvent state while π^{*H}, α^H, and β^H are values in the solute state.

Often additional parameters are used for various applications. For example, Diorazio et al. (2016) have developed a solvent selection tool that combines several descriptors of 272 compounds to suggest prospective solvents for green chemistry and pharmaceutical applications. Another example is the prediction of the Henry's law constant H_v^{px} for a solute (2) in ambient water (1) at 25 °C (Sherman et al., 1996):

$$\ln(H_v^{px}) = -0.536\log\left(L_2^{16}\right) - 5.508\ \pi_2^{*H} - 8.251\ \alpha_2^H - 10.54\ \beta_2^H - 1.598\left[\ln\left(\left(\frac{V_2}{V_1}\right)^{0.75}\right) + 1 - \left(\frac{V_2}{V_1}\right)^{0.75}\right] + 16.10 \tag{9.2-11}$$

where L^{16} is the hexadecane-air partition coefficient, usually measured by gas chromatography retention on a hexadecane column (Table 9.2), and V is the molar volume (Table 9.2). Hexadecane is a convenient solvent for characterizing solutes in that it is easy to run as a stationary phase in a gas chromatograph, and it has no polar or hydrogen-bonding interactions. L^{16} gives a good measure of the cavity term plus the dispersive interactions. Then values for γ^∞ may be found from

$$\ln\left(\gamma_2^\infty\right) = \ln\left(H_v^{px}\right) - \ln\left(f_2^o\right) \tag{9.2-12}$$

where f_2^o is the reference-state fugacity.

Example 9.1 Inference of IDACs from solvatochromic parameters.

To illustrate application of this technique, we calculate γ_2^∞ for benzene in water (1) at 25 °C.

Solution. For water, $V_1 = 18$ and from Table 9.2, for benzene

$$L^{16} = 2.786;\ \pi_2^{*H} = 0.52;\ \alpha_2^H = 0;\ \beta_2^H = 0.14;\ V_2 = 89.4$$

Then, $H_v^{px} = 174 \times 10^3$ torr. Using the vapor pressure at 25 °C, 95.14 torr, for f_2^o, we get $\gamma_2^\infty = 1830$. The experimental value is 2495 (Li et al., 1993).

TABLE 9.2 Solvatochromic parameters for sample solvents and solutes (Abraham, 1993).

Name	RI	V	π^{*KT}	π^{*H}	α^{KT}	α^{H}	β^{KT}	β^{H}	log L^{16}
n-hexane	1.37226	131.6	–0.04	0.00	0.00	0.00	0.00	0.00	2.668
n-heptane	1.38511	147.5	–0.01	0.00	0.00	0.00	0.00	0.00	3.173
n-octane	1.39505	163.5	0.01	0.00	0.00	0.00	0.00	0.00	3.677
2,2,4-trimethylpentane	1.38898	166.1	–0.04	0.00	0.00	0.00	0.00	0.00	3.106
diethylamine	1.38250	104.3	0.24	0.30	0.03	0.08	0.70	0.68	2.395
triethylamine	1.39800	140.0	0.14	0.15	0.00	0.00	0.71	0.79	3.040
benzene	1.49792	89.4	0.59	0.52	0.00	0.00	0.10	0.14	2.786
toluene	1.49413	106.9	0.54	0.52	0.00	0.00	0.11	0.14	3.325
1,4-dimethylbenzene	1.49325	123.9	0.43	0.52	0.00	0.00	0.12	0.16	3.839
ethylbenzene	1.49320	123.1	0.53	0.51	0.00	0.00	0.12	0.15	3.778
phenol	1.55090	87.8	0.72	0.89	1.65	0.60	0.30	0.31	3.766
cyclohexane	1.42354	108.8	0.00	0.10	0.00	0.00	0.00	0.00	2.964
tetrahydrofuran	1.40496	81.9	0.58	0.52	0.00	0.00	0.55	0.48	2.636
diethylether	1.34954	104.7	0.27	0.25	0.00	0.00	0.47	0.45	2.015
methylacetate	1.35890	79.8	0.60	0.64	0.00	0.00	0.42	0.45	1.911
ethylacetate	1.36978	98.5	0.55	0.62	0.00	0.00	0.45	0.45	2.314
butylacetate	1.39180	132.6	0.46	0.60	0.00	0.00	0.45	0.45	3.353
dichloromethane	1.42115	64.5	0.82	0.57	0.13	0.10	0.10	0.05	2.019
chloroform	1.44293	80.7	0.58	0.49	0.20	0.15	0.10	0.02	2.480
chlorobenzene	1.52185	102.3	0.71	0.65	0.00	0.00	0.07	0.07	3.657
acetone	1.35596	74.1	0.71	0.70	0.08	0.04	0.43	0.51	1.696
2-butanone	1.37685	90.2	0.67	0.70	0.06	0.00	0.48	0.51	2.287
butyraldehyde	1.37660	90.5	0.63	0.65	0.00	0.00	0.41	0.45	2.270
acetonitrile	1.34163	52.9	0.75	0.90	0.19	0.04	0.40	0.33	1.739
propionitrile	1.36360	70.9	0.71	0.90	0.00	0.02	0.39	0.36	2.082
butyronitrile	1.38200	87.9	0.71	0.90	0.00	0.00	0.40	0.36	2.548
ethanol	1.35941	58.7	0.54	0.42	0.86	0.37	0.75	0.48	1.485
1-propanol	1.38370	75.2	0.52	0.42	0.84	0.37	0.90	0.48	2.031
1-butanol	1.39741	92.0	0.47	0.42	0.84	0.37	0.84	0.48	2.601
1-octanol	1.42760	158.5	0.40	0.42	0.77	0.37	0.81	0.48	4.619
2-propanol	1.37520	76.9	0.48	0.36	0.76	0.33	0.84	0.56	1.764
methanol	1.32652	40.8	0.60	0.44	0.98	0.43	0.66	0.47	0.970
carbontetrachloride	1.45739	97.1	0.28	0.38	0.00	0.00	0.10	0.00	2.823
1,4-dioxane	1.42025	85.7	0.55	0.75	0.00	0.00	0.37	0.64	2.892

9.2.2 The MOSCED Method

The solvatochromic technique has been coupled with modifications of the Hildebrand solubility parameter to estimate IDACs for nonionic liquids (other than water) near 25 °C. These methods differ substantially from the UNIFAC methods as they are not made up of group contributions but rather reflect measurements or estimates of molecular properties. In other words, these methods sum contributions to the cohesive energy density by different types of contributions. Examples of this approach have been presented by Hansen (1967), Hait et al. (1993), and Lazzaroni et al. (2005). The publication of Lazzaroni et al. pertains to a method called MOSCED which originated with work by Eckert and coworkers in the 1980s. The publication by Hait et al. pertains to a method called SPACE developed in the 1990s, also by Eckert and coworkers. The SPACE method is designed to use fewer adjustable parameters than MOSCED. Both methods are similar in that they relate the IDAC to dispersion, polarity, acidity, and basicity parameters. The SPACE method relates these parameters to refractive index and the Kamlet-Taft parameters, providing correlations for several homologous series. The MOSCED method is somewhat simpler to apply because it takes the parameters as tabulated whereas SPACE adds several steps to the computations. A significant limitation of the MOSCED method is that Lazzaroni et al. (2005) tabulated parameters for only 132 compounds. The SPACE method can be viewed as one method of correlating those parameters to expand the number of compounds covered, although there are small differences that preclude an exact correspondence. Additionally, Paluch and coworkers have presented methods for the estimation of MOSCED parameters from quantum DFT (cf. Diaz-Rodrigues et al., 2016). Similarly, Gnap and Elliott (2019) presented correlations for MOSCED parameters from the COSMO σ^c-profiles of Mullins et al. (2006). Supplementary material of Gnap and Elliott provides predictive estimates of MOSCED parameters for over 1300 compounds. We describe the MOSCED method below.

The MOSCED formulation for γ_2^∞ in solvent 1 is

$$\ln\left(\gamma_2^\infty\right) = \frac{V_2}{RT}\left[\left(\lambda_2 - \lambda_1\right)^2 + q_1^2 q_2^2 \frac{\left(\tau_2^T - \tau_1^T\right)^2}{\psi_1} + \frac{\Delta\alpha\Delta\beta}{\xi_1}\right] + d_{12} \tag{9.2-13}$$

$$\Delta\alpha\Delta\beta = \left(\alpha_2^T - \alpha_1^T\right)\left(\beta_2^T - \beta_1^T\right);\ \alpha_i^T = \alpha_i^M\left(\frac{293}{T(K)}\right)^{0.8};\ \beta_i^T = \beta_i^M\left(\frac{293}{T(K)}\right)^{0.8};\ \tau_i^T = \tau_i^M\left(\frac{293}{T(K)}\right)^{0.4}$$

$$\psi_1 = POL + 0.002629\alpha_1^T\beta_1^T;\ POL = 1 + 1.15q_1^4\left[1 - \exp\left(-0.002337\left(\tau_1^T\right)^3\right)\right]$$

$$\xi_1 = 0.68(POL - 1) + \left[3.24 - 2.4\exp\left(-0.002687\left(\alpha_1\beta_1\right)^{1.5}\right)\right]^{\left(\frac{293}{T}\right)^2}$$

$$d_{12} = 1 - \left(\frac{V_2}{V_1}\right)^{aa} + aa\ln\left(\frac{V_2}{V_1}\right);\ aa = 0.953 - 0.002314\left(\left(\tau_2^T\right)^2 + \alpha_2^T\beta_2^T\right)$$

where $R = 8.314$ J/mole-K, T is the temperature in Kelvins, V_i is the liquid molar volume at 25 °C, λ_i^M is the dispersion parameter, τ_i^M is the polarity parameter, α_i^M is the acidity parameter, β_i^M is the basicity parameter, q_i^M is a factor ranging between 0.9 and 1.

The functional form of the MOSCED model derives from Flory-Huggins theory. The key quantity is the one involving $\Delta\alpha\Delta\beta/\xi_1$ in Eq. (9.2-13). When $(\alpha_2^T - \alpha_1^T)(\beta_2^T - \beta_1^T)$ is negative, dispersion and polarity differences are often moderate, and the activity coefficient may be less than 1. The contribution of ξ_1 is also very important, as it impacts the activity coefficient quite significantly at alternative ends of the composition range. Although $\Delta\alpha\Delta\beta/\xi_1$ comprises an explicit term for hydrogen bonding, its form is not consistent with the form expected from TPT1 or QCT as described in Sec. 8.6. In TPT1 and QCT, the bonding energy appears in an exponential. Also, the functional form of the IDAC changes significantly when one component is associating and the other is not. Thus, the functional form of the MOSCED model is phenomenological. Insights from various applications could be improved if this model for IDAC could be put on a consistent basis with the most advanced models.

Example 9.2 Inference of IDACs using the MOSCED method.

Use MOSCED to calculate the IDAC at 25 °C for methanol (1) in 2-nitropropane (2).

Solution. Referring to Lazzaroni et al. (2005), the required parameters are tabulated below:

Comp	V^L	λ^M	τ^M	α^M	β^M	q^M	τ^T	α^T	β^T
1	40.6	14.43	2.53	17.43	14.49	1	2.51	17.2	14.3
2	90.6	14.6	8.3	0.55	3.43	1	8.24	0.54	3.38

Where τ^T, α^T, and β^T have been computed with their temperature dependence. Further applying the defining equations below Eq. (9.2-13),

aa	d_{12}	POL	y_2	x_2	$\ln(\gamma_1^\infty)$	γ_1^∞
0.370	–0.040	1.840	1.844	1.432	2.340	10.30

Thomas et al. (1982) report an experimental value at 20 °C of 8.35. The correction for a temperature difference of 5 °C is likely to be small, so the error is roughly 22%LD.

When parameters are available, the MOSCED method performs quite well in our evaluations for binary mixtures of organic compounds. Figure 9.1 shows a parity plot comparing MOSCED with original UNIFAC for nonaqueous systems. Gnap and Elliott (2019) reported that average deviations with UNIFAC are about double those with MOSCED, and deviations with COSMO-RS/SAC are roughly four times larger. Similar deviations are observed in Sec. 9.5.

9.2.3 IDACs for Aqueous Systems

Aqueous systems are quite common in chemical processing. Typically, the values of interest are near ambient conditions, making the Yaws compilations especially relevant. While Yaws provides correlations for estimating IDACs when not experimentally available, the correlations are based on broad classes of compounds with individual parameters for each class, like the correlation of Pierotti et al. Owing to ambiguities for multi-functional compounds and similar issues, these kinds of correlations are difficult to automate reproducibly. The Yaws correlations are further complicated by their dependence on knowledge of the normal boiling temperature. When experimental values of the boiling temperature are unavailable, they must be predicted using methods from Chap. 4 and this leads to increased uncertainty. Several alternatives are available that are related to activity models like MOSCED, UNIFAC, and COSMO-RS/SAC. With the advent of machine learning, methods based on semiempirical descriptors like zero-point energy and dipole moment have become feasible and more common.

Interest in IDACs of organic compounds in water is motivated by environmental considerations for obvious reasons, but also by drug solubility. Aqueous solubility is a primary target for optimization in drug discovery (Wang et al., 2009). Substantial literature has developed to address this problem. Although the biological literature favors expressing solubility in terms of $\log S \equiv \log_{10}(\text{mol/L})$, conversion to consistent units is straightforward. The biochemistry literature also favors fundamental descriptors over the kinds of group contribution descriptors more familiar to chemical engineers. Altogether, the biochemistry literature can provide a rich resource for methods and data relevant to IDACs. Unfortunately, the accuracy of measurements from the biochemistry literature raises questions. According to Boobier et al. (2020), "...the expected level of noise in the training data [is] ($\log_{[10]} S \pm 0.7$)..." This level of "noise" corresponds to a factor of 5, or a

%AALD of 161%. Admittedly, aqueous IDACs are difficult to measure, but that level of noise is difficult to interpret meaningfully.

More problematic is ambiguity in the literature regarding LLE versus SLE in defining the solubility. Melting temperatures are not reported in the Yaws databases, so it is difficult to know if the solubility of the reported IDACs pertain to LLE or SLE. Similarly, the publication by Boobier et al. (2020) includes a water solubility database for about 900 compounds, a conveniently small number. For those compounds where we could cross-reference a melting temperature, roughly half had melting temperatures below 300 K, suggesting solubilities based on LLE. On the other hand, roughly 20% of the compounds had melting temperatures above 400 K, suggesting SLE. This ambiguity often makes the solubility literature difficult to interpret. Even an ideal solution would have low solubility if the compound was 100 K below its melting temperature.

Additional care must be exercised when referring to the Yaws databases because many of the reported values are based on predictions. Since the predictions are based on correlations using the normal boiling point, there may be an extra layer of prediction for that quantity. The prevalence of predicted values in the Yaws databases increases with carbon number. For example, 50% of values for C# = 5 are predicted while 70% are predicted for C# = 8. To compile databases for evaluating aqueous IDACs, only values indicating an experimental source were considered. Yaws's tables of IDACs were supplemented with IDACs computed from his solubility tables when the solute was less than 2 wt%, using Eq. (9.2-7). These tabulations were then cross-referenced against the TDE database and the database of Moine et al. (2017) as described above.

Evaluations of the various methods show that correlating IDACs for aqueous systems is indeed a challenging task, even when the databases have been carefully curated. For example, the original UNIFAC method yields roughly 90%AALD in correlating the IDACs of organic compounds in water. The best available methods yield roughly 70%AALD, and these require customization.

As an example of customization, Dhakal and Paluch (2018, MOSCED18) found that careful attention to the MOSCED water parameters was required when considering IDACs for water in an organic solvent versus IDACs for organic compounds in water as the solvent. They recommended revising the MOSCED parameters of Lazzaroni et al. (2005, MOSCED05) for water as given by MOSCED18 in Table 9.3. Their recommendation was based on a database of IDACs from the Yaws compilation selected without consideration of the source. We have reconsidered their recommendation in our evaluations by comparing to our curated databases. Furthermore, the problems of water as solute versus water as solvent are sufficiently distinct that separate parameters for each can be justified. Dilute water in a hydrocarbon, for example, has nothing to hydrogen bond with and has properties like a polar methane molecule. Concentrated water can hydrogen-bond strongly in three dimensions, and evidence shows that hydrogen bonds after the first are stronger than the first bond. This is called a cooperativity effect. The enhanced α^M value of the MOSCED21 model in Table 9.3 is consistent with stronger hydrogen bonding when water is concentrated. Adjusting α^M in this way gives every advantage to the MOSCED21 method with the rationale of seeking a method that comes close to matching the reliability of the experimental data. Even then, the MOSCED21 method yields 63%AALD when water is the solvent and 38%AALD when water is the solute. Further detailed comparisons, including comparison to other activity models, are presented in Sec. 9.5.

For design of water-pollution abatement processes, it is often necessary to estimate the activity coefficient of a pollutant dilute in aqueous solution. When developing stripping processes to remove pollutants, temperatures

TABLE 9.3 MOSCED water parameters for aqueous IDACs.

Name	V^L	λ^M	τ^M	α^M	β^M	q	Source
MOSCED05	36.0	10.58	10.48	52.78	15.86	1	Lazzaroni et al. (2005)
MOSCED18	26.6	6.53	14.49	45.34	12.81	1	Dhakal and Paluch (2018)
MOSCED21	27.8	7.87	8.95	45.53	15.86	1	Water solvent (2021)
MOSCED21	27.8	7.87	8.95	30.17	15.86	1	Water solute (2021)

closer to 100 °C may be more relevant. Hwang et al. (1992) proposed an empirical correlation based on the molecular structure of the organic pollutants:

$$\begin{aligned}\log_{10}\left(K_1^{\infty}\right) = {} & 3.097 + 0.386 n_{\text{satC}} + 0.323 n_{=C} + 0.097 n_{\equiv C} + 0.145 n_{\text{aroC}} \\ & - 0.013 n_{\text{C}}^2 + 0.366 n_{\text{F}} + 0.096 n_{\text{cl}} - 0.496 n_{\text{BrI}} \\ & - 1.954 n_{-\text{O}-} - 2.528 n_{=\text{O}} - 3.464 n_{\text{OH}} + 0.331 n_{\text{O}}^2 \\ & - 2.674 n_{\text{N}} - 2.364 n_{=\text{N}} - 1.947 n_{\text{NO}_2} - 1.010 n_s \end{aligned} \tag{9.2-14}$$

where subscript 1 denotes the organic solute; the distribution coefficient K_1^{∞} (100 °C) at infinite dilution is defined as the ratio of the mole fraction of the solute in the vapor phase to that of the solute in the liquid phase at 100 °C; n denotes the number of atoms or groups specified in the subscript. Atoms or groups in the subscripts represent the categories in PGL5ed. Although Hwang et al. compare their correlation to a database of 404 compounds, most of their raw data pertain to ambient temperatures and the temperature effects are estimated. They do not provide a correlation for IDACs near ambient temperatures.

Example 9.3 Inference of IDACs for aqueous systems.

In this example, we use MOSCED correlations to estimate IDACs in water at 20 °C for the following six solutes: benzene, toluene, chlorobenzene, phenol, aniline, and nitrobenzene. (a) Use the MOSCED18 parameters of Dhakal and Paluch. (b) Use the updated MOSCED21 parameters of Table 9.3.

Solution. To use the MOSCED model, we must tabulate the MOSCED parameters for all compounds of interest.

							MOSCED	Version	ln(IDAC)
Name	V^L	λ^M(J/cc)$^{1/2}$	τ^M	α^M	β^M	q	**_18**	**_21**	**expt***
Benzene	89.5	16.71	3.95	0.63	2.24	0.9	8.10	7.44	7.83
Toluene	106.7	16.61	3.22	0.57	2.23	0.9	9.50	8.74	9.08
Chlorobenzene	102.3	16.72	4.17	0	2.5	0.89	9.02	8.26	9.28
Phenol	88.9	16.66	4.5	25.14	5.35	0.9	5.58	4.04	3.90
Aniline	91.6	16.51	9.41	6.51	6.34	0.9	5.87	4.86	5.07
Nitrobenzene	102.7	16.06	8.23	0.98	3.29	0.9	7.78	7.32	8.06

*Cf. Moine et al. (2017).

We illustrate the detailed calculation for phenol with the MOSCED21 model. Since water is the (pure) solvent, the POL, ψ, and ξ parameters are constant.

POL	ψ	ξ
2.1490	4.4592	4.0213

The aa parameter is computed from the properties of the solvent:

$$aa = 0.953 - 0.002314(\tau^2 + \alpha\beta) = 0.953 - 0.002314[4.50^2 + 25.14(5.35)] = 0.9136$$

The d_{12} parameter is computed from the volume ratio:

$$d_{12} = 1 - (88.9/26.6)^{0.9136} + 0.9136 \ln(88.9/26.6) = -0.3312$$

Finally, the IDAC is computed as

$$\ln(\text{IDAC}) = -0.3312 + [88.8/(8.314*293)]*$$
$$[(16.66 - 11.88)^2 + 0.9^2(4.50 - 14.45)^2/4.4592 + (25.14 - 55.97)(5.35 - 15.70)/4.0213] = 4.04$$

9.3 LIQUID-LIQUID EQUILIBRIA

We are concerned with a liquid mixture that, at temperature T and pressure P, is in equilibrium with another liquid mixture at the same temperature and pressure. The quantities of interest are typically the temperature and the compositions of both phases. Pressure is occasionally unspecified in LLE or SLE measurements because liquids and solids are relatively insensitive to pressure changes. In those cases, it usually suffices to specify the bubble pressure by the ideal solution approximation. Given some of these quantities, our task is to calculate the others.

Many liquids are only partially miscible, and in some cases, e.g., mercury and hexane at normal temperatures, the mutual solubilities are so small that, for practical purposes, the liquids may be considered immiscible. Partial miscibility is observed not only in binary mixtures but also in ternary (and higher) systems. This observation is the basis of liquid extraction as a separation method. This section introduces some useful thermodynamic relations which, in conjunction with limited experimental data, can be used to obtain quantitative estimates of phase compositions in liquid-liquid systems.

At ordinary temperatures and pressures, it is straightforward to obtain experimentally the compositions of two coexisting liquid phases and, as a result, the technical literature is rich in experimental results for a variety of binary and ternary systems near 25 °C and near atmospheric pressure. However, as temperature and pressure deviate appreciably from ambient conditions, the availability of experimental data falls rapidly.

Partial miscibility in liquids is often called *phase splitting*. The thermodynamic criteria which indicate phase splitting are well understood regardless of the number of components (Tester and Modell, 1977), but most thermodynamics texts confine discussion to binary systems. Stability analysis shows that, for a binary system, phase splitting occurs when

$$\left(\frac{\partial^2 g}{\partial x_1^2}\right)_{T,P} < 0 \tag{9.3-1}$$

where g is the molar excess Gibbs energy of the binary mixture. Briefly, downward concavity implies that Gibbs energy is minimized by a molar average between tangent points rather than the mixed phase curve (Elliott and Lira, 2012, Chap. 14). Substituting the ideal solution definition provides an alternative form that may be more convenient,

$$\left(\frac{\partial^2 g^E}{\partial x_1^2}\right)_{T,P} + RT\left(\frac{1}{x_1} + \frac{1}{x_2}\right) < 0 \tag{9.3-2}$$

where g^E is the molar excess Gibbs energy of the binary mixture. To illustrate Eq. (9.3-1), consider the simplest nontrivial case. Let

$$g^E = Ax_1x_2 \tag{9.3-3}$$

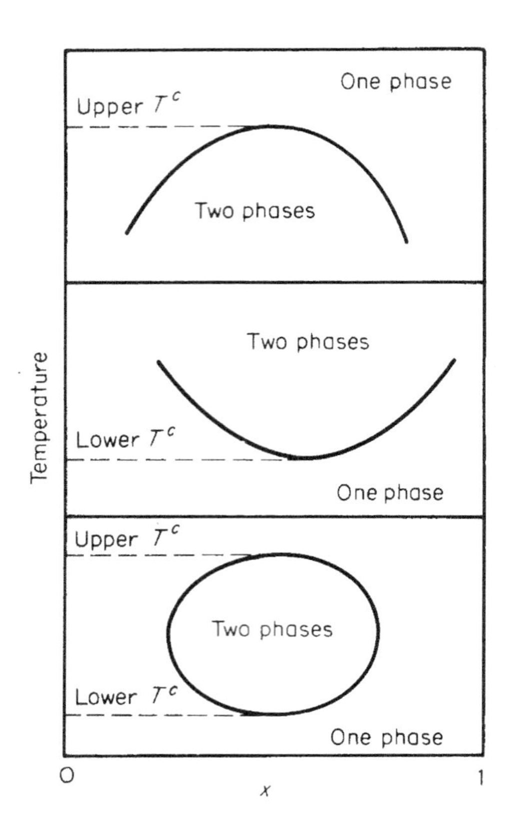

FIGURE 9.2
Phase stability in three binary mixtures (Prausnitz et al., 1999).

where A is an empirical coefficient characteristic of the binary mixture. Substituting into Eq. (9.3-1), we find that phase splitting occurs if

$$A > 2RT \tag{9.3-4}$$

In other words, if $A < 2\,RT$, the two components 1 and 2 are completely miscible; there is only one liquid phase. However, if $A > 2\,RT$, two liquid phases form because components 1 and 2 are only partially miscible.

The condition when $A = 2\,RT$ is called *incipient instability,* and the temperature corresponding to that condition is called the *consolute temperature,* designated by T^c. Since Eq. (9.3-3) is symmetric in mole fractions x_1 and x_2, the composition at the consolute or critical point is $x_1^c = x_2^c = 0.5$. In a typical binary mixture, the coefficient A is a function of temperature, and therefore it is possible to have either an upper consolute temperature or a lower consolute temperature, or both, as indicated in Figs. 9.2 and 9.3. Upper consolute temperatures are more common than lower consolute temperatures. Except for those containing polymers, and surfactant systems, systems with both upper and lower consolute temperatures are rare.[1] Note that the behavior of the bottom figure in Fig. 9.3 would require quadratic temperature dependence for A/RT if the Margules 1-parameter model was used. To the extent that Eq. (9.3-3) represents a simplification of the Scatchard (1931)-Hildebrand (1929) model for equal sized molecules, there is no theoretical basis for quadratic temperature dependence of Eq. (9.3-3). Only upper consolute temperatures are expected in that context.

[1]Although Eq. (9.3-4) is based on the simple two-suffix (one-parameter) Margules equation, similar calculations can be made using other expression for g^E. See, for example, Shain and Prausnitz (1963).

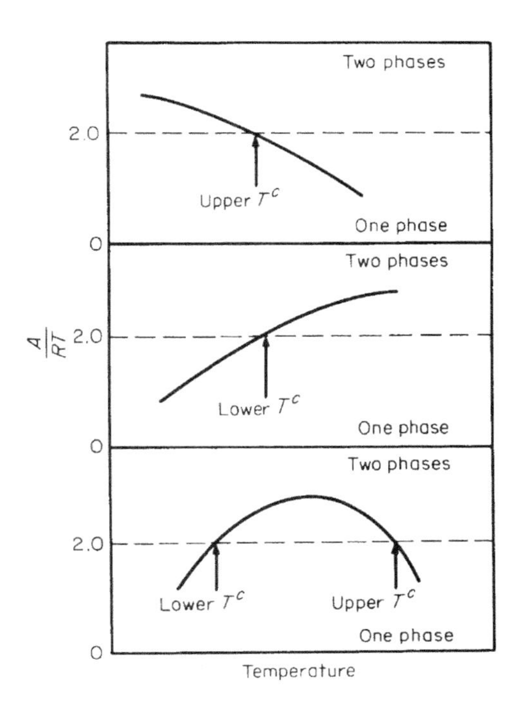

FIGURE 9.3
Phase stability in three binary mixtures whose excess Gibbs energy is given by the one-parameter Margules model (Prausnitz et al., 1999).

Stability analysis for ternary (and higher) systems is, in principle, like that for binary systems, although the mathematical complexity rises with the number of components (see, for example, Beegle and Modell, 1974). The expressions for a ternary system are

$$\left(\frac{\partial \mu_1}{\partial n_1}\right)_{n_2,n_3} > 0; \left(\frac{\partial \mu_2}{\partial n_2}\right)_{n_1,n_3} > 0$$

$$\left(\frac{\partial \mu_1}{\partial n_1}\right)_{n_2,n_3}\left(\frac{\partial \mu_2}{\partial n_2}\right)_{n_1,n_3} - \left(\frac{\partial \mu_1}{\partial n_2}\right)_{n_1,n_3} \geq 0 \tag{9.3-5}$$

However, it is important to recognize that stability analysis can tell us only whether a system can or cannot *somewhere* exhibit phase splitting at a given temperature. That is, if we have an expression for g^E at a particular temperature, stability analysis can determine whether there is *some* range of composition where two liquids exist. It does *not* tell us what that composition range is. To find the range of compositions within which two liquid phases exist at equilibrium requires a more elaborate calculation. To illustrate, consider again a simple binary mixture whose excess Gibbs energy is given by Eq. (9.3-3). If $A > 2\,RT$, we can calculate the compositions of the two coexisting equations by solving the two equations of phase equilibrium

$$(\gamma_1 x_1)' = (\gamma_1 x_1)'' \quad \text{and} \quad (\gamma_2 x_2)' = (\gamma_2 x_2)'' \tag{9.3-6}$$

where the prime and double prime designate, respectively, the two liquid phases.

From Eq. (9.3-3), we have

$$\ln \gamma_1 = \frac{A}{RT} x_2^2 \tag{9.3-7}$$

$$\ln \gamma_2 = \frac{A}{RT} x_1^2 \tag{9.3-8}$$

Substituting into the equation of equilibrium and noting that $x_1' + x_2' = 1$ and $x_1'' + x_2'' = 1$ we obtain

$$x_1' \exp \frac{A(1 - x_1')^2}{RT} = x_1'' \exp \frac{A(1 - x_1'')^2}{RT} \tag{9.3-9}$$

$$x_2' \exp \frac{A(1 - x_2')^2}{RT} = x_2'' \exp \frac{A(1 - x_2'')^2}{RT} \tag{9.3-10}$$

Equations (9.3-9) and (9.3-10) contain two unknowns $(x_1'$ and $x_1'')$ that can be found by iteration. Mathematically, several solutions of these two equations can be obtained. However, to be physically meaningful, it is necessary that $0 < x_1' < 1$ and $0 < x_1'' < 1$.

Similar calculations can be performed for ternary (or higher) mixtures. For a ternary system, the three equations of equilibrium are

$$(\gamma_1 x_1)' = (\gamma_1 x_1)''; \quad (\gamma_2 x_2)' = (\gamma_2 x_2)''; \quad (\gamma_3 x_3)' = (\gamma_3 x_3)'' \tag{9.3-11}$$

If we have an equation relating the excess molar Gibbs energy g^E of the mixture to the overall composition (x_1, x_2, x_3), we can obtain corresponding expressions for the activity coefficients γ_1, γ_2, and γ_3. The equations of equilibrium coupled with material-balance relations form the basis of a flash calculation equivalent to the flash problem discussed in Chap. 8. LLE K-values can be defined for each component analogous to the VLE K-values.

$$K_i^{\mathrm{LL}} \equiv \frac{x_i''}{x_i'} = \frac{\gamma_i'}{\gamma_i''} \tag{9.3-12}$$

For problems where the K^{LL} values are far from unity, convergence can usually be achieved by Picard iteration. For more difficult systems, the methods of Michelsen and Mollerup (2007) are applicable. Balder and Prausnitz (1966) provide guidelines for fine tuning activity model parameters to accommodate modeling of ternary systems.

Although the thermodynamics of multicomponent LLE is, in principle, straightforward, it is difficult to obtain an expression for g^E that is sufficiently accurate to yield reliable results. LLE are much more sensitive to small changes in activity coefficients than VLE. In the latter, activity coefficients play a role which is secondary to the all-important pure-component vapor pressures. In LLE, however, the activity coefficients are dominant; pure-component vapor pressures play no role at all. Therefore, it has often been observed that good estimates of VLE can be made for many systems by using only approximate activity coefficients, provided the pure-component vapor pressures are accurately known. However, in calculating LLE, small inaccuracies in activity coefficients can lead to serious errors. Regardless of which equation is used to model activity coefficients, much care must be exercised in determining parameters from experimental data and more care is required when applying the parameters to compute phase equilibria. Whenever possible, such parameters should come from mutual solubility data.

When parameters are obtained from reduction of VLE data, there is always some ambiguity. Unless the experimental data and the thermodynamic model are of very high accuracy, it is usually not possible to obtain a truly unique set of parameters. In a typical case, there is a range of parameter sets such that any set in that range can equally well reproduce the experimental data. When multicomponent LLE are calculated, results are more sensitive to the choice of binary parameters than VLE is. Therefore, it is difficult to establish reliable ternary (or higher) LLE by using only binary parameters obtained from binary LLE and binary VLE data. For reliable results with current models, it is usually necessary to utilize at least some multicomponent LLE data.

Experience in this field is not yet plentiful, but all indications are that it is always best to use binary data for calculating binary parameters. Since it often happens that binary parameter sets cannot be determined uniquely, ternary (or higher) data should then be used to fix the best binary sets from the ranges obtained from the binary data. It is, of course, always possible to add ternary (or higher) terms to the expression for the excess Gibbs energy and thereby introduce ternary (or higher) constants. This is sometimes justified, but it is meaningful only if the multicomponent data are plentiful and of high accuracy. In calculating multicomponent equilibria, the general rule is to use binary data first then use multicomponent data for fine-tuning.

9.3.1 Caveats of Common Thermodynamic Models Applied to LLE

At the outset, we remind readers of the constraints implied by Sec. 8.2. Evaluations in PGL6ed favor the temperature dependence assigned to the models by their original authors. This constraint can cause more drastic effects for LLE characterizations than for VLE. If readers choose to deviate from this constraint, they are advised to do so cautiously, with careful attention to impacts that extrapolation of their parameters may cause. Subject to this constraint, neither the UNIQUAC nor NRTL models can characterize lower consolute temperature behavior. In principle, models that include representations of hydrogen bonding could have that capability, as demonstrated by Paricaud et al. (2003) and Barker (1952) for example. For the general models available for our evaluations, however, we have not identified any cases of lower consolute temperature behavior being correctly characterized.

An additional constraint pertains to the NRTL model. As shown by Heidemann and Mandhane (1973), unusual results can be calculated for LLE if the α_N parameter is not chosen with care. The basic problem is illustrated in Fig. 9.4 for the *n*-pentane(1)+water(2) system at 25 °C. IDACs at this temperature are roughly

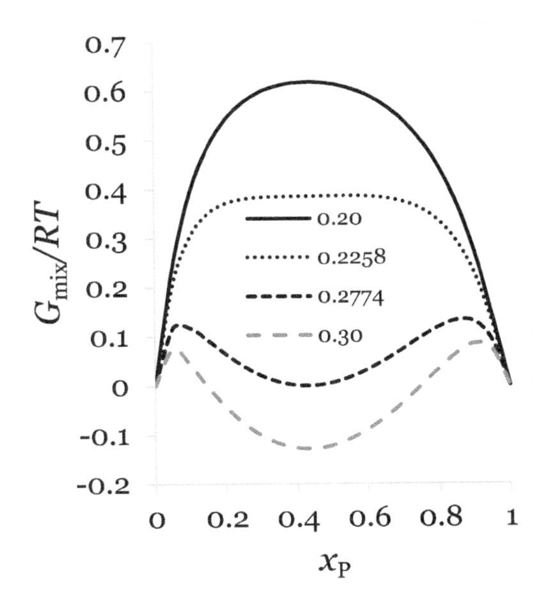

FIGURE 9.4
Artificial phase behavior when the NRTL α_N parameter is chosen imprudently.

$\gamma_1^\infty = 90909$ and $\gamma_2^\infty = 2475$. These IDACs suffice to specify τ_{12} and τ_{21} for any given value of α_N. As shown in Fig. 9.4, four liquid phases are indicated when $\alpha_N = 0.3$, which is highly artificial of course. Note that the minima near $x_P = 0$ and $x_P = 1$ are difficult to discern on the scale of this graph, but they do exist. When $\alpha_N = 0.2774$, the artificial roots barely disappear, but the reliability of extending such narrowly permissible parameters to multicomponent mixtures would be questionable. Hence the recommendation by Heidemann and Mandhane (1973) is to select a value of $\alpha_N < 0.2258$ for this system, in which case the curve for g is practically always concave down between the two liquid roots. Heidemann and Mandhane provide graphical guidelines for selecting α_N depending on either τ_{ij} parameters or IDACs but we prefer equations. When IDACs are available,

$$\alpha_N^{max} = \min\left(0.43, \frac{0.52}{\left(\gamma_1^\infty\gamma_2^\infty\right)^{0.06}} + 0.06\right) \tag{9.3-13}$$

This correlation sets a cap on α_N^{max} at 0.43. If given τ_{ij} parameters, γ^∞ can be estimated from

$$\ln(\gamma^\infty) = \langle\tau\rangle[1 + \exp(-0.3\langle\tau\rangle)] \tag{9.3-14}$$

where $\langle\tau\rangle \equiv (\tau_{12}\ \tau_{21})^{1/2}$. Our evaluations of the NRTL model for LLE correlation apply this rule consistently, effectively transforming NRTL into a two-parameter model.

9.3.2 Sample Calculations

Sample calculations can be useful in clarifying the principles and procedures for implementing LLE phase behavior modeling. We illustrate with two examples. The first example applies the UNIQUAC model to ternary LLE at a single temperature. The second example applies several EOS models to binary LLE over a range of temperatures.

Example 9.4 LLE Determination by an isothermal flash.

Acetonitrile (1) is used to extract benzene (2) from a mixture of benzene and *n*-heptane (3) at 45 °C.

(a) 0.5148 mol of acetonitrile is added to a mixture containing 0.0265 mol of benezene and 0.4587 mol of *n*-heptane to form 1 mol of feed.
(b) 0.4873 mol of acetonitrile is added to a mixture containing 0.1564 mol of benzene and 0.3563 mol of *n*-heptane to form 1 mol of feed.

For (*a*) and for (*b*), find the composition of the extract phase E, the composition of the raffinate phase R and L_E, the fraction of feed in the extract phase. Assume the UNIQUAC model with parameters in Table 9.4. Note that the binary parameters relate to the temperature dependence given by

$$\tau_{ij} = \exp\left(-\frac{a_{ij}}{T}\right) \qquad \tau_{ji} = \exp\left(-\frac{a_{ji}}{T}\right) \tag{9.3-15}$$

TABLE 9.4 UNIQUAC parameters for the system acetonitrile(1)+benzene(2)+*n*-heptane(3).

	Pure-Component Parameters		Components		Binary Parameters	
Comp.	***r***	***q***	***i***	***j***	**a_{ij}, K**	**a_{ji}, K**
1	1.87	1.72	1	2	60.3	89.6
2	3.19	2.4	1	3	23.7	545.8
3	5.17	4.4	2	3	−135.9	245.4

Solution. To find the desired quantities, we must solve an isothermal flash problem in which 1 mol of feed separates into L_E mol of extract and $1 - L_E$ mol of raffinate

There are five unknowns: two mole fractions in E, two mole fractions in R, and L_E. To find these five unknowns, we require five independent equations. They are three equations of phase equilibrium

$$(\gamma_i x_i)^E = (\gamma_i x_i)^R \qquad i = 1, 2, 3$$

and two material balances

$$z_i = x_i^E L_E + x_i^R(1 - L_E) \qquad \text{for any two components}$$

Here z_i is the mole fraction of component i in the feed; x^E and x^R are, respectively, mole fractions in E and in R, and γ is the activity coefficient.

To solve five equations simultaneously, we use an iterative procedure based on the Newton-Raphson method as described, for example, by Prausnitz et al. (1980). The objective function is

$$F^{\text{obj}}\left(x^R, x^E, L_E\right) = \sum_{i=1}^{3} \frac{(K_i - 1)z_i}{(K_i - 1)\, L_E + 1} \rightarrow 0$$

where

$$K_i = \frac{x_i^E}{x_i^R} = \frac{\gamma_i^R}{\gamma_i^E}$$

In the accompanying table calculated results are compared with experimental data.[2]

LLE in the system acetonitrile (1)-benzene (2)-*n*-heptane (3) at 45 °C.

			$100x_i^R$			$100x_i^E$	
	i	γ_i^R	Calc.	Exp.	γ_i^E	Calc.	Exp.
(*a*)	1	7.15	13.11	11.67	1.03	91.18	91.29
	2	1.25	3.30	3.41	2.09	1.98	1.88
	3	1.06	83.59	84.92	12.96	6.84	6.83
(*b*)	1	3.38	25.63	27.23	1.17	73.96	70.25
	2	1.01	18.08	17.71	1.41	12.97	13.56
	3	1.35	56.29	55.06	5.80	13.09	16.19

For (*a*), the calculated $L_E = 0.4915$; for (*b*), it is 0.4781. When experimental data are substituted into the material balance, $L_E = 0.5$ for both (*a*) and (*b*). In this case, there is good agreement between calculated and experimental results because the binary parameters were selected by using binary and ternary data.

Example 9.5 Binary LLE by EOS methods.

While activity models are commonly used for LLE, EOS formulations can also be used. For example, Tsonopoulos and Wilson (1983) correlated the solubility of water in hydrocarbons using a variation of the Redlich-Kwong EOS that was proposed by Zudkevitch and Joffe (1970). Their correlation uses different binary parameters, k_{12}, according to the phase of computation and other elaborate empiricisms. Results for the mutual solubilities of benzene(1) + water (2) in the range of 0 to 200 °C were reported in PGL5ed. As an alternative, repeat the calculations at 40 °C using (a) the ESD model with $k_{12} = 0.1020$, Elliott et al. (1990) (b) the PC-SAFT model with $k_{12} = -0.0964$ (Gross and Sadowski, 2002) and (c) the tcPR-GE model with $a_{12} = 1631$, $a_{21} = 25000$ (cf. Pina-Martinez et al., 2021).

[2]Palmer and Smith (1972).

Solution

(a) The ESD parameters for water were reported by Suresh and Elliott (1992). For benzene, critical properties suffice.

NAME	c	ε/k(K)	B	Nd	K^{AD}	ε^{AD}/k (K)
BENZENE	1.767	336.72	29.61	0	0	0
WATER	1.005	427.25	9.411	1	0.00156	5.14

We present the calculation for the final iteration and show that the "guessed" compositions are consistent with the computed compositions. The converged value for water solubility in the benzene-rich phase is $x_2' = 0.003078$. The converged value for benzene solubility in the water-rich phase is $x_1'' = 0.000564$.

Phase	b	η	$(\alpha_1^{AD})^{1/2}$	F^{AD}	X_1^A	$qYb\rho$	Z^{rep}	Z^{att}	Z^{chem}
benzene-rich′	29.543	0.3204	10.034	0.0247	0.8011	1.471	5.778	−6.773	−0.00156
water-rich″	9.422	0.3977	25.045	0.9800	0.0392	1.145	6.542	−3.613	−3.92819

The component-specific properties are:

	Benzene-rich′				Water-rich″				
	$\ln(\varphi^{rep})$	$\ln(\varphi^{att})$	$\ln(\varphi^{chem})$	$\ln(\varphi')$	$\ln(\varphi^{rep})$	$\ln(\varphi^{att})$	$\ln(\varphi^{chem})$	$\ln(\varphi'')$	K^{LL}
(1)	9.2807	−16.325	−0.0010	−1.402	25.794	−17.401	−9.325	6.0705	1759.6
(2)	3.8258	−5.7521	−0.4438	3.273	9.514	−9.571	−9.445	−2.4991	0.00311

According to the binary flash equation:

$$x_1'' = \frac{1 - K_2}{K_1 - K_2};\ x_1' = K_1 x_1'' \tag{9.3-16}$$

Substituting the K^{LL} values, $x_1'' = (1 - 0.00311)/(1759.6 - 0.00311) = 0.000567$ and $x_2' = 0.003098$. These values differ slightly from the guessed values owing to the number of significant figures retained. A more precise calculation would provide convergence to five significant figures.

(b) The EOS parameters for water and benzene are given in pcsaft_pure_parametersGross.txt.

NAME	m	σ(nm)	ε/k(K)	μ	q	N^d	K^{AD}	ε^{AD}/k(K)
BENZENE	2.291	0.3756	294.06	0	5.5907	0	0	0
WATER	1.271	0.2820	281.94	1.855	3.4353	2	0.004354	952.66

We present the calculation for the final iteration and show that the "guessed" compositions are consistent with the computed compositions. The converged value for water solubility in the benzene-rich phase is $x_2' = 0.03380$. The converged value for benzene solubility in the water-rich phase is $x_1'' = 0.0002514$.

	η	Z^{HS}	Z^{HC}	Z_1/T	$Z_2/T2$	Z^{chem}	Z^{DD}	Z^{QQ}	Z^{DQ}
′	0.4064	13.9402	−2.2605	−12.6948	0.5114	−0.0005	−0.0023	−0.4514	−0.0386
″	0.4990	67.6085	−60.8989	−7.1738	−0.4642	20.2539	−26.6609	−3.5033	−9.9103

The component-specific properties are:

	Benzene-rich′				Water-rich″				
	$\ln(\varphi^{rep})$	$\ln(\varphi^{att})$	$\ln(\varphi^{chem})$	$\ln(\varphi')$	$\ln(\varphi^{rep})$	$\ln(\varphi^{att})$	$\ln(\varphi^{chem})$	$\ln(\varphi'')$	K^{LL}
(1)	17.6100	−24.6965	−0.0003	−1.4174	67.6085	−60.8989	−7.1738	6.8360	3840.3
(2)	6.0952	−10.9781	−0.0114	0.7752	20.2539	−26.6609	−3.5033	−2.6102	0.03387

Substituting the K^{LL} values, $x_1'' = (1 - 0.03387)/(3840.3 - 0.03387) = 0.000252$ and $x_1' = 0.03386$. Once again, a more precise calculation would provide convergence to five significant figures.

(c) The equilibrium compositions for tcPR-GE EOS are $x_1' = 0.0002315$, $x_2'' = 0.001710$. From Sec. 6.6,

$$Z = 1 + \frac{\eta^P}{1-\eta^P} - \left(\frac{a_c\alpha^S}{bRT}\right)\frac{\eta^P}{(1-r_1\eta^P)(1-r_2\eta^P)} \tag{9.3-17}$$

where $r_1 = -(2^{½} + 1)$ and $r_2 = (2^{½} - 1)$.

The pure-component parameters are:

Component	T_c (K)	P_c (bar)	L^{tc}	M^{tc}	N^{tc}	a^{PR}	b^{PR}	c^{VT}
BENZENE	562.05	48.95	0.1349	0.8482	2.579	2827052	74.27	−1.439
WATER	647.10	220.64	0.3872	0.8720	1.9669	948723	18.97	−1.439

Where a_i^{PR} [=] J-cm³/mol² and b_i^{PR} [=] c_i^{VT} [=] cm³/mol.

The mixing rules of Sec. 6.6 give the mixture parameters a_m and b_m as

$$b_m = \left(x_1^2 b_{11} + 2x_1x_2b_{12} + x_2^2 b_{22}\right);\ b_{ij}^{2/3} = \frac{\left(b_{ii}^{2/3} + b_{jj}^{2/3}\right)}{2} \tag{9.3-18}$$

$$a_m = b_m\left(\frac{g^E}{C^{GE}} + x_1\frac{a_{11}}{b_1} + x_2\frac{a_{22}}{b_2}\right) \tag{9.3-19}$$

where g^E is the excess Gibbs energy at the given composition and $C^{GE} = -0.623$. With the Huron-Vidal (1979) mixing rules, we can use any convenient model for g^E; here we use the Wilson (1964) expression, which for a volume translated binary mixture is

$$\frac{g^E}{RT} = -x_1\ln\left(\Phi_1^{VT} + \Phi_2^{VT}\Lambda_{21}\right) - x_2\ln\left(\Phi_2^{VT} + \Phi_1^{VT}\Lambda_{12}\right) \tag{9.3-20}$$

with $\Lambda_{ij} = \exp(-a_{ij}/T)$, $a_{12} = 1631$, $a_{21} = 25000$, $\Phi_j^{VT} = x_jb_j^{VT}/b_m^{VT}$, $b_j^{VT} = b_j^{PR} - c_j^{VT}$, $b_m^{VT} = \Sigma x_ib_i^{VT}$. We use the Wilson activity model here because it is most consistent with EOS models in modeling the athermal contribution. Although the Wilson model fails to model LLE when directly used for activities, LLE modeling is feasible when using it as a mixing rule. Pina-Martinez et al. (2021) have evaluated several activity models for adaptation in this manner and all were similar. Substitution at $x_1'' = 0.0003143$ shows that $b_m = 18.99$ cm³/mol, $b_m^{VT} = 13.72$ cm³/mol, $\Sigma x_ia_i/b_iRT = 19.21$, $g^E/RT = -0.00527$, $a_m = 949212$ J-cm³/mol². Solving gives $Z = 0.0082596$, $\eta^P = 0.88284$. Substitution at $x_2' = 0.001170$ shows that $b_m = 74.198$ cm³/mol, $b_m^{VT} = 75.636$ cm³/mol, $\Sigma x_ia_i/b_iRT = 14.61$, $g^E/RT = -0.0162$, $a_m = 2822365$ J-cm³/mol². Solving gives $Z = 0.003396$, $\eta^P = 0.8390$.

The component-specific properties are:

	Benzene-rich′			Water-rich″			
	$\ln(\varphi^{rep})$	$\ln(\varphi^{att})$	$\ln(\varphi')$	$\ln(\varphi^{rep})$	$\ln(\varphi^{att})$	$\ln(\varphi'')$	K^{LL}
(1)	1.8266	−8.9266	−1.4150	2.1443	−2.5959	6.6473	3172.6
(2)	1.8266	−3.9442	3.5674	2.1443	−11.8420	−2.5990	0.00210

Checking the phase compositions implied by these K^{LL} values, $x_1'' = 0.000315$, $x_2' = 0.00209$.

9.4 SOLUBILITIES OF SOLIDS IN LIQUIDS

The solubility of a solid in a liquid is determined not only by the intermolecular forces between solute and solvent but also by packing within the solid as indicated by the melting point and the enthalpy of fusion of the solute. For example, at 25 °C, the solid aromatic hydrocarbon phenanthrene is highly soluble in benzene; its solubility is 20.7 mole percent. By contrast, the solid aromatic hydrocarbon anthracene, an isomer of phenanthrene, is only slightly soluble in benzene at 25 °C; its solubility is 0.81 mole percent. For both solutes, intermolecular forces between solute and benzene are essentially identical with activity coefficients near 1.0. However, the melting points of the solutes are significantly different: phenanthrene melts at 100 °C and anthracene at 217 °C. In general, it can be shown that, when other factors are held constant, the solute with the higher melting point has the lower solubility. Also, when other factors are held constant, the solute with the higher enthalpy of fusion has the lower solubility.

These qualitative conclusions follow from a quantitative thermodynamic analysis given in numerous texts. (See, for example, Prigogine and Defay, 1954; Prausnitz et al., 1999; Elliott and Lira, 2012, Chap. 14). In a binary system, let subscript 1 stand for solvent and subscript 2 for solute. Assume that the solid phase is pure. At temperature T, the solubility (mole fraction) x_2 is given by

$$\ln(\gamma_2 x_2) = -\frac{\Delta H_{fus}}{RT}\left(1 - \frac{T}{T_t}\right) + \frac{\Delta C_P}{R}\left(\frac{T_t - T}{T}\right) - \frac{\Delta C_P}{R}\ln\left(\frac{T_t}{T}\right) \tag{9.4-1}$$

where ΔH_{fus} is the enthalpy change for melting the solute at the triple-point temperature T_t and ΔC_P is given by the molar heat capacity of the pure solute:

$$\Delta C_P = C_P(\text{subcooled liquid solute}) - C_P(\text{solid solute}) \tag{9.4-2}$$

The standard state for activity coefficient γ_2 is pure (subcooled) liquid 2 at system temperature T. Tsonopoulos and Prausnitz (1971) discuss the significance of ΔC_P. For example, when $T_m/T = 1.75$, neglecting ΔC_P causes a 20% error in x_2 relative to using the best estimate of ΔC_P.

Another source of uncertainty can be the heat of fusion at the temperature of interest. For example, ethane goes through several phase transitions between its melting temperature and a few degrees lower. The estimate of H_{fus} for ethane tabulated in Appendix A reflects the sum of those transition energies, assuming that most SLE calculations of interest will be substantially below the melting temperature. Choi and Mclaughlin (1983) discuss this issue in more detail.

To a good approximation, we can substitute normal melting temperature T_m for triple-point temperature T_t, and we can assume that ΔH_{fus} is essentially the same at the two temperatures. In Eq. (9.4-1), the first term on the right-hand side is much more important than the remaining two terms, and therefore a simplified form of that equation is

$$\ln(\gamma_2 x_2) = \frac{-\Delta H_{fus}}{R}\left(\frac{1}{T} - \frac{1}{T_m}\right) \tag{9.4-3}$$

If we substitute

$$\Delta S_{\text{fus}} = \frac{\Delta H_{\text{fus}}}{T_{\text{m}}} \tag{9.4-4}$$

we obtain an alternative simplified form

$$\ln(\gamma_2 x_2) = -\frac{\Delta S_{\text{fus}}}{R}\left(\frac{T_{\text{m}}}{T} - 1\right) \tag{9.4-5}$$

where ΔS_{fus} is the entropy of fusion.

If we let $\gamma_2 = 1$, x_2 is known as the "ideal solubility" at temperature T, knowing only the solute's melting temperature and its enthalpy (or entropy) of fusion. This ideal solubility depends only on properties of the solute; it is independent of the solvent's properties. Since ΔH_{fus}, T_{m}, and ΔS_{fus} are constants, an Arrhenius plot of the ideal solubility is obviously linear.

The effects of intermolecular forces between molten solute and solvent are reflected in activity coefficient γ_2. To describe γ_2, we can use any of the expressions for the excess Gibbs energy. However, since γ_2 depends on the mole fraction x_2, solution of Eq. (9.4-3) requires iteration, initializing x_2 with the ideal solubility or including the IDAC for γ_2 in Eq. (9.4-6). Including the IDAC is valuable if the IDAC > 1 and solubility is high. Convergence is generally achieved in 10 to 20 Picard iterations with $\alpha_{\text{P}} \approx 0.5$.

$$\ln(x_2) = \frac{-\Delta H_{\text{fus}}}{R}\left(\frac{1}{T} - \frac{1}{T_{\text{m}}}\right) - \ln(\gamma_2) \tag{9.4-6}$$

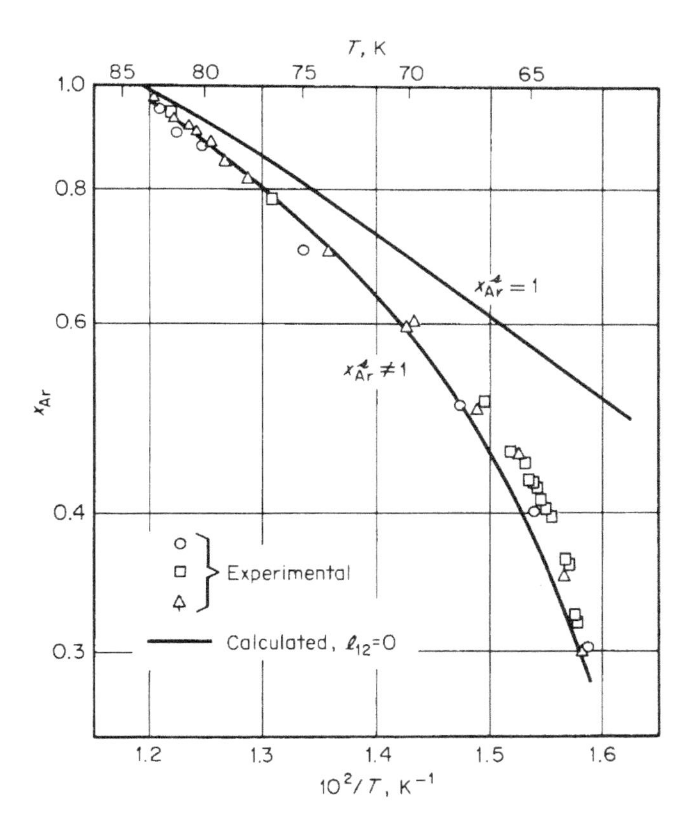

FIGURE 9.5
Solubility of argon in nitrogen: effect of solid-phase composition (From Preston and Prausnitz, 1970).

Guidelines for choosing an activity model are like those for VLE and LLE. As shown by Preston and Prausnitz (1970), regular solution theory is useful for calculating solubilities in nonpolar systems, especially when the geometric-mean assumption is relaxed through introduction of an empirical correction k_{12}. Unfortunately, *some* mixture data point is needed to estimate k_{12}. In a few fortunate cases one freezing point datum, e.g., the eutectic point, may be available to fix k_{12}. In many cases, it is possible to use UNIFAC for estimating solubilities of solids, as discussed by Gmehling et al. (1978). The performance of various models for SLE correlation and prediction is summarized in Sec. 9.5.

It is important to remember that the calculations outlined above rest on the assumption that the solid phase is pure, i.e., that there is no solubility of the solvent in the solid phase. This assumption is often a good one, especially if the two components differ appreciably in molecular size and shape. However, in many known cases, the two components are at least partially miscible in the solid phase, and in that event, it is necessary to correct for solubility and nonideality in the solid phase as well as in the liquid phase. This complicates the thermodynamic description, but, more important, solubility in the solid phase may significantly affect the phase diagram. Figure 9.5 shows results for the solubility of solid argon in liquid nitrogen. The top line presents calculated results if x^S (argon) = 1, where superscript ^{S} denotes the solid phase. The bottom line considers the experimentally known solubility of nitrogen in solid argon [x^S (argon) $\neq$ 1]. In this case, serious error is introduced by neglecting solubility of the solvent in the solid phase. Wax formation is a common situation in which solutes of various molar masses may contribute to the solid phase (Elliott and Lira, 2012, Example 14.12).

Variations of the UNIQUAC equation have been used to correlate experimental solubility data for polar solid organic solutes in water. Examples 9.6 and 9.7 illustrate such correlations for solubilities of sugars and amino acids.

Example 9.6 SLE calculations with the UNIQUAC model.

Peres and Macedo (1997) proposed a modified UNIQUAC model to describe VLE and SLE for mixtures containing D-glucose, D-fructose, sucrose, and water.

Use the modified UNIQUAC model to calculate the SLE composition phase diagram for the ternary D-glucose (1)-sucrose (2)-water (3) at 70 °C.

Solution. If each sugar is assumed to exist as a pure solid phase, the two equations at equilibrium are

$$f_i^{\mathrm{L}} = f_{i,\mathrm{p}}^{\mathrm{S}} \qquad (i = 1, 2) \tag{9.4-7}$$

or equivalently,

$$x_i\gamma_i = f_{i,\mathrm{p}}^{S}/f_{i,\mathrm{p}}^{\mathrm{L}} \qquad (i = 1, 2) \tag{9.4-8}$$

where subscript i denotes sugar i and subscript p denotes pure. Superscripts L and S stand for liquid and solid; f is fugacity; x is the liquid-phase mole fraction; γ is the liquid-phase activity coefficient.

The standard state for each sugar is chosen as the pure subcooled liquid at system temperature; further, the difference between the heat capacity of pure liquid sugar and that of pure solid sugar $\Delta C_{P,i}$ is assumed to be linearly dependent on temperature:

$$\Delta C_{P,i} = \Delta A_i - \Delta B_i(T - T_o) \qquad (i = 1, 2) \tag{9.4-9}$$

where ΔA_i (J mol^{-1} K^{-1}) and ΔB_i (J mol^{-1} K^{-2}) are constants for each sugar; T = 343.15 K; T_o is the reference temperature 298.15 K.

The ratio $f_{i,\mathrm{p}}^{\mathrm{S}}/f_{i,\mathrm{p}}^{\mathrm{L}}$ [Eq. 9.4-6] is given by Gabas and Laguerie (1993), Raemy and Schweizer (1983), and Roos (1993):

$$\ln(x_i\gamma_i) = \left[\frac{-\Delta H_{\mathrm{fus},i}}{RT_{\mathrm{m},i}} + \frac{\Delta A_i - \Delta B_i T_{\mathrm{o}}}{RT_{\mathrm{m},i}} + 1 + \frac{\Delta B_i}{2R}T_{\mathrm{m},i}\right]\left(\frac{T_{\mathrm{m},i}}{T} - 1\right) + \frac{\Delta A_i - \Delta B_i T_{\mathrm{o}}}{R}\ln\left(\frac{T}{T_{\mathrm{m},i}}\right) + \frac{\Delta B_i}{2R}(T - T_{\mathrm{m},i}) \tag{9.4-10}$$

where R is the universal gas constant; $T_{\mathrm{m},i}$ is the melting temperature of pure sugar i; ΔH_{fus} is the enthalpy change of melting at $T_{\mathrm{m},i}$; ΔA_i, ΔB_i, T, and T_{o} are those in Eq. (9.4-9).

Substituting Eq. (9.4-10) into (9.4-8), we obtain the specific correlation of Eq. (9.4-1) to be used in this case

$$\ln(x_i\gamma_i) = \left[\begin{matrix}\dfrac{-\Delta H_{\mathrm{fus},i}}{R} + \dfrac{\Delta A_i - \Delta B_i T_{\mathrm{o}}}{R} \\ + T_{\mathrm{m},i} + \dfrac{\Delta B_i}{2R}T_{\mathrm{m},i}^2\end{matrix}\right]\left(\frac{1}{T} - \frac{1}{T_{\mathrm{m},i}}\right) + \frac{\Delta A_i - \Delta B_i T_{\mathrm{o}}}{R}\ln\left(\frac{T}{T_{\mathrm{m},i}}\right) + \frac{\Delta B_i}{2R}(T - T_{\mathrm{m},i}) \tag{9.4-11}$$

For the two sugars of interest, physical properties used in Eq. (9.4-11) are given by Peres and Macedo (1997):

	D-glucose	**Sucrose**
$T_{\mathrm{m},i}$ (K)	423.15	459.15
$\Delta H_{\mathrm{m},i}$ (J mol^{-1})	32432	46187
$\Delta A_i'$ (J mol^{-1} K^{-1})	139.58	316.12
$\Delta B_i'$ (J mol^{-1} K^{-2})	0	−1.15

At fixed T, the right side of Eq. (9.4-11) can be solved simultaneously for $i = 1$ and $i = 2$ to yield x_1 and x_2 (x_3 is given by mass balance $x_1 + x_2 + x_3 = 1$). Activity coefficients are given by a modified form of UNIQUAC as described in Example 8.9.

For D-glucose, sucrose, and water, volume and surface parameters are:

Component	r_i	q_i
D-glucose	8.1528	7.920
Sucrose	14.5496	13.764
Water	0.9200	1.400

Peres and Macedo (1997) set interaction parameters between the two sugars to zero ($a_{12} = a_{21} = 0$) while they assume that interaction parameters between water and sugars (a_{i3} and a_{3i}, $i = 1, 2$) are linearly dependent on temperature:

$$a_{i3} = a_{i3}^{(\mathrm{o})} + a_{i3}^{(1)}(T - T_{\mathrm{o}})$$
$$a_{i3} = a_{3i}^{(\mathrm{o})} + a_{3i}^{(1)}(T - T_{\mathrm{o}}) \tag{9.4-12}$$

TABLE 9.5 Solid-liquid equilibria for the ternary D-glucose (1)–sucrose (2)–water (3) at 70 °C.

Calculated		Experimental	
x_1	x_2	x_1	x_2
0.0000	0.1442	0.0000	0.1462
0.0538	0.1333	0.0545	0.1333
0.1236	0.1193	0.1233	0.1182
0.1777	0.1132	0.1972	0.1060
0.2104	0.0662	0.2131	0.0614
0.2173	0.0304	0.2292	0.0304
0.2475	0.0059	0.2531	0.0059
0.2550	0.0000	0.2589	0.0000

Using experimental binary water-sugar data from 0 to 100 °C and from very dilute to saturated concentration, Peres and Macedo (1997) give

	$a_{i3}^{(o)}$ (K)	$a_{3i}^{(o)}$ (K)	$a_{i3}^{(1)}$	$a_{3i}^{(1)}$
$i = 1$(D-glucose)	−68.6157	96.5267	−0.0690	0.2770
$i = 2$(sucrose)	−89.3391	−89.3391	0.3280	−0.3410

Table 9.5 shows x_1 and x_2 from solving simultaneously two Eqs. (9.4-11) once for $i = 1$ and once for $i = 2$. Corresponding experimental data are from Abed et al. (1992).

Figure 9.6 suggests that the UNIQUAC equation is useful for describing SLE for this aqueous system where the two solutes (D-glucose and sucrose) are chemically similar enough that the solute-solute parameters

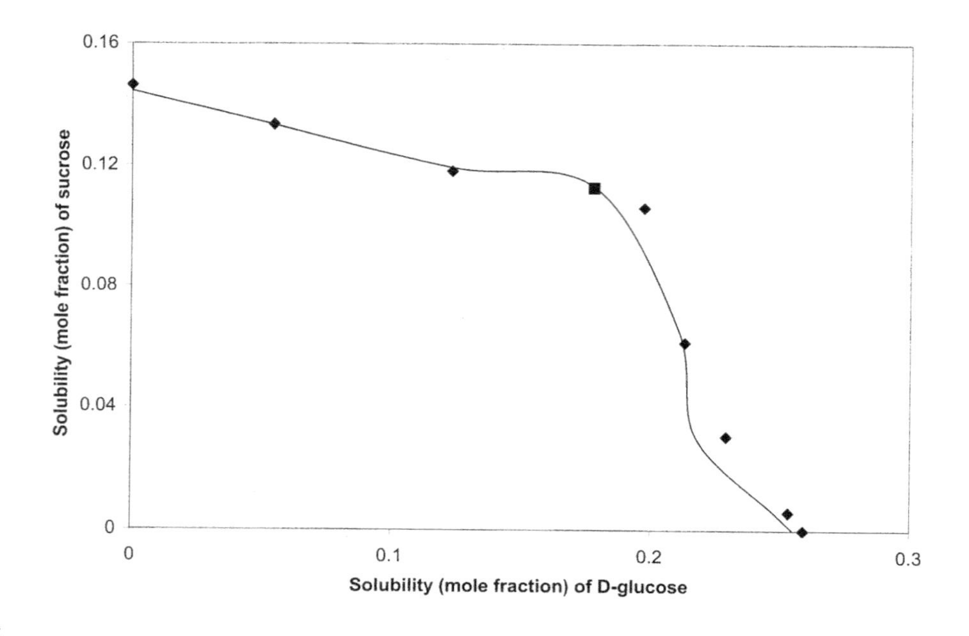

FIGURE 9.6
Solubility plot for D-glucose (1)–sucrose (2)–water (3) at 70 °C.
▪ Experimental, —calculated, ■ calculated three-phase point.

a_{12} and a_{21} can be set to zero, yet structurally dissimilar enough that there is no significant mutual solubility in the solid phase.

Example 9.7 SLE computations for aqueous amino acids.

Kuramochi et al. (1996) studied solid-liquid equilibria (SLE) for mixtures containing DL-alanine, DL-serine, DL-valine, and water. Use the modified UNIFAC model of Larsen et al. (1987) to calculate the solubility diagrams at 25 °C for the ternaries.

(a) DL-alanine (1); DL-valine (2); water (3) (b) DL-alanine (1); DL-serine (2); water (3)

Solution. The procedure to solve this problem is like that of Example 9.6. The activity-coefficient prediction uses the UNIFAC method established for liquid-liquid systems rather than the method developed for VLE. If each amino acid is assumed to form a pure solid phase, the two equations at equilibrium are Eqs. (9.4-7) and (9.4-8) of Example 9.6. In this case, however, since most amino acids decompose before reaching their melting temperatures, the quantities $T_{m,i}$, $\Delta H_{fus,i}$, and $\Delta C_{p,i}$ are not available from experiment.

Therefore, Kuramochi et al. assumed,

$$\ln\left(f_{i,p}^{S}/f_{i,p}^{L}\right) = A_i - B_i/T + C_i \ln(T) \quad (i = 1, 2) \tag{9.4-13}$$

The final form of Eq. (9.4-1) for this case is then

$$\ln(x_i\gamma_i) = A_i - B_i/T + C_i \ln(T) \quad (i = 1, 2) \tag{9.4-14}$$

Values for γ_i in the amino acid-water binaries and ternaries are calculated from UNIFAC at the same temperatures as those of the experimental data. To use UNIFAC, Kuramochi et al. (1996) needed to introduce five new groups: α-CH (α indicates adjacent to an NH_2 group), sc-CH (sc stands for side chain), α-CH_2, sc-CH_2, and CONH. The constitutive groups for the three amino acids and water are as follows:

Compound	Constitutive Groups
DL-alanine [CH_3 $CH(NH_2)COOH$]	NH_2 + COOH + α-CH + CH_3
DL-valine [$(CH_3)_2CHCH(NH_2)COOH$]	NH_2 + COOH + α-CH + sc-CH + 2CH_3
DL-serine [$OHCH_2CH(NH_2)COOH$]	NH_2 + COOH + α-CH + sc-CH_2 + OH
Water [H_2O]	H_2O

Group-volume (R_i) and surface-area (Q_i) parameters are given as:

Group	R_i	Q_i
CH_3	0.9011	0.848
sc-CH_2	0.6744	0.540
sc-CH	0.4469	0.228
α-CH	0.4469	0.228
NH_2	0.6948	1.150
COOH	1.3013	1.224
OH	1.000	1.200
H_2O	0.9200	1.400

For interactions involving the newly assigned groups, Kuramochi et al. (1996) calculated interaction parameters a_{ij} using experimental osmotic coefficients for the three amino acid-water binaries at 25 °C. Other interaction a_{ij}

values parameters are taken from the LLE UNIFAC table of Larsen et al. (1987). Pertinent interaction parameters (K) are:

Group	CH_X*	α–XH	NH_2	COOH	OH	H_2O
CH_X*	0	−896.5	218.6	1554	707.2	49.97
α–XH	−167.3	0	−573.2	−960.5	−983.1	−401.4
NH	1360	921.8	0	867.7	−92.21	86.44
COOH	3085	−603.4	−489	0	−173.7	−244.5
OH	1674	−1936	61.78	−176.5	0	155.6
H_2O	85.7	−1385	8.62	−66.39	−47.15	0

*CH_x is shorthand for CH_3/sc-CH_2/sc-CH

To obtain constants A, B, and C, for the amino acids, Kuramochi, et al. minimized an objective function F^{obj}, defined by

$$F^{obj} \equiv \sum_{data}^{NPTS} \frac{\left| x_i^{calc} - x_i^{expt} \right|}{x_i^{expt}} \tag{9.4-15}$$

where the summation is over all binary experimental data points from 273 to 373 K; x_i^{expt} and x_i^{calc} are experimental and calculated solubilities of the amino acid (i)-water binary. Simultaneously solving Eq. (9.4-14) and optimizing (9.4-15) for the three amino acid-water binaries. Kuramochi et al. (1996) give constants A, B, and C for each amino acid:

	A	$B(K^{-1})$	C
DL-alanine	77.052	−2668.6	11.082
DL-valine	−5236.3	−5236.3	17.455
DL-serine	−28.939	−318.35	4.062

For the ternary systems [amino acids (1) and (2) and water (3)] calculated solubilities x_1 and x_2 are found by solving simultaneously two Eqs. (9.4-14) once for each amino acid. Results (in terms of molality m) are shown in Tables 9.6 and 9.7. Experimental ternary data are from Kuramochi et al. (1996).

Figures 9.7 and 9.8 compare calculated and experimental data.

TABLE 9.6 Solubilities for DL-alanine (1)–DL-valine (2)–water (3) at 25 °C.

Experimental		Calculated	
m_1 (mol/Kg water)	m_2 (mol/Kg water)	m_1 (mol/Kg water)	m_2 (mol/Kg water)
0.0000	0.6099	0.0000	0.6078
0.4545	0.5894	0.4487	0.5862
0.7704	0.5817	0.7693	0.5706
1.1292	0.5623	1.1276	0.5504
1.4962	0.5449	1.4907	0.5331
1.7765	0.5269	1.7862	0.5348
1.7920	0.5010	1.7932	0.5244
1.8350	0.3434	1.8382	0.3402
1.8674	0.1712	1.8634	0.1737
1.8830	0.0000	1.8849	0.0000

TABLE 9.7 Solubilities for DL-alanine (1)–DL-serine (2)–water (3) at 25 °C.

Experimental		Calculated	
m_1 (mol/Kg water)	m_2 (mol/Kg water)	m_1 (mol/Kg water)	m_2 (mol/Kg water)
0.0000	0.4802	0.0000	0.4853
0.4056	0.4929	0.4021	0.4968
0.7983	0.5053	0.7077	0.5051
1.1703	0.5179	1.1644	0.5188
1.4691	0.5193	1.4690	0.5206
1.9027	0.5234	1.9062	0.5255
1.9012	0.4116	1.9038	0.4187
1.8959	0.3120	1.9014	0.3136
1.8928	0.1993	1.8952	0.2009
1.8887	0.1074	1.8904	0.1123
1.8830	0.0000	1.8849	0.0000

9.5 EVALUATIONS

Evaluations for IDACs, LLE, and SLE are consolidated in this section. The models of interest have been described in preceding chapters, with a few added variations on the MOSCED model, and clarifications of details like the Heidemann-Mandhane constraint on the NRTL α^N parameter as discussed in this chapter. There is considerable overlap among these three properties because LLE and SLE systems are usually more dilute than VLE systems. A trend in common with VLE is that aqueous systems tend to exhibit larger deviations for all methods, so we break out the metrics for those systems once again as we did in Chapter 8.

It is not obvious how to evaluate model reliability such that all properties are equally respected. For example, IDACs can be matched exactly by two parameter values if both IDACs are available at the same temperature, so all deviations would be zero if that was the protocol. On the other hand, IDACs, LLE, and SLE tend to be much more sensitive to BIP values than VLE is, so constraining the BIPs to be independent of temperature while comparison data span a substantial temperature range can provide a meaningful comparison (see Sec. 8.2).

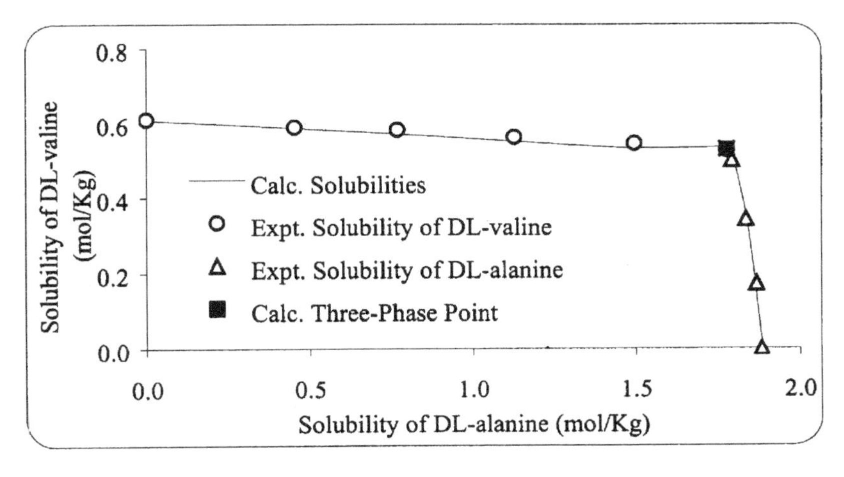

FIGURE 9.7
Solubility plot for DL-alanine (1)–DL-valine (2)–water (3) at 25 °C.

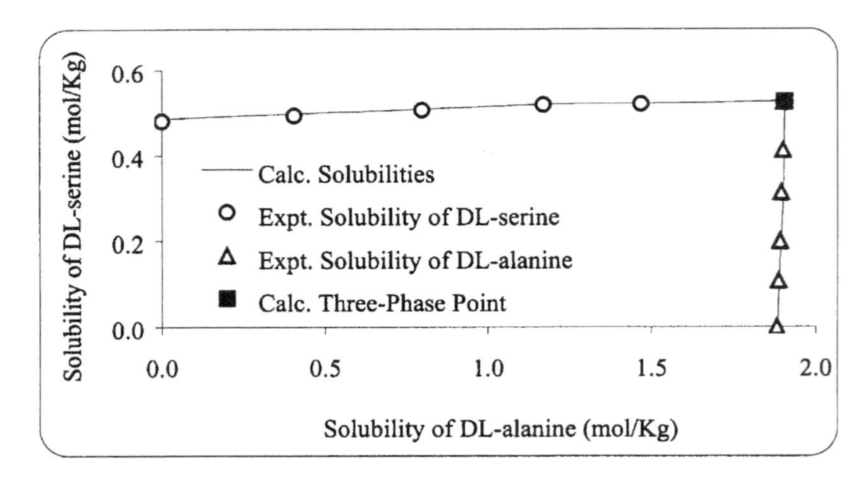

FIGURE 9.8
Solubility plot for DL-alanine (1)–DL-serine (2)–water (3) at 25 °C.

Challenging evaluations would be to constrain all properties to be correlated by VLE data with a single BIP (or two for local composition models), but predicting these properties based on BIPs characterized from VLE data is inadvisable. Predicting VLE from BIPs characterized based on IDAC, LLE, or SLE might seem more reasonable. Unfortunately, the uncertainty of IDACs reported in the literature is highly variable. Furthermore, the temperature ranges for IDACs, LLE, and SLE tend to be lower than those for VLE. With these considerations, VLE deviations using BIPs characterized from IDAC data may be misleading. In conclusion, current models are not quite ready for such a challenge.

Therefore, our evaluations in PGL6ed are based on individual correlations for each property without attempting to simultaneously correlate all forms of phase behavior with a single BIP. Models are improving and new molecular insights continue to accrue. Let us hope that more stringent challenges will be more attainable in the future.

9.5.1 Evaluations of Methods for Estimating IDACs

The MOSCED model is only one of the many models that can be applied to estimate IDACs. In this section we compare predictions for several models, including MOSCED. We begin with comparisons for nonaqueous systems. Our database for nonaqueous systems is adapted from the supplementary information of Lazzaroni et al. (2005). It comprises 4327 values with 3048 solute-solvent combinations, the difference representing temperature variations, with 87 distinct solvents. Table 9.8 presents our comparisons of MOSCED with implementations of UNIFAC and COSMO-RS for nonaqueous systems. It is not surprising that the MOSCED model shows the smallest deviations. It was developed exclusively for IDACs. Furthermore, the evaluation database for nonaqueous systems was developed by Lazzaroni et al. specifically for training MOSCED. Although the comparison is not entirely fair, it suggests that the MOSCED method should be included when considering methods to estimate IDACs. It would be desirable to extend the coverage of MOSCED to more compounds. While PMOSCED provides such an extension, its accuracy is no better than the COSMO-RS/SAC approach. The open source GAMESS implementation of COSMO-RS/SAC by Ferrarini et al. (2018) appears to be an improvement over the Va Tech (VT) implementation by Mullins et al. (2005), but deviations for both exceed what can be achieved with UNIFAC. Another alternative might combine functional groups like those in UNIFAC with a COSMO-RS/SAC implementation for IDAC prediction. Soares et al. (2013) have reported %AALD of around 11% for such a method (F-SAC), but for a much smaller database. Silveira and Salau (2018) compared the F-SAC implementation to UNIFAC for a database of roughly 1800 IDACs and found F-SAC to be superior. Fingerhut et al. (2018) have performed a more extensive comparison of UNIFAC and another COSMO-RS/SAC

TABLE 9.8 Evaluation of IDAC methods for nonaqueous systems.

Lazzaroni DB	%AALD	NPTS
MOSCED	10.6%	4260
Orig UNIFAC	18.3%	3538
NIST-mod-UNIFAC	19.1%	3848
UNIFAC(Do)	23.1%	3372
PMOSCED	31.8%	4260
COSMO-RS/SAC(VT)	44.4%	4056
COSMO-RS/GAMESS(HB2)	32.4%	4010
COSMO-RS/F-SAC	9.6%	2062

implementation (SAC10), using a database of roughly 25,000 IDACs, 18% of which were aqueous. They found that SAC10 was slightly less accurate than UNIFAC for nonaqueous systems overall, but this conclusion was dependent on the interactions of molecular classes, for which they provided greater resolution than our relatively simple evaluation. Our independent evaluation in Table 9.8 reaches a conclusion like that of Silveira and Salau (2018), with a qualification: the number of values covered by F-SAC is much smaller than other methods. Since Soares et al. (2013) focused on IDACs in their training set, the accuracy in Table 9.8 may reflect a large degree of overlap between the training set and the values for which the method worked. If the F-SAC method could be extended to more compounds, it might be a promising method for general application.

We compare results for aqueous systems in Tables 9.9 and 9.10. The database for water as solvent comprises 3272 values for 544 solutes. The database for water as solute comprises 519 values for 282 solvents. All databases are available through the PGL6ed website.

Table 9.9 shows that MOSCED21 and MOSCED05 provide similar performance when water is the solvent. Upon reflection, this observation is not surprising because Lazzaroni et al. only considered systems with water as a solvent when developing their parameters. Referring to Table 9.3, the parameters for MOSCED21 and MOSCED05 are similar. The molar volume of MOSCED21 was constrained to match the value from MOSCED18 and the variations in the other MOSCED21 parameters appear to cancel that change relative to the MOSCED05 model. Either the MOSCED21 or MOSCED05 parameters could be used with equal "reliability," noting that 70%AALD does not suggest a high degree of reliability. The GAMESS implementation of COSMO-RS/SAC is slightly more accurate than the UNIFAC methods, and the F-SAC method is superior to the "trained" method of MOSCED. This observation raises doubts about the viability of the MOSCED method for aqueous systems. Possibly, the more accurate treatment of strong interactions afforded by the SS-QCT basis

TABLE 9.9 Evaluation of IDAC methods for systems with water as solvent.

Water as Solvent	COMPDS	NPTS	%AALD	%BIASLD
MOSCED21	101	1681	65.6	−13.1
MOSCED18	101	1681	67.1	12.0
MOSCED05	101	1681	63.9	−30.5
PMOSCED	436	3047	163.6	134.0
PMOSCED*	436	3047	109.2	−2.1
Original UNIFAC-VLE	372	3006	91.0	−53.6
NIST-mod-UNIFAC	384	3028	91.3	−62.5
COSMO-RS/GAMESS(HB2)	410	2915	83.8	−9.6
COSMO-RS/F-SAC2	205	1831	52.9	−15.6
UNIFAC(Do)	335	2571	108.1	−60.7

TABLE 9.10 Evaluation of IDAC methods for systems with water as solute.

Water as Solute	COMPDS	NPTS	%AALD	%BIASLD
MOSCED21	73	278	30.8	0.0
MOSCED18	73	278	90.4	90.4
MOSCED05	73	278	364.1	364.1
PMOSCED	232	465	72.9	60.9
PMOSCED*	232	465	56.4	13.9
Original UNIFAC-VLE	219	496	44.4	–16.1
NIST-mod-UNIFAC	213	489	98.0	–58.7
COSMO-RS/GAMESS(HB2)	215	449	35.4	–21.7
COSMO-RS/F-SAC	111	273	35.1	–9.6
UNIFAC(Do)	181	397	112.7	–55.3

in COSMO-RS/SAC methods explains this superior performance. Further exploration of this prospect would be desirable. The PMOSCED approach does not provide reliable accuracy for aqueous systems. It is possible to improve its predictions by correcting for its large bias. A divisor of 3.9 for all compounds reduces the bias of the PMOSCED method, as indicated by PMOSCED* in Table 9.9. Nevertheless, the corrected results are still worse than the UNIFAC methods.

Table 9.10 shows that MOSCED05 is not useful when water is the solute, as suggested by Dhakal and Paluch (2018). Figure 9.9b shows the wide disparity of MOSCED05 when water is the solute. In this case, MOSCED21 and MOSCED18 provide similar results and referring to Table 9.3 shows their parameters are also similar. If it was necessary to always apply a single set of parameters, the MOSCED18 values should suffice. The alternative models exhibit larger deviations, but not as large as when water is the solvent. As shown by the scales of the axes in Fig. 9.9, the magnitudes of the IDACs are much smaller when water is the solute. The role of solute molar volume clearly plays a significant role in this effect. The UNIFAC models exhibit a significant negative bias, and the PMOSCED model exhibits a strong positive bias. Fingerhut et al. (2018) also observed a strong positive bias for the COSMO-RS/SAC10 method when water was the solute. In the case of PMOSCED,

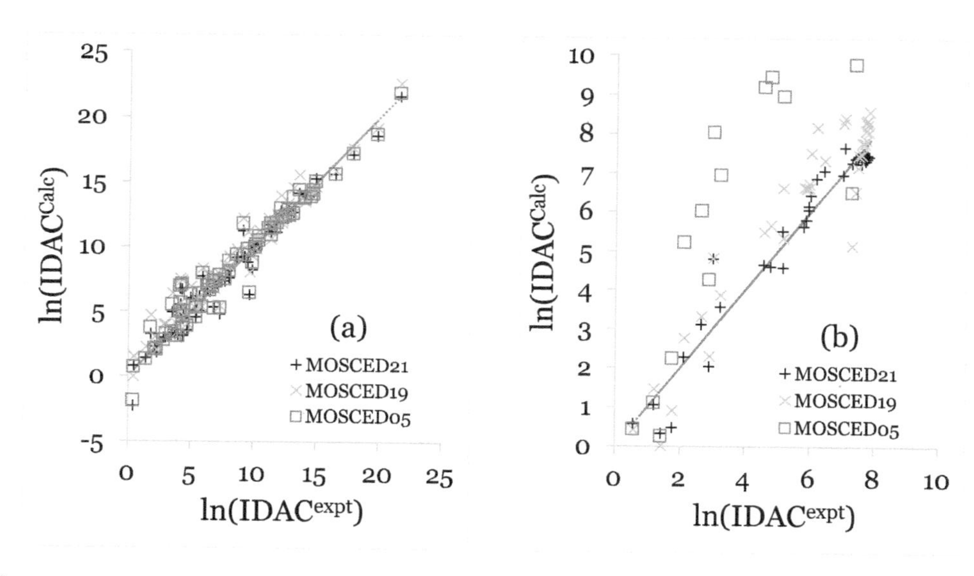

FIGURE 9.9
Parity plots of MOSCED implementations for aqueous systems. (a) water as solvent. (b) water as solute.

the bias can be traced to normal and heavy compounds. The divisor for PMOSCED* is 1.6 for all compounds, but it would be desirable to develop a class-specific correction. Of course, it would be preferable if fundamental analysis could eliminate the need for an ad hoc correction. The MOSCED methods stand out as a reasonable choice for correlating organic IDACs when sufficient experimental data are available. For example, if IDACs have been measured for a homologous series, the trends in the MOSCED parameters could be extrapolated. If no data are available, the COSMO-RS/GAMESS method compares quite favorably to the "trained" MOSCED method. The COSMO-RS/F-SAC method also shows promise, but the number of working compounds was again quite small.

Clearly, there is much room for improvement when considering methods to predict or correlate IDACs. The situation for nonaqueous systems appears to be less daunting than for aqueous systems, but the database developed by Lazzaroni et al. might be obscuring some details. We hope to expand on this database in the future. For aqueous systems, the available methods demonstrate a need for new insights. Could there be cooperative effects of hydrogen bonding of water when one site or more is solvated? When water is the solute, do its bonding sites behave in a very different manner than when water is the solvent? It is tempting to suggest that TPT1 might provide answers to these questions, but it is not so straightforward. The majority of IDACs are available near ambient temperature, so the improved temperature dependence of TPT1 could be difficult to appreciate. Implementation of TPT1 is hindered because it depends on bonding volumes and bonding energies that often vary from one compound to another in complex manners when using current methods of characterization. Furthermore, the bonding energies for TPT1 need to be correlated to individual acidity and basicity tendencies to provide predictive capability.

The problems for aqueous systems are not limited to the availability of viable methods. Viable experimental data are also in dire need. Uncertainties of 160%AALD in the experimental data cannot provide a basis for meaningful method development, and uncertainties of 40%AALD cannot be sustained if methods are to be improved from their current status. The large deviations and inconsistencies when water is the solvent suggest that greater care is required when analyzing these systems. Could some form of aggregation like micellization in these systems explain the wide variability? What new experimental techniques could be applied to reduce this variability? Noting the significance of water solubility in bio/pharma applications, these questions merit serious scientific inquiry. This is certainly one area where advancement of the field requires more substantial resources.

9.5.2 Evaluating Methods for LLE Modeling

For LLE and SLE evaluations, we have relied entirely on the TDE database. Common activity models like NRTL and UNIQUAC provide less accurate correlations of LLE behavior when temperature-dependence is not included in their parameters, relative to VLE and SLE correlations. When temperature dependence is added, a two-parameter model can have four or more parameters, making it difficult to justify using the result for anything other than interpolation of the specific data used for correlation. Extension to multicomponent application or beyond the regressed temperature range would be unreliable. Hence, we reiterate the protocol of Sec. 8.2.

The LLE database comprises 96 binary mixtures, 47 of which are aqueous. Most systems exhibit upper consolute temperature behavior. Among the nonaqueous systems, 28 involve normal compounds with associating compounds and 18 involve normal compounds with polar compounds. These proportions are roughly representative of the complete TDE database for LLE. Coverage has been restricted to the compounds listed in Appendix A. For further details and analysis, the complete database (LLeDbPGL6ed96b.txt) is linked to the PGL6ed website.

Table 9.11 summarizes the performance of methods for correlating LLE. As expected, the local composition models usually provide the greatest accuracy and breadth of coverage. Accuracy for LLE is expressed in terms of the %AALDS defined by

$$\%\text{AALDS} \equiv \sum_{\text{data}}^{\text{NPTS}} \left| \ln\left(\frac{x_i^{\text{calc}}}{x_i^{\text{expt}}} \right) \right| \times \frac{100}{\text{NPTS}} \tag{9.5-1}$$

TABLE 9.11 Evaluation of methods for correlating LLE.

	Nonaqueous			Aqueous			Overall	
	Systems	Points	% AALDS	Systems	Points	% AALDS	Systems	% AALDS
ESD96	49	4355	21.3	47	7638	44.7	96	36.2
PC-SAFT	28	3060	37.4	44	6650	97.7	72	78.7
SPEADMD	43	4017	30.3	46	6940	65.9	89	52.8
tcPR-GE(W)	49	4219	39.7	47	7326	56.7	96	50.5
UNIQUAC	49	4455	34.1	43	7660	35.3	92	34.8
LSG	42	4174	34.1	43	7660	34.6	85	34.4
NRTL	49	4455	37.8	43	7660	40.3	92	39.4

where the x_i values represent the mole fraction of the dilute component. For example, the ESD model for a mixture of benzene+water at 40 °C gives mutual solubilities of $x''_B = 0.000566$ and $x'_B = 0.99690$, or $x'_W = 0.003096$. Tsonopoulos and Wilson (1983) report $x''_B = 0.0004435$ and $x'_W = 0.00501$. So, these mutual solubilities would provide two deviation points with $dev_B = 24.4\%$ and $dev_W = 48.1\%$.

One small surprise is that the ESD model performs as well as it does. The ESD model provides the highest accuracy of any model in correlating LLE for nonaqueous systems (21%AALDS), and its accuracy for aqueous systems (45%) is competitive with the results for UNIQUAC (35%) and NRTL (40%). It was expected that some implementation of TPT1 could provide improved representation of temperature effects, owing to the stronger theoretical basis of explicitly treating strong interactions as well as dispersion interactions. On the other hand, the PC-SAFT and SPEADMD models are more sophisticated than the ESD model, so they should perform better, in principle. The PC-SAFT model includes explicit treatment of polar interactions as well as having the capability to treat strong solvation interactions. The SPEADMD model includes greater molecular detail than the ESD model along with explicit treatment of strong solvation interactions. The ESD implementation here ignores strong solvation interactions, applying the geometric combining rule to all solvation interactions in favor of computational speed. It should be feasible to refine the PC-SAFT and SPEADMD models to perform at least as well as the ESD model, but that would be beyond the scope of our current effort.

Another small surprise was that the UNIQUAC model performed better than the NRTL model in all cases. Most literature seems to favor NRTL in comparisons of this sort. The difference between the two models is small, however. Consistent application of the Heidemann and Mandhane (1973) constraint is essential in our opinion, but not always applied in practice. Alternative implementations of the NRTL model may be neglecting this constraint in a way that (artificially) reduces their %AALDS.

Table 9.12 summarizes results for predictions of LLE. The surprise here was that the original UNIFAC method performed relatively well compared to the NIST/KT method, especially for aqueous systems.

TABLE 9.12 Evaluation of methods predicting LLE.

	Nonaqueous			Aqueous			Overall	
	Systems	Points	%AALDS	Systems	Points	%AALDS	Systems	%AALDS
UNIFAC-VLE	47	4334	86.3	42	7576	62.1	89	70.9
UNIFAC-NIST/KT	42	3982	79.8	42	7576	124.6	84	109.2
UNIFAC-NIST/Mod	41	3911	66.9	41	7537	66.9	82	60.5
UNIFAC-Dortmund	43	5063	65.1	36	5962	68.2	79	66.8
COSMO-RS/FSAC2	25	3419	124.4	27	4376	61.0	52	88.8
COSMO-RS/GAMESS	40	4204	139.9	36	5905	55.3	76	90.5

The NIST/KT implementation includes LLE data in its original development and incorporates a temperature dependence in the binary group interaction parameters. Overall, the NIST/mod implementation performs best, but this is not surprising because its development was like that of NIST/KT. Altogether, none of these predictive methods is very satisfactory. Keep in mind that 69%AALDS corresponds to a factor of 2 discrepancy between the calculated and experimental solubility values.

9.5.3 Evaluating Methods for SLE Modeling

For SLE evaluations, NIST/TRC has contributed a database comprising 171 binary mixtures, 18 of which are aqueous. Coverage has been restricted to the compounds listed in Appendix A. Among the nonaqueous systems, 41 involve normal compounds with associating compounds and 32 involve normal compounds with normal compounds, and other mixture classes were fairly evenly distributed. These proportions are roughly representative of the complete TDE database for SLE. For further details and analysis, the complete database (SLeDbPGL6ed.txt) is linked to the PGL6ed website.

Table 9.13 summarizes the results of correlating SLE. The NRTL model is capable of correlating SLE data with a high degree of accuracy. The deviations observed for the NRTL model reflect uncertainty in the experimental data as much as correlation error. The deviations with the Wilson model are close to those with the NRTL model, but clearly inferior, especially for aqueous systems. The tcPR-GE model performs well for nonaqueous systems but surprisingly poorly for aqueous systems. A similar problem was evident during the LLE evaluations. The problem is made more mysterious because the Wilson model forms the basis of the Huron-Vidal mixing rule in this specific implementation of the tcPR-GE model. Communications with the authors of the method were unable to resolve the issue. Among the TPT1 models, the ESD model performs surprisingly well again. Although deviations for aqueous systems with the ESD model (29%AALDS) were substantially larger than those with the NRTL model (9%), they were substantially smaller than with the PC-SAFT model (62%). The deviations of the SPEADMD model are small enough to be promising, but only half of the aqueous systems were represented by the SLE compounds that had been characterized in the SPEADMD database. For completeness, we included the tcPRq model with quadratic mixing. As expected, it performed poorly for aqueous systems. On the other hand, it performed reasonably well for nonaqueous systems. Overall, SLE appears to be a property that can be characterized within the uncertainty of the experimental data. The primary caveat is to note that the experimental uncertainty includes uncertainty in H_{fus} and T_m as well as in measured solubilities. The examples of anthracene and phenanthrene cited in Sec. 9.4 should not be overlooked. The propagation of uncertainty discussed in Chap. 2 should also be kept in mind.

TABLE 9.13 Evaluation of methods for correlating SLE.

	Nonaqueous			Aqueous			Overall	
	Systems	Points	%AALDS	Systems	Points	%AALDS	Systems	%AALDS
NRTL	153	5776	6.4	18	1405	9.0	171	7.2
UNIQUAC	144	5604	8.2	17	1401	11.2	161	9.0
LSG	144	5604	8.6	17	1401	12.1	161	8.6
tcPR-GE	151	5642	8.7	18	1138	47.1	169	10.6
Wilson	153	5775	11.4	18	1405	17.1	171	13.2
ESD96	150	5499	10.6	18	1279	28.5	168	14.4
PC-SAFT	122	4586	11.9	13	903	61.6	135	23.4
SPEADMD	100	3412	12.6	12	709	16.1	112	13.1
tcPRq	131	4850	14.1	15	1086	73.8	146	26.1

TABLE 9.14 Evaluation of methods for predicting SLE.

	Nonaqueous			Aqueous			Overall	
	Systems	Points	%AALDS	Systems	Points	%AALDS	Systems	%AALDS
Id Soln	152	5793	42.5	18	1403	506.9	170	133.0
UNIFAC-VLE	141	5502	40.3	18	1403	52.6	159	42.8
UNIFAC-NIST/KT	137	5377	37.2	16	1244	57.8	153	41.1
UNIFAC-NIST/Mod	141	5450	28.1	16	1221	64.9	157	34.8
UNIFAC-Dortmund	133	5870	50.0	9	590	69.6	142	51.8
COSMO-RS/FSAC2	74	2992	20.5	4	247	62.0	78	23.6
COSMO-RS/GAMESS	136	5813	54.9	9	582	134.4	145	62.2

Table 9.14 presents SLE results for several predictive models, including the ideal solution model as a kind of null hypothesis. In this case, the original UNIFAC method is disappointing for nonaqueous systems but surprisingly good for aqueous systems. The COSMO-RS/FSAC2 method is promising for nonaqueous systems, although its coverage is relatively narrow at present. The large deviations of the COSMO-RS/FSAC2 method for aqueous systems reflect a large bias like that for aqueous IDACs. Overall, the NIST/mod version of UNIFAC performs the most reliably, providing consistently superior performance for a broad range of compounds and all forms of phase equilibria and excess enthalpy.

Altogether, the performance of SLE modeling was much like VLE modeling. Correlations to the uncertainty of experimental data are feasible. Whereas vapor pressure tends to mask impacts of the activity model in the case of VLE, the ideal solubility (with uncertainty in T_m and H_{fus}) masks the activity model in SLE. Like VLE, deviations for aqueous SLE systems are much higher than for nonaqueous systems, especially for predictive models. On the other hand, average deviations are three to four times higher for SLE than for VLE. That result could derive from uncertainty in the experimental data or it could be due to greater sensitivity to details of the activity model.

9.6 CONCLUDING REMARKS

Multiple deliberate efforts will be required to improve modeling capabilities for IDACs and LLE. These properties are more sensitive to details of the activity models, especially the implied temperature dependence. But requisite effort is not limited to modeling of these properties; the experimental measurements require improvement and the databases require careful curation as well.

The problems are particularly challenging for aqueous systems. Aqueous systems are especially important for environmental and pharmaceutical applications. These applications are very important but the economic motivations are quite different from those that motivated the petrochemical industry to collect large amounts of accurate data. These economics do not encourage optimism for the prospects of more plentiful and accurate experimental data for these properties. Nevertheless, it is helpful to have identified some of the issues.

The situation with SLE is more encouraging. It appears that better than 10% accuracy can be achieved in correlating SLE data if accurate H_{fus} and T_m measurements are available. Like the situation with VLE, deficiencies in the activity model are masked somewhat by the underlying influence of the pure component properties. Deficiencies in the available activity models are more apparent for predictive models, especially for aqueous systems.

A common theme throughout our phase equilibrium evaluations continues to be poor modeling of aqueous systems. Modeling efforts for IDACs and LLE may be hampered by poor experimental data, but the models could be improved substantially before their deviations matched the uncertainty of the experimental data.

For example, fundamental qualitative behaviors like lower critical solution behavior were lacking in all models evaluated here. The challenges of aqueous systems have echoed over several decades and new ideas continue to be advanced. We look forward to learning more as these ideas progress and to putting them into context with the methods that are currently practiced.

9.7 NOTATION NEW TO CHAPTER 9*

a_i activity of component i
ΔH_{fus} molar enthalpy of fusion
f_i fugacity of component i
F_i nonideality factor
g^E molar excess Gibbs energy
G^E extensive excess Gibbs energy
$H_{2,1}$ Henry's constant for solute (2) in solvent (1)
h^E molar excess enthalpy
h_i partial molar enthalpy of component i
k_{12} binary interaction parameter for dispersion energies in EOS models
K_i^{LL} liquid-liquid molar distribution coefficient
L^{16} hexadecane-air partition coefficient
M_k molar mass of solute
M_s molar mass of solvent
N_C number of components
n_i number of moles of component i
P total pressure
p_i partial pressure of the ith component
q_i^M MOSCED polarity prefactor, Eq. (9.2-13)
R_{gas} gas constant
RI index of refraction
s^E molar excess entropy
s_j number of size groups in molecule j
T absolute temperature, K
t temperature, °C
T_m melting point temperature
T_t triple-point temperature
u_{ij} empirical constant in the UNIQUAC activity model
V molar volume
w_k weight fraction of component k
x_i liquid phase mole fraction of component i
X_k group mole fraction for group k
x_M mole fraction of monomer in a chemical-physical treatment of strong interactions.
y_i vapor phase mole fraction of component i
Z compressibility factor
z coordination number
z_i overall mole fraction

*See also Chap. 8 for common notation on phase equilibrium modeling

Greek

α^{KT} Kamlet-Taft acidity parameter, Table 9.2
α^{M} MOSCED acidity parameter, Eq. (9.2-13)
α_{N} NRTL nonrandomness parameter
β^{KT} Kamlet-Taft basicity parameter, Table 9.2
β^{M} MOSCED basicity parameter, Eq. (9.2-13)
λ^{M} MOSCED nonpolar solubility parameter or dispersion parameter, Eq. (9.2-13)
π^{*} Kamlet-Taft dipolarity/polarized scale, Table 9.2
τ^{M} MOSCED polar solubility or polar parameter, Eq. (9.2-13)
φ_{i} fugacity coefficient of component i
Φ_{i} volume fraction of component i

Superscripts

c consolute
calc calculated
chem chemical-physical contributions for strong interactions like hydrogen bonding
expt experimental
E excess
FHA Flory-Huggins athermal contribution
GSA Guggenheim-Staverman athermal contribution, Table 8.2
H Kamlet-Taft solute parameter, Table 9.2
KT Kamlet-Taft solvent parameter, Table 9.2
liq liquid phase
$\circ$ standard state as in f_i°, estimated true value
res Residual
sat saturation
sub sublimation
vap vapor phase
∞ infinite dilution

9.8 REFERENCES

Abed, Y., N. Gabas, M. L. Delia, T. Bounahmidi, *Fluid Phase Equil.,* **73:** 175 (1992).
Abraham, M. H., *Chem. Soc. Rev.,* **22:** 73 (1993).
Abrams, D. S., J. M. Prausnitz, *AIChE J.,* **21:** 116 (1975).
Balder, J. R., J. M. Prausnitz, *Ind. Eng. Chem. Fund.,* **5:** 449–454 (1966).
Beegle, B. L., M. Modell, R. C. Reid, *AIChE J.,* **20:** 1200 (1974).
Bender, E., U. Block, *Verfahrenstechnic,* **9:** 106 (1975).
Boobier, S., D. R. J. Hose, A. J. Blacker, B. N. Nguyen, *Nat. Comm.,* **11:** 5753 (2020).
Boublik, T., *J. Chem. Phys.,* **53:** 471 (1970).
Carr, P. W., *Microchem. J.,* **48:** 1 (1993).
Chapman, W. G., K. E. Gubbins, G. Jackson, M. Radosz, *Ind. Eng. Chem. Res.* **29:** 1709 (1990).
Choi, P. B., E. Mclaughlin, *AIChE J.,* **29:**150–153 (1983).
Cooling, M. R. B. Khalfaoui, B., D. M. T. Newsham, *Fluid Phase Equilib.,* **81:** 217–229 (1992).
Diaz-Rodriguez, S., S. M. Bozada, J. R. Phifer, A. S. Paluch, *J. Comput. Aided Mol. Des.,* **30:** 1007–1017 (2016).
Dhakal, P., A. S. Paluch, *Ind. Eng. Chem. Res.,* **57:** 1689–1695 (2018).
Diorazio, L. J., D. R. J. Hose, N. K. Adlington, *Org. Process Res. Dev.,* **20:** 760–773 (2016).
Elliott, J. R., Suresh, S. J., Donohue, *Ind. Eng. Chem. Res.,* **29:** 1624–1629 (1990).

Elliott, J. R., C. T. Lira, *Introductory Chemical Engineering Thermodynamics,* 2nd ed., Prentice-Hall, Englewood Cliffs, NJ, 2012.

Elliott, J. R., *Fluid Phase Equilib.,* **416:** 27–41 (2016).

Gabas, N., C. Laguerie, *J. Crystal Growth,* **128:** 1245 (1993).

Gmehling, J., T. F. Anderson, J. M. Prausnitz, *Ind. Eng. Chem. Fundam.,* **17:** 269 (1978).

Gnap, M., Elliott, J. R., *Fluid Phase Equil.,* **470:** 241–248 (2018).

Gross, J., G. Sadowski, *Ind. Eng. Chem. Res.,* **41:** 5510–5515 (2002).

Guggenheim, E. A., "Mixtures," Clarendon, Oxford (1952).

Hait, M. J., C. L. Liotta, C. A. Eckert, D. L. Bergmann, A. M. Karachewski, A. J. Dallas, D.Eikens, J. J. Li, P. W. Carr, R. B. Poe, S. C. Rutan, *Ind. Eng. Chem. Res.,* **32:** 2905 (1993).

Hansen, C. M., *J. Paint Technol.,* **39:** 104, 505 (1967).

Hao, Y., C.-C. Chen, *AIChE J.,* **67:** e17061 (2021).

Heidemann, R. A., J. M. Mandhane, *Chem. Eng. Sci.,* **28:** 1213–1221 (1973).

Hildebrand, J. H., *J. Am. Chem. Soc.,* **51:** 66–80 (1929).

Hwang, Y.-L., G. E. Keller, J. D. Olson: *Ind. Eng. Chem. Res.,* **31:** 1753 and 1759 (1992).

Kamlet, M. J., R. W. Taft, *J. Am. Chem. Soc.,* **98:** 377 (1976).

Kamlet, M. J., J.-L. M. Aboud, M. H. Abraham, R. W. Taft, *J. Org. Chem.,* **48:** 2877 (1983).

Kamlet, M. J., M. H. Abraham, P. W. Carr, R. M. Doherty, R. W. Taft, *J. Chem. Soc., Perkins Trans.,* **2:** 2087 (1988).

Klamt, A., *J. Phys. Chem.,* **99:** 2224–2235 (1995).

Klamt, A., V. Jonas, T. Burger, J. C. W. Lohrenz, *J. Phys. Chem.,* **102:** 5074–5085 (1998).

Klamt, A., G. J. P. Krooshof, R. Taylor, *AIChE J.,* **48:** 2332–2349 (2002).

Kuramochi, H., H. Noritomi, D. Hoshino, K. Nagahama, *Biotech. Progr.,* **12**(3): 371 (1996).

Larsen, B. L., P. Rasmussen, and Aa. Fredenslund, *Ind. Eng. Chem. Res.,* **26:** 2274 (1987).

Lazzaroni, M. J., D. Bush, C. A. Eckert, T. C. Frank, S. Gupta, J. D. Olson, *Ind. Eng. Chem. Res.,* **44:** 4075 (2005).

Le Guennec, Y., R. Privat, J.-N. Jaubert, *Fluid Phase Equil.,* **429:** 301–312 (2016).

Li, J., A. J. Dallas, J. I. Eikens, P. W. Carr, D. L. Bergmann, M. J. Hait, C. A. Eckert, *Anal. Chem.,* **65:** 3312 (1993).

Mansoori, G. A., N. F. Carnahan, K. E. Starling, and T. W. Leland, Jr., *J. Chem. Phys.* **54:** 1523 (1971).

Michelsen M. L., J. Mollerup, *Thermodynamic Modelling: Fundamentals and Computational Aspects*, 2nd ed. Tie-Line Publications, 2007.

Moine, E., R. Privat, B. Sirjean, J. N. Jaubert, *J. Phys. Chem. Ref. Data,* **46:** 033102 (2017).

Mullins, E., R. Oldland, Y. A. Liu, S. Wang, S. I. Sandler, C.-C. Chen, M. Zwolak, K. C. Seavey, *Ind. Eng. Chem. Res.,* **45:** 4389–4415 (2006).

Nikolaidis, I. G. Economou, R. Privat, *Ind. Eng. Chem. Res.,* **59:** 14981–15027 (2020)

Palmer, D. A., B. D. Smith, *J. Chem. Eng. Data,* **17:** 71 (1972).

Paricaud, P., A. Galindo, G. Jackson, *Mol. Phys.,* **101:** 2575–2600 (2003).

Peng, D. Y., D. B. Robinson, *Ind. Eng. Chem. Fundam.,* **15:** 59 (1976).

Peres, A. M., E. A. Macedo, *Fluid Phase Equil.,* **139:** 47 (1997).

Pierotti, G. J., C. H. Deal, E. L. Derr, *Ind. Eng. Chem.,* **51:** 95 (1959).

Pina-Martinez, A., R. Privat, I. K. Nikolaidis, I. G. Economou, J.-N. Jaubert, *Ind. Eng. Chem. Res.,* **60:** 17228–17247 (2021).

Preston, G. T., J. M. Prausnitz, *Ind. Eng. Chem. Process Design Develop.,* **9:** 264 (1970).

Prigogine, I., R. Defay, *Chemical Thermodynamics,* Longmans, London, 1954.

Prausnitz, J. M., R. N. Lichtenthaler, E. G. Azevedo, *Molecular Thermodynamics of Fluid-Phase Equilibria,* 3rd ed., Prentice Hall, Englewood Cliffs, NJ, 1999.

Raemy, A., T. F. Schweizer, *J. Thermal Anal.,* **28:** 95 (1983).

Renon, H., J. M. Prausnitz, *Ind. Eng. Chem. Process Design Develop.,* **8:** 413 (1969).

Roos, Y., *Carbohydr. Res.,* **238:** 39 (1993).

Scatchard, G., *Chem. Rev.,* **8:** 321–333 (1931).

Schreiber, L. B., C. A. Eckert, *Ind. Eng. Chem. Process Design Develop.,* **10:** 572 (1971).

Shain, S. A., J. M. Prausnitz, *Chem. Eng. Sci.,* **18:** 244 (1963).

Sherman, S. R., D. B. Trampe, D. M. Bush, M. Schiller, C. A. Eckert, A. J. Dallas, J. Li, P. W. Carr, *Ind. Eng. Chem. Res.,* **35:** 1044 (1996).

Soares, R. D. P., R. P. Gerber, L. F. K. Possani, P. B. Staudt, *Ind. Eng. Chem. Res.,* **52:** 11172–11181 (2013).

Soave, G., *Chem. Eng. Sci.,* **27:** 1197 (1972).

Suresh, S. J., J. R. Elliott, *Ind. Eng. Chem. Res.,* **31:** 2783–2794 (1992).

Taft, R. W., J. M. Abboud, M. J. Kamlet, M. H. Abraham, *J. Soln. Chem.,* **3:** 153 (1985).
Tester, J. W., M. Modell, *Thermodynamics and Its Applications,* 3rd ed., Prentice Hall, Englewood Cliffs, NJ, 1997.
Thomas, E. R., B. A. Newman, G. L. Nicolaides, C. A. Eckert, *J. Chem. Eng. Data,* **27:** 233 (1982).
Treybal, R. E., *Liquid Extraction,* 2nd ed., McGraw-Hill, New York, 1963.
Tsonopoulos, C., J. M. Prausnitz, *Ind. Eng. Chem. Fund.,* **10:** 593–600 (1971).
Tsonopoulos, C., G. M. Wilson, *AIChE J.,* **29:** 990 (1983).
Unlu, O., N. H. Gray, Z. N. Gerek, J. R. Elliott, *Ind. Eng. Chem. Res.,* **43:** 1788–1793 (2004).
van Konynenburg, P. H., Scott, R. L. *Phil. Trans. Roy. Soc. London, Ser. A,* **298**(1442): 495–540 (1980).
Vera, J. H., S. G. Sayegh, G. A. Ratcliff, *Fluid Phase Equil.,* **1:** 113–135 (1977).
Wang, J., T. Hu, X. Xu, *J. Chem. Inf. Model.,* **49:** 571–581 (2009).
Wertheim, M. S., *J. Stat. Phys.* **35:** 19 (1984), *ibid.* **35:** 35 (1984).
Wilson, G. M., *J. Am. Chem. Soc.,* **86:** 127, 133 (1964).
Yaws, C. L., *Yaws' Handbook of Properties for Aqueous Systems,* Knovel, Norwich, NY, 2012.
Zudkevitch, D., J. Joffe, *AIChE J.,* **16:** 112 (1970).

10

Viscosity

10.1 SCOPE

The first part of this chapter deals with the viscosity of gases and the second with the viscosity of liquids. In each part, methods are recommended for: (1) correlating viscosities with temperature; (2) estimating viscosities when no experimental data are available; (3) estimating the effect of pressure on viscosity; and (4) estimating the viscosities of mixtures. The molecular theory of viscosity is considered briefly.

10.2 DEFINITIONS OF UNITS OF VISCOSITY

If a shearing stress is applied to any portion of a confined fluid, the fluid will move with a velocity gradient with its maximum velocity at the point where the stress is applied. If the local shear stress per unit area at any point is divided by the velocity gradient, the ratio obtained is defined as the viscosity of the medium. Thus, viscosity is a measure of the internal fluid friction, which tends to oppose any dynamic change in the fluid motion. An applied shearing force will result in a large velocity gradient at low viscosity. Increased viscosity causes each fluid layer to exert a larger frictional drag on adjacent layers which in turn decreases the velocity gradient.

Viscosity differs in one important respect from the properties discussed previously in this book; namely, viscosity can only be measured in a nonequilibrium experiment. This is unlike density which can be found in a static apparatus and so is an equilibrium property. On the microscale, however, both properties reflect the effects of molecular motion and interaction. Thus, even though viscosity is ordinarily referred to as a nonequilibrium property, it is, like density, a function of the thermodynamic state of the fluid; in fact, it may even be used to define the state of the material. Brulé and Starling (1984) have emphasized the desirability of using both viscosity and thermodynamic data to characterize complex fluids and to develop correlations.

This discussion is limited to Newtonian fluids, i.e., fluids in which the viscosity, as defined, is independent of either the magnitude of the shearing stress or velocity gradient (rate of shear). For polymer solutions which are non-Newtonian, the reader is referred to Ferry (1980) or Larson (1999).

The mechanisms and molecular theory of gas viscosity have been reasonably well clarified by nonequilibrium statistical mechanics and the kinetic theory of gases (Millat et al., 1996). Brief summaries of both theories will be presented. The theory of liquid viscosity is less well developed.

Since viscosity is defined as a shearing stress per unit area divided by a velocity gradient, it should have the dimensions of (force) (time)/(length)^2 or mass/(length) (time). Both dimensional groups are used, although for most scientific work, viscosities are expressed in poises, centipoises, micropoises, etc. A poise (P) denotes a viscosity of 0.1 N · s/m^2 and 1.0 cP = 0.01 P. Centipoise (cP) is a popular unit because the viscosity of liquid water at room temperature is approximately 1 cP. The following conversion factors apply to viscosity units:

$1\ \text{P} = 100\ \text{cP} = 10^6\ \mu\text{P} = 0.1\ \text{N}\cdot\text{s/m}^2 = 1\ \text{g/(cm}\cdot\text{s)} = 0.1\ \text{Pa}\cdot\text{s}$

$1\ \text{P} = 0.067197\ \text{lb}_\text{m}/(\text{ft}\cdot\text{s}) = 241.91\ \text{lb}_\text{m}/(\text{ft}\cdot\text{h})$

$1\ \text{cP} = 1\ \text{mPa}\cdot\text{s}$

The *kinematic viscosity* is the ratio of the viscosity to the density. With viscosity in poises and the density of grams per cubic centimeter, the unit of kinematic viscosity is the *stokes,* with the units square centimeters per second. In the SI system of units, viscosities are expressed in N · s/m^2 (or Pa · s) and kinematic viscosities in either m^2/s or cm^2/s.

10.3 THEORY OF GAS TRANSPORT PROPERTIES

The theory of gas transport properties is simply stated, but it is quite complex to express in equations that can be used directly to calculate viscosities. In simple terms, when a gas undergoes a shearing stress so that there is some bulk motion, the molecules at any one point have the bulk velocity vector added to their own random velocity vector. Molecular collisions cause an interchange of momentum throughout the fluid, and this bulk motion velocity (or momentum) becomes distributed. Near the source of the applied stress, the bulk velocity vector is large, but as the molecules move away from the source, they are "slowed down" due to random molecular collisions. This random, molecular momentum interchange is the predominant cause of gaseous viscosity.

10.3.1 Elementary Kinetic Theory

If the gas is modeled in the simplest manner, it is possible to show the general relations among viscosity, temperature, pressure, and molecular size. More rigorous treatments yield similar relations but with important correction factors. The elementary gas model assumes all molecules to be nonattracting rigid spheres of diameter σ (with mass m) moving randomly at a mean velocity v. The molar density is n molecules in a unit volume while the mass density is the mass in a unit volume. Molecules move in the gas and collide transferring momentum in a velocity gradient and energy in a temperature gradient. The motion also transfers molecular species in a concentration gradient. The net flux of momentum, energy, or component mass between two layers is assumed proportional to the momentum, energy, or mass density gradient, i.e.,

$$\text{Flux} \propto -\frac{d\rho'}{dz} \tag{10.3-1}$$

where the density ρ' decreases in the $+z$ direction and ρ' may be ρ_i (mass density), nmv_y (momentum density), or $C_v nT$ (energy density). The coefficient of proportionality for all these fluxes is given by elementary kinetic theory as $vL/3$, where L is the mean free path.

Equation (10.3-1) is also used to define the transport coefficients of diffusivity D_{ab}, viscosity η, and thermal conductivity λ; that is,

$$\text{Mass Flux} = -D_{ab}m\frac{dn_i}{dz} = -\frac{vL}{3}\frac{d\rho_i}{dz} \tag{10.3-2}$$

$$\text{Momentum Flux} = -\eta\frac{dv_y}{dz} = -\frac{vL}{3}mn\frac{dv_y}{dz} \tag{10.3-3}$$

$$\text{Momentum Flux} = -\lambda\frac{dT}{dz} = -\frac{vL}{3}C_v n\frac{dT}{dz} \tag{10.3-4}$$

Equations (10.3-2) to (10.3-4) define the transport coefficients D_{ab}, η, and λ. If the average speed is proportional to $\left(\frac{RT}{M}\right)^{\frac{1}{2}}$, and the mean free path is proportional to $(n\sigma^2)^{-1}$,

$$D_{ab} = \frac{vL}{3} = (\text{const})\frac{T^{\frac{3}{2}}}{M^{\frac{1}{2}}P\sigma^2} \tag{10.3-5}$$

$$\eta = \frac{m\rho vL}{3} = (\text{const})\frac{T^{\frac{1}{2}}M^{\frac{1}{2}}}{\sigma^2} \tag{10.3-6}$$

$$\lambda = \frac{vLC_v n}{3} = (\text{const})\frac{T^{\frac{1}{2}}}{M^{\frac{1}{2}}\sigma^2} \tag{10.3-7}$$

The constant multipliers in Eqs. (9.3-5) to (9.3-7) are different in each case; the interesting fact to note from these results is the dependency of the various transfer coefficients on T, P, M, and σ. A similar treatment for rigid, nonattracting spheres having a Maxwellian velocity distribution yields the same final equations but with slightly different numerical constants.

The viscosity relation [Eq. (9.3-6)] for a rigid, nonattracting sphere model is (see Hirschfelder et al., 1954, p. 14.)

$$\eta = 26.693\frac{(MT)^{\frac{1}{2}}}{\sigma^2} \tag{10.3-8}$$

where η is viscosity in μP, M is molecular weight in g/mol, T is temperature in K, and σ is the hard-sphere diameter in Å. Analogous equations for λ and D_{ab} are given in Chaps. 11 and 12.

10.3.2 Effect of Intermolecular Forces

If the molecules attract or repel one another by virtue of intermolecular forces, the theory of Chapman and Enskog is normally employed (Chapman and Cowling, 1939; Hirschfelder et al., 1954). There are four important assumptions in this development: (1) the gas is sufficiently dilute for only binary collisions to occur; (2) the motion of the molecules during a collision can be described by classical mechanics; (3) only elastic collisions occur; and (4) the intermolecular forces act only between fixed centers of the molecules; i.e., the intermolecular potential function is spherically symmetric. With these restrictions, it would appear that the resulting theory should be applicable only to low-pressure, high-temperature monatomic gases. However, the pressure and temperature restrictions are valid for polyatomic gases and, except for thermal conductivity (see Chap. 10), are adequate for most modeling purposes.

The Chapman-Enskog treatment develops integral relations for the transport properties when the interactions between colliding molecules are described by a potential energy function $\psi(r)$. The equations require complex numerical solution for each choice of intermolecular potential model. In general terms, the first-order solution for viscosity can be written as

$$\eta = \frac{(26.693)(MT)^{\frac{1}{2}}}{\sigma^2\Omega_v} \tag{10.3-9}$$

where the temperature dependence of the collision integral, Ω_v, is different for each $\psi(r)$, and all symbols and units are as defined in Eq. (10.3-8). Ω_v is unity if the molecules do not attract each other. Corrections can be found in Chapman and Cowling (1939) and Hirschfelder et al. (1954). The use of Ω_v from the Lennard-Jones (12-6) potential function is illustrated in Sec. 10.4.

10.4 ESTIMATION OF LOW-PRESSURE GAS VISCOSITY

Essentially all gas viscosity estimation techniques are based on either the Chapman-Enskog theory or the law of corresponding states. Both approaches are discussed below, and recommendations are presented at the end of the section. Experimental values of low-pressure gas viscosities are compiled in Landolt-Bornstein (1955),

Stephan and Lucas (1979), and Vargaftik et al. (1996). Literature references for a number of substances along with equations with which to calculate gas viscosities based on critically evaluated data may be found in the DIPPR 801 database (Wilding et al., 2017; DIPPR 2021) and ThermoData Engine from the Thermodynamics Research Center (Diky et al., 2021). Gas phase viscosity information can also be found in Speight (2016), Rumble (2021), Green and Southard (2018), and Yaws (1995, 1995a). This information should be used with caution in those cases where constants in equations have been determined from estimated rather than experimental viscosities.

10.4.1 Theoretical Approach

The first-order Chapman-Enskog viscosity equation was given as Eq. (10.3-9). To use this relation to estimate viscosities, the collision diameter σ and the collision integral Ω_v must be found. In the derivation of Eq. (10.3-9), Ω_v is obtained as a function of a dimensionless temperature T^* which depends upon the intermolecular potential chosen. For any potential curve, the dimensionless temperature T^* is related to ε by

$$T^* = \frac{kT}{\varepsilon} \tag{10.4-1}$$

where k is Boltzmann's constant and ε is the minimum of the pair-potential energy. The working equation for η must have as many parameters as were used to define the original $\psi(r)$ relation. While many potential models have been proposed (Hirschfelder et al., 1954), the Lennard-Jones 12-6 model was the first and has most often been applied for ideal gas viscosity.

$$\psi(r) = 4\varepsilon\left[\left(\frac{\sigma}{r}\right)^{12} - \left(\frac{\sigma}{r}\right)^{6}\right] \tag{10.4-2}$$

In Eq. (10.4-2), σ is like a molecular diameter and is the value of r that causes $\psi(r)$ to be zero. With this potential, the collision integral has been determined by a number of investigators (Barker et al., 1964; Hirschfelder et al., 1954; Itean et al., 1961; Klein and Smith, 1968; Monchick and Mason, 1961; O'Connell and Prausnitz, 1965). Neufeld et al. (1972) proposed an empirical equation which is convenient for computer application:

$$\Omega_v = \left[A(T^*)^{-B}\right] + C\left[\exp(-DT^*)\right] + E\left[\exp(-FT^*)\right] \tag{10.4-3}$$

where $T^* = \frac{kT}{\varepsilon}$, $A = 1.16145$, $B = 0.14874$, $C = 0.52487$, $D = 0.77320$, $E = 2.16178$, and $F = 2.43787$. Equation (10.4-3) is applicable from $0.3 \le T^* \le 100$ with an average deviation of only 0.064%. A graph of $\log \Omega_v$ as a function of $\log T^*$ is shown in Fig. 10.1.

With values of Ω_v as a function of T^*, several investigators have used Eq. (10.3-9) and regressed experimental viscosity-temperature data to find the best values of $\frac{\varepsilon}{k}$ and σ for many substances. Appendix B lists a number of such sets as reported by Svehla (1962). It should be noted, however, that there appears to be a number of other quite satisfactory *sets* of $\frac{\varepsilon}{k}$ and σ for any given compound. For example, with *n*-butane, Svehla suggested $\frac{\varepsilon}{k} = 513.4$ K, $\sigma = 4.730$ Å, whereas Flynn and Thodos (1962) recommend $\frac{\varepsilon}{k} = 208$ K and $\sigma = 5.869$ Å.

Both sets, when used to calculate viscosities, yield almost exactly the same values as shown in Fig. 10.2. This interesting paradox has been resolved by Reichenberg (1971), who suggested that $\log \Omega_v$ is essentially a linear function of $\log T^*$ (see Fig. 10.1).

$$\Omega_v = a(T^*)^n \tag{10.4-4}$$

Kim and Ross (1967) do, in fact, propose that:

$$\Omega_v = 1.604(T^*)^{-0.5} \tag{10.4-5}$$

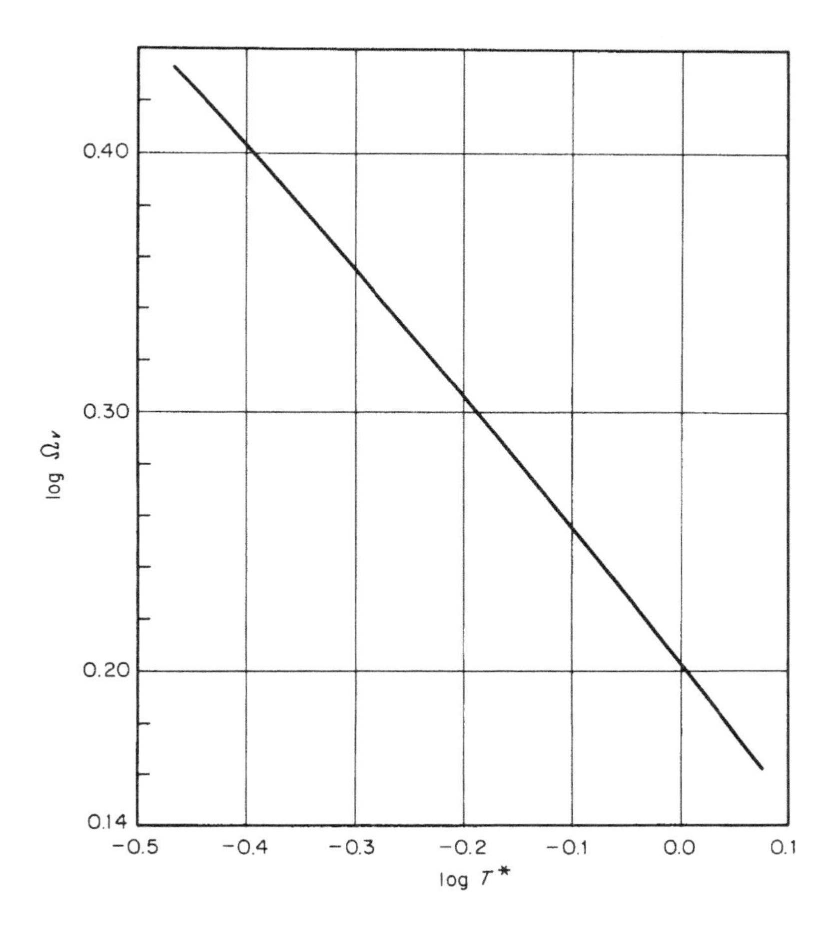

FIGURE 10.1
Effect of temperature on the Lennard-Jones viscosity collision integral.

where $0.4 \leq T^* \leq 1.4$. They note a maximum error of only 0.7%. Substitution of Eq. (10.4-5) into Eq. (10.3-9) leads to

$$\eta = \frac{16.64(M)^{\frac{1}{2}}T}{\left(\frac{\varepsilon}{k}\right)^{\frac{1}{2}}\sigma^2} \tag{10.4-6}$$

where the units are the same as in Eq. (10.3-9). Here the parameters σ and $\frac{\varepsilon}{k}$ are combined as a *single* term $\left(\frac{\varepsilon}{k}\right)^{\frac{1}{2}}\sigma^2$. There is then no way of delineating individual values of $\frac{\varepsilon}{k}$ and σ by using experimental viscosity data, at least over the range where Eq. (10.4-5) applies. Equation (10.4-6) suggests that when reduced temperature $T_r \leq 1.4$, low-pressure gas viscosities are essentially proportional to the absolute temperature.

The conclusion to be drawn from this discussion is that Eq. (10.3-9) can be used to calculate gas viscosity, although the chosen set of $\frac{\varepsilon}{k}$ and σ may have little relation to molecular properties. There will be an infinite number of acceptable sets as long as the temperature range is not too broad, e.g., if one limits the estimation to the range of reduced temperatures from about 0.3 to 1.2. Of course, when using published values of $\frac{\varepsilon}{k}$ and σ for a fluid of interest, the two values from the same set must be used—never $\frac{\varepsilon}{k}$ from one set and σ from another.

The difficulty in obtaining a priori meaningful values of $\frac{\varepsilon}{k}$ and σ has led most authors to specify rules which relate $\frac{\varepsilon}{k}$ and σ to macroscopic parameters, such as the critical constants. One such method is shown below.

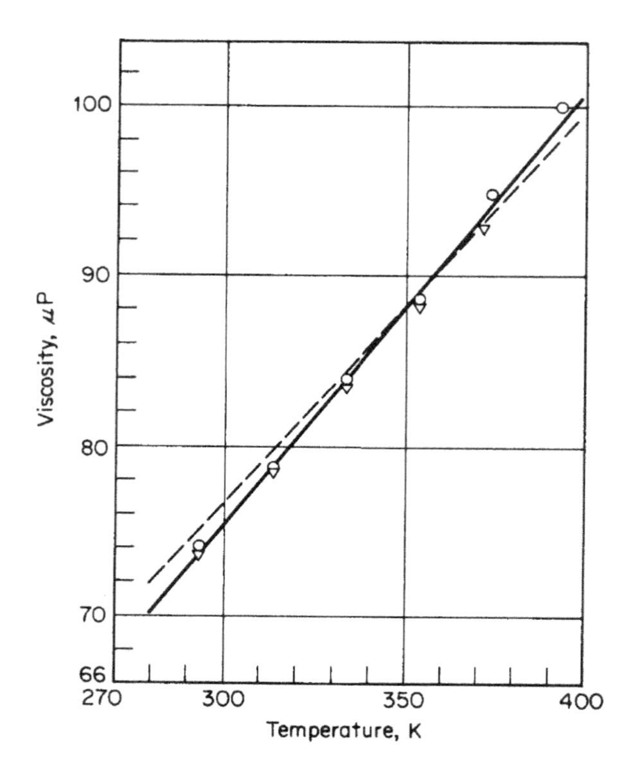

FIGURE 10.2
Comparison of calculated and experimental low-pressure gas viscosity of *n*-butane with Eq. (10.3-9) and the Lennard-Jones Potential: (dashed line) Flynn and Thodos (1962) with $\sigma = 5.869$ Å and $\frac{\varepsilon}{k} = 208$ K; (solid line) Svehla (1962) with $\sigma = 4.730$ Å and $\frac{\varepsilon}{k} = 513.4$ K; (o) Titani (1929); (∇) Wobster and Mueller (1941).

10.4.2 Method of Chung et al. (1984, 1988)

These authors have employed Eq. (10.3-9) with

$$\frac{\varepsilon}{k} = \frac{T_c}{1.2593} \tag{10.4-7}$$

$$\sigma = 0.809 V_c^{\frac{1}{3}} \tag{10.4-8}$$

where T_c is the critical temperature in K, $\frac{\varepsilon}{k}$ is in K, σ is in Å, and V_c is the critical volume in cm³/mol. Then, using Eqs. (10.4-1) and (10.4-7),

$$T^* = 1.2593 T_r \tag{10.4-9}$$

Ω_v in Eq. (10.3-9) is found from Eq. (10.4-3) with T^* defined by Eq. (10.4-9). Chung et al. also multiply the right-hand side of Eq. (10.3-9) by a factor F_c to account for molecular shapes and polarities of dilute gases. Their final result may be expressed as:

$$\eta = 40.785 \frac{F_c (MT)^{\frac{1}{2}}}{V_c^{\frac{2}{3}} \Omega_v} \tag{10.4-10}$$

TABLE 10.1 The association factor κ in Eq. (10.4-11) to predict gas viscosity using the method of Chung et al. (1988).

Compound	κ	Compound	κ
Methanol	0.215	*n*-Pentanol	0.122
Ethanol	0.175	*n*-Hexanol	0.114
n-Propanol	0.143	*n*-Heptanol	0.109
i-Propanol	0.143	Acetic acid	0.0916
n-Butanol	0.132	Water	0.076
i-Butanol	0.132		

where η is viscosity in μP, M is molecular weight in g/mol, T is temperature in K, V_c is critical volume in cm^3/mol, and Ω_v is the viscosity collision integral from Eq. (10.4-3) with T^* from Eq. (10.4-9). The molecular shape and correction factor is given by:

$$F_c = 1 - 0.2756\omega + 0.059035\mu_r^4 + \kappa \qquad (10.4\text{-}11)$$

Here, ω is the acentric factor (see Chap. 3), μ_r is the reduced dipole moment (discussed later), and κ is a special correction for highly polar substances, such as alcohols and acids. Values of κ for a few such materials are shown in Table 10.1. Chung et al. (1984) suggest that for alcohols not shown in Table 10.1,

$$\kappa = 0.0682 + 4.703203\left[\frac{N_{OH}}{M}\right] \qquad (10.4\text{-}12)$$

where N_{OH} is the number of —OH groups in a molecule and M is molecular weight in g/mol.

The term μ_r is a dimensionless dipole moment. Various methods to nondimensionalize dipole moment have been proposed. Chung et al. (1984) use a definition based on ε and σ which, combined with Eqs. (10.4-7) and (10.4-8), yields the following:

$$\mu_r = 131.3\frac{\mu}{(V_c T_c)^{\frac{1}{2}}} \qquad (10.4\text{-}13)$$

where V_c is in cm^3/mol, T_c is in K, and μ is in Debye (D).

Example 10.1 Estimate the viscosity of sulfur dioxide gas at atmospheric pressure and 300 °C by using the Chung et al. method. The experimental viscosity is 246 μP (Landolt-Bornstein, 1955).

Solution. From Appendix A, T_c = 430.75 K, V_c = 122 cm^3/mole, M = 64.0638 g/mol, μ = 1.63091 D, and ω = 0.245381. The method only gives values for κ for alcohols, so for sulfur dioxide κ = 0. From Eq. (10.4-13),

$$\mu_r = (131.3)(1.63091)/[(122)(430.75)]^{0.5} = 0.9341$$

and with Eqs. (10.4-11) and (10.4-9),

$$F_c = 1 - (0.2756)(0.245381) + (0.059035)(0.9341)^4 = 0.9773$$

$$T^* = 1.2593(573.15/430.75) = 1.6756$$

Then, with Eq. (10.4-3), $\Omega_v = 1.2557$. The viscosity is determined from Eq. (10.4-10).

$$\eta = (40.785)(0.9773)[(64.0638)(573.15)]^{0.5}/[(122)^{2/3}(1.2557)] = 247.3\ \mu\text{P}$$

$$\eta \text{ Difference} = 247.3 - 246 = 1.3\ \mu\text{P or } 0.53\%$$

10.4.3 Corresponding States Methods

A dimensionless viscosity can be defined by starting from an equation such as (10.3-9) and recognizing that σ^3 is directly related to V_c [see Eq. (10.4-8)]. By assuming V_c is proportional to $\frac{RT_c}{P_c}$, the reduced viscosity can be defined as

$$\eta_r = \zeta\eta = f(T_r) \tag{10.4-14}$$

$$\zeta = \left[\frac{RT_c N_A^2}{M^3 P_c^4}\right]^{\frac{1}{6}} \tag{10.4-15}$$

In SI units, if $R = 8314$ J/(kmol · K) and N_A (Avogadro's number) $= 6.023 \times 10^{26}$ kmol^{-1} and with T_c in K, M in kg/kmol, and P_c (critical pressure) is in Pa, ζ has the units of m^2/(N · s) or inverse viscosity. In more convenient units,

$$\zeta = 0.176\left(\frac{T_c}{M^3 P_c^4}\right)^{\frac{1}{6}} \tag{10.4-16}$$

where ζ is the inverse viscosity in μP^{-1}, T_c is in K, M is in g/mol, and P_c is in bar.

Equation (10.4-14) has been recommended by several authors (Flynn and Thodos, 1961; Golubev, 1959; Malek and Stiel, 1972; Mathur and Thodos, 1963; Trautz, 1931; Yoon and Thodos, 1970). The specific form suggested by Lucas (Lucas, 1980; Lucas, 1983; and Lucas, 1984a) is illustrated below.

$$\eta\zeta = \left[0.807T_r^{0.618} - 0.357\exp(-0.449T_r) + 0.340\exp(-4.058T_r) + 0.018\right]F_P^\circ F_Q^\circ \tag{10.4-17}$$

Here, ζ is defined by Eq. (10.4-16), η is in μP, T_r is the reduced temperature, and F_P° and F_Q° are correction factors to account for polarity or quantum effects. To obtain F_P°, a reduced dipole moment is required. Lucas defines this quantity as

$$\mu_r = 52.46\frac{\mu^2 P_c}{T_c^2} \tag{10.4-18}$$

where μ is in Debye, P_c is in bar, and T_c is in K. Then F_P° values are found as

$$\begin{aligned} F_P^\circ &= 1 & 0 \le \mu_r < 0.022 \\ F_P^\circ &= 1 + 30.55(0.292 - Z_c)^{1.72} & 0.022 \le \mu_r < 0.075 \\ F_P^\circ &= 1 + 30.55(0.292 - Z_c)^{1.72}\left|0.96 + 0.1(T_r - 0.7)\right| & 0.075 \le \mu_r \end{aligned} \tag{10.4-19}$$

The factor F_Q° is used only for the quantum gases He, H_2, and D_2.

$$F_Q^\circ = 1.22Q^{0.15}\left\{1 + 0.00385\left[(T_r - 12)^2\right]^{\frac{1}{M}}\text{sign}(T_r - 12)\right\} \tag{10.4-20}$$

where $Q = 1.38$ for He, $Q = 0.76$ for H_2, $Q = 0.52$ for D_2 and sign( ) indicates that either +1 or −1 should be used depending on whether the value of the argument ( ) is greater than or less than zero. It is interesting to note that if $T_r < 1$, the $f(T_r)$ in Eq. (10.4-17) is closely approximated by $0.606T_r$, that is,

$$\eta\zeta \approx 0.606T_r F_P^\circ F_Q^\circ \qquad T_r \le 1 \tag{10.4-21}$$

As mentioned above, several authors have proposed corresponding states prediction methods for gas viscosities based on the form of Eq. (10.4-14). The expression of Yoon and Thodos (1970) to obtain η is

$$\eta\zeta 10^5 = 46.10T_r^{0.618} - 20.4\exp(-0.449T_r) + 19.4\exp(-4.058T_r) + 1 \quad (10.4\text{-}22)$$

where

$$\zeta = \left(\frac{T_c}{M^3 P_c^4}\right)^{\frac{1}{6}} \quad (10.4\text{-}23)$$

Here, T_c is in K, M is in g/mol, P_c is in atm, and η is in cP. This expression was specifically designed for non-quantum, nonpolar gases but is evaluated for all classes of compounds later.

The method of Lucas is illustrated in Example 10.2 while that of Yoon and Thodos is demonstrated in Example 10.3.

Example 10.2 Estimate the viscosity of methanol vapor at 550 K and 1 bar using the method of Lucas. The experimental value is 181 μP (Stephan and Lucas, 1979).

Solution. From Appendix A, $T_c = 512.5$ K, $P_c = 80.84$ bar, $Z_c = 0.222$, $M = 32.04186$ g/mol, and $\mu = 1.69987$ D. With these values, $T_r = 550/512.5 = 1.0732$.

With Eq. (10.4-18),

$$\mu_r = 52.46[(1.69987)^2(80.84)/(512.5)^2] = 4.6667 \times 10^{-2}.$$

From Eq. (10.4-19),

$$F_P^\circ = 1 + (30.55)(0.292 - 0.222)^{1.72} = 1.3152$$

With Eq. (10.4-16),

$$\zeta = 0.176\{512.5/[(32.04186)^3(80.84)^4]\}^{1/6} = 4.7046 \times 10^{-3}\ \mu\text{P}^{-1}$$

Then, with Eq. (10.4-17)

$$\eta\zeta = \{(0.807)(1.0732)^{0.618} - 0.357\exp[-(0.449)(1.0732)] + 0.340\exp[-(4.058)(1.0732)] + 0.018\}(1.3152) = 0.84815$$

Yielding

$$\eta = 0.84815/4.7046 \times 10^{-3} = 180\ \mu\text{P}$$

$$\eta \text{ Difference} = 180 - 181 = -1\ \mu\text{P or } 0.55\%$$

Example 10.3 Estimate the viscosity of neopentane vapor at 305 K and 1 bar using the method of Yoon and Thodos. The experimental value is 74.0 μP (McCoubrey and Singh, 1963).

Solution. From Appendix A, $T_c = 433.8$ K, $P_c = 31.96$ bar $= 31.542$ atm, and $M = 72.149$ g/mol. With these values, $T_r = 305/433.8 = 0.7031$.

With Eq. (10.4-23),

$$\zeta = [433.8/(72.149^3 \cdot 31.542^4)]^{1/6} = 0.032447$$

Then, with Eq. (10.4-22),

$$\eta\zeta = \{(46.10)(0.7031)^{0.618} - 20.4\exp[(-0.449)(0.7031)] + 19.4\exp[(-4.058)(0.7031)] + 1\} = 2.4323 \times 10^{-4}$$

Yielding

$$\eta = 2.4323 \times 10^{-4}/0.032447 = 0.007496 \text{ cP} = 75.0\ \mu\text{P}$$

$$\eta \text{ Difference} = 75 - 74 = 1\ \mu\text{P or } 1.35\%$$

Reichenberg (1971, 1979) added a group contribution element to a corresponding states approach for low-pressure gas viscosity of organic compounds. In this formalism,

$$\eta = \frac{M^{\frac{1}{2}}T}{a^*\left[1 + \dfrac{4}{T_c}\right]\left[1 + 0.36T_r(T_r - 1)\right]^{\frac{1}{6}}} \frac{T_r(1 + 270\mu_r^4)}{T_r + 270\mu_r^4} \tag{10.4-24}$$

where M is the molecular weight in g/mol, T is the temperature in K, T_c is the critical temperature in K, T_r is the reduced temperature ($T_r = \frac{T}{T_c}$), μ_r is the reduced dipole moment given by Eq. (10.4-18), and η is in units of μP. The parameter a^* is defined as

$$a^* = \sum_i N_i \left(\Delta_{a^*}\right)_i \tag{10.4-25}$$

where $\left(\Delta_{a^*}\right)_i$ are the contributions of group i and N_i is the number of times group i is found in the compound. The values of $\left(\Delta_{a^*}\right)_i$ are contained in Table 10.2. The term $\left[1 + \frac{4}{T_c}\right]$ in the denominator of Eq. (10.4-24) may be neglected except for treating quantum gases with low values of T_c. The method is illustrated in Example 10.4.

Example 10.4 Estimate the viscosity of ethyl acetate vapor at 398.15 K and 1 bar using the method of Reichenberg. The experimental value is reported to be 101 μP (Landolt-Bornstein Tabellen, 1955).

Solution. From Appendix A, $T_c = 523.3$ K, $P_c = 38.80$ bar, $M = 88.105$ g/mol, and $\mu = 1.78081$ D. With these values, $T_r = 398.15/523.3 = 0.7608$. With Eq. (10.4-18),

$$\mu_r = (52.46)(1.78081)^2(38.80)/523.3^2 = 0.02327$$

Ethyl acetate is composed of two —CH_3 groups, one —CH_2— group, and one —COO— (ester) group. Using Eq. (10.4-25) with the values in Table 10.2,

$$a^* = 2(9.04) + 6.47 + 13.41 = 37.96$$

With Eq. (10.4-24),

$$\eta = (88.105)^{1/2}(398.15)(0.7608)[1 + (270)(0.02327)^4]/ \{(37.96)[1 + (0.36)(0.7608)(0.7608 - 1)]^{1/6}[0.7608 + (270)(0.02327)^4]\} = 99.6\ \mu\text{P}.$$

$$\eta \text{ Difference} = 99.6 - 101 = -1.4\ \mu\text{P or } -1.4\%$$

TABLE 10.2 Group contributions for the Reichenberg (1971, 1979) method for low-pressure gas viscosities to use in Eq. (10.4-25). Groups with atoms in rings are identified by the letter "r" following the single-letter element symbol (e.g. Cr, Or, Sr, and Nr). Do not confuse these ring designations with symbols for elements such as chromium or strontium.

Group	Δ_{a^*}
Nonring Carbon Increments	
—CH_3	9.04
—CH_2—	6.47
>CH—	2.67
>C<	−1.53
=CH_2	7.68
=CH—	5.53
=C<	1.78
≡CH	7.41
≡C—	5.24
Ring Carbon Increments	
—CrH_2—	6.91
>CrH—	1.16
>Cr<	0.23
=CrH—	5.90
=Cr<	3.59
Halogen Increments	
—F	4.46
—Cl	10.06
—Br	12.83
Oxygen Increments	
—OH (alcohol)	7.96
—O— (nonring)	3.59
>C=O (noring ketone)	12.02
—HC=O (aldehyde)	14.02
—COOH (acid)	18.65
—COO— (ester) or HCOO— (formate)	13.41
Nitrogen Increments	
—NH_2	9.71
—NH— (nonring)	3.68
=Nr— (ring)	4.97
—C≡N	18.15
Sulfur Increments	
—Sr— (ring)	8.86

10.4.4 Recommendations for Estimating Low-Pressure Viscosities of Pure Gases

A test database was created to evaluate the performance of pure component low-pressure vapor viscosity prediction methods. This database consisted of all reliable, experimental vapor viscosity data from the October 2021 release of the DIPPR database. Temperatures ranged from T_m to T_c and included over 2600 total points (N_{pt}) from 159 unique compounds (N_c). The techniques tested were: the theoretical approach of Eq. (10.4-10) by Chung (1984, 1988) (abbreviated CH), the corresponding states approaches of Eq. (10.4-17) by Lucas (1980, 1983, 1984a) (abbreviated LU), Eq. (10.4-22) by Yoon and Thodos (1970) (abbreviated YT), and the group contribution method of Reichenberg (1971, 1979) (abbreviated RE).

To identify if a method was better for one class of compounds than other methods, each chemical in the test database was assigned a *class* according to the scheme below.

- **Normal:** Compounds containing only C, H, F, Br, and/or I with molecular weights below 213 g/mol.
- **Heavy:** Compounds containing only C, H, F, Br, and/or I with molecular weights greater than or equal to 213 g/mol. Also compounds containing only C, H, F, Br, I and/or metal atoms (e.g., Fe, As, Al, etc.) regardless of MW.
- **Polar:** Compounds containing atoms other than C, H, F, Br, metals (e.g., O, N, P, S, etc.) that cannot form hydrogen bonds (e.g., esters, ketones, aldehydes, thioesters, sulfonyl, etc.).
- **Associating:** Compounds containing atoms other than C, H, F, Br, metals (e.g., O, N, P, S, etc.) that can form hydrogen bond (e.g., acids, amines, alcohols, etc.).

Results were quantified using the average absolute relative deviation ($AARD$) defined as

$$AARD = \frac{1}{N_{pt}} \sum_{i=1}^{N_{pt}} \left| \frac{X_{pred_i} - X_{exp_i}}{X_{exp_i}} \right| \tag{10.4-26}$$

where the index is for point i in the test set, X_{pred_i} is the predicted value for property X for point i, and X_{exp_i} is the experimental value for property X for point i.

Table 10.3 contains the average absolute relative deviation [see Eq. (10.4-26)] of each method according to compound class. Also listed are the number of points tested in each class (N_{pt}) and number of compounds (N_c). Overall, each of the methods perform approximately equally well. The CH and YT methods appear slightly better for normal compounds. Surprisingly, the YT method has the best performance for polar compounds even though it was not developed for such. However, it is significantly worse than all other methods for associating compounds. The RE technique produces significantly better results for associating compounds, while the LU approach was most consistent with an AARD of 6.6% or less for all classes. The maximum AARD for the other methods was between 7.6% and 9.8%.

TABLE 10.3 Performance of prediction techniques for low-pressure vapor η against the test database.

	Compound Class											
	Normal			Heavy			Polar			Associating		
Method	N_{pt}	N_c	AARD	N_{pt}	N_c	AARD	N_{pt}	N_c	AARD	N_{pt}	N_c	AARD
CH	1938	109	3.8%	41	4	5.1%	320	19	7.6%	363	27	5.2%
LU	1938	109	4.7%	41	4	6.6%	320	19	6.4%	363	27	5.0%
YT	1938	109	3.9%	41	4	6.5%	320	19	5.3%	363	27	9.8%
RE	1516	89	4.5%	32	3	5.8%	126	12	8.5%	223	17	3.9%

10.5 VISCOSITIES OF GAS MIXTURES AT LOW PRESSURES

The rigorous kinetic theory of Chapman and Enskog can be extended to determine the viscosity of low-pressure multicomponent mixtures (Brokaw, 1964, 1965, 1968, 1965; Chapman and Cowling, 1939; Hirschfelder et al., 1954; Kestin et al., 1976). The final expressions are quite complicated and are rarely used to estimate mixture viscosities. Three simplifications of the rigorous theoretical expressions are described below. Reichenberg's equations are the most complex, but, as shown later, the most consistently accurate. Wilke's method is simpler, and that of Herning and Zipperer is even more so. All these methods are essentially interpolative; i.e., the viscosity values for the pure components must be available. The methods then lead to estimations showing how the mixture viscosity varies with composition. Later in this section, two corresponding states methods are described; they do not require pure component values as inputs. A compilation of references dealing with gas mixture viscosities (low and high pressure) has been prepared by Sutton (1976).

10.5.1 Method of Reichenberg (1974, 1977, 1979)

In this technique, Reichenberg has incorporated elements of the kinetic theory approach of Hirschfelder et al. (1954) with corresponding states methodology to obtain desired parameters. In addition, a polar correction has been included. The general, multicomponent mixture viscosity equation is

$$\eta_{\mathrm{m}} = \sum_{i=1}^{n} K_i \left(1 + 2\sum_{j=1}^{i-1} H_{ij} K_j + \sum_{j=1 \neq i}^{n} \sum_{k=1 \neq i}^{n} H_{ij} H_{ik} K_j K_k \right) \tag{10.5-1}$$

where η_m is the mixture viscosity and n is the number of components. With η_i the viscosity of pure i, M_i the molecular weight of i, and y_i the mole fraction of i in the mixture,

$$K_i = y_i \eta_i \left(y_i + \eta_i \sum_{k=1 \neq i}^{n} y_k H_{ik} \left[3 + \frac{2M_k}{M_i} \right] \right)^{-1} \tag{10.5-2}$$

Two other component properties used are

$$U_i = \frac{[1 + 0.36 T_{r_i} (T_{r_i} - 1)]^{\frac{1}{6}} F_{R_i}}{(T_{r_i})^{\frac{1}{2}}} \tag{10.5-3}$$

$$C_i = \frac{M_i^{\frac{1}{4}}}{(\eta_i U_i)^{\frac{1}{2}}} \tag{10.5-4}$$

where $T_{r_i} = \frac{T}{T_{c_i}}$ and F_{R_i} is a polar correction given by

$$F_{R_i} = \frac{T_{r_i}^{3.5} + (10\mu_{ri})^7}{T_{r_i}^{3.5} [1 + (10\mu_{ri})]^7} \tag{10.5-5}$$

Here μ_{r_i} is the reduced dipole moment of i and is calculated as shown earlier in Eq. (10.4-18). For Eq. (10.5-1), $H_{ij} = H_{ji}$ and

$$H_{ij} = \left[\frac{M_i M_j}{32(M_i + M_j)^3} \right]^{\frac{1}{2}} (C_i + C_j)^2 \frac{\left[1 + 0.36 T_{r_{ij}} (T_{r_{ij}} - 1) \right]^{\frac{1}{6}} F_{R_{ij}}}{\left(T_{r_{ij}} \right)^{\frac{1}{2}}} \tag{10.5-6}$$

with

$$T_{r_{ij}} = \frac{T}{(T_{c_i} + T_{c_j})^{\frac{1}{2}}} \tag{10.5-7}$$

$F_{R_{ij}}$ is found from Eq. (10.5-5) with T_{r_i} replaced by $T_{r_{ij}}$ and μ_{r_i} by $\mu_{r_{ij}} = (\mu_{r_i}\mu_{r_j})^{\frac{1}{2}}$.

For a binary gas mixture of 1 and 2, these equations may be written as

$$\eta_m = K_1(1 + H_{12}^2 K_2^2) + K_2(1 + 2H_{12}K_1 + H_{12}^2 K_1^2) \tag{10.5-8}$$

$$K_1 = \frac{y_1\eta_1}{y_1 + \eta_1\left\{y_2 H_{12}\left[3 + \frac{2M_2}{M_1}\right]\right\}} \tag{10.5-9}$$

$$K_2 = \frac{y_2\eta_2}{y_2 + \eta_2\left\{y_1 H_{12}\left[3 + \frac{2M_1}{M_2}\right]\right\}} \tag{10.5-10}$$

$$U_1 = \frac{\left[1 + 0.36T_{r_1}(T_{r_1} - 1)\right]^{\frac{1}{6}}}{(T_{r_1})^{\frac{1}{2}}} \frac{T_{r_1}^{3.5} + (10\mu_{r_1})^7}{T_{r_1}^{3.5}\left[1 + (10\mu_{r_1})\right]^7} \tag{10.5-11}$$

and a comparable expression for U_2. The meaning of C_1 and C_2 is clear from Eq. (10.5-4). Finally, with

$$T_{r_{ij}} = \frac{T}{(T_{c_1} + T_{c_2})^{\frac{1}{2}}} \quad \text{and} \quad \mu_{12} = (\mu_{r_1}\mu_{r_2})^{\frac{1}{2}}$$

$$H_{12} = \left[\frac{M_1 M_2}{32(M_1 + M_2)^3}\right]^{\frac{1}{2}} (C_1 + C_2)^2 \frac{\left[1 + 0.36T_{r_{12}}(T_{r_{12}} - 1)\right]^{\frac{1}{6}}}{(T_{r_{12}})^{\frac{1}{2}}} \frac{T_{r_{12}}^{3.5} + (10\mu_{r_{12}})^7}{T_{r_{12}}^{3.5}[1 + (10\mu_{r_{12}})]^7} \tag{10.5-12}$$

To employ Reichenberg's method, one needs the pure gas viscosity for each component at the system temperature as well as the molecular weight, dipole moment, critical temperature, and critical pressure. The temperature and composition are state variables. The method is illustrated in Example 10.5. A comparison of experimental and calculated gas-mixture viscosities is shown in Table 10.4.

Example 10.5 Use Reichenberg's method to estimate the viscosity of a nitrogen-chlorodifluoromethane (R-22) mixture at 50 °C and atmospheric pressure. The mole fraction of nitrogen is 0.286. The experimental viscosity is 145 μP (Tanaka et al., 1977).

Solution. The following pure component properties are used:

	N_2	$CHClF_2$
T_c, K	126.2	369.28
P_c, bar	33.98	49.86
M, g/mol	28.014	86.468
μ, D	0	1.4
η, 50 °C, μP	188	134

TABLE 10.4 Comparison of calculated and experimental low-pressure gas mixture viscosities.

System	T (K)	Mole Fraction First Component	η_m (exp.) (μP)	Ref.*	Percent Deviation† Calculated by Method of: Reichenberg Eq. (10.5-1)	Wilke Eqs. (10.5-16) & (10.5-14)	Herning and Zipperer Eq. (10.5-16) & (10.5-17)	Lucas Eq. (10.4-17) and Eqs. (10.5-18) to (10.5-23)	Chung et al. Eq. (10.5-24)
Nitrogen-hydrogen	373	0.0	104.2	6, 11	—	—	—	0.8	−11
		0.2	152.3		4.3	12	2.0	2.1	−23
		0.51	190.3		1.8	5.6	−1.0	−2.0	−11
		0.80	205.8		0.1	1.4	−1.2	3.6	−3.3
		1.0	210.1		—	—	—	0.4	0
Methane-propane	298	0.0	81.0	1	—	—	—	3.5	1.3
		0.2	85.0		0.2	−0.3	−0.2	4.6	1.7
		0.4	89.9		0.1	−0.8	−0.6	5.0	1.7
		0.6	95.0		0.6	−0.4	−0.2	5.4	1.9
		0.8	102.0		0.2	−0.6	−0.5	3.7	1.0
		1.0	110.0		—	—	—	0.3	1.0
	498	0.0	131.0	1	—	—	—	4.0	2.3
		0.2	136.0		0.4	0.0	−0.2	5.2	2.7
		0.4	142.0		0.6	0.0	−0.5	5.6	2.6
		0.6	149.0		0.7	0.0	−0.6	5.2	2.0
		0.8	157.0		0.7	0.0	−0.3	3.7	1.1
		1.0	167.0		—	—	—	−0.2	0.2
Carbon tetrafluoride-sulfur hexafluoride	303	0.0	159.0	8	—	—	—	6.6	0.7
		0.257	159.9		2.0	2.0		9.3	3.8
		0.491	161.5		3.4	3.4		11.0	6.2
		0.754	164.3		4.6	4.6		13.0	8.4
		1.0	176.7		—	—	—	7.4	3.6

(Continued)

TABLE 10.4 Comparison of calculated and experimental low-pressure gas mixture viscosities. *(Continued)*

System	T (K)	Mole Fraction First Component	η_m (exp.) (μP)	Ref.*	Percent Deviation† Calculated by Method of: Reichenberg Eq. (10.5-1)	Wilke Eqs. (10.5-16) & (10.5-14)	Herning and Zipperer Eq. (10.5-16) & (10.5-17)	Lucas Eq. (10.4-17) and Eqs. (10.5-18) to (10.5-23)	Chung et al. Eq. (10.5-24)
Nitrogen-carbon dioxide	293	0.0	146.6	5	—	—	—	1.6	−1.2
		0.213	153.5		0.5	−1.3	−1.0	0.4	−0.3
		0.495	161.8		0.4	−1.8	−1.5	−0.2	0.7
		0.767	172.1		−2.0	−2.8	−2.5	−1.7	−0.7
		1.0	175.8		—	—	—	0.4	−0.2
Ammonia-hydrogen	306	0.0	90.6	6	—	—	—	2.4	−9.9
		0.195	118.4		−4.0	−11	−18	−2.7	2.1
		0.399	123.8		−4.6	−12	−19	−3.0	10.0
		0.536	122.4		−4.5	−11	−16	−2.7	10.0
		0.677	120.0		−4.8	−9.7	−14	−3.1	7.1
		1.0	105.9		—	—	—	1.3	0.9
Hydrogen sulfide-ethyl ether	331	0.0	84.5	7	—	—	—	−1.7	−4.0
		0.204	87		−2.9	−3.2	0.2	2.3	0.4
		0.500	97		−2.2	−2.8	3.2	3.4	1.7
		0.802	116		0.0	−0.4	4.2	0.6	−0.7
		1.0	137		—	—	—	−3.0	−4.1
Ammonia-methylamine	423	0.0	130.0	2	—	—	—	−2.1	−8.0
		0.25	134.5		−0.8	−0.3	−0.6	−1.5	−7.5
		0.75	142.2		−1.0	−0.3	−0.7	0.1	−3.4
		1.0	146.0		—	—	—	1.1	1.1
	673	0.0	204.8	2	—	—	—	−4.6	−11
		0.25	212.8		−2.6	−0.7	−0.9	−4.9	−11
		0.75	228.3		−3.1	−0.7	−0.9	−4.7	−9.3
		1.0	235.0		—	—	—	−4.3	−5.4

†Percent deviation = [(calc. − exp.)/(exp.)] × 100.

*References: 1, Bircher (1943); 2, Burch and Raw (1967); 3, Carmichael and Sage (1966); 4, Chakraborti and Gray (1965); 5, Kestin and Leidenfrost (1959); 6, Pal and Baruna (1967); 7, Pal and Bhattacharyya (1969); 8, Raw and Tang (1963); 9, Stephan and Lucas (1979); 10, Tanka et al. (1977); 11, Trautz and Baumann (1929).

With $T = 50$ °C, T_r (N_2) = 2.56, and T_r ($CHClF_2$) = 0.875,

$$T_{r_{12}} = (50 + 273.2)/[(126.2)(369.3)]^{1/2} = 1.497$$

μ_r (N_2) = 0, and from Eq. (10.4-18),

$$\mu_r\,(CHClF_2) = (52.46)(1.4)^2(49.86)/369.28^2 = 0.0376$$

Since $\mu_{r_{12}} = (\mu_{r_1}\mu_{r_1})^{\frac{1}{2}}$, then for this mixture, $\mu_{r_{12}} = 0$. With Eq. (10.5-11), for $CHC1F_2$,

$$U\,(CHClF_2) = [1 + (0.36)(0.875)(0.875 - 1)]^{1/6}\,[(0.875)^{3.5} + (10)^7(0.0376)^7]/ \{(0.875)^{0.5}(0.875)^{3.5}[1 + (10)^7(0.0376)^7]\} = 1.063$$

and $U(N_2) = 0.725$. Then, from Eq. (10.5-4),

$$C\,(N_2) = (28.014)^{1/4}/[(188)(0.725)]^{1/2} = 0.197$$

and C ($CHClF_2$) = 0.256. Next, with Eq. (10.5-6) followed by Eq. (10.5-2),

$$H\,(N_2\text{—}CHClF_2) = (0.197 + 0.256)^2[(28.014)(86.468)]^{1/2}\,[1 + (0.36)(1.497)(1.497 - 1)]^{1/6}/ \{[32(28.014 + 86.468)^3]^{1/2}\,(1.497)^{1/2}\}(1.0) = 1.239 \times 10^{-3}$$

$$K\,(N_2) = (0.286)(188)/\{0.286 + (188)(0.714)(1.239 \times 10^{-3})[3 + (2)(86.469)/28.014]\} = 29.68$$

and K ($CHClF_2$) = 107.9. Substituting into Eq. (10.5-8),

$$\eta_m = (29.71)[1 + (1.239 \times 10^{-3})^2(107.9)^2] + (107.9)[1 + (2)(1.239 \times 10^{-3})(29.68) + (1.239 \times 10^{-3})^2(29.68)^2)$$
$$= 146.2\ \mu P$$

η_m Difference = 146.2 − 145 = 1.2 μP or 0.83%

10.5.2 Method of Wilke (1950)

In a further simplification of the kinetic theory approach, Wilke (1950) neglected second-order effects and proposed:

$$\eta_m = \sum_{i=1}^{n} \frac{y_i \eta_i}{\sum_{j=1}^{n} y_j \phi_{ij}} \tag{10.5-13}$$

where

$$\phi_{ij} = \frac{\left[1 + \left(\frac{\eta_i}{\eta_j}\right)^{\frac{1}{2}} \left(\frac{M_j}{M_i}\right)^{\frac{1}{4}}\right]^2}{\left[8\left(1 + \frac{M_i}{M_j}\right)\right]^{\frac{1}{2}}} \tag{10.5-14}$$

ϕ_{ji} is found by interchanging subscripts or by

$$\phi_{ji} = \frac{\eta_j}{\eta_i}\frac{M_i}{M_j}\phi_{ij} \tag{10.5-15}$$

For a binary system of 1 and 2, with Eqs. (10.5-13) to (10.5-15),

$$\eta_m = \frac{y_1\eta_1}{y_1 + y_2\phi_{12}} + \frac{y_2\eta_2}{y_2 + y_1\phi_{21}} \tag{10.5-16}$$

where η_m is the viscosity of the mixture, η_1 and η_2 are the pure component viscosities, y_1 and y_2 are the mole fractions and

$$\phi_{12} = \frac{\left[1 + \left(\frac{\eta_1}{\eta_2}\right)^{\frac{1}{2}}\left(\frac{M_2}{M_1}\right)^{\frac{1}{4}}\right]^2}{\left[8\left(1 + \frac{M_1}{M_2}\right)\right]^{\frac{1}{2}}}$$

$$\phi_{21} = \phi_{12}\frac{\eta_2 M_1}{\eta_1 M_2}$$

Equation (10.5-13), with ϕ_{ij} from Eq. (10.5-14), has been extensively tested. Wilke (1950) compared values with data on 17 binary systems and reported an average deviation of less than 1%; several cases in which η_m passed through a maximum were included. Many other investigators have tested this method (Amdur and Mason, 1958; Bromley and Wilke, 1951; Cheung, 1958; Dahler, 1959; Gandhi and Saxena, 1964; Ranz and Brodowsky, 1962; Saxena and Gambhir, 1963, 1963a; Strunk et al., 1964; Vanderslice et al. 1962; Wright and Gray, 1962). In most cases, only nonpolar mixtures were compared, and very good results obtained. For some systems containing hydrogen as one component, less satisfactory agreement was noted. In Table 10.4, Wilke's method predicted mixture viscosities that were larger than experimental for the H_2—N_2 system, but for H_2—NH_3, it underestimated the viscosities. Gururaja et al. (1967) found that this method also overpredicted in the H_2—O_2 case but was quite accurate for the H_2—CO_2 system. Wilke's approximation has proved reliable even for polar-polar gas mixtures of aliphatic alcohols (Reid and Belenyessy, 1960). The principal reservation appears to lie in those cases where $M_i >> M_j$ and $\eta_i >> \eta_j$.

Example 10.6 Kestin and Yata (1968) report that the viscosity of a mixture of methane and *n*-butane is 93.35 μP at 293 K when the mole fraction of *n*-butane is 0.303. Compare this result with the value estimated by Wilke's method. For pure methane and *n*-butane, these same authors report viscosities of 109.4 and 72.74 μP.

Solution. Let 1 refer to methane and 2 to *n*-butane. M_1 = 16.043 g/mol and M_2 = 58.123 g/mol.

$$\phi_{12} = [1 + (109.4/72.74)^{1/2}(58.123/16.043)^{1/4}]^2/\{8[1 + (16.043/58.123)]\}^{1/2} = 2.268$$

$$\phi_{21} = (2.268)(72.74/109.4)(16.043/58.123) = 0.416$$

$$\eta_m = (0.697)(109.4)/[0.697 + (0.303)(2.268)]$$
$$+ (0.303)(72.74)/[0.303 + (0.697)(0.416)] = 92.26\ \mu\text{P}$$

$$\eta_m \text{ Difference} = 92.26 - 93.35 = -1.09\ \mu\text{P or } -1.2\%$$

10.5.3 Herning and Zipperer (1936) Approximation of ϕ_{ij}

As an approximate expression for ϕ_{ij} of Eq. (10.5-14), the following is proposed (Herning and Zipperer, 1936):

$$\phi_{ij} = \left(\frac{M_j}{M_i}\right)^{\frac{1}{2}} = \phi_{ji}^{-1} \tag{10.5-17}$$

When Eq. (10.5-17) is used with Eq. (10.5-16) to estimate low-pressure binary gas mixture viscosities, quite reasonable predictions are obtained (Table 10.4) except for systems such as H_2—NH_3. The technique is illustrated in Example 10.7. Note that Example 10.6 and Example 10.7 treat the same problem; each provides a viscosity estimate close to the experimental value. But the ϕ_{12} and ϕ_{21} values employed in the two cases are quite different. Apparently, multiple sets of ϕ_{ij} and ϕ_{ji} work satisfactorily in Eq. (10.5-13).

Example 10.7 Repeat Example 10.6 by using the Herning and Zipperer approximation for ϕ_{ij}.

Solution. As before, with 1 as methane and 2 as *n*-butane

$$\phi_{12} = (58.123/16.043)^{1/2} = 1.903$$

$$\phi_{21} = 1.903^{-1} = 0.525$$

$$\eta_m = (0.697)(109.4)/[0.697 + (0.303)(1.903)] + (0.303)(72.74)/[0.303 + (0.697)(0.525)] = 92.82\ \mu P$$

$$\eta_m \text{ Difference} = 92.82 - 93.35 = -0.53\ \mu P \text{ or } -0.6\%$$

10.5.4 Corresponding States Methods

In this approach, one estimates pseudocritical and other mixture properties (see Sec. 5.3) from pure-component properties, the composition of the mixture, and appropriate combining and mixing rules.

10.5.4.1 Lucas Rules (1980, 1983, 1984a)

Lucas (1980, 1983, 1984a) defined mixture properties as shown below for use in Eqs. (10.4-17) through (10.4-22).

$$T_{c_m} = \sum_{i=1}^{n} y_i T_{c_i} \tag{10.5-18}$$

$$P_{c_m} = RT_{c_m} \frac{\sum_{i=1}^{n} y_i Z_{c_i}}{\sum_{i=1}^{n} y_i V_{c_i}} \tag{10.5-19}$$

$$M_m = \sum_{i=1}^{n} y_i M_i \tag{10.5-20}$$

$$F^{\circ}_{P_m} = \sum_{i=1}^{n} y_i F^{\circ}_{P_i} \tag{10.5-21}$$

$$F^{\circ}_{Q_m} = \left(\sum_{i=1}^{n} y_i F^{\circ}_{Q_i}\right) A \tag{10.5-22}$$

and, letting the subscript H denote the mixture component of highest molecular weight and L the component of lowest molecular weight,

$$A = 1 - 0.01\left(\frac{M_H}{M_L}\right)^{0.87} \quad \text{for } \frac{M_H}{M_L} > 9 \text{ and } 0.05 < y_H < 0.7 \qquad (10.5\text{-}23)$$

$$A = 1 \qquad \text{otherwise}$$

The method of Lucas does not necessarily lead to the pure-component viscosity η_i when all $y_i = 0$ except $y_i = 1$. Thus, the method is not interpolative in the same way as are the techniques of Reichenberg, Wilke, and Herning and Zipperer. Nevertheless, as seen in Table 10.4, the method provides reasonable estimates of η_m in most test cases.

Example 10.8 Estimate the viscosity of a binary mixture of ammonia and hydrogen at 33 °C and low pressure by using the Lucas corresponding states method.

Solution. Let us illustrate the method for a mixture containing 67.7 mole% ammonia. We use the following pure-component values:

	Ammonia	Hydrogen
T_c, K	405.50	33.2
P_c, bar	113.5	13.0
V_c, cm^3/mol	72.5	64.3
Z_c	0.244	0.306
M, g/mol	17.031	2.016
μ, D	1.47	0
T_r	0.755	9.223

Using Eqs. (10.5-18) to (10.5-20), $T_{c_m} = 285.2$ K, $P_{c_m} = 89.6$ bar, and $M_m = 12.18$ g/mol. From these values and Eq. (10.4-16), $\zeta_m = 6.46 \times 10^{-3}\ \mu P^{-1}$. With Eq. (10.4-18), $\mu_r(NH_3) = 7.825 \times 10^{-2}$ and $\mu_r(H_2) = 0$. Then, with Eq. (10.4-19),

$$F_P^\circ(NH_3) = 1 + 30.55(0.292 - 0.244)^{1.72}\,|0.96 + 0.1(0.755 - 0.7)| = 1.159$$

$$F_P^\circ(H_2) = 1.0$$

$$F_{P_m}^\circ = (1.159)(0.677) + (1.0)(0.323) = 1.108$$

For the quantum correction, with Eq. (10.5-23), since $\frac{M_H}{M_L} = 17.031/2.016 = 8.4 < 9$, then $A = 1$.

$$F_Q^\circ(NH_3) = 1.0$$

And with Eq. (10.4-20),

$$F_Q^\circ = (1.22)(0.76)^{0.15}\,\{1 + 0.00385[(9.223 - 12)^2]^{1/2.016} \times \text{sign}(9.223 - 12)\}$$

$$= (1.171)[1 + (0.01061)(-1)] = 1.159$$

And from Eq. (10.5-23),

$$F^{\circ}_{Q_m} = (1.159)(0.323) + (1.0)(0.677) = 1.051$$

Next, from Eq. (10.4-17) with $T_{r_m} = (33 + 273.15)/285.3 = 1.073$

$$\eta_m \zeta_m = (0.645)(1.107)(1.051) = 0.750$$

$$\eta_m = 0.750/6.46 \times 10^{-3} = 116.1\ \mu\text{P}$$

The experimental value is 120.0 μP; thus

$$\eta_m \text{ Difference} = 116.1 - 120.0 = -3.9\ \mu\text{P or } -3.3\%.$$

The viscosity of the ammonia-hydrogen mixture at 33 °C is line 4 in Fig. 10.3.

No.	System	T (K)	Reference
1	Hydrogen sulfide-ethyl ether	331	Pal and Bhattacharyya (1969)
2	Ammonia-ethyl ether	331	Pal and Bhattacharyya (1969)
3	Methane-*n*-butane	293	Kestin and Yata (1968)
4	Ammonia-hydrogen	306	Pal and Barua (1967)
5	Ammonia-methyl amine	423	Burch and Raw (1967)
6	Ethylene-ammonia	293	Trautz and Heberling (1931)

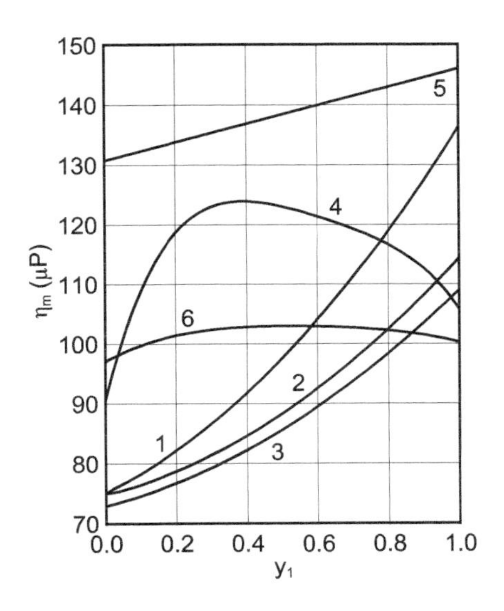

FIGURE 10.3
Gas mixture viscosities.

10.5.4.2 Chung et al. Rules (1984, 1988)

In this case, Eq. (10.3-9) is employed to estimate the mixture viscosity with, however, a factor F_{c_m} as used in Eq. (10.4-10) to correct for shape and polarity.

$$\eta_m = 26.69 \frac{F_{c_m}(M_m T)^{\frac{1}{2}}}{\sigma_m^2 \Omega_v} \tag{10.5-24}$$

where $\Omega_v = f(T_m^*)$. In the Chung et al. approach, the mixing rules are:

$$\sigma_m^3 = \sum_i \sum_j y_i y_j \sigma_{ij}^3 \tag{10.5-25}$$

$$T_m^* = \frac{T}{\left(\frac{\varepsilon}{k}\right)_m} \tag{10.5-26}$$

$$\left(\frac{\varepsilon}{k}\right)_m = \frac{\sum_i \sum_j y_i y_j \left(\frac{\varepsilon_{ij}}{k}\right) \sigma_{ij}^3}{\sigma_m^3} \tag{10.5-27}$$

$$M_m = \left[\frac{\sum_i \sum_j y_i y_j \left(\frac{\varepsilon_{ij}}{k}\right) \sigma_{ij}^2 M_{ij}^{\frac{1}{2}}}{\left(\frac{\varepsilon}{k}\right)_m \sigma_m^2}\right]^2 \tag{10.5-28}$$

$$\omega_m = \frac{\sum_i \sum_j y_i y_j \omega_{ij} \sigma_{ij}^3}{\sigma_m^3} \tag{10.5-29}$$

$$\mu_m^4 = \sigma_m^3 \sum_i \sum_j \frac{y_i y_j \mu_i^2 \mu_j^2}{\sigma_{ij}^3} \tag{10.5-30}$$

$$\kappa_m = \sum_i \sum_j y_i y_j \kappa_{ij} \tag{10.5-31}$$

and the combining rules are:

$$\sigma_{ii} = \sigma_i = 0.809 V_{c_i}^{\frac{1}{3}} \tag{10.5-32}$$

$$\sigma_{ij} = \xi_{ij} (\sigma_i \sigma_j)^{\frac{1}{2}} \tag{10.5-33}$$

$$\frac{\varepsilon_{ii}}{k} = \frac{\varepsilon_i}{k} = \frac{T_{c_i}}{1.2593} \tag{10.5-34}$$

$$\frac{\varepsilon_{ij}}{k} = \zeta_{ij} \left(\frac{\varepsilon_i}{k} \frac{\varepsilon_j}{k}\right)^{\frac{1}{2}} \tag{10.5-35}$$

$$\omega_{ii} = \omega_i \tag{10.5-36}$$

$$\omega_{ij} = \frac{\omega_i + \omega_j}{2} \tag{10.5-37}$$

$$\kappa_{ii} = \kappa_i \tag{10.5-38}$$

$$\kappa_{ij} = (\kappa_i \kappa_j)^{\frac{1}{2}} \tag{10.5-39}$$

$$M_{ij} = \frac{2 M_i M_j}{M_i + M_j} \tag{10.5-40}$$

Here, ξ_{ij} and ζ_{ij} are binary interaction parameters which are normally set equal to unity in the absence of experimental data from which they may be regressed. The F_{c_m} term in Eq. (10.5-24) is defined as in Eq. (10.4-11).

$$F_{c_m} = 1 - 0.275\omega_m + 0.059035\mu_{r_m}^4 + \kappa_m \tag{10.5-41}$$

where μ_{r_m} is as in Eq. (10.4-13)

$$\mu_{r_m} = \frac{131.3\mu_m}{(V_{c_m}T_{c_m})^{\frac{1}{2}}} \tag{10.5-42}$$

$$V_{c_m} = \left(\frac{\sigma_m}{0.809}\right)^3 \tag{10.5-43}$$

$$T_{c_m} = 1.2593\left(\frac{\varepsilon}{k}\right)_m \tag{10.5-44}$$

In these equations, T_{c_m} is in K, V_{c_m} in cm^3/mol, and μ in D.

The rules suggested by Chung et al. are illustrated for a binary gas mixture in Example 10.9. As with the Lucas approach, the technique is not interpolative between pure component viscosities. Some calculated binary gas mixture viscosities are compared with experimental values in Table 10.4. Errors vary, but they are usually less than about ±5%.

Example 10.9 Use the Chung et al. method to estimate the low-pressure gas viscosity of a binary of hydrogen sulfide (H_2S) and ethyl ether (EE) containing 20.4 mole% H_2S. The temperature is 331 K.

Solution. The properties listed below are used for this example. See Table 10.1 for κ values.

	Hydrogen Sulfide	Ethyl Ether
T_c, K	373.4	466.70
V_c, cm^3/mol	98	280
ω	0.090	0.281
μ, D	0.9	1.3
κ	0	0
M, g/mol	34.082	74.123
y	0.204	0.796

From Eqs. (10.5-32) and (10.5-33) and assuming $\xi_{ij} = 1$,

$$\sigma(H_2S) = (0.809)(98)^{1/3} = 3.730\ \text{Å}$$

$$\sigma(EE) = (0.809)(280)^{1/3} = 5.293\ \text{Å}$$

$$\sigma(H_2S\text{-}EE) = [(3.730)(5.293)]^{1/2} = 4.443\ \text{Å}$$

Then, with Eq. (10.5-25),

$$\sigma_m^3 = (0.204)^2(3.730)^3 + (0.796)^2(5.293)^3 + (2)(0.204)(0.796)(4.443)^3 = 124.60\ \text{Å}^3$$

From Eqs. (10.5-34) and (10.5-35) and assuming $\zeta_{ij} = 1$,

$$\frac{\varepsilon}{k}(\mathrm{H_2S}) = 373.4/1.2593 = 296.5\ \mathrm{K}$$

$$\frac{\varepsilon}{k}(\mathrm{EE}) = 466.70/1.2593 = 370.6\ \mathrm{K}$$

$$\frac{\varepsilon}{k}(\mathrm{H_2S\text{-}EE}) = [(296.5)(370.6)]^{1/2} = 331.5\ \mathrm{K}$$

Then, with Eq. (10.5-27),

$$\left(\frac{\varepsilon}{k}\right)_{\mathrm{m}} = [(0.204)^2(296.5)(3.730)^3 + (0.796)^2(370.6)(5.293)^3 + (2)(0.204)(0.796)(331.5)(4.443)^3]/124.60 = 360.4\ \mathrm{K}$$

With Eqs. (10.5-28) and (10.5-40),

$$M_{\mathrm{m}} = (\{(0.204)^2(296.5)(3.730)^2(34.082)^{1/2} + (0.796)^2(370.6)(5.293)^2(74.123)^{1/2} + (2)(0.204)(0.796)(331.5)(4.443)^2[(2)(34.082)(74.123)/(34.082 + 74.123)]^{1/2}\}/[(360.4)(124.60)^{2/3}])^2 = 64.42\ \mathrm{g/mol}$$

With Eq. (10.5-29),

$$\omega_{\mathrm{m}} = \{(0.204)^2(0.090)(3.730)^3 + (0.796)^2(0.281)(5.293)^3 + (2)(0.204)(0.796)[(0.090 + 0.281)/2](4.443)^3\}/124.60 = 0.256$$

and with Eq. (10.5-30),

$$\mu_{\mathrm{m}}^4 = \{[(0.204)^2(0.9)^4/(3.730)^3] + [(0.796)^2(1.3)^4/(5.293)^3] + [(2)(0.204)(0.796)(0.9)^2(1.3)^2/(4.443)^3]\}(124.60) = 2.217$$

$$\mu_{\mathrm{m}} = 1.22\ \mathrm{D}$$

so, with Eqs. (10.5-42) to (10.5-44),

$$V_{c_{\mathrm{m}}} = 124.60/(0.809)^3 = 235.3\ \mathrm{cm^3/mol}$$

$$T_{c_{\mathrm{m}}} = (1.2593)(360.4) = 453.9\ \mathrm{K}$$

$$\mu_{\mathrm{m}} = (131.3)(1.22)/[(235.3)(453.9)]^{1/2} = 0.490$$

Since $\kappa_{\mathrm{m}} = 0$, with Eq. (10.5-41),

$$F_{c_{\mathrm{m}}} = 1 - (0.275)(0.256) + (0.059035)(0.490)^4 = 0.933$$

Using Eq. (10.5-26),

$$T_{\mathrm{m}}^* = 331/360.4 = 0.918$$

and with Eq. (10.4-3),

$$\Omega_v = (1.16156)(0.918^{-0.14874}) + 0.52487\exp[-(0.77320)(0.918)] + 2.16178\ \exp[-(2.43787)(0.918)] = 1.665$$

Finally, with Eq. (10.5-24),

$$\eta_m = (26.69)(0.933)[(64.42)(331)]^{1/2}/[(124.60)^{2/3}(1.665)] = 87.5\ \mu P$$

The experimental value is 87 μP (Table 10.4).

$$\eta_m \text{ Difference} = 87.5 - 87 = 0.5\ \mu P \text{ or } 0.6\%$$

10.5.5 Discussion and Recommendations to Estimate the Low-Pressure Viscosity of Gas Mixtures

As is obvious from the estimation methods discussed in this section, the viscosity of a gas mixture can be a complex function of composition. This is evident from Fig. 10.3. There may be a maximum in mixture viscosity in some cases, e.g., System 4: ammonia-hydrogen. However, cases of a viscosity minimum have also been reported. Behavior similar to that of the ammonia-hydrogen case occurs most often in polar-nonpolar mixtures in which the pure component viscosities are not greatly different (Hirschfelder et al., 1960; Rutherford et al., 1960). Maxima are more pronounced as the molecular weight ratio differs from unity.

Of the five estimation methods described in this section, three (Herning and Zipperer, Wilke, and Reichenberg) use the kinetic theory approach and yield interpolative equations between the pure-component viscosities. Reichenberg's method is most consistently accurate, but it is the most complex. To use Reichenberg's procedure, one needs, in addition to temperature and composition, the viscosity, critical temperature, critical pressure, molecular weight, and dipole moment of each constituent. Wilke's and Herning and Zipperer's methods require only the pure-component viscosities and molecular weights; these latter two yield reasonably accurate predictions of the mixture viscosity.

Arguing that it is rare to have available the pure-gas viscosities at the temperature of interest, both Lucas and Chung et al. provide estimation methods to cover the entire range of composition. At the end points where only pure components exist, their methods reduce to those described earlier in Sec. 10.3. Although the errors from these two methods are, on the average, slightly higher than those of the interpolative techniques, they are usually less than ±5% as seen from Table 10.4. Such errors could be reduced even further if pure-component viscosity data were available and were employed in a simple linear correction scheme. For example, if the pure-component viscosity predictions are too high, the mixture prediction would be improved if it were lowered by composition-averaged error of the pure-component predictions.

It is recommended that Reichenberg's method [Eq. (10.5-8)] be used to calculate η_m if pure-component viscosity values are available. Otherwise, either the Lucas method [Eq. (10.4-17)] or the Chung et al. method [Eq. (10.5-24)] can be employed if critical properties are available for all components.

10.6 EFFECT OF PRESSURE ON THE VISCOSITY OF PURE GASES

Figure 10.4 shows the viscosity of carbon dioxide ($T_c = 304.1$ K and $P_c = 73.8$ bar) as a function of temperature and pressure. In some ranges ($T_r > 1.5$ and $P_r < 2$), pressure has little effect on viscosity. But when $1 < T_r < 1.5$ and when $P > P_c$, pressure has a strong effect on viscosity as can be seen by the nearly vertical isobars in this region of Fig. 10.4. Figure 10.4 shows isobars as a function of temperature, while Fig. 10.5 shows isotherms as a function of pressure for nitrogen ($T_c = 77.4$ K, $P_c = 33.9$ bar). Lucas (1981, 1983) has generalized the viscosity phase diagrams (for nonpolar gases) as shown in Fig. 10.6. In this case, the ordinate is $\eta\zeta$ and the temperatures and pressures are reduced values. ζ is the inverse reduced viscosity defined earlier in Eq. (10.4-16).

At the critical point, the viscosity diverges so that its value is larger than would otherwise be expected. However, this effect is much smaller for viscosity than for thermal conductivity (see Fig. 11.5). Whereas the thermal conductivity can increase by a factor of two near the critical point, the increase in viscosity is on the

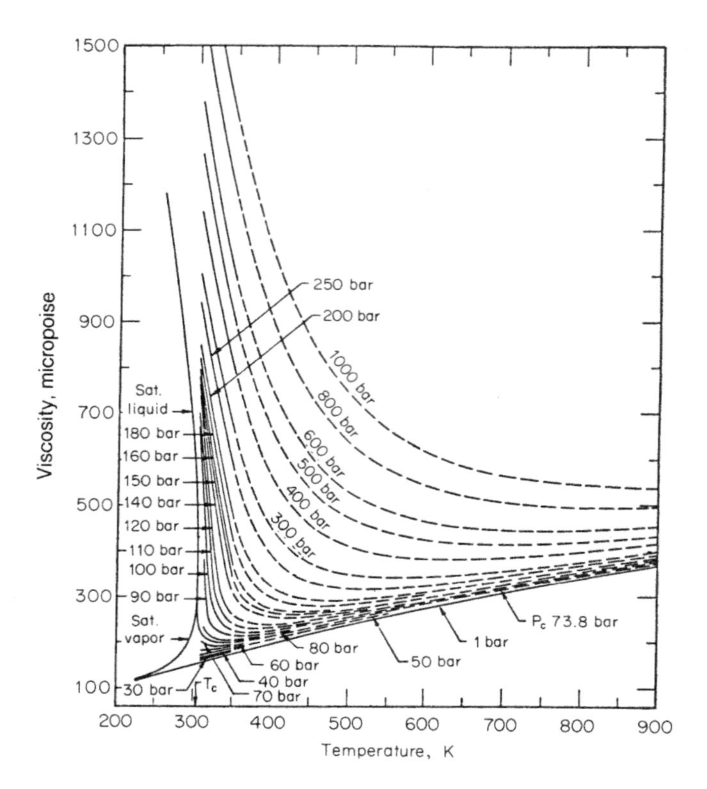

FIGURE 10.4
Viscosity of carbon dioxide. (Stephan and Lucas, 1979)

order of 1%. In fact, Vesovic et al. (1990) state that for carbon dioxide, "the viscosity enhancement is less than 1% at densities and temperatures outside the range bounded approximately by 300 K $< T <$ 310 K and 300 kg/m^3 $<$ $\rho <$ 600 kg/m^3."

In Fig. 10.6, the lower limit of the P_r curves would be indicative of the dilute gas state, as described in Sec. 10.4. In such a state, η increases with temperature. At high reduced pressures, we see there is a wide range of temperatures where η decreases with temperature. In this region the viscosity behavior more closely resembles a liquid state, and, as will be shown in Sec. 10.10, an increase in temperature results in a decrease in viscosity. Finally, at very high-reduced temperatures, a condition again is found where pressure has little effect and viscosities increase with temperature.

The temperature–pressure region in Fig. 10.4 where viscosity changes rapidly with pressure is the very region where density also changes rapidly with pressure. Figure 10.7 shows a plot of the residual viscosity as a function of density for *n*-butane. A smooth curve results even though values over a range of temperatures are shown. This suggests that density is an important variable when describing viscosity behavior at high pressures, and several of the correlations presented in this section take advantage of this importance.

10.6.1 Enskog Dense-Gas Theory

One of the very few theoretical efforts to predict the effect of pressure on the viscosity of gases is due to Enskog and is treated in detail by Chapman and Cowling (1939). The theory has also been applied to dense gas diffusion coefficients, bulk viscosities, and, for monatomic gases, thermal conductivities. The assumption is made that the gas consists of dense, hard spheres and behaves like a low-density hard-sphere system except that all events

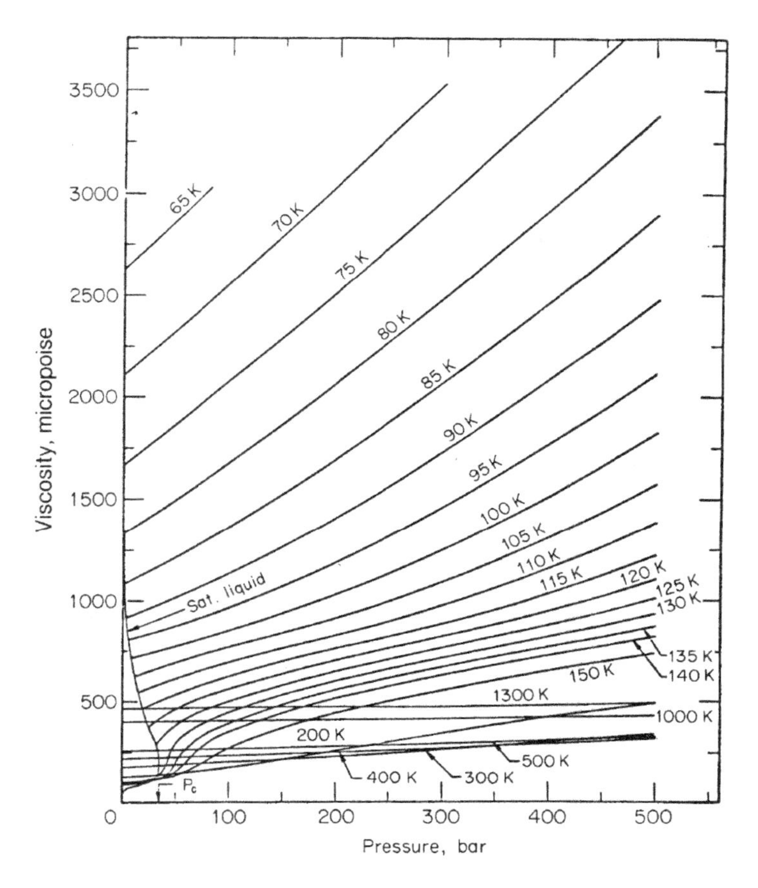

FIGURE 10.5
Viscosity of nitrogen. (Stephan and Lucas, 1979)

occur at a faster rate due to the higher rates of collision (Alder, 1966; Alder and Dymond, 1966). The increase in collision rate is proportional to the radial distribution function Ψ. The Enskog equation for shear viscosity is

$$\frac{\eta}{\eta^{\circ}} = \Psi_c^{-1} + 0.8b_0\rho + 0.761\Psi_c(b_0\rho)^2 \tag{10.6-1}$$

where η is the viscosity in μP, η° is the low-pressure viscosity in μP, $b_0 = \frac{2}{3}\pi N_A\sigma^3$ is the excluded volume in cm^3/mol, N_A is Avogadro's number, σ is the hard-sphere diameter in Å, and ρ is the molar density in mol/cm^3. Ψ_c is the radial distribution function at contact and can be related to an equation of state by

$$\Psi_c = \frac{Z-1}{\rho b_0} \tag{10.6-2}$$

where Z is the compressibility factor.

Dymond, among others (Assael et al., 1996; Dymond and Assael, 1996), has continued efforts to modify the hard sphere approach in order to predict transport properties and has shown that viscosities of dense fluids can be correlated by the universal equation

$$\log_{10}\eta_r = 1.0945 - \frac{9.26324}{V_r} + \frac{71.038}{V_r^2} - \frac{301.9012}{V_r^3} + \frac{797.69}{V_r^4} - \frac{1221.977}{V_r^5} + \frac{987.5574}{V_r^6} - \frac{319.4636}{V_r^7} \tag{10.6-3}$$

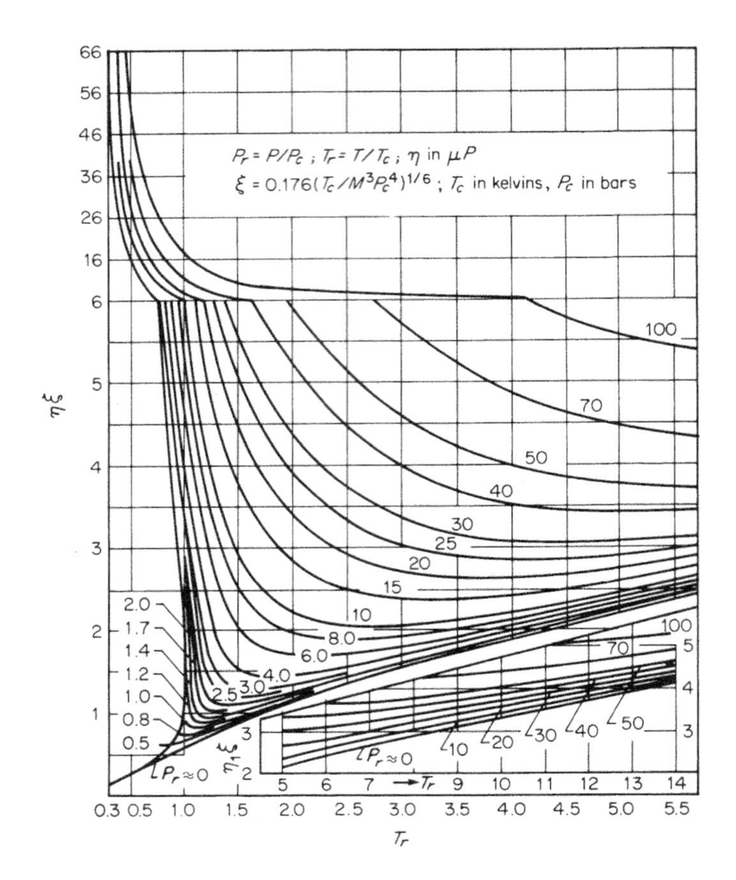

FIGURE 10.6
Generalized phase diagram for gas viscosity. (Lucas, 1981; 1983)

where $V_r = \frac{V}{V_0}$ and V_0 is a close-packed volume. η_r is a reduced viscosity defined by

$$\eta_r = 6.619 \times 10^5 \frac{\eta V^{\frac{2}{3}}}{R_\eta (MT)^{\frac{1}{2}}} \tag{10.6-4}$$

where η is in Pa · s, V is in cm^3/mol, M is g/mol and T is in K. R_η is a parameter that accounts for deviations from smooth hard spheres. The two parameters, V_0 and R_η are compound specific and are not functions of density. V_0 is a function of temperature as is R_η for n-alcohols (Assael et al., 1994). For n-alkanes (Assael et al., 1992a, 1992b), aromatic hydrocarbons (Assael et al., 1992c), refrigerants (Assael et al., 1995), and a number of other compounds (Assael et al. 1992; Bleazard and Teja, 1996), R_η has been found to be independent of temperature. In theory, the two parameters V_0 and R_η could be set with two experimental viscosity-density data, but in practice, Eq. (10.6-3) has been used only for systems for which extensive data are available. It has been applied to densities above the critical density and applicability to temperatures down to $T_r \approx 0.6$ has been claimed. Values of V_0 and R_η at 298 K and 350 K (as well as R_λ which is discussed in Chap. 11) for 16 fluids have been calculated with equations given in the above references and are shown in Table 10.5.

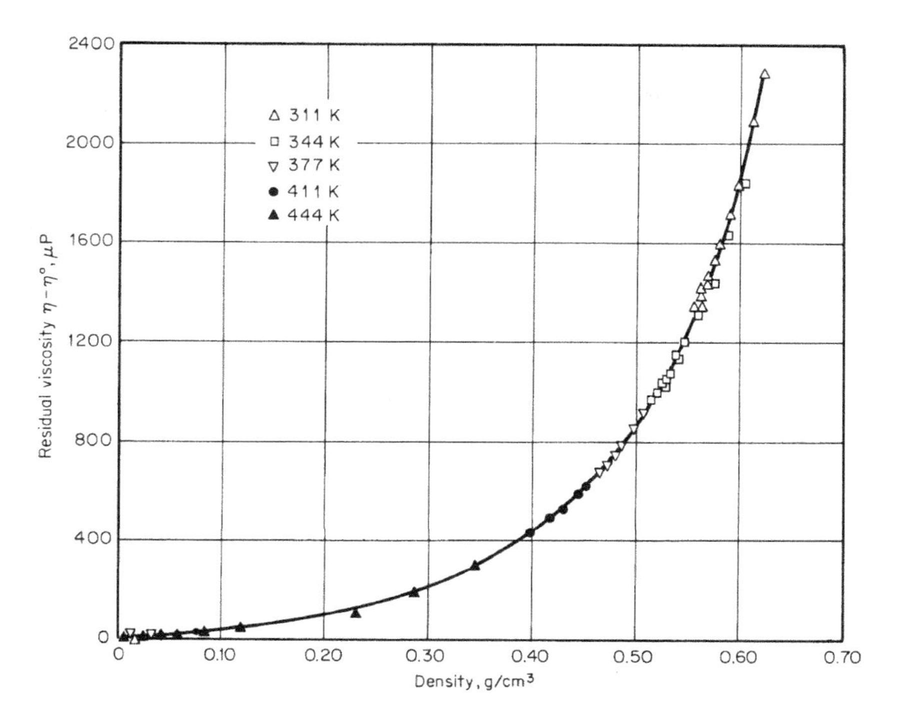

FIGURE 10.7
Residual *n*-butane viscosity as a function of density. (Dolan et al., 1963)

TABLE 10.5 Typical values of V_0, R_η, and R_λ.

	V_0/cm³/mol		R_η		R_λ	
	298 K	**350 K**	**298 K**	**350 K**	**298 K**	**350 K**
Methane	17.9	17.3	1.00	1.00	1.16	1.16
n-Butane	84.1	84.3	1.08	1.08	1.67	1.67
n-Decane	134.3	130.1	1.53	1.53	3.13	3.13
Cyclohexane	77.4	75.7	0.93	0.93	1.35	1.35
CCl_4	65.4	63.8	1.07	1.07	1.57	1.57
R134a	43.7	41.7	1.10	1.10	1.61	1.82
Ethanol	33.8	32.7	4.62	2.80	1.42	1.42
n-Hexanol	91.9	84.4	3.00	3.00	1.95	1.95
Acetic acid	41.3	41.2	0.76	0.76	0.92	0.98
Butyl ethanoate	89.8	88.3	1.10	1.10	2.20	2.25
2-Ethoxyethanol	70.7	68.0	1.33	1.33	1.65	1.78
1,3-Propanediol	62.2	59.1	1.04	1.04	1.16	1.32
Diethylene glycol	78.1	73.6	1.83	1.83	1.65	1.94
Diethanolamine	87.9	80.6	2.38	2.38	1.37	1.72
Triethylamine	73.8	66.4	2.48	2.48	3.03	3.47
Dimethyl disulfide	58.3	57.0	1.12	1.12	1.79	1.79

Xiang et al. (1999) have extended Eq. (10.6-1) to cover the entire fluid range by introducing a crossover function between the low-pressure limit and the high-pressure limit. With a single equation, they fit low-pressure viscosities to within 4% and liquid and high-pressure viscosities generally to within 10% for 18 pure fluids. Their equation requires the density, critical properties, the acentric factor, and values for σ and $\frac{\varepsilon}{k}$.

10.6.2 Reichenberg Method (1971, 1975, 1979)

In this case, the viscosity ratio $\frac{\eta}{\eta^\circ}$ is given by Eq. (10.6-5)

$$\frac{\eta}{\eta^\circ} = 1 + Q\frac{AP_r^{\frac{3}{2}}}{BP_r + \left(1 + CP_r^{D}\right)^{-1}} \tag{10.6-5}$$

The constants, A, B, C, and D are functions of the reduced temperature T_r as shown below, and η° is the viscosity of the gas at the same T and low pressure.

$$A = \frac{1.9824 \times 10^{-3}}{T_r}\exp 5.2683T_r^{-0.5767} \tag{10.6-6}$$

$$B = A(1.6552T_r - 1.2760) \tag{10.6-7}$$

$$C = \frac{0.1319}{T_r}\exp 3.7035T_r^{-79.8678} \tag{10.6-8}$$

$$D = \frac{2.9496}{T_r}\exp 2.9190T_r^{-16.6169} \tag{10.6-9}$$

$$Q = (1 - 5.655\mu_r) \tag{10.6-10}$$

Here, μ_r is defined in Eq. (10.4-18). For nonpolar materials, $Q = 1.0$. Example 10.10 illustrates the application of Eq. (10.6-5), and, in Table 10.7, experimental dense gas viscosities are compared to the viscosities estimated with this method. Errors are generally only a few percent; the poor results for ammonia at 420 K seem to be an anomaly.

Example 10.10 Use Reichenberg's method to estimate the viscosity of *n*-pentane vapor at 500 K and 101 bar. The experimental value is 546 μP (Stephan and Lucas, 1979).

Solution. Whereas one could estimate the low-pressure viscosity of *n*-pentane at 500 K by using the methods described in Sec. 10.4, an experimental value is available (114 μP) (Stephan and Lucas, 1979) and will be used. The dipole moment of *n*-pentane is zero, so $Q = 1.0$. From Appendix A, $T_c = 469.7$ K and $P_c = 33.7$ bar. Thus $T_r = (500/469.7) = 1.0645$ and $P_r = (101/33.7) = 2.9970$. From Eqs. (10.6-6) to (10.6-10), $A = 0.2999$, $B = 0.1458$, $C = 0.1271$, and $D = 7.7846$. With Eq. (10.6-5),

$$\frac{\eta}{\eta^\circ} = 1 + (0.2999)(2.9970)^{3/2}/\{(0.1458)(2.9970) + [1 + (0.1271)(2.9970)^{7.7846}]^{-1}\} = 4.550$$

$$\eta = (4.550)(114) = 519\ \mu\text{P}$$

$$\eta \text{ Difference} = 519 - 546 = -27\ \mu\text{P or } -4.9\%$$

If one refers back to Fig. 10.6, at $T_r = 1.0$ and $P_r = 3.0$, the viscosity is changing rapidly with both temperature and pressure. Thus, an error of only 5% is quite remarkable.

10.6.3 Lucas Method (1980, 1981, 1983)

In a technique which, in some aspects, is similar to Reichenberg's, Lucas (1980, 1981, 1983) recommends the following procedure. For the reduced temperature of interest, first calculate a parameter Z_1 with Eq. (10.6-11).

$$Z_1 = \eta^\circ\zeta = [0.807T_r^{0.618} - 0.357\exp(-0.449T_r) + 0.340\exp(-4.058T_r) + 0.018]F_P^\circ F_Q^\circ \tag{10.6-11}$$

where η° refers to the low-pressure viscosity. Next calculate Z_2. If $T_r < 1.0$ and $P_r < \left(\frac{P^{vp}}{P_c}\right)$, then

$$\begin{aligned} Z_2 &= 0.600 + 0.760P_r^\alpha + \left(6.990P_r^\beta - 0.6\right)(1 - T_r) \\ \alpha &= 3.262 + 14.98P_r^{5.508} \\ \beta &= 1.390 + 5.746P_r \end{aligned} \tag{10.6-12}$$

If $(1 < T_r < 40)$ and $(0 < P_r \leq 100)$, then

$$Z_2 = \eta^\circ\zeta\left[1 + \frac{aP_r^{1.3088}}{bP_r^f + \left(1 + cP_r^d\right)^{-1}}\right] \tag{10.6-13}$$

where $\eta^\circ\zeta$ is found from Eq. (10.6-11). The term multiplying this group is identical to the pressure correction term in Reichenberg's method, Eq. (10.6-5), but the values of the constants are different.

$$a = \frac{1.245 \times 10^{-3}}{T_r}\exp 5.1726T_r^{-0.3286} \tag{10.6-14}$$

$$b = a(1.6553T_r - 1.2723) \tag{10.6-15}$$

$$c = \frac{0.4489}{T_r}\exp 3.0578T_r^{-37.7332} \tag{10.6-16}$$

$$d = \frac{1.7368}{T_r}\exp 2.2310T_r^{-7.6351} \tag{10.6-17}$$

$$f = 0.9425\exp\left[-0.1853T_r^{0.4489}\right] \tag{10.6-18}$$

After computing Z_1 and Z_2, we define

$$Y = \frac{Z_2}{Z_1} \tag{10.6-19}$$

and the correction factors F_P and P_Q,

$$F_P = \frac{1 + (F_P^\circ - 1)Y^{-3}}{F_P^\circ} \tag{10.6-20}$$

$$F_Q = \frac{1 + (F_Q^\circ - 1)[Y^{-1} - (0.007)(\ln Y)^4]}{F_Q^\circ} \tag{10.6-21}$$

where F_P° and F_Q° are low-pressure polarity and quantum factors determined as shown in Eqs. (10.4-19) and (10.4-20). Finally, the dense gas viscosity is calculated as

$$\eta = \frac{Z_2F_PF_Q}{\zeta} \tag{10.6-22}$$

where ζ is defined in Eq. (10.4-16). At low pressures, Y is essentially unity, and $F_P = 1$, $F_Q = 1$. Also, Z_2 then equals $\eta^\circ\zeta$ so $\eta \rightarrow \eta^\circ$, as expected.

The Lucas method is illustrated in Example 10.11, and calculated dense gas viscosities are compared with experimental data in Table 10.7. In Example 10.11 and Table 10.7, the low-pressure viscosity η° was not obtained from experimental data but was estimated by the Lucas method from Sec. 10.4. Except in a few cases, the error was found to be less than 5%. The critical temperature, critical pressure, critical compressibility factor, and dipole moment are required, as well as the system temperature and pressure.

Example 10.11 Estimate the viscosity of ammonia gas at 420 K and 300 bar by using Lucas's method. The experimental values of η and η° are 571 and 146 μP (Stephan and Lucas, 1979).

Solution. For ammonia we use $M = 17.031$ g/mol, $Z_c = 0.244$, $T_c = 405.50$ K, $P_c = 113.53$ bar, and $\mu = 1.47$ D. Thus, $T_r = 420/405.50 = 1.036$ and $P_r = 300/113.53 = 2.643$. From Eq. (10.4-16),

$$\zeta = 0.176\{405.50/[(17.031)^3(113.53)^4]\}^{1/6} = 0.004949\ \mu\text{P}^{-1}$$

with Eq. (10.4-18),

$$\mu_r = 52.46[(1.47)^2(113.53)/(405.52)^2] = 0.07827$$

Since ammonia is not a quantum gas, $F_Q^\circ = 1$, and with Eq. (10.4-19),

$$F_P^\circ = 1 + 30.55(0.292 - 0.244)^{1.72}\,|\,0.96 + 0.1(1.036 - 0.7)\,| = 1.164$$

From Eq. (10.6-11),

$$\begin{aligned} Z_1 = \eta^\circ\zeta &= \{0.807(1.036)^{0.618} - 0.357\exp[(-0.449)(1.036)] \\ &\quad + 0.340\exp[(-4.058)(1.036)] + 0.018\}(1.164) = 0.7260 \\ \eta^\circ &= 0.7260/0.004949 = 147\ \mu\text{P} \\ \eta^\circ\ \text{Difference} &= 147 - 146 = 1\ \mu\text{P or } 0.7\% \end{aligned}$$

The estimation of the low-pressure viscosity of ammonia agrees very well with the experimental value.

Since $T_r > 1.0$, we use Eq. (10.6-13) to determine Z_2. The values of the coefficients from Eqs. (10.6-14) to (10.6-18) are $a = 0.1997$, $b = 8.839 \times 10^{-2}$, $c = 0.9692$, $d = 9.205$, and $f = 0.7808$. Then,

$$Z_2 = (0.7260)[1 + (0.1997)(2.643)^{1.3088}/\{(0.08839)(2.643)^{0.7808} + [1 + (0.9692)(2.643)^{9.205}]^{-1}\}] = 3.464$$

with Eqs. (10.6-19) to (10.6-21),

$$\begin{aligned} Y &= 3.464/0.7258 = 4.771 \\ F_P &= [1 + (1.164 - 1)(4.771)^{-3}]/1.164 = 0.860 \\ F_Q &= 1 \end{aligned}$$

and, with Eq. (10.6-22),

$$\begin{aligned} \eta &= (3.464)(0.860)(1.0)/0.004949 = 602\ \mu\text{P} \\ \eta\ \text{Difference} &= 602 - 571 = 31\ \mu\text{P or } 5.4\% \end{aligned}$$

The Reichenberg and Lucas methods employ temperature and pressure as the state variables. In most other dense gas viscosity correlations, however, the temperature and density (or specific volume) are used. In those cases, one must have accurate volumetric data or an applicable equation of state to determine the dense gas viscosity. Three different methods are illustrated below.

10.6.4 Method of Jossi, Stiel, and Thodos (Jossi et al., 1962; Stiel and Thodos, 1964)

In this case, the residual viscosity $\eta - \eta^{\circ}$ is correlated with fluid density. All temperature effects are incorporated in the η° term. To illustrate the behavior of the $\eta - \eta^{\circ}$ function, consider Fig. 10.7, which shows $\eta - \eta^{\circ}$ for *n*-butane graphed as a function of density (Dolan et al., 1963). Note that there does not appear to be any specific effect of temperature over the range shown. At the highest density, 0.6 g/cm^3, the reduced density ρ/ρ_c is 2.63. Similar plots for many other substances are available, for example, He, air, O_2, N_2, CH_4 (Kestin and Leidenfrost, 1959); ammonia (Carmichael et al., 1963; Shimotake and Thodos, 1963); rare gases (Shimotake and Thodos, 1958); diatomic gases (Brebach and Thodos, 1958); sulfur dioxide (Shimotake and Thodos, 1963a); CO_2 (Kennedy and Thodos, 1961; Vesovic et al., 1990); steam (Kestin and Moszynski, 1959); and various hydrocarbons (Carmichael and Sage, 1963; Eakin and Ellington, 1963; Giddings, 1963; Starling et al., 1960; Starling and Ellington, 1964). Other authors have also shown the applicability of a residual viscosity-density correlation (Golubev 1959; Hanley et al., 1969; Kestin and Moszynski, 1959; Rogers and Brickwedde 1965; Starling, 1960, 1962).

In the Jossi, Stiel, and Thodos method, separate residual viscosity expressions are given for nonpolar and polar gases, but no quantitative criterion is presented to distinguish these classes.

10.6.4.1 Nonpolar Gases (Jossi et al., 1962)

The basic relation is

$$[(\eta - \eta^{\circ})\xi_T + 1]^{\frac{1}{4}} = 1.0230 + 0.23364\rho_r + 0.58533\rho_r^2 - 0.40758\rho_r^3 + 0.093324\rho_r^4 \qquad (10.6\text{-}23)$$

where η is the dense gas viscosity in μP, η° is the low-pressure gas viscosity in μP, $\rho_r = \frac{\rho}{\rho_c} = \frac{V_c}{V}$ is the reduced gas density, and M is molecular weight in g/mol. Also,

$$\xi_T = \left(\frac{T_c}{M^3 P_c^4}\right)^{\frac{1}{6}} \qquad (10.6\text{-}24)$$

with T_c in K and P_c in atm. This relation is reported by Jossi et al. to be applicable in the range $0.1 \le \rho_r < 3$.

10.6.4.2 Polar Gases (Stiel and Thodos, 1964)

The relation to be used depends on the reduced density:

$$(\eta - \eta^{\circ})\xi_T = 1.656\rho_r^{1.111} \qquad \rho_r \le 0.1 \qquad (10.6\text{-}25)$$

$$(\eta - \eta^{\circ})\xi_T = 0.0607(9.045\rho_r + 0.63)^{1.739} \qquad 0.1 < \rho_r \le 0.9 \qquad (10.6\text{-}26)$$

$$\log\{4 - \log[(\eta - \eta^{\circ})\xi_T]\} = 0.6439 - 0.1005\rho_r - \Delta \qquad 0.9 < \rho_r \le 2.6 \qquad (10.6\text{-}27)$$

where

$$\Delta = 0 \qquad 0.9 < \rho_r \le 2.2$$

$$\Delta = 4.75 \times 10^{-4}\left(\rho_r^3 - 10.65\right)^2 \qquad 2.2 < \rho_r \le 2.6 \qquad (10.6\text{-}28)$$

The ρ_r range over which the method may be used can also be extended by interpolation since $(\eta - \eta^{\circ})\xi_T = 90.0$ and 250 at $\rho_r = 2.8$ and 3.0, respectively. The notation used in Eqs. (10.6-25) to (10.6-27) is defined under Eq. (10.6-23). Note that the parameter ξ_T is *not* the same as ζ defined earlier in Eq. (10.4-16). An example of the Jossi et al. method is shown below, and calculated dense gas viscosities are compared with experimental values in Table 10.7.

Example 10.12 Use the Jossi, Stiel, and Thodos method to estimate the viscosity of isobutane at 500 K and 100 bar. The experimental viscosity is 261 μP (Stephan and Lucas, 1979) and the specific volume is 243.8 cm³/mol (Waxman and Gallagher, 1983). At low pressure and 500 K, $\eta° = 120\ \mu$P.

Solution. Since isobutane is nonpolar, Eq. (10.6-23) is used. From Appendix A, $T_c = 407.8$ K, $P_c = 36.4$ bar = 35.9 atm, $V_c = 259$ cm³/mol, and $M = 58.122$. Then

$$\xi_T = \{407.8/[(58.122)^3(35.9)^4]\}^{1/6} = 0.03282\ \mu\text{P}^{-1}$$

The reduced density is $\rho_r = 259/243.8 = 1.062$. With Eq. (10.6-23),

$$[(\eta - 120)(0.03282) + 1]^{1/4} = 1.0230 + (0.23364)(1.062) + (0.58533)(1.062)^2 - (0.40758)(1.062)^2 + (0.093324)(1.062)^4 = 1.562$$

$$\eta = 271\ \mu\text{P}$$

$$\eta \text{ Difference} = 271 - 261 = 10\ \mu\text{P or } 3.8\%$$

10.6.5 Chung et al. Method (1988)

In an extension of the Chung et al. technique to estimate low-pressure gas viscosities, the authors began with Eq. (10.3-9) and employed empirical correction factors to account for the fact that the fluid has a high density. Their relations are shown below:

$$\eta = \eta^* \frac{36.344(MT_c)^{\frac{1}{2}}}{V_c^{\frac{2}{3}}} \tag{10.6-29}$$

where η is the viscosity in μP, M is the molecular weight in g/mol, T_c is the critical temperature in K, V_c is the critical volume in cm³/mol, and

$$\eta^* = \frac{(T^*)^{\frac{1}{2}}}{\Omega_v}\{F_c[G_2^{-1} + E_6 y]\} + \eta^{**} \tag{10.6-30}$$

Here, T^* and F_c are defined as in Eqs. (10.4-9) and (10.4-11). Ω_v is found with Eq. (10.4-3) as a function of T^*. With ρ in mol/cm³,

$$y = \frac{\rho V_c}{6} \tag{10.6-31}$$

$$G_1 = \frac{1 - 0.5y}{(1-y)^3} \tag{10.6-32}$$

$$G_2 = \frac{E_1 \dfrac{1 - \exp(-E_4 y)}{y} + E_2 G_1 \exp(E_5 y) + E_3 G_1}{E_1 E_4 + E_2 + E_3} \tag{10.6-33}$$

$$\eta^{**} = E_7 y^2 G_2 \exp\left[E_8 + \frac{E_9}{T^*} + \frac{E_{10}}{(T^*)^2}\right] \tag{10.6-34}$$

TABLE 10.6 Chung et al. coefficients to calculate $E_i = a_i + b_i\omega + c_i\mu_r^4 + d_i\kappa$.

i	a_i	b_i	c_i	d_i
1	6.324	50.412	−51.680	1189.0
2	1.210×10^{-3}	-1.154×10^{-3}	-6.257×10^{-3}	0.03728
3	5.283	254.209	−168.48	3898.0
4	6.623	38.096	−8.464	31.42
5	19.745	7.630	−14.354	31.53
6	−1.900	−12.537	4.985	−18.15
7	24.275	3.450	−11.291	69.35
8	0.7972	1.117	0.01235	−4.117
9	−0.2382	0.06770	−0.8163	4.025
10	0.06863	0.3479	0.5926	−0.727

and the parameters E_1 to E_{10} are given by

$$E_i = a_i + b_i\omega + c_i\mu_r^4 + d_i\kappa \tag{10.6-35}$$

where the coefficients a_i through d_i are in Table 10.6, ω is the acentric factor, μ_r is the reduced dipole moment as defined in Eq. (10.4-13), and κ is the association factor (see Table 10.1). At very low densities, y approaches zero, G_1 and G_2 approach unity, and η^{**} is negligible. At these limiting conditions, combining Eqs. (10.6-29), (10.6-30), and (10.4-9) leads to Eq. (10.4-10), which then applies for estimating $\eta°$.

The application of the Chung et al. method is shown in Example 10.13. Some calculated values of η are compared with experimental results in Table 10.7. The errors usually are below 5%.

Example 10.13 Estimate the viscosity of ammonia at 520 K and 600 bar with the Chung et al. method. The experimental value of η is 466 μP (Stephan and Lucas, 1979). At this temperature, $\eta° = 182\ \mu$P. The specific volume of ammonia at 520 K and 600 bar is 48.2 cm^3/mol (Haar and Gallagher, 1978).

Solution. The following constants are used for ammonia. $T_c = 405.50$ K, $V_c = 72.4$ cm^3/mol, $\omega = 0.256$, $M = 17.031$ g/mol, and $\mu = 1.47$ D. Thus $T_r = 520/405.50 = 1.282$ and $\rho = 1/48.2 = 0.0207$ mol/cm^3. With Eq. (10.4-13),

$$\mu_r = (131.3)(1.47)/[(72.4)(405.50)]^{1/2} = 1.126$$

and with Eqs. (10.4-11) and (10.4-9), and assuming $\kappa = 0$ since it is not listed in Table 10.1,

$$F_c = 1 - (0.2756)(0.256) + (0.059035)(1.126)^4 = 1.024$$

$$T^* = (1.2593)(1.282) = 1.614$$

Using Eq. (10.4-3), $\Omega_V = 1.2746$, and with Eqs. (10.6-31) and (10.6-32),

$$y = (0.0207)(72.4)/6 = 0.250$$

$$G_1 = [1 - (0.5)(0.250)]/(1 - 0.250)^3 = 2.074$$

TABLE 10.7 Comparison of experimental and calculated dense gas viscosities.

Compound	T (K)	P (bar)	V (cm^3/mol)	Ref.*	η (μP)	$\eta°$(μP)	Percent Deviation† Calculated by Method of: Reichenberg Eq. (10.6-5)	Lucas Eq. (10.6-22)	Jossi et al. Eq. (10.6-23)	Chung et al. Eq. (10.6-29) & Table 10.6	Brulé and Starling Eq. (10.6-29) & Table 10.8	Trapp
Oxygen	300	30.4	806.1	6	212.8	207.2	−1.0	−1.6	0.6	−1.5	0.2	−0.2
		81.0	295.3		225.7		−1.2	−1.1	−0.6	−1.9	0.8	−0.4
		152.0	155.3		250.3		−0.3	−0.2	−0.8	−0.2	1.6	−0.3
		304.0	81.4		319.3		3.6	0.8	2.8	3.9	4.6	0.6
Methane	200	40.0	282.0	3	90	78.0	7.0	0.6	5.9	3.5	1.7	6.8
		100.0	60.2		296		10.0	8.2	5.1	3.1	14	7.8
		200.0	51.1		415		3.8	5.0	−0.5	−2.2	16	8.1
	500	40	1039.0	3	180	177	−0.4	−5.6	0.9	−5.3	−3.8	0.1
		100	417.7		187		−0.6	−5.1	0.3	−7.2	−2.3	0.0
		200	213.7		204		−1.0	−5.0	−0.3	−2.9	−1.0	−1.0
		500	98.9		263		1.5	−3.3	3.3	2.5	5.3	−1.1
Isobutane	500	20	2396.0	7	127	120	0.9	6.3	0.2	0.8	2.2	−0.3
		50	620.0		146		5.7	12.0	4.5	9.3	12	5.9
		100	244.0		261		−5.2	3.8	5.4	5.5	8.6	2.3
		200	159.0		506		−11	2.3	−9.0	−7.2	−5.0	−6.1
		400	130.0		794		−19	−8.2	−16	−10	−9.9	−11

Ammonia	420	50	588.1	4	149	146	3.0	–2.4	1.7	3.1	–5.2	4.9
		150	61.9		349		–17	–6.5	–15	–13	5.1	14
		300	39.8		571		–21	5.2	3.6	–4.0	22	60
		600	34.3		752		–24	7.8	11	–1.3	31	84
	520	50	807.6	4	185	182	0.7	–5.8	0.5	–0.1	–9.2	2.0
		150	229.6		196		4.5	0.9	2.3	4.0	1.6	9.4
		300	90.7		296		–1.4	5.3	–2.3	0.7	13	5.0
		600	48.2		466		–13	5.8	–3.2	–3.4	12	36
Carbon dioxide	360	50	514.6	1	190	177	3.0	3.1	1.9	1.1	2.4	2.6
		100	211.2		230		2.1	3.3	0.8	3.6	6.1	1.2
		400	55.0		730		1.3	7.7	–3.5	–0.8	1.2	6.3
		800	45.8		1104		–7.0	1.1	–9.6	–2.2	–1.3	4.0
	500	50	802.8	1	243	235	0.0	3.0	1.3	0.7	1.6	0.7
		100	389.2		254		1.7	5.1	1.4	3.3	5.3	2.0
		400	97.1		411		9.8	7.4	2.6	3.6	9.4	2.1
		800	62.9		636		10	9.6	0.9	–3.2	2.8	3.0
n–pentane	600	20.3	2240	2	143	134	0.0	1.4	0.0	1.2	2.1	0.2
		81.1	418.3		242		–7.5	–5.3	–11	–4.6	–1.0	–8.5
		152	237.5		383		0.9	2.3	–7.9	–7.0	–3.7	–1.3

†Percent deviation = [(calc. – exp.)/exp.] × 100.

*References: 1, Angus et al. (1976); 2, Das et al. (1977); 3, Goodwin (1973); 4, Haar and Gallaghee (1978); 5, Stephan and Lucas (1979); 6, Stewart (1966). Ideal gas values from Stephan and Lucas (1979).

From Eq. (10.6-35) and Table 10.6, the following coefficients were computed: $E_1 = -63.85$, $E_2 = -0.009154$, $E_3 = -200.5$, $E_4 = 2.770$, $E_5 = -1.376$, $E_6 = 2.904$, $E_7 = 7.008$, $E_8 = 1.103$, $E_9 = -1.533$, and $E_{10} = 1.110$. Then, with Eq. (10.6-33), $G_2 = 1.440$ and, from Eq. (10.6-34), $\eta^{**} = 1.126$. Finally, using Eqs. (10.6-30) and (10.6-29),

$$\eta^* = (1.614)^{1/2}(1.2746)^{-1}(1.024)[(1.440)^{-1} + (2.904)(0.250)] + 1.126 = 2.576$$

$$\eta = (2.576)(36.344)[(17.031)(405.50)]^{1/2}/(72.4)^{2/3} = 448\ \mu\text{P}$$

$$\eta \text{ Difference} = 448 - 466 = -18\ \mu\text{P or } -3.9\%$$

10.6.6 TRAPP Method (Huber, 1996)

The transport property prediction (TRAPP) method is a corresponding states method to calculate viscosities (η) and thermal conductivities (λ) of pure fluids and mixtures. In its original version (Ely, 1981; Ely and Hanley, 1981), it was used to estimate *low pressure* values of λ and η and employed methane as a reference fluid. In the most recent version, presented below for pure fluids and later in Sec. 10.7 for mixtures, the high-pressure values are estimated from low-pressure values—the latter of which are estimated by one of the methods presented earlier in this chapter. The most recent version also uses propane as the reference fluid, and shape factors are no longer functions of density as they were in the initial method. Other reference fluids could be chosen and, in fact, Huber and Ely (1992) use R134a as the reference fluid to describe the viscosity behavior of refrigerants. The TRAPP method was originally developed only for nonpolar compounds, but there have been efforts to extend the method to polar compounds as well (Hwang and Whiting, 1987).

In the TRAPP method, the residual viscosity of a pure fluid is related to the residual viscosity of the reference fluid, propane:

$$\eta - \eta^\circ = F_\eta[\eta_R - \eta_R^\circ] \tag{10.6-36}$$

The reference fluid values (denoted by subscript R) are evaluated at T_η and density ρ_η (described later), not T and ρ. In Eq. (10.6-36), η° is the viscosity at low pressure, η_R is the true viscosity of the reference fluid (propane) at temperature T_η and density ρ_η, η_R° is the low-pressure value for propane at temperature T_η, and F_η is the factor that relates the reference fluid to the fluid of interest and is discussed below.

Younglove and Ely (1987) report an expression for the viscosity of propane across a wide range of conditions. When applied at T_η and ρ_η (Zaytsev and Aseyev, 1992), the residual viscosity for the reference fluid in Eq. (10.6-36) is given by

$$\eta_R - \eta_R^\circ = G_1 \exp\left[\rho_\eta^{0.1} G_2 + \rho_\eta^{0.5}\left(\rho_{\eta_r} - 1\right) G_3\right] - G_1 \tag{10.6-37}$$

where ρ_η is in mol/dm^3, $\eta_R - \eta_R^\circ$ is in μPa·s $\rho_{\eta_r} = \rho_\eta/\rho_{c_R}$, ρ_{c_R} is the critical density of the reference fluid (propane), and

$$G_1 = \exp\left(E_1 + \frac{E_2}{T_\eta}\right) \tag{10.6-38}$$

$$G_2 = E_3 + \frac{E_4}{T_\eta^{1.5}} \tag{10.6-39}$$

$$G_3 = E_5 + \frac{E_6}{T_\eta} + \frac{E_7}{T_\eta^2} \tag{10.6-40}$$

Here, T_η is in K, $E_1 = -14.113294896$, $E_2 = 968.22940153$, $E_3 = 13.686545032$, $E_4 = -12511.628378$, $E_5 = 0.0168910864$, $E_6 = 43.527109444$, and $E_7 = 7659.4543472$. T_η, ρ_η, and F_η are calculated by

$$T_\eta = \frac{T}{f} \tag{10.6-41}$$

$$\rho_\eta = \rho h \tag{10.6-42}$$

$$F_\eta = \left(\frac{M}{44.094} f\right)^{\frac{1}{2}} h^{-\frac{2}{3}} \tag{10.6-43}$$

where M is the molecular weight in g/mol and f and h are equivalent substance reducing ratios and are determined as described below.

If vapor pressure and liquid density information are available for the substance of interest, and if $T < T_c$, it is recommended that f be obtained from the equation

$$\frac{P^{\text{vp}}}{\rho^{\text{S}}} = f \frac{P_{\text{R}}^{\text{vp}}(T_\eta)}{\rho_{\text{R}}^{\text{S}}(T_\eta)} \tag{10.6-44}$$

where P^{vp} and ρ^{S} are the vapor pressure and saturated liquid density at temperature T and $P_{\text{R}}^{\text{vp}}(T_\eta)$ and $\rho_{\text{R}}^{\text{S}}(T_\eta)$ are the vapor pressure and saturated liquid density for the reference fluid propane evaluated at T_η. As indicated in Eq. (10.6-41), T_η is a function of f, so Eq. (10.6-44) must be solved iteratively to find f. Once f is found, h is determined from

$$h = \frac{\rho_{\text{R}}^{\text{S}}(T_\eta)}{\rho^{\text{S}}} \tag{10.6-45}$$

If $T > T_c$, or if vapor pressure and saturated liquid density information are not available, h and f can be calculated by

$$f = \frac{T_c}{T_{c_{\text{R}}}}[1 + (\omega - \omega_{\text{R}})(0.05203 - 0.7498 \ln T_r)] \tag{10.6-46}$$

$$h = \frac{\rho_{c_{\text{R}}}}{\rho_c}\frac{Z_{c_{\text{R}}}}{Z_c}[1 - (\omega - \omega_{\text{R}})(0.1436 - 0.2822 \ln T_r)] \tag{10.6-47}$$

where T_c, ρ_c, and ω are the critical temperature, critical density, and acentric factor of the fluid of interest, respectively; T_{c_R}, ρ_{c_R}, and ω_{R} are the same properties but for the reference fluid (propane); and $T_r = T/T_c$.

The application of the TRAPP method is shown in Example 10.14. Some calculated values of η are compared with experimental results in Table 10.7. Huber (1996) gives results of additional comparisons and suggests methods to improve predictions if some experimental data are available.

Example 10.14 Use the TRAPP method to estimate the viscosity of isobutane at 500 K and 100 bar. The experimental viscosity is 261 μP (Stephan and Lucas, 1979) and the specific volume is 243.8 cm³/mol (Waxman and Gallagher, 1983). At low pressure and 500 K, $\eta° = 120\ \mu$P.

Solution. From Appendix A, for the reference fluid, propane, $T_c = 369.83$ K, $V_c = 200$ cm³/mol, $Z_c = 0.276$, and $\omega = 0.152291$. For isobutane, $T_c = 407.8$ K, $V_c = 259$ cm³/mol, $Z_c = 0.278$, $\omega = 0.183521$, and $M = 58.122$ g/mol. At the conditions given, $T_r = 500/407.8 = 1.226$. Since $T > T_c$, Eqs. (10.6-46) and (10.6-47) should be used to calculate f and h which give

$$f = (407.8)(369.83)^{-1}\{1 + [0.183521 - 0.152291][0.05203 - 0.7498 \ln (1.226)]\} = 1.099$$

$$h = (259/200)(0.276/0.278)\{1 - [0.183521 - 0.152291][0.1436\text{-}0.2822 \ln (1.226)]\} = 1.282$$

Using Eqs. (10.6-41) through (10.6-43) yields

$$T_\eta = 500/1.099 = 455.0 \text{ K}$$
$$\rho_\eta = 1.282/243.8 = 0.005258 \text{ mol/cm}^3 = 5.258 \text{ mol/dm}^3$$
$$F_\eta = [(58.122)(1.099)/44.094]^{1/2}(1.282)^{-2/3} = 1.020$$

The value of $\eta_R - \eta_R^\circ$ is now needed. Using Eqs. (10.6-38) to (10.6-40) gives

$$G_1 = \exp(-14.113294896 + 968.22940153/455.0) = 6.235 \times 10^{-6}$$
$$G_2 = 13.686545032 - 12511.628378/455.0^{1.5} = 12.397$$
$$G_3 = 0.0168910864 + 43.527109444/455.0 + 7659.4543472/455.0^2 = 0.150$$

With Eq. (10.6-37),

$$\rho_{\eta_r} = (0.005258)(200) = 1.052$$
$$\eta_R - \eta_R^\circ = 6.235 \times 10^{-6} \exp[(5.258)^{0.1}(12.397) + (5.258)^{0.5}(1.052 - 1)(0.150)] - 6.235 \times 10^{-6} = 14.41\ \mu\text{Pa}\cdot\text{s} = 144.1\ \mu\text{P}$$

Eq. (10.6-36) gives

$$\eta = 120 + (1.020)(144.1) = 267.0\ \mu\text{P}$$
$$\eta \text{ Difference} = 267 - 261 = 6\ \mu\text{P or } 2.3\%$$

10.6.7 Other Corresponding States Methods

In a manner identical in form with that of Chung et al., Brulé and Starling (1984) proposed a different set of coefficients for E_1 to E_{10} to be used instead of those in Table 10.6. These are shown in Table 10.8. Note that no polarity terms are included and the *orientation* parameter γ has replaced the acentric factor ω. If values of γ are not available, the acentric factor may be substituted.

TABLE 10.8 Brulé and Starling coefficients to calculate $E_i = a_i + b_i\gamma$.

i	a_i	b_i
1	17.450	34.063
2	-9.611×10^{-4}	7.235×10^{-3}
3	51.043	169.46
4	−0.6059	71.174
5	21.382	−2.110
6	4.668	−39.941
7	3.762	56.623
8	1.004	3.140
9	-7.774×10^{-2}	−3.584
10	0.3175	1.1600

The Brulé and Starling technique was developed to be more applicable for heavy hydrocarbons rather than for simple molecules as tested in Table 10.7. Okeson and Rowley (1991) have developed a four-parameter corresponding states method for polar compounds at high pressures but did not test their method for mixtures.

10.6.8 Discussion and Recommendations for Estimating Dense Gas Viscosities

Six estimation techniques were discussed in this section. Two (Reichenberg and Lucas) were developed to use temperature and pressure as the input variables to estimate the viscosity. The other four require temperature and density; thus, an equation of state would normally be required to obtain the necessary volumetric data if not directly available. In systems developed to estimate many types of properties, it would not be difficult to couple the *PVT* and viscosity programs to provide densities when needed. In fact, the Brulé and Starling method (Brulé and Starling, 1984) is predicated on combining thermodynamic and transport analyses to obtain the characterization parameters most suitable for both types of estimations.

Another difference to be recognized among the methods noted in this section is that Reichenberg's, Jossi et al.'s, and the TRAPP methods require a low-pressure viscosity at the same temperature. The other techniques bypass this requirement and have imbedded into the methods a low-pressure estimation method; i.e., at low densities they reduce to techniques as described in Sec. 10.4. If the Lucas, Chung et al., or Brulé-Starling methods were selected, no special low-pressure estimation method would have to be included in a property estimation package.

With these few remarks, along with the testing in Table 10.7 and evaluations by authors of the methods, we recommend that either the Lucas or Chung et al. procedure be used to estimate dense gas viscosities of both polar and nonpolar compounds. The Brulé-Starling method is, however, preferable when complex hydrocarbons are of interest, but even for those materials, the Chung et al. procedure should be used at low reduced temperatures ($T_r < 0.5$). For nonpolar compounds, we recommend the TRAPP method as well as the Lucas or Chung et al. methods.

Except when one is working in temperature and pressure ranges in which viscosities are strong functions of these variables (see Fig. 10.6), errors for the recommended methods are usually only a few percent. Near the critical point, and in regions where the fluid density is approaching that of a liquid, higher errors may be encountered.

10.7 VISCOSITY OF GAS MIXTURES AT HIGH PRESSURES

The most convenient method to estimate the viscosity of dense gas mixtures is to combine, where possible, techniques given previously in Secs. 10.5 and 10.6.

10.7.1 Lucas Approach (Lucas 1980, 1981, 1983)

In the (pure) dense gas viscosity approach suggested by Lucas, Eqs. (10.6-11) to (10.6-22) were used. To apply this technique to mixtures, rules must be chosen to obtain T_c, P_c, M, and μ as functions of composition. For T_c, P_c, and M of the mixture, Eqs. (10.5-18) to (10.5-20) should be used. The polarity (and quantum) corrections are introduced by using Eqs. (10.6-20) and (10.6-21), where F_P° and F_Q° refer to mixture values from Eqs. (10.5-21) and (10.5-22). The parameter Y in Eqs. (10.6-20) and (10.6-21) must be based on T_{c_m} and P_{c_m}. F_P° and F_Q°, for the pure components, were defined in Eqs. (10.4-19) and (10.4-20).

10.7.2 Chung et al. (1988) Approach

To use this method for dense gas mixtures, Eqs. (10.6-29) to (10.6-34) are used. The parameters T_c, V_c, ω, M, μ, and κ in these equations are given as functions of composition in Sec. 10.5. That is,

Parameter	Equation to Use
T_{cm}	(10.5-44)
V_{cm}	(10.5-43)
ω_m	(10.5-29)
M_m	(10.5-28)
μ_m	(10.5-30)
κ_m	(10.5-31)

10.7.3 TRAPP Method (Huber, 1996)

For gas mixtures at high pressure, the viscosity is determined by a combination of the techniques introduced for high-pressure gases (Sec. 10.6) with appropriate mixing rules. The viscosity of the mixture is given by

$$\eta_m - \eta_m^\circ = F_{\eta_m}[\eta_R - \eta_R^\circ] + \Delta\eta^{\text{ENSKOG}} \tag{10.7-1}$$

The quantity $\eta_R - \eta_R^\circ$ that appears in Eq. (10.7-1) is for the reference fluid propane and is evaluated with Eq. (10.6-37) at T_{η_m} and ρ_{η_m}. The following mixing rules are used to determine F_{η_m}, T_{η_m}, and ρ_{η_m}.

$$h_m = \sum_i \sum_j y_i y_j h_{ij} \tag{10.7-2}$$

$$f_m h_m = \sum_i \sum_j y_i y_j f_{ij} h_{ij} \tag{10.7-3}$$

$$h_{ij} = \frac{\left[h_i^{\frac{1}{3}} + h_j^{\frac{1}{3}}\right]^3}{8} \tag{10.7-4}$$

$$f_{ij} = (f_i f_j)^{\frac{1}{2}} \tag{10.7-5}$$

f_i and h_i are determined as in Sec. 10.6. T_{η_m} and ρ_{η_m} are calculated by equations similar to Eqs. (10.6-41) and (10.6-42), namely

$$T_{\eta_m} = \frac{T}{f_m} \tag{10.7-6}$$

$$\rho_{\eta_m} = \rho h_m = \frac{h_m}{V} \tag{10.7-7}$$

Finally,

$$F_{\eta_m} = \frac{1}{\sqrt{44.094}\, h_m^2} \sum_i \sum_j y_i y_j (f_{ij} M_{ij})^{\frac{1}{2}} (h_{ij})^{\frac{4}{3}} \tag{10.7-8}$$

where

$$M_{ij} = \frac{2 M_i M_j}{M_i + M_j} \tag{10.7-9}$$

The term, $\Delta\eta^{\text{ENSKOG}}$, accounts for size differences (Ely, 1981) and is calculated by

$$\Delta\eta^{\text{ENSKOG}} = \eta_{\text{m}}^{\text{ENSKOG}} - \eta_{\text{x}}^{\text{ENSKOG}} \tag{10.7-10}$$

where $\eta_{\text{x}}^{\text{ENSKOG}}$ is discussed later and

$$\eta_{\text{m}}^{\text{ENSKOG}} = \sum_i \beta_i Y_i + \alpha\rho^2 \sum_i \sum_j y_i y_j \sigma_{ij}^6 \eta_{ij}^{\circ} g_{ij} \tag{10.7-11}$$

Here, ρ is density in mol/L, σ is in Å, and η° and $\eta_{\text{m}}^{\text{ENSKOG}}$ are in μP. Also,

$$\alpha = \frac{48}{25\pi}\left[\frac{2\pi}{3}(6.023 \times 10^{-4})\right]^2 = 9.725 \times 10^{-7}$$

$$\sigma_i = 4.771 h_i^{\frac{1}{3}} \tag{10.7-12}$$

$$\sigma_{ij} = \frac{\sigma_i + \sigma_j}{2} \tag{10.7-13}$$

Because $\Delta\eta^{\text{ENSKOG}}$ is a correction based on a hard sphere assumption, Eq. (10.3-8) is used to calculate η_{ij}°. The radial distribution function, g_{ij}, is calculated (Tham and Gubbins, 1971) by

$$g_{ij} = (1-\xi)^{-1} + \frac{3\xi}{(1-\xi)^2}\Theta_{ij} + \frac{2\xi^2}{(1-\xi)^3}\Theta_{ij}^2 \tag{10.7-14}$$

$$\Theta_{ij} = \frac{\sigma_i \sigma_j}{2\sigma_{ij}} \frac{\sum_k y_k \sigma_k^2}{\sum_k y_k \sigma_k^3} \tag{10.7-15}$$

$$\xi = (6.023 \times 10^{-4})\frac{\pi}{6}\rho \sum_i y_i \sigma_i^3 \tag{10.7-16}$$

$$Y_i = y_i\left[1 + \frac{8\pi}{15}(6.023 \times 10^{-4})\rho \sum_j y_j \left(\frac{M_j}{M_i + M_j}\right)\sigma_{ij}^3 g_{ij}\right] \tag{10.7-17}$$

where ρ is in mol/L and the n values of β_i are obtained by solving the n linear equations of the form

$$\sum_j B_{ij}\beta_j = Y_i \tag{10.7-18}$$

where

$$B_{ij} = 2\sum_k y_i y_k \frac{g_{ik}}{\eta_{ik}^{\circ}}\left(\frac{M_k}{M_i + M_k}\right)^2\left[\left(1 + \frac{5}{3}\frac{M_i}{M_k}\right)\delta_{ij} - \frac{2M_i}{3M_k}\delta_{jk}\right] \tag{10.7-19}$$

In Eq. (10.7-19), δ_{ij} is the Kronecker delta function, 1 if $i = j$, and 0 if $i \neq j$. The quantity $\eta_{\text{x}}^{\text{ENSKOG}}$ that appears in Eq. (10.7-10) is for a pure hypothetical fluid with the same density as the mixture and is determined with Eq. (10.7-11) with σ_{x} defined by

$$\sigma_{\text{x}} = \left(\sum_i \sum_j y_i y_j \sigma_{ij}^3\right)^{\frac{1}{3}} \tag{10.7-20}$$

$$M_{\text{x}} = \left[\sum_i \sum_j y_i y_j M_{ij}^{\frac{1}{2}} \sigma_{ij}^4\right]^2 \sigma_{\text{x}}^{-8} \tag{10.7-21}$$

M_{ij} and σ_{ij} are defined in Eqs. (10.7-9) and (10.7-13). Huber (1996) tested the TRAPP method on a number of binary hydrocarbon mixtures over a wide range of densities and reports an average absolute error of about 5%, although, in some cases, significantly larger deviations were found. The method is illustrated in Example 10.15.

Example 10.15 Use the TRAPP method to estimate the viscosity of a mixture of 80 mol% methane (1) and 20 mol% *n*-decane at 377.6 K and 413.7 bar. Lee et al. (1966) report at these conditions, $\rho = 0.4484$ g/cm³ and $\eta = 126$ μPa·s, although this value is considerably higher than values reported by Knapstad et al. (1990) at similar conditions.

Solution. The following properties are used:

	M	T_c, K	V_c, cm³/mol	Z_c	ω
CH_4	16.043	190.56	98.6	0.286	0.011
$C_{10}H_{22}$	142.285	617.7	624	0.256	0.490

With Eqs. (10.7-2) to (10.7-9), the procedure illustrated in Example 10.14 leads to $f_m = 0.9819$, $h_m = 0.8664$, $T_\eta = 384.6$ K, $\rho_\eta = 9.408$ mol/L, and $\eta_R - \eta_R^\circ = 51.72$ μPa·s. Equation (10.7-8) gives $F_{\eta m} = 1.260$. Calculation of $\Delta\eta^{\text{ENSKONG}}$ requires the application of the method described in Eqs. (10.7-10) through (10.7-21). Intermediate results include $\rho = 10.86$ mol/L and $\xi = 0.3668$ with other values shown below.

ij	σ_{ij} (Å)	η_{ij}° (μP)	g_{ij}	$B_{ij} \times 10^4$ (μP)$^{-1}$
11	3.716	150	2.696	210.7
12	5.314	98.6	3.085	−6.076
22	6.913	130	3.873	28.18

When $Y_1 = 2.014$ and $Y_2 = 0.5627$ are used in Eq. (10.7-18), this equation is written for each of the two components and solved to give $\beta_1 = 102.0$ and $\beta_2 = 221.6$. Finally, Eq. (10.7-11) gives $\eta_m^{\text{ENSKOG}} = 911$ μP. From Eqs. (10.7-20) and (10.7-21), $\sigma_x = 4.548$ Å and $M_x = 47.50$. Then, for the hypothetical pure fluid, Eq. (10.7-16) gives

$$\xi = (6.023 \times 10^{-4})(3.14159)(10.86)(4.548)^3/6 = 0.3222$$

Eq. (10.7-14) with $\Theta_{xx} = ½$ gives

$$g_{xx} = 1/(1 - 0.3222) + (3/2)(0.3222)/(1 - 0.3222)^2 + (2/4)(0.3222)^2/(1 - 0.3222)^3 = 2.694$$

Eq. (10.7-17) gives

$$Y_x = 1 + (8/15)(3.14159)(6.023 \times 10^{-4})(10.86)(4.548)^3(2.694)/2 = 2.389$$

From Eq. (10.3-8)

$$\eta_x^\circ = (26.69)[(47.50)(377.6)]^{1/2}/(4.548)^2 = 172.8\ \mu\text{P}$$

For a pure component, Eq. (10.7-19) reduces to $B_{xx} = \frac{g_{xx}}{\eta_x^\circ} B_{xx}$, so that $B_{xx} = 2.694/172.8 = 0.01559$. Then Eq. (10.7-18) gives $\beta_x = 2.389/0.01559 = 153.2$. Applying Eq. (10.7-11) to a pure fluid gives

$$\eta^{\text{ENSKOG}} = (153.2)(2.389) + (9.725 \times 10^{-7})(10.86)^2(4.548)^6(172.8)(2.694) = 839\ \mu\text{P}$$

With Eq. (9-7.10)

$$\Delta\eta^{\text{ENSKOG}} = 911 - 839 = 72\ \mu\text{P} = 7.2\ \mu\text{Pa}\cdot\text{s}$$

Then using Eq. (10.7-1)

$$\eta_m = 10.2 + (1.260)(51.72) + 7.2 = 82.6\ \mu\text{Pa}\cdot\text{s}$$

$$\eta_m \text{ Difference} = 82.6 - 126 = -43.4\ \mu\text{Pa}\cdot\text{s or } -34\%$$

In the above example, for the low pressure contribution (η_m°) of 10.2 μPa · s, Eq. (10.3-9) was used for methane, Eq. (10.4-17) was used for decane, and Eq. (10.5-16) was used for the mixture. Although the TRAPP prediction in Example 10.15 was lower than the experimental value reported by Lee et al. (1966), TRAPP predictions are considerably higher than the experimental values reported in Knapstad et al. (1990) for the same system at similar conditions. High-quality data for mixtures with different size molecules at high pressures are limited. This in turn limits the ability to evaluate models for this case.

10.7.4 Discussion

Both the Lucas and Chung et al. methods use the relations for the estimation of dense gas viscosity and apply a one-fluid approximation to relate the component parameters to composition. The TRAPP method uses the term $\Delta\eta^{\text{ENSKOG}}$ to improve the one-fluid approximation. In the Lucas method, the state variables are T, P, and composition, whereas in the TRAPP and Chung et al. procedures, T, ρ, and composition are used.

The accuracy of the Lucas and Chung et al. forms is somewhat less than when applied to pure, dense gases. Also, as noted at the end of Sec. 10.6, the accuracy is often poor when working in the critical region or at densities approaching those of a liquid at the same temperature. The TRAPP procedure can be extended into the liquid region. The paucity of accurate high-pressure gas mixture viscosity data has limited the testing that could be done, but Chung et al. (1988) report absolute average deviations of 8 to 9% for both polar and nonpolar dense gas mixtures. A comparable error would be expected from the Lucas form. The TRAPP method gives similar deviations for nonpolar mixtures but has not been tested for polar mixtures. Tilly et al. (1994) recommended a variation of the TRAPP method to correlate viscosities of supercritical fluid mixtures in which various solutes were dissolved in supercritical carbon dioxide.

As a final comment to the first half of this chapter, if one were planning a property estimation system for use on a computer, it is recommended that the Lucas, Chung et al., or Brulé and Starling method be used in the dense gas mixture viscosity correlations. Then, at low pressures or for pure components, the relations simplify directly to those described in Secs. 10.4 to 10.6. In other words, it is not necessary, when using these particular methods, to program separate relations for low-pressure pure gases, low-pressure gas mixtures, and high-pressure pure gases. One program is sufficient to cover all those cases as well as high-pressure gas mixtures.

10.8 LIQUID VISCOSITY

Most gas and gas mixture estimation techniques for viscosity are modifications of theoretical expressions described briefly in Secs. 10.3 and 10.5. There is no comparable theoretical basis for the estimation of liquid viscosities. Thus, it is particularly desirable to determine liquid viscosities from experimental data when such data exist. Viswanath and Natarajan (1989) have published a compilation of liquid viscosity data for over 900 compounds and list constants that correlate these data. Liquid viscosity data can also be found in Gammon et al. (1993–1998), Riddick et al. (1986), Stephan and Lucas (1979), Stephen and Hildwein (1987), Stephan and Heckenberger (1988), Timmermans (1965), and Vargaftik et al. (1996). Data for aqueous electrolyte solutions may be found in Kestin and Shankland (1981), Lobo (1990), and Zaytsev and Aseyev (1992). Critically evaluated tabulations of data and constants have been published in the DIPPR 801 database (Wilding et al., 2017; DIPPR 2021) and ThermoData Engine from the Thermodynamics Research Center (Diky et al., 2021). Tabulations of constants are also found in Duhne (1979), van Velzen et al. (1972), Yaws et al. (1976), and Yaws (1995, 1995a) that allow estimations of liquid viscosities. When these constants are derived from experimental data they can be used with confidence, but sometimes (Yaws 1995, 1995a) they are based on estimated viscosities, and in such instances, they should be used with caution. Liquid phase viscosity values can also be found in Speight (2016), Rumble (2021), and Green and Southard (2018).

The viscosities of liquids are larger than those of gases at the same temperature. As an example, in Fig. 10.8, the viscosities of liquid and vapor benzene are plotted as functions of temperature. Near the normal boiling point (353.24 K), the liquid viscosity is about 36 times the vapor viscosity, and at lower temperatures, this ratio increases even further. Two vapor viscosities are shown in Fig. 10.8. The low-pressure gas line would correspond to vapor at about 1 bar. As noted earlier in Eq. (10.4-21), below T_c, low-pressure gas viscosities vary in a nearly linear manner with temperature. The curve noted as saturated vapor reflects the effect of the increase in vapor pressure at higher temperatures. The viscosity of the saturated vapor should equal that of the saturated liquid at the critical temperature (for benzene, $T_c = 562.05$ K).

Much of the curvature in the liquid viscosity-temperature curve may be eliminated if the logarithm of the viscosity is plotted as a function of reciprocal (absolute) temperature. This change is illustrated in Fig. 10.9 for four saturated liquids: ethanol, benzene, n-heptane, and nitrogen. (To allow for variations in the temperature

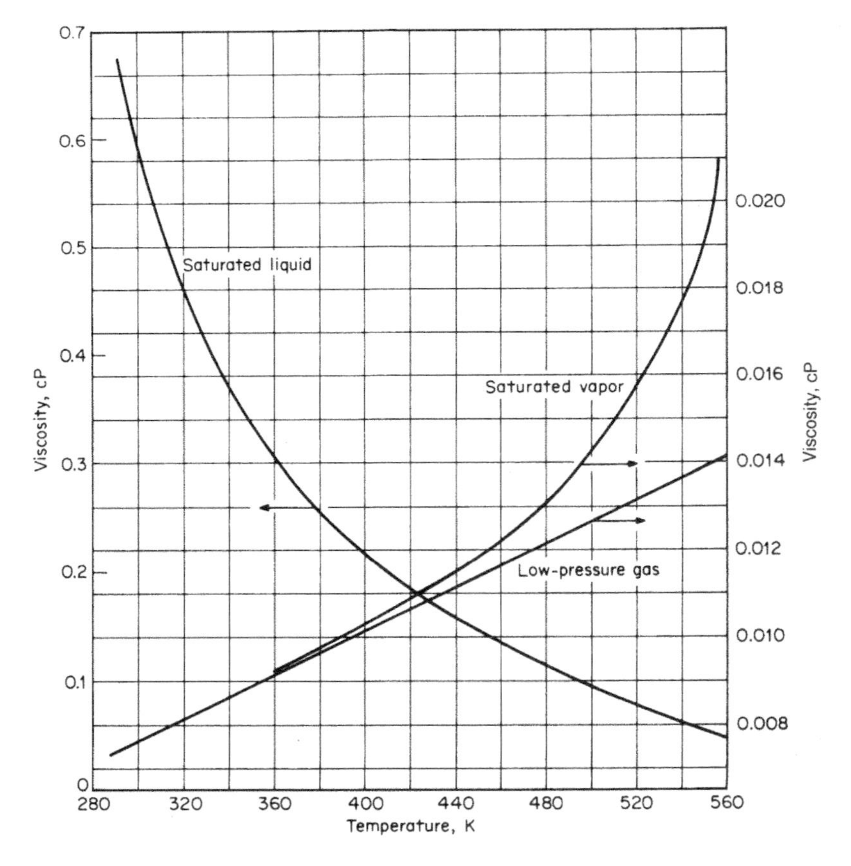

FIGURE 10.8
Viscosities of liquid and vapor benzene ($T_b = 353.24$ K; $T_c = 562.05$ K).

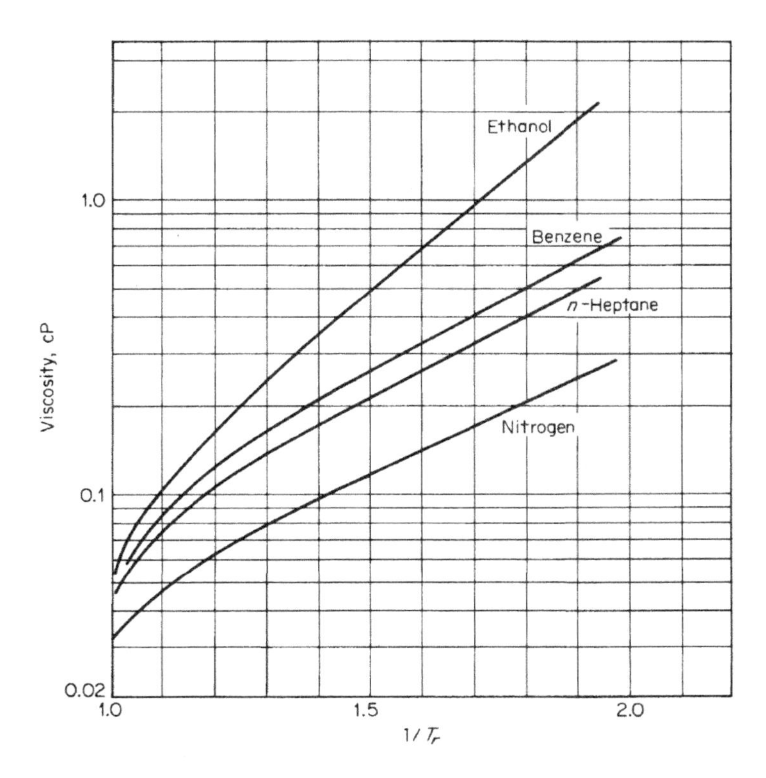

FIGURE 10.9
Viscosities of various liquids as functions of temperature. (Stephan and Lucas, 1979)

range, the reciprocal of the reduced temperature is employed.) Typically, the normal boiling point would be at a value of $T_r^{-1} \approx 1.5$. For temperatures below the normal boiling point ($T_r^{-1} > 1.5$), the logarithm of the viscosity varies linearly with T_r^{-1}. Above the normal boiling point, this no longer holds. In the nonlinear region, several corresponding states estimation methods have been suggested, and they are covered in Sec. 10.12. In the linear region, most corresponding states methods have not been found to be accurate, and many estimation techniques employ a group contribution approach to emphasize the effects of the chemical structure on viscosity. The curves in Fig. 10.9 suggest that, at comparable reduced temperatures, viscosities of polar fluids are higher than those of nonpolar liquids such as hydrocarbons, which themselves are larger than those of simple molecules such as nitrogen. If one attempts to replot Fig. 10.9 by using a nondimensional viscosity such as $\eta\zeta$ [see, for example, Eqs. (10.4-14) to (10.4-16)] as a function of T_r, the separation between curves diminishes, especially at $T_r > 0.7$. However, at lower values of T_r, there are still significant differences between the example compounds.

In the use of viscosity in engineering calculations, one is often interested not in the dynamic viscosity, but, rather, in the ratio of the dynamic viscosity to the density. This quantity, called the *kinematic viscosity,* would normally be expressed in m^2/s or in stokes. One stokes (St) is equivalent to 10^{-4} m^2/s. The kinematic viscosity ν decreases with increasing temperature in a manner such that ln ν is nearly linear in temperature for both the saturated liquid and vapor as illustrated in Fig. 10.10 for benzene. As with the dynamic viscosity, the kinematic viscosities of the saturated vapor and liquid become equal at the critical point.

The behavior of the kinematic viscosity with temperature has led to several correlation schemes to estimate ν rather than η. However, in most instances, ln ν is related to T^{-1} rather than T. If Fig. 10.10 is replotted by using T^{-1}, again there is a nearly linear correlation with some curvature near the critical point (as there is in Fig. 10.9).

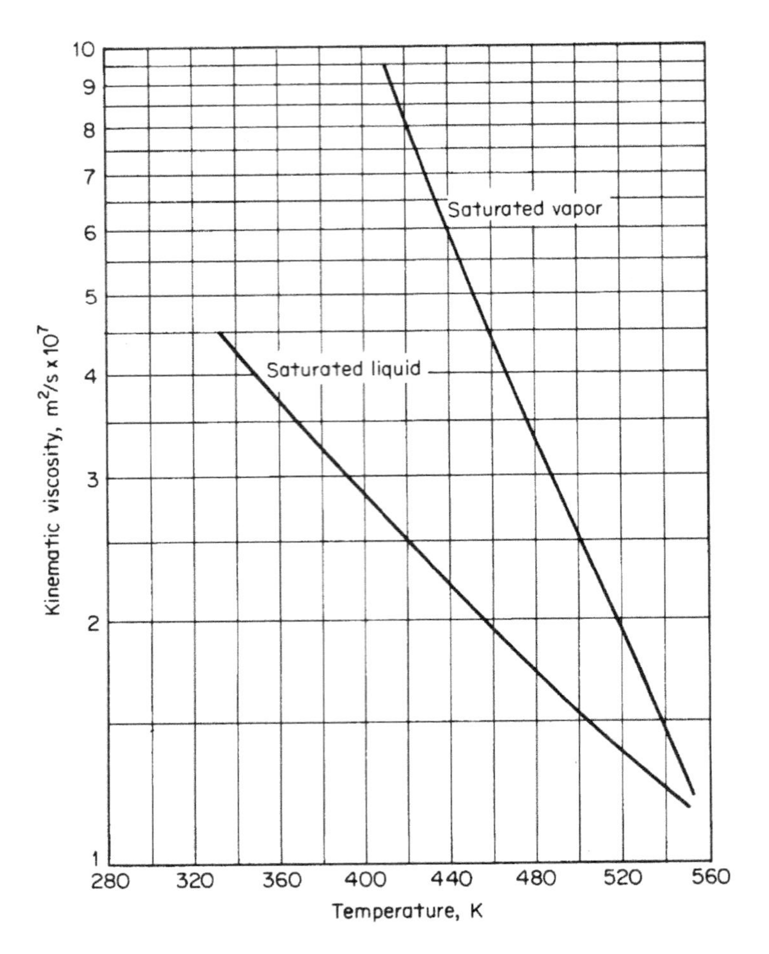

FIGURE 10.10
Kinematic viscosities of saturated liquid and vapor benzene ($T_b = 353.24$ K; $T_c = 562.05$ K).

In summary, pure-liquid viscosities at high reduced temperatures are usually correlated with some variation of the law of corresponding states (Sec. 10.12). At lower temperatures, most methods are empirical and involve a group contribution approach (Sec. 10.11). Current liquid *mixture* correlations are essentially mixing rules relating pure-component viscosities to composition (Sec. 10.13). Little theory has been shown to be applicable to estimating liquid viscosities (Andrade, 1954; Brokaw et al., 1965; Brush, 1962; Gemant, 1941; Hirschfelder et al., 1954).

10.9 EFFECT OF HIGH PRESSURE ON LIQUID VISCOSITY

Increasing the pressure over a liquid results in an increase in viscosity. Lucas (1981) has suggested that the change may be estimated from Eq. (10.9-1)

$$\frac{\eta}{\eta^{SL}} = \frac{1 + D\left(\dfrac{\Delta P_r}{2.118}\right)^A}{1 + C\omega\Delta P_r} \tag{10.9-1}$$

where η is the viscosity of the liquid at pressure P, η^{SL} is the viscosity of the saturated liquid at P^{vp}, $\Delta P_r = \frac{(P-P^{VP})}{P_c}$, ω is acentric factor, and A, C, and D are coefficients that are dependent on reduced temperature according to

$$A = 0.9991 - \left[\frac{4.674 \times 10^{-4}}{1.0523 T_r^{-0.03877} - 1.0513}\right] \quad (10.9\text{-}2)$$

$$\begin{aligned} C = &-0.07921 + 2.1616 T_r - 13.4040 T_r^2 + 44.1706 T_r^3 - 84.8291 T_r^4 \\ &+ 96.1209 T_r^5 - 59.8127 T_r^6 + 15.6719 T_r^7 \end{aligned} \quad (10.9\text{-}3)$$

$$D = \left[\frac{0.3257}{\left(1.0039 - T_r^{2.573}\right)^{0.2906}}\right] - 0.2086 \quad (10.9\text{-}4)$$

In a test with 55 liquids, polar and nonpolar, Lucas found errors in the calculated viscosities of less than 10%. To illustrate the predicted values of Eq. (10.9-1), Figs. 10.11 and 10.12 were prepared. In both, η/η^{SL} was plotted as a function of ΔP_r for various reduced temperatures. In Fig. 10.11, $\omega = 0$, and in Fig. 10.12, $\omega = 0.2$. Except at high values of T_r, η/η^{SL} is approximately proportional to ΔP_r. The effect of pressure is more important at the high reduced temperatures. As the acentric factor increases, there is a somewhat smaller effect of pressure. The method is illustrated in Example 10.16.

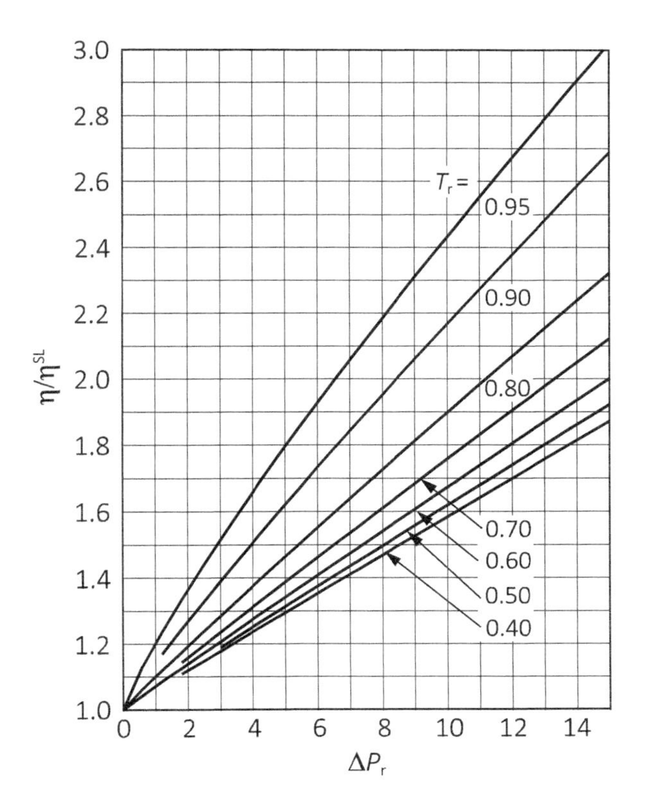

FIGURE 10.11
Effect of pressure on the viscosity of liquids, $\omega = 0$.

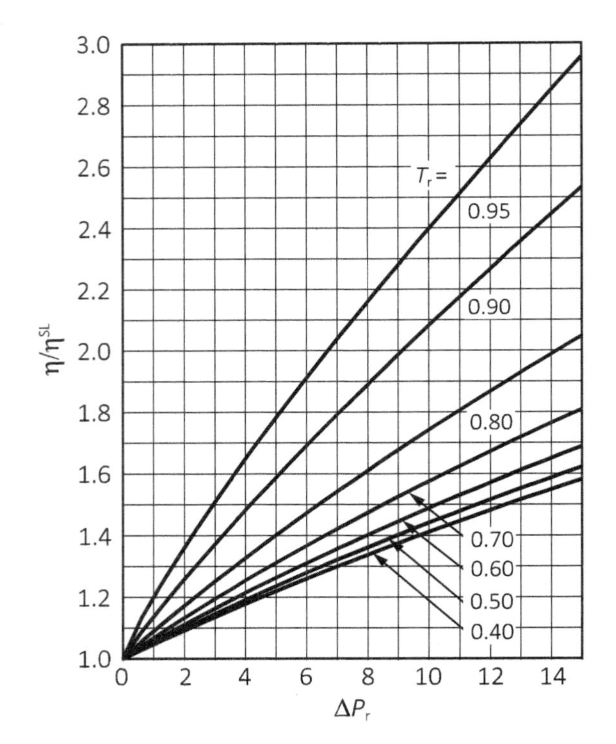

FIGURE 10.12
Effect of pressure on the viscosity of liquids, $\omega = 0.2$.

Example 10.16 Estimate the viscosity of liquid methylcyclohexane at 300 K and 500 bar. The viscosity of the saturated liquid at 300 K is 0.68 cP, and the vapor pressure is less than 1 bar.

Solution. From Appendix A, $T_c = 572.1$ K, $P_c = 34.80$ bar, and $\omega = 0.236055$. Thus $T_r = 300/572.1 = 0.5244$ and $\Delta P_r = 500/34.80 = 14.37$. ($P^{vp}$ was neglected.) Then

$$A = 0.9991 - 4.674 \cdot 10^{-4}/[(1.0523)(0.5244)^{-0.03877} - 1.0513] = 0.9822$$

$$C = -0.07921 + (2.1616)(0.5244) - (13.4040)(0.5244)^2 + (44.1706)(0.5244)^3$$
$$- (84.8291)(0.5244)^4 + (96.1209)(0.5244)^5 - (59.8127)(0.5244)^6$$
$$+ (15.6719)(0.5244)^7 = 0.06191$$

$$D = 0.3257/[1.0039 - (0.5244)^{2.573}]^{0.2906} - 0.2086 = 0.1372$$

With Eq. (10.9-1),

$$\frac{\eta}{\eta^{SL}} = [1 + (0.1372)(14.37/2.118)^{0.9822}]/[1 + (0.6191)(0.236044)(14.37)] = 1.570$$

$$\eta = (1.570)(0.68) = 1.07 \text{ cP}$$

The experimental value of η at 300 K and 500 bar is 1.09 cP (Titani, 1929).

$$\eta \text{ Difference} = 1.07 - 1.09 = 0.02 \text{ cP or } -1.8\%$$

Whereas the correlation by Lucas would encompass most pressure ranges, at pressures over several thousand bar the data of Bridgman suggest that the logarithm of the viscosity is proportional to pressure and that the structural complexity of the molecule becomes important. Those who are interested in such high-pressure regions should consult the original publications of Bridgman (1926) or the work of Dymond and Assael (see Sec. 10.6, Assael et al., 1996, or Dymond and Assael, 1996).

10.10 EFFECT OF TEMPERATURE ON LIQUID VISCOSITY

The viscosities of liquids decrease with increasing temperature either under isobaric conditions or as saturated liquids. This behavior can be seen in Fig. 10.9, where, for example, the viscosity of saturated liquid benzene is graphed as a function of temperature. Also, as noted in Sec. 10.8 and as illustrated in Fig. 10.10, for a temperature range from the freezing point to somewhere around the normal boiling temperature, it is often a good approximation to assume $\ln \eta$ is linear in reciprocal absolute temperature; i.e.,

$$\ln \eta = A + \frac{B}{T} \tag{10.10-1}$$

This simple form was apparently first proposed by de Guzman (1913) (O'Loane, 1979), but it is more commonly referred to as the Andrade equation (1930, 1934). Variations of Eq. (10.10-1) have been proposed to improve upon its correlation accuracy; many include some function of the liquid molar volume in either the A or B parameter (Bingham and Stookey, 1939; Cornelissen and Waterman, 1955; Eversteijn et al., 1960; Girifalco, 1955; Gutman and Simmons, 1952; Innes, 1956; Marschalko and Barna, 1957; Medani and Hasan, 1977; Miller, 1963, 1963a; Telang, 1945; and van Wyk et al., 1940). Another variation involves the use of a third constant to obtain the Vogel equation (1921),

$$\ln \eta = A + \frac{B}{T + C} \tag{10.10-2}$$

Goletz and Tassios (1977) have used this form (for the kinematic viscosity) and report values of A, B, and C for many pure liquids. More complex equations are needed when viscosity is correlated over large temperature ranges. For example, DIPPR (Wilding et al., 2017) reports liquid viscosity correlations from T_m to $0.8T_c$ using the modified Reidel equation (see Sec. 5.5), namely,

$$\ln \eta = A + \frac{B}{T} + C \ln T + DT^E \tag{10.10-3}$$

Equation (10.10-1) requires at least two viscosity-temperature data points to determine the two constants. If only one data point is available, one of the few ways to extrapolate this value is to employ the approximate Lewis-Squires chart (1934), which is based on the empirical fact that the sensitivity of viscosity to temperature variations appears to depend primarily upon the value of the viscosity. This chart, shown in Fig. 10.13, can be used by locating the known value of viscosity on the ordinate and then extending the abscissa by the required number of degrees to find the new viscosity. Figure 10.13 can be expressed in an equation form as

$$\eta_T^{-0.2661} = \eta_K^{-0.2661} + \frac{T - T_K}{233} \tag{10.10-4}$$

where η_T is liquid viscosity in cP at temperature T, η_K is a known value of liquid viscosity in cP at known temperature T_K. T and T_K may be expressed in either °C or K. Thus, given a value of η at T_K, one can estimate values of η at other temperatures. Equation (10.10-4) or Fig. 10.13 is only approximate, and errors of 5 to 15% (or greater) may be expected. This method should not be used if the temperature is much above the normal boiling point.

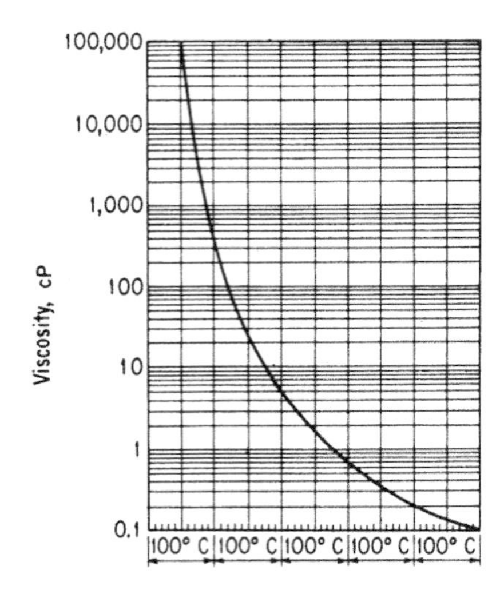

FIGURE 10.13
Lewis and Squires liquid viscosity-temperature correlation. (Lewis and Squires, 1934 as adapted in Gambill, 1959)

Example 10.17 The viscosity of acetone at 30 °C is 0.292 cP. Estimate the viscosities at –90 °C, –60 °C, 0 °C, and 60 °C.

Solution. At –90 °C, with Eq. (10.10-4),

$$\eta^{-0.2661} = 0.292^{-0.2661} + (-90 - 30)/233 = 0.8726$$

$$\eta^{0.2661} = 1/0.8726 = 1.146$$

$$\eta = 1.146^{1/0.2661} = 1.7 \text{ cP}$$

For all the cases,

T, °C	η, cP Eq. (10.10-4)	η, cP Experimental	Precent error
–90	1.7	2.1	–19
–60	0.99	0.98	1
0	0.42	0.39	8
60	0.21	0.23	–9

In summary, Eq. (10.10-3) provides robust fitting across a wide range of temperatures, from the melting point to approaching the critical temperature, when experimental data are available. Below the normal boiling point, simpler expressions such as those of Eqs. (10.10-1) and (10.10-2) may provide satisfactory correlations. At least two data points are required for Eq. (10.10-1). If only one point is known, a rough approximation of the viscosity at other temperatures can be obtained from Eq. (10.10-4) or Fig. 10.13.

10.11 ESTIMATION OF LOW-TEMPERATURE LIQUID VISCOSITY

Estimation methods for low-temperature liquid viscosity often employ structurally sensitive parameters which are valid only for certain homologous series or are found from group contributions. Techniques employing corresponding states concepts have also been developed. Most methods are limited to reduced temperatures less than about 0.75. As will be shown later, none are particularly reliable.

10.11.1 Orrick and Erbar (1974) Method

This method employs a group contribution technique to find the coefficients A and B in Eq. (10.11-1).

$$\ln \frac{\eta}{\rho M} = A + \frac{B}{T} \tag{10.11-1}$$

Here, η is liquid viscosity in cP, ρ is liquid density at 20 °C in g/cm^3, M is molecular weight in g/mol, and T is temperature in K. Table 10.9 contains the groups to find A and B. Notice that all primary (—CH_3) and secondary (—CH_2—) carbons are accounted for with the algebraic expression found in the first entry of the table. This contribution is added to other contributions found in the compound. All other portions of the molecule are accounted for by typical group summations of the form $\Sigma_i N_i \Delta_i$ where N_i is the number of times group i is found in the compound, and Δ_i is the contribution in the last two columns of Table 10.9 for all groups but those pertaining to the first entry. For liquids that have a normal boiling point below 20 °C, use the value of ρ at 20 °C; for liquids whose freezing point is above 20 °C, ρ at the melting point should be employed. Compounds containing nitrogen or sulfur cannot be treated.

Orrick and Erbar tested this method for 188 organic liquids. The errors varied widely, but they reported an average deviation of 15%. The method is evaluated along with other techniques later. Notice that since ρ is evaluated at a specific temperature other than T, Eq. (10.11-1) follows the form of Eq. (10.10-1).

TABLE 10.9 Orrick and Erbar (1974) group contributions for A and B in Eq. (10.11-1).

Algebraic Contributions	***A***	***B***
Carbon atoms not part of any group below[†]	$-(6.95 + 0.21N'_C)$	$275 + 99N'_C$
Additive Group Contributions		
>C— (bonds not to H's)	−0.15	35
>C< (bonds not to H's)	−1.20	400
Double bond (nonaromatic)	0.24	−90
Five-membered nonaromatic ring	0.10	32
Six-membered nonaromatic ring	−0.45	250
Aromatic ring	0	20
Ortho substitution	−0.12	100
Meta substitution	0.05	−34
Para substitution	−0.01	−5
—Cl	−0.61	220
—Br	−1.25	365
—I	−1.75	400
—OH (alcohol)	−3.00	1600
—COO— (ester, formate)	−1.00	420
—O— (ether)	−0.38	140
>C(=O (aldehyde, ketone)	−0.50	350
—COOH (acid)	−0.90	770

[†]N'_C = number of carbon atoms not including those described by Additive Group Contributions

Example 10.18 Estimate the viscosity of liquid 2-methyl-2-propanol (tert-butanol) at 70 °C with the Orrick-Erbar method. Bravo-Sanchez et al. (2013) report an experimental value of 0.756 cP.

Solution. 2-methyl-2-propanol is composed of $N = 3$ carbon atoms accounted for using the first entry of Table 10.9, one >C< group and one —OH group. The contributions of these to A and B are outlined below.

Algebraic Contribution		N'_C	A	B
Carbon atoms not part of any group below		3	−7.58	572
Group Contributions		N	A	B
1	>C<	1	−1.20	400
2	—OH	1	−3.00	1600
Total			−11.78	2572

At 20 °C, Kumar et al. (2012) report the density to be 0.7885 g/cm^3. With $M = 74.1216$ g/mol, Eq. (10.11-1) gives,

$$\eta = (0.7885)(74.1216)\exp[-11.78 + 2572/(70 + 273.15)] = 0.805 \text{ cP}$$

$$\eta \text{ Difference} = 0.805 - 0.756 = 0.049 \text{ cP or } 6.5\%$$

10.11.2 Sastri-Rao Method (1992)

In this method, the pure-liquid viscosity is calculated with the equation

$$\eta = \eta_b \left(P_{SR}^{vp}\right)^{A_{SR}} \tag{10.11-2}$$

Here, P_{SR}^{vp} is the vapor pressure in atm predicted from an expression given by Sastri and Rao (see below), η_b is the viscosity at the normal boiling point (T_b) in mPa·s and is obtained from group contributions as discussed below, and A_{SR} is obtained from group contributions as is also described later. Below T_b, Sastri and Rao determine P_{SR}^{vp} with the equation

$$\ln P_{SR}^{vp} = (4.5398 + 1.0309 \ln T_b)\left[1 - \frac{\left(3 - 2\dfrac{T}{T_b}\right)^{0.19}}{\dfrac{T}{T_b}} - 0.38\left(\left(3 - 2\frac{T}{T_b}\right)^{0.19} \ln\left(\frac{T}{T_b}\right)\right)\right] \tag{10.11-3}$$

Equation (10.11-3) should be used only when $T < T_b$. This expression is not necessarily the most accurate equation for vapor pressure predictions but should be used with Eq. (10.11-2) because the group contributions used to estimate η_b and A_{SR} were determined when P_{SR}^{vp} was calculated with Eq. (10.11-3). η_b is found with the equation

$$\eta_b = \sum_i N_i \left(\Delta_{\eta_b}\right)_i + \sum_j \left(\Delta_{\eta_b^{cor}}\right)_j \tag{10.11-4}$$

where $(\Delta_{\eta_b})_i$ is the contribution from group i, $(\Delta_{\eta_b^{cor}})_j$ is the contribution for group j which represent *corrections* due to certain functional groups, and N_i is the number of times group i is found in the molecule. The value of A_{SR} in Eq. (10.11-2) is determined from

$$A_{SR} = 0.2 + \sum_i (\Delta_{A_{SR}})_i + \sum_j (\Delta_{A_{SR}^{cor}})_j \qquad (10.11\text{-}5)$$

where $(\Delta_{A_{SR}})_i$ and $(\Delta_{A_{SR}^{cor}})_j$ are the contributions for groups i and corrections j, respectively. Values for contributions to determine the summations in Eqs. (10.11-4) and (10.11-5) are given in the file CH10SastriRaoLiqVisc.docx file on the PGLed6 website. Table 10.10 contains example groups for the method. The contributions of the functional groups i to η_b are generally cumulative (one is added for each identical group found in the molecule) as indicated by the presence of N_i in the first term in Eq. (10.11-4). However, if the compound contains more than one identical functional group, its contributions for A_{SR} should be taken only once unless otherwise mentioned as indicated by the absence of N_i in first term of Eq. (10.11-5). Thus, for branched hydrocarbons with multiple >CH— groups (e.g., 2,3-dimethylbutane), A_{SR} is 0.2 + 0.05 = 0.25. In Table 10.10, the term alicyclic means cycloparaffins and cycloolefins and excludes aromatics and heterocyclics. In the contributions of halogen groups, "others" means aromatics, alicyclics, and heterocyclics while the carbon groups listed are meant for aliphatic compounds. Also for halogens, the values of $\Delta_{A_{SR}}$ for aliphatic, alicyclics, and aromatics are not used if other nonhydrocarbon groups are present in the cyclic compound (see footnote b in the halogen section of Table 10.10). For example, the corrections for halogenated pyridines and anilines are given in footnote b and are not to be used in conjunction with the corrections listed under "aliphatic, alicyclics, and aromatics." Calculation of η_b and A_{SR} is illustrated in Example 10.19. The method is evaluated along with others later in this section.

Example 10.19 Determine the values of η_b and A_{SR} to be used in Eq. (10.11-2) for *o*-xylene, ethanol, ethylbenzene, 2,3-dimethylbutane, and *o*-chlorophenol.

Solution.

o-xylene

o-xylene has four =CH— (ring, not alicyclic), 2 =C< (ring, not alicyclic) and two —CH_3 groups. There is a correction $(\Delta_{\eta_b^{cor}})$ to η_b of 0.070 for multiple substitution and a correction $(\Delta_{A_{SR}^{cor}})$ to A_{SR} of 0.050. With values from Table 10.10,

$$\eta_b = 4(0.05) + 2(-0.120) + 2(0.105) + 0.070 = 0.24 \text{ mPa}\cdot\text{s}$$
$$A_{SR} = 0.2 + 1(0.000) + 1(0.000) + 1(0.000) + 0.050 = 0.25$$

Notice in the calculation for A_{SR} that each group contribution i is zero and is only counted once regardless of the number of times it appears. Also notice that in both calculations the corrections are only counted once.

ethanol

Ethanol has 1 —CH_3, 1 —CH_2— (non-ring), and 1 —OH. With values and expressions from Table 10.10 (notice that the Δ_{η_b} contribution for the —OH group is an algebraic expression based on the number of carbon atoms (2) in the molecule),

$$\eta_b = 1(0.105) + 1(0.000) + [0.615 - (0.092)(2) + (0.004)(2)^2 - 10^{(-0.58)(2)}] = 0.483 \text{ mPa}\cdot\text{s}$$
$$A_{SR} = 0.2 + 1(0.000) + 1(0.000) + 0.15 = 0.35$$

ethylbenzene

Ethylbenzene has 5 =CH— (ring, not alicyclic), 1 =C< (ring not alicyclic), 1 —CH_3, and 1 —CH_2— (nonring). There is a branching correction to $\Delta_{A_{SR}^{cor}}$ of 0.025. With values from Table 10.10,

$$\eta_b = 5(0.05) + 1(-0.12) + 1(0.105) + 1(0.000) = 0.235 \text{ mPa}\cdot\text{s}$$
$$A_{SR} = 0.2 + 1(0.000) + 1(0.000) + 1(0.000) + 0.025 = 0.225$$

TABLE 10.10 Example groups to calculate η_b and A_{SR} in Eq. (10.11-2) for the method of Sastri and Rao (1992) for liquid viscosity. See the file CH10SastriRaoLiqVisc.docx on the PGLed6 website for the complete table.

Group		Δ_{η_b}(mPa·s)	$\Delta_{A_{SR}}$	Remarks and Examples
				Nonring Hydrocarbon Groups
$—CH_3$		0.105	0.000	For *n*-alkanes, *n*-alkenes or *n*-alkynes with C > 8 $\Delta_{A_{SR}} = 0.050$ counted once
$—CH_2—$		0.000	0.000	
>CH—		–0.110	0.050	(i) If both >CH— and >C< groups are present $\Delta_{A_{SR}} = 0.050$ only (ii) $\Delta_{A_{SR}}$ values applicable only for aliphatic hydrocarbons and halogenated derivatives of aliphatic compounds (e.g., 2,2,4 trimethyl pentane, chloroform, bromal) in other cases $\Delta_{A_{SR}} = 0.000$
>C<		–0.180	0.100	
				Ring Hydrocarbon Groups
=CH—	Alicylic	0.040	0.000	
=CH—	Others	0.050	0.000	
=C<	Alicylic	–0.100	0.000	
=C<	Others	–0.120	0.000	

Correction Contributions ($\Delta_{A_{SR}^{cor}}$) of Ring Structure and Hydrocarbon Chains for A_{SR}

Structure	$\Delta_{A_{SR}^{cor}}$	Remarks and Examples
Monoalkyl benzenes with $C_{br} > 1$	0.025	ethylbenzene
Correction for multiple substitution in aromatics by hydrocarbon groups		
ortho	0.050	*o*-xylene, *o*-nitrotoluene
meta and para	0.000	*p*-xylene

Correction Contributions ($\Delta_{\eta_b^{cor}}$) of Ring Structure and Hydrocarbon Chains for η_b

Structure	$\Delta_{\eta_b^{cor}}$ (mPa·s)	Remarks
multiple substitution in aromatics by hydrocarbon groups	0.070	Counted once if more than one substitution is found regardless of the number of substitutions

Contribution of Halogen Groups

	Δ_{η_b} (mPa·s) for Halogen Attached to Carbon in						$\Delta_{A_{SR}}$ in Halogenated Hydrocarbons with No Other Functional Groups[b]		
	Aliphatic Compounds					Others			
Group	—CH_3 or —CH_2—[a]	>CH—	>C<[a]	=CH—	=C<		**Alicyclics**	**Aromatics**	**Others**
—F[c]	0.185	0.155	0.115	n.d.	n.d.	0.185	0.075	0.025	0.00
—Cl[a]	0.185	0.170	0.170	0.180	0.150	0.170	0.075	0.025	0.00

[a]Special configurations/function group structure combination	$\Delta_{A_{SR}^{cor}}$	Remarks and examples
(1) X—$(CH_2)_n$—X where X is halogen	0.050	1,3-dichloropropane
[b]Case of nonhydrocarbon group present in cyclic compounds		
(1) Halogen attached to ring carbons in compounds containing		
(A) —NH_2 or phenolic —OH	–0.075	2-chloro-6-methyl aniline
(B) oxygen-containing groups other than OH	0.050	2-chlorophenylmethyl ether
(C) other nonoxygen functional groups	–0.050	2-chloropyridine
(2) Halogen attached to nonhydrocarbon functional group	–0.050	benzoylbromide

[c]Fluorine groups in perfluorocompounds

Group	Δ_{η_b} **(mPa·s)**	$\Delta_{A_{SR}}$
Non-ring		
—CF_3	0.210	$\Delta_{A_{SR}} = 0.150$ for all perfluoro *n*-compounds
—CF_2—	0.000	
>CF—	–0.080	$\Delta_{A_{SR}} = 0.200$ for all isocompounds
Ring		
—CF_2—	0.145	$\Delta_{A_{SR}} = 0.200$ for all cyclic compounds
>CF—	–0.170	

(Continued)

TABLE 10.10 Example groups to calculate η_b and A_{SR} in Eq. (10.11-2) for the method of Sastri and Rao (1992) for liquid viscosity. See the file CH10SastriRaoLiqVisc.docx on the PGLed6 website for the complete table. (*Continued*)

Group	Δ_{η_b}(mPa·s)	$\Delta_{A_{SR}}$	Remarks and Examples
Contribution of Hydroxyl Groups			
Structure	Δ_{η_b} **(mPa·s)**	$\Delta_{A_{SR}}$	**Remarks and Examples**
—OH in aliphatics saturated primary	$0.615 - 0.092C + 0.004C^2 - 10^{-0.58C}$ for $C \le 10$	0.3 for $2 < C < 12$	(i) In compounds containing —O— group special value for the combination, $\Delta_{A_{SR}} = 0.100$ (2-methoxyethanol)
	0.095 for $C > 10$	0.15 for others	(ii) In compounds containing >NH group, special value for the combination, $\Delta_{A_{SR}} = 0.300$ (aminoethyl ethanolamine)
In cyclic alcohols	0.270	0.150	
Phenolic	0.270	0.200	(i) In compounds containing —NH_2 or —CHO groups in ortho position, special value for the combination, $\Delta_{A_{SR}} =$ 0.075 (2- nitrophenol, salicylaldehyde) (ii) In compounds containing —O— $\Delta_{A_{SR}}^{cor} = 0.050$ (4-methoxphenol)

2,3-dimethylbutane

2,3-dimethylbutane has 4 —CH_3 and 2 >CH— (nonring). With values from Table 10.10,

$$\eta_b = 4(0.105) + 2(-0.110) = 0.20 \text{ mPa} \cdot \text{s}$$
$$A_{SR} = 0.2 + 1(0.000) + 1(0.050) = 0.25$$

Notice that $\Delta_{A_{SR}}$ for the >CH— groups is applied only once when calculating A_{SR}.

o-chlorophenol

o-chlorophenol has 4 =CH— (ring, not alicyclic), 2 =C< (ring, not alicyclic), 1 —Cl attached to an "other," and 1 —OH (phenolic). Note that the —Cl contribution to $\Delta_{A_{SR}}$ of 0.025 is not used. Footnote b in the halogen section of Table 10.10 applies because of the presence of the nonhydrocarbon —OH group. With values from Table 10.10,

$$\eta_b = 4(0.050) + 2(-0.120) + 1(0.17) + 1(0.270) = 0.40 \text{ mPa} \cdot \text{s}$$
$$A_{SR} = 0.2 + 1(0.000) + 1(0.000) + 1(-0.075) + 1(0.200) = 0.325$$

10.11.3 Przezdziecki and Sridhar (1985) Method

In this technique, the authors propose using the Hildebrand-modified Batschinski equation (Batschinski, 1913; Hildebrand, 1971; Vogel and Weiss, 1981)

$$\eta = \frac{V_{PS}}{E_{PS}(V - V_{PS})} \tag{10.11-6}$$

where η is liquid viscosity in cP, V is liquid molar volume in cm^3/mol, and the parameters E_{PS} and V_{PS} are defined below.

$$E_{PS} = -1.12 + \frac{V_c}{12.94 + 0.10M - 0.23P_c + 0.0424T_m - 11.58(T_m/T_c)} \tag{10.11-7}$$

$$V_{PS} = 0.0085\omega T_c - 2.02 + \frac{V_m}{0.342(T_m/T_c) + 0.894} \tag{10.11-8}$$

Here, T_c is critical temperature in K, P_c is critical pressure in bar, V_c is critical volume in cm^3/mol, M is molecular weight in g/mol, T_m is melting point in K, ω is acentric factor, and V_m is liquid molar volume at T_m in cm^3/mol. The authors recommend that V_m and V be estimated from T_m and T by the Gunn-Yamada (1971) method. In the Gunn-Yamada method, one accurate value of V is required in the temperature range of applicability of Eq. (10.11-6). We define this datum point as V^R at T^R. Then, at any other temperature T,

$$V(T) = \frac{f(T)}{f(T^R)} V^R \tag{10.11-9}$$

where

$$f(T) = H_1(1 - \omega H_2) \tag{10.11-10}$$

$$H_1 = 0.33593 - 0.33953T_r + 1.51941T_r^2 - 2.02512T_r^3 + 1.11422T_r^4 \tag{10.11-11}$$

$$H_2 = 0.29607 - 0.09045T_r - 0.04842T_r^2 \tag{10.11-12}$$

Luckas and Lucas (1986) note that this prediction method often produces larger errors at low temperatures—an outcome that is expected from the form of Eq. (10.11-6). That is, because V_{PR} is of the order of the volume at the freezing point and $\eta \propto (V - V_{PR})^{-1}$, the estimated value of η becomes exceedingly sensitive to the choice of V. Luckas and Lucas (1986) suggest that Eq. (10.11-6) should not be used below T_r values of about 0.55.

Example 10.20 Use the Przezdziecki and Sridhar correlation to estimate the liquid viscosity of toluene at 383 K. The experimental value is 0.249 cP (Vargaftik et al., 1996).

Solution. From Appendix A, $T_c = 591.75$ K, $P_c = 41.08$ bar, $V_c = 316$ cm^3/mol, $T_m = 178.18$ K, $M = 92.13842$ gm/mol, and $\omega = 0.264012$. The liquid molar volume for the compound at 298.15 K is also given in Appendix A, so $V^R = 106.652$ cm^3/mol at $T^R = 298.15$ K and $T_r^R = 298.15/591.75 = 0.504$. The molar volumes at $T = 383$ K and $T_m = 178.18$ K are needed for the prediction method. Each of these are calculated from the reference point from Eq. (10.11-9). For the reference temperature, Eqs. (10.11-10) to (10.11-12) give

$$H_1(T_r^R) = 0.33593 - (0.33953)(0.504) + (1.51941)(0.504)^2 - (2.02512)(0.504)^3 + (1.11422)(0.504)^4 = 0.363$$
$$H_2 = 0.29607 - (0.09045)(0.504) - (0.04842)(0.504)^2 = 0.238$$
$$f = 0.363[1 - (0.264012)(0.238)] = 0.340$$

Similarly for the other two temperatures,

	T, *K*	T_r	H_1	H_2	$f(T)$
T_m	178.18	0.301	0.325	0.264	0.302
T	383	0.647	0.399	0.217	0.376

Using Eq. (10.11-9),

$$V_m = (0.302/0.340)(106.652) = 94.73 \text{ cm}^3/\text{mol}$$
$$V = (0.376/0.340)(106.652) = 117.94 \text{ cm}^3/\text{mol}$$

This value for V agrees with that given in Vargaftik et al. (1996). With Eqs. (10.11-7) and (10.11-8),

$$E_{PS} = -1.12 + 316/[12.94 + (0.10)(92.13842) - (0.23)(41.08) + (0.0424)(178.18) - 11.58(178.18/591.75)] = 17.72$$
$$V_{PS} = (0.0085)(0.264012)(591.75) - 2.02 + 94.73/[0.342(178.18/591.75) + 0.894] = 94.33 \text{ cm}^3/\text{mol}$$

Then, with Eq. (10.11-6),

$$\eta = 94.33/[17.72(117.94 - 94.33)] = 0.225 \text{ cP}$$

$$\eta \text{ Difference} = 0.225 - 0.249 = -0.024 \text{ cP or } -9.5\%$$

10.11.4 Bhethanabotla (1983) Method

Bhethanabotla (1983) proposed a method to predict liquid viscosity using the same correlating form as Orrick and Erbar. In this method,

$$\ln \frac{\eta}{\rho(T)M} = A_{Bh} + \frac{B_{Bh}}{T} \tag{10.11-13}$$

where M is molecular weight in g/mol, $\rho(T)$ is the density in g/cm^3 at temperature T, and T is in K. This last fact differs from the method of Orrick and Erbar in which the density was always evaluated at 20 °C. The coefficients are obtained by

$$A_{Bh} = \sum_i N_i \left(\Delta_{A_{Bh}}\right)_i \tag{10.11-14}$$

$$B_{Bh} = \sum_i N_i \left(\Delta_{B_{Bh}}\right)_i \tag{10.11-15}$$

TABLE 10.11 Example groups to calculate A_{Bh} and B_{Bh} in Eq. (10.11-13) for the Bhethanabotla method (1983) for liquid viscosity. See the file CH10BhethLiqVisc.xlsx on the PGLed6 website for the complete list. The symbol "Cd" means =C, and the symbol "Cb" means an aromatic carbon.

Group	$\Delta_{A_{Bh}}$	$\Delta_{B_{Bh}}$
C—(C)(H)$_3$	–0.3000	100.00
C—(C)$_2$(H)$_2$	–0.2000	85.00
Cd—(H)$_2$	–0.3500	90.00
Cd—(C)(H)	–0.2800	75.00
Cb—(H)	–0.3855	145.00
Cb—(C)	0.0300	–20.00
C—(Cb)(C)(H)$_2$	–0.3814	140.00
Correction Groups		
Alkene cis corr.	–0.0440	6.50
Cyclopentane ring corr.	–0.6500	210.00
Cyclohexane ring corr.	–0.9500	290.00
Family Corrections*		
Saturated hydrocarbons	–6.4000	250.00
Alcohols	–6.0000	1500.00
Straight chain aliphatic acids	–(5.667 + 0.2166C)	111C – 292
Ethanolamines	–6.0000	450 + 775C
All other classes of compounds	–6.0000	200.00

*C = number of carbon atoms in the compound

where $(\Delta_{A_{Bh}})_i$ and $(\Delta_{B_{Bh}})_i$ are the contributions for group i and N_i is the number of times group i is found in the molecule. Table 10.11 contains examples of the groups for the method, and the complete set is found in the file CH10BhethLiqVisc.xlsx on the PGLed6 website. Notice that the method has "Correction groups" for certain structures such as rings and cis orientations. There are also family corrections which must be added for every compound.

Example 10.21 Use the Bhethanabotla method to estimate the liquid viscosity of *n*-propylbenzene at 298.15 K. The experimental value is 0.792 cP (Sirbu et al., 2019).

Solution. *n*-propylbenzene contains 1 C—(C)(H)3, 1 C—(C)2(H)2, 5 Cb—(H), 1 Cb—(C), and 1 C—(Cb)(C)(H). No Correction groups are applicable, but the "all other classes of compounds" family correction is used. The group summations become

	Group	N	$N\Delta_{A_{Bh}}$	$N\Delta_{B_{Bh}}$
1	C—(C)(H)$_3$	1	–0.3000	100.00
2	C—(C)2(H)2	1	–0.2000	85.00
3	Cb—(H)	5	–1.9275	725.00
4	Cb—(C)	1	0.0300	–20.00
5	C—(Cb)(C)(H)$_2$	1	–0.3814	140.00
6	"all other classes of compounds"	1	–6.0000	200.00
$\sum_{k=1}^{6} N_k F_k$			–8.7789	1230

From Appendix A, $M = 120.19158$ g/mol and the liquid molar volume at 298.15 K is 139.969 cm^3/mol. Thus,

$$\rho = (120.19158)/(139.969) = 0.85873 \text{ g/cm}^3$$

With Eq. (10.11-13), the liquid viscosity is

$$\eta = (0.85873)(120.19158)\exp[-8.7789 + 1230/298.15] = 0.983 \text{ cP}$$

$$\eta \text{ Difference} = 0.983 - 0.792 = 0.191 \text{ cP or } 24.1\%$$

10.11.5 Hsu (2002) Method

Hsu et al. (2002) developed a group contribution method for liquid viscosity at atmospheric pressure and for $T_r < 0.75$. After trying several different correlating equations, they settled on the following expression:

$$\ln \eta = \sum_i N_i \left[(\Delta_{A_{\text{Hsu}}})_i + (\Delta_{B_{\text{Hsu}}})_i T + \frac{(\Delta_{C_{\text{Hsu}}})_i}{T^2} + (\Delta_{D_{\text{Hsu}}})_i \ln P_c \right] \quad (10.11\text{-}16)$$

Here, η is liquid viscosity in mPa·s, T is temperature in K, P_c is critical pressure in bar, N_i is the number of times group i is found in the molecule, and $(\Delta_{A_{\text{Hsu}}})_i$, $(\Delta_{B_{\text{Hsu}}})_i$, $(\Delta_{C_{\text{Hsu}}})_i$, and $(\Delta_{D_{\text{Hsu}}})_i$ are the contributions for group i. This method has more types of groups than previous approaches, and the authors claim superior performance for sulfur-containing, highly branched, and heterocyclic compounds. Table 10.12 lists example groups for the technique, and the complete list is found in the file C10HsuSheuTuLiqVisc.xlsx on the PGLed6 website.

TABLE 10.12 Example Groups for the Hsu et al. Method (2002) for Liquid Viscosity. See the File CH10HsuSheuTuLiqVisc.xlsx on the PGLed6 Website for the Complete List.

Group	$\Delta_{A_{\text{Hsu}}}$	$\Delta_{B_{\text{Hsu}}}$	$\Delta_{C_{\text{Hsu}}}$	$\Delta_{D_{\text{Hsu}}}$
$—CH_3$	0.0570	−0.002383	7556	−0.1765
$—CH_2—$	−0.1497	0.000060	14157	0.0751
—S—	−3.2767	0.000779	44123	0.9549

Example 10.22 Use the Hsu et al. method to estimate the liquid viscosity of methyl ethyl sulfide at 298.15 K. The experimental value is 0.354 mPa·s (Haines et al., 1954).

Solution. Methyl ethyl sulfide is composed of the 2 $—CH_3$, 1 $—CH_2—$, and 1 —S—. The group summations become

	Group	N	$N\Delta_{A_{\text{Bh}}}$	$N\Delta_{B_{\text{Bh}}}$	$\Delta_{C_{\text{Hsu}}}$	$\Delta_{D_{\text{Hsu}}}$
1	$—CH_3$	2	0.1140	−0.004766	15112	−0.3530
2	$—CH_2—$	1	−0.1497	0.000060	14157	0.0751
3	—S—	1	−3.2767	0.000779	44123	0.9549
$\sum_{k=1}^{3} N_k F_k$			−3.3124	−0.003927	73392	0.6770

The critical pressure from Appendix A for this compound is 42.60 bar. With this value, and the group summations, the viscosity from Eq. (10.11-16) is

$$\eta = \exp[-3.3124 - (0.003927)(298.15) + 73392/298.15^2 + 0.6770 \ln(42.60)]$$
$$= 0.327 \text{ mPa} \cdot \text{s}$$
$$\eta \text{ Difference} = 0.327 - 0.354 = -0.027 \text{ mPa} \cdot \text{s or } -7.6\%$$

10.11.6 Other Correlations

Other viscosity-correlating methods have been proposed, and a number of these are summarized in Mehrotra et al. (1996) and Monnery et al. (1995). Other recent correlations are given in Mehrotra (1991), and the earlier literature was reviewed in the 4th edition of this book.

10.11.7 Recommendations for Estimating Low-Temperature Liquid Viscosities

A test database was created to evaluate the performance of pure-component low-temperature ($T_r \lesssim 0.75$) liquid viscosity prediction methods. This database consisted of all reliable, experimental liquid viscosity data from the October 2021 release of the DIPPR database. Temperatures ranged from T_m to approximately T_b and included almost 13,000 total points (N_{pt}) from 978 unique compounds (N_c). The techniques tested included those from: Orrick and Erbar (1974) Eq. (10.11-1) (abbreviated OE), Sastri and Rao (1992) Eq. (10.11-2) (abbreviated SR), Przezdziecki and Sridhar (1985) Eq. (10.11-6) (abbreviated PS), Bhethanabotla (1983) Eq. (10.11-13) (abbreviated BH), and Hsu et al. (2002) Eq. (10.11-16) (abbreviated HS). The evaluation also includes the approach from van Velzen et al. (1972) (abbreviated VV) described in the 4th edition of this book and the technique of Thomas (1946) (abbreviated TH) outlined in the 3rd edition of this book. To identify if a method was better at one class of compounds than another, each chemical in the test database was assigned a *class* as described at the end of Sec. 10.4. Results were quantified using the average absolute relative deviation (*AARD*) defined in Eq. (10.4-26).

The results are found in Table 10.13 where the *AARD* is listed for each method according to compound class. Also listed are the number of points tested in each class (N_{pt}) and the number of compounds (N_c). Compared with other properties presented in this book, accurate prediction of liquid viscosity is difficult. Only a few classes are predicted with better than 20% *AARD* (the SR and TH methods for heavy compounds, the OE and TH methods

TABLE 10.13 Performance of prediction techniques for low-temperature ($T_r \lesssim 0.7$) liquid η against the test database.

	Compound Class											
	Normal			Heavy			Polar			Associating		
Method	N_{pt}	N_c	AARD	N_{pt}	N_c	AARD	N_{pt}	N_c	AARD	N_{pt}	N_c	AARD
OE	2977	244	27%	404	31	26%	1120	105	15%	614	104	30%
SR	4491	351	22%	472	42	16%	1449	132	24%	1324	181	29%
PS	4898	393	27%	471	41	48%	1620	157	20%	1471	191	76%
BH	2743	222	48%	446	38	65%	978	83	22%	662	106	23%
HS	4023	330	50%	414	39	24%	1467	135	42%	1407	188	101%
VV	4654	365	45%	472	42	121%	1397	136	20%	1309	178	35%
TH	2679	208	31%	399	29	12%	1185	104	11%	350	27	11%

for polar compounds, and the TH method for associating compounds), and none of these are better than 10%. Strikingly, many of the *AARD*s are over 50%, and some exceed 100%. The OE and TH approaches have consistently lower *AARD*s for all classes, but they are also the most limited in terms of applicability (generally lower values of N_{pt} and N_c). Among the techniques which can be used across a larger number of compounds, the SR method performs consistently across all classes of compounds and has lower, or nearly equal, *AARD*s compared to PS, BH, HS, or VV. The PS approach works well for normal and polar compounds but performs poorly for heavy and associating chemicals. The accuracy of the VV method is only satisfactory for polar compounds, while the BH technique is good for polar and associating substances.

All the methods can fail spectacularly and unexpectedly as each produced errors greater than 100% in certain instances. There was no discernable pattern to these extreme outliers, and the problematic compounds were different for each technique. For this reason, it is highly suggested that the family analysis approach, described in Chap. 3 and demonstrated in Example 10.23, be followed to select the best method to use when predicting liquid viscosity.

Example 10.23 Determine which prediction method will yield the most accurate value for liquid η of 3-ethylthiophene.

Solution. 3-ethylthiophene is a member of the *Sulfides/Thiophenes* family in the DIPPR database. The database contains 62 compounds in this family, but only 22 have experimental values for liquid η. These data consist of 191 points at different temperatures for the 22 compounds. Determining which of the prediction methods in this chapter is likely to give the best estimate of liquid η for 3-ethylthiophene was done automatically using DIPPR's DIADEM program in a process described by Rowley et al. (2007) and summarized in Example 3.16. Using this process, the prediction methods discussed in this chapter reproduce the experimental values for the 22 compounds according to the statistics found in the table below.

Method	Number of Compounds	Ave. Abs. Rel. Dev.	Min. Abs. Rel. Dev.	Max. Abs. Rel. Dev.
OE	0	—	—	—
SR	21	11%	0.9%	76%
PS	22	22%	2%	54%
BH	3	19%	12%	28%
HS	20	60%	0.4%	348%
VV	22	23%	0.2%	79%
TH	17	10%	0.5%	83%

Notice that groups are not available in some methods to predict a value for all 22 compounds. The OE approach is not applicable to the family, and the BH can only be used for four members. The TH and SR methods perform the best on average for the family. The data in the tables do not give convincing evidence to choose one over the other, but the TH approach does not have the needed groups to describe 3-ethylthiophene. Thus, the SR technique is the preferred method to predict liquid η for this compound.

10.12 ESTIMATION OF LIQUID VISCOSITY AT HIGH TEMPERATURES

Below a reduced temperature of about 0.7, ln η for liquids is usually assumed to follow a linear relationship with reciprocal absolute temperature (see Sec. 10.10). Above a reduced temperature of about 0.7, this relation is no longer valid, as illustrated in Fig. 10.10. In the region from about $T_r = 0.7$ to near the critical point, many estimation methods are of a corresponding states type that resemble or are identical with those used in the first sections of this chapter to treat gases. For this temperature range, Sastri (1998) recommends

$$\ln \eta = \left[\frac{\ln \eta_b}{\ln(\alpha \eta_b)}\right]^{\phi} \ln(\alpha \eta_b) \tag{10.12-1}$$

where η is in mPa · s, η_b is viscosity at T_b in mPa · s, $\alpha = 0.1175$ for alcohols and 0.248 for other compounds and

$$\phi = \frac{1 - T_r}{1 - T_{b_r}} \tag{10.12-2}$$

where $T_r = T/T_c$ and $T_{b_r} = T_b/T_c$. Sastri reports average deviations of 10% for $T_r > 0.9$ and 6% for $T_{b_r} < T_r < 0.9$.

Example 10.24 Estimate the saturated liquid viscosity of *n*-propanol at 433.2 K by using Eq. (10.12-1). The experimental value is 0.188 cP.

Solution. From Appendix A, $T_b = 370.35$ K and $T_c = 536.8$ K. With contributions from Table 10.10 for 1 —CH_3, 2 —CH_2—, and 1 primary —OH (the algebraic contribution with C = 3)

$$\eta_b = 1(0.105) + 2(0.000) + 1[0.615 - 0.092(3) + 0.004(3)^2 - 10^{-(0.58)(3)}]$$
$$= 0.462 \text{ mPa} \cdot \text{s} = 462\ \mu\text{Pa} \cdot \text{s}$$

From Eq. (10.12-2),

$$\phi = (1 - 433.2/536.8)/(1 - 370.35/536.8) = 0.623$$

With $\alpha = 0.1175$, Eq. (10.12-1) gives

$$\ln \eta = \ln[(0.1175)(462)]\{\ln(462)/\ln[(0.1175)(462)]\}^{0.624} = 5.221$$
$$\eta = \exp(5.221) = 185\ \mu\text{Pa} \cdot \text{s} = 0.185 \text{ cP}$$

$$\eta \text{ Difference} = 0.185 - 0.188 = -0.003 \text{ cP or } -1.6\%$$

A more general estimation method would logically involve the extension of the high-pressure gas viscosity correlations described in Sec. 10.6 into the liquid region. Two techniques have, in fact, been rather widely tested and found reasonably accurate for reduced temperatures above about 0.5. These methods are those of Chung et al. (1988) and Brulé and Starling (1984). Both methods use Eq. (10.6-27), but they have slightly different coefficients to compute some of the parameters. The Chung et al. form is preferable for simple molecules and will treat polar as well as nonpolar compounds. The Brulé and Starling relation was developed primarily for complex hydrocarbons, and the authors report their predictions are within 10% of experimental values in the majority of cases. The Chung et al. method has a similar accuracy for most nonpolar compounds, but significantly higher errors can occur with polar, halogenated, or high-molecular weight compounds. In both cases, one needs accurate liquid density data, and the reliability of the methods decreases significantly for T_r less than about 0.5. The liquids need not be saturated; subcooled compressed liquid states simply reflect a higher liquid density. The Chung et al. technique was illustrated for dense gas ammonia in Example 10.13. The procedure is identical when applied to high-temperature liquids.

10.12.1 Discussion

The quantity of accurate liquid viscosity data at temperatures much above the normal boiling point is not large. In addition, to test estimation methods such as those of Chung et al. or Brulé and Starling, one needs accurate liquid density data under the same conditions which apply to the viscosity data. This matching makes it somewhat difficult to test the methods with many compounds. However, Brulé and Starling developed their technique so that they would be coupled to a separate computation program using a modified BWR equation of state to provide densities. They report relatively low errors, and this fact appears to confirm the general approach (see also Brulé and Starling, 1984). Hwang et al. (1982) have proposed viscosity (as well as density and surface tension) correlations for coal liquids.

Regardless of what high-temperature estimation method is chosen, there is the problem of joining both high- and low-temperature estimated viscosities should that be necessary.

10.13 LIQUID MIXTURE VISCOSITY

Essentially all correlations for liquid mixture viscosity refer to solutions of liquids below or only slightly above their normal boiling points; i.e., they are restricted to reduced temperatures (of the pure components) below about 0.7. The bulk of the discussion below is limited to that temperature range. At the end of the section, however, we suggest approximate methods to treat high-pressure, high-temperature liquid mixture viscosity.

At temperatures below $T_r \approx 0.7$, liquid viscosities are very sensitive to the structure of the constituent molecules (see Sec. 10.11). This generality is also true for liquid mixtures, and even mild association effects between components can often significantly affect the viscosity. For a mixture of liquids, the shape of the curve of viscosity as a function of composition can be nearly linear for so-called ideal mixtures. But systems that contain alcohols and/or water often exhibit a maximum or a minimum and sometimes both (Irving, 1977a).

Almost all methods to estimate or correlate liquid mixture viscosities assume that values of the viscosities of the pure components are available. Thus the methods are interpolative. Nevertheless, there is no agreement on the best way to carry out the interpolation. Irving (1977) surveyed more than 50 equations for binary liquid viscosities and classified them by type. He points out that only very few do not have some adjustable constant that must be determined from experimental mixture data, and the few that do not require such a parameter are applicable only to systems of similar components with comparable viscosities. In a companion report from the National Engineering Laboratory, Irving (1977a) has also evaluated 25 of the more promising equations with experimental data from the literature. He recommends the one-constant Grunberg-Nissan (1949) equation [see Eq. (10.13-1)] as being widely applicable yet reasonably accurate except for aqueous solutions. This NEL report is also an excellent source of viscosity data tabulated from the literature. Other data and literature sources for data may be found in Aasen et al. (1990), Aucejo et al. (1995), supplementary material of Cao et al. (1993), Franjo et al. (1995), Kouris and Panayiotou (1989), Krishnan et al. (1995, 1995a), Kumagai and Takahashi (1995), Petrino et al. (1995), Stephan and Hildwein (1987), Stephan and Heckenberger (1988), Teja et al., (1985), and Wu et al. (1998).

10.13.1 Method of Grunberg and Nissan (1949)

In this procedure, the low-temperature liquid viscosity for mixtures is given as

$$\ln \eta_m = \sum_{i=1}^{n} x_i \ln \eta_i + \frac{1}{2}\sum_{i=1}^{n}\sum_{j=1}^{n} x_i x_j G_{ij} \tag{10.13-1}$$

where n is the number of components in the mixture, x_i is the liquid mole fraction of component i, η_i is the liquid viscosity of pure component i, and G_{ij} is an interaction parameter (which is a function of the components i and j, temperature, and, in some cases, composition). This relation has probably been more extensively examined than any other liquid mixture viscosity correlation. For a binary of 1 and 2, Eq. (10.13-1) reduces to

$$\ln \eta_m = x_i \ln \eta_1 + x_2 \ln \eta_2 + x_1 x_2 G_{12} \tag{10.13-2}$$

since $G_{ii} = 0$.

Isdale (1979) presents the results of a very detailed testing using more than 2000 experimental mixture data points. When the interaction parameter was regressed from experimental data, nonassociated mixtures and many mixtures containing alcohols, carboxylic acids, and ketones were fitted satisfactorily. The overall root mean square deviation for the mixtures tested was 1.6%. Isdale et al. (1985) later proposed a group contribution method to estimate the binary interaction parameter G_{ij} at 298 K. The procedure is:

1. For a binary of i and j, select i (the first component) using the following priority rules. (j then becomes the second component.)

   a. i = an alcohol, if present
   b. i = an acid, if present
   c. i = the component with the most carbon atoms
   d. i = the component with the most hydrogen atoms
   e. i = the component with the most —CH_3 groups

   If none of these rules establish a priority, $G_{ij} = 0$, and the process is complete for the two components. If priority is established, proceed to Step 2.

2. Determine the parameter W.

   a. If either i or j contains atoms other than carbon and hydrogen, set $W = 0$.
   b. Else,

$$W = \frac{0.3161\left(N_{C_i} - N_{C_j}\right)^2}{N_{C_i} + N_{C_j}} - 0.1188\left(N_{C_i} - N_{C_j}\right) \tag{10.13-3}$$

   where N_{C_i} and N_{C_j} are the number of carbon atoms in i and j, respectively.

3. Calculate G_{ij} from

$$G_{ij} = \sum_k N_k(\Delta_i)_k + \sum_k N_k(\Delta_j)_k + W \tag{10.13-4}$$

   where $(\Delta_i)_k$ is the contribution for group k in component i, $(\Delta_j)_k$ is the contribution for group k in component j, and N_k is the number of times group k is found in the component for the summation for which N_k pertains. Table 10.14 lists the values for the group contributions for the method.

G_{ij} is sometimes a function of temperature. However, existing data suggest that, for alkane-alkane solutions, or for mixtures of an associated component with an unassociated one, G_{ij} is independent of temperature. However, for mixtures of nonassociated compounds (but not of only alkanes) or for mixtures of associating compounds, G_{ij} is a mild function of temperature. Isdale et al. (1985) suggest for these latter two cases,

$$G_{ij}(T) = 1 - [1 - G_{ij}(298\text{ K})]\frac{573 - T}{275} \tag{10.13-5}$$

where T is in K and $G_{ij}(298\text{ K})$ is the binary interaction parameters at 298 K.

Example 10.25 Estimate the viscosity of a mixture of acetic acid and acetone at 323 K (50 °C) that contains 70 mole% acetic acid. Isdale et al. (1985) quote the experimental value to be 0.587 cP, and, at 50 °C, the viscosities of pure acetic acid and acetone are 0.798 and 0.241 cP, respectively.

TABLE 10.14 Group contributions for G_{ij} at 298 K in the method of Isdale et al. (1985) for determining η_m. Use in Eq. (10.13-4) for both components i and j.

Group	Notes	Δ_x (x is either i or j)
$—CH_3$		−0.100
$—CH_2—$		0.096
>CH—		0.204
>C<		0.433
Benzene ring		0.766
Benze ring substitutions:		
Ortho		0.174
Meta		—
Para		0.154
Cyclohexane ring		0.887
—OH	Methanol	0.887
	Ethanol	−0.023
	Higher aliphatic alcohols	−0.443
>C═O	Ketones	1.046
—Cl		$0.653 - 0.161\ N_{Cl}$
—Br		−0.116
—COOH	Acid with:	
	Nonassociated liquids	$-0.411 + 0.06074\ N_C$
	Ketones	1.130
	Formic acid with ketones	0.167

N_{Cl} = number of chlorine atoms in the molecule.
N_C = *total* number of carbon atoms in both compounds.

Solution. First we must estimate G_{ij} at 298 K. Component i is acetic acid (priority rule b). Since the mixture contains atoms other than carbon and hydrogen (i.e., oxygen), $W = 0$. With the groups in Table 10.14, acetic acid is composed of 1 $—CH_3$ and 1 —COOH (with ketone), and acetone has 2 $—CH_3$ and 1 >C═O. The groups summations in Eq. (10.13-4) become

$$\sum_k N_k(\Delta_i)_k \ (i \text{ is acetic acid}) = 1(-0.100) + 1(1.130) = 1.030$$

$$\sum_k N_k(\Delta_j)_k \ (j \text{ is acetone}) = 2(-0.100) + 1(1.046) = 0.846$$

With Eq. (10.13-4),

$$G_{ij} \text{ (at 298 K)} = 1.030 - 0.846 + 0 = 0.184$$

G_{ij} at 50 °C = 323 K is determined with Eq. (10.13-5).

$$G_{ij} \text{ (at 323 K)} = 1 - [1 - 0.184](573 - 323)/275 = 0.258$$

Then, using Eq. (10.13-2),

$$\ln \eta_m = 0.7\ \ln(0.798) + 0.3\ \ln(0.241) + (0.7)(0.3)(0.258) = -0.531$$
$$\eta_m = 0.588\ \text{cP}$$

This estimated value is essentially identical with the experimental result of 0.587 cP.

To summarize the Isdale modification of the Grunberg-Nissan equation, for each possible binary pair in the mixture, first decide which component is to be labeled i and which j using the priority rules. Determine $\Sigma_k N_k(\Delta_i)_k$ and $\Sigma_k N_k(\Delta_j)_k$ using groups in Table 10.14 and W from Eq. (10.13-3), if necessary. Use Eq. (10.13-4) to calculate G_{ij}. Correct for temperatures other than 298 K, if necessary, with Eq. (10.13-5). With the values of G_{ij} so determined, use either Eq. (10.13-1) [or Eq. (10.13-2) for a binary system] to determine the viscosity of the liquid mixture. This technique yields acceptable estimates of low-temperature liquid mixture viscosities for many systems, but Table 10.14 is limited in the number of classes of compounds that can be treated. Also, the method does not cover aqueous mixtures.

10.13.2 UNIFAC-VISCO Method (Chevalier et al., 1988; Gaston-Bonhomme et al., 1994)

Gaston-Bonhomme, Petrino, and Chevalier have modified the UNIFAC activity coefficient method (described in Chap. 8) to predict viscosities. In this method, mixture viscosity (η_m) is calculated by

$$\ln \eta_m = \sum_i x_i \ln(\eta_i V_i) - \ln V_m + \left(\frac{g^E}{RT}\right)^{\text{comb}} + \left(\frac{g^E}{RT}\right)^{\text{res}} \tag{10.13-6}$$

where x_i is the mole fraction of component i, η_i is the viscosity of pure compound i in cP, V_i is the molar volume of pure compound i in cm^3/mol, V_m is the mixture molar volume in cm^3/mol, and $\left(\frac{g^E}{RT}\right)^{\text{comb}}$ and $\left(\frac{g^E}{RT}\right)^{\text{res}}$ are the combinatorial and residual UNIQUAC terms (see Table 8.2). The mixture molar volume is determined by

$$V_m = \frac{1}{\rho_m}\sum_i x_i M_i \tag{10.13-7}$$

Here, ρ_m is the mixture density in g/cm^3 and M_i is the molecular weight of compound i in g/mol. The combinatorial term is calculated by

$$\left(\frac{g^E}{RT}\right)^{\text{comb}} = \sum_i x_i \ln\frac{\phi_i}{x_i} + \frac{z}{2}\sum_i q_i x_i \ln\frac{\theta_i}{\phi_i} \tag{10.13-8}$$

where the summations over all components i in the system, z is the coordination number, equal to 10, and θ_i and ϕ_i are the molecular surface area fraction and molecular volume fraction, respectively, given by

$$\theta_i = \frac{x_i q_i}{\sum_j x_j q_j} \tag{10.13-9}$$

and

$$\phi_i = \frac{x_i r_i}{\sum_j x_j r_j} \tag{10.13-10}$$

where q_i, the van der Waals' surface area, and r_i, the van der Waals' volume of component i, are found by summation of the corresponding group contributions. Thus, if $N_k^{(i)}$ is the number of groups of type k in the molecule i,

$$q_i = \sum_k N_k^{(i)} Q_k \tag{10.13-11}$$

$$r_i = \sum_k N_k^{(i)} R_k \tag{10.13-12}$$

where Q_k is the contribution of group k to the surface area parameter for compound i, and R_k is the contribution of group k to the size parameter for compound i. The group contributions are given in Table 10.15.

The residual term in Eq. (10.13-6) is calculated by

$$\left(\frac{g^{\mathrm{E}}}{RT}\right)^{\mathrm{res}} = -\sum_i x_i \ln \gamma_i^{\mathrm{res}} \tag{10.13-13}$$

where

$$\ln \gamma_i^{\mathrm{res}} = \sum_k N_k^{(i)} \left[\ln \gamma_k - \ln \gamma_k^{(i)}\right] \tag{10.13-14}$$

and

$$\ln \gamma_k = Q_k \left[1 - \ln\left(\sum_m \Theta_m \Psi_{mk}\right) - \sum_m \frac{\Theta_m \Psi_{mk}}{\sum_n \Theta_n \Psi_{nm}}\right] \tag{10.13-15}$$

where m and n are over all the distinct groups in the first column of Table 10.15 found in the system and Θ_{m} is the surface area fraction in the mixture of groups given by

$$\Theta_{\mathrm{m}} = \frac{Q_{\mathrm{m}} X_{\mathrm{m}}}{\sum_j X_j Q_j} \tag{10.13-16}$$

TABLE 10.15 Group contributions for to determine the surface area (q_i) and volume (r_i) parameters for the UNIFAC-VISCO model for determining η_{m}. Use in Eqs. (10.13-11) and (10.13-12).

Group	R	Q
CH_2, CH_{2cy}	0.6744	0.540
CH_3	0.9011	0.848
CH_{ar}	0.5313	0.400
Cl	0.7910	0.724
CO	0.7713	0.640
COO	1.0020	0.880
OH	1.0000	1.200
CH_3OH	1.4311	1.432

Here, j is over all the distinct groups in the first column of Table 10.15 found in the system [the same as m and n in Eq. (10.13-15)], and X_m is the mole fraction in the mixture of groups found through

$$X_{\mathrm{m}} = \frac{\sum_i N_{\mathrm{m}}^{(i)} x_i}{\sum_j \sum_i N_j^{(i)} x_i} \tag{10.13-17}$$

where i is over the compounds in the system and j is over the distinct groups given in the first column of Table 10.15 *found in compound i*. Except for the minus sign in Eq. (10.13-13), these last equations are identical to those in the UNIFAC method described in Chap. 8. However, the groups are chosen differently, and the interaction parameters are different and are calculated by

$$\Psi_{\mathrm{nm}} = \exp\left(-\frac{a_{\mathrm{nm}}}{298}\right) \tag{10.13-18}$$

Values of a_{nm} are given in Table 10.16. γ_k is the activity coefficient of group k in a mixture of groups in the actual mixture, and $\gamma_k^{(i)}$ is the activity coefficient of group k in a mixture of groups formed from the groups in pure component i. Groups in branched hydrocarbons and substituted cyclic and aromatic hydrocarbons are chosen as follows:

Type of Compound	Group	Representation
Branched cyclic	>CH—CH_3	2 CH_2 groups
	>CH_{cy}—CH_3	1 CH_{2cy} + 1 CH_2
	>C_{cy}(—CH_3)$_2$	1 CH_{acy} + 2 CH_2
Aromatic	>C_{ar}—CH_3	1 CH_{ar} + 1 CH_2

Table 10.17 compares results calculated with the UNIFAC-VISCO method to experimental values. Of all the methods evaluated, the UNIFAC-VISCO method was the only one that demonstrated any success in predicting viscosities of mixtures of compounds with large size differences. The method has also been successfully applied to ternary and quaternary alkane systems. The average absolute deviation for 13 ternary alkane systems was 2.6%, while for four quaternary systems it was 3.6%. The method is illustrated in Example 10.26.

TABLE 10.16 UNIFAC-VISCO group interaction parameters (a_{nm}). Use in Eq. (10.13-18).

n/m	CH_2	CH_3	CH_{2cy}	CH_{ar}	Cl	CO	COO	OH	CH_3OH
CH_2	0	66.53	224.9	406.7	60.30	859.5	1172.0	498.6	−219.7
CH_3	−709.5	0	−130.7	−119.5	82.41	11.86	−172.4	594.4	−228.7
CH_{2cy}	−538.1	187.3	0	8.958	215.4	−125.4	−165.7	694.4	−381.53
CH_{ar}	−623.7	237.2	50.89	0	177.2	128.4	−49.85	419.3	−88.81
Cl	−710.3	375.3	−163.3	−139.8	0	−404.3	−525.4	960.2	−165.4
CO	586.2	−21.56	740.6	−117.9	−4.145	0	29.20	221.5	55.52
COO	541.6	−44.25	416.2	−36.17	240.5	22.92	0	186.8	69.62
OH	−634.5	1209.0	−138	197.7	195.7	664.1	68.35	0	416.4
CH_3OH	−526.1	653.1	751.3	51.31	−140.9	−22.59	−286.2	−23.91	0

TABLE 10.17 Performance of the UNIFAC-VISCO prediction method for viscosity of liquid mixtures.

1st Component	2nd Component	x_1	T, K	η_{exp}, mPa·s	Ref*	η_{calc}, mPa·s	% Deviation
n-$C_{10}H_{22}$	n-$C_{60}H_{122}$	0.749	384.1	3.075	1	2.309	−25
		0.749	446.4	1.423	1	1.275	−10
n-$C_{10}H_{22}$	n-$C_{44}H_{99}$	0.354	368.8	5.286	1	5.256	−0.6
		0.354	464.1	1.465	1	1.654	−13
		0.695	374.1	2.318	1	1.960	−15
butane	squalane	0.839	293.1	1.060	2	0.8812	−17
ethanol	benzene	0.5113	298.1	0.681	3	0.6403	−6.0
acetone	benzene	0.3321	298.1	0.4599	4	0.4553	−1.0
acetone	ethanol	0.3472	298.1	0.5133	5	0.4860	−5.3

*References: 1, Aasen et al. (1990); 2, Kumagai and Takahashi (1995); 3, Kouris and Panayiotou (1989); 4, Petrino et al. (1995); 5, Wei et al. (1985).

Example 10.26 Use the UNIFAC-VISCO method to estimate the viscosity of a mixture of 35.4 mole% *n*-decane (1) and 64.6 mole% *n*-tetratetracontane, $C_{44}H_{90}$ (2) at 397.49 K. The experimental viscosity and density (Aasen et al., 1990) are $\eta_m = 3.278$ cP and $\rho_m = 0.7447$ g/cm^3.

Solution. From Aasen et al. (1990), $\eta_1 = 0.2938$ cP, $\eta_2 = 4.937$ cP, $V_1 = 220$ cm^3/mol, $V_2 = 815.5$ cm^3/mol, $M_1 = 142.28$ g/mol, and $M_2 = 619.16$ g/mol. With Eq. (10.13-7),

$$V_m = [(0.354)(142.28) + (0.646)(619.16)]/0.7447 = 604.7 \text{ cm}^3/\text{mol}$$

In *n*-decane, there are 8 CH_2 groups and 2 CH_3 groups. In tetratetracontane, there are 42 CH_2 groups and 2 CH_3 groups. Equations (10.13-11) and (10.13-12), with the group values in Table 10.15, give

$$r_1 = 8(0.6744) + 2(0.9011) = 7.1974$$
$$r_2 = 42(0.6744) + 2(0.9011) = 30.127$$
$$q_1 = 8(0.540) + 2(0.848) = 6.016$$
$$q_2 = 42(0.540) + 2(0.848) = 24.376$$

Equations (10.13-9) and (10.13-10) give

$$\theta_1 = [(0.354)(6.016)]/[(0.354)(0.6016) + (0.646)(24.376)] = 0.1191$$
$$\theta_2 = [(0.646)(24.376)]/[(0.354)(0.6016) + (0.646)(24.376)] = 0.8809$$
$$\phi_1 = [(0.354)(7.1974)]/[(0.354)(7.1974) + (0.646)(30.127)] = 0.1158$$
$$\phi_2 = [(0.646)(30.127)]/[(0.354)(7.1974) + (0.646)(30.127)] = 0.8842$$

Equation (10.13-8) is now used to calculate the combinatorial contribution.

$$\left(\frac{g^E}{RT}\right)^{\text{comb}} = 0.354 \ln(0.1158/0.354) + 0.646 \ln(0.8842/0.646) + 5[(0.354)(6.016) \ln(0.1191/0.1158) + (0.646)(24.376) \ln(0.8809/0.8842)] = -0.1880$$

Two distinct groups are found in the system. In this mixture of groups, with CH_2 designated by subscript 1 and CH_3 by subscript 2, Eq. (10.13-17) yields

$$X_1 = [8(0.354) + 42(0.646)]/[8(0.354) + 42(0.646) + 2(0.354) + 2(0.646)] = 0.9374$$
$$X_2 = [2(0.354) + 2(0.646)]/[8(0.354) + 42(0.646) + 2(0.354) + 2(0.646)] = 0.0626$$

Equation (10.13-16) written for these same distinct groups gives

$$\Theta_1 = (0.54)(0.9374)/[(0.54)(0.9374) + (0.848)(0.0626)] = 0.9051$$
$$\Theta_2 = (0.848)(0.0626)/[(0.54)(0.9374) + (0.848)(0.0626)] = 0.0949$$

From Table 10.16, for the two types of groups found in the system [(1) CH_2 and (2) CH_3], $a_{12} = 66.53$ and $a_{21} = -709.5$. Equation (10.13-18), with these interaction parameters, gives

$$\Psi_{12} = \exp(-66.53/298) = 0.7999$$
$$\Psi_{21} = \exp(709.5/298) = 10.81$$

Equation (10.13-15) for groups 1 and 2 gives

$$\ln \gamma_1 = 0.54\{1 - \ln[0.9051 + (0.0949)(10.81)] - 0.9051/[0.9051 + (0.0949)(10.81)] - (0.0949)(0.7999)/[(0.9051)(0.7999) + 0.0949]\} = -0.1185$$
$$\ln \gamma_2 = 0.848\{1 - \ln[(0.9051)(0.7999) + (0.0949)] - (0.9051)(10.81)/[0.9051 + (0.0949)(10.81)] - (0.0949)/[(0.9051)(0.7999) + 0.0949]\} = -3.3776$$

$\ln \gamma_k^{(i)}$ is now needed. These quantities are determined from the same equations as $\ln \gamma_i$ except only the groups in each single component i are considered. In the mixture of groups from pure component 1, recognizing the $x_2 = 0$ for pure component 1, Equation (10.13-17) yields $X_1^{(1)} = 0.8$ and $X_2^{(1)} = 0.2$ where the subscripts refer to groups (1) CH_2 and (2) CH_3 and the superscript. Using Eq. (10.13-16) for pure component 1 then gives

$$\Theta_1^{(1)} = (0.8)(0.54)/[(0.8)(0.54) + (0.2)(0.848)] = 0.7181$$
$$\Theta_2^{(1)} = (0.2)(0.848)/[(0.8)(0.54) + (0.2)(0.848)] = 0.2819$$

With Eq. (10.13-15),

$$\ln \gamma_1^{(1)} = 0.54\{1 - \ln[0.7181 + (0.2819)(10.81)] - 0.7181/[0.7181 + (0.2819)(10.81)] - (0.2819)(0.7999)/[(0.7181)(0.7999) + 0.2819] = -0.4211$$

Similarly, $\ln \gamma_2^{(1)} = -1.0478$. In pure component 2, the results are $\Theta_1^{(2)} = 0.9304$, $\Theta_2^{(2)} = 0.0696$, $\ln \gamma_1^{(2)} = -0.07654$, and $\ln \gamma_2^{(2)} = -4.1182$. Equation (10.13-14) gives

$$\ln \gamma_1^{res} = 8(-0.1185 + 0.4211) + 2(-3.3776 + 1.0478) = -2.239$$
$$\ln \gamma_2^{res} = 42(-0.1185 + 0.07654) + 2(-3.3776 + 4.1182) = -0.2811$$

Finally, the residual contribution is calculated with Eq. (10.13-13)

$$\left(\frac{g^E}{RT}\right)^{res} = -[(0.354)(-2.239) + (0.646)(-0.2811)] = 0.9742$$

Equation (10.13-6) is now used to calculate the mixture viscosity.

$$\ln \eta_m = 0.354 \ln[(0.2938)(220)] + 0.646 \ln[(4.937)(815.5)] - \ln(604.7) - 0.1880 + 0.9742 = 1.219$$
$$\eta_m = \exp(1.219) = 3.384 \text{ cP}$$

$$\eta_m \text{ Difference} = 3.384 - 3.278 = 0.106 \text{ cP or } 3.22\%$$

10.13.3 Method of Teja and Rice (1981, 1981a)

Based on a corresponding-states treatment for mixture compressibility factors (Teja 1980; Teja and Sandler, 1980), Teja and Rice proposed an analogous form for liquid mixture viscosity.

$$\ln(\eta_m \varepsilon_m) = \ln(\eta\varepsilon)^{(R1)} + \left[\ln(\eta\varepsilon)^{(R2)} - \ln(\eta\varepsilon)^{(R1)}\right]\frac{\omega_m - \omega^{(R1)}}{\omega^{(R2)} - \omega^{(R1)}} \tag{10.13-19}$$

where the superscripts (R1) and (R2) refer to two reference fluids, η_m is the mixture viscosity, η is the viscosity of the indicated reference fluid, ω is the acentric factor of the indicated reference fluid, ω_m is the mixture acentric factor (described below), ε is a parameter similar to ζ in Eq. (10.4-16) but defined here as

$$\varepsilon = \frac{V_c^{\frac{2}{3}}}{(T_c M)^{\frac{1}{2}}} \tag{10.13-20}$$

and ε_m is calculated using Eq. (10.13-20) with the mixture properties discussed shortly. In Eq. (10.13-20), V_c is in cm³/mol, T_c is in K, and M is in g/mol. The variable of composition is introduced in three places: the definitions of ω_m, T_{c_m}, and M_m. The rules suggested by the authors to compute these mixture parameters are

$$V_{c_m} = \sum_i \sum_j x_i x_j V_{c_{ij}} \tag{10.13-21}$$

$$T_{c_m} = \frac{1}{V_{c_m}} \sum_i \sum_j x_i x_j T_{c_{ij}} V_{c_{ij}} \tag{10.13-22}$$

$$M_m = \sum_i x_i M_i \tag{10.13-23}$$

$$\omega_m = \sum_i x_i \omega_i \tag{10.13-24}$$

$$V_{c_{ij}} = \frac{\left(V_{c_i}^{\frac{1}{3}} + V_{c_j}^{\frac{1}{3}}\right)^3}{8} \tag{10.13-25}$$

$$T_{c_{ij}} V_{c_{ij}} = \psi_{ij}\left(T_{c_i} T_{c_j} V_{c_i} V_{c_j}\right)^{\frac{1}{2}} \tag{10.13-26}$$

ψ_{ij} is an interaction parameter of order unity which must be found from experimental data.

When using Eq. (10.13-19) for a given mixture at a specified temperature, the viscosity values for the two reference fluids $\eta^{(R1)}$ and $\eta^{(R2)}$ are to be obtained *not at* T, but at a temperature equal to $T\left[\frac{(T_c)^{(R1)}}{T_{c_m}}\right]$ for (R1) and $T\left[\frac{(T_c)^{(R2)}}{T_{c_m}}\right]$ for (R2) with T_{c_m} given by Eq. (10.13-22).

Whereas the reference fluids (R1) and (R2) may be chosen as different from the actual components in the mixture, it is normally advantageous to select them from the principal components in the mixture. In fact, for a binary of 1 and 2, if (R1) is selected as component 1 and (R2) as component 2, then, by virtue of Eq. (10.13-24), Eq. (10.13-19) simplifies to

$$\ln(\eta_m \varepsilon_m) = x_1 \ln(\eta\varepsilon)_1 + x_2 \ln(\eta\varepsilon)_2 \tag{10.13-27}$$

but, as noted above, η_1 is evaluated at $T\left[\frac{T_{c1}}{T_{c_m}}\right]$ and η_2 at $T\left[\frac{T_{c2}}{T_{c_m}}\right]$.

Our further discussion of this method will be essentially limited to Eq. (10.13-27), since that is the form most often used for binary liquid mixtures and, by this choice, one is assured that the relation gives correct results when $x_1 = 0$ or 1. In addition, the assumption is made that the interaction parameter ψ_{ij} is not a function of temperature or composition.

The authors claim good results for many mixtures ranging from strictly nonpolar to highly polar aqueous-organic systems. For nonpolar mixtures, errors averaged about 1%. For nonpolar-polar and polar-polar mixtures, the average rose to about 2.5%, whereas for systems containing water, an average error of about 9% was reported.

In comparison with the Grunberg-Nissan correlation [Eq. (10.13-1)], with G_{ij} found by regressing data, Teja and Rice show that about the same accuracy is achieved for both methods for nonpolar-nonpolar and nonpolar-polar systems, but their technique was significantly more accurate for polar-polar mixtures, and particularly for aqueous solutions for which Grunberg and Nissan's form should not be used.

Example 10.27 Estimate the viscosity of a liquid mixture of water and 1,4-dioxane at 60 °C when the mole fraction water is 0.83. For this very nonideal solution, Teja and Rice suggest an interaction parameter $\psi_{ij} = 1.37$. This value was determined by regressing data at 20 °C.

Solution. From Appendix A, for water, $T_c = 674.095$ K, $V_c = 55.9472$ cm^3/mol, and $M = 18.01528$ g/mol; for 1,4-dioxane, $T_c = 587.0$ K, $V_c = 238$ cm^3/mol, and $M = 88.10512$. Let 1 be water and 2 be 1,4-dioxane. With Eq. (10.13-20),

$$\varepsilon_1 = 55.9472^{2/3}/[(647.095)(18.01528)]^{1/2} = 0.135$$

$$\varepsilon_2 = 238^{2/3}/[(587.0)(88.10512)]^{1/2} = 0.169$$

From Eq. (10.13-21),

$$V_{c_m} = (0.83)^2(55.9472) + (0.17)^2(238) + (2)(0.83)(0.17)[55.9472^{1/3} + 238^{1/3}]^3/8$$
$$= 80.93 \text{ cm}^3\text{/mol}$$

and with Eqs. (10.13-22) and (10.13-23),

$$T_{c_m} = \{(0.83)^2(647.095)(55.9472) + (0.17)^2(587)(238)$$
$$+ (2)(0.83)(0.17)(1.37)[(647.095)(55.9472)(587)(238)]^{1/2}\}/80.93 = 697.8 \text{ K}$$

$$M_m = (0.83)(18.01528) + (0.17)(88.10512) = 29.9306 \text{ g/mol}$$

So, with Eq. (10.13-20)

$$\varepsilon_m = (80.93)^{2/3}/[(697.8)(29.9306)]^{1/2} = 0.1295$$

Next, the viscosity of water is required at $T\left[\frac{T_{c1}}{T_{cm}}\right] = (333.2)(647.095)/(697.9) = 308.9$ K, not $T = 333.2$ K (60 °C). This value is 0.712 cP (Irving, 1977a). [Note that, at 60 °C, $\eta_{water} = 0.468$ cP.] For 1,4-dioxane, the reference temperature is (333.2)(587)/697.9 = 280.3 K (7.1 °C), and at that temperature $\eta = 1.63$ cP (Irving, 1977a). Again, this value is quite different from the viscosity of 1,4-dioxane at 60 °C, which is 0.715 cP. Finally, with Eq. (10.13-27),

$$\ln \eta_m = 0.83 \ln[(0.712)(0.135)] + 0.17 \ln[(1.63)(0.169)] - \ln(0.1295) = -0.1191$$

$$\eta_m = \exp(-0.1191) = 0.888 \text{ cP}$$

This is essentially equal to the experimental viscosity of 0.89 cP.

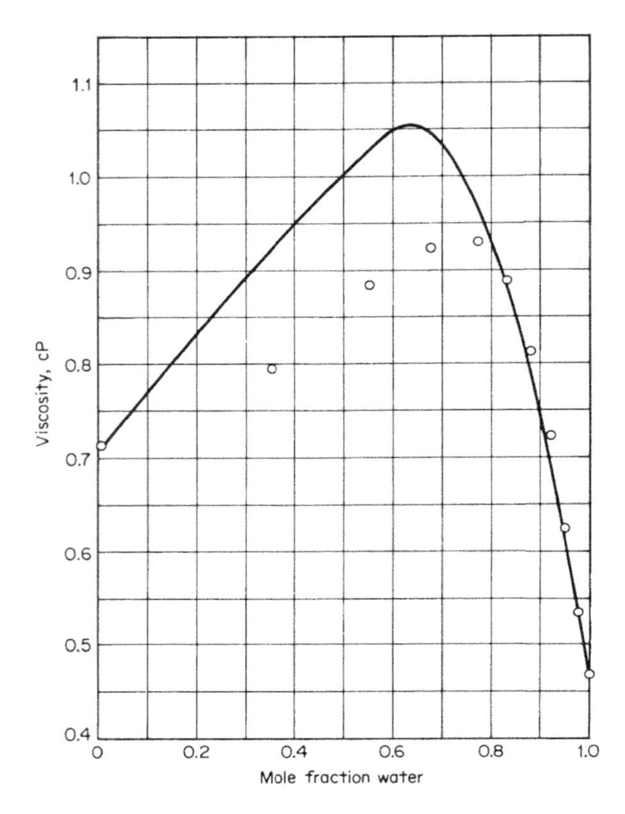

FIGURE 10.14
Viscosity of water and 1,4-dioxane at 333 K. Line is Eq. (10.13-27) with ψ_{ij} = 1.37; symbols are experimental data. (Irving, 1977a)

Although the agreement between the experimental and estimated viscosity in Example 10.27 is excellent, higher errors occur in other composition ranges. Figure 10.14 shows the estimated and experimental values of the mixture viscosity over the entire range of composition. From a mole fraction water of about 0.8 (weight fraction = 0.45) to unity, the method provides an excellent fit to experimental results. At smaller concentrations of water, the technique overpredicts η_m. Still, for such a nonideal aqueous mixture, the general fit should be considered good.

10.13.4 Discussion

Three methods have been introduced to estimate the viscosity of liquid mixtures: the Grunberg-Nissan relation [Eq. (10.13-1)], the UNIFAC-VISCO method [Eq. (10.13-6)], and the Teja-Rice form [Eq. (10.13-19)]. The Grunberg-Nissan and Teja-Rice forms contain one adjustable parameter per binary pair in the mixture. The UNIFAC-VISCO method is predictive but limited in the types of compounds to which it can be applied. The method correctly predicts the behavior of the methanol-toluene system which demonstrates both a maximum and minimum in the viscosity versus concentration curve (Hammond et al., 1958). An approximate technique is available to estimate the Grunberg-Nissan parameter G_{ij} as a function of temperature [Eq. (10.13-5)] for many types of systems. Teja and Rice suggest that their parameter ψ_{ij} is independent of temperature—at least over reasonable temperature ranges. This latter technique seems better for highly polar systems, especially if water is one of the components, and it has also been applied to undefined mixtures of coal liquids (Teja et al., 1985; Thurner, 1984) with the introduction of reference components [see Eq. (10.13-18)]. The UNIFAC-VISCO method has been successfully applied to ternary and quaternary alkane mixtures (Chevalier et al., 1988) but otherwise, evaluation of the above methods for multicomponent mixtures has been limited.

The above three methods are by no means a complete list of available methods. For example, Twu (1985, 1986) presents an equation to estimate the viscosity of petroleum fractions based on the specific gravity and boiling point. This method is particularly useful for cases where the exact chemical composition of a mixture is unknown. Allan and Teja (1991) have also presented a method applicable to petroleum fractions, and Chhabra (1992) presents a method for mixtures of hydrocarbons. Chhabra and Sridhar (1989) extend Eq. (10.11-6) to mixtures. For the treatment of electrolyte solutions, the reader is referred to Lencka et al. (1998). Cao et al. (1993) presented a UNIFAC-based method, but our testing did not reproduce their excellent results in a number of cases. Other mixture correlations are reviewed in Monnery et al. (1995) as well as the 4th edition of this book. For an example of gases dissolved in liquids under pressure, see Tilly et al. (1994).

An equation developed by McAllister (1960) has been used successfully to correlate data for binary as well as multicomponent mixtures (Aminabhavi et al., 1982; Aucejo et al, 1995; Dizechi and Marschall, 1982a; Noda et al., 1982). For binaries, the McAllister (1960) equation has been written to contain either two or three adjustable parameters. For ternary mixtures, the equation has been used with one (Dizechi and Marschall, 1982a) or three (Noda et al., 1982) ternary parameters in addition to the binary parameters. Dizechi and Marschall (1982) have extended the equation to mixtures containing alcohols and water, and Asfour et al. (1991) have developed a method to estimate the parameters in the McAllister equation from pure-component properties. Because of the variable number of parameters that can be introduced into the McAllister equation, it has had considerable success in the correlation of mixture viscosity behavior.

Lee et al. (1999) used an equation of state method to successfully correlate the behavior of both binary and multicomponent mixtures. Nonaqueous mixtures required one parameter per binary while aqueous mixtures required two parameters per binary. One drawback of their method is the nonsymmetrical mixing rule used for multicomponent aqueous mixtures (Michelsen and Kistenmacher, 1990). The equation of state structure allowed the method to be successfully applied to liquid mixtures at high pressure.

To finish this section, we again reiterate that the methods proposed should be limited to situations in which the reduced temperatures of the components comprising the mixture are less than about 0.7, although the exact temperature range of the Teja-Rice procedure is as yet undefined.

Should one desire the viscosity of liquid mixtures at high pressures and temperatures, it is possible to employ the Chung et al. (1988) method described in Sec. 10.7 to estimate high-pressure gas mixture viscosities. This recommendation is tempered by the fact that such a procedure has been only slightly tested, and usually with rather simple systems where experimental data exist.

10.13.5 Recommendations to Estimate the Viscosities of Liquid Mixtures

To estimate low-temperature liquid mixture viscosities, either the Grunberg-Nissan equation [Eq. (10.13-1) or (10.13-2)], the UNIFAC-VISCO method [Eq. (10.13-6)], or the Teja-Rice relation [Eq. (10.13-19) or (10.13-27)] may be used. The Grunberg-Nissan and Teja-Rice methods require some experimental data to establish the value of an interaction parameter specific for each binary pair in the mixture. In the absence of experimental data, the UNIFAC-VISCO method is recommended if group interaction parameters are available. The UNIFAC-VISCO method is particularly recommended for mixtures in which the components vary greatly in size. It is possible to estimate the Grunberg-Nissan interaction parameter G_{ij} by a group contribution technique, and this technique can be applied to more compounds than can the UNIFAC-VISCO method. All three methods are essentially interpolative in nature, so viscosities of the pure components comprising the mixture must be known (or in the Teja-Rice procedure, one may instead use reference fluids of similar structure rather than the actual mixture components). The errors to be expected range from a few percent for nonpolar or slightly polar mixtures to 5 to 10% for polar mixtures. With aqueous solutions, neither the Grunberg-Nissan form nor the UNIFAC-VISCO method is recommended.

10.14 NOTATION

$A, a, B, b, C, c, \ldots$	parameters in correlating equations or prediction methods
$A_i, a_i, B_i, b_i, C_i, c_i, \ldots$	parameters in correlating equations or prediction methods
a^*	parameter in the Reichenberg (1971, 1979) method for predicting low pressure gas viscosity, see Eq. (10.4-24)
a_{nm}	group interaction parameter in the UNIFAC-VISCO method to predict liquid viscosities of mixtures
A_{SR}	parameter in the Sastri and Rao (1992) method for predicting liquid viscosity
A_{Bh}	parameter in the Bhethanabotla (1983) method for predicting liquid viscosity
$AARD$	average absolute relative deviation
b_0	excluded volume $\left(\frac{2}{3}\pi N_A \sigma^3\right)$
B_{Bh}	parameter in the Bhethanabotla (1983) method for predicting liquid viscosity
B_{ij}	parameter to calculate radial distribution function g_{ij}, see Eq. (10.7-19)
C_i	parameters in the method of Reichenberg (1974, 1977, 1979) to predict low-pressure gas viscosity of mixtures
C_v	heat capacity at constant volume
D_{ab}	diffusivity
E_{PS}	parameter in the Przezdziecki and Sridhar (1985) method to predict liquid viscosity
f, f_m, f_i, f_{ij}	parameters in the TRAPP (Huber, 1996) method to predict viscosity of gas mixtures
F_c	factor that accounts for molecular shapes and polarities in the method of Chung et al. (1984, 1988) for gas viscosities, see Eq. (10.4-10)
F_p, F_P°	correction factor accounting for polarity effects in multiple methods
$F_{P_m}^\circ$	correction factor accounting for polarity effects in a mixture, see Eq. (10.5-21)
F_Q, F_Q°	correction factor accounting for quantum effects in multiple methods
$F_{Q_m}^\circ$	correction factor accounting for quantum effects in a mixture, see Eq. (10.5-22)
F_{R_i}	parameters in the method of Reichenberg (1974, 1977, 1979) to predict low-pressure gas viscosity of mixtures
F_η, F_{η_m}	parameters in the TRAPP (Huber, 1996) method to predict viscosity of pure gases and gas mixtures
g_{ij}	radial distribution function, see Eq. (10.7-14)
G_{ij}	interaction parameter between components i and j needed to calculate liquid viscosity of a mixture using the Grunberg and Nissan (1949) method
$\left(\frac{g^E}{RT}\right)^{comb}$	combinatorial term in the UNIFAC-VISCO method to predict liquid viscosities of mixtures
$\left(\frac{g^E}{RT}\right)^{res}$	residual term in the UNIFAC-VISCO method to predict liquid viscosities of mixtures
h, h_m, h_i, h_{ij}	parameters in the TRAPP (Huber, 1996) method to predict viscosity of pure gases and gas mixtures
H_1, H_2	parameters in the method of Gunn and Yamada to calculate V
H_{ij}	parameters in the method of Reichenberg (1974, 1977, 1979) to predict low-pressure gas viscosity of mixtures

k	Boltzmann constant (1.380649×10^{-23} J/K)
K_i	parameters in the method of Reichenberg (1974, 1977, 1979) to predict low pressure gas viscosity of mixtures
L	mean free path
m	mass of molecule
M	molecular weight
M_H	molecular weight of the component with the highest molecular weight in a mixture, see Eq. (10.5-23)
M_i	molecular weight of pure component i in a mixture
M_{ij}	cross interaction molecular weight between components i and j in a mixture
M_L	molecular weight of the component with the lowest molecular weight in a mixture, see Eq. (10.5-23)
M_m	molecular weight of a mixture
M_x	parameter to calculate η_x^{ENSKOG}, see Eq. (10.7-21)
n	number density of molecules; number of components in a mixture; parameter in correlating equations or prediction methods
N_A	Avogadro's number ($6.02214076 \times 10^{23}$ mol^{-1})
N_c	total number of compounds in a test database of experimental data used to evaluate prediction methods
N_{C_i}	number of carbon atoms in component i
N'_C	number of carbon atoms not including those described by additive group contributions in the Orrick and Erbar (1974) prediction method for liquid viscosity
N_i	number of times group i is found in a compound
$N_m^{(i)}$	number of times group m is found in compound i
N_{OH}	number of —OH groups in a molecule
N_{pt}	total number of points in a test database of experimental data used to evaluate prediction methods
P	pressure
P_c	critical pressure
P_{c_m}	critical pressure of a mixture
P_r	reduced pressure $\left(\dfrac{P}{P_c}\right)$
P^{vp}	saturated liquid vapor pressure
$P_R^{vp}(T_\eta)$	saturated liquid vapor pressure of reference fluid at temperature T_η, used in the TRAPP (Huber, 1996) method to predict viscosity of pure gases and gas mixtures
P_{SR}^{vp}	vapor pressure parameter in the Sastri and Rao (1992) method for predicting liquid viscosity
q_i	van der Waals surface area of component i
Q	parameter in the Reichenberg method (1971, 1975, 1979) to calculate gas viscosity at high pressure, see Eq. (10.6-10)
Q_k	contribution of group k to the van der Waals surface area for compound i, see Eq. (10.13-11) and Table 10.15
r_i	van der Waals volume of component i
R	gas constant [8.314462618 J/(mol·K)]
R_k	contribution of group k to the van der Waals volume for compound i, see Eq. (10.13-12) and Table 10.15

R_η	parameter in Enskog Dense-gas Theory, see Table 10.5
T	temperature
T^*	dimensionless temperature $\left(\frac{kT}{\varepsilon}\right)$; parameter in the Chung et al. (1984, 1988) method to calculate gas viscosity
T^R	reference temperature
T_b	normal ($P = 101325$ Pa) boiling point
T_{b_r}	reduced normal ($P = 101325$ Pa) boiling point temperature $\left(\frac{T_b}{T_c}\right)$
T_c	critical temperature
T_{c_i}	critical temperature of pure component i in a mixture
$T_{c_{ij}}$	cross interaction critical temperature between components i and j in a mixture
T_{c_m}	critical temperature of a mixture
T_{c_R}	critical temperature of reference fluid, used in the TRAPP (Huber, 1996) method to predict viscosity of pure gases and gas mixtures
$(T_c)^{(R1)}$, $(T_c)^{(R2)}$	critical temperatures for reference fluids 1 (R1) and 2 (R2) in the Teja and Rice (1981, 1981a) method to predict liquid viscosity of mixtures
T_m	normal ($P = 101325$ Pa) melting temperature
T_r	reduced temperature $\left(\frac{T}{T_c}\right)$
T_{r_i}	reduced temperature $\left(\frac{T}{T_{c_i}}\right)$ of pure component i in a mixture
$T_{r_{ij}}$	cross interaction reduced temperature between components i and j in a mixture
T_η, T_{η_m}	parameters in the TRAPP (Huber, 1996) method to predict viscosity of pure gases and gas mixtures
U_i	parameters in the method of Reichenberg (1974, 1977, 1979) to predict low-pressure gas viscosity of mixtures
v	molecular velocity
v_y	molecular velocity in the y direction
V	molar volume
V^R	molar volume of reference fluid
V_c	critical volume
V_0	close-packed volume in Enskog Dense-gas Theory, see Table 10.5
V_{c_i}	critical volume of component i in a mixture
$V_{c_{ij}}$	cross interaction critical volume between components i and j in a mixture
V_{c_m}	critical volume of a mixture
V_m	liquid molar volume at T_m
V_{PS}	parameter in the Przezdziecki and Sridhar (1985) method to predict liquid viscosity
V_r	reduced volume $\left(\frac{V}{V_0}\right)$
W	parameter in the Isdale (1979) method to calculate liquid viscosities of mixtures, see Eq. (10.13-13)

x_i	mole fraction of component i in a liquid mixture
X_m	parameter in the UNIFAC-VISCO prediction method for liquid viscosity of mixtures
X_{exp_i}	value of experimental data point i for property X
X_{pred_i}	value of predicted point i for property X
y	parameter in the Chung et al. (1988) method to calculate gas viscosities at high pressures
y_i	mole fraction of component i in a gas mixture
Y	parameter in the Lucas (1980, 1981, 1983) method to calculate gas viscosity at high pressure, see Eq. (10.6-19)
Y_i	parameter in the TRAPP (Huber, 1996) method to predict viscosity of pure gases and gas mixtures
z	z direction
Z	compressibility factor
Z_c	critical compressibility factor
Z_{c_i}	critical compressibility factor of component i in a mixture
Z_{c_m}	critical compressibility factor of a mixture
Z_{c_R}	critical compressibility factor of reference fluid, used in the TRAPP (Huber, 1996) method to predict viscosity of pure gases and gas mixtures
Z_1, Z_2	parameters in the Lucas (1980, 1981, 1983) method to calculate gas viscosity at high pressure, see Eqs. (10.6-11) and (10.6-12)

Greek

α	parameter in the Lucas (1980, 1981, 1983) method to calculate gas viscosity at high pressure, see Eq. (10.6-12); parameter in the TRAPP (Huber, 1996) method to predict viscosity of pure gases and gas mixtures; parameter in the Sastri (1998) method to predict liquid viscosity at high temperatures
β	parameter in the Lucas (1980, 1981, 1983) method to calculate gas viscosity at high pressure
β_i	parameter in the TRAPP (Huber, 1996) method to predict viscosity of pure gases and gas mixtures
β_{ij}	parameter to calculate radial distribution function g_{ij}, see Eq. (10.7-18)
γ	orientation parameter, see Table 10.8
γ_i	activity coefficient of component i
$\gamma_k^{(i)}$	parameter for pure component i in the UNIFAC-VISCO prediction method for liquid viscosity of mixtures
γ_i^{res}	residual activity coefficient of component i in the UNIFAC-VISCO prediction method for liquid viscosity of mixtures
δ_{ij}	Kronecker delta function
Δ	parameter in the Jossi et al. (1962) and Stiel and Thodos (1964) methods to predict gas viscosity
$(\Delta_{a^*})_i$	contribution for group i to obtain a^* in the prediction method of Reichenberg (1971, 1979) for low-pressure gas viscosity
$(\Delta_i)_k$, $(\Delta_j)_k$	contributions for group k for the prediction method of Isdale (1979) for liquid viscosity of mixtures
$(\Delta_{\eta_b})_i$, $(\Delta_{\eta_b^{cor}})_i$, $(\Delta_{A_{SR}})_i$, $(\Delta_{A_{SR}^{cor}})_i$	contributions for group i for the prediction method of Sastri and Rao (1992) for liquid viscosity
$(\Delta_{A_{Bh}})_i$, $(\Delta_{B_{Bh}})_i$	contributions for group i for the prediction method of Bhethanabotla (1983) for liquid viscosity

$(\Delta_{A_{Hsu}})_i$, $(\Delta_{B_{Hsu}})_i$, $(\Delta_{C_{Hsu}})_i$, $(\Delta_{D_{Hsu}})_i$	contributions for group i for the prediction method of Hsu et al. (2002) for liquid viscosity
ΔP_r	$\dfrac{(P - P^{VP})}{P_c}$
$\Delta\eta^{ENSKOG}$	parameter in the TRAPP (Huber, 1996) method to predict viscosity of pure gases and gas mixtures
ε	parameter in the potential energy function describing the depth of the attractive well
ε_i	energy parameter ε for component i in a mixture
ε_{ij}	cross interaction energy parameter ε between components i and j in a mixture
ε_m	parameter in the Teja and Rice (1981, 1981a) method to predict liquid viscosity of mixtures
ζ	inverse viscosity used to reduce (nondimensionalize) viscosity, see Eq. (10.4-15)
ζ_{ij}	binary interaction parameter, see Eq. (10.5-35)
ζ_m	mixture inverse viscosity used to reduce (nondimensionalize) viscosity, see Example 10.8
η	viscosity
η°	viscosity at low pressure (about 1 bar)
η^*, η^{**}	parameter in the Chung et al. (1988) method to calculate gas viscosities at high pressures
η^{SL}	viscosity of saturated liquid at P^{vp}
η_b	viscosity at the normal boiling point obtained from Eq. (10.11-4) needed for the prediction method of Sastri and Rao (1992) for liquid viscosity.
η_i	viscosity of pure component i in a mixture
η_{ij}°	cross interaction viscosity at low pressure (about 1 bar) between components i and j in a mixture
η_K	known liquid viscosity needed to use Eq. (10.10-4)
η_m	viscosity of a mixture
η_m°	viscosity of a mixture at low pressure (about 1 bar)
η_r	reduced viscosity
η_R	viscosity of reference fluid at elevated pressure, used in the TRAPP (Huber, 1996) method to predict viscosity of gas mixtures
η_R°	viscosity of reference fluid at low pressure, used in the TRAPP (Huber, 1996) method to predict viscosity of gas mixtures
η_T	unknown liquid viscosity to be predicted, at temperature T, using Eq. (10.10-4)
η_m^{ENSKOG}, η_x^{ENSKOG}	parameters in the TRAPP (Huber, 1996) method to predict viscosity of gas mixtures
$(\eta\varepsilon)^{(R1)}$, $(\eta\varepsilon)^{(R2)}$	parameter for reference fluids 1 (R1) and 2 (R2) in the Teja and Rice (1981, 1981a) method to predict liquid viscosity of mixtures
θ_i	molecular surface area fraction in the UNIFAC-VISCO prediction method for liquid viscosities of mixtures.
Θ_m	parameter in the UNIFAC-VISCO prediction method for liquid viscosity of mixtures
Θ_{ij}	parameter to calculate radial distribution function g_{ij}, see Eq. (10.7-15)
κ	polarity correction factor in the method Chung et al. (1984, 1988) for gas viscosities
κ_i	polarity correction factor κ for component i in a mixture
κ_{ij}	cross interaction polarity correction factor κ between components i and j in a mixture
κ_m	polarity correction factor κ of a mixture
λ	thermal conductivity
μ	dipole moment

μ_m	dipole moment of a mixture
μ_r	reduced dipole moment
μ_{r_i}	reduced dipole moment of pure component i in a mixture
$\mu_{r_{ij}}$	cross interaction reduced dipole moment between components i and j in a mixture
μ_{r_m}	reduced dipole moment of a mixture
v	kinematic viscosity
ξ	parameter to calculate radial distribution function g_{ij}, see Eq. (10.7-16)
ξ_{ij}	binary interaction parameter, see Eq. (10.5-33)
ξ_T	parameter in the Jossi et al. (1962) and Stiel and Thodos (1964) methods to predict gas viscosity
ρ	density
ρ'	general density representing either mass density (ρ_i), momentum density (nmv_y), or energy density ($C_v nT$)
ρ_r	reduced density $\left(\frac{\rho}{\rho_c}\right)$
ρ^*	parameter in the Chung et al. (1988) method to calculate gas viscosities at high pressures
ρ^S	saturated liquid density
ρ_c	critical density
ρ_{c_R}	critical density of reference fluid, used in the TRAPP (Huber, 1996) method to predict viscosity of pure gases and gas mixtures
$\rho_R^S(T_\eta)$	saturated liquid density of reference fluid at temperature T_η, used in the TRAPP (Huber, 1996) method to predict viscosity of pure gases and gas mixtures
ρ_η, ρ_{η_m}	parameters in the TRAPP (Huber, 1996) method to predict viscosity of pure gases and gas mixtures
σ	molecular diameter
σ_i	molecular diameter of component i in a mixture
σ_{ij}	cross interaction molecular diameter between components i and j in a mixture
σ_m	molecular diameter parameter in a mixture
σ_x	parameter to calculate η_x^{ENSKOG}, see Eq. (10.7-20)
ϕ	parameter in the Sastri (1998) method to predict liquid viscosity at high temperatures
ϕ_i	molecular volume area fraction in the UNIFAC-VISCO prediction method for liquid viscosities of mixtures.
ϕ_{ij}	cross interaction parameter between components i and j in the prediction method of Wilke (1950) for low-pressure gas viscosity of a mixture
$\psi(r)$	potential energy function between pairs of molecules as a function of distance r
ψ_{ij}	interaction parameter in the Teja and Rice (1981, 1981a) prediction method for liquid viscosities of mixtures
Ψ_c	radial distribution function at contact
Ψ_{km}	parameter in the UNIFAC-VISCO prediction method for liquid viscosity of mixtures
ω	acentric factor
ω_i	acentric factor of component i in a mixture
ω_{ij}	cross interaction acentric factor between components i and j in a mixture
ω_m	acentric factor of a mixture

ω_R	acentric factor of reference fluid, used in the TRAPP (Huber, 1996) method to predict viscosity of pure gases and gas mixtures
$\omega^{(R1)}$, $\omega^{(R2)}$	acentric factors for reference fluids 1 (R1) and 2 (R2) in the Teja and Rice (1981, 1981a) method to predict liquid viscosity of pure gases and mixtures
Ω_v	collision integral for viscosity

10.15 REFERENCES

Aasen, E., E. Rytter, and H. A. Øye: *Ind. Eng. Chem. Res.,* **29:** 1635 (1990).

Alder, B. J.: "Prediction of Transport Properties of Dense Gases and Liquids," UCRL 14891-T, University of California, Berkeley, Calif., May 1966.

Alder, B. J., and J. H. Dymond: "Van der Waals Theory of Transport in Dense Fluids," UCRL 14870-T, University of California, Berkeley, Calif., April 1966.

Allan, J. M., and A. S. Teja: *Can. J. Chem. Eng.,* **69:** 986 (1991).

Amdur, I., and E. A. Mason: *Phys. Fluids,* **1:** 370 (1958).

American Petroleum Institute, *Selected Values of Physical and Thermodynamic Properties of Hydrocarbons and Related Compounds,* Project 44, Carnegie Press, Pittsburgh, Pa., 1953, and supplements.

Aminabhavi, T. M., R. C. Patel, and K. Bridger: *J. Chem. Eng. Data,* **27:** 125 (1982).

Andrade, E. N. da C.: *Nature,* **125:** 309 (1930).

Andrade, E. N. da C.: *Phil. Mag.,* **17:** 497, 698 (1934).

Andrade, E. N. da C.: *Endeavour,* **13:** 117 (1954).

Angus, S., B. Armstrong, and K. M. deReuck: *International Thermodynamic Tables of the Fluid State-Carbon Dioxide,* Pergamon, New York, 1976.

Asfour, A.-F. A., E. F. Cooper, J. Wu, and R. R. Zahran: *Ind. Eng. Chem. Res.,* **30:** 1669 (1991).

Assael, M. J., J. P. M. Trusler, and T. F. Tsolakis: *Thermophysical Properties of Fluids, An Introduction to their Prediction,* Imperial College Press, London, 1996.

Assael, M. J., J. H. Dymond, M. Papadaki, and P. M Patterson: *Fluid Phase Equil.,* **75:** 245 (1992).

Assael, M. J., J. H. Dymond, M. Papadaki, and P. M Patterson: *Int. J. Thermophys.,* **13:** 269 (1992a).

Assael, M. J., J. H. Dymond, M. Papadaki, and P. M Patterson: *Intern. J. Thermophys.,* **13:** 659 (1992b).

Assael, M. J., J. H. Dymond, M. Papadaki, and P. M Patterson: *Intern. J. Thermophys.,* **13:** 895 (1992c).

Assael, M. J., J. H. Dymond, and S. K. Polimatidou: *Intern. J. Thermophys.,* **15:** 189 (1994).

Assael, M. J., J. H. Dymond, and S. K. Polimatidou: *Intern. J. Thermophys.,* **16:** 761 (1995).

Aucejo, A., M. C. Burget, R. Mun˜oz, and J. L. Marques: *J. Chem. Eng. Data,* **40:** 141 (1995).

Barker, J. A., W. Fock, and F. Smith: *Phys. Fluids,* **7:** 897 (1964).

Batschinski, A. J.: *Z. Physik. Chim.,* **84:** 643 (1913).

Bhethanabotla, V. R.: *A Group Contribution Method For Liquid Viscosity*, M.S. Thesis, The Pennsylvania State University, University Park, Penn., 1983.

Bingham, E. C., and S. D. Stookey: *J. Am. Chem. Soc.,* **61:** 1625 (1939).

Bircher, L. B.: Ph.D. thesis, University of Michigan, Ann Arbor, Mich., 1943.

Bleazard, J. G., and A. S. Teja: *Ind. Eng. Chem. Res.,* **35:** 2453 (1996).

Bravo-Sanchez, M. G., G. A. Iglesia-Silva, and A. Estrada-Baltazar: *J. Chem. Eng. Data*, **58:** 2538 (2013).

Brebach, W. J., and G. Thodos: *Ind. Eng. Chem.,* **50:** 1095 (1958).

Bridgman, P. W.: *Proc. Am. Acad. Arts Sci.,* **61:** 57 (1926).

Brokaw, R. S.: *NASA Tech. Note D-2502,* November 1964.

Brokaw, R. S.: *J. Chem. Phys.,* **42:** 1140 (1965).

Brokaw, R. S.: *NASA Tech. Note D-4496,* April 1968

Brokaw, R. S., R. A. Svehla, and C. E. Baker: *NASA Tech. Note D-2580,* January 1965.

Bromley, L. A., and C. R. Wilke: *Ind. Eng. Chem.,* **43:** 1641(1951).

Brulé, M. R., and K. E. Starling: *Ind. Eng. Chem. Process Design Develop.,* **23:** 833 (1984).

Brush, S. G.: *Chem. Rev.,* **62:** 513 (1962).

Burch, L. G., and C. J. G. Raw: *J. Chem. Phys.,* **47:** 2798 (1967).

Cao, W., K. Knudsen, A. Fredenslund, and P. Rasmussen: *Ind. Eng. Chem. Res.,* **32:** 2088 (1993).

Carmichael, L. T., H. H. Reamer, and B. H. Sage: *J. Chem. Eng. Data,* **8:** 400 (1963).
Carmichael, L. T., and B. H. Sage: *J. Chem. Eng. Data,* **8:** 94 (1963).
Carmichael, L. T., and B. H. Sage: *AIChE J.,* **12:** 559 (1966).
Chakraborti, P. K., and P. Gray: *Trans. Faraday Soc.,* **61:** 2422 (1965).
Chapman, S., and T. G. Cowling: *The Mathematical Theory of Nonuniform Gases,* Cambridge, New York, 1939.
Cheung, H.: *UCRL Report 8230,* University of California, Berkeley, Calif., April 1958.
Chevalier, J. L., P. Petrino, and Y. Gaston-Bonhomme: *Chem. Eng. Sci.,* **43:** 1303 (1988).
Chhabra, R. P.: *AIChE J.,* **38:** 1657 (1992).
Chhabra, R. P. and T. Sridhar: *Chem. Eng. J.,* **40:** 39 (1989).
Chung, T.-H.: Ph.D. thesis, University of Oklahoma, Norman, Okla., 1980.
Chung, T.-H., M. Ajlan, L. L. Lee, and K. E. Starling: *Ind. Eng. Chem. Res.,* **27:** 671 (1988).
Chung, T.-H., L. L. Lee, and K. E. Starling: *Ind. Eng. Chem. Fundam.,* **23:** 8 (1984).
Cornelissen, J., and H. I. Waterman: *Chem. Eng. Sci.,* **4:** 238 (1955).
Dahler, J. S.: *Thermodynamic and Transport Properties of Gases, Liquids, and Solids,* McGraw-Hill, New York, 1959, pp. 14–24.
Das, T. R., C. O. Reed, Jr., and P. T. Eubank: *J. Chem. Eng. Data,* **22:** 3 (1977).
Davidson, T. A.: *A Simple and Accurate Method for Calculating Viscosity of Gaseous Mixtures.* U.S. Bureau of Mines, RI9456, 1993.
de Guzman, J.: *Anales Soc. Espan. Fiz. y Quim.,* **11:** 353 (1913).
Dean, J. A.: *Lange's Handbook of Chemistry,* 15th ed., McGraw-Hill, New York, 1999.
Dizechi, M., and E. Marschall: *Ind. Eng. Chem. Process Design Develop.;* **21:** 282 (1982).
Diky, V., R. D. Chirico, M. Frenkel, A. Bazyleva, J. W. Magee, E. Paulechka, A. Kazakov, E. W. Lemmon, C. D. Muzny, A. Y. Smolyanitsky, S. Townsend, and K. Kroenlein. *NIST ThermoData Engine, NIST Standard Reference Database 103 a/b,* National Institute of Standards and Technology, USA (2021): https://www.nist.gov/mml/acmd/trc/thermodata-engine
DIPPR, Design Institute for Physical Properties, American Institute of Chemical Engineers, 120 Wall St. FL 23, New York, NY 10005-4020 USA (2021): https://www.aiche.org/dippr
Dizechi, M., and E. Marschall: *J. Chem. Eng. Data,* **27:** 358 (1982a).
Dolan, J. P., K. E. Starling, A. L. Lee, B. E. Eakin, and R. T. Ellington: *J. Chem Eng. Data,* **8:** 396 (1963).
Duhne, C. R.: *Chem. Eng.,* **86**(15)**:** 83 (1979).
Dymond, J. H., and M. J. Assael: Chap 10 in *Transport Properties of Fluids, Their Correlation, Prediction and Estimation,* J. Millat, J. H. Dymond, and C. A. Nieto de Castro (eds.), IUPAC, Cambridge Univ. Press, Cambridge, 1996.
Eakin, B. E., and R. T. Ellington: *J. Petrol. Technol.,* **14:** 210 (1963).
Ely, J. F.: *J. Res. Natl. Bur. Stand.,* **86:** 597 (1981).
Ely, J. F., and H. J. M. Hanley: *Ind. Eng. Chem. Fundam.,* **20:** 323 (1981).
Eversteijn, F. C., J. M. Stevens, and H. I. Waterman: *Chem. Eng. Sci.,* **11:** 267 (1960).
Ferry, J. D.: *Viscoelastic Properties of Polymers,* Wiley, New York (1980).
Flynn, L. W., and G. Thodos: *J. Chem. Eng. Data,* **6:** 457 (1961).
Flynn, L. W., and G. Thodos: *AIChE J.,* **8:** 362 (1962).
Franjo, C., E. Jiménez, T. P. Iglesias, J. L. Legido, and M. I. Paz Andrade: *J. Chem. Eng. Data,* **40:** 68 (1995).
Gambill, W. R.: *Chem. Eng.,* **66**(3)**:** 123 (1959).
Gammon, B. E., K. N. Marsh, and A. K. R. Dewan: *Transport Properties and Related Thermodynamic Data of Binary Mixtures, Parts 1–5,* Design Institute of Physical Property Data (DIPPR), New York, 1993–1998.
Gandhi, J. M., and S. C. Saxena: *Indian J. Pure Appl. Phys.,* **2:** 83 (1964).
Gaston-Bonhomme, Y., P. Petrino, and J. L. Chevalier: *Chem. Eng. Sci.,* **49:** 1799 (1994).
Gemant, A. J.: *Appl. Phys.,* **12:** 827 (1941).
Giddings, J. D.: Ph.D. thesis, Rice University, Houston, Texas, 1963.
Girifalco, L. A.: *J. Chem. Phys.,* **23:** 2446 (1955).
Goletz, E., and D. Tassios: *Ind. Eng. Chem. Process Design Develop.,* **16:** 75 (1977).
Golubev, I. F.: "Viscosity of Gases and Gas Mixtures: A Handbook," *Natl. Tech. Inf. Serv., TT 70 50022,* 1959.
Goodwin, R. D.: "The Thermophysical Properties of Methane from 90 to 500 K at Pressures up to 700 bar," *NBSIR 93-342, Natl. Bur. Stand.,* October 1973.
Green, D. W., and M. Z. Southard (eds.): *Chemical Engineers' Handbook,* 9th ed., McGraw-Hill, New York, 2019.
Grunberg, L., and A. H. Nissan: *Nature,* **164:** 799 (1949).
Gunn, R. D., and T. Yamada: *AIChE J.,* **17:** 1341 (1971).
Gururaja, G. J., M. A. Tirunarayanan, and A. Ramachandran: *J. Chem. Eng. Data,* **12:** 562 (1967).
Gutman, F., and L. M. Simmons: *J. Appl. Phys.,* **23:** 977 (1952).

Haar, L., and J. S. Gallagher: *J. Phys. Chem. Ref. Data,* **7:** 635 (1978).
Haines, W. E., R. V. Helm, C. W. Bailey, and J. S. Ball, *J. Phys. Chem.*, **58:** 270 (1954).
Hammond, L. W., K. S. Howard, and R. A. McAllister: *J. Phys. Chem.*, **62:** 637 (1958).
Hanley, H. J. M., R. D. McCarty, and J. V. Sengers: *J. Chem. Phys.,* **50:** 857 (1969).
Herning, F., and L. Zipperer: *Gas Wasserfach,* **79:** 49 (1936).
Hildebrand, J. H.: *Science,* **174:** 490 (1971).
Hirschfelder, J. O., C. F. Curtiss, and R. B. Bird: *Molecular Theory of Gases and Liquids,* Wiley, New York, 1954.
Hirschfelder, J. O., M. H. Taylor, and T. Kihara: *Univ. Wisconsin Theoret. Chem, Lab., WISOOR-29,* Madison, Wis., July 8, 1960.
Hsu, H-C., Y-W Sheu, and C-H Tu: *Chem. Eng. J.*, **88:** 27 (2002).
Huber, M. L., and J. F. Ely: *Fluid Phase Equil.*, **80:** 239 (1992).
Huber, M. L.: Chap 12 in *Transport Properties of Fluids, Their Correlation, Prediction and Estimation,* J. Millat, J. H. Dymond, and C. A. Nieto de Castro (eds.), IUPAC, Cambridge Univ. Press, Cambridge, 1996.
Hwang, M.-J., and W. B. Whiting: *Ind. Eng. Chem. Res.,* **26:** 1758 (1987).
Hwang, S. C., C. Tsonopoulos, J. R. Cunningham, and G. M. Wilson: *Ind. Eng. Chem. Proc. Des. Dev.,* **21:** 127 (1982).
Innes, K. K.: *J. Phys. Chem.,* **60:** 817 (1956).
Irving, J. B.: "Viscosities of Binary Liquid Mixtures: A Survey of Mixture Equations," *Natl. Eng. Lab., Rept. 630,* East Kilbride, Glasgow, Scotland, February 1977.
Irving, J. B.: "Viscosities of Binary Liquid Mixtures: The Effectiveness of Mixture Equations," *Natl. Eng. Lab., Rept. 631,* East Kilbride, Glasgow, Scotland, February 1977a.
Isdale, J. D.: Symp. Transp. Prop. Fluids and Fluid Mixtures, *Natl. Eng. Lab.,* East Kilbride, Glasgow, Scotland, 1979.
Isdale, J. D., J. C. MacGillivray, and G. Cartwright: "Prediction of Viscosity of Organic Liquid Mixtures by a Group Contribution Method," *Natl. Eng. Lab. Rept.,* East Kilbride, Glasgow, Scotland, 1985.
Itean, E. C., A. R. Glueck, and R. A. Svehla: *NASA Lewis Research Center,* TND 481, Cleveland, Ohio, 1961.
Jossi, J. A., L. I. Stiel, and G. Thodos: *AIChE J.,* **8:** 59 (1962).
Kennedy, J. T., and G. Thodos: *AIChE J.,* **7:** 625 (1961).
Kestin, J., H. E. Khalifa, and W. A. Wakeham: *J. Chem. Phys.,* **65:** 5186 (1976).
Kestin, J., and W. Leidenfrost: *Physica,* **25:** 525 (1959).
Kestin, J., and W. Leidenfrost: in Y. S. Touloukian (ed.), *Thernodynamic and Transport Properties of Gases, Liquids, and Solids,* ASME and McGraw-Hill, New York, 1959, pp. 321–338.
Kestin, J., and J. R. Moszynski: in Y. S. Touloukian (ed.), *Thermodynamic and Transport Properties of Gases, Liquids, and Solids,* ASME and McGraw-Hill, New York, 1959, pp. 70–77.
Kestin, J., and I. R. Shankland: in J. V. Sengers (ed.), *Proc. 8th Symp. Thermophys. Prop., II,* ASME, New York, 1981, p. 352.
Kestin, J., and J. Yata: *J. Chem. Phys.,* **49:** 4780 (1968).
Kim, S. K., and J. Ross: *J. Chem. Phys.,* **46:** 818 (1967).
Klein, M., and F. J. Smith: *J. Res. Natl. Bur. Stand.,* **72A:** 359 (1968).
Knapstad, B., P. A. Skølsvik, and H. A. Øye: *Ber. Bunsenges. Phys. Chem.,* **94:** 1156 (1990).
Kouris, S., and C. Panayiotou: *J. Chem. Eng. Data,* **34:** 200 (1989).
Krishnan, K. M., K. Ramababu, D. Ramachandran, P. Venkateswarlu, and G. K. Raman: *Fluid Phase Equil.,* **105:** 109 (1995).
Krishnan, K. M., K. Ramababu, P. Venkateswarlu, and G. K. Raman: *J. Chem. Eng. Data,* **40:** 132 (1995a).
Kumagai, A., and S. Takahashi: *Intern. J. Thermophys.,* **16:** 773 (1995).
Kumar, M. P., S. Tasleem, and G. R. Kumar: *J. Chem. Eng. Data,* **57:** 3109–3113 (2012).
Landolt-Bornstein Tabellen, vol. 4, pt. 1, Springer-Verlag, Berlin, 1955.
Larson, R. G.: *The Structure and Rheology of Complex Fluids,* Oxford Univ. Press, New York, 1999.
Lee, A. L., M. H. Gonzalez, and B. E. Eakin: *J. Chem. Eng. Data,* **11:** 281 (1966).
Lee, M.-J., J.-Y. Chiu, S.-M. Hwang, and H.-M. Lin: *Ind. Eng. Chem. Res.,* **38:** 2867 (1999).
Lencka, M. M., A. Anderko, and R. D. Young: *Intern. J. Thermophys.,* **19:** 367 (1998).
Lewis, W. K., and L. Squires: *Refiner Nat. Gasoline Manuf.,* **13**(12): 448 (1934).
Lobo, V. M. M.: "Handbook of Electrolyte Solutions," *Phys. Sci. Data Ser.,* Nos. 41a & 41b, Elsevier, 1990.
Lucas, K.: *Phase Equilibria and Fluid Properties in the Chemical Industry,* Dechema, Frankfurt, 1980, p. 573.
Lucas, K.: *Chem. Ing. Tech.,* **53:** 959 (1981).
Lucas, K.: personal communications, August 1983, September 1984.

Lucas, K.: VDI-Warmeatlas, Abschnitt DA, "Berechnungsmethoden für Stoffeigenschaften," *Verein Deutscher Ingenieure*, Düsseldorf, 1984a.
Luckas, M., and K. Lucas: *AIChE J.*, **32:** 139 (1986).
McAllister, R. A.: *AIChE J.*, **6:** 427 (1960).
Malek, K. R., and L. I. Stiel: *Can. J. Chem. Eng.*, **50:** 491 (1972).
Marschalko, B., and J. Barna: *Acta Tech. Acad. Sci. Hung.*, **19:** 85 (1957).
Mathur, G. P., and G. Thodos: *AIChE J.*, **9:** 596 (1963).
McCoubrey, J. C., and N. M. Singh: *J. Phys. Chem.*, **67:** 517 (1963).
Medani, M. S., and M. A. Hasan: *Can. J. Chem. Eng.*, **55:** 203 (1977).
Mehrotra, A. K.: *Ind. Eng. Chem. Res.*, **30:** 420, 1367 (1991).
Mehrotra, A. K., W. D. Monnery, and W. Y. Scrcek: *Fluid Phase Equil.*, **117:** 344 (1996).
Michelsen, M. L., and H. Kistenmacher: *Fluid Phase Equil.*, **58:** 229 (1990).
Millat, J., V. Vesovic, and W. A. Wakeham: Chap. 4 in J. Millat, J. H. Dymond, and C. A. Nieto de Castro (eds.), *Transport Properties of Fluids, Their Correlation, Prediction and Estimation*, IUPAC, Cambridge Univ. Press, Cambridge, 1996.
Miller, A. A.: *J. Chem. Phys.*, **38:** 1568 (1963).
Miller, A. A.: *J. Phys. Chem.*, **67:** 1031, 2809 (1963a).
Monchick, L., and E. A. Mason: *J. Chem. Phys.*, **35:** 1676 (1961).
Monnery, W. D., W. S. Svrcek, and A. K. Mehrotra: *Can. J. Chem. Eng.*, **73:** 3 (1995).
Neufeld, P. D., A. R. Janzen, and R. A. Aziz: *J. Chem. Phys.*, **57:** 1100 (1972).
Noda, K., M. Ohashi, and K. Ishida: *J. Chem. Eng. Data*, **27:** 326 (1982).
O'Connell, J. P., and J. M. Prausnitz: "Applications of the Kihara Potential to Thermodynamic and Transport Properties of Gases," in S. Gratch (ed.), *Advances in Thermophysical Properties at Extreme Temperatures and Pressures*, ASME, New York, 1965, pp. 19–31.
Okeson, K. J., and R. L. Rowley: *Intern. J. Thermophys.*, **12:** 119 (1991).
O'Loane, J. K.: personal communication, June 1979.
Orrick, C., and J. H. Erbar: personal communication, December 1974.
Pal, A. K., and A. K. Barua: *J. Chem. Phys.*, **47:** 216 (1967).
Pal, A. K., and P. K. Bhattacharyya: *J. Chem. Phys.*, **51:** 828 (1969).
Petrino, P. J., Y. H. Gaston-Bonhomme, and J. L. E. Chevalier: *J. Chem. Eng. Data*, **40:** 136 (1995).
Przezdziecki, J. W., and T. Sridhar: *AIChE J.*, **31:** 333 (1985).
Ranz, W. E., and H. A. Brodowsky: Univ. Minn. *OOR Proj. 2340 Tech. Rept. 1*, Minneapolis, Minn., March 15, 1962.
Raw, C. J. G., and H. Tang: *J. Chem. Phys.*, **39:** 2616 (1963).
Reichenberg, D.: "The Viscosities of Gas Mixtures at Moderate Pressures," *NPL Rept. Chem. 29*, National Physical Laboratory, Teddington, England, May, 1974.
Reichenberg, D.: (a) *DCS report 11*, National Physical Laboratory, Teddington, England, August 1971; (b) *AIChE J.*, **19:** 854 (1973); (c) *ibid.*, **21:** 181 (1975).
Reichenberg, D.: "The Viscosities of Pure Gases at High Pressures," *Natl. Eng. Lab., Rept. Chem. 38*, East Kilbride, Glasgow, Scotland, August 1975.
Reichenberg, D.: "New Simplified Methods for the Estimation of the Viscosities of Gas Mixtures at Moderate Pressures," *Natl. Eng. Lab. Rept. Chem. 53*, East Kilbride, Glasgow, Scotland, May 1977.
Reichenberg, D.: *Symp. Transp. Prop. Fluids and Fluid Mixtures, Natl. Eng. Lab.*, East Kilbride, Glasgow, Scotland, 1979.
Reid, R. C., and L. I. Belenyessy: *J. Chem. Eng. Data*, **5:** 150 (1960).
Riddick, J. A., W. B. Bunger, and T. K. Sakano: *Organic Solvents Physical Properties and Methods of Purification*, 4th ed., Wiley, New York.
Rogers, J. D., and F. G. Brickwedde: *AIChE J.*, **11:** 304 (1965).
Rowley, J. R., W. V. Wilding, J. L. Oscarson, and R. L. Rowley: *Int. J. Thermophys.*, **28:** 824–834 (2007).
Rumble, J. R. (ed.): *Handbook of Chemistry and Physics*, 102nd ed., CRC Press, Boca Raton, 2021.
Rutherford, R., M. H. Taylor, and J. O. Hirschfelder: *Univ. Wisconsin Theoret. Chem. Lab., WIS-OOR-29a*, Madison, Wis., August 23, 1960.
Sastri, S. R. S., and K. K. Rao: *Chem. Eng. J.*, **50:** 9 (1992).
Sastri, S. R. S.: personal communication, Regional Research Laboratory, Bhubaneswar, India, 1998.
Saxena, S. C., and R. S. Gambhir: *Brit. J. Appl. Phys.*, **14:** 436 (1963).
Saxena, S. C., and R. S. Gambhir: *Proc. Phys. Sec. London*, **81:** 788 (1963a).
Shimotake, H., and G. Thodos: *AIChE J.*, **4:** 257 (1958).
Shimotake, H., and G. Thodos: *AIChE J.*, **9:** 68 (1963).

Shimotake, H., and G. Thodos: *J. Chem. Eng. Data,* **8:** 88 (1963a).
Sirbu, F., D. Dragoescu, A. Shchamialiou, and T. Khasanshin: *J. Chem. Thermo,* **128**: 383 (2019).
Speight, J.: Lange's *Handbook of Chemistry,* 17th ed., McGraw-Hill, New York, 2016.
Starling, K. E.: M.S. thesis, Illinois Institute of Technology, Chicago, Illinois, 1960.
Starling, K. E.: Ph.D. thesis, Illinois Institute of Technology, Chicago, Illinois, 1962.
Starling, K. E., B. E. Eakin, and R. T. Ellington: *AIChE J.,* **6:** 438 (1960).
Starling, K. E., and R. T. Ellington: *AIChE J.,* **10:** 11 (1964).
Stephan, K., and K. Lucas: *Viscosity of Dense Fluids,* Plenum, New York, 1979.
Stephan, K., and H. Hildwein: Recommended Data of Selected Compounds and Binary Mixtures, *Chemistry Data Series,* vol. IV, parts 1 & 2, DECHEMA, Frankfurt, 1987.
Stephan, K., and T. Heckenberger: Thermal Conductivity and Viscosity Data of Fluid Mixtures, *Chemistry Data Series,* vol. X, part 1, DECHEMA, Frankfurt, 1988.
Stewart, R. B.: Ph.D. thesis, University of Iowa, Iowa City, Iowa, June 1966.
Stiel, L. I., and G. Thodos: *AIChE J.,* **10:** 275 (1964).
Strunk, M. R., W. G. Custead, and G. L. Stevenson: *AIChE J.,* **10:** 483 (1964).
Sutton, J. R.: "References to Experimental Data on Viscosity of Gas Mixtures," *Natl. Eng. Lab. Rept. 613,* East Kilbride, Glasgow, Scotland, May 1976.
Svehla, R. A.: "Estimated Viscosities and Thermal Conductivities at High Temperatures," *NASA TRR-132,* 1962.
Tanaka, Y., H. Kubota, T. Makita, and H. Okazaki: *J. Chem. Eng. Japan,* **10:** 83 (1977).
Teja, A. S.: *AIChE J.,* **26:** 337 (1980).
Teja, A. S., and P. Rice: *Chem. Eng. Sci.,* **36:** 7 (1981).
Teja, A. S., and P. Rice: *Ind. Eng. Chem. Fundam.,* **20:** 77 (1981a).
Teja, A. S., and S. I. Sandler: *AIChE J.,* **26:** 341 (1980).
Teja, A. S., P. A. Thurner, and B. Pasumarti: paper presented at the *Annual AIChE Mtg., Washington, D.C., 1983; Ind. Eng. Chem. Process Design Develop.,* **24:** 344 (1985).
Telang, M. S.: *J. Phys. Chem.,* **49:** 579 (1945); **50:** 373 (1946).
Tham, M. K. and K. E. Gubbins: *J. Chem. Phys.,* **55:** 268 (1971).
Thomas, L. H.: *J. Chem. Soc.*, 573 (1946).
Thurner, P. A.: S.M. thesis, Georgia Institute of Technology, Atlanta, Ga., 1984.
Tilly, K. D., N. R. Foster, S. J. Macnaughton, and D. L. Tomasko: *Ind. Eng. Chem. Res.,* **33:** 681 (1994).
Timmermans, J.: *Physico-Chemical Constants of Pure Organic Compounds,* vol. 2, Elsevier, Amsterdam, 1965.
Titani, T.: *Bull. Inst. Phys. Chem. Res. Tokyo,* **8:** 433 (1929).
Trautz, M., and P. B. Baumann: *Ann. Phys.,* **5:** 733 (1929).
Trautz, M., and R. Heberling: *Ann. Phys.,* **10:** 155 (1931).
Trautz, M.: *Ann. Phys.,* **11:** 190 (1931).
Twu, C. H.: *Ind. Eng. Chem. Process Des. Dev.,* **24:** 1287 (1985).
Twu, C. H.: *AIChE J.,* **32:** 2091 (1986).
van Velzen, D., R. L. Cardozo, and H. Langenkamp: "Liquid Viscosity and Chemical Constitution of Organic Compounds: A New Correlation and a Compilation of Literature Data," *Euratom,* 4735e, Joint Nuclear Research Centre, Ispra Establishment, Italy, 1972.
van Wyk, W. R., J. H. van der Veen, H. C. Brinkman, and W. A. Seeder: *Physica,* **7:** 45 (1940).
Vanderslice, J. T., S. Weissman, E. A. Mason, and R. J. Fallon: *Phys. Fluids,* **5:** 155 (1962).
Vargaftik, N. B., Y. K. Vinogradov, and V. S. Yargin: *Handbook of Physical Properties of Liquids and Gases,* Begell House, New York, 1996.
Vesovic, V., W. A. Wakeham, G. A. Olchowy, J. V. Sengers, J. T. R. Watson, and J. Millat: *J. Phys. Chem. Ref. Data,* **19:** 763 (1990).
Viswanath, D. S., and G. Natarajan: *Data Book on the Viscosity of Liquids,* Hemisphere Pub. Co., New York, 1989.
Vogel, H.: *Physik Z.,* **22:** 645 (1921).
Vogel. E., and A. Weiss: *Ber. Bunsenges. Phys. Chem.,* **85:** 539 (1981).
Waxman, M., and J. S. Gallagher: *J. Chem. Eng. Data,* **28:** 224 (1983).
Wei, I.-Chien, and R. Rowley: *J. Chem. Eng. Data,* **29:** 332, 336 (1984); *Chem. Eng. Sci.,* **40:** 401 (1985).
Wilding, W. V., T. A. Knotts, N. F. Giles, R. L. Rowley, DIPPR® Data Compilation of Pure Chemical Properties, Design Institute for Physical Properties, AIChE, New York, NY (2017).
Wilke, C. R.: *J. Chem. Phys.,* **18:** 517 (1950).
Wobster, R., and F. Mueller: *Kolloid Beih.,* **52:** 165 (1941).

Wright, P. G., and P. Gray: *Trans. Faraday Soc.,* **58:** 1 (1962).
Wu, J., A. Shan, and A.F. A. Asfour: *Fluid Phase Equil.,* **143:** 263 (1998)
Xiang, H. W., W. A. Wakeham, and J. P. M. Trusler: "Correlation and Prediction of Viscosity Consistently from the Dilute Vapor to Highly Compressed Liquid State," submitted to *High Temp.-High Press* (1999).
Yaws, C. L., J. W. Miller, P. N. Shah, G. R. Schorr, and P. M. Patel: *Chem. Eng.,* **83**(25)**:** 153 (1976).
Yaws, C. L.: *Handbook of Viscosity,* Gulf Pub., Houston, 1995.
Yaws, C. L.: *Handbook of Transport Property Data,* Gulf Pub., Houston, 1995a.
Yoon, P., and G. Thodos: *AIChE J.,* **16:** 300 (1970).
Younglove, B. A. and J. F. Ely: *J. Phys. Chem. Ref. Data,* **16:** 577 (1987).
Zaytsev, I. D., and G. G. Aseyev: *Properties of Aqueous Electrolyte Solutions,* CRC Press, Boca Raton, 1992.

11

Thermal Conductivity

11.1 SCOPE

Thermal conductivity of gases and liquids is considered in this chapter. Thermal conductivity is the transport of energy across a temperature gradient due to random molecular motion. The basic theory behind the property is discussed first. This is followed by a discussion of estimation techniques, including the effects of temperature and pressure, for the thermal conductivity of pure gases. Similar topics for pure liquids are then presented. The thermal conductivities of nonreacting gas and liquid mixtures are also discussed. Thermal conductivities of reacting gas mixtures are not covered but are reviewed in Curtiss et al. (1982).

The units used for thermal conductivity are W/(m·K). Conversions to English or cgs units can be accomplished with the following factors:

$$\text{W/(m·K)} \times 0.5778 = \text{BTU/(hr·ft·°R)}$$

$$\text{W/(m·K)} \times 8.604 \times 10^{-3} = \text{kcal/(cm·hr·K)}$$

$$\text{W/(m·K)} \times 2.390 \times 10^{-3} = \text{cal/(cm·s·K)}$$

or

$$\text{BTU/(hr·ft·°R)} \times 1.731 = \text{W/(m·K)}$$

$$\text{kcal/(cm·hr·K)} \times 116.2 = \text{W/(m·K)}$$

$$\text{cal/(cm·s·K)} \times 418.4 = \text{W/(m·K)}$$

11.2 THEORY OF THERMAL CONDUCTIVITY

In Sec. 10.3, through rather elementary arguments, the thermal conductivity (λ) of an ideal gas was found to be equal to $\frac{1}{3}vLC'_vn$ [Eq. (10.3-7)], where v is the average molecular velocity, L is the mean free path, C'_v is the constant volume heat capacity *per molecule* (the prime being used to distinguish this quantity from the more common *per mole* property), and n is the *number density* of molecules. Similar relations have been derived for the viscosity and diffusion coefficients of gases. In the case of the last two properties, this elementary approach yields approximate but reasonable values. For thermal conductivity, it is quite inaccurate. A more detailed treatment is necessary to account for the effect of having a wide spectrum of molecular velocities; also, molecules may store energy in forms other than translational. For monatomic gases, which have no rotational or vibrational degrees of freedom, a more rigorous analysis yields

$$\lambda = \frac{25}{32}(\pi m_m kT)^{\frac{1}{2}} \frac{\frac{C'_v}{m_m}}{\pi\sigma^2\Omega_v} \tag{11.2-1}$$

where m_m is the mass of the molecule, k is Boltzmann's constant, T is temperature, σ is the characteristic size of the molecule, and Ω_v is the collision integral. Written for computational ease, with $C_v = \frac{3}{2}k$ and converting to a per mole rather than a per molecule basis,

$$\lambda = 2.632 \times 10^{-23} \frac{\left(\frac{T}{M'}\right)^{\frac{1}{2}}}{\sigma^2\Omega_v} \tag{11.2-2}$$

where M' is the molecular weight of the molecule, and the other variables are as described above. This equation returns λ in units of W/(m·K) by requiring T in K, M' in kg/mol (the prime being used as a reminder of the units), and σ in m. Ω_v is dimensionless.

Values of λ from Eq. (11.2-2) for xenon and helium at 300 K are 0.008 and 0.1 W/(m·K), respectively. For a hard-sphere molecule, Ω_v is unity; normally, however, it is a function of temperature, and the exact dependence is related to the intermolecular force law chosen. If the Lennard-Jones 12-6 potential [Eq. (10.4-2)] is selected, Ω_v is given by Eq. (10.4-3).

A useful dimensionless group, which forms the basis of several prediction methods for λ, is obtained by dividing Eq. (11.2-1) by Eq. (10.3-9). Converting the *per molecule* masses in Eq. (11.2-1) to molecular weight using Avogadro's number, this division yields

$$\frac{\lambda M'}{\eta C_v} = 2.5 \tag{11.2-3}$$

where λ is the thermal conductivity in W/(m·K), M' is molecular weight in kg/mol, η is the viscosity in Pa·s, and C_v is the *per mole* constant volume heat capacity in J/(mol·K).

With $\gamma = \frac{C_p}{C_v}$, the Prandtl number N_{Pr} is

$$N_{Pr} = \frac{C_p\eta}{\lambda M'} = \frac{\gamma}{2.5} \tag{11.2-4}$$

where the units are the same as those described for Eq. (11.2-3). Since γ for monatomic gases is close to $\frac{5}{3}$ except at very low temperatures, Eq. (11.2-4) would indicate that $N_{Pr} \approx \frac{2}{3}$, a value close to that found experimentally. To obtain Eq. (11.2-3), the terms σ^2 and Ω_v in Eq. (11.2-2) cancel, and the result is essentially independent of the intermolecular potential law chosen.

Our discussion so far has considered only energy associated with translational motion. Since heat capacities of polyatomic molecules exceed those for monatomic gases, a substantial fraction of molecular energy resides in modes other than translational. This has a much greater effect on the thermal conductivity than on viscosity or the diffusion coefficient. The next section describes many prediction techniques for polyatomic gases. Some are based on the dimensionless group $\frac{\lambda M'}{\eta C_v}$ which is known as the *Eucken factor.* As per Eq. (11.2-4), it is close to 2.5 for monatomic gases, but it is significantly less for polyatomic gases.

11.3 THERMAL CONDUCTIVITIES OF POLYATOMIC GASES

11.3.1 Eucken, Modified Eucken, and Related Models

Eucken proposed that Eq. (11.2-3) be modified for polyatomic gases by separating the contributions due to translational (tr) and internal (int) degrees of freedom into separate terms

$$\frac{\lambda M'}{\eta C_v} = f_{\text{tr}}\left(\frac{C_{\text{tr}}}{C_v}\right) + f_{\text{int}}\left(\frac{C_{\text{int}}}{C_v}\right) \tag{11.3-1}$$

where f_{tr} and f_{int} are factors which scale the impact of the individual phenomena on the Eucken factor and C_{tr} and C_{int} are the translation and internal heat capacity contributions, respectively. Thus the contribution due to translational degrees of freedom has been decoupled from that due to internal degrees of freedom (Cottrell and McCoubrey, 1961; Lambert and Bates, 1962; Mason and Monchick, 1962; O'Neal and Brokaw, 1962; Saxena, et al., 1964; Srivastava and Srivastava, 1959; Vines, 1958; Vines and Bennett, 1954), although the validity of this step has been questioned (Hirschfelder, 1957; Saxena and Agrawal, 1961; Svehla, 1962). Invariably, f_{tr}, is set equal to 2.5 to force Eq. (11.3-1) to reduce to Eq. (11.2-3) for a monatomic ideal gas. C_{tr} is set equal to the classical value of $1.5R$, and C_{int} is conveniently expressed as $C_v - C_{\text{tr}}$. Then

$$\frac{\lambda M'}{\eta C_v} = \frac{\frac{15}{4}}{\frac{C_v}{R}} + f_{\text{int}}\left(1 - \frac{\frac{3}{2}}{\frac{C_v}{R}}\right)$$

$$= \frac{\frac{15}{4}}{\frac{C_p}{R} - 1} + f_{\text{int}}\left(1 - \frac{\frac{3}{2}}{\frac{C_p}{R} - 1}\right) \tag{11.3-2}$$

where the ideal-gas heat capacity relation ($C_p - C_v = R$) has been used.

Eucken chose $f_{\text{int}} = 1.0$, whereby Eq. (11.3-2) reduces to

$$\frac{\lambda M'}{\eta C_v} = 1 + \frac{\frac{9}{4}}{\frac{C_v}{R}} = 1 + \frac{\frac{9}{4}}{\frac{C_p}{R} - 1} \tag{11.3-3}$$

the well-known Eucken correlation for polyatomic gases.

Many of the assumptions leading to Eq. (11.3-3) are open to question, in particular, the choice of $f_{\text{int}} = 1.0$. Ubbelohde (1935), Chapman and Cowling (1961), Hirschfelder (1957), and Schafer (1943) have suggested that molecules with excited internal energy states could be regarded as separate chemical species, and the transfer of internal energy is then analogous to a diffusional process. This concept leads to a result that

$$f_{\text{int}} = \frac{M'\rho D}{\eta} \tag{11.3-4}$$

where M' is the molecular weight in kg/mol, η is the viscosity in Pa·s, ρ is the molar density in mol/m^3, and D is the diffusion coefficient in m^2/s. Most early theories selected D to be equivalent to the molecular self-diffusion coefficient, and f_{int} is then the reciprocal of the Schmidt number. With Eqs. (10.3-9) and (12.4-2), it can be shown that $f_{\text{int}} \approx 1.32$ and is almost independent of temperature. With this formulation, Eq. (11.3-2) becomes

$$\frac{\lambda M'}{\eta C_v} = 1.32 + \frac{1.77}{\frac{C_v}{R}} = 1.32 + \frac{1.77}{\frac{C_p}{R} - 1} \tag{11.3-5}$$

Equation (11.3-5), often referred to as the modified Eucken correlation, was used by Svehla (1962) in his compilation of high-temperature gas properties.

The modified Eucken relation [Eq. (11.3-5)] predicts larger values of λ than the Eucken form [Eq. (11.3-3)], and the difference becomes greater as C_v increases above the monatomic gas value of about 12.6 J/(mol·K). Both yield Eq. (11.2-3) when $C_v = \frac{3}{2}R$. Usually, experimental values of λ lie between those calculated by the two Eucken forms except for polar gases, when both predict λ values that are too high. For nonpolar gases, Stiel and Thodos (1964) suggested a compromise between Eqs. (11.3-3) and (11.3-5) as

$$\frac{\lambda M'}{\eta C_v} = 1.15 + \frac{2.03}{\frac{C_v}{R}} = 1.15 + \frac{2.03}{\frac{C_p}{R} - 1} \tag{11.3-6}$$

Equations (11.3-3), (11.3-5), and (11.3-6) indicate that the Eucken factor, $\frac{\lambda M'}{\eta C_v}$, should decrease with increasing temperature as the heat capacity rises, but experimental data indicate that the Eucken factor is often remarkably constant or increases slightly with temperature. In Fig. 11.1 we illustrate the case for ethyl chloride, where the data of Vines and Bennett show the Eucken factor increases from only about 1.41 to 1.48 from 40 to 140 °C. On this same graph, the predictions of Eqs. (11.3-3), (11.3-5), and (11.3-6) are plotted and, as noted earlier, all predict a small decrease in the Eucken factor as temperature increases. In Fig. 11.2 we have graphed the experimental Eucken factor as a function of reduced temperature for 13 diverse low-pressure gases. Except for ethane, all show a small rise with an increase in temperature.

11.3.2 Method of Misic and Thodos (1961)

Misic and Thodos (1961), citing some of the questionable assumptions of the Eucken factor, particularly for polyatomic gases, proposed a prediction method based on dimensional analysis. They recognized that thermal conductivity (λ) was directly related to heat capacity (C_p) and used this as the basis of the procedure by proposing that

$$\lambda = \beta M^a T_c^b T^c P_c^d V_c^e C_p^f R^g \tag{11.3-7}$$

where β and a through g are constants, M is molecular weight, T_c is critical temperature, P_c is the critical pressure, V_c is the critical volume, and R is the gas constant. Using fundamental dimensions of mass, length, time, and temperature, the dimensional analysis yielded

$$\lambda \frac{M^{\frac{1}{2}} T_c^{\frac{1}{6}}}{P_c^{\frac{2}{3}}} = \lambda\Lambda = \beta T_r^c C_p^f Z_c^{\frac{5}{6}-f-g} \tag{11.3-8}$$

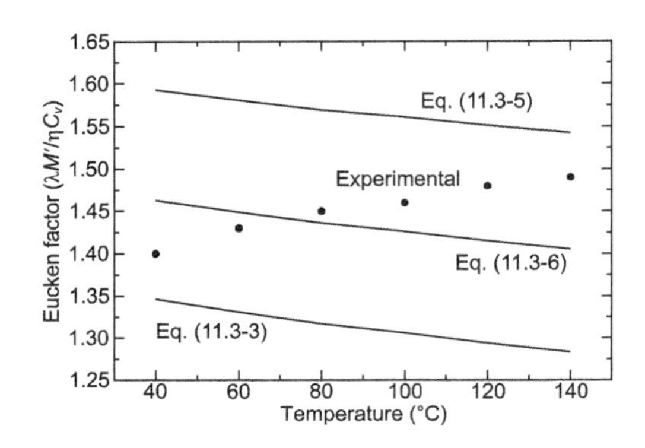

FIGURE 11.1
Eucken factor for ethyl chloride at low pressure. (Data from Vines and Bennett, 1954)

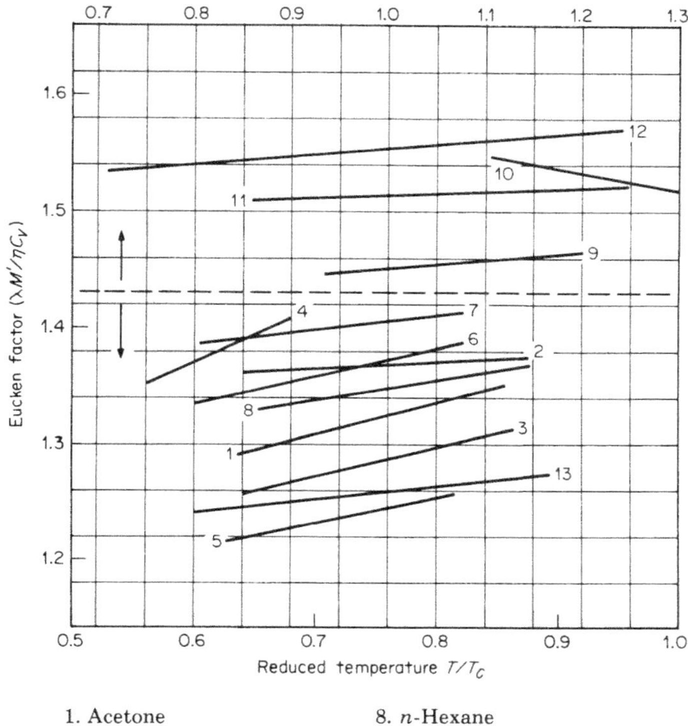

1. Acetone
2. Methyl alcohol
3. Acetaldehyde
4. Water
5. Acetonitrile
6. Cyclohexane
7. Benzene
8. n-Hexane
9. Propane
10. Ethane
11. Acetylene
12. Dichlorodifluoromethane
13. Ethyl acetate

FIGURE 11.2
Variation of the Eucken factor with temperature. (Data primarily from Stiel and Thodos, 1964)

where $Z_c = \frac{P_c V_c}{RT_c}$ is the critical compressibility factor, the group $\Lambda = \frac{M^{\frac{1}{2}} T_c^{\frac{1}{6}}}{P_c^{\frac{2}{3}}}$ is a constant for a given substance, and the values of β, c, f, and g must be obtained by fitting to experimental heat capacity and thermal conductivity data.

Misic and Thodos performed the analysis to obtain the values of β, c, f, and g *only for hydrocarbons*. This work showed that $\frac{\lambda\Lambda}{C_p}$ is a constant (meaning $f = 1$) and independent of Z_c (meaning the power on $Z_c = 0$ or $g = -\frac{1}{6}$) for this class of molecules. The values of β and c appeared dependent on degree of internal rotations and vibrations present in the molecule. The thermal conductivity of methane, cycloalkanes, and aromatics (hydrocarbons with lower degrees of flexibility) below $T_r = 1.0$ is given by

$$\frac{\lambda\Lambda}{C_p} = 0.445 \times 10^{-5} T_r \tag{11.3-9}$$

while that for all other aliphatic hydrocarbons (more flexible hydrocarbons) from $T_r = 0.5 - 3.0$ is

$$\frac{\lambda\Lambda}{C_p} \times 10^6 = (14.52 T_r - 5.14)^{\frac{2}{3}} \tag{11.3-10}$$

Using the last two equations requires M in g/mol, T in K, T_c in K, P_c in atm, and C_p in cal/(mol·K). They return λ in units of cal/(s·cm·K).

11.3.3 Method of Chung et al. (1984, 1988)

Chung et al. employed an approach similar to that of Mason and Monchick (1962) to obtain a relation for λ. By using their form and a similar one for low-pressure viscosity [Eq. (10.4-10)], one obtains

$$\frac{\lambda M'}{\eta C_v} = \frac{3.75\Psi}{\dfrac{C_v}{R}} \qquad (11.3\text{-}11)$$

where λ is thermal conductivity in W/(m·K), M' is molecular weight in kg/mol, η is the low-pressure gas viscosity in Pa·s, C_v is the heat capacity at constant volume in J/(mol·K), R is the gas constant in J/(mol·K), and Ψ is obtained with the following relationships:

$$\Psi = 1 + \alpha \frac{0.215 + 0.28288\alpha - 1.061\beta + 0.26665Z}{0.6366 + \beta Z + 1.061\alpha\beta} \qquad (11.3\text{-}12)$$

$$\alpha = \frac{C_v}{R} - \frac{3}{2} \qquad \beta = 0.7862 - 0.7109\,\omega + 1.3168\,\omega^2 \qquad Z = 2.0 + 10.5T_r^2 \qquad (11.3\text{-}13)$$

The β term is an empirical correlation for $(f_{\text{int}})^{-1}$ [Eq. (11.3-4)] and is said to apply only for nonpolar materials. For polar materials, β is specific for each compound, and Chung et al. (1984) list values for a few materials. If the compound is polar and β is not available, use a default value of $(1.32)^{-1} = 0.758$.

Z represents the number of collisions required to interchange a quantum of rotational energy with translational energy. For large values of Z, Ψ reduces to

$$\Psi = 1 + 0.2665\frac{\alpha}{\beta} \qquad \text{for large } Z \qquad (11.3\text{-}14)$$

If Eq. (11.3-14) is used in Eq. (11.3-11), the Eucken correlation [Eq. (11.3-3)] is obtained when β is set equal to unity. If $\beta = (1.32)^{-1}$, the modified Eucken relation [Eq. (11.3-5)] is recovered. The method is illustrated in Example 11.1.

Example 11.1 Use the above methods to estimate the thermal conductivity of 2-methylbutane (isopentane) vapor at 1 bar and 100 °C. The value tabulated in Bretsznajder (1971) is 0.022 W/(m·K).

Solution. From Appendix A, $T_c = 460.4$ K, $P_c = 33.80$ bar = 33.36 atm, $V_c = 306$ cm^3/mol, $Z_c = 0.27$, $\omega = 0.227875$, and $M = 72.149$ g/mol ($M' = 72.149 \times 10^{-3}$ kg/mol).

The prediction methods require the Eucken factor, so the viscosity (η) of the vapor and the ideal gas constant volume heat capacity (C_v) are needed. McCoubrey and Singh (1963) report experimental measurements for η from 309–464 K for a value of 8.876×10^{-6} Pa·s at 373.15 K. (One of the prediction methods discussed in Chap. 10 can be used if experimental data for η are unavailable.) The correlation in Appendix A for the ideal gas isobaric heat capacity (C_p) yields a value of 145.5 J/(mol·K) = 34.78 cal/(mol·K) meaning C_v (from $C_v = C_p - R$) is $C_v = 145.5 - 8.31 = 137.2$ J/(mol·K)

EUCKEN METHOD, Eq. (11.3-3)

$$\lambda = [(8.876 \cdot 10^{-6})(137.2)/0.072149][1 + (9/4)/(137.2/8.314)] = 0.0192 \text{ W/(m·K)}$$

$$\lambda \text{ Difference} = 0.0192 - 0.022 = -0.0028 \text{ W/(m·K) or } -12.7\%$$

MODIFIED EUCKEN METHOD, Eq. (11.3-5)

$$\lambda = [(8.876 \cdot 10^{-6})(137.2)/0.072149][1.32 + 1.77/(137.2/8.314)] = 0.0241 \text{ W/(m·K)}$$

$$\lambda \text{ Difference} = 0.0241 - 0.022 = 0.0021 \text{ W/(m·K) or } 9.5\%$$

STIEL AND THODOS METHOD, Eq. (11.3-6)

$$\lambda = [(8.876 \cdot 10^{-6})(137.2)/0.072149][1.15 + 2.03/(137.2/8.314)] = 0.0215 \text{ W/(m·K)}$$

$$\lambda \text{ Difference} = 0.0215 - 0.022 = -0.0005 \text{ W/(m·K) or } -2.3\%$$

MISIC AND THODOS METHOD, Eq. (11.3-10)

$$\Lambda = [(72.149)^{1/2}(460.4)^{1/6}/33.36^{2/3}] = 2.278$$

$$\lambda = (34.78/2.278)[14.52(373.15/460.4) - 5.14]^{2/3}(10^{-6}) = 5.387 \times 10^{-5} \text{ cal/(s·cm·K)}$$

$$= 0.0226 \text{ W/(m·K)}$$

$$\lambda \text{ Difference} = 0.0225 - 0.022 = 0.0005 \text{ W/(m·K) or } 2.3\%$$

CHUNG ET AL. METHOD, Eq. (11.3-11)

As defined under Eq. (11.3-13),

$$\alpha = 137.2/8.314 - 3/2 = 15.00$$

$$\beta = 0.7862 - (0.7109)(0.227875) + (1.3168)(0.227875)^2 = 0.693$$

$$Z = 2 + 10.5(373.15/460.4)^2 = 8.897$$

$$\Psi = 1 + (15.00)[0.215 + (0.28288)(15.00) - (1.061)(0.693) + (0.26665)(8.897)]/$$
$$[0.6366 + (0.693)(8.897) + (1.061)(15.00)(0.693)] = 6.127$$

Using Eq. (11.3-11),

$$\lambda = [(8.876 \cdot 10^{-6})(137.2)/0.072149][(3.76)(6.127)/(137.2/8.314)] = 0.0236 \text{ W/(m·K)}$$

$$\lambda \text{ Difference} = 0.0236 - 0.022 = 0.0016 \text{ W/(m·K) or } 7.3\%$$

11.3.4 Discussion and Recommendations

A test database was created to evaluate the performance of pure-component, low-pressure vapor thermal conductivity prediction methods. This database consisted of all reliable, experimental vapor thermal conductivity data from the October 2021 release of the DIPPR database and included over 2000 total points (N_{pt}) from 173 unique compounds (N_c). The techniques tested were: Euken Eq. (11.3-3) (abbreviated EU), Modified Euken

Eq. (11.3-5) (abbreviated ME), Stiel and Thodos Eq. (11.3-6) (abbreviated ST), Misic and Thodos Eq. (11.3-10) (abbreviated MT), and Chung et al. Eq. (11.3-11) (abbreviated CH).

To identify if a method was better at one class of compounds than another, each chemical in the test database was assigned a *class* according to the scheme below.

- **Normal:** Compounds containing only C, H, F, Br, and/or I with molecular weights below 213 g/mol.
- **Heavy:** Compounds containing only C, H, F, Br, and/or I with molecular weights greater than or equal to 213 g/mol. Also compounds containing only C, H, F, Br, I and/or metal atoms (e.g., Fe, As, Al, etc.) regardless of MW.
- **Polar:** Compounds containing atoms other than C, H, F, Br, metals (e.g., O, N, P, S, etc.) that cannot form hydrogen bonds (e.g., esters, ketones, aldehydes, thioesters, sulfonyl, etc.).
- **Associating:** Compounds containing atoms other than C, H, F, Br, metals (e.g., O, N, P, S, etc.) that can form hydrogen bond (e.g., acids, amines, alcohols, etc.).

Results were quantified using the average absolute relative deviation (*AARD*) defined as

$$AARD = \frac{1}{N_{\mathrm{pt}}} \sum_{i=1}^{N_{\mathrm{pt}}} \left| \frac{X_{\mathrm{pred}_i} - X_{\mathrm{exp}_i}}{X_{\mathrm{exp}_i}} \right| \tag{11.3-15}$$

where the index is for point i in the test set, X_{pred_i} is the predicted value for property X for point i, and X_{exp_i} is the experimental value for property X for point i.

Table 11.1 contains the average absolute relative deviation [see Eq. (11.3-15)] of each method according to compound class. Also listed are the number of points tested in each class (N_{pt}) and number of compounds (N_c). Each of the methods perform relatively poorly for associating compounds with all the AARDs for the class being between 21% and 27%. However, errors of less than 25% can be expected in this case if the CH method is used. The MT method is limited to just hydrocarbons, but it does well for these compounds with an AARD of just 4.1% for the normal class (the best of all methods) and 8.9% for heavy. Thus, in cases where this method is applicable, errors 5 to 10% can be assumed reasonably. The ST approach performs consistently around 10% AARD for all but the associating class, while the ME technique appears to be the best selection for heavy compounds and is also the second best for the normal family. The EU method should be avoided. Overall, for compounds other than associating, all methods but EU produce about 10% error on average.

Thermal conductivity data for pure gases are compiled in Tsederberg (1965), Vargaftik et al. (1994, 1996), the DIPPR 801 database (Wilding et al., 2017; DIPPR 2021), and the Thermodynamics Research Center (Diky et al., 2021). Constants that may be used to calculate pure gas thermal conductivities at different temperatures are tabulated in Miller et al. (1976a), Yaws (1995, 1995a), the DIPPR 801 database (Wilding et al., 2017; DIPPR 2021), and ThermoData Engine from the Thermodynamics Research Center (Diky et al., 2021). The constants in

TABLE 11.1 Performance of prediction techniques for low-pressure vapor λ against the test database.

	Compound Class											
	Normal			Heavy			Polar			Associating		
Method	N_{pt}	N_c	AARD	N_{pt}	N_c	AARD	N_{pt}	N_c	AARD	N_{pt}	N_c	AARD
EU	1556	114	15.2%	20	4	21.2%	196	23	12.9%	306	32	26.1%
ME	1556	114	6.9%	20	4	2.7%	196	23	11.0%	306	32	23.6%
ST	1556	114	9.0%	20	4	10.1%	196	23	9.3%	306	32	23.9%
MT	697	48	4.1%	20	4	8.9%	—	—	—	—	—	—
CH	1556	114	7.2%	20	4	13.7%	196	23	9.5%	306	32	21.3%

the last three sources are based on critically evaluated data while those in the other references are not. Thermal conductivity values for common gases are tabulated at different temperatures in Speight (2016), Rumble (2021), and Green and Southard (2019).

11.4 EFFECT OF TEMPERATURE ON THE LOW-PRESSURE THERMAL CONDUCTIVITIES OF GASES

Thermal conductivities of low-pressure gases increase with temperature. The exact dependence of λ on T is difficult to judge from the λ-estimation methods in Sec. 11.3 because other temperature-dependent parameters (e.g., heat capacities and viscosities) are incorporated in the correlations. Generally, $d\lambda/dT$ ranges from 4×10^{-5} to 1.2×10^{-4} W/(m·K^2), with the more complex and polar molecules having the larger values. Several power laws relating λ with T have been proposed (Correla et al., 1968; Missenard, 1972), but they are not particularly accurate. To illustrate the trends, Fig. 11.3 has been drawn to show λ as a function of temperature for a few selected gases.

11.5 EFFECT OF PRESSURE ON THE THERMAL CONDUCTIVITIES OF GASES

The thermal conductivities of all gases increase with pressure, although the effect is relatively small at low and moderate pressures. Three pressure regions in which the effect of pressure is distinctly different are discussed below.

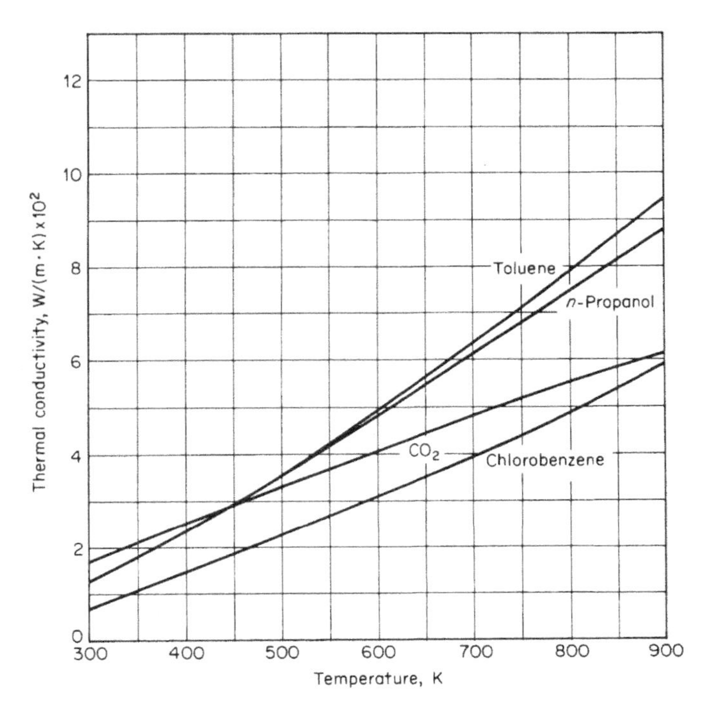

FIGURE 11.3
Effect of temperature on the thermal conductivity of some low-pressure gases.

11.5.1 Very Low Pressure

Below pressures of about 10^{-3} bar, the mean free path of the molecules is large compared to typical dimensions of a measuring cell, and there λ is almost proportional to pressure. This region is called the *Knudsen domain.* In reported thermal conductivity data, the term *zero-pressure value* is often used; however, it refers to values extrapolated from higher pressures (above 10^{-3} bar) and not to measured values in the very-low-pressure domain.

11.5.2 Low Pressure

This region extends from approximately 10^{-3} to 10 bar and includes the domain discussed in Secs. 11.3 and 11.4. The thermal conductivity increases about 1% or less per bar (Kannuliuk and Donald, 1950; Vines, 1953, 1958; Vines and Bennett, 1954). Such increases are often ignored in the literature, and either the 1-bar value or the "zero-pressure" extrapolated value may be referred to as the low-pressure conductivity.

11.5.3 High Pressure

Figure 11.4 shows the thermal conductivity of propane over a wide range of pressures and temperatures (Holland et al., 1979). The high-pressure gas domain would be represented by the curves on the right-hand side of the graph above the critical temperature (369.8 K). Increasing pressure raises the thermal conductivity, with the region around the critical point being particularly sensitive. Increasing temperature at low pressures results in a larger thermal conductivity, but at high pressure the opposite effect is noted. Similar behavior is shown for the

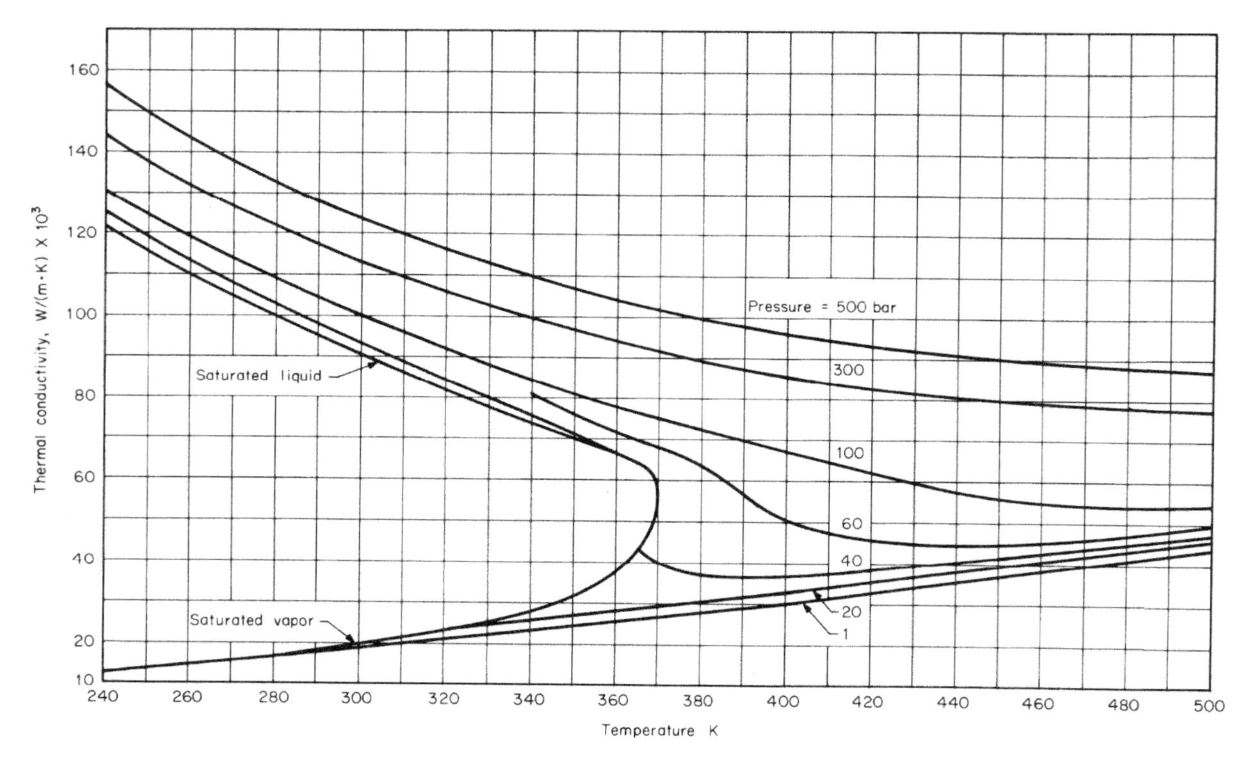

FIGURE 11.4
Thermal conductivity of propane. (Data from Holland et al., 1979)

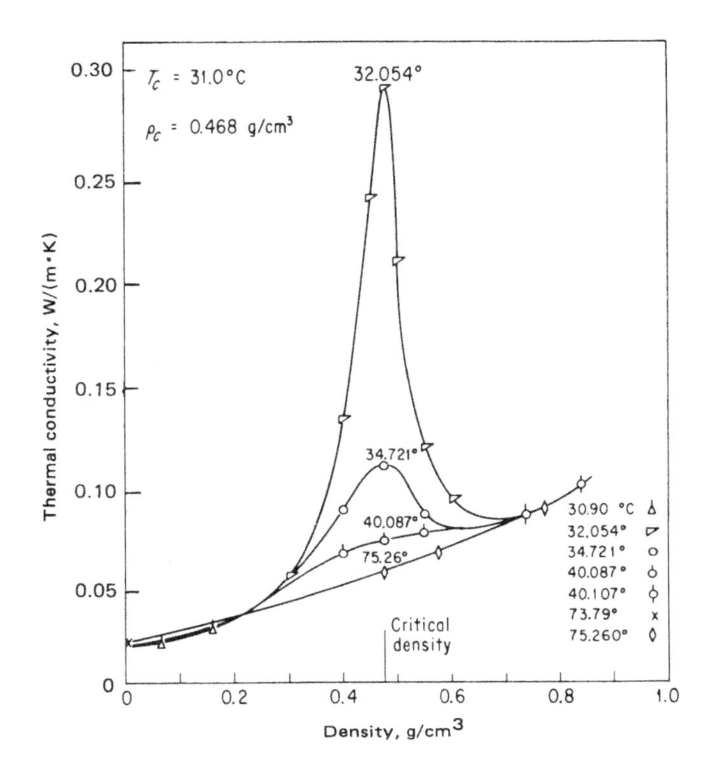

FIGURE 11.5
Thermal conductivity of carbon dioxide near the critical point. (Data from Guildner, 1958)

region below T_c, where λ for liquids decreases with temperature, whereas for gases (see Sec. 11.4), there is an increase of λ with T. Pressure effects (except at very high pressures) are small below T_c. Not shown in Fig. 11.4 is the unusual behavior of λ near the critical point. In this region, the thermal conductivity is quite sensitive to both temperature and pressure (Basu and Sengers, 1983). Figure 11.5 shows a plot of λ for CO_2 near the critical point (Guildner, 1958). It can be seen from Fig 11.5 that near the critical point, λ can increase by a factor of 6 due to the enhancement that occurs in the critical region. This enhancement is much bigger for thermal conductivity than for viscosity. The region for which this enhancement increases λ for CO_2 by at least 1% is bounded approximately by $240 < T < 450$ K ($0.79 < T_r < 1.48$) and 25 kg/m³ $< \rho <$ 1000 kg/m³ ($0.53 < \rho_r < 2.1$). See Krauss et al. (1993), Sengers and Luettmer-Strathmann (1996), and Vesocic et al. (1990) for additional discussion of the critical region enhancement and examples of how to describe the enhancement mathematically. When generalized charts of the effect of pressure on λ are drawn, the enhancement around T_c and P_c is usually smoothed out and not shown.

11.5.4 Excess Thermal Conductivity Correlations

Many investigators have adopted the suggestion of Vargaftik (1951, 1958) that the excess thermal conductivity, $\lambda - \lambda^\circ$, be correlated as a function of the PVT properties of the system in a corresponding states manner. (Here λ° is the low-pressure thermal conductivity of the gas at the same temperature.) In its simplest form,

$$\lambda - \lambda^\circ = f(\rho) \qquad (11.5\text{-}1)$$

where ρ is the fluid density. The correlation has been shown to be applicable to ammonia (Groenier and Thodos, 1961; Richter and Sage, 1964), ethane (Carmichael et al., 1963), n-butane (Carmichael and Sage, 1964; Kramer and Comings, 1960), nitrous oxide (Richter and Sage, 1963), ethylene (Owens and Thodos, 1960), methane (Carmichael et al., 1966; Mani and Venart, 1973; Owens and Thodos, 1957a), diatomic gases (Misic and Thodos, 1965; Schafer and Thodos, 1959), hydrogen (Schafer and Thodos, 1958), inert gases (Owens and Thodos, 1957), and carbon dioxide (Kennedy and Thodos, 1961). Temperature and pressure do not enter explicitly, but their effects are included in the parameters $\lambda°$ (temperature only) and ρ.

11.5.4.1 Method of Stiel and Thodos (1964)

Stiel and Thodos (1964) have generalized Eq. (11.5-1) by assuming that $f(\rho)$ depends only on T_c, P_c, V_c, M (molecular weight), and ρ. By dimensional analysis they obtain a correlation between $\lambda - \lambda°$, Z_c, Γ, and ρ, where Γ is a reduced thermal conductivity discussed below. From data on 20 nonpolar substances, including inert gases, diatomic gases, CO_2, and hydrocarbons, they established the approximate analytical expressions:

$$(\lambda - \lambda°)\Gamma Z_c^5 = 1.22 \times 10^{-2}[\exp(0.535\rho_r) - 1] \qquad \rho_r < 0.5 \tag{11.5-2}$$

$$(\lambda - \lambda°)\Gamma Z_c^5 = 1.14 \times 10^{-2}[\exp(0.67\rho_r) - 1.069] \qquad 0.5 < \rho_r < 2.0 \tag{11.5-3}$$

$$(\lambda - \lambda°)\Gamma Z_c^5 = 2.60 \times 10^{-3}[\exp(1.155\rho_r) - 2.016] \qquad 2.0 < \rho_r < 2.8 \tag{11.5-4}$$

where λ is in W/(m·K), Z_c is the critical compressibility, and $\rho_r = \dfrac{\rho}{\rho_c} = \dfrac{V_c}{V}$ is the reduced density.

In the same way that the viscosity was nondimensionalized in Eqs. (10.4-14) and (10.4-15), a reduced thermal conductivity may be expressed as

$$\lambda_r = \lambda\Gamma \tag{11.5-5}$$

$$\Gamma = \left[\frac{T_c (M')^3 N_A^2}{R^5 P_c^4}\right]^{\frac{1}{6}} \tag{11.5-6}$$

In SI units, if $R = 8314$ J/(kmol·K), N_A (Avogadro's number) $= 6.023 \times 10^{26}$ kmol^{-1}, and with T_c in K, M' in kg/kmol, and P_c in Pa, Γ has the units of m·K/W or inverse thermal conductivity. In more convenient units,

$$\Gamma = 210\left(\frac{T_c M^3}{P_c^4}\right)^{\frac{1}{6}} \tag{11.5-7}$$

where Γ is the reduced inverse thermal conductivity in m·K/W, T_c is in K, M is in g/mol, and P_c is in bar.

Equations (11.5-2) to (11.5-4) should not be used for polar substances or for hydrogen or helium. The general accuracy is in doubt, and errors of ±10 to 20% or more are possible. The method is illustrated in Example 11.2.

Example 11.2 Estimate the thermal conductivity of nitrous oxide at 105 °C and 138 bar. At this temperature and pressure, the experimental value is 3.90×10^{-2} W/(m·K) (Richter and Sage, 1963). At 1 bar and 105 °C, $\lambda° = 2.34 \times 10^{-2}$ W/(m·K) (Richter and Sage, 1963). From Couch and Dobe (1961), $T_c = 309.6$ K, $P_c = 72.54$ bar, $V_c = 97.27$ cm³/mol, $Z_c = 0.274$, and $M = 44.013$ g/mol. At 105 °C and 138 bar, Z for N_2O is 0.63 (Couch and Dobe, 1961).

Solution. With Eq. (11.5-7),

$$\Gamma = 210[(309.6)(44.013)^3/72.54^4]^{1/6} = 208.3$$
$$V = (0.63)(8.314)(378.15)/(138 \times 10^5) \times 10^6 = 143.5 \text{ cm}^3/\text{mol}$$
$$\rho_r = 97.27/143.5 = 0.678$$

Then, with Eq. (11.5-3),

$$\lambda - \lambda^\circ = (0.0114)\{\exp[(0.67)(0.678)] - 1.069\}/[(208.3)(0.274)^5] = 0.0179 \text{ W/(m·K)}$$
$$\lambda = 0.0234 + 0.0179 = 0.0413 \text{ W/(m·K)}$$
$$\lambda \text{ Difference} = 0.0413 - 0.0390 = 0.0023 \text{ or } 5.9\%$$

11.5.4.2 Method of Chung et al. (1984, 1988)

The low-pressure estimation procedure for pure-component thermal conductivities developed by these authors [Eq. (11.3-11)] is modified to treat materials at high pressures (or densities).

$$\lambda = \frac{31.2\eta^\circ \Psi}{M'}(G_2^{-1} + B_6 y) + qB_7 y^2 T_r^{\frac{1}{2}} G_2 \tag{11.5-8}$$

Here, λ is thermal conductivity in W/(m·K), η° is *low-pressure*, pure-gas viscosity in Pa·s, M' is molecular weight in kg/mol, $T_r = \frac{T}{T_c}$ is reduced temperature, T is temperature in K, T_c is critical temperature in K, Ψ is defined in Eq. (11.3-12), and

$$q = 3.586 \times 10^{-3}\left(\frac{T_c}{M'}\right)^{\frac{1}{2}} V_c^{\frac{2}{3}} \tag{11.5-9}$$

$$y = \frac{V_c}{6V} \tag{11.5-10}$$

$$G_1 = \frac{1 - 0.5y}{(1 - y)^3} \tag{11.5-11}$$

$$G_2 = \frac{\frac{B_1}{y}[1 - \exp(-B_4 y)] + B_2 G_1 \exp(B_5 y) + B_3 G_1}{B_1 B_4 + B_2 + B_3} \tag{11.5-12}$$

$$B_i = a_i + b_i\omega + c_i\mu_r^4 + d_i\kappa \tag{11.5-13}$$

$$\mu_r = 131.3\frac{\mu}{(V_c T_c)^{\frac{1}{2}}} \tag{11.5-14}$$

where V_c is the critical volume in cm^3/mol, ω is the acentric factor, κ is the association factor (see Table 10.1), u is the dipole moment in Debye, and a_i, b_i, c_i, and d_i are given in Table 11.2.

The relation for high-pressure thermal conductivities is similar to the Chung et al. form for high-pressure viscosities [Eqs. (10.6-29) through (10.6-35)].

In Eq. (11.5-10), if V becomes large, y then approaches zero. In such a case, both G_1 and G_2 are essentially unity and Eq. (11.5-8) reduces to Eq. (11.3-11), the relation for λ at low pressures. To use Eq. (11.5-8), η° can be obtained from experimental values or estimated by the techniques given in Sec. 11-4.

Chung et al. tested Eq. (11.5-8) with data from a large range of hydrocarbon types and from data for simple gases. Deviations over a wide pressure range were usually less than 5 to 8%. For highly polar materials, the correlation for β as given under Eq. (11.3-13) is not accurate, and, at present, no predictive technique to apply to

TABLE 11.2 Constants to calculate B_i for the Chung et al. (1984, 1988) method to predict vapor thermal conductivities at high pressures. Use in Eq. (11.5-13).

i	a_i	b_i	c_i	d_i
1	2.4166	0.74824	-0.91858	121.72
2	−0.50924	−1.5094	−49.991	69.983
3	6.6107	5.6207	64.760	27.039
4	14.543	−8.9139	−5.6379	74.344
5	0.79274	0.82019	−0.69369	6.3173
6	−5.8634	12.801	9.5893	65.529
7	91.089	128.11	−54.217	523.81

such compounds is available. [See the discussion dealing with polar materials under Eq. (11.3-13).] The high-pressure Chung et al. method is illustrated in Example 11.3.

11.5.4.3 TRAPP Method (Huber, 1996)

The transport property prediction (TRAPP) method is a corresponding states method to calculate viscosities and thermal conductivities of pure fluids and mixtures. In its original version (Ely and Hanley, 1983; Hanley, 1976), it was also used to estimate low-pressure values of λ and η and employed methane as a reference fluid. In the most recent version, presented below for pure fluids and later in Sec. 11.7 for mixtures, low-pressure values are estimated by one of the methods presented earlier in the chapter, propane is the reference fluid, and shape factors are no longer functions of density.

In this method, the excess thermal conductivity of a pure fluid is related to the excess thermal conductivity of the reference fluid, propane:

$$\lambda - \lambda^\circ = F_\lambda X_\lambda [\lambda_R - \lambda_R^\circ] \tag{11.5-15}$$

The reference fluid values (denoted by subscript R) are evaluated at T_λ and density ρ_λ, not T and ρ. In Eq. (11.5-15), λ° is the thermal conductivity at low pressure, λ_R is the true thermal conductivity of the reference fluid (propane) at temperature T_λ and density ρ_λ, λ_R° is the low-pressure value for propane at temperature T_λ, and F_λ and X_λ are factors that relate the reference fluid to the fluid of interest and are discussed below. Equation (11.3-5) leads to

$$\lambda^\circ = (1.32 C_p^\circ + 3.741)\frac{\eta^\circ}{M'} \tag{11.5-16}$$

where η° is the low-pressure viscosity in Pa·s, M' is molecular weight in kg/mol, λ° is in W/(m·K), and C_p° is ideal gas isobaric heat capacity in J/(mol·K). Ely (Huber, 1998) has found for propane that

$$\lambda_R - \lambda_R^\circ = C_1 \rho_{r_R} + C_2 (\rho_{r_R})^3 + \left(C_3 + \frac{C_4}{T_{r_R}}\right)(\rho_{r_R})^4 + \left(C_5 + \frac{C_6}{T_{r_R}}\right)(\rho_{r_R})^5 \tag{11.5-17}$$

where $T_{r_R} = \frac{T_\lambda}{T_{c_R}}$, $\rho_{r_R} = \frac{\rho_\lambda}{\rho_{c_R}}$, T_{cR} is the critical temperature of the reference fluid (propane), ρ_{cR} is the critical density of the reference fluid (propane), $\lambda_R - \lambda_R^\circ$ is in mW/(m·K), $C_1 = 15.2583985944$, $C_2 = 5.29917319127$, $C_3 = -3.05330414748$, $C_4 = 0.450477583739$, $C_5 = 1.03144050679$, and $C_6 = -0.185480417707$. X_λ, T_λ, ρ_λ, and F_λ are calculated by

$$T_\lambda = \frac{T}{f} \tag{11.5-18}$$

$$\rho_\lambda = \rho h \tag{11.5-19}$$

$$F_\lambda = \left(\frac{0.044094}{M'} f\right)^{\frac{1}{2}} h^{-\frac{2}{3}} \tag{11.5-20}$$

$$X_\lambda = \left[1 + \frac{2.1866(\omega - \omega_R)}{1 - 0.505(\omega - \omega_R)}\right]^{\frac{1}{2}} \tag{11.5-21}$$

where ω is the acentric factor of the compound of interest, ω_R is the acentric factor of the reference fluid (propane), and f and h are equivalent substance reducing ratios and are determined as described below.

If vapor pressure and liquid density information are available for the substance of interest, and if $T < T_c$, it is recommended that f be obtained from the equation

$$\frac{P^{vp}}{\rho^S} = f \frac{P_R^{vp}(T_\lambda)}{\rho_R^S(T_\lambda)} \tag{11.5-22}$$

where P^{vp} and ρ^S are the vapor pressure and saturated liquid density at temperature T and $P_R^{vp}(T_\lambda)$ and $\rho_R^S(T_\lambda)$ are for the reference fluid propane evaluated at T_λ. As indicated in Eq. (11.5-18), T_λ is a function of f, so Eq. (11.5-22) must be solved iteratively to find f. Once f is found, h is determined from

$$h = \frac{\rho_R^S(T_\lambda)}{\rho^S} \tag{11.5-23}$$

If $T > T_c$, or if vapor pressure and saturated liquid density information are not available, h and f can be calculated by

$$f = \frac{T_c}{T_{c_R}}[1 + (\omega - \omega_R)(0.05203 - 0.7498 \ln T_r)] \tag{11.5-24}$$

$$h = \frac{\rho_{c_R}}{\rho_c}\frac{Z_{c_R}}{Z_c}[1 - (\omega - \omega_R)(0.1436 - 0.2822 \ln T_r)] \tag{11.5-25}$$

where $T_r = T/T_c$, and the other variables are defined above.

Example 11.3 Estimate the thermal conductivity of propylene at 473 K and 150 bar by using the (a) Chung et al. and (b) TRAPP methods. Under these conditions, Vargaftik et al. (1996) report $\lambda = 6.64 \times 10^{-2}$ W/(m·K) and $V = 172.1$ cm^3/mol. Also, these same authors list the low-pressure viscosity and thermal conductivity of propylene at 473 K as $\eta° = 134 \times 10^{-7}$ N·s/m^2 and $\lambda° = 3.89 \times 10^{-2}$ W/(m·K).

Solution. From Appendix A for propylene, $T_c = 364.85$ K, $P_c = 46.0$ bar, $V_c = 185$ cm^3/mol, $Z_c = 0.281$, $\omega = 0.137588$, $M = 42.080$ g/mol, $M' = 0.042080$ kg/mol, and $\mu = 0.366$ D. Since propylene is not found in Table 10.1, the association factor in Chung et al.'s method $\kappa = 0$. Also, from the coefficients in Appendix A, $C_p° = 90.655$ J/(mol·K), so $C_v = 90.655 - 8.314 = 82.341$ J/(mol·K).

METHOD OF CHUNG ET AL. With Eq. (11.3-13),

$$\alpha = 82.341/8.314 - 3/2 = 8.404$$

$$\beta = 0.7862 - (0.7109)(0.137588) + (1.3168)(0.137588)^2 = 0.7133$$

$$Z = 2.0 + (10.5)(473/364.85)^2 = 19.65$$

Then, by Eq. (11.3-12),

$$\Psi = 1 + (8.404)[0.215 + (0.28288)(8.404) - (1.061)(0.7133) + (0.26665)(19.65)]/$$
$$[0.6366 + (0.7133)(19.65) + (1.061)(8.404)(0.7133)] = 3.830$$

From Eq. (11.5-10),

$$y = 185/[(6)(172.1)] = 0.1792$$

And with Eq. (11.5-14),

$$\mu_r^4 = \{(131.3)(0.366)/[(185)(365.85)]^{1/2}\}^4 = 0.001171$$

The values of B_i are found from Eq. (11.5-13) with the coefficients from Table 11.2. The first value becomes

$$B_1 = 2.4166 + (0.74824)(0.137588) - (0.91858)(0.001171) = 2.5185$$

and the others are $B_2 = -0.77545$, $B_3 = 7.4599$, $B_4 = 13.310$, $B_5 = 0.90478$, $B_6 = -4.0909$, and $B_7 = 108.65$. With Eqs. (11.5-11), (11.5-12), and (11.5-9),

$$G_1 = [1 - (0.5)(0.1792)]/(1 - 0.1792)^3 = 1.646$$
$$G_2 = \{(2.5185/0.1792)\{1 - \exp[(-13.310)(0.1792)]\}$$
$$+ (-0.77545)(1.646)\exp[(0.90478)(0.1792)] + (7.4599)(1.646)\}/$$
$$[(2.5185)(13.310) - 0.77545 + 7.4599] = 0.5854$$
$$q = 0.003586(364.85/0.042080)^{1/2}/185^{2/3} = 0.01028$$

Finally, using Eq. (11.5-8),

$$\lambda = [(31.2)(134 \times 10^{-7})(3.830)/0.042080][(0.5854)^{-1} - (4.0909)(0.1792)]$$
$$+ (0.01028)(108.65)(0.1792)^2(473/364.85)^{1/2}(0.5854) = 0.0610\ \text{W/(m·K)}$$
$$\lambda\ \text{Difference} = 0.0610 - 0.0664 = -0.0054\ \text{W/(m·K) or } -8.1\%$$

TRAPP METHOD. From Appendix A, the properties for the reference fluid propane are $T_{c_R} = 369.83$ K, $V_{c_R} = 200$ cm³/mol, $Z_{c_R} = 0.276$, and $\omega_R = 0.152291$.

Equations (11.5-24) and (11.5-25), with $T_r = 473/364.85 = 1.2964$, give

$$f = (364.85/369.83)[1 + (0.137588 - 0.152291)(0.05203 - 0.7498\ln(1.2964)] = 0.9886$$
$$h = (185/200)(0.279/0.281)[1 - (0.137588 - 0.152291)(0.1436 - 0.2822\ln(1.2964)]$$
$$= 0.9194$$

Equations (11.5-18) and (11.5-19) give

$$T_\lambda = 473/0.9886 = 478.5\ \text{K}$$
$$\rho_\lambda = 0.9194/172.1 = 0.005342\ \text{cm}^3/\text{mol}$$

Equation (11.5-17) with $T_{r_R} = 478.5/369.83 = 1.294$ and $\rho_{r_R} = (0.005342)(200) = 1.068$ gives

$$\lambda_R - \lambda_R^\circ = (15.2583985944)(1.068) + (5.29917319127)(1.068)^3$$
$$+ (-3.05330414748 + 0.450477583739/1.294)(1.068)^4$$
$$+ (1.03144050679 - 0.185480417707/1.294)(1.068)^5$$
$$= 20.47 \text{ mW/(m·K)} = 0.02047 \text{ W/(m·K)}$$

Equations (11.5-20) and (11.5-21) give

$$F_\lambda = [(0.044094/0.042080)(0.9886)]^{1/2}(0.9194)^{-2/3} = 1.076$$
$$X_\lambda = \{1 + [2.1866(0.137588 - 0.152291)]/[1 - 0.505(0.137588 - 0.152291)]\}^{1/2}$$
$$= 0.9839$$

Equation (11.5-16) yields

$$\lambda^\circ = [(1.32)(90.655) + 3.741](134 \times 10^{-7}/0.042080) = 0.0393 \text{ W/(m·K)}$$

Finally, using Eq. (11.5-15),

$$\lambda = 0.0393 + (1.076)(0.9839)(0.02047) = 0.0610 \text{ W/(m·K)}$$
$$\lambda \text{ Difference} = 0.0610 - 0.0664 = -0.0054 \text{ W/(m·K) or } -8.1\%$$

11.5.4.4 Discussion

Three methods for estimating the thermal conductivity of pure materials in the dense gas region were presented. All use the fluid density rather than pressure as a system variable. The low-density thermal conductivity is required in the Stiel and Thodos [Eqs. (11.5-2) to (11.5-4)] and TRAPP [Eqs. (11.5-15) to (11.5-25)] methods, but it is calculated as a part of the procedure in the Chung et al. [Eq. (11.5-8)] method. None of the techniques are applicable for polar gases, and even for nonpolar materials, errors can be large. The Chung et al. and TRAPP procedures are reported to be applicable over a wide density domain even into the liquid phase. No one of the methods appears to have a clear superiority over the others.

Other methods for dense-fluid thermal conductivity have been proposed. Assael et al. (1996) and Dymond and Assael (1996) have correlated thermal conductivities of dense fluids by the universal equation

$$\log_{10} \lambda_r = 1.0655 - \frac{3.538}{V_r} + \frac{12.12}{V_r^2} - \frac{12.469}{V_r^3} + \frac{4.562}{V_r^4} \tag{11.5-26}$$

where $V_r = \frac{V}{V_0}$ and V_0 is the same close-packed volume that appears in Eq. (10.6-3). λ_r is a reduced thermal conductivity defined by

$$\lambda_r = 21.23 \frac{\lambda V^{\frac{2}{3}}}{R_\lambda}\left(\frac{M}{T}\right)^{\frac{1}{2}} \tag{11.5-27}$$

where λ is in W/(m·K), V is in cm^3/mol, M is g/mol, and T is in K. R_λ is a parameter that accounts for deviations from smooth hard spheres. The two parameters V_0 and R_λ are compound-specific and are not functions of density. V_0 is a function of temperature as is R_λ for n-alcohols, refrigerants, and a number of other polar compounds (Assael et al., 1994, 1995; Bleazard and Teja, 1996). For n-alkanes (Assael et al, 1992, 1992a), aromatic hydrocarbons (Assael et al., 1992b), and other simple molecular fluids (Assael et al., 1992), R_λ has been found to be independent of temperature. In theory, the two parameters V_0 and R_λ could be set with two experimental viscosity-density data, but in fact Eq. (11.5-26) has been used only for systems for which extensive data are available. It has been applied to densities above the critical density and applicability to temperatures down to $T_r \approx 0.6$ has been claimed. Values of V_0 and R_λ at 298 K and 350 K for 16 fluids are shown in Table 10.5.

Thermal conductivity data for selected fluids appear in the references above as well as in Fleeter et al. (1980), Le Neindre (1987), Prasad and Venart (1981), Tufeu and Neindre (1981), Yorizane et al. (1983), and Zheng et al. (1984).

11.6 THERMAL CONDUCTIVITIES OF LOW-PRESSURE GAS MIXTURES

The thermal conductivity of a gas mixture is not usually a linear function of mole fraction. Generally, if the constituent molecules differ greatly in polarity, the mixture thermal conductivity is larger than would be predicted from a mole fraction average; for nonpolar molecules, the opposite trend is noted and is more pronounced the greater the difference in molecular weights or sizes of the constituents (Gray et al., 1970; Misic and Thodos, 1961). Some of these trends are evident in Fig. 11.6, which shows experimental thermal

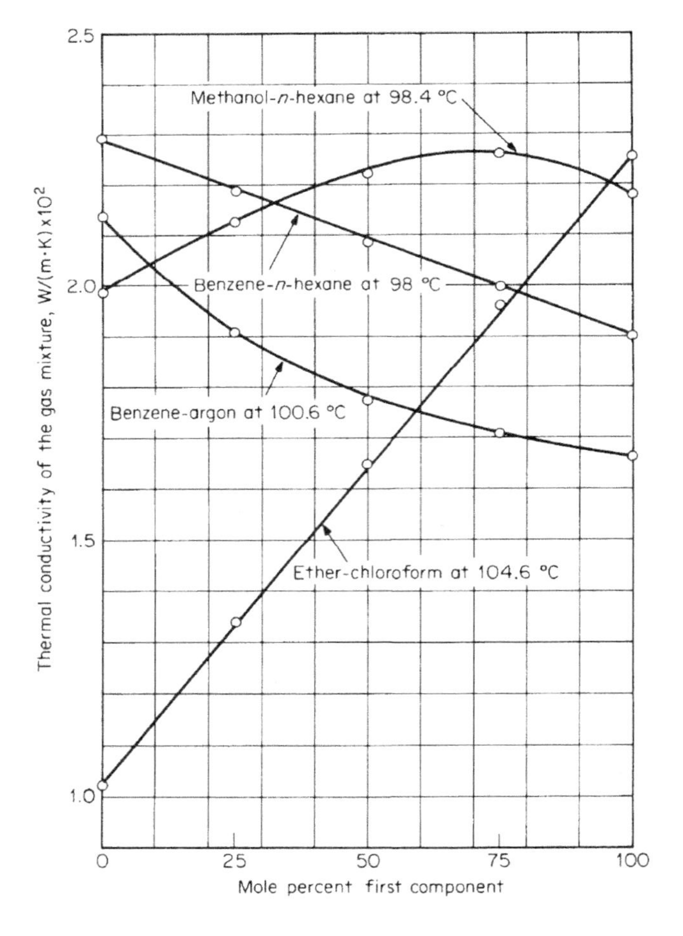

FIGURE 11.6
Typical gas-mixture thermal conductivities. (From Bennett and Vines, 1955)

conductivities for four systems. The argon-benzene system typifies a nonpolar case with different molecular sizes, and the methanol-*n*-hexane system is a case representing a significant difference in polarity. The linear systems benzene-*n*-hexane and ether-chloroform represent a balance between the effects of size and polarity.

Many theoretical papers discussing the problems, approximations, and limitations of the various methods also have appeared. The theory for calculating the conductivity for rare-gas mixtures has been worked out in detail (Brokaw, 1958; Hirschfelder et al., 1954; Mason 1958; Mason and Saxena, 1959; Mason and von Ubisch, 1960; Muckenfuss, 1958). The more difficult problem, however, is to modify monatomic mixture correlations to apply to polyatomic molecules.

Many techniques have been proposed; all are essentially empirical, and most reduce to some form of the Wassiljewa equation. Corresponding states methods for low-pressure thermal conductivities have also been adapted for mixtures, but the results obtained in testing several were not encouraging.

11.6.1 Wassiljewa Equation

In a form analogous to the theoretical relation for mixture viscosity, Eq. (10.5-13), Wassiljewa (1904) proposed the empirical relationship

$$\lambda_{\mathrm{m}} = \sum_{i=1}^{n} \frac{y_i \lambda_i}{\sum_{j=1}^{n} y_j A_{ij}} \tag{11.6-1}$$

where λ_{m} is the thermal conductivity of the gas mixture, λ_i is the thermal conductivity of pure i, y_i is the mole fraction of component i, A_{ij} is a function discussed below, and $A_{ii} = 1.0$.

11.6.2 Mason and Saxena Modification

Mason and Saxena (1958) suggested that A_{ij} in Eq. (11.6-1) could be expressed as

$$A_{ij} = \frac{\varepsilon\left[1 + \left(\dfrac{\lambda_{\mathrm{tr}_i}}{\lambda_{\mathrm{tr}_j}}\right)^{\frac{1}{2}} \left(\dfrac{M_i}{M_j}\right)^{\frac{1}{4}}\right]^2}{\left[8\left(1 + \dfrac{M_i}{M_j}\right)\right]^{\frac{1}{2}}} \tag{11.6-2}$$

where M is the molecular weight in g/mol, λ_{tr_i} is the monatomic (translational) value of the thermal conductivity for component i proposed by Roy (1967), and ε is a numerical constant near unity. Mason and Saxena proposed a value of 1.065 for ε, and Tandon and Saxena (1965) later suggested 0.85. As used here, $\varepsilon = 1.0$.

From Eq. (11.2-3), noting for monatomic gases that $C_v = C_{\mathrm{tr}} = \frac{3}{2} R$,

$$\frac{\lambda_{\mathrm{tr}_i}}{\lambda_{\mathrm{tr}_j}} = \frac{\eta_i}{\eta_j} \frac{M_j}{M_i} \tag{11.6-3}$$

Substituting Eq. (11.6-3) into Eq. (11.6-2) and comparing with Eq. (10.5-14) gives

$$A_{ij} = \phi_{ij} \tag{11.6-4}$$

where ϕ_{ij} is the interaction parameter for gas-mixture viscosity. Thus the relation for estimating mixture viscosities is also applicable to thermal conductivities by simply substituting λ for η. In this approximation, to determine λ_{m} one needs data giving the pure-component thermal conductivities and viscosities.

An alternative way to proceed is to use Eqs. (11.6-1) and (11.6-2) but obtain the ratio of translational thermal conductivities using the expression for λ_{tr} given by Roy (1967; Roy and Thodos 1968, 1970). This ratio is

$$\frac{\lambda_{tr_i}}{\lambda_{tr_j}} = \frac{\Gamma_j[\exp(0.0464T_{r_i}) - \exp(-0.2412T_{r_i})]}{\Gamma_i[\exp(0.0464T_{r_j}) - \exp(-0.2412T_{r_j})]} \tag{11.6-5}$$

where Γ is defined by Eq. (11.5-7). With Eq. (11.6-5), values of A_{ij} become functions of the reduced temperatures of both i and j. However, with this latter approach, pure-gas viscosities are not required. Both techniques are illustrated in Example 11.4.

Lindsay and Bromley (1950) have also proposed a technique to estimate A_{ij}. It is slightly more complex than Eq. (11.6-2), and the results obtained do not differ significantly from the Mason-Saxena approach.

The Wassiljewa equation can represent low-pressure mixture thermal conductivities with either a maximum or minimum as composition is varied. As Gray et al. (1970) have shown, if $\lambda_1 < \lambda_2$,

- $\frac{\lambda_1}{\lambda_2} < A_{12}A_{21} < \frac{\lambda_2}{\lambda_1}$ — λ_m varies monotonically with composition
- $A_{12}A_{21} \geq \frac{\lambda_2}{\lambda_1}$ — λ_m has a minimum value below λ_1
- $\frac{\lambda_1}{\lambda_2} \geq A_{12}A_{21}$ — λ_m has a maximum value above λ_2

11.6.3 Corresponding States Methods

The Chung et al. (1984, 1988) method for estimating low-pressure thermal conductivities [Eqs. (11.3-11) and (11.3-14)] has been adapted to handle mixtures. The emphasis of these authors, however, was to treat systems at high pressure and, if possible, as liquids. When their method is used for low-pressure gas mixtures, the accuracy away from the pure components is often not particularly high. However, in their favor is the fact that pure-component thermal conductivities are not required as input; the method generates its own values of pure-component conductivities.

To use the Chung et al. form for mixtures, we need to have rules to obtain M', η, C_v, ω, and T_c for the mixture. η_m is found from Eq. (10.5-24), and in using this relation, one also obtains M_m, ω_m, and T_{c_m} [Eqs. (10.5-28), (10.5-29), and (10.5-44)]; $M'_m = \frac{M_m}{1000}$. For C_{v_m} a mole fraction average rule is used, i.e.,

$$C_{v_m} = \sum_i y_i C_{v_i} \tag{11.6-6}$$

With these mixture values, the procedure to compute λ_m is identical with that used for the pure-component conductivity (see Example 11.1). The method is also illustrated for a mixture in Example 11.4.

11.6.4 Discussion

Three techniques were suggested to estimate the thermal conductivity of a gas mixture at low pressure. Two employ the Wassiljewa formulation [Eq. (11.6-1)] and differ only in the manner $\frac{\lambda_{tr_i}}{\lambda_{tr_j}}$ is calculated. The third method (Chung et al.) uses a corresponding states approach. It is the least accurate, but it has the advantage that pure-component thermal conductivities do not have to be known. The other two methods require either experimental or estimated values of λ for all pure components. All three methods are illustrated in Example 11.4.

For nonpolar gas mixtures, we recommend the Wassiljewa equation with the Mason-Saxena relation for A_{ij}, where $\frac{\lambda_{tr_i}}{\lambda_{tr_j}}$ is calculated from Eq. (11.6-5). Errors will generally be less than 3 to 4%. For nonpolar-polar and polar-polar gas mixtures, none of the techniques examined were found to be particularly accurate. As an example, in Fig. 11.6, none predicted the maximum in λ_m for the methanol-*n*-hexane system. Thus, in such cases, errors greater than 5 to 8% may be expected when one employs the procedures recommended for nonpolar gas mixtures. For mixtures in which the sizes and polarities of the constituent molecules are not greatly different, λ_m can be estimated satisfactorily by a mole fraction average of the pure-component conductivities (e.g., the benzene-*n*-hexane and ether-chloroform cases in Fig. 11.6).

Example 11.4 Estimate the thermal conductivity of a gas mixture containing 25 mole% benzene and 75 mole% argon at 100.6 °C and about 1 bar. The experimental value is 0.0192 W/(m·K) (Bennett and Vines, 1955).

Solution. The following pure component properties are used:

	Benzene (1)	Argon (2)
T_c, K	562.05	150.86
P_c, bar	48.95	48.98
V_c, cm^3/mol	256.0	74.57
ω	0.210	−0.002
Z_c	0.268	0.291
M, g/mol	78.114	39.948
M', kg/mol	0.078114	0.039948
C_p°(373.8 K), J/(mol·K)	104.5	20.8
C_v°(373.8 K) = C_p°(373.8 K) − R, J/(mol·K)	96.2	12.5
$\eta \times 10^7$, Pa·s	92.5	271.0
$\lambda \times 10^2$, W/(m·K)	1.66	2.14

Here, the ideal gas heat capacities are used due to the low pressure involved, and R is approximated as 8.3 J/(mol·K).

MASON AND SAXENA. Equation (11.6-1) is used with $A_{12} = \phi_{12}$ and $A_{21} = \phi_{21}$ from Eqs. (10.5-14) and (10.5-15).

$$A_{12} = [1 + (92.5/271)^{1/2}(39.948/78.114)^{1/4}]^2/[8(1 + 78.114/39.948)]^{1/2} = 0.459$$

$$A_{21} = 0.459(271/92.5)(78.114/39.948) = 2.630$$

With Eq. (11.6-1)

$$\lambda_m = \{[(0.25)(1.66)]/[0.25 + (0.75)(0.459)]$$

$$+ [(0.75)(2.14)]/[(0.25)(2.630) + 0.75]\}/100 = 0.0184 \text{ W/(m·K)}$$

$$\lambda_m \text{ Difference} = 0.0184 - 0.0192 = -0.0008 \text{ W/(m·K) or } -4.2\%$$

MASON AND SAXENA Form with Eq. (11.6-5). In this case, $\frac{\lambda_{tr_i}}{\lambda_{tr_j}}$ is obtained from Eq. (11.6-5). Γ is determined with Eq. (11.5-7).

$$\Gamma_1 = 210\{[(562.05)(78.114)^3]/48.95^4\}^{1/6} = 398.5$$

$$\Gamma_2 = 210\{[(150.86)(39.948)^3]/48.7^4\}^{1/6} = 229.6$$

At 373.8 K, $T_{r_1} = 373.8/562.05 = 0.665$; $T_{r_2} = 373.8/150.86 = 2.478$. Then,

$$\frac{\lambda_{tr_1}}{\lambda_{tr_2}} = (229.6/398.5)\{\exp[(0.0464)(0.665)] - \exp[(-0.2412)(0.665)]\}/$$
$$\{\exp[(0.0464)(2.478)] - \exp[(-0.2412)(2.478)]\} = 0.1809$$

$$\frac{\lambda_{tr_2}}{\lambda_{tr_1}} = (0.1890)^{-1} = 5.528$$

Inserting these values into Eq. (11.6-2) with $\varepsilon = 1.0$ gives

$$A_{12} = [1 + (0.1809)^{1/2}(78.114/39.948)^{1/4}]^2/\{8[1 + (78.114/39.948)]\}^{1/2} = 0.4646$$

$$A_{21} = [1 + (5.528)^{1/2}(39.948/78.114)^{1/4}]^2/\{8[1 + (39.948/78.114)]\}^{1/2} = 2.568$$

Then, using Eq. (11.6-1),

$$\lambda_m = 10^{-2}\{[(0.25)(1.66)]/[0.25 + (0.75)(0.4646)]$$
$$+ [(0.75)(2.14)]/[(0.25)(2.568) + (0.75)]\} = 0.0185 \text{ W/(m·K)}$$

$$\lambda_m = 0.0185 - 0.0192 = -0.0007 \text{ W/(m·K) or } -3.6\%$$

CHUNG ET AL. With this method, we use the relations described in Chap. 10 to determine the mixture properties η_m, M'_m, ω_m, and T_{c_m}. (See Example 10.9.) In this case, at 25 mole% benzene, $\eta_m = 182.2$ μP $= 182.2 \times 10^{-7}$ Pa·s, $M'_m = 0.04631$ kg/mol, $\omega_m = 0.0788$, and $T_{c_m} = 276.9$ K. From Eq. (11.6-6),

$$C_{v_m} = (0.25)(96.2) + (0.75)(12.5) = 33.4 \text{ J/(mol·K)}$$

The mixture thermal conductivity is then found from Eq. (11.3-11) with the parameters given by Eqs. (11.3-12) and (11.3-13) calculated using the mixture properties as given below.

$$\alpha_m \frac{C_{v\,m}}{R} - \frac{3}{2} = 33.4/8.31 - 3/2 = 2.52$$

$$\beta_m = 0.7862 - 0.7109\omega_m + 1.3168\omega_m^2$$
$$= 0.7862 - 0.7109(0.0788) + 1.3168(0.0788)^2 = 0.7384$$

$$T_{r_m} = \frac{T}{T_{c_m}} = 373.8/276.9 = 1.350$$

$$Z_m = 2.0 + 10.5T_{c_m}^2 = 2.0 + (10.5)(1.348)^2 = 21.13$$

$$\Psi_m = 1 + \alpha_m \frac{0.215 + 0.28288\alpha_m - 1.061\beta_m + 0.26665Z_m}{0.6366 + \beta_m Z_m + 1.061\alpha_m\beta_m}$$
$$= 1 + \{2.52[0.215 + (0.28288)(2.52) - (1.061)(0.7384) + (0.26665)(21.13)]\}/$$
$$[0.6366 + (0.7384)(21.13) + (1.061)(2.52)(0.7384)] = 1.800$$

$$\lambda_m = \frac{\left(\frac{\eta_m C_{v_m}}{M'_m}\right)(3.75\Psi_m)}{\frac{C_{v_m}}{R}} = \frac{\eta_m R}{M'_m}(3.75\Psi_m)$$

$$= [(182.2 \times 10^{-7})(8.314)/0.0461][(3.75)(1.800)] = 0.0221 \text{ W/(m·K)}$$

λ_m Difference = 0.0221 − 0.0192 = 0.0029 W/(m·K) or 15.1%

11.7 THERMAL CONDUCTIVITIES OF GAS MIXTURES AT HIGH PRESSURES

There are few experimental data for gas mixtures at high pressures and, even here, most studies are limited to simple gases and light hydrocarbons. The nitrogen-carbon dioxide system was studied by Keyes (1951), and Comings and his colleagues reported on ethylene mixtures with nitrogen and carbon dioxide (Junk and Comings, 1953), rare gases (Peterson et al., 1971), and binaries containing carbon dioxide, nitrogen, and ethane (Gilmore and Comings, 1966). Rosenbaum and Thodos (1967, 1969) investigated methane-carbon dioxide and methane-carbon tetrafluoride binaries. Binaries containing methane, ethane, nitrogen, and carbon dioxide were also reported by Christensen and Fredenslund (1979), and data for systems containing nitrogen, oxygen, argon, methane, ethylene, and carbon dioxide were published by Zheng et al. (1984) and Yorizane et al. (1983). Data for selected systems are summarized in Stephan and Hildwein (1987), Stephan and Heckenberger (1988), and Sutton (1976).

We present below three estimation methods. All are modifications of procedures developed earlier for low- and high-pressure pure-gas thermal conductivities. For the extension of Eqs. (11.5-26) and (11.5-27) to mixtures, see Assael et al. (1992a, 1996).

11.7.1 Stiel and Thodos Modification

Equations (11.5-2) to (11.5-4) were suggested as a way to estimate the high-pressure thermal conductivity of a pure gas. This procedure may be adapted for mixtures if mixing and combining rules are available to determine T_{c_m}, P_{c_m}, V_{c_m}, Z_{c_m}, and M_m. Yorizane et al. (1983a) have studied this approach and recommend the following:

$$T_{c_m} = \frac{1}{V_{c_m}} \sum_i \sum_j y_i y_j V_{c_{ij}} T_{c_{ij}} \tag{11.7-1}$$

$$V_{c_m} = \sum_i \sum_j y_i y_j V_{c_{ij}} V_{c_{ij}} \tag{11.7-2}$$

$$\omega_m = \sum_i y_i \omega_i \tag{11.7-3}$$

$$Z_{c_m} = 0.291 - 0.08\omega_m \tag{11.7-4}$$

$$P_{c_m} = \frac{Z_{c_m} R T_{c_m}}{V_{c_m}} \tag{11.7-5}$$

$$M_m = \sum_i y_i M_i \tag{11.7-6}$$

$$T_{c_{ii}} = T_{c_i} \quad (11.7\text{-}7)$$

$$T_{c_{ij}} = \left(T_{c_i} T_{c_j}\right)^{\frac{1}{2}} \quad (11.7\text{-}8)$$

$$V_{c_{ii}} = V_{c_i} \quad (11.7\text{-}9)$$

$$V_{c_{ij}} = \frac{\left[\left(V_{c_i}\right)^{\frac{1}{3}} + \left(V_{c_j}\right)^{\frac{1}{3}}\right]^3}{8} \quad (11.7\text{-}10)$$

Using these simple rules, they found they could correlate their high-pressure thermal conductivity data for CO_2—CH_4 and CO_2—Ar systems quite well. In Fig. 11.7, we show a plot of λ_m for the CO_2—Ar system at 298 K. This case is interesting because the temperature is slightly below the critical temperature of CO_2 and, at high pressure, λ for carbon dioxide increases more rapidly than that for argon. The net result is that the λ_m composition curves are quite nonlinear. Still, the Stiel-Thodos method, with Eqs. (11.7-1) through (11.7-10), appears to give a quite satisfactory fit to the data. We illustrate the approach in Example 11.5.

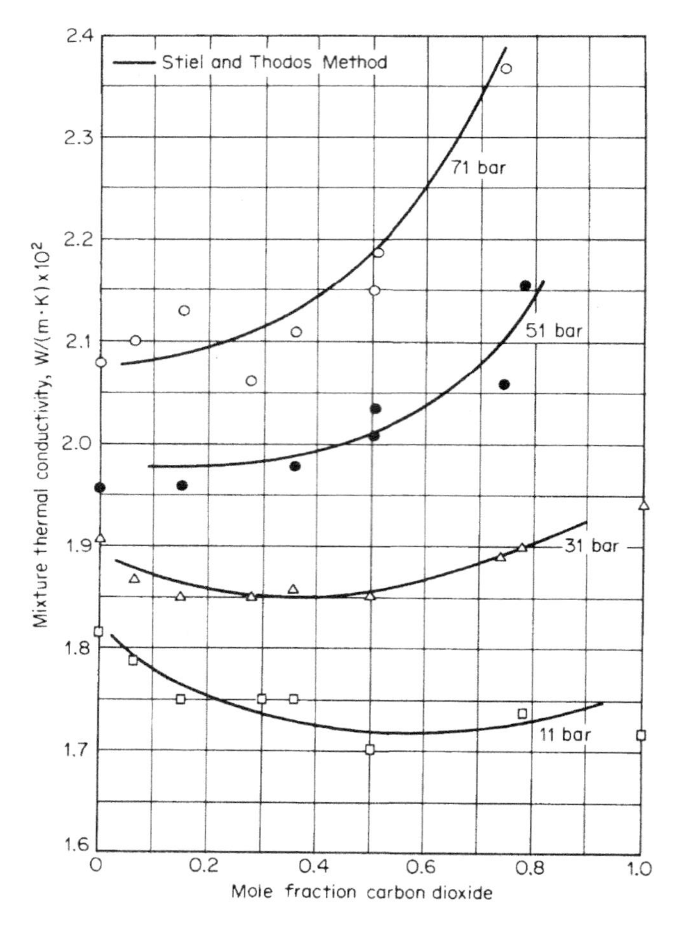

FIGURE 11.7
High-pressure thermal conductivities of the argon–carbon dioxide system. (Data from Yorizane et al., 1983a)

Example 11.5 Estimate the thermal conductivity of a methane (1)-carbon dioxide (2) mixture containing 75.5 mole% methane at 370.8 K and 174.8 bar. Rosenbaum and Thodos (1969) show an experimental value of 5.08×10^{-2} W/(m·K); these same investigators report experimental values for the mixture of $V = 159$ cm³/mol and, at 1 bar, $\lambda_m^\circ = 3.77 \times 10^{-2}$ W/(m·K).

Solution. The pure-component constants for methane and carbon dioxide given below will be used in this example and in Example 11.6 and Example 11.7.

	CH_4 (1)	CO_2 (2)
T_c, K	190.56	304.12
P_c, bar	45.99	73.74
V_c, cm³/mol	98.6	94.07
Z_c	0.286	0.274
ω	0.011	0.225
μ, debye	0	0
C_p(370.8 K), J/(mol·K)	39.09	40.04
C_v(370.8 K), J/(mol·K)	30.78	31.73
M, g/mol	16.043	44.010
M', kg/mol	0.01604	0.04401

With Eqs. (11.7-1) through (11.7-10),

$$T_{c_{12}} = [(190.56)(304.12)]^{1/2} = 240.73 \text{ K}$$

$$V_{c_{12}} = (1/8)[(98.6)^{1/3} + (94.07)^{1/3}]^3 = 96.3 \text{ cm}^3/\text{mol}$$

$$V_{c_m} = (0.755)^2(98.6) + (0.245)^2(94.07) + (2)(0.755)(0.245)(96.3) = 97.9 \text{ cm}^3/\text{mol}$$

$$T_{c_m} = [(0.755)^2(190.56)(98.6) + (0.245)^2(304.12)(94.07)]$$
$$+ (2)(0.755)(0.245)(240.73)(96.3)]/97.5 = 215.4 \text{ K}$$

$$\omega_m = (0.755)(0.011) + (0.245)(0.225) = 0.063$$

$$Z_{c_m} = 0.291 - (0.08)(0.063) = 0.286$$

$$P_{c_m} = (0.286)(8.314)(215.4)/97.5 \times 10^{-6} = 5.24 \times 10^6 \text{ Pa} = 52.4 \text{ bar}$$

$$M_m = (0.755)(16.04) + (0.245)(44.01) = 22.9 \text{ g/mol}$$

With Eq. (11.5-7),

$$\Gamma = 210[(215.4)(22.9)/(52.4)^4]^{1/6} = 176$$

and

$$\rho_{r_m} = 97.5/159 = 0.613$$

Using Eq. (11.5-3),

$$\lambda_m - \lambda_m^\circ = (0.0114)\{\exp[(0.67)(0.613)] - 1.069\}[(176)(0.285)^5]^{-1}$$

$$= 0.0151 \text{ W/(m·K)}$$

$$\lambda_m = 0.0151 + 0.0377 = 0.0528 \text{ W/(m·K)}$$

$$\lambda_m \text{ Difference} = 0.0528 - 0.0508 = 0.002 \text{ W/(m·K) or } 3.9\%$$

11.7.2 Chung et al. Method (1984, 1988)

To apply this method to estimate the thermal conductivities of high-pressure gas mixtures, one must combine the high-pressure pure component relations with the mixing rules given in Secs. 11.6 and 10.5. To be specific, Eq. (11.5-5) is employed with all variables subscripted with m to denote them as mixture properties. Example 11.6 illustrates the procedure in detail.

Example 11.6 Repeat Example 11.5 by using the Chung et al. approach.

Solution. For the methane (1)-carbon dioxide (2) system, the required pure-component properties were given in Example 11.5.

To use Eq. (11.5-5), let us first estimate η_m° with the procedures in Chap. 10. From Eqs. (10.4-7) and (10.4-8),

$$\left(\frac{\varepsilon}{k}\right)_1 = 190.56/1.2593 = 151.3 \text{ K}$$

$$\left(\frac{\varepsilon}{k}\right)_2 = 304.13/1.2593 = 241.5 \text{ K}$$

Interaction values are then found from Eqs. (10.5-33), (10.5-35), (10.5-37), and (10.5-40).

$$\sigma_{12} = [(3.737)(3.769)]^{1/2} = 3.711 \text{ Å}$$

$$\left(\frac{\varepsilon}{k}\right)_{12} = [(151.3)(241.5)]^{1/2} = 191.1 \text{ K}$$

$$\omega_{12} = (0.011 + 0.225)/2 = 0.118$$

$$M_{12} = (2)(16.04)(44.01)/(16.04 + 44.01) = 23.51$$

With $y_1 = 0.755$ and $y_2 = 0.245$, using Eqs. (10.5-25) to (10.5-29) and Eq. (10.5-41),

$$\sigma_m = 3.723 \text{ Å} \qquad \left(\frac{\varepsilon}{k}\right)_m = 171.0 \text{ K} \qquad T_m^* = 2.168$$

$$M_m = 20.91 \qquad \omega_m = 0.063 \qquad F_{c_m} = 0.983$$

So, with Eq. (10.4-3), $\Omega_v = 1.144$. Then, with Eq. (10.5-24),

$$\eta_m^\circ = (26.69)(0.983)[(20.91)(370.8)]^{1/2}/[(3.723)^2(1.144)] = 145.7 \ \mu\text{P} = 145.7 \times 10^{-7} \text{ Pa·s}$$

With Eqs. (10.5-43) and (10.5-44),

$$V_{c_m} = (3.723/0.809)^3 = 97.46 \text{ cm}^3/\text{mol}$$

$$T_{c_m} = (1.2593)(171.0) = 215.3 \text{ K}$$

$$T_{r_m} = 370.8/215.3 = 1.722$$

C_{v_m} is found with Eq. (11.6-6) to be 31.47 J/(mol·K), and Ψ_m is determined with Eqs. (11.3-12) and (11.3-13) as

$$\alpha_m = 31.47/8.314 - 1.5 = 2.285$$

$$\beta_m = 0.7862 - (0.7109)(0.063) + (1.3168)(0.063)^2 = 0.747$$

$$Z_m = 2.0 + (10.5)(1.722)^2 = 33.14$$

$$\Psi_m = 1.748$$

and

$$M'_m = M_m/1000 = 0.02091 \text{ kg/mol}$$

With Eqs. (11.5-9) to (11.5-13) and Table 11.2,

$$B_1 = 2.464 \qquad B_2 = -0.6043 \qquad B_3 = 6.965 \qquad B_4 = 13.98$$

$$B_5 = 0.8444 \qquad B_6 = -5.057 \qquad B_7 = 99.16$$

$$y_m = 0.1022 \qquad G_{1_m} = 1.311 \qquad G_{2_m} = 0.6519$$

$$q_m = (3.586 \times 10^{-3})(215.3/0.02091)^{1/2}/(97.46)^{2/3} = 0.01718$$

Finally, substituting these values into Eq. (11.5-8),

$$\lambda_m = [(31.2)(145.7 \times 10^{-7})(1.748)/0.02091][(0.6519)^{-1} - (5.057)(0.1022)]$$
$$+ (0.01718)(99.16)(0.1022)^2(1.722)^{1/2}(0.6519) = 0.0539 \text{ W/(m·K)}$$

$$\lambda_m \text{ Difference} = 0.0539 - 0.0508 = 0.0031 \text{ W/(m·K) or } 6.1\%$$

If the pressure were reduced to 1 bar, y_m would become quite small and G_2 would be essentially unity. In that case, $\lambda^\circ_m = 0.0380$ W/(m·K), very close to the experimental value of 0.0377 W/(m·K). The error found in Example 11.6 is typical for this method when simple gas mixtures are treated. As noted before, the Chung et al. method should not be used for polar gases. Its accuracy for nonpolar gas mixtures containing other than simple gases or light hydrocarbons has not been tested.

11.7.3 TRAPP Method (Huber, 1996)

For gas mixtures at high pressure, the thermal conductivity is determined by a combination of the techniques introduced for high-pressure gases (Sec. 11.5) with appropriate mixing rules. The thermal conductivity of the mixture is given by

$$\lambda_m = \lambda^\circ_m + F_{\lambda_m} X_{\lambda_m} [\lambda_R - \lambda^\circ_R] \tag{11.7-11}$$

where λ_m° is the mixture value at low pressure and may be determined by methods described earlier in the chapter. The quantity $\lambda_R - \lambda_R^\circ$ that appears in Eq. (11.7-11) is for the reference fluid propane and is evaluated with Eq. (11.5-17) at T_{λ_m} and ρ_{λ_m}. The following mixing rules are used to determine F_{λ_m}, T_{λ_m}, and ρ_{λ_m}.

$$h_m = \sum_i \sum_j y_i y_j h_{ij} \tag{11.7-12}$$

$$f_m h_m = \sum_i \sum_j y_i y_j f_{ij} h_{ij} \tag{11.7-13}$$

$$h_{ij} = \frac{\left[(h_i)^{\frac{1}{3}} + (h_j)^{\frac{1}{3}}\right]^3}{8} \tag{11.7-14}$$

$$f_{ij} = (f_i f_j)^{\frac{1}{2}} \tag{11.7-15}$$

f_i and h_i are determined by the method described in Sec. 11.5. T_{λ_m} and ρ_{λ_m} are calculated by equations similar to Eqs. (11.5-18) and (11.5-19):

$$T_{\lambda_m} = \frac{T}{f_m} \tag{11.7-16}$$

$$\rho_{\lambda_m} = \rho h_m = \frac{h_m}{V} \tag{11.7-17}$$

Finally,

$$F_{\lambda_m} = (44.094)^{\frac{1}{2}}(h_m)^{-2} \sum_i \sum_j y_i y_j \left(\frac{f_{ij}}{M_{ij}}\right)^{\frac{1}{2}} (h_{ij})^{\frac{4}{3}} \tag{11.7-18}$$

$$X_{\lambda_m} = \left[1 + \frac{2.1866(\omega_m - \omega_R)}{1 - 0.505(\omega_m - \omega_R)}\right]^{\frac{1}{2}} \tag{11.7-19}$$

where

$$M_{ij} = \left(\frac{1}{2M_i} + \frac{1}{2M_j}\right)^{-1} \tag{11.7-20}$$

$$\omega_m = \sum_i y_i \omega_i \tag{11.7-21}$$

Huber (1996) tested the TRAPP method on several binary hydrocarbon mixtures and one ternary hydrocarbon mixture over a wide range of densities and reports an average absolute error of about 5%, although, in some cases, significantly larger deviations were found. The technique is illustrated in Example 11.7.

Example 11.7 Repeat Example 11.6 by using the TRAPP procedure.

Solution. The pure-component properties for both components of the methane (1), carbon dioxide (2) binary are given at the beginning of the solution of Example 11.5. Pure-component values of h_i and f_i are determined from Eqs. (11.5-24) and (11.5-25) using $T_{r_1} = 370.8/190.56 = 1.946$ and $T_{r_2} = 370.8/304.12 = 1.219$.

$$f_1 = (190.564/369.83)[1 + (0.011 - 0.152)(0.05203) - 0.7498 \ln(1.946)] = 0.5478$$

$$h_1 = (98.60/200.0)(0.276/0.286)[1 - (0.011 - 0.152)(0.1436 - 0.2822 \ln(1.946))]$$

$$= 0.4728$$

Similarly, f_2 and h_2 are 0.8165 and 0.4708, respectively. Mixture values along with T_{λ_m} and ρ_{λ_m} are determined with Eqs. (11.7-12) to (11.7-21).

$$f_{12} = [(0.5478)(0.8165)]^{1/2} = 0.6688$$

$$h_{12} = (1/8)[(0.4728)^{1/3} + (0.4708)^{1/3}]^3 = 0.4718$$

$$h_m = (0.755)^2(0.4728) + 2(0.755)(0.245)(0.4718) + (0.245)^2(0.4708) = 0.4723$$

$$f_m h_m = (0.755)^2(0.5478)(0.4728) + 2(0.755)(0.245)(0.6688)(0.4718)$$
$$+ (0.245)^2(0.8165)(0.4708) = 0.2874$$

$$f_m = 0.2874/0.4723 = 0.6086$$

$$T_{\lambda_m} = 370.8/0.6086 = 609.3$$

$$\rho_{\lambda_m} = 0.4723/159 = 0.002970$$

$$M_{12} = [(0.5)(1/16.043) + (0.5)(1/44.010)]^{-1} = 23.5143$$

$$F_{\lambda_m} = [(44.094)^{1/2}/(0.4723)^2][(0.755)^2(0.5478/16.043)^{1/2}(0.4728)^{4/3}$$
$$+ 2(0.755)(0.245)(0.6688/23.5143)^{1/2}(0.4718)^{4/3}$$
$$+ (0.245)^2(0.8165/44.010)^{1/2}(0.4708)^{4/3}] = 1.926$$

Equations (11.7-21) and (11.7-19) give

$$\omega_m = (0.755)(0.011) + (0.245)(0.2276) = 0.063$$

$$X_{\lambda_m} = \{1 + [2.1866(0.063 - 0.152)]/[1 - 0.505(0.063 - 0.152)]\}^{1/2} = 0.9026$$

With $T_{r_R} = 609.3/369.83 = 1.648$ and $\rho_{r_R} = (0.002970)(200) = 0.5941$, Eq. (11.7-17) gives

$$\lambda_R - \lambda_R^\circ = 9.898 \text{ mW/(m·K)} = 0.009898 \text{ W/(m·K)}$$

Using Eq. (11.7-11)

$$\lambda_m = 0.0377 + (1.926)(0.9026)(0.009898) = 0.0549 \text{ W/(m·K)}$$

$$\lambda_m \text{ Difference} = 0.0549 - 0.0508 = 0.0041 \text{ W/(m·K) or } 8.1\%$$

11.7.4 Discussion

Of the three methods presented to estimate the thermal conductivity of high-pressure (or high-density) gas mixtures, all have been tested on available data and shown to be reasonably reliable with errors averaging about 5 to 7%. However, the database used for testing is small and primarily comprises permanent gases and light hydrocarbons. None are believed to be applicable to polar fluid mixtures. The Chung et al. and TRAPP methods have also been tested on more complex (hydrocarbon) systems at densities which are in the liquid range with quite encouraging results.

For simple hand calculation of one or a few values of λ_m, the Stiel and Thodos and TRAPP methods are certainly the simplest. If many values are to be determined, the somewhat more complex, but probably more accurate, method of Chung et al. or the TRAPP method should be programmed and used.

11.8 THERMAL CONDUCTIVITIES OF LIQUIDS

For many simple organic liquids, the thermal conductivity is between 10 and 100 times larger than that of the low-pressure gases at the same temperature. There is little effect of pressure, and raising the temperature usually decreases the thermal conductivity. These characteristics are similar to those noted for liquid viscosity, although the temperature dependence of the latter is pronounced and nearly exponential, whereas that for thermal conductivity is weak and nearly linear.

Values of λ_L for most common organic liquids range between 0.09 and 0.19 W/(m·K) at temperatures below the normal boiling point, but water, ammonia, and other highly polar molecules have values several times as large. Also, in many cases the dimensionless ratio $\frac{M\lambda}{R\eta}$ is nearly constant (for nonpolar liquids) between values of 2 and 3, so that viscous liquids have a correspondingly larger thermal conductivity. Liquid metals and some organosilicon compounds have large values of λ_L; the former often are 100 times larger than those for normal organic liquids. The solid thermal conductivity at the melting point is approximately 20 to 40% larger than that of the liquid.

The difference between transport property values in the gas phase and values in the liquid phase indicates a distinct change in mechanism of energy (or momentum or mass) transfer, i.e.,

$$\frac{\lambda_L}{\lambda_G} \cong 10 - 100 \qquad \frac{\eta_L}{\eta_G} \cong 10 - 100 \qquad \frac{D_L}{D_G} \cong 10^{-4}$$

In the gas phase, the molecules are relatively free to move about and transfer momentum and energy by a collisional mechanism. The intermolecular force fields, though not insignificant, do not drastically affect the value of λ, η, or D. That is, the attractive intermolecular forces are reflected solely in the collision integral terms Ω_v and Ω_D, which are *really* ratios of collision integrals for a real force field and an artificial case in which the molecules are rigid, nonattracting spheres. The variation of Ω_v, or Ω_D from unity then yields a rough quantitative measure of the importance of attractive intermolecular forces in affecting gas phase transport coefficients. Reference to Eq. (10.4-3) (for Ω_v) or Eq. (12.4-6) (for Ω_D) shows that Ω values are often near unity. One then concludes that a rigid, nonattracting spherical molecular model yields a low-pressure transport coefficient λ, η, or D not greatly different from that computed when intermolecular forces are included.

In the liquid, however, the close proximity of molecules to one another emphasizes strongly the intermolecular forces of attraction. There is little wandering of the individual molecules, as evidenced by the low value of liquid diffusion coefficients, and often a liquid is modeled as a lattice with each molecule caged by its nearest neighbors. Energy and momentum are primarily exchanged by oscillations of molecules in the shared force fields surrounding each molecule. McLaughlin (1964) discusses in more detail the differences in transport mechanisms between a dense gas or liquid and a low-pressure gas.

To date, theory has not been successful in formulating useful and accurate expressions to calculate liquid thermal conductivities, although Eqs. (11.5-26) and (11.5-27) are theory-based and have been used for a number of liquids. Generally, approximate techniques must be employed for engineering applications.

Only relatively simple organic liquids are considered in the sections to follow. Ho et al. (1972) have presented a comprehensive review covering the thermal conductivity of the elements, and Ewing et al. (1957) and Gambill (1959) consider, respectively, molten metals and molten salt mixtures. Cryogenic liquids are discussed by Preston et al. (1967) and Mo and Gubbins (1974).

Liquid thermal conductivity data have been compiled in Jamieson et al. (1975), Le Neindre (1987), Liley et al. (1988), Nieto de Castro et al. (1986), Stephan and Hildwein (1987), Vargaftik et al. (1994), the DIPPR 801 database (Wilding et al., 2017; DIPPR 2021), and the Thermodynamics Research Center (Diky et al., 2021). Data for mixtures and electrolyte solutions are also given in Jamieson et al. (1975) and Zaytsev and Aseyev (1992). Other sources include Jamieson and Tudhope (1964), Tsederberg (1965), and Vargaftik et al. (1996). New liquid thermal conductivity data were reported for alcohols (Cai et al., 1993; Jamieson, 1979; Jamieson and Cartwright, 1980; Ogiwara et al., 1982), alkyl amines (Jamieson, 1979; Jamieson and Cartwright, 1980), hydrocarbons (Nieto de Castro et al., 1981; Ogiwara, et al., 1980), nitroalkanes (Jamieson and Cartwright, 1981), ethanolamines (DiGuilio et al, 1992), glycols (DiGuilio and Teja, 1990), and other organic compounds (Cai et al. 1993; Qun-Fung et al., 1997; Venart and Prasad, 1980).

Constants that may be used to calculate thermal conductivities for pure liquids at different temperatures are tabulated in Miller et al. (1976a), Yaws (1995, 1995a), the DIPPR 801 database (Wilding et al., 2017; DIPPR 2021), and ThermoData Engine from the Thermodynamics Research Center (Diky et al., 2021). The constants in the last three sources are based on critically evaluated data while those in the other references are not. Values are tabulated at various temperatures for common fluids in Speight (2016), Rumble (2021), and Green and Southard (2019).

11.9 ESTIMATION OF THE THERMAL CONDUCTIVITIES OF PURE LIQUIDS

All estimation techniques for the thermal conductivity of pure liquids are empirical. As noted earlier, however, below the normal boiling point, the thermal conductivities of most organic, nonpolar liquids lie between 0.09 and 0.19 W/(m·K). With this fact in mind, it is not too difficult to devise various schemes for estimating λ_L within this limited domain. Four estimation methods that were tested are described below. Others that were considered are noted briefly at the end of the section.

11.9.1 Missenard Method

One method to predict thermal conductivity as a function of temperature is based on an approach suggested by Missenard (1965, 1965a). First, the liquid thermal conductivity at 0 °C is obtained from (Missenard 1965a)

$$\lambda_{L_0} = \frac{84 \times 10^{-6} (T_b \rho_0)^{\frac{1}{2}} C_{p_0}}{M^{\frac{1}{2}} N^{\frac{1}{4}}} \tag{11.9-1}$$

where λ_{L_0} is the liquid thermal conductivity at 0 °C (273.15 K) in cal/(cm·s·K), T_b is the normal (101325 Pa) boiling point in K, ρ_0 is the liquid density at 0 °C in mol/cm^3, C_{p_0} is the isobaric liquid heat capacity at 0 °C in cal/(mol·K), M is the molecular weight in g/mol, and N is the number of atoms in the molecule. Values of λ_L at any temperature T may then be obtained from

$$\lambda_L = \lambda_{L_0} \frac{3 + 20(1 - T_r)^{\frac{2}{3}}}{3 + 20\left(1 - \frac{273}{T_c}\right)^{\frac{2}{3}}} \tag{11.9-2}$$

Here, T_c is the critical temperature in K and $T_r = \frac{T}{T_c}$. The approach is demonstrated in Example 11.9.

11.9.2 Latini et al. Method

In an examination of the thermal conductivity of many diverse liquids, Latini and his coworkers (Baroncini et al., 1980, 1981, 1981a, 1983, 1983a, 1984; Latini and Pacetti, 1977) suggest a correlation of the form

$$\lambda_L = \frac{A_\lambda (1 - T_r)^{0.38}}{T_r^{\frac{1}{6}}} \tag{11.9-3}$$

$$A_\lambda = \frac{A^* T_b^\alpha}{M^\beta T_c^\gamma} \tag{11.9-4}$$

where λ_L is the thermal conductivity of the liquid in W/(m·K), $T_r = \frac{T}{T_c}$, T_c is the critical temperature in K, T_b is the normal (at 1 atm) boiling temperature in K, M is the molecular weight in g/mol, and the parameters A^*, α, β, and γ are shown in Table 11.3 for various classes of organic compounds. Specific values of A_λ are given for many compounds by Baroncini et al. (1981). Equation (11.9-4) is only an approximation of the regressed value of A_λ, and this simplification introduces significant error unless $50 < M < 250$. More recently Latini et al. (1989, 1996) have suggested a different form than Eq. (11.9-3) for alkanes, aromatics, and refrigerants.

Many types of compounds (e.g., nitrogen or sulfur-containing materials and aldehydes) cannot be treated with the method, and problems arise if the compound may be fitted into two families. *m*-cresol is an example. It could be considered an aromatic compound or an alcohol. The method is demonstrated in Example 11.9.

11.9.3 Sastri and Rao Method

Sastri and Rao (1999) recommend

$$\lambda_L = \lambda_b a_{SR}^{1-\left(\frac{1-T_r}{1-T_{b_r}}\right)^{n_{SR}}} \tag{11.9-5}$$

where $T_r = \frac{T}{T_c}$, $T_{b_r} = \frac{T_b}{T_c}$, T_c is the critical temperature, T_b is the normal (at 1 atm) boiling temperature, and the other parameters are as follows. For alcohols and phenols, $a_{SR} = 0.856$ and $n_{SR} = 1.23$. For other compounds, $a_{SR} = 0.16$ and $n_{SR} = 0.2$. The thermal conductivity at the normal boiling point, λ_b, is determined with the group contribution values according to

$$\lambda_b = \sum_i N_i \left(\Delta_{\lambda_b}\right)_i + \sum_j N_j^{corr} \left(\Delta_{\lambda_b}^{corr}\right)_j \tag{11.9-6}$$

TABLE 11.3 Latini et al. Correlation Parameters for Eq. (11.9-4).

Family	A^*	α	β	γ
Saturated hydrocarbons	0.00350	1.2	0.5	0.167
Olefins	0.0361	1.2	1.0	0.167
Cycloparaffins	0.0310	1.2	1.0	0.167
Aromatics	0.0346	1.2	1.0	0.167
Alcohols	0.00339	1.2	0.5	0.167
Acids (organic)	0.00319	1.2	0.5	0.167
Ketones	0.00383	1.2	0.5	0.167
Esters	0.0415	1.2	1.0	0.167
Ethers	0.0385	1.2	1.0	0.167
Refrigerants				
R20, R21, R22, R23	0.562	0.0	0.5	−0.167
Others	0.494	0.0	0.5	−0.167

where N_i is the number of times group i is found in the compound, $(\Delta_{\lambda_b})_i$ is the contribution of group i as found in Table 11.4, N_j^{corr} is the number of times correction j is found in the compound, and $(\Delta_{\lambda_b}^{corr})_j$ is the contribution from correction j as found in Table 11.5. Examples of calculating λ_b using Eq. (11.9-6) are found in Example 11.8. The complete method is demonstrated in Example 11.9.

TABLE 11.4 Group contributions, in W/(m·K), to calculate λ_b in the Sastri and Rao (1998) method for λ_L. Use in Eq. (11.9-6).

Group	Δ_{λ_b}
Hydrocarbon Groups	
$—CH_3$	0.0545
$—CH_2—$	−0.0008
>CH—	−0.0600
>C<	−0.1230
$=CH_2$	0.0545
=CH—	0.0020
=C<	−0.0630
=C=	0.1200
Ring[1]	0.1130
Nonhydrocarbon Groups	
—O—	0.0100
—OH[2]	0.0830
—OH[3]	0.0680
>CO (ketone)	0.0175
>CHO (aldehyde)	0.0730
—COO— (ester)	0.0070
—COOH (acid)	0.0650
$—NH_2$	0.0880
—NH—	0.0065
—NH— (ring)	0.0450
>N—	−0.0605
N (ring)	0.0135
—CN	0.0645
$—NO_2$	0.0700
—S—	0.0100
—F[4]	0.0568
—F[5]	0.0510
—Cl	0.0550
—Br	0.0415
—I	0.0245
—H[6]	0.0675
3-member ring	0.1500
Ring (other)[7]	0.1100

[1]In polycyclic compounds, all rings are treated as separate rings.
[2]In aliphatic primary alcohols and phenols with no branch chains.
[3]In all alcohols except as described in 2 above.
[4]In perfluoro carbons.
[5]In all cases except as described in 4 above.
[6]This contribution is used for methane, formic acid, and formates.
[7]In polycyclic nonhydrocarbon compounds, all rings are considered as nonhydrocarbon rings.

TABLE 11.5 Corrections, in W/(m·K), to calculate λ_b in the Sastri and Rao (1998) method for λ_L. Use in Eq. (11.9-6).

Description of Correction	$\Delta_{\lambda_b}^{corr}$
Hydrocarbons when number of carbon atoms, C, is less than 5	0.0150(5-C)
Compounds with single CH_3 group, no other hydrocarbon groups, and nonhydrocarbon groups other than COOH, Br, or I (e.g., CH_3Cl)[1]	0.0600
Compounds with 2 hydrocarbon groups (2 CH_3, CH_3CH_2, or $CH_2{=\!=}CH$) and nonhydrocarbon groups other than COOH, Br, or I[1]	0.0285
Unsaturated aliphatic compounds with 3 hydrocarbon groups (e.g., allylamine or vinyl acetate)	0.0285
Special groups $Cl(CH_2)_nCl$	0.0350
Compounds with more than one nonhydrocarbon group and at least one hydrocarbon group (e.g., propyl formate or furfural)[1]	0.0095
Compounds with nonhydrocarbon groups only (e.g., formic acid)	0.1165

[1]Nonhydrocarbons with more than one type of nonhydrocarbon group and either (i) one or two methyl groups, or (ii) one methyl group require both correction factors (e.g., the correction for methylformate, with one methyl group and two nonhydrocarbon groups is 0.0600 + 0.0095 = 0.0695).

Example 11.8 Determine the value of λ_b from Eq. (11.9-6) to be used in Eq. (11.9-5) for tetrahydrofuran, carbon tetrachloride, phenol, and acetonitrile.

Solution.

(i) As per Table 11.4, tetrahydrofuran is a nonhydrocarbon with four —CH_2— groups, one —O—, and one nonhydrocarbon ring correction. Thus, $\lambda_b = 4(-0.0008) + 0.0100 + 0.1100 = 0.1168$ W/(m·K).

(ii) As per Table 11.4, carbon tetrachloride has one >C< and four —Cl groups. Thus, $\lambda_b = -0.1230 + 4(0.0550) = 0.097$ W/(m·K).

(iii) As per Table 11.4, phenol has five =CH—, one =C<, one —OH, and one nonhydrocarbon ring correction. Thus, $\lambda_L = 5(0.0020) - 0.0630 + 0.0830 + 0.1100 = 0.14$ W/(m·K).

(iv) As per Table 11.4, acetonitrile has one —CH_3 and one —CN. It also has one correction for a compound with one hydrocarbon group that is —CH_3 and one nonhydrocarbon group other than COOH, Br, or I as listed in Table 11.5. Thus, $\lambda_b = 0.0545 + 0.0645 + 0.0600 = 0.179$ W/(m·K).

Example 11.9 Estimate the thermal conductivity of carbon tetrachloride at 293 K. At this temperature, Jamieson and Tudhope (1964) list 11 values. Six are given a ranking of A and are considered reliable. They range from 0.102 to 0.107 W/(m·K). Most, however, are close to 0.103 W/(m·K).

Solution. From Appendix A, $T_c = 556.35$ K, $T_b = 349.79$ K, and $M = 153.8227$ g/mol, and $\rho(273.15 \text{ K}) = \rho_0 = 0.01059$ mol/cm³. The isobaric liquid heat capacity from Touloukian and Ho (1970) at 0 °C is $C_p(273.15 \text{ K}) = C_{p_0} = 31.21$ cal/(mol·K).

MISSENARD The number of atoms in CCl_4 is $N = 5$. With Eq. (11.9-1),

$$\lambda_{L_0} = (84 \times 10^{-6})[(349.79)(0.01059)]^{0.5}(31.21)/[(153.8227)^{0.5}(5)^{0.25}]$$

$$= 2.721 \times 10^{-4} \text{ cal/(cm·s·K)}$$

Then, with $T_r = 293/556.35 = 0.5266$ and Eq. (11.9-2),

$$\lambda_L = 2.721 \times 10^{-4}[3 + 20(1 - 0.5266)^{2/3}]/[3 + 20(1 - 273/556.35)^{2/3}]$$
$$= 2.616 \times 10^{-4} \text{ cal/(cm·s·K)} = 0.109 \text{ W/(m·K)}$$
$$\lambda_L \text{ Difference} = 0.109 - 0.103 = 0.006 \text{ W/(m·K) or } 5.8\%$$

LATINI ET AL. Assuming CCl_4 to be a refrigerant, by Eq. (11.9-4) and Table 11.3,

$$A_\lambda = (0.494)(349.79)^0/[(153.8227)^{0.5}(556.35)^{-0.167}] = 0.1145$$

Then, with $T_r = 293/556.35 = 0.5266$ and Eq. (11.9-4),

$$\lambda_L = (0.1145)(1 - 0.5266)^{0.38}/(0.5266)^{1/6} = 0.0959 \text{ W/(m·K)}$$
$$\lambda_L \text{ Difference} = 0.0959 - 0.103 = -0.0071 \text{ W/(m·K) or } -6.9\%$$

SASTRI AND RAO Using Eq. (11.9-5) and the value of λ_b from Example 11.8,

$$\left[1 - \left(\frac{1 - T_r}{1 - T_{b_r}}\right)\right]^{n_{SR}} = 1 - [(1 - 293/556.35)/(1 - 349.79/556.35)]^{0.2} = -0.0498$$
$$\lambda_L = (0.097)(0.16)^{-0.0498} = 0.0106$$
$$\lambda_L \text{ Difference} = 0.0106 - 0.0103 = 0.0003 \text{ W/(m·K) or } 2.9\%$$

11.9.4 Danner and Daubert Method

Danner and Daubert (1997) report a technique to predict the liquid thermal conductivity for hydrocarbons. The approach is based on ideas reported by Pachaiyappan et al. (1967) and Riedel (1949). The method is stated to be applicable at pressures below 500 psia (34 atm) and $0.25 < T_r < 0.8$. In this technique,

$$\lambda_L = \frac{C_{DD} M^{n_{DD}}}{V}\left[\frac{3 + 20(1 - T_r)^{\frac{2}{3}}}{3 + 20\left(1 - \frac{527.67}{T_c}\right)^{\frac{2}{3}}}\right] \tag{11.9-7}$$

where λ_L is the liquid thermal conductivity in BTU/(hr·ft·°R), M is the molecular weight in lb_m/lbmol (or equivalently g/mol), V is the molar volume of the compound at 68 °F (293.15 K) in ft^3/lbmol, $T_r = \frac{T}{T_c}$ is the reduced temperature, T is temperature in °R, and T_c is the critical temperature in °R. The constants C_{DD} and n_{DD} depend on the type of hydrocarbon. Specifically, $C_{DD} = 1.676 \times 10^{-3}$ and $n_{DD} = 1.001$ *for unbranched, straight chain hydrocarbons*, and $C_{DD} = 4.079 \times 10^{-3}$ and $n_{DD} = 0.7717$ *for branched and cyclic hydrocarbons*. The method is demonstrated in Example 11.10.

Example 11.10 Use the Daubert and Danner (1997) method to estimate the liquid thermal conductivity of *n*-butylbenzene at 1 atm and 333.15 K = 599.67 °R. The experimental value reported by Rastorguev et al. (1970) is 0.118 W/(m·K).

Solution. From Appendix A, $T_c = 660.5$ K = 1188.9 °R and $M = 134.218$ g/mol = 134.218 lb_m/lbmol. Luning Prak et al. (2019) report the liquid density at 68 °F (293.15 K) to be 860.69 kg/m^3 = 0.40033 lbmol/ft^3. With these values, $T_r = 599.67/1188.9 = 0.5044$ and $V = 1/0.40033 = 2.4979$ ft^3/lbmol. *n*-butylbenzene is a cyclic hydrocarbon so $C_{DD} = 4.079 \times 10^{-3}$ and $n_{DD} = 0.7717$. Using Eq. (11.9-7),

$$\lambda_L = [(4.079 \times 10^{-3})(134.218)^{0.7717}/2.4979]\{[3 + 20(1 - 0.5044)^{2/3}]/$$

$$[3 + 20(1 - 527.67/1188.9)^{2/3}]\} = 0.06728 \text{ BTU/(hr·ft·°R)} = 0.116 \text{ W/(m·K)}$$

$$\lambda_L \text{ Difference} = 0.116 - 0.118 = -0.002 \text{ W/(m·K) or } -1.7\%$$

11.9.5 Other Liquid Thermal Conductivity Estimation Techniques

For nonpolar materials, the estimation procedures in Sec. 11.5 may be employed to obtain λ_L when temperatures are well above the normal boiling point and accurate fluid densities are available. In particular, the Chung et al. and TRAPP methods were specifically devised to treat liquid systems at high reduced temperatures as well as high-pressure gases.

Teja and Rice (1981, 1982) have suggested that, in some cases, values of λ_L are available for compounds similar to the one of interest, and these data could be employed in an interpolative scheme as follows. Two liquids, similar chemically and with acentric factors bracketing the liquid of interest, are selected. The liquid thermal conductivities of these reference liquids should be known over the range of reduced temperatures of interest. We denote the properties of one reference fluid by a prime and the other by a double prime. Defining

$$\phi = \frac{V_c^{\frac{2}{3}} M^{\frac{1}{2}}}{T_c^{\frac{1}{2}}} \tag{11.9-8}$$

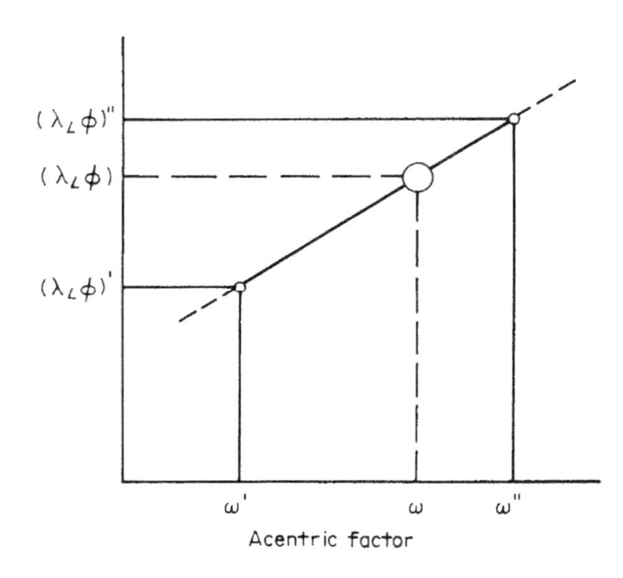

FIGURE 11.8
Schematic representation of the Teja and Rice interpolation procedure. At the circle, $\lambda_L\phi = (\lambda_L\phi)' + \frac{\omega - \omega'}{\omega'' - \omega'}[(\lambda_L\phi)'' - (\lambda_L\phi)']$.

with V_c in cm^3/mol, M in g/mol, and T_c in K, then $\lambda_L \phi$ is found by an interpolation based on the acentric factor ω as shown in Fig. 11.8.

$$\lambda_L \phi = (\lambda_L \phi)' + \frac{\omega - \omega'}{\omega'' - \omega'}[(\lambda_L \phi)'' - (\lambda_L \phi)'] \tag{11.9-9}$$

In Eq. (11.9-9) when one selects λ_L' and λ_L'', they should be evaluated at the same reduced temperature as for the compound of interest. The procedure is illustrated in Example 11.11.

Example 11.11 Using the Teja and Rice scheme, estimate the thermal conductivity of liquid 2-methyl-2-propanol (*tert*-butyl alcohol) at 318 K. Ogiwara et al. (1982) report the experimental value of 2-methyl-2-propanol at 318 K to be 0.128 W/(m·K). They also give the thermal conductivities of *n*-propanol and *n*-hexanol as shown below.

n-propanol: $$\frac{\lambda_L}{\text{W m}^{-1}\text{ K}^{-1}} = 0.202 - 1.76 \times 10^{-4}\frac{T}{\text{K}}$$

n-hexanol: $$\frac{\lambda_L}{\text{W m}^{-1}\text{ K}^{-1}} = 0.190 - 1.36 \times 10^{-4}\frac{T}{\text{K}}$$

Solution. From Appendix A:

Compound	V_c cm^3/mol	T_c K	M g/mol	ω
n-propanol	219	536.8	60.10	0.621
n-hexanol	382	611.3	102.17	0.559
2-methyl-2-propanol	275	506.2	74.12	0.615

Thus, with Eq. (11.9-8),

$$\phi\,(n\text{-propanol}) = (219)^{2/3}(60.10)^{1/2}/(536.8)^{1/2} = 12.16$$

$$\phi\,(n\text{-hexanol}) = (382)^{2/3}(102.17)^{1/2}/(611.3)^{1/2} = 21.52$$

$$\phi\,(\text{2-methyl-2-propanol}) = (275)^{2/3}(74.12)^{1/2}/(506.2)^{1/2} = 16.18$$

At 318 K, for 2-methyl-2-propanol, $T_r = 318/506.2 = 0.628$. At this reduced temperature, the appropriate temperature to use for n-propanol is (0.628)(536.8) = 337.11 K, and for n-hexanol it is (0.628)(611.3) = 383.9 K. With these, using the Ogiwara et al. correlations,

$$\lambda_L\,(n\text{-propanol}) = 0.202 - (1.76 \times 10^{-4})(337.11) = 0.143 \text{ W/(m·K)}$$

$$\lambda_L\,(n\text{-hexanol}) = 0.190 - (1.36 \times 10^{-4})(383.9) = 0.138 \text{ W/(m·K)}$$

Then, using Eq. (11.9-9) with n-propanol as the $'$ reference and n-hexanol as $''$ reference,

$$\lambda_L\,(16.18) = (0.143)(12.16) + [(0.615 - 0.621)/(0.559 - 0.621)][(0.138)(21.52) - (0.143)(12.16)] = 1.86$$

$$\lambda_L = 1.86/16.18 = 0.115 \text{ W/(m·K)}$$

$$\lambda_L \text{ Difference} = 0.115 - 0.128 = -0.013 \text{ W/(m·K) or } -10.3\%$$

Arikol and Gürbüz (1992) developed a correlation for the thermal conductivity of aliphatic and aromatic hydrocarbons and for ethers, aldehydes, ketones, and esters with more than four carbon atoms. Their equation requires the temperature, critical temperature, critical pressure, molecular weight, and normal boiling point and gives reported average deviations of 2%. Ogiwara et al. (1982) suggested a general estimation relation for λ_L for aliphatic alcohols. Jamieson (1979) and Jamieson and Cartwright (1980) proposed a general equation to correlate λ_L over a wide temperature range (see Sec. 11.10), and they discuss how the constants in their equation vary with structure and molecular size. Other methods can be found in Assael et al. (1989), Bleazard and Teja (1996), Dymond and Assael (1996), Klass and Viswanath (1998), Lakshmi and Prasad (1992), Sastri and Rao (1993), and Teja and Tardieu (1988). If critical properties are not available, the methods in Lakshmi and Prasad (1992) or Sastri and Rao (1993) may be used, or the critical properties may be estimated by the methods of Chap. 3. The methods in Klass and Viswanath (1998) and Teja and Tardieu (1988) require an experimental value of the thermal conductivity.

11.9.6 Discussion and Recommendations

A test database was created to evaluate the performance of pure-component liquid thermal conductivity prediction methods. This database consisted of all reliable, experimental liquid thermal conductivity data from the October 2021 release of the DIPPR database and included over 5600 total points (N_{pt}) from 431 unique compounds (N_c). The techniques tested were: Missenard Eq. (11.9-2) (abbreviated MI), Latini et al. Eq. (11.9-3) (abbreviated LA), Sastri and Rao Eq. (11.9-5) (abbreviated SR), and Danner and Daubert Eq. (11.9-7) (abbreviated DD). To identify if a method was better at one class of compounds than another, each chemical in the test database was assigned a *class* according to the scheme described at the end of Sec. 11.3. Results were quantified using the average absolute relative deviation (*AARD*) defined by Eq. (11.3-15).

Two tables of data are presented. The first, Table 11.6 contains the average absolute relative deviation [see Eq. (11.3-15)] of each method according to compound class for all applicable compounds for a method. Also listed are the number of points tested in each class (N_{pt}) and number of compounds (N_c). The second, Table 11.7, is restricted to only hydrocarbons—compounds for which the DD approach is applicable. For polar compounds, the results indicate that MI performs better on average than SR or LA, while for associating compounds LA is slightly better than the other two. The errors for polar compounds with the MI method and those in associating compounds with the LA technique can reasonably be expected to be below 10%. The DD approach works well for hydrocarbons with molecular weights below 213 g/mol (the normal category), and errors are expected to be better than 5%. The superiority of the DD method for this class is maintained even when all methods are evaluated with only hydrocarbons (see Table 11.7). For heavier hydrocarbons, both the DD and the MI techniques work well, with MI having a slightly better AARD (Table 11.7). For nonhydrocarbon compounds in the normal or heavy classes (halogenated hydrocarbons), the LA technique should be used as

TABLE 11.6 Performance of prediction techniques for liquid λ against the test database for all compounds that can be predicted for the listed method.

	Compound Class											
	Normal			Heavy			Polar			Associating		
Method	N_{pt}	N_c	AARD	N_{pt}	N_c	AARD	N_{pt}	N_c	AARD	N_{pt}	N_c	AARD
MI	2942	198	11.8%	249	25	12.8%	868	82	8.1%	1612	126	10.9%
LA	2764	173	7.4%	244	24	11.1%	741	58	13.0%	727	60	9.8%
SR	2882	184	8.7%	243	24	15.9%	827	70	11.0%	1593	122	11.6%
DD	1651	85	4.0%	152	13	5.9%	—	—	—	—	—	—

TABLE 11.7 Performance of prediction techniques for liquid λ against only the hydrocarbons in the test database.

	Compound Class					
	Normal			Heavy		
Method	N_{pt}	N_c	AARD	N_{pt}	N_c	AARD
MI	1651	85	7.7%	152	13	5.4%
LA	1651	85	7.2%	152	13	12.5%
SR	1651	85	6.7%	152	13	8.7%
DD	1651	85	4.0%	152	13	5.9%

its performance between the entire test database (Table 11.6) and hydrocarbons only (Table 11.7) is the most consistent among the available options. Errors for such compounds in the normal class can be expected to be better than 10%, with those in the heavy category being slightly worse. In many instances, the experimental data are not believed to be particularly reliable, and the estimation errors are in the same range as the experimental uncertainty. This is clear from the careful survey of liquid thermal conductivity data provided by the National Engineering Laboratory (Jamieson and Tudhope, 1964; Jamieson, 1979; Jamieson and Cartwright, 1980). None of the procedures predict the large increase in λ_L near the critical point.

11.10 EFFECT OF TEMPERATURE ON THE THERMAL CONDUCTIVITIES OF LIQUIDS

Except for aqueous solutions, water, and some multihydroxy and multiamine molecules, the thermal conductivities of most liquids decrease with temperature. The major effect causing this is believed to be the decrease in liquid density that occurs as temperature increases, which means the molecules are farther apart and less able to transfer energy between neighbors. Below or near the normal boiling point, the decrease is nearly linear and is often represented over small temperature ranges by

$$\lambda_L = A - BT \tag{11.10-1}$$

where A and B are constants and B generally is in the range of 1 to 3×10^{-4} W/(m·K^2). Figure 11.9 shows the temperature effect on λ_L for a few liquids. For wider temperature ranges, the equations in Sec. 11.9, or the following correlation suggested by Riedel (1951), may be used.

$$\lambda_L = B\left[3 + 20(1 - T_r)^{\frac{2}{3}}\right] \tag{11.10-2}$$

Although not suited for water, glycerol, glycols, hydrogen, or helium, Jamieson (1971) indicates that Eq. (11.10-2) represented well the variation of λ_L with temperature for a wide range of compounds. Although, as noted earlier, few data for λ_L exist over the temperature range from near the melting point to near the critical point, for those that are available, Jamieson (1979) has found that neither Eq. (11.10-1) nor Eq. (11.10-2) is suitable, and he recommends

$$\lambda_L = A\left(1 + B_\tau^{\frac{1}{3}} + C\tau^{\frac{2}{3}} + D\tau\right) \tag{11.10-3}$$

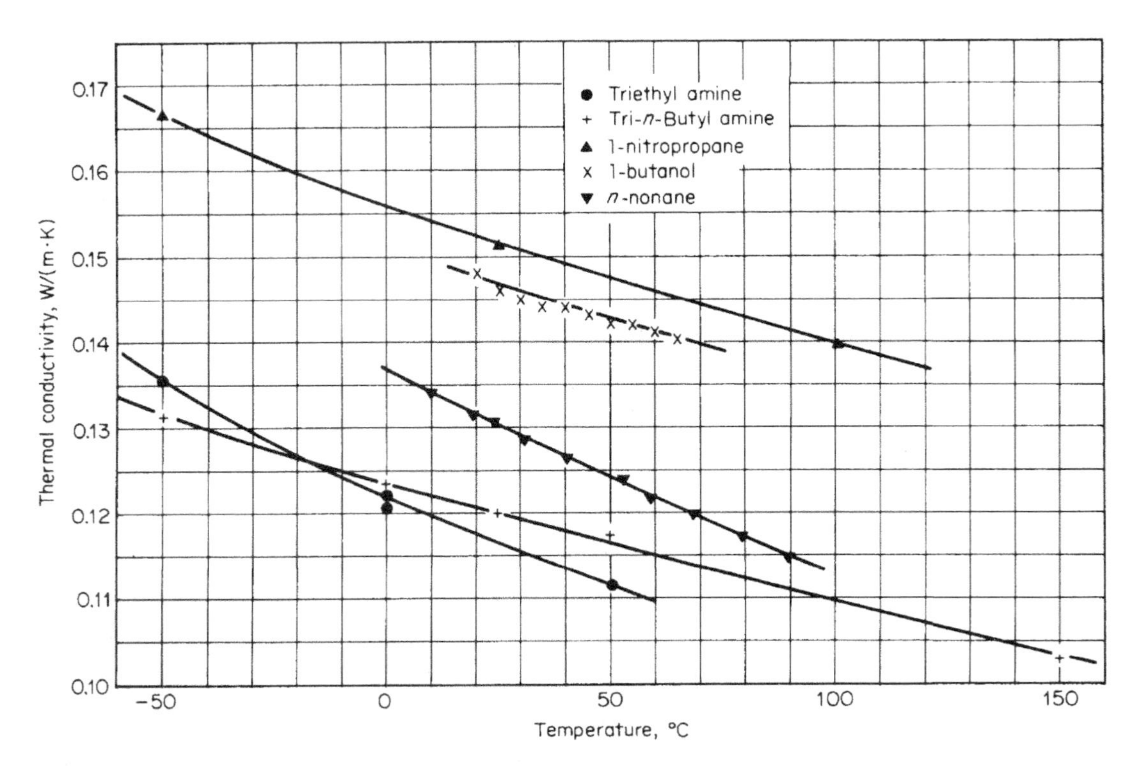

FIGURE 11.9
Thermal conductivity of a few organic liquids as functions of temperature. (Data from Jamieson and Cartwright, 1980, 1981; Nagata, 1973; Ogiwara et al., 1982)

where A, B, C, and D are constants and $\tau = 1 - T_r$. For nonassociating liquids, $C = 1 - 3B$ and $D = 3B$. With these simplifications, Eq. (11.10-3) becomes

$$\lambda_L = A\left[1 + \tau^{\frac{2}{3}} + B\left(\tau^{\frac{1}{3}} - 3\tau^{\frac{2}{3}} + 3\tau\right)\right] \tag{11.10-4}$$

As an example, in Fig. 11.9, if one fits the data for tributyl amine (a polar, but nonassociating, liquid), to Eq. (11.10-4), approximate values of A and B are $A = 0.0590$ W/(m·K) and $B = 0.875$. Using them, one can show by differentiating Eq. (11.10-4) that $\frac{d\lambda_L}{dT}$ decreases with increasing temperature, although, as is obvious from Fig. 11.9, the change in slope is not large in the temperature region shown. For other materials for which data are available over a quite wide temperature range, Eq. (11.10-4) is clearly preferable to Eq. (11.10-1) or (11.10-2) (Jamieson, 1984).

For associated liquids, $C = 1 - 2.6B$ and $D \approx 6.5$ for alcohols and 6.0 for alkyd and dialkyd amines. Correlations for C and D for other types of associated molecules are not available. The constants A and B have been correlated, approximately, with carbon number for several homologous series (Jamieson, 1979; Jamieson and Cartwright, 1980, 1981).

For saturated liquids at high pressure, variations of λ_L with temperature should probably be determined by using the high-pressure correlations in Sec. 11.5.

11.11 EFFECT OF PRESSURE ON THE THERMAL CONDUCTIVITIES OF LIQUIDS

At moderate pressures, up to 50 to 60 bar, the effect of pressure on the thermal conductivity of liquids is usually neglected, except near the critical point, where the liquid behaves more like a dense gas than a liquid (see Sec. 11.5). At lower temperatures, λ_L increases with pressure. Data showing the effect of pressure on several organic liquids are available in Bridgman (1923) and Jamieson et al. (1975).

A convenient way of estimating the effect of pressure on λ_L is by Eq. (11.11-1)

$$\frac{\lambda_2}{\lambda_1} = \frac{L_2}{L_1} \tag{11.11-1}$$

where λ_2 and λ_1 refer to liquid thermal conductivities at T and pressures P_2 and P_1 and L_2 and L_1 are functions of the reduced temperature and pressure, as shown in Fig. 11.10. This correlation was devised by Lenoir (1957). Testing with data for 12 liquids, both polar and nonpolar, showed errors of only 2 to 4%. The use of Eq. (11.11-1) and Fig. 11.10 is illustrated in Example 11.12 with liquid NO_2, a material *not* used in developing the correlation.

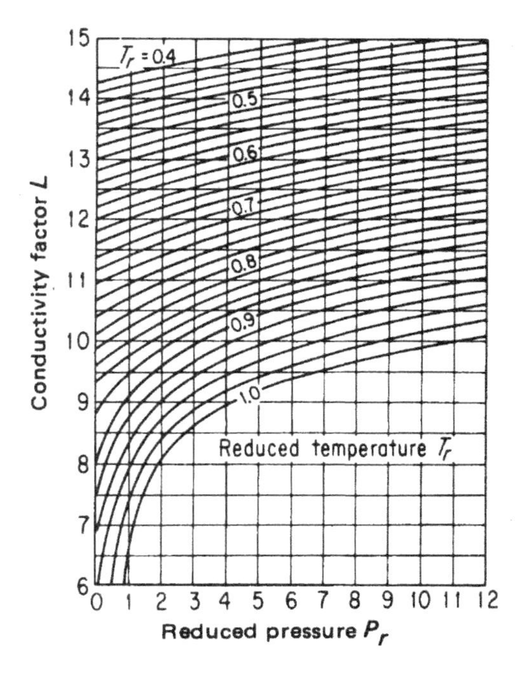

FIGURE 11.10
Effect of pressure on liquid thermal conductivities. (From Le Neindre, 1987)

Example 11.12 Estimate the thermal conductivity of nitrogen dioxide at 311 K and 276 bar. The experimental value quoted is 0.134 W/(m·K) (Richter and Sage, 1957). The experimental value of λ_L for the saturated liquid at 311 K and 2.1 bar is 0.124 W/(m·K) (Richter and Sage, 1957).

Solution. From Daubert et al. (1997), $T_c = 431.35$ K, $P_c = 101.33$ bar; thus $T_r = 311/431.35 = 0.721$, $P_{r_1} = 2.1/101.33 = 0.021$, and $P_{r_2} = 276/101.33 = 2.72$. From Fig. 11.10, $L_2 = 11.75$ and $L_1 = 11.17$. With Eq. (11.11-1),

$$\lambda_L(276 \text{ bar}) = (0.124)(11.75/11.17) = 0.130 \text{ W/(m·K)}$$

$$\lambda_L \text{ Difference} = 0.130 - 0.134 = -0.004 \text{ W/(m·K) or } -3.0\%$$

Missenard (1970) has proposed a simple correlation for λ_L that extends to much higher pressures. In analytical form

$$\frac{\lambda_L(P_r)}{\lambda_L(\text{low pressure})} = 1 + QP_r^{0.7} \qquad (11.11\text{-}2)$$

where $\lambda_L(P_r)$ and λ_L(low pressure) refer to liquid thermal conductivities at high and low, i.e., near saturation, pressure, both at the same temperature. Q is a parameter given in Table 11.8. The correlation is shown in Fig. 11.11. The approaches of Missenard and Lenoir agree up to a reduced pressure of 12, the maximum value shown for the Lenoir form.

TABLE 11.8 Values of Q in Eq. (11.11-2).

	Reduced Pressure					
T_r	1	5	10	50	100	200
0.8	0.036	0.038	0.038	(0.038)	(0.038)	(0.038)
0.7	0.018	0.025	0.027	0.031	0.032	0.032
0.6	0.015	0.020	0.022	0.024	(0.025	0.025
0.5	0.012	0.0165	0.017	0.019	0.020	0.020

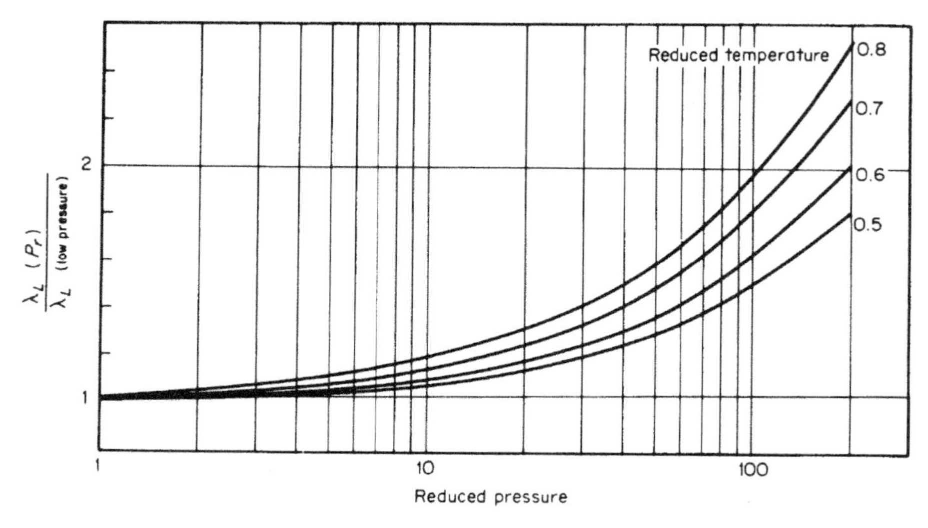

FIGURE 11.11
Missenard (1970) correlation for liquid thermal conductivities at high pressures.

Example 11.13 Estimate the thermal conductivity of liquid toluene at 6330 bar and 304 K. The experimental value at this high pressure is 0.228 W/(m·K) (Kandiyoti, et al. 1973). At 1 bar and 304 K, $\lambda_L = 0.129$ W/(m·K) (Kandiyoti et al., 1973).

Solution. From Appendix A, $T_c = 591.75$ K and $P_c = 41.08$ bar. Therefore, $T_r = 304/591.75 = 0.514$ and $P_r = 6330/41.08 = 154$. From Table 11.7, $Q = 0.0205$. Then, using Eq. (11.11-2),

$$\lambda_L(P_r) = (0.129)[1 + (0.0205)(154)^{0.7}] = 0.219 \text{ W/(m·K)}$$

$$\lambda_L(P_r) \text{ Difference} = 0.219 - 0.228 = -0.009 \text{ W/(m·K) or } -3.9\%$$

Latini and Baroncini (1983) correlated the effect of pressure on liquid thermal conductivity by using Eq. (11.9-3), but they expressed the A parameter as

$$A = A_0 + A_1 P_r \tag{11.11-3}$$

Thus, A_0 would represent the appropriate A parameter at low pressures, as described in Sec. 11.9 and given by Eq. (11.9-4). Values of A_1 were found to range from 6×10^{-3} to 6×10^{-4} W/(m·K); thus the term A_1P_r is negligibly small except at quite high values of P_r. The authors have generalized the parameter A_1 for hydrocarbons as

$$A_1 = \frac{0.0673}{M^{0.84}} \qquad \text{saturated hydrocarbons} \tag{11.11-4}$$

$$A_1 = \frac{102.50}{M^{2.4}} \qquad \text{aromatics} \tag{11.11-5}$$

For hydrocarbons, the authors found average errors usually less than 6% with maximum errors of 10 to 15%. The method should not be used for reduced pressures exceeding 50. More recently, Latini et al. (1989) have extended Eq. (11.9-3) to higher pressures by replacing the exponent of 0.38 with a pressure-dependent expression.

Example 11.14 Rastorguev et al. (1968), as quoted in Jamieson et al. (1975), show the liquid thermal conductivity of *n*-heptane at 313 K to be 0.115 W/(m·K) at 1 bar. Estimate the thermal conductivity of the compressed liquid at the same temperature and 490 bar. The experimental value is 0.136 W/(m·K) (Rastorguev et al., 1968).

Solution. From Appendix A, $T_c = 540.2$ K, $P_c = 27.4$ bar, and $M = 100.202$ g/mol. Since we know the low-pressure value of λ_L, we can estimate A with Eq. (11.9-3). With $T_r = 313/540.2 = 0.580$,

$$A = (0.115)(0.580)^{1/6}/(1 - 0.580)^{0.38} = 0.146 \text{ W/(m·K)}$$

This value of A then becomes A_0 in Eq. (11.11-3). Using Eq. (11.11-4),

$$A_1 = 0.0673/100.202^{0.84} = 0.00140$$

Then, using Eqs. (11.9-3) and (11.11-3) with $P_r = 490/27.4 = 17.9$,

$$\frac{\lambda_L(P_r = 17.9)}{\lambda_L(\text{low pressure})} = \frac{A_0 + A_1P_r}{A_0} = [0.146 + (0.00140)(17.9)]/0.146 = 1.17$$

$$\lambda_L(P_r = 17.9) = (0.115)(1.17) = 0.135 \text{ W/(m·K)}$$

$$\lambda_L \text{ Difference} = 0.135 - 0.136 = -0.001 \text{ W/(m·K) or } -0.7\%$$

11.12 THERMAL CONDUCTIVITIES OF LIQUID MIXTURES

The thermal conductivities of most mixtures of organic liquids are usually less than those predicted by either a mole or weight fraction average, although the deviations are often small. We show data for several binaries in Fig. 11.12 to illustrate this point.

Many correlation methods for λ_m have been proposed (Arikol and Gürbüz, 1992; Assael et al., 1992a, 1996; Bleazard and Teja, 1996; Fareleira et al., 1990). Five were selected for presentation in this section. They are described separately and evaluated later when examples are presented to illustrate the methodology in using each of the methods.

There is a surprisingly large amount of experimental mixture data (Baroncini et al., 1984; Cai et al., 1993; DiGuilio, 1990; Gaitonde et al., 1978; Jamieson et al., 1969, 1973; Jamieson and Irving, 1973; Ogiwara et al.,1980, 1982; Qun-Fung et al., 1997; Rabenovish, 1971; Shroff, 1968; Stephan and Hildwein, 1987; Teja and Tardieu, 1988; Usmanov and Salikov, 1977; Vesovic and Wakeham, 1991), although most are for temperatures near ambient.

11.12.1 Filippov Equation

The Filippov equation (Filippov, 1955; Filippov and Novoselova, 1955) is

$$\lambda_m = w_1\lambda_1 + w_2\lambda_2 - 0.72w_1w_2(\lambda_2 - \lambda_1) \tag{11.12-1}$$

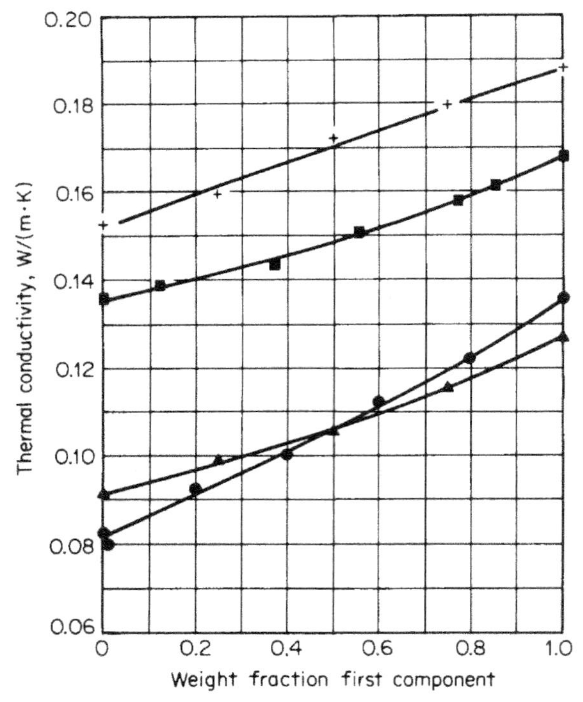

FIGURE 11.12
Thermal conductivities of liquid mixtures. (Data from Filippovand Novoselova, 1955, and Jamieson and Irving, 1973)

where w_1 and w_2 are the weight fractions of components 1 and 2, and λ_1 and λ_2 are the pure-component thermal conductivities. The components were so chosen that $\lambda_2 \geq \lambda_1$. The constant 0.72 may be replaced by an adjustable parameter if binary mixture data are available. The technique is not suitable for multicomponent mixtures but has been extensively tested for binary mixtures.

11.12.2 Jamieson et al. Correlation (1975)

Research and data evaluation at the National Engineering Laboratory has suggested, for binary mixtures,

$$\lambda_m = w_1\lambda_1 + w_2\lambda_2 - \alpha(\lambda_2 - \lambda_1)\left[1 - w_2^{\frac{1}{2}}\right]w_2 \tag{11.12-2}$$

where w_1 and w_2 are weight fractions and, as in the Filippov method, the components are so selected that $\lambda_2 \geq \lambda_1$. α is an adjustable parameter that is set equal to unity if mixture data are unavailable for regression purposes. The authors indicate that Eq. (11.12-2) enables one to estimate λ_m within about 7% (with a 95% confidence limit) for all types of binary mixtures with or without water. It cannot, however, be extended to multicomponent mixtures.

11.12.3 Baroncini et al. (1981a, 1983, 1984) Correlation

The Latini et al. method to estimate pure-liquid thermal conductivities [Eq. (11.9-3)] has been adapted to treat binary liquid mixtures as shown in Eq. (11.12-3)

$$\lambda_m = \left[x_1^2 A_1 + x_2^2 A_2 + 2.2\left(\frac{A_1^3}{A_2}\right)^{\frac{1}{2}} x_1 x_2\right]\frac{\left(1 - T_{r_m}\right)^{0.38}}{T_{r_m}^{\frac{1}{6}}} \tag{11.12-3}$$

where x_1 and x_2 are the mole fractions of components 1 and 2. The A parameters, introduced in Eq. (11.9-3), can be estimated from Eq. (11.9-4) and Table 11.3, or they can be calculated from pure-component thermal conductivities (see Example 11.14). The reduced temperature of the mixture is $T_{r_m} = \frac{T}{T_{c_m}}$ where

$$T_{c_m} = x_1 T_{c_1} + x_2 T_{c_2} \tag{11.12-4}$$

with T_{c_1} and T_{c_2} the pure-component critical temperatures. The choice of which component is number 1 is made with criterion $A_1 \leq A_2$.

This correlation was tested (Baroncini et al., 1984) with over 600 datum points on 50 binary systems including those with highly polar components. The average error found was about 3%. The method is not suitable for multicomponent mixtures.

11.12.4 Method of Rowley (1988)

In this procedure, the liquid phase is modeled by using a two-liquid theory wherein the energetics of the mixture are assumed to favor local variations in composition. The basic relation assumed by Rowley is

$$\lambda_m = \sum_{i=1}^{n} w_i \sum_{j=1}^{n} w_{ji}\lambda_{ji} \tag{11.12-5}$$

where λ_m is the liquid mixture thermal conductivity in W/(m·K), w_i is the weight fraction of component i, w_{ji} is the local weight fraction of component j relative to a central molecule of component i, and λ_{ji} is the characteristic parameter for the thermal conductivity that expresses the interactions between j and i in W/(m·K).

Mass fractions were selected instead of mole fractions in Eq. (11.12-5) because it was found that the excess mixture thermal conductivity

$$\lambda_{\mathrm{m}}^{\mathrm{ex}} = \lambda_{\mathrm{m}} - \sum_{i=1}^{n} w_i \lambda_i \tag{11.12-6}$$

was more symmetrical when weight fractions were employed.

The two-liquid (or local composition) theory was developed in Chap. 8 to derive several of the liquid activity coefficient-composition models. Rowley develops expressions for w_{ji} and relates this quantity to parameters in the NRTL equation (see Chap. 8). In his treatment, he was able to show that Eq. (11.12-5) could be expressed as

$$\lambda_{\mathrm{m}} = \sum_{i=1}^{n} w_i \frac{\sum_{j=1}^{n} w_j G_{ji} \lambda_{ji}}{\sum_{k=1}^{n} w_k G_{ki}} \tag{11.12-7}$$

where G_{ji} and G_{ij} (or G_{ki} and G_{ik}) are the same NRTL parameters as used in activity coefficient correlations for the system of interest.

To obtain λ_{ji} (= λ_{ij}), Rowley makes the important assumption that for any binary, say 1 and 2, $\lambda_{\mathrm{m}} = \lambda_{12} = \lambda_{21}$ when the local *mole fractions* are equal, that is, $x_{12} = x_{21}$. Then, after some algebra, the final correlation is obtained.

$$\lambda_{\mathrm{m}} = \sum_{i=1}^{n} w_i \lambda_i \frac{\sum_{j=1}^{n} w_j G_{ji} (\lambda_{ji} - \lambda_i)}{\sum_{k=1}^{n} w_k G_{ki}} \tag{11.12-8}$$

In Eq. (11.12-8), $\lambda_{ii} = \lambda_i$, $\lambda_{ij} = \lambda_{ji}$. In the original formulation (Rowley, 1982), λ_{ij} was given by

$$\lambda_{ij} = \frac{(w_i^*)^2 (w_j^* + w_i^* G_{ij}) \lambda_i + (w_j^*)^2 (w_i^* + w_j^* G_{ji}) \lambda_j}{(w_i^*)^2 (w_j^* + w_i^* G_{ij}) + (w_j^*)^2 (w_i^* + w_j^* G_{ji})} \tag{11.12-9}$$

$$w_i^* = \frac{M_i (G_{ji})^{\frac{1}{2}}}{M_i (G_{ji})^{\frac{1}{2}} + M_j (G_{ij})^{\frac{1}{2}}} \tag{11.12-10}$$

$$w_j^* = 1 - w_i^* \tag{11.12-11}$$

Note that $G_{ii} = G_{jj} = 1$. w_i^* is that weight fraction in a binary $i - j$ mixture such that $x_{12} = x_{21}$. More recently, Rowley (1988) recommended for systems not containing water, Eq. (11.12-9) be replaced by

$$\lambda_{ij} = \frac{M_i (w_i^*)^2 (w_j^* + w_i^* G_{ij}) \lambda_i + M_j (w_j^*)^2 (w_i^* + w_j^* G_{ji}) \lambda_j}{M_i (w_i^*)^2 (w_j^* + w_i^* G_{ij}) + M_j (w_j^*)^2 (w_i^* + w_j^* G_{ji})} \tag{11.12-12}$$

For a binary system of 1 and 2, Eq. (11.12-8) becomes

$$\lambda_{\mathrm{m}} = w_1 \lambda_1 + w_2 \lambda_2 + w_1 w_2 \left[\frac{G_{21}(\lambda_{12} - \lambda_1)}{w_1 + w_2 G_{21}} + \frac{G_{12}(\lambda_{12} - \lambda_2)}{w_1 G_{12} + w_2} \right] \tag{11.12-13}$$

λ_{12} is found from Eq. (11.12-9) if water is one of the components or Eq. (11.12-12) if water is absent. Using Eq. (11.12-9), with $i = 1$ and $j = 2$ gives

$$\lambda_{ij} = \frac{(w_1^*)^2 (w_2^* + w_1^* G_{12}) \lambda_1 + (w_2^*)^2 (w_1^* + w_2^* G_{21}) \lambda_2}{(w_1^*)^2 (w_2^* + w_1^* G_{12}) + (w_2^*)^2 (w_1^* + w_2^* G_{21})} \tag{11.12-14}$$

Equations (11.12-10) and (11.12-11) give

$$w_1^* = \frac{M_1(G_{21})^{\frac{1}{2}}}{M_1(G_{21})^{\frac{1}{2}} + M_2(G_{12})^{\frac{1}{2}}} \tag{11.12-15}$$

$$w_2^* = 1 - w_1^* \tag{11.12-16}$$

With some algebra, it can be shown that the quantity in brackets in Eq. (11.12-13) is equal to $(\lambda_2 - \lambda_1)\Phi$ where

$$\Phi = \frac{G_{21}}{(w_1 + w_2 G_{21})(1 + \Psi)} - \frac{G_{12}}{(w_2 + w_1 G_{12})(1 + \Psi^{-1})} \tag{11.12-17}$$

$$\Psi = \left(\frac{w_1^*}{w_2^*}\right)^2 \frac{w_2^* + w_1^* G_{12}}{w_1^* + w_2^* G_{21}} \tag{11.12-18}$$

Thus, the entire nonideal effect is included in the Φ parameter, and the form is quite similar to the Filippov and Jamieson et al. relations described earlier.

To employ this technique, values for the liquid thermal conductivities of all pure components are required. In addition, the NRTL parameters G_{ij} and G_{ji} must be found from data sources or from regressing vapor-liquid equilibrium data. When tested with data on 18 ternary mixtures, Rowley (1988) found an average absolute deviation of 1.86%. The concept of relating transport and thermodynamic properties is an interesting one and bears further study; Brulé and Starling (1984) also advocated such an approach.

11.12.5 Power Law Method

Following Vredeveld (1973), the following equation may be used for nonaqueous systems in which the ratio of component thermal conductivities does not exceed 2:

$$\lambda_m = \left(\sum_i w_i \lambda_i^{-2}\right)^{-\frac{1}{2}} \tag{11.12-19}$$

where w_i is the weight fraction of component i and λ_i is the thermal conductivity of pure i. Equation (11.12-19) has been used successfully for both binary (Carmichael et al., 1963) and ternary systems (Rowley, 1988). Attempts to use Eq. (11.12-19) for aqueous binaries have been unsuccessful and typically led to deviations as high as 10% (Rowley, 1988).

11.12.6 Discussion

All five methods for estimating λ_m described in this section have been extensively tested by using binary mixture data. All require the thermal conductivities of the pure components making up the system (or an estimate of the values), and thus they are interpolative in nature. The Filippov, Jamieson et al., and power-law procedures require no additional information other than the weight fractions and pure-component values of λ_L. The Baroncini et al. method also needs pure-component critical properties. Rowley's correlation requires the NRTL parameters G_{ij} and G_{ji} from phase equilibrium data. Only the power-law and Rowley's methods will treat multicomponent mixtures. The power-law method should not be used if water is present or if the ratio of pure-component λ values exceeds 2. Figure 11.13 shows some measurements of Usmanov and Salikov (1977) for very polar systems and illustrates how well Filippov's relation (11.12-1) fits these data. Methods described in this section other than the power-law method would have been equally satisfactory. Gaitonde et al. (1978)

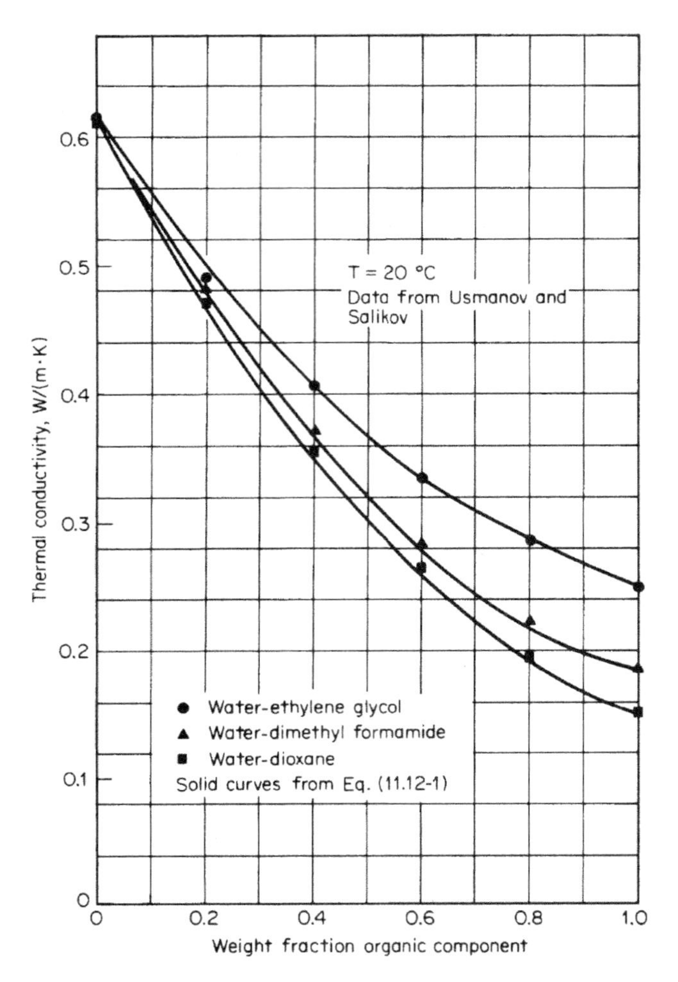

FIGURE 11.13
Filippov correlation of liquid mixture thermal conductivity. (Data from Osmanov and Salikov, 1977)

measured λ_m for liquid mixtures of alkanes and silicone oils to study systems with large differences in the molecular sizes of the components. They found the Filippov and Jamieson et al. correlations provide a good fit to the data, but they recommended the general form of McLaughlin (1964) with the inclusion of an adjustable binary parameter. Teja and Tardieu (1988) have used a method based on effective carbon number to estimate thermal conductivities of crude oil fractions.

In summary, one can use any of the relations described in this section to estimate λ_m with the expectation that errors will rarely exceed about 5%.

In the case of aqueous (*dilute*) solutions containing electrolytes, the mixture thermal conductivity usually decreases with an increase in the concentration of the dissolved salts. To estimate the thermal conductivity of such mixtures, Jamieson and Tudhope (1964) recommend the use of an equation proposed originally by Riedel (1951) and tested by Vargaftik and Os'minin (1956). At 293 K:

$$\lambda_m = \lambda_{H_2O} + \sum_i \sigma_i C_i \qquad (11.12\text{-}20)$$

where λ_m is the thermal conductivity of the ionic solution at 293 K in W/(m·K), λ_{H_2O} is the thermal conductivity of water at 293 K in W/(m·K), C_i is the concentration of the electrolyte in mol/L, and σ_i is a coefficient that is characteristic for each ion. Values of σ_i are shown in Table 11.9. To obtain λ_m at other temperatures T,

$$\lambda_m(T) = \lambda_m(293\ \text{K}) \frac{\lambda_{H_2O}(T)}{\lambda_{H_2O}(293\ \text{K})} \tag{11.12-21}$$

Except for strong acids and bases at high concentrations, Eqs. (11.12-20) and (11.12-21) are usually accurate to within ±5%.

TABLE 11.9 Values of for anions and cations in Eq. (11.12-20) (Jamieson and Tudhope, 1964).

Anion	$\sigma_i \times 10^5$	Cation	$\sigma_i \times 10^5$
OH^-	20.934	H^+	−9.071
F^-	2.0934	L^+	−3.489
Cl^-	−5.466	Na^+	0.000
Br^-	−17.445	K^+	−7.560
I^-	−27.447	NH_4	−11.63
NO_2^-	−4.652	Mg^{2+}	−9.304
NO_3^-	−6.978	Ca^{2+}	−0.5815
NO_4^-	−14.189	Sr^{2+}	−3.954
ClO_4^-	−17.445	Ba^{2+}	−7.676
BrO_3^-	−14.189	Ag^+	−10.47
CO_3^{2-}	−7.560	Cu^{2+}	−16.28
SiO_3^{2-}	−9.300	Zn^{2+}	−16.28
SO_3^{2-}	−2.326	Pb^{2+}	−9.304
SO_4^{2-}	1.163	Co^{2+}	−11.63
$S_2O_3^{2-}$	8.141	Al^{3+}	−32.56
CrO_4^{2-}	−1.163	Th^{4+}	−43.61
$Cr_2O_7^{2-}$	15.93		
PO_4^{3-}	−20.93		
$Fe(CN)_6^{4-}$	18.61		
Acetate$^-$	−22.91		
Oxalate^{2-}	−3.489		

Example 11.15 Using Filippov's and Jamieson et al.'s methods, estimate the thermal conductivity of a liquid mixture of methanol and benzene at 273 K. The weight fraction methanol is 0.4. At this temperature, Jamieson et al. (1969) report the thermal conductivities of pure benzene and methanol as 0.152 and 0.210 W/(m·K), respectively. The experimental mixture value (Jamieson et al., 1969) is 0.170 W/(m·K).

Solution. FILIPPOV METHOD We use Eq. (11.12-1). Here methanol is component 2, since λ (methanol) > λ (benzene). Thus,

$$\lambda_m = (0.6)(0.152) + (0.4)(0.210) - (0.72)(0.6)(0.4)(0.210 - 0.152)$$
$$= 0.165 \text{ W/(m·K)}$$
$$\lambda_m \text{ Difference} = 0.165 - 0.170 = -0.005 \text{ W/(m·K) or } -2.9\%$$

JAMESON ET AL. METHOD. Again, methanol is chosen as component 2. With Eq. (11.12-2) and $\alpha = 1$,

$$\lambda_m = (0.6)(0.152) + (0.4)(0.210) - (0.210 - 0.152)[1 - (0.4)^{1/2}](0.4)$$
$$= 0.167 \text{ W/(m·K)}$$
$$\lambda_m \text{ Difference} = 0.167 - 0.170 = -0.003 \text{ W/(m·K) or } -1.8\%$$

Example 11.16 Estimate the liquid thermal conductivity of a mixture of benzene (1) and methyl formate (2) at 323 K by using the method of Baroncini et al. At this temperature, the values of λ_L for the pure components are $\lambda_1 = 0.138$ and $\lambda_2 = 0.179$ W/(m·K) (Baroncini et al., 1984).

Solution. We will estimate the values of λ_m at 0.25, 0.50, and 0.75 weight fraction benzene. First, however, we need to determine A_1 and A_2. Although Eq. (11.9-4) and Table 11.3 could be used, it is more convenient to employ the pure-component values of λ_L with Eq. (11.9-3). From Appendix A, $T_{c_1} = 562.05$ K and $T_{c_2} = 487.2$ K, so $T_{r_1} = 323/562.05 = 0.575$ and $T_{r_2} = 323/487.2 = 0.663$. Then, with Eq. (11.9-3), for benzene,

$$A_1 = (0.138)(0.575)^{1/6}/(1 - 0.575)^{0.38} = 0.174$$

Similarly, $A_2 = 0.252$. [Note that, if Eq. (11.9-4) and Table 11.3 had been used, we would have $A_1 = 0.176$ and $A_2 = 0.236$.] We have selected components 1 and 2 to agree with the criterion $A_1 \leq A_2$.

Considering first a mixture containing 0.25 weight fraction benzene, i.e., $w_1 = 0.25$ and $w_2 = 0.75$, the mole fractions are $x_1 = 0.204$ and $x_2 = 0.796$. Thus,

$$T_{c_m} = (0.204)(562.05) + (0.796)(487.2) = 502.5 \text{ K}$$

$$T_{r_m} = 323/502.5 = 0.643$$

With Eq. (11.12-3),

$$\lambda_m = \{(0.204)^2(0.174) + (0.796)^2(0.252)$$
$$+ (2.2)[(0.174)^3/0.252]^{1/2}(0.204)(0.769)\}[(1 - 0.643)^{0.38}/(0.643)^{1/6}]$$
$$= 0.159 \text{ W/(m·K)}$$

Calculated results for this and other compositions are shown below with the experimental values (Baroncini et al., 1984) and percent errors.

Results for benzene (1) – methyl formate mixtures at $T = 323$ K.

w_1	x_1	T_{c_m}, K	λ_m, calc, W/(m·K)	λ_m, exp, W/(m·K)	Percent error
0.25	0.204	502.5	0.159	0.158	0.6
0.50	0.435	519.8	0.143	0.151	–5.3
0.75	0.698	539.6	0.135	0.140	–3.6

Example 11.17 Use Rowley's method to estimate the thermal conductivity of a liquid mixture of acetone (1) and chloroform (2) that contains 66.1 weight % of the former. The temperature is 298 K. As quoted by Jamieson et al. (1975), Rodriguez (1962) reports λ_1 = 0.161 W/(m·K), λ_2 = 0.119 W/(m·K), and for the mixture, λ_m = 0.143 W/(m·K).

Solution. First, we need the NRTL parameters for this binary at 298 K. Nagata (1973) suggests $G_{12} = 1.360$ and $G_{21} = 0.910$. From Appendix A, $M_1 = 58.08$ and $M_2 = 119.38$ g/mol. Using Eqs. (11.12-15) and (11.12-16),

$$w_1^* = (58.08)(0.910)^{1/2}/[(58.08)(0.910)^{1/2} + (119.38)(1.360)^{1/2}] = 0.285$$

$$w_2^* = 1 - 0.285 = 0.715$$

$$w_1^* = \frac{(58.08)(0.910)^{1/2}}{(58.08)(0.910)^{1/2} + (119.38)(1.360)^{1/2}} = 0.285$$

$$w_2^* = 1 - 0.285 = 0.715$$

With Eq. (11.12-14),

$$\lambda_{12} = \{(0.285)^2[0.715 + (0.285)(1.360)](0.161) + (0.715)^2[0.285 + (0.715)(0.910)](0.119)\}/ \{(0.285)^2[0.715 + (0.285)(1.360)] + (0.715)^2[0.285 + (0.715)(0.910)]\} = 0.126$$

Then, with Eq. (11.12-13),

$$\lambda_m = (0.661)(0.161) + (0.339)(0.119) + (0.661)(0.339)\{[(0.910)(0.126 - 0.161)]/[0.661 + (0.339)(0.910)] + [(1.360)(0.126 - 0.119)]/[(0.661)(1.360) + 0.339]\} = 0.141 \text{ W/(m·K)}$$

$$\lambda_m \text{ Difference} = 0.141 - 0.143 = -0.002 \text{ W/(m·K) or } -1.4\%$$

Using Eq. (11.12-12) instead of Eq. (11.12-14) for λ_{12} gives $\lambda_m = 0.140$ for an error of –2%.

11.13 NOTATION

$A, a, B, b, C, c, \ldots$	parameters in correlating equations or prediction methods
$A_i, a_i, B_i, b_i, C_i, c_i, \ldots$	parameters in correlating equations or prediction methods
a_{SR}	parameter in the Sastri and Rao (1999) method for predicting liquid thermal conductivity
A^*, A_λ	parameters in the Latini et al. (1977) method for predicting liquid thermal conductivity
A_{ij}	interaction parameter between components i and j in a mixture, see Eq. (11.6-1)
$AARD$	average absolute relative deviation, see Eq. (11.3-15)
C_{DD}	parameter in the Danner and Daubert (1997) method for predicting liquid thermal conductivity
C_i	concentration of the electrolyte in an aqueous solution
C_p	heat capacity at constant pressure per mole
C_p°	idea gas heat capacity at constant pressure per mole
C_{p0}	isobaric liquid heat capacity at 0 °C used in the Missenard (1965, 1965a) method to predict thermal conductivity of liquids
C_{tr}, C_{int}	translational and internal contributions to the heat capacity
C_v	heat capacity at constant volume per mole
C_v'	heat capacity at constant volume per molecule
C_{v_i}	heat capacity at constant volume per mole of component i in a mixture
C_{vm}	heat capacity at constant volume per mole of a mixture
D_{ab}	diffusivity
f, f_m, f_i, f_{ij}	parameters in the TRAPP (Huber, 1996) method to predict thermal conductivity of gases and gas mixtures
f_{tr}, f_{int}	scaling factors for the translation and internal heat capacity contributions to the Eucken factor
F_λ, F_{λ_m}	parameters in the TRAPP (Huber, 1996) method to predict thermal conductivity of gases and gas mixtures
G_{ij}	NRTL activity coefficient parameter, see Chap. 8
h, h_m, h_i, h_{ij}	parameters in the TRAPP (Huber, 1996) method to predict thermal conductivity of gases and gas mixtures
k	Boltzmann constant (1.380649×10^{-23} J/K)
L	mean free path
L_1, L_2	functions of reduced temperature and pressure in the correlation of Lenoir (1957)
m_m	mass of molecule
M	molecular weight (g/mol)
M'	molecular weight (kg/mol)
M_i	molecular weight of pure component i in a mixture
M_{ij}	cross interaction molecular weight between components i and j in a mixture
M_m	molecular weight of a mixture
n	number density of molecules; number of components in a mixture; parameter in correlating equations or prediction methods
n_{SR}	parameter in the Sastri and Rao (1999) method for predicting liquid thermal conductivity
N	number of atoms in a molecule
N_A	Avogadro's number ($6.02214076 \times 10^{23}$ mol^{-1})

N_c	total number of compounds in a test database of experimental data used to evaluate prediction methods
n_{DD}	parameter in the Danner and Daubert (1997) method for predicting liquid thermal conductivity
N_i	number of times group i is found in a compound
N_j^{corr}	number of times correction j is found in a compound
N_{Pr}	Prandtl number
N_{pt}	total number of points in a test database of experimental data used to evaluate prediction methods
P	pressure
P_c	critical pressure
P_{c_m}	critical pressure of a mixture
P_r	reduced pressure $\left(\frac{P}{P_c}\right)$
P^{vp}	saturated liquid vapor pressure
$P_R^{vp}(T_\lambda)$	saturated liquid vapor pressure of reference fluid at temperature T_λ, used in the TRAPP (Huber, 1996) method to predict thermal conductivity of gases and gas mixtures
q	parameter in the Chung et al. (1984, 1988) method for predicting thermal conductivity of gases
Q	parameter in the Missenard method (1970) to calculate liquid viscosity at high pressure, see Eq. (11.11-2)
R	gas constant [8.314462618 J/(mol·K)]
R_λ	parameter in Enskog Dense-gas Theory, see Eq. (11.5-27) and Table 10.5
T	temperature
T_b	normal (P = 101325 Pa) boiling point
T_{b_r}	reduced normal (P = 101325 Pa) boiling point temperature $\left(\frac{T_b}{T_c}\right)$
T_c	critical temperature
T_{c_i}	critical temperature of pure component i in a mixture
$T_{c_{ij}}$	cross interaction critical temperature between components i and j in a mixture
T_{c_m}	critical temperature of a mixture
T_{c_R}	critical temperature of reference fluid, used in the TRAPP (Huber, 1996) method to predict thermal conductivity of gases and gas mixtures
T_r	reduced temperature $\left(\frac{T}{T_c}\right)$
T_{r_i}	reduced temperature $\left(\frac{T}{T_{c_i}}\right)$ of pure component i in a mixture
T_{r_m}	reduced temperature of a mixture
T_{r_R}	reduced temperature $\left(\frac{T_\lambda}{T_{c_R}}\right)$ of the reference fluid in the TRAPP (Huber, 1996) method to predict thermal conductivity of gases and gas mixtures
T_λ, T_{λ_m}	parameters in the TRAPP (Huber, 1996) method to predict thermal conductivity of gases and gas mixtures
v	molecular velocity

V	molar volume
V_c	critical volume
V_0	close-packed volume in Enskog Dense-gas Theory, see Table 10.5
V_{c_i}	critical volume of component i in a mixture
$V_{c_{ij}}$	cross interaction critical volume between components i and j in a mixture
V_{c_m}	critical volume of a mixture
V_r	reduced volume $\left(\dfrac{V}{V_0}\right)$
w_i	weight fraction of component i in a liquid mixture
w_{ji}	local weight fraction of component j relative to a central molecule of component i in the Rowley (1988) method to calculate liquid thermal conductivity of mixtures
w_i^*, w_j^*	parameters in the Rowley (1988) method to calculate liquid thermal conductivity of mixtures
x_i	mole fraction of component i in a liquid mixture
$X_{\exp_i}$	value of experimental data point i for property X
X_{pred_i}	value of predicted point i for property X
X_λ, X_{λ_m}	parameters in the TRAPP (Huber, 1996) method to predict thermal conductivity of gases and gas mixtures
y	parameter in the Chung et al. (1984, 1988) method for predicting thermal conductivity of gases
y_i	mole fraction of component i in a gas mixture
Z	compressibility factor; number of collisions required to interchange a quantum of rotational energy with translational energy in the Chung et al. (1984, 1988) prediction method for gas thermal conductivity, see Eq. (11.3-12)
Z_c	critical compressibility factor
Z_{c_i}	critical compressibility factor of component i in a mixture
Z_{c_m}	critical compressibility factor of a mixture
Z_{c_R}	critical compressibility factor of reference fluid, used in the TRAPP (Huber, 1996) method to predict thermal conductivity of gases and gas mixtures

Greek

α	parameter in the Chung et al. (1984, 1988) prediction method for gas thermal conductivity, see Eq. (11.3-13); parameter in the Latini et al. (1977) method for predicting liquid thermal conductivity; parameter in the Jamieson et al. (1975) method for predicting liquid thermal conductivity of mixtures, see Eq. (11.12-2)
β	parameter in the Misic and Thodos (1961) method to calculate gas thermal conductivity, see Eq. (11.3-7); parameter in the Chung et al. (1984, 1988) prediction method for gas thermal conductivity, see Eq. (11.3-13); parameter in the Latini et al. (1977) method for predicting liquid thermal conductivity
γ	$\dfrac{C_p}{C_v}$; parameter in the Latini et al. (1977) method for predicting liquid thermal conductivity
Γ	inverse thermal conductivity used for nondimensionalization of λ, see Eq. (11.5-6)
Γ_i	inverse thermal conductivity (Γ) of component i in a mixture
$(\Delta_{\lambda_b})_i$, $(\Delta_{\lambda_b}^{\text{corr}})_i$	contributions for group i for the prediction method of Sastri and Rao (1999) for liquid viscosity

ε	numerical constant in the Mason and Saxena (1958) method to predict thermal conductivity of mixtures, see Eq. (11.6-2)
η	viscosity
η°	viscosity at low pressure (about 1 bar)
η_i	viscosity of pure component i in a mixture
η_m	viscosity of a mixture
κ	polarity correction factor in the method Chung et al. (1984, 1988) for gas thermal conductivities
λ	thermal conductivity
λ°	thermal conductivity at low pressure (about 1 bar)
λ_b	parameter in the Sastri and Rao (1999) method for predicting liquid thermal conductivity
λ_i	thermal conductivity of pure component i in a mixture
λ_{ji}	parameter in the Rowley (1988) method to calculate liquid thermal conductivity of mixtures
λ_{H_2O}	thermal conductivity of water
λ_L	thermal conductivity of a liquid
λ_{L_0}	thermal conductivity of a liquid at 0 °C used in the Missenard (1965, 1965a) method to predict thermal conductivity of liquids
λ_m	thermal conductivity of a mixture
λ_m°	thermal conductivity of a mixture at low pressure (about 1 bar)
λ_m^{ex}	excess mixture thermal conductivity, see Eq. (11.12-16)
λ_r	reduced thermal conductivity
λ_R	thermal conductivity of reference fluid at elevated pressure, used in the TRAPP (Huber, 1996) method to predict thermal conductivity of gases and gas mixtures
λ_R°	thermal conductivity of reference fluid at low pressure, used in the TRAPP (Huber, 1996) method to predict thermal conductivity of gases and gas mixtures
λ_{tr_i}	monatomic (translation) value of the thermal conductivity for component i in a mixture
$\dfrac{\lambda M'}{\eta C_v}$	Eucken factor
Λ	parameter in the Misic and Thodos (1961) method to calculate gas thermal conductivity $\left(\dfrac{M^{\frac{1}{2}} T_c^{\frac{1}{6}}}{P_c^{\frac{2}{3}}}\right)$
μ	dipole moment
μ_r	reduced dipole moment
ρ	density
ρ_r	reduced density $\left(\dfrac{\rho}{\rho_c}\right)$
ρ_{r_R}	reduced density $\left(\dfrac{\rho_\lambda}{\rho_{c_R}}\right)$ of the reference fluid in the TRAPP (Huber, 1996) method to predict thermal conductivity of gases and gas mixtures
ρ^S	saturated liquid density
ρ_c	critical density

ρ_{c_R}	critical density of reference fluid, used in the TRAPP (Huber, 1996) method to predict thermal conductivity of gases and gas mixtures
$\rho_R^S(T_\lambda)$	saturated liquid density of reference fluid at temperature T_λ, used in the TRAPP (Huber, 1996) method to predict thermal conductivity of gases and gas mixtures
ρ_λ, ρ_{λ_m}	parameters in the TRAPP (Huber, 1996) method to predict thermal conductivity of gases and gas mixtures
ρ_0	liquid density at 0 °C used in the Missenard (1965, 1965a) method to predict thermal conductivity of liquids
σ	molecular diameter
σ_i	parameter to calculate the thermal conductivity of aqueous solutions, see Eq. (11.12-20) and Table 11.9
τ	$1 - T_r$
ϕ, ϕ', ϕ''	parameters in the Teja and Rice (1981, 1982) method to predict liquid thermal conductivity
ϕ_{ij}	cross interaction parameter between components i and j in the prediction method of Wilke (1950) for low pressure gas viscosity of a mixture, see Chap. 10
Φ	parameter in the Rowley (1988) method to calculate liquid thermal conductivity of mixtures
Ψ	parameter in the Chung et al. (1984, 1988) prediction method for gas thermal conductivity; parameter in the Rowley (1988) method to calculate liquid thermal conductivity of mixtures
ω	acentric factor
ω_i	acentric factor of component i in a mixture
ω_{ij}	cross interaction acentric factor between components i and j in a mixture
ω_m	acentric factor of a mixture
ω_R	acentric factor of reference fluid, used in the TRAPP (Huber, 1996) method to predict thermal conductivity of gases and gas mixtures
ω', ω''	acentric factors for reference fluids 1 and 2 in the Teja and Rice (1981, 1982) method to predict liquid thermal conductivity
Ω_v	collision integral for viscosity and thermal conductivity
Ω_D	collision integral for diffusivity

11.14 REFERENCES

Arikol, M., and H. Gürbüz: *Can. J. Chem. Eng.*, **70:** 1157 (1992).

Assael, M. J., J. P. M. Trusler, and T. F. Tsolakis: *Thermophysical Properties of Fluids, An Introduction to Their Prediction*, Imperial College Press, London, 1996.

Assael, M. J., J. H. Dymond, M. Papadaki, and P. M Patterson: *Fluid Phase Equil.*, **75:** 245 (1992).

Assael, M. J., E. Charitidou, and W. A. Wakeham: *Intern. J. Thermophys.*, **10:** 779 (1989).

Assael, M. J., J. H. Dymond, M. Papadaki, and P. M Patterson: *Intern. J. Thermophys.*, **13:** 269 (1992).

Assael, M. J., J. H. Dymond, M. Papadaki, and P. M Patterson: *Intern. J. Thermophys.*, **13:** 659 (1992a).

Assael, M. J., J. H. Dymond, M. Papadaki, and P. M Patterson: *Intern. J. Thermophys.*, **13:** 895 (1992b).

Assael, M. J., J. H. Dymond, and S. K. Polimatidou: *Intern. J. Thermophys.*, **15:** 189 (1994).

Assael, M. J., J. H. Dymond, and S. K. Polimatidou: *Intern. J. Thermophys.*, **16:** 761 (1995).

Baroncini, C., P. Di Filippo, G. Latini, and M. Pacetti: *Intern. J. Thermophys.*, **1**(2)**:** 159 (1980).

Baroncini, C., P. Di Filippo, G. Latini, and M. Pacetti: *Intern. J. Thermophys.*, **2**(1)**:** 21 (1981).

Baroncini, C., P. Di Filippo, G. Latini, and M. Pacetti: *Thermal Cond.*, 1981a (pub. 1983), 17th, Plenum Pub. Co., p. 285.

Baroncini, C., P. Di Filippo, and G. Latini: "Comparison Between Predicted and Experimental Thermal Conductivity Values for the Liquid Substances and the Liquid Mixtures at Different Temperatures and Pressures," paper presented at the *Workshop on Thermal Conductivity Measurement, IMEKO, Budapest,* March 14–16, 1983.

Baroncini, C., P. Di Filippo, and G. Latini: *Intern. J. Refrig.,* **6**(1)**:** 60 (1983a).
Baroncini, C., G. Latini, and P. Pierpaoli: *Intern. J. Thermophys.,* **5**(4)**:** 387 (1984).
Basu, R. S., and J. V. Sengers: "Thermal Conductivity of Fluids in the Critical Region," in D. C. Larson (ed.), *Thermal Conductivity,* Plenum, New York, 1983, p. 591.
Bennett, L. A., and R. G. Vines: *J. Chem. Phys.,* **23:** 1587 (1955).
Bleazard, J. G., and A. S. Teja: *Ind. Eng. Chem. Res.,* **35:** 2453 (1996).
Bretsznajder, S.: *Prediction of Transport and Other Physical Properties of Fluids,* trans. by J. Bandrowski, Pergamon, New York, 1971, p. 251.
Bridgman, P. W.: *Proc. Am. Acad. Art Sci.,* **59:** 154 (1923).
Brokaw, R. S.: *J. Chem. Phys.,* **29:** 391 (1958).
Bromley, L. A.: "Thermal Conductivity of Gases at Moderate Pressures," *Univ. California Rad. Lab. UCRL-1852,* Berkeley, Calif., June 1952.
Brulé, M. R., and K. E. Starling: *Ind. Eng. Chem. Proc. Des. Dev.,* **23:** 833 (1984).
Cai, G., H. Zong, Q. Yu, and R. Lin: *J. Chem. Eng. Data,* **38:** 332 (1993).
Carmichael, L. T., V. Berry, and B. H. Sage: *J. Chem. Eng. Data,* **8:** 281 (1963).
Carmichael, L. T., H. H. Reamer, and B. H. Sage: *J. Chem. Eng. Data,* **11:** 52 (1966).
Carmichael, L. T., and B. H. Sage: *J. Chem. Eng. Data,* **9:** 511 (1964).
Chapman, S., and T. G. Cowling: *The Mathematical Theory of Non-uniform Gases,* Cambridge, New York, 1961.
Christensen, P. L., and A. Fredenslund: *J. Chem. Eng. Data,* **24:** 281 (1979).
Chung, T.-H., M. Ajlan, L. L. Lee, and K. E. Starling: *Ind. Eng. Chem. Res.* **27:** 671 (1988).
Chung, T.-H., L. L. Lee, and K. E. Starling: *Ind. Eng. Chem. Fundam.,* **23:** 8 (1984).
Cohen, Y., and S. I. Sandler: *Ind. Eng. Chem. Fundam.,* **19:** 186 (1980).
Correla, F. von, B. Schramm, and K. Schaefer: *Ber. Bunsenges. Phys. Chem.,* **72**(3)**:** 393 (1968).
Cottrell, T. L., and J. C. McCoubrey: *Molecular Energy Transfer in Gases,* Butterworth, London, 1961.
Couch, E. J., and K. A. Dobe: *J. Chem. Eng. Data,* **6:** 229 (1961).
Curtiss, L. A., D. J. Frurip, and M. Blander: *J. Phys. Chem.,* **86:** 1120 (1982).
Daubert, T. E. and R. P. Danner, R.P: *Technical Data Book—Petroleum Refining,* 6th ed., American Petroleum Institute, Washington, DC 1997.
Speight, J.: Lange's *Handbook of Chemistry,* 17th ed., McGraw-Hill, New York, 2016.
DiGuilio, R. M., W. L. McGregor, and A. S. Teja: *J. Chem. Eng. Data,* **37:** 342 (1992).
DiGuilio, R. M., and A. S. Teja: *J. Chem. Eng. Data,* **35:** 117 (1990)
Diky, V., R. D. Chirico, M. Frenkel, A. Bazyleva, J. W. Magee, E. Paulechka, A. Kazakov, E. W. Lemmon, C. D. Muzny, A. Y. Smolyanitsky, S. Townsend, and K. Kroenlein. *NIST ThermoData Engine, NIST Standard Reference Database 103 a/b,* National Institute of Standards and Technology, USA (2021). https://www.nist.gov/mml/acmd/trc/thermodata-engine
DIPPR, Design Institute for Physical Properties, American Institute of Chemical Engineers, 120 Wall St. FL 23, New York, NY 10005-4020 USA (2021). https://www.aiche.org/dippr
Donaldson, A. B.: *Ind. Eng. Chem. Fundam.,* **14:** 325 (1975).
Dymond, J. H., and M. J. Assael: Chap. 10 in J. Millat, J. H. Dymond, and C. A. Nieto de Castro (eds.), *Transport Properties of Fluids, Their Correlation, Prediction and Estimation,* IUPAC, Cambridge Univ. Press, Cambridge, 1996.
Ehya, H., F. M. Faubert, and G. S. Springer: *J. Heat Transfer,* **94:** 262 (1972).
Ely, J. F., and H. J. M. Hanley: *Ind. Eng. Chem. Fundam.,* **22:** 90 (1983).
Ewing, C. T., B. E. Walker, J. A. Grand, and R. R. Miller*: Chem. Eng. Progr. Symp. Ser.,* **53**(20)**:** 19 (1957).
Fareleira, J. M. N., C. A. Nieto de Castro, and A. A. H. Pa´dua: *Ber. Bunsenges. Phys. Chem.,* **94:** 553 (1990).
Filippov, L. P.: *Vest. Mosk. Univ., Ser. Fiz. Mat. Estestv. Nauk,* (8)**10**(5)**:** 67–69 (1955); *Chem. Abstr.,* **50:** 8276 (1956).
Filippov, L. P., and N. S. Novoselova: *Vestn. Mosk. Univ., Ser. Fiz. Mat. Estestv Nauk,* (3)**10**(2)**:** 37–40 (1955); *Chem. Abstr.,* **49:** 11366 (1955).
Fleeter, R. J., Kestin, and W. A. Wakeham: *Physica A (Amsterdam),* **103A:** 521 (1980).
Gaitonde, U. N., D. D. Deshpande, and S. P. Sukhatme: *Ind. Eng. Chem. Fundam.,* **17:** 321 (1978).
Gambill, W. R.: *Chem. Eng.,* **66**(16)**:** 129 (1959).
Gilmore, T. F., and E. W. Comings: *AIChE J.,* **12:** 1172 (1966).
Gray, P., S. Holland, and A. O. S. Maczek: *Trans. Faraday Soc.,* **66:** 107 (1970).
Green, D. W., and M. Z. Southard (eds.): *Chemical Engineers' Handbook,* 9th ed., McGraw-Hill, New York, 2019.
Groenier, W. S., and G. Thodos: *J. Chem. Eng. Data,* **5:** 285 (1960).
Groenier, W. S., and G. Thodos: *J. Chem. Eng. Data,* **6:** 240 (1961).
Guildner, L. A.: *Proc. Natl. Acad. Sci.,* **44:** 1149 (1958).
Hanley, H. J. M.: *Cryogenics,* **16**(11)**:** 643 (1976).

Hirschfelder, J. O.: *J. Chem. Phys.,* **26:** 282 (1957).
Hirschfelder, J. O., C. F. Curtiss, and R. B. Bird: *Molecular Theory of Gases and Liquids,* Wiley, New York, 1954.
Ho, C. Y., R. W. Powell, and P. E. Liley: *J. Phys. Chem. Ref. Data,* **1:** 279 (1972).
Holland, P. M., H. J. M. Hanley, K. E. Gubbins, and J. M. Haile: *J. Phys. Chem. Ref. Data,* **8:** 559 (1979).
Huber, M. L.: Chap. 12 in J. Millat, J. H. Dymond, and C. A. Nieto de Castro (eds.), *Transport Properties of Fluids, Their Correlation, Prediction and Estimation,* IUPAC, Cambridge Univ. Press, Cambridge, 1996.
Huber, M. L.: personal communication, NIST, Boulder, 1998.
Jamieson, D. T.: personal communication, National Engineering Laboratory, East Kilbride, Glasgow, March 1971.
Jamieson, D. T.: *J. Chem. Eng. Data,* **24:** 244 (1979).
Jamieson, D. T.: personal communication, National Engineering Laboratory, East Kilbride, Scotland, October 1984.
Jamieson, D. T., and G. Cartwright: *J. Chem. Eng. Data,* **25:** 199 (1980).
Jamieson, D. T., and G. Cartwright: *Proc. 8th Symp. Thermophys. Prop.,* Vol. 1, *Thermo-physical Properties of Fluids,* J. V. Sengers (ed.), ASME, New York, 1981, p. 260.
Jamieson, D. T., and E. H. Hastings: in C. Y. Ho and R. E. Taylor (eds.), *Proc. 8th Conf. Thermal Conductivity,* Plenum, New York, 1969, p. 631.
Jamieson, D. T., and J. B. Irving: paper presented in *13th Intern. Thermal Conductivity Conf., Univ. Missouri,* 1973.
Jamieson, D. T., and J. S. Tudhope: *Natl. Eng. Lab. Glasgow Rep.* 137, March 1964.
Jamieson, D. T., J. B. Irving, and J. S. Tudhope: *Liquid Thermal Conductivity. A Data Survey to 1973,* H. M. Stationary Office, Edinburgh, 1975.
Junk, W. A., and E. W. Comings: *Chem. Eng. Progr.,* **49:** 263 (1953).
Kandiyoti, R., E. McLaughlin, and J. F. T. Pittman: *Chem. Soc. (London), Faraday Trans.,* **69:** 1953 (1973).
Kannuliuk, W. G., and H. B. Donald: *Aust. J. Sci. Res.,* **3A:** 417 (1950).
Kennedy, J. T., and G. Thodos: *AIChE J.,* **7:** 625 (1961).
Keyes, F. G.: *Trans. ASME,* **73:** 597 (1951).
Klass, D. M., and D. S. Viswanath: *Ind. Eng. Chem. Res.,* **37:** 2064 (1998).
Kramer, F. R., and E. W. Comings: *J. Chem. Eng. Data,* **5:** 462 (1960).
Krauss, R., J. Luettmer-Strathmann, J. V. Sengers, and K. Stephan: *Intern. J. Thermophys.,* **14:** 951 (1993).
Lakshmi, D. S., and D. H. L. Prasad: *Chem. Eng. J.,* **48:** 211 (1992).
Lambert, J. D.: in D. R. Bates (ed.), *Atomic and Molecular Processes,* Academic Press, New York, 1962.
Latini, G., and C. Baroncini: *High Temp.-High Press.,* **15:** 407 (1983).
Latini, G., and M. Pacetti: *Therm. Conduct.,* **15:** 245 (1977); pub. 1978.
Latini, G., G. Passerini, and F. Polonara: *Intern. J. Thermophys,* **17:** 85 (1996).
Latini, G., F. Marcotullio, P. Pierpaoli, and A. Ponticiello: *Thermal Conductivity,* vol. 20, p. 205, Plenum Pr., New York, 1989.
Le Neindre: *Recommended Reference Materials for the Realization of Physicochemical Properties,* K. N. Marsh (ed.), IUPAC, Blackwell Sci. Pub., Oxford, 1987.
Lenoir, J. M.: *Petrol. Refiner,* **36**(8)**:** 162 (1957).
Luning Prak, D. J., J. M. Fries, R. T. Gober, R. Vozka, G. Kilaz, T. R. Johnson, S. L. Graft, P. C. Trulove, and J. S. Cowart: *J. Chem. Eng. Data,* **64:** 1725–1745 (2019).
Rumble, J. R. (ed.): *Handbook of Chemistry and Physics,* 102 ed., CRC Press, Boca Raton, 2021.
Liley, P. E., T. Makita, and Y. Tanaka: in C. Y. Ho (ed.), *Cindas Data Series on Material Properties, Properties of Inorganic and Organic Fluids,* Hemisphere, New York, 1988.
Lindsay, A. L., and L. A. Bromley: *Ind. Eng. Chem.,* **42:** 1508 (1950).
McLaughlin, E.: *Chem. Rev.,* **64:** 389 (1964).
Mani, N., and J. E. S. Venart: *Advan. Cryog. Eng.,* **18:** 280 (1973).
Mason, E. A.: *J. Chem. Phys.,* **28:** 1000 (1958).
Mason, E. A., and L. Monchick: *J. Chem. Phys.,* **36:** 1622 (1962).
Mason, E. A., and S. C. Saxena: *Phys. Fluids,* **1:** 361 (1958).
Mason, E. A., and S. C. Saxena: *J. Chem. Phys.,* **31:** 511 (1959).
Mason, E. A., and H. von Ubisch: *Phys. Fluids,* **3:** 355 (1960).
McCoubrey, J. C. and N. M. Singh: *J. Phys. Chem.,* **67:** 517 (1963).
Miller, J. W., J. J. McGinley, and C. L. Yaws: *Chem. Eng.,* **83**(23)**:** 133 (1976).
Miller, J. W., P. N. Shah, and C. L. Yaws: *Chem. Eng.,* **83**(25)**:** 153 (1976a).
Misic, D., and G. Thodos: *AIChE J.,* **7:** 264 (1961).
Misic, D., and G. Thodos: *AIChE J.,* **11:** 650 (1965).
Missenard, A.: *Conductivite Thermique des Solides, Liquides, Gaz et de Leurs Melanges,* Editions Eyrolles, Paris, 5 (1965).

Missenard, A.: *Comptes Rendus*, **260:** 5521 (1965a).
Missenard, A.: *Rev. Gen. Thermodyn.*, **101**(5)**:** 649 (1970).
Missenard, A.: *Rev. Gen. Thermodyn.*, **11:** 9 (1972).
Mo, K. C., and K. E. Gubbins: *Chem. Eng. Comm.*, **1:** 281 (1974).
Muckenfuss, C., and C. F. Curtiss; *J. Chem. Phys.*, **29:** 1273 (1958).
Nagata, I.: *J. Chem. Eng. Japan*, **6:** 18 (1973).
Nieto de Castro, C. A., J. M. N. A. Fareleira, J. C. G. Calado, and W. A. Wakeham: *Proc. 8th Symp. Thermophys. Prop.*, Vol. 1, *Thermophysical Prop. of Fluids*, J. V. Sengers (ed.), ASME, New York, 1981, p. 247.
Nieto de Castro, C. A., S. F. Y. Li, A. Nagashima, R. D. Trengove, and W. A. Wakeham: *J. Phys. Chem. Ref. Data*, **15:** 1073 (1986).
O'Neal, C. Jr., and R. S. Brokaw: *Phys. Fluids*, **5:** 567 (1962).
Ogiwara, K., Y. Arai, and S. Saito: *Ind. Eng. Chem. Fundam.*, **19:** 295 (1980).
Ogiwara, K., Y. Arai, and S. Saito: *J. Chem. Eng. Japan*, **15:** 335 (1982).
Owens, E. J., and G. Thodos: *AIChE J.*, **3:** 454 (1957).
Owens, E. J., and G. Thodos: *Proc. Joint Conf. Thermodyn. Transport Prop. Fluids, London*, July 1957a, pp. 163–68, Inst. Mech. Engrs., London, 1958.
Owens, E. J., and G. Thodos: *AIChE J.*, **6:** 676 (1960).
Pachaiyappan, V., S. H. Ibrahim, and N. R. Kuloor: *Chem. Eng.*, **74:** 140 (1967).
Peterson, J. N., T. F. Hahn, and E. W. Comings: *AIChE J.*, **17:** 289 (1971).
Prasad, R. C., and J. E. S. Venart: *Proc. 8th Symp. Thermophys. Prop.*, Vol. 1, *Themophysical Prop. of Fluids*, J. V. Sengers (ed.), ASME, New York, 1981, p. 263.
Preston, G. T., T. W. Chapman, and J. M. Prausnitz: *Cryogenics*, **7**(5)**:** 274 (1967).
Qun-Fung, L., L. Rui-Sen, and N. Dan-Yan: *J. Chem. Eng. Data*, **42:** 971, (1997).
Rabenovish, B. A.: *Thermophysical Properties of Substances and Materials*, 3rd ed., Standards, Moscow, 1971.
Rastorguev, Yu. L., G. F. Bogatov, and B. A. Grigor'ev: *Isv. vyssh. ucheb. Zaved., Neft'i Gaz.*, **11**(12)**:** 59 (1968).
Richter, G. N., and B. H. Sage: *J. Chem. Eng. Data*, **2:** 61 (1957).
Richter, G. N., and B. H. Sage: *J. Chem. Eng. Data*, **8:** 221 (1963).
Richter, G. N., and B. H. Sage: *J. Chem. Eng. Data*, **9:** 75 (1964).
Riedel, L.: *Chem. Ing. Tech.*, **21:** 349 (1949); **23:** 59, 321, 465 (1951).
Rodriguez, H. V.: Ph.D. thesis, Louisiana State Univ., Baton Rouge, La., 1962.
Rosenbaum, B. M., and G. Thodos: *Physica*, **37:** 442 (1967).
Rosenbaum, B. M., and G. Thodos: *J. Chem. Phys.*, **51:** 1361 (1969).
Rowley, R. L.: *Chem. Eng. Sci.*, **37:** 897 (1982).
Rowley, R. L.: *Chem. Eng. Sci.*, **43:** 361 (1988).
Roy, D.: M.S. thesis, Northwestern University, Evanston, Ill., 1967.
Roy, D., and G. Thodos: *Ind. Eng. Chem. Fundam.*, **7:** 529 (1968).
Roy, D., and G. Thodos: *Ind. Eng. Chem. Fundam.*, **9:** 71 (1970).
Sastri, S. R. S.: personal communication, Regional Research Laboratory, Bhubaneswar (1998).
Sastri, S. R. S., and K. K. Rao: *Chem. Eng.*, **106:** Aug. 1993.
Saxena, S. C., and J. P. Agrawal: *J. Chem. Phys.*, **35:** 2107 (1961).
Saxena, S. C., M. P. Saksena, and R. S. Gambhir: *Brit. J. Appl. Phys.*, **15:** 843 (1964).
Schaefer, C. A., and G. Thodos: *Ind. Eng. Chem*, **50:** 1585 (1958).
Schaefer, C. A., and G. Thodos: *AIChE J.*, **5:** 367 (1959).
Schafer, K.: *Z. Phys. Chem.*, **B53:** 149 (1943).
Sengers, J. V., and J. Luettmer-Strathmann: Chap. 6 in J. Millat, J. H. Dymond, and C. A. Nieto de Castro (eds.), *Transport Properties of Fluids, Their Correlation, Prediction and Estimation, IUPAC*, Cambridge Univ. Press, Cambridge, 1996.
Shroff, G. H.: Ph.D. thesis, University of New Brunswick, Fredericton, 1968. Srivastava, B. N., and R. C. Srivastava: *J. Chem. Phys.*, **30:** 1200 (1959).
Stephan, K. and T. Heckenberger: Thermal Conductivity and Viscosity Data of Fluid Mixtures, *Chem. Data Ser.*, vol. X, Part 1, DECHEMA, Frankfurt, 1988.
Stephan, K., and H. Hildwein: Recommended Data of Selected Compounds and Binary Mixtures, *Chem. Data Ser.*, vol. IV, Parts 1 + 2, DECHEMA, Frankfurt, 1987.
Stiel, L. I., and G. Thodos: *AIChE J.*, **10:** 26 (1964).
Sutton, J. R.: "References to Experimental Data on Thermal Conductivity of Gas Mixtures," *Natl. Eng. Lab., Rept. 612*, East Kilbride, Glasgow, Scotland, May 1976.

Svehla, R. A.: "Estimated Viscosities and Thermal Conductivities of Gases at High Temperatures," *NASA Tech. Rept. R-132,* Lewis Research Center, Cleveland, Ohio, 1962.

Tandon, P. K., and S. C. Saxena: *Appl. Sci. Res.,* **19:** 163 (1965).

Teja, A. S., and P. Rice: *Chem. Eng. Sci.,* **36:** 417 (1981).

Teja, A. S., and P. Rice: *Chem. Eng. Sci.,* **37:** 790 (1982).

Teja, A. S., and G. Tardieu: *Can. J. Chem. Eng.,* **66:** 980 (1988).

Touloukian, Y. S., and C. Y. Ho: *Thermophysical Properties of Matter*, 13 Vols, IFI/Plenum, New York (1970).

Tsederberg, N. V.: *Thermal Conductivity of Gases and Liquids,* The M.I.T. Press Cambridge, Mass., 1965.

Tufeu, R., and B. L. Neindre: *High Temp.-High Press.,* **13:** 31 (1981).

Ubbelohde, A. R.: *J. Chem. Phys.,* **3:** 219 (1935).

Usmanov, I. U., and A. S. Salikov: *Russ. J. Phys. Chem.,* **51**(10)**:** 1488 (1977).

Vargaftik, N. B.: "Thermal Conductivities of Compressed Gases and Steam at High Pressures," *Izu. Vses. Telpotelzh. Inst.,* Nov. 7, 1951; personal communication, Prof. N. V. Tsederberg, Moscow Energetics Institute.

Vargaftik, N. B.: *Proc. Joint Conf. Thermodyn. Transport Prop. Fluids, London,* July 1957, p. 142, Inst. Mech. Engrs., London, 1958.

Vargaftik, N. B., and Y. P. Os'minin: *Teploenergetika,* **3**(7)**:** 11 (1956).

Vargaftik, N. B., L. P. Filippov, A. A. Tarzimanov, and E. E. Totskii: *Handbook of Thermal Conductivity of Liquids and Gases,* CRC Press, Boca Raton, 1994.

Vargaftik, N. B., Y. K. Vinogradov, and V. S. Yargin: *Handbook of Physical Properties of Liquids and Gases,* Begell House, New York, 1996.

Venart, J. E. S., and R. C. Prasad: *J. Chem. Eng. Data,* **25:** 198 (1980).

Vesovic, V., W. A. Wakeham, G. A. Olchowy, J. V. Sengers, J. T. R. Watson, and J. Millat: *J. Phys. Chem. Ref. Data,* **19:** 763 (1990).

Vesovic, V., and W. A. Wakeham: *High Temp.-High Press.,* **23:** 179 (1991).

Vines, R. G.: *Aust. J. Chem.,* **6:** 1 (1953).

Vines, R. G.: *Proc. Joint Conf. Thermodyn. Transport Prop. Fluids, London,* July 1957, Inst. Mech. Engrs., London, 1958, pp. 120–123.

Vines, R. G., and L. A. Bennett: *J. Chem. Phys.,* **22:** 360 (1954).

Vredeveld, D.: personal communication, 1973.

Wassiljewa, A.: *Physik. Z.,* **5:** 737 (1904).

Wilding, W. V., T. A. Knotts, N. F. Giles, R. L. Rowley, and J. L. Oscarson, *DIPPR® Data Compilation of Pure Chemical Properties*, Design Institute for Physical Properties, AIChE, New York, NY (2017).

Yaws, C. L.: *Handbook of Thermal Conductivity,* Gulf Pub., Houston, 1995.

Yaws, C. L.: *Handbook of Transport Property Data: Viscosity, Thermal Conductivity, and Diffusion Coefficients of Liquids and Gases,* Gulf Pub., Houston, 1995a.

Yorizane, M., S. Yoshimura, H. Masuoka, and H. Yoshida: *Ind. Eng. Chem. Fundam.,* **22:** 454 (1983).

Yorizane, M., S. Yoshimura, H. Masuoka, and H. Yoshida: *Ind. Eng. Chem. Fundam.,* **22:** 458 (1983a).

Zaytsev, I. D., and G. G. Aseyev: *Properties of Aqueous Solutions of Electrolytes,* CRC Press, Boca Raton, 1992.

Zheng, X.-Y., S. Yamamoto, H. Yoshida, H. Masuoka, and M. Yorizane: *J. Chem. Eng. Japan,* **17:** 237 (1984).

12

Diffusion

12.1 SCOPE

In Sec. 12.2, we discuss briefly several frames of reference from which diffusion can be related and define the diffusion coefficient. Most recent progress in diffusivity research has focused on self-diffusivity, as reviewed in Sec. 12.3. Low-pressure binary gas diffusion coefficients are treated in Secs. 12.4 and 12.5. The pressure and temperature effects on gas-phase diffusion coefficients are covered in Secs. 12.6 and 12.7, respectively. The theory for liquid diffusion coefficients is introduced in Sec. 12.9, and estimation methods for binary liquid diffusion coefficients at infinite dilution are described in Sec. 12.10. Concentration effects are considered in Sec. 12.11 and temperature and pressure effects in Sec. 12.12. Brief comments on multicomponent mixtures are made in Secs. 12.8 (gases) and 12.13 (liquids); ionic solutions are covered in Sec. 12.14.

12.2 BASIC CONCEPTS AND DEFINITIONS

The extensive use of the term "diffusion" in the chemical engineering literature is based on an intuitive feel for the concept, i.e., diffusion refers to the net transport of material within a single phase in the absence of mixing (by mechanical means or by convection). Both experiment and theory have shown that diffusion can result from pressure gradients (pressure diffusion), temperature gradients (thermal diffusion), external force fields (forced diffusion), and concentration gradients. Only the last type is considered in this chapter; i.e., the discussion is limited to diffusion in isothermal, isobaric systems with no external force field gradients.

Even with this limitation, confusion can easily arise unless care is taken to clearly define diffusion fluxes and diffusion potentials, e.g., driving forces. The proportionality constant between the flux and potential is the *diffusion coefficient,* or *diffusivity.*

12.2.1 Diffusion Fluxes

A detailed discussion of diffusion fluxes has been given by Bird et al. (1960) and Cussler (1997). Various types originate because different reference frames are employed. The most obvious reference plane is fixed on the equipment in which diffusion is occurring. This plane is designated by RR' in Fig. 12.1. Suppose in a binary mixture of A and B, that A is diffusing to the left and B to the right. If the diffusion rates of these species are not

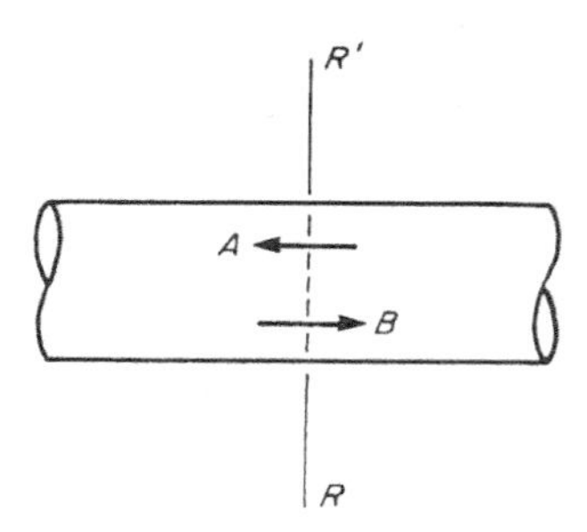

FIGURE 12.1
Diffusion across plane RR'.

identical, there will be a net depletion on one side and an accumulation of molecules on the other side of RR'. To maintain the requirements of an isobaric, isothermal system, bulk motion of the mixture occurs. Net movement of A (as measured in the fixed reference frame RR') then results from both diffusion and bulk flow.

Although many reference planes can be delineated, a plane of *no net mole flow* is normally used to define a diffusion coefficient in binary mixtures. If J_A^M represents a mole flux in a mixture of A and B, J_A^M is then the net mole flow of A across the boundaries of a hypothetical (moving) plane such that the total moles of A and B are invariant on both sides of the plane. J_A^M can be related to fluxes across RR' by

$$J_A^M = N_A - x_A(N_A + N_B) \tag{12.2-1}$$

where N_A and N_B are the fluxes of A and B across RR' (relative to the fixed plane) and x_A is the mole fraction of A at RR'. Note that J_A^M, N_A, and N_B are vectorial quantities, and a sign convention must be assigned to denote flow directions. Equation (12.2-1) shows that the net flow of A across RR' is due to a diffusion contribution J_A^M and a bulk flow contribution $x_A(N_A + N_B)$. For equimolar counterdiffusion, $N_A + N_B = 0$ and $J_A^M = N_A$.

One other flux is extensively used, i.e., one relative to the plane of *no net volume flow*. This plane is less readily visualized. By definition,

$$J_A^M + J_B^M = 0 \tag{12.2-2}$$

and if J_A^V and J_B^V are vectorial molar fluxes of A and B relative to the plane of no net volume flow, then, by definition,

$$J_A^V \overline{V}_A + J_B^V \overline{V}_B = 0 \tag{12.2-3}$$

where $\overline{V}_A$ and $\overline{V}_B$ are the partial molar volumes of A and B in the mixture. It can be shown that

$$J_A^V = \frac{\overline{V}_B}{V} J_A^M \quad \text{and} \quad J_B^V = \frac{\overline{V}_A}{V} J_B^M \tag{12.2-4}$$

where V is the volume per mole of mixture. Obviously, if $\overline{V}_A = \overline{V}_B = V$, as in an ideal mixture, then $J_A^V = J_A^M$.

12.2.2 Diffusion Coefficients

Diffusion coefficients for a binary mixture of A and B are commonly defined by

$$J_A^M = -cD_{AB}\frac{dx_A}{dz} \tag{12.2-5}$$

$$J_B^M = -cD_{BA}\frac{dx_B}{dz} \tag{12.2-6}$$

where c is the total molar concentration ($= V^{-1}$) and diffusion is in the z direction. With Eq. (12.2-2), since $(dx_A/dz) + (dx_B/dz) = 0$, we have $D_{AB} = D_{BA}$. The diffusion coefficient then represents the proportionality between the flux of A relative to a plane of no net molar flow and the gradient $c(dx_A/dz)$. From Eqs. (12.2-4) to (12.2-6) and the definition of a partial molar volume it can be shown that, for an isothermal, isobaric binary system,

$$J_A^V = -D_{AB}\frac{dc_A}{dz} \text{ and } J_B^V = -D_{AB}\frac{dc_B}{dz} \tag{12.2-7}$$

When fluxes are expressed in relation to a plane of no net volume flow, the potential is the concentration gradient. D_{AB} in Eq. (12.2-7) is identical with that defined in Eq. (12.2-5), when $\overline{V}_A \approx \overline{V}_B \approx V =$ as in ideal gases, $J_A^V \approx J_A^M$, $J_B^V \approx J_B^M$.

12.2.3 Mutual, Self-, and Tracer Diffusion Coefficients

The diffusion coefficient D_{AB} introduced above is termed the *mutual diffusion coefficient*, and it refers to the diffusion of one constituent in a binary system. A similar coefficient D_{1m} would imply the diffusivity of component 1 in a mixture (see Secs. 12.8 and 12.13).

Tracer diffusion coefficients (sometimes referred to as *intradiffusion coefficients*) relate to the diffusion of a labeled component within a *homogeneous* mixture. Like mutual diffusion coefficients, tracer diffusion coefficients can be a function of composition. If D_A^* is the tracer diffusivity of A in a mixture of A and B, then as $x_A \rightarrow 1.0$, $D_A^* \rightarrow D_{AA}$, where D_{AA} is the *self-diffusion coefficient* of A in pure A.

In Fig. 12.2, the various diffusion coefficients noted above are shown for a binary liquid mixture of n-octane and n-dodecane at 60 °C (Van Geet and Adamson, 1964). In this case, the mutual diffusion of these two hydrocarbons increases as the mixture becomes richer in n-octane. With A as n-octane and B as n-dodecane, as $x_A \rightarrow 1.0$, $D_{AB} = D_{BA} \rightarrow D_{BA}^\circ$, where this notation signifies that this limiting diffusivity represents the diffusion of B in a medium consisting essentially of A, that is, n-dodecane molecules diffusing through almost pure n-octane. Similarly, D_{AB}° is the diffusivity of A in essentially pure B. Except in the case of infinite dilution, tracer diffusion coefficients differ from binary-diffusion coefficients, and there is no way to relate the two coefficients (Cussler, 1997). Similarly, there is no relation between quantities such as D_{BB} and D_{AB}° or D_{AA} and D_{BA}°. In this chapter, correlation techniques for D_{ij} (or D_{ij}°) are the primary focus; methods for correlating D_{ii} are briefly reviewed, however, in Sec. 12.3.

12.2.4 Chemical Potential Driving Force

The mutual diffusion coefficient D_{AB} in Eq. (12.2-7) indicates that the flux of a diffusing component is proportional to the concentration gradient. Diffusion is, however, affected by more than just the gradient in concentration, e.g., the intermolecular interactions of the molecules (Dullien, 1974; Turner, 1975) which can give complex composition dependence of the behavior in addition to that from temperature and pressure. Thus, fluxes may not be linear in the concentration gradient; it is even possible that species can diffuse opposite to their concentration gradient.

Modern theories of diffusion (Ghai et al., 1973) have adopted the premise that if one perturbs the equilibrium composition of a binary system, the subsequent diffusive flow required to attain a new equilibrium state is proportional to the gradient in chemical potential $(d\mu_A/dz)$. Since the diffusion coefficient was defined in Eqs. (12.2-5) and (12.2-6) in terms of a mole fraction gradient instead of a chemical potential gradient, it is argued that one

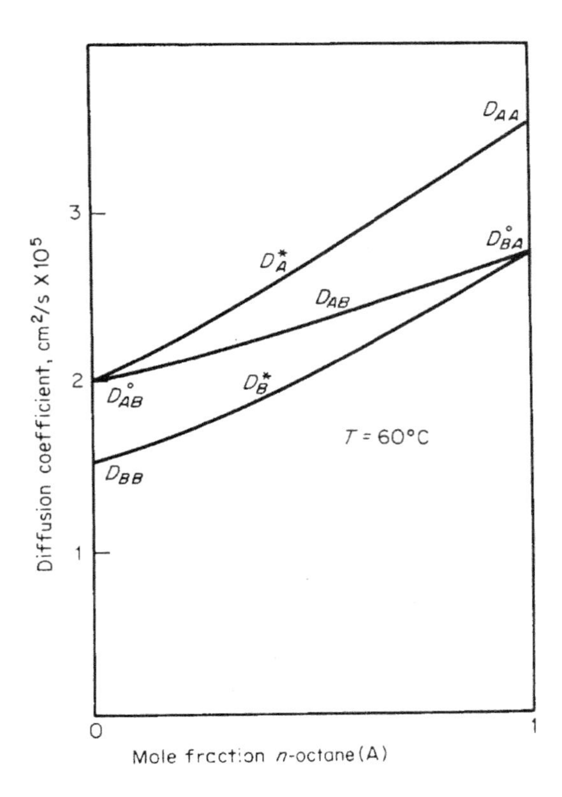

FIGURE 12.2
Mutual, self-, and tracer diffusion coefficients in a binary mixture of n-octane and n-dodecane. (van Geet and Adamson, 1964)

should include a thermodynamic correction in any correlation for D_{AB} based on ideal solution considerations. This correction is

$$\alpha^D = \left[\frac{(\partial \ln(a_A))}{(\partial \ln(x_A))} \right]_{T,P} \tag{12.2-8}$$

where the activity $a_A = x_A \gamma_A$ and γ_A, the activity coefficient is described in Chap. 8. By virtue of the Gibbs-Duhem equation, α^D is the same regardless of whether activities and mole fractions of either A or B are used in Eq. (12.2-8). For gases, α^D is almost always close to unity (except at high pressures), and this correction is seldom used. For liquid mixtures, however, it is widely adopted, as will be illustrated in Sec. 12.11.

12.3 PROGRESS IN SELF-DIFFUSIVITY CORRELATION

Let one of the components, say A, in a homogenous multicomponent system at uniform temperature be partly replaced by a labeled otherwise equivalent species, say A^* that has negligible chemical difference from A with respect to diffusion in the system. Both A and A^* will have equimolal counter diffusion, known as intradiffusion (Albright and Mills, 1965). In the absence of chemical species other than A and A^* in the mixture, the observed diffusivity coefficient will be the self-diffusivity of A in pure A, i.e., D_{AA}.

Self-diffusivity is the simplest transport property to estimate by molecular simulation, but it was a tedious and expensive transport property to measure experimentally when isotope tracer experiments were the only means. Fortunately, nuclear magnetic resonance (NMR) spin echo (SE) methods have been developed to greatly facilitate these measurements (Carr and Purcell, 1954; Hahn, 1950; McCall et al., 1963; Stejskal and Tanner, 1965; Stilbs, 1987). The self-diffusivity establishes how a molecule diffuses through Brownian motion in an isotropic system. Both the experiments and the molecular dynamics simulations employ Einstein's equation to relate self-diffusivity in three-dimensional space to the mean squared displacement (MSD), ***r***. (Maginn et al., 2018)

$$D^S = \lim_{t\to\infty} \frac{\langle [r(t) - r(t_0)]^2 \rangle}{6(t - t_0)} \tag{12.3-1}$$

where t is time and t_0 is an initial reference time, $\boldsymbol{r}$ is the molecular center of mass position.

The MSD of a molecule relative to its own previous position can pertain to a pure fluid or to a component in a mixture. When $x_A \to 0$ in a binary mixture, $D_A^S(x_B \approx 1) = D_{AB}^\circ$; on the other hand, when $x_A \to 1$, $D_A^S(x_A = 1) = D_{AA}$. Pulsed Gradient Spin Echo (PGSE) methods provide a rich spectrum of accurate measurements over a wide range of conditions and applications. With Fourier-Transforms (FT-PGSE) multicomponent self-diffusivities for all but heavy molecules can be resolved in a few minutes (Stilbs, 1987). Self-diffusivities of heavy molecules generally require a broader field with stronger amplitude, undermining FT-PGSE, but individual component diffusivities can still be resolved if they are very different, as in a polymer mixed with a solvent (von Meerwall, 1991). Such analysis can provide valuable molecular scale insight into phenomena like association and aggregation. For example, FT-PGSE has been applied to resolve the states of aggregation and microstructure in microemulsions over the entire range of external phases and cosolvent compositions (Stilbs, 1987).

Methods of correlating self-diffusivity are like those for mutual diffusivity. Most methods are closely related to the Chapman-Enskog equation for hard spheres. Lee and Thodos (1983), Murad (1981), Liu et al. (1998), and Zabaloy et al. (2006) present correlations based on the hard sphere perspective. In this approach, it is normal to represent large molecules like n-hexadecane as large spheres; the cube of the effective spherical diameter is then proportional to the molecular volume. Yu and Gao (1999) were the first to point out that the hard sphere perspective implies the self-diffusivity should scale as molecular volume to the $-1/6$ power. Gerek and Elliott (2010) showed that $-1/6$ scaling is inconsistent with the Rouse scaling well-known in the polymer literature. Rouse scaling suggests that the self-diffusivity should decrease as the reciprocal of molecular volume. Molecular simulations of fused sphere chain molecules follow Rouse scaling at all densities. Figure 12.3a shows how self-diffusivity of n-alkanes behaves in the low-density, high-temperature limit, clarifying the inconsistency with the hard sphere perspective. Figure 12.3b shows how hard fused-sphere chains behave over all densities. The behavior in Fig. 12.3b reflects the packing fraction reduced relative to the estimated packing fraction at the glass transition (η_g).

The self-diffusivity drops dramatically when the molecules are jammed at the glass transition, regardless of whether the molecules are large or small. Small (spherical) molecules reach their glass transition density at lower packing fractions than long chain molecules do. When plotted with mass density on the abscissa, the various curves cross in a confusing manner, as in Figure 2 of Gerek and Elliott. Plotting the reduced density on the abscissa makes the behavior appear to be much simpler, as in Fig. 12.3b. Gerek and Elliott (2010) were able to correlate the fused hard sphere chain simulations with

$$\frac{\rho D_C}{\rho D_o^{mon}} = C(m^{Eff})\left(1 - \frac{\eta^P}{\eta^g}\right)\left(1 + d_2(m^{Eff})\left(\frac{\eta^P}{\eta^g}\right)^2 - d_4(m^{Eff})\left(\frac{\eta^P}{\eta^g}\right)^4\right) \tag{12.3-2}$$

$$\eta^g = 0.57 + 0.43\,\Delta_m;\ \Delta_m \equiv (m^{Eff} - 1)/m^{Eff} \tag{12.3-3}$$

$$C(m^{Eff}) = \frac{1 + C_1^{FS}\Delta_m + C_2^{FS}\Delta_m^2}{m^{Eff}} \tag{12.3-4}$$

$$d_i = d_i^{HS} - d_i^{FS}\Delta_m \tag{12.3-5}$$

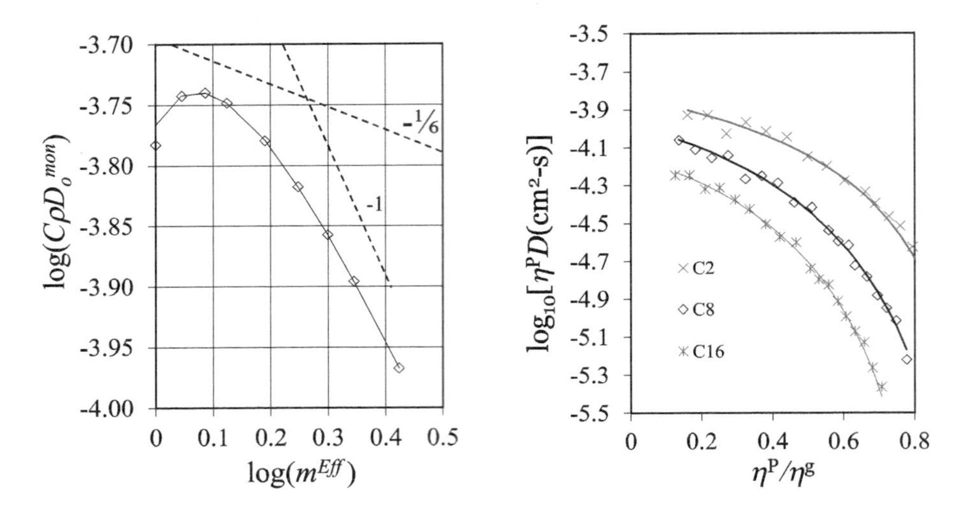

FIGURE 12.3
Self-diffusivity of fused-hard-sphere chain models of n-alkanes from molecular dynamics simulation.

where η^P is the packing fraction. The correlation constants are: $\rho D_o^{mon} = 0.000171$ g/cm-s, $d_2^{HS} = 0.4740$, $d_4^{HS} = 1.1657$, $d_2^{FS} = 3.733$, $d_4^{FS} = 2.619$, $c_1^{FS} = 1.662$, $c_2^{FS} = -1.005$, and m^{Eff} is the effective number of hard tangent sphere segments that comprise the chain molecule. For n-alkanes,

$$m^{Eff} = 1 + 0.11(N^C - 1) \tag{12.3-6}$$

where N^C is the carbon number. For non-alkanes, m^{Eff} and ρCD_o^{mon} are treated as parameters for correlating the molecular simulation results. Thus far, there may seem to be too many correlating parameters, but keep in mind that these are correlations for interpolating the simulation results of the reference fluid. No experimental data have been applied yet. One additional correction must be applied to the reference simulation results to correct for the softness of the repulsive potential at high temperatures.

$$\frac{V^{soft}(T)}{V^{Eff}} = 1 - \frac{V^{mon}}{V^{Eff}}\left[1 - \left(\frac{d^{EHS}}{\sigma^{LJ}}\right)^3\right] \tag{12.3-7}$$

Where $V^{mon} = 0.02443$ nm^3 and V^{Eff} is determined from the molecular simulation. $\eta^P = V^{soft}/V$.

$$\frac{d^{EHS}}{\sigma^{LJ}} = 2^{1/6}(0.6288(T^*)^2 + 11T^* + 1)^{\frac{-1}{25}} \tag{12.3-8}$$

where $T^* = T(\mathrm{K})/79$. To account for the effects of attractive forces on self-diffusivity, Gerek and Elliott (2010) proposed a simple mean field correlation,

$$\ln\left(\frac{\rho D^{ref}}{\rho D}\right) = \left(\frac{\eta^P}{1 - \dfrac{\eta^P}{\eta^g}}\right)^2 \frac{\varepsilon^D}{kT} \tag{12.3-9}$$

For n-alkanes,

$$\frac{\varepsilon^{D}}{k} = 0.6T_c(0.56\Delta_m + 1.26\Delta_m^2) \tag{12.3-10}$$

For non-alkanes, ε^D is the single parameter fit to experimental data. Since non-alkanes may contain branches, rings, polar, or associating functionality, treating all molecules with the same $\{c_i\}$ and $\{d_i\}$ is imperfect, but it simplifies the correlation of molecular simulation data if non-alkanes can be characterized as n-alkanes with a regressed value of m^{Eff}. This is usually successful, except in the cases of fluoroform and pyridine, where customized values of η_g were necessary.

A representative example of a more traditional approach to self-diffusivity modeling was provided by Silva et al. (1998). Silva et al. used a correlation of the self-diffusivity of Lennard-Jones spheres [Eq. (12.3-11)] as the basis for correlating self-diffusivity of any molecule, including a special correction for hydrogen bonding molecules.

$$D^{LJ}\left(\frac{cm^2}{s}\right) = \frac{11.08\sigma_{eff}^{BMD}A^D}{\eta^P}\left(\frac{1000RT}{MW}\right)^{1/2}\exp\left(\frac{-1.1379\eta^P}{1-\eta^P/\eta^g} - \frac{E^D}{RT}\right) \tag{12.3-11}$$

where $\eta^P = N_A\rho(\sigma_{eff}^{BMD})^3\pi/6$, ρ is the molar density, $\eta^g = 0.6591$, and

$$\sigma_{eff}^{BMD} = \sigma^{LJ}2^{1/6}\left[1 + \left(\frac{T}{T^D}\right)^{1/2}\right]^{-1/6} \tag{12.3-12}$$

Silva et al. (1998) implemented this correlation in four formats, referred to here as LJ0 to LJ4. For the LJ0, LJ1, and LJ2 models: $A^D = 1$, $T^D = 0.7559\varepsilon^{LJ}/k$, and $E^D/R = 0.27862\varepsilon^{LJ}/k$. Further specifications are: (1) in the LJ0 model, $\varepsilon^{LJ} = 0.744\ kT_c$ and σ^{LJ} is given by Eq. (12.3-13) below; (2) in the LJ1 model, $\varepsilon^{LJ} = 0.744\ kT_c$ and σ^{LJ} is an adjustable parameter; (3) in the LJ2 model, both σ^{LJ} and ε^{LJ} are adjustable parameters; (4) in the LJ4 model, σ^{LJ}, ε^{LJ}, A^D, and T^D are adjustable parameters

$$\sigma^{LJ} = 0.17791 + 11.779\left(\frac{T_c}{P_c}\right) - 0.049029\left(\frac{T_c}{P_c}\right)^2 \tag{12.3-13}$$

Our evaluations begin by comparing the LJ0 correlation with the generalized correlation of Gerek and Elliott for n-alkanes. Both models are based on correlations of measured compounds but provide predictions for unmeasured compounds. Table 12.1 summarizes the comparisons. The LJ0 method does perform remarkably well, even including n-hexadecane, but the extrapolation to n-triacontane is problematic. The DIPPR compilation does not provide critical properties for n-hexacontane (C60), so they were estimated from the correlation of Emami et al. (2009) as $T_c = 961$K and $P_c = 0.443$ MPa. This led to an estimate of $\sigma^{LJ} = 0.628$ nm for hexacontane compared to 0.887 nm for triacontane so the LJ0 correlation was abandoned for hexacontane. Larger values of T_c/P_c lead to negative values of $(\sigma^{LJ})^3$, so this appears to be a problem with the quadratic form of Eq. (12.3-13). The Gerek-Elliott method shows remarkable accuracy for n-alkanes heavier than n-butane, but inferior accuracy for the short alkanes. The Gerek-Elliott predictions for triacontane and hexacontane lend support to their scaling arguments for both the repulsive and attractive contributions.

Table 12.1 also includes comparisons for n-alkanes with the LJ1 and LJ2 methods. Overall, the LJ1 method provides accuracy nearly equivalent to the LJ2 method, despite having half as many adjustable parameters. This reflects the sensitivity of Eq. (12.3-11) to value of σ^{LJ}. A 1% change in the value σ^{LJ} results in a change of the %AAD from 5% to 15%, whereas a 1% change in ε^{LJ} leads to an insignificant change in %AAD.

Table 12.2 presents the results for non-alkanes. The values for the LJ models are taken from the reports of Liu et al. (1998) and Silva et al. (1998) because the databases are very similar for these compounds. For these

TABLE 12.1 Evaluation of self-diffusivity correlations for n-alkanes.

Compound	NPTS	T_{min}(K)	T_{max}	P_{min} (MPa)	P_{max}	ρ_{min}– (g/cm³)	ρ_{max}	%AAD G&E	LJ0	LJ1	LJ2
Methane	296	110	454	0.77	221.6	0.008	0.528	11	6.4	6.2	6.2
Ethane	71	136	454	25	200	0.238	0.672	10.3	5.2	3.5	3.5
Propane	57	154	453	0.1	200	0.37	0.756	12.4	7.6	5.9	5.9
n-Butane	18	177	451	0.1	200	0.518	0.741	16.7	15.3	13.9	13.9
n-hexane	63	223	333	0.1	393.8	0.614	0.819	6.4	20.2	3.1	3.1
n-heptane	27	173	373	0.1	0.1	0.613	0.778	6.4	18.0	5.2	5.2
n-octane	50	248	348	0.1	360.8	0.658	0.834	4.6	22.1	4.3	4.3
n-nonane	12	236	423	0.1	0.1	0.609	0.764	3.8	11.2	4.2	4.2
n-decane	24	250	440	0.1	60	0.61	0.764	6.2	14.8	4.8	4.8
n-dodecane	16	263	433	0.1	60	0.644	0.787	7.3	13.6	4.4	4.4
n-hexadecane	25	298	348	0.1	279.2	0.736	0.853	14.3	19.4	20.3	20.3
n-triacontane	6	343	443	0.1	0.1	0.722	0.781	10.8	2779.8	51.7	35.4
n-hexacontane	4	383	443	0.1	0.1	0.753	0.785	6.7	NA	49.6	29.2
Overall	659							10.0	36.0	5.7	6.8

non-alkanes, all the LJ models perform better than the Gerek and Elliott (2010) model. The most direct comparison is between the Gerek-Elliott model and the LJ1 model, in which case each model has one adjustable parameter. While the Gerek-Elliott model adjusts the strength of attraction, the LJ1 model adjusts the strength of repulsion through σ^{LJ}. As noted for n-alkanes, the %AAD is much more sensitive to repulsive characterization, giving the advantage to the LJ1 model. Note that all the non-alkanes studied here are roughly spherical, so it is not surprising that the results here are like those for n-alkanes smaller than n-pentane. Comparing the LJ1 and the LJ2 models, the advantage of the LJ2 model is small, as observed for n-alkanes. For nonassociating compounds, the LJ1 model is reasonably reliable. For associating compounds, however, the LJ4 model offers a significant advantage. For completeness, Table 12.2 includes results for the Lee-Thodos (1983) (LT) model as reported by Liu et al. (1998). Noting its four parameters, the Lee-Thodos model provides only a small advantage over the LJ1 model. The complete database of Gerek and Elliott (2010) is available through the PGL6ed website.

12.3.1 Discussion

In summary, the LJ1 model of Silva et al. (1998) provides the most promising means of correlating a few experimentally measured data and extrapolating them carefully. For associating compounds, it may be necessary to incorporate the LJ4 method, requiring substantially more data to perform reliably. For compounds less spherical than n-butane, like polymers or heavy hydrocarbons, the Gerek-Elliott model should be considered. Looking to the future, the observations for the Gerek-Elliott model suggest that molecular simulation models of self-diffusivity may provide a reasonable basis for predictions, but they also suggest a caveat. Since the repulsive character of the molecular model is typically fixed by fitting vapor pressure and liquid density data, characterizations of self-diffusivity may run into the problem of misrepresenting the effect of repulsive forces on self-diffusivity. This may be a case where transport data should be included with thermodynamic data when developing the potential model. Finally, we note that we have not evaluated recent advances in modeling self-diffusivity with residual entropy (Bell, 2020; Rosenfeld, 1977). Gerek and Elliott (2010) briefly considered that approach for predicting trends with chain length, but not for correlating temperature effects on a single molecule. It seems the latter approach would be more promising, but further evaluation must await future analysis.

TABLE 12.2 Evaluation of self-diffusivity correlations for non-alkanes.

	%AAD				
Compound	**GE**	**LJ1**	**LJ2**	**LJ4**	**LT**
Argon	14.0	5.7	5.7	4.0	6.8
Hydrogen	71.3	6.7	5.5	5.3	4.0
Carbondisulphide	4.6	1.6	1.1	1.1	0.9
Carbondioxide	27.0	9.9	6.4	6.2	5.3
Ethylene	24.6	6.6	5.8	5.8	4.5
Cyclohexane	5.5	3.7	2.6	2.2	1.7
Benzene	12.8	5.9	5.2	4.6	5.2
Toluene	13.5	4.5	3.0	1.6	1.4
Chlorobenzene	19.1	8.1	1.0	0.6	0.1
Chloromethane	9.5	3.8	3.1	2.5	10.7
Dichloromethane	33.2	6.6	6.5	3.3	8.0
Chloroform	6.2	9.8	9.7	8.7	9.2
Chlorotrifluoromethane	8.4	13.4	6.5	6.1	9.6
Carbontetrachloride	64.4	3.5	3.5	3.4	2.4
Trifluoromethane	55.1	8.6	7.9	6.9	8.5
Carbontetrafluoride	12.4	7.0	6.5	5.6	5.5
Sulfurhexafluoride	10.8	8.6	8.5	6.3	18.4
Perfluorocyclobutane	34.7	7.7	3.8	3.3	3.9
Ammonia	6.5	10.6	1.4	0.9	1.5
Acetone	9.5	2.1	1.0	0.8	0.5
Pyridine	13.0	9.8	9.0	2.7	7.0
Methanol	46.2	37.9	35.5	6.2	47.4
Ethanol	39.9	43.4	29.5	3.5	41.0
Water	17.8	29.3	19.1	16.2	29.5
Overall	23.3	14.9	11.0	5.6	13.9

12.4 DIFFUSION COEFFICIENTS FOR BINARY GAS SYSTEMS AT LOW PRESSURES: PREDICTION FROM THEORY

The theory describing diffusion in binary gas mixtures at low to moderate pressures has been well developed. As noted earlier in Chaps. 10 (Viscosity) and 11 (Thermal Conductivity), the theory results from solving the Boltzmann equation, and the results are usually credited to both Chapman and Enskog, who independently derived the working equation

$$D_{AB} = \frac{3}{16}\frac{(4\pi kT/M_{AB})^{1/2}}{n\pi\sigma_{AB}^2\Omega_D} f_D \tag{12.4-1}$$

where M_A, M_B = molecular weights of A and B
$M_{AB} = 2[(1/M_A) + (1/M_B)]^{-1}$
n = number density of molecules in the mixture
k = Boltzmann's constant
T = absolute temperature

Ω_D, the collision integral for diffusion, is a function of temperature; it depends upon the choice of the intermolecular force law between colliding molecules. σ_{AB} is the characteristic length of the intermolecular force law. Finally, f_D is a correction term, which is of the order of unity. If M_A is of the same order as M_B; f_D lies between 1.0 and 1.02 regardless of composition or intermolecular forces. Only if the molecular masses are very unequal and the light component is present in trace amounts, the value of f_D is significantly different from unity, and even in such cases, f_D is usually between 1.0 and 1.1 (Marrero and Mason, 1972).

If f_D is chosen as unity and n is expressed by the ideal-gas law, Eq. (12.4-1) may be written as

$$D_{AB} = \frac{0.00266T^{3/2}}{PM_{AB}^{1/2}\sigma_{AB}^2\Omega_D} \tag{12.4-2}$$

where D_{AB} = diffusion coefficient, cm²/s
T = temperature, K
P = pressure, bar
σ_{AB} = characteristic length, Å
Ω_D = diffusion collision integral, dimensionless

and M_{AB} is defined under Eq. (12.4-1). The key to the use of Eq. (12.4-2) is the selection of an intermolecular force law and the evaluation of σ_{AB} and Ω_D.

12.4.1 Lennard-Jones 12-6 Potential

As noted in Chap. 10, a popular correlation relating the intermolecular energy ψ, between two molecules to the distance of separation r, is given by

$$\psi^{LJ} = 4\varepsilon^{LJ}\left[\left(\frac{\sigma^{LJ}}{r}\right)^{12} - \left(\frac{\sigma^{LJ}}{r}\right)^{6}\right] \tag{12.4-3}$$

with ε^{LJ} and σ^{LJ} as the characteristic Lennard-Jones energy and length, respectively. Application of the Chapman-Enskog theory to the viscosity of pure gases has led to the determination of many values of ε^{LJ} and σ^{LJ}, sample values are given on the PGL6ed website.

To use Eq. (12.4-2), some rule must be chosen to obtain the interaction value σ_{AB} from σ_A and σ_B. Also, it can be shown that Ω_D is a function only of kT/ε_{AB}, where again some rule must be selected to relate ε_{AB} to ε_A and ε_B. The simple rules shown below are usually employed:

$$\varepsilon_{AB} = (\varepsilon_A\varepsilon_B)^{1/2} \tag{12.4-4}$$

$$\sigma_{AB} = \frac{\sigma_A + \sigma_B}{2} \tag{12.4-5}$$

Ω_D is tabulated as a function of kT/ε for the Lennard-Jones 12-6 potential (Hirschfelder et al., 1954), and various analytical approximations also are available (Hattikudur and Thodos, 1970; Johnson and Colver, 1969; Kestin et al., 1977; Neufeld et al., 1972). The accurate relation of Neufield et al. (1972) is

$$\Omega_D = \frac{A}{(T^*)^B} + \frac{C}{\exp(DT^*)} + \frac{E}{\exp(FT^*)} + \frac{G}{\exp(HT^*)} \tag{12.4-6}$$

where $T^* = kT/\varepsilon_{AB}$ $\quad A = 1.06036 \quad B = 0.15610$
$C = 0.19300 \quad D = 0.47635 \quad E = 1.03587$
$F = 1.52996 \quad G = 1.76474 \quad H = 3.89411$

Example 12.1 Low-pressure diffusion coefficients

Estimate the diffusion coefficient for the system N_2-CO_2 at 590 K and 1 bar. The experimental value reported by Ellis and Holsen (1969) is 0.583 cm^2/s.

Solution. To use Eq. (12.4-2), values of $\sigma(CO_2)$, $\sigma(N_2)$, $\varepsilon(CO_2)$, and $\varepsilon(N_2)$ must be obtained. Using the values in Appendix B of PGL5ed with Eqs. (12.4-4) and (12.4-5) gives $\sigma(CO_2) = 3.941$ Å; $\sigma(N_2) = 3.798$ Å; $\sigma(CO_2\text{-}N_2) = (3.941 + 3.798)/2 = 3.8695$ Å; $\varepsilon(CO_2)/k = 195.2$ K; $\varepsilon(N_2)/k = 71.4$ K; $\varepsilon(CO_2\text{-}N_2)/k = [(195.2)(71.4)]^{1}/_{2} = 118$ K. Then $T^* = kT/\varepsilon\,(CO_2\text{-}N_2) = 590/118 = 5.0$. With Eq. (12.4-6), $\Omega_D = 0.842$. Since $M\,(CO_2) = 44.0$ and $M\,(N_2) = 28.0$, $M_{AB} = (2)[(1/44.0) + (1/28.0)]^{-1} = 34.22$. With Eq. (12.4-2),

$$D(CO_2 - N_2) = \frac{(0.00266)(560)^{3/2}}{(1)(34.22)^{1/2}(3.8695)^2(0.842)} = 0.52 \text{ cm}^2/\text{s}$$

The error is 11%. Ellis and Holsen recommend values of $\varepsilon\,(CO_2\text{-}N_2) = 134$ K and $\sigma(CO_2\text{-}N_2) = 3.660$ Å. With these parameters, they predicted D to be 0.56 cm^2/s, a value closer to that found experimentally.

Equation (12.4-2) is derived for dilute gases consisting of nonpolar, spherical, monatomic molecules; and the potential function (12.4-3) is essentially empirical, as are the combining rules [Eqs. (12.4-4) and (12.4-5)]. Yet Eq. (12.4-2) gives good results over a wide range of temperatures and provides useful approximate values of D_{AB} (Gotch et al., 1974; Gotoh et al., 1973). The general nature of the errors to be expected from this estimation procedure is indicated by the comparison of calculated and experimental values discussed in Table 12.4.

The calculated value of D_{AB} is relatively insensitive to the value of ε_{AB} employed and even to the form of the assumed potential function, especially if values of ε and σ are obtained from viscosity measurements.

No effect of composition is predicted. A more detailed treatment does indicate that there may be a small effect for cases in which M_A and M_B differ significantly. In a specific study of this effect (Yabsley et al., 1973), the low-pressure binary diffusion coefficient for the system He-$CClF_3$ did vary from about 0.416 to 0.430 cm^2/s over the extremes of composition. In another study (Mrazek et al., 1968), no effect of concentration was noted for the methyl alcohol-air system, but a small change was observed with chloroform-air.

12.4.2 Low-pressure Diffusion Coefficients from Viscosity Data

Since the equations for low-pressure gas viscosity [Eq. (10.3-9)] and diffusion [Eq. (12.4-2)] have a common basis in the Chapman-Enskog theory, they can be combined to relate the two gas properties. Experimental data on viscosity as a function of composition at constant temperature are required as a basis for calculating the *binary* diffusion coefficient D_{AB} (Di Pippo et al., 1967; Gupta and Saxena, 1968; Hirschfelder et al., 1954; Kestin et al., 1977; Kestin and Wakeham, 1983). Weissman and Mason (1962) and Weissman (1964) compared the method with a large collection of experimental viscosity and diffusion data and find excellent agreement.

12.4.3 Polar Gases

If one or both components of a gas mixture are polar, a modified Lennard-Jones relation such as the Stockmayer potential is often used. A different collision integral relation [rather than Eq. (12.4-6)] is then necessary and Lennard-Jones σ^{LJ} and ε^{LJ} values are not sufficient.

Brokaw (1969) has suggested an alternative method for estimating diffusion coefficients for binary mixtures containing polar components. Equation (12.4-1) is still used, but the collision integral Ω_D is now given as

$$\Omega'_D = \Omega_D(\text{Eq. } 12.46) + \frac{0.198\,\delta_{AB}^2}{T^*} \qquad (12.4\text{-}7)$$

where $T^* = \dfrac{kT}{\varepsilon_{AB}}$

$$\delta = \frac{1.94 \times 10^3 \; \mu_p^2}{V_b T_b} \tag{12.4-8}$$

μ_p = dipole moment, debyes
V_b = liquid molar volume at the normal boiling point, cm³/mol
T_b = normal boiling point (1 atm), K

$$\frac{\varepsilon}{k} = 1.18(1 + 1.38\delta^2)T_b \tag{12.4-9}$$

$$\sigma = \left(\frac{1.585 V_b}{1 + 1.3\delta^2}\right)^{1/3} \tag{12.4-10}$$

$$\delta_{AB} = (\delta_A \delta_B)^{1/2} \tag{12.4-11}$$

$$\frac{\varepsilon_{AB}}{k} = \left(\frac{\varepsilon_A}{k}\frac{\varepsilon_B}{k}\right)^{1/2} \tag{12.4-12}$$

$$\sigma_{AB} = (\sigma_A \sigma_B)^{1/2} \tag{12.4-13}$$

Note that the polarity effect is related exclusively to the dipole moment; this may not always be a satisfactory assumption (Byrne et al., 1967).

Example 12.2 Polar diffusion coefficients

Estimate the diffusion coefficient for a mixture of methyl chloride (MC) and sulfur dioxide (SD) at 1 bar and 323 K. The data required to use Brokaw's relation from Appendix A are shown below:

	Methyl Chloride (MC)	**Sulfur Dioxide (SD)**
Dipole moment, debyes	1.9	1.6
Liquid molar volume at T_b, cm³/mol	50.1	44.03
Normal boiling temperature, K	248.95	263.13

Solution. With Eqs. (12.4-8) and (12.4-11),

$$\delta(\text{MC}) = \frac{(1.94 \times 10^3)(1.9)^2}{(50.1)(248.95)} = 0.56$$

$$\delta(\text{SD}) = \frac{(1.94 \times 10^3)(1.6)^2}{(44.03)(263.1)} = 0.43$$

$$\delta(\text{MC-SD}) = [(0.55(0.43)]^{1/2} = 0.49$$

Also, with Eqs. (12.4-9) and (12.4-12),

$$\frac{\varepsilon(\text{MC})}{k} = 1.18[1 + 1.3(0.56)^2](248.95) = 414 \text{ K}$$

$$\frac{\varepsilon(\text{SD})}{k} = 1.18[1 + 1.3(0.43)^2](263.1) = 385 \text{ K}$$

$$\frac{\varepsilon(\text{MC-SD})}{k} = [(414)(385)]^{1/2} = 399 \text{ K}$$

Then, with Eqs. (12.4-10) and (12.4-13),

$$\sigma(\text{MC}) = \left[\frac{(1.585)(50.1)}{1 + (1.3)(0.56)^2}\right]^{1/3} = 3.84 \text{ Å}$$

$$\sigma(\text{SD}) = \left[\frac{(1.585)(44.03)}{1 + (1.3)(0.43)^2}\right]^{1/3} = 3.83 \text{ Å}$$

$$\sigma(\text{MC-SD}) = [(3.84)(3.83)]^{1/2} = 3.84 \text{ Å}$$

To determine Ω_D, $T^* = kT/\varepsilon$ (MC-SD) $= 323/399 = 0.810$. With Eq. (12.4-6), $\Omega_D = 1.60$. Then with Eq. (12.4-7),

$$\Omega_D = 1.6 + \frac{(0.19)(0.49)^2}{(0.810)} = 1.66$$

With Eq. (12.4-2) and M (MC) = 50.49, M (SD) = 64.06, and $M_{AB} = (2)[1/50.49) + (1/64.06)]^{-1} = 56.47$

$$D_{\text{MC-SD}} = \frac{(0.00266)(323)^{3/2}}{(1)(56.47)^{1/2}(3.84)^2(1.66)} = 0.084 \text{ cm}^2/\text{s}$$

The experimental value is 0.078 cm^2/s (Brokaw, 1969) and the error is 8%.

12.4.4 Discussion

A comprehensive review of the theory and experimental data for gas diffusion coefficients is available (Marrero and Mason, 1972). There have been many studies covering wide temperature ranges, and the applicability of Eq. (12.4-1) is well verified. Most investigators select the Lennard-Jones potential for its convenience and simplicity. The difficult task is to locate appropriate values of σ and ε. Brokaw suggests other relations, e.g., Eqs. (12.4-9) and (12.4-10). Even after the pure-component values of σ and ε have been selected, a combination rule is necessary to obtain σ_{AB} and ε_{AB}. Most studies have employed Eqs. (12.4-4) and (12.4-5) because they are simple and theory suggests no particularly better alternatives. Ravindran et al. (1979) have used Eq. (12.4-2) to correlate diffusivities of low-volatile organics in light gases.

It is important to employ values of σ and ε obtained from the same source. Published values of these parameters differ considerably, but σ and ε from a single source often lead to the same result as the use of a quite different pair from another source.

The estimation equations described in this section were used to calculate diffusion coefficients for a number of different gases, and the results are shown in Table 12.4. The accuracy of the theoretical relations is discussed in Sec. 12.5 after some empirical correlations for the diffusion coefficient have been described.

12.5 DIFFUSION COEFFICIENTS FOR BINARY GAS SYSTEMS AT LOW PRESSURES: EMPIRICAL CORRELATIONS

Several proposed methods for estimating D_{AB} in low-pressure binary gas systems retain the general form of Eq. (12.4-2), with empirical constants based on experimental data. These include the equations proposed by Arnold (1930), Gilliland (1934), Wilke and Lee (1955), Slattery and Bird (1958), Bailey (1975), Chen and Othmer (1962), Othmer and Chen (1962), and Fuller et al. (1965, 1966, 1969). Values of D_{AB} estimated by these equations generally agree with experimental values to within 5 to 10%, although discrepancies of more than 20% are possible. We illustrate two methods which have been shown to be quite general and reliable.

12.5.1 Wilke and Lee (1955)

Equation (12.4-2) is rewritten as

$$D_{AB} = \frac{[3.03 - (0.98/M_{AB}^{1/2})](10^{-3})T^{3/2}}{PM_{AB}^{1/2}\sigma_{AB}^{2}\Omega_D} \tag{12.5-1}$$

where D_{AB} = binary diffusion coefficient, cm²/s
T = temperature, K
M_A, M_B = molecular weights of A and B, g/mol
$M_{AB} = 2[(1/M_A) + (1/M_B)]^{-1}$
P = pressure, bar

The scale parameter σ_{AB} is given by Eq. (12.4-5) where, for each component,

$$\sigma = 1.18V_b^{1/3} \tag{12.5-2}$$

and V_b is the liquid molar volume at the normal boiling temperature, cm³/mol, found from experimental data or estimated by the methods in Chap. 4. Ω_D is determined from Eq. (12.4-6) with $(\varepsilon/k)_{AB}$ from Eq. (12.4-4) and, for each component,

$$\frac{\varepsilon}{k} = 1.15T_b \tag{12.5-3}$$

with T_b as the normal boiling point (at 1 atm) in kelvins. Note, for systems in which one component is air, σ (air) = 3.62 Å and ε/k (air) = 97.0 K. Equations (12.5-2) and (12.5-3) should not be used for hydrogen or helium. We illustrate this method in Example 12.3.

12.5.2 Fuller et al. (1965, 1966, 1969)

These authors modified Eq. (12.4-2) to

$$D_{AB} = \frac{0.00143T^{1.75}}{PM_{AB}^{1/2}\left[\left(\sum_{\upsilon}\right)_A^{1/3} + \left(\sum_{\upsilon}\right)_B^{1/3}\right]^2} \tag{12.5-4}$$

where the terms have been defined under Eq. (12.5-1) and Σ_v is found for each component by summing atomic diffusion volumes in Table 12.3 (Fuller et al., 1969). These atomic parameters were determined by a regression analysis of many experimental data, and the authors report an average absolute error of about 4% when using Eq. (12.5-4). The technique is illustrated in Example 12.3.

TABLE 12.3 Atomic diffusion volumes.

Atomic and Structural Diffusion Volume Increments			
C	15.9	F	14.7
H	2.31	Cl	21.0
O	6.11	Br	21.9
N	4.54	I	29.8
Aromatic Ring	−18.3	S	22.9
Heterocyclic ring	−18.3		
Diffusion Volumes of Simple Molecules			
He	2.67	CO	18.0
Ne	5.98	CO_2	26.9
Ar	16.2	N_2O	35.9
Kr	24.5	NH_3	20.7
Xe	32.7	H_2O	13.1
H_2	6.12	SF_6	71.3
D_2	6.84	Cl_2	38.4
N_2	18.5	Br_2	69.0
O_2	16.3	SO_2	41.8
Air	19.7		

Example 12.3 Diffusion coefficients in air

Estimate the diffusion coefficient of allyl chloride (AC) in air at 298 K and 1 bar. The experimental value reported by Lugg (1968) is 0.098 cm²/s.

Solution

WILKE AND LEE METHOD. As suggested in the text, for air $\sigma = 3.62$ Å and $\varepsilon/k = 97.0$ K. For allyl chloride, from Daubert et al. (1997), $T_b = 318.3$ K and $V_b = 84.7$ cm³/mol. Thus, using Eqs. (12.5-2) and (12.5-3),

$$\sigma(\text{AC}) = (1.18)(84.7)^{1/3} = 5.18 \text{ Å}$$

$$\varepsilon(\text{AC})/k = (1.15)(318.3) = 366 \text{ K}$$

Then, with Eqs. (12.4-4) and (12.4-5),

$$\varepsilon(\text{AC-air})/k = [(366)(97.0)]^{1/2} = 188 \text{ K}$$

$$\sigma(\text{AC-air}) = (5.18 + 3.62)/2 = 4.40 \text{ Å}$$

$$T^* = \frac{T}{\varepsilon(\text{AC-air})/k} = \frac{298}{188} = 1.59$$

and, with Eq. (12.4-6), $\Omega_D = 1.17$. With M (AC) = 76.5 and M (air) = 29.0, $M_{AB} = (2)[(1/76.5) + (1/29.0)]^{-1} = 42.0$. Finally, with Eq. (12.5-1) when $P = 1$ bar,

$$D = \frac{\{3.03 - [0.98/(42.0)^{1/2}]\}(10^{-3})(298)^{3/2}}{(1)(42.0)^{1/2}(4.40)^2(1.17)} = 0.10 \text{ cm}^2/\text{s}$$

$$\text{Error} = \frac{0.10 - 0.098}{0.098} \times 100 = 2\%$$

FULLER ET AL. METHOD. Equation (12.5-4) is used. $P = 1$ bar; M_{AB} was shown above to be equal to 42.0; and $T = 298$ K. For air $(\Sigma_v) = 19.7$, and for allyl chloride, C_3H_5Cl, with Table 12.3, $(\Sigma_v) = (3)(15.9) + (5)(2.31) + 21 = 80.25$. Thus,

$$D = \frac{(0.00143)(298)^{1.75}}{(1)(42.0)^{1/2}\,[(19.7)^{1/3} + (80.25)^{1/3}]^2} = 0.096 \text{ cm}^2/\text{s}$$

$$\text{Error} = \frac{0.096 - 0.098}{0.098} \times 100 = -2\%$$

12.5.3 Discussion

In Table 12.4, we show experimental diffusion coefficients for a number of binary systems and note the errors found when estimating D_{AB} for (a) the basic theoretical Eq. (12.4-2), (b) Brokaw's method [Eqs. (12.4-2) and (12.4-7)], (c) Wilke and Lee's method [Eq. 12.5-1], and (d) Fuller et al.'s method [Eq. (12.5-4)]. For (a), no calculations were made if σ and ε/k were not available in Appendix B of PGL5ed. For (b), calculations were done for systems in which at least one of the species had a nonzero dipole moment. For hydrogen and helium, σ and ε/k were used from Appendix B of PGL5ed. For all other compounds, Eqs. (12.4-9) and (12.4-10) were used. For systems in which at least one of the components was polar, Brokaw's method usually, but not always, gave a more accurate prediction than did Eq. (12.4-2). For the 26 cases in Table 12.4 for which predictions are given for both methods, the average absolute percent deviation for Brokaw's method was about 1% less than the predictions of Eq. (12.4-2).

For all methods, there were always a few systems for which large errors were found. These differences may be due to inadequacies of the method or to inaccurate data. In general, however, the Fuller et al. procedure [Eq. (12.5-4) and Table 12.3] yielded the smallest average error, and it is the method recommended for use. Other evaluations (Elliott and Watts, 1972; Gotoh et al., 1973; Gotoh et al., 1974; Lugg, 1968; Pathak et al., 1981) have shown both the Fuller et al. and the Wilke-Lee forms to be reliable.

Reviews of experimental data of binary diffusion coefficients are available (Gordon, 1977; Marrero and Mason, 1972) and Massman (1998) presents a review of diffusivities of components commonly found in air.

12.6 THE EFFECT OF PRESSURE ON THE BINARY DIFFUSION COEFFICIENTS OF GASES

At low to moderate pressures, binary diffusion coefficients vary inversely with pressure or density as suggested by Eqs. (12.4-1) and (12.4-2). At high pressures, the product DP or $D\rho$ is no longer constant but decreases with an increase in either P or ρ. Note that it is possible to have a different behavior in the products DP and $D\rho$ as the pressure is raised, since ρ is proportional to pressure only at low pressures, and gas nonidealities with their concomitant effect on the system density may become important. Also, as indicated earlier, at low pressures, the binary diffusion coefficient is essentially independent of composition. At high pressures, where the gas phase may deviate significantly from an ideal gas, small, but finite effects of composition have been noted, e.g., Takahaski and Hongo (1982).

With the paucity of reliable data, it is not surprising that few estimation methods have been proposed. Takahashi (1974) has suggested a very simple corresponding states method that is satisfactory for the limited database available. His correlation is

$$\frac{D_{AB}P}{(D_{AB}P)^+} = f(T_r,\ P_r) \tag{12.6-1}$$

where D_{AB} = diffusion coefficient, cm^2/s
P = pressure, bar

TABLE 12.4 Comparison of methods for estimating gas diffusion coefficients at low pressures.

				Errors as Percent of Experimental Values			
System	**T, K**	**$D_{AB}P$ (exp.), (cm^2/s) bar**	**Ref.***	**Theory**	**Brokaw**	**Wilke-Lee**	**Fuller et al.**
Air-carbon dioxide	276	0.144	9	−6		2	−3
	317	0.179		−2		6	−1
Air-ethanol	313	0.147	13	−10	−16	−11	−8
Air-helium	276	0.632	9	0		1	−5
	346	0.914		0		2	−2
Air-*n*-hexane	294	0.081	5	−6		−4	−7
	328	0.094		−1		1	−2
Air-2-methylfuran	334	0.107	1		2	9	8
Air-naphthalene	303	0.087	4			−18	−20
Air-water	313	0.292	5	−18	−15	−16	−5
Ammonia-diethyl ether	288	0.101	20	−24	−12	−15	2
	337	0.139		−24	−12	−15	−2
Argon-ammonia	255	0.152	19	3	5	4	13
	333	0.256		3	5	2	7
Argon-benzene	323	0.085	12	9		14	15
	373	0.112		9		13	13
Argon-helium	276	0.655	9	−2		−5	−1
	418	1.417	6	−9		−12	−6
Argon-hexafluorobenzene	323	0.082	12			−5	−18
	373	0.095				8	−9
Argon-hydrogen	295	0.84	22	−9		−16	−4
	628	3.25		−15		−22	−7
	1068	8.21		−19		−25	−7
Argon-krypton	273	0.121	18	−1		3	0
Argon-methane	298	0.205	6	5		4	5
Argon-sulfur dioxide	263	0.078	13	18		24	25
Argon-xenon	195	0.052	6	−2		5	9
	378	0.18		−3		3	0
Carbon dioxide-helium	298	0.62	17	−3		0	−5
	498	1.433		−1		2	1
Carbon dioxide-nitrogen	298	0.169	21	−7		−3	−3
Cargon dioxide-nitrous oxide	313	0.13	3	−6	−4	3	−3
Carbon dioxide-sulfur dioxide	473	0.198	13	7	14	18	15
Carbon dioxide-tetrafluoromethane	298	0.087	11	0		11	−12
	673	0.385		−3		9	−17
Carbon dioxide-water	307	0.201	7	−21	−12	−13	11
Carbon monoxide-nitrogen	373	0.322	2	−6	−13	−8	−4
Ethylene-water	328	0.236	15	−25	−16	−20	−3
Helium-benzene	423	0.618	17	9		8	−4
Helium-bromobenzene	427	0.55	8		28	8	−2
Helium-2-chlorobutane	429	0.568	8		33	12	−2

(Continued)

TABLE 12.4 Comparison of methods for estimating gas diffusion coefficients at low pressures. (*Continued*)

				Errors as Percent of Experimental Values			
System	T, K	$D_{AB}P$ (exp.), (cm²/s) bar	Ref.*	Theory	Brokaw	Wilke-Lee	Fuller et al.
Helium-*n*-butanol	423	0.595	17		28	10	−2
Helium-1-iodobutane	428	0.524	8			11	1
Helium-methanol	432	1.046	17	11	8	−3	2
Helium-nitrogen	298	0.696	17	1		−3	2
Helium-water	352	1.136	15	1	8	−11	0
Hydrogen-acetone	296	0.43	15	−1	11	−9	2
Hydrogen-ammonia	263	0.58	13	3	8	−7	7
	358	1.11	13	−4	0	−15	−4
	473	1.89		−6	−3	−19	−9
Hydrogen-cyclohexane	289	0.323	10	−4	14	−5	−2
Hydrogen-naphthalene	303	0.305	4			−7	−1
Hydrogen-nitrobenzene	493	0.831	14	16		−10	4
Hydrogen-nitrogen	294	0.773	16	−5		−14	−1
	573	2.449		−8		−15	1
Hydrogen-pyridine	318	0.443	10		9	−8	5
Hydrogen-water	307	0.927	7	−12	−7	−21	4
Methane-water	352	0.361	15	−19	−13	−18	−2
Nitrogen-ammonia	298	0.233	13	−5	−4	−7	−2
	358	0.332		−6	−5	−9	−5
Nitrogen-aniline	473	0.182	14		4	7	9
Nitrogen-sulfur dioxide	263	0.105	13	−3	−4	0	0
Nitrogen-water	308	0.259	15	−11	−8	−12	7
	352	0.364		−18	−15	−20	−4
Nitrogen-sulfur hexafluoride	378	0.146	11	6		12	3
Oxygen-benzene	311	0.102	10	−9		−5	−3
Oxygen-carbon tetrachloride	296	0.076	13	−6		6	6
Oxygen-cyclohexane	289	0.076	10	−7		2	−1
Oxygen-water	352	0.357	15	−17	−12	−16	0
Average absolute error				7.9		9.6	5.4

*References: 1, Alvarez et al. (1983); 2, Amdur and Shuler (1963); 3, Amdur et al. (1952); 4, Caldwell (1984); 5, Carmichael et al. (1955); 6, Carswell and Stryland (1963); 7, Crider (1956); 8, Fuller et al. (1969); 9, Holson and Strunk (1964); 10, Hudson et al. (1960); 11, Kestin et al. (1977); 12, Maczek and Edwards (1979); 13, Mason and Monschick (1962); 14, Pathak et al. (1981); 15, Schwartz and Brow (1951); 16, Scott and Cox (1960); 17, Seager et al. (1963); 18, Srivastava and Srivastava (1959); 19, Srivastava and Srivastava (1962); 20, Srivastava and Srivastava (1963); 21, Walker and Westenberg (1958); 22, Westenberg and Frazier (1962).

The superscript $^+$ indicates that low-pressure values are to be used. The function $f(T_r, P_r)$ is shown in Fig. 12.4, and to obtain pseudocritical properties from which to calculate the reduced temperatures and pressures, Eqs. (12.6-2) to (12.6-5) are used.

$$T_r = \frac{T}{T_c} \tag{12.6-2}$$

$$T_c = y_A T_{cA} + y_B T_{cB} \tag{12.6-3}$$

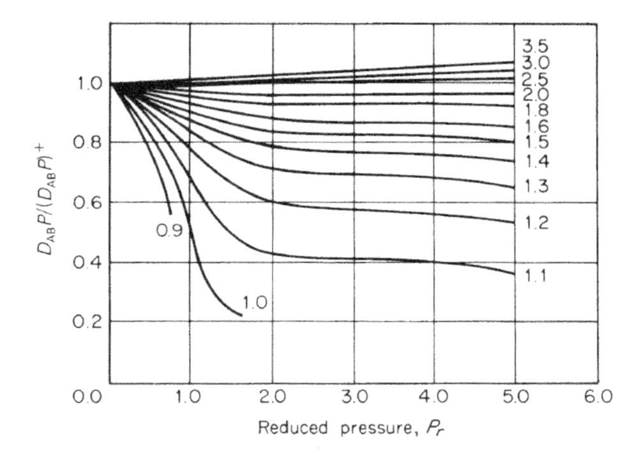

FIGURE 12.4
Takahashi correlation for the effect of pressure and temperature on the binary diffusion coefficient. Lines are at constant reduced temperature.

$$P_r = \frac{P}{P_c} \tag{12.6-4}$$

$$P_c = y_A P_{cA} + y_B P_{cB} \tag{12.6-5}$$

As an illustration of this technique, in Fig. 12.5 we have plotted the data of Takahashi and Hongo (1982) for the system carbon dioxide-ethylene. Two cases are considered: one with a very low concentration of ethylene and the other with a very low concentration of carbon dioxide. Up to about 80 bar, the two limiting diffusion coefficients are essentially identical. Above that pressure, D_{AB} for the trace CO_2 system is significantly higher. Plotted as solid curves on this graph are the predicted values of D_{AB} in Fig. 12.4 and Eq. (12.6-1) using the $(D_{AB}P)^+$ product at low pressure to be 0.149 (cm^2/s) bar as found by Takahashi and Hongo. Also, the dashed curve has been drawn to indicate the estimated value of D_{AB} if one had assumed that $D_{AB}P$ was a constant. Clearly, this assumption is in error above a pressure of about 10 to 15 bar.

Riazi and Whitson (1993) propose Eq. (12.6-6)

$$\frac{\rho D_{AB}}{(\rho D_{AB})^+} = 1.07 \left(\frac{\mu}{\mu^o} \right)^{b + cP_r} \tag{12.6-6}$$

where $b = -0.27 - 0.38\ \omega$
$c = -0.05 + 0.1\ \omega$
μ^0 is the viscosity at low pressure
ω is the acentric factor
$P_r = P/P_c$

$$P_c = y_A P_{cA} + y_B P_{cB} \tag{12.6-7}$$

$$\omega = y_A \omega_A + y_B \omega_B \tag{12.6-8}$$

As in Eq. (12.6-1), the superscript $^+$ represents low-pressure values. Riazi and Whitson (1997) claim that Eq. (12.6-6) can represent high-pressure liquid behavior as well as high-pressure gas behavior. Equation (12.6-6) gives a slightly worse description of the systems shown in Fig. 12.5 than does Eq. (12.6-1). When Eq. (12.6-6) is compared to the data in Fig. 12.5, the average absolute deviation is 14% while the maximum deviation is 30%.

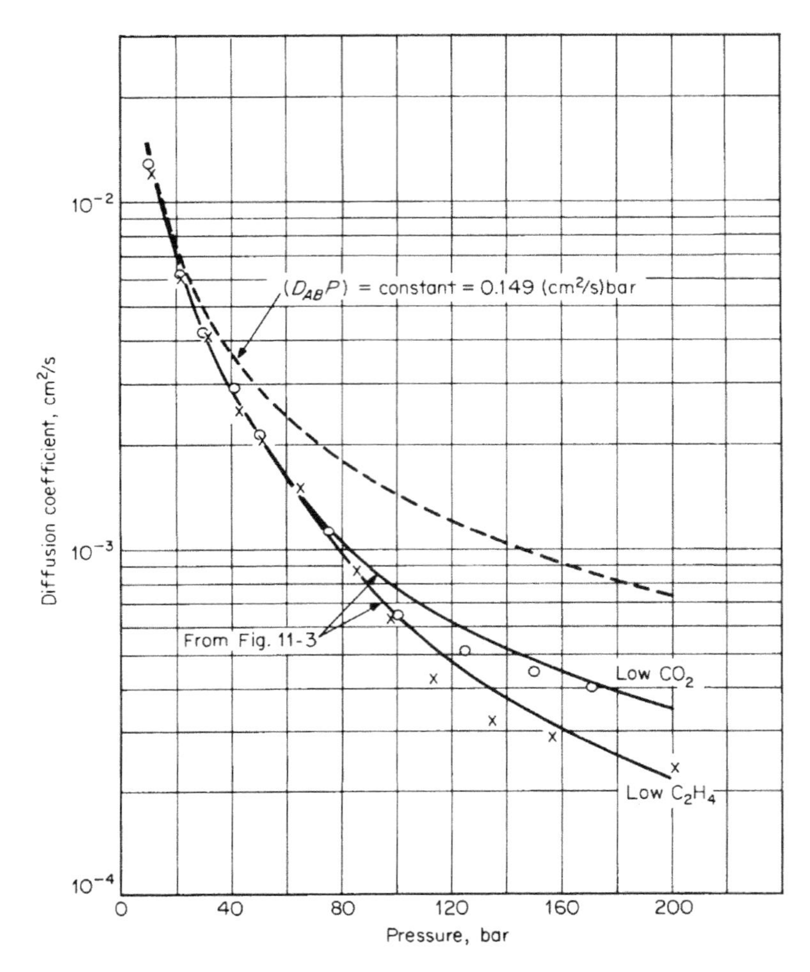

FIGURE 12.5
Effect of pressure and composition on the binary diffusion coefficient in the CO_2-C_2H_4 system at 323.2 K.

Neither Eq. (12.6-1) nor Eq. (12.6-6) is entirely satisfactory. The former requires that values be read from Fig. 12.4. The latter requires viscosity information and does not reproduce the correct value in the limit of low pressure.

Many of the more recent data for diffusion coefficients at high pressure involve a trace solute in a supercritical fluid. To illustrate some data for the diffusion coefficient of complex solutes in supercritical fluids, we show Fig. 12.6 (Debenedetti and Reid, 1986). There the diffusion coefficient is given as a function of reduced pressure from the ideal-gas range to reduced pressures up to about 6. The solutes are relatively complex molecules, and the solvent gases are CO_2 and ethylene. No temperature dependence is shown, since the temperatures studied (see legend) were such that all the reduced temperatures were similar and were, in most cases, in the range of 1 to 1.05. Since the concentrations of the solutes were quite low, the pressure and temperature were reduced by P_c and T_c of the pure solvent. Up to about half the critical pressure, $D_{AB}P$ is essentially constant. Above that pressure, the data show the product $D_{AB}P$ decreasing, and at reduced pressures of about 2, $D_{AB} \propto P_r^{1/2}$. As supercritical extractions are often carried out in a reduced temperature range of about 1.1 to 1.2 and in a reduced pressure range of 2 to 4, this plot would indicate that $D_{AB} \cong 10^{-4}$ cm²/s, a value much less than for a low-pressure gas but still significantly higher than for a typical liquid (see Sec. 12.10).

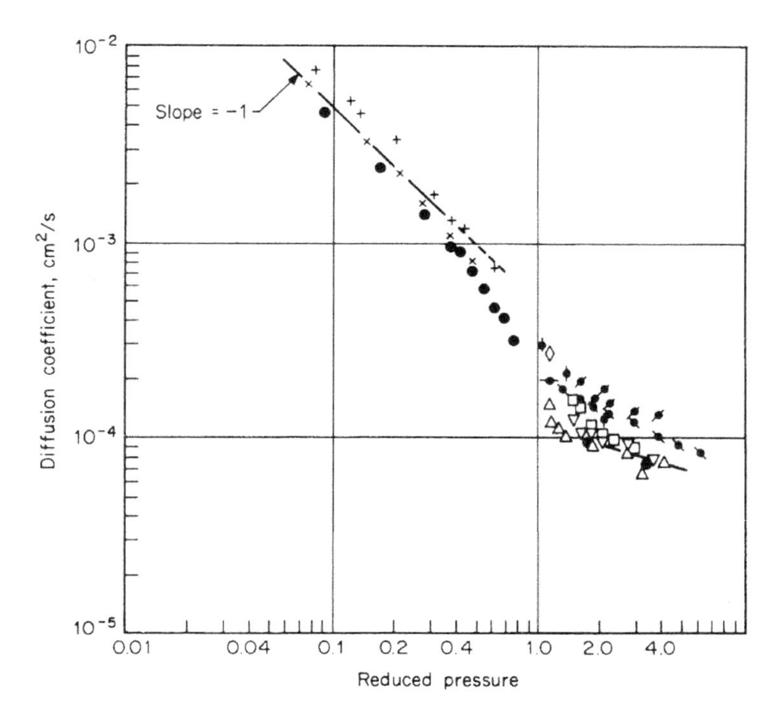

FIGURE 12.6
Diffusion coefficients in supercritical fluids.

Key	System	Ref.*
●, ×, Δ, +, ∇	CO_2-naphthalene at 20, 30, 35, 40, and 55 °C	1, 2, 3, 4
(symbols), (symbols)	Ethylene-naphthalene at 12 and 35 °C	3, 4
(symbol)	CO_2-benzene at 40 °C	5
(symbol)	CO_2-propylbenzene at 40 °C	6
0	CO_2-1,2,3-trimethylbenzene at 40 °C	6

*References: 1, Morozov and Vinkler (1975); 2, Vinkler and Morozov (1975); 3, Iomtev and Tsekhanskaya (1964); 4, Tsekhanskaya (1971); 5, Schneider (1978); 6, Swaid and Schneider (1979).

Several correlations have been presented for diffusion coefficients of solutes in supercritical fluids. One of the simplest is that of He and Yu (1998) as shown in Eq. (12.6-9)

$$D_{AB} = \alpha \times 10^{-5}\left(\frac{T}{M_A}\right)^{1/2} \exp\left(-\frac{0.3887}{V_{rB} - 0.23}\right) \tag{12.6-9}$$

$$\alpha = 14.882 + 0.005908\,\frac{T_{cB}\,V_{cB}}{M_B} + 2.0821 \times 10^{-6}\left(\frac{T_{cB}\,V_{cB}}{M_B}\right)^2 \tag{12.6-10}$$

where D_{AB} = diffusion coefficient of solute A in solvent B, cm^2/s
$V_{rB} = V_B/V_{cB}$
V_{cB} = critical volume of the solvent in cm^3/mol
M = molecular weight in g/mol
T_{cB} = critical temperature of the solvent in K

When Eqs. (12.6-9) and (12.6-10) were tested on 1300 data points involving 11 different solvents, the authors found an average error of 8%, which is remarkable for such a simple equation. The cases examined included solvents that were high-temperature liquids as well as supercritical fluids and covered temperature and density ranges of $0.66 < T_r < 1.78$ and $0.22 < \rho_r < 2.62$. Because of its simplicity, Eq. (12.6-9) should likely not be used for conditions outside the range of fit. Furthermore, Eq. (12.6-9) does not include any effect of solvent viscosity or solute density and for solutes or solvents for which these properties are dramatically different than those tested, Eq. (12.6-9) would likely give higher errors. For example, the solutes for the 1300 data points tested were most often organic compounds for which the pure-component liquid density is typically 0.8 to 0.9 g/cm^3. The errors for chloroform and iodine (densities of 1.5 and 4.9 g/cm^3, respectively) in carbon dioxide were −20% and −38%, respectively.

There are methods to estimate diffusion coefficients of solutes in supercritical fluids other than Eq. (12.6-9) for which slightly improved accuracy is claimed but which are also more involved. Funazukuri et al. (1992) propose a method that uses the ratio of the Schmidt number to its value at low pressure. Liu and Ruckenstein (1997) have developed a correlation that uses the Peng-Robinson equation of state to calculate a thermodynamic factor (see Sec. 12.2). References to many of the data on solutes in supercritical fluids are summarized in these latter two references as well as Catchpole and King (1994) and He and Ye (1998).

Example 12.4 High-pressure diffusion coefficients

Estimate the diffusion coefficient of vitamin K_1 in supercritical carbon dioxide at 313 K and 160 bar. The experimental value is reported to be 5.43×10^{-5} cm^2/s (Funazukuri et al., 1992).

Solution. For vitamin K_1, $M = 450.7$ g/mol. From Appendix A, for CO_2, $T_c = 304.12$ K, $P_c = 73.74$ bar, $V_c = 94.07$ cm^3/mol, and $\omega = 0.225$. With these values, the Lee-Kesler equation of state gives $V = 54.97$ cm^3/mol for pure CO_2 at $T = 313$ K and $P = 160$ bar. Thus, $V_{rB} = 54.97/94.07 = 0.584$. From Eqs. (12.6-9) and (12.6-10),

$$\frac{T_{cB}\ V_{cB}}{M_B} = \frac{(304.12)(94.07)}{44.01} = 650.$$

$$\alpha = 14.822 + (0.005908)(650) + 2.0821 \times 10^{-6}(650)^2 = 19.60$$

$$D_{AB} = 19.60\left(\frac{313}{450.7}\right)^{1/2} \exp\left(-\frac{0.3887}{0.584 - 0.23}\right) = 5.45 \times 10^{-5}\ \text{cm}^2/\text{s}$$

$$\text{Error} = \frac{5.45 - 5.43}{5.43} \times 100 = 0.4\%$$

12.7 THE EFFECT OF TEMPERATURE ON DIFFUSION IN GASES

At pressures where the ideal-gas law approximation is valid, it is seen from Eq. (12.4-2) that

$$D_{AB}\,\alpha\,\frac{T^{3/2}}{\Omega_D(T)} \tag{12.7-1}$$

or

$$\left(\frac{\partial \ln D_{AB}}{\partial \ln T}\right)_P = \frac{3}{2} - \frac{d \ln \Omega_D}{d \ln T} \tag{12.7-2}$$

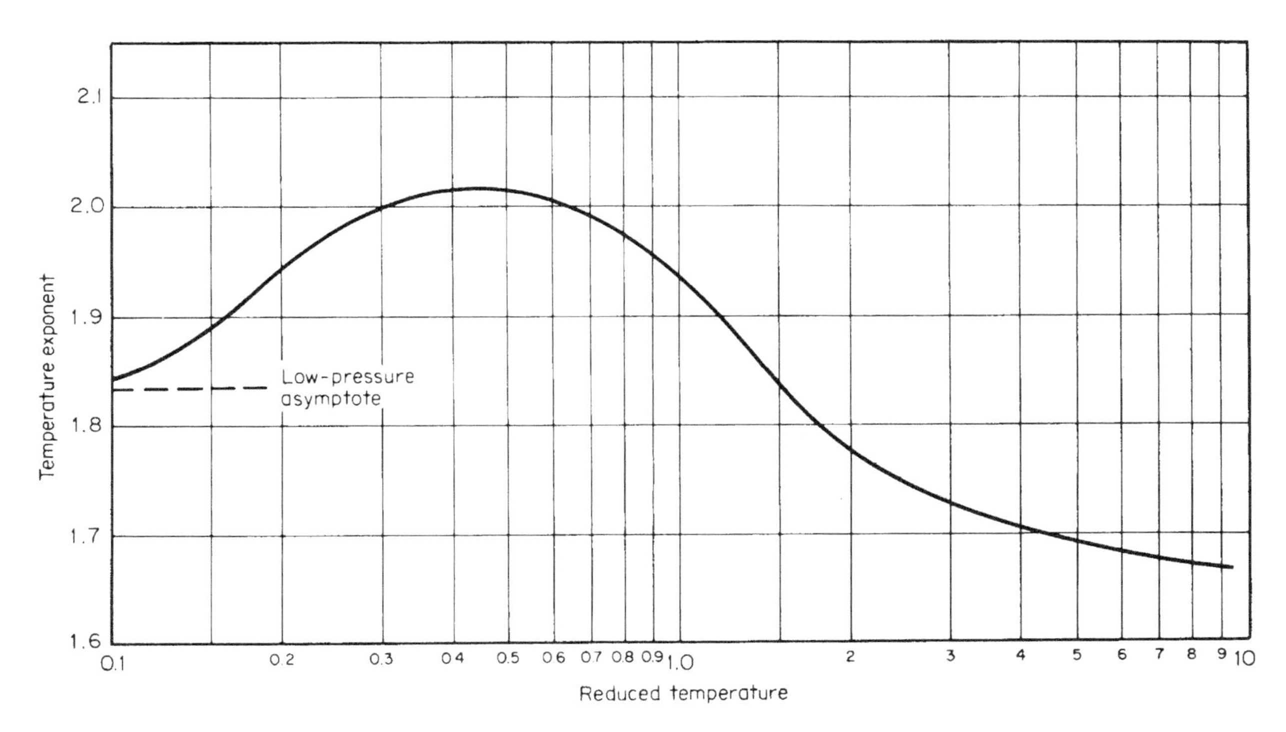

FIGURE 12.7
Exponent of temperature for diffusion in gases. [Adapted from Marrero and Mason (1972) with the approximation that $\varepsilon/k \approx \mathbf{0.75}T_c$]

Marrero and Mason (1972) indicate that, in most cases, the term $d \ln \Omega_D / d \ln T$ varies from 0 to –½. Thus D_{AB} varies as $T^{3/2}$ to T^2. This result agrees with the empirical estimation methods referred to in Sec. 12.5, e.g., in the Fuller et al. method, $D \propto T^{1.75}$. Over wide temperature ranges, however, the exponent on temperature changes. Figure 12.7 shows the approximate variation of this exponent with reduced temperature. The very fact that the temperature exponent increases and then decreases indicates that empirical estimation techniques with a constant exponent will be limited in their range of applicability. The theoretical and the Wilke-Lee methods are therefore preferable if wide temperature ranges are to be covered. Dunlop and Bignell (1997) relate the temperature dependence to the thermal diffusion factor.

12.8 DIFFUSION IN MULTICOMPONENT GAS MIXTURES

A few general concepts of diffusion in multicomponent liquid mixtures presented later (in Sec. 12.13) are applicable for gas mixtures also. One of the problems with diffusion in liquids is that even the binary diffusion coefficients are often very composition dependent. For multicomponent liquid mixtures, therefore, it is difficult to obtain numerical values of the diffusion coefficients relating fluxes to concentration gradients.

In gases, D_{AB} is normally assumed independent of composition. With this approximation, multicomponent diffusion in gases can be described by the Stefan-Maxwell equation

$$\frac{dx_i}{dz} = \sum_{j=1}^{n} \frac{c_i c_j}{c^2 D_{ij}} \left(\frac{J_j}{c_j} - \frac{J_i}{c_i} \right) \tag{12.8-1}$$

where c_i = concentration of i
c = mixture concentration.
J_i, J_j = flux of i, j
D_{ij} = Binary diffusion coefficient of the ij system
(dx_i/dz) = gradient in mole fraction of i in the z direction

This relation is different from the basic binary diffusion relation (12.2-5), but the employment of common binary diffusion coefficients is particularly desirable. Marrero and Mason (1972) discuss many of the assumptions behind Eq. (12.8-1).

Few attempts have been made by engineers to calculate fluxes in multicomponent systems. However, one important and simple limiting case is often cited. If a dilute component i diffuses into a *homogeneous* mixture, then $J_j \approx 0$. With $c_j/c = x_j$, Eq. (12.8-1) reduces to

$$\frac{dx_i}{dz} = -J_i \sum_{\substack{j=1 \\ j \neq i}}^{n} \frac{x_j}{cD_{ij}} \tag{12.8-2}$$

Defining

$$D_{im} = \frac{-J_i}{dx_i/d_z} \tag{12.8-3}$$

gives

$$D_{im} = \left(\sum_{\substack{j=1 \\ j \neq i}}^{n} \frac{x_j}{D_{ij}} \right)^{-1} \tag{12.8-4}$$

This simple relation is sometimes called Blanc's law (Blanc, 1908; Marrero and Mason, 1972). It was shown to apply to several ternary cases in which i was a trace component (Mather and Saxena, 1966). Deviations from Blanc's law are discussed by Sandler and Mason (1968).

The general theory of diffusion in multicomponent gas systems is covered by Cussler (1997) and by Hirschfelder et al. (1954). The problem of diffusion in three-component gas systems has been generalized by Toor (1957) and verified by Fairbanks and Wilke (1950), Walker et al. (1960), and Duncan and Toor (1962).

12.9 DIFFUSION IN LIQUIDS: THEORY

Binary liquid diffusion coefficients are defined by Eq. (12.2-5) or (12.2-7). Since molecules in liquids are densely packed and strongly affected by force fields of neighboring molecules, values of D_{AB} for liquids are much smaller than for low-pressure gases. That does not mean that diffusion rates are necessarily low, since concentration gradients can be large.

Liquid state theories for calculating diffusion coefficients are quite idealized, and none is satisfactory in providing relations for calculating D_{AB}. In several cases, however, the form of a theoretical equation has provided the framework for useful prediction methods. A case in point involves the analysis of large spherical molecules diffusing in a dilute solution. Hydrodynamic theory (Bird et al., 1960; Gainer and Metzner, 1965) then indicates that

$$D_{AB} = \frac{RT}{6\pi\eta_B r_A} \tag{12.9-1}$$

where η_B is the viscosity of the solvent and r_A is the radius of the "spherical" solute. Equation (12.9-1) is the Stokes-Einstein equation which strictly applies to macroscopic systems. However, many authors have used the form as a starting point in developing correlations for molecular diffusion.

Other theories for modeling diffusion in liquids have been based on kinetic theory (Anderson, 1973; Bearman, 1961; Carman, 1973; Carman and Miller, 1959; Darken, 1948; Dullien, 1961; Hartley and Crank, 1949; Kett and Anderson, 1969; Miller and Carman, 1961), absolute-rate theory (Cullinan and Cusick, 1967; Eyring and Ree, 1961; Gainer and Metzner, 1965; Glasstone et al., 1941; Leffler and Cullinan, 1970; Li and Chang, 1955; Olander, 1963; Ree et al., 1958), statistical mechanics (Bearman, 1960; Bearman, 1961; Kamal and Canjar, 1962), and other concepts (Albright et al., 1983; Brunet and Doan, 1970; Horrocks and Mclaughlin, 1962; Kuznetsova and Rashidova, 1980; Raina, 1980). Several reviews are available for further consideration (Dullien, 1963; Ghai et al., 1973; Ghai et al., 1974; Himmelblau, 1964; Loflin and McLaughlin, 1969).

Diffusion in liquid metals is not treated, although estimation techniques are available (Pasternak and Olander, 1967).

12.10 ESTIMATION OF BINARY LIQUID DIFFUSION COEFFICIENTS AT INFINITE DILUTION

For a binary mixture of solute A in solvent B, the diffusion coefficient D^o_{AB} of A diffusing in an infinitely dilute solution of A in B implies that each A molecule is in an environment of essentially pure B. In engineering work, however, D^o_{AB} is assumed to be a representative diffusion coefficient even for concentrations of A of 5 to 10 mole%. In this section, several estimation methods for D^o_{AB} are introduced; the effect of concentration for mutual diffusion coefficients is covered in Sec. 12.11.

12 10.1 Wilke-Chang Estimation Method (Wilke and Chang, 1955)

An older but still widely used correlation for D^o_{AB}, the Wilke-Chang technique is, in essence, an empirical modification of the Stokes-Einstein relation (12.9-1):

$$D^o_{AB} = \frac{7.4 \times 10^{-8}\ (\phi M_B)^{1/2} T}{\eta_B V_A^{0.6}} \tag{12.10-1}$$

where D^o_{AB} = mutual diffusion coefficient of *solute* A at very low concentrations in *solvent* B, cm^2/s
M_B = molecular weight of *solvent* B, g/mol
T = temperature, K
η_B = viscosity of *solvent* B, cP
V_A = molar volume of *solute* A at its normal boiling temperature, cm^2/mol
ϕ = association factor of *solvent* B, dimensionless

If experimental data to obtain V_A at T_{Ab} do not exist, estimation methods from Chap. 4 may be used.

Wilke and Chang recommend that ϕ be chosen as 2.6 if the solvent is water, 1.9 if it is methanol, 1.5 if it is ethanol, and 1.0 if it is unassociated. When 251 solute-solvent systems were tested by these authors, an average error of about 10% was noted. Figure 12.8 is a graphical representation of Eq. (12.10-1) with the dashed line representing Eq. (12.9-1); the latter is assumed to represent the maximum value of the ordinate for any value of V_A.

A number of authors have suggested modifications of Eq. (12.10-1) particularly to improve its accuracy for systems where water is the solute and the solvent is an organic liquid (Amourdam and Laddha, 1967; Caldwell and Babb, 1956; Hayduk and Buckley, 1972; Hayduk et al., 1973; Hayduk and Laudie, 1974; Lees and Sarram, 1971; Lusis, 1971; Lusis and Ratcliff, 1971; Olander, 1961; Scheibel, 1954; Shrier, 1967; Wise and Houghton, 1966; Witherspoon and Bonoli, 1969). However, none of these suggestions have been widely accepted. In

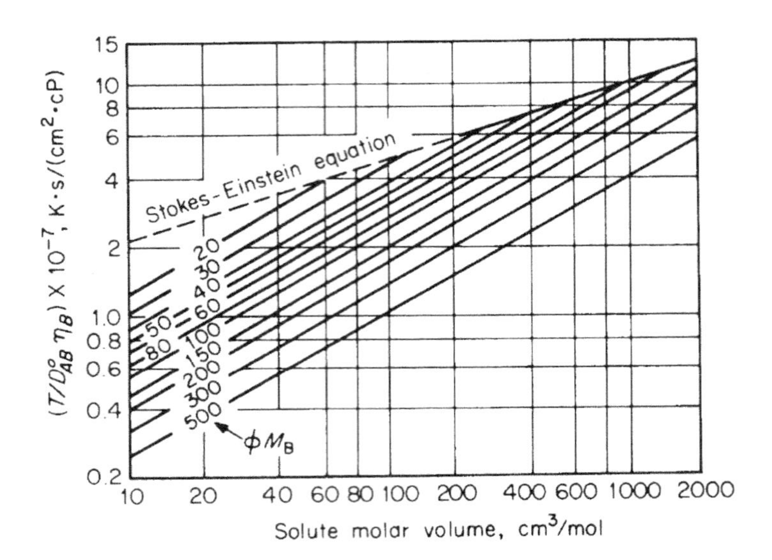

FIGURE 12.8
Graphical representation of Wilke-Chang correlation of diffusion coefficients in dilute solutions. (Wilke and Chang, 1955)

Table 12.7, we show a comparison of estimated and experimental values of D°_{AB}. The errors vary so greatly that the concept of an *average* error is meaningless. The method should not be used when water is the *solute*.

Example 12.5 Wilke-Chang estimation for diffusion in liquid water

Use the Wilke-Chang correlation to estimate D°_{AB} for ethylbenzene diffusing into water at 293 K. The viscosity of water at this temperature is essentially 1.0 cP. The experimental value of D°_{AB} is 0.81×10^{-5} cm²/s (Witherspoon and Bonoli, 1969).

Solution. The normal boiling point of ethylbenzene is 409.36 K (Appendix A). At that temperature, the density is 0.761 g/cm³ (Vargaftik et al., 1996), so with $M_A = 106.17$, $V_A = 106.17/0.761 = 139.5$ cm³/mol. Then, using Eq. (12.10-1) with $\phi = 2.6$ and $M_B = 18.0$ for water,

$$D^{\circ}_{AB} = 7.4 \times 10^{-8} \frac{[(2.6)(18.0)]^{1/2}(293)}{(1.0)(139.5)^{0.6}} = 0.77 \times 10^{-5} \text{ cm}^2/\text{s}$$

$$\text{Error} = \frac{0.77 - 0.81}{0.81} \times 100 = -5\%$$

12.10.2 Tyn and Calus Method (Tyn and Calus, 1975)

These authors have proposed that D°_{AB} be estimated by the relation

$$D^{\circ}_{AB} = 8.93 \times 10^{-8} \left(\frac{V_A}{V_B^2}\right)^{1/6} \left(\frac{P_B}{P_A}\right)^{0.6} \frac{T}{\eta_B} \tag{12.10-2}$$

where V_B = molar volume of the solvent at the normal boiling temperature, cm^3/mol, $\mathbf{P}_A$ and $\mathbf{P}_B$ are parachors for the solute and solvent, and the other terms are defined under Eq. (12.10-1).

The parachor is related to the liquid surface tension (see Chap. 13) as

$$\mathbf{P} = V\sigma^{1/4} \tag{12.10-3}$$

where σ is the surface tension in dyn/cm = $g/s^2 = 10^{-3}$ N/m^2 and V is the molar volume, cm^3/mol, both measured at the same temperature. Thus the units of $\mathbf{P}$ are $cm^3 \cdot g^{1/4}/(s^{1/2} \cdot mol)$. Quayle (1953) has tabulated values of $\mathbf{P}$ for many chemicals; alternatively, $\mathbf{P}$ may be estimated from additive group contributions as shown in Table 12.5. Over moderate temperature ranges, $\mathbf{P}$ is essentially a constant.

When using the correlation shown in Eq. (12.10-2), the authors note several restrictions:

1. The method should not be used for diffusion in viscous solvents. Values of η_B above about 20 to 30 cP would classify the solvent as viscous.
2. If the solute is water, a dimer value of V_A and $\mathbf{P}_A$ should be used. In the calculations for Table 12.7, we used $V_A = V_w = 37.4$ cm^3/mol and $\mathbf{P}_A = \mathbf{P}_w = 105.2$ $cm^3 \cdot g^{1/4}/(s^{1/2} \cdot mol)$.

TABLE 12.5 Structural contributions for calculating the parachor*.

Group	Contribution	Group	Contribution
Carbon-hydrogen:		R—[—CO—]—R′ (for the —CO— in ketones)	
C	9.0	R + R′ = 2	51.3
H	15.5	R + R′ = 3	49.0
CH_3	55.5	R + R′ = 4	47.5
CH_2 in $—(CH_2)_n—$		R + R′ = 5	46.3
$n < 12$	40.0	R + R′ = 6	45.3
$n > 12$	40.3	R + R′ = 7	44.1
Alkyl groups		—CHO	66
1-Methylethyl	133.3	O (if not noted above)	20
1-Methylpropyl	171.9	N (if not noted above)	17.5
1-Methylbutyl	211.7	S	49.1
2-Methylpropyl	173.3	P	40.5
1-Ethylpropyl	209.5	F	26.1
1,1-Dimethylethyl	170.4	Cl	55.2
1,1-Dimethylpropyl	207.5	Br	68.0
1,2-Dimethylpropyl	207.9	I	90.3
1,1,2-Trimethylpropyl	243.5	Ethylenic bonds (=C<)	19.1
C_6H_5	189.6	Terminal	
Special groups:		2,3-position	17.7
—COO— (esters)	63.8	3,4-position	16.3
—COOH (acids)	73.8	Triple bond	40.6
—OH	29.8	Ring closure:	
$—NH_2$	42.5	Three-membered	12
—O—	20.0	Four-membered	6.0
$—NO_2$	74	Five-membered	3.0
$—NO_3$ (nitrate)	93	Six-membered	0.8
$—CO(NH_2)$	91.7		

*As modified from Quayle (1953).

3. If the solute is an organic acid and the solvent is other than water, methanol, or butanol, the acid should be considered a dimer with twice the expected values of V_A and $\mathbf{P}_A$.
4. For nonpolar solutes diffusing into monohydroxy alcohols, the values of V_B and $\mathbf{P}_B$ should be multiplied by a factor equal to $8\eta_B$, where η_B is the solvent viscosity in cP.

By using Eq. (12.10-2) with the restrictions noted above, values of D°_{AB} were estimated for several systems. The results are shown in Table 12.7, along with experimentally reported results. In most cases, quite reasonable estimates of D°_{AB} were found and errors normally were less than 10%.

To use the Tyn-Calus form, however, the parachors of both the solute and the solvent must be known. Although the compilation of Quayle (1953) is of value, it is still incomplete. The structural contributions given in Table 12.5 also are incomplete, and many functional groups are not represented.

A modified form of Eq. (12.10-2) may be developed by combining Eqs. (12.10-2) and (12.10-3) to give

$$D^\circ_{AB} = 8.93 \times 10^{-8} \frac{V_B^{0.267}}{V_A^{0.433}} \frac{T}{\eta_B} \left(\frac{\sigma_B}{\sigma_A} \right)^{0.15} \tag{12.10-4}$$

The definitions of the terms are the same as before except, when substituting Eq. (12.10-3), we must define V and σ at T_b. Thus σ_B and σ_A in Eq. (12.10-4) refer to surface tensions at T_b. Note also the very low exponent on this ratio of surface tensions. Since most organic liquids at T_b have similar surface tensions, one might choose to approximate this ratio as equal to unity. (For example, $0.80^{0.15} = 0.97$ and $1.2^{0.15} = 1.03$.) Then,

$$D^\circ_{AB} = 8.93 \times 10^{-8} \frac{V_B^{0.267}}{V_A^{0.433}} \frac{T}{\eta_B} \tag{12.10-5}$$

Alternatively, an approximation to the σ_B/σ_A ratio may be developed by using one of the correlations shown in Chap. 12. For example, if the Brock and Bird corresponding states method [Eqs. (12.4-3) and (12.4-4)] were used, then

$$\sigma = P_c^{2/3} T_c^{1/3} (0.132\alpha_c - 0.279)(1 - T_{br})^{11/9} \tag{12.10-6}$$

with P_c in bars and T_b and T_c in kelvins, $T_{br} = T_b/T_c$, and

$$\alpha_c = 0.9076 \left[1 + \frac{T_{br} \ln(P_c/1.013)}{1 - T_{br}} \right] \tag{12.10-7}$$

Equation (12.10-6) is only approximate, but it may be satisfactory when used to develop the *ratio* (σ_A/σ_B). Also, considering the low power (0.15) to which the ratio is raised, estimates of $(\sigma_B/\sigma_A)^{0.15}$ should be quite reasonable.

When Eq. (12.10-5) was employed to estimate D°_{AB} for the systems shown in Table 12.7, the results, as expected, were very similar to those found from the original Tyn and Calus form Eq. (12.10-2), except when σ_B differed appreciably from σ_A, for example, in the case of water and an organic liquid. In such situations, however, Eq. (12.10-4) with Eqs. (12.10-6) and (12.10-7) still led to results not significantly different from those with Eq. (12.10-2). The various forms of the Tyn-Calus correlation are illustrated in Example 12.6.

12.10.3 Hayduk and Minhas (1982) Correlation

These authors considered many correlations for the infinite dilution binary diffusion coefficient. By regression analysis, they proposed several depending on the type of solute-solvent system.

For *normal paraffin solutions:*

$$D^\circ_{AB} = 13.3 \times 10^{-8} \frac{T^{1.47} \eta_B^{\varepsilon}}{V_A^{0.71}} \tag{12.10-8}$$

where $\varepsilon = (10.2/V_A) - 0.791$ and the other notation is the same as in Eq. (12.10-1). Equation (12.10-8) was developed from data on solutes ranging from C_5 to C_{32} in normal paraffin solvents encompassing C_5 to C_{16}. An average error of only 3.4% was reported.

For *solutes in aqueous solutions:*

$$D^o_{AB} = 1.25 \times 10^{-8}(V_A^{-0.19} - 0.292)T^{1.52}\,\eta_w^{\varepsilon^*} \tag{12.10-9}$$

with $\varepsilon^* = (9.58/V_A) - 1.12$. The rest of the terms are defined in the same manner as under Eq. (12.10-1) except that the subscript w refers to the solvent, water. The authors report that this relation predicted D^o_{AB} values with an average deviation of slightly less than 10%.

For *nonaqueous (nonelectrolyte) solutions:*

$$D^o_{AB} = 1.55 \times 10^{-8} \frac{T^{1.29}}{\eta_B^{0.92}\,V_B^{0.23}} \frac{\mathbf{P}_B^{0.5}}{\mathbf{P}_A^{0.42}} \tag{12.10-10}$$

The notation is the same as in Eq. (12.10-2).

The appropriate equation in the set of Eqs. (12.10-8) to (12.10-10) was used in computing the errors shown in Table 12.7.

It is important to note that, when using the Hayduk-Minhas correlations, the same restrictions apply as in the Tyn-Calus equations.

If Eqs. (12.10-3) and (12.10-4) are used in Eq. (12.10-10) to eliminate the parachors, one obtains

$$D^o_{AB} = 1.55 \times 10^{-8} \frac{V_B^{0.27}}{V_A^{0.42}} \frac{T^{1.29}}{\eta_B^{0.92}} \frac{\sigma_B^{0.125}}{\sigma_A^{0.105}} \tag{12.10-11}$$

This relation is remarkably similar to the modified Tyn-Calus Eq. (12.10-4) except for the larger exponent on temperature. As before, when σ_A and σ_B are not greatly different, the surface tension ratio may be set equal to unity as was done to obtain Eq. (12.10-5), or if σ_A and σ_B differ appreciably, Eqs. (12.10-6) and (12.10-7) may be employed.

Example 12.6 Infinite dilution diffusion coefficient

Estimate the infinitely dilute diffusion coefficient of acetic acid into acetone at 313 K. The experimental value is 4.04×10^{-5} cm²/s (Wilke and Chang, 1955).

Solution. The data, from Appendix A, Quayle (1953) and Vargaftik et al. (1996), are:

	Acetic Acid (Solute) A	Acetone (Solvent) B
T_b, K	391.0	329.2
T_c, K	594.45	508.1
P_c, bar	57.9	47.0
ρ (at T_b), g/cm³	0.939	0.749
M, g/mol	60.05	58.08
P, cm g$^{1/4}$/(s$^{1/2}$ mol)	129	162
η_B, cP		0.270

TYN-CALUS, Eq. (12.10-2). By rule 3, acetic acid should be treated as a dimer; thus, $V = (2)(64.0) = 128$ cm^3/mol and $\mathbf{P} = (2)(129) = 258$ cm^{3}g$^{1/4}$/(s$^{1/2}$ mol).

$$D^{\circ}_{AB} = 8.93 \times 10^{-8}\left(\frac{128}{(77.5)^2}\right)^{1/6}\left(\frac{162}{258}\right)^{0.6}\frac{313}{0.270} = 4.12 \times 10^{-5} \text{ cm}^2/\text{s}$$

$$\text{Error} = \frac{4.12 - 4.04}{4.04} \times 100 = 2\%$$

MODIFIED TYN-CALUS, Eq. (12.10-5)

$$D^{\circ}_{AB} = 8.93 \times 10^{-8}\frac{(77.5)^{0.267}}{(128)^{0.433}}\frac{313}{0.270} = 4.04 \times 10^{-5} \text{ cm}^2/\text{s}$$

$$\text{Error} = 0\%$$

MODIFIED TYN-CALUS, Eqs. (12.10-4), (12.10-6), and (12.10-7). For acetic acid, $T_{br} = 391.1/594.45 = 0.658$. With Eq. (12.10-7),

$$\alpha_c = 0.9076\left\{1 + (0.658)\left[\frac{\ln(57.9/1.013)}{1 - 0.658}\right]\right\} = 7.972$$

Similarly, α_c for acetone = 7.316. Then, with Eq. (12.10-6)

$$\sigma_A = (57.9)^{2/3}(594.45)^{1/3}[(0.132)(7.972) - 0.278](1 - 0.658)^{11/9} = 26.3 \text{ erg/cm}^2$$

For acetone, $\sigma_B = 19.9$ erg/cm^2 and $(\sigma_B/\sigma_A)^{0.15} = 0.959$; thus,

$$D^{\circ}_{AB} = (4.04 \times 10^{-5})(0.959) = 3.87 \times 10^{-5} \text{ cm}^2/\text{s}$$

$$\text{Error} = \frac{3.87 - 4.04}{4.04} \times 100 = -4\%$$

In this case, the use of the $(\sigma_B/\sigma_A)^{0.15}$ factor actually increased the error; in most other cases, however, errors were less when it was employed.

HAYDUK-MINHAS, Eq. (12.10-10)

$$D^{\circ}_{AB} = 1.55 \times 10^{-8}(313)^{1.29}\frac{(162)^{0.5}/(258)^{0.42}}{(0.270)^{0.92}(77.5)^{0.23}} = 3.89 \times 10^{-5} \text{ cm}^2/\text{s}$$

$$\text{Error} = \frac{3.89 - 4.04}{4.04} \times 100 = -4\%$$

12.10.3 Nakanishi (1978) Correlation

In this method, empirical parameters were introduced to account for specific interactions between the solvent and the (infinitely dilute) solute. As originally proposed, the scheme was applicable only at 298.2 K. We have scaled the equation assuming $D^{\circ}_{AB}\,\eta_B/T$ to be constant.

$$D^{\circ}_{AB} = \left[\frac{9.97 \times 10^{-8}}{(I_A V_A)^{1/3}} + \frac{2.40 \times 10^{-8}\, A_B S_B V_B}{I_A S_A V_A}\right]\frac{T}{\eta_B} \qquad (12.10\text{-}12)$$

TABLE 12.6 Nakanishi parameter values for liquid diffusion coefficients.

	As Solutes (A)†		As Solvents (B)	
Compound(s)	I_A	S_A	A_B	S_B
Water	2.8 (1.8)‡	1	2.8	1
Methanol	2.2 (1.5)	1	2.0	1
Ethanol	2.5 (1.5)	1	2.0	1
Other monohydric alcohols	1.5	1	1.8	1
Glycols, organic acids, and other associated compounds	2.0	1	2.0	1
Highly polar materials	1.5	1	1.0	1
Paraffins ($5 \leq n \leq 12$)	1.0	0.7	1.0	0.7
Other substances	1.0	1	1.0	1

†If the solute is He, H_2, D_2, or Ne, the values of V_A should be multiplied by $[1 + (0.85)A^2]$, where A = 3.08 for He^3, 2.67 for He^4, 1.73 for H_2, 1.22 for D_2, and 0.59 for Ne.
‡The values in parentheses are for cases in which these solutes are dissolved in a solvent which is more polar.

where D^o_{AB} is the diffusion coefficient of solute A in solvent B at low concentrations, cm³/s. V_A and V_B are the liquid molar volumes of A and B at 298 K, cm³/mol, and the factors I_A, S_A, S_B, and A_B are given in Table 12.6. η_B is the solvent viscosity, in cP, at the system temperature T.

Should the solute (pure) not be a liquid at 298 K, it is recommended that the liquid molar volume at the boiling point be obtained either from data or from correlations in Chap. 4. Then,

$$V_A(298 \text{ K}) = \beta V_A(T_B) \tag{12.10-13}$$

where $\beta = 0.894$ for compounds that are solid at 298 K and $\beta = 1.065$ for compounds that are normally gases at 298 K (and 1 bar). For example, if oxygen is the solute, then, at the normal boiling point of 90.2 K, the molar liquid volume is 27.9 cm³/mol (Appendix A). With Eq. (12.10-13), $V_A = (1.065)(27.9) = 29.7$ cm³/mol.

Values of D^o_{AB} were estimated for a number of solute-solvent systems and the results were compared with experimental values in Table 12.7. In this tabulation, V_A for water was set equal to the dimer value of 37.4 cm³/mol to obtain more reasonable results. The poorest estimates were obtained with dissolved gases and with solutes in the more viscous solvents such as n-butanol. The use of definite values of I_A to account for solute polarity may cause problems, since it is often difficult to decide whether a compound should be counted as polar ($I_A = 1.5$) or not ($I_A = 1.0$). It might be better to select an average $I_A \cong 1.25$ if there is doubt about the molecular polarity. In Table 12.7, I_A was set equal to 1.5 for pyridine, aniline, nitrobenzene, iodine, and ketones.

Example 12.7 Estimating diffusion from viscosity

Estimate the value of D^o_{AB} for CCl_4 diffusing into ethanol at 298 K. At this temperature, the viscosity of ethanol is 1.08 cP (Riddick et al., 1986). The experimental value of D^o_{AB} is 1.50×10^{-5} cm²/s (Lusis and Ratcliff, 1971).

Solution. For this system with CCl_4 as solute A and ethanol as solvent B, from Table 12.6, $I_A = 1$, $S_A = 1$, $A_B = 2$, and $S_B = 1$. From Appendix A, for CCl_4 at 298 K, $V_A = 97.07$ cm³/mol and for ethanol at 298 K, $V_B = 58.68$ cm³/mol. Then, with Eq. (12.10-12),

$$D^o_{AB} = \left[\frac{9.97 \times 10^{-8}}{(97.07)^{1/3}} + \frac{(2.40 \times 10^{-8})(2)(58.68)}{97.07}\right]\frac{298}{1.08} = 1.40 \times 10^{-5} \text{ cm}^2/\text{s}$$

$$\text{Error} = \frac{1.40 - 1.50}{1.50} \times 100 = -6.7\%$$

TABLE 12.7 Diffusion coefficients in liquids at infinite dilution.

Solute A	Solvent B	T(K)	Expt. cm^2/s	Ref.**	Wilke-Chang	Tyn-Calus	Hayduk-Minhas	Nakanishi
Acetone	Chloroform	298	2.5	5	42	8.2	5.8	−8.8
		313	2.9		38	5.1	3.2	−11
Benzene		288	2.51	10	1.3	−21	−22	−15
		328	4.25	20	−0.7	−23	−23	−17
Ethanol		288	2.2	10	47	12	8.3	−29
Ethyl ether		298	2.13	20	29	7.4	4.5	4.4
Ethyl acetate		298	2.02	18	36	12	9.4	15
Methyl ethyl ketone		298	2.13	18	37	9.3	6.9	−11
Acetic acid	Benzene	298	2.09	4	28	−9.4	−10	−7
Aniline		298	1.96	17	0.4	0.1	0.1	−11
Benzoic acid		298	1.38	4	28	0.6	0.8	−10
Bromobenzene		281	1.45	22	−8.8	−6.4	−6	2.4
Cyclohexane		298	2.09	20	−11	−6.9	−7.5	−3.5
		333	3.45		−8.1	−4.2	−5	−0.7
Ethanol		288	2.25	10	−1.8	−5.2	−7	−39
Formic acid		298	2.28	4	53	−4.2	−4.2	13
n-Heptane		298	2.1	3	−27	−16	−17	−6.2
		353	4.25		−20	−7.1	−8.9	3.6
Methyl ethyl ketone		303	2.09	1	8.7	10	9	−8.5
Naphthalene		281	1.19	22	−2.1	5.2	5.6	18
Toluene		298	1.85	20	0.1	4.1	3.5	10
1,2,4-Trichlorobenzene		281	1.34	22	−13	−8.5	−7.7	−0.8
Vinyl chloride		281	1.77	22	8.7	0.1	−0.2	12
Acetic Acid	Acetone	288	2.92	2	35	3.2	−3.2	4.2
		313	4.04	25	33	2.1	−3.7	3.1
Benzoic acid		298	2.62	4	13	−3.5	−8.2	−13
Formic acid		298	3.77	4	56	5.5	0.1	21
Nitrobenzene	Acetone	293	2.94	19	−2.4	8.7	3	−6.2
Water		298	4.56	16		5.6	3.7	−19
Bromobenzene	*n*-Hexane	281	2.6	25	16	17	12	26
Carbon tetrachloride		298	3.7	6	12	8.6	4.6	18
Dodecane		298	2.73	23	−17	8.1	−6.7	7.9
n-Hexane		298	4.21	15	−18	−9	−4	1.3
Methyl ethyl ketone		303	3.74	1	23	23	17	−0.2
Propane		298	4.87	7	3.4	0.6	20	1.3
Toluene		298	4.21	4	−9	−6.3	−11	−3.1
Allyl alchohol	Ethanol	293	0.98	9	22	9.7	14	30
Isoamyl alcohol		293	0.81	9	9.8	10	14	15
Benzene		298	1.81	14	−39	−3	2.8	−18
Iodine		298	1.32	4	2.4			12
Oxygen		303	2.64	11	−3.2	32	34	47

(Continued)

TABLE 12.7 Diffusion coefficients in liquids at infinite dilution. (*Continued*)

Solute A	Solvent B	T(K)	Expt. cm^2/s	Ref.**	Wilke-Chang	Tyn-Calus	Hayduk-Minhas	Nakanishi
Pyridine		293	1.1	9	−1	−14	−9.1	3.3
Water		298	1.24	12		1.7	11	6.9
Carbon tetrachloride		298	1.5	14	−30	11	18	−6.7
Adipic Acid		303	0.4	1	1.1	16	29	8
Benzene		298	1	14	−52	5.3	20	−23
Butyric Acid		303	0.51	1	1.5	7.4	19	2.5
p-Dichlorobenzene		298	0.82	14	−52	13	29	−22
Methanol		303	0.59	14	51	37	50	51
Olelic acid		303	0.25	1	−8.4	26	41	−4.4
Propane		298	1.57	2	−65	−23	−13	−47
Water		298	0.56	14		7.2	26	22
Benzene	*n*-Heptane	298	3.4	3	8.4	1.2	−1.2	13
		372	8.4		1.6	−5	−5.9	6
Acetic acid	Ethyl acetate	293	2.18	21	69	12	8.5	21
Acetone		293	3.18	21	3.2	−6.8	−9.6	−15
Ethyl benzoate		293	1.85	21	9	16	13	−3.9
Methyl ethyl ketone		303	2.93	1	14	8.1	4.8	−5.4
Nitrobenzene		293	2.25	21	10	6.4	4.2	−2.8
Water		298	3.2	12		16	17	−4.2
Methane	Water	275	0.85	26	10	−3.6	0	14
		333	3.55		15	0.7	−2.6	19
Carbon dioxide		298	2	24	1.6	−22	−13	−19
Propylene		298	1.44	24	−7.7	−13	−13	−6.2
Methanol		288	1.26	10	5.4	−8.7	−5.4	−9.6
Ethanol		288	1	9	5.3	−1.6	−2.7	−8.7
Allyl alcohol		288	0.9	9	5.5	0.5	−2	−7.4
Acetic acid		293	1.19	13	2.6	−5	−4.7	−24
Ethyl acetate		293	1	13	−10	−9.4	−16	−0.9
Aniline		293	0.92	13	−2.5	−5.9	−8.9	−10
Diethylamine		293	0.97	13	−8.6	−7.3	−15	−21
Pyridine		288	0.58	9	49	37	38	31
Ethylbenzene		293	0.81	26	−8.9	−0.2	−18	8
Methylcylopentane		275	0.48	26	−2.5	0.3	−14	7
		293	0.85		−1.7	1.1	−9	7.8
		333	1.92		6.3	9.4	8.5	17
Vinyl chloride		298	1.34	8	3.6	−7.6	−3.3	4.3
		348	3.67		4.2	−7.1	3	4.9
ave. abs. % dev.					17	9	11	13

*Percent error = [(calc. −exp.)/exp.] × 100.

**References: 1, Amourdam and Laddha (1967); 2, Bidlack and Anderson (1964); 3, Calus and Tyn (1973); 4, Chang and Wilke (1955); 5, Haluska and Colver (1971); 6, Hammond and Stokes (1955); 7, Hayduk et al. (1973); 8, Hayduk and Laudie (1974); 9, Int. Critical Tables (1926); 10, Johnson and Babb (1956); 11, Krieger et al. (1967); 12, Lees and Sarram (1971); 13, Lewis (1955); 14, Lusis and Ratcliff (1971); 15, McCall and Douglas (1959); 16, Olander (1961) 17, Rao and Bennett (1971); 18, Ratcliff and Lusis (1971); 19; Reddy and Doraiswamy (1967); 20, Sanni and Hutchinson (1973); 21, Sitaraman et al. (1963); 22, Stearn et al. (1940); 23, Vadoic and Colver (1973); 24, Vivian and King (1964); 25, Wilke and Chang (1955); 26, Witherspoon and Bonoli (1969).

Other infinite dilution correlations for diffusion coefficients have been proposed, but after evaluation they were judged either less accurate or less general than the ones noted above (Akgerman, 1976; Akgerman and Gainer, 1972; Albright et al., 1983; Brunet and Doan, 1970; Chen, 1984; Faghri and Riazi, 1983; Fedors, 1979; Gainer, 1966; Hayduk and Laudie, 1974; King et al., 1965; Kuznetsova and Rashidova, 1980; Lusis and Ratcliff, 1968; Othmer and Thakar, 1953; Raina, 1980; Reddy and Doraiswamy, 1967; Siddiqi and Lucas, 1986; Sridhar and Potter, 1977; Teja, 1982; Umesi and Danner, 1981; Vadovic and Colver, 1973).

12.10.4 Effect of Solvent Viscosity

Most of the estimation techniques introduced in this section have assumed that D°_{AB} varies inversely with the viscosity of the solvent. This inverse dependence originated from the Stokes-Einstein relation for a large (spherical) molecule diffusing through a continuum solvent (small molecules). If, however, the solvent is viscous, one may question whether this simple relation is applicable. Davies et al. (1967) found for CO_2 that in various solvents, $D^{\circ}_{AB}\eta_B^{0.45} \cong$ constant for solvents ranging in viscosity from 1 to 27 cP. These authors also noted that Arnold (1930a) had proposed an empirical estimation scheme by which $D^{\circ}_{AB} \propto \eta_B^{0.45}$. Oosting et al. (1985) noted that, for the diffusion of 1-hexanol and 2-butanone in malto-dextrin solutions, the viscosity exponent was close to −0.5 over a range of temperatures and concentrations.

Hayduk and Cheng (1971) investigated the effect of solvent viscosity more extensively and proposed that, for nonaqueous systems,

$$D^{\circ}_{AB} = Q\eta_B^{q}, \tag{12.10-14}$$

where the constants Q and q are particular for a given solute; some values are listed by these authors. In Fig. 12.9, CO_2 diffusion coefficients in various solvents are shown. The solvent viscosity range is reasonably large, and the correlation for organic solvents is satisfactory. In contrast, the data for water as a solvent also are shown (Himmelblau, 1964). These data fall well below the organic solvent curve and have a slope close to −1. Hiss and Cussler (1973) measured diffusion coefficients of n-hexane and naphthalene in hydrocarbons with viscosities ranging from 0.5 to 5000 cP and report that $D^{\circ}_{AB} \propto \eta_B^{-2/3}$, whereas Hayduk et al. (1973) found that, for methane, ethane, and propane, D°_{AB} was proportional to $\eta_B^{-0.545}$.

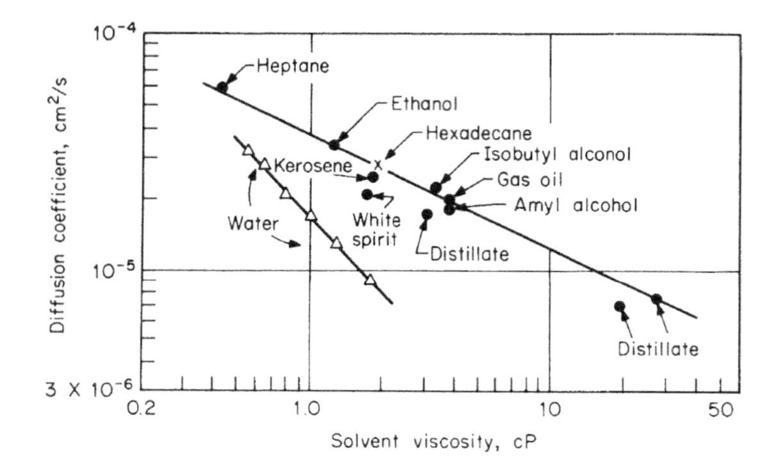

FIGURE 12.9
Diffusion coefficients of carbon dioxide in various solvents. • Davies et al. (1967); × Hayduk and Cheng (1971); Δ Himmelblau (1964).

These studies and others (Gainer and Metzner, 1965; Lusis, 1974; Way, 1971) clearly show that, over wide temperature or solvent viscosity ranges, simple empirical correlations, as presented earlier, are inadequate. The diffusion coefficient does not decrease in proportion to an increase in solvent viscosity, but $D^o_{AB} \propto \eta^q_B$, where q varies, usually from −0.5 to −1.

12.10.5 Discussion

Four estimation techniques were described to estimate the infinite dilution diffusion coefficient of a solute A in a solvent B. In Table 12.7, we show comparisons between calculated and experimental values of D^o_{AB} for a number of binary systems. Several comments are pertinent when analyzing the results. First, the temperature range covered is small; thus, any conclusions based upon this sample may not hold at much higher (or lower) temperatures. Second, while D^o_{AB} (exp.) is reported to three significant figures, the true accuracy is probably much less because diffusion coefficients are difficult to measure with high precision. Third, all estimation schemes tested showed wide fluctuations in the percent errors. These "failures" may be due to inadequacies in the correlation or to poor data. However, with such wide error ranges, the value of a single average percent error is in doubt.

With these caveats, it is clearly seen that, in general, the Tyn-Calus and the Hayduk-Minhas correlations usually yield the lowest errors; they are, therefore, recommended for calculating D^o_{AB}. Both require values of the solute and solvent parachors, but this is obviated with modifications such as Eq. (12.10-5) when $\sigma_A \approx \sigma_B$, or Eq. (12.10-4) [or Eq. (12.10-10)] with, say, Eqs. (12.10-6) and (12.10-7) when σ_A differs much from σ_B.

In special situations such as diffusion in n-paraffin solutions, Eq. (12.10-8) is recommended. We did not find a clear advantage for Eq. (12.10-9) over Eq. (12.10-2) for solutes diffusing into water, but the former would be more convenient to use.

New experimental data include the systems H_2S-H_2O (Halmour and Sandall, 1984), SO_2-H_2O (Leaist, 1984), CO_2 in binary mixtures (Takahashi et al., 1982), normal paraffin solutions (Hayduk and Ioakimidis, 1976), hydrocarbons in n-hexane (Dymond, 1981; Dymond and Woolf, 1982), and rare gases in water (Verhallen et al., 1984). Baldauf and Knapp (1983) studied a wide variety of polar and nonpolar systems at different temperatures and compositions. Mohan and Srinivasan (1984) and McKeigue and Gulari (1989) discuss the reduction of D^o_{AB} due to association of alcohols in nonpolar solvents (benzene, carbon tetrachloride).

12.11 CONCENTRATION DEPENDENCE OF BINARY LIQUID DIFFUSION COEFFICIENTS

The concentration dependence of binary diffusion coefficients is not simple. In some cases, it varies linearly between the two limiting diffusion coefficients, while in others strong positive or negative deviations from linearity are observed. In Sec. 12.2, it was suggested that the diffusion coefficient D_{AB} in a binary mixture may be proportional to a thermodynamic correction $\alpha = [(\partial \ln a_A/\partial \ln x_A)]_{T,P}$; a_A and x_A are the activity and mole fraction of species A respectively. From the Gibbs-Duhem equation, the derivative $(\partial \ln a_A/\partial \ln x_A)$ is the same whether written for A or B.

Several liquid models purport to relate D_{AB} to composition, e.g., the Darken equation (Darken, 1948; Ghai and Dullien, 1976) predicts that

$$D_{AB} = (D^*_A x_A + D^*_B x_B)\alpha \tag{12.11-1}$$

where D^*_A and D^*_B are tracer diffusion coefficients at x_A and x_B and α is evaluated at the same composition. Equation (12.11-1) was originally proposed to describe diffusion in metals, but it has been used for organic liquid mixtures by a number of investigators (Carman, 1967; Carman and Miller, 1959; Ghai and Dullien 1976; McCall and Douglas, 1967; Miller and Carman, 1961; Tyn and Calus, 1975; Vignes, 1966) with reasonable success except for mixtures in which the components may solvate (Hardt et al., 1959). The unavailability of tracer diffusion coefficients in most instances has led to a modification of Eq. (12.11-1) as

$$D_{AB} = (D_{BA}^{o} x_A + D_{AB}^{o} x_B)\alpha = [x_A(D_{BA}^{o} - D_{AB}^{o}) + D_{AB}^{o}]\alpha \tag{12.11-2}$$

That is, D_{AB} is a linear function of composition (see Fig. 12.2) corrected by the thermodynamic factor α. Equation (12.11-2) is easier to use because the infinitely dilute diffusion coefficients D_{BA}^{o} and D_{AB}^{o} may be estimated by techniques shown in Sec. 12.10. The thermodynamic term in Eq. (12.11-2) often overcorrects D_{AB}. Rathbun and Babb (1966) suggest α be raised to a fractional power; for associated systems, the exponent chosen was 0.6 unless there were *negative* deviations from Raoult's law when an exponent of 0.3 was recommended. Siddiqi and Lucas (1986) also recommend an exponent of 0.6 when one component is polar and the other nonpolar. When both A and B are polar, Siddiqi and Lucas (1986) recommend using Eq. (12.11-2) with the mole fraction replaced with the volume fraction. It is interesting to note (Sanchez and Clifton, 1977) that curves showing α and D_{AB} as a function of x_A tend to have the same curvature, thus providing some credence to the use of α as a correction factor.

Sanchez and Clifton (1977) found they could correlate D_{AB} with composition for a wide variety of binary systems by using a modification of Eq. (12.11-2):

$$D_{AB} = (D_{BA}^{o} x_A + D_{AB}^{o} x_B)(1 - m + m\alpha) \tag{12.11-3}$$

where the parameter m is to be found from one mixture datum point, preferably in the mid-compositional range. m varies from system to system and may be either greater or less than unity. When $m = 1$, Eq. (12.11-3) reduces to Eq. (12.11-2). Interestingly, for several highly associated systems, m was found to be between 0.8 and 0.9. The temperature dependence of m is not known.

Another theory predicts that the group $D_{AB}\eta/\alpha$ should be a linear function of mole fraction (Anderson and Babb, 1961; Bidlack and Anderson, 1964; Byers and King, 1966). Vignes (1966) shows graphs indicating this is not even approximately true for the systems acetone-water and acetone-chloroform. Rao and Bennett (1971) studied several very nonideal mixtures and found that, while the group $D_{AB}\eta/\alpha$ did not vary appreciably with composition, no definite trends could be discerned. One of the systems studied (aniline-carbon tetrachloride) is shown in Fig. 12.10. In this case, D_{AB}, η, α, and $D_{AB}\eta$ varied widely; the group $D_{AB}\eta/\alpha$ also showed an unusual variation with composition. Carman and Stein (1956) stated that $D_{AB}\eta/\alpha$ is a linear function of x_A for the nearly ideal system benzene-carbon tetrachloride and for the nonideal system acetone-chloroform but not for ethyl alcohol-water. Vignes (1966) suggested a convenient way of correlating the composition effect on the liquid diffusion coefficient:

$$D_{AB} = [(D_{AB}^{o})^{x_B}(D_{BA}^{o})^{x_A}]\alpha \tag{12.11-4}$$

and, therefore, a plot of log (D_{AB}/α) versus mole fraction should be linear. He illustrated this relation with many systems, and, except for strongly associated mixtures, excellent results.

Figure 12.11 shows the same aniline-carbon tetrachloride system plotted earlier in Fig. 12.10. Although not perfect, there is a good agreement with Eq. (12.11-4).

Dullien (1971) carried out a statistical test of the Vignes correlation. It was found to fit experimental data extremely well for ideal or nearly ideal mixtures, but there were several instances when it was not particularly accurate for nonideal, nonassociating solutions. Other authors report that the Vignes correlation is satisfactory for benzene and n-heptane (Calus and Tyn, 1973) and toluene and methylcyclo-hexane (Haluska and Colver, 1970), but not for benzene and cyclohexane (Loflin and McLaughlin, 1969).

The Vignes relation can be derived from absolute rate theory, and a logical modification of this equation is found to be (Leffler and Cullinan, 1970)

$$D_{AB}\eta = [(D_{AB}^{o}\ \eta_B)^{x_B}(D_{BA}^{o}\ \eta_A)^{x_A}]\alpha \tag{12.11-5}$$

A test of 11 systems showed that this latter form was marginally better in fitting experimental data. In Fig. 12.12, we have plotted both log $(D_{AB}\eta/\alpha)$ and log (D_{AB}/α) as a function of composition for the

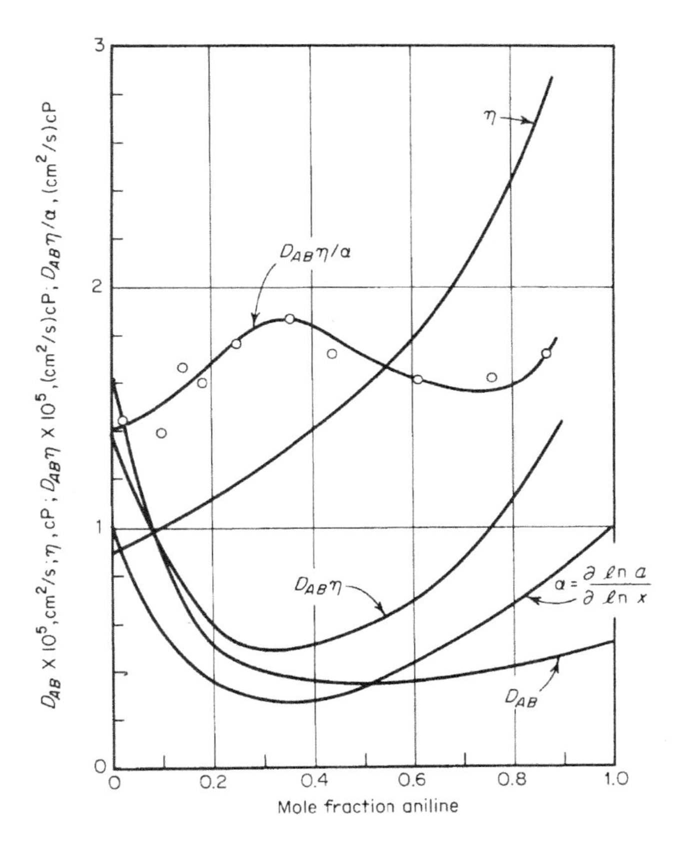

FIGURE 12.10
Diffusion coefficients for the system aniline-carbon tetrachloride at 298 K. (Rao and Bennett, 1971)

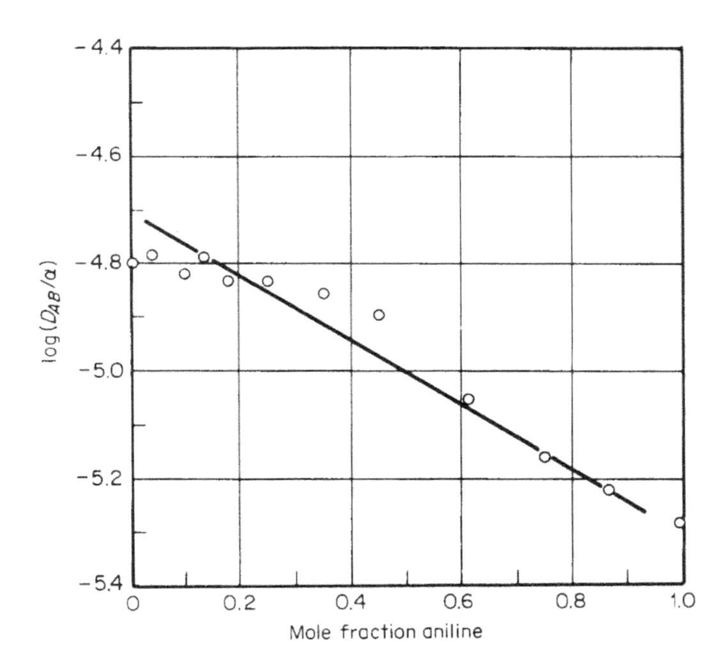

FIGURE 12.11
Vignes plot for the system aniline-carbon tetrachloride at 298 K.

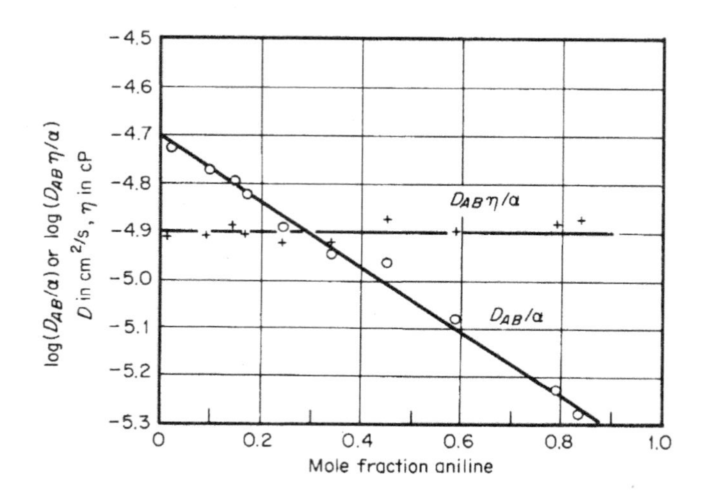

FIGURE 12.12
Vignes plot for the system aniline-benzene at 298 K. (Rao and Bennett, 1971)

aniline-benzene system. The original Vignes equation fits the data well, but so does Eq. (12.11-5); in fact, for the latter $D_{AB}\eta/\alpha$ is essentially constant.

Tyn and Calus (1975a) measured the binary diffusion coefficient for several associating systems (ethanol-water, acetone-water, and acetone-chloroform) and found that Eq. (12.11-4) was, generally, preferable to Eq. (12.11-5), although the mean deviation for the Vignes relation was about 14% for the three systems studied.

Other correlation methods have been proposed (Anderson et al., 1958; Cram and Adamson, 1960; Cullinan, 1971; Gainer, 1970; Haase and Jansen, 1980; Haluska and Colver, 1971; Ratcliff and Holdcroft, 1963; Teja, 1982), but they are either less accurate or less general than those discussed above.

Baldauf and Knapp (1983) present an exceptionally complete data set for 11 binary liquid mixtures giving D_{AB}, η_m, ρ_m, and the refractive index as a function of composition.

In summary, no single correlation is always satisfactory for estimating the concentration effect on liquid diffusion coefficients. The Vignes method [Eq. (12.11-4)] is recommended here as a well-tested and generally accurate correlation. It is also the easiest to apply, and no mixture viscosities are necessary. The thermodynamic correction factor α must, however, be known. If constants are available for a particular activity coefficient model, these constants along with the appropriate equations can be used to calculate α. For example, for the NRTL equation, α is given by the equation

$$\alpha = 1 - 2x_A x_B \left[\frac{\tau_A \, G_{BA}^2}{(x_A + x_B G_{BA})^3} + \frac{\tau_B \, G_{AB}^2}{(x_B + x_A G_{AB})^3} \right] \tag{12.11-6}$$

where the symbols are defined in Table 8.2.

In an approach that does not involve the thermodynamic correction factor α, Hsu and Chen (1998) used absolute reaction rate theory to develop Eq. (12.11-7). This equation contains two adjustable parameters and has been effective for the correlation of binary diffusion coefficients with composition.

$$\begin{aligned} \ln D_{AB} = {} & x_B \ln D_{AB}^{o} + x_A D_{BA}^{o} + 2\left(x_A \ln \frac{x_A}{\phi_A} + x_B \ln \frac{x_B}{\phi_B} \right) \\ & + 2x_A x_B \left\{ \frac{\phi_A}{x_A}\left(1 - \frac{\lambda_A}{\lambda_B}\right) + \frac{\phi_B}{x_B}\left(1 - \frac{\lambda_B}{\lambda_A}\right) \right\} \\ & + x_B q_A[(1 - \theta_{BA}^2) \ln \tau_{BA} + (1 - \theta_{BB}^2)\tau_{AB} \ln \tau_{AB}] \\ & + x_A q_B[(1 - \theta_{AB}^2) \ln \tau_{AB} + (1 - \theta_{AA}^2)\tau_{BA} \ln \tau_{BA}] \end{aligned} \tag{12.11-7}$$

$$\theta_{ji} = \frac{\theta_j \tau_{ji}}{\sum_l \theta_l \tau_{li}} \tag{12.11-8}$$

$$\theta_j = \frac{x_j q_j}{\sum_l x_l q_l} \tag{12.11-9}$$

$$\tau_{ji} = \exp\left(-\frac{a_{ji}}{T}\right) \tag{12.11-10}$$

$$\phi_i = \frac{x_i \lambda_i}{\sum_i x_l \lambda_l} \tag{12.11-11}$$

where a_{12}, a_{21} = adjustable parameters
$\lambda_i = (r_i)^{1/3}$
q_i, r_i = UNIFAC volume and surface parameters respectively (see Chap. 8).

Equation (12.11-7) expresses the excess part of the diffusion coefficient in a form similar to that of a UNIQUAC equation. Hsu and Chen give values of a_{12} and a_{21} for 49 systems. For n-alkane systems Hsu and Chen took a_{12} and a_{21} to be zero. For 13 n-alkane systems, they found an average error of 1.6%, while for all systems tested they found an average error of 2.3%.

Example 12.8 Diffusion in a liquid mixture

Calculate the diffusion coefficient for methanol (A) water (B) at 313.13 K when $x_A = 0.25$ with the Hsu and Chen method. The experimental value is 1.33×10^{-5} cm²/s (Lee and Li, 1991).

Solution. From Hsu and Chen (1998), $a_{BA} = 194.5302$ and $a_{AB} = -10.7575$. From Chap. 8, $r_A = 1.4311$, $r_B = 0.92$, $q_A = 1.432$, and $q_B = 1.4$. λ_A and λ_B are 1.127 and 0.973, respectively. From Lee and Li (1991), $D^\circ_{AB} = 2.1 \times 10^{-5}$ cm²/s and 2.67×10^{-5} cm²/s. Substitution into Eqs. (12.10-8) to (12.11-11) leads to $\theta_A = 0.254$, $\theta_B = 0.721$, $\phi_A = 0.279$, $\phi_B = 0.746$, $\theta_{BA} = 0.612$, $\theta_{AB} = 0.261$, $\theta_{AA} = 0.388$, $\theta_{BB} = 0.739$, $\tau_{AB} = 1.035$, and $\tau_{BA} = 0.5373$. Equation (12.10-7) then gives

$$\begin{aligned}
\ln D_{AB} &= 0.75 \ln (2.10 \times 10^{-5} + 0.25 \ln (2.67 \times 10^{-5}) \\
&\quad + 2(0.25)(0.75)\left[\frac{0.279}{0.025}\left(1 - \frac{1.127}{0.973}\right) + \frac{0.721}{0.75}\left(1 - \frac{0.973}{1.127}\right)\right] \\
&\quad + 2\left(0.25 \ln \frac{0.25}{0.279} + 0.75 \ln \frac{0.75}{0.721}\right) \\
&\quad + (0.75)(1.432)[1 - (0.612)^2) \ln(0.5373) + (1 - (0.739)^2)(1.035) \ln(1.035)] \\
&\quad + (0.25)(1.4)[1 - (0.261)^2) \ln(1.035) + (1 - (0.388)^2)(0.5373) \ln(0.5373)] \\
D_{AB} &= 1.351 \times 10^{-5} \text{ cm}^2/\text{s} \\
\text{Error} &= \frac{1.351 - 1.33}{1.33} \times 100 = 1.6\%
\end{aligned}$$

When the predictive Eqs. (12.11-2), (12.11-4), and (12.11-5) were used to work the problem in Example 12.8, the errors were 21%, 21%, and –32%, respectively. Furthermore, these equations required a value for α which itself could not be determined with confidence. For example, Gmehling and Onken (1977) list three sets of activity coefficient data for the methanol-water system at 313 K. The three different sets of NRTL constants along with Eq. (12.11-6) gave α values that ranged from 0.72 to 0.84.

12.12 THE EFFECTS OF TEMPERATURE AND PRESSURE ON DIFFUSION IN LIQUIDS

For the Wilke-Chang and Tyn-Calus correlations for D°_{AB} in Sec. 12.10, the effect of temperature was accounted for by assuming

$$\frac{D^{\circ}_{AB}\ \eta_B}{T} = \text{constant} \tag{12.12-1}$$

In the Hayduk-Minhas method, the (absolute) temperature was raised to a power > 1, and the viscosity parameter was a function of solute volume. While these approximations may be valid over small temperature ranges, it is usually preferable (Sanchez et al., 1977) to assume that

$$D_{AB}\ (\text{or } D^{\circ}_{AB}) = A \exp\frac{-B}{T} \tag{12.12-2}$$

Equation (12.12-2) has been employed by a number of investigators (Innes and Albright, 1957; McCall et al., 1961; Robinson et al., 1966; Tyn, 1975). We illustrate its applicability in Fig. 12.13 with the system ethanol-water from about 298 K to 453 K for both infinitely dilute diffusion coefficients and D_{AB} for a 20 mole% solution (Kircher et al., 1981). Note that we have not included the thermodynamic correction factor α because it is assumed to be contained in the A and B parameters. Actually, since the viscosity of liquids is an exponentially decreasing function of temperature, below reduced temperatures of about 0.7, the product $D_{AB}\eta$ shows much less variation with temperature than do D_{AB} or η individually. The energies of activation for diffusion and viscosity are opposite in sign and often of the same magnitude numerically (Robinson et al., 1966).

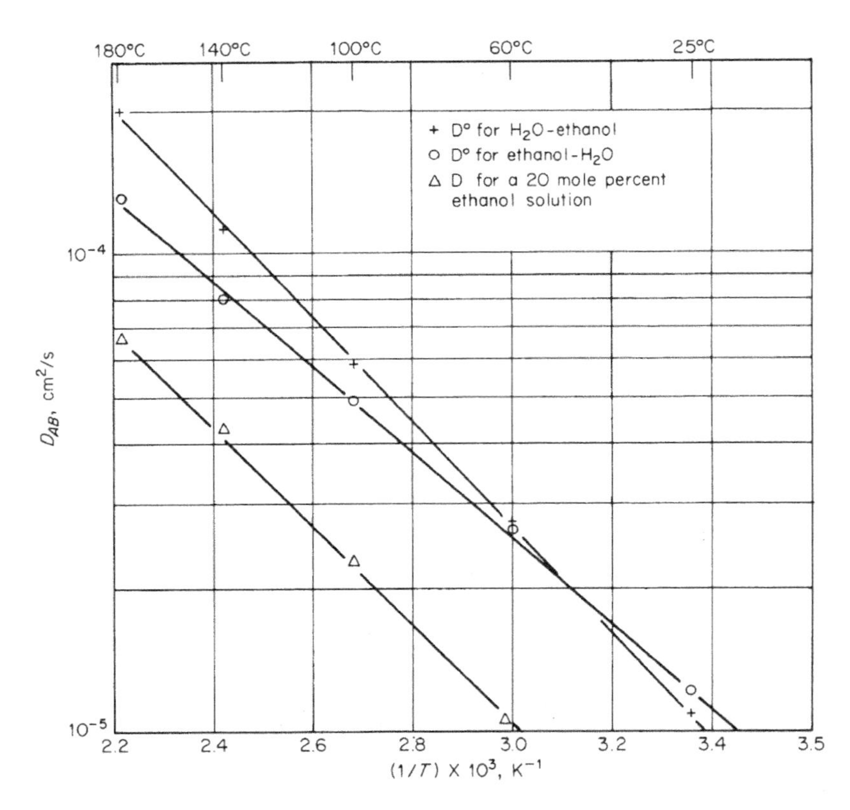

FIGURE 12.13
Variation of diffusion coefficients with temperature. [Data from Kircher et al. (1981)]

Tyn (1976, 1981) reviewed the various proposed techniques to correlate infinitely dilute binary (and also self-)diffusion coefficients with temperature. He suggested that

$$\frac{D^{o}_{AB}\ (T_2)}{D^{o}_{AB}\ (T_1)} = \left(\frac{T_c - T_1}{T_c - T_2}\right)^n \tag{12.12-3}$$

where T_c is the critical temperature of the solvent B. T_c, T_1, and T_2 are in kelvins. The parameter n was related to the heat of vaporization of the solvent at T_b (solvent) as follows:

n	$\Delta H_v(T_b)$, J/mol
3	7900 to 30,000
4	30,000 to 39,700
6	39,700 to 46,000
8	46,000 to 50,000
10	> 50,000

Typical compounds falling into these categories would be n = 3, n-pentane, acetone, cyclohexane, chloroform; n = 4, benzene, toluene, chlorobenzene, n-octane, carbon tetrachloride; n = 6, cyclohexane, propanol, butanol, water; n = 8, heptanol; and n = 10, ethylene and propylene glycols. See Chap. 7 for more information about $\Delta H_v(T_b)$.

Equation (12.12-3), which does not require mixture viscosity data, was tested with a large number of binary systems, and an error of about 9% was found. When Eq. (12.12-1) also was examined, Tyn reported an error of about 10%. The temperature ranges for Eq. (12.12-3) are about 10 K above the freezing point to about 10 K below the normal boiling point. Larger errors were noted if these ranges were exceeded.

The effect of pressure on liquid diffusion coefficients has received little attention. Easteal (1984) attempted to correlate tracer or self-diffusion coefficients with pressure and suggested

$$\ln D^*_j = a + bP^{0.75} \tag{12.12-4}$$

where D^*_j is a tracer or self-diffusion coefficient and a and b are constants for a given solute, but they do vary significantly with temperature. b is a negative number, and thus D^*_j decreases with an increase in pressure. As an example, the self-diffusion coefficient for n-hexane decreases from about 4.2×10^{-5} cm²/s at 1 bar to about 0.7×10^{-5} cm²/s at 3500 bar at a temperature of 298 K.

From Eq. (12.12-1), at a given temperature, it can be inferred that

$$D^{o}_{AB}\ \eta_B = \text{constant} \tag{12.12-5}$$

If solvent-liquid viscosity data are available at high pressures, or are estimated with the methods in Chap. 10, it should then be possible to employ Eq. (12.12-5) to estimate D^{o}_{AB} at the elevated pressure from low-pressure diffusion coefficient data. Dymond and Woolf (1982) show, however, that this proportionality is only approximate for tracer-diffusion coefficients, but they indicate it may be satisfactory for binaries with large solute molecules.

12.13 DIFFUSION IN MULTICOMPONENT LIQUID MIXTURES

In a binary liquid mixture, as indicated in Secs. 12.2 and 12.9, a single diffusion coefficient was sufficient to express the proportionality between the flux and concentration gradient. In multicomponent systems, the situation is considerably more complex, and the flux of a given component depends upon the gradient of $n - 1$ components in the mixture. For example, in a ternary system of A, B, and C, the flux of A can be expressed as

$$J_A = \mathbf{D}_{AA} \frac{dc_A}{dz} + \mathbf{D}_{AB} \frac{dc_B}{dz} \tag{12.13-1}$$

Similar relations can be written for J_B and J_C. The coefficients $\mathbf{D}_{AA}$ and $\mathbf{D}_{BB}$ are called main coefficients; they are not self-diffusion coefficients. $\mathbf{D}_{AB}$, $\mathbf{D}_{BA}$, etc., are cross-coefficients, because they relate the flux of a component i to a gradient in j. $\mathbf{D}_{ij}$ is normally not equal to $\mathbf{D}_{ji}$ for multicomponent systems.

One important case of multicomponent diffusion results when a solute diffuses through a homogeneous solution of mixed solvents. When the solute is dilute, there are no concentration gradients for the solvent species and one can speak of a single solute diffusivity with respect to the mixture D°_{Am}. This problem has been discussed by several authors (Cullinan and Cusick, 1967a; Holmes et al., 1962; Perkins and Geankoplis, 1969; Tang and Himmelblau, 1965) and empirical relations for D°_{Am} have been proposed. Perkins and Geankoplis (1969) evaluated several methods and suggested

$$D^{\circ}_{Am}\eta_m^{0.8} = \sum_{\substack{j=1 \\ j \neq A}}^{n} x_j D^{\circ}_{Aj}\, n_j^{0.8} \tag{12.13-2}$$

where D°_{Am} = effective diffusion coefficient for a dilute solute A into the mixture, cm²/s
D°_{Aj} = infinite dilution binary diffusion coefficient of solute A into solvent j, cm²/s
x_j = mole fraction of j
η_m = mixture viscosity, cP
η_j = pure-component viscosity, cP

When tested with data for eight ternary systems, errors were normally less than 20%, except for cases involving CO_2. These same authors also suggested that the Wilke-Chang equation (12.10-1) might be modified to include the mixed solvent case, i.e.,

$$D^{\circ}_{Am} = 7.4 \times 10^{-8} \frac{(\phi M)^{1/2} T}{\eta_m V_A^{0.6}} \tag{12.13-3}$$

$$\phi M = \sum_{\substack{j=1 \\ j \neq A}}^{n} x_j \phi_j M_j \tag{12.13-4}$$

Although not extensively tested, Eq. (12.13-3) provides a rapid, reasonably accurate estimation method.

For CO_2 as a solute diffusing into mixed solvents, Takahashi et al. (1982) recommend

$$D^{\circ}_{CO_2m}\left(\frac{\eta_m}{V_m}\right)^{1/3} = \sum_{\substack{j=1 \\ j \neq CO_2}}^{n} x_j D^{\circ}_{CO_2 j}\left(\frac{\eta_j}{V_j}\right)^{1/3} \tag{12.13-5}$$

where V_m is the molar volume, cm³/mol, for the mixture at T and V_j applies to the pure component. Tests with several ternary systems involving CO_2 led to deviations from experimental values usually less than 4%.

Example 12.9 Diffusion in a mixed solvent

Estimate the diffusion coefficient of acetic acid diffusing into a mixed solvent containing 20.7 mole% ethyl alcohol in water. The acetic acid concentration is small. Assume the temperature is 298 K. The experimental value reported by Perkins and Geankoplis (1969) is 0.571×10^{-5} cm²/s.

Data Let E = ethyl alcohol, W = water, and A = acetic acid. At 298 K, Perkins and Geankoplis (1969) give $\eta_E = 1.10$ cP, $\eta_W = 0.894$ cP, $D_{AE} = 1.03 \times 10^{-5}$ cm²/s, $D_{AW} = 1.30 \times 10^{-5}$ cm²/s, and for the solvent mixture under consideration, $\eta_m = 2.35$ cP.

Solution. From Eq. (12.13-2),

$$D^{\circ}_{Am} = (2.35)^{-0.8} [(0.207)(1.03 \times 10^{-5})(1.10)^{0.8} + (0.793)(1.30 \times 10^{-5})(0.894)^{0.8}]$$
$$= 0.59 \times 10^{-5} \text{ cm}^2/\text{s}$$

$$\text{Error} = \frac{0.59 - 0.571}{0.571} \times 100 = 3.3\%$$

Note that this diffusion coefficient is significantly below the two limiting binary values. The decrease in the mixture diffusivity appears to be closely related to the increase in solvent mixture viscosity relative to the pure components. Had the modified Wilke-Chang equation been used, with $V_A = 64.1$ cm³/mol (Chap. 3), $\phi_E = 1.5$, and $\phi_W = 2.6$, and with $M_E = 46$ and $M_W = 18$, using Eqs. (12.13-3) and (12.13-4),

$$\phi M = (0.207)(1.5)(46) + (0.793)(2.6)(18) = 51.39$$

$$D^{\circ}_{Am} = \frac{(7.4 \times 10^{-8})(51.39)^{1/2}(298)}{(2.35)(64.1)^{0.6}} = 0.55 \times 10^{-5} \text{ cm}^2/\text{s}$$

$$\text{Error} = \frac{0.55 - 0.571}{0.571} \times 100 = -3.7\%$$

Example 12.10 Gas diffusion in a mixed liquid solvent

Estimate the diffusion coefficient of CO_2 (D) into a mixed solvent of *n*-octanol (L) and carbon tetrachloride (C) containing 60 mole % *n*-octanol. The temperature is 298 K. The experimental value (Takahashi et al., 1982) is 1.96×10^{-5} cm²/s.

Data From Takahashi et al. (1982), $D^{\circ}_{DL} = 1.53 \times 10^{-5}$ cm²/s, $D^{\circ}_{DC} = 3.17 \times 10^{-5}$ cm²/s, $\eta_m = 3.55$ cP, $\eta_L = 7.35$ cP, and $\eta_C = 0.88$ cP. From Appendix A, $V_L = 158.37$ cm³/mol and $V_C = 97.07$ cm³/mol at 298 K. The mixture volume is not known. If we assume that the mole fraction of CO_2 in the liquid mixture is small and that *n*-octanol and carbon tetrachloride form ideal solutions,

$$V_m \approx (0.6)(158.37) + (0.4)(97.07) = 133.8 \text{ cm}^3/\text{mol}$$

Solution. With Eq. (12.13-5)

$$D(CO_2\text{-m}) = (3.55/133.8)^{-1/3}[(0.6)(1.53 \times 10^{-5})(7.35/158.37)^{1/3}]$$
$$+ (0.4)(3.17 \times 10^{-5})(0.88/97.07)^{1/3} = 1.99 \times 10^{-5} \text{ cm}^2/\text{s}.$$

$$\text{Error} = \frac{1.99 - 1.96}{1.96} \times 100 = 1.5\%$$

When dealing with the general case of multicomponent diffusion coefficients, there are no convenient and simple estimation methods. Kooijman and Taylor (1991) have discussed extension of the Vignes correlation [Eq. (12.13-4)] to ternary systems; the reader is also referred to page 570 of Bird et al. (1960) or to Curtiss and Bird (1999) for discussion of this problem. Kett and Anderson (1969, 1969a) apply hydrodynamic theory to estimate ternary diffusion coefficients. Although some success was achieved, the method requires extensive data on activities, pure component and mixture volumes, and viscosities, as well as tracer and binary diffusion coefficients. Bandrowski and Kubaczka (1982) suggest using the mixture critical volume as a correlating parameter to estimate D_{Am} for multicomponent mixtures.

12.14 DIFFUSION IN ELECTROLYTE SOLUTIONS

When a salt dissociates in solution, ions rather than molecules diffuse. In the absence of an electric potential, however, the diffusion of a single salt may be treated as molecular diffusion.

The theory of diffusion of salts at low concentrations is well developed. At concentrations encountered in most industrial processes, one normally resorts to empirical correlations, with a concomitant loss in generality and accuracy. A comprehensive discussion of this subject is available (Newman and Tobias, 1967).

For dilute solutions of a single salt, the diffusion coefficient is given by the Nernst-Haskell equation

$$D^{\circ}_{AB} = \frac{RT[1/z_+] + (1/z_-)]}{F^2[(1/\lambda^{\circ}_+) + (1/\lambda^{\circ}_-)]} \tag{12.14-1}$$

where D°_{AB} = diffusion coefficient at infinite dilution, based on molecular concentration, cm^2/s
T = temperature, K
R = gas constant, 8.314 J/(mol·K)
λ°_+, λ°_- = limiting (zero concentration) ionic conductances, (A/cm^2) (V/cm) $(g\text{-equiv}/cm^3)$
z_+, z_- = valences of cation and anion, respectively
F = faraday = 96,500 C/g-equiv

Values of λ°_+ and λ°_- can be obtained for many ionic species at 298 K from Table 12.8 or from alternative sources (Moelwyn-Hughes, 1957; Robinson and Stokes, 1959). If values of λ°_+ and λ°_- at other temperatures are needed, an approximate correction factor is $T/334\ \eta_W$, where η_W is the viscosity of water at T in centipoises.

TABLE 12.8 Limiting ionic conductances in water at 298 K (*Harned and Owen*, 1950) $(A/cm^2)(V/cm)(g\text{-equiv}/cm^2)$.

Anion	λ°_-	Cation	λ°_+
OH^-	197.6	H^+	349.8
Cl^-	76.3	Li^+	38.7
Br^-	78.3	Na^+	50.1
I^-	76.8	K^+	73.5
NO_3^-	71.4	NH^+	73.4
ClO_4^-	68.0	Ag^+	61.9
HCO_3^-	44.5	Tl^+	74.7
HCO_2^-	54.6	$(1/2)Mg^{2+}$	53.1
$CH_3CO_2^-$	40.9	$(1/2)Ca^{2+}$	59.5
$ClCH_2CO_2^-$	39.8	$(1/2)Sr^{2+}$	50.5
$CNCH_2CO_2^-$	41.8	$(1/2)Ba^{2+}$	63.6
$CH_3CH_2CO_2^-$	35.8	$(1/2)Cu^{2+}$	54
$CH_3(CH_2)_2CO_2^-$	32.6	$(1/2)Zn^{2+}$	53
$C_6H_5CO_2^-$	32.3	$(1/3)La^{3+}$	69.5
$HC_2O_4^-$	40.2	$(1/3)Co(NH_3)_6^{3+}$	102
$(1/2)C_2O_4^{2-}$	74.2		
$(1/2)SO_4^{2-}$	80		
$(1/3)\ Fe(CN)_6^{3-}$	101		
$(1/4)Fe(CN)_6^{4-}$	111		

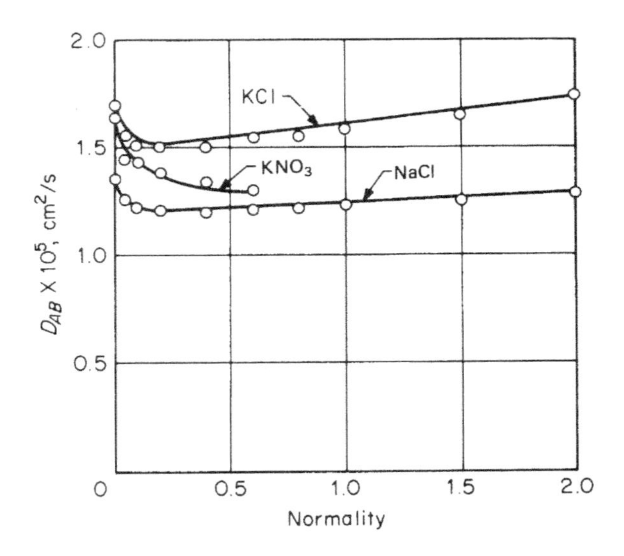

FIGURE 12.14
Effect of concentration on diffusivity of electrolytes in aqueous solution at 18.5 °C. Solid lines calculated by using Eq. (12.14-3). [Data from Gordon (1937)]

As the salt concentration becomes finite and increases, the diffusion coefficient decreases rapidly and then usually rises, often becoming greater than D°_{AB} at high normalities. Figure 12.14 illustrates the typical trend for three simple salts. The initial decrease at low concentrations is proportional to the square root of the concentration, but deviations from this trend are usually significant above 0.1 N.

Figure 12.15 shows the behavior for the rather atypical system HCl in water at 298 K. This system illustrates not only the minimum illustrated in Fig. 12.14 but shows a maximum as well.

Also shown in Fig. 12.15 is the behavior of $D^{\circ}_{AB}/\alpha^{D}$, where the thermodynamic factor α^{D} is given by

$$\alpha^{D} = 1 + m\frac{\partial \ln \gamma_{\pm}}{\partial m} \tag{12.14-2}$$

m = molality of the solute, mol/kg solvent
$\gamma_{\pm}$ = mean ionic activity coefficient of the solute

From Fig. 12.15, the quantity D°_{AB}/α varies more smoothly with concentration than does D°_{AB}. While D°_{AB}/α for HCl drops smoothly over the whole concentration range, Rizzo et al. (1997) observe that for most salts, D°_{AB}/α shows a maximum at low concentrations.

The behavior shown in Fig. 12.15 is consistent with an empirical equation proposed by Gordon (1937) that has been applied to systems at concentrations up to 2 N:

$$D_{AB} = D^{\circ}_{AB}\frac{\eta_s}{\eta}(\rho_s\bar{V}_s)^{-1}\alpha \tag{12.14-3}$$

where α is defined in Eq. (12.14-2) and
D°_{AB} = diffusion coefficient at infinite dilution, [Eq. (12.14-1)], cm^2/s
ρ_s = molar density of the solvent, mol/cm^3
$\bar{V}_s$ = partial molar volume of the solvent, cm^3/mol
η_s = viscosity of the solvent, cP
η = viscosity of the solution, cP

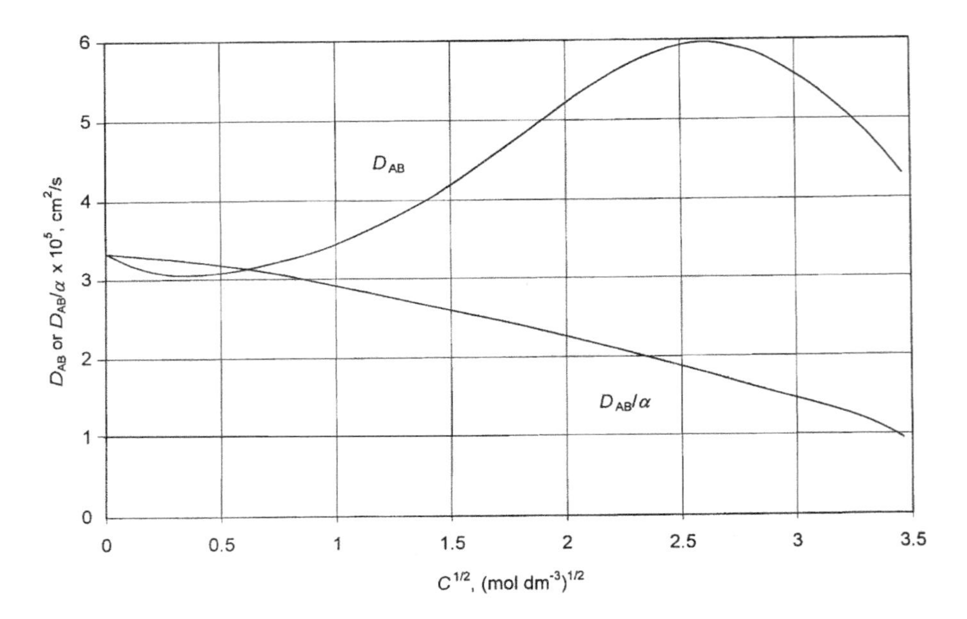

FIGURE 12.15
Effect of concentration on diffusivity of HCl in aqueous solution at 25 °C. [Curves are through data from Rizzo et al. (1997) and Stokes (1950)]

In many cases, the product $\rho_s \bar{V}_s$ is close to unity, as is the viscosity ratio η_s/η, so that Gordon's relation provides an activity correction to the diffusion coefficient at infinite dilution. Though Harned and Owen (1950) tabulate $\gamma_\pm$ as a function of m for many aqueous solutions, there now exist several semiempirical correlation techniques to relate $\gamma_\pm$ to concentration. These correlations are discussed in detail in Prausnitz et al. (1999).

Data on the diffusion of CO_2 into electrolyte solutions have been reported by Ratcliff and Holdcroft (1963). The diffusion coefficient was found to decrease linearly with an increase in salt concentration.

In summary, for very dilute solutions of electrolytes, employ Eq. (12.14-1). When values of the limiting ionic conductances in water are not available at the desired temperature, use those in Table 12.8 for 298 K and multiply D°_{AB} at 298 K by $T/334\ \eta_w$, where η_w is the viscosity of water at T in centipoises.

For concentrated solutions, use Eq. (12.14-3). If values of $\gamma_\pm$ and λ° are not available at T, calculate D°_{AB} at 298 K and multiply it by $(T/298)[(\eta \text{ at } 298)/(\eta \text{ at } T)]$. If necessary, the ratio of the solution viscosity at 298 K to that at T may be assumed to be the same as the corresponding ratio for water.

Example 12.11 Electrolyte diffusion

Estimate the diffusion coefficient of NaOH in a 2 N aqueous solution at 288 K.

Solution. From data on densities of aqueous solutions of NaOH, it is evident that, up to 12 weight% NaOH (about 3 N), the density increases almost exactly in inverse proportion to the weight fraction of water; i.e., the ratio of moles of water per liter is essentially constant at 55.5. Thus both V/n and $\bar{V}_1$ are very nearly 55.5 and cancel in Eq. (12.14-3). In this case, the molality m is essentially identical with the normality. Values of $\gamma_\pm$ for NaOH at 298 K are given in Harned and Owen (1950). The value at $m = 2$ is 0.698. When the values are plotted versus molality m, the slope at 2 m is approximately 0.047. Hence,

$$m \frac{\partial \ln \gamma_\pm}{\partial m} = \frac{m}{\gamma_\pm}\frac{\partial \gamma_\pm}{\partial m} = \frac{2}{0.698}(0.047) = 0.135$$

The viscosities of water and 2 N NaOH solution at 298 K are 0.894 and 1.42 cP, respectively. Substituting in Eqs. (12.14-1) and (12.14-3) gives

$$D^{\circ}_{AB} = \frac{(2)(8.314)(298)}{[(1/50) + (1/198)](96,500)^2}$$
$$= 2.12 \times 10^{-5} \text{ cm}^2/\text{s}$$

$$D_{AB} = (2.12 \times 10^{-5}) \frac{0.894}{1.42} \frac{55.5}{55.5} [1 + (2)(0.135)]$$
$$= 1.70 \times 10^{-5} \text{ cm}^2/\text{s}$$

At 288 K, the viscosity of water is 1.144 cP, and so the estimated value of D_{AB} at 288 K is

$$1.70 \times 10^{-5} \frac{(288)}{(334)(1.144)} = 1.28 \times 10^{-5} \text{ cm}^2/\text{s}$$

which may be compared with the International Critical Tables (1926) value of 1.36×10^{-5} cm^2/s for an error of −5.9%.

In a system of mixed electrolytes, such as in the simultaneous diffusion of HCl and NaCl in water, the faster-moving H^+ ion may move ahead of its Cl^- partner, the electric current being maintained at zero by the lagging behind of the slower-moving Na^+ ions. In such systems, the unidirectional diffusion of each ion species results from a combination of electric and concentration gradients:

$$N_+ = \frac{\lambda_+}{F^2}\left(-RT\frac{\partial c_+}{\partial z} - Fc_+\frac{\partial E}{\partial z}\right) \quad (12.14\text{-}4)$$

$$N_- = \frac{\lambda_-}{F^2}\left(-RT\frac{\partial c_-}{\partial z} - Fc_-\frac{\partial E}{\partial z}\right) \quad (12.14\text{-}5)$$

where N_+, N_- = diffusion flux densities of the cation and anion, respectively, g-equiv/cm^2·s
c_+, c_- = corresponding ion concentrations, g-equiv/cm^3
$\partial E/\partial z$ = gradient in electric potential
λ_+, λ_- = ionic equivalent conductances

Collision effects, ion complexes, and activity corrections are neglected. The electric field gradient may be imposed externally but is present in the ionic solution even if, owing to the small separation of charges that results from diffusion itself, there is no external electrostatic field.

One equation for each cation and one for each anion can be combined with the requirement of zero current at any z to give $\Sigma N_+ = \Sigma N_-$. Solving for the unidirectional flux densities (Ratcliff and Lusis, 1971),

$$n_+N_+ = -\frac{RT\lambda_+}{F^2 n_+}(G_+ - n_+c_+Y) \quad (12.14\text{-}6)$$

$$n_-N_- = -\frac{RT\lambda_-}{F^2 n_-}(G_- - n_-c_-Y) \quad (12.14\text{-}7)$$

$$Y = \frac{(\Sigma\gamma_+G_+/n_+) - (\Sigma\gamma_-G_-/n_-)}{\Sigma\gamma_+c_+ + \Sigma\gamma_-c_-} \quad (12.14\text{-}8)$$

where G_+ and G_- are the concentration gradients $\partial c/\partial z$ in the direction of diffusion.

Vinograd and McBain (1941) have used these relations to represent their data on diffusion in multi-ion solutions. D_{AB} for the hydrogen ion was found to decrease from 12.2 to 4.9×10^{-5} cm^2/s in a solution of HCl and $BaCl_2$ when the ratio of H^+ to Ba^{2+} was increased from zero to 1.3; D_{AB} at the same temperature is 9.03×10^{-5} for the free H^+ ion, and 3.3×10^{-5} for HCl in water. The presence of the slow-moving Ba^{2+} accelerates the H^+ ion, the potential existing with zero current causing it to move in dilute solution even faster than it would as a free ion with no other cation present. That is, electrical neutrality is maintained by the hydrogen ions moving ahead of the chlorine faster than they would as free ions, while the barium diffuses more slowly than as a free ion.

The interaction of ions in a multi-ion system is important when the several ion conductances differ greatly, as they do when H^+ or OH^- is diffusing. When the diffusion of one of these two ions is not involved, no great error is introduced by the use of "molecular" diffusion coefficients for the salt species present.

The relations proposed by Vinograd and McBain, Eqs. (12.14-5) to (12.14-7), are not adequate to represent a ternary system, in which four independent diffusion coefficients must be known to predict fluxes. The subject of ion diffusion in multicomponent systems is covered in detail by Cussler (1976) and in the papers by Wendt (1965) and Miller (1966, 1967) in which it is demonstrated how one can obtain multicomponent ion diffusion coefficients, although the data required are usually not available. Anderko and Lencka (1998) have used self-diffusion coefficients to model diffusion in multicomponent electrolyte systems; Mills and Lobo (1989) present a compilation of self-diffusion coefficient values.

12.15 NOTATION

a_j — activity of component $j = x_j \gamma_j$

A_B — parameter in Table 12.6

c — concentration, mol/cm³; c_j, for component j; c_+, c_-, ion concentrations

D_{AB} — binary diffusion coefficient of A diffusing into B, cm²/s; D°_{AB}, at infinite dilution of A in B; D_{AB}, cross-coefficient in multicomponent mixtures; D_{im} of i into a homogeneous mixture; D^+_{AB}, at a low pressure

D^*_A — tracer-diffusion coefficient of A, cm²/s

D_{AA} — self-diffusion coefficient of A, cm²/s; D_{AA}, main coefficient for A in multicomponent diffusion

E — electric potential

f_D — correction term in Sec. 12.4

F — faraday = 96,500 C/g-equiv

G_+, G_- $\partial c_+/\partial z$ and $\partial c_-/\partial z$, Sec. 12.14

I_A — parameter in Table 12.6

J_A — flux of A, mol/(cm² · s); J^M_A, flux relative to a plane of no net mole flow; J^V_A flux relative to a plane of no net volume flow

k — Boltzmann's constant

m — molality of solute, mol/kg solvent; parameter in Sec. 12.11

M_A — molecular weight of A, g/mol; M_{AB}, $2[(1/M_A) + (1/M_B)]^{-1}$

n — number density of molecules; parameter in Sec. 12.14

N_A — flux of A relative to a fixed coordinate plane, mol/(cm² · s), Sec. 12.3

N_+, N_- — diffusion flux of cation and anion, respectively, Sec. 12.14

P — pressure, bar; P_c, critical pressure; P_r, reduced pressure, P/P_c

$\mathbf{P}_j$ — parachor of component j

q, Q — parameters in Sec. 12.10; q_i, UNIFAC surface parameter

r — distance of separation between molecules, Å

r_A — molecular radius in the Stokes-Einstein equation, Sec. 12.9; r_i, UNIFAC volume parameter

R — gas constant, Chap. 6

S_A, S_B — parameters in Table 12.6

T	temperature, kelvins; T_b, at the normal boiling point (at 1 atm); T_c critical temperature; T_r, reduced temperature, T/T_c
V	volume, cm^3/mol; V_b, at T_b; V_A, partial molar volume of A
V_j	molar volume of component j at either T_b or T, cm^3/mol
x_j	mole fraction of j, usually liquid
y_j	mole fraction of j, usually vapor
z	direction coordinate for diffusion
z_+, z_-	valences of cation and anion, respectively

Greek

α	$\partial \ln a_A/\partial \ln x_A$, $\partial \ln a/\partial \ln m$
β	$1/kT$ or parameter in Sec. 12.10
γ	activity coefficient; $\gamma_\pm$, mean ionic activity coefficient
δ	polar parameter defined in Sec. 12.4
ε	characteristic energy parameter; ε_A, for pure A; ε_{AB}, for an A-B interaction
ε, ε^*	parameters in Sec. 12.10
η	viscosity, cP; η_A, for pure A; η_m for a mixture
η^P	packing fraction, Sec. 12.3
η^g	packing fraction at glassy (jamming) transition, Sec. 12.3
λ_+^o, λ_-^o	limiting (zero concentration) ionic conductances, (A/cm^2)(V/cm)(g-equiv/cm^3)
μ_A	chemical potential of A, J/mol
μ_p	dipole moment, debyes, Chap. 3
ρ	density, g/cm^3
σ	characteristic length parameter, Å; σ_A, for pure A; σ_{AB}, for an A-B interaction; surface tension
Σ_v	Fuller et al. volume parameter, Table 12.3
ϕ	association parameter for the solvent
ψ	intermolecular potential energy of interaction
ω	acentric factor
Ω_D	collision integral for diffusion

Superscripts

$^\circ$	infinite dilution
*	tracer value
+	low pressure

Subscripts

A, B	components A and B; usually B is the solvent
m	mixture
w	water
s	solvent

12.16 REFERENCES

Akgerman, A.: *Ind. Eng. Chem. Fundam.,* **15:** 78 (1976).
Akgerman, A., and J. L. Gainer: *J. Chem. Eng. Data,* **17:** 372 (1972).
Albright, J.G., R. Mills, *J. Phys. Chem.,* **69:** 3120 (1965).
Albright, J. G., A. Vernon, J. Edge, and R. Mills: *J. Chem. Soc., Faraday Trans.* 1, **79:** 1327 (1983).
Alvarez, R., I. Medlina, J. L. Bueno, and J. Coca: *J. Chem. Eng. Data,* **28:** 155 (1983).
Amdur, I., and L. M. Shuler: *J. Chem. Phys.,* **38:** 188 (1963).
Amdur, I., J. Ross, and E. A. Mason: *J. Chem. Phys.,* **20:** 1620 (1952).
Amourdam, M. J., and G. S. Laddha: *J. Chem. Eng. Data,* **12:** 389 (1967).
Anderko, A., and M. M. Lencka: *Ind. Eng. Chem. Res.,* **37:** 2878 (1998).
Anderson, D. K., and A. L. Babb: *J. Phys. Chem.,* **65:** 1281 (1961).
Anderson, D. K., J. R. Hall, and A. L. Babb: *J. Phys. Chem.,* **62:** 404 (1958).
Anderson, J. L.: *Ind. Eng. Chem. Fundam.,* **12:** 490 (1973).
Arnold, J. H.: *Ind. Eng. Chem.,* **22:** 1091 (1930).
Arnold, J. H.: *J. Am. Chem. Soc.,* **52:** 3937 (1930a).
Bailey, R. G.: *Chem. Eng.,* **82**(6): 86 (1975).
Baldauf, W., and H. Knapp: *Ber. Bunsenges. Phys. Chem.,* **87:** 304 (1983).
Bandrowski, J., and A. Kubaczka: *Chem. Eng. Sci.,* **37:** 1309 (1982).
Bearman, R. J.: *J. Chem. Phys.,* **32:** 1308 (1960).
Bearman, R. J.: *J. Phys. Chem.,* **65:** 1961 (1961).
Bell, I. H., R. Hellmann, and A. H. Harvey, *J. Chem. Eng. Data,* **65:** 1038–1050 (2020).
Bidlack, D. L., and D. K. Anderson: *J. Phys. Chem.,* **68:** 3790 (1964).
Bird, R. B., W. E. Stewart, and E. N. Lightfoot: *Transport Phenomena,* Wiley, New York, 1960, Chap. 16.
Blanc, A.: *J. Phys.,* **7:** 825 (1908).
Brokaw, R. S.: *Ind. Eng. Chem. Process Design Develop.,* **8:** 240 (1969).
Brunet, J., and M. H. Doan: *Can. J. Chem. Eng.,* **48:** 441 (1970).
Byers, C. H., and C. J. King: *J. Phys. Chem.,* **70:** 2499 (1966).
Byrne, J. J., D. Maguire, and J. K. A. Clarke: *J. Phys. Chem.,* **71:** 3051 (1967).
Caldwell, L.: *J. Chem. Eng. Data,* **29:** 60 (1984).
Caldwell, C. S., and A. L. Babb: *J. Phys. Chem.,* **60:** 14, 56 (1956).
Calus, W. F., and M. T. Tyn: *J. Chem. Eng. Data,* **18:** 377 (1973).
Carman, P. C.: *J. Phys. Chem.,* **71:** 2565 (1967).
Carman, P. C.: *Ind. Eng. Chem. Fundam.,* **12:** 484 (1973).
Carman, P. C., and L. Miller: *Trans. Faraday Soc.,* **55:** 1838 (1959).
Carman, P. C., and L. H. Stein: *Trans. Faraday Soc.,* **52:** 619 (1956).
Carmichael, L. T., B. H. Sage, and W. N. Lacey: *AIChE J.,* **1:** 385 (1955).
Carr, H. Y., E. M. Purcell, *Phys. Rev.,* **94**:640 (1954).
Carswell, A. J., and J. C. Stryland: *Can. J. Phys.,* **41:** 708 (1963).
Catchpole, O. J., and M. B. King: *Ind. Eng. Chem. Res.,* **33:** 1828 (1994).
Chang, P., and C. R. Wilke: *J. Phys. Chem.,* **59:** 592 (1955).
Chen, N. H., and D. P. Othmer: *J. Chem. Eng. Data,* **7:** 37 (1962).
Chen, S.-H.: *AIChE J.,* **30:** 481 (1984).
Cram, R. R., and A. W. Adamson: *J. Phys. Chem.,* **64:** 199 (1960).
Crider, W. L.: *J. Am. Chem. Soc.* **78:** 924 (1956).
Cullinan, H. T., Jr.: *Can. J. Chem. Eng.,* **49:** 130 (1971).
Cullinan, H. T., Jr., and M. R. Cusick: *Ind. Eng. Chem. Fundam.,* **6:** 72 (1967).
Cullinan, H. T., Jr., and M. R. Cusick: *AIChE J.,* **13:** 1171 (1967a).
Curtiss, C. F., and R. B. Bird: *Ind. Eng. Chem. Res.,* **38:** 2515 (1999).
Cussler, E. L.: *Multicomponent Diffusion,* Elsevier, New York, 1976.
Cussler, E. L.: *Diffusion: Mass Transfer in Fluid Systems,* 2d ed. Cambridge, 1997, Chaps. 3, 7.
Darken, L. S.: *Trans. Am. Inst. Mining Metall. Eng.,* **175:** 184 (1948).
Daubert, T. E., R. P. Danner, H. M. Sibel, and C. C. Stebbins: *Physical and Thermodynamic Properties of Pure Chemicals: Data Compilation,* Taylor & Francis, Washington, D.C., 1997.
Davies, G. A., A. B. Ponter, and K. Craine: *Can. J. Chem. Eng.,* **45:** 372 (1967).

Debenedetti, P., and R. C. Reid: *AIChE J.,* **12:** 2034 (1986).
Di Pippo, R., J. Kestin, and K. Oguchi: *J. Chem. Phys.,* **46:** 4986 (1967).
Dullien, F. A. L.: *Nature,* **190:** 526 (1961).
Dullien, F. A. L.: *Trans. Faraday Soc.,* **59:** 856 (1963).
Dullien, F. A. L.: *Ind. Eng. Chem. Fundam.,* **10:** 41(1971).
Dullien, F. A. L.: personal communication, January, 1974.
Duncan, J. B., and H. L. Toor: *AIChE J.,* **8:** 38 (1962).
Dunlop, P. J., and C. M. Bignell: *Intern. J. Thermophys.,* **18:** 939 (1997).
Dymond, J. H.: *J. Phys. Chem,* **85:** 3291 (1981).
Dymond, J. H., and L. A. Woolf: *J. Chem. Soc., Faraday Trans.* 1, **78:** 991 (1982).
Easteal, A. J.: *AIChE J.,* **30:** 641 (1984).
Elliott, R. W., and H. Watts: *Can. J. Chem.,* **50:** 31 (1972).
Ellis, C. S., and J. N. Holsen: *Ind. Eng. Chem. Fundam.,* **8:** 787 (1969).
Emami, F. S., A. Vahid, and J. R. Elliott, *J. Chem. Thermo.,* **41:** 530–537 (2009).
Eyring, H., and T. Ree: *Proc. Natl. Acad. Sci.,* **47:** 526 (1961).
Faghri, A., and M.-R. Riazi: *Intern. Comm. Heat Mass Transfer,* **10:** 385 (1983).
Fairbanks, D. F., and C. R. Wilke: *Ind. Eng. Chem.,* **42:** 471 (1950).
Fedors, R. F.: *AIChE J.,* **25:** 200, 716 (1979).
Fuller, E. N., and J. C. Giddings: *J. Gas Chromatogr.,* **3:** 222 (1965).
Fuller, E. N., P. D. Schettler, and J. C. Giddings: *Ind. Eng. Chem.,* **58**(5)**:** 18 (1966).
Fuller, E. N., K. Ensley, and J. C. Giddings: *J. Phys. Chem.,* **73:** 3679 (1969).
Funazukuri, T., Y. Ishiwata, and N. Wakao: *AIChE J.,* **38:** 1761 (1992).
Gainer, J. L.: *Ind. Eng. Chem. Fundam.,* **5:** 436 (1966).
Gainer, J. L.: *Ind. Eng. Chem. Fundam.,* **9:** 381 (1970).
Gainer, J. L., and A. B. Metzner: *AIChE-Chem E Symp. Ser.,* no. 6, 1965, p. 74.
Gerek, Z. N., J. R. Elliott, *Ind. Eng. Chem. Res.,* **49:** 3411–3423 (2010).
Ghai, R. K., and F. A. L. Dullien: personal communication, 1976.
Ghai, R. K., H. Ertl, and F. A. L. Dullien: *AIChE J.,* **19:** 881 (1973).
Ghai, R. K., H. Ertl, and F. A. L. Dullien: *AIChE J.,* **20:** 1 (1974).
Gilliland, E. R.: *Ind. Eng. Chem.,* **26:** 681 (1934).
Glasstone, S., K. J. Laidler, and H. Eyring: *The Theory of Rate Processes,* McGraw-Hill, New York, 1941, Chap. 9.
Gmehling, J., and U. Onken: *Vapor-Liquid Equilibrium Data Collection,* Chem.Data Ser., Vol. I, Parts 1 and 1b, DECHEMA, Frankfurt, 1977.
Gordon, A. R.: *J. Chem. Phys.,* **5:** 522 (1937).
Gordon, M.: "References to Experimental Data on Diffusion Coefficients of Binary Gas Mixtures," *Natl. Eng. Lab. Rept.,* Glasgow, Scotland, 1977.
Gotoh, S., M. Manner, J. P. Sørensen, and W. E. Stewart: *Ind. Eng. Chem.,* **12:** 119 (1973).
Gotoh, S., M. Manner, J. P. Sørensen, and W. E. Stewart: *J. Chem. Eng. Data,* **19:** 169, 172 (1974).
Gupta, G. P., and S. C. Saxena: *AIChE J.,* **14:** 519 (1968).
Haase, R., and H.-J. Jansen: *Z. Naturforsch.,* **35A:** 1116 (1980).
Hahn, E.L., *Phys. Rev.,* **80:** 580 (1950).
Halmour, N., and O. C. Sandall: *J. Chem. Eng. Data,* **29:** 20 (1984).
Haluska, J. L., and C. P. Colver: *AIChE J.,* **16:** 691 (1970).
Haluska, J. L., and C. P. Colver: *Ind. Eng. Chem. Fundam.,* **10:** 610 (1971).
Hammond, B. R., and R. H. Stokes: *Trans. Faraday Soc.,* **51:** 1641 (1955).
Hardt, A. P., D. K. Anderson, R. Rathbun, B. W. Mar, and A. L. Babb: *J. Phys. Chem.,* **63:** 2059 (1959).
Harned, H. S., and B. B. Owen: "The Physical Chemistry of Electrolytic Solutions," *ACS Monogr.* **95,** 1950.
Hartley, G. S., and J. Crank: *Trans. Faraday Soc.,* **45:** 801 (1949).
Hattikudur, U. R., and G. Thodos: *J. Chem. Phys.,* **52:** 4313 (1970).
Hayduk, W., and W. D. Buckley: *Chem. Eng. Sci.,* **27:** 1997 (1972).
Hayduk, W., R. Casteneda, H. Bromfield, and R. R. Perras: *AIChE J.,* **19:** 859 (1973).
Hayduk, W., and S. C. Cheng: *Chem. Eng. Sci.,* **26:** 635 (1971).
Hayduk, W., and S. Ioakimidis: *J. Chem. Eng. Data,* **21:** 255 (1976).
Hayduk, W., and H. Laudie: *AIChE J.,* **20:** 611 (1974).
Hayduk, W., and B. S. Minhas: *Can. J. Chem. Eng.,* **60:** 295 (1982).

He, C.-H., and Y.-S. Yu: *Ind. Eng. Chem. Res.,* **37:** 3793 (1998).
Himmelblau, D. M.: *Chem. Rev.,* **64:** 527 (1964).
Hirschfelder, J. O., C. F. Curtiss, and R. B. Bird: *Molecular Theory of Gases and Liquids,* Wiley, New York, 1954.
Hiss, T. G., and E. L. Cussler: *AIChE J.,* **19:** 698 (1973).
Holmes, J. T., D. R. Olander, and C. R. Wilke: *AIChE J.,* **8:** 646 (1962).
Holson, J. N., and M. R. Strunk: *Ind. Eng. Chem. Fundam.,* **3:** 163 (1964).
Horrocks, J. K., and E. McLaughlin: *Trans. Faraday Soc.,* **58:** 1367 (1962).
Hsu, Y.-D., and Y.-P. Chen: *Fluid Phase Equil.,* **152:** 149 (1998).
Hudson, G. H., J. C. McCoubrey, and A. R. Ubbelohde: *Trans. Faraday Soc.,* **56:** 1144 (1960).
Innes, K. K., and L. F. Albright: *Ind. Eng. Chem.,* **49:** 1793 (1957).
International Critical Tables, McGraw-Hill, New York, 1926–1930.
Iomtev, M. B., and Y. V. Tsekhanskaya: *Russ. J. Phys. Chem.,* **38:** 485 (1964).
Johnson, D. W., and C. P. Colver: *Hydrocarbon Process. Petrol. Refiner,* **48**(3)**:** 113 (1969).
Johnson, P. A., and A. L. Babb: *Chem. Rev.,* **56:** 387 (1956).
Kamal, M. R., and L. N. Canjar: *AIChE J.,* **8:** 329 (1962).
Kestin, J., H. E. Khalifa, S. T. Ro, and W. A. Wakeham: *Physica,* **88A:** 242 (1977).
Kestin, J., and W. A. Wakeham: *Ber. Bunsenges. Phys. Chem.,* **87:** 309 (1983).
Kett, T. K., and D. K. Anderson: *J. Phys. Chem.,* **73:** 1262 (1969).
Kett, T. K., and D. K. Anderson: *J. Phys. Chem.,* **73:** 1268 (1969a).
King, C. J., L. Hsueh, and K. W. Mao: *J. Chem. Eng. Data,* **10:** 348 (1965).
Kircher, K., A. Schaber, and E. Obermeier: *Proc. 8th Symp. Thermophys. Prop.,* Vol. 1, ASME, 1981, p. 297.
Kooijman, H. A., and R. Taylor: *Ind. Eng. Chem. Res.,* **30:** 1217 (1991).
Krieger, I. M., G. W. Mulholland, and C. S. Dickey: *J. Phys. Chem.,* **71:** 1123 (1967).
Kuznetsova, E. M., and D. Sh. Rashidova: *Russ. J. Phys. Chem.,* **54:** 1332, 1339 (1980).
Leaist, D. G.: *J. Chem. Eng. Data,* **29:** 281 (1984).
Lee, H., and G. Thodos: *Ind. Eng. Chem. Fundam.,* **22:** 17 (1983).
Lee, Y. E., and F. Y Li: *J. Chem. Eng. Data,* **36:** 240 (1991).
Lees, F. P., and P. Sarram: *J. Chem. Eng. Data,* **16:** 41 (1971).
Leffler, J., and H. T. Cullinan, Jr.: *Ind. Eng. Chem. Fundam.,* **9:** 84, 88 (1970).
Lewis, J. B.: *J. Appl. Chem. London,* **5:** 228 (1955).
Li, J. C. M., and P. Chang: *J. Chem. Phys.,* **23:** 518 (1955).
Liu, H., and E. Ruckenstein: *Ind. Eng. Chem. Res.,* **36:** 888 (1997).
Liu, H.; Silva, C. M.; Macedo, E. A., *Chem. Eng. Sci.,* **53:** 2403 (1998).
Loflin, T., and E. McLaughlin: *J. Phys. Chem.,* **73:** 186 (1969).
Lugg, G. A.: *Anal. Chem.,* **40:** 1072 (1968).
Lusis, M. A.: *Chem. Proc. Eng.,* **5:** 27 (May 1971).
Lusis, M. A.: *Chem. Ind. Devel. Bombay,* January 1972, **48;** *AIChE J.,* **20:** 207 (1974).
Lusis, M. A., and G. A. Ratcliff: *Can. J. Chem. Eng.,* **46:** 385 (1968).
Lusis, M. A., and G. A. Ratcliff: *AIChE J.,* **17:** 1492 (1971).
Maczek, A. O. S., and C. J. C. Edwards: "The Viscosity and Binary Diffusion Coefficients of Some Gaseous Hydrocarbons, Fluorocarbons and Siloxanes," *Symp. Transport Prop. Fluids and Fluid Mixtures, Natl. Eng. Lab.,* East Kilbride, Glasgow, Scotland, April 1979.
Maginn, E. J., R. A. Messerly, D. J. Carlson, D. R. Roe, J.R. Elliott, *Living J. Comp. Mol. Sci.,* **1:** 6324 (2018).
McCall, D. W., and D. C. Douglas: *Phys. Fluids,* **2:** 87 (1959).
McCall, D. W., D. C. Douglass, and E. W. Anderson: *Bunsenges. Phys. Chem.,* **67:** 336 (1963).
McCall, D. W., and D. C. Douglas: *J. Phys. Chem.,* **71:** 987 (1967).
McCall, D. W., D. C. Douglas, and E. W. Anderson: *Phys. Fluids,* **4:** 162 (1961).
Marrero, T. R., and E. A. Mason: *J. Phys. Chem. Ref. Data,* **1:** 3 (1972).
Mason, E. A., and L. Monchick: *J. Chem. Phys.,* **36:** 2746 (1962).
Massman, W. J.: *Atmos. Environ.,* **32:** 1111 (1998).
Mather, G. P., and S. C. Saxena: *Ind. J. Pure Appl. Phys.,* **4:** 266 (1966).
McKeigue, K., and E. Gulari: *AIChE J.,* **35:** 300 (1989).
Miller, D. G.: *J. Phys. Chem.,* **70:** 2639 (1966); **71:** 616 (1967).
Miller, L., and P. C. Carman: *Trans. Faraday Soc.* **57:** 2143 (1961).
Mills, R., and V. M. M. Lobo: *Self-diffusion in Electrolyte Solutions, A Critical Examination of Data Compiled from the Literature,* Phys. Sci. Data 36, Elsevier, Amsterdam, 1989.

Moelwyn-Hughes, E. A.: *Physical Chemistry,* Pergamon, London, 1957.
Mohan, V., and D. Srinivasan: *Chem. Eng. Comm.,* **29:** 27 (1984).
Morozov, V. S., and E. G. Vinkler: *Russ. J. Phys. Chem.,* **49:** 1404 (1975).
Mrazek, R. V., C. E. Wicks, and K. N. S. Prabhu: *J. Chem. Eng. Data,* **13:** 508 (1968).
Murad, S.: *Chem. Eng. Sci.,* **36:** 1867 (1981).
Nakanishi, K.: *Ind. Eng. Chem. Fundam.,* **17:** 253 (1978).
Neufeld, P. D., A. R. Janzen, and R. A. Aziz: *J. Chem. Phys.,* **57:** 1100 (1972).
Newman, J. S.: in C. W. Tobias (ed.), *Advances in Electrochemistry and Electrochemical Engineering,* Vol. 5, Interscience, New York, 1967.
Olander, D. R.: *AIChE J.,* **7:** 175 (1961).
Olander, D. R.: *AIChE J.,* **9:** 207 (1963).
Oosting, E. M., J. I. Gray, and E. A. Grulke: *AIChE J.,* **31:** 773 (1985).
Othmer, D. F., and T. T. Chen: *Ind. Eng. Chem. Process Design Develop.,* **1:** 249 (1962).
Othmer, D. F., and M. S. Thakar: *Ind. Eng. Chem.,* **45:** 589 (1953).
Pasternak, A. D., and D. R. Olander: *AIChE J.,* **13:** 1052 (1967).
Pathak, B. K., V. N. Singh, and P. C. Singh: *Can. J. Chem. Eng.,* **59:** 362 (1981).
Perkins, L. R., and C. J. Geankoplis: *Chem. Eng. Sci.,* **24:** 1035 (1969).
Prausnitz, J. M., R. N. Lichtenthaler, and E. G. de Azevedo: *Molecular Thermodynamics of Fluid-Phase Equilibria,* 3rd ed., Prentice Hall, Upper Saddle River, NJ, 1999.
Quayle, O. R.: *Chem. Rev.,* **53:** 439 (1953).
Raina, G. K.: *AIChE J.,* **26:** 1046 (1980).
Rao, S. S., and C. O. Bennett: *AIChE J.,* **17:** 75 (1971).
Ratcliff, G. A., and J. G. Holdcroft: *Trans. Inst. Chem. Eng. London,* **41:** 315 (1963).
Ratcliff, G. A., and M. A. Lusis: *Ind. Eng. Chem. Fundam.,* **10:** 474 (1971).
Rathbun, R. E., and A. L. Babb: *Ind. Eng. Chem. Process Design Develop.,* **5:** 273 (1966).
Ravindran, P., E. J. Davis, and A. K. Ray: *AIChE J.,* **25:** 966 (1979).
Reddy, K. A., and L. K. Doraiswamy: *Ind. Eng. Chem. Fundam.,* **6:** 77 (1967).
Ree, F. H., T. Ree, and H. Eyring: *Ind. Eng. Chem.,* **50:** 1036 (1958).
Riazi, M. R., and C. H. Whitson: *Ind. Eng. Chem. Res.,* **32:** 3081 (1993).
Riddick, J. A., W. B. Bunger, and T. K. Sakano: *Organic Solvents, Physical Properties and Methods of Purification,* 4th ed., Wiley-Interscience, New York, 1986.
Rizzo, R., J. G. Albright, and D. G. Miller: *J. Chem. Eng. Data,* **42:** 623 (1997).
Robinson, R. A., and R. H. Stokes: *Electrolyte Solutions,* 2nd ed., Academic, New York, 1959.
Robinson, R. L., Jr., W. C. Edmister, and F. A. L. Dullien: *Ind. Eng. Chem. Fundam.,* **5:** 74 (1966).
Rosenfeld, Y., *Phys. Rev. A,* **15:** 2545 (1977).
Sanchez, V., and M. Clifton: *Ind. Eng. Chem. Fundam.,* **16:** 318 (1977).
Sanchez, V., H. Oftadeh, C. Durou, and J.-P. Hot: *J. Chem. Eng. Data,* **22:** 123 (1977).
Sandler, S., and E. A. Mason: *J. Chem. Phys.,* **48:** 2873 (1968).
Sanni, S. A., and P. Hutchinson: *J. Chem. Eng. Data,* **18:** 317 (1973).
Scheibel, E. G.: *Ind. Eng. Chem.,* **46:** 2007 (1954).
Schneider, G. M.: *Angew. Chem. Intern. Ed. English,* **17:** 716 (1978).
Schwartz, F. A., and J. E. Brow: *J. Chem. Phys.,* **19:** 640 (1951).
Scott, D. S., and K. E. Cox: *Can. J. Chem. Eng.,* **38:** 201 (1960).
Seager, S. L., L. R. Geertson, and J. C. Giddings: *J. Chem. Eng. Data,* **8:** l68 (1963).
Shrier, A. L.: *Chem. Eng. Sci.,* **22:** 1391 (1967).
Siddiqi, M. A., and K. Lucas: *Can. J. Chem. Eng.,* **64:** 839 (1986).
Silva, C. M., Liu, H. and Macedo, E. A. *Chem. Eng. Sci.,* **53:** 2423–2429 (1998).
Sitaraman, R., S. H. Ibrahim, and N. R. Kuloor: J. Chem. Eng. Data, 8: 198 (1963).
Slattery, J. C., and R. B. Bird: *AIChE J.,* **4:** 137 (1958).
Sridhar, T., and O. E. Potter: *AIChE J.,* **23:** 590, 946 (1977).
Srivastava, B. N., and K. P. Srivastava: *J. Chem. Phys.,* **30:** 984 (1959).
Srivastava, B. N., and I. B. Srivastava: *J. Chem. Phys.,* **36:** 2616 (1962).
Srivastava, B. N., and I. B. Srivastava: *J. Chem. Phys.,* **38:** 1183 (1963).
Stejskal, E. O., J. E. Tanner, *J. Chem. Phys.,* **42**:288 (1965).
Stearn, A. E., E. M. Irish, and H. Eyring: *J. Phys. Chem.,* **44:** 981 (1940).
Stilbs, P., *Progr. NMR Spectrosc.* **19**: 1–45 (1987).

Stokes, R. H.: *J. Am. Chem. Soc.,* **72:** 2243 (1950).
Swaid, I., and G. M. Schneider: *Ber. Bunsenges. Phys. Chem.,* **83:** 969 (1979).
Takahashi, S.: *J. Chem. Eng. Japan,* **7:** 417 (1974).
Takahashi, S., and M. Hongo: *J. Chem. Eng. Japan,* **15:** 57 (1982).
Takahashi, M., Y. Kobayashi, and H. Takeuchi: *J. Chem. Eng. Data,* **27:** 328 (1982).
Tang, Y. P., and D. M. Himmelblau: *AIChE J.,* **11:** 54 (1965).
Teja, A. S.: personal communication, 1982.
Toor, H. L.: *AIChE J.,* **3:** 198 (1957).
Tsekhanskaya, Y. V.: *Russ. J. Phys. Chem.,* **45:** 744 (1971).
Turner, J. C. R.: *Chem. Eng. Sci.,* **30:** 151 (1975).
Tyn, M. T.: *Chem. Eng.,* **82**(12)**:** 106 (1975).
Tyn, M. T.: *Chem. Eng. J.,* **12:** 149 (1976).
Tyn, M. T.: *Trans. Inst. Chem. Engrs.,* **59**(2)**:** 112 (1981).
Tyn, M. T., and W. F. Calus: *J. Chem. Eng. Data,* **20:** 106 (1975).
Tyn, M. T., and W. F. Calus: *J. Chem. Eng. Data,* **20:** 310 (1975a).
Umesi, N. O., and R. P. Danner: *Ind. Eng. Chem. Process Design Develop.,* **20:** 662 (1981).
Vadovic, C. J., and C. P. Colver: *AIChE J.,* **19:** 546 (1973).
Van Geet, A. L., and A. W. Adamson: *J. Phys. Chem.,* **68:** 238 (1964).
Vargaftik, N. B., Y. K. Vinogradov, and V. S. Yargin: *Handbook of Physical Properties of Liquids and Gases,* Begell House, New York, 1996.
Verhallen, P. T. H. M., L. J. O. Oomen, A. J. J. M. v. d. Elsen, and A. J. Kruger: *Chem. Eng. Sci.,* **39:** 1535 (1984).
Vignes, A.: *Ind. Eng. Chem. Fundam.,* **5:** 189 (1966).
Vinkler, E. G., and V. S. Morozov: *Russ. J. Phys. Chem.,* **49:** 1405 (1975).
Vinograd, J. R., and J. W. McBain: *J. Am. Chem. Soc.,* **63:** 2008 (1941).
Vivian, J. E., and C. J. King: *AIChE J.,* **10:** 220 (1964).
von Meerwall, E. D., *J. Non-Cryst. Solids,* **131–133:** 735–741 (1991).
Walker, R. E., N. de Haas, and A. A. Westenberg: *J. Chem. Phys.,* **32:** 1314 (1960).
Walker, R. E., and A. A. Westenberg: *J. Chem. Phys.,* **29:** 1139 (1958).
Way, P.: Ph.D. thesis, Massachusetts Institute of Technology, Cambridge, Mass. 1971.
Weissman, S.: *J. Chem. Phys.,* **40:** 3397 (1964).
Weissman, S., and E. A. Mason: *J. Chem. Phys.,* **37:** 1289 (1962).
Wendt, R. P.: *J. Phys. Chem.,* **69:** 1227 (1965).
Westenberg, A. A., and G. Frazier: *J. Chem. Phys.,* **36:** 3499 (1962).
Wilke, C. R., and P. Chang: *AIChE J.,* **1:** 264 (1955).
Wilke, C. R., and C. Y. Lee: *Ind. Eng. Chem.,* **47:** 1253 (1955).
Wise, D. L., and G. Houghton: *Chem. Eng. Sci.,* **21:** 999 (1966).
Witherspoon, P. A., and L. Bonoli: *Ind. Eng. Chem. Fundam.,* **8:** 589 (1969).
Yabsley, M. A., P. J. Carlson, and P. J. Dunlop: *J. Phys. Chem.,* **77:** 703 (1973).
Yu, Y.-X., and G.-H. Gao: *Fluid Phase Equilib.* **166:** 111–124 (1999).
Zabaloy, M. S., V. R. Vasquez, and E. A. Macedo: *Fluid Phase Equilib.,* **242:** 43–56 (2006).

13

Surface Tension

13.1 SCOPE

The surface tension of pure liquids and liquid mixtures is considered in this chapter. For the former, methods based on the law of corresponding states and the parachor are explained and evaluated. For mixtures, extensions of the pure-component methods are presented, as is a method based upon a thermodynamic analysis of the system. Interfacial tensions for liquid-solid systems are not included. Also, in recent years, considerable progress has been made with molecular theories of surface tension like classical density functional theory and square gradient theory (Lawrence et al., 2020; Mejia et al., 2021). Unfortunately, model development with such approaches is either too narrow in scope or too recently available for inclusion in this edition of the book.

13.2 INTRODUCTION

The boundary between a liquid phase and a gas phase can be considered a third phase with properties distinct from those of the liquid and gas. A qualitative picture of the microscopic surface layer shows that there are unequal forces acting upon the molecules; i.e., at low gas densities, the surface molecules are attracted sidewise and toward the bulk liquid but experience little attraction in the direction of the bulk gas. Thus, the surface layer is in tension and tends to contract to the smallest area compatible with the mass of material, container restraints, and external forces, e.g., gravity.

This tension can be presented in various quantitative ways; the most common is the surface tension σ, defined as the force exerted in the plane of the surface per unit length. We can consider a reversible isothermal process whereby surface area A is increased by pulling the surface apart and allowing molecules from the bulk to enter at constant temperature and pressure. The differential reversible work is σdA. Since it is also the differential Gibbs energy change, σ is the surface Gibbs energy per unit of area. At equilibrium, systems tend to a state of minimum Gibbs energy at fixed T and P, and the product σA also tends to a minimum. For a fixed σ, equilibrium is a state of minimum area consistent with the system conditions. Analogously, the boundary between two liquid phases may also be considered a third phase which is characterized by the interfacial tension.

Surface tension and interfacial tension are usually expressed in units of dynes per centimeter, which is the same as ergs per square centimeter. With relation to SI units, 1 dyn/cm = 1 erg/cm^2 = 1 mJ/m^2 = 1 mN/m.

The thermodynamics of surface layers furnishes a fascinating subject for study. Guggenheim (1959), Gibbs (1957), and Tester and Modell (1997) have formulated treatments that differ considerably but reduce to similar equations relating macroscopically measurable quantities. In addition to the thermodynamic perspective, treatments of the physics and chemistry of surfaces have been published (Adamson, 1982; Aveyard and Haden, 1973; Barbulescu, 1974; Brown, 1974; Chattoraj and Birdi, 1984; Evans and Wennerström, 1999; Everett, 1988; Ross, 1965; Rowlinson and Widom, 1982). These subjects are not covered here; instead, the emphasis is placed upon the few reliable methods available to estimate σ from either semitheoretical or empirical equations.

13.3 ESTIMATION OF PURE-LIQUID SURFACE TENSION

As the temperature is raised, the surface tension of a liquid in equilibrium with its own vapor decreases and becomes zero at the critical point (Rowlinson and Widom, 1982). In the reduced-temperature range 0.45 to 0.65, σ for most organic liquids ranges from 20 to 40 dyn/cm, but for some low-molecular-weight dense liquids such as formamide, $\sigma > 50$ dyn/cm. For water $\sigma = 72.8$ dyn/cm at 293 K, and for liquid metals σ is between 300 and 600 dyn/cm; e.g., mercury at 293 K has a value of about 476 dyn/cm.

A thorough critical evaluation of experimental surface tensions has been prepared by Jasper (1972). Additional data and tabulations of data are given in Körösi and Kováts (1981), Riddick et al. (1986), Timmermans (1965), and Vargaftik (1996). Daubert et al. (1997) list references to original data and Gray et al. (1983) have correlated surface tensions of coal liquid fractions. The DIPPR 801 Database (Wilding et al. 2017; DIPPR 2021) also has data and correlations for surface tension.

Essentially all useful estimation techniques for the surface tension of a liquid are empirical. Several are discussed below, and others are briefly noted at the end of this section.

13.3.1 Parachor-Based Methods

Sugden (1924), based on a similar expression by Macleod (1923), suggested a relation between σ, a structural parameter called the parachor ($\boldsymbol{P}$), and the liquid (ρ_L) and vapor densities (ρ_V). The expression is

$$\sigma^{\frac{1}{4}} = \boldsymbol{P}(\rho_L - \rho_V) \tag{13.3-1}$$

where the densities must be given in units of mol/cm^3 and the calculated surface tension is in units of dyn/cm. Sugden calculated the parachor for 167 compounds using this equation and experimental data available at the time and suggested an additive scheme to calculate $\boldsymbol{P}$ based on atomic, ring, and bond patterns in the molecule. Mumford and Phillips (1929) found Sugden's emphasis on atoms unreliable for branched molecules and proposed new rules to calculate $\boldsymbol{P}$ by including bonding patterns using CH_x groups. Quayle (1953) expanded on these efforts with more compounds; however, even this scheme is incomplete in terms of the functional groups represented. Knotts et al. (2001) proposed an approach to standardize and expand the groups to alleviate this limitation. They created two training sets from available experimental surface tension data to regress the group contributions: one from data with errors < 1% and one from data with errors < 5% (the latter encompassing the former). Increasing the error limit when crafting the training set was done to increase the applicability of the method to compounds with multiple functional groups for which experimental data are less accurate and less abundant. Table 13.1 lists group values (Δ_{P_i}) for both the 1% and 5% training sets. The authors recommend using the 5% set for alcohols, aldehydes, epoxides, halides, and sulfur-containing compounds or if the compound is composed of two or more functional groups from different nonhydrocarbon families. The 1% set should be used in other situations. With the group values in Table 13.1, $\boldsymbol{P}$ is determined by

$$\boldsymbol{P} = \sum_i N_i \Delta_{P_i} \tag{13.3-2}$$

TABLE 13.1 Group contribution parameters for the Knotts et al. (2001) method to predict parachor (P).

Group	Δ_{P_i} (1%)	Δ_{P_i} (5%)
Nonring Carbon		
—CH_3	55.24	55.25
>CH_2 (n = 1–11)	39.90	39.92
>CH_2 (n = 12–20)	40.11	40.11
>CH_2 (n > 20)	40.11	40.51
>CH—	28.88	28.90
>C<	15.65	15.76
=CH_2	49.87	49.76
=CH—	34.61	34.57
=C<	24.46	24.50
=C=	24.53	24.76
≡CH	43.66	43.64
≡C—	28.66	28.64
per branch correction	−6.02	−6.02
sec-sec adjacency correction	−2.75	−2.73
sec-tert adjacency correction	−3.72	−3.61
tert-tert adjacency correction	−6.19	−6.10
Nonaromatic Ring Carbon		
—CH_2—	39.53	39.21
>CH—	22.06	23.94
>C<	5.11	7.19
=CH—	33.33	34.07
=C<	24.82	18.85
>CH— (fused ring)	20.57	22.05
three-member ring correction	13.12	12.67
four-member ring correction	15.00	15.76
five-member ring correction	7.74	7.04
six-member ring correction	5.42	5.19
seven-member ring correction	0.79	3.00
Aromatic Ring Carbon		
>CH	34.37	34.36
>C—	16.08	16.07
—C— (fused arom/arom)	19.73	19.73
—C— (fused arom/aliph)	14.41	14.41
ortho correction	−0.60	−0.60
para correction	3.40	3.40
meta correction	2.24	2.24
substituted naphthalene correction	−7.07	−7.07

(*Continued*)

TABLE 13.1 Group contribution parameters for the Knotts et al. (2001) method to predict parachor (*P*). (*Continued*)

Group	Δ_{P_i} (1%)	Δ_{P_i} (5%)
	Oxygen	
—OH (primary alcohol)	30.20	31.42
—OH (secondary alcohol)	22.60	22.68
—OH (tertiary alcohol)	18.93	20.66
—OH (phenol)	19.25	30.32
—O— (nonring)	20.72	20.61
—O— (nonaromatic ring)	20.97	21.67
—O— (aromatic ring)	23.43	23.54
>C═O (nonring)	46.92	47.02
>C═O (ring)	49.22	50.04
O═CH— (aldehyde)	65.96	66.06
HCOOH (formic acid)	93.93	94.01
—COOH (acid)	74.48	74.57
—OCHO (formate)	82.42	82.29
—COO— (ester)	64.96	64.97
—COOCO— (acid anhydride)	115.11	115.07
—OC(═O)O (ring)	84.10	84.05
	Halogen	
—F (nonaromatic)	19.98	21.81
—Cl (nonaromatic)	50.98	26.24
—Br (nonaromatic)	65.73	51.16
—I (nonaromatic)	90.82	54.56
—F (aromatic)	27.29	66.30
—Cl (aromatic)	54.07	70.39
—Br (aromatic)	72.07	90.84
—I (aromatic)	92.08	92.04
	Nitrogen	
R—NH_2 (primary R)	45.40	44.98
R—NH_2 (secondary R)	45.85	44.63
R—NH_2 (tertiary R)	46.40	46.44
A-NH_2 (attached to aromatic ring)	43.90	46.53
>NH (nonring)	29.54	29.04
>NH (nonaromatic ring)	33.49	31.97
>NH (aromatic ring; e.g. pyrrole)	34.12	33.92
>N— (nonring)	8.03	10.77
>N— (nonaromatic ring)	16.05	15.71
—N═ (nonring)	24.44	23.24

(*Continued*)

TABLE 13.1 Group contribution parameters for the Knotts et al. (2001) method to predict parachor (*P*). (*Continued*)

Group	Δ_{P_i} (1%)	Δ_{P_i} (5%)
Nitrogen		
>N (aromatic ring; e.g., pyridine)	26.46	26.49
HC≡N (hydrogen cyanide)	80.94	80.94
R—C≡N (nonaromatic R)	66.15	65.23
R—C≡N (aromatic R)	67.42	67.54
Nitrogen and Oxygen		
—C(=O)NH_2 (primary amide)	93.44	93.43
—C(=O)NH- (secondary amide)	73.65	73.64
—C(=O)N< (tertiary amide)	56.33	57.05
—NHC(=O)H (monosub. formamide)	91.69	91.69
>NC(=O)H (disub. formamide)	77.14	77.12
—N=O	64.49	64.32
R—NO_2 (nonaromatic R)	72.31	73.86
R—NO_2 (aromatic R)	74.17	75.05
Sulfur		
R—SH (primary R)	66.87	66.89
R—SH (secondary R)	63.37	63.34
R—SH (tertiary R)	65.37	65.33
R—SH (aromatic R)	68.24	68.30
—S— (nonring)	51.29	51.37
—S— (nonaromatic ring)	50.27	51.75
—S— (aromatic ring)	52.70	51.47
>S=O (nonring)	72.22	72.21
>SO_2 (nonring)	93.53	93.20
>SO_2 (nonaromatic ring)	88.82	91.03
Silicon		
SiH_4	105.11	105.11
>SiH—	55.01	54.50
>Si<	44.07	44.93
>Si< (nonaromatic ring)	29.44	28.64
Other Inorganics		
>PO_4^-	115.67	115.59
>P—	49.35	48.84
>B—	28.19	22.65
>Al—	25.15	25.06
—ClO_3	107.87	106.03

where N_i is the number of times group i is found in the compound. The parachor method for predicting surface tension is illustrated in Example 13.1.

Example 13.1 Use Eq. (13.3-1), the Macleod-Sugden relationship, to estimate the surface tension of isobutyric acid at 333 K. The experimental value quoted by Jasper (1972) is 21.36 dyn/cm.

Solution. DIPPR (Wilding et al., 2017; DIPPR, 2021) reports the liquid density (ρ_L) at 333 K to be 10.342 kmol/m^3 = 0.010342 mol/cm^3. With the correlation in Appendix A the vapor pressure is 1860 Pa. The ideal gas equation of state is valid at such low pressures giving

$$\rho_V = 1860/8.3145/333 = 0.671829 \text{ mol/m}^3 = 6.71829 \times 10^{-7} \text{ mol/cm}^3$$

which is negligible compared to ρ_L and is thus ignored.

For the groups defined in Table 13.1, isobutyric acid consists of one —COOH, two nonring —CH_3, one nonring >CH—, and one chain branching correction. Only one functional group family is present, the carboxyl group, which is not one of the families for which the 5% parameters are recommended, so the 1% parameters in the table should be used. These give

$$\boldsymbol{P} = 1(74.48) + 2(55.24) + 1(28.88) + 1(-6.02) = 207.82$$

Then, with Eq. (13.3-1),

$$\sigma = [(207.82)(0.010342)]^4 = 21.34 \text{ dyn/cm}$$

$$\text{Error} = 21.34 - 21.36 = -0.02 \text{ dyn/cm or } -0.094\%$$

Though not recommended by Knotts et al. (2001), if the 5% parameters are used

$$\boldsymbol{P} = 1(74.57) + 2(55.25) + 1(28.90) + 1(-6.02) = 207.95$$

and

$$\sigma = [(207.95)(0.010342)]^4 = 21.39 \text{ dyn/cm}$$

$$\text{Error} = 21.39 - 21.36 = 0.03 \text{ dyn/cm or } -0.15\%$$

The difference between the 1% and 5% sets is negligible for this compound.

Since σ is proportional to $(\boldsymbol{P}\rho_L)^4$, Eq. (13.3-1) is very sensitive to the values of the parachor and liquid density chosen. The accuracy of the approach is thus remarkable.

Corresponding States Methods

Instead of correlating surface tensions with densities, several prediction methods have been developed using the corresponding states principle. The group $\dfrac{\sigma}{P_c^{\frac{2}{3}} T_c^{\frac{1}{3}}}$ is dimensionless except for a numerical constant which depends upon the units of σ, P_c (critical pressure), and T_c (critical temperature). Van der Waals (1894) suggested that the group could be correlated with $1 - T_r$ (T_r is the reduced temperature). Brock and Bird (1955) developed this idea for nonpolar liquids and proposed that

$$\frac{\sigma}{P_c^{\frac{2}{3}} T_c^{\frac{1}{3}}} = (0.132\alpha_c - 0.279)(1 - T_r)^{\frac{11}{9}} \tag{13.3-3}$$

where α_c is the Riedel (1954) parameter at the critical point and α is defined as $\dfrac{d \ln P_r^{vp}}{d \ln T_r}$ as per Eq. (5.7-1).

Miller (1963) suggested the following to relate α_c to T_{b_r} (the reduced normal boiling point, $\frac{T_b}{T_c}$) and P_c.

$$\alpha_c = 0.9076\left[1 + \frac{T_{b_r}\ln\left(\dfrac{P_c}{1.01325}\right)}{1 - T_{b_r}}\right] \tag{13.3-4}$$

Using this relationship, it can be shown that

$$\sigma = P_c^{\frac{2}{3}} T_c^{\frac{1}{3}} Q (1 - T_r)^{\frac{11}{9}} \tag{13.3-5}$$

$$Q = 0.1196\left[1 + \frac{T_{b_r}\ln\left(\dfrac{P_c}{1.01325}\right)}{1 - T_{b_r}}\right] - 0.279 \tag{13.3-6}$$

Pressures are in units of bar, temperatures in units of kelvin, and surface tension in units of dyn/cm in Eqs. (13.3-4) to (13.3-6).

Pitzer (Curl and Pitzer, 1958; Pitzer, 1995) gives a different series of relations for σ in terms of P_c, T_c, and ω (acentric factor) that together lead to the following corresponding states relation

$$\sigma = P_c^{\frac{2}{3}} T_c^{\frac{1}{3}} \frac{1.86 + 1.18\omega}{19.05}\left[\frac{3.75 + 0.91\omega}{0.291 - 0.08\omega}\right]^{\frac{2}{3}} (1 - T_r)^{\frac{11}{9}} \tag{13.3-7}$$

Here, as with Eqs. (13.3-4) to (13.3-6), pressures are in units of bar, temperatures in units of kelvin, and surface tension in units of dyn/cm. It is stated that a deviation of more than 5% from this relation for a substance "appears to indicate significant *abnormality*" in the sense that the three-parameter corresponding states principles don't apply. This means that Eq. (13.3-7) can be used as a test for whether a fluid can be considered a "normal" (nonpolar, nonassociating) fluid (see Sec. 3.6).

Zuo and Stenby (1997) used a two reference fluid corresponding states approach patterned after work done by Rice and Teja (1982) to estimate surface tensions. Unlike Rice and Teja who used T_c and V_c as reducing parameters, Zuo and Stenby used T_c and P_c according to:

$$\sigma_r = \ln\left[1 + \frac{\sigma}{T_c^{\frac{1}{3}} P_c^{\frac{2}{3}}}\right] \tag{13.3-8}$$

The units in Eq. (13.3-8) are bar, kelvin, and dyn/cm. To use this method, σ_r for the fluid of interest (with acentric factor ω) is related to σ_r for two reference fluids—(1) methane (with acentric factor $\omega^{(1)}$) and (2) *n*-octane (with acentric factor $\omega^{(2)}$) —by

$$\sigma_r = \sigma_r^{(1)} + \frac{\omega - \omega^{(1)}}{\omega^{(2)} - \omega^{(1)}}\left(\sigma_r^{(2)} - \sigma_r^{(1)}\right) \tag{13.3-9}$$

For methane,

$$\sigma^{(1)} = 40.520(1 - T_r)^{1.287} \tag{13.3-10}$$

TABLE 13.2 Values of constants for the Sastri and Rao (1995) method to predict σ using Eq. (13.3-12).

Family	K	x	y	z	m
Alcohols	2.280	0.175	0.25	0	0.8
Acids	0.125	0.350	0.50	−1.85	11/9
All Others	0.158	0.350	0.50	−1.85	11/9

and for n-octane,

$$\sigma^{(2)} = 52.095(1 - T_r)^{1.21548} \tag{13.3-11}$$

The procedure to calculate surface tensions with Eqs. (13.3-8) to (13.3-11) is illustrated in Example 13.2.

While the corresponding states methods described above are satisfactory for nonpolar liquids, they are not satisfactory for compounds that exhibit strong hydrogen-bonding (alcohols, acids). To deal with these compounds, Sastri and Rao (1995) present the following modification of the above equations

$$\sigma = K T_b^x P_c^y T_{b_r}^z \left[\frac{1 - T_r}{1 - T_{b_r}}\right]^m \tag{13.3-12}$$

where T_b is the normal boiling point, and $T_{b_r} = \dfrac{T_b}{T_c}$. The units in Eq. (13.3-12) are bar, kelvin, and dyn/cm. Values for the constants are given in Table 13.2.

The Sastri and Rao method is illustrated in Example 13.2.

Example 13.2 Estimate the surface tension of ethyl mercaptan at 303 K with the corresponding states methods of Brock and Bird (1955), Pitzer (1995), Zuo and Stenby (1997), and Sastri and Rao (1995). The experimental value is 22.68 dyn/cm (Jasper, 1972).

Solution. From Appendix A, for ethyl mercaptan, $T_c = 499.15$ K, $T_b = 308.153$ K, $P_c = 54.90$ bar, and $\omega = 0.187751$. Thus $T_{b_r} = 308.153/499.15 = 0.6174$ and $T_r = 303/499.15 = 0.6070$.

BROCK AND BIRD, Eq. (13.3-5).

With Eq. (13.3-6)

$$Q = 0.1196[1 + (0.6174)\ln(54.90/1.01325)/(1 - 0.6174)] - 0.279 = 0.6111$$

$$\sigma = (54.90)^{2/3}(499.15)^{1/3}(0.6111)(1 - 0.6070)^{11/9} = 22.36 \text{ dyn/cm}$$

$$\text{Error} = 22.31 - 22.68 = -0.32 \text{ dyn/cm or } -1.41\%$$

PITZER, EQ. (13.3-7)

$$\sigma = (54.90)^{2/3}(499.15)^{1/3}[(1.86 + 1.18 \times 0.187751)/19.05][(3.75 + 0.91 \times 0.187751)$$
$$/(0.291 - 0.08 \times 0.187751)]^{2/3}(1 - 0.6070)^{11/9} = 23.45 \text{ dyn/cm}$$

$$\text{Error} = 23.45 - 22.68 = 0.77 \text{ dyn/cm or } 3.40\%$$

ZUO AND STENBY, Eqs. (13.3-8) to (13.3-11)

For methane, from Appendix A, $T_c = 190.564$ K, $P_c = 45.99$ bar, and $\omega = 0.0115478$.

For n-octane, $T_c = 568.7$ K, $P_c = 24.90$ bar, and $\omega = 0.399552$. With Eqs. (13.3-8) and (13.3-10):

$$\sigma^{(1)} = 40.520(1 - 0.6070)^{1.287} = 12.180 \text{ dyn/cm}$$

$$\sigma_r^{(1)} = \ln[1 + 12.180/(190.564^{1/3} \times 45.99^{2/3})] = 0.1526$$

Similarly for reference fluid 2, n-octane, $\sigma^{(2)} = 16.7413$ dyn/cm, and $\sigma_r^{(2)} = 0.2127$. With Eq. (13.3-9):

$$\sigma_r = 0.1526 + [(0.187751 - 0.0115478)/(0.399552 - 0.0115478)](0.2127 - 0.1526) = 0.1799$$

Finally, Eq. (13.3-8) may be solved for the desired value of σ to give:

$$\sigma = (499.15)^{1/3}(54.90)^{2/3}[\exp(0.1799) - 1] = 22.58 \text{ dyn/cm}$$

$$\text{Error} = 22.58 - 22.68 = -0.10 \text{ dyn/cm or } -0.44\%$$

SASTRI AND RAO, Eq. (13.3.12)
From Table 13.2, $K = 0.158$, $x = 0.350$, $y = 0.50$, $z = -1.85$, and $m = 11/9$.

$$\sigma = (0.158)(308.153)^{0.350}(54.90)^{0.50}(0.6174)^{-1.85}[(1 - 0.6070)/(1 - 0.6174)]^{11/9} = 21.94 \text{ dyn/cm}$$

$$\text{Error} = 21.94 - 22.68 = -0.74 \text{ dyn/cm or } -3.26\%$$

13.3.2 Discussion

Other approaches to estimate surface tension include a method by Escobedo and Mansoori (1996) in which the parachor is related to the refractive index, and a method by Hugill and van Welsenes (1986) which employs a corresponding states type expression for the parachor. Both these methods work well for nonpolar fluids but can lead to large errors for polar compounds.

Another approach to the prediction of surface tensions of pure fluids is to change the reference fluids used in Eq. (13.3-9). For example, Rice and Teja (1982) predicted surface tensions of six alcohols generally to within 5% when ethanol and pentanol were chosen as the reference fluids which is much better than when the reference fluids are methane and octane.

A test database was created to evaluate the performance of pure-component surface tension prediction methods. This database consisted of all reliable, experimental surface tension data from the October 2021 release of the DIPPR database. Temperatures ranged from T_m to T_c and included over 12,500 total points (N_{pt}) from more than 1000 unique compounds (N_c). The techniques tested were: the parachor approach of Eq. (13.3-1) with the Knotts et al. 1% and 5% groups from Table 13.1 (abbreviated K1 and K5), Brock and Bird Eqs. (13.3-5) to (13.3-6) (abbreviated BB), Pitzer Eq. (13.3-7) (abbreviated PI), Zuo and Stenby Eqs. (13.3-8) to (13.3-11) (abbreviated ZS), and Sastri and Rao Eq. (13.3-12) (abbreviated SR).

To identify if a method was better at one class of compounds than another, each chemical in the test database was assigned a *class* according to the scheme below.

- **Normal:** Compounds containing only C, H, F, Br, and/or I with molecular weights below 213 g/mol.
- **Heavy:** Compounds containing only C, H, F, Br, and/or I with molecular weights greater than or equal to 213 g/mol. Also compounds containing only C, H, F, Br, I and/or metal atoms (e.g., Fe, As, Al, etc.) regardless of MW.
- **Polar:** Compounds containing atoms other than C, H, F, Br, metals (e.g., O, N, P, S, etc.) that cannot form hydrogen bonds (e.g., esters, ketones, aldehydes, thioesters, sulfonyl, etc.).
- **Associating:** Compounds containing atoms other than C, H, F, Br, metals (e.g., O, N, P, S, etc.) that can form hydrogen bond (e.g., acids, amines, alcohols, etc.).

TABLE 13.3 Performance of prediction techniques for σ against the test database.

	Compound Class											
	Normal			Heavy			Polar			Associating		
Method	N_{pt}	N_c	AARD	N_{pt}	N_c	AARD	N_{pt}	N_c	AARD	N_{pt}	N_c	AARD
K1	6188	443	10.4%	498	48	10.9%	2617	225	6.7%	3032	309	10.3%
K5	6188	443	25.1%	498	48	10.7%	2617	225	12.7%	3032	309	10.7%
BB	6275	454	4.2%	498	48	8.1%	2638	225	8.0%	3177	311	21.6%
PI	6332	456	6.5%	498	48	7.1%	2638	225	11.4%	3177	311	28.2%
ZS	6332	456	4.0%	498	48	7.2%	2638	225	8.2%	3177	311	21.8%
SR	5883	431	4.7%	498	48	7.3%	2397	214	18.4%	3043	302	14.0%

Results were quantified using the average absolute relative deviation ($AARD$) defined as

$$AARD = \frac{1}{N_{pt}} \sum_{i=1}^{N_{pt}} \left| \frac{X_{pred\,i} - X_{exp\,i}}{X_{exp\,i}} \right| \tag{13.3-13}$$

where the index is for point i in the test set, X_{pred_i} is the predicted value for property X for point i, and X_{exp_i} is the experimental value for property X for point i.

Table 13.3 contains the average absolute relative deviation [see Eq. (13.3-13)] of each method according to compound class. Also listed are the number of points tested in each class (N_{pt}) and number of compounds (N_c). As with other properties (see Chap. 3), the corresponding states methods work better for nonpolar (normal and heavy) compounds and become increasingly worse as polarity increases. The methods are worse for associating compounds (those with hydrogen bonding potential) with none of these CSP methods being better than 14% AARD and the worse being more than 28% AARD. The parachor approaches work much better for associating compounds with both K1 and K5 being less than 11%. The K1 method is most consistent, with AARDs for all classes of compounds being less than 11%.

13.3.3 Recommendations

The best source of surface tension values are correlations provided in evaluated databases or data compilations such as those of Jasper (1972), NIST TRC, or DIPPR (Wilding et al., 2017, DIPPR 2021). When prediction is needed, the CSP methods perform better for nonpolar compounds with the Brock and Bird (1955), Zuo and Stenby (1997), and Sastri and Rao (1995) approaches being slightly better than the Pitzer (1995) technique. The parachor method with the 1% groups from Knotts et al. (2001) should be used for polar and associating compounds.

As with other properties, the best way to select a prediction method is to perform a family analysis first described in Chap. 3. Example 13.3 illustrates this approach for surface tension.

Example 13.3 Determine which prediction method will yield the most accurate value for the surface tension of 3-ethylthiophene.

Solution. 3-ethylthiophene is a member of the *Sulfides/Thiophenes* family in the DIPPR database. The database contains 62 compounds in this family, but only 31 of these have experimental values for σ at various temperatures for a total of 258 data points. Determining which of the prediction methods in this chapter is likely to give the best estimate of σ for 3-ethylthiophene was done automatically using DIPPR's DIADEM program in a process described by Rowley et al. (2007) and summarized in Example 3.16. Using this process, the prediction methods discussed in this chapter reproduce the 258 experimental points according to the statistics found in the table below.

Method	Number of Compounds	Number of Points	Ave. Abs. Rel. Dev. (%)	Min. Abs. Rel. Dev. (%)	Max. Abs. Rel. Dev. (%)
K1	31	258	3.8	0.07	12.9
K5	31	258	4.1	0.17	16.0
BB	31	258	5.1	0.15	22.9
PI	31	258	8.5	0.59	29.4
ZS	31	258	5.6	1.7	24.0
SR	31	258	4.9	0.18	20.1

Notice that all the methods examined can be applied to all 31 compounds. For this property, the parachor method with the 1% group values (K1) gives the lowest average, minimum, and maximum deviations. The K1 method is thus the preferred approach to predict the surface tension of 3-ethylthiophene.

13.4 TEMPERATURE DEPENDENCE OF PURE-LIQUID SURFACE TENSION

Equations in the previous section indicate that

$$\sigma \propto (1 - T_r)^n \tag{13.4-1}$$

where n varies from 0.8 for alcohols in the Sastri-Rao method to 1.22, or 11/9 for other compounds. In Fig. 13.1, log σ is plotted against $\log(1 - T_r)$ with experimental data for acetic acid, diethyl ether, and ethyl acetate. For the latter two, the slope is close to 1.25; for acetic acid, the slope is 1.16. For most organic liquids except alcohols, this encompasses the range normally found for n.

For values of T_r between 0.4 and 0.7, Eq. (13.4-1) indicates that $\frac{d\sigma}{dT}$ is almost constant, and often the surface tension-temperature relation is represented by a linear equation

$$\sigma = a + bT \tag{13.4-2}$$

As an example, for nitrobenzene between 313 and 473 K, data from Jasper (1972) are plotted in Fig. 13.2. The linear approximation is satisfactory. Jasper lists values of a and b for many materials.

Extrapolation to wider temperature ranges is possible as the surface tension must become zero at the critical temperature. However, this behavior deviates from the linear relationship explained above. Surface tension from T_m to T_c can be reliably correlated using the following expression.

$$\sigma = A(1 - T_r)^{(B+CT_r+DT_r^2+ET_r^3)} \tag{13.4-3}$$

where $A - E$ are fitting constants. For reliable extrapolations, DIPPR (Wilding et al., 2017; DIPPR, 2021) policy is that $B + C + D$ should be approximately 1.2 and $C + D$ should be less than B.

Figure 13.3 illustrates the deficiencies of Eq. (13.4-2) for wide temperature ranges. Experimental data for the surface tension of ethanol, reported by Muratov (1980), are depicted as circles and the critical point as a square. The dashed line is Eq. (13.4-2) with a and b from Jasper (1972) which is based on data from approximately 283 K to 343 K. This correlation does not extrapolate well as temperature increases and does not yield $\sigma(T_c) = 0$. The solid line is the fit of the data to Eq. (13.4-3) with $A = 82.713085$, $B = 3.737437$, $C = -5.580000$, $D = 3.071600$, and $E = 0$. This correlation displays the proper limiting behavior as T approaches T_c and captures the variations in the data very well with a coefficient of determination for the fit of 0.99999 and a bias of 0.00. Notice that the coefficients satisfy the limits listed above, specifically, $B + C + D = 1.229 \approx 1.2$ and $C + D = -1.254$ is less than $B = 3.737$.

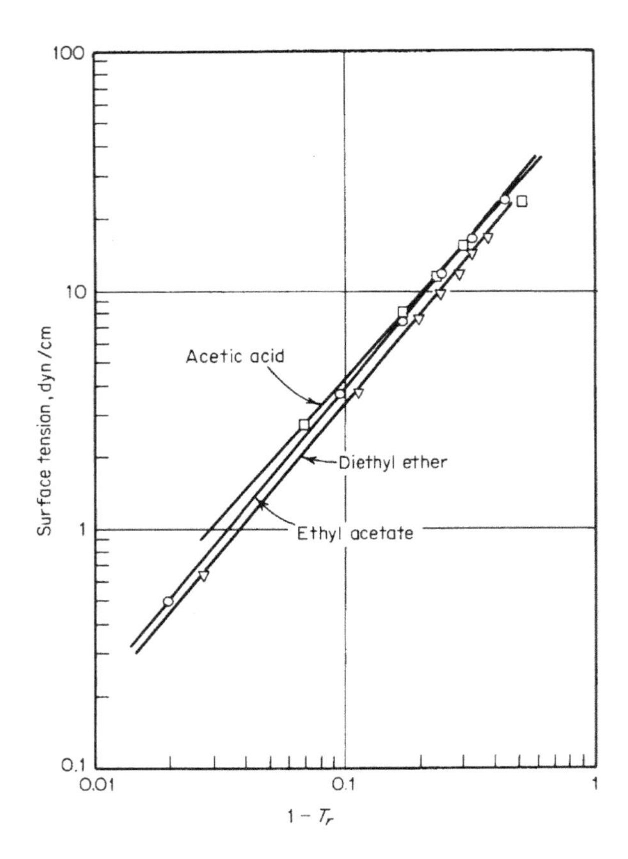

FIGURE 13.1
Variation of surface tension with temperature. (Data from Macleod, 1923)

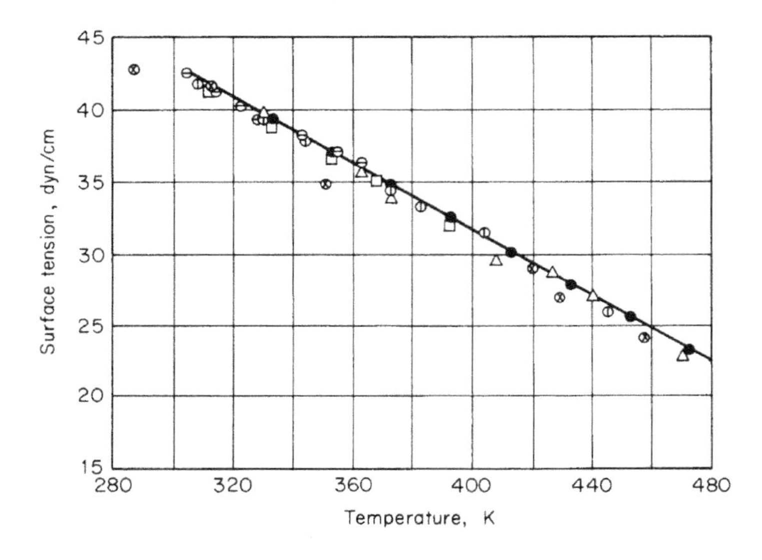

FIGURE 13.2
Surface tension of nitrobenzene, from Jasper (1972) which presents similar graphs for 56 compounds along with references to the original data.

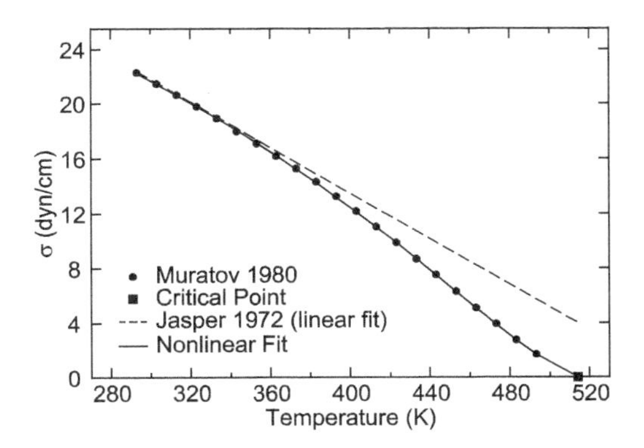

FIGURE 13.3
Surface tension data for ethanol with low-temperature, linear fit of Jasper et al. to Eq. (13.4-2) and a nonlinear fit of the entire temperature range to Eq. (13.4-3).

13.5 SURFACE TENSIONS OF MIXTURES

The surface tension of a liquid mixture (σ_{m}) is not a simple function of the surface tensions of the pure components because, in a mixture, the composition of the surface is not the same as that of the bulk. In a typical situation, we know the bulk composition but not the surface composition. The derivative $\frac{d\sigma_{\mathrm{m}}}{dx_i}$, where x_i is the mole fraction of the component in the mixture with the largest pure-component surface tension, typically increases as x_i increases. Usually, the component with the lowest surface tension concentrates in the surface phase so that the surface tension of a mixture σ_{m} is usually, but not always (Agarwal et al., 1979; Zihao and Jufu, 1982), less than that calculated from a mole fraction average of the surface tensions of the pure components (the excess surface tension is usually negative). This behavior is illustrated in Fig. 13.4 which shows mixture surface tensions for several systems. All illustrate the usual trend with nonlinearity of the σ_{m} versus x_i relation to different degrees. The surface tension of the acetophenone-benzene system is almost linear in composition, the nitromethane-benzene and nitrobenzene-carbon tetrachloride systems are decidedly nonlinear, and the diethyl ether-benzene case is intermediate. Systems become more nonlinear as the difference in pure-component surface tension increases, or as the system becomes more nonideal. For the systems shown in Fig. 13.4, System 1 deviates the most from ideal solution behavior while System 4 has the greatest difference between pure-component surface tensions. These are also the two systems with the greatest deviation from linearity.

The techniques suggested for estimating σ_{m} can be divided into two categories: those based on empirical relations suggested earlier for pure liquids and those derived from thermodynamics. The empirical relations can be used when the pure component surface tensions do not differ greatly and when deviations from ideal solution behavior are not large. In practice, these relations have been used most often to correlate existing data, and in this role, they have been successful at both low and high pressures. Thermodynamic-based methods require more involved calculations but lead to more reliable results.

13.5.1 Parachor-based Approach

Applying Eq. (13.3-1) to mixtures gives

$$\sigma_{\mathrm{m}} = \left(\boldsymbol{P}_{\mathrm{L_m}} \rho_{\mathrm{L_m}} - \boldsymbol{P}_{\mathrm{V_m}} \rho_{\mathrm{V_m}} \right)^n \tag{13.5-1}$$

where σ_{m} is the surface tension of mixture, $\boldsymbol{P}_{\mathrm{L_m}}$ is the parachor of the liquid mixture, $\boldsymbol{P}_{\mathrm{V_m}}$ is the parachor of the vapor mixture, $\rho_{\mathrm{L_m}}$ is the density of the liquid mixture, and $\rho_{\mathrm{V_m}}$ is the density of the vapor mixture, and n is a parameter discussed more below. The units for this equation are dyn/cm and mol/cm^3.

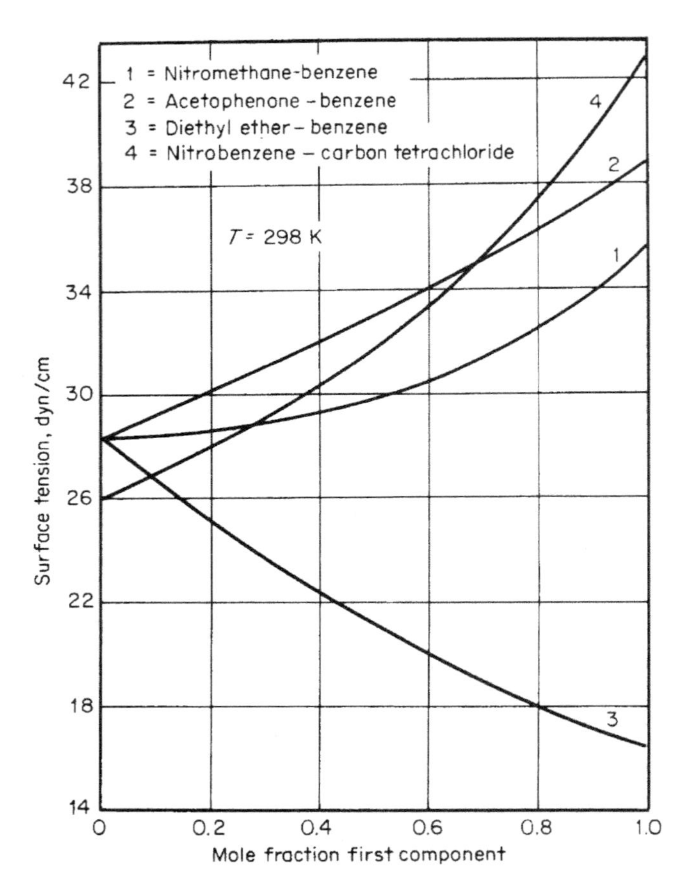

FIGURE 13.4
Mixture surface tensions. (Hammick and Andrew, 1979)

Hugill and van Welsenes (1986) recommend

$$\boldsymbol{P}_{\mathrm{L_m}} = \sum_i \sum_j x_i x_j \boldsymbol{P}_{ij} \tag{13.5-2}$$

$$\boldsymbol{P}_{\mathrm{V_m}} = \sum_i \sum_j y_i y_j \boldsymbol{P}_{ij} \tag{13.5-3}$$

where x_i is the mole fraction of component i in the liquid and y_i is the mole fraction of component i in the vapor. In Eqs. (13.5-2) and (13.5-3)

$$\boldsymbol{P}_{ij} = \lambda_{ij} \frac{\boldsymbol{P}_i + \boldsymbol{P}_j}{2} \tag{13.5-4}$$

where $\boldsymbol{P}_i$ parachor of pure component i, and λ_{ij} is a binary interaction coefficient determined from experimental data. In the absence of experimental data, λ_{ij} may be set equal to one, and if n in Eq. (13.5-1) is set equal to 4, Eq. (13.5-1) reduces to the Weinaug-Katz (1943) equation. Some researchers (Gasem et al. 1989; Zuo and Stenby 1997) who fitted experimental data to Eq. (13.5-1) recommend a value of 3.6 for n.

At low pressures, the term involving the vapor density may be neglected; when this simplification is possible, Eq. (13.5-1) has been employed to correlate mixture surface tensions for a wide variety of organic liquids with reasonably good results (Bowden and Butler, 1939; Gambill, 1958; Hammick and Andrew, 1929; Meissner and Michaels, 1949; Riedel, 1955). Many authors, however, do not obtain $\boldsymbol{P}_i$ from general group contribution methods or from pure-component density and surface tension behavior; instead, they regress mixture data to obtain the best value of $\boldsymbol{P}_i$ for each component in the mixture. This procedure leads to an improved description of the mixture data but may not reproduce the pure-component behavior. Application of Eq. (13.5-1) is illustrated in Example 13.4.

For gas-liquid systems under high pressure, the vapor term in Eq. (13.5-1) becomes significant. Weinaug and Katz (1943) showed that Eqs. (13.5-1) and (13.5-4) with $n = 4$ and all $\lambda_{ij} = 1.0$ correlate methane-propane surface tensions from 258 to 363 K and from 2.7 to 103 bar. Deam and Maddox (1970) also employed these same equations for the methane-nonane mixture from 239 to 297 K and 1 to 101 bar. Some smoothed data are shown in Fig. 13.5. At the conditions of the data shown in the figure, the vapor phase is essentially pure methane, and the liquid composition varies due to changes in the solubility of methane with temperature and pressure. At any temperature, σ_m decreases with increasing pressure as more methane dissolves in the liquid phase. The effect of temperature is more unusual; instead of decreasing with rising temperature, σ_m increases, except at the lowest pressures. This phenomenon illustrates the fact that at the lower temperatures methane is more soluble in nonane and the effect of liquid composition is more important than the effect of temperature in determining σ_m.

Gasem et al. (1989) have used Eqs. (13.5-1) to (13.5-4) to correlate the behavior of mixtures of carbon dioxide and ethane in various hydrocarbon solvents including butane, decane, tetradecane, cyclohexane, benzene, and trans-decalin. The measurements range from about 10 bar to the critical point of each system. They recommended a value of $n = 3.6$. When values of $\boldsymbol{P}_i$ were regressed and λ_{ij} was set to unity, the average absolute deviations for the ethane and CO_2 systems were 5% and 9%, respectively. When λ_{ij} was also regressed, there was only marginal improvement in the description of the ethane systems while the average deviation in the CO_2 systems decreased to about 5%. Other systems for which Eq. (13.5-1) has been used to correlate high-pressure surface tension data include methane-pentane and methane-decane (Stegemeier, 1959), nitrogen-butane and

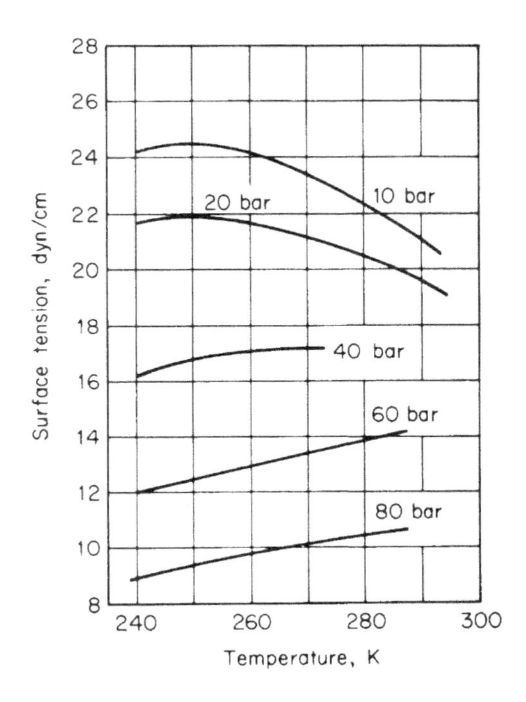

FIGURE 13.5
Surface tension for the system methane-nonane.

nitrogen-heptane (Reno and Katz, 1943), and the effect of pressure of N_2 and H_2 on the surface tension of liquid ammonia (Lefrancois and Bourgeois, 1972).

When the Macleod-Sugden correlation is used, errors at low pressures rarely exceed 5 to 10% and can be much less if P_i values are obtained from experimental data. It is desirable that mixture liquid and vapor densities and compositions be known accurately. However, Zuo and Stenby (1997) have correlated the behavior of a number of systems including petroleum fractions by calculating densities with the Soave equation of state, even though this equation does not predict accurate liquid densities. They then fit P to surface and interfacial tension data so the error in liquid density is compensated for in the correlation for the parachor. This emphasizes the fact that the parachor is a calculated quantity. Parachor values calculated by Eq. (13.5-1) with an exponent of 4 should obviously not be used in a mixture equation in which the exponent is some other value.

Example 13.4 Use Eq. (13.5-1) to estimate the interfacial tension of a carbon dioxide (1) – *n*-decane (2) mixture at 344.3 K, 11380 kPa, and with $x_1 = 0.775$. At these conditions, Nagarajan and Robinson (1986) report $y_1 = 0.986$, $\rho_{L_m} = 0.7120$ g/cm³, $\rho_{V_m} = 0.3429$ g/cm³, and $\sigma_m = 1.29$ mN/m.

Solution. Gasem et al. (1989) recommend $P_1 = 73.5$ and $P_2 = 446.2$ when $n = 3.6$ and $\lambda_{ij} = 1$. From Appendix A, the molecular weights (M_i) are $M_1 = 44.010$ g/mol and $M_2 = 142.285$ g/mol.

With Eqs. (13.5-2) to (13.5-4),

$$P_{L_m} = (0.775)^2(73.5) + 2(0.775)(0.225)[(73.5 + 446.2)/2] + (0.225)^2(446.2) = 157.4$$

$$P_{V_m} = (0.986)^2(73.5) = 2(0.986)(0.014)\ [(73.5 + 446.2)/2] + (0.014)^2(446.2) = 78.7$$

Converting density to a molar density,

$$\rho_{L_m} = 0.7120/[(0.775)(44.01) + (0.225)(142.285)] = 0.01077 \text{ mol/cm}^3$$

$$\rho_{V_m} = 0.3429/[(0.986)(44.01) + (0.014)(142.285)] = 0.00756 \text{ mol/cm}^3$$

With Eq. (13.5-1),

$$\sigma_m = [(157.4)(0.01077) - (78.7)(0.00756)]^{3.6} = 1.41 \text{ dyn/cm} = \text{mN/m}$$

$$\text{Error} = 1.41 - 1.29 = 0.12 \text{ mN/m or } 9.3\%.$$

In Example 13.4, the value used for P_2 of 446.2 was determined by a fit to the data set of Nagarajan and Robinson for which the carbon dioxide liquid phase mole fractions ranged from 0.5 to 0.9. Using the surface tension and density of pure decane to determine P_2 leads to a value of 465. Using this value in Example 13.4 leads to an error of 25%. In other words, Eq. (13.5-1) does not describe the behavior of the CO_2–decane system over the entire composition range for the temperature in Example 13.4.

13.5.2 Discussion

Often, when only approximate estimates of σ_m are necessary, one may choose the general form

$$\sigma_m^r = \sum_i^n x_i \sigma_i^r \tag{13.5-5}$$

Hadden (1966) recommends $r = 1$ for most hydrocarbon mixtures, which would predict linear behavior in surface tension versus composition. For the nonlinear behavior as shown in Fig. 13.4, closer agreement is found if $r = -1$ to -3.

Zuo and Stenby (1997) have extended Eqs. (13.3-8) to (13.3-11) to mixtures with success at low to moderate pressures by using a pseudocritical temperature and pressure calculated from the Soave equation of state by applying Eq. (6.6-6) to the mixture EoS. For mixtures containing only hydrocarbons, no interaction parameter was required, but for mixtures containing CO_2 or methane, an interaction parameter was fit to experimental data. Because the pseudocritical point differs from the true critical point, this method breaks down as the true critical point of the mixture is approached. For this case, Eq. (13.5-1) has led to better results because the equation necessarily predicts that σ_m goes to zero as the true critical point is approached. In addition to the work of Gasem et al. already described, both Hugill and van Welsenes (1986) and Zuo and Stenby (1997) have developed equations for $\boldsymbol{P}_i$ in terms of T_{c_i}, P_{c_i}, and ω_i. These two sets of investigators used different values for n, calculated densities with different equations of state, and ended up with two different equations for $\boldsymbol{P}_i$, one predicting that $\boldsymbol{P}_i$ goes up with ω_i, while the other predicts that $\boldsymbol{P}_i$ goes down with ω_i. This illustrates the importance of documenting how one obtains phase densities and compositions and illustrates the empirical nature of the parachor approach. As explained in the 5th edition of this book, using these equations to predict pure-component surface tensions results in higher deviations than for the methods evaluated in Table 13.3.

13.5.3 Surface Tensions of Aqueous Systems

Whereas for nonaqueous solutions the mixture surface tension in some cases can be approximated by a linear dependence on mole fraction, aqueous solutions show pronounced nonlinear characteristics. A typical case is shown in Fig. 13.6 for acetone-water at 353 K. The surface tension of the mixture is represented by an approximately straight line on semilogarithmic coordinates. This behavior is typical of organic-aqueous systems, in which small concentrations of the organic material may significantly affect the mixture surface tension. The hydrocarbon portion of the organic molecule behaves like a hydrophobic material and tends to be rejected from the water phase by preferentially concentrating at the surface. In such a case, the bulk concentration is very different from the surface concentration. Unfortunately, the latter is not easily measured. Meissner and Michaels (1949) show graphs similar to Fig. 13.6 for a variety of dilute solutions of organic materials in water and suggest that the general behavior is approximated by the Szyszkowski equation, which they modify to the form

$$\frac{\sigma_m}{\sigma_W} = 1 - 0.411 \log\left(1 + \frac{x}{a}\right) \tag{13.5-6}$$

where σ_W is the surface tension of pure water, x is the mole fraction of the organic material, and a is a constant characteristic of the organic material. Values of a are listed in Table 13.4 for a few compounds. This equation

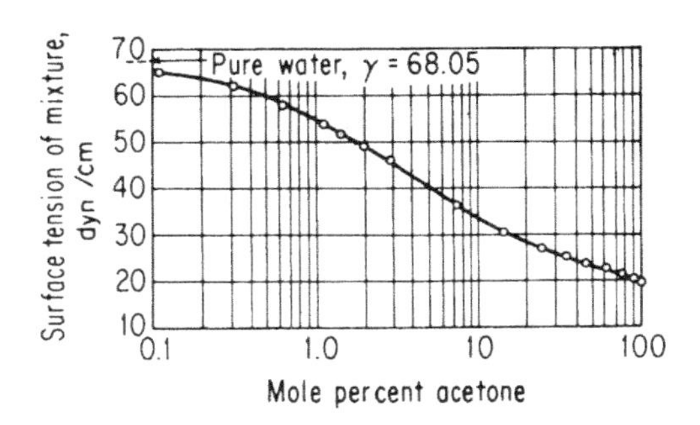

FIGURE 13.6
Surface tensions of water-acetone solutions at 353 K. (McAllister and Howard, 1957)

TABLE 13.4 Constants for the Szyszkowski equation (13.5-6) (Meissner and Michaels, 1949).

Compound	$a \times 10^4$	Compound	$a \times 10^4$
propionic acid	26	diethyl ketone	8.5
n-propyl alcohol	26	ethyl propionate	3.1
isopropyl alcohol	26	propyl acetate	3.1
methyl acetate	26	*n*-valeric acid	1.7
n-propyl amine	19	isovaleric acid	1.7
methyl ethyl ketone	19	*n*-amyl alcohol	1.7
n-butyric acid	7.0	isoamyl alcohol	1.7
isobutyric acid	7.0	propyl propionate	1.0
n-butyl alcohol	7.0	*n*-caproic acid	0.75
isobutyl alcohol	7.0	*n*-heptanoic acid	0.17
propyl formate	8.5	*n*-octanoic acid	0.034
ethyl acetate	8.5	*n*-decanoic acid	0.0025
methyl propionate	8.5		

should not be used if the mole fraction of the organic solute exceeds 0.01. For some substances this is well below the solubility limit.

The method of Tamura et al. (1955) may be used to estimate surface tensions of aqueous binary mixtures over wide concentration ranges of the dissolved organic material and for both low- and high-molecular weight organic-aqueous systems. Equation (13.5-1) is assumed as a starting point, but the significant densities and concentrations are taken to be those characteristic of the surface layer, that is, $(V^\sigma)^{-1}$ replaces ρ_{L_m}, where V^σ is a hypothetical molar volume of the surface layer. V^σ is estimated with

$$V^\sigma = \sum_j x_j^\sigma V_j \tag{13.5-7}$$

where x_j^σ is the mole fraction of compound j in the surface layer. V_j, however, is chosen as the pure-liquid molar volume of compound j. Then, with Eq. (13.5-1), assuming $\rho_L >> \rho_V$,

$$V^\sigma \sigma_m^{\frac{1}{4}} = x_W^\sigma \boldsymbol{P}_W + x_O^\sigma \boldsymbol{P}_O \tag{13.5-8}$$

where the subscripts W and O represent water and the organic component, respectively. To eliminate the parachor, however, Tamura et al. introduce Eq. (13.3-1); the result is

$$\sigma_m^{\frac{1}{4}} = \psi_W^\sigma \sigma_W^{\frac{1}{4}} + \psi_O^\sigma \sigma_O^{\frac{1}{4}} \tag{13.5-9}$$

In Eq. (13.5-9), ψ_W^σ is the superficial volume fraction water in the surface layer

$$\psi_W^\sigma = \frac{x_W^\sigma V_W}{V^\sigma} \tag{13.5-10}$$

and similarly for ψ_O^σ.

Equation (13.5-9) is the final correlation. To obtain values of the superficial surface volume fractions ψ_W^σ and ψ_O^σ, equilibrium is assumed between the surface and bulk phases. Tamura's equation is complex, and after rearrangement it can be written in the following set of equations.

$$B = \log \frac{\psi_W^q}{\psi_O} \tag{13.5-11}$$

$$W = 0.441 \frac{q}{T} \left(\frac{\sigma_O V_O^{\frac{2}{3}}}{q} - \sigma_W V_W^{\frac{2}{3}} \right) \tag{13.5-12}$$

$$C = B + W \tag{13.5-13}$$

$$\frac{(\psi_W^\sigma)^q}{\psi_O^\sigma} = 10^C \tag{13.5-14}$$

Here ψ_W^σ is defined by Eq. (13.5-10), V_W is the molar volume of pure water, V_O is the molar volume of pure organic component, σ_W is the surface tension of pure water, σ_O is the surface tension of pure organic component, T is temperature in K, and q is a constant depending upon type and size of organic constituent. Also, ψ_W and ψ_O are the superficial bulk volume fractions of water and organic material determined as

$$\psi_W = \frac{x_W V_W}{x_W V_W + x_O V_O} \qquad \psi_O = \frac{x_O V_O}{x_W V_W + x_O V_O} \tag{13.5-15}$$

where x_W is the bulk mole fraction of pure water, and x_O is the bulk mole fraction pure organic component. Equation (13.5-14), along with the fact that $\psi_W^\sigma + \psi_O^\sigma = 1$, allows values of ψ_W^σ and ψ_O^σ to be determined so that σ_m can be found from Eq. (13.5-9).

The method is illustrated in Example 13.5. Tamura et al. (1955) tested the method with some 14 aqueous systems and two alcohol-alcohol systems; the percentage errors are less than 10% when q is less than 5 and within 20% for q greater than 5. The method cannot be applied to multicomponent mixtures. For nonaqueous mixtures comprising polar molecules, the method is unchanged except that q = ratio of molar volumes of the solute to solvent.

Materials	q	Example
Fatty acids, alcohols	Number of carbon atoms	Acetic acid, $q = 2$
Ketones	One less than the number of carbon atoms	Acetone, $q = 2$
Halogen derivatives of fatty acids	Number of carbons times ratio of molar volume of halogen derivative to parent fatty acid	$q = 2\dfrac{V_b(\text{chloroacetic acids})}{V_b(\text{acetic acid})}$

Example 13.5 Estimate the surface tension of a mixture of methanol and water at 303 K when the alcohol mole fraction is 0.122. The experimental value is 46.1 dyn/cm (Tamura et al., 1955).

Solution. At 303 K (O represents methanol, W water), $\sigma_W = 71.18$ dyn/cm, $\sigma_O = 21.75$ dyn/cm, $V_W = 18$ cm^3/mol, $V_O = 41$ cm^3/mol, and q = number of carbon atoms = 1. From Eq. (13.5-15),

$$\frac{\psi_W}{\psi_O} = \frac{(0.878)(18)}{(0.122)(41)} = 3.16$$

and from Eq. (13.5-11),

$$B = \log 3.16 = 0.50$$

and from Eq. (13.5-12),

$$W = (0.144)(303)^{-1}[(21.75)(41)^{2/3} - (71.18)(18)^{2/3}] = -0.34$$

Hence, from Eq. (13.5-13),

$$C = 0.50 - 0.34 = 0.16$$

From Eq. (13.5-14), with $q = 1$,

$$\frac{\psi_W^\sigma}{\psi_O^\sigma} = 10^{0.16} = 1.45$$

Using $\psi_W^\sigma + \psi_O^\sigma = 1$ yields

$$\frac{\psi_W^\sigma}{1 - \psi_W^\sigma} = 1.45$$

$$\psi_W^\sigma = 0.59 \quad \psi_O^\sigma = 0.41$$

Finally, from Eq. (13.5-9)

$$\sigma_m = [(0.59)(71.18)^{1/4} + (0.41)(21.75)^{1/4}]^4 = 46 \text{ dyn/cm}$$

$$\text{Error} = 46 - 46.1 = -0.1 \text{ dyn/cm or } -0.2\%$$

13.5.4 Thermodynamic-Based Relations

The estimation procedures introduced earlier in this section are empirical; all except the Tamura et al. method employ the bulk liquid (and sometimes vapor) composition to characterize a mixture. However, the "surface phase" usually differs in composition from that of the bulk phases, and it is reasonable to suppose that, in mixture surface tension relations, surface compositions are more important than bulk compositions. The fact that σ_m is almost always less than the bulk mole fraction average is interpreted as indicating that the component or components with the lower pure component values of σ preferentially concentrate in the surface phase.

The assumptions that the bulk and surface phases are in equilibrium, and the partial molar area of component i is the same as the molar area of i, leads to the following equations (Sprow and Prausnitz 1966a):

$$\sigma_m = \sigma_i + \frac{RT}{A_i} \ln \frac{x_i^\sigma \gamma_i^\sigma}{x_i \gamma_i} \quad (i = 1, 2, \dots N) \tag{13.5-16}$$

$$\sum_i x_i^\sigma = 1 \tag{13.5-17}$$

where σ_m is the mixture surface tension in units of dyn/cm, σ_i is the surface tension of pure component i in dyn/cm, $R = 8.314 \times 10^7$ dyn·cm/(mol·K), A_i is the surface area of component i, cm²/mol (see Table 13.5), T is temperature in K, x_i is the mole fraction of component i in the bulk phase, x_i^σ is the mole fraction of component i in the surface phase, γ_i is the activity coefficient of component i in the bulk phase, and γ_i^σ is the activity coefficient of component i in the surface phase.

When γ_i^σ is related to the surface composition and γ_i to the bulk liquid composition, Eqs. (13.5-4) and (13.5-5) represent $N + 1$ equations in the $N + 1$ unknowns, σ_m and the N values of x_i^σ. Hildebrand and Scott (1964) have examined the case where $\gamma_i^\sigma = 1$, and Eckert and Prausnitz (1964) and Sprow and Prausnitz (1966, 1966a) have used regular solution theory for γ_i^σ. None of these versions, however, was particularly successful for aqueous mixtures. Suarez et al. (1989) have used a version of the UNIFAC model by Larsen et al. (1987) to determine the surface and bulk phase activity coefficients. Larsen's UNIFAC model differs from the one

TABLE 13.5 Values of A_i to be used in Eq. (13.5-16).

Component	$A_i \times 10^8$ cm^2 mol^{-1}
Water	0.7225
Methanol	3.987
Ethanol	8.052
1-Propanol	17.41
2-Propanol	20.68
Ethylene glycol	4.123
Glycerol	3.580
1,2-Propanediol	6.969
1,3-Propanediol	8.829
1,3-Butanediol	9.314
1,4-Butanediol	8.736
Acetonitrile	6.058
Acetic acid	6.433
1,4-Dioxane	12.27
Acetone	8.917
Methyl ethyl ketone	12.52
n-Hexane	11.99
Benzene	9.867
Toluene	9.552
Pyridine	10.35

presented in Chap. 8 in that ln γ^c is determined from Eq. (13.5-18) and the a_{mn} parameters of Eq. (8.11-4) are functions of temperature.

$$\ln \gamma_i^c = \ln \frac{\phi_i}{x_i} + 1 - \frac{\phi_i}{x_i} \tag{13.5-18}$$

$$\phi_i = \frac{x_i r_i^{\frac{2}{3}}}{\sum_j x_j r_j^{\frac{2}{3}}} \tag{13.5-19}$$

where x_i is the mole fraction of component i in the bulk phase and r_i is the UNIFAC volume parameter of molecule i determined by method in Chap. 8. For nonaqueous mixtures with a difference between pure-component surface tensions not exceeding around 20 dyn/cm, Suarez et al. (1989) claim that σ_m values are predicted with an average error of 3.5% when pure-component areas A_i are calculated by

$$A_i = 1.021 \times 10^8 V_c^{\frac{6}{15}} V_b^{\frac{4}{15}} \tag{13.5-20}$$

where V_c and V_b are in cm^3/mol and A_i is in cm^2/mol. Suarez et al. claim improved results when pure-component areas shown in Table 13.5 are used. Table 13.5 values should be used if values of A_i are tabulated for all components. Otherwise, Eq. (13.5-20) should be used for all values. Values from Eq. (13.5-20) and Table 13.5 should not be mixed. Suarez et al. report average deviations for binary systems including aqueous systems of 3% and deviations of 4% for ternary systems. Zhibao et al. (1990) have also used UNIFAC to predict surface tensions and report results similar to those of Suarez et al. The Suarez method is illustrated in Example 13.6.

Example 13.6 Use the Suarez method, Eqs. (13.5-16) and (13.5-17) along with Larsen's (1987) UNIFAC method, to estimate σ_m for a mixture of 5 weight % *n*-propanol(1) and 95 weight % H_2O(2) at 298 K. Vázquez et al. (1995) report an experimental valve of 41.83 dyn/cm. They also give $\sigma_1 = 23.28$ dyn/cm and $\sigma_2 = 72.01$ dyn/cm.

Solution. There are four UNIFAC groups, —CH_3, >CH_2, —OH, and H_2O. R_i and Q_i for these groups are

	—CH_3	>CH_2	—OH	H_2O
R_i	0.9011	0.6744	1.0	0.92
Q_i	0.848	0.54	1.2	1.4

With these group values, the UNIFAC volume parameter for *n*-propanol is

$$r_i = 0.9011 + (2)(0.6744) + 1.0 = 3.2499$$

Similarly, $r_2 = 0.92$, $q_1 = 3.128$, and $q_2 = 1.4$

a_{mn} values at 298 K from Larsen et al. (1987) are

	—CH_3	>CH_2	—OH	H_2O
—CH_3	0	0	972.8	1857
>CH_2	0	0	972.8	1857
—OH	637.5	637.5	0	155.6
H_2O	410.7	410.7	–47.15	0

From Table 13.5, $A_1 = 17.41 \times 10^8$ cm²/mol and $A_2 = 0.7225 \times 10^8$ cm²/mol. Equations (13.5-18) and (13.5-19) are used for the combinatorial contribution to γ and Eqs. (8.11-1) to (8.11-4) are used for the residual contribution. The bulk composition of 5 weight % *n*-propanol corresponds to $x_1 = 0.01553$ and $x_2 = 0.98447$. At this composition, $\gamma_1 = 10.015$ and $\gamma_2 = 1.002$.

Using Eq. (13.5-16) for component 1

$$\begin{aligned}\sigma_m &= \sigma_1 + \frac{(8.314 \times 10^7)(298)}{8.052 \times 10^8} \ln \frac{x_1^\sigma \gamma_1^\sigma}{(0.01553)(10.015)} \\ &= 23.28 + 30.77 \ln(6.4295 x_1^\sigma \gamma_1^\sigma)\end{aligned}$$

similarly for component 2,

$$\sigma_m = 72.01 + 342.9 \ln(1.0136 x_2^\sigma \gamma_2^\sigma)$$

These two equations plus the condition, $x_1^\sigma + x_2^\sigma = 1$ (Eq. 13.5-17), along with the UNIFAC relations for γ_1^σ and γ_2^σ must be solved iteratively. The solution is

$$x_1^\sigma = 0.269, \; x_2^\sigma = 0.731, \; \gamma_1^\sigma = 2.05, \; \gamma_2^\sigma = 1.234, \text{ and } \sigma_m = 41.29 \text{ dyn/cm}$$

$$\text{Error} = 41.29 - 41.83 = -0.54 \text{ dyn/cm or } -1.3\%$$

Note that this model predicts that the component with the lower surface tension, *n*-propanol, is 17 times more concentrated in the surface than the bulk. Using the UNIFAC method in Chap. 8 (instead of that of Larsen et al. (1987) as specified in the Suarez method) to calculate activity coefficients at both the bulk and surface concentrations would have predicted a value of σ_m of 43.40 dyn/cm for an error of 3.8%. If one just takes both bulk and surface activity coefficients equal to 1 in Example 13.6, the error is 13%. The Suarez method gives an error of –0.4% for the case of Example 13.5.

13.5.5 Recommendations

For estimating the surface tensions of mixtures, the Suarez method [Eqs. (13.5-16)–(13.5-19) and Example 13.6] is generally recommended. However, in certain circumstances, other methods might be preferred. Near mixture critical points, the Macleod-Sugden correlation [Eq. (13.5-1) and Example 13.4] should be used because the form of the equation necessarily gives the correct limit that σ goes to zero at the critical point. For nonpolar mixtures, extension of the corresponding-states method of Zuo and Stenby [Eqs. (13.3-8) to (13.3-11)] to mixtures gives results as reliable as the Suarez method and the calculational procedure is simpler.

For estimating the surface tensions of binary organic-aqueous mixtures, use either the Suarez method or the method of Tamura et al. as given by Eqs. (13.5-9) to (13.5-15) and illustrated in Example 13.5. For multicomponent mixtures with water as one component, the Suarez method should be used. If the solubility of the organic compound in water is low, the Szyszkowski equation (13.5-6), as developed by Meissner and Michaels, may be used. Of these three methods, the Suarez approach is most broadly applicable but the most complex. The Szyszkowski method is the simplest but should be used only when the solute mole fraction is less than 0.01. Furthermore, a value for constant a must be available in Table 13.4.

13.5.6 Interfacial Tensions in Liquid-Liquid Binary Systems

Li and Fu (1991) have presented a UNIQUAC-based equation to predict interfacial tensions in systems with two liquid phases and two components. Unlike the empirical methods presented earlier for interfacial tensions at high pressure (Sec. 13.5), the Li-Fu equation is for highly nonideal systems at low (near atmospheric) pressure. Li and Fu propose

$$\sigma_m = 3.14 \times 10^{-9}(1 - k_{12})W_{12}(\phi_1^I - \phi_1^{II})^2 \tag{13.5-21}$$

where k_{12} is an empirical parameter discussed below, ϕ_i^J is the volume fraction of i in phase J (where $J = 1$ or II) and is calculated by

$$\phi_1^I = \frac{x_1^I r_1}{x_1^I + x_2^I r_2} \tag{13.5-22}$$

and

$$W_{12} = \frac{R(\Delta U_{21} + \Delta U_{21})}{z} \tag{13.5-23}$$

In these equations, x_1^I is the mole fraction of component 1 in the phase rich in component 1, x_1^{II} is the mole fraction of component 1 in the phase rich in component 2, r_i is the UNIQUAC volume parameter for component i (see Chap. 8), z the coordination number (taken as 10), R is the gas constant (taken as 8.314×10^7 dyn·cm/(mol·K)), and ΔU_{12} and ΔU_{21} are UNIQUAC parameters. The UNIQUAC parameters, along with solubility data required in Eq. (13.5-22), have been tabulated for many binary systems in Sørensen and Arlt (1979). In Eq. (13.5-21), the constant 3.14×10^{-9} has units mol/cm^2 and σ is in dyn/cm when the value of R given above is used.

Li and Fu suggest that the parameter k_{12} accounts for orientation effects of molecules at the interface and recommend the empirical equation

$$k_{12} = 0.467 - 0.185X + 0.016X^2 \tag{13.5-24}$$

where

$$X = -\ln\left(x_1^{II} + x_2^I\right) \tag{13.5-25}$$

For 48 binary systems, Li and Fu claim an average absolute percent deviation of 8.8% with Eq. (13.5-21). Other methods (Hecht, 1979; Li and Fu, 1989) give slightly lower deviations but require either numerical integration or a numerical solution of a set of nonlinear equations.

Example 13.7 Use Eq. (13.5-21) to estimate the interfacial tension of the benzene (1)—water (2) system at 20 °C. The experimental value (Fu et al., 1986) is 33.9 dyn/cm. Also, from Page 341 of Sørensen and Arlt (1979), $x_1^{II} = 2.52 \times 10^{-3}$, $x_2^{I} = 4.00 \times 10^{-4}$, $\Delta U_{12} = 882.10$ K, and $\Delta U_{21} = 362.50$ K.

Solution. From Eqs. (13.5-24) and (13.5-25),

$$X = -\ln(2.52 \times 10^{-3} + 4.00 \times 10^{-4}) = 5.836$$

$$k_{12} = 0.467 - (0.185)(5.836) + (0.016)(5.836)^2 = -0.0677$$

From Eq. (12.5-23)

$$W_{12} = (8.314 \times 10^7/10)(882.10 + 362.50) = 1.035 \times 10^{10} \text{ dyn·cm/mol}$$

Using the Sørensen and Arlt (1979) data, the volume fractions [Eq. (13.5-22)] are determined from

$$x_1^{I} = 1 - 4.00 \times 10^{-4} = 0.9996$$

$$x_1^{II} = 1 - 2.52 \times 10^{-3}\ 0.99748$$

$$r_1 = 3.1878;\ r_2 = 0.92$$

$$\phi_1^{I} = (0.9996)(3.1878)[(0.9996)(3.1878) + (4.00 \times 10^{-4})(0.92)]^{-1} = 0.9999$$

$$\phi_1^{II} = (2.52 \times 10^{-3})(3.1878)[(2.52 \times 10^{-3})(3.1878) + (0.99748)(0.92)]^{-1} = 0.00868$$

With these values, the surface tension predicted by Eq. (13.5-21) is

$$\sigma_m = (3.14 \times 10^{-9})(1 + 0.677)(1.035 \times 10^{10})(0.9999 - 0.00868)^2 = 34.1 \text{ dyn/cm}$$

$$\text{Error} = 34.1 - 33.9 = 0.02 \text{ dyn/cm or } 0.6\%$$

13.6 NOTATION

A	surface area
A_i	surface area of component i
a	compound-specific parameter for prediction of σ_m for binary aqueous mixtures using the Szyszkowski as reported by Meissner and Michaels (1949)
a_{mn}	parameter in UNIFAC model for activity coefficients
AARD	absolute average relative deviation
B	parameter in Tamura et al. (1955) prediction method for σ_m of binary aqueous mixtures
C	parameter in Tamura et al. (1955) prediction method for σ_m of binary aqueous mixtures
K	parameter in Sastri and Rao (1995) prediction method for σ
k_{ij}	parameter in the Liu and Fu prediction method for the interfacial tension of a liquid-liquid binary mixture

m	parameter in Sastri and Rao (1995) prediction method for σ
N_c	number of compounds in test set database
n	parameter in predictions of σ and σ_m
N_i	number of times group i is found in a compound
N_{pt}	number of points in test set database
P	system pressure
P	parachor
P_i	parachor of component i
P_{ij}	cross interaction parachor in a mixture
P_{L_m}	parachor of liquid mixture
P_O	parachor of the organic compound in an aqueous mixture
P_{V_m}	parachor of vapor mixture
P_W	parachor of water in an aqueous mixture
P_c	critical pressure
P_{c_i}	critical pressure of component i
P^{vp}	saturated liquid vapor pressure
P_r^{vp}	reduced saturated liquid vapor pressure; $\frac{P^{vp}}{P_c}$
Q	parameter in Brock and Bird (1955) prediction method of σ
Q_i	parameter in UNIFAC model for activity coefficients
q	parameter in Tamura et al. (1955) prediction method for σ_m of binary aqueous mixtures
R	gas constant (8.314462618 J/(mol·K))
R_i	parameter in UNIFAC model for activity coefficients
r (superscript)	power describing nonlinear behavior of surface tensions of mixtures
r_i	volume parameter in UNIFAC/UNIQUAC model for activity coefficients
T	system temperature
T_b	normal (101325 Pa) boiling temperature
T_{br}	reduced normal boiling temperature; $\frac{T_b}{T_c}$
T_c	critical temperature
T_{c_i}	critical temperature of component i
T_r	reduced temperature; $\frac{T}{T_c}$
T_m	melting temperature
V^σ	hypothetical molar volume of the surface layer in a mixture
V_b	liquid molar volume at T_b
V_i	pure liquid molar volume of component i
V_O	pure liquid molar volume of the organic compound in an aqueous mixture
V_W	pure liquid molar volume of water in an aqueous mixture
W	parameter in Tamura et al. (1955) prediction method for σ_m of binary aqueous mixtures
W_{12}	parameter in the Liu and Fu prediction method for the interfacial tension of a liquid-liquid binary mixture
x	parameter in Sastri and Rao (1995) prediction method for σ
x_i	liquid mole fraction of component i

x_i^σ	mole fraction of component i in the surface layer
x_O^σ	mole fraction of the organic compound in the surface layer of an aqueous mixture
x_W^σ	mole fraction of water in the surface layer of an aqueous mixture
x_1^I	mole fraction of component 1 in the phase rich in component 1 in a liquid/liquid binary mixture
x_1^{II}	mole fraction of component 1 in the phase rich in component 2 in a liquid/liquid binary mixture
x_2^I	mole fraction of component 2 in the phase rich in component 1 in a liquid/liquid binary mixture
x_2^{II}	mole fraction of component 2 in the phase rich in component 2 in a liquid/liquid binary mixture
X	parameter in the Liu and Fu prediction method for the interfacial tension of a liquid-liquid binary mixture
y	parameter in Sastri and Rao (1995) prediction method for σ
y_i	vapor mole fraction of component i
z	(1) parameter in Sastri and Rao (1995) prediction method for σ; (2) coordination number in the Liu and Fu (1991) prediction method for the interfacial tension of a liquid-liquid binary mixture

Greek

α_c	parameter in Riedel-based predictions for P^{vp} used in Brock and Bird (1955) prediction of σ
γ_i	activity coefficient of component i in the bulk
γ_i^σ	activity coefficient of component i in the surface layer
$\Delta U_{12}, \Delta U_{21}$	parameters in UNIQUAC model for activity coefficients
Δ_{P_i}	contribution for group i for prediction of parachor (***P***)
λ_{ij}	binary interaction parameter
ρ_L	liquid density
ρ_{L_m}	liquid density of mixture
ρ_V	vapor density
ρ_{V_m}	vapor density of mixture
σ	surface tension
σ_i	surface tension component i
σ_m	surface tension of a mixture
σ_O	surface tension of organic compound in an aqueous mixture
σ_r	reduced surface tension
$\sigma_r^{(1)}, \sigma_r^{(2)}$	reduced surface tensions of reference fluids 1 and 2 in Zuo and Stenby (1997) prediction of σ
σ_W	surface tension of water
ϕ_i	parameter in UNIFAC model for activity coefficients
ϕ_i^J	volume fraction of component i in phase J ($J = I$ or II) used in the Liu and Fu prediction method for the interfacial tension of a liquid-liquid binary mixture
ψ_O	superficial bulk volume fraction of the organic compound in an aqueous mixture
ψ_O^σ	superficial volume fraction of the organic compound in the surface layer of an aqueous mixture
ψ_W	superficial bulk volume fraction of water in an aqueous mixture
ψ_W^σ	superficial volume fraction of water in the surface layer of an aqueous mixture
ω	acentric factor

$\omega^{(1)}, \omega^{(2)}$ acentric factors of reference fluids 1 and 2 in Zuo and Stenby (1997) prediction of σ
ω_i acentric factor of component i

13.7 REFERENCES

Adamson, A. W.: *Physical Chemistry of Surfaces,* 4th ed., Wiley, New York, 1982.
Agarwal, D. K., R. Gopal, and S. Agarwal: *J. Chem. Eng. Data,* **24:** 181 (1979).
Aveyard, R., and D. A. Haden: *An Introduction to Principles of Surface Chemistry,* Cambridge, London, 1973.
Barbulescu, N.: *Rev. Roum. Chem.,* **19:** 169 (1974).
Bowden, S. T., and E. T. Butler: *J. Chem. Soc.,* **1939:** 79.
Brock, J. R., and R. B. Bird: *AIChE J.,* **1:** 174 (1955).
Brown, R. C.: *Contemp. Phys.,* **15:** 301 (1974).
Chattoraj, D. K., and K. S. Birdi: *Adsorption and the Gibbs Surface Excess,* Plenum Press, New York, 1984.
Curl, R. F., Jr., and K. S. Pitzer: *Ind. Eng. Chem.,* **50:** 265 (1958).
Daubert, T. E., R. P. Danner, H. M. Sibel, and C. C. Stebbins, *Physical and Thermodynamic Properties of Pure Chemicals: Data Compilation,* Taylor & Francis, Washington, DC, 1997.
Deam, J. R., and R. N. Maddox: *J. Chem. Eng. Data,* **15:** 216 (1970).
DIPPR, Design Institute for Physical Properties, American Institute of Chemical Engineers, 120 Wall St. FL 23, New York, NY 10005-4020 USA (2021). https://www.aiche.org/dippr
Eckert, C. A., and J. M. Prausnitz: *AIChE J.,* **10:** 677 (1964).
Escobedo, J., and G. A. Mansoori: *AIChE J.,* **42:** 1425 (1996).
Evans, D. F., and J. Wennerstöm: The Colloidal Domain, Wiley-VCH, New York, 1999.
Everett, D. H.: *Basic Principles of Colloid Science,* Royal Society of Chemistry, London, 1988.
Fu, J., B. Li, and Z. Wang: *Chem. Eng. Sci.,* **41:** 2673 (1986).
Gambill, W. R.: *Chem. Eng.,* **64**(5)**:** 143 (1958).
Gasem, K. A. M., P. B. Dulcamara, Jr., K. B. Dickson, and R. L. Robinson, Jr.: *Fluid Phase Equil.,* **53:** 39 (1989).
Gibbs, J. W.: *The Collected Works of J. Willard Gibbs,* vol. I, *Thermodynamics,* Yale University Press, New Haven, Conn., 1957.
Gray, J. A., C. J. Brady, J. R. Cunningham, J. R. Freeman, and G. M. Wilson: *Ind. Eng. Chem. Process Des. Dev.,* **22:** 410 (1983).
Guggenheim, E. A.: *Thermodynamics,* 4th ed., North-Holland, Amsterdam, 1959.
Hadden, S. T.: *Hydrocarbon Process Petrol. Refiner,* **45**(10)**:** 161 (1966).
Hammick, D. L., and L. W. Andrew: *J. Chem. Soc.,* **1929:** 754.
Hecht, G.: *Chem. Technol.,* **31:** 143 (1979).
Hildebrand, J. H., and R. L. Scott: *The Solubility of Nonelectrolytes,* 3rd ed., Dover New York, 1964, chap. 21.
Hugill, J. A., and A. J. van Welsenes: *Fluid Phase Equil.,* **29:** 383 (1986).
International Critical Tables, vol. III, McGraw-Hill, New York, 1928, p. 28.
Jasper, J. J.: *J. Phys. Chem. Ref. Data,* **1:** 841 (1972).
Knotts, T. A., W. V. Wilding, J. L. Oscarson, and R. L. Rowley: *J. Chem. Eng. Data*, **46:** 1007–1012 (2001).
Körösi, G., and E. sz. Kováts: *J. Chem. Eng. Data,* **26:** 323 (1981).
Larsen, B. L., P. Rasmussen, and A. Fredenslund: *Ind. Eng. Chem. Res.,* **26:** 2274 (1987). Lefrançois, H., and Y. Bourgeois: *Chim. Ind. Genie Chim.,* **105**(15)**:** 989 (1972).
Lawrence, A. N., L. Wang, W. G. Chapman, S. Gupta, Surface Tension Models for Industrial Applications, Presented at AIChE Annual Meeting, Virtual (2020).
Li, B., and J. Fu: *Chem. Eng. Sci.,* **44:** 1519 (1989).
Li, B., and J. Fu: *Fluid Phase Equil,* **64:** 129 (1991).
Macleod, D. B.: *Trans. Faraday Soc.,* **19:** 38 (1923).
McAllister, R. A., and K. S. Howard: *AIChE J.,* **3:** 325 (1957).
Meissner, H. P., and A. S. Michaels: *Ind. Eng. Chem.,* **41:** 2782 (1949).
Mejía, A., E. A. Müller, G. Chaparro-Maldonado, *J. Chem. Inf. Model.,* **61:** 1244–1250 (2021).
Miller, D. G.: *Ind. Eng. Chem. Fundam.,* **2:** 78 (1963).
Mumford, S. A. and J. W. C. Phillips: *J. Chem. Soc*., 2112–2133 (1929).
Muratov, G. N.: *Zh. Fiz. Khim.*, **54:** 2088–2089 (1980).
Nagarajan, N., and R. L. Robinson, Jr.: *J. Chem. Eng. Data,* **31:** 168 (1986).

Pitzer, K. S.: *Thermodynamics,* 3rd ed., New York, McGraw-Hill, 1995, p. 521.
Quayle, O. R.: *Chem. Rev.,* **53:** 439 (1953).
Reno, G. J., and D. L. Katz: *Ind. Eng. Chem.,* **35:** 1091 (1943).
Rice, O. K.: *J. Phys. Chem.,* **64:** 976 (1960).
Rice, P., and A. S. Teja: *J. Colloid. Interface Sci.,* **86:** 158 (1982).
Riddick, J. A., W. B. Bunger, and T. K. Sakano: *Organic Solvents, Physical Properties and Methods of Purification,* 4th ed., Techniques of Chemistry, Vol. II, Wiley, New York, 1986.
Riedel, L.: *Chem. Ing. Tech.,* **26:** 83 (1954).
Riedel, L.: *Chem. Ing. Tech.,* **27:** 209 (1955).
Ross, S. (Chairman): *Chemistry and Physics of Interfaces,* American Chemical Society, Wash-ington, DC, 1965.
Rowley, J. R., W. V. Wilding, J. L. Oscarson, and R. L. Rowley: *Int. J. Thermophys.,* **28:** 824–834 (2007).
Rowlinson, J. S., and G. Widom: *Molecular Theory of Capillarity,* Oxford University Press, New York, 1982.
Sastri, S. R. S., and K. K. Rao: *Chem. Eng. J.,* **59:** 181 (1995).
Sørensen, J. M., and W. Arlt: *Liquid-Liquid Equilibrium Data Collection,* Vol. 1, DECHEMA, 1979.
Sprow, F. B., and J. M. Prausnitz: *Trans. Faraday Soc.,* **62:** 1097 (1966).
Sprow, F. B., and J. M. Prausnitz: *Trans. Faraday Soc.,* **62:** 1105 (1966a); *Can. J. Chem. Eng.,* **45:** 25 (1967).
Stegemeier, G. L.: Ph.D. dissertation, University of Texas, Austin, Tex., 1959.
Suarez, J. T., C. Torres-Marchal, and P. Rasmussen: *Chem. Eng. Sci.,* **44:** 782 (1989).
Sugden, S.: *J. Chem. Soc.,* **32:** 1177 (1924).
Tamura, M., M. Kurata, and H. Odani: *Bull. Chem. Soc. Japan,* **28:** 83 (1955).
Tester, J. W., and M. Modell: *Thermodynamics and Its Applications,* 3rd ed., Prentice Hall, Englewood Cliffs, NJ, 1997.
Timmermans, J.: *Physico-Chemical Constants of Pure Organic Compounds,* Vol. 2, Elsevier, Amsterdam, 1965.
van der Waals, J. D.: *Z. Phys. Chem.,* **13:** 716 (1894).
Vargaftik, N. B.: *Handbook of Physical Properties of Liquids and Gases: Pure Substances and Mixtures,* Begell House, New York, 1996.
Vázquez, G., E. Alvarez, and J. M. Navaza: *J. Chem. Eng. Data,* **40:** 611 (1995).
Weinaug, C. F., and D. L. Katz: *Ind. Eng. Chem.,* **35:** 239 (1943).
Wilding, W. V., T. A. Knotts, N. F. Giles, R. L. Rowley, DIPPR® Data Compilation of Pure Chemical Properties, Design Institute for Physical Properties, AIChE, New York, NY (2017).
Zhibao, L., S. Shiquan, S. Meiren, and S. Jun: *Themochem. Acta,* **169:** 231 (1990).
Zihao, W., and F. Jufu: "An Equation for Estimating Surface Tension of Liquid Mixtures," *Proc. Jt. Mtg. of Chem. Ind. and Eng. Soc. of China and AIChE,* Beijing, China, September 1982.
Zuo, Y.-X., and E. H. Stenby: *Can. J. Chem. Eng.,* **75:** 1130 (1997).

A

Property Data Bank

This appendix contains selected property values and correlations for 470 pure compounds and elements from the May 2022 public release of the DIPPR database (Wilding et al., 2017). These are the values recommended by the DIPPR 801 project after critically reviewing the available literature and evaluating the thermodynamic picture of the entire compound in a procedure outlined by Bloxham et al. (2021). In a few cases, newer data than the values found in the property data bank are available. These newer data are not listed, but their existence is indicated by an asterisk. They are not included because they have not gone through the rigorous DIPPR review process and may not be more accurate than the values listed.

Most property values are accompanied by an expected *error* identified by a column labelled "Er" immediately following the property value column. The *error* codes are given as a percentage of the value indicating the likely uncertainty. Additionally, the font style of the entry (e.g., italics (I), bold (B), underline (U), bold/underline (BU), etc.) identifies the source/type of the property value according to the list below.

Source Type	Font Styling	Font Example	Explanation
Experimental	normal	12345	Data from direct experimentation
Smoothed	U	<u>12345</u>	Value obtained from fitting data from direct experimentation to an appropriate correlation
Defined	BU	**<u>12345</u>**	A constant value defined by an equation using other properties (e.g., ω) or by convention (e.g., ideal gas $\Delta_f H^\circ = 0$ for argon)
Derived	IBU	***<u>12345</u>***	Value calculated from well-established thermodynamic relationships using other properties (e.g., $\Delta_{vap}H$ from the Clapeyron equation)
Experimental & Predicted	B	**12345**	Temperature dependent properties whose correlation is based on both experimental and predicted values
Predicted	I	*12345*	Value or correlation obtained using a prediction method
Not specified	IU	*<u>12345</u>*	Data of unknown origin

For example, the melting point of *n*-pentane is listed as 143.42 K with an *error* ("Er") of "0.2" and is typeset in upright font *without* an underline, bolding, or italics. This means the value came from experimental data and has an uncertainty of <0.2% or ± 0.29 K. The exact sources of experimental data, or the methods used for predicted values, are available in the DIPPR database. Some *error* fields are left blank which indicates that either the uncertainty could not be estimated at the time the compound was evaluated for entry into the database or that the error is identically 0 by definition such as the ideal gas enthalpy of formation for elemental gases in their natural state (e.g., Ar, O_2, etc.).

"Smoothed" data, as mentioned above, are experimental values across multiple temperatures, or multiple values for members of a family, that have been fit to an appropriate equation to reduce errors in any one data point. For example, many authors of older papers reporting measurements of vapor pressure only list the results as coefficients to an Antione equation and not the individual data points. Such data are classified as smoothed. Another class of smoothed data are illustrated in Fig. A.1 which depicts the critical temperature of the *n*-alcohols from C3 to C13. Over 114 experimental points (the symbols) from many authors are found on the figure representing many accurate, independent, slightly different, numerical measurements for each compound in the family. Rather than picking any one point as the accepted critical temperature for a compound, all the data for the family are fit to a quadratic as shown in the solid line, and the accepted values are obtained from this correlation. The data in this case follow a quadratic model very well as the resulting equation has a coefficient of determination of $R^2 = 0.9998$.

Some property entries are blank which indicates one of two things. The first is that the property has not been measured experimentally and no reliable prediction technique exists to obtain an estimated value. The other is that the property is not applicable to the compound. For example, the normal boiling point of CO_2 is blank as the compound does not exist as a liquid at 101325 Pa.

For some compounds, solid-solid transitions exist so near to the melting temperature (T_m) that they are difficult to separate from the solid-liquid transition experimentally. The enthalpies of melting in such cases thus consist of contributions from both transitions. Values for $\Delta_m H(T_m)$ where this occurs, or is suspected to occur, are identified by dagger symbols (†) next to the numerical value.

Compounds are listed with multiple identifiers. The *DIPPR ID* number, *Formula*, and *Name* are consistent across all four Appendix A tables with each sorted by the *Formula* entry using standard Hill notation (Hill, 1900).

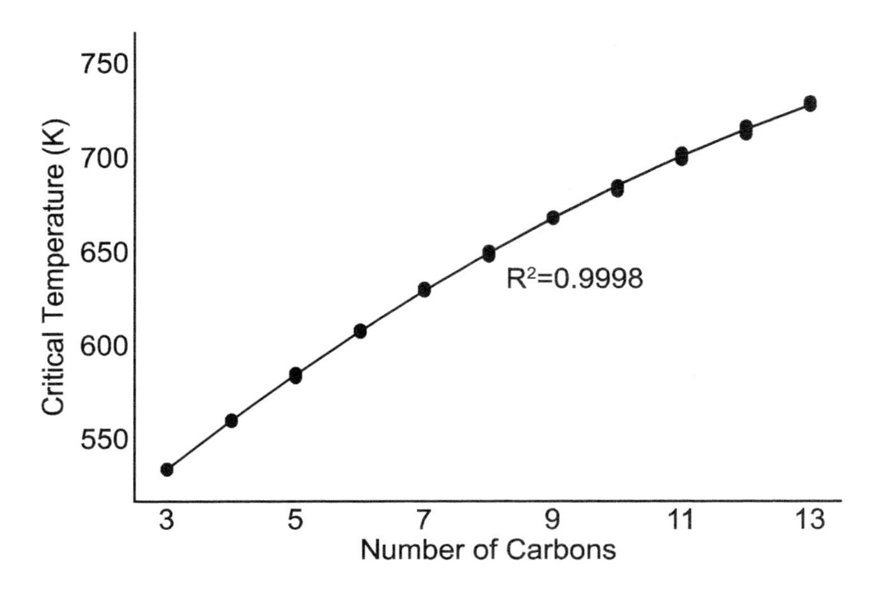

FIGURE A.1
Example of the "smoothed" data type using the critical temperatures of the *n*-alcohol family from C3 to C13. Points: 114 independent experimental data from various authors; Line: fit of the data to a quadratic equation.

Specifically, substances are listed by the total number of carbons, followed by the number of hydrogens and, further, by additional atoms in alphabetical order. The *Name* listed is that which is commonly found in industry. Each species' *CAS RN*® is also provided in Table A.1 for further compound identification.

Ideal gas isobaric heat capacities, C_p°, as a function of temperature are correlated to one of the following equations:

$$C_p^\circ(T) = A + BT + CT^2 + DT^3 + ET^4 \tag{A-1}$$

$$C_p^\circ(T) = A + B\left[\frac{\left(\frac{C}{T}\right)}{\sinh\left(\frac{C}{T}\right)}\right]^2 + D\left[\frac{\left(\frac{E}{T}\right)}{\cosh\left(\frac{E}{T}\right)}\right]^2 \tag{A-2}$$

$$C_p^\circ(T) = A + B\left[\frac{\left(\frac{C}{T}\right)^2 \exp\left(\frac{C}{T}\right)}{\left(\exp\left(\frac{C}{T}\right) - 1\right)^2}\right] + D\left[\frac{\left(\frac{E}{T}\right)^2 \exp\left(\frac{E}{T}\right)}{\left(\exp\left(\frac{E}{T}\right) - 1\right)^2}\right] + F\left[\frac{\left(\frac{G}{T}\right)^2 \exp\left(\frac{G}{T}\right)}{\left(\exp\left(\frac{G}{T}\right) - 1\right)^2}\right] \tag{A-3}$$

Here, T is temperature in K, $C_p^\circ(T)$ is in J/(kmol·K), and $(A, B, C, \cdots)$ are the regression coefficients listed in Table A.3. The equation to use for each compound is also listed in this table under the column heading C_p° Eq. Blank entries for any regression coefficient indicate a value of 0 or that the data were regressed with fewer parameters than the total possible for the indicated equation.

Liquid vapor pressures, P^{vp}, as a function of temperature are correlated to the modified Riedel equation given below (see Chap. 5).

$$P^{vp} = \exp\left[A + \frac{B}{T} + C\ln T + DT^E\right] \tag{A-4}$$

The regression coefficients in this equation (A, B, C, D, E), found in Table A.4, require temperature T in K, and return P^{vp} in Pa. Two temperature ranges are also found in this table. The first, denoted by the column headings Eq. T_{min} and Eq. T_{max}, identify the temperature range over which the correlation may be reliably used. This range for most compounds is from melting temperature T_m to critical temperature T_c. The second temperature range found in the table, delineated by column headings Data T_{min} and Data T_{max}, identify the limits of the experimental data used when regressing the correlating equation. This range is usually smaller than that over which the correlation is deemed accurate. As described in Sec. 5.6, vapor pressure correlations may be reliably extrapolated beyond the limits of the experimental data, and significant evaluation has gone into creating the vapor pressure correlation listed in Table A.4. This includes considering available critical temperature, critical pressure, liquid density, enthalpy of vaporization, liquid heat capacity, and sublimation pressure data as well as ensuring the correlation adheres to expected vapor pressure trends in the process described in Sec. 5.6. For most used cases, Eq. T_{min} and Eq. T_{max} constitute the range of interest and describe the temperature limits over which the correlation is expected to be reliable. The range quantified by Data T_{min} and Data T_{max} is provided as additional information. The error ("Er") listed in this table is for the *equation* range and accounts for the fact that some extrapolation has been done outside the range of the data where applicable.

The standard state Gibbs energy and enthalpy of formation, Table A.2, are for the species in the ideal gas state at 298.15 K and 1 bar. The reference states for the elements are as follows:

- Ideal gases at 298.15 K and 1 bar: Ar, Cl_2, D_2, F_2, He, H_2, Kr, Ne, O_2, Rn, T_2, Xe.
- Crystalline solid at 298.15 K and 1 bar: Al, As, B, C, I_2, P, S, Se, Si, Ti, U
- Saturated liquid at 298.15 K: Br_2, Hg

The column heading symbols for each property, their location in the subsequent tables, and an explanation are given below.

Table	Column Heading	Explanation
All	DIPPR ID	DIPPR ID number for the compound in this database
All	Formula	listing of chemical formula following Hill notation
All	Name	common name
All	Er	error (see above) in the property in the preceding column or for the given temperature-dependent correlation
A.1	CAS RN®	CAS Registry Number
A.1	MW	molecular weight, g/mol; IUPAC 2009 (Wieser and Copen, 2011)
A.1	T_m	normal (P = 101325 Pa) melting temperature, K
A.1	T_b	normal (P = 101325 Pa) boiling point, K
A.1	T_c	critical temperature, K
A.1	P_c	critical pressure, kPa
A.1	V_c	critical volume, cm^3/mol
A.1	ω	Pitzer acentric factor (calculated from Eq. (3.3-1) from the properties listed in this databank)
A.2	$\Delta_f H^\circ(298)$	ideal gas standard (P = 1 bar) state enthalpy of formation at 298.15 K (see Chap. 4), kJ/mol
A.2	$\Delta_f G^\circ(298)$	ideal gas standard (P = 1 bar) state Gibbs energy of formation at 298.15 K (see Chap. 4), kJ/mol
A.2	$\Delta_{vap} H(T_b)$	enthalpy change of boiling (vaporization) at the normal (P = 101325 Pa) boiling point T_b, kJ/mol
A.2	$\Delta_m H(T_m)$	enthalpy change of melting at normal (P = 101325 Pa) melting temperature T_m, kJ/mol
A.2	V_L	liquid molar volume, cm^3/mol, at the conditions indicated below depending on the standard state of the compound • $T_c > 298.15$ K and $T_m < 298.15$ K: T = 298.15 K, P = max[P^{vp}(298.15 K), 101325 Pa] • $T_c < 298.15$ K: T = T_b, P = 101325 Pa • TPT > 298.15 K: T = T_m, P = $P^{vp}(T_m)$
A.2	μ	molecular dipole moment, Debye
A.3	C_p° Eq	equation number for calculating ideal gas isobaric heat capacity, J/(kmol·K); see Eqs. (A-1) to (A-3)
A.3	T_{min}	minimum temperature for which the equation for ideal gas isobaric heat capacity should be used, K
A.3	T_{max}	maximum temperature for which the equation for ideal gas isobaric heat capacity should be used, K
A.3	A, B, C, D, E, F, G	parameters for equation indicated by C_p° Eq to calculate ideal gas isobaric heat capacity in units of J/(kmol·K)
A.3	$C_p^{liq}(298)$	value of the isobaric heat capacity of the liquid at 298.15 K, J/(mol·K)
A.4	Eq. T_{min}	minimum temperature for which the correlation for liquid vapor pressure may be used reliably, K

(*Continued*)

Table	Column Heading	Explanation
A.4	Eq. T_{max}	maximum temperature for which the correlation for liquid vapor pressure may be used reliably, K
A.4	Data T_{min}	minimum temperature of the experimental data sets used when regressing the vapor pressure correlation, K
A.4	Data T_{max}	maximum temperature of the experimental data sets used when regressing the vapor pressure correlation, K
A.4	A, B, C, D, E	parameters for the modified Riedel equation [Eq. (A-4)] to calculate liquid vapor pressure in units of Pa

TABLE A.1 Basic Constants 1

DIPPR ID	Formula	Name	CAS RN®	*MW* (g/mol)	T_m (K)	Er (%)	T_b (K)	Er (%)	T_c (K)	Er (%)	P_c (kPa)	Er (%)	V_c (cm^3/mol)	Er (%)	ω
914	Ar	argon	7440-37-1	39.948	83.78	1	87.28	0.2	150.86	3	4898	3	74.59	5	**−0.004**
922	Br_2	bromine	7726-95-6	159.808	265.90	1	331.90	1	584.15	3	10300	10	*135.00*	10	**0.129**
1906	BrH	hydrogen bromide	10035-10-6	80.912	186.34	1	206.45	1	363.15	3	8552	5	100.00	10	**0.073**
2686	$CBrClF_2$	bromochlorodifluoromethane	353-59-3	165.365	113.65	1	269.14	1	426.15	3	4254	5	246.00	10	**0.187**
2687	$CBrF_3$	bromotrifluoromethane	75-63-8	148.910	105.15	3	215.26	1	340.15	1	3970	3	200.00	5	**0.170**
2688	CBr_2F_2	dibromodifluoromethane	75-61-6	209.816	163.05	3	295.94	3	*478.00*	5	*4070*	10	*249.00*	25	**0.108**
1606	$CClF_3$	chlorotrifluoromethane	75-72-9	104.459	92.15	1	191.74	1	302.00	1	3870	3	180.28	5	**0.172**
1601	CCl_2F_2	dichlorodifluoromethane	75-71-8	120.914	115.15	1	243.36	3	384.95	1	4125	3	217.00	5	**0.180**
1894	CCl_2O	phosgene	75-44-5	98.916	145.37	3	280.71	3	455.00	3	5674	5	190.00	10	**0.201**
1602	CCl_3F	trichlorofluoromethane	75-69-4	137.368	162.04	1	296.97	3	471.20	1	4408	3	248.00	5	**0.189**
1501	CCl_4	carbon tetrachloride	56-23-5	153.823	250.33	1	349.79	1	556.35	1	4560	3	276.00	5	**0.193**
1616	CF_4	carbon tetrafluoride	75-73-0	88.004	89.40	0.2	145.10	1	227.51	3	3745	3	*143.00*	25	**0.179**
1634	$CHBrF_2$	bromodifluoromethane	1511-62-2	130.919	128.00	3	257.67	1	411.98	3	*5132*	10	167.00	10	**0.205**
1604	$CHClF_2$	chlorodifluoromethane	75-45-6	86.468	115.73	3	232.32	1	369.30	1	4971	3	166.00	5	**0.219**
1696	$CHCl_2F$	dichlorofluoromethane	75-43-4	102.923	138.15	1	282.05	1	451.58	1	5184	3	196.00	5	**0.205**
1521	$CHCl_3$	chloroform	67-66-3	119.378	209.63	1	334.33	1	536.40	1	5472	5	239.00	5	**0.222**
1615	CHF_3	trifluoromethane	75-46-7	70.014	117.97	1	191.09	1	299.01	1	4816	3	132.00	5	**0.264**
1771	CHN	hydrogen cyanide	74-90-8	27.025	259.86	0.2	298.80	0.2	457.00	1	5400	3	135.00	5	**0.404**
1511	CH_2Cl_2	dichloromethane	75-09-2	84.933	178.01	1	312.90	3	510.00	1	6080	3	*185.00*	5	**0.199**
1614	CH_2F_2	difluoromethane	75-10-5	52.023	137.00	3	221.50	1	351.25	0.2	5784	0.2	123.00	3	**0.277**
1001	CH_2O	formaldehyde	50-00-0	30.026	155.15	1	253.85	1	420.00	1	6590	3	*85.10*	25	**0.168**
1251	CH_2O_2	formic acid	64-18-6	46.026	281.55	1	373.71	1	588.00	1	*5810*	10	125.00	25	**0.313**
1641	CH_3Br	bromomethane	74-83-9	94.939	179.44	0.2	276.66	1	464.00	5	*6929*	3	*152.00*	25	**0.153**
1502	CH_3Cl	methyl chloride	74-87-3	50.487	175.43	1	248.95	1	416.25	3	6680	3	*141.00*	5	**0.151**
1613	CH_3F	methyl fluoride	593-53-3	34.033	131.35	1	194.82	1	317.42	1	5875	3	113.00	5	**0.195**
1681	CH_3I	methyl iodide	74-88-4	141.939	206.70	5	315.60	5	528.20	5	6600	5	175.00	3	**0.157**
2851	CH_3NO	formamide	75-12-7	45.041	275.70	3	*493.00*	5	*771.00*	5	*7800*	3	*163.00*	10	**0.412**
1760	CH_3NO_2	nitromethane	75-52-5	61.040	244.60	1	374.35	1	588.15	1	6310	3	173.00	5	**0.348**
1	CH_4	methane	74-82-8	16.043	*90.69*	0.2	111.67	0.2	190.56	0.2	4599	0.2	98.60	1	**0.012**
1101	CH_4O	methanol	67-56-1	32.042	175.47	1	337.85	1	512.50	0.2	8084	1	117.00	1	**0.566**
1801	CH_4S	methyl mercaptan	74-93-1	48.107	150.18	1	279.11	1	469.95	1	7230	3	145.00	5	**0.158**
1701	CH_5N	methylamine	74-89-5	31.057	179.69	1	266.82	1	430.80	0.2	7620	1	*154.00*	3	**0.290**
908	CO	carbon monoxide	630-08-0	28.010	68.15	1	81.70	1	132.92	1	3499	3	94.40	5	**0.048**
1893	COS	carbonyl sulfide	463-58-1	60.075	134.34	1	222.87	1	378.80	1	6349	3	135.10	5	**0.094**
909	CO_2	carbon dioxide	124-38-9	44.010	216.58	1			304.21	1	7383	3	94.00	5	**0.224**
1938	CS_2	carbon disulfide	75-15-0	76.141	161.11	0.2	319.65	0.2	552.49	1	7329	3	173.00	3	**0.084**

TABLE A.1 Basic Constants 1 (*Continued*)

DIPPR ID	Formula	Name	CAS RN®	MW (g/mol)	T_m (K)	Er (%)	T_b (K)	Er (%)	T_c (K)	Er (%)	P_c (kPa)	Er (%)	V_c (cm³/mol)	Er (%)	ω
2692	C_2ClF_5	chloropentafluoroethane	76-15-3	154.466	173.71	1	234.04	1	353.15	1	3157	3	252.00	5	**0.251**
1611	$C_2Br_2F_4$	1,2-dibromotetrafluoroethane	124-73-2	259.823	163.00	3	320.31	1	487.80	1	3393	3	341.00	5	**0.243**
1693	$C_2Cl_2F_4$	1,1-dichlorotetrafluoroethane	374-07-2	170.921	216.58	3	276.17	1	418.55	3	3300	5	294.00	10	**0.253**
1609	$C_2Cl_2F_4$	1,2-dichlorotetrafluoroethane	76-14-2	170.921	180.62	1	276.92	1	418.85	1	3260	3	294.00	5	**0.252**
2655	$C_2Cl_3F_3$	1,1,2-trichlorotrifluoroethane	76-13-1	187.376	236.92	1	320.75	1	487.25	1	3410	3	325.00	5	**0.252**
1542	C_2Cl_4	tetrachloroethylene	127-18-4	165.833	250.80	1	394.40	3	*620.00*	5	*4490*	10	*248.00*	10	**0.214**
1630	C_2F_4	tetrafluoroethylene	116-14-3	100.015	*142.00*	3	197.51	1	306.45	1	3944	3	172.00	5	**0.226**
2693	C_2F_6	hexafluoroethane	76-16-4	138.012	172.15	3	*194.95*	1	293.03	3	3043	3	*224.00*	10	**0.255**
2640	$C_2HBrClF_3$	halothane	151-67-7	197.382	157.40	1	323.35	1	*496.00*	5	*3920*	10	*296.00*	25	**0.275**
2648	C_2HClF_4	2-chloro-1,1,1,2-tetrafluoroethane	2837-89-0	136.476	*74.00*	10	261.05	1	395.65	1	3660	3	243.81	3	**0.288**
1694	$C_2HCl_2F_3$	2,2-dichloro-1,1,1-trifluoroethane	306-83-2	152.931	166*	3	301.05	1	456.94	1	3676	1	278.00	3	**0.282**
2647	$C_2HCl_2F_3$	1,2-dichloro-1,1,2-trifluoroethane	354-23-4	152.931	195.15	3	301.15	1	461.60	5	*3610*	10	*278.00*	25	**0.241**
1646	C_2HF_5	pentafluoroethane	354-33-6	120.021	170.15	1	225.04	1	339.17	0.2	3620	1	211.30	3	**0.305**
3605	C_2HF_5O	difluoromethyl trifluoromethyl ether	3822-68-2	136.021	116.00	1	237.90	0.2	354.49	0.2	*3376*	0.2	*212.00*	25	**0.325**
401	C_2H_2	acetylene	74-86-2	26.037	192.40	1			308.30	0.2	6138	0.2	112.00	3	**0.191**
1581	$C_2H_2Cl_2$	trans-1,2-dichloroethylene	156-60-5	96.943	223.35	1	320.85	1	516.50	3	5510	5	*224.00*	10	**0.223**
1529	$C_2H_2Cl_4$	1,1,2,2-tetrachloroethane	79-34-5	167.849	229.35	1	418.25	1	*645.00*	3	*4090*	5	*325.00*	10	**0.247**
1629	$C_2H_2F_2$	1,1-difluoroethylene	75-38-7	64.034	129.15	3	187.50	1	302.80	1	4460	3	154.00	5	**0.136**
2650	$C_2H_2F_4$	1,1,1,2-tetrafluoroethane	811-97-2	102.031	172.00	1	247.08	1	374.18	0.2	4056	1	198.80	3	**0.327**
2660	$C_2H_2F_4$	1,1,2,2-tetrafluoroethane	359-35-3	102.031	*172.00*	25	250.15	3	391.80	3	*4640*	10	*191.00*	25	**0.266**
1504	C_2H_3Cl	vinyl chloride	75-01-4	62.498	119.36	1	259.25	1	*432.00*	3	*5670*	5	*179.00*	10	**0.100**
2695	$C_2H_3ClF_2$	1-chloro-1,1-difluoroethane	75-68-3	100.495	142.35	1	263.95	1	410.29	1	4041	1	225.00	3	**0.231**
2649	$C_2H_3Cl_2F$	1,1-dichloro-1-fluoroethane	1717-00-6	116.950	169.65	3	305.15	1	478.85	1	4340	3	253.00	3	**0.221**
1527	$C_2H_3Cl_3$	1,1,1-trichloroethane	71-55-6	133.404	*243.10*	1	347.23	1	545.00	1	4300	3	*281.00*	10	**0.218**
1524	$C_2H_3Cl_3$	1,1,2-trichloroethane	79-00-5	133.404	236.50	1	387.00	1	*602.00*	3	*4480*	5	*281.00*	10	**0.259**
1619	$C_2H_3F_3$	1,1,1-trifluoroethane	420-46-2	84.040	*161.82*	1	225.81	1	345.88	0.2	3764	1	195.00	5	**0.261**
2619	$C_2H_3F_3$	1,1,2-trifluoroethane	430-66-0	84.040	189.15	1	278.15	1	*429.80*	5	*5241*	25	*179.00*	25	**0.326**
1772	C_2H_3N	acetonitrile	75-05-8	41.052	229.31	0.2	354.81	0.2	545.50	0.2	4850	1	*193.00*	25	**0.342**
201	C_2H_4	ethylene	74-85-1	28.053	104.00	1	169.41	1	282.34	0.2	5041	0.2	131.00	1	**0.086**
1673	$C_2H_4Br_2$	1,2-dibromoethane	106-93-4	187.861	282.94	1	404.51	1	650.15	3	5477	5	261.60	10	**0.207**
1522	$C_2H_4Cl_2$	1,1-dichloroethane	75-34-3	98.959	176.19	1	330.45	1	523.00	1	5070	3	240.00	5	**0.234**
1523	$C_2H_4Cl_2$	1,2-dichloroethane	107-06-2	98.959	237.49	1	356.59	1	561.60	0.2	5370	5	220.00	5	**0.287**
1640	$C_2H_4F_2$	1,1-difluoroethane	75-37-6	66.050	156.15	1	249.13	1	386.44	0.2	4520	1	179.00	5	**0.275**
1002	C_2H_4O	acetaldehyde	75-07-0	44.053	149.78	1	294.15	1	466.00	1	5570	3	*154.00*	25	**0.262**
1441	C_2H_4O	ethylene oxide	75-21-8	44.053	160.65	1	283.60	1	469.15	1	7190	3	140.30	5	**0.197**

(*Continued*)

TABLE A.1 Basic Constants 1 (*Continued*)

DIPPR ID	Formula	Name	CAS RN®	*MW* (g/mol)	T_m (K)	Er (%)	T_b (K)	Er (%)	T_c (K)	Er (%)	P_c (kPa)	Er (%)	V_c (cm³/mol)	Er (%)	ω
1252	$C_2H_4O_2$	acetic acid	64-19-7	60.052	289.81	1	391.05	1	591.95	1	5786	3	177.00	5	**0.467**
1301	$C_2H_4O_2$	methyl formate	107-31-3	60.052	174.15	1	304.90	1	487.20	1	6000	3	172.00	5	**0.256**
1645	C_2H_5Br	bromoethane	74-96-4	108.965	154.25	1	311.49	1	503.80	1	*5565*	3	*204.00*	25	**0.205**
1503	C_2H_5Cl	ethyl chloride	75-00-3	64.514	136.75	3	285.45	1	460.35	3	5270	3	*192.00*	5	**0.189**
1617	C_2H_5F	ethyl fluoride	353-36-6	48.059	129.95	1	235.45	1	375.31	1	5028	3	159.00	5	**0.218**
2	C_2H_6	ethane	74-84-0	30.069	90.35	0.2	184.57	0.2	305.32	0.2	4872	1	145.50	3	**0.099**
1102	C_2H_6O	ethanol	64-17-5	46.068	159.05	1	351.44	1	514.00	0.2	6137	1	168.00	1	**0.644**
1401	C_2H_6O	dimethyl ether	115-10-6	46.068	131.66	1	248.31	1	400.10	1	5370	3	170.00	5	**0.200**
1201	$C_2H_6O_2$	ethylene glycol	107-21-1	62.068	260.15	1	470.38	0.2	719.00	0.2	8200	1	*187.00*	1	**0.521**
1802	C_2H_6S	ethyl mercaptan	75-08-1	62.134	125.26	1	308.15	1	499.15	3	5490	5	207.00	10	**0.188**
1820	C_2H_6S	dimethyl sulfide	75-18-3	62.134	174.88	1	310.48	1	503.04	1	5530	3	201.00	5	**0.194**
1704	C_2H_7N	ethylamine	75-04-7	45.084	192.15	1	289.73	1	456.15	1	5620	3	*206.00*	3	**0.288**
1702	C_2H_7N	dimethylamine	124-40-3	45.084	180.96	1	280.03	3	437.20	1	*5340*	3	*180.00*	5	**0.300**
1723	C_2H_7NO	monoethanolamine	141-43-5	61.083	283.65	1	443.15	1	678.20	5	*7124*	10	*225.00*	25	**0.447**
1741	$C_2H_8N_2$	ethylenediamine	107-15-3	60.098	284.29	0.2	390.41	1	613.10	0.2	6707	0.2	*202.00*	3	**0.343**
2652	C_3F_8	octafluoropropane	76-19-7	188.019	*125.46*	3	236.45	1	345.05	1	2680	3	299.00	5	**0.328**
2630	C_3HF_7	1,1,1,2,3,3,3-heptafluoropropane	431-89-0	170.029	142.00	1	256.79	1	374.83	3	*2912*	5	274.00	10	**0.355**
2631	$C_3H_2F_6$	1,1,1,2,3,3-hexafluoropropane	431-63-0	152.038	*170**	5	279.40	1	412.38	1	3412	3	*270.00*	3	**0.369**
1620	$C_3H_2F_6$	1,1,1,3,3,3-hexafluoropropane	690-39-1	152.038	179.75	3	272.45	3	398.07	3	3219	3	*266.00*	25	**0.390**
2629	$C_3H_3F_5$	1,1,1,2,2-pentafluoropropane	1814-88-6	134.048	*134.00*	10	255.50	3	380.11	3	3137	5	273.20	10	**0.305**
2627	$C_3H_3F_5$	1,1,1,3,3-pentafluoropropane	460-73-1	134.048	*		288.45	3	427.20	0.2	*3640*		259.00	3	**0.375**
2628	$C_3H_3F_5$	1,1,2,2,3-pentafluoropropane	679-86-7	134.048	191.15*	5	298.40	1	447.57	0.2	*3925*	1	256.00	3	**0.353**
2868	$C_3H_3F_5O$	2-(difluoromethoxy)-1,1,1-trifluoroethane	1885-48-9	150.047	*150*	10	302.15	3	*444.03*	1	*3420*	3	*289.00*	25	**0.389**
1774	C_3H_3N	acrylonitrile	107-13-1	53.063	189.63	1	350.45	1	540.00	1	4660	1	*216.00*	25	**0.311**
5892	C_3H_3NO	isoxazole	288-14-2	69.062	206.05	3	367.90	1	590.00	5	6100	5	190.94	5	**0.257**
402	C_3H_4	methylacetylene	74-99-7	40.064	170.45	1	249.94	1	402.40	0.2	5630	1	164.00	3	**0.212**
301	C_3H_4	propadiene	463-49-0	40.064	136.87	1	238.65	1	394.00	1	5250	3	*165.00*	25	**0.104**
1773	C_3H_5N	propionitrile	107-12-0	55.078	180.37	0.2	370.25	0.2	561.30	0.2	4260	3	*242.00*	25	**0.350**
202	C_3H_6	propylene	115-07-1	42.080	87.9	1	225.45	0.2	364.85	0.2	4600	1	185.00	3	**0.138**
101	C_3H_6	cyclopropane	75-19-4	42.080	145.59	1	240.37	1	398.00	0.2	5540	1	162.00	3	**0.128**
1526	$C_3H_6Cl_2$	1,2-dichloropropane	78-87-5	112.986	172.71	0.2	369.52	0.2	578.00	0.2	***4650***	3	289.71	1	**0.258**
1167	C_3H_6O	allyl alcohol	107-18-6	58.079	144.15	1	370.23	1	545.10	1	*5620*	25	*208.00*	25	**0.570**
1051	C_3H_6O	acetone	67-64-1	58.079	178.45	1	329.28	0.2	508.10	0.2	4700	1	213.00	3	**0.307**
1003	C_3H_6O	propanal	123-38-6	58.079	165*	1	322.15	1	503.60	3	5038	3	*204.00*	25	**0.281**
1442	C_3H_6O	1,2-propylene oxide	75-56-9	58.079	161.22	1	307.65	1	482.25	1	4920	5	186.00	5	**0.268**
1253	$C_3H_6O_2$	propionic acid	79-09-4	74.079	252.45	1	414.32	1	600.81	1	4668	1	235.00	5	**0.580**

TABLE A.1 Basic Constants 1 (*Continued*)

DIPPR ID	Formula	Name	CAS RN®	*MW* (g/mol)	T_m (K)	Er (%)	T_b (K)	Er (%)	T_c (K)	Er (%)	P_c (kPa)	Er (%)	V_c (cm^3/mol)	Er (%)	ω
1312	$C_3H_6O_2$	methyl acetate	79-20-9	74.079	175.15	1	330.09	1	506.55	1	4750	3	228.00	5	**0.331**
1302	$C_3H_6O_2$	ethyl formate	109-94-4	74.079	193.55	1	327.46	1	508.40	1	4740	3	229.00	5	**0.285**
2391	$C_3H_6O_3$	dimethyl carbonate	616-38-6	90.078	273.15*	3	363.40	1	*548.00*	5	*4500*	10	*251.00*	25	**0.385**
1585	C_3H_7Cl	propyl chloride	540-54-5	78.541	150.35	1	319.67	1	503.15	3	*4425*	10	*243.00*	5	**0.215**
1876	C_3H_7NO	*N,N*-dimethylformamide	68-12-2	73.094	212.72	1	425.15	5	649.60	0.2	*4420*	25	261.99	10	**0.318**
3	C_3H_8	propane	74-98-6	44.096	85.47	0.2	231.04	0.2	369.83	0.2	4248	1	200.00	3	**0.152**
1103	C_3H_8O	1-propanol	71-23-8	60.095	146.95	0.2	370.35	0.2	536.80	0.2	5169	1	219.00	3	**0.621**
1104	C_3H_8O	isopropanol	67-63-0	60.095	185.26	1	355.30	1	508.30	0.2	4765	1	222.00	3	**0.663**
1407	C_3H_8O	methyl ethyl ether	540-67-0	60.095	*160*	25	280.50	1	437.80	1	4400	3	221.00	5	**0.231**
1431	$C_3H_8O_2$	methylal	109-87-5	76.094	168.35	1	315.00	1	480.6*	3	3950	10	213.00	25	**0.286**
1212	$C_3H_8O_2$	1,3-propylene glycol	504-63-2	76.094	245.5	1	487.37	1	718.20	1	6550	3	*239.00*	1	**0.624**
1231	$C_3H_8O_3$	glycerol	56-81-5	92.094	291.33	1	*561.00*	1	850.00	3	7500	5	*264.00*	10	**0.513**
1813	C_3H_8S	methyl ethyl sulfide	624-89-5	76.161	167.23	0.2	339.80	1	533.00	3	4260	5	*254.00*	25	**0.209**
1711	C_3H_9N	*n*-propylamine	107-10-8	59.110	190.15	1	321.70	1	496.95	1	4740	3	*258.00*	5	**0.310**
1719	C_3H_9N	isopropylamine	75-31-0	59.110	177.95	1	304.92	1	471.85	1	4540	3	221.00	10	**0.276**
1703	C_3H_9N	trimethylamine	75-50-3	59.110	156.08	1	276.02	3	433.25	1	4070	3	254.00	5	**0.206**
6862	C_3H_9NO	methylethanolamine	109-83-1	75.110	268.65	1	431.15		*630.00*	10	*5220*	10	*253.00*	25	**0.585**
2654	C_4F_8	octafluorocyclobutane	115-25-3	200.030	*232.96*	1	267.17	1	388.37	1	2778	3	324.80	5	**0.356**
1478	C_4H_4O	furan	110-00-9	68.074	187.55	0.2	304.50	3	490.15	1	5500	3	218.00	5	**0.202**
1821	C_4H_4S	thiophene	110-02-1	84.140	234.94	1	357.31	1	579.35	1	5690	3	219.00	5	**0.197**
1721	C_4H_5N	pyrrole	109-97-7	67.089	249.74	3	403.00	5	639.75	1	*6530*	5	*230.00*	10	**0.298**
403	C_4H_6	ethylacetylene	107-00-6	54.090	147.43	0.2	281.22	1	440.00	1	4600	5	208.00	25	**0.247**
302	C_4H_6	1,2-butadiene	590-19-2	54.090	136.95	1	284.00	1	*452.00*	5	*4360*	10	*220.00*	25	**0.166**
303	C_4H_6	1,3-butadiene	106-99-0	54.090	164.25	1	268.74	1	425.00	1	4320	3	221.00	5	**0.195**
1321	$C_4H_6O_2$	vinyl acetate	108-05-4	86.089	180.35	3	345.65	5	519.13	3	*3958*	5	*270.00*	10	**0.351**
1291	$C_4H_6O_3$	acetic anhydride	108-24-7	102.089	200.15	0.2	412.70	1	606.00	3	4000	5	*304.00*	25	**0.455**
1782	C_4H_7N	butyronitrile	109-74-0	69.105	161.3	0.2	390.74	0.2	585.40	1	3880	3	*291.00*	25	**0.360**
102	C_4H_8	cyclobutane	287-23-0	56.106	*182.48*	1	285.66	1	*459.93*	1	*4980*	1	*210.00*	5	**0.185**
204	C_4H_8	1-butene	106-98-9	56.106	87.8	1	266.91	1	419.50	0.2	4020	1	241.00	3	**0.184**
206	C_4H_8	trans-2-butene	624-64-6	56.106	167.62	1	274.03	1	428.60	0.2	4100	1	238.00		**0.218**
205	C_4H_8	cis-2-butene	590-18-1	56.106	*134.26*	0.2	276.87	1	435.50	0.2	4210	3	234.00	5	**0.202**
207	C_4H_8	isobutene	115-11-7	56.106	132.81	1	266.25	1	417.90	0.2	4000	1	239.00	3	**0.195**
1005	C_4H_8O	butanal	123-72-8	72.106	176.8	1	347.94	1	537.20	3	4410	3	*258.00*	25	**0.283**
1052	C_4H_8O	methyl ethyl ketone	78-93-3	72.106	186.46	1	352.73	0.2	536.70	0.2	4207	1	267.00	3	**0.324**
1479	C_4H_8O	tetrahydrofuran	109-99-9	72.106	164.65	1	339.12	1	540.15	1	5190	3	224.00	5	**0.225**

(*Continued*)

TABLE A.1 Basic Constants 1 (*Continued*)

DIPPR ID	Formula	Name	CAS RN®	*MW* (g/mol)	T_m (K)	Er (%)	T_b (K)	Er (%)	T_c (K)	Er (%)	P_c (kPa)	Er (%)	V_c (cm³/mol)	Er (%)	ω
1256	$C_4H_8O_2$	*n*-butyric acid	107-92-6	88.105	267.95	1	436.42	1	615.70	1	4060	1	293.00	3	**0.675**
1260	$C_4H_8O_2$	isobutyric acid	79-31-2	88.105	227.15	1	427.65	1	605.00	0.2	3700	1	292.00	5	**0.614**
2422	$C_4H_8O_2$	1,3-dioxane	505-22-6	88.105	228.15	3	378.15	1	*590.00*	10	*5150*	10	*239.00*	25	**0.289**
1421	$C_4H_8O_2$	1,4-dioxane	123-91-1	88.105	284.95	1	374.47	1	587.00	1	5208	3	238.00	5	**0.279**
1322	$C_4H_8O_2$	methyl propionate	554-12-1	88.105	185.65	1	352.60	1	530.60	1	4004	3	282.00	5	**0.347**
1313	$C_4H_8O_2$	ethyl acetate	141-78-6	88.105	189.6	1	350.21	1	523.30	1	3880	3	286.00	5	**0.366**
1303	$C_4H_8O_2$	*n*-propyl formate	110-74-7	88.105	180.25	1	353.97	1	538.00	1	*4020*	3	285.00	5	**0.309**
1587	C_4H_9Cl	sec-butyl chloride	78-86-4	92.567	141.85	1	341.25	1	520.60	1	*3900*	5	*300.00*	10	**0.291**
1766	C_4H_9N	pyrrolidine	123-75-1	71.121	215.31	0.2	359.72	1	568.55	1	5610	3	249.00	5	**0.267**
5	C_4H_{10}	*n*-butane	106-97-8	58.122	*134.86*	0.2	272.66	0.2	425.12	0.2	3796	1	255.00	3	**0.200**
4	C_4H_{10}	isobutane	75-28-5	58.122	113.54	1	261.43	1	407.80	0.2	3640	3	259.00	3	**0.184**
2752	$C_4H_{10}N_2$	piperazine	110-85-0	86.136	384.6*	3	421.72	0.2	656.3*	1	5420	5	267.51	3	**0.313**
1105	$C_4H_{10}O$	1-butanol	71-36-3	74.122	183.85	0.2	391.90	0.2	563.10	0.2	4414	1	273.00	3	**0.588**
1106	$C_4H_{10}O$	2-methyl-1-propanol	78-83-1	74.122	165.15*	3	380.81	1	547.80	0.2	4295	1	274.00	3	**0.586**
1108	$C_4H_{10}O$	2-methyl-2-propanol	75-65-0	74.122	*298.97*	1	355.57	1	506.20	0.2	3972	1	275.00	3	**0.615**
1107	$C_4H_{10}O$	2-butanol	78-92-2	74.122	*158.45**	1	372.90	1	535.90	0.2	4188	1	270.00	3	**0.581**
1402	$C_4H_{10}O$	diethyl ether	60-29-7	74.122	156.85	1	307.58	1	466.70	1	3640	3	280.00	5	**0.281**
1455	$C_4H_{10}O_2$	1,2-dimethoxyethane	110-71-4	90.121	215.15	1	357.75	1	536.15	1	3871	3	270.64	25	**0.347**
1220	$C_4H_{10}O_2$	1,2-butanediol	584-03-2	90.121	*220*	5	469.57	1	680.00	3	*5210**	10	303.00	10	**0.630**
1221	$C_4H_{10}O_2$	1,3-butanediol	107-88-0	90.121	196.15	3	481.38	1	676.00	3	4020	5	305.00	5	**0.704**
1202	$C_4H_{10}O_3$	diethylene glycol	111-46-6	106.120	264.15	1	518.15	0.2	753.00	0.2	4770	5	*327.00*	3	**0.573**
1818	$C_4H_{10}S$	diethyl sulfide	352-93-2	90.187	169.2	1	365.25	1	557.15	1	3960	3	318.00	5	**0.290**
1712	$C_4H_{11}N$	*n*-butylamine	109-73-9	73.137	224.05	1	351.20	1	531.90	1	4150	3	*309.00*	5	**0.336**
1710	$C_4H_{11}N$	diethylamine	109-89-7	73.137	223.35	1	328.60	1	496.60	1	3710	3	301.00	5	**0.304**
1714	$C_4H_{11}N$	isobutylamine	78-81-9	73.137	188.55	1	340.88	1	*513.73*	3	*4215*	5	*312.60*	10	**0.363**
1727	$C_4H_{11}N$	*tert*-butylamine	75-64-9	73.137	206.19	1	317.55	3	483.90	3	3840	5	293.00	10	**0.275**
1724	$C_4H_{11}NO_2$	diethanolamine	111-42-2	105.136	301.15	1	541.54	3	736.60	5	*4270*	10	*349.00*	10	**0.953**
2623	C_5F_{12}	perfluoro-*n*-pentane	678-26-2	288.034	147.51	1	302.40	0.2	423.10	1	2047	3	*461.00*	25	**0.402**
1889	$C_5H_4O_2$	furfural	98-01-1	96.084	236.65	3	434.85	1	670.15	3	5660	5	*252.00*	10	**0.368**
1791	C_5H_5N	pyridine	110-86-1	79.100	231.53	1	388.41	1	619.95	1	5630	3	254.00	3	**0.239**
405	C_5H_8	1-pentyne	627-19-0	68.117	167.45	1	313.33	1	*481.20*	3	*4170*	10	*277.00*	25	**0.290**
269	C_5H_8	cyclopentene	142-29-0	68.117	*138.13*	0.2	317.38	1	507.00	0.2	4800	3	245.00	5	**0.196**
1079	C_5H_8O	cyclopentanone	120-92-3	84.116	221.85	3	403.80	1	624.50	3	4600	10	*258.00*	25	**0.288**
1783	C_5H_9N	valeronitrile	110-59-8	83.132	176.92	0.2	414.71	0.2	610.30	0.2	3580	3	*342.00*	25	**0.407**
1071	C_5H_9NO	*N*-methyl-2-pyrrolidone	872-50-4	99.131	249.15	1	477.42	1	721.60	0.2	4520	1	310.00	3	**0.373**
104	C_5H_{10}	cyclopentane	287-92-3	70.133	179.31	1	322.40	1	511.70	0.2	4510	1	260.00	3	**0.195**

TABLE A.1 Basic Constants 1 (*Continued*)

DIPPR ID	Formula	Name	CAS RN®	*MW* (g/mol)	T_m (K)	Er (%)	T_b (K)	Er (%)	T_c (K)	Er (%)	P_c (kPa)	Er (%)	V_c (cm^3/mol)	Er (%)	ω
209	C_5H_{10}	1-pentene	109-67-1	70.133	*108.02*	1	303.22	1	464.80	0.2	3560	3	293.40	3	**0.237**
210	C_5H_{10}	*cis*-2-pentene	627-20-3	70.133	121.75	1	310.08	1	475.00	1	3640	3	292.00	5	**0.245**
211	C_5H_{10}	*trans*-2-pentene	646-04-8	70.133	132.89	1	309.49	1	474.20	1	3660	1	293.00	5	**0.248**
212	C_5H_{10}	2-methyl-1-butene	563-46-2	70.133	135.58	1	304.31	1	465.00	1	3447	3	292.00	1	**0.234**
214	C_5H_{10}	2-methyl-2-butene	513-35-9	70.133	139.39	1	311.70	1	470.00	1	3420	3	292.00	3	**0.287**
213	C_5H_{10}	3-methyl-1-butene	563-45-1	70.133	104.66	1	293.20	1	452.70	0.2	3530	1	305.00	5	**0.210**
3112	$C_5H_{10}O$	cyclopentanol	96-41-3	86.132	255.6	1	413.95	0.2	619.50	0.2	4900	3	288.00	5	**0.427**
1060	$C_5H_{10}O$	2-pentanone	107-87-9	86.132	196.27	1	375.41	0.2	561.19	0.2	3672	1	321.00	3	**0.345**
1053	$C_5H_{10}O$	3-pentanone	96-22-0	86.132	234.18	1	375.14	1	560.95	1	3740	3	336.00	5	**0.345**
1061	$C_5H_{10}O$	methyl isopropyl ketone	563-80-4	86.132	181.15	1	367.48	0.2	552.80	0.2	3800	3	310.00	3	**0.333**
1400	$C_5H_{10}O$	2-methyltetrahydrofuran	96-47-9	86.132	137.15	1	352.94	0.2	537.00	1	3758	3	267.01	10	**0.282**
1258	$C_5H_{10}O_2$	*n*-pentanoic acid	109-52-4	102.132	239.15	1	458.95	1	639.16	1	3630	1	350.00	3	**0.707**
1261	$C_5H_{10}O_2$	isovaleric acid	503-74-2	102.132	243.85	1	449.68	1	629.09	1	3688	3	311.60	3	**0.682**
1332	$C_5H_{10}O_2$	methyl *n*-butyrate	623-42-7	102.132	187.35	1	375.90	1	554.50	1	3473	3	340.00	5	**0.378**
1323	$C_5H_{10}O_2$	ethyl propionate	105-37-3	102.132	199.25	1	372.25	1	546.00	1	3362	3	345.00	5	**0.394**
2332	$C_5H_{10}O_2$	methyl isobutyrate	547-63-7	102.132	188.45	1	365.45	1	540.70	1	3432	3	339.00	5	**0.364**
1314	$C_5H_{10}O_2$	*n*-propyl acetate	109-60-4	102.132	178.15	1	374.65	1	549.73	1	3360	3	345.00	5	**0.389**
1305	$C_5H_{10}O_2$	isobutyl formate	542-55-2	102.132	177.35	1	371.22	1	551.35	1	3881	5	352.00	10	**0.390**
1319	$C_5H_{10}O_2$	isopropyl acetate	108-21-4	102.132	199.75	1	361.65	1	532.00	1	3290	1	*336.00*	5	**0.368**
1588	$C_5H_{11}Cl$	1-chloropentane	543-59-9	106.594	174.15	1	380.65	0.2	571.20	0.2	*3352*	10	*356.00*	10	**0.295**
7	C_5H_{12}	*n*-pentane	109-66-0	72.149	143.42	0.2	309.21	0.2	469.70	0.2	3370	1	313.00	3	**0.252**
8	C_5H_{12}	isopentane	78-78-4	72.149	113.25	1	300.99	1	460.40	0.2	3380	3	306.00	1	**0.228**
9	C_5H_{12}	neopentane	463-82-1	72.149	256.6	1	282.65	1	433.80	0.2	3196	1	307.00	3	**0.196**
1109	$C_5H_{12}O$	1-pentanol	71-41-0	88.148	*195.56*	0.2	410.90	0.2	588.10	0.2	3897	1	326.00	3	**0.575**
1110	$C_5H_{12}O$	2-pentanol	6032-29-7	88.148	200	3	392.20	1	561.00	0.2	3700	1	326.00	10	**0.555**
1112	$C_5H_{12}O$	2-methyl-1-butanol	137-32-6	88.148	*195*	10	401.85	0.2	575.40	0.2	3940	1	*329.00*	25	**0.588**
1111	$C_5H_{12}O$	2-methyl-2-butanol	75-85-4	88.148	264.15	3	375.20	1	543.70	0.2	3710	1	*324.00*	25	**0.482**
1123	$C_5H_{12}O$	3-methyl-1-butanol	123-51-3	88.148	155.95	3	404.15	0.2	577.20	0.2	3930	1	*329.00*	25	**0.590**
1124	$C_5H_{12}O$	3-methyl-2-butanol	598-75-4	88.148			383.88	1	556.10	0.2	3870	1	*327.00*	25	**0.502**
1415	$C_5H_{12}O$	ethyl propyl ether	628-32-0	88.148	*145.65*	1	337.01	1	500.23		3370	1	339.00	5	**0.347**
1405	$C_5H_{12}O$	methyl *tert*-butyl ether	1634-04-4	88.148	164.55	1	328.20	0.2	497.10	0.2	*3286*	25	*329.00*	25	**0.247**
1120	$C_5H_{12}O$	3-pentanol	584-02-1	88.148	204.15	3	388.45	1	559.60	0.2	*3714*	10	325.00	10	**0.514**
1864	C_6F_6	hexafluorobenzene	392-56-3	186.055	278.25	3	353.41	3	516.73	1	3273	3	335.00	5	**0.395**
1623	C_6F_{14}	perfluoro-*n*-hexane	355-42-0	338.042	185	1	330.30	0.2	449.55	1	*1802*	5	*563.00*	5	**0.497**
1573	$C_6H_4Cl_2$	*m*-dichlorobenzene	541-73-1	147.002	248.39	1	446.23	1	683.95	3	*4070*	10	*351.00*	10	**0.279**

(*Continued*)

TABLE A.1 Basic Constants 1 (*Continued*)

DIPPR ID	Formula	Name	CAS RN®	*MW* (g/mol)	T_m (K)	Er (%)	T_b (K)	Er (%)	T_c (K)	Er (%)	P_c (kPa)	Er (%)	V_c (cm^3/mol)	Er (%)	ω
1572	$C_6H_4Cl_2$	*o*-dichlorobenzene	95-50-1	147.002	256.15	1	453.57	1	*705.00*	3	*4070*	5	*351.00*	10	**0.219**
1571	C_6H_5Cl	monochlorobenzene	108-90-7	112.557	227.95	1	404.87	5	632.35	1	4519	5	308.00	5	**0.250**
1691	C_6H_5I	iodobenzene	591-50-4	204.008	241.83	1	461.60	1	721.15	1	4519	3	351.00	5	**0.247**
1886	$C_6H_5NO_2$	nitrobenzene	98-95-3	123.109	278.91	3	483.95	5	*719.00*	3	*4400*	5	*349.00*	10	**0.449**
501	C_6H_6	benzene	71-43-2	78.112	278.68	1	353.24	1	562.05	0.2	4895	0.2	256.00	3	**0.210**
1181	C_6H_6O	phenol	108-95-2	94.111	*314.06*	1	454.99	1	694.25	1	6130	3	229.00	5	**0.443**
1792	C_6H_7N	aniline	62-53-3	93.126	267.12	0.2	457.25	0.2	705.00	1	5630	1	293.00	5	**0.365**
1797	C_6H_7N	2-methylpyridine	109-06-8	93.126	*206.44*	1	402.55	1	621.00	1	4600	3	335.00	10	**0.299**
2797	C_6H_7N	3-methylpyridine	108-99-6	93.126	255.01	1	417.29	1	645.00	1	*4660*	10	*320.00*	25	**0.298**
2799	C_6H_7N	4-methylpyridine	108-89-4	93.126	276.8	1	418.50	1	646.15	3	4660	3	326.00	5	**0.301**
2088	C_6H_8O	2-cyclohexene-1-one	930-68-7	96.127	*220.2*	1	*445.15*	1	*684.20*	5	*4535*	10	*315.00*	25	**0.306**
270	C_6H_{10}	cyclohexene	110-83-8	82.144	*169.67*	0.2	356.12	1	560.40	0.2	*4350*	5	*291.00*	25	**0.212**
1080	$C_6H_{10}O$	cyclohexanone	108-94-1	98.143	242	1	428.58	1	653.00	3	4000	5	*311.00*	10	**0.299**
1065	$C_6H_{10}O$	mesityl oxide	141-79-7	98.143	220.15	3	402.95	1	*600.00*	10	*3410*	10	*355.00*	10	**0.325**
137	C_6H_{12}	cyclohexane	110-82-7	84.159	279.69	1	353.87	1	553.80	0.2	4080	1	308.00	1	**0.208**
105	C_6H_{12}	methylcyclopentane	96-37-7	84.159	*130.73*	0.2	344.96	1	532.70	0.2	3790	3	319.00	3	**0.229**
216	C_6H_{12}	1-hexene	592-41-6	84.159	*133.39*	0.2	336.63	0.2	504.00	0.2	3210	1	348.00	3	**0.285**
223	C_6H_{12}	4-methyl-1-pentene	691-37-2	84.159	119.51	1	327.01	1	*496.00*	5	*3220*	5	*345.00*	10	**0.239**
1151	$C_6H_{12}O$	cyclohexanol	108-93-0	100.159	296.6	3	434.00	1	650.10	1	4260	3	*322.00*	10	**0.369**
1062	$C_6H_{12}O$	2-hexanone	591-78-6	100.159	217.4	1	400.70	0.2	587.10	0.2	3300	3	377.00	3	**0.390**
1059	$C_6H_{12}O$	3-hexanone	589-38-8	100.159	217.5	1	396.65	1	582.82	1	3320	3	378.00	5	**0.380**
1054	$C_6H_{12}O$	methyl isobutyl ketone	108-10-1	100.159	189.15	1	389.15	0.2	574.60	1	3270	3	374.00	3	**0.353**
1447	$C_6H_{12}O$	butyl vinyl ether	111-34-2	100.159	181.15	3	366.97	3	*536.00*	3	*3120*	5	*364.00*	10	**0.380**
1262	$C_6H_{12}O_2$	*n*-hexanoic acid	142-62-1	116.158	269.25	1	478.85	1	660.20	1	3308	3	408.00	5	**0.733**
1333	$C_6H_{12}O_2$	ethyl *n*-butyrate	105-54-4	116.158	175.15	3	394.65	1	571.00	1	*2950*	5	*403.00*	10	**0.401**
1324	$C_6H_{12}O_2$	*n*-propyl propionate	106-36-5	116.158	197.25	1	395.65	1	568.60	1	3060	3	*389.00*	10	**0.449**
2337	$C_6H_{12}O_2$	ethyl isobutyrate	97-62-1	116.158	185	3	383.00	3	553.15	3	3040	5	*410.00*	10	**0.426**
1315	$C_6H_{12}O_2$	*n*-butyl acetate	123-86-4	116.158	199.65	3	399.26	1	575.40	1	3090	1	*389.00*	10	**0.439**
1316	$C_6H_{12}O_2$	isobutyl acetate	110-19-0	116.158	174.3	1	389.80	1	560.80	0.2	3010	1	*389.00*	10	**0.434**
1306	$C_6H_{12}O_2$	*n*-pentyl formate	638-49-3	116.158	199.65	1	405.45	1	576.00	3	3125	5	*389.00*	25	**0.515**
5884	$C_6H_{12}O_3$	2-ethoxyethyl acetate	111-15-9	132.158	211.45	1	429.74	1	607.30	1	3166	1	453.00	5	**0.552**
11	C_6H_{14}	*n*-hexane	110-54-3	86.175	177.83	0.2	341.86	0.2	507.60	0.2	3025	1	371.00	3	**0.301**
12	C_6H_{14}	2-methylpentane	107-83-5	86.175	*119.55*	0.2	333.41	1	497.70	0.2	3040	1	368.00	3	**0.279**
13	C_6H_{14}	3-methylpentane	96-14-0	86.175	110.25	1	336.42	1	504.60	0.2	3120	3	368.00	3	**0.270**
14	C_6H_{14}	2,2-dimethylbutane	75-83-2	86.175	*174.28*	0.2	322.88	1	489.00	0.2	3100	1	358.00	3	**0.234**
15	C_6H_{14}	2,3-dimethylbutane	79-29-8	86.175	*145.19*	0.2	331.13	1	500.00	0.2	3150	1	361.00	3	**0.249**
1114	$C_6H_{14}O$	1-hexanol	111-27-3	102.175	228.55	0.2	429.90	0.2	611.30	0.2	3446	1	382.00	3	**0.559**

TABLE A.1 Basic Constants 1 (*Continued*)

DIPPR ID	Formula	Name	CAS RN®	*MW* (g/mol)	T_m (K)	Er (%)	T_b (K)	Er (%)	T_c (K)	Er (%)	P_c (kPa)	Er (%)	V_c (cm³/mol)	Er (%)	ω
1115	$C_6H_{14}O$	2-hexanol	626-93-7	102.175	*223*	3	412.40	1	585.30	0.2	3311	1	385.00	3	**0.553**
1116	$C_6H_{14}O$	3-hexanol	623-37-0	102.175	*208.5*	10	406.15	1	582.40	0.2	3360	1	383.00	3	**0.498**
1117	$C_6H_{14}O$	2-methyl-1-pentanol	105-30-6	102.175	*208*	10	421.15	1	604.40	0.2	3450	1	*380.00*	10	**0.510**
1130	$C_6H_{14}O$	4-methyl-2-pentanol	108-11-2	102.175			404.85	0.2	574.40	0.2	*3470*	10	*380.00*	10	**0.573**
1403	$C_6H_{14}O$	diisopropyl ether	108-20-3	102.175	187.65	1	341.45	1	500.05	1	2880	3	386.00	5	**0.339**
1446	$C_6H_{14}O$	di-*n*-propyl ether	111-43-3	102.175	149.95	1	363.23	3	530.60	1	3028	3	*382.00*	10	**0.369**
1432	$C_6H_{14}O_2$	acetal	105-57-7	118.174	173.15	3	376.65	1	539.70	0.2	3220	3	*402.00*	10	**0.487**
1706	$C_6H_{15}N$	triethylamine	121-44-8	101.190	158.45	1	361.92	3	535.15	1	3040	3	390.00	5	**0.316**
1707	$C_6H_{15}N$	di-*n*-propylamine	142-84-7	101.190	210.15*	1	382.00	3	550.00	3	3140	3	*402.00*	5	**0.450**
1725	$C_6H_{15}NO_3$	triethanolamine	102-71-6	149.188	294.35	1	608.54	3	772.10	5	*2743*	10	*472.00*	10	**1.284**
2626	C_7F_{16}	perfluoro-*n*-heptane	335-57-9	388.049	221.86	1	355.66	0.2	475.65	1	1610	1	657.70	5	**0.543**
1874	$C_7H_5ClO_2$	*o*-chlorobenzoic acid	118-91-2	156.566	415.15	3	560.15	5	*792.00*	3	*4030*	10	*383.00*	10	**0.664**
1790	C_7H_5N	benzonitrile	100-47-0	103.121	260.28	1	464.15	1	702.30	1	4215	5	313.20	10	**0.343**
1041	C_7H_6O	benzaldehyde	100-52-7	106.122	*216.02*	0.2	451.90	1	695.00	1	*4650*	10	*324.00*	10	**0.313**
1578	C_7H_7Cl	*p*-chlorotoluene	106-43-4	126.583	280.65	1	435.65	1	*660.00*	3	*3910*	10	*360.00*	10	**0.312**
502	C_7H_8	toluene	108-88-3	92.138	178.18	1	383.78	1	591.75	0.2	4108	1	316.00	3	**0.264**
1461	C_7H_8O	anisole	100-66-3	108.138	235.65	1	426.73	1	645.60	0.2	4250	3	*337.00*	10	**0.350**
1180	C_7H_8O	benzyl alcohol	100-51-6	108.138	257.85	0.2	478.60	0.2	713.00	1	4330	3	*335.00*	3	**0.408**
1182	C_7H_8O	*o*-cresol	95-48-7	108.138	*304.19*	1	464.15	1	697.55	1	5010*	3	282.00	5	**0.434**
1183	C_7H_8O	*m*-cresol	108-39-4	108.138	*285.39*	1	475.43	1	705.85	1	4560	3	312.00	5	**0.448**
1184	C_7H_8O	*p*-cresol	106-44-5	108.138	*307.93*	1	475.13	1	704.65	1	5150	3	277.00	5	**0.507**
2796	C_7H_9N	2,6-dimethylpyridine	108-48-5	107.153	267	1	417.20	1	623.75	3	*3750*	10	*373.00*	25	**0.345**
1344	$C_7H_{12}O_2$	*n*-butyl acrylate	141-32-2	128.169	208.55	3	420.55	3	*598.00*	3	*2910*	3	*428.00*	5	**0.482**
159	C_7H_{14}	cycloheptane	291-64-5	98.186	*265.12*	1	391.94	1	604.20	0.2	3820	3	353.00	3	**0.241**
138	C_7H_{14}	methylcyclohexane	108-87-2	98.186	*146.58*	1	374.08	1	572.10	0.2	3480	1	369.00	1	**0.236**
107	C_7H_{14}	ethylcyclopentane	1640-89-7	98.186	*134.71*	1	376.62	1	569.50	0.2	3400	3	375.00	3	**0.270**
111	C_7H_{14}	*cis*-1,3-dimethylcyclopentane	2532-58-3	98.186	139.45	1	363.92	1	*551.00*	1	*3445*	1	*360.00*	5	**0.274**
112	C_7H_{14}	*trans*-1,3-dimethylcyclopentane	1759-58-6	98.186	139.18	1	364.88	1	*553.00*	1	*3445*	1	*360.00*	5	**0.270**
234	C_7H_{14}	1-heptene	592-76-7	98.186	154.12	0.2	366.79	0.2	537.40	0.2	2920	3	402.00	3	**0.343**
1326	$C_7H_{14}O_2$	*n*-butyl propionate	590-01-2	130.185	183.63	1	418.26	1	594.60	1	*2800*	3	*442.00*	10	**0.464**
2261	$C_7H_{14}O_2$	*n*-heptanoic acid	111-14-8	130.185	*265.83*	1	496.15	1	677.30	1	3043	3	466.00	5	**0.760**
1347	$C_7H_{14}O_2$	ethyl isovalerate	108-64-5	130.185	173.85	3	407.45	1	587.95	1	*2840*	5	*442.00*	10	**0.400**
1327	$C_7H_{14}O_2$	*n*-propyl *n*-butyrate	105-66-8	130.185	177.95	1	415.85	1	593.70	1	2720	3	*442.00*	25	**0.433**
1336	$C_7H_{14}O_2$	*n*-propyl isobutyrate	644-49-5	130.185	*200*	10	408.65	1	579.40	1	*2840*	10	*442.00*	25	**0.484**
1317	$C_7H_{14}O_2$	isopentyl acetate	123-92-2	130.185	194.65	1	415.15	1	586.10	3	2760	1	*442.00*	10	**0.489**

(*Continued*)

TABLE A.1 Basic Constants 1 (*Continued*)

DIPPR ID	Formula	Name	CAS RN®	*MW* (g/mol)	T_m (K)	Er (%)	T_b (K)	Er (%)	T_c (K)	Er (%)	P_c (kPa)	Er (%)	V_c (cm³/mol)	Er (%)	ω
17	C_7H_{16}	*n*-heptane	142-82-5	100.202	182.57	0.2	371.53	0.2	540.20	0.2	2740	3	428.00	5	**0.349**
18	C_7H_{16}	2-methylhexane	591-76-4	100.202	*154.9*	1	363.20	1	530.40	0.2	2740	1	421.00	3	**0.328**
19	C_7H_{16}	3-methylhexane	589-34-4	100.202	153.75	1	365.00	1	535.20	0.2	2810	3	404.00	3	**0.320**
20	C_7H_{16}	3-ethylpentane	617-78-7	100.202	154.55	1	366.62	1	540.60	0.2	2890	3	416.00	3	**0.309**
21	C_7H_{16}	2,2-dimethylpentane	590-35-2	100.202	149.34	1	352.34	1	520.50	0.2	2770	3	416.00	3	**0.287**
22	C_7H_{16}	2,3-dimethylpentane	565-59-3	100.202	*141.23*	10	362.93	1	537.30	0.2	2910	3	393.00	3	**0.296**
23	C_7H_{16}	2,4-dimethylpentane	108-08-7	100.202	153.91	1	353.64	1	519.80	0.2	2740	3	418.00	3	**0.302**
24	C_7H_{16}	3,3-dimethylpentane	562-49-2	100.202	138.7	1	359.21	1	536.40	0.2	2950	3	414.00	3	**0.268**
25	C_7H_{16}	2,2,3-trimethylbutane	464-06-2	100.202	*248.57*	1	354.03	1	531.10	0.2	2950	3	398.00	3	**0.250**
1125	$C_7H_{16}O$	1-heptanol	111-70-6	116.201	239.15	0.2	448.60	0.2	632.30	0.2	3085	1	444.00	3	**0.562**
1126	$C_7H_{16}O$	2-heptanol	543-49-7	116.201	*220*	10	432.90	1	608.30	0.2	3000	1	447.00	3	**0.568**
1625	C_8F_{18}	perfluoro-*n*-octane	307-34-6	438.057	250	1	376.45	1	502.20	1	1478	5	*718.41*	5	**0.508**
601	C_8H_8	styrene	100-42-5	104.149	242.54	1	418.31	1	635.20	1	3870	3	*352.00*	1	**0.302**
1090	C_8H_8O	acetophenone	98-86-2	120.149	292.81	1	475.26	1	709.60	1	4010	1	386.00	3	**0.383**
1283	$C_8H_8O_2$	*p*-toluic acid	99-94-5	136.148	452.34	0.2	547.65	0.2	775.00	3	3800	3	*394.00*	25	**0.629**
504	C_8H_{10}	ethylbenzene	100-41-4	106.165	178.2	0.2	409.31	0.2	617.26	0.2	3616	0.2	372.00	1	**0.303**
505	C_8H_{10}	*o*-xylene	95-47-6	106.165	247.98	0.2	417.44	0.2	630.30	0.2	3763	1	370.00	1	**0.313**
506	C_8H_{10}	*m*-xylene	108-38-3	106.165	225.3	0.2	412.22	0.2	617.00	0.2	3541	1	*376.00*	0.2	**0.324**
507	C_8H_{10}	*p*-xylene	106-42-3	106.165	286.4	0.2	411.43	0.2	616.20	0.2	3531	1	*379.00*	1	**0.323**
1185	$C_8H_{10}O$	*o*-ethylphenol	90-00-6	122.164	269.84	1	477.67	1	703.00	3	*4290*	10	*351.00*	25	**0.476**
2112	$C_8H_{10}O$	*m*-ethylphenol	620-17-7	122.164	269.15	1	491.57	1	716.45	1	*4290*	10	*387.00*	25	**0.519**
1187	$C_8H_{10}O$	*p*-ethylphenol	123-07-9	122.164	318.23	5	491.14	1	716.45	1	*4290*	10	*374.00*	25	**0.515**
1170	$C_8H_{10}O$	2,3-xylenol	526-75-0	122.164	345.71	1	490.07	1	722.95	1	*4900*	5	*360.00*	10	**0.516**
1172	$C_8H_{10}O$	2,4-xylenol	105-67-9	122.164	297.68	1	484.13	1	707.65	1	*4400*	5	*390.00*	5	**0.514**
1174	$C_8H_{10}O$	2,5-xylenol	95-87-4	122.164	347.99	1	484.33	1	707.05	1	*4900*	5	*350.00*	10	**0.564**
1176	$C_8H_{10}O$	2,6-xylenol	576-26-1	122.164	318.76	1	474.22	1	701.05	1	*4300*	5	*390.00*	10	**0.455**
1177	$C_8H_{10}O$	3,4-xylenol	95-65-8	122.164	338.25	1	500.15	1	729.95	1	*5000*	5	*350.00*	10	**0.576**
1178	$C_8H_{10}O$	3,5-xylenol	108-68-9	122.164	336.59	1	494.89	1	715.65	1	*3600*	5	*480.00*	10	**0.485**
160	C_8H_{16}	cyclooctane	292-64-8	112.213	287.95	1	424.29	1	647.20	0.2	3560	3	410.00	3	**0.252**
147	C_8H_{16}	*trans*-1,4-dimethylcyclohexane	2207-04-7	112.213	236.21	1	392.51	1	587.70	0.2	2940	3	*450.00*	10	**0.255**
250	C_8H_{16}	1-octene	111-66-0	112.213	*171.45*	1	394.41	1	*566.90*	1	*2663*	3	464.00	5	**0.392**
1265	$C_8H_{16}O_2$	*n*-octanoic acid	124-07-2	144.211	289.65	1	512.85	1	694.26	1	2779	3	523.00	5	**0.773**
2260	$C_8H_{16}O_2$	2-ethyl hexanoic acid	149-57-5	144.211	155.15	1	500.66	1	674.60	0.2	2778	0.2	528.00	10	**0.801**
1385	$C_8H_{16}O_2$	*n*-butyl *n*-butyrate	109-21-7	144.211	181.15	1	438.15	1	*616.00*	3	*2540*	10	*494.00*	25	**0.482**
1360	$C_8H_{16}O_2$	isobutyl isobutyrate	97-85-8	144.211	192.5	0.2	420.45	1	601.89	0.2	*2610*	10	*502.00*	25	**0.401**
27	C_8H_{18}	*n*-octane	111-65-9	114.229	216.38	0.2	398.77	0.2	568.70	0.2	2490	3	*486.00*	5	**0.400**

TABLE A.1 Basic Constants 1 (*Continued*)

DIPPR ID	Formula	Name	CAS RN®	*MW* (g/mol)	T_m (K)	Er (%)	T_b (K)	Er (%)	T_c (K)	Er (%)	P_c (kPa)	Er (%)	V_c (cm³/mol)	Er (%)	ω
28	C_8H_{18}	2-methylheptane	592-27-8	114.229	164.16	1	390.80	1	559.70	0.2	2500	1	488.00	1	**0.380**
29	C_8H_{18}	3-methylheptane	589-81-1	114.229	152.6	1	392.08	1	563.60	0.2	2550	3	464.00	3	**0.372**
30	C_8H_{18}	4-methylheptane	589-53-7	114.229	152.2	1	390.86	1	561.70	0.2	2540	3	476.00	3	**0.371**
31	C_8H_{18}	3-ethylhexane	619-99-8	114.229	*173.6*	10	391.69	1	565.50	0.2	2610	3	455.00	3	**0.361**
32	C_8H_{18}	2,2-dimethylhexane	590-73-8	114.229	151.97	1	379.99	1	549.80	0.2	2530	3	478.00	3	**0.338**
33	C_8H_{18}	2,3-dimethylhexane	584-94-1	114.229	*162.67*	10	388.76	1	563.50	0.2	2630	3	468.00	3	**0.346**
34	C_8H_{18}	2,4-dimethylhexane	589-43-5	114.229	*153.02*	10	382.58	1	553.50	0.2	2560	3	472.00	3	**0.344**
35	C_8H_{18}	2,5-dimethylhexane	592-13-2	114.229	182	1	382.26	1	550.00	0.2	2490	1	482.00	1	**0.358**
36	C_8H_{18}	3,3-dimethylhexane	563-16-6	114.229	147.05	1	385.12	1	562.00	0.2	2650	3	443.00	3	**0.320**
37	C_8H_{18}	3,4-dimethylhexane	583-48-2	114.229	*161.9*	10	390.88	1	568.80	0.2	2690	3	466.00	3	**0.338**
38	C_8H_{18}	2-methyl-3-ethylpentane	609-26-7	114.229	158.2	1	388.80	1	567.10	0.2	2700	3	442.00	3	**0.330**
39	C_8H_{18}	3-methyl-3-ethylpentane	1067-08-9	114.229	182.28	1	391.42	1	576.50	0.2	2810	3	455.00	3	**0.305**
40	C_8H_{18}	2,2,3-trimethylpentane	564-02-3	114.229	160.89	1	383.00	1	563.50	0.2	2730	3	436.00	3	**0.297**
41	C_8H_{18}	2,2,4-trimethylpentane	540-84-1	114.229	165.78	1	372.39	1	543.80	0.2	2570	1	468.00	1	**0.303**
42	C_8H_{18}	2,3,3-trimethylpentane	560-21-4	114.229	172.22	3	387.92	1	573.50	0.2	2820	3	455.00	3	**0.290**
43	C_8H_{18}	2,3,4-trimethylpentane	565-75-3	114.229	163.95	1	386.62	1	566.40	0.2	2730	3	460.00	3	**0.315**
44	C_8H_{18}	2,2,3,3-tetramethylbutane	594-82-1	114.229	373.96	1	379.44	1	*568.00*	5	*2870*	5	*461.00*	10	**0.245**
1132	$C_8H_{18}O$	1-octanol	111-87-5	130.228	257.65	0.2	467.10	0.2	652.30	0.2	2783	1	509.00	3	**0.570**
1133	$C_8H_{18}O$	2-octanol	123-96-6	130.228	*241.55*	3	452.90	0.2	629.80	0.2	2749	1	512.00	3	**0.588**
1121	$C_8H_{18}O$	2-ethyl-1-hexanol	104-76-7	130.228	203.15	1	457.60	1	640.40	0.2	2756	3	*492.00*	5	**0.550**
1404	$C_8H_{18}O$	di-*n*-butyl ether	142-96-1	130.228	175.3	1	414.15	1	584.10	1	*2460*	5	*487.00*	10	**0.448**
2708	$C_8H_{19}N$	*n*-octylamine	111-86-4	129.243	273.37	1	452.75	1	641*	1	*2673*	5	*543.00*	5	**0.466**
1744	$C_8H_{19}N$	di-*n*-butylamine	111-92-2	129.243	211.15	1	432.00	1	602.30	1	2570	3	*512.00*	5	**0.530**
1626	C_9F_{20}	perfluoro-*n*-nonane	375-96-2	488.064	257.15	1	398.45	0.2	*523.90*	1	*1296**	10	*823.00*	25	**0.525**
1748	C_9H_7N	quinoline	91-22-5	129.159	*258.37*	0.2	510.31	1	782.15	1	*4860*	5	*371.00*	10	**0.346**
2785	C_9H_7N	isoquinoline	119-65-3	129.159	*299.62*	0.2	516.37	1	803.15	1	*5100*	10	*374.00*	1	**0.303**
820	C_9H_{10}	indane	496-11-7	118.176	221.74	1	451.12	1	684.90	0.2	3950	3	*396.00*	25	**0.309**
509	C_9H_{12}	*n*-propylbenzene	103-65-1	120.192	*173.55*	1	432.39	1	638.35	0.2	3200	1	440.00	3	**0.344**
510	C_9H_{12}	cumene	98-82-8	120.192	177.14	1	425.56	1	631.00	0.2	3209	3	434.00	5	**0.327**
513	C_9H_{12}	*p*-ethyltoluene	622-96-8	120.192	210.83	1	435.16	1	640.20	0.2	3230	5	*427.00*	10	**0.367**
514	C_9H_{12}	1,2,3-trimethylbenzene	526-73-8	120.192	247.79	1	449.27	1	664.50	0.2	3454	3	*414.00*	25	**0.367**
515	C_9H_{12}	1,2,4-trimethylbenzene	95-63-6	120.192	229.33	1	442.53	1	649.10	0.2	3232	3	430.00	5	**0.379**
516	C_9H_{12}	mesitylene	108-67-8	120.192	228.42	1	437.89	1	637.30	0.2	3127	3	433.00	5	**0.399**
259	C_9H_{18}	1-nonene	124-11-8	126.239	*191.91*	0.2	420.02	1	593.10	1	*2428*	3	524.00	5	**0.437**
1259	$C_9H_{18}O_2$	*n*-nonanoic acid	112-05-0	158.238	285.55	1	528.75	1	710.70	1	2514	3	*584.00*	10	**0.779**

(*Continued*)

TABLE A.1 Basic Constants 1 (*Continued*)

DIPPR ID	Formula	Name	CAS RN®	*MW* (g/mol)	T_m (K)	Er (%)	T_b (K)	Er (%)	T_c (K)	Er (%)	P_c (kPa)	Er (%)	V_c (cm³/mol)	Er (%)	*ω*
3318	$C_9H_{18}O_2$	isopentyl butyrate	106-27-4	158.238	*189*	5	450.65	1	618.83	1	*2355*	10	*570.00*	10	**0.573**
46	C_9H_{20}	*n*-nonane	111-84-2	128.255	219.66	1	423.91	0.2	594.60	0.2	2290	3	*551.00*	5	**0.443**
91	C_9H_{20}	2-methyloctane	3221-61-2	128.255	193.05	1	416.45	1	582.87	0.2	2310	1	*541.00*	5	**0.460**
96	C_9H_{20}	2,2-dimethylheptane	1071-26-7	128.255	160.15	1	405.84	1	576.70	0.2	2350	3	*519.00*	5	**0.391**
47	C_9H_{20}	2,2,5-trimethylhexane	3522-94-9	128.255	167.39	1	397.24	1	569.80	1	*2330*	3	*519.00*	5	**0.345**
51	C_9H_{20}	2,2,3,3-tetramethylpentane	7154-79-2	128.255	263.26	1	413.44	1	607.50	0.2	2741	3	478.00	5	**0.305**
52	C_9H_{20}	2,2,3,4-tetramethylpentane	1186-53-4	128.255	152.06	1	406.18	1	592.60	0.2	2600	3	490.00	5	**0.314**
53	C_9H_{20}	2,2,4,4-tetramethylpentane	1070-87-7	128.255	206.95	1	395.44	1	574.60	0.2	2490	3	504.00	5	**0.314**
54	C_9H_{20}	2,3,3,4-tetramethylpentane	16747-38-9	128.255	171.03	1	414.70	1	607.50	0.2	2720	3	493.00	5	**0.314**
1134	$C_9H_{20}O$	1-nonanol	143-08-8	144.255	268.15	0.2	485.20	0.2	670.90	0.2	2527	1	576.00	3	**0.584**
1135	$C_9H_{20}O$	2-nonanol	628-99-9	144.255	*238.15*	1	471.70	0.2	649.50	0.2	2541	1	577.00	3	**0.609**
1627	$C_{10}F_{22}$	perfluoro-*n*-decane	307-45-9	538.072	309.15	0.2	417.35	1	542.35	1	*1172**	5	893.20	3	**0.513**
1381	$C_{10}H_{10}O_4$	dimethyl terephthalate	120-61-6	194.184	413.8	0.2	559.20	1	*777.40*	3	2760	10	*529.00*	25	**0.581**
701	$C_{10}H_8$	naphthalene	91-20-3	128.171	353.43	1	491.14	1	748.40	0.2	4050	3	407.00	3	**0.302**
706	$C_{10}H_{12}$	1,2,3,4-tetrahydronaphthalene	119-64-2	132.202	237.4	1	480.77	1	720.00	0.2	3650	3	408.00	5	**0.335**
503	$C_{10}H_{12}$	1-methylindan	767-58-8	132.202	*182.1*	10	463.75	1	*694.10*	5	*3437*	10	*441.00*	5	**0.314**
508	$C_{10}H_{12}$	5-methylindan	874-35-1	132.202	*213.3*	10	*475.15*	1	*711.20*	5	*3522*	10	*440.00*	5	**0.326**
518	$C_{10}H_{14}$	*n*-butylbenzene	104-51-8	134.218	*185.3*	1	456.45	1	660.50	0.2	2890	3	497.00	5	**0.394**
519	$C_{10}H_{14}$	isobutylbenzene	538-93-2	134.218	221.7	1	445.94	1	650.00	1	3050	3	*478.00*	10	**0.382**
527	$C_{10}H_{14}$	*p*-diethylbenzene	105-05-5	134.218	230.32	1	456.94	1	657.90	0.2	2803	3	*497.00*	10	**0.403**
524	$C_{10}H_{14}$	*p*-cymene	99-87-6	134.218	205.25	1	450.28	1	652.00	1	2800	3	497.00	10	**0.374**
532	$C_{10}H_{14}$	1,2,4,5-tetramethylbenzene	95-93-2	134.218	352.38	1	469.99	1	676.00	1	2900	3	*482.00*	10	**0.422**
520	$C_{10}H_{14}$	*sec*-butylbenzene	135-98-8	134.218	197.72	1	446.48	1	664.54	3	2950	5	*497.00*	10	**0.279**
153	$C_{10}H_{18}$	*cis*-decahydronaphthalene	493-01-6	138.250	*230.2*	1	468.96	1	703.60	1	3200	10	*480.00*	10	**0.279**
154	$C_{10}H_{18}$	*trans*-decahydronaphthalene	493-02-7	138.250	*242.79*	1	460.46	1	687.00	1	*3200*	10	*480.00*	10	**0.299**
260	$C_{10}H_{20}$	1-decene	872-05-9	140.266	*206.9*	0.2	443.75	1	616.60	1	2223	1	584.00	5	**0.480**
1254	$C_{10}H_{20}O_2$	*n*-decanoic acid	334-48-5	172.265	304.75	1	543.15	1	722.10	1	2280	10	*639.00*	10	**0.814**
56	$C_{10}H_{22}$	*n*-decane	124-18-5	142.282	243.51	1	447.27	0.2	617.70	0.2	2110	3	*617.00*	5	**0.492**
48	$C_{10}H_{22}$	3,3,5-trimethylheptane	7154-80-5	142.282	*165*	5	428.83	1	609.50	0.2	2320	3	*577.00*	25	**0.385**
57	$C_{10}H_{22}$	2,2,3,3-tetramethylhexane	13475-81-5	142.282	219.15	1	433.46	1	623.00	0.2	2510	3	*557.00*	25	**0.366**
58	$C_{10}H_{22}$	2,2,5,5-tetramethylhexane	1071-81-4	142.282	260.55	1	410.61	1	581.40	0.2	2190	3	*584.00*	10	**0.377**
1136	$C_{10}H_{22}O$	1-decanol	112-30-1	158.281	280.05	0.2	503.00	0.2	688.00	0.2	2308	1	645.00	3	**0.607**
702	$C_{11}H_{10}$	1-methylnaphthalene	90-12-0	142.197	242.67	1	517.83	1	772.00	0.2	3600	3	*465.00*	10	**0.342**
703	$C_{11}H_{10}$	2-methylnaphthalene	91-57-6	142.197	307.73	1	514.26	1	761.00	0.2	3500	3	*465.00*	10	**0.378**
63	$C_{11}H_{24}$	*n*-undecane	1120-21-4	156.308	247.57	1	469.05	0.2	639.00	0.2	1950	5	*685.00*	5	**0.530**
1137	$C_{11}H_{24}O$	1-undecanol	112-42-5	172.308	288.45	0.2	520.30	0.2	703.90	0.2	2119	1	715.00	3	**0.624**

TABLE A.1 Basic Constants 1 (*Continued*)

DIPPR ID	Formula	Name	CAS RN®	*MW* (g/mol)	T_m (K)	Er (%)	T_b (K)	Er (%)	T_c (K)	Er (%)	P_c (kPa)	Er (%)	V_c (cm³/mol)	Er (%)	ω
558	$C_{12}H_{10}$	biphenyl	92-52-4	154.208	*342.2*	1	528.15	1	773.00	1	3380	3	497.00	5	**0.403**
715	$C_{12}H_{12}$	2,7-dimethylnaphthalene	582-16-1	156.224	*368.81*	1	536.15	1	775.00	1	3230	10	600.00	25	**0.448**
3545	$C_{12}H_{18}$	1,3,5-triethylbenzene	102-25-0	162.271	206.74	0.2	489.05	1	*678.00*	5	*2300*	10	*593.00*	25	**0.508**
814	$C_{12}H_{20}$	1,3-dimethyladamantane	702-79-4	164.287	247.59	3	476.44	0.2	*708.00*	3	*3000*	3	570.00	3	**0.293**
262	$C_{12}H_{24}$	1-dodecene	112-41-4	168.319	237.95	1	486.15	1	*657.10*	1	*1888*	3	*725.00*	10	**0.562**
1269	$C_{12}H_{24}O_2$	*n*-dodecanoic acid	143-07-7	200.318	316.98	1	571.85	1	743.00	1	1930	3	767.00	25	**0.898**
64	$C_{12}H_{26}$	*n*-dodecane	112-40-3	170.335	263.57	1	489.45	0.2	658.00	0.2	1820	10	*755.00*	5	**0.576**
1140	$C_{12}H_{26}O$	1-dodecanol	112-53-8	186.334	296.95	0.2	537.10	0.2	718.70	0.2	1954	1	787.00	3	**0.656**
563	$C_{13}H_{12}$	diphenylmethane	101-81-5	168.234	298.37	1	537.42	1	778.00	3	*2923*	10	561.60	1	**0.395**
65	$C_{13}H_{28}$	*n*-tridecane	629-50-5	184.361	267.76	0.2	508.63	0.2	675.00	0.2	1680	10	*826.00*	10	**0.617**
1141	$C_{13}H_{28}O$	1-tridecanol	112-70-9	200.361	303.75	0.2	553.40	0.2	732.40	0.2	1810	1	*860.00*	5	**0.688**
805	$C_{14}H_{10}$	phenanthrene	85-01-8	178.229	372.38	1	612.70	1	893.00	1	3250	3	*554.00*	3	**0.410**
804	$C_{14}H_{10}$	anthracene	120-12-7	178.229	*488.93*	1	615.18	1	873.00	1	*2900*	10	*554.00*	10	**0.486**
534	$C_{14}H_{22}$	1,4-di-*tert*-butylbenzene	1012-72-2	190.324	350.76	0.2	510.43	1	708.00	0.2	*2300*	5	732.00	10	**0.502**
66	$C_{14}H_{30}$	*n*-tetradecane	629-59-4	198.388	*279.01*	0.2	526.74	0.2	693.00	1	1570	25	*897.00*	10	**0.643**
1142	$C_{14}H_{30}O$	1-tetradecanol	112-72-1	214.387	310.65	0.2	569.00	1	745.30	0.2	1682	1	933.00	5	**0.717**
1198	$C_{15}H_{16}O_2$	bisphenol A	80-05-7	228.286	430.15	3	633.65	5	*849.00*	3	*2930*	10	*677.00*	25	**0.945**
67	$C_{15}H_{32}$	*n*-pentadecane	629-62-9	212.415	283.07	0.2	543.91	0.2	708.00	1	1480	25	*969.00*	10	**0.686**
1143	$C_{15}H_{32}O$	1-pentadecanol	629-76-5	228.414	317.05	0.2	*583.90*	1	757.30	3	1569	5	*1010.00*	5	**0.751**
68	$C_{16}H_{34}$	*n*-hexadecane	544-76-3	226.441	291.31	0.2	560.20	1	722.00	1	1373	3	1040.00	3	**0.708**
90	$C_{16}H_{34}$	2,2,4,4,6,8,8-heptamethylnonane	4390-04-9	226.441			*519.50*	3	692.00	1	*1570*	5	*863.00*	25	**0.548**
1144	$C_{16}H_{34}O$	1-hexadecanol	36653-82-4	242.441	322.35	0.2	*598.00*	1	768.60	1	1468	3	*1080.00*	5	**0.779**
69	$C_{17}H_{36}$	*n*-heptadecane	629-78-7	240.468	295.13	0.2	575.70	1	735.30	1	1288	3	1120.00	3	**0.739**
1145	$C_{17}H_{36}O$	1-heptadecanol	1454-85-9	256.467	327.05	0.2	*611.30*	1	779.20	1	1377	3	*1160.00*	5	**0.812**
561	$C_{18}H_{14}$	*o*-terphenyl	84-15-1	230.304	329.35	1	609.15	1	857.00	1	2990	25	731.00	25	**0.551**
560	$C_{18}H_{14}$	*m*-terphenyl	92-06-8	230.304	360	1	648.15	3	883.00	3	2480	25	724.00	25	**0.651**
559	$C_{18}H_{14}$	*p*-terphenyl	92-94-4	230.304	485	1	655.15	3	908.00	3	2990	25	729.00	25	**0.643**
1276	$C_{18}H_{36}O_2$	*n*-octadecanoic acid	57-11-4	284.477	342.75	1	*647.15*	1	803.00	1	*1330*	3	1140.00	25	**1.039**
70	$C_{18}H_{38}$	*n*-octadecane	593-45-3	254.494	301.31	0.2	590.40	1	747.80	1	1212	3	1189.23	3	**0.769**
1146	$C_{18}H_{38}O$	1-octadecanol	112-92-5	270.494	331.05	0.2	*623.60*	3	789.30	1	1295	1	*1230.00*	5	**0.832**
71	$C_{19}H_{40}$	*n*-nonadecane	629-92-5	268.521	305.04	0.2	604.50	1	*759.50*	1	1143	3	1260.00	3	**0.800**
1149	$C_{19}H_{40}O$	1-nonadecanol	1454-84-8	284.520	334.85	0.2	*635.10*	3	798.80	1	*1220*	3	1310.00	10	**0.858**
73	$C_{20}H_{42}$	*n*-eicosane	112-95-8	282.547	309.58	0.2	617.80	1	*770.60*	1	1081	3	1340.00	3	**0.834**
1148	$C_{20}H_{42}O$	1-eicosanol	629-96-9	298.547	338.55	0.2	645.50	3	807.70	0.2	1154	3	1380.00	10	**0.879**
74	$C_{21}H_{44}$	*n*-heneicosane	629-94-7	296.574	313.35	0.2	630.60	1	*781.00*	1	1026	3	1410.00	3	**0.862**

(*Continued*)

TABLE A.1 Basic Constants 1 (*Continued*)

DIPPR ID	Formula	Name	CAS RN®	*MW* (g/mol)	T_m (K)	Er (%)	T_b (K)	Er (%)	T_c (K)	Er (%)	P_c (kPa)	Er (%)	V_c (cm^3/mol)	Er (%)	ω
75	$C_{22}H_{46}$	*n*-docosane	629-97-0	310.601	317.15	0.2	642.80	1	*790.90*	1	975	3	1490.00	3	**0.894**
76	$C_{23}H_{48}$	*n*-tricosane	638-67-5	324.627	320.65	0.2	654.50	1	*800.30*	1	929	3	1570.00	3	**0.921**
77	$C_{24}H_{50}$	*n*-tetracosane	646-31-1	338.654	323.75	0.2	665.70	1	*809.30*	1	887	3	1650.00	3	**0.943**
1987	$ClFO_3$	perchloryl fluoride	7616-94-6	102.450	125.41	1	226.49	1	368.40	3	5370	5	161.00	10	**0.173**
1904	ClH	hydrogen chloride	7647-01-0	36.461	158.97	1	188.15	1	324.65	3	8310	5	81.00	10	**0.132**
2928	ClH_4N	ammonium chloride	12125-02-9	53.491					*1155.20*	3	*164000*	3	*111.00*	25	**0.441**
1939	ClNa	sodium chloride	7647-14-5	58.443	1073.95	1	1738.15	5	*3400.00*	25	*35500*	50	*266.00*	50	**0.189**
1986	ClNO	nitrosyl chloride	2696-92-6	65.459	213.55	1	267.77	1	*440.65*	1	9120	3	139.00	3	**0.300**
918	Cl_2	chlorine	7782-50-5	70.906	172.12	0.2	239.12	1	417.15	3	7710	5	124.00	10	**0.069**
1937	Cl_4Si	tetrachlorosilane	10026-04-7	169.898	204.3	1	330.00	1	507.00	3	3590	5	326.00	10	**0.232**
925	D_2	deuterium	7782-39-0	4.032	*18.73*	0.2	23.65	3	38.35	3	1662	5	60.26		**–0.145**
1997	D_2O	deuterium oxide	7789-20-0	20.028	276.96	0.2	374.57	0.2	643.89	0.2	21671	1	56.30	3	**0.366**
1905	FH	hydrogen fluoride	7664-39-3	20.006	189.79	1	292.67	1	461.15	3	6480	5	69.00	10	**0.382**
917	F_2	fluorine	7782-41-4	37.997	53.54	0.2	84.95	3	*144.12*	3	*5172*	5	*66.55*		**0.053**
1972	F_3N	nitrogen trifluoride	7783-54-2	71.002	*66.46*	1	144.09	1	234.00	1	4461	3	118.75	5	**0.120**
1989	F_4N_2	tetrafluorohydrazine	10036-47-2	104.007	111.65	1	198.95	3	309.35	3	3710	5	*213.00*	50	**0.223**
1940	F_6S	sulfur hexafluoride	2551-62-4	146.055	222.45	3			318.69	3	3760	5	198.52	10	**0.215**
1907	HI	hydrogen iodide	10034-85-2	127.912	222.38	1	237.55	1	423.85	3	*8310*	5	*121.90*	10	**0.038**
1912	HNaO	sodium hydroxide	1310-73-2	39.997	596	1	*1830.00*		*2820.00*	50	*25000*	50	*200.00*	50	
902	H_2	hydrogen	1333-74-0	2.016	13.95	1	20.39	1	33.19	3	1313	10	64.15	5	**–0.216**
1921	H_2O	water	7732-18-5	18.015	273.15	0.2	373.15	0.2	647.10	0.2	22064	0.2	55.95	0.2	**0.345**
1901	H_2O_4S	sulfuric acid	7664-93-9	98.078	283.46	0.2	*610.00*	25	924.00	5	*6400*	25	177.00	10	
1922	H_2S	hydrogen sulfide	7783-06-4	34.081	187.68	1	212.80	3	373.53	1	8963	5	98.50	10	**0.094**
3951	H_2Se	hydrogen selenide	7783-07-5	80.976	207.45	1	231.15	3	411.15	3	8920	5	*120.00*	10	**0.054**
1911	H_3N	ammonia	7664-41-7	17.031	195.41	1	239.72	1	405.65	3	11280	5	72.47	10	**0.253**
1981	H_3P	phosphine	7803-51-2	33.998	139.37	1	185.41	1	324.75	3	6540	5	*113.00*	10	**0.045**
1717	H_4N_2	hydrazine	302-01-2	32.045	274.69	1	386.65	1	653.15	1	14700	3	*158.00*	25	**0.314**
1982	H_4Si	silane	7803-62-5	32.117	*88.48*	1	161.00	1	269.70	3	4840	5	*132.70*	25	**0.094**
913	He	helium	7440-59-7	4.003	1.76	3	4.22	1	5.20	1	228	3	57.30	5	**–0.390**
1998	I_2	iodine	7553-56-2	253.809	386.75	1	457.56	1	819.15	3	11654	5	155.00	10	**0.111**
920	Kr	krypton	7439-90-9	83.798	115.78	3	119.80	3	209.35	3	5502	5	91.20	10	**–0.001**
912	NO	nitric oxide	10102-43-9	30.006	112.15	5	121.38	1	180.15	3	6480	5	58*	10	**0.583**
905	N_2	nitrogen	7727-37-9	28.013	63.15	1	77.34	1	126.20	3	3400	5	89.21	10	**0.038**
899	N_2O	nitrous oxide	10024-97-2	44.013	182.33	3	184.67	3	309.57	3	7245	5	97.40	10	**0.141**
906	N_2O_4	nitrogen tetroxide	10544-72-6	92.011	261.92	0.2	294.30	0.2	431.35	1	*9976*	25	*137.00*	25	**0.843**
919	Ne	neon	7440-01-9	20.180	24.55	1	27.09	1	44.40	1	2653	3	41.70	5	**–0.040**

TABLE A.1 Basic Constants 1 (*Continued*)

DIPPR ID	Formula	Name	CAS RN®	*MW* (g/mol)	T_m (K)	Er (%)	T_b (K)	Er (%)	T_c (K)	Er (%)	P_c (kPa)	Er (%)	V_c (cm³/mol)	Er (%)	ω
901	O_2	oxygen	7782-44-7	31.999	54.36	1	90.19	1	154.58	3	5043	5	73.40	10	**0.022**
910	O_2S	sulfur dioxide	7446-09-5	64.064	200	3	263.13	1	430.75	3	7884	5	122.00	10	**0.245**
924	O_3	ozone	10028-15-6	47.998	80.15	1	161.85	1	261.00	3	5570	5	89.00	10	**0.212**
911	O_3S	sulfur trioxide	7446-11-9	80.063	289.95	3	317.90	3	490.85	3	8210	5	127.00	10	**0.424**
970	Rn	radon	10043-92-2	222.018	202	0.2	211.00	1	377.50	1	6280	1	*137.65*	5	**–0.052**
1923	S	sulfur	7704-34-9	32.065	388.36	0.2	717.82	10	*1313.00*	3	*18208*	25	*158.00*	25	**0.246**
959	Xe	xenon	7440-63-3	131.293	161.36	1	165.03	3	289.74	3	5840	5	118.00	10	**0.012**

TABLE A.2 Basic Constants 2

DIPPR IDC	Formula	Name	$\Delta_f H°(298)$ (kJ/mol)	Er (%)	$\Delta_f G°(298)$ (kJ/mol)	Er (%)	$\Delta_{vap} H(T_b)$ (kJ/mol)	Er (%)	$\Delta_m H(T_m)$ (kJ/mol)	Er (%)	V_L (cm³/mol)	Er (%)	μ (Debye)	Er (%)
914	Ar	argon	**0.00**	0.2	**0.00**	0.2	**6.43**	1	1.18	1	28.62	1	**0.00**	
922	Br_2	bromine	30.91	1	**3.14**	1	*29.71*	3	10.57	3	51.48	3	**0.00**	
1906	BrH	hydrogen bromide	−36.29	1	**−53.34**	1	**17.73**	3	2.41	1	46.22	3	0.82	3
2686	$CBrClF_2$	bromochlorodifluoromethane	−431.37	3	**−408.28**	3	*22.39*	10	*3.98*	25	91.37	3	*0.49*	10
2687	$CBrF_3$	bromotrifluoromethane	−648.98	1	**−622.69**	1	*17.61*	5			96.93	1	0.65	10
2688	CBr_2F_2	dibromodifluoromethane	−386.60	3	**−376.00**	3	*23.92*	10			*91.93*	1	0.66	5
1606	$CClF_3$	chlorotrifluoromethane	−707.93	1	**−667.38**	1	*15.61*	3	*3.17*	25	125.15	1	0.51	3
1601	CCl_2F_2	dichlorodifluoromethane	−491.62	3	**−452.71**	3	*20.17*	5	4.14		92.38	1	0.51	10
1894	CCl_2O	phosgene	−218.90	1	**−204.79**	1	*24.44*	5	5.74	3	72.6	1	1.17	1
1602	CCl_3F	trichlorofluoromethane	−288.70	3	−249.40	3	**24.99**	5	6.90		92.96	1	0.45	10
1501	CCl_4	carbon tetrachloride	−95.81	3	**−53.54**	3	**29.71**	1	2.53	1	97.14	1	**0.00**	
1616	CF_4	carbon tetrafluoride	−922.10	1	**−877.60**	1	**11.81**	5	0.71	3	54.86	3	**0.00**	
1634	$CHBrF_2$	bromodifluoromethane	−424.90	1	**−408.59**	3	*24.08*	5	*4.64*	25	72.54	5	1.50	
1604	$CHClF_2$	chlorodifluoromethane	−481.60	3	**−450.50**	3	**20.25**	3	4.12	3	72.59	1	1.46	1
1696	$CHCl_2F$	dichlorofluoromethane	−283.30	5	**−252.80**	5	*24.72*	3	*5.12*	25	75.24	1	1.29	3
1521	$CHCl_3$	chloroform	−102.90	1	**−70.10**	1	**29.36**	3	9.54		80.5	1	1.01	1
1615	CHF_3	trifluoromethane	−697.05	1	**−662.61**	1	**16.76**	3	4.06	3	102.03	3	1.65	1
1771	CHN	hydrogen cyanide	***135.14***	5	**124.30**	10	**26.93**	10	8.41	3	39.73	0.2	2.99	3
1511	CH_2Cl_2	dichloromethane	−95.52	1	**−68.96**	3	**28.16**	3	4.60		64.43	1	1.60	3
1614	CH_2F_2	difluoromethane	−452.30	0.2	**−424.75**	0.2	**20.02**	3	*5.16*	50	54.22	3	1.98	1
1001	CH_2O	formaldehyde	−108.60	1	**−102.60**	3	**23.03**	5	*5.44*	25	*39.21*	5	2.33	3
1251	CH_2O_2	formic acid	−378.80	1	**−351.10**	5	22.06	10	12.70	1	37.92	1	1.42	0.2
1641	CH_3Br	bromomethane	−37.70	3	**−27.04**	10	**24.06**	10	5.98	0.2	57.1	3	1.84	3
1502	CH_3Cl	methyl chloride	−85.70	3	**−62.09**	10	**21.78**	5	6.43	1	55	1	1.87	3
1613	CH_3F	methyl fluoride	−234.30	25	**−210.30**	25	*16.97*	10	*4.34*	25	*59.21*	10	1.86	3
1681	CH_3I	methyl iodide	14.40	5	**16.11**	5	27.23	5	*6.54*	5	62.71	1	1.59	5
2851	CH_3NO	formamide	−192.20	3	**−147.10**	3	**51.30**	10	7.98	25	39.89	1	3.72	3
1760	CH_3NO_2	nitromethane	−74.70	3	**−6.93**	5	**34.58**	5	9.70	1	54.08	1	3.45	3
1	CH_4	methane	−74.52	1	**−50.49**	1	**8.17**	1	0.94	1	37.97	1	**0.00**	
1101	CH_4O	methanol	−200.94	1	**−162.32**	1	**35.26**	3	3.21	0.2	40.58	1	1.70	1
1801	CH_4S	methyl mercaptan	−22.90	3	**−9.80**	10	**24.42**	3	5.90		55.8	1	1.52	5
1701	CH_5N	methylamine	−22.97	1	**32.17**	3	**26.05**	10	6.13	3	47.33	1	1.31	1
908	CO	carbon monoxide	−110.53	0.2	**−137.15**	1	*6.00*	1	0.84	3	35.44	1	0.11	1
1893	COS	carbonyl sulfide	−142.40	1	**−169.30**	3	**18.70**	5	4.73	0.2	62.01	1	0.71	1
909	CO_2	carbon dioxide	−393.51	0.2	**−394.37**				9.02	5	61.68	1	**0.00**	
1938	CS_2	carbon disulfide	***117.40***	1	**67.33**	3	**26.88**	5	4.39	3	60.61	0.2	**0.00**	

TABLE A.2 Basic Constants 2 (*Continued*)

DIPPR IDC	Formula	Name	$\Delta_f H^\circ(298)$ (kJ/mol)	Er (%)	$\Delta_f G^\circ(298)$ (kJ/mol)	Er (%)	$\Delta_{vap}H(T_b)$ (kJ/mol)	Er (%)	$\Delta_m H(T_m)$ (kJ/mol)	Er (%)	V_L (cm^3/mol)	Er (%)	μ (Debye)	Er (%)
2692	C_2ClF_5	chloropentafluoroethane	–1123.00	1	**–1042.00**	3	**19.45**	3	1.88		120.23	3	0.52	3
1611	$C_2Br_2F_4$	1,2-dibromotetrafluoroethane	–789.10	1	**–735.00**	25	**26.97**	3	7.04	3	120.15	3	*0.88*	10
1693	$C_2Cl_2F_4$	1,1-dichlorotetrafluoroethane	–926.80	3	**–845.50**	3	*22.98*	3	*8.52*	25	117.16	1	*0.66*	10
1609	$C_2Cl_2F_4$	1,2-dichlorotetrafluoroethane	–916.30	1	**–834.10**	3	**23.19**	3	1.51	0.2	117.43	1	0.67	0.2
2655	$C_2Cl_3F_3$	1,1,2-trichlorotrifluoroethane	–759.29	1	**–680.80**	3	**27.20**	3	2.47	0.2	119.78	1	*0.49*	10
1542	C_2Cl_4	tetrachloroethylene	–12.13	25	**22.62**	25	**34.36**	3	10.46	3	102.81	1	**0.00**	
1630	C_2F_4	tetrafluoroethylene	–658.56	1	**–623.69**	1	*16.83*	3	7.71	3	109.39	3	**0.00**	
2693	C_2F_6	hexafluoroethane	–1343.90	1	**–1260.00**	3	**16.13**	5	2.79	5	87.1	1	**0.00**	
2640	$C_2HBrClF_3$	halothane	–690.40	1	**–630.06**	25	**27.78**	5	4.84	1	*105.82*	5	*1.31*	10
2648	C_2HClF_4	2-chloro-1,1,1,2-tetrafluoroethane	–924.70	3	**–851.90**	3	*22.43*	5	*2.81*	25	*100.8*	3	1.47	1
1694	$C_2HCl_2F_3$	2,2-dichloro-1,1,1-trifluoroethane	–743.90	3	**–669.00**	3	**25.95**	3	*6.54*	25	104.45	1	1.36	3
2647	$C_2HCl_2F_3$	1,2-dichloro-1,1,2-trifluoroethane	–710.00	3	**–639.70**	3	*25.33*	5	*7.68*	25	104.39	1	1.30	1
1646	C_2HF_5	pentafluoroethane	–1100.40	1	**–1025.57**	3	*19.73*	5	*5.64**	25	101	3	1.54	5
3605	C_2HF_5O	difluoromethyl trifluoromethyl ether	*–1316.00*	1	**–1212.00**	3	*21.04*	10	*4.36*	25	106.27	1	1.50	3
401	C_2H_2	acetylene	228.20	1	210.68	1			3.77		68.99	1	**0.00**	
1581	$C_2H_2Cl_2$	trans-1,2-dichloroethylene	6.15	50	**28.55**	50	**28.13**	5	*11.98*		77.63	1	**0.00**	
1529	$C_2H_2Cl_4$	1,1,2,2-tetrachloroethane	–149.40	5	**–82.24**	5	**38.27**	10	9.17	1	105.75	1	1.29	1
1629	$C_2H_2F_2$	1,1-difluoroethylene	–328.96	3	**–305.50**	3	*15.66*	10	*5.40*	50	*105*	3	1.38	3
2650	$C_2H_2F_4$	1,1,1,2-tetrafluoroethane	–895.79	3	**–826.90**	3	*22.22*	5	*6.48*	50	84.64	1	2.06	1
2660	$C_2H_2F_4$	1,1,2,2-tetrafluoroethane	–892.40	3	**–824.60**	3	*22.51*	5			79.02	1	0.99	1
1504	C_2H_3Cl	vinyl chloride	28.45	3	**41.95**	5	*22.40*	5	4.74		69.22	1	1.45	3
2695	$C_2H_3ClF_2$	1-chloro-1,1-difluoroethane	–529.70	3	**–465.80**	3	*22.44*	3	2.69		90.54	1	2.14	
2649	$C_2H_3Cl_2F$	1,1-dichloro-1-fluoroethane	–339.70	3	**–276.20**	3	*26.06*	3	*6.87*	25	94.86	1	2.01	1
1527	$C_2H_3Cl_3$	1,1,1-trichloroethane	–142.30	3	**–76.21**	3	**29.65**	5	2.35	1	100.31	1	1.78	3
1524	$C_2H_3Cl_3$	1,1,2-trichloroethane	–142.00	3	**–80.97**	3	**34.75**	5	11.30	3	92.99	1	1.25	1
1619	$C_2H_3F_3$	1,1,1-trifluoroethane	–736.40	3	**–667.40**	3	*19.20*	3	6.19	3	90.54	3	2.32	3
2619	$C_2H_3F_3$	1,1,2-trifluoroethane	–730.70	0.2	**–671.00**	3	*26.49*	5	*10.80*	25	69.09	3	*1.68*	1
1772	C_2H_3N	acetonitrile	*64.67*	5	**82.41**	3	**31.07**	5	8.17	1	52.79	1	3.30	5
201	C_2H_4	ethylene	52.51	3	**68.44**	3	**13.49**	1	3.35	1	49.32	1	**0.00**	
1673	$C_2H_4Br_2$	1,2-dibromoethane	–38.90	5	**–10.54**	5	35.66	5	10.96	1	86.61	1	1.01	
1522	$C_2H_4Cl_2$	1,1-dichloroethane	–129.41	3	**–72.59**	3	**29.22**	3	7.87	0.2	84.72	1	2.06	3
1523	$C_2H_4Cl_2$	1,2-dichloroethane	–129.79	3	**–73.94**	3	**31.85**	3	8.83	3	79.39	1	1.44	25
1640	$C_2H_4F_2$	1,1-difluoroethane	–497.00	3	**–439.48**	3	*21.89*	3	*5.87**	50	73.46	3	2.26	1
1002	C_2H_4O	acetaldehyde	–171.00	3	**–137.80**	5	**25.94**	5	2.31	3	56.88	1	2.69	3
1441	C_2H_4O	ethylene oxide	–52.63	3	**–13.23**	3	**25.58**	3	5.17	3	50.83	1	1.89	1

(*Continued*)

TABLE A.2 Basic Constants 2 (*Continued*)

DIPPR IDC	Formula	Name	$\Delta_f H°(298)$ (kJ/mol)	Er (%)	$\Delta_f G°(298)$ (kJ/mol)	Er (%)	$\Delta_{vap}H(T_b)$ (kJ/mol)	Er (%)	$\Delta_m H(T_m)$ (kJ/mol)	Er (%)	V_L (cm³/mol)	Er (%)	μ (Debye)	Er (%)
1252	$C_2H_4O_2$	acetic acid	–432.80	1	**–374.50**	3	23.38	10	11.73	1	57.63	1	1.74	25
1301	$C_2H_4O_2$	methyl formate	–352.40	1	–295.00	1	**28.06**	3	7.53	3	62.1	1	1.77	3
1645	C_2H_5Br	bromoethane	–63.60	3	**–25.74**	10	**26.94**	5	5.86	1	75.17	3	2.03	1
1503	C_2H_5Cl	ethyl chloride	–112.30	1	**–60.45**	10	**24.85**	5	4.45	5	73	1	2.05	1
1617	C_2H_5F	ethyl fluoride	*–264.40*	1	**–212.30**	1	**20.44**	3	*5.52*	25	*69*	3	1.94	3
2	C_2H_6	ethane	–83.82	1	**–31.92**	1	**14.68**	1	2.86†	1	95.36	1	**0.00**	
1102	C_2H_6O	ethanol	–234.95	1	**–167.85**	1	**39.18**	1	4.93	3	58.62	1	1.69	
1401	C_2H_6O	dimethyl ether	–184.10	0.2	–112.80	3	**21.51**	5	4.94	1	70.23	1	1.30	0.2
1201	$C_2H_6O_2$	ethylene glycol	***–392.70***	3	**–305.80**	3	**53.09**	3	9.96	5	55.91	0.2	2.41	3
1802	C_2H_6S	ethyl mercaptan	–46.30	3	**–4.81**	3	**26.77**	3	4.97		74.61	1	1.58	3
1820	C_2H_6S	dimethyl sulfide	–37.24	10	**7.30**	3	**26.95**	3	7.98	1	73.74	1	1.50	3
1704	C_2H_7N	ethylamine	–47.15	1	**36.29**	3	**27.25**	5	*8.45*	10	66.64	0.2	1.22	5
1702	C_2H_7N	dimethylamine	–18.45	3	**68.39**	3	**26.50**	3	5.94	3	69.34	1	1.03	1
1723	C_2H_7NO	monoethanolamine	–206.70	3	**–103.30**	5	***50.23***	10	*20.50*		60.34	1	2.37	10
1741	$C_2H_8N_2$	ethylenediamine	***–17.96***	5	**102.50**	3	**39.43**	1	22.58	0.2	67.27	0.2	1.89	3
2652	C_3F_8	octafluoropropane	–1784.70	1	**–1661.00**	3	***20.02***	10	*0.48*	5	*144.23*	3	**0.00**	
2630	C_3HF_7	1,1,1,2,3,3,3-heptafluoropropane	*–1552.00*	3	**–1433.00**	3	***22.56***	5	*5.67*	25	122.29	3	1.46	3
2631	$C_3H_2F_6$	1,1,1,2,3,3-hexafluoropropane	*–1333.00*	3	**–1222.00**	3	***25.23***	3	*8.71*	25	106.57	3	1.17	3
1620	$C_3H_2F_6$	1,1,1,3,3,3-hexafluoropropane	*–1368.00*	3	**–1255.00**	5	**25.03**	10	8.58	5	*111.8*	3	1.98	3
2629	$C_3H_3F_5$	1,1,1,2,2-pentafluoropropane	*–1111.00*	3	**–1004.00**	3	***21.98***	5			113.89	1	*2.39*	10
2627	$C_3H_3F_5$	1,1,1,3,3-pentafluoropropane	*–1174.00*	10	**–1029.00**	10	***27.16***	1			*100.1*	1	1.56	3
2628	$C_3H_3F_5$	1,1,2,2,3-pentafluoropropane	*–564.00*	3	**–464.30**	3	***27.37***	10	*8.78*	25	*96.55*	3	1.74	3
2868	$C_3H_3F_5O$	2-(difluoromethoxy)-1,1,1-trifluoroethane					*27.67*	5	*7.06*	25	108.24	1	1.63	
1774	C_3H_3N	acrylonitrile	179.70	1	**189.20**	5	***31.71***	10	6.23	0.2	66.27	1	3.87	3
5892	C_3H_3NO	isoxazole	82.02	3	**93.77**	3	33.68	10	*7.63*	25	64.5	1	2.90	10
402	C_3H_4	methylacetylene	184.90	1	**193.84**	1	***22.19***	3	*5.35*	25	66.05	1	0.78	3
301	C_3H_4	propadiene	190.50	1	**200.80**	3	**20.24**	5	4.40		69.18	1	**0.00**	
1773	C_3H_5N	propionitrile	*51.55*	5	**96.88**	5	**32.75**	5	5.05	0.2	70.76	1	3.39	3
202	C_3H_6	propylene	***20.23***	1	**62.64**	3	***18.73***	3	2.94	1	83.24	1	0.37	1
101	C_3H_6	cyclopropane	53.30	1	**104.40**	1	**19.97**	5	5.44	1	*69.9*	5	**0.00**	
1526	$C_3H_6Cl_2$	1,2-dichloropropane	***–162.70***	1	**–79.95**	5	**32.28**	1	6.40	1	98.33	0.2	1.85	10
1167	C_3H_6O	allyl alcohol	–124.50	3	**–63.72**	3	**39.83**	5	*5.12*	25	68.53	1	1.60	5
1051	C_3H_6O	acetone	***–215.70***	1	**–151.30**	3	**29.71**	3	5.77	3	73.93	0.2	2.88	3
1003	C_3H_6O	propanal	*–184.90*	3	**–123.70**	25	**28.86**	3	8.59	5	73.4	3	2.52	3
1442	C_3H_6O	1,2-propylene oxide	–93.70	3	**–26.80**	5	**27.21**	5	6.53	1	70.55	3	2.01	1
1253	$C_3H_6O_2$	propionic acid	–453.50	1	**–358.20**	3	31.11	10	10.66	1	74.97	1	1.75	3

TABLE A.2 Basic Constants 2 (*Continued*)

DIPPR IDC	Formula	Name	$\Delta_f H°(298)$ (kJ/mol)	Er (%)	$\Delta_f G°(298)$ (kJ/mol)	Er (%)	$\Delta_{vap}H(T_b)$ (kJ/mol)	Er (%)	$\Delta_m H(T_m)$ (kJ/mol)	Er (%)	V_L (cm³/mol)	Er (%)	μ (Debye)	Er (%)
1312	$C_3H_6O_2$	methyl acetate	–411.90	1	**–324.20**	3	**30.46**	1	*7.97*	10	79.82	1	1.68	5
1302	$C_3H_6O_2$	ethyl formate	–388.30	3	–303.10	3	**29.80**	3	9.20		80.91	1	1.93	
2391	$C_3H_6O_3$	dimethyl carbonate	*–570.10*	5	**–452.40**	5	**33.70**	3	*12.00*	25	84.72	1	0.90	
1585	C_3H_7Cl	propyl chloride	–133.20	3	**–52.51**	10	**27.62**	5	5.54	3	89	1	2.15	5
1876	C_3H_7NO	*N,N*-dimethylformamide	–191.70	1	**–88.40**	5	**39.54**	5	16.15*		77.39	1	3.81	3
3	C_3H_8	propane	–104.68	1	**–24.39**	1	**18.75**	3	3.52	1	89.8	1	**0.00**	
1103	C_3H_8O	1-propanol	–254.60	1	**–159.90**	1	**41.59**	10	7.00	5	75.17	3	1.68	10
1104	C_3H_8O	isopropanol	–272.10	1	–175.20	1	40.53	3	5.41	1	76.86	1	1.66	3
1407	C_3H_8O	methyl ethyl ether	–216.40	1	**–117.10**	3	**24.01**	5	*7.98*	10	86.75	1	1.23	5
1431	$C_3H_8O_2$	methylal	–348.20	1	**–226.30**	3	**28.21**	5	8.33	0.2	89.13	3	0.74	
1212	$C_3H_8O_2$	1,3-propylene glycol	–408.00	3	**–295.80**	3	**56.10**	3	7.10	5	72.46	0.2	2.55	3
1231	$C_3H_8O_3$	glycerol	–577.90	0.2	**–447.10**	3	**66.41**	10	18.28	3	73.2	1	2.68	
1813	C_3H_8S	methyl ethyl sulfide	–59.60	1	**11.47**	1	**29.45**	3	9.76	1	91.02	1	1.56	
1711	C_3H_9N	*n*-propylamine	***–70.52***	3	**41.97**	3	**29.32**	3	10.97	10	82.93	0.2	1.17	5
1719	C_3H_9N	isopropylamine	–83.80	1	**31.92**	3	**27.93**	5	7.32	0.2	86.43	1	1.45	
1703	C_3H_9N	trimethylamine	–24.31	3	**98.99**	3	**22.87**	5	6.54	1	94.07	1	0.61	3
6862	C_3H_9NO	methylethanolamine	*–198.00*	25	**–61.10**	25	***46.22***	10	*8.76*	50	80.45	3	2.16	3
2654	C_4F_8	octafluorocyclobutane	–1528.00	1	**–1398.80**	3	23.23	3	2.77	1	133.67	1	**0.00**	
1478	C_4H_4O	furan	–34.80	3	**0.82**	3	**27.14**	3	3.80	1	73.11	1	0.66	1
1821	C_4H_4S	thiophene	115.44	1	**126.70**	1	**31.48**	3	5.09	1	79.48	0.2	0.54	5
1721	C_4H_5N	pyrrole	108.30	1	**160.40**	1	**38.99**	3	7.91		69.5	1	1.84	5
403	C_4H_6	ethylacetylene	165.20	1	**202.25**	1	**24.68**	1	6.03	0.2	83.51	1	0.81	
302	C_4H_6	1,2-butadiene	162.30	1	**198.60**	3	**24.07**	5	6.96	0.2	83.82	1	0.40	3
303	C_4H_6	1,3-butadiene	109.24	3	**149.72**	3	**22.37**	1	7.98	1	87.88	1	**0.00**	
1321	$C_4H_6O_2$	vinyl acetate	–314.90	3	**–227.90**	3	***32.01***	5	*5.37*	25	92.95	1	1.79	
1291	$C_4H_6O_3$	acetic anhydride	***–575.50***	3	**–476.00**	3	**40.14**	10	10.50	10	95.04	1	3.12	10
1782	C_4H_7N	butyronitrile	*33.42*	3	**105.70**	3	**34.81**	5	5.02	1	*87.76*	1	3.47	3
102	C_4H_8	cyclobutane	28.50	3	**112.20**	3	**23.96**	3	1.09	1	*81.42*	1	**0.00**	
204	C_4H_8	1-butene	*–0.50*	3	**70.41**	3	**22.05**	3	3.85	1	95.52	1	0.34	25
206	C_4H_8	*trans*-2-butene	–11.00	10	**63.20**	10	**22.78**	3	9.76	1	93.61	1	**0.00**	
205	C_4H_8	*cis*-2-butene	–7.40	25	**65.36**	10	**23.28**	3	7.31	1	90.89	1	0.30	50
207	C_4H_8	isobutene	–17.10	10	**58.08**	10	***22.04***	3	5.93	1	95.36	1	0.50	1
1005	C_4H_8O	butanal	***–206.20***	5	**–114.80**	25	**30.99**	5	11.10	5	90.41	3	2.72	5
1052	C_4H_8O	methyl ethyl ketone	***–238.70***	3	**–146.10**	5	**31.52**	3	8.38	1	90.13	0.2	2.78	3
1479	C_4H_8O	tetrahydrofuran	–184.18	1	**–79.69**	3	**29.85**	3	8.54	1	81.94	1	1.63	5

(*Continued*)

TABLE A.2 Basic Constants 2 (*Continued*)

DIPPR IDC	Formula	Name	$\Delta_f H°(298)$ (kJ/mol)	Er (%)	$\Delta_f G°(298)$ (kJ/mol)	Er (%)	$\Delta_{vap}H(T_b)$ (kJ/mol)	Er (%)	$\Delta_m H(T_m)$ (kJ/mol)	Er (%)	V_L (cm³/mol)	Er (%)	μ (Debye)	Er (%)
1256	$C_4H_8O_2$	*n*-butyric acid	–475.80	1	**–360.00**	5	35.55	10	11.59	5	92.59	1	1.65	10
1260	$C_4H_8O_2$	isobutyric acid	*–484.10*	10	**–362.10**	10	34.52	10	5.02		93.33	1	1.09	
2422	$C_4H_8O_2$	1,3-dioxane	–342.30	1	**–209.00**	3	**34.84**	10	*10.30*	5	85.76	1	2.06	1
1421	$C_4H_8O_2$	1,4-dioxane	–315.80	1	**–181.60**	3	**34.37**	3	12.84	3	85.66	1	**0.00**	
1322	$C_4H_8O_2$	methyl propionate	–427.50	3	–311.00	3	**32.31**	3	*10.10*	10	96.96	1	1.70	0.2
1313	$C_4H_8O_2$	ethyl acetate	–444.50	1	**–328.00**	1	**32.15**	1	10.48	3	98.59	1	1.78	5
1303	$C_4H_8O_2$	*n*-propyl formate	–407.60	3	–293.60	3	**32.02**	3	*13.20*	10	97.94	1	1.91	1
1587	C_4H_9Cl	sec-butyl chloride	–165.69	10	**–55.19**	10	**29.11**	5	*6.14*	25	106.63	1	2.04	5
1766	C_4H_9N	pyrrolidine	–3.60	25	114.70	25	**33.45**	3	8.58	0.2	82.74	3	1.58	1
5	C_4H_{10}	*n*-butane	–125.79	1	**–16.70**	1	**22.40**	3	4.66	1	101.39	1	**0.00**	
4	C_4H_{10}	isobutane	–134.99	1	–21.44	3	**21.23**	3	4.54	1	105.35	1	0.13	10
2752	$C_4H_{10}N_2$	piperazine	***28.80***	25	**221.40**	3	***40.42***	10	26.7*	1	*95.48*	10	1.47	3
1105	$C_4H_{10}O$	1-butanol	–275.10	1	**–150.70**	1	**43.00**	10	9.37	1	92.19	1	1.67	10
1106	$C_4H_{10}O$	2-methyl-1-propanol	–283.20	1	**–154.90**	10	41.67	5	6.32	0.2	92.91	1	1.64	3
1108	$C_4H_{10}O$	2-methyl-2-propanol	–312.40	1	–177.60	1	39.03	3	6.70	1	94.74	1	1.67	
1107	$C_4H_{10}O$	2-butanol	–292.90	1	**–167.00**	1	**40.75**	3	5.97	1	92.39	1	1.66	10
1402	$C_4H_{10}O$	diethyl ether	–252.10	1	**–122.10**	3	**26.58**	3	7.19	3	104.69	1	1.15	1
1455	$C_4H_{10}O_2$	1,2-dimethoxyethane	*–346.00*		**–198.16**		**32.88**	5	12.56		104.89	1	1.71	
1220	$C_4H_{10}O_2$	1,2-butanediol	–445.80	3	**–304.40**	3	**52.59**	10	*9.72*	25	90.19	1	2.18	
1221	$C_4H_{10}O_2$	1,3-butanediol	–433.20	0.2	**–291.80**	3	**54.83**	10	*8.32*	25	89.98	1	2.52	
1202	$C_4H_{10}O_3$	diethylene glycol	***–547.70***	1	**–385.60**	3	***56.62***	5	13.48*	3	95.27	0.2	2.52	3
1818	$C_4H_{10}S$	diethyl sulfide	–83.56	1	**17.74**	1	**31.79**	3	11.90	1	108.4	1	1.54	5
1712	$C_4H_{11}N$	*n*-butylamine	***–92.35***	1	**49.12**	5	**31.77**	3	*14.80*	10	99.75	0.2	1.39	10
1710	$C_4H_{11}N$	diethylamine	–71.42	3	**73.08**	3	29.33	5	*11.40*	25	104.07	1	0.92	5
1714	$C_4H_{11}N$	isobutylamine	–98.80	3	**46.10**	3	**31.13**	3	*9.99*	25	100.25	1	1.27	
1727	$C_4H_{11}N$	*tert*-butylamine	–119.87	1	28.87		**28.09**	5	0.88	0.2	106.29	1	1.29	
1724	$C_4H_{11}NO_2$	diethanolamine	–408.47	3	**–225.74**	5	***64.40***	5	25.10		*96.24*	1	2.79	10
2623	C_5F_{12}	perfluoro-*n*-pentane	–2570.00	3	**–2366.00**	5	***26.12***	10	6.80	10	*180*	3	**0.00**	
1889	$C_5H_4O_2$	furfural	–151.00	3	**–102.80**	3	***41.95***	5	14.40	3	83.22	1	3.60	
1791	C_5H_5N	pyridine	140.37	1	**190.49**	3	**35.18**	3	8.28	1	80.88	1	2.19	3
405	C_5H_8	1-pentyne	144.40	3	**210.30**	3	***27.06***	5			99	1	0.81	1
269	C_5H_8	cyclopentene	32.30	5	**110.50**	5	***26.98***	3	3.36	1	88.87	1	0.20	3
1079	C_5H_8O	cyclopentanone	–194.10	1	**–93.35**	3	**36.34**	5	7.74*	3	89.09	1	3.24	1
1783	C_5H_9N	valeronitrile	*11.00*	3	**112.30**	3	**37.90**	5	4.73	1	104.9	1	3.54	3
1071	C_5H_9NO	*N*-methyl-2-pyrrolidone	–210.85	3	**–71.60**	10	**44.65**	5	*9.11*		96.61	1	4.08	1
104	C_5H_{10}	cyclopentane	–77.03	3	**38.85**	3	**27.20**	3	0.61	1	94.61	1	**0.00**	

TABLE A.2 Basic Constants 2 (*Continued*)

DIPPR IDC	Formula	Name	$\Delta_f H°(298)$ (kJ/mol)	Er (%)	$\Delta_f G°(298)$ (kJ/mol)	Er (%)	$\Delta_{vap}H(T_b)$ (kJ/mol)	Er (%)	$\Delta_m H(T_m)$ (kJ/mol)	Er (%)	V_L (cm³/mol)	Er (%)	μ (Debye)	Er (%)
209	C_5H_{10}	1-pentene	***−21.62***	3	**78.37**	3	**25.11**	3	5.94	1	*110.3*	1	0.51	25
210	C_5H_{10}	*cis*-2-pentene	*−26.30*	5	**73.70**	5	***26.31***	1	7.11	1	107.92	1	*0.30*	
211	C_5H_{10}	*trans*-2-pentene	*−31.10*	5	**69.80**	5	***26.13***	3	8.35	1	109.1	1	**0.00**	
212	C_5H_{10}	2-methyl-1-butene	−35.30	3	**66.68**	3	**25.54**	3	7.91	1	108.72	1	0.51	
214	C_5H_{10}	2-methyl-2-butene	−41.80	3	**60.45**	3	**26.24**	3	7.60	1	107.19	1	0.34	
213	C_5H_{10}	3-methyl-1-butene	−27.60	5	**76.05**	5	**24.00**	3	5.36	1	112.81	1	0.32	3
3112	$C_5H_{10}O$	cyclopentanol	***−243.00***	3	**−111.30**	5	**44.23**	10	1.54	10	91.38	0.2	1.72	5
1060	$C_5H_{10}O$	2-pentanone	***−259.10***	3	**−137.60**	5	**33.57**	5	10.62	1	107.2	0.2	2.77	3
1053	$C_5H_{10}O$	3-pentanone	−257.90	1	**−134.40**	3	**33.53**	3	11.59		106.41	1	2.82	
1061	$C_5H_{10}O$	methyl isopropyl ketone	***−263.20***	3	**−139.30**	5	**32.54**	5	9.34	3	107.5	0.2	2.76	3
1400	$C_5H_{10}O$	2-methyltetrahydrofuran	*−215.30*	3	**−79.77**	10	**30.81**	5	*7.95*	25	101.5	1	*1.38*	10
1258	$C_5H_{10}O_2$	*n*-pentanoic acid	−491.30	3	**−347.00**	3	42.23	10	14.16	1	109.5	1	1.61	10
1261	$C_5H_{10}O_2$	isovaleric acid	−514.70	3	**−367.00**	3	39.70	10	7.32	3	110.26	3	0.63	3
1332	$C_5H_{10}O_2$	methyl *n*-butyrate	−450.70	3	**−305.30**	3	**34.18**	3	*11.50*	10	114.38	1	1.72	3
1323	$C_5H_{10}O_2$	ethyl propionate	−463.60	1	−319.30	3	**33.73**	3	*12.30*	10	115.62	1	1.75	3
2332	$C_5H_{10}O_2$	methyl isobutyrate	*−464.00*	3	**−316.50**	3	**33.11**	5	*11.50*	25	115.58	1	1.80	
1314	$C_5H_{10}O_2$	*n*-propyl acetate	−464.80	3	−320.40	3	**33.99**	3	*11.20*	10	115.76	1	1.79	1
1305	$C_5H_{10}O_2$	isobutyl formate	−436.30		−293.20		**34.07**	3	*11.20*	10	116.76	1	1.89	1
1319	$C_5H_{10}O_2$	isopropyl acetate	−481.70	1	**−333.70**	3	**33.07**	5	*8.88**	10	117.58	1	1.75	5
1588	$C_5H_{11}Cl$	1-chloropentane	−175.00	3	**−37.26**	25	**32.90**	3	*12.80*	25	121	1	*2.37*	10
7	C_5H_{12}	*n*-pentane	−146.76	1	**−8.81**	1	**25.76**	3	8.40	1	116.05	1	**0.00**	
8	C_5H_{12}	isopentane	−153.70	1	**−14.05**	3	**24.72**	3	5.15	1	117.06	1	0.13	50
9	C_5H_{12}	neopentane	−168.07	1	**−17.14**	3	**22.68**	1	3.15	1	123.26	1	**0.00**	
1109	$C_5H_{12}O$	1-pentanol	−295.70	1	**−142.30**	1	**43.87**	10	12.00	5	108.5	1	1.70	10
1110	$C_5H_{12}O$	2-pentanol	−313.70	1	**−158.80**	3	***42.98***	10	*8.48*	25	109.5	1	*1.67*	10
1112	$C_5H_{12}O$	2-methyl-1-butanol	−302.09	1	**−144.50**	5	**43.45**	5	*7.43*	25	*108.3*	1	1.88	5
1111	$C_5H_{12}O$	2-methyl-2-butanol	−330.80	1	**−166.10**	3	**39.85**	5	4.46	1	109.5	3	1.70	25
1123	$C_5H_{12}O$	3-methyl-1-butanol	*−303.00*	1	**−145.40**	3	**44.51**	5	*6.61*	25	109.2	1	1.85	5
1124	$C_5H_{12}O$	3-methyl-2-butanol	−313.50	1	**−153.70**	5	**41.40**	5			108.3	10	*1.72*	10
1415	$C_5H_{12}O$	ethyl propyl ether	−272.20	3	−115.20	3	*27.58*	5	8.39		121.74	3	1.16	10
1405	$C_5H_{12}O$	methyl *tert*-butyl ether	−283.00	1	*−117.00*	10	**28.20**	3	7.60	3	119.8	1	1.36	3
1120	$C_5H_{12}O$	3-pentanol	−315.40	1	**−156.40**	1	**42.29**	10	*7.78*	25	107.9	5	1.64	5
1864	C_6F_6	hexafluorobenzene	−956.00	1	−878.40		31.97	3	11.59		115.84	1	**0.00**	
1623	C_6F_{14}	perfluoro-*n*-hexane	*−2993.00*	3	**−2747.00**	5	*28.79*	10	6.84	5	*202.5*	3	**0.00**	
1573	$C_6H_4Cl_2$	*m*-dichlorobenzene	25.70	10	**77.90**	10	**39.39**	5	12.59	1	114.53	1	1.72	5

(*Continued*)

TABLE A.2 Basic Constants 2 (*Continued*)

DIPPR IDC	Formula	Name	$\Delta_f H°(298)$ (kJ/mol)	Er (%)	$\Delta_f G°(298)$ (kJ/mol)	Er (%)	$\Delta_{vap}H(T_b)$ (kJ/mol)	Er (%)	$\Delta_m H(T_m)$ (kJ/mol)	Er (%)	V_L (cm³/mol)	Er (%)	μ (Debye)	Er (%)
1572	$C_6H_4Cl_2$	*o*-dichlorobenzene	30.20	10	**82.90**	10	**40.15**	5	12.66	3	112.97	1	2.50	3
1571	C_6H_5Cl	monochlorobenzene	51.09	3	**98.29**	3	**35.57**	5	9.56		102.29	1	1.69	1
1691	C_6H_5I	iodobenzene	164.90	3	**190.20**	3	**40.52**	5	9.75		111.98	1	1.70	5
1886	$C_6H_5NO_2$	nitrobenzene	67.50	1	**163.00**	3	**44.08**	5	11.60	3	102.72	1	4.23	3
501	C_6H_6	benzene	82.88	3	**129.60**	3	**30.74**	1	9.87	1	89.48	1	**0.00**	
1181	C_6H_6O	phenol	–96.40	1	**–32.64**	1	*45.87*	3	11.51	3	*88.94*	1	1.45	10
1792	C_6H_7N	aniline	*86.93*	3	**166.80**	3	*44.06*	3	10.54	1	91.56	0.2	1.53	0.2
1797	C_6H_7N	2-methylpyridine	98.95	3	177.07	5	**36.27**	3	9.72	1	99.08	1	2.04	
2797	C_6H_7N	3-methylpyridine	106.15	3	184.26	5	**37.51**	3	14.18	1	97.8	3	2.40	
2799	C_6H_7N	4-methylpyridine	104.00	1	**183.80**	5	**37.68**	3	11.57	1	98.03	1	2.58	
2088	C_6H_8O	2-cyclohexene-1-one	*–122.20*	3	*–17.33*	5	*40.69*	10	*9.36*	25	*97.4*	10	3.63	3
270	C_6H_{10}	cyclohexene	–4.60	3	**107.70**	3	**30.32**	5	3.29	0.2	101.88	1	0.33	1
1080	$C_6H_{10}O$	cyclohexanone	–226.10	1	**–90.28**	5	**37.86**	3	1.19	10	104.11	3	3.09	
1065	$C_6H_{10}O$	mesityl oxide	*–195.60*	5	**–78.81**	5	**36.13**	5			115.17	1	3.21	
137	C_6H_{12}	cyclohexane	–123.30	3	**31.91**	3	**29.90**	3	2.74	3	108.86	1	**0.00**	
105	C_6H_{12}	methylcyclopentane	–106.20	1	**36.30**	3	**29.05**	3	6.93	1	113.05	1	**0.00**	
216	C_6H_{12}	1-hexene	***–41.67***	3	**87.00**	3	*29.16*	10	9.35	1	*125.8*	1	0.45	25
223	C_6H_{12}	4-methyl-1-pentene	*–51.20*	3	**82.90**	3	*27.07*	3	3.56		127.69	3	*0.41*	10
1151	$C_6H_{12}O$	cyclohexanol	–286.20	1	**–109.50**	3	**45.01**	5	1.78	5	105.75	1	1.86	
1062	$C_6H_{12}O$	2-hexanone	***–279.20***	1	**–129.20**	5	**36.12**	3	14.90	1	124	0.2	2.68	3
1059	$C_6H_{12}O$	3-hexanone	–277.60	1	–126.00	3	**35.52**	3	13.49	0.2	123.63	1	*2.85*	10
1054	$C_6H_{12}O$	methyl isobutyl ketone	–286.40	3	**–134.90**	5	**34.52**	5	*9.71*	25	125.7	0.2	2.69	5
1447	$C_6H_{12}O$	butyl vinyl ether	*–183.00*	25	**–36.20**	25	*32.81*	5	*13.80*	25	129.4	3	1.25	
1262	$C_6H_{12}O_2$	*n*-hexanoic acid	–511.90	1	**–338.00**	3	45.86	10	15.40	1	126.4	3	1.57	5
1333	$C_6H_{12}O_2$	ethyl *n*-butyrate	–485.50	3	**–312.20**	5	**36.10**	5	*12.50*	10	132.92	1	1.81	3
1324	$C_6H_{12}O_2$	*n*-propyl propionate	–483.10	3	**–309.90**	5	**36.35**	5	*15.00*	10	132.47	1	1.79	3
2337	$C_6H_{12}O_2$	ethyl isobutyrate	–499.60	3	–324.00	5	**34.73**	3			*134.57*	5	2.07	
1315	$C_6H_{12}O_2$	*n*-butyl acetate	–485.60	0.2	**–312.60**	1	**36.44**	3	14.59		132.61	3	1.84	
1316	$C_6H_{12}O_2$	isobutyl acetate	–494.70	3	**–321.20**	10	**36.00**	5	*12.40*	10	133.81	1	1.87	1
1306	$C_6H_{12}O_2$	*n*-pentyl formate	*–448.20*	3	**–276.00**	3	*36.64*	10	*19.00*	25	131.81	1	1.90	5
5884	$C_6H_{12}O_3$	2-ethoxyethyl acetate	*–611.00*	25	**–422.00**	25	**40.97**	3	*17.20*	25	136.55	5	2.25	
11	C_6H_{14}	*n*-hexane	–166.94	1	**–0.07**	1	**28.80**	3	13.08	1	131.36	1	**0.00**	
12	C_6H_{14}	2-methylpentane	–174.55	0.2	**–5.34**	5	**27.86**	3	6.27	1	132.92	1	**0.00**	
13	C_6H_{14}	3-methylpentane	–172.00	0.2	**–3.42**	5	**28.06**	3	5.30	0.2	130.45	1	**0.00**	
14	C_6H_{14}	2,2-dimethylbutane	–184.68	1	**–8.74**	5	**26.20**	3	0.58	1	133.71	1	**0.00**	
15	C_6H_{14}	2,3-dimethylbutane	–176.80	1	**–3.12**	5	**27.32**	3	0.80	1	131.15	1	**0.00**	

TABLE A.2 Basic Constants 2 (*Continued*)

DIPPR IDC	Formula	Name	$\Delta_f H°(298)$ (kJ/mol)	Er (%)	$\Delta_f G°(298)$ (kJ/mol)	Er (%)	$\Delta_{vap} H(T_b)$ (kJ/mol)	Er (%)	$\Delta_m H(T_m)$ (kJ/mol)	Er (%)	V_L (cm³/mol)	Er (%)	μ (Debye)	Er (%)
1114	$C_6H_{14}O$	1-hexanol	–316.20	1	**–133.90**	1	**45.46**	10	15.40	1	*125.2*	0.2	1.65	10
1115	$C_6H_{14}O$	2-hexanol	–334.60	3	**–150.60**	3	**44.25**	5	*10.30*	25	126.1	1	*1.66*	10
1116	$C_6H_{14}O$	3-hexanol	*–331.80*	3	**–147.10**	10	***43.02***	10	*10.70*	25	125.7	1	*1.63*	10
1117	$C_6H_{14}O$	2-methyl-1-pentanol	*–322.90*	25	**–138.70**	25	***44.51***	10	*10.40*	25	124.59	3	*1.48*	10
1130	$C_6H_{14}O$	4-methyl-2-pentanol	–344.90	3	**–158.20**	3	***41.54***	5			127.18	1	1.70	
1403	$C_6H_{14}O$	diisopropyl ether	–319.20	1	**–124.80**	3	29.62	3	11.05	1	141.78	1	1.13	10
1446	$C_6H_{14}O$	di-*n*-propyl ether	–292.90	1	**–105.50**	3	**31.49**	5	10.77		137.65	1	1.21	5
1432	$C_6H_{14}O_2$	acetal	–453.50	1	**–245.00**	3	**33.88**	5	*10.90*	25	143.86	3	1.38	25
1706	$C_6H_{15}N$	triethylamine	–95.80	3	**114.10**	3	**31.06**	5	*10.70*	25	139.67	1	0.66	5
1707	$C_6H_{15}N$	di-*n*-propylamine	–116.00	3	**119.60**	3	34.76	10	*14.50*	25	137.26	1	1.07	
1725	$C_6H_{15}NO_3$	triethanolamine	–561.45	3	**–299.40**	3	**73.55**	10	*27.19*		132.99	1	1.08	
2626	C_7F_{16}	perfluoro-*n*-heptane	–3385.40	1	**–3102.00**	3	***31.42***	10	6.95	1	224.7	1	**0.00**	
1874	$C_7H_5ClO_2$	*o*-chlorobenzoic acid	–325.00	1	**–240.53**	5	*79.5**	5	25.76	3	112.66*	10	2.45	
1790	C_7H_5N	benzonitrile	215.70	1	**258.00**	3	**43.07**	10	10.87	0.2	*103.1*	1	4.17	3
1041	C_7H_6O	benzaldehyde	–36.80	10	**22.39**	10	**41.32**	5	9.32		101.89	1	3.21	1
1578	C_7H_7Cl	*p*-chlorotoluene	***26.90***	10	**106.00**	10	**38.57**	5	12.98	3	119.1	3	2.21	3
502	C_7H_8	toluene	50.17	3	**122.20**	3	**33.28**	3	6.64	1	106.65	1	0.36	25
1461	C_7H_8O	anisole	–67.90	3	**22.70**	10	**39.14**	3	*9.60**	25	109.17	1	1.36	
1180	C_7H_8O	benzyl alcohol	***–94.16***	3	**–5.53**	100	**48.69**	3	8.97	10	103.8	0.2	1.70	3
1182	C_7H_8O	*o*-cresol	–128.57	1	**–35.43**	3	***45.53***	5	15.82	1	*104.44*	1	1.45	
1183	C_7H_8O	*m*-cresol	–132.30	1	**–40.19**	3	***47.73***	3	10.71	1	105	1	1.59	5
1184	C_7H_8O	*p*-cresol	–125.35	3	–31.66	3	***47.67***	5	12.71	1	*105.83*	1	1.56	
2796	C_7H_9N	2,6-dimethylpyridine	58.10	3	**167.70**	5	**37.98**	3	10.04*	3	116.75	3	1.68	
1344	$C_7H_{12}O_2$	*n*-butyl acrylate	*–385.00*	25	**–223.00**	25	**38.62**	10	*		143.22	1	1.93	
159	C_7H_{14}	cycloheptane	–119.03	3	63.39	3	**33.28**	3	1.88	1	121.7	3	**0.00**	
138	C_7H_{14}	methylcyclohexane	–154.80	1	**27.33**	3	**31.09**	3	6.75	1	128.2	1	**0.00**	
107	C_7H_{14}	ethylcyclopentane	–126.90	1	**44.80**	3	**31.94**	3	6.87	1	128.75	1	**0.00**	
111	C_7H_{14}	*cis*-1,3-dimethylcyclopentane	–135.85	3	**39.27**	3	**30.93**	3	7.40		132.65	1	**0.00**	
112	C_7H_{14}	*trans*-1,3-dimethylcyclopentane	–133.60	3	**41.52**	3	**30.66**	5	7.40		131.92	1	**0.00**	
234	C_7H_{14}	1-heptene	–62.89	3	**94.82**	3	***31.39***	3	12.64	5	141.4	1	*0.63*	5
1326	$C_7H_{14}O_2$	*n*-butyl propionate	–502.60	3	–300.50	5	**38.93**	5	*17.20*	25	149.37	3	1.80	3
2261	$C_7H_{14}O_2$	*n*-heptanoic acid	–536.20	1	**–334.00**	3	***51.70***	10	15.44	3	142.4	3	*1.68*	10
1347	$C_7H_{14}O_2$	ethyl isovalerate	–527.00	3	**–325.00**	3	***37.63***	5	*10.80*	25	151.19	1	1.97	
1327	$C_7H_{14}O_2$	*n*-propyl *n*-butyrate	–505.30	5	–303.20	5	**37.93**	5	*17.80*	10	150.01	3	1.75	
1336	$C_7H_{14}O_2$	*n*-propyl isobutyrate	*–518.10*	3	**–314.10**	3	***38.22***	5	*13.30*	25	151.42	5	1.89	

(*Continued*)

TABLE A.2 Basic Constants 2 (*Continued*)

DIPPR IDC	Formula	Name	$\Delta_f H°(298)$ (kJ/mol)	Er (%)	$\Delta_f G°(298)$ (kJ/mol)	Er (%)	$\Delta_{vap} H(T_b)$ (kJ/mol)	Er (%)	$\Delta_m H(T_m)$ (kJ/mol)	Er (%)	V_L (cm³/mol)	Er (%)	μ (Debye)	Er (%)
1317	$C_7H_{14}O_2$	isopentyl acetate	*−511.90*	3	**−308.30**	3	**36.60**	10	*11.90*	25	149.94	1	1.80	
17	C_7H_{16}	*n*-heptane	−187.65	1	**8.16**	1	**31.74**	3	14.05	1	147.02	1	**0.00**	
18	C_7H_{16}	2-methylhexane	−194.60	1	**3.47**	3	**30.83**	3	9.18	1	148.69	1	**0.00**	
19	C_7H_{16}	3-methylhexane	−191.30	1	**5.12**	3	**30.96**	3	*9.46*	25	146.41	1	**0.00**	
20	C_7H_{16}	3-ethylpentane	−189.33	3	11.38	5	**31.10**	3	9.55	1	144.11	1	**0.00**	
21	C_7H_{16}	2,2-dimethylpentane	−205.81	3	0.55	3	**29.16**	3	5.82	1	148.89	1	**0.00**	
22	C_7H_{16}	2,3-dimethylpentane	−194.10	3	**5.72**	3	**30.48**	3			144.95	1	**0.00**	
23	C_7H_{16}	2,4-dimethylpentane	−201.67	1	**3.41**	3	**29.53**	3	6.84	1	150.04	1	**0.00**	
24	C_7H_{16}	3,3-dimethylpentane	−199.79	3	4.91	3	***29.68***	3	7.07	1	145.86	1	**0.00**	
25	C_7H_{16}	2,2,3-trimethylbutane	−204.43	3	4.68	3	***28.96***	3	2.26	0.2	145.86	1	**0.00**	
1125	$C_7H_{16}O$	1-heptanol	−336.80	1	**−125.50**	1	**46.79**	10	18.60	1	141.5	1	1.74	10
1126	$C_7H_{16}O$	2-heptanol	*−353.00*	3	**−137.00**	5	***45.30***	10	*15.50*	25	142.7	1	*1.65*	10
1625	C_8F_{18}	perfluoro-*n*-octane	*−3816.00*	3	**−3491.00**	5	33.68	5	9.58	1	248.1	1	**0.00**	
601	C_8H_8	styrene	***147.20***	1	**213.80**	3	**36.80**	3	10.95	1	115.5	0.2	0.13	3
1090	C_8H_8O	acetophenone	−86.70	3	**−1.36**	5	**43.99**	3	*10.30*	25	117.39	3	3.03	3
1283	$C_8H_8O_2$	*p*-toluic acid	−330.00	1	**−212.00**	5	*58.68*	10	22.48	1	*127.2*	10	1.17	10
504	C_8H_{10}	ethylbenzene	29.92	3	**130.90**	1	***35.84***	100	9.17	0.2	123.1	0.2	0.60	10
505	C_8H_{10}	*o*-xylene	19.11	1	**123.00**	1	**36.62**	1	13.61	0.2	121.3	0.2	0.63	10
506	C_8H_{10}	*m*-xylene	***17.36***	1	**119.20**	1	**36.19**	1	11.64	1	123.3	0.2	0.30	1
507	C_8H_{10}	*p*-xylene	18.03	1	**121.50**	1	**35.86**	1	17.11	1	123.9	0.2	**0.00**	
1185	$C_8H_{10}O$	*o*-ethylphenol	−145.20	3	*−23.30*	3	***46.03***	5	13.94	1	120.36	5	1.57	
2112	$C_8H_{10}O$	*m*-ethylphenol	−146.10	3	**−25.32**	3	***49.35***	5	*13.3**	25	121.38	3	1.69	
1187	$C_8H_{10}O$	*p*-ethylphenol	−144.05	1	−21.58	1	***50.87***	10	11.89	3	*123.16*	10	1.78	
1170	$C_8H_{10}O$	2,3-xylenol	−157.19	1	**−34.59**	1	***48.47***	5	21.02	1	*129.69*	5	1.25	1
1172	$C_8H_{10}O$	2,4-xylenol	−162.88	1	**−42.55**	1	**47.60**	5	12.84	1	120.4	1	1.39	1
1174	$C_8H_{10}O$	2,5-xylenol	−161.63	1	**−40.72**	1	***48.74***	5	23.38	1	126.28	1	1.43	1
1176	$C_8H_{10}O$	2,6-xylenol	−161.75	1	−39.04	1	***44.92***	5	18.90	1	*137.06*	5	1.41	3
1177	$C_8H_{10}O$	3,4-xylenol	−156.56	1	−34.28	1	***51.24***	5	18.13	1	123.17	1	1.77	3
1178	$C_8H_{10}O$	3,5-xylenol	−161.54	1	−39.41	1	***49.43***	5	18.00	1	124.72	1	1.76	3
160	C_8H_{16}	cyclooctane	−125.80	3	**89.97**	3	**36.15**	3	2.41	0.2	134.91	1	**0.00**	
147	C_8H_{16}	*trans*-1,4-dimethylcyclohexane	−184.60	3	31.71	3	**32.50**	3	12.33	1	147.92	1	**0.00**	
250	C_8H_{16}	1-octene	***−81.94***	3	**105.70**	3	**34.00**	3	15.31	1	157.4	1	*0.42*	25
1265	$C_8H_{16}O_2$	*n*-octanoic acid	−556.00	1	**−325.00**	3	***53.50***	10	21.35	1	159.5	1	1.70	5
2260	$C_8H_{16}O_2$	2-ethyl hexanoic acid	−559.50	1	**−324.90**	3	***54.10***	10	*18.30*	25	**159.9**	1	0.81	10
1385	$C_8H_{16}O_2$	*n*-butyl *n*-butyrate	−524.90	3	**−294.10**	5	***39.67***	5	14.93		166.66	1	2.05	
1360	$C_8H_{16}O_2$	isobutyl isobutyrate	***−538.50***	3	**−304.00**	10	**38.03**	5	*11.40*	25	169.4	1	*1.95*	5

TABLE A.2 Basic Constants 2 (*Continued*)

DIPPR IDC	Formula	Name	$\Delta_f H°(298)$ (kJ/mol)	Er (%)	$\Delta_f G°(298)$ (kJ/mol)	Er (%)	$\Delta_{vap} H(T_b)$ (kJ/mol)	Er (%)	$\Delta_m H(T_m)$ (kJ/mol)	Er (%)	V_L (cm³/mol)	Er (%)	μ (Debye)	Er (%)
27	C_8H_{18}	*n*-octane	–208.75	1	**16.00**	1	**34.58**	3	20.74	1	162.56	1	**0.00**	
28	C_8H_{18}	2-methylheptane	–215.35	1	**11.69**	1	**33.62**	3	11.88	1	164.4	1	**0.00**	
29	C_8H_{18}	3-methylheptane	–212.51	1	**12.76**	1	**33.70**	3	11.63	1	162.88	1	**0.00**	
30	C_8H_{18}	4-methylheptane	–211.96	1	**15.72**	1	**33.61**	3	10.80	1	162.8	1	**0.00**	
31	C_8H_{18}	3-ethylhexane	–210.71	3	16.83	3	***33.58***	3			160.97	1	**0.00**	
32	C_8H_{18}	2,2-dimethylhexane	–224.60	3	10.44	3	***32.03***	3	6.78	1	165.12	1	**0.00**	
33	C_8H_{18}	2,3-dimethylhexane	–213.80	3	15.49	3	***33.05***	3			161.28	1	**0.00**	
34	C_8H_{18}	2,4-dimethylhexane	–219.24	3	11.35	3	***32.39***	3			164.8	1	**0.00**	
35	C_8H_{18}	2,5-dimethylhexane	–222.51	3	9.66	3	***32.42***	3	12.95	1	165.56	1	**0.00**	
36	C_8H_{18}	3,3-dimethylhexane	–219.99	3	13.39	3	***32.30***	3	7.11	1	161.6	1	**0.00**	
37	C_8H_{18}	3,4-dimethylhexane	–212.67	3	16.74	3	***33.21***	3			159.47	1	**0.00**	
38	C_8H_{18}	2-methyl-3-ethylpentane	–212.80	3	18.94	3	***32.91***	3	11.34	1	159.43	1	**0.00**	
39	C_8H_{18}	3-methyl-3-ethylpentane	–214.85	3	**22.58**	3	**32.71**	3	10.80		157.85	1	**0.00**	
40	C_8H_{18}	2,2,3-trimethylpentane	–219.95	1	**17.73**	3	**31.87**	3	8.62	1	160.34	1	**0.00**	
41	C_8H_{18}	2,2,4-trimethylpentane	–224.01	1	**13.94**	3	**30.73**	3	9.20	1	165.48	1	**0.00**	
42	C_8H_{18}	2,3,3-trimethylpentane	–218.45	3	18.28	5	***32.21***	3	0.86	1	158.16	1	**0.00**	
43	C_8H_{18}	2,3,4-trimethylpentane	–217.32	3	19.00	3	***32.46***	3	9.27	1	159.5	1	**0.00**	
44	C_8H_{18}	2,2,3,3-tetramethylbutane	–225.60	1	**22.39**	1	**31.40**	5	7.54		*174.7*	1	**0.00**	
1132	$C_8H_{18}O$	1-octanol	–357.30	1	**–117.00**	1	**47.90**	10	22.60	1	157.7	0.2	1.65	10
1133	$C_8H_{18}O$	2-octanol	–376.20	1	**–134.30**	5	***46.24***	10	*22.00*	25	159.6	3	1.65	10
1121	$C_8H_{18}O$	2-ethyl-1-hexanol	–363.00	1	**–117.40**	5	**46.08**	10	*13.90*	25	157	0.2	1.74	3
1404	$C_8H_{18}O$	di-*n*-butyl ether	–333.40	1	**–88.27**	3	**36.97**	5	*16.30*	25	170.41	3	1.17	10
2708	$C_8H_{19}N$	*n*-octylamine	***–173.50***	3	**83.92**	5	***40.68***	10	*27.50**	25	165.8	1	1.42	10
1744	$C_8H_{19}N$	di-*n*-butylamine	–156.60	1	**104.00**	3	**39.86**	5	*19.00*	25	170.71	1	0.98	
1626	C_9F_{20}	perfluoro-*n*-nonane	*–4227.00*	3	**–3862.00**	5	*36.08*	10	*15.70*	25	272.2	1	**0.00**	
1748	C_9H_7N	quinoline	200.50	5	**271.70**	5	*46.59*	3	10.66	1	118.55	1	2.29	5
2785	C_9H_7N	isoquinoline	208.40	3	279.70	5	*47.24*	3	13.54	0.2	*118.21*	1	2.61	10
820	C_9H_{10}	indane	60.70	3	**166.80**	5	39.62	5	8.60	0.2	123.13	1	0.54	
509	C_9H_{12}	*n*-propylbenzene	7.90	10	**137.60**	10	**37.88**	3	9.27	1	139.97	1	0.37	
510	C_9H_{12}	cumene	4.00	3	**137.90**	3	**37.12**	3	7.33	1	139.91	1	0.39	
513	C_9H_{12}	*p*-ethyltoluene	–3.20	50	**126.80**	25	***38.09***	3	13.36	1	140.28	1	**0.00**	
514	C_9H_{12}	1,2,3-trimethylbenzene	–9.50	3	126.10	3	**40.18**	3	8.18	1	134.94	1	0.56	
515	C_9H_{12}	1,2,4-trimethylbenzene	–13.80	3	**117.10**	3	**39.33**	3	13.19	1	137.81	1	0.30	
516	C_9H_{12}	mesitylene	–15.90	3	**118.10**	3	**39.16**	3	9.51	1	139.53	1	**0.00**	
259	C_9H_{18}	1-nonene	*–103.50*	3	**112.30**	3	***37.07***	3	19.36	1	173.81	5	0.60	25

(*Continued*)

TABLE A.2 Basic Constants 2 (*Continued*)

DIPPR IDC	Formula	Name	$\Delta_f H°(298)$ (kJ/mol)	Er (%)	$\Delta_f G°(298)$ (kJ/mol)	Er (%)	$\Delta_{vap}H(T_b)$ (kJ/mol)	Er (%)	$\Delta_m H(T_m)$ (kJ/mol)	Er (%)	V_L (cm^3/mol)	Er (%)	μ (Debye)	Er (%)
1259	$C_9H_{18}O_2$	*n*-nonanoic acid	–577.30	0.2	**–317.00**	3	*56.61*	10	19.82	3	175.5	1	*1.68*	10
3318	$C_9H_{18}O_2$	isopentyl butyrate	*–550.10*	3	**–290.10**	10	***41.55***	25	*15.10*	25	183.9	1	*1.68*	5
46	C_9H_{20}	*n*-nonane	–228.74	1	**24.98**	1	**37.26**	3	15.47	1	178.89	1	**0.00**	
91	C_9H_{20}	2-methyloctane	*–234.80*	1	**22.10**	1	**37.07**	3	18.00	3	180.7	3	**0.00**	
96	C_9H_{20}	2,2-dimethylheptane	–246.10	3	**17.90**	3	***34.37***	5	8.95	3	181.51	1	**0.00**	
47	C_9H_{20}	2,2,5-trimethylhexane	–253.10	1	**13.94**	3	**33.40**	3	6.20	1	182.25	1	**0.00**	
51	C_9H_{20}	2,2,3,3-tetramethylpentane	–237.10	3	37.40	3	***34.43***	3	2.30	1	170.33	3	**0.00**	
52	C_9H_{20}	2,2,3,4-tetramethylpentane	–235.00	3	35.20	10	***33.79***	5	0.50	5	174.45	3	**0.00**	
53	C_9H_{20}	2,2,4,4-tetramethylpentane	–242.30	3	34.10	3	**32.51**	3	9.70	1	179.23	3	**0.00**	
54	C_9H_{20}	2,3,3,4-tetramethylpentane	–236.10	1	**38.34**	3	**34.68**	5	8.95	1	170.76	1	**0.00**	
1134	$C_9H_{20}O$	1-nonanol	–377.90	1	**–108.60**	1	**48.65**	10	26.8*	1	175	1	1.61	10
1135	$C_9H_{20}O$	2-nonanol	–397.10	1	**–126.10**	5	***47.26***	10	*20.50*	25	175.4	3	*1.55*	10
1627	$C_{10}F_{22}$	perfluoro-*n*-decane	*–4639.00*	3	**–4234.00**	5	***38.32****	10	*20.00*	25	*297.5*	3	**0.00**	
1381	$C_{10}H_{10}O_4$	dimethyl terephthalate	–627.42	0.2	**–419.70**	3	***53.55***	10	31.63	3	*177.3*	1	2.30	3
701	$C_{10}H_8$	naphthalene	150.58	3	**224.08**	3	43.43	3	18.98	1	131.02	1	**0.00**	
706	$C_{10}H_{12}$	1,2,3,4-tetrahydronaphthalene	26.61	3	167.10	3	42.01	5	12.45	0.2	136.73	1	0.22	
503	$C_{10}H_{12}$	1-methylindan	*37.08*	3	**172.10**	3	***40.73***	10	*9.33*	25	141.2	0.2	*0.46*	10
508	$C_{10}H_{12}$	5-methylindan	*31.72*	3	**164.80**	3	***41.33***	10	*10.80*	25	139.9	0.2	*0.27*	10
518	$C_{10}H_{14}$	*n*-butylbenzene	–13.14	10	**145.40**	10	**40.27**	3	11.22	1	156.61	1	0.37	
519	$C_{10}H_{14}$	isobutylbenzene	–21.55	25	**138.80**	10	***39.43***	3	12.50	1	158.08	1	0.31	
527	$C_{10}H_{14}$	*p*-diethylbenzene	***–22.00***	50	**138.50**	25	***40.24***	3	10.60	1	156.44	1	**0.00**	
524	$C_{10}H_{14}$	*p*-cymene	–29.00	10	**133.52**	10	***39.18***	3	9.66	1	157.46	1	**0.00**	
532	$C_{10}H_{14}$	1,2,4,5-tetramethylbenzene	–47.10	10	**117.70**	10	***42.46***	3	21.00	1	159.94	1	**0.00**	
520	$C_{10}H_{14}$	sec-butylbenzene	–16.90	3	**145.23**	3	**39.40**	5	9.83	1	156.5	1	0.39	
153	$C_{10}H_{18}$	cis-decahydronaphthalene	–169.24	1	**85.62**	3	**40.05**	5	9.49	1	154.66	1	**0.00**	
154	$C_{10}H_{18}$	trans-decahydronaphthalene	–182.17	1	73.55	3	38.59	5	14.41	1	159.52	1	**0.00**	
260	$C_{10}H_{20}$	1-decene	–124.70	3	**122.70**	3	***39.60***	3	13.81	0.2	190	1	0.42	5
1254	$C_{10}H_{20}O_2$	*n*-decanoic acid	–594.30	3	**–305.00**	3	***58.21***	10	27.80	0.2	*193*	1	1.68	5
56	$C_{10}H_{22}$	*n*-decane	–249.46	1	**33.18**	1	**39.43**	3	28.71	1	195.83	1	**0.00**	
48	$C_{10}H_{22}$	3,3,5-trimethylheptane	*–258.80*	1	**33.46**	1	***36.14***	3	*14.00*	5	192.51	5	**0.00**	
57	$C_{10}H_{22}$	2,2,3,3-tetramethylhexane	*–258.30*	1	**46.68**	1	***36.31***	5	12.40	1	187.06	3	**0.00**	
58	$C_{10}H_{22}$	2,2,5,5-tetramethylhexane	*–283.80*	1	**21.40**	1	***35.10***	3	9.80	1	198.72	5	**0.00**	
1136	$C_{10}H_{22}O$	1-decanol	–398.50	1	**–100.20**	1	**49.85**	10	31.20	1	192.8	1	1.62	10
702	$C_{11}H_{10}$	1-methylnaphthalene	116.90	3	**217.90**	3	**46.99**	3	6.94	3	139.9	1	0.51	
703	$C_{11}H_{10}$	2-methylnaphthalene	116.10	3	**216.30**	3	**45.93**	3	12.12	1	*143.09*	1	0.42	
63	$C_{11}H_{24}$	*n*-undecane	–270.43	1	**41.16**	1	**41.90**	3	22.18	1	212.24	1	**0.00**	

TABLE A.2 Basic Constants 2 (*Continued*)

DIPPR IDC	Formula	Name	$\Delta_f H^\circ(298)$ (kJ/mol)	Er (%)	$\Delta_f G^\circ(298)$ (kJ/mol)	Er (%)	$\Delta_{vap}H(T_b)$ (kJ/mol)	Er (%)	$\Delta_m H(T_m)$ (kJ/mol)	Er (%)	V_L (cm³/mol)	Er (%)	μ (Debye)	Er (%)
1137	$C_{11}H_{24}O$	1-undecanol	–419.00	1	**–91.77**	1	***49.48***	10	35.7*	1	207.5	1	1.67	10
558	$C_{12}H_{10}$	biphenyl	178.49	3	**276.30**	3	**48.43**	3	18.58	1	*155.64*	1	**0.00**	
715	$C_{12}H_{12}$	2,7-dimethylnaphthalene	***79.90***	10	**211.00**	10	***49.54***	5	23.35	0.2	*166.08*	10	0.41	
3545	$C_{12}H_{18}$	1,3,5-triethylbenzene	–74.73	3	**144.80**	3	***44.35***	10	*14.80*	25	*188.94*	5	0.10	
814	$C_{12}H_{20}$	1,3-dimethyladamantane	–213.00	1	**82.97**	3	**40.23**	3	1.58	5	182.9	3	**0.00**	
262	$C_{12}H_{24}$	1-dodecene	*–165.40*	1	**136.10**	1	***44.20***	3	19.91	0.2	*222.8*	1	0.52	3
1269	$C_{12}H_{24}O_2$	*n*-dodecanoic acid	*–640.00*	1	**–293.00**	3	***61.47***	5	36.30	1	*229.2*	1	1.64	5
64	$C_{12}H_{26}$	*n*-dodecane	–290.72	1	**49.81**	1	**44.13**	3	36.84	1	228.6	1	**0.00**	
1140	$C_{12}H_{26}O$	1-dodecanol	–439.60	1	**–83.35**	1	**51.10**	10	40.20	1	226.3	1	1.69	10
563	$C_{13}H_{12}$	diphenylmethane	157.20	3	**283.00**	3	**48.72**	3	19.01	0.2	168	1	0.77	10
65	$C_{13}H_{28}$	*n*-tridecane	–311.77	1	**57.71**	1	**46.31**	3	28.50	1	245.63	1	**0.00**	
1141	$C_{13}H_{28}O$	1-tridecanol	–460.10	1	**–74.92**	1	***52.50***	10	44.80	3	241.8	1	*1.65*	10
805	$C_{14}H_{10}$	phenanthrene	***201.80***	3	**303.00**	3	***54.93***	3	16.46	1	167	1	**0.00**	
804	$C_{14}H_{10}$	anthracene	230.10	1	331.70	3	***54.90***	3	29.37	1	183.29	1	**0.00**	
534	$C_{14}H_{22}$	1,4-di-*tert*-butylbenzene	*–141.80*	3	**159.22**	3	*45.42*	5	22.48†	3	231.56	25	0.58	
66	$C_{14}H_{30}$	*n*-tetradecane	–332.44	1	**65.99**	1	**48.15**	3	45.07	1	261.27	1	**0.00**	
1142	$C_{14}H_{30}O$	1-tetradecanol	–480.70	1	**–66.49**	1	**53.48**	10	49.50	1	260.6	1	1.55	10
1198	$C_{15}H_{16}O_2$	bisphenol A	***–191.80***	10	**44.20**	10	***80.77***	25	29.29		*215.37*	10	*0.91*	10
67	$C_{15}H_{32}$	*n*-pentadecane	–353.11	1	**74.26**	1	**48.94**	5	34.59	1	277.78	1	**0.00**	
1143	$C_{15}H_{32}O$	1-pentadecanol	–501.20	1	**–58.07**	1	***54.82***	10	54.2†	1	*277.4*	10	*1.65*	10
68	$C_{16}H_{34}$	*n*-hexadecane	–374.17	1	**82.71**	5	**50.96**	5	53.36	1	294.3	0.2	**0.00**	
90	$C_{16}H_{34}$	2,2,4,4,6,8,8-heptamethylnonane	*–413.20*	3	**73.96**	3	*44.04*	10			289.71	5	**0.00**	
1144	$C_{16}H_{34}O$	1-hexadecanol	–521.80	1	**–49.64**	1	**56.05**	10	59.00†	1	*297.2*	1	1.67	10
69	$C_{17}H_{36}$	*n*-heptadecane	–394.45	1	**91.41**	5	**52.46**	5	40.16	1	311	0.2	**0.00**	
1145	$C_{17}H_{36}O$	1-heptadecanol	–542.30	1	**–41.22**	1	*58.03*	10	63.7†	1	*313*	3	*1.65*	10
561	$C_{18}H_{14}$	*o*-terphenyl	*276.60*	10	**423.00**	10	**57.62**	5	17.19	1	219.66	1	*0.10*	10
560	$C_{18}H_{14}$	*m*-terphenyl	*276.60*	10	**423.00**	10	**63.35**	10	24.00	5	221.28	1	0.20	
559	$C_{18}H_{14}$	*p*-terphenyl	*276.60*	10	**424.00**	10	*65.99*	5	33.70	5	*239.93*	1	0.60	
1276	$C_{18}H_{36}O_2$	*n*-octadecanoic acid	–764.00	1	**–244.00**	3	***65.39***	10	61.21	1	*336.2*	1	1.67	5
70	$C_{18}H_{38}$	*n*-octadecane	–415.12	1	**99.74**	5	***53.87***	10	61.71	5	329	1	**0.00**	
1146	$C_{18}H_{38}O$	1-octadecanol	–562.90	1	**–32.79**	1	***58.58***	10	68.5†	1	333.1	3	1.66	10
71	$C_{19}H_{40}$	*n*-nonadecane	–435.79	1	**108.00**	5	***55.29***	10	45.81	1	346.1	0.2	**0.00**	
1149	$C_{19}H_{40}O$	1-nonadecanol	–583.50	1	**–24.37**	1	*59.86*	10	73.30	1	*349.1*	10	*1.65*	10
73	$C_{20}H_{42}$	*n*-eicosane	–456.46	1	**116.30**	5	***56.76***	10	69.87	5	363.7	0.2	**0.00**	
1148	$C_{20}H_{42}O$	1-eicosanol	–604.00	1	**–15.94**	1	*61.31*	10	78.1†	10	*368.2*	1	*1.65*	10

(*Continued*)

TABLE A.2 Basic Constants 2 (*Continued*)

DIPPR IDC	Formula	Name	$\Delta_f H°(298)$ (kJ/mol)	Er (%)	$\Delta_f G°(298)$ (kJ/mol)	Er (%)	$\Delta_{vap}H(T_b)$ (kJ/mol)	Er (%)	$\Delta_m H(T_m)$ (kJ/mol)	Er (%)	V_L (cm³/mol)	Er (%)	μ (Debye)	Er (%)
74	$C_{21}H_{44}$	*n*-heneicosane	*-477.86*	3	**123.60**	10	***57.97***	10	47.70	3	381.6	1	**0.00**	
75	$C_{22}H_{46}$	*n*-docosane	*-498.50*	3	**131.50**	10	***59.30***	10	48.95	3	399.2	0.2	**0.00**	
76	$C_{23}H_{48}$	*n*-tricosane	*-519.20*	3	**139.80**	10	***60.44***	10	53.97	3	417.4	0.2	**0.00**	
77	$C_{24}H_{50}$	*n*-tetracosane	*-540.00*	3	**148.10**	10	***61.38***	10	54.89	3	435.5	0.2	**0.00**	
1987	$ClFO_3$	perchloryl fluoride	-23.80	25	**48.20**	25	19.31	1	3.83	0.2	*72.8*	1	0.02	1
1904	ClH	hydrogen chloride	-92.31	0.2	**-95.30**	0.2	**16.20**	3	2.00	3	45.72	3	1.08	3
2928	ClH_4N	ammonium chloride	*-69.10*	1	**15.78**	3			*25.30*	25	*41.43*	5		
1939	$ClNa$	sodium chloride	-181.40	3	**-201.35**	3	*170.71*	25	28.20	1	37.71	1	8.99	1
1986	$ClNO$	nitrosyl chloride	51.76	3	66.11	3	**24.87**	3	5.98	5	*51.87*	10	1.90	25
918	Cl_2	chlorine	**0.00**		**0.00**		*20.41*	1	6.41	0.2	50.87	1	**0.00**	
1937	Cl_4Si	tetrachlorosilane	-662.75	0.2	**-622.83**	0.2	***28.01***	10	7.66	3	115.58	1	**0.00**	
925	D_2	deuterium	0.00		0.00		***1.19***	5	0.20		*25.11**	5	**0.00**	
1997	D_2O	deuterium oxide	-249.20	3	**-234.58**	3	41.45	5	6.37	1	18.13	0.2	1.78	3
1905	FH	hydrogen fluoride	-273.30	1	-275.40	1	7.52	10	4.58		*20.7*	3	1.82	1
917	F_2	fluorine	**0.00**		**0.00**		**6.58**	5	0.51	1	25.28	1	**0.00**	
1972	F_3N	nitrogen trifluoride	-132.09	1	**-90.63**	1	**11.58**	3	0.40	1	*45.66*	5	0.24	5
1989	F_4N_2	tetrafluorohydrazine	*-8.37*	100+	*79.80*	100+	***15.13***	3			*117.1*	25	0.26	3
1940	F_6S	sulfur hexafluoride	-1220.47	1	-1116.53	1			5.02	3	*110.5*	5	**0.00**	
1907	HI	hydrogen iodide	26.50	1	**1.72**	1	**20.02**	3	2.87	3	50.76	3	0.45	5
1912	$HNaO$	sodium hydroxide	-197.76	10	**-200.51**	10	**0.00**	3	6.61	25	22.44	3	*6.54*	10
902	H_2	hydrogen	**0.00**		**0.00**		0.90	3	0.12	3	28.57	3	**0.00**	
1921	H_2O	water	-241.82	0.2	**-228.57**	0.2	**40.69**	1	6.00	0.2	18.07	0.2	1.85	1
1901	H_2O_4S	sulfuric acid	-735.20	3	**-653.50**	3	*58.16*	25	10.71	3	53.64	1	2.73	
1922	H_2S	hydrogen sulfide	-20.63	3	-33.44	3	*18.72*	3	2.38	3	43.8	1	0.97	0.2
3951	H_2Se	hydrogen selenide	29.70	3	**16.05**	25	***20.48***	5	2.51		*46.38*	1	0.24	10
1911	H_3N	ammonia	-45.90	1	**-16.40**	3	23.33	1	5.66	0.2	28.29	1	1.47	1
1981	H_3P	phosphine	5.40	10	**13.40**	10	**14.55**	5	*1.13*	1	*69.2*	3	0.57	1
1717	H_4N_2	hydrazine	95.35	1	**159.17**	1	***40.84***	1	12.66	3	31.93	5	1.75	5
1982	H_4Si	silane	34.31	10	**56.79**	10	***11.95***	10	0.67	3	*55.13*	5	**0.00**	
913	He	helium	**0.00**	0.2	**0.00**	0.2	**0.08**	10	0.05		32.28	1	**0.00**	
1998	I_2	iodine	62.42	0.2	**19.38**	0.2	**41.81**	5	15.52	3	63.83	3	**0.00**	
920	Kr	krypton	**0.00**	0.2	**0.00**	0.2	*9.08*	3	1.64	3	34.65	1	**0.00**	
912	NO	nitric oxide	90.25	1	86.57	1	**13.56**	3	2.30	3	*23.43*	5	0.15	1
905	N_2	nitrogen	**0.00**		**0.00**		5.57	1	0.72		34.67	1	**0.00**	
899	N_2O	nitrous oxide	82.05	1	**104.16**	1	16.54	3	6.54	3	*59.28*	3	0.17	1
906	N_2O_4	nitrogen tetroxide	9.08	3	**97.81**	1	***32.85***	10	14.65	0.2	64.01	1	**0.00**	

TABLE A.2 Basic Constants 2 (*Continued*)

DIPPR IDC	Formula	Name	$\Delta_f H°(298)$ (kJ/mol)	Er (%)	$\Delta_f G°(298)$ (kJ/mol)	Er (%)	$\Delta_{vap} H(T_b)$ (kJ/mol)	Er (%)	$\Delta_m H(T_m)$ (kJ/mol)	Er (%)	V_L (cm^3/mol)	Er (%)	μ (Debye)	Er (%)
919	Ne	neon	**0.00**	0.2	**0.00**	0.2	**1.71**	3	0.33		*16.76*	1	**0.00**	
901	O_2	oxygen	**0.00**		**0.00**		**6.79**	1	0.44	3	28.02	1	**0.00**	
910	O_2S	sulfur dioxide	–296.84	0.2	**–300.12**	0.2	**25.34**	3	7.40	3	46.88	1	1.63	1
924	O_3	ozone	142.67	1	**163.16**	3	***14.01***	5	*2.00*		35.58	1	0.54	3
911	O_3S	sulfur trioxide	–395.72	1	–370.95	1	**40.72**	5	7.53		42.1	3	**0.00**	
970	Rn	radon	**0.00**		**0.00**		***17.01***	10	*2.95*	3	*62.17*	1	**0.00**	
1923	S	sulfur	277.17	0.2	**236.72**	1	**9.01**	25	1.73	0.2	17.87	5	**0.00**	
959	Xe	xenon	**0.00**	0.2	**0.00**	0.2	*12.62*	3	2.29		44.45	1	**0.00**	

†likely a combination of two close transtions that cannot be separated experimentally.

TABLE A.3 Ideal gas heat capacity correlation parameters and liquid heat capacity at 298.15 K values.

DIPPR ID	Formula	Name	C_p° Eq	T_{min} (K)	T_{max} (K)	A	B	C	D	E	F	G	Er (%)	C_p^{liq} (298) (J/(mol·K))	Er (%)
914	Ar	argon	1	100	1500	*20786*							1		
922	Br_2	bromine	2	100	1500	*30113*	*8009*	*751.4*	*10780*	*314.6*			1	75.63	3
1906	BrH	hydrogen bromide	2	50	1500	*29120*	*9530*	*2142*	*1570*	*1400*			1		
2686	$CBrClF_2$	bromochlorodifluoromethane	2	100	1500	*37467*	*70523*	*656.3*	*48200*	*290.8*			1	128.05	5
2687	$CBrF_3$	bromotrifluoromethane	2	100	1500	*36417*	*71594*	*728.36*	*45711*	*324.83*			1		
2688	CBr_2F_2	dibromodifluoromethane	2	100	1500	*39722*	*68255*	*644.7*	*49140*	*279*			1	*121.51*	25
1606	$CClF_3$	chlorotrifluoromethane	2	100	1500	**34671**	**73358**	**751.56**	**45633**	**346.45**			1		
1601	CCl_2F_2	dichlorodifluoromethane	2	99.82	1500	*34210*	*73840*	*620.8*	*44000*	*276*			1	**117.86**	3
1894	CCl_2O	phosgene	2	100	1500	*35513*	*47190*	*887.2*	*41170*	*388.9*			1		
1602	CCl_3F	trichlorofluoromethane	2	88.71	1500	*35600*	*72620*	*572.6*	*48230*	*257.6*			1	121.39	3
1501	CCl_4	carbon tetrachloride	2	100	1500	*37582*	*70540*	*512.1*	*48500*	*236.1*			1	**131.90**	3
1616	CF_4	carbon tetrafluoride	3	50	1500	*33260*	*–2847*	*2586*	*30660*	*1908*	*46960*	*825.56*	1		
1634	$CHBrF_2$	bromodifluoromethane	2	100	1500	**37907**	**67070**	**945.24**	**36166**	**418.71**			3	*129.73*	25
1604	$CHClF_2$	chlorodifluoromethane	2	100	1500	*34574*	*70901*	*995.73*	*46085*	*442.97*			1		
1696	$CHCl_2F$	dichlorofluoromethane	2	100	1500	**36822**	**68464**	**943.48**	**45526**	**404.65**			1	108.38	5
1521	$CHCl_3$	chloroform	2	100	1500	*39420*	*65730*	*928*	*49300*	*399.6*			1	113.67	3
1615	CHF_3	trifluoromethane	2	50	1500	*33616*	*71905*	*1026.4*	*36091*	*467.9*			1		
1771	CHN	hydrogen cyanide	3	20	1500	*33258*	*17215*	*1498.4*	*15297*	*4533.7*	*0*	*1*	10	**70.86**	10
1511	CH_2Cl_2	dichloromethane	2	100	1500	*36280*	*68040*	*1256*	*42750*	*548*			1	100.88	5
1614	CH_2F_2	difluoromethane	3	20	1500	*33258*	*9705*	*796.54*	*46514*	*1794.67*	*18311.9*	*4087.45*	1		
1001	CH_2O	formaldehyde	3	20	1500	*33258*	*27337*	*1848.8*	*22022*	*3657.7*	*0*	*1*	10		
1251	CH_2O_2	formic acid	2	50	1500	*33810*	*75930*	*1192.5*	*31800*	*550*			3	99.39	3
1641	CH_3Br	bromomethane	3	20	1500	33258	41641	1382.4	31238	3834.5	0	1	5		
1502	CH_3Cl	methyl chloride	3	20	1500	33258	32089	1314.5	36070	3022.10	0	1	3	81.30	3
1613	CH_3F	methyl fluoride	3	20	6000	33258	*–16916*	*1791.6*	*63588*	*1791.6*	*28130*	*4149.9*	5		
1681	CH_3I	methyl iodide	3	20	1500	33258	34415110	6477.9	–34405890	6480.95	49987.7	1487.39	5	82.40	5
2851	CH_3NO	formamide	2	150	1500	*38220*	*93000*	*1845*	*69000*	*850*			3	**108.30**	10
1760	CH_3NO_2	nitromethane	3	20	1500	*33258*	*6173.3*	*117.55*	*56569*	*1173.34*	*53318.5*	*2900.66*	1	106.60	3
1	CH_4	methane	2	50	1500	*33298*	*79933*	*2086.9*	*41602*	*991.96*			1		
1101	CH_4O	methanol	3	20	1500	*33258*	*36199*	*1205.7*	*15373000*	*3212.2*	*–1.5318E+07*	*3212.2*	1	81.12	1
1801	CH_4S	methyl mercaptan	3	20	1500	*33258*	*10382*	*314.11*	*43301*	*1483.28*	*38798.2*	*3599.23*	1		
1701	CH_5N	methylamine	3	20	1500	*33258*	*25368*	*712.74*	*79861*	*2505*	*0*	*1*	3	**106.92**	3
908	CO	carbon monoxide	2	60	1500	*29108*	*8773*	*3085.1*	*8455.3*	*1538.2*			1		
1893	COS	carbonyl sulfide	3	20	6000	33258	22624	1105.7	15350000	3423.4	–15341000	3423.2	3	**77.77**	5
909	CO_2	carbon dioxide	2	50	5000	*29370*	*34540*	*1428*	*26400*	*588*			1		
1938	CS_2	carbon disulfide	3	20	1500	33258	24111	858.23	6315	3228.3	0	1	10	**78.98**	3

TABLE A.3 Ideal gas heat capacity correlation parameters and liquid heat capacity at 298.15 K values. ***(Continued)***

DIPPR ID	Formula	Name	C_p° Eq	T_{min} (K)	T_{max} (K)	A	B	C	D	E	F	G	Er (%)	C_p^{liq} (298) (J/(mol·K))	Er (%)
2692	C_2ClF_5	chloropentafluoroethane	2	200	1500	*80270*	*99580*	*1084.8*	*81960*	*581.6*			1	179.59	3
1611	$C_2Br_2F_4$	1,2-dibromotetrafluoroethane	2	298.15	1200	*144600*	*40000*	*933*	*–34700000*	*7.223*			25	170.24	1
1693	$C_2Cl_2F_4$	1,1-dichlorotetrafluoroethane	2	50	1500	*41436*	*139040*	*571.03*	*80848*	*229.84*			1	*181.22*	10
1609	$C_2Cl_2F_4$	1,2-dichlorotetrafluoroethane	2	273.15	1500	*87120*	*93062*	*1062*	*84800*	*567*			1	174.68	3
2655	$C_2Cl_3F_3$	1,1,2-trichlorotrifluoroethane	2	200	1500	*101720*	*82104*	*594.51*	*300000*	*0.00003*			1	172.96	3
1542	C_2Cl_4	tetrachloroethylene	2	100	1500.1	*51830*	*80920*	*723.8*	*66980*	*312.73*			1	140.99	10
1630	C_2F_4	tetrafluoroethylene	2	100	1500	*41691*	*90688*	*854.37*	*66399*	*361.51*			1		
2693	C_2F_6	hexafluoroethane	3	50	1500	**33258**	**81604**	**495.21**	**31759000**	**2041.7**	**–31694000**	**2042.9**	5		
2640	$C_2HBrClF_3$	halothane	2	100	1500	*81960*	*99361*	*1358.5*	*89584*	*649.85*			3	156.54	3
2648	C_2HClF_4	2-chloro-1,1,1,2-tetrafluoroethane	2	100	1500	*49134*	*126920*	*771.79*	*69884*	*333.78*			1	*150.34*	5
1694	$C_2HCl_2F_3$	2,2-dichloro-1,1,1-trifluoroethane	2	100	1500	**51054**	**126650**	**755.14**	**70565**	**325.48**			1	**158.12**	3
2647	$C_2HCl_2F_3$	1,2-dichloro-1,1,2-trifluoroethane	2	100	1500	*49971*	*126730*	*728.75*	*72269*	*316.6*			1	*162.38*	10
1646	C_2HF_5	pentafluoroethane	3	20	1500	**33258**	**19307**	**193.07**	**50573**	**656.33**	**72930.69**	**1683.23**	1	161.54	3
3605	C_2HF_5O	difluoromethyl trifluoromethyl ether	2	260	1500	**79881**	**114860**	**1132.8**	**81671**	**610.47**			3	**173.75**	10
401	C_2H_2	acetylene	3	20	1500	*33258*	*71101530*	*1601.8*	*–71077610*	*1602.12*	*31821.29*	*4056.64*	1		
1581	$C_2H_2Cl_2$	*trans*-1,2-dichloroethylene	2	200	1500	*49990*	*80550*	*1414.5*	*62300*	*625.15*			1	**113.90**	10
1529	$C_2H_2Cl_4$	1,1,2,2-tetrachloroethane	2	298.15	1500	*76894*	*105420*	*1420*	*91253*	*652.78*			1	**166.03**	10
1629	$C_2H_2F_2$	1,1-difluoroethylene	2	100	1500	*39060*	*91390*	*1384.4*	*71500*	*611.25*			3		
2650	$C_2H_2F_4$	1,1,1,2-tetrafluoroethane	3	20	1500	*33258*	*42636*	*413.07*	*85068*	*1459.74*	*16896.78*	*3859.35*	1	144.82	3
2660	$C_2H_2F_4$	1,1,2,2-tetrafluoroethane	3	20	1500	*33258*	*18585*	*193.09*	*57705*	*701.77*	*65023.75*	*2183.29*	1		
1504	C_2H_3Cl	vinyl chloride	2	200	1500	*42364*	*87350*	*1649.2*	*65560*	*739.07*			1	*85.92*	25
2695	$C_2H_3ClF_2$	1-chloro-1,1-difluoroethane	3	20	1500	*33258*	*44463*	*484.07*	*68317*	*1396.55*	*30751.50*	*3656.44*	1	131.40	5
2649	$C_2H_3Cl_2F$	1,1-dichloro-1-fluoroethane	3	20	1500	*33258*	*51336*	*463.14*	*65140*	*1419.63*	*27359.23*	*3779.56*	1	133.55	3
1527	$C_2H_3Cl_3$	1,1,1-trichloroethane	3	20	1500	*33258*	*55303*	*437.62*	*61758*	*1399.38*	*26878.20*	*3692.90*	1	144.16	5
1524	$C_2H_3Cl_3$	1,1,2-trichloroethane	2	298.15	1500	*66554*	*112570*	*1545.4*	*97196*	*717.04*			3	150.85	5
1619	$C_2H_3F_3$	1,1,1-trifluoroethane	3	20	1500	**33258**	**39618**	**503.05**	**72706**	**1451.27**	**31344.43**	**3746.54**	1		
2619	$C_2H_3F_3$	1,1,2-trifluoroethane	2	200	3000	*57425*	*120550*	*1580.3*	*97285*	*709.56*			5	*149.49*	25
1772	C_2H_3N	acetonitrile	3	20	1500	**33258**	**105320**	**993.6**	**50689**	**3028.9**	**–61642**	**992.83**	10	91.28	3
201	C_2H_4	ethylene	2	60	1500	*33380*	*94790*	*1596*	*55100*	*740.8*			1		
1673	$C_2H_4Br_2$	1,2-dibromoethane	2	200	1500	*74906*	*127250*	*1981*	*94370*	*845.2*			3	135.70	5
1522	$C_2H_4Cl_2$	1,1-dichloroethane	3	20	1500	*33258*	*36601*	*442.12*	*69076*	*1466.22*	*39200.00*	*3907.88*	1	126.57	3
1523	$C_2H_4Cl_2$	1,2-dichloroethane	2	200	1500	*65271*	*112540*	*1737.6*	*87800*	*795.45*			3	**129.25**	3
1640	$C_2H_4F_2$	1,1-difluoroethane	3	20	1500	**33258**	**29008**	**484.85**	**77874**	**1572.41**	**37868.76**	**4045.61**	1	117.31	1
1002	C_2H_4O	acetaldehyde	3	20	1500	*33258*	*114840*	*1056*	*61246*	*3047*	*–60366*	*1056*	10		

(Continued)

TABLE A.3 Ideal gas heat capacity correlation parameters and liquid heat capacity at 298.15 K values. (*Continued*)

DIPPR ID	Formula	Name	C_p° Eq	T_{min} (K)	T_{max} (K)	A	B	C	D	E	F	G	Er (%)	C_p^{liq} (298) (J/(mol·K))	Er (%)
1441	C_2H_4O	ethylene oxide	2	50	1500	*33460*	*121160*	*1608.4*	*82410*	*737.3*			1		
1252	$C_2H_4O_2$	acetic acid	2	50	1500	40200	136750	1262	70030	569.7			1	123.86	3
1301	$C_2H_4O_2$	methyl formate	2	250	1500	*50600*	*121900*	*1637*	*89400*	*743*			3	**119.69**	10
1645	C_2H_5Br	bromoethane	3	20	1500	*33258*	*84503*	*1110.4*	*66044*	*3765.4*	*0*	*1*	10	99.73	5
1503	C_2H_5Cl	ethyl chloride	3	20	1500	**33258**	**61292**	**2550.6**	**63689**	**958.81**	**57358.82**	**6954.27**	3		
1617	C_2H_5F	ethyl fluoride	3	20	1500	*33258*	*16970*	*431.03*	*78420*	*1556*	*48580*	*3922*	1	**110.38**	25
2	C_2H_6	ethane	3	20	1500	*33258*	*70254*	*1605.8*	*60623*	*3784.68*	*12107.57*	*489.58*	1		
1102	C_2H_6O	ethanol	3	20	1500	*33258*	*79290*	*1029.3*	*15385000*	*3123.9*	*−15306000*	*3123.9*	1	112.34	3
1401	C_2H_6O	dimethyl ether	3	20	1500	*33258*	*25893*	*383.55*	*73861*	*1606.47*	*64223.38*	*3686.56*	1		
1201	$C_2H_6O_2$	ethylene glycol	3	20	1500	*33258*	*103580*	*1008.6*	*91778*	*3812.6*	*0*	*1*	10	**149.83**	1
1802	C_2H_6S	ethyl mercaptan	3	20	1500	*33258*	*29571*	*329.93*	*74832*	*1482.34*	*56862.99*	*3577.37*	1	117.82	1
1820	C_2H_6S	dimethyl sulfide	2	200	1500	*60370*	*137470*	*1641*	*79880*	*743.5*			3	118.15	3
1704	C_2H_7N	ethylamine	3	20	1500	*33258*	*89803*	*983.37*	*92039*	*3136.7*	*0*	*1*	3	**133.33**	10
1702	C_2H_7N	dimethylamine	2	200	1500	*55650*	*163840*	*1734.1*	*108990*	*793.04*			3		
1723	C_2H_7NO	monoethanolamine	2	298.15	1500	*72140*	*181500*	*2030*	*131400*	*860*			25	*161.29*	25
1741	$C_2H_8N_2$	ethylenediamine	3	20	1500	*33258*	*119170*	*914.5*	*110260*	*3314.3*	*0*	*1*	10	**172.86**	3
2652	C_3F_8	octafluoropropane	3	50	1500	*33258*	*143910*	*577.1*	*13264000*	*2380.4*	*−13194000*	*2385.6*	10		
2630	C_3HF_7	1,1,1,2,3,3,3-heptafluoropropane	2	298.15	1000	*100100*	*147460*	*1073.6*	*101210*	*573.16*			3		
2631	$C_3H_2F_6$	1,1,1,2,3,3-hexafluoropropane	2	298.15	1000	*91034*	*149780*	*1058.1*	*90155*	*565.2*			3	*208.06*	25
1620	$C_3H_2F_6$	1,1,1,3,3,3-hexafluoropropane	2	298.15	1500	**105150**	**148240**	**1858.2**	**167760**	**820.12**			10		
2629	$C_3H_3F_5$	1,1,1,2,2-pentafluoropropane	2	298	1200	*102720*	*138610*	*808.03*	*150000*	*17000*			5		
2627	$C_3H_3F_5$	1,1,1,3,3-pentafluoropropane	2	290	1500	**97947**	**163590**	**2009.8**	**166010**	**846.71**			3		
2628	$C_3H_3F_5$	1,1,2,2,3-pentafluoropropane	2	298.15	1500	**36050**	**213460**	**1179**	**201560**	**478.64**			10	*175.03*	25
2868	$C_3H_3F_5O$	2-(difluoromethoxy)-1,1,1-trifluoroethane	2	277.61	1500	*44257*	*220770*	*939.03*	*171390*	*396.44*			10	*207.85*	25
1774	C_3H_3N	acrylonitrile	3	20	1500	*33258*	*70696*	*989.83*	*15372000*	*3344.3*	*−15319000*	*3344.3*	10	108.70	5
5892	C_3H_3NO	isoxazole	2	100	1500	**33909**	**144270**	**1380**	**105240**	**645.05**			1	**106.03**	10
402	C_3H_4	methylacetylene	3	20	1500	*33258*	*23922*	*530.62*	*52471*	*1493.21*	*46387.30*	*3810.04*	1		
301	C_3H_4	propadiene	3	20	1500	*33258*	*19318*	*535.61*	*61686*	*1476.66*	*42452.60*	*3941.22*	1		
1773	C_3H_5N	propionitrile	3	20	1500	*33258*	*88674*	*963.68*	*15384000*	*2983.9*	*−15308000*	*2982.9*	10	119.61	3
202	C_3H_6	propylene	2	130	1500	**43852**	**150600**	**1398.8**	**74754**	**616.46**			3		
101	C_3H_6	cyclopropane	2	100	1500	*33800*	*168940*	*1613.5*	*117680*	*722.8*			1		
1526	$C_3H_6Cl_2$	1,2-dichloropropane	3	20	1500	*33258*	*48729*	*351.26*	*104170*	*1417.2*	*67563*	*3558.5*	10	156.43	1
1167	C_3H_6O	allyl alcohol	2	200	1500	*58414*	*162170*	*1731.2*	*116200*	*773.05*			1	*150.57*	25
1051	C_3H_6O	acetone	3	20	1500	**33258**	**31740**	**309.1**	**76072**	**1374.7**	**78900**	**3334.7**	5	**125.13**	3
1003	C_3H_6O	propanal	3	20	1500	**33258**	**199990**	**880.88**	**129230**	**3217.8**	**−108490**	**880.88**	10	*132.05*	5

TABLE A.3 **Ideal gas heat capacity correlation parameters and liquid heat capacity at 298.15 K values. (*Continued*)**

DIPPR ID	Formula	Name	C_p° Eq	T_{min} (K)	T_{max} (K)	A	B	C	D	E	F	G	Er (%)	C_p^{liq} (298) (J/(mol·K))	Er (%)
1442	C_3H_6O	1,2-propylene oxide	2	200	1500	*49450*	*174450*	*1563*	*114580*	*702.2*			1	120.40	3
1253	$C_3H_6O_2$	propionic acid	2	298.15	1500	*69590*	*177780*	*1709.8*	*126540*	*763.78*			10	151.76	3
1312	$C_3H_6O_2$	methyl acetate	2	298	1500	*55500*	*178200*	*1260*	*85300*	*562*			3	**142.03**	10
1302	$C_3H_6O_2$	ethyl formate	2	100	1500	*53700*	*188600*	*1207*	*86400*	*496*			3	**146.67**	10
2391	$C_3H_6O_3$	dimethyl carbonate	2	298.15	1500	*71662*	*189620*	*1423.2*	*122710*	*671.1*			10	170.43	10
1585	C_3H_7Cl	propyl chloride	3	20	1500	33258	95809	853.12	15395170	2600.53	−15295700	2600.53	1	**131.91**	3
1876	C_3H_7NO	*N,N*-dimethylformamide	2	200	1500	*72200*	*178300*	*1532*	*131000*	*762*			3	**150.43**	3
3	C_3H_8	propane	3	20	1500	33258	79306	3837.3	28074	413.45	106269.64	1543.58	1	120.08	1
1103	C_3H_8O	1-propanol	3	20	1500	*33258*	*118200*	*978.61*	*−67036000*	*3135.6*	*67145000*	*3135.6*	1	144.47	3
1104	C_3H_8O	isopropanol	3	50	1500	*33260*	*796300*	*1001*	*101100*	*3286*	*−667100*	*1001*	1	156.83	3
1407	C_3H_8O	methyl ethyl ether	3	20	1500	*33258*	*43831*	*328.85*	*94342*	*1401.59*	*97133.39*	*3383.37*	1		
1431	$C_3H_8O_2$	methylal	2	298.15	1000.2	*74976*	*161660*	*862.87*	*789640*	*4671.8*			5	161.50	1
1212	$C_3H_8O_2$	1,3-propylene glycol	3	20	1500	*33258*	*141670*	*936.36*	*118770*	*3547.6*	*0*	*1*	10	**175.35**	3
1231	$C_3H_8O_3$	glycerol	2	298.15	1200.2	*96490*	*151870*	*821.2*	*182800*	*3272*			25	221.79	3
1813	C_3H_8S	methyl ethyl sulfide	2	273.16	1500	*75083*	*195770*	*1642.4*	*119490*	*749.19*			3	144.69	3
1711	C_3H_9N	*n*-propylamine	3	20	1500	*33258*	*36661*	*300.51*	*121910*	*1415.9*	*100830*	*3546.4*	5	**162.72**	5
1719	C_3H_9N	isopropylamine	3	20	1500	*33258*	*41602*	*307.74*	*116664*	*1352.98*	*96658.20*	*3560.68*	1	163.81	3
1703	C_3H_9N	trimethylamine	2	200	1500	*71070*	*150510*	*796.62*	*84537*	*2187.6*			3		
6862	C_3H_9NO	methylethanolamine	2	298.15	1200	*75020*	*229940*	*1593.2*	*152540*	*721.2*			25		
2654	C_4F_8	octafluorocyclobutane	2	200	1500	*70208*	*211160*	*750.7*	*117140*	*325.2*			3		
1478	C_4H_4O	furan	3	20	1500	*33258*	*56573*	*1015.7*	*81674*	*1704.19*	*35937.12*	*4367.27*	1	114.68	1
1821	C_4H_4S	thiophene	3	20	1500	*33258*	*49461*	*846.2*	*86757*	*1547.49*	*37444.98*	*4174.46*	1	123.98	1
1721	C_4H_5N	pyrrole	2	200	1500	*40055*	*186480*	*1520.9*	*144660*	*680.63*			3	127.83	1
403	C_4H_6	ethylacetylene	3	20	1500	*33258*	*33007*	*399.0*	*86509*	*1350.64*	*71885.81*	*3613.72*	1		
302	C_4H_6	1,2-butadiene	3	20	1500	*33258*	*31365*	*342.8*	*84925*	*1381.00*	*73431.16*	*3422.76*	1		
303	C_4H_6	1,3-butadiene	2	200	1500	*50950*	*170500*	*1532.4*	*133700*	*685.6*			1	123.60	1
1321	$C_4H_6O_2$	vinyl acetate	2	100	1500	*53600*	*211900*	*1198*	*114700*	*510*			1	**171.47**	5
1291	$C_4H_6O_3$	acetic anhydride	3	20	1500	*33258*	*95397*	*628.05*	*122300*	*1822.1*	*56573*	*3743.5*	5	**189.56**	5
1782	C_4H_7N	butyronitrile	3	20	1500	*33258*	*83628*	*637.71*	*51881*	*3627.5*	*92547*	*1946.6*	10	*144.31*	5
102	C_4H_8	cyclobutane	3	20	1500	*33258*	*15112*	*390.9*	*81862*	*3764.15*	*146243.15*	*1508.58*	1		
204	C_4H_8	1-butene	2	250	1500	**64257**	**206180**	**1676.8**	**133240**	**757.06**			1	128.30	3
206	C_4H_8	*trans*-2-butene	3	20	1500	*33258*	*42303*	*354.4*	*104949*	*1525.80*	*89927.85*	*3608.22*	1		
205	C_4H_8	*cis*-2-butene	3	20	1500	*33258*	*112805*	*1375.4*	*98776*	*3409.20*	*23233.48*	*152.22*	1	127.45	1
207	C_4H_8	isobutene	3	20	1500	*33258*	*41250*	*397.5*	*107862*	*1475.17*	*88331.99*	*3687.58*	1	131.10	3
1005	C_4H_8O	butanal	3	20	1500	*33258*	*210840*	*855.11*	*150670*	*3100.7*	*−80439*	*855.11*	10	162.43	3

(*Continued*)

TABLE A.3 Ideal gas heat capacity correlation parameters and liquid heat capacity at 298.15 K values. (*Continued*)

DIPPR ID	Formula	Name	C_p° Eq	T_{min} (K)	T_{max} (K)	A	B	C	D	E	F	G	Er (%)	C_p^{liq} (298) (J/(mol·K))	Er (%)
1052	C_4H_8O	methyl ethyl ketone	3	20	1500	**33258**	**47651**	**295.36**	**118050**	**1432.9**	**101960**	**3308.2**	5	**158.86**	3
1479	C_4H_8O	tetrahydrofuran	3	20	1500	33258	12024	206.7	154100	1397.70	99717.73	3368.09	1	123.98	1
1256	$C_4H_8O_2$	*n*-butyric acid	2	298.15	1500	148800	135220	1146	−67800000	6.98			10	177.75	3
1260	$C_4H_8O_2$	isobutyric acid	2	298.15	1200	74694	243560	1715	184840	757.75			10	181.72	10
2422	$C_4H_8O_2$	1,3-dioxane	3	20	1500	33258	40264	496.4	164300	1566.38	90448.74	3564.23	1	**144.88**	5
1421	$C_4H_8O_2$	1,4-dioxane	3	20	1500	33258	41306	483.49	166045	1542.70	88633.44	3629.14	1	153.65	5
1322	$C_4H_8O_2$	methyl propionate	2	298.15	1200	77650	244200	1714	181800	716			5		
1313	$C_4H_8O_2$	ethyl acetate	2	200	1500	99810	209310	2022.6	180300	928.05			1	170.80	1
1303	$C_4H_8O_2$	*n*-propyl formate	2	298.15	1500	87100	244700	1925.4	188800	821.3			3	**172.93**	10
1587	C_4H_9Cl	sec-butyl chloride	2	150	1500	78230	233900	1557.5	152500	700			1	159.27	10
1766	C_4H_9N	pyrrolidine	2	200	1500	53150	272480	1730.7	200890	793.4			1	156.45	3
5	C_4H_{10}	*n*-butane	3	20	1500	33258	47101	306.73	98932	3782.21	137464.89	1521.87	1	140.50	1
4	C_4H_{10}	isobutane	3	20	1500	33258	48701	411.26	141643	1544.73	93408.55	3850.62	1	141.43	3
2752	$C_4H_{10}N_2$	piperazine	3	20	1500	33258	115610	3640.4	223410	1087.6	0	1	10		
1105	$C_4H_{10}O$	1-butanol	3	50	1500	33258	160070	941.26	15414000	3102.5	−15277000	3102.5	1	177.87	3
1106	$C_4H_{10}O$	2-methyl-1-propanol	2	298.15	1200	87940	241600	1718	165400	798.7			5	181.78	3
1108	$C_4H_{10}O$	2-methyl-2-propanol	3	20	1500	33258	65702	404.47	141906	1499.43	94354.38	3836.80	1		
1107	$C_4H_{10}O$	2-butanol	3	50	1500	33260	134100	3269	−165700	939.653	332500	939.653	10	197.48	3
1402	$C_4H_{10}O$	diethyl ether	3	20	1500	33258	151488	2907.1	102000	1085.39	49009.59	236.171	3	175.63	3
1455	$C_4H_{10}O_2$	1,2-dimethoxyethane	2	298.15	1500	175440	164880	1325.1	−74185000	8.193			25	**195.18**	5
1220	$C_4H_{10}O_2$	1,2-butanediol	2	298.15	1500.1	104780	254900	1877.6	187500	833			10	231.03	3
1221	$C_4H_{10}O_2$	1,3-butanediol	2	298.15	1500.2	106600	257500	1967	195100	860.5			5	209.19	3
1202	$C_4H_{10}O_3$	diethylene glycol	3	20	1500	33258	145990	2704.9	174500	797.26	0	1	10	**244.34**	3
1818	$C_4H_{10}S$	diethyl sulfide	2	200	1500	91273	241000	1668.6	165200	771.08			3	171.27	0.2
1712	$C_4H_{11}N$	*n*-butylamine	3	20	1500	33258	148800	806.68	153110	2678.8	0	1	5	**191.94**	10
1710	$C_4H_{11}N$	diethylamine	2	200	1500	91020	267400	1719	179260	794.94			3	**173.83**	10
1714	$C_4H_{11}N$	isobutylamine	2	298.15	1500	86855	260410	1587.6	166580	734.47			3	**183.19**	25
1727	$C_4H_{11}N$	*tert*-butylamine	2	200	1500	91545	264480	1712.9	193390	779.4			3	192.10	3
1724	$C_4H_{11}NO_2$	diethanolamine	2	298.15	1500.1	120800	306600	2089	234300	891			25		
2623	C_5F_{12}	perfluoro-*n*-pentane	3	50	1500	33258	−20736000	2453.1	241120	559.16	20844000	2446.9	10	347.79	25
1889	$C_5H_4O_2$	furfural	2	100	1500	47300	198300	1040.6	109000	472.6			1	157.93	5
1791	C_5H_5N	pyridine	3	20	1500	33258	36589	678.29	131940	1531.62	53751.76	3848.07	1	132.65	1
405	C_5H_8	1-pentyne	2	200	1500	75300	209050	1530.7	137800	672.8			1	**162.71**	10
269	C_5H_8	cyclopentene	2	150	1500	48074	251590	1580.3	174540	718.37			1	122.72	3
1079	C_5H_8O	cyclopentanone	3	20	1500	33258	23200	258.41	156955	1283.17	111582.62	3171.62	1	**149.83**	10
1783	C_5H_9N	valeronitrile	3	20	1500	33258	169040	885.95	15412000	2983.9	−15279000	2983.1	10	174.10	5

TABLE A.3 Ideal gas heat capacity correlation parameters and liquid heat capacity at 298.15 K values. (*Continued*)

DIPPR ID	Formula	Name	C_p° Eq	T_{min} (K)	T_{max} (K)	A	B	C	D	E	F	G	Er (%)	C_p^{liq} (298) (J/(mol·K))	Er (%)
1071	C_5H_9NO	*N*-methyl-2-pyrrolidone	2	298.15	1200.1	*73440*	*346700*	*1874.5*	*249100*	*821.5*			25		
104	C_5H_{10}	cyclopentane	2	100	1500	*41600*	*301400*	*1461.7*	*180950*	*668.8*			1	127.23	3
209	C_5H_{10}	1-pentene	2	298.15	1500	**82523**	**259430**	**1729.1**	**176800**	**778.7**			3	155.29	1
210	C_5H_{10}	*cis*-2-pentene	2	200	1500	*76600*	*266300*	*1800.4*	*184520*	*826.65*			1	152.12	1
211	C_5H_{10}	*trans*-2-pentene	2	200	1500	*86080*	*256230*	*1781.1*	*172300*	*808.58*			1	156.46	1
212	C_5H_{10}	2-methyl-1-butene	2	200	1500	*87026*	*255560*	*1775.7*	*176360*	*807.82*			1	157.33	3
214	C_5H_{10}	2-methyl-2-butene	2	200	1500	*81924*	*260380*	*1759.3*	*171950*	*800.93*			1	152.83	3
213	C_5H_{10}	3-methyl-1-butene	2	200	1500	*91600*	*250030*	*1708.7*	*166970*	*755.31*			1	156.14	3
3112	$C_5H_{10}O$	cyclopentanol	3	20	1500	33258	28091	197.18	217610	1317.8	140440	4960.6	10	**183.35**	10
1060	$C_5H_{10}O$	2-pentanone	3	20	1500	*33258*	*64732*	*282.92*	*165330*	*1500.9*	*115260*	*3610.4*	10	**184.39**	3
1053	$C_5H_{10}O$	3-pentanone	2	200	1500	*96896*	*249070*	*1417.7*	*130100*	*646.7*			3	190.81	3
1061	$C_5H_{10}O$	methyl isopropyl ketone	3	20	1500	**33258**	**64448**	**307.97**	**119540**	**1277.9**	**153140**	**2966.2**	5	**180.50**	5
1400	$C_5H_{10}O$	2-methyltetrahydrofuran	3	20	1500	*33258*	*123760*	*3255.7*	*201080*	*1103*	*0*	*1*	10	**155.80**	5
1258	$C_5H_{10}O_2$	*n*-pentanoic acid	2	298.15	1500	*283600*	*108000*	*2107*	*–356000*	*283*			10	**210.14**	3
1261	$C_5H_{10}O_2$	isovaleric acid	2	298.15	1500	*237880*	*132300*	*1511.7*	*–314900*	*217.3*			3	**200.76**	10
1332	$C_5H_{10}O_2$	methyl *n*-butyrate	2	298	1200	*89400*	*291000*	*1570*	*207300*	*678.3*			5	**196.99**	10
1323	$C_5H_{10}O_2$	ethyl propionate	2	298.15	1200	*93700*	*282900*	*1648*	*215500*	*724.7*			5	**195.62**	10
2332	$C_5H_{10}O_2$	methyl isobutyrate	2	298.15	1200	*106310*	*330400*	*2184.6*	*251120*	*891.13*			5	*192.86*	10
1314	$C_5H_{10}O_2$	*n*-propyl acetate	2	298.15	1500	*179940*	*175300*	*1196*	*–412000*	*108.2*			5	**197.92**	10
1305	$C_5H_{10}O_2$	isobutyl formate	2	298.15	1500	*299400*	*114000*	*2382*	*–394500*	*298.3*			3	**216.71**	25
1319	$C_5H_{10}O_2$	isopropyl acetate	2	298.15	1200.2	*94879*	*297760*	*1771.8*	*233790*	*763.26*			5	**194.67**	10
1588	$C_5H_{11}Cl$	1-chloropentane	3	20	1500	*33258*	*122832*	*605.35*	*157121*	*1932.74*	*96037.42*	*4851.22*	10	**191.28**	3
7	C_5H_{12}	*n*-pentane	2	200	1500	**88050**	**301100**	**1650.2**	**189200**	**747.6**			1	166.91	1
8	C_5H_{12}	isopentane	2	200	1500	**74600**	**326500**	**1545**	**192300**	**666.7**			1	165.89	3
9	C_5H_{12}	neopentane	2	200	1500	**66200**	**368700**	**1555**	**212000**	**632.9**			1		
1109	$C_5H_{12}O$	1-pentanol	3	20	1500	*33258*	*197720*	*907.77*	*15425000*	*2991.2*	*–15266000*	*2991.2*	1	209.25	3
1110	$C_5H_{12}O$	2-pentanol	3	50	1500	*33260*	*427000*	*895.49*	*166000*	*3183*	*–224100*	*895.489*	10	**229.52**	5
1112	$C_5H_{12}O$	2-methyl-1-butanol	2	298.15	1500	*107410*	*316130*	*1936*	*229000*	*849.69*			10	**216.35**	5
1111	$C_5H_{12}O$	2-methyl-2-butanol	2	298	1500	**121900**	**300020**	**1784.8**	**205190**	**807.74**			5	248.56	10
1123	$C_5H_{12}O$	3-methyl-1-butanol	2	298.15	1500	**92139**	**333710**	**1836.1**	**246440**	**757.83**			10	**209.89**	3
1124	$C_5H_{12}O$	3-methyl-2-butanol	3	50	1500	*33258*	*100300*	*4915*	*118300*	*575.135*	*175000*	*1860*	10	245.44	3
1415	$C_5H_{12}O$	ethyl propyl ether	2	298.15	1500	*113200*	*294000*	*1827*	*205500*	*852*			5	197.24	3
1405	$C_5H_{12}O$	methyl *tert*-butyl ether	3	50	1500	*33258*	*52830*	*5673*	*182300*	*824.925*	*151500*	*2577*	10	188.12	3
1120	$C_5H_{12}O$	3-pentanol	2	298.15	1500	*108660*	*307440*	*1869.7*	*222900*	*826.53*			10	247.72	5
1864	C_6F_6	hexafluorobenzene	2	200	1500	*123700*	*213100*	*1627.4*	*161430*	*714.4*			1	221.38	1

(*Continued*)

TABLE A.3 Ideal gas heat capacity correlation parameters and liquid heat capacity at 298.15 K values. (*Continued*)

DIPPR ID	Formula	Name	C_p° Eq	T_{min} (K)	T_{max} (K)	A	B	C	D	E	F	G	Er (%)	C_p^{liq} (298) (J/(mol·K))	Er (%)
1623	C_6F_{14}	perfluoro-*n*-hexane	3	50	1500	33258	−25296000	2464.9	289200	553.24	25423000	2458.7	10	380.63	10
1573	$C_6H_4Cl_2$	*m*-dichlorobenzene	2	200	1500	70000	207460	1366.4	159830	620.16			1	170.71	5
1572	$C_6H_4Cl_2$	*o*-dichlorobenzene	2	200	1500	69480	208040	1363.2	159400	619.2			1	**168.51**	10
1571	C_6H_5Cl	monochlorobenzene	2	200	1500	80110	231000	2157	204600	897.6			1	150.12	3
1691	C_6H_5I	iodobenzene	2	200	1500	71193	206080	1581.6	164920	730.64			3	158.94	3
1886	$C_6H_5NO_2$	nitrobenzene	2	298.15	1500	113900	218000	2122	210700	948			25	181.30	3
501	C_6H_6	benzene	3	20	1500	33258	51445	761.09	139737	1616.91	56829.10	4111.40	1	136.49	3
1181	C_6H_6O	phenol	2	100	1500	43400	244500	1152	151200	507			3		
1792	C_6H_7N	aniline	3	20	1500	33258	143190	2354.1	131370	786.82	0	1	3	191.37	1
1797	C_6H_7N	2-methylpyridine	2	200	1500	59510	262150	1482.7	182000	675.75			1	158.61	1
2797	C_6H_7N	3-methylpyridine	2	200	1500	60130	265300	1504	183500	684.3			1	158.83	3
2799	C_6H_7N	4-methylpyridine	2	200	1500	59000	262700	1481.5	181870	675.55			1	159.01	3
2088	C_6H_8O	2-cyclohexene-1-one	2	298	1500	87077	258493	797.32	−90517	962.718			10	169.46	5
270	C_6H_{10}	cyclohexene	2	150	1500	58171	317170	1543.5	212730	701.62			1	148.41	1
1080	$C_6H_{10}O$	cyclohexanone	3	20	1500	33258	33591	278.82	167943	1231.48	156006.56	2636.12	3	**185.28**	25
1065	$C_6H_{10}O$	mesityl oxide	2	298.15	1000.2	106500	282370	1466.2	167280	637.32			5	**212.30**	10
137	C_6H_{12}	cyclohexane	2	100	1500	43200	373500	1192	163500	530.1			1	154.85	3
105	C_6H_{12}	methylcyclopentane	3	20	1500	33258	210049	1366.4	141164	3388.82	31675.64	209.203	1	158.60	1
216	C_6H_{12}	1-hexene	2	298	1500	104340	307490	1745.9	207280	793.53			1	182.98	3
223	C_6H_{12}	4-methyl-1-pentene	2	200	1500	93390	307330	1650.3	204600	755.8			1	**178.02**	10
1151	$C_6H_{12}O$	cyclohexanol	2	200	1500	90430	257710	788.2	130680	1952.2			3	214.32	5
1062	$C_6H_{12}O$	2-hexanone	3	20	1500	33258	80354	270.42	201730	1503.6	137400	3610.9	10	**213.21**	3
1059	$C_6H_{12}O$	3-hexanone	2	150	1500	112370	293600	1401	160100	650.5			1	216.62	3
1054	$C_6H_{12}O$	methyl isobutyl ketone	3	20	1500	**33258**	**77849**	**289.91**	**187990**	**1359.2**	**155270**	**3540.1**	5	**214.26**	5
1447	$C_6H_{12}O$	butyl vinyl ether	2	298.15	1500	106620	201200	729.3	150300	2004			25	**214.26**	25
1262	$C_6H_{12}O_2$	*n*-hexanoic acid	2	298.15	1500	116220	207080	686.61	153550	1932.5			10	**238.16**	25
1333	$C_6H_{12}O_2$	ethyl *n*-butyrate	2	298	1200	111500	339100	1670.5	251800	733.6			5	**227.05**	10
1324	$C_6H_{12}O_2$	*n*-propyl propionate	2	298.15	1200	107590	348870	1687.3	261810	725.32			5	**227.00**	10
2337	$C_6H_{12}O_2$	ethyl isobutyrate	2	298.15	1500	104880	251180	749	144200	2495			5	212.10	10
1315	$C_6H_{12}O_2$	*n*-butyl acetate	2	298.15	1200	116840	376900	1956	281800	811.2			5	**226.49**	10
1316	$C_6H_{12}O_2$	isobutyl acetate	2	298.15	1400.2	121910	312000	1653.9	232490	780.35			5	**227.49**	10
1306	$C_6H_{12}O_2$	*n*-pentyl formate	2	298.15	1500	119500	348800	1813.3	260200	788			3	240.73	10
5884	$C_6H_{12}O_3$	2-ethoxyethyl acetate	2	298.15	1500	106120	240330	658.9	149500	1967			25		
11	C_6H_{14}	*n*-hexane	2	200	1500	104400	352300	1694.6	236900	761.6			1	196.20	1
12	C_6H_{14}	2-methylpentane	2	200	1500	90300	380100	1602	245300	691.6			1	**193.65**	1
13	C_6H_{14}	3-methylpentane	2	200	1500	86300	379100	1542	234600	669.6			1	190.61	1

TABLE A.3 Ideal gas heat capacity correlation parameters and liquid heat capacity at 298.15 K values. (*Continued*)

DIPPR ID	Formula	Name	C_p° Eq	T_{min} (K)	T_{max} (K)	A	B	C	D	E	F	G	Er (%)	C_p^{liq} (298) (J/(mol·K))	Er (%)
14	C_6H_{14}	2,2-dimethylbutane	2	200	1500	*91460*	*412800*	*1651.5*	*250000*	*706.4*			1	188.96	3
15	C_6H_{14}	2,3-dimethylbutane	2	200	1500	*77720*	*403200*	*1544*	*250800*	*649.95*			1	189.01	1
1114	$C_6H_{14}O$	1-hexanol	3	20	1500	*33258*	*236090*	*886.57*	*15436000*	*2937.5*	*−15254000*	*2937.5*	1	242.17	3
1115	$C_6H_{14}O$	2-hexanol	3	50	1500	*33260*	*190200*	*2021*	*103100*	*4892*	*166400*	*646.096*	10	**250.41**	10
1116	$C_6H_{14}O$	3-hexanol	3	50	1500	*33258*	*240200*	*871.74*	*15450000*	*3132.58*	*−15250000*	*3132.58*	10	**286.02**	5
1117	$C_6H_{14}O$	2-methyl-1-pentanol	2	298.15	1500	*120050*	*241100*	*793.6*	*146300*	*2292*			5		
1130	$C_6H_{14}O$	4-methyl-2-pentanol	2	298.15	1500	*124300*	*241200*	*794*	*142200*	*2348*			5	**272.84**	10
1403	$C_6H_{14}O$	diisopropyl ether	2	298.15	1500	*109300*	*368300*	*1605.7*	*234200*	*699*			3	216.77	1
1446	$C_6H_{14}O$	di-*n*-propyl ether	2	200	1500	*120300*	*341230*	*1545.5*	*204800*	*723.2*			3	**221.90**	3
1432	$C_6H_{14}O_2$	acetal	2	298.15	1500	*110740*	*370020*	*1668.2*	*285230*	*758.75*			3	*207.06*	25
1706	$C_6H_{15}N$	triethylamine	2	200	1500	*127660*	*255590*	*809.37*	*148290*	*2231.7*			3	**221.24**	10
1707	$C_6H_{15}N$	di-*n*-propylamine	2	298.15	1500	*121140*	*261270*	*789.56*	*169030*	*2394.4*			3	*216.75*	10
1725	$C_6H_{15}NO_3$	triethanolamine	2	298.15	1200.2	*157400*	*446000*	*1974*	*340800*	*830*			25	*363.96*	25
2626	C_7F_{16}	perfluoro-*n*-heptane	3	50	1500	*33258*	*−75628000*	*2480.5*	*338650*	*552.37*	*75774000*	*2478*	10	417.50	3
1874	$C_7H_5ClO_2$	*o*-chlorobenzoic acid	2	298.15	1000.2	*93302*	*207370*	*806.62*	*60609*	*2688.9*			25		
1790	C_7H_5N	benzonitrile	3	50	1500	*33258*	*162200*	*944.03*	*15400000*	*2824*	*−15290000*	*2824*	10	166.33	5
1041	C_7H_6O	benzaldehyde	2	200	1500	*77147*	*248090*	*1608.3*	*199760*	*741.12*			1	172.15	3
1578	C_7H_7Cl	*p*-chlorotoluene	2	298.15	1000.2	*90049*	*267890*	*1730.4*	*215920*	*784.57*			10		
502	C_7H_8	toluene	2	200	1500	*58140*	*286300*	*1440.6*	*189800*	*650.43*			1	**156.51**	3
1461	C_7H_8O	anisole	2	298.15	1200	*76370*	*293770*	*1605.1*	*217000*	*751.2*			10	*199.78*	10
1180	C_7H_8O	benzyl alcohol	3	20	1500	*33258*	*140500*	*3306.6*	*207710*	*1074.8*	*0*	*1*	10	**216.17**	3
1182	C_7H_8O	*o*-cresol	2	200	1500	*79880*	*285300*	*1476.5*	*204200*	*664.7*			1		
1183	C_7H_8O	*m*-cresol	3	20	1500	*33258*	*51926*	*367.23*	*178818*	*1286.99*	*101326.35*	*3336.38*	1	224.97	5
1184	C_7H_8O	*p*-cresol	3	20	1500	*33258*	*50938*	*354.26*	*175045*	*1266.23*	*105712.39*	*3278.06*	1		
2796	C_7H_9N	2,6-dimethylpyridine	2	200	1500	*74120*	*316600*	*1478.7*	*209500*	*670.83*			1	*184.95*	5
1344	$C_7H_{12}O_2$	*n*-butyl acrylate	2	298.15	1200.2	*106640*	*346200*	*1179.8*	*199620*	*510.04*			25	*252.80*	10
159	C_7H_{14}	cycloheptane	2	150	1500	*74877*	*423110*	*1542.2*	*284280*	*705.09*			1	180.74	1
138	C_7H_{14}	methylcyclohexane	2	200	1500	*92270*	*411500*	*1650.4*	*290060*	*779.48*			1	184.75	1
107	C_7H_{14}	ethylcyclopentane	3	20	1500	*33258*	*239148*	*1354.0*	*164196*	*3359.55*	*47957.70*	*198.546*	1	186.12	1
111	C_7H_{14}	*cis*-1,3-dimethylcyclopentane	2	200	1500	*97394*	*401460*	*1805.6*	*302100*	*820.25*			1	**190.14**	25
112	C_7H_{14}	*trans*-1,3-dimethylcyclopentane	2	200	1500	*97394*	*401460*	*1805.6*	*302100*	*820.25*			1	**188.19**	25
234	C_7H_{14}	1-heptene	2	298.15	1500	*118510*	*363620*	*1735.9*	*250480*	*785.73*			3	211.82	1
1326	$C_7H_{14}O_2$	*n*-butyl propionate	2	298.15	1200.2	*87800*	*232500*	*518.5*	*231000*	*1627*			5	**254.20**	10
2261	$C_7H_{14}O_2$	*n*-heptanoic acid	2	298.15	1500	*131350*	*233170*	*675.67*	*182400*	*1846*			10	266.23	3
1347	$C_7H_{14}O_2$	ethyl isovalerate	2	298.15	1200	*130350*	*266960*	*729.2*	*154000*	*2209*			5	**253.77**	5

(*Continued*)

TABLE A.3 Ideal gas heat capacity correlation parameters and liquid heat capacity at 298.15 K values. (*Continued*)

DIPPR ID	Formula	Name	C_p° Eq	T_{min} (K)	T_{max} (K)	A	B	C	D	E	F	G	Er (%)	C_p^{liq} (298) (J/(mol·K))	Er (%)
1327	$C_7H_{14}O_2$	*n*-propyl *n*-butyrate	2	298	1200	*119200*	*404400*	*1627*	*294700*	*702.2*			5	**254.14**	25
1336	$C_7H_{14}O_2$	*n*-propyl isobutyrate	2	298.15	1200	*136980*	*430570*	*2035.5*	*334220*	*855.41*			5	**253.93**	10
1317	$C_7H_{14}O_2$	isopentyl acetate	2	298.15	1500	*119100*	*274400*	*708*	*162000*	*2160*			5	**252.90**	10
17	C_7H_{16}	*n*-heptane	2	200	1500	**120150**	**400100**	**1676.6**	**274000**	**756.4**			1	224.37	1
18	C_7H_{16}	2-methylhexane	2	200	1500	**106000**	**425070**	**1577**	**275800**	**690.8**			1	222.94	3
19	C_7H_{16}	3-methylhexane	2	200	1500	**101300**	**429100**	**1557**	**279700**	**677.8**			1	219.78	3
20	C_7H_{16}	3-ethylpentane	2	200	1500	**115800**	**414800**	**1624**	**266600**	**722.49**			1	219.60	1
21	C_7H_{16}	2,2-dimethylpentane	2	200	1500	**104100**	**467300**	**1657**	**306300**	**700.4**			1	221.21	0.2
22	C_7H_{16}	2,3-dimethylpentane	2	200	1500	*85438*	*457720*	*1518.1*	*297400*	*641.01*			1	217.76	3
23	C_7H_{16}	2,4-dimethylpentane	2	200	1500	**96500**	**447800**	**1540**	**297500**	**645.2**			1	224.26	3
24	C_7H_{16}	3,3-dimethylpentane	2	200	1500	**98430**	**470400**	**1596.6**	**299000**	**674.3**			1	214.45	1
25	C_7H_{16}	2,2,3-trimethylbutane	2	200	1500	**95300**	**480400**	**1573**	**291400**	**664.3**			1	213.51	1
1125	$C_7H_{16}O$	1-heptanol	3	20	1500	*33258*	*273290*	*868.74*	*15449000*	*2883.5*	*−15241000*	*2883.5*	1	278.99	3
1126	$C_7H_{16}O$	2-heptanol	3	50	1500	*33300*	*145000*	*471.3*	*239000*	*1670*	*121000*	*4059.7*	10	**295.14**	5
1625	C_8F_{18}	perfluoro-*n*-octane	3	50	1500	*33258*	*−75619000*	*2503.4*	*387470*	*549.52*	*75783000*	*2500.5*	10	*454.70*	10
601	C_8H_8	styrene	3	20	1500	*33258*	*108320*	*663.64*	*162880*	*1748.9*	*73212*	*4243.6*	3	**182.89**	3
1090	C_8H_8O	acetophenone	2	298.15	1500	*85400*	*233400*	*831*	*77300*	*2227*			5		
1283	$C_8H_8O_2$	*p*-toluic acid	3	20	1500	*33258*	*109380*	*592.68*	*218990*	*1935.2*	*84575*	*5771.8*	10		
504	C_8H_{10}	ethylbenzene	3	20	1500	*33258*	*43028*	*233.7*	*204910*	*1273.6*	*132890*	*3260.8*	5	**185.66**	1
505	C_8H_{10}	*o*-xylene	3	20	1500	*33258*	*61769*	*307.78*	*182620*	*1333.2*	*137140*	*3264.6*	3	**187.70**	1
506	C_8H_{10}	*m*-xylene	3	20	1500	33258	44147	256.51	191240	1262.9	144870	3172.9	3	181.96	1
507	C_8H_{10}	*p*-xylene	3	20	1500	*33258*	*109660*	*568.47*	*247380*	*2071.5*	*0*	*1*	3	181.25	1
1185	$C_8H_{10}O$	*o*-ethylphenol	2	298.15	1500	*117330*	*310380*	*1718.8*	*251150*	*794.01*			5	*251.87*	5
2112	$C_8H_{10}O$	*m*-ethylphenol	2	298.15	1500	*106460*	*319260*	*1628.9*	*253220*	*748.02*			5	*222.31*	5
1187	$C_8H_{10}O$	*p*-ethylphenol	2	200	1500	*110450*	*245260*	*765.7*	*104470*	*2476*			1	*251.87*	5
1170	$C_8H_{10}O$	2,3-xylenol	2	200	1500	**106530**	**334950**	**1444.5**	**227200**	**640.8**			1		
1172	$C_8H_{10}O$	2,4-xylenol	2	200	1500	*98640*	*343700*	*1468.6*	*239100*	*655.2*			1		
1174	$C_8H_{10}O$	2,5-xylenol	2	200	1500	*102250*	*340200*	*1480.4*	*235900*	*664.6*			1		
1176	$C_8H_{10}O$	2,6-xylenol	2	200	1500	*98970*	*343000*	*1458*	*233300*	*651.1*			1		
1177	$C_8H_{10}O$	3,4-xylenol	2	200	1500	*102020*	*339400*	*1437.1*	*235560*	*631.7*			1		
1178	$C_8H_{10}O$	3,5-xylenol	2	200	1500	*93130*	*348840*	*1451.7*	*241700*	*648.62*			1		
160	C_8H_{16}	cyclooctane	2	200	1500	*89059*	*483540*	*1585.4*	*314110*	*730.72*			1	215.37	3
147	C_8H_{16}	*trans*-1,4-dimethylcyclohexane	2	200	1500	*109580*	*464680*	*1670.6*	*338770*	*787.66*			1	210.30	3
250	C_8H_{16}	1-octene	2	298.15	1500	*135990*	*416050*	*1731.7*	*286750*	*784.47*			3	**241.16**	3
1265	$C_8H_{16}O_2$	*n*-octanoic acid	2	298.15	1500	*140820*	*434360*	*1466.2*	*276870*	*659.38*			10	**298.11**	10
2260	$C_8H_{16}O_2$	2-ethyl hexanoic acid	2	298.15	1500	*157770*	*440170*	*1749.4*	*323780*	*792.34*			10	**295.61**	10

TABLE A.3 Ideal gas heat capacity correlation parameters and liquid heat capacity at 298.15 K values. (*Continued*)

DIPPR ID	Formula	Name	C_p° Eq	T_{min} (K)	T_{max} (K)	A	B	C	D	E	F	G	Er (%)	C_p^{liq} (298) (J/(mol·K))	Er (%)
1385	$C_8H_{16}O_2$	*n*-butyl *n*-butyrate	2	298	1200	144030	451920	1693.6	331900	738.93			5	**280.44**	10
1360	$C_8H_{16}O_2$	isobutyl isobutyrate	3	20	1500	33258	217330	3274.5	339090	945.73	0	1	10	**280.24**	3
27	C_8H_{18}	*n*-octane	2	200	1500	135540	443100	1635.6	305400	746.4			1	254.43	1
28	C_8H_{18}	2-methylheptane	2	150	1500	122150	472700	1586.6	316500	699.6			1	251.43	1
29	C_8H_{18}	3-methylheptane	2	150	1500	118100	476800	1572	317500	690			1	249.67	1
30	C_8H_{18}	4-methylheptane	2	150	1500	116000	476300	1552	321600	680.3			1	250.65	1
31	C_8H_{18}	3-ethylhexane	2	200	1500	129850	461700	1585.5	300300	707.7			1	**251.63**	10
32	C_8H_{18}	2,2-dimethylhexane	2	200	1500	120900	509500	1625.4	331300	701.7			1	**250.36**	5
33	C_8H_{18}	2,3-dimethylhexane	2	200	1500	101150	507100	1530.5	338200	650.7			1	**246.94**	10
34	C_8H_{18}	2,4-dimethylhexane	2	200	1500	107100	497300	1497.5	327800	632.8			1	**252.96**	10
35	C_8H_{18}	2,5-dimethylhexane	2	200	1500	107200	501300	1541.5	329500	660			1	249.28	3
36	C_8H_{18}	3,3-dimethylhexane	2	200	1500	109000	515300	1543.9	341500	656.6			1	246.57	3
37	C_8H_{18}	3,4-dimethylhexane	2	200	1500	93770	514600	1517	347800	639.7			1	**244.27**	10
38	C_8H_{18}	2-methyl-3-ethylpentane	2	200	1500	106330	468240	1330.2	269450	577.6			1	**250.75**	25
39	C_8H_{18}	3-methyl-3-ethylpentane	2	150	1500	108800	516000	1544	332400	659			1	**247.97**	25
40	C_8H_{18}	2,2,3-trimethylpentane	2	200	1500	101700	542300	1580.5	353400	659			1	**245.36**	5
41	C_8H_{18}	2,2,4-trimethylpentane	2	200	1500	113900	528600	1594	335100	677.94			1	238.24	3
42	C_8H_{18}	2,3,3-trimethylpentane	2	200	1500	98200	540200	1531	349300	639.9			1	245.55	1
43	C_8H_{18}	2,3,4-trimethylpentane	2	200	1500	96700	522700	1493.6	343500	615.2			1	248.65	1
44	C_8H_{18}	2,2,3,3-tetramethylbutane	2	200	1500	113520	563310	1621.1	338290	681.9			3		
1132	$C_8H_{18}O$	1-octanol	3	20	1500	33258	310830	856.26	15460000	2840.8	−15229000	2840.8	1	308.39	3
1133	$C_8H_{18}O$	2-octanol	3	50	1500	33258	295020	807.09	247310	2688.7	185520	10040	10	**330.39**	5
1121	$C_8H_{18}O$	2-ethyl-1-hexanol	3	20	1500	**33258**	**273940**	**2937.8**	**296440**	**811.21**	**0**	**1**	10	318.79	3
1404	$C_8H_{18}O$	di-*n*-butyl ether	2	200	1500	161220	447770	1683.1	291800	781.6			3	**278.08**	3
2708	$C_8H_{19}N$	*n*-octylamine	3	20	1500	33258	281990	2625.3	283240	748.49	0	1	10	**309.17**	3
1744	$C_8H_{19}N$	di-*n*-butylamine	2	298.15	1500	158240	499530	1823.8	356660	810.96			3		
1626	C_9F_{20}	perfluoro-*n*-nonane	3	50	1500	33258	−75609000	2497.2	436260	548.8	75793000	2494	10	500.50	10
1748	C_9H_7N	quinoline	2	200	1500	82530	320820	745.95	−89940	895.5			1	195.18	3
2785	C_9H_7N	isoquinoline	2	200	1500	83990	320580	760.7	−88010	926.3			1		
820	C_9H_{10}	indane	3	20	1500	33258	41834	305.83	223729	1272.42	147666.04	3104.05	1	190.65	3
509	C_9H_{12}	*n*-propylbenzene	3	20	1500	33258	145080	3202.3	66499	240.82	229699.38	1308.43	1	214.25	1
510	C_9H_{12}	cumene	2	200	1500	108100	379320	1750.5	300270	794.8			1	209.25	3
513	C_9H_{12}	*p*-ethyltoluene	3	20	1500	33258	54736	205.16	175250	3029.27	216183.92	1217.85	3	209.43	10
514	C_9H_{12}	1,2,3-trimethylbenzene	2	200	1500	105200	379000	1481.4	233100	667.3			1	216.21	3
515	C_9H_{12}	1,2,4-trimethylbenzene	3	20	1500	33258	56782	197.30	182272	1089.52	204886.57	2845.70	1	214.94	3

(*Continued*)

TABLE A.3 Ideal gas heat capacity correlation parameters and liquid heat capacity at 298.15 K values. (*Continued*)

DIPPR ID	Formula	Name	C_p° Eq	T_{min} (K)	T_{max} (K)	A	B	C	D	E	F	G	Er (%)	C_p^{liq} (298) (J/(mol·K))	Er (%)
516	C_9H_{12}	mesitylene	3	20	1500	33258	185862	3028.7	52750	180.79	208508.59	1194.24	3	209.27	3
259	C_9H_{18}	1-nonene	2	298.15	1500	153520	468440	1728.8	323040	783.67			3	269.69	3
1259	$C_9H_{18}O_2$	*n*-nonanoic acid	2	298.15	1500	12660	601100	1081.5	459460	418.2			10	**326.39**	5
3318	$C_9H_{18}O_2$	isopentyl butyrate	3	20	1500	33258	293760	3127.8	376690	846.57	0	1	10	297.25	25
46	C_9H_{20}	*n*-nonane	2	200	1500	151750	491500	1644.8	347000	749.6			1	284.16	1
91	C_9H_{20}	2-methyloctane	2	200	1500	162790	510070	831.25	–203090	973.17			1	282.69	3
96	C_9H_{20}	2,2-dimethylheptane	2	150	1500	136000	553000	1599	365200	698.3			1	275.57	5
47	C_9H_{20}	2,2,5-trimethylhexane	2	150	1500	123000	582700	1591.4	381600	681			3	**276.28**	5
51	C_9H_{20}	2,2,3,3-tetramethylpentane	2	200	1500	114400	627300	1581	401300	652.2			1	271.56	1
52	C_9H_{20}	2,2,3,4-tetramethylpentane	2	150	1500	110600	605400	1540.6	387100	645			1	**272.10**	5
53	C_9H_{20}	2,2,4,4-tetramethylpentane	2	150	1500	138200	607000	1653	375000	702			1	266.19	3
54	C_9H_{20}	2,3,3,4-tetramethylpentane	2	200	1500	110090	604510	1507.7	394880	617.6			3	272.80	5
1134	$C_9H_{20}O$	1-nonanol	3	20	1500	33258	348060	845.5	15473000	2813.7	–15217000	2813.7	1		
1135	$C_9H_{20}O$	2-nonanol	3	50	1500	33300	145000	4890	276000	2040	256000	630	10	348.23	5
1627	$C_{10}F_{22}$	perfluoro-*n*-decane	3	50	1500	33258	–75600000	2517	485250	547.05	75802000	2513.3	10		
1381	$C_{10}H_{10}O_4$	dimethyl terephthalate	2	298.15	1500	114025	536801	2088.6	413440	809.837			10		
701	$C_{10}H_8$	naphthalene	3	20	1500	33258	56715	443.50	227394	1352.55	107771.2042	3387.67	1		
706	$C_{10}H_{12}$	1,2,3,4-tetrahydronaphthalene	3	20	1500	33258	51771	291.03	265753	1260.38	169167.83	3183.18	1	217.53	5
503	$C_{10}H_{12}$	1-methylindan	3	20	1500	33258	208280	2839.8	264250	1013.5	0	1	10	208.93	25
508	$C_{10}H_{12}$	5-methylindan	3	20	1500	33258	224820	2769.1	245440	958.83	0	1	10	213.33	25
518	$C_{10}H_{14}$	*n*-butylbenzene	2	200	1500	113800	445400	1550.7	304970	708.86			1	243.12	1
519	$C_{10}H_{14}$	isobutylbenzene	2	298.15	1500	124660	427650	1650.9	313790	761.61			3	**241.29**	10
527	$C_{10}H_{14}$	*p*-diethylbenzene	2	200	1500	128500	419760	1626.5	296370	757.44			3	241.58	10
524	$C_{10}H_{14}$	*p*-cymene	2	200	1500	131860	430360	1773.4	325700	811.9			1	236.37	3
532	$C_{10}H_{14}$	1,2,4,5-tetramethylbenzene	2	200	1500	126400	426900	1503	255570	678.3			1		
520	$C_{10}H_{14}$	*sec*-butylbenzene	2	200	1500	130420	423080	1713.2	319310	786.84			3	230.13	10
153	$C_{10}H_{18}$	*cis*-decahydronaphthalene	2	200	1500	105800	559320	1646.4	404310	774.5			3	232.52	1
154	$C_{10}H_{18}$	*trans*-decahydronaphthalene	2	200	1500	86840	554800	1357.6	320650	651.1			3	229.13	1
260	$C_{10}H_{20}$	1-decene	2	298.15	1500	171010	520890	1726.5	359350	782.92			1	299.81	3
1254	$C_{10}H_{20}O_2$	*n*-decanoic acid	2	298.15	1500	24457	654600	1089.9	486420	424			10		
56	$C_{10}H_{22}$	*n*-decane	2	200	1500	167200	535300	1614.1	378200	742			1	315.06	1
48	$C_{10}H_{22}$	3,3,5-trimethylheptane	2	200	1500	130750	631570	1516.5	413050	648.95			3	303.63	5
57	$C_{10}H_{22}$	2,2,3,3-tetramethylhexane	2	200	1500	124580	673990	1540.6	441440	639.38			3	296.51	5
58	$C_{10}H_{22}$	2,2,5,5-tetramethylhexane	2	200	1500	138480	670110	1629.3	426200	692.18			1	296.52	5
1136	$C_{10}H_{22}O$	1-decanol	3	20	1500	33258	383380	834.03	15485000	2767.6	–15204000	2767.6	1	370.87	3
702	$C_{11}H_{10}$	1-methylnaphthalene	3	20	1500	33258	105403	489.36	239297	1492.53	120692.7	3646.23	1	224.36	1

TABLE A.3 Ideal gas heat capacity correlation parameters and liquid heat capacity at 298.15 K values. ***(Continued)***

DIPPR ID	Formula	Name	C_p° Eq	T_{min} (K)	T_{max} (K)	A	B	C	D	E	F	G	Er (%)	C_p^{liq} (298) (J/(mol·K))	Er (%)
703	$C_{11}H_{10}$	2-methylnaphthalene	2	298.15	1500	111060	389230	1651.1	300690	756.1			1		
63	$C_{11}H_{24}$	*n*-undecane	2	200	1500	195290	609980	1708.7	413020	775.4			1	345.87	1
1137	$C_{11}H_{24}O$	1-undecanol	3	20	1500	33258	421010	827.92	15497000	2751.8	−15191000	2751.8	1	**400.40**	3
558	$C_{12}H_{10}$	biphenyl	2	200	1500	**107590**	**421050**	**1904.1**	**417850**	**828.81**			5		
715	$C_{12}H_{12}$	2,7-dimethylnaphthalene	2	298.15	1500	135680	433640	1657.7	322050	759.3			3		
3545	$C_{12}H_{18}$	1,3,5-triethylbenzene	2	298.15	1500	160840	517950	1590	359900	737.1			3	288.63	5
814	$C_{12}H_{20}$	1,3-dimethyladamantane	2	298.15	1500	86327	615920	811.27	179250	2109.3			3	232.26	3
262	$C_{12}H_{24}$	1-dodecene	2	298.15	1500	216430	427460	815.94	264900	2417.3			3	360.23	1
1269	$C_{12}H_{24}O_2$	*n*-dodecanoic acid	2	298.15	1500	210520	634390	1492.3	402900	681.9			5		
64	$C_{12}H_{26}$	*n*-dodecane	2	200	1500	212950	663300	1715.5	451610	777.5			1	375.84	3
1140	$C_{12}H_{26}O$	1-dodecanol	3	20	1500	33258	456260	819.49	15510000	2718.3	−15178000	2718.3	1	430.39	3
563	$C_{13}H_{12}$	diphenylmethane	3	20	1500	33258	325910	957.53	219830	2856.6	0	1	10		
65	$C_{13}H_{28}$	*n*-tridecane	2	200	1500	214960	730450	1669.5	499980	741.02			1	**408.05**	3
1141	$C_{13}H_{28}O$	1-tridecanol	3	20	1500	33258	494340	816.38	15522000	2711.5	−15166000	2711.5	1		
805	$C_{14}H_{10}$	phenanthrene	3	20	1500	33258	71445	338.17	293621	1192.69	169977	2994.74	3		
804	$C_{14}H_{10}$	anthracene	3	20	1500	33258	70438	305.65	297666	1213.10	166807.4031	3003.71	1		
534	$C_{14}H_{22}$	1,4-di-*tert*-butylbenzene	2	298.15	1500	175640	641770	1542.5	472790	712.62			5		
66	$C_{14}H_{30}$	*n*-tetradecane	2	200	1500	230820	786780	1682.3	544860	743.1			1	**438.38**	3
1142	$C_{14}H_{30}O$	1-tetradecanol	3	20	1500	33258	582570	2685.3	529120	808.86	−199810	2685.3	1		
1198	$C_{15}H_{16}O_2$	bisphenol A	2	298.15	1600.2	185490	583520	1634.7	484860	734.17			3		
67	$C_{15}H_{32}$	*n*-pentadecane	2	200	1500	**246790**	**842120**	**1686.5**	**585370**	**743.6**			1	**470.71**	3
1143	$C_{15}H_{32}O$	1-pentadecanol	3	20	1500	33260	1371000	828.68	389500	2709	−790400	828.678	1		
68	$C_{16}H_{34}$	*n*-hexadecane	3	20	1500	33258	391480	3498.7	222830	241.74	562810	1463.8	5	**500.92**	3
90	$C_{16}H_{34}$	2,2,4,4,6,8,8-heptamethylnonane	2	298.15	1500	279680	869160	1720.6	659540	785.5			1	464.69	5
1144	$C_{16}H_{34}O$	1-hexadecanol	3	20	1500	33260	595500	800.19	890400	2580	−463100	2580	1		
69	$C_{17}H_{36}$	*n*-heptadecane	3	20	1500	33258	237970	240.42	597670	1462.7	415060	3495.7	5	**532.83**	3
1145	$C_{17}H_{36}O$	1-heptadecanol	3	20	1500	33258	604760	753.54	−2305400000	2534.7	2305900000	2534.7	1		
561	$C_{18}H_{14}$	*o*-terphenyl	2	298.15	1500	**207190**	**626680**	**2404.4**	**634500**	**967.71**			10		
560	$C_{18}H_{14}$	*m*-terphenyl	2	298.15	1500	163970	601250	1690.2	513140	757.5			10		
559	$C_{18}H_{14}$	*p*-terphenyl	2	298.15	1500	210840	670120	2608	666320	992.52			10		
1276	$C_{18}H_{36}O_2$	*n*-octadecanoic acid	2	298.15	1500	326200	947300	1626	641540	743.1			5		
70	$C_{18}H_{38}$	*n*-octadecane	3	20	1500	33258	253130	239.11	632560	1461.7	438620	3493.5	5		
1146	$C_{18}H_{38}O$	1-octadecanol	3	20	1500	33260	1935000	734.18	532100	2433	−1319000	734.179	1		
71	$C_{19}H_{40}$	*n*-nonadecane	3	20	1500	33258	268250	238.03	667440	1460.7	462220	3491.3	5		
1149	$C_{19}H_{40}O$	1-nonadecanol	3	20	1500	33260	894500	829.95	474200	2729	−150500	829.954	1		

(Continued)

TABLE A.3 Ideal gas heat capacity correlation parameters and liquid heat capacity at 298.15 K values. (*Continued*)

DIPPR ID	Formula	Name	C_p° Eq	T_{min} (K)	T_{max} (K)	A	B	C	D	E	F	G	Er (%)	C_p^{liq} (298) (J/(mol·K))	Er (%)
73	$C_{20}H_{42}$	*n*-eicosane	3	20	1500	*33258*	*283420*	*237.14*	*702220*	*1459.8*	*485710*	*3488.4*	5		
1148	$C_{20}H_{42}O$	1-eicosanol	3	20	1500	*33258*	*896580*	*790.81*	*534540*	*2622.1*	*–148640*	*790.81*	1		
74	$C_{21}H_{44}$	*n*-heneicosane	3	20	1500	*33258*	*298590*	*236.53*	*737290*	*1459.3*	*509280*	*3488*	5		
75	$C_{22}H_{46}$	*n*-docosane	3	20	1500	*33258*	*313730*	*235.54*	*771990*	*1458.4*	*532840*	*3485*	5		
76	$C_{23}H_{48}$	*n*-tricosane	3	20	1500	*33258*	*328880*	*234.89*	*807090*	*1457.9*	*556300*	*3485.1*	5		
77	$C_{24}H_{50}$	*n*-tetracosane	3	20	1500	*33258*	*344000*	*233.74*	*841720*	*1457.1*	*580050*	*3482.2*	5		
1987	$ClFO_3$	perchloryl fluoride	2	100	1500	*34005*	*74023*	*806.93*	*48496*	*380*			1	113.39	5
1904	ClH	hydrogen chloride	2	50	1500	*29157*	*9048*	*2093.8*	*–107*	*120*			1		
2928	ClH_4N	ammonium chloride	3	20	1500	*31200*	*10689*	*1784.7*	*18524*	*4457.9*	*261.64*	*427.15*	5		
1939	ClNa	sodium chloride	2	100	1500	*29820*	*8750*	*831*	*11350*	*340.3*			1		
1986	ClNO	nitrosyl chloride	3	20	4500	*33258*	*–106850*	*810.58*	*127760*	*810.58*	*10835*	*5085.3*	1		
918	Cl_2	chlorine	2	50	1500	*29142*	*9176*	*949*	*10030*	*425*			1		
1937	Cl_4Si	tetrachlorosilane	3	20	1500	*33260*	*43750*	*281.53*	*31060*	*852.77*	*0*	*1*	1	145.33	10
925	D_2	deuterium	2	100	1500	*30290*	*9750*	*2515*	*–2750*	*368*			1		
1997	D_2O	deuterium oxide	2	100	6000	33633	27839	2497	15953	1143.5			1	84.31	1
1905	FH	hydrogen fluoride	2	50	1500	*29134*	*9325.2*	*2905*	*195*	*1326*			1		
917	F_2	fluorine	2	50	1500	*29122*	*10132*	*1453*	*9410.1*	*662.91*			1		
1972	F_3N	nitrogen trifluoride	2	100	1500	33284	49837	709.3	23264	372.91			1		
1989	F_4N_2	tetrafluorohydrazine	2	100	1500	*39645*	*93385*	*684.61*	*43753*	*344.95*			3		
1940	F_6S	sulfur hexafluoride	2	100	1500	*35256*	*122700*	*679.38*	*78407*	*351.27*			1		
1907	HI	hydrogen iodide	2	100	1500	*29117*	*9201.3*	*1689.4*	*451.4*	*980*			1		
1912	HNaO	sodium hydroxide	2	100	1500	*32400*	*25830*	*1033*	*34500*	*379*			1		
902	H_2	hydrogen	2	250	1500	*27617*	*9560*	*2466*	*3760*	*567.6*			1		
1921	H_2O	water	2	100	2273.1	**33363**	**26790**	**2610.5**	**8896**	**1169**			1	75.38	1
1901	H_2O_4S	sulfuric acid	2	100	1500	*40240*	*109500*	*943*	*83700*	*393.8*			1	**139.09**	3
1922	H_2S	hydrogen sulfide	2	100	1500	*33288*	*26086*	*913.4*	*–17979*	*949.4*			1	73.61	3
3951	H_2Se	hydrogen selenide	2	100	1500	*33235*	*24070*	*1535*	*7480*	*718*			1		
1911	H_3N	ammonia	2	100	1500	*33427*	*48980*	*2036*	*22560*	*882*			1	79.97	3
1981	H_3P	phosphine	2	100	1500	33266	49816	1605.3	24764	771			1		
1717	H_4N_2	hydrazine	3	20	1500	*33258*	*12508*	*593.03*	*47101*	*1598.82*	*34634.6*	*4457.73*	1	98.86	3
1982	H_4Si	silane	3	20	1500	*33260*	*41420*	*1345*	*33400*	*3143*	*0*	*1*	1		
913	He	helium	1	100	1500	*20786*							1		
1998	I_2	iodine	3	35	1500	*33258*	*1002.6*	*4229.4*	*42609*	*11453.94*	*4555.43*	*500.154*	1		
920	Kr	krypton	1	100	1500	*20786*							1		
912	NO	nitric oxide	1	100	1500	*34980*	*–35.32*	*0.07729*	*–5.7357E–05*	*1.4526E–08*			3		
905	N_2	nitrogen	2	50	1500	**29105**	**8614.9**	**1701.6**	**103.47**	**909.79**			1		

TABLE A.3 Ideal gas heat capacity correlation parameters and liquid heat capacity at 298.15 K values. (*Continued*)

DIPPR ID	Formula	Name	C_p° Eq	T_{min} (K)	T_{max} (K)	A	B	C	D	E	F	G	Er (%)	C_p^{liq} (298) (J/(mol·K))	Er (%)
899	N_2O	nitrous oxide	2	100	1500	*29338*	*32360*	*1123.8*	*21770*	*479.4*			1		
906	N_2O_4	nitrogen tetroxide	3	20	1500	**33258**	**13293**	**134.5**	**45856**	**700.69**	**38866**	**1982.5**	10	144.04	3
919	Ne	neon	1	100	1500	*20786*							1		
901	O_2	oxygen	2	50	1500	**29103**	**10040**	**2526.5**	**9356**	**1153.8**			1		
910	O_2S	sulfur dioxide	2	100	1500	*33375*	*25864*	*932.8*	*10880*	*423.7*			1	87.46	3
924	O_3	ozone	2	100	1500	*33483*	*29577*	*1521.7*	*27151*	*680.35*			1		
911	O_3S	sulfur trioxide	2	100	1500	*33408*	*49677*	*873.22*	*28563*	*393.74*			3		
970	Rn	radon	3	20	1500	*20786*		*1*		*1*	*0*	*1*	1	*53.18*	10
1923	S	sulfur	1	273.15	1500	*25639*	*–7.987*	*0.004786*	*–9.57E–07*				3		
959	Xe	xenon	1	50	1500	*20786*							1		

TABLE A.4 Vapor pressure correlation parameters.

DIPPR ID	Formula	Name	Eq. T_{min} (K)	Eq. T_{max} (K)	Data T_{min} (K)	Data T_{max} (K)	A	B	C	D	E	Er (%)
914	Ar	argon	83.78	150.86	83.8	150.86	4.2127E+01	−1.0931E+03	−4.1425E+00	5.7254E-05	2	1
922	Br_2	bromine	265.85	584.15	267.95	383.15	1.0826E+02	−6.5920E+03	−1.4160E+01	1.6043E-02	1	3
1906	BrH	hydrogen bromide	185.15	363.15	195.15	343.75	**2.9315E+01**	**−2.4245E+03**	**−1.1354E+00**	**2.3806E-18**	**6**	3
2686	$CBrClF_2$	bromochlorodifluoromethane	113.65	426.15	177.81	418.15	6.7852E+01	−4.3699E+03	−7.2955E+00	1.0089E-05	2	3
2687	$CBrF_3$	bromotrifluoromethane	105.15	340.15	156.15	338.71	4.0365E+01	−2.8290E+03	−2.9248E+00	1.2241E-16	6	3
2688	CBr_2F_2	dibromodifluoromethane	163.05	478	221.75	296.78	**1.2039E+02**	**−6.0757E+03**	**−1.6626E+01**	**2.1162E-02**	**1**	10
1606	$CClF_3$	chlorotrifluoromethane	92.15	302	140.54	302	6.1908E+01	−3.0055E+03	−6.7276E+00	1.7855E-05	2	3
1601	CCl_2F_2	dichlorodifluoromethane	115.15	384.95	170.93	384.95	9.4110E+01	−4.4117E+03	−1.2613E+01	1.9914E-02	1	3
1894	CCl_2O	phosgene	145.37	455	180.25	455	1.1232E+02	−5.6774E+03	−1.5351E+01	2.1250E-02	1	3
1602	CCl_3F	trichlorofluoromethane	162.04	471.2	162.04	471.15	7.3689E+01	−5.0236E+03	−8.0908E+00	9.3489E-06	2	3
1501	CCl_4	carbon tetrachloride	250.33	556.35	252.65	556.3	7.8441E+01	−6.1281E+03	−8.5766E+00	6.8465E-06	2	1
1616	CF_4	carbon tetrafluoride	89.56	227.51	89.56	227.5	6.1890E+01	−2.2963E+03	−7.0860E+00	3.4687E-05	2	1
1634	$CHBrF_2$	bromodifluoromethane	128	411.98	194.65	411.85	4.4482E+01	−3.6659E+03	−3.3749E+00	3.9225E-17	6	3
1604	$CHClF_2$	chlorodifluoromethane	115.73	369.3	162.95	369.3	1.1606E+02	−4.8884E+03	−1.6486E+01	2.7234E-02	1	3
1696	$CHCl_2F$	dichlorofluoromethane	138.15	451.58	181.85	451.4	8.1939E+01	−5.2218E+03	−9.3544E+00	1.1125E-05	2	3
1521	$CHCl_3$	chloroform	207.15	536.4	207.15	431	1.4643E+02	−7.7923E+03	−2.0614E+01	2.4578E-02	1	3
1615	CHF_3	trifluoromethane	117.97	299.01	130	298.15	6.6043E+01	−3.2440E+03	−7.2767E+00	1.8665E-05	2	3
1771	CHN	hydrogen cyanide	200	457	256.75	456.75	3.0476E+01	−3.6814E+03	−1.1661E+00	2.4564E-17	6	3
1511	CH_2Cl_2	dichloromethane	178.01	510	195.15	358.03	1.0160E+02	−6.5416E+03	−1.2247E+01	1.2311E-05	2	3
1614	CH_2F_2	difluoromethane	136.95	351.26	191.19	351.23	6.9132E+01	−3.8477E+03	−7.5868E+00	1.5065E-05	2	3
1001	CH_2O	formaldehyde	155.15	420	163.75	250.85	4.9363E+01	−3.8479E+03	−4.0983E+00	4.6363E-17	6	5
1251	CH_2O_2	formic acid	281.45	588	285.65	398.25	4.3807E+01	−5.1310E+03	−3.1878E+00	2.3782E-06	2	3
1641	CH_3Br	bromomethane	179.44	464	179.48	464	4.4764E+01	−3.9078E+03	−3.4016E+00	2.9499E-17	6	5
1502	CH_3Cl	methyl chloride	175.45	416.25	175.45	416.25	4.4555E+01	−3.5213E+03	−3.4258E+00	5.6312E-17	6	5
1613	CH_3F	methyl fluoride	131.35	317.42	131.35	317.25	4.1274E+01	−2.6767E+03	−3.0391E+00	2.4491E-16	6	5
1681	CH_3I	methyl iodide	206.7	528.2	206.71	528	7.9533E+01	−5.5531E+03	−8.9235E+00	9.4206E-06	2	5
2851	CH_3NO	formamide	275.6	771	391.01	465.53	**1.0030E+02**	**−1.0763E+04**	**−1.0946E+01**	**3.8503E-06**	**2**	10
1760	CH_3NO_2	nitromethane	244.6	588.15	265.25	463.05	5.7278E+01	−6.0890E+03	−4.9821E+00	1.2154E-17	6	5
1	CH_4	methane	90.69	190.56	90.69	190.56	**3.9205E+01**	**−1.3244E+03**	**−3.4366E+00**	**3.1019E-05**	**2**	1
1101	CH_4O	methanol	175.47	512.5	257.7	503.15	8.2718E+01	−6.9045E+03	−8.8622E+00	7.4664E-06	2	1
1801	CH_4S	methyl mercaptan	150.18	469.95	202.85	297.84	**5.4150E+01**	**−4.3377E+03**	**−4.8127E+00**	**4.5000E-17**	**6**	3
1701	CH_5N	methylamine	179.69	430.8	179.69	417.75	7.0068E+01	−4.9952E+03	−7.1470E+00	2.1040E-11	4	3
908	CO	carbon monoxide	68.15	132.92	68.15	132.92	**4.5698E+01**	**−1.0766E+03**	**−4.8814E+00**	**7.5673E-05**	**2**	1
1893	COS	carbonyl sulfide	134.34	378.8	140.15	373	4.6878E+01	−3.1705E+03	−3.9104E+00	1.2569E-16	6	3
909	CO_2	carbon dioxide	216.58	304.21	216.58	304.21	4.7017E+01	−2.8390E+03	−3.8639E+00	2.8112E-16	6	1

TABLE A.4 Vapor pressure correlation parameters. (*Continued*)

DIPPR ID	Formula	Name	Eq. T_{min} (K)	Eq. T_{max} (K)	Data T_{min} (K)	Data T_{max} (K)	A	B	C	D	E	Er (%)
1938	CS_2	carbon disulfide	161.11	552.49	247.15	552.49	3.2308E+01	–3.8132E+03	–1.5356E+00	3.4360E-18	6	3
2692	C_2ClF_5	chloropentafluoroethane	173.71	353.15	175	353.15	4.9516E+01	–3.4325E+03	–4.2790E+00	1.3631E-16	6	3
1611	$C_2Br_2F_4$	1,2-dibromotetrafluoroethane	163	487.8	293.4	483.15	9.0238E+01	–6.0942E+03	–1.0530E+01	1.0363E-05	2	3
1693	$C_2Cl_2F_4$	1,1-dichlorotetrafluoroethane	216.58	418.55	216.91	413.15	**8.0816E+01**	**–4.9295E+03**	**–9.3293E+00**	**1.3098E-05**	**2**	3
1609	$C_2Cl_2F_4$	1,2-dichlorotetrafluoroethane	180.62	418.85	202.59	418.71	5.1933E+01	–4.1445E+03	–4.5269E+00	5.2767E-17	6	3
2655	$C_2Cl_3F_3$	1,1,2-trichlorotrifluoroethane	236.92	487.25	237.84	377.59	5.3074E+01	–4.8687E+03	–4.5729E+00	1.9505E-17	6	3
1542	C_2Cl_4	tetrachloroethylene	250.8	620	252.55	393.95	5.8764E+01	–6.1912E+03	–5.3312E+00	2.1269E-06	2	3
1630	C_2F_4	tetrafluoroethylene	142	306.45	142	206.99	7.5184E+01	–3.5219E+03	–8.8314E+00	2.1903E-05	2	3
2693	C_2F_6	hexafluoroethane	172.15	293.03	173.15	291.2	4.0279E+01	–2.5850E+03	–2.9414E+00	2.8049E-16	6	3
2640	$C_2HBrClF_3$	halothane	157.4	496	222.45	344.91	7.4250E+01	–5.8523E+03	–7.7324E+00	4.8624E-17	6	5
2648	C_2HClF_4	2-chloro-1,1,1,2-tetrafluoroethane	74	395.65	173.15	394.5	8.8895E+01	–5.0607E+03	–1.0614E+01	1.5867E-05	2	3
1694	$C_2HCl_2F_3$	2,2-dichloro-1,1,1-trifluoroethane	166	456.94	256.4	456.74	**8.5414E+01**	**–5.6519E+03**	**–9.8284E+00**	**1.0843E-05**	**2**	1
2647	$C_2HCl_2F_3$	1,2-dichloro-1,1,2-trifluoroethane	195.15	461.6	223.15	322.15	**7.5221E+01**	**–5.1882E+03**	**–8.2865E+00**	**9.1647E-06**	**2**	5
1646	C_2HF_5	pentafluoroethane	170.15	339.17	205.04	339.12	**8.2644E+01**	**–4.2543E+03**	**–9.8199E+00**	**1.9244E-05**	**2**	3
3605	C_2HF_5O	difluoromethyl trifluoromethyl ether	116	354.49	216.45	353.6	5.9064E+01	–3.9477E+03	–5.6609E+00	1.7017E-16	6	0.2
401	C_2H_2	acetylene	192.4	308.3	192.59	293.35	3.9630E+01	–2.5522E+03	–2.7800E+00	2.3930E-16	6	1
1581	$C_2H_2Cl_2$	trans-1,2-dichloroethylene	223.35	516.5	234.96	357.99	1.1001E+02	–6.8205E+03	–1.3647E+01	1.4887E-05	2	5
1529	$C_2H_2Cl_4$	1,1,2,2-tetrachloroethane	229.35	645	298.15	445.15	5.3334E+01	–6.5703E+03	–4.3313E+00	2.3072E-07	2	5
1629	$C_2H_2F_2$	1,1-difluoroethylene	129.15	302.8	133.15	302	**2.0327E+02**	**–5.7766E+03**	**–3.3235E+01**	**6.9341E-02**	**1**	3
2650	$C_2H_2F_4$	1,1,1,2-tetrafluoroethane	169.85	374.18	210.96	374.15	8.1808E+01	–4.6766E+03	–9.4881E+00	1.5122E-05	2	3
2660	$C_2H_2F_4$	1,1,2,2-tetrafluoroethane	172	391.8	185.15	390.13	1.3279E+02	–5.4010E+03	–1.9595E+01	3.4077E-02	1	5
1504	C_2H_3Cl	vinyl chloride	119.36	432	167.5	333.49	9.1432E+01	–5.1417E+03	–1.0981E+01	1.4318E-05	2	3
2695	$C_2H_3ClF_2$	1-chloro-1,1-difluoroethane	142.35	410.29	212.51	410.18	7.4090E+01	–4.5732E+03	–8.2593E+00	1.1647E-05	2	3
2649	$C_2H_3Cl_2F$	1,1-dichloro-1-fluoroethane	169.65	478.85	250	450.01	7.9319E+01	–5.4469E+03	–8.8835E+00	9.4125E-06	2	3
1527	$C_2H_3Cl_3$	1,1,1-trichloroethane	243.1	545	251.25	371.35	8.3221E+01	–6.2815E+03	–9.3256E+00	7.8717E-06	2	5
1524	$C_2H_3Cl_3$	1,1,2-trichloroethane	236.5	602	285.15	412.15	**5.4153E+01**	**–6.0418E+03**	**–4.5383E+00**	**4.9833E-18**	**6**	3
1619	$C_2H_3F_3$	1,1,1-trifluoroethane	161.82	345.88	173.6	345.75	7.1776E+01	–3.8701E+03	–8.1092E+00	1.6372E-05	2	3
2619	$C_2H_3F_3$	1,1,2-trifluoroethane	189.15	429.8	314.02	400.96	**5.7131E+01**	**–4.6925E+03**	**–5.1067E+00**	**3.4841E-17**	**6**	5
1772	C_2H_3N	acetonitrile	229.32	545.5	277.92	373.17	4.6735E+01	–5.1262E+03	–3.5406E+00	1.3995E-17	6	3
201	C_2H_4	ethylene	104	282.34	110	282.34	5.3963E+01	–2.4430E+03	–5.5643E+00	1.9079E-05	2	1
1673	$C_2H_4Br_2$	1,2-dibromoethane	282.85	650.15	291.75	404.65	**4.3751E+01**	**–5.5877E+03**	**–3.0891E+00**	**8.2664E-07**	**2**	3
1522	$C_2H_4Cl_2$	1,1-dichloroethane	176.19	523	212.45	505.2	6.6611E+01	–5.4931E+03	–6.7301E+00	5.3579E-06	2	3
1523	$C_2H_4Cl_2$	1,2-dichloroethane	237.49	561.6	239.15	560.3	9.2355E+01	–6.9204E+03	–1.0651E+01	9.1426E-06	2	1
1640	$C_2H_4F_2$	1,1-difluoroethane	154.56	386.44	154.56	386.43	7.3491E+01	–4.3859E+03	–8.1851E+00	1.2978E-05	2	3
1002	C_2H_4O	acetaldehyde	149.78	466	191.7	382.9	5.2911E+01	–4.6431E+03	–4.5068E+00	2.7028E-17	6	3

(*Continued*)

TABLE A.4 Vapor pressure correlation parameters. (*Continued*)

DIPPR ID	Formula	Name	Eq. T_{min} (K)	Eq. T_{max} (K)	Data T_{min} (K)	Data T_{max} (K)	A	B	C	D	E	Er (%)
1441	C_2H_4O	ethylene oxide	160.65	469.15	160.65	469	9.1944E+01	–5.2934E+03	–1.1682E+01	1.4902E-02	1	3
1252	$C_2H_4O_2$	acetic acid	289.81	591.95	293.15	573.15	5.3270E+01	–6.3045E+03	–4.2985E+00	8.8865E-18	6	1
1301	$C_2H_4O_2$	methyl formate	174.15	487.2	198.95	487.15	7.7184E+01	–5.6061E+03	–8.3920E+00	7.8468E-06	2	1
1645	C_2H_5Br	bromoethane	154.25	503.8	160.49	503	5.7324E+01	–4.9312E+03	–5.2244E+00	3.0761E-17	6	5
1503	C_2H_5Cl	ethyl chloride	136.75	460.35	211.86	422.65	4.4677E+01	–4.0260E+03	–3.3710E+00	2.2730E-17	6	3
1617	C_2H_5F	ethyl fluoride	129.95	375.31	169.55	375.31	3.8593E+01	–3.1233E+03	–2.5301E+00	5.2994E-17	6	3
2	C_2H_6	ethane	90.35	305.32	92	304	**5.1857E+01**	**–2.5987E+03**	**–5.1283E+00**	**1.4913E-05**	**2**	1
1102	C_2H_6O	ethanol	159.05	514	243.15	513.92	7.3304E+01	–7.1223E+03	–7.1424E+00	2.8853E-06	2	1
1401	C_2H_6O	dimethyl ether	131.65	400.1	178.2	400.05	4.4704E+01	–3.5256E+03	–3.4444E+00	5.4574E-17	6	5
1201	$C_2H_6O_2$	ethylene glycol	260.15	719	282.83	570.35	9.1594E+01	–1.0811E+04	–9.2821E+00	3.0231E-18	6	10
1802	C_2H_6S	ethyl mercaptan	125.26	499.15	196.45	339.26	6.5551E+01	–5.0274E+03	–6.6853E+00	6.3208E-06	2	3
1820	C_2H_6S	dimethyl sulfide	174.88	503.04	225.75	331.47	8.4390E+01	–5.7406E+03	–9.6454E+00	1.0073E-05	2	3
1704	C_2H_7N	ethylamine	192.15	456.15	192.15	456.15	8.7598E+01	–5.9041E+03	–9.8894E+00	1.5152E-08	3	5
1702	C_2H_7N	dimethylamine	180.96	437.2	200.95	435.75	7.1738E+01	–5.3020E+03	–7.3324E+00	6.4200E-17	6	3
1723	C_2H_7NO	monoethanolamine	283.65	678.2	285	623.17	9.2624E+01	–1.0367E+04	–9.4699E+00	1.9000E-18	6	10
1741	$C_2H_8N_2$	ethylenediamine	284.29	613.1	284.29	486.65	7.0822E+01	–7.4411E+03	–6.7467E+00	6.3477E-18	6	3
2652	C_3F_8	octafluoropropane	125.46	345.05	173.15	345.05	3.5913E+01	–3.0313E+03	–2.1186E+00	2.7190E-17	6	3
2630	C_3HF_7	1,1,1,2,3,3,3-heptafluoropropane	142	374.83	237.65	373.19	**5.9773E+01**	**–4.2780E+03**	**–5.6986E+00**	**1.0808E-16**	**6**	5
2631	$C_3H_2F_6$	1,1,1,2,3,3-hexafluoropropane	170	412.38	242.58	410	9.8062E+01	–5.8287E+03	–1.1871E+01	1.5279E-05	2	3
1620	$C_3H_2F_6$	1,1,1,3,3,3-hexafluoropropane	179.75	398.07	247.99	360.42	6.7322E+01	–4.9237E+03	–6.7337E+00	8.6307E-17	6	5
2629	$C_3H_3F_5$	1,1,1,2,2-pentafluoropropane	134	380.11	232.25	380.11	**5.5993E+01**	**–4.0633E+03**	**–5.1581E+00**	**9.8084E-17**	**6**	5
2627	$C_3H_3F_5$	1,1,1,3,3-pentafluoropropane	193.15	427.2	293.25	426.12	7.2783E+01	–5.4945E+03	–7.4573E+00	5.8777E-17	6	3
2628	$C_3H_3F_5$	1,1,2,2,3-pentafluoropropane	191.15	447.57	243.05	391.86	5.7434E+01	–4.9192E+03	–5.1675E+00	3.4974E-17	6	1
2868	$C_3H_3F_5O$	2-(difluoromethoxy)-1,1,1-trifluoroethane	150	444.03			*6.4815E+01*	*–5.3191E+03*	*–6.2541E+00*	*4.3493E-17*	6	5
1774	C_3H_3N	acrylonitrile	189.63	540	273.15	528.2	5.7316E+01	–5.6622E+03	–5.0622E+00	1.5068E-17	6	10
5892	C_3H_3NO	isoxazole	206.05	590	298.15	403.959	1.2025E+02	–8.4728E+03	–1.4765E+01	1.1323E-05	2	5
402	C_3H_4	methylacetylene	170.45	402.4	186.87	393.15	5.0242E+01	–3.8119E+03	–4.2526E+00	6.5326E-17	6	3
301	C_3H_4	propadiene	136.87	394	138	292	5.7069E+01	–3.6827E+03	–5.5662E+00	6.5133E-06	2	5
1773	C_3H_5N	propionitrile	180.37	561.3	188.49	393.89	5.9996E+01	–6.0062E+03	–5.4600E+00	1.7041E-17	6	3
202	C_3H_6	propylene	87.89	364.85	99.46	364.5	4.3905E+01	–3.0978E+03	–3.4425E+00	9.9989E-17	6	5
101	C_3H_6	cyclopropane	145.59	398	179.65	393.15	4.0608E+01	–3.1796E+03	–2.8937E+00	5.6131E-17	6	5
1526	$C_3H_6Cl_2$	1,2-dichloropropane	172.71	578	293.673	406.461	7.3869E+01	–6.4770E+03	–7.6351E+00	6.4479E-09	3	3
1167	C_3H_6O	allyl alcohol	144.15	545.1	273.15	393.15	**8.4739E+01**	**–8.0576E+03**	**–8.7051E+00**	**1.6596E-17**	**6**	5
1051	C_3H_6O	acetone	178.45	508.1	215.15	508.1	5.7947E+01	–5.3553E+03	–5.2106E+00	1.2449E-14	5	3
1003	C_3H_6O	propanal	165	503.6	231.15	373.15	5.0877E+01	–4.9310E+03	–4.1667E+00	1.6674E-17	6	3

TABLE A.4 Vapor pressure correlation parameters. (*Continued*)

DIPPR ID	Formula	Name	Eq. T_{min} (K)	Eq. T_{max} (K)	Data T_{min} (K)	Data T_{max} (K)	A	B	C	D	E	Er (%)
1442	C_3H_6O	1,2-propylene oxide	161.22	482.25	198.15	348.15	9.1037E+01	–5.9763E+03	–1.0686E+01	1.1993E-05	2	5
1253	$C_3H_6O_2$	propionic acid	252.45	600.81	298.18	438.05	5.4552E+01	–7.1494E+03	–4.2769E+00	1.1843E-18	6	3
1312	$C_3H_6O_2$	methyl acetate	175.15	506.55	215.95	503.15	6.1267E+01	–5.6186E+03	–5.6473E+00	2.1080E-17	6	1
1302	$C_3H_6O_2$	ethyl formate	193.55	508.4	212.65	508.4	7.3833E+01	–5.8170E+03	–7.8090E+00	6.3200E-06	2	1
2391	$C_3H_6O_3$	dimethyl carbonate	273.15	548	287.67	370.56	5.8033E+01	–5.9913E+03	–5.0971E+00	1.3402E-17	6	3
1585	C_3H_7Cl	propyl chloride	150.35	503.15	248.05	367.05	5.8359E+01	–5.1113E+03	–5.3526E+00	2.4673E-17	6	3
1876	C_3H_7NO	*N,N*-dimethylformamide	212.72	649.6	293.15	363.15	**8.2762E+01**	**–7.9555E+03**	**–8.8038E+00**	**4.2431E-06**	**2**	10
3	C_3H_8	propane	85.47	369.83	85.47	360	5.9078E+01	–3.4926E+03	–6.0669E+00	1.0919E-05	2	3
1103	C_3H_8O	1-propanol	146.95	536.8	273.15	536.71	8.4664E+01	–8.3072E+03	–8.5767E+00	7.5091E-18	6	3
1104	C_3H_8O	isopropanol	185.26	508.3	275	508.3	1.1072E+02	–9.0400E+03	–1.2676E+01	5.5380E-06	2	3
1407	C_3H_8O	methyl ethyl ether	160	437.8	182.15	437.8	7.8586E+01	–5.1763E+03	–8.7501E+00	9.1727E-06	2	5
1431	$C_3H_8O_2$	methylal	168.35	480.6	185.45	315.2	9.1640E+01	–6.2791E+03	–1.0631E+01	9.7948E-06	2	5
1212	$C_3H_8O_2$	1,3-propylene glycol	245.5	718.2	293.5	716.6	1.1020E+02	–1.2513E+04	–1.1804E+01	3.9867E-18	6	5
1231	$C_3H_8O_3$	glycerol	291.33	850	298.75	535	9.9986E+01	–1.3808E+04	–1.0088E+01	3.5712E-19	6	10
1813	C_3H_8S	methyl ethyl sulfide	167.23	533	247.55	373.97	7.9070E+01	–6.1141E+03	–8.6310E+00	6.5333E-06	2	3
1711	C_3H_9N	*n*-propylamine	190.15	496.95	237.15	487.65	1.1199E+02	–7.1031E+03	–1.3836E+01	1.4469E-05	2	5
1719	C_3H_9N	isopropylamine	177.95	471.85	181.84	463.14	1.3666E+02	–7.2015E+03	–1.8934E+01	2.2255E-02	1	3
1703	C_3H_9N	trimethylamine	156.08	433.25	176.05	363.15	1.3468E+02	–6.0558E+03	–1.9415E+01	2.8619E-02	1	5
6862	C_3H_9NO	methylethanolamine	268.65	630	325.15	328.85	**8.8254E+01**	**–9.5067E+03**	**–9.0217E+00**	**7.3345E-18**	**6**	10
2654	C_4F_8	octafluorocyclobutane	232.96	388.37	233.15	388.37	5.6891E+01	–4.3166E+03	–5.2321E+00	7.3846E-17	6	3
1478	C_4H_4O	furan	187.55	490.15	228.25	483.15	7.4738E+01	–5.4170E+03	–8.0636E+00	7.4700E-06	2	3
1821	C_4H_4S	thiophene	234.94	579.35	243.15	579.35	9.3193E+01	–7.0015E+03	–1.0738E+01	8.2308E-06	2	3
1721	C_4H_5N	pyrrole	249.74	639.75	298.15	615.15	1.2502E+02	–9.0926E+03	–1.6079E+01	1.3712E-02	1	3
403	C_4H_6	ethylacetylene	147.43	440	204.23	353.15	7.7004E+01	–5.0545E+03	–8.5665E+00	1.0161E-05	2	1
302	C_4H_6	1,2-butadiene	136.95	452	204.03	303.55	**3.9714E+01**	**–3.7699E+03**	**–2.6407E+00**	**6.9379E-18**	**6**	3
303	C_4H_6	1,3-butadiene	164.25	425	193.1	425	7.5572E+01	–4.6219E+03	–8.5323E+00	1.2269E-05	2	1
1321	$C_4H_6O_2$	vinyl acetate	180.35	519.13	273	515.13	**5.7406E+01**	**–5.7028E+03**	**–5.0307E+00**	**1.1042E-17**	**6**	5
1291	$C_4H_6O_3$	acetic anhydride	200.15	606	293.15	429.3	6.7182E+01	–7.4635E+03	–6.2439E+00	6.8593E-18	6	5
1782	C_4H_7N	butyronitrile	161.3	585.4	278.15	393.373	6.0658E+01	–6.4043E+03	–5.4929E+00	1.1329E-17	6	3
102	C_4H_8	cyclobutane	182.48	459.93	204.95	453.15	8.5899E+01	–4.8844E+03	–1.0883E+01	1.4934E-02	1	3
204	C_4H_8	1-butene	87.8	419.5	125.85	419.5	5.1836E+01	–4.0192E+03	–4.5229E+00	4.8833E-17	6	3
206	C_4H_8	trans-2-butene	167.62	428.6	196.76	413.15	7.1704E+01	–4.5631E+03	–7.9053E+00	1.1319E-05	2	3
205	C_4H_8	cis-2-butene	134.26	435.5	199.66	423.15	7.2541E+01	–4.6912E+03	–7.9776E+00	1.0368E-05	2	3
207	C_4H_8	isobutene	132.81	417.9	190.85	417.9	7.8010E+01	–4.6341E+03	–8.9575E+00	1.3413E-05	2	3
1005	C_4H_8O	butanal	176.8	537.2	303.86	353.15	5.1648E+01	–5.3014E+03	–4.2559E+00	1.1406E-17	6	3

(*Continued*)

TABLE A.4 Vapor pressure correlation parameters. (*Continued*)

DIPPR ID	Formula	Name	Eq. T_{min} (K)	Eq. T_{max} (K)	Data T_{min} (K)	Data T_{max} (K)	A	B	C	D	E	Er (%)
1052	C_4H_8O	methyl ethyl ketone	186.46	536.7	265.15	536.7	8.4530E+01	−6.7872E+03	−9.2336E+00	9.0891E-09	3	1
1479	C_4H_8O	tetrahydrofuran	164.65	540.15	253.15	540.15	5.4898E+01	−5.3054E+03	−4.7627E+00	1.4291E-17	6	1
1256	$C_4H_8O_2$	*n*-butyric acid	267.95	615.7	278.2	493.15	7.8117E+01	−8.9244E+03	−7.5993E+00	7.3908E-18	6	5
1260	$C_4H_8O_2$	isobutyric acid	227.15	605	326.85	445.6	1.1038E+02	−1.0540E+04	−1.2262E+01	1.4310E-17	6	5
2422	$C_4H_8O_2$	1,3-dioxane	228.15	590	374.15	378.15	**5.8260E+01**	**−6.2286E+03**	**−5.1013E+00**	**7.0585E-18**	**6**	10
1421	$C_4H_8O_2$	1,4-dioxane	284.95	587	285.15	583.15	4.4494E+01	−5.4067E+03	−3.1287E+00	2.8913E-18	6	1
1322	$C_4H_8O_2$	methyl propionate	185.65	530.6	253.15	530.55	7.0717E+01	−6.4397E+03	−6.9845E+00	2.0129E-17	6	1
1313	$C_4H_8O_2$	ethyl acetate	189.6	523.3	229.75	523.25	6.6824E+01	−6.2276E+03	−6.4100E+00	1.7914E-17	6	1
1303	$C_4H_8O_2$	*n*-propyl formate	180.25	538	262.05	538	1.0408E+02	−7.5359E+03	−1.2348E+01	9.6020E-06	2	3
1587	C_4H_9Cl	sec-butyl chloride	141.85	520.6	252.65	367.05	6.8867E+01	−5.5641E+03	−7.1727E+00	6.8734E-06	2	3
1766	C_4H_9N	pyrrolidine	215.31	568.55	273.15	566.48	8.5194E+01	−6.9777E+03	−9.3533E+00	6.0256E-06	2	1
5	C_4H_{10}	*n*-butane	134.86	425.12	140	420	6.6343E+01	−4.3632E+03	−7.0460E+00	9.4509E-06	2	3
4	C_4H_{10}	isobutane	113.54	407.8	238.98	406.87	**1.0843E+02**	**−5.0399E+03**	**−1.5012E+01**	**2.2725E-02**	**1**	3
2752	$C_4H_{10}N_2$	piperazine	384.6	656.3	418	655	6.5472E+01	−7.4769E+03	−5.9955E+00	3.9600E-18	6	3
1105	$C_4H_{10}O$	1-butanol	183.85	563.1	295.75	562.98	1.0629E+02	−9.8664E+03	−1.1655E+01	1.0832E-17	6	3
1106	$C_4H_{10}O$	2-methyl-1-propanol	165.15	547.8	297.85	547.71	**1.2178E+02**	**−1.0504E+04**	**−1.3921E+01**	**1.6898E-17**	**6**	3
1108	$C_4H_{10}O$	2-methyl-2-propanol	298.97	506.2	300.85	506.15	1.7227E+02	−1.1589E+04	−2.2113E+01	1.3703E-05	2	3
1107	$C_4H_{10}O$	2-butanol	158.45	535.9	278.15	535.9	1.2255E+02	−1.0236E+04	−1.4125E+01	2.3559E-17	6	3
1402	$C_4H_{10}O$	diethyl ether	156.85	466.7	198.85	466.7	1.3690E+02	−6.9543E+03	−1.9254E+01	2.4508E-02	1	3
1455	$C_4H_{10}O_2$	1,2-dimethoxyethane	215.15	536.15	264.25	533.15	6.1814E+01	−6.1029E+03	−5.6547E+00	1.1802E-17	6	5
1220	$C_4H_{10}O_2$	1,2-butanediol	220	680	329.15	506.39	1.0328E+02	−1.1548E+04	−1.0925E+01	4.2560E-18	6	10
1221	$C_4H_{10}O_2$	1,3-butanediol	196.15	676	319.99	512.05	**1.2322E+02**	**−1.2620E+04**	**−1.3986E+01**	**3.9260E-06**	**2**	10
1202	$C_4H_{10}O_3$	diethylene glycol	264.15	753	293.15	538.91	1.2186E+02	−1.3799E+04	−1.3403E+01	3.4123E-18	6	5
1818	$C_4H_{10}S$	diethyl sulfide	169.2	557.15	283.15	395.58	4.6705E+01	−5.1774E+03	−3.5985E+00	1.7147E-06	2	3
1712	$C_4H_{11}N$	*n*-butylamine	224.05	531.9	259.15	373.15	1.2182E+02	−8.1176E+03	−1.5157E+01	1.3486E-05	2	5
1710	$C_4H_{11}N$	diethylamine	223.35	496.6	240.15	493.15	4.9314E+01	−4.9490E+03	−3.9256E+00	9.1978E-18	6	3
1714	$C_4H_{11}N$	isobutylamine	188.55	513.73	223.15	373.76	6.6734E+01	−6.0301E+03	−6.4406E+00	2.4888E-17	6	3
1727	$C_4H_{11}N$	*tert*-butylamine	206.19	483.9	292.47	348.36	6.0529E+01	−4.9518E+03	−6.0687E+00	4.9204E-03	1	5
1724	$C_4H_{11}NO_2$	diethanolamine	301.15	736.6	330	645	1.0638E+02	−1.3714E+04	−1.1060E+01	3.2645E-18	6	5
2623	C_5F_{12}	perfluoro-*n*-pentane	147.51	423.1	221.17	421.85	7.2801E+01	−5.5390E+03	−7.5271E+00	6.1648E-17	6	3
1889	$C_5H_4O_2$	furfural	236.65	670.15	298.15	526.95	**9.4570E+01**	**−8.3721E+03**	**−1.1130E+01**	**8.8150E-03**	**1**	5
1791	C_5H_5N	pyridine	231.51	619.95	253.15	619.95	8.2154E+01	−7.2113E+03	−8.8646E+00	5.2528E-06	2	1
405	C_5H_8	1-pentyne	167.45	481.2	233.15	334.15	**8.2805E+01**	**−5.6838E+03**	**−9.4301E+00**	**1.0767E-05**	**2**	3
269	C_5H_8	cyclopentene	138.13	507	222.35	393.15	6.7952E+01	−5.1875E+03	−7.0785E+00	6.8165E-06	2	3
1079	C_5H_8O	cyclopentanone	221.85	624.5	273.1	621.55	5.6405E+01	−6.4445E+03	−4.8222E+00	4.8774E-18	6	5

TABLE A.4 Vapor pressure correlation parameters. (*Continued*)

DIPPR ID	Formula	Name	Eq. T_{min} (K)	Eq. T_{max} (K)	Data T_{min} (K)	Data T_{max} (K)	A	B	C	D	E	Er (%)
1783	C_5H_9N	valeronitrile	176.92	610.3	267.15	413.95	5.8645E+01	−6.6581E+03	−5.1611E+00	8.8787E-18	6	5
1071	C_5H_9NO	*N*-methyl-2-pyrrolidone	249.15	721.6	279.15	522.13	6.8476E+01	−8.4679E+03	−6.3622E+00	3.2235E-18	6	3
104	C_5H_{10}	cyclopentane	179.28	511.7	221.15	344.69	6.6341E+01	−5.1985E+03	−6.8103E+00	6.1930E-06	2	3
209	C_5H_{10}	1-pentene	108.02	464.8	272.99	463.4	4.6994E+01	−4.2895E+03	−3.7345E+00	2.5424E-17	6	3
210	C_5H_{10}	*cis*-2-pentene	121.75	475	230.17	343.15	5.1101E+01	−4.6165E+03	−4.3069E+00	2.3505E-17	6	3
211	C_5H_{10}	*trans*-2-pentene	132.89	474.2	223.75	343.15	7.7480E+01	−5.4132E+03	−8.6036E+00	9.1643E-06	2	1
212	C_5H_{10}	2-methyl-1-butene	135.58	465	145.15	446.93	**9.3131E+01**	**−5.5254E+03**	**−1.1852E+01**	**1.4205E-02**	**1**	3
214	C_5H_{10}	2-methyl-2-butene	139.39	470	146.15	433.15	8.3927E+01	−5.6405E+03	−9.6453E+00	1.1121E-05	2	3
213	C_5H_{10}	3-methyl-1-butene	104.66	452.7	133.15	452.69	7.4855E+01	−4.9617E+03	−8.3216E+00	1.0099E-05	2	3
3112	$C_5H_{10}O$	cyclopentanol	255.6	619.5	283.3	437.51	1.1077E+02	−1.0479E+04	−1.2275E+01	8.3127E-18	6	3
1060	$C_5H_{10}O$	2-pentanone	196.27	561.19	268.15	560	9.1259E+01	−7.5157E+03	−1.0148E+01	8.4195E-09	3	3
1053	$C_5H_{10}O$	3-pentanone	234.18	560.95	283.15	555.37	4.4286E+01	−5.4151E+03	−3.0913E+00	1.8580E-18	6	3
1061	$C_5H_{10}O$	methyl isopropyl ketone	181.15	552.8	276.25	552.8	8.6109E+01	−7.0910E+03	−9.4310E+00	8.4371E-09	3	3
1400	$C_5H_{10}O$	2-methyltetrahydrofuran	137.15	537	263.44	533.15	5.3628E+01	−5.4269E+03	−4.5596E+00	1.1636E-17	6	3
1258	$C_5H_{10}O_2$	*n*-pentanoic acid	239.15	639.16	283.2	573.25	9.3208E+01	−1.0471E+04	−9.6135E+00	5.6156E-18	6	3
1261	$C_5H_{10}O_2$	isovaleric acid	243.85	629.09	244.76	464.44	**9.1787E+01**	**−1.0140E+04**	**−9.4547E+00**	**6.0472E-18**	**6**	5
1332	$C_5H_{10}O_2$	methyl *n*-butyrate	187.35	554.5	263.15	554.45	7.1870E+01	−6.8857E+03	−7.0944E+00	1.4903E-17	6	3
1323	$C_5H_{10}O_2$	ethyl propionate	199.25	546	245.15	546	1.0564E+02	−8.0070E+03	−1.2477E+01	9.0000E-06	2	1
2332	$C_5H_{10}O_2$	methyl isobutyrate	188.45	540.7	260.15	540.7	6.8739E+01	−6.5413E+03	−6.6696E+00	1.5079E-17	6	5
1314	$C_5H_{10}O_2$	*n*-propyl acetate	178.15	549.73	263.15	549.35	1.1516E+02	−8.4339E+03	−1.3934E+01	1.0346E-05	2	3
1305	$C_5H_{10}O_2$	isobutyl formate	177.35	551.35	240.45	551.15	4.3097E+01	−5.3627E+03	−2.9026E+00	3.8186E-07	2	3
1319	$C_5H_{10}O_2$	isopropyl acetate	199.75	532	234.85	385.26	4.9754E+01	−5.5639E+03	−3.8789E+00	2.4755E-18	6	3
1588	$C_5H_{11}Cl$	1-chloropentane	174.15	571.2	283.72	390.65	6.2352E+01	−6.2620E+03	−5.7910E+00	1.1390E-17	6	3
7	C_5H_{12}	*n*-pentane	143.42	469.7	223.05	469.65	7.8741E+01	−5.4203E+03	−8.8253E+00	9.6171E-06	2	3
8	C_5H_{12}	isopentane	113.25	460.4	217.19	448.15	**7.1308E+01**	**−4.9760E+03**	**−7.7169E+00**	**8.7271E-06**	**2**	3
9	C_5H_{12}	neopentane	256.6	433.8	258.02	433.78	4.2256E+01	−3.7232E+03	−3.1133E+00	3.1540E-17	6	1
1109	$C_5H_{12}O$	1-pentanol	195.56	588.1	286.75	573.15	**1.1475E+02**	**−1.0643E+04**	**−1.2858E+01**	**1.2491E-17**	**6**	3
1110	$C_5H_{12}O$	2-pentanol	200	561	293.15	393.43	1.1683E+02	−1.0453E+04	−1.3177E+01	1.0712E-17	6	3
1112	$C_5H_{12}O$	2-methyl-1-butanol	195	575.4	248	410.65	1.1924E+02	−1.0738E+04	−1.3522E+01	1.4271E-17	6	3
1111	$C_5H_{12}O$	2-methyl-2-butanol	264.15	543.7	274.3	375.49	1.1578E+02	−9.8601E+03	−1.3162E+01	1.4681E-17	6	3
1123	$C_5H_{12}O$	3-methyl-1-butanol	155.95	577.2	298.13	404.55	**1.1707E+02**	**−1.0743E+04**	**−1.3165E+01**	**1.1670E-17**	**6**	10
1124	$C_5H_{12}O$	3-methyl-2-butanol	188	556.1	280.3	385.15	1.1226E+02	−9.9257E+03	−1.2591E+01	1.1433E-17	6	5
1415	$C_5H_{12}O$	ethyl propyl ether	145.65	500.23	246.54	359.96	8.6898E+01	−6.6464E+03	−9.5758E+00	5.9615E-17	6	5
1405	$C_5H_{12}O$	methyl *tert*-butyl ether	164.55	497.1	273.15	496.4	5.7130E+01	−5.2007E+03	−5.1398E+00	1.6513E-17	6	3
1120	$C_5H_{12}O$	3-pentanol	204.15	559.6	245.05	409.15	9.5581E+01	−9.2258E+03	−1.0115E+01	1.0753E-18	6	3

(*Continued*)

TABLE A.4 Vapor pressure correlation parameters. (*Continued*)

DIPPR ID	Formula	Name	Eq. T_{min} (K)	Eq. T_{max} (K)	Data T_{min} (K)	Data T_{max} (K)	A	B	C	D	E	Er (%)
1864	C_6F_6	hexafluorobenzene	278.25	516.73	278.35	516.67	6.4988E+01	–6.1417E+03	–6.1557E+00	1.8827E-17	6	3
1623	C_6F_{14}	perfluoro-*n*-hexane	185	449.55	256.43	448.77	6.3705E+01	–5.6667E+03	–6.0445E+00	2.7396E-17	6	3
1573	$C_6H_4Cl_2$	*m*-dichlorobenzene	248.39	683.95	285.25	475.62	5.3187E+01	–6.8275E+03	–4.3233E+00	2.3112E-18	6	3
1572	$C_6H_4Cl_2$	*o*-dichlorobenzene	256.15	705	293.15	483.42	7.7105E+01	–8.1111E+03	–7.8886E+00	2.7267E-06	2	3
1571	C_6H_5Cl	monochlorobenzene	227.95	632.35	260.15	632.35	**5.4144E+01**	**–6.2444E+03**	**–4.5343E+00**	**4.7030E-18**	**6**	5
1691	C_6H_5I	iodobenzene	241.83	721.15	297.25	721.15	8.0266E+01	–8.2487E+03	–8.4197E+00	3.6633E-06	2	3
1886	$C_6H_5NO_2$	nitrobenzene	278.87	719	283.15	564.15	8.5828E+01	–9.4938E+03	–8.8595E+00	6.8912E-18	6	5
501	C_6H_6	benzene	278.68	562.05	283.1	553.15	8.3107E+01	–6.4862E+03	–9.2194E+00	6.9844E-06	2	1
1181	C_6H_6O	phenol	314.06	694.25	315	690	9.5444E+01	–1.0113E+04	–1.0090E+01	6.7603E-18	6	3
1792	C_6H_7N	aniline	267.12	705	268.11	699	6.5632E+01	–8.1661E+03	–5.9216E+00	2.6871E-18	6	3
1797	C_6H_7N	2-methylpyridine	206.44	621	273.15	620	9.1273E+01	–7.8412E+03	–1.0216E+01	6.2148E-06	2	3
2797	C_6H_7N	3-methylpyridine	255.01	645	273.15	645	9.3075E+01	–8.1722E+03	–1.0444E+01	6.0393E-06	2	3
2799	C_6H_7N	4-methylpyridine	276.82	646.15	280	645	9.0839E+01	–8.1036E+03	–1.0096E+01	5.7026E-06	2	3
2088	C_6H_8O	2-cyclohexene-1-one	220.2	684.2	316.65	481.36	**5.8337E+01**	**–7.2418E+03**	**–5.0120E+00**	**2.8659E-18**	**6**	10
270	C_6H_{10}	cyclohexene	169.67	560.4	213.15	423.15	8.8184E+01	–6.6249E+03	–1.0059E+01	8.2566E-06	2	3
1080	$C_6H_{10}O$	cyclohexanone	242	653	242	458.34	8.5424E+01	–7.9444E+03	–9.2862E+00	4.9957E-06	2	3
1065	$C_6H_{10}O$	mesityl oxide	220.15	600	287.15	403.05	6.2679E+01	–6.8150E+03	–5.7103E+00	5.3452E-18	6	5
137	C_6H_{12}	cyclohexane	279.69	553.8	280.05	553.5	5.1087E+01	–5.2264E+03	–4.2278E+00	9.7554E-18	6	1
105	C_6H_{12}	methylcyclopentane	130.73	532.7	255.06	527.59	5.5368E+01	–5.1498E+03	–5.0136E+00	3.2220E-06	2	3
216	C_6H_{12}	1-hexene	133.39	504	273.15	503.52	**5.1977E+01**	**–5.1047E+03**	**–4.3484E+00**	**1.1716E-17**	**6**	10
223	C_6H_{12}	4-methyl-1-pentene	119.51	496	235.65	349.9	1.0799E+02	–6.3009E+03	–1.4283E+01	1.6832E-02	1	3
1151	$C_6H_{12}O$	cyclohexanol	296.6	650.1	350.8	456.54	1.8919E+02	–1.4337E+04	–2.4148E+01	1.0740E-05	2	3
1062	$C_6H_{12}O$	2-hexanone	217.4	587.1	294.59	585	9.7675E+01	–8.3351E+03	–1.0987E+01	7.7951E-09	3	3
1059	$C_6H_{12}O$	3-hexanone	217.5	582.82	292.85	422.25	7.3155E+01	–7.2429E+03	–7.2569E+00	1.2741E-17	6	3
1054	$C_6H_{12}O$	methyl isobutyl ketone	189.15	574.6	283.15	415.82	9.4870E+01	–7.9175E+03	–1.0643E+01	8.0701E-09	3	3
1447	$C_6H_{12}O$	butyl vinyl ether	181.15	536	293.15	298.15	**6.8365E+01**	**–6.5015E+03**	**–6.6315E+00**	**1.6639E-17**	**6**	5
1262	$C_6H_{12}O_2$	*n*-hexanoic acid	269.25	660.2	297.2	533.15	9.8377E+01	–1.1394E+04	–1.0224E+01	3.2948E-18	6	10
1333	$C_6H_{12}O_2$	ethyl *n*-butyrate	175.15	571	254.75	422.65	5.7661E+01	–6.3465E+03	–5.0320E+00	8.2534E-18	6	3
1324	$C_6H_{12}O_2$	*n*-propyl propionate	197.25	568.6	258.95	420.35	7.8318E+01	–7.2569E+03	–8.2280E+00	4.8584E-06	2	3
2337	$C_6H_{12}O_2$	ethyl isobutyrate	185	553.15	248.85	553.15	5.4972E+01	–6.0999E+03	–4.6324E+00	8.5966E-18	6	3
1315	$C_6H_{12}O_2$	*n*-butyl acetate	199.65	575.4	326.19	410.04	1.2282E+02	–9.2532E+03	–1.4990E+01	1.0470E-05	2	3
1316	$C_6H_{12}O_2$	isobutyl acetate	174.3	560.8	288.15	523.15	7.2310E+01	–6.9443E+03	–7.2980E+00	3.7892E-06	2	3
1306	$C_6H_{12}O_2$	*n*-pentyl formate	199.65	576	230	490	8.6931E+01	–8.1124E+03	–9.2418E+00	2.3290E-17	6	5
5884	$C_6H_{12}O_3$	2-ethoxyethyl acetate	211.45	607.3	303.15	468.71	8.8865E+01	–8.9786E+03	–9.3229E+00	1.2714E-17	6	3
11	C_6H_{14}	*n*-hexane	177.83	507.6	196.15	503.15	1.0465E+02	–6.9955E+03	–1.2702E+01	1.2381E-05	2	3

TABLE A.4 Vapor pressure correlation parameters. (*Continued*)

DIPPR ID	Formula	Name	Eq. T_{min} (K)	Eq. T_{max} (K)	Data T_{min} (K)	Data T_{max} (K)	A	B	C	D	E	Er (%)
12	C_6H_{14}	2-methylpentane	119.55	497.7	240.85	497.5	5.3579E+01	−5.0412E+03	−4.6404E+00	1.9443E-17	6	3
13	C_6H_{14}	3-methylpentane	110.25	504.6	243.15	504.4	5.7090E+01	−5.2397E+03	−5.1592E+00	2.1702E-17	6	3
14	C_6H_{14}	2,2-dimethylbutane	174.28	489	199.15	488.78	9.3196E+01	−6.0852E+03	−1.1092E+01	1.2087E-05	2	3
15	C_6H_{14}	2,3-dimethylbutane	145.19	500	238.25	499.98	7.7161E+01	−5.6911E+03	−8.5010E+00	8.0325E-06	2	3
1114	$C_6H_{14}O$	1-hexanol	228.55	611.3	310	603.15	**1.3542E+02**	**−1.2288E+04**	**−1.5732E+01**	**1.2701E-17**	**6**	5
1115	$C_6H_{14}O$	2-hexanol	223	585.3	240	585	1.2270E+02	−1.0870E+04	−1.4192E+01	3.8710E-06	2	5
1116	$C_6H_{14}O$	3-hexanol	208.5	582.4	263.14	411.15	1.2446E+02	−1.1067E+04	−1.4273E+01	1.1583E-17	6	10
1117	$C_6H_{14}O$	2-methyl-1-pentanol	223	604.4	261	428.15	2.8435E+02	−1.6344E+04	−4.1270E+01	3.6496E-02	1	5
1130	$C_6H_{14}O$	4-methyl-2-pentanol	183	574.4	293.15	427.15	7.8617E+01	−8.2868E+03	−7.7706E+00	6.6663E-18	6	5
1403	$C_6H_{14}O$	diisopropyl ether	187.65	500.05	284.78	494.26	4.1631E+01	−4.6687E+03	−2.8551E+00	6.3693E-04	1	3
1446	$C_6H_{14}O$	di-*n*-propyl ether	149.95	530.6	200	530	6.5935E+01	−6.1985E+03	−6.3423E+00	2.0909E-17	6	5
1432	$C_6H_{14}O_2$	acetal	173.15	539.7	273.15	377.35	1.0287E+02	−7.6647E+03	−1.2245E+01	1.1529E-05	2	5
1706	$C_6H_{15}N$	triethylamine	158.45	535.15	283.15	367.75	5.6550E+01	−5.6819E+03	−4.9815E+00	1.2363E-17	6	5
1707	$C_6H_{15}N$	di-*n*-propylamine	210.15	550	273.15	381.15	**5.4000E+01**	**−6.0185E+03**	**−4.4981E+00**	**9.9684E-18**	**6**	10
1725	$C_6H_{15}NO_3$	triethanolamine	294.35	772.1	420	645	1.2832E+02	−1.6855E+04	−1.3935E+01	4.6601E-18	6	10
2626	C_7F_{16}	perfluoro-*n*-heptane	221.87	475.65	271.26	475.65	1.0610E+02	−7.6148E+03	−1.2686E+01	1.0733E-05	2	3
1874	$C_7H_5ClO_2$	*o*-chlorobenzoic acid	415.15	792			*9.6025E+01*	*−1.2774E+04*	*−9.7594E+00*	*1.8357E-18*	*6*	10
1790	C_7H_5N	benzonitrile	260.28	702.3	325.15	508.98	5.5040E+01	−7.3638E+03	−4.5061E+00	1.9494E-18	6	10
1041	C_7H_6O	benzaldehyde	216.02	695	262.3	695	1.1628E+02	−9.3312E+03	−1.4639E+01	1.1932E-02	1	5
1578	C_7H_7Cl	*p*-chlorotoluene	280.65	660	322.05	435.55	**1.4740E+02**	**−1.0076E+04**	**−1.9840E+01**	**1.7958E-02**	**1**	5
502	C_7H_8	toluene	178.18	591.75	247.15	591.7	7.6945E+01	−6.7298E+03	−8.1790E+00	5.3017E-06	2	3
1461	C_7H_8O	anisole	235.65	645.6	242.11	640.41	1.2806E+02	−9.3077E+03	−1.6693E+01	1.4919E-02	1	3
1180	C_7H_8O	benzyl alcohol	257.85	713	282.2	675	9.9097E+01	−1.0982E+04	−1.0479E+01	3.2564E-18	6	3
1182	C_7H_8O	*o*-cresol	304.19	697.55	313.15	693.15	2.1088E+02	−1.3928E+04	−2.9483E+01	2.5182E-02	1	1
1183	C_7H_8O	*m*-cresol	285.39	705.85	288.15	705.85	9.5403E+01	−1.0581E+04	−1.0004E+01	4.3032E-18	6	3
1184	C_7H_8O	*p*-cresol	307.93	704.65	313.15	704.65	1.1853E+02	−1.1957E+04	−1.3293E+01	8.6988E-18	6	3
2796	C_7H_9N	2,6-dimethylpyridine	267	623.75	273.35	600	7.8801E+01	−7.6378E+03	−8.2293E+00	3.9754E-06	2	3
1344	$C_7H_{12}O_2$	*n*-butyl acrylate	208.55	598	272.65	420.55	8.1598E+01	−7.6129E+03	−8.7765E+00	5.9550E-06	2	3
159	C_7H_{14}	cycloheptane	265.12	604.2	283.05	593.15	8.4578E+01	−7.1115E+03	−9.3545E+00	6.1711E-06	2	3
138	C_7H_{14}	methylcyclohexane	146.58	572.1	259.98	566.48	9.2684E+01	−7.0808E+03	−1.0695E+01	8.1366E-06	2	3
107	C_7H_{14}	ethylcyclopentane	134.71	569.5	240.95	566.48	8.8671E+01	−7.0127E+03	−1.0045E+01	7.4578E-06	2	3
111	C_7H_{14}	*cis*-1,3-dimethylcyclopentane	139.45	551	263.15	389.15	8.6463E+01	−6.2828E+03	−1.0444E+01	1.0732E-02	1	3
112	C_7H_{14}	*trans*-1,3-dimethylcyclopentane	139.18	553	263.95	390.15	**1.2737E+02**	**−7.6263E+03**	**−1.7277E+01**	**1.9136E-02**	**1**	3
234	C_7H_{14}	1-heptene	154.12	537.4	255.46	374.15	6.5922E+01	−6.1890E+03	−6.3629E+00	2.0091E-17	6	3
1326	$C_7H_{14}O_2$	*n*-butyl propionate	183.63	594.6	305.55	431.46	**7.1228E+01**	**−7.7098E+03**	**−6.8418E+00**	**6.3588E-18**	**6**	5

(*Continued*)

TABLE A.4 Vapor pressure correlation parameters. (*Continued*)

DIPPR ID	Formula	Name	Eq. T_{min} (K)	Eq. T_{max} (K)	Data T_{min} (K)	Data T_{max} (K)	A	B	C	D	E	Er (%)
2261	$C_7H_{14}O_2$	*n*-heptanoic acid	265.83	677.3	283.2	496.15	1.1237E+02	−1.2660E+04	−1.2147E+01	4.3880E-18	6	5
1347	$C_7H_{14}O_2$	ethyl isovalerate	173.85	587.95	267.05	407.45	**5.4702E+01**	**−6.5988E+03**	**−4.4931E+00**	**9.1346E-08**	**2**	5
1327	$C_7H_{14}O_2$	*n*-propyl *n*-butyrate	177.95	593.7	271.55	445.15	5.8395E+01	−6.8763E+03	−5.0335E+00	3.2675E-18	6	5
1336	$C_7H_{14}O_2$	*n*-propyl isobutyrate	200	579.4	407.15	407.15	**7.5263E+01**	**−7.7764E+03**	**−7.4416E+00**	**9.5324E-18**	**6**	5
1317	$C_7H_{14}O_2$	isopentyl acetate	194.65	586.1	293.15	415.15	9.9558E+01	−8.8768E+03	−1.1075E+01	2.4723E-17	6	5
17	C_7H_{16}	*n*-heptane	182.57	540.2	194.15	540	8.7829E+01	−6.9964E+03	−9.8802E+00	7.2099E-06	2	3
18	C_7H_{16}	2-methylhexane	154.9	530.4	270.55	530.3	5.9531E+01	−5.8264E+03	−5.4269E+00	1.4542E-17	6	3
19	C_7H_{16}	3-methylhexane	153.75	535.2	271.85	535.19	5.7485E+01	−5.7566E+03	−5.1207E+00	1.2330E-17	6	3
20	C_7H_{16}	3-ethylpentane	154.55	540.6	266.35	540.57	9.0196E+01	−6.9235E+03	−1.0309E+01	8.0770E-06	2	3
21	C_7H_{16}	2,2-dimethylpentane	149.34	520.5	260.85	513.15	8.7305E+01	−6.4698E+03	−9.9730E+00	8.6229E-06	2	3
22	C_7H_{16}	2,3-dimethylpentane	160	537.3	204.75	523.15	7.8335E+01	−6.3487E+03	−8.5105E+00	6.4311E-06	2	3
23	C_7H_{16}	2,4-dimethylpentane	153.91	519.8	262.45	519.73	8.6089E+01	−6.4919E+03	−9.7506E+00	8.1176E-06	2	3
24	C_7H_{16}	3,3-dimethylpentane	138.7	536.4	213.14	536.34	7.4562E+01	−6.0587E+03	−7.9839E+00	6.2561E-06	2	3
25	C_7H_{16}	2,2,3-trimethylbutane	248.57	531.1	254.35	527.78	7.9383E+01	−6.1271E+03	−8.7696E+00	7.3637E-06	2	3
1125	$C_7H_{16}O$	1-heptanol	239.15	632.3	315.55	603.15	**1.4741E+02**	**−1.3466E+04**	**−1.7353E+01**	**1.1284E-17**	**6**	3
1126	$C_7H_{16}O$	2-heptanol	220	608.3	243.95	433.15	1.5309E+02	−1.2619E+04	−1.8748E+01	7.4507E-06	2	3
1625	C_8F_{18}	perfluoro-*n*-octane	251.4	502.2			*7.5601E+01*	*−6.9036E+03*	*−7.7230E+00*	*2.3803E-17*	*6*	10
601	C_8H_8	styrene	242.54	635.2	265	635.2	8.5914E+01	−7.9681E+03	−9.2338E+00	5.4384E-09	3	3
1090	C_8H_8O	acetophenone	292.81	709.6	298.15	628.55	6.4239E+01	−8.1739E+03	−5.7673E+00	2.6743E-18	6	3
1283	$C_8H_8O_2$	*p*-toluic acid	452.34	775			*1.0349E+02*	*−1.3019E+04*	*−1.0823E+01*	*2.1536E-18*	*6*	10
504	C_8H_{10}	ethylbenzene	178.2	617.26	306.244	617.26	7.6626E+01	−7.3303E+03	−7.9000E+00	4.7230E-09	3	1
505	C_8H_{10}	*o*-xylene	247.98	630.3	262.35	630.29	6.0616E+01	−6.8148E+03	−5.4351E+00	5.9397E-18	6	1
506	C_8H_{10}	*m*-xylene	225.3	617	226.65	617	6.0117E+01	−6.7518E+03	−5.3539E+00	5.5096E-18	6	1
507	C_8H_{10}	*p*-xylene	286.4	616.2	286.4	616.2	5.9356E+01	−6.6310E+03	−5.2735E+00	6.5192E-18	6	1
1185	$C_8H_{10}O$	*o*-ethylphenol	269.84	703	271.95	506.61	1.6119E+02	−1.3145E+04	−2.0181E+01	1.0283E-05	2	5
2112	$C_8H_{10}O$	*m*-ethylphenol	269.15	716.45	273.95	520.46	1.5968E+02	−1.3714E+04	−1.9751E+01	8.9216E-06	2	5
1187	$C_8H_{10}O$	*p*-ethylphenol	318.23	716.45	374.15	520.01	6.9902E+01	−9.4194E+03	−6.3292E+00	9.4022E-19	6	10
1170	$C_8H_{10}O$	2,3-xylenol	345.71	722.95	373.15	722.95	1.2669E+02	−1.1755E+04	−1.4992E+01	7.0404E-06	2	3
1172	$C_8H_{10}O$	2,4-xylenol	297.68	707.65	320.32	707.65	1.4436E+02	−1.2558E+04	−1.7611E+01	8.4952E-06	2	3
1174	$C_8H_{10}O$	2,5-xylenol	347.99	707.05	373.15	707.05	6.2262E+01	−8.5800E+03	−5.3452E+00	2.8244E-18	6	3
1176	$C_8H_{10}O$	2,6-xylenol	318.76	701.05	323.15	701.05	1.2933E+02	−1.1290E+04	−1.5546E+01	7.9759E-06	2	3
1177	$C_8H_{10}O$	3,4-xylenol	338.25	729.95	393.15	729.95	1.3988E+02	−1.2871E+04	−1.6820E+01	7.6585E-06	2	3
1178	$C_8H_{10}O$	3,5-xylenol	336.59	715.65	353.15	715.65	1.0868E+02	−1.1822E+04	−1.1820E+01	4.7309E-18	6	3
160	C_8H_{16}	cyclooctane	287.95	647.2	290.96	467.55	8.6649E+01	−7.8144E+03	−9.5263E+00	5.1845E-06	2	3
147	C_8H_{16}	*trans*-1,4-dimethylcyclohexane	236.21	587.7	290.39	419.91	8.1510E+01	−6.9275E+03	−8.9113E+00	5.7659E-06	2	3
250	C_8H_{16}	1-octene	171.45	566.9	264.15	395.37	**7.4936E+01**	**−7.1559E+03**	**−7.5843E+00**	**1.7106E-17**	**6**	3

TABLE A.4 Vapor pressure correlation parameters. (*Continued*)

DIPPR ID	Formula	Name	Eq. T_{min} (K)	Eq. T_{max} (K)	Data T_{min} (K)	Data T_{max} (K)	A	B	C	D	E	Er (%)
1265	$C_8H_{16}O_2$	*n*-octanoic acid	289.65	694.26	297.4	513.58	1.1648E+02	−1.3300E+04	−1.2675E+01	3.9834E-18	6	5
2260	$C_8H_{16}O_2$	2-ethyl hexanoic acid	155.15	674.6	356.15	513.69	1.2236E+02	−1.3309E+04	−1.3571E+01	6.4180E-18	6	10
1385	$C_8H_{16}O_2$	*n*-butyl *n*-butyrate	181.15	616	250	610	1.0918E+02	−9.3840E+03	−1.2770E+01	7.4704E-06	2	5
1360	$C_8H_{16}O_2$	isobutyl isobutyrate	192.5	601.89	274.2	441.25	7.7684E+01	−7.9957E+03	−7.8100E+00	7.5670E-18	6	1
27	C_8H_{18}	*n*-octane	216.38	568.7	228.15	568.65	9.6084E+01	−7.9002E+03	−1.1003E+01	7.1802E-06	2	3
28	C_8H_{18}	2-methylheptane	164.16	559.7	222.15	559.6	9.0847E+01	−7.4757E+03	−1.0262E+01	6.9106E-06	2	3
29	C_8H_{18}	3-methylheptane	152.6	563.6	223.15	563.6	9.6070E+01	−7.7009E+03	−1.1067E+01	7.6863E-06	2	3
30	C_8H_{18}	4-methylheptane	152.2	561.7	292.55	561.7	9.8617E+01	−7.7983E+03	−1.1454E+01	8.0052E-06	2	3
31	C_8H_{18}	3-ethylhexane	173.6	565.5	285.95	565.42	9.7343E+01	−7.7432E+03	−1.1264E+01	7.8528E-06	2	3
32	C_8H_{18}	2,2-dimethylhexane	151.97	549.8	276.25	549.8	9.4157E+01	−7.3316E+03	−1.0857E+01	8.0124E-06	2	3
33	C_8H_{18}	2,3-dimethylhexane	162.67	563.5	283.05	563.42	9.4189E+01	−7.5127E+03	−1.0816E+01	7.6731E-06	2	3
34	C_8H_{18}	2,4-dimethylhexane	153.02	553.5	278.35	553.45	1.3450E+02	−8.3411E+03	−1.8225E+01	1.8850E-02	1	1
35	C_8H_{18}	2,5-dimethylhexane	182	550	278.45	549.99	9.7056E+01	−7.5172E+03	−1.1282E+01	8.3354E-06	2	3
36	C_8H_{18}	3,3-dimethylhexane	147.05	562	279.25	561.95	9.2555E+01	−7.3075E+03	−1.0617E+01	7.7867E-06	2	3
37	C_8H_{18}	3,4-dimethylhexane	161.9	568.8	284.45	568.78	9.3800E+01	−7.5259E+03	−1.0752E+01	7.5458E-06	2	1
38	C_8H_{18}	2-methyl-3-ethylpentane	158.2	567.1	282.65	559.97	9.2154E+01	−7.4045E+03	−1.0515E+01	7.3988E-06	2	3
39	C_8H_{18}	3-methyl-3-ethylpentane	182.28	576.5	249	576	8.6063E+01	−7.1084E+03	−9.6172E+00	6.7639E-06	2	3
40	C_8H_{18}	2,2,3-trimethylpentane	160.89	563.5	283.95	563.5	8.7680E+01	−7.0388E+03	−9.8896E+00	7.1519E-06	2	1
41	C_8H_{18}	2,2,4-trimethylpentane	165.78	543.8	199.15	543.8	8.4912E+01	−6.7222E+03	−9.5157E+00	7.2244E-06	2	3
42	C_8H_{18}	2,3,3-trimethylpentane	172.22	573.5	280.05	573.49	8.3105E+01	−6.9037E+03	−9.1858E+00	6.4703E-06	2	3
43	C_8H_{18}	2,3,4-trimethylpentane	163.95	566.4	222.8	566.34	8.3530E+01	−6.9609E+03	−9.2248E+00	6.4027E-06	2	3
44	C_8H_{18}	2,2,3,3-tetramethylbutane	373.96	568	374.65	406.15	**5.7963E+01**	**−5.9015E+03**	**−5.2048E+00**	**9.1301E-18**	**6**	5
1132	$C_8H_{18}O$	1-octanol	257.65	652.3	293.15	549	1.4411E+02	−1.3667E+04	−1.6826E+01	9.3666E-18	6	3
1133	$C_8H_{18}O$	2-octanol	241.55	629.8	283.15	480.76	1.8583E+02	−1.4520E+04	−2.3624E+01	1.0885E-05	2	3
1121	$C_8H_{18}O$	2-ethyl-1-hexanol	203.15	640.4	293.8	478.08	2.2338E+02	−1.6439E+04	−2.9206E+01	1.4261E-05	2	3
1404	$C_8H_{18}O$	di-*n*-butyl ether	175.3	584.1	337.01	413.21	**7.2227E+01**	**−7.5376E+03**	**−7.0596E+00**	**9.1442E-18**	**6**	5
2708	$C_8H_{19}N$	*n*-octylamine	273.37	641	275	635	1.4626E+02	−1.1405E+04	−1.8276E+01	1.0826E-05	2	5
1744	$C_8H_{19}N$	di-*n*-butylamine	211.15	602.3	233.67	520.12	7.8056E+01	−8.1482E+03	−7.8692E+00	1.2426E-17	6	3
1626	C_9F_{20}	perfluoro-*n*-nonane	257.15	523.9			*7.7319E+01*	*−7.3306E+03*	*−7.9282E+00*	*1.8797E-17*	*6*	10
1748	C_9H_7N	quinoline	258.37	782.15	285.77	644.2	9.6846E+01	−1.0118E+04	−1.0687E+01	4.3897E-06	2	3
2785	C_9H_7N	isoquinoline	299.62	803.15	313.15	565.8	9.3736E+01	−1.0105E+04	−1.0189E+01	3.7859E-06	2	3
820	C_9H_{10}	indane	221.74	684.9	273.15	684.9	9.5488E+01	−8.7935E+03	−1.0735E+01	5.6169E-06	2	3
509	C_9H_{12}	*n*-propylbenzene	173.55	638.35	266.35	461.02	9.1379E+01	−8.2768E+03	−1.0176E+01	5.6240E-06	2	3
510	C_9H_{12}	cumene	177.14	631	276.05	553.15	1.0281E+02	−8.6746E+03	−1.1922E+01	7.0048E-06	2	3
513	C_9H_{12}	*p*-ethyltoluene	210.83	640.2	319.26	640.2	1.0907E+02	−9.1001E+03	−1.2858E+01	7.8533E-06	2	3

(*Continued*)

TABLE A.4 Vapor pressure correlation parameters. (*Continued*)

DIPPR ID	Formula	Name	Eq. T_{min} (K)	Eq. T_{max} (K)	Data T_{min} (K)	Data T_{max} (K)	A	B	C	D	E	Er (%)
514	C_9H_{12}	1,2,3-trimethylbenzene	247.79	664.5	323.13	664.47	7.8341E+01	−8.0198E+03	−8.1458E+00	3.8971E-06	2	3
515	C_9H_{12}	1,2,4-trimethylbenzene	229.33	649.1	313.13	643.15	8.5301E+01	−8.2159E+03	−9.2166E+00	4.7979E-06	2	3
516	C_9H_{12}	mesitylene	228.42	637.3	298.15	637.25	8.8697E+01	−8.3170E+03	−9.7330E+00	5.3187E-06	2	3
259	C_9H_{18}	1-nonene	191.91	593.1	306.65	421.01	6.3313E+01	−7.0404E+03	−5.8055E+00	7.5753E-18	6	3
1259	$C_9H_{18}O_2$	*n*-nonanoic acid	285.55	710.7	291.99	528.75	1.2337E+02	−1.4215E+04	−1.3561E+01	3.1658E-18	6	10
3318	$C_9H_{18}O_2$	isopentyl butyrate	189	618.83	273.15	452.61	7.4241E+01	−8.3441E+03	−7.2449E+00	8.6136E-18	6	25
46	C_9H_{20}	*n*-nonane	219.66	594.6	310.93	510.93	1.0935E+02	−9.0304E+03	−1.2882E+01	7.8544E-06	2	3
91	C_9H_{20}	2-methyloctane	193.05	582.87	195	581.4	**6.8954E+01**	**−7.3542E+03**	**−6.6010E+00**	**8.9378E-18**	**6**	10
96	C_9H_{20}	2,2-dimethylheptane	160.15	576.7	303.15	433.15	**7.0767E+01**	**−7.0721E+03**	**−6.9725E+00**	**1.3412E-17**	**6**	5
47	C_9H_{20}	2,2,5-trimethylhexane	167.39	569.8	206.45	424.25	1.3938E+02	−8.8173E+03	−1.8903E+01	1.8789E-02	1	3
51	C_9H_{20}	2,2,3,3-tetramethylpentane	263.26	607.5	306.65	442.05	6.2855E+01	−6.6877E+03	−5.8419E+00	8.3613E-18	6	3
52	C_9H_{20}	2,2,3,4-tetramethylpentane	152.06	592.6	301.35	434.25	**6.2907E+01**	**−6.5816E+03**	**−5.8632E+00**	**9.3233E-18**	**6**	5
53	C_9H_{20}	2,2,4,4-tetramethylpentane	206.95	574.6	292.75	423.05	6.2641E+01	−6.3690E+03	−5.8616E+00	1.1507E-17	6	3
54	C_9H_{20}	2,3,3,4-tetramethylpentane	171.03	607.5	307.81	443.27	1.2325E+02	−8.4765E+03	−1.6212E+01	1.5517E-02	1	5
1134	$C_9H_{20}O$	1-nonanol	268.15	670.9	364.85	613.15	1.6285E+02	−1.5205E+04	−1.9424E+01	1.0722E-17	6	5
1135	$C_9H_{20}O$	2-nonanol	238.15	649.5	263.09	364.15	2.1307E+02	−1.6246E+04	−2.7620E+01	1.3183E-05	2	3
1627	$C_{10}F_{22}$	perfluoro-*n*-decane	309.15	542.35	405.95	542.35	**3.7042E+01**	**−5.1358E+03**	**−2.1974E+00**	**9.3028E-18**	**6**	10
1381	$C_{10}H_{10}O_4$	dimethyl terephthalate	413.79	777.4	413.79	771.2	6.6180E+01	−9.8704E+03	−5.8560E+00	1.4718E-18	6	5
701	$C_{10}H_8$	naphthalene	353.43	748.4	353.43	693.15	6.2964E+01	−8.1375E+03	−5.6317E+00	2.2675E-18	6	1
706	$C_{10}H_{12}$	1,2,3,4-tetrahydronaphthalene	237.38	720	273.15	713.15	1.3723E+02	−1.0620E+04	−1.7908E+01	1.4506E-02	1	3
503	$C_{10}H_{12}$	1-methylindan	182.1	694.1	333.15	463.15	5.8059E+01	−7.3651E+03	−4.9969E+00	2.6359E-18	6	5
508	$C_{10}H_{12}$	5-methylindan	213.3	711.2	347.15	474.25	4.9682E+01	−6.9358E+03	−3.8261E+00	2.0934E-18	6	5
518	$C_{10}H_{14}$	*n*-butylbenzene	185.3	660.5	268.45	523.15	1.0122E+02	−9.2554E+03	−1.1538E+01	5.9208E-06	2	3
519	$C_{10}H_{14}$	isobutylbenzene	221.7	650	287.25	475.6	7.1645E+01	−7.6180E+03	−7.1624E+00	3.2995E-06	2	3
527	$C_{10}H_{14}$	*p*-diethylbenzene	230.32	657.9	335.98	486.74	1.0790E+02	−9.5823E+03	−1.2537E+01	6.6241E-06	2	3
524	$C_{10}H_{14}$	*p*-cymene	205.25	652	267.65	633.15	1.0545E+02	−9.3000E+03	−1.2204E+01	6.4251E-06	2	3
532	$C_{10}H_{14}$	1,2,4,5-tetramethylbenzene	352.38	676	353	500.15	**8.8899E+01**	**−9.0130E+03**	**−9.6101E+00**	**4.2346E-06**	**2**	3
520	$C_{10}H_{14}$	*sec*-butylbenzene	197.72	664.54	270.15	474.82	1.7987E+02	−1.1606E+04	−2.4932E+01	2.1872E-02	1	3
153	$C_{10}H_{18}$	*cis*-decahydronaphthalene	230.2	703.6	263.15	523.15	1.0753E+02	−9.6002E+03	−1.2502E+01	6.1923E-06	2	5
154	$C_{10}H_{18}$	*trans*-decahydronaphthalene	242.79	687	273.15	632.7	1.0094E+02	−9.0555E+03	−1.1585E+01	6.1499E-06	2	5
260	$C_{10}H_{20}$	1-decene	206.89	616.6	359.92	444.75	6.8401E+01	−7.7769E+03	−6.4637E+00	6.3750E-18	6	3
1254	$C_{10}H_{20}O_2$	*n*-decanoic acid	304.55	722.1	307.49	543.15	1.2641E+02	−1.4865E+04	−1.3907E+01	2.5132E-18	6	10
56	$C_{10}H_{22}$	*n*-decane	243.51	617.7	268.15	490.29	1.1273E+02	−9.7496E+03	−1.3245E+01	7.1266E-06	2	3
48	$C_{10}H_{22}$	3,3,5-trimethylheptane	165	609.5	313.15	458.15	**1.2067E+02**	**−9.4261E+03**	**−1.4668E+01**	**9.4541E-06**	**2**	3
57	$C_{10}H_{22}$	2,2,3,3-tetramethylhexane	219.15	623	314.15	463.15	**6.4908E+01**	**−7.1366E+03**	**−6.0888E+00**	**7.8905E-18**	**6**	3
58	$C_{10}H_{22}$	2,2,5,5-tetramethylhexane	260.55	581.4	300.15	438.15	8.9478E+01	−7.8393E+03	−9.9327E+00	5.4151E-06	2	3

TABLE A.4 Vapor pressure correlation parameters. (*Continued*)

DIPPR ID	Formula	Name	Eq. T_{min} (K)	Eq. T_{max} (K)	Data T_{min} (K)	Data T_{max} (K)	A	B	C	D	E	Er (%)
1136	$C_{10}H_{22}O$	1-decanol	280.05	688	336	528.32	1.5624E+02	−1.5212E+04	−1.8424E+01	8.5006E-18	6	5
702	$C_{11}H_{10}$	1-methylnaphthalene	242.67	772	370.25	755.37	6.7566E+01	−8.7370E+03	−6.3362E+00	1.6377E-06	2	3
703	$C_{11}H_{10}$	2-methylnaphthalene	307.73	761	377.15	707.15	8.8401E+01	−1.0133E+04	−9.1720E+00	4.2657E-18	6	3
63	$C_{11}H_{24}$	*n*-undecane	247.57	639	348.25	498.95	1.3100E+02	−1.1143E+04	−1.5855E+01	8.1871E-06	2	3
1137	$C_{11}H_{24}O$	1-undecanol	288.45	703.9	293.18	406.65	1.8257E+02	−1.7112E+04	−2.2125E+01	1.1284E-17	6	5
558	$C_{12}H_{10}$	biphenyl	342.2	773	347.8	672.04	7.7314E+01	−9.9104E+03	−7.5079E+00	2.2385E-18	6	3
715	$C_{12}H_{12}$	2,7-dimethylnaphthalene	368.81	775	369.15	536.15	**5.0867E+01**	**−8.2747E+03**	**−3.8078E+00**	**5.9002E-19**	**6**	5
3545	$C_{12}H_{18}$	1,3,5-triethylbenzene	206.74	678	363.15	371.75	**1.1808E+02**	**−1.0965E+04**	**−1.3827E+01**	**6.2556E-06**	**2**	10
814	$C_{12}H_{20}$	1,3-dimethyladamantane	247.59	708	350.65	526.208	5.3334E+01	−7.0898E+03	−4.3710E+00	2.2085E-18	6	10
262	$C_{12}H_{24}$	1-dodecene	237.95	657.1	362	487.62	8.2022E+01	−9.4856E+03	−8.2517E+00	4.9813E-18	6	3
1269	$C_{12}H_{24}O_2$	*n*-dodecanoic acid	316.98	743	319.07	555.65	1.4388E+02	−1.6742E+04	−1.6254E+01	3.4422E-18	6	10
64	$C_{12}H_{26}$	*n*-dodecane	263.57	658	289.15	520.23	1.3747E+02	−1.1976E+04	−1.6698E+01	8.0906E-06	2	3
1140	$C_{12}H_{26}O$	1-dodecanol	296.95	718.7	303.15	618.9	1.8699E+02	−1.7927E+04	−2.2640E+01	9.8439E-18	6	5
563	$C_{13}H_{12}$	diphenylmethane	298.37	778	303.4	647.25	8.6886E+01	−1.0715E+04	−8.8245E+00	2.3329E-18	6	3
65	$C_{13}H_{28}$	*n*-tridecane	267.76	675	290.15	540.19	1.3745E+02	−1.2549E+04	−1.6543E+01	7.1275E-06	2	1
1141	$C_{13}H_{28}O$	1-tridecanol	303.75	732.4	335	730	1.7771E+02	−1.7736E+04	−2.1273E+01	8.0501E-18	6	5
805	$C_{14}H_{10}$	phenanthrene	372.38	893	372.38	667.74	9.4249E+01	−1.2064E+04	−9.8971E+00	2.1100E-09	3	5
804	$C_{14}H_{10}$	anthracene	488.93	873	488.93	673.15	**6.5069E+01**	**−1.0251E+04**	**−5.7509E+00**	**1.1238E-18**	**6**	3
534	$C_{14}H_{22}$	1,4-di-*tert*-butylbenzene	350.76	708	384.15	559.06	9.8734E+01	−1.0899E+04	−1.0577E+01	5.7444E-18	6	5
66	$C_{14}H_{30}$	*n*-tetradecane	279.01	693	279.55	559.15	1.4047E+02	−1.3231E+04	−1.6859E+01	6.5877E-06	2	3
1142	$C_{14}H_{30}O$	1-tetradecanol	310.65	745.3	333.19	745	1.8085E+02	−1.8318E+04	−2.1656E+01	7.5637E-18	6	10
1198	$C_{15}H_{16}O_2$	bisphenol A	430.15	849	493.15	633.65	**2.0096E+02**	**−2.4713E+04**	**−2.3329E+01**	**1.0329E-18**	**6**	25
67	$C_{15}H_{32}$	*n*-pentadecane	283.07	708	345.99	576.95	1.3557E+02	−1.3478E+04	−1.6022E+01	5.6136E-06	2	3
1143	$C_{15}H_{32}O$	1-pentadecanol	317.05	757.3	343.1	755	1.8511E+02	−1.9068E+04	−2.2167E+01	6.8889E-18	6	5
68	$C_{16}H_{34}$	*n*-hexadecane	291.31	722	294.95	563.13	**1.6264E+02**	**−1.5734E+04**	**−1.9620E+01**	**6.4337E-09**	**3**	5
90	$C_{16}H_{34}$	2,2,4,4,6,8,8-heptamethylnonane	163	692	423.61	545.42	**1.2831E+02**	**−1.1839E+04**	**−1.5333E+01**	**6.9676E-06**	**2**	5
1144	$C_{16}H_{34}O$	1-hexadecanol	322.35	768.6	362.95	569.05	1.8619E+02	−1.9523E+04	−2.2258E+01	6.3145E-18	6	5
69	$C_{17}H_{36}$	*n*-heptadecane	295.13	735.3	298.15	558.06	**1.6973E+02**	**−1.6613E+04**	**−2.0542E+01**	**6.3144E-09**	**3**	5
1145	$C_{17}H_{36}O$	1-heptadecanol	327.05	779.2			*1.7269E+02*	*−1.9015E+04*	*−2.0312E+01*	*4.8665E-18*	6	10
561	$C_{18}H_{14}$	*o*-terphenyl	329.35	857	343.47	613	1.1052E+02	−1.4045E+04	−1.1861E+01	2.2121E-18	6	5
560	$C_{18}H_{14}$	*m*-terphenyl	360	883	533	653	**8.2803E+01**	**−1.2948E+04**	**−7.9324E+00**	**8.3072E-19**	**6**	10
559	$C_{18}H_{14}$	*p*-terphenyl	485	908	543	653	**9.6294E+01**	**−1.4393E+04**	**−9.6956E+00**	**9.1365E-19**	**6**	5
1276	$C_{18}H_{36}O_2$	*n*-octadecanoic acid	342.75	803	365.58	628.35	2.1473E+02	−2.3660E+04	−2.5807E+01	5.3801E-18	6	10
70	$C_{18}H_{38}$	*n*-octadecane	301.31	747.8	318.15	589.65	**1.7667E+02**	**−1.7476E+04**	**−2.1442E+01**	**6.2096E-09**	**3**	5
1146	$C_{18}H_{38}O$	1-octadecanol	331.05	789.3	433.15	574.3	1.8464E+02	−2.0220E+04	−2.1906E+01	4.9355E-18	6	5

(*Continued*)

TABLE A.4 Vapor pressure correlation parameters. (*Continued*)

DIPPR ID	Formula	Name	Eq. T_{min} (K)	Eq. T_{max} (K)	Data T_{min} (K)	Data T_{max} (K)	A	B	C	D	E	Er (%)
71	$C_{19}H_{40}$	*n*-nonadecane	305.04	759.5	306.15	588.13	**1.8488E+02**	**−1.8440E+04**	**−2.2517E+01**	**6.1575E-09**	**3**	5
1149	$C_{19}H_{40}O$	1-nonadecanol	334.85	798.8			*1.9849E+02*	*−2.1711E+04*	*−2.3721E+01*	*4.7579E-18*	*6*	10
73	$C_{20}H_{42}$	*n*-eicosane	309.58	770.6	312.76	625.99	**2.0033E+02**	**−1.9958E+04**	**−2.4584E+01**	**6.2887E-09**	**3**	3
1148	$C_{20}H_{42}O$	1-eicosanol	338.55	807.7			*1.8667E+02*	*−2.1260E+04*	*−2.2021E+01*	*3.6754E-18*	*6*	10
74	$C_{21}H_{44}$	*n*-heneicosane	313.35	781	351.54	488.15	**2.0466E+02**	**−2.0628E+04**	**−2.5125E+01**	**6.1604E-09**	**3**	5
75	$C_{22}H_{46}$	*n*-docosane	317.15	790.9	379.15	573.13	**2.1897E+02**	**−2.2078E+04**	**−2.7028E+01**	**6.2694E-09**	**3**	5
76	$C_{23}H_{48}$	*n*-tricosane	320.65	800.3	412.18	472.65	**2.2636E+02**	**−2.2986E+04**	**−2.7986E+01**	**6.2261E-09**	**3**	10
77	$C_{24}H_{50}$	*n*-tetracosane	323.75	809.3	333.09	588.13	**2.2726E+02**	**−2.3366E+04**	**−2.8061E+01**	**6.0521E-09**	**3**	10
1987	$ClFO_3$	perchloryl fluoride	125.41	368.4	133.15	353.15	9.8154E+01	−4.2699E+03	−1.3458E+01	2.2978E-02	1	3
1904	ClH	hydrogen chloride	158.97	324.65	163.61	201	**1.0427E+02**	**−3.7312E+03**	**−1.5047E+01**	**3.1340E-02**	**1**	1
2928	ClH_4N	ammonium chloride	793.15	1155.2	793.15	1123.15	1.4891E+02	−1.9547E+04	−1.6585E+01	2.9072E-06	2	5
1939	ClNa	sodium chloride	1073.9	3400	1138.15	1738.15	**8.5587E+01**	**−3.1057E+04**	**−7.6371E+00**	**2.6231E-07**	**2**	5
1986	ClNO	nitrosyl chloride	213.55	440.65	213.55	294.25	3.7327E+01	−3.5303E+03	−2.3064E+00	3.8777E-06	2	3
918	Cl_2	chlorine	172.12	417.15	179.62	400.2	7.1334E+01	−3.8550E+03	−8.5171E+00	1.2378E-02	1	1
1937	Cl_4Si	tetrachlorosilane	204.3	507	209.75	504.9	7.0402E+01	−5.5056E+03	−7.3965E+00	6.3480E-06	2	3
925	D_2	deuterium	18.73	38.35	18.73	38.34	1.8947E+01	−1.5447E+02	−5.7226E−01	3.8899E-02	1	3
1997	D_2O	deuterium oxide	257.74	643.89	257.74	643.89	7.8433E+01	−7.6012E+03	−7.9714E+00	4.3816E-06	2	0.2
1905	FH	hydrogen fluoride	189.79	461.15	198.45	312.85	**5.9544E+01**	**−4.1438E+03**	**−6.1764E+00**	**1.4161E-05**	**2**	3
917	F_2	fluorine	53.48	144.12	53.48	140	4.2393E+01	−1.1033E+03	−4.1203E+00	5.7815E-05	2	3
1972	F_3N	nitrogen trifluoride	66.46	234	89.33	234	6.8149E+01	−2.2579E+03	−8.9118E+00	2.3233E-02	1	3
1989	F_4N_2	tetrafluorohydrazine	111.65	309.35	111.95	259.02	4.2764E+01	−2.2194E+03	−4.3123E+00	1.3787E-02	1	3
1940	F_6S	sulfur hexafluoride	223.15	318.69	223.15	318.65	2.9160E+01	−2.3836E+03	−1.1342E+00			
1907	HI	hydrogen iodide	222.38	423.85	223.45	400.65	**5.4233E+01**	**−3.3421E+03**	**−5.5756E+00**	**7.8006E-03**	**1**	3
1912	HNaO	sodium hydroxide	596	1830	1283.15	1681.15	−1.0627E+02	−4.4532E+03	1.7000E+01	−2.0150E-06	2	25
902	H_2	hydrogen	13.95	33.19	15.1	33.18	**1.2690E+01**	**−9.4896E+01**	**1.1125E+00**	**3.2915E-04**	**2**	3
1921	H_2O	water	273.16	647.1	273.16	647.1	7.3649E+01	−7.2582E+03	−7.3037E+00	4.1653E-06	2	0.2
1901	H_2O_4S	sulfuric acid	283.15	603.15			*1.4422E+01*	*−9.7577E+03*	*2.3632E+00*	*3.2700E-19*	*6*	10
1922	H_2S	hydrogen sulfide	187.68	373.53	190	373.4	8.5584E+01	−3.8399E+03	−1.1199E+01	1.8848E-02	1	3
3951	H_2Se	hydrogen selenide	207.45	411.15	210.55	411.15	9.1056E+01	−4.6519E+03	−1.1090E+01	1.7849E-05	2	3
1911	H_3N	ammonia	195.41	405.65	196.56	403.15	9.0483E+01	−4.6697E+03	−1.1607E+01	1.7194E-02	1	1
1981	H_3P	phosphine	139.37	324.75	140	199.46	**5.6954E+01**	**−2.6894E+03**	**−6.0466E+00**	**1.8861E-05**	**2**	5
1717	H_4N_2	hydrazine	274.69	653.15	288.15	653.15	7.6858E+01	−7.2452E+03	−8.2200E+00	6.1557E-03	1	1
1982	H_4Si	silane	88.48	269.7	113.15	269.65	7.2385E+01	−2.5842E+03	−9.0452E+00	4.4324E-05	2	5
913	He	helium	1.76	5.2	1.8	5.2	1.1533E+01	−8.9900E+00	6.7240E−01	2.7430E-01	1	1
1998	I_2	iodine	386.75	819.15	386.75	487.55	8.6593E+01	−8.6299E+03	−9.3114E+00	3.9884E-06	2	3
920	Kr	krypton	115.78	209.35	116.65	129.23	**2.6224E+01**	**−1.2262E+03**	**−9.3273E−01**	**1.6350E-15**	**6**	3

TABLE A.4 Vapor pressure correlation parameters. (*Continued*)

DIPPR ID	Formula	Name	Eq. T_{min} (K)	Eq. T_{max} (K)	Data T_{min} (K)	Data T_{max} (K)	A	B	C	D	E	Er (%)
912	NO	nitric oxide	109.5	180.15	116.09	178.35	**7.2974E+01**	**−2.6500E+03**	**−8.2610E+00**	**9.7000E-15**	**6**	3
905	N_2	nitrogen	63.15	126.2	63.15	116	5.8282E+01	−1.0841E+03	−8.3144E+00	4.4127E-02	1	1
899	N_2O	nitrous oxide	182.3	309.57	183.31	309.55	9.6512E+01	−4.0450E+03	−1.2277E+01	2.8860E-05	2	3
906	N_2O_4	nitrogen tetroxide	261.85	431.35	262.35	430.55	2.5943E+00	−3.0193E+03	3.3750E+00	7.0500E-18	6	5
919	Ne	neon	24.56	44.4	24.56	44.4	2.9755E+01	−2.7106E+02	−2.6081E+00	5.2700E-04	2	1
901	O_2	oxygen	54.36	154.58	62.5	97.2	**5.1245E+01**	**−1.2002E+03**	**−6.4361E+00**	**2.8405E-02**	**1**	1
910	O_2S	sulfur dioxide	197.67	430.75	203.13	414.85	**4.7365E+01**	**−4.0845E+03**	**−3.6469E+00**	**1.7990E-17**	**6**	1
924	O_3	ozone	80.15	261	85	173.15	4.0067E+01	−2.2048E+03	−2.9351E+00	7.7520E-16	6	3
911	O_3S	sulfur trioxide	289.95	490.85	289.95	471.15	1.8099E+02	−1.2060E+04	−2.2839E+01	7.2350E-17	6	3
970	Rn	radon	202	377.5	263	377.5	4.2114E+01	−2.7840E+03	−3.2513E+00	7.0786E-17	6	5
1923	S	sulfur	514.65	1313	514.65	1273	1.2949E+02	−1.3691E+04	−1.6208E+01	1.0714E-02	1	10
959	Xe	xenon	161.36	289.74	163.6	280	4.0935E+01	−1.8659E+03	−3.9013E+00	1.1049E-02	1	3

References

Bloxham, J. C., M. E. Redd, N. F. Giles, T. A Knotts IV, and W. V. Wilding: *J. Chem. Eng. Data*, **66:** 3–10 (2021).
Hill, E. A.: *J. Am. Chem. Soc.*, **22:** 478–494 (1900).
Wieser, M. E., and T. B. Coplen: *Pure Appl. Chem.* **83:** 359–396 (2011).
Wilding, W. V., T. A. Knotts, N. F. Giles, R. L. Rowley: DIPPR® Data Compilation of Pure Chemical Properties, Design Institute for Physical Properties, AIChE, New York, NY (2017).

B

Lennard-Jones Potentials as Determined from Viscosity Data[1]

	Substance	b_0,* cm^3/g-mol	σ, Å	ε/k, K
Ar	Argon	56.08	3.542	93.3
He	Helium	20.95	2.551†	10.22
Kr	Krypton	61.62	3.655	178.9
Ne	Neon	28.30	2.820	32.8
Xe	Xenon	83.66	4.047	231.0
Air	Air	64.50	3.711	78.6
AsH_3	Arsine	89.88	4.145	259.8
BCl_3	Boron chloride	170.1	5.127	337.7
BF_3	Boron fluoride	93.35	4.198	186.3
$B(OCH_3)_3$	Methyl borate	210.3	5.503	396.7
Br_2	Bromine	100.1	4.296	507.9
CCl_4	Carbon tetrachloride	265.5	5.947	322.7
CF_4	Carbon tetrafluoride	127.9	4.662	134.0
$CHCl_3$	Chloroform	197.5	5.389	340.2
CH_2Cl_2	Methylene chloride	148.3	4.898	356.3
CH_3Br	Methyl bromide	88.14	4.118	449.2
CH_3Cl	Methyl chloride	92.31	4.182	350.0
CH_3OH	Methanol	60.17	3.626	481.8
CH_4	Methane	66.98	3.758	148.6
CO	Carbon monoxide	63.41	3.690	91.7

(Continued)

[1] R. A. Svehla, *NASA Tech. Rep.* R-132, Lewis Research Center, Cleveland, Ohio, 1962.

	Substance	b_0,* cm³/g-mol	σ, Å	ε/k, K
COS	Carbonyl sulfide	88.91	4.130	336.0
CO_2	Carbon dioxide	77.25	3.941	195.2
CS_2	Carbon disulfide	113.7	4.483	467.0
C_2H_2	Acetylene	82.79	4.033	231.8
C_2H_4	Ethylene	91.06	4.163	224.7
C_2H_6	Ethane	110.7	4.443	215.7
C_2H_5Cl	Ethyl chloride	148.3	4.898	300.0
C_2H_5OH	Ethanol	117.3	4.530	362.6
C_2N_2	Cyanogen	104.7	4.361	348.6
CH_3OCH_3	Methyl ether	100.9	4.307	395.0
CH_2CHCH_3	Propylene	129.2	4.678	298.9
CH_3CCH	Methylacetylene	136.2	4.761	251.8
C_3H_6	Cyclopropane	140.2	4.807	248.9
C_3H_8	Propane	169.2	5.118	237.1
n-C_3H_7OH	n-Propyl alcohol	118.8	4.549	576.7
CH_3COCH_3	Acetone	122.8	4.600	560.2
CH_3COOCH_3	Methyl acetate	151.8	4.936	469.8
n-C_4H_{10}	n-Butane	130.0	4.687	531.4
iso-C_4H_{10}	Isobutane	185.6	5.278	330.1
$C_2H_5OC_2H_5$	Ethyl ether	231.0	5.678	313.8
$CH_3COOC_2H_5$	Ethyl acetate	178.0	5.205	521.3
n-C_5H_{12}	n-Pentane	244.2	5.784	341.1
$C(CH_3)_4$	2,2-Dimethylpropane	340.9	6.464	193.4
C_6H_6	Benzene	193.2	5.349	412.3
C_6H_{12}	Cyclohexane	298.2	6.182	297.1
n-C_6H_{14}	n-Hexane	265.7	5.949	399.3
Cl_2	Chlorine	94.65	4.217	316.0
F_2	Fluorine	47.75	3.357	112.6
HBr	Hydrogen bromide	47.58	3.353	449.0
HCN	Hydrogen cyanide	60.37	3.630	569.1
HCl	Hydrogen chloride	46.98	3.339	344.7
HF	Hydrogen fluoride	39.37	3.148	330.0
HI	Hydrogen iodide	94.24	4.211	288.7
H_2	Hydrogen	28.51	2.827	59.7
H_2O	Water	23.25	2.641	809.1
H_2O_2	Hydrogen peroxide	93.24	4.196	289.3
H_2S	Hydrogen sulfide	60.02	3.623	301.1
Hg	Mercury	33.03	2.969	750.0
$HgBr_2$	Mercuric bromide	165.5	5.080	686.2
$HgCl_2$	Mercuric chloride	118.9	4.550	750.0
HgI_2	Mercuric iodide	224.6	5.625	695.6
I_2	Iodine	173.4	5.160	474.2
NH_3	Ammonia	30.78	2.900	558.3

(*Continued*)

	Substance	b_0,* cm^3/g-mol	σ, Å	ε/k, K
NO	Nitric oxide	53.74	3.492	116.7
NOCl	Nitrosyl chloride	87.75	4.112	395.3
N_2	Nitrogen	69.14	3.798	71.4
N_2O	Nitrous oxide	70.80	3.828	232.4
O_2	Oxygen	52.60	3.467	106.7
PH_3	Phosphine	79.63	3.981	251.5
SF_6	Sulfur hexafluoride	170.2	5.128	222.1
SO_2	Sulfur dioxide	87.75	4.112	335.4
SiF_4	Silicon tetrafluoride	146.7	4.880	171.9
SiH_4	Silicon hydride	85.97	4.084	207.6
$SnBr_4$	Stannic bromide	329.0	6.388	563.7
UF_6	Uranium hexafluoride	268.1	5.967	236.8

*$b_0 = \frac{2}{3}\pi N_A \sigma^3$, where N_A is Avogadro's number.
†The parameter σ was determined by quantum-mechanical formulas.

Index

Note: Page numbers followed by *f* indicate figures; page numbers followed by *t* indicate tables.

1A approach, 269
1C site type, 269
12-term Span-Wagner EOS, 26, 27, 27*f*, 242, 272
28-Carboxylic acids, 268–269
2016 local correlation calculation (localcc=2016), 123

A^{chain}, 225
A^{chem}, 225
%AAD, 8
%AADo, 9
%AADP, 337
%AALD, 9
%AALDS, 447, 449
AARD. *See* Average absolute relative deviation (AARD)
Abbott et al. (1975), 371
Absolute entropy, 99, 100*t*
Absolute reaction rate theory, 640, 642
Acentric factor, 66
Acetonitrile(1)+Benzene(2)+*n*-Heptane(3), 432*t*
Activity coefficients, 341–343, 374–389
Activity modeling, 288, 344–362
  activity coefficients, 341–343, 374–389
  chemical-physical interactions: association and strong solvation, 347–349, 354–357
  common activity models, 343–344
  correlating azeotropic data, 367–369
  correlating multiple measurements, 363–367
  lattice fluid hydrogen bonding (LFHB) theory, 354
  linear free energy relations (LFER), 362
  local composition theory, 357–361
  local surface Guggenheim (LSG) model, 359
  preliminary fundamentals, 345–346
  QCT. *See* Quasichemical theory (QCT)
  quadratic mixing rules, 347
  separation of cohesive energy density (SCED), 361–362
Adiabatic flash, 397
Allan and Teja (1991) method, 531
α. *See* Thermodynamic correction factor α
Ambrose (1979) method, 85–92
Andrade equation, 505
Anisotropic UA (AUA) model, 271
Antoine's equation, 160–161, 162, 178, 382, 384
APACT. *See* Associated perturbed anisotropic chain theory (APACT)
Aqueous amino acids, 441–443
Aqueous solubility, 424
Aqueous systems:
  IDACs, 424–427, 450, 451
  surface tensions of mixtures, 675–678
Arikol and Gürbüz (1992) correlation, 582
Artificial intelligence (AI), 91–92
Aspen process simulator, 274, 333
Aspen-TDE-SAFT, 274, 275*t*, 277*t*, 278*t*, 279*t*, 280*t*
Associated perturbed anisotropic chain theory (APACT), 249, 302
Associating compounds, 4
  ideal gases, 134
  low-pressure viscosities of pure gases, 466
  pure component constants, 86
  pure liquid surface tension, 667
  saturated liquid density, 155
  thermal conductivity of polyatomic gases, 552
Association, 347–349
Asymmetric mixture, 298
Athermal mixing rules, 297–300
Atom contributions, 36
ATOMIC, 119
ATOMIC/ATOMIC(hc) technique, 134*t*, 135–141
Attractive mixing rules, 300–301
AUA. *See* Anisotropic UA (AUA) model
Average absolute relative deviation (AARD), 86, 134, 462, 517, 582
Avogadro's number, 462, 481, 546
Azeotropic data, 367–369
Azeotropy, 367

B3LYP/6-311+G(3df,2p):
  enthalpy of sublimation, 197
  ideal gases, 117, 118*t*, 123, 125*t*, 126, 130*t*, 139
Bakowies (2009, 2020) ATOMIC/ATOMIC(hc) technique, 134*t*, 135–141
Barker-Henderson (BH), 254, 255, 265
Barker's method, 363n
Baroncini et al. method, 589, 591, 594–595
Benedict-Webb-Rubin (BWR), 222, 239
Benson group indistinguishabilities, 105*t*

Benson method, 102–107, 134*t*, 135–141
Benzene(1)+methanol(2), 346, 346*f*, 349
BH. *See* Barker-Henderson (BH)
Bhethanabotla (1983) method, 514–516, 517–518
%*BIAS*, 8
Binary interaction parameters (BIPs), 288, 291, 334
Binary liquid diffusion coefficients, 628
Binary LLE by EOS methods, 433–436
Binary mixture, 20–22, 28–30
Binary mixtures, 405–406
Binodal composition determined by titration, 22
BIPs. *See* Binary interaction parameters (BIPs)
Blanc's law, 628
BMCSL, 294, 295, 346, 348, 409
Boiling and melting points, 66–85
  Constantinou and Gani (1994) method, 66–68
  definitions, 66
  Emami et al. (2009) method, 81
  estimation methods, compared, 85–92
  Joback (1984) method, 66
  Nannoolal et al. (2004) method, 69–80
  Stein and Brown (1994) method, 81–85
Boiling point (T_b), 86, 89, 89*t*. *See also* Boiling and melting points
Boltzmann constant, 546, 613
Boltzmann equation, 613
Bond contributions, 36
Bondi (1968) methods, 180–184, 196
Bonding arrangements, 250*t*
Bonding volume, 348
Book. *See Properties of Gases and Liquids* (Elliott et al.)
"Bottom of the well" energy, 115
Boublik (1970)-Mansoori-Carnahan-Starling-Leland (1971). *See* BMCSL
Brock and Bird (1955) method, 664–665, 666, 667, 668*t*, 669*t*
Brokaw's method, 616, 620
Brownian motion, 609
Brulé and Starling method:
  liquid viscosity at high temperatures, 519, 520
  pressure and viscosity of pure gases, 490–491*t*, 494–495
Bubble-point *T* problem, 397
Bubble-point temperature, 374
Bubble pressure calculation:
  ESD model, 399–400
  gaseous compounds, 402–403
  PPC-SAFT, 400–401
  SPEADMD, 401–402
  ternary systems, 403–404
Bunsen coefficient, 390
BWR. *See* Benedict-Webb-Rubin (BWR)

Calibrant property values, 20
Carnahan-Starling equation:
  CPA model, 252
  hard sphere EOS, 223
  molecular simulation models, 270
  PC-SAFT, 225
  SAFT-VR, 254
  SAFTγ-WCA, 261, 262
  Wertheim's theory of association, 246
CBS. *See* Complete basic set (CBS) approaches
ccCA. *See* Correlation consistent composite approach (ccCA)
CCT. *See* Coupled cluster theory (CCT)
Centipose (cP), 456
CH04Benson-CHETAHdHSCp.xlsx, 102, 104, 106*t*
CH04DomalskiHearingHSCp.xlsx, 107, 108*t*
CH05RuzickaDomalskiCp. xlsx, 175, 176*f*
CH06SPREOS.xlsx, 239
CH06SRKEOS.xlsx, 239
CH08ActCoeff.xlsx, 364, 368
CH10HsuSheuTuLiqVisc.xlsx, 516, 516*t*
CH10SastriRaoLiqVisc.docx, 509, 510–512*t*
Chapman-Enskog theory, 457, 609, 615
Chapman-Enskog viscosity equation, 458
Chemical potential driving force, 607–608
Chemical theory EOS, 267–269
Chemical+Wertheim theory, 268–269
CHETAH program, 102, 104
Chickos and Acree (2009) method, 186*t*, 187, 187*t*, 188–191*t*, 191, 192–193*t*, 193
Chickos et al. method, 184–194, 196
Chueh and Prausnitz (1967) method, 208
Chung et al. method:
  high-pressure gas mixture viscosities, 496
  high-temperature liquid viscosity, 519
  liquid mixture viscosity, 531
  low-pressure gas mixture viscosities, 469–470*t*, 476–479, 496
  low-pressure gas viscosity, 460–462
  pressure and thermal conductivities of gases, 557–558
  pressure and viscosity of pure gases, 488–492, 490–491*t*, 495
  thermal conductivity of high-pressure gas mixtures, 570–571, 574
  thermal conductivity of low-pressure gas mixtures, 564, 566–567
  thermal conductivity of polyatomic gases, 550, 551, 552, 552*t*
  viscosity of liquid mixtures at high pressures and temperatures, 531
Clapeyron equation, 156, 165, 178
Classical density functional theory, 659
"Classical" region, 266
Clausius-Clapeyron equation, 156, 196, 199
CODESSA Pro, 91
Cohesive energy density, 361
Combined standard uncertainty, 15
Comparison of estimation methods for pure-component constants, 85–92. *See also* Boiling and melting points; Vapor-liquid critical properties
  abbreviations, 86*t*
  boiling point, 89, 89*t*
  critical pressure, 87–88
  critical temperature, 86–87
  critical volume, 88
  factor analysis, 90
  Kazakov et al. (2010) method, 92
  machine learning/artificial intelligence techniques, 91–92
  melting point, 88–89
  molecular descriptors, 90–91
  philosophy, 85–86
  QSPR and other methods, 90–92
  selecting the best prediction method, 89–90
  Turner et al. (1998) method, 91
Complete basic set (CBS) approaches, 119
Composite method, 119
Compound classifications. *See* Associating compounds; Heavy compounds; Normal compounds; Polar compounds
Compressibility factor *(Z)*, 218, 219, 481
Compressibility-factor Z constant, 28
Conceptuals, 230
Conformer-Rotamer Ensemble Sampling Tool (CREST), 128
Consolute temperature, 428
Constantinou and Gani (1994) method, 42–48, 66–68, 134*t*, 135–141
Contact counting, 352
CoolProp compounds, 317
Cooperativity effect, 425
Correlation consistent composite approach (ccCA), 119, 137
Corresponding states methods:
  liquid heat capacity, 177, 179*t*
  low-pressure gas mixture viscosities, 473–479
  low-pressure gas viscosity, 462–465
  pressure and viscosity of pure gases, 494–495
  pure liquid surface tension, 664–665
  thermal conductivity of low-pressure gas mixtures, 564

Corresponding states principle (CSP):
compound classifications, 4
defined, 219
dipole moments, 93
EOS parameter values, 236–237
extended corresponding states (ECS) model, 237
Lee-Kesler model, 237
liquid heat capacity, 177, 179*t*
liquid vapor pressure, 168
modifications of the law, 3
nonpolar and weakly polar molecules, 4
PVT properties of methane and nitrogen, 3, 3*f*
strongly polar and associating molecules, 4–5
TRAPP corresponding states method, 3
two-parameter and three-parameter CSP, 316
what is it?, 3, 219
COSMO-based models, 25
COSMO-RS, 353, 409
COSMO-RS/FSAC, 324, 327*t*, 328*t*, 445*t*, 447
COSMO-RS/FSAC2, 445*t*, 448*t*, 450, 450*t*
COSMO-RS/SAC, 2, 386–389, 406, 408, 445, 445*t*, 446*t*
COSMO-RS/SAC-GAMESS:
IDACs, 445*t*, 446*t*, 447
liquid-liquid equilibria (LLE), 448*t*
mixtures, 324, 327*t*, 328*t*, 405*t*, 406
solid-liquid-equilibria (SLE), 450*t*
COSMO-RS/SAC-Phi, 324, 326*t*, 327*t*, 328*t*, 405*t*, 406
COSMO-RS/Therm(3ds), 405*t*, 406, 408
COSMOtherm, 327*t*
COSMOtherm (3ds), 328*t*
Coupled cluster theory (CCT), 5
Cox chart, 161
cP. *See* Centipose (cP)
CPA model. *See* Cubic plus association (CPA) model
CREST. *See* Conformer-Rotamer Ensemble Sampling Tool (CREST)
Critical compressibility factor (Z_c), 4
Critical evaluated data, 11, 22–23
Critical pressure (P_c), 87–88. *See also* Vapor-liquid critical properties
Critical region, 226–229, 266
Critical temperature (T_c), 86–87. *See also* Vapor-liquid critical properties
Critical volume (V_c), 88. *See also* Vapor-liquid critical properties
Cross coefficients, 318
CSP. *See* Corresponding states principle (CSP)
Cubic EOS, 235–239, 277
Cubic plus association (CPA) model:
1A approach, 269
assumptions, 246
compressed liquid density deviations, 278*t*
critical region density deviations, 279*t*
generally, 252
measures of compressibility deviations, 279*t*
measures of thermal properties deviations, 280*t*
saturated liquid density deviations, 277*t*
vapor density deviations, 278*t*
vapor pressure deviations, 275*t*
Curtiss et al. (2007) G4 method, 119–121

D-Glucose(1)-Sucrose(2)-Water(3), 440*f*, 440*t*
Dannenfelser and Yalkowsky (1996) method, 195, 196
Danner and Daubert (1997) method, 579–580, 582, 582*t*, 583*t*
Danner-Gess database, 8, 405, 405*t*
Darken equation, 639
Daubert et al. (1997) equation, 148, 151, 151*f*, 152
De Santis and Grande (1979) method, 208
Deam and Maddox (1970) method, 673
DECHEMA, 254, 333
Departure functions, 230, 232, 233–234
Design Institute for Physical Property Data (DIPPR), 13, 23, 256, 505
DIADEM software, 35, 93, 141, 518
Diffusion, 605–658
chemical potential driving force, 607–608
coefficients, 605, 606–607
defined, 605
electrolyte solutions, 648–652
fluxes, 605–606
gases. *See* Gas diffusion
intradiffusion, 608
liquids. *See* Liquid diffusion
notation (acronyms/symbols/Greek), 652–653
self-diffusivity, 608–613
supercritical fluids, 625, 625*f*, 626
Diffusion coefficients, 606–607
mutual, 607, 608*f*
polar, 616–617
self-, 607, 608*f*
supercritical fluids, 625*f*
tracer, 607, 608*f*
Diffusion fluxes, 605–606
Diffusivity, 605
Dilution effect, 294
Dipole moments, 92–93
DIPPR. *See* Design Institute for Physical Property Data (DIPPR)
DIPPR database, 2, 7, 23, 24, 35, 668
DIPPR 801 database, 91, 93, 141, 458, 500, 552, 660
DIPPR 2021 database, 517, 667, 668
Direct simulation, 270–271
DL-Alanine(1)-DL-Valine(2)-Water(3), 442*t*, 443*f*, 443*t*, 444*f*
Do-mod UNIFAC. *See* Dortmund modified (Do-mod) UNIFAC
Domalski and Hearing (1993, 1994) method, 107–111, 134*t*, 135–141
Domalski and Hearing (1996) method, 180*t*
Dortmund data base, 377
Dortmund modified (Do-mod) UNIFAC, 371, 376, 377, 448*t*, 450*t*
Dragon 7, 91
Dreisbach (1995, 1959) method, 180*t*
Dyn/cm, 659

EA model. *See* Explicit atom (EA) model
ECS. *See* Extended corresponding states (ECS) model
Efficient ESD formula, 273
EGC-ESD:
mixtures, 326*t*, 327*t*, 328*t*
pure gases and liquids, 275*t*, 277*t*, 278*t*, 279*t*, 280*t*
EGC-ESD(T_b):
mixtures, 326*t*, 327*t*, 328*t*
pure gases and liquids, 254, 275*t*, 277*t*, 278*t*, 279*t*, 280*t*
EGC-PC-SAFT:
mixtures, 326*t*, 327*t*
pure gases and liquids, 275*t*, 277*t*, 278*t*, 279*t*, 280*t*
EGC-PC-SAFT(T_b), 254, 275*t*, 277*t*, 278*t*, 279*t*, 280*t*
Einstein's equation, 609
Elbro et al. (1991) method, 152–154, 155*t*
Electrolyte diffusion, 648–652
Elementary kinetic theory, 456–457
"elements," 350
Elliott, J. Richard (jelliott@uakron.edu), 6. *See also Properties of Gases and Liquids* (Elliott et al.)
Elliott-Suresh-Donohue (ESD) model:
assumptions, 246
benzene(1)+methanol(2), 346, 346*f*
compressed liquid density deviations, 278*t*
critical region density deviations, 279*t*
efficient ESD formula, 273
ESD96, 324, 405*t*, 448*t*, 449*t*
measures of compressibility deviations, 279*t*
measures of thermal properties deviations, 280*t*
mixtures, 297, 305–306, 326*t*, 327*t*, 328*t*

Elliott-Suresh-Donohue (ESD) model (*Cont.*):
original formulation, 302
pure gases and liquids, 249–252
residual functions for dispersion interactions, 272*t*, 323*t*
saturated liquid density deviations, 277*t*
vapor density deviations, 278*t*
vapor pressure deviations, 275*t*
Emami et al. method, 256–257
Emami et al. (2009) method, 59–60, 61–62*t*, 81
Emami method, 256–257
Enantiomer, 105
Enskog dense-gas theory, 480–484
Enthalpy, 97–99
Enthalpy of combustion, 131–133
Enthalpy of formation, 97, 98, 117–124, 131–133
Enthalpy of melting, 180–196
Bondi (1968) methods, 180–184, 196
Chickos et al. method, 184–194, 196
Dannenfelser and Yalkowsky (1996) method, 195, 196
discussion and recommendations, 195–196
Enthalpy of sublimation, 132, 196–199
Enthalpy of vaporization, 99, 132
basic theory, 156–157
correlation, 157
extrapolation of vapor pressure, enthalpy of vaporization and heat capacity, 165–167
predicting, 174–175
Entropy, 99, 100*t*
Entropy of fusion, 437
EOS. *See* Equations of state (EOS)
Equations of state (EOS), 217–286
"adjustments," 217
applications, 396
binary LLE by EOS methods, 433–436
chemical theory EOS, 267–269
comparisons of EOS models. *See* Evaluations of equations of state
compressibility factor *(Z)*, 218, 219
critical region, 226–229
cubic EOS, 235–239
customized parameters. *See* Perturbation models with customized parameters
defined, 217
departure functions, 230, 232, 233–234
fundamental form of EOS, 218
high pressure region, 229–230
Lee-Kesler EOS, 220–221*f*, 239–240, 273
metastable region, 230
modified Benedict-Webb-Rubin (MBWR) EOS, 239–241
molecular simulation models, 270–272
multiparameter EOS, 223, 239–242
near-critical region, 310
notation (acronyms/symbols/Greek), 281–282
PC-SAFT. *See* PC-SAFT
residual functions, 232, 233–234
residual functions for evaluated models, 272–273
robustness, 229
SAFT EOS family, 225, 243
thermodynamic consistency, 25–27
thermodynamic perturbation theory (TPT), 223–226
transferable parameters. *See* Perturbation models with transferable parameters
vapor and supercritical fugacities, 333
virial EOS, 219–222, 234–235
VLE, 396–404
volume translation, 237–238
Wagner models, 239, 241–242
Ergs/cm^2, 659
Escobedo and Mansoori (1996) method, 667
ESD model. *See* Elliott-Suresh-Donohue (ESD) model
ESD96, 324, 405*t*, 448*t*, 449*t*
Estimation of physical properties, 1–2
ideal system for estimation of a physical property, 2
non-traditional estimation methods, 5
objectives of estimation, 2–3
traditional estimation methods, 2–5
Ethanol(1)+water(2), 367*f*
Eucken factor, 546, 548, 548*f*, 549*f*
Eucken method, 547, 551, 552, 552*t*
Evaluations of equations of state, 273–280
compressed liquid density deviations, 278*t*
critical region density deviations, 279*t*
measures of compressibility deviations, 279*t*
measures of thermal properties deviations, 280*t*
saturated liquid density deviations, 277*t*
shorthand notation, 274
vapor density deviations, 278*t*
vapor pressure deviations, 275*t*
Excess Gibbs energy (g^E), 335–336*t*, 342, 343*t*, 371, 420
Excess Gibbs energy mixing rules, 269, 311–315
Expanded uncertainty, 15
Experimental accuracy, 114, 137
Explicit atom (EA) model, 270, 271
Extended corresponding states (ECS) model, 237

F-SAC, 444, 445
F90 code, 274
Factor analysis, 60, 90
Fedors (1979) method, 85–92
FH. *See also* Flory-Huggins model
defined, 347, 409
IDACs, 374
methanol+benzene, 335*t*
FHA. *See* Flory-Huggins athermal term (FHA)
Filippov method, 588–589, 591, 592*f*, 594
First-order Chapman-Enskog viscosity equation, 458
First-order QCT, 351
First-order thermodynamic theory (TPT1), 244
"Five Famous Fugacity Formulae," 339
"Flash" computations, 397, 398
Flores and Amador (2004) method, 194
Flory-Huggins athermal term (FHA), 409
Flory-Huggins model. *See also* FH
defined, 347
EOSs with quadratic mixing rules, 362
MOSCED, 423
QCT, compared, 357
van der Waals EOS, 345
Flory-Huggins parameter, 347, 419
Fluctuation, 266
Fluid mixtures. *See* Mixtures
Flux, 456
Focal point method, 119
Force field, 255, 270
Fourier-Transform PGSE (FT-PGSE), 609
Frequency calculation, 115, 126*t*
FSAC-Phi EOS, 288
FT-PGSE. *See* Fourier-Transform PGSE (FT-PGSE)
Fugacity:
defined in absolute terms, 231
"Five Famous Fugacity Formulae," 339
pure liquid, 339–340
standard-state, 373
vapor, 396
Fuller et al. method, 618, 620

g^E. *See* Excess Gibbs energy (g^E)
G^E. *See* Total excess Gibbs energy (G^E)
G4 method of Curtiss et al. (2007), 119–121
Gas diffusion, 613–628
  binary gas systems at low pressure, 613–622
  mixed liquid solvent, 645
  multicomponent gas mixtures, 627–628
  pressure, 620–626
  temperature, 626–627
Gas solubility, 391, 391*t*, 395
Gas viscosity, 456–499. *See also* Viscosity
  elementary kinetic theory, 456–457
  high-pressure gas mixtures. *See* High-pressure gas mixture viscosities
  intermolecular forces, 457
  low-pressure. *See* Low-pressure gas viscosity
  low-pressure gas mixtures. *See* Low-pressure gas mixture viscosities
  phase diagram, 482*f*
  pressure and pure gases. *See* Pressure and viscosity of pure gases
Gasem et al. (1989) method, 673, 674
Gaussian, 91
Gaussian 09, 119
Gaussian extrapolation, 227, 228, 228*f*
GaussView 5.0, 119
GC-PPC-SAFT, 247, 296
GC-PPC-SAFT(IFP), 274, 275*t*, 277*t*, 278*t*, 279*t*, 280*t*
GCLF. *See* Group contribution lattice fluid (GCLF)
GCR. *See* Geometric combining rule (GCR)
Genetic cross algorithm, 128
Geometric combining rule (GCR), 247, 303
Geometry optimization, 115, 126*t*, 127, 128
Gerek-Elliott method, 609–612
GERG-2008 compounds, 317, 318
Ghobadi and Elliott Gaussian extrapolation, 227, 228, 299*f*
Gibbs-Duhem equation, 28, 30, 341–343, 371, 608, 639
Gibbs energy (conformers), 128
Gibbs energy change of reaction, 99
Gibbs energy of formation, 97
Gibbs excess mixing rules, 269, 311–315
Gibbs excess models, 407
Gibbs phase rule, 15, 25
D-Glucose(1)-Sucrose(2)-Water(3), 440*f*, 440*t*
Gn methods, 119
Goodman et al. (2004) method, 196–199
Gordon (1937) method, 649, 649*f*, 650
GPEC, 397
GQCT. *See* Guggenheim's (1935) QCT (GQCT)
Gross, Joachim, 243, 274
Group additivity, 5
Group contribution lattice fluid (GCLF), 299*f*, 300
Group contributions, 36
Grunberg and Nissan equation, 520, 529, 530, 531
GSA:
  athermal excess entropy, 299, 299*f*
  defined, 409
  g^E, 336*t*
  multicomponent g^E expressions, 342
  Panayiotou and Vera (1980), 351
Guggenheim-Staverman athermal term. *See* GSA
Guggenheim's (1935) QCT (GQCT), 352
Guide to the Expression of Uncertainty in Measurement (GUM), 13, 14, 17, 31
Guidelines for Evaluating and Expressing the Uncertainty of NIST Measurement Results (NIST Technical Note 1297), 13–14, 15
GUM. *See Guide to the Expression of Uncertainty in Measurement* (GUM)
Gunn-Yamada (1971) method, 513

Hansen's method, 362
Hard-sphere (HS) potential, 205
Hayden and O'Connell (1975) correlation, 373
Hayduk-Minhas (1982) correlation, 632–633, 634, 639, 644
HBT correlation, 149, 155*t*
He and Yu (1998) correlation, 625, 626
HEAT, 119
Heat capacity:
  ideal gas, 112–114, 126–127, 139–141
  isobaric ideal gas, 125
  liquid. *See* Liquid heat capacity
Heavy compounds, 4
  ideal gases, 134
  low-pressure viscosities of pure gases, 466
  pure component constants, 86
  pure liquid surface tension, 667
  saturated liquid density, 155
  thermal conductivity of polyatomic gases, 552
Heidemann-Mandhane constraint, 443
Heidemann-Prausnitz (1976) method, 4, 302
Heisenberg Uncertainty Principle, 115
Helmholtz energy equations of state, 100
Henry's constant, 392, 394, 395, 396, 421
Henry's law, 390–396
Henry's law volatility constant, 392
Henry's volatility, 392, 395
Herington test, 30
Herning and Zipper approximation, 469–470*t*, 472–473, 479
Hexadecane, 421
HF. *See* Hydrofluoric acid (HF)
HF/6-31G*:
  enthalpy of sublimation, 197
  ideal gases, 141
High-pressure gas mixture viscosities, 495–499
  Chung et al. method, 496
  discussion and recommendations, 499
  Lucas method, 495, 499
  TRAPP method, 496–499
High pressure region, 229–230
High-temperature liquid viscosity
Higher order virial coefficients, 208
Hildebrand-modified Batschinski equation, 513
Hildebrand solubility parameter, 422
Hogge et al (2017) method, 171–173, 179*t*
Homodesmic, 119
Homodesmotic, 119
Homodesmotic reaction, 119
Homomorph, 180
HR-SAFT, 249
HS potential. *See* Hard-sphere (HS) potential
Hsu and Chen (1998) method, 642–643
Hsu (2002) method, 516–517, 517–518
Hugill and van Welsenes (1986) method, 667, 672, 675
Huron-Vidal mixing rules, 238, 311–312, 314–315, 449
Hydrodynamic theory, 628
Hydrofluoric acid (HF), 267
Hyperhomodesmotic reaction, 119
Hypohomodesmotic reaction, 119

IAPWS EOS, 266
Id Soln, 450*t*
IDACs. *See* Infinite dilution activity coefficients (IDACs)
Ideal gas. *See* Thermodynamic properties of ideal gases
Ideal gas absolute entropy, 124–126, 137–139
Ideal gas heat capacity, 112–114, 126–127, 139–141
Ideal gas Helmholtz energy, 100
Ideal gas law, 614
Ideal gas model, 2

Ideal gas standard state enthalpy of formation, 117–124
Ideal gas standard state Gibbs energy of formation, 139
Ideal solubility, 394, 437
Ideal Solution model, 405*t*
IEM. *See* Infinite equilibrium model (IEM)
IFPSC. *See* Industrial Fluid Properties Simulation Challenge (IFPSC)
Iglesias-Silva and Hall (2001) method, 204–205, 207*t*
Incipient instability, 428
Industrial Fluid Properties Simulation Challenge (IFPSC), 270
Infinite dilution activity coefficients (IDACs), 418–427
  aqueous systems, 424–427, 450, 451
  benzene(1)+methanol(2), 349
  conversion of solvation free energy to IDAC, 419*t*
  correlation for IDACs in binary systems, 419–420
  discussion and recommendations, 8, 444–447
  experimental methods, 418
  FH theory combined with SCED perspective, 374
  Kamlet-Taft method, 420–421
  M1 model, 419
  MOSCED method, 361, 422–424, 444
  reported and compiled by numerous sources, 418
  solvatochromic method, 420–422
  TPT1, 447
  usefulness, 417, 418
Infinite dilution diffusion coefficients. *See* Liquid diffusion coefficients at infinite dilution
Infinite Dilution test, 30
Infinite equilibrium model (IEM), 268
Infinite order approximation, 227
Infinite pressure matching, 294
Interfacial tensions in liquid-liquid binary systems, 681–682
International Union of Pure and Applied Chemistry (IUPAC) values, 36
International vocabulary of metrology—Basic and general concepts and associated terms (VIM 3), 14
Intradiffusion, 608
Invariance, 294
Ion diffusion, 651–652
Irving (1977) survey, 520
Isdale modification of Grunberg-Nissan equation, 521–523
ISO/TAG 4, 31n3
Isobaric ideal gas heat capacity, 125
Isobaric VLE data, 365–367, 372–374
Isodesmic reaction, 117
Isothermal flash, 397, 432–433
Isothermal VLE data, 364–365
IUPAC Guidelines for Reporting of Phase Equilibrium Measurements, 17, 23
IUPAC values. *See* International Union of Pure and Applied Chemistry (IUPAC) values

Jamieson et al. (1975) method, 589, 594
Jamieson et al. (1969) report, 594
Jaubert, Jean-Noel, 274
Jaubert et al. database, 8
Joback and Reid (1987). *See* Joback (1984) method
Joback (1984) method, 40–42, 66, 134*t*, 135–141
Jossi, Stiel, and Thodos method, 487–488, 490–491*t*, 495
Joule-Thompson coefficient, 274
Just's (1901) rules, 394

K values, 25
K^{LL} values, 428
Kamlet-Taft (1976) multiparameter approach, 420–421
Kamlet-Taft parameters, 362
Kay's rules (Kay, 1936), 316
Kazakov et al. (2010) method, 92
Kestin and Yata (1968) report, 472
Kinematic viscosity, 456
Kinetic theory of gases, 455, 456–459
Klamt's self-consistency relation, 353
Knotts et al. (2001) method, 660, 661–663*t*, 667, 668, 668*t*, 669*t*
Kronecker delta, 353, 497
KT-modified-UNIFAC, 384
KT-UNIFAC, 375

Larsen's UNIFAC model, 678
Latini and Baroncini (1983) correlation, 587
Latini et al. method, 576, 576*t*, 579, 582, 582*t*, 583*t*, 587, 589
Lattice fluid hydrogen bonding (LFHB) theory, 354
Lattice surface Guggenheim model. *See* LSG
Law of mass action, 399
Layout of book. *See Properties of Gases and Liquids* (Elliott et al.)
Lazzaroni et al. database, 8, 419
LCCSD(T)/aug-cc-pVQZ model chemistry, 123
LCMV model, 313
lcorthr=tight. *See* "Tight" accuracy of local correlation calculations (lcorthr=tight)
LCT. *See* Lower consolute temperature (LCT)
Lee and Kesler EOS:
  compressed liquid density deviations, 278*t*
  corresponding states principle (CSP), 237
  critical region, 273
  critical region density deviations, 279*t*
  diminishment of advantages of Lee-Kesler model, 230
  illustration of Lee-Kesler correlation, 219, 220*f*, 239–240
  MBWR correlation, 239–240
  measures of compressibility deviations, 279*t*
  measures of thermal properties deviations, 280*t*
  methane and *n*-Butane, 229*t*
  saturated liquid density deviations, 277*t*
  superior accuracy, 273
  vapor density deviations, 278*t*
  vapor pressure deviations, 275*t*
Lee and Kesler (1975) method, 173, 179*t*
Lee et al. (199) method, 531
Lee et al. (1999) method, 529, 530, 531
Lee-Kesler EOS. *See* Lee and Kesler EOS
Lee-Kesler method. *See* Lee and Kesler (1975) method
Lennard-Jones 12-6 potential, 458, 546, 614
Lennard-Jones (LJ) potential, 206, 221, 222*f*, 253, 258*f*, 460*f*
Lennard-Jones viscosity collision integral, 459*f*
Level of confidence, 12, 12*f*
Level of theory (LT), 114
Lewis-Squires chart, 505, 506*f*
LFER. *See* Linear free energy relations (LFER)
LFHB. *See* Lattice fluid hydrogen bonding (LFHB) theory
Li and Fu (1991) method, 681–682
Linear free energy relations (LFER), 362, 421
Liquid diffusion, 628–652
  binary liquid diffusion coefficients, 628
  concentration dependence of binary diffusion coefficients, 639–643
  electrolyte solutions, 648–652
  infinite dilution diffusion coefficients. *See* Liquid diffusion coefficients at infinite dilution
  liquid mixture, 643
  multicomponent liquid mixtures, 645–647
  pressure, 644–645
  temperature, 644–645
Liquid diffusion coefficients at infinite dilution, 629–639
  discussion and recommendations, 639
  estimating diffusion from viscosity, 635, 638
  Hayduk-Minhas (1982) correlation, 632–633, 634
  modified Tyn-Calus method, 634

Liquid diffusion coefficients at infinite dilution (*Cont.*):
Nakanishi (1978) correlation, 634–635
solvent viscosity, 638–639
Tyn-Calus (1975) method, 630–632, 634, 639
Wilke-Chang (1955) method, 629–630
Liquid heat capacity:
basic theory, 157–158
correlations, 164
corresponding state methods, 177, 179*t*
discussion and recommendations, 178–180
extrapolation of vapor pressure, enthalpy of vaporization and heat capacity, 165–167
predicting, 175–177
Ruzicka and Domalski (1993) method, 175–177, 179*t*
Liquid-liquid equilibria (LLE), 30, 337, 427–436
binary flash computation, 417
binary LLE by EOS methods, 433–436
common thermodynamic models, 431–432
consolute temperature, 428
discussion and recommendations, 447–449
isothermal flash, 432–433
K^{LL} values, 428
phase splitting, 427, 428
wax formation, 438
Liquid mixture viscosity, 520–531
Allan and Teja (1991) method, 531
Chung et al. method, 531
discussion and recommendations, 530–531
Grunberg and Nissan equation, 520, 531
Isdale modification of Grunberg-Nissan equation, 521–523
Lee et al. (199) method, 531
McAllister equation, 531
Teja and Rice method, 528–530, 530–531, 531
Twu (1985, 1986) equation, 531
UNIFAC-VISCO method, 523–527, 530, 531
Liquid vapor pressure:
Antoine equation, 160–161, 162
basic theory, 156–157
correlations, 158–164
discussion and recommendations, 178–180
extrapolation of vapor pressure, enthalpy of vaporization and heat capacity, 165–167
fitting vapor pressure data, 162–164
Hogge et al (2017) method, 171–173, 179*t*
Lee and Kesler (1975) method, 173, 179*t*
predicting, 168–174
Riedel (1954) method, 168–171, 179*t*
shifted-rotated plot, 162–164
Vetere (1991) method, 169–171, 179*t*
Wagner equation, 161–162
Liquid viscosity, 499–531
high temperature, 518–520
low temperature. *See* Low-temperature liquid viscosity
mixtures. *See* Liquid mixture viscosity
pressure, effect of, 502–505
temperature, effect of, 505–506
viscosities of various liquids, 501*f*
Liu and Ruckenstein (1997) correlation, 626
LJ potential. *See* Lennard-Jones (LJ) potential
LJ0 to LJ4, 611, 612
LLE. *See* Liquid-liquid equilibria (LLE)
LLeDbPGL6ed96b.txt, 447
Local composition theory, 293, 357–361, 362
Local coupled-cluster "all5" method of Paulechka and Kazakov (2018), 121–122, 129–130
Local mole fraction, 358
Local surface Guggenheim (LSG) model, 359
localcc+2016. *See* 2016 local correlation calculation (localcc=2016)
Low-pressure diffusion coefficients, 615
Low-pressure gas mixture viscosities, 467–479
Chung et al. method, 469–470*t*, 476–479
compilation of references, 467
corresponding states methods, 473–479
discussion and recommendations, 479
Herning and Zipper approximation, 469–470*t*, 472–473, 479
Lucas method, 469–470*t*, 473–475, 479
Reichenberg method, 467–471, 479
Wilke (1950) method, 469–470*t*, 471–472, 479
Low-pressure gas viscosity, 457–466
Chung et al. method, 460–462
corresponding states methods, 462–465
discussion and recommendations, 466, 466*t*
Lucas method, 462, 463
Reichenberg method, 464, 465*t*
Yoon and Thodos (1970) method, 463–464
Low-temperature liquid viscosity, 506–518
Bhethanabotla (1983) method, 514–516, 517–518
discussion and recommendations, 517–518
Hsu (2002) method, 516–517, 517–518
Orrick and Erbar (1974) method, 507–508, 517–518
other correlations, 517
Przezdziecki and Sridhar (1985) method, 513–514, 517–518
Sastri-Rao (1992) method, 508–513, 517–518
Thomas (1946) method, 517–518
van Velzen et al. (1972) method, 517–518
Lower consolute temperature (LCT), 364, 428
LSG:
binary VLE models, 405*t*
defined, 359, 409
evaluation of, 408
g^E, 336*t*
liquid-liquid equilibria (LLE), 448*t*
solid-liquid equilibria (SLE), 449*t*
LT. *See* Level of theory (LT)
Lucas method:
high-pressure gas mixture viscosities, 495, 499
low-pressure gas mixture viscosities, 469–470*t*, 473–475, 479, 495
low-pressure gas viscosity, 462, 463
pressure and viscosity of pure gases, 485–486, 490–491*t*, 495
Lydersen (1955) method, 37–39
Lyons, John W., 12

M1 model, 419. *See also* Margules one-parameter model (Margules1)
Machine learning (ML), 6, 91–92, 396
Macleod-Sugden correlation:
pure liquid surface tension, 660, 664
surface tensions of mixtures, 674, 681
Macroscopic properties, 5
Margules models, 358, 371
Margules one-parameter model (Margules1). *See also* M1 model
g^E, 335*t*
G^E, 342
IDACs, 417
LLE, 428, 429*f*
quadratic temperature dependence, 428
Margules three-parameter model (Margules3)
g^E, 335*t*
nonideal carbons, 344
Margules two-parameter model (Margules2), 334
g^E, 335*t*
methanol+benzene, 335*t*
nonideal carbons, 344
Mason and Saxena approach, 563–564, 565, 566
Mass flux, 456
Maxwellian velocity distribution, 457
Maxwell's relations, 233

MBAR, 271
MBWR. *See* Modified Benedict-Webb-Rubin (MBWR)
MC. *See* Monte Carlo (MC)
McAllister equation, 531
McCann and Danner (1984) method, 207, 207*t*
MD. *See* Molecular dynamics (MD)
Mean mixture model, 371
Mean spherical approximation (MSA), 360, 360*f*
Mean squared displacement (MSD), 609
Mean value theorem (MVT), 263
Meissner and Michaels (1949) method, 675–676, 676*t*, 681
Melting point (T_m), 66, 88–89. *See also* Boiling and melting points
Meta-dynamics simulation, 128
Metastable region, 230
Methanol+benzene, 334*f*, 335*t*, 368, 369*f*
Method developers, 6
MHV1 and MHV2 mixing rules, 313
Michelsen and Mollerup (2007) monograph, 397
Michelson's simplification, 246
Mie potential, 206, 253
Milano Chemometrics and QSAR Research Group, 91
Miller (1963) method, 665
Miscibility, 427
Miscibility gap, 344
Misic and Thodos (1961) method, 548–550, 551, 552, 552*t*
Missenard method:
  pressure and thermal conductivities of liquids, 586, 586*f*
  thermal conductivity of pure liquids, 575, 578–579, 582, 582*t*, 583*t*
Mixtures, 287–332
  activity modeling, 288
  asymmetric, 298
  athermal mixing rules, 297–300
  attractive mixing rules, 300–301
  composition variations, 289–290
  cubic EOSs, 292
  diffusion, 627–628, 643, 645–647
  dilution effect, 294
  empirical correlations, 324
  energy equation, 291, 292
  errors in *PVTy* relations, 289
  evaluations and recommendations, 324–328
  excess Gibbs energy mixing rules, 311–315
  gas mixtures viscosities, 467–479, 495–499
  Huron-Vidal, 311–312, 314–315
  invariance, 294
  liquid viscosity. *See* Liquid mixture viscosity
  local composition theory, 293
  low- vs. high-density phases, 289
  MHV1 and MHV2, 313
  mixing and combining rules, 290–291
  molar volume estimation, 316–317
  molecular simulations, 288
  multiparameter EOSs, 315–318
  notation (acronyms/symbols/Greek), 328–330
  one-fluid, 290
  perturbation. *See* Perturbation models
  phase behavior. *See* Specialized phase behavior in mixtures
  programming errors, 294
  quadratic mixing rules, 347
  REFPROP and related models, 317–318
  residual functions for dispersion interactions, 323*t*
  separation operations, 333
  "solution of concepts" concept, 288
  surface tension. *See* Surface tensions of mixtures
  theory of mixture modeling, 291–294
  thermal conductivities of high-pressure gas mixtures, 567–574
  thermal conductivities of liquid mixtures, 588–595
  thermal conductivities of low-pressure gas mixtures, 562–567
Mixtures (*Cont.*):
  thermodynamic consistency, 294
  tradeoffs (Gibbs + Huron-Vidal vs. perturbation models), 288
  two-parameter and three-parameter CSP, 316
  UNIQUAC, 292–294
  van der Waals activity models, 292
  virial equations of state, 318–322
  VLE. *See* Vapor-liquid equilibria in mixtures
  Wong-Sandler, 312–313, 313–314
mod-GSA, 336*t*
Model chemistry, 114
Modeling of temperature effects, 338
Modern uncertainty assessment procedures, 22–30
  article content analysis, 23
  critical evaluated data, 11, 22–23
  DIPPR v. NIST/TRC evaluations, 23
  endpoints, 24
  equations of state (EOS), 25–27
  Gibbs phase rule, 25
  IUPAC guidelines, 23
  liquid-liquid equilibria (LLE) data, 30
  literature comparisons, 23–24
  phase equilibrium studies on binary mixtures, 28–30
  property prediction, 24–25
  reference books, 23
  single-property regression and visualization, 25
  solid-liquid equilibria (SLE) data, 30
  thermodynamic consistency, 25–30
  trends in a series of compounds, 30
  true value, 22, 23*f*
  vapor-liquid equilibria (VLE) data, 25, 28–30
Modified Benedict-Webb-Rubin (MBWR), 239–241, 520
Modified Eucken method, 547–548, 551, 552, 552*t*
Modified Joback method, 112–114, 134*t*, 135–141
Modified Lennard-Jones relation, 615
Modified Raoult's law, 341, 342
Modified Riedel equation, 161, 505
Modified Sanchez-Lacombe, 26
Modified SCED. *See* MOSCED method
Modified Tyn-Calus method, 634
Moine et al. database, 8, 419
Molar excess Gibbs energy. *See* Excess Gibbs energy (g^E)
Molecular descriptors, 36, 90–91
Molecular dynamics (MD), 128
Molecular QCT (MQCT), 357
Molecular simulation:
  athermal mixing rules, 297–300
  attractive mixing rules, 300–301
  direct simulation, 270–271
  fused sphere chain molecules, 609
  larger molecules, 128
  local composition models, 362
  phase equilibrium predictions, 288
  simulation surrogates, 271–272
Molecular simulation EOS, 270–272
Momentum flux, 456
Monte Carlo (MC), 128
MOPAC, 91
MOSCED, 445*t*
MOSCED method, 361, 422–424, 444
MOSCED05, 425, 425*t*, 445, 445*t*, 446, 446*t*
MOSCED18, 425, 425*t*, 426, 445*t*, 446, 446*t*
MOSCED21, 425, 425*t*, 426, 445, 445*t*, 446, 446*t*
MQCT. *See* Molecular QCT (MQCT)
MRCC, 123
MSD. *See* Mean squared displacement (MSD)
Multi-ion system and diffusion, 651–652
Multicomponent gas mixtures, 627–628

Multicomponent g^E expressions, 342
Multicomponent isobaric VLE data, 372–374
Multicomponent liquid mixtures, 645–647
Multiion system and diffusion, 651–652
Multiparameter EOSs:
  mixtures, 315–318
  pure gases and liquids, 223, 239–242
Multiproperty optimization, 166
Mutual diffusion coefficient, 607, 608*f*
MVT. *See* Mean value theorem (MVT)

Nakanishi (1978) correlation, 634–635
Nannoolal et al. (2004) method, 69–80
Nannoolal et al. (2007) method, 48–59
National Engineering Laboratory (NEL) report, 520
National Institute of Standards and Technology (NIST), 13, 333
Near-critical region, 266–267, 310
Neighbor swapping, 352
NEL report. *See* National Engineering Laboratory (NEL) report
Nernst-Haskell equation, 648
Newton-Raphson method, 373, 433
Newtonian fluids, 455
Next-nearest neighbor interactions, 102
NIST. *See* National Institute of Standards and Technology (NIST)
NIST Guide for the Expression of Uncertainty (Technical Note 1297), 13–14, 15
NIST JANAF Thermochemical Tables, 99, 100*t*
NIST-KT-UNIFAC:
  liquid-liquid equilibria (LLE), 448*t*
  mixtures, 324, 327*t*, 328*t*, 377, 405*t*, 406
  solid-liquid equilibria (SLE), 450*t*
NIST-mod-UNIFAC:
  IDACs, 445*t*, 446*t*
  liquid-liquid equilibria (LLE), 448*t*
  mixtures, 324, 327*t*, 328*t*, 377, 405*t*
  solid-liquid equilibria (SLE), 450*t*
NIST REFPROP, 242
NIST/SEMATECH e-Handbook of Statistical Methods, 16
NIST-TDE, 222*f*, 242, 256, 333
NIST ThermoData Engine (TDE). *See* ThermoData Engine (TDE)
NIST Thermodynamics Research Center. *See* Thermodynamics Research Center (TRC)
NIST ThermoLit website, 333
NIST/TRC, 7, 8, 23, 24, 35, 449, 668. *See also* Thermodynamics Research Center (TRC)
NIST Uncertainty Machine, 17
Non-random two-liquid theory. *See* NRTL
Non-traditional estimation methods, 5
Nonequilibrium statistical mechanics, 455
Normal compounds, 4
  ideal gases, 134
  low-pressure viscosities of pure gases, 466
  pure component constants, 86
  pure liquid surface tension, 667
  saturated liquid density, 155
  thermal conductivity of polyatomic gases, 552
NRTL:
  azeotropic data, 368–369
  binary VLE models, 405*t*
  consistently reliable, 364
  defined, 409
  g^E, 336*t*
  isobaric VLE data, 365–367
  isothermal VLE data, 364–365
  limitations, 420
  liquid-liquid equilibria (LLE), 431, 447, 448, 448*t*
  mean mixture model, 371
  methanol+benzene, 335*t*
NRTL (*Cont.*):
  multicomponent g^E expressions, 342
  solid-liquid equilibria (SLE), 449, 449*t*
  ternary VLE, 407
  UNIQUAC model, 358
  VLE and LLE, 344
NRTL with TPT1, 360, 407, 408

Ogiwara et al (1982) method, 582
Okeson and Rowley (1991) method, 495
One-constant Grunberg-Nissan (1949) equation, 520
One-fluid mixture, 290
One-parameter Margules. *See* Margules one-parameter model (Margules1)
Open-source codes, 8
OPLS-UA model, 271
Orbey and Vera (1983) method, 208
Original UNIFAC method. *See* UNIFAC-Original
Ornstein-Zernike equation, 360
Orrick and Erbar (1974) method, 507–508, 517–518
Ostwald coefficient, 390
Overview of book. *See Properties of Gases and Liquids* (Elliott et al.)

P. *See* Poise (P)
P_c. *See* Critical pressure (P_c)
P_{VLE}. *See* Vapor-liquid equilibrium (VLE) pressures (P_{VLE})
P-T-x-y data, 366, 367
P-x-T data, 363
Parachor-based methods:
  pure liquid surface tension, 660–667
  surface tensions of mixtures, 671–675
Parity plot, 420*f*, 446*f*
PC-SAFT, 26, 225
  caveats, 243–244
  compressed liquid density deviations, 278*t*
  critical region density deviations, 279*t*
  defined, 409
  HR-SAFT, compared, 249
  liquid-liquid equilibria (LLE), 448, 448*t*
  measures of compressibility deviations, 279*t*
  measures of thermal properties deviations, 280*t*
  methanol+benzene, 335*t*
  mixtures, 296, 310, 324, 326*t*, 327*t*, 328*t*
  perturbation models, 296, 310
  repulsive terms, 346
  residual functions for dispersion interactions, 272*t*, 273, 323*t*
  saturated liquid density deviations, 277*t*
  solid-liquid equilibria (SLE), 449, 449*t*
  TPT contributions, 225, 227*f*
  vapor density deviations, 278*t*
  vapor pressure deviations, 275*t*
PCC-SAFT, 310, 400–401
Peng-Robinson (1976). *See* PR76
Peng-Robinson (PR), 4, 225
  compressed liquid density deviations, 278*t*
  critical point, 256
  critical region density deviations, 279*t*
  cubic EOS, 235
  gas diffusion, 626
  measures of compressibility deviations, 279*t*
  measures of thermal properties deviations, 280*t*
  methane and *n*-Butane, 229*t*
  mixtures, 323*t*
  saturated liquid density deviations, 277*t*
  vapor density deviations, 278*t*
  vapor pressure deviations, 275*t*
Peng-Robinson with Translation, 229*t*
Percentage absolute average logarithmic deviation (&AALD), 9
Percentage average absolute deviation (%AAD), 8

Perturbation models, 294–310
  athermal mixing rules, 297–300
  attractive mixing rules, 300–301
  chemical-physical contributions, 301–308
  customized parameters. *See* Perturbation models with customized parameters
  ESD EOS, 297, 305–306
  fundamental form, 294
  geometric combining rule (GCR), 303
  near-critical region, 310
  PC-SAFT, 296, 310
  polarity contributions, 308–310
  reference and dispersion contributions, 294
  SPEADMD, 306–308
  transferable parameters. *See* Perturbation models with transferable parameters
  underlying concept, 243
  user added model for process simulation, 302
Perturbation models with customized parameters, 243–256. *See also* Perturbation models
  cubic plus association (CPA) model, 252, 269
  Elliott-Suresh-Donohue (ESD) model, 249–252
  HR-SAFT, 249
  PC-SAFT, 243–244
  polar contributions, 248–249
  SAFT-VR, 253–256
  Soft-SAFT. 253
  Wertheim's theory of association, 244–248
Perturbation models with transferable parameters, 256–267. *See also* Perturbation models
  crossover EOS models for near-critical region, 266–267
  Emami method, 256–257
  GC-PPC-SAFT, 247
  SAFT for branched interactions, 266
  SAFTγ, 259–266
  SPREADMD, 257–259
  Tihic method, 256
PGL6ed. *See Properties of Gases and Liquids* (Elliott et al.)
PGSE. *See* Pulsed gradient spin echo (PGSE)
Phase behavior. *See* Specialized phase behavior in mixtures
Phase equilibria, 337, 390–396, 408. *See also* Vapor-liquid equilibria in mixtures
Phase splitting, 427, 428
"physical basis," 345
Picard iterations, 430, 437
Pitzer (1995) method, 665, 666, 667, 668*t*, 669*t*
PM3 (Stewart, 1989) method, 91
PMOSCED, 444, 445*t*, 446, 446*t*, 447
Point test, 30
Poise (P), 456
Polar compounds, 4
  ideal gases, 134
  low-pressure viscosities of pure gases, 466
  pure component constants, 86
  pure liquid surface tension, 667
  saturated liquid density, 155
  thermal conductivity of polyatomic gases, 552
Polar diffusion coefficient, 615–617
Polar PC-SAFT (PCC-SAFT), 310, 400–401
Polymer/solvent volume ratio, 299*f*
Potoff-Mie model, 271
Power law method, 591
Poynting factor, 339, 340, 366, 392
PR. *See* Peng-Robinson (PR)
PR76:
  binary VLE models, 405*t*
  quadratic mixing, 405
  representative of evaluated models, 404
Prandtl number, 546
Prausnitz et al (1999) method, 200
Pressure:
  bubble. *See* Bubble pressure calculation
  critical pressure (P_c), 87–88
  gas diffusion, 620–626
  high-pressure gas mixture viscosity. *See* High-pressure gas mixture viscosities
  liquid diffusion, 644–645
  liquid viscosity, 502–505
  low-pressure gas mixture viscosity. *See* Low-pressure gas mixture viscosities
  low-pressure gas viscosity. *See* Low-pressure gas viscosity
  sublimation, 199–201
  thermal conductivity of gases, 553–562
  thermal conductivity of liquids, 585–587
  vapor. *See* Liquid vapor pressure
  vapor-liquid equilibrium (VLE) pressures (P_{VLE}), 21, 21*f*
  vapor pressure correlations, 158–164
  viscosity of pure gases, 479–495
Pressure and thermal conductivities of gases, 553–562
  Chung et al. method, 557–558
  discussion and recommendations, 561–562
  excess thermal conductivity correlations, 555–562
  high pressure, 554–555
  low pressure, 554
  Stiel and Thodos (1964) method, 556
  TRAPP method, 558–561
  very low pressure, 554
Pressure and viscosity of pure gases, 479–495
  Brulé and Starling method, 490–491*t*, 494–495
  Chung et al. method, 488–492, 490–491*t*, 495
  Enskog dense-gas theory, 480–484
  evaluations and recommendations, 490–491*t*, 495
  Jossi, Stiel, and Thodos method, 487–488, 490–491*t*, 495
  Lucas method, 485–486, 490–491*t*, 495
  Reichenberg method, 484, 490–491*t*, 495
  TRAPP method, 490–491*t*, 492–494, 495
Principle of corresponding states. *See* Corresponding states principle (CSP)
PRLorraine module, 324
Process simulation software developers, 225
Properties of Gases and Liquids (Elliott et al.):
  contact information, 6
  database development, 7–8
  evaluation metrics, 8–9
  experimental data, 13
  focus of book is nonelectrolytes, 333
  IUPAC values, 36
  major focus/primary objective, 35, 85
  method developers, 6
  mission, 2, 345
  molecular simulation, 6
  molecular weights, 36
  objective and comprehensive set of metrics, 7
  omissions, 9
  open-source codes, 8
  organization of book, 9
  predictive methods, 8
  property evaluations, 24
  QSPR and machine learning, 5, 6
  quantum mechanical methods, 6
  reproducibility by broad range of readers, 6, 7
  spirit of "molecular engineering," 1
  structure, bonding, and group additivity, 5
  test database, 35, 85
  TPT1 implementations, 337
  working hypothesis, 7

Property prediction, 24–25
PRSV EOS, 313–314, 403
Przezdziecki and Sridhar (1985) method, 513–514, 517–518
Pseudocomponent properties, 36
Psi4, 117, 118, 123, 125, 126
PSRK, 324, 327*t*, 328*t*, 405*t*
Pulsed gradient spin echo (PGSE), 609
Pure-component constants, 35–96
  T_c, P_c, V_c. *See* Vapor-liquid critical properties
  T_m, T_b. *See* Boiling and melting points
  acentric factor, 66
  computer programs, 93
  database, 93
  dipole moments, 92–93
  estimation methods, compared, 85–92. *See also* Comparison of estimation methods for pure-component constants
  group/bond/atom contribution methods, 36
  molecular descriptors, 36
  notation (acronyms/symbols/Greek), 93–94
  pseudocomponent properties, 36
  uses, 35
Pure fluid thermodynamic properties - temperature, 147–215
  Clapeyron equation, 156, 165, 178
  Clausius-Clapeyron equation, 156
  enthalpy of melting. *See* Enthalpy of melting
  enthalpy of sublimation, 196–199
  enthalpy of vaporization. *See* Enthalpy of vaporization
  Goodman et al. (2004) method, 196–199
  heat capacity. *See* Liquid heat capacity
  notation (acronyms/symbols/Greek), 208–212
  saturated liquid density. *See* Saturated liquid density
  solid vapor pressure, 199–201
  sublimation pressure, 199–201
  vapor pressure. *See* Liquid vapor pressure
  virial coefficients. *See* Virial coefficients
Pure gases and liquids, 217–286. *See also* Equations of state (EOS)
Pure liquid surface tension, 660–669
  Brock and Bird (1955) method, 664–665, 666, 667, 668*t*, 669*t*
  corresponding states methods, 664–665
  discussion and recommendations, 667–669
  Knotts et al. (2001) method, 660, 661–663*t*, 667, 668, 668*t*, 669*t*
  Macleod-Sugden relationship, 660, 664
  parachor-based methods, 660–667
  Pitzer (1995) method, 665, 666, 667, 668*t*, 669*t*
  Sastri and Rao method, 666, 667, 668*t*, 669*t*
  temperature, 669, 670*f*, 671*f*
  Zuo and Stenby (1997) method, 665–666, 666–667, 668*t*, 669*t*
PvT relations, 217

QCT. *See* Quasichemical theory (QCT)
QDFT. *See* Quantum density functional theory (QDFT)
QM-ATOMIC. *See* Quantum mechanical ATOMIC/ATOMIC(hc) technique of Bakowies (2009, 2020)
QM-LCC. *See* Quantum mechanical local coupled-cluster method of Paulechka and Kazakov (2017)
QM methods. *See* Quantum mechanical methods
QSPR. *See* Quantitative structure property relationship (QSPR)
Quadratic mixing rules, 347
Quantitative structure property relationship (QSPR), 5, 6, 25, 90–91
Quantum density functional theory (QDFT), 1, 5, 352
Quantum mechanical ATOMIC/ATOMIC(hc) technique of Bakowies (2009, 2020), 134*t*, 135–141
Quantum mechanical local coupled-cluster method of Paulechka and Kazakov (2017), 134*t*, 135–141
Quantum mechanical methods, 5, 6, 114–130
  "all5" method of Paulechka and Kazakov (2018), 121–122, 129–130
  basic set (BS), 114
  estimation methods compared, 134*t*, 135–141
Quantum mechanical methods (*Cont.*):
  flexible molecules and QM approaches, 127–130
  frequency scaling factors in CCCBDB, 116, 116*t*
  G4 method of Curtiss et al. (2007), 119–121
  geometry optimization, 127, 128
  ideal gas absolute entropy, 124–126
  ideal gas heat capacity, 126–127
  ideal gas standard state enthalpy of formation, 117–124
  isodesmic reaction, 117
  level of theory (LT), 114
  RRHO approximation, 124, 126
  zero-point energy (ZPE), 115, 116
Quasichemical theory (QCT), 349–352
  bonding energy, 423
  first-order QCT, 351
  formulations of QCT relations, 352
  Guggenheim's (1935) QCT (GQCT), 352
  local composition theory, 357–361, 362
  methanol+benzene, 335*t*
  molecular QCT (MQCT), 357
  reactionary stoichiometry, 357
  SS-QCT, 352–354
  strong solvation, 356*f*
  uses, 349

R. *See* Universal gas constant *(R)*
Rackett equation, 148–152, 155*t*
Raoult's law, 340, 341, 640
Ratcliff and Lusis (1971) method, 651
Redlich-Kister, 344
Redlich-Kwong EOS, 229*f*
Reference books, 23
Reference fugacities, 339
Reference materials, 16
REFPROP:
  binary VLE models, 405*t*
  compressed liquid density deviations, 278*t*
  critical region density deviations, 279*t*
  Huber et al., 275
  ideal gas heat capacity, 139
  measures of compressibility deviations, 279*t*
  measures of thermal properties deviations, 280*t*
  mixtures, 317–318, 324, 326*t*, 327*t*, 328*t*
  other EOS, compared, 273, 275–277
  saturated liquid density deviations, 277*t*
  superior accuracy, 273
  TRAPP model, 241
  vapor density deviations, 278*t*
  vapor pressure deviations, 275*t*
Regular solution theory, 438
Reichenberg method:
  low-pressure gas mixture viscosities, 467–471, 479
  low-pressure gas viscosity, 464, 465*t*
  pressure and viscosity of pure gases, 484, 490–491*t*, 495
Reliable uncertainties, 12
Repeatability, 14
Reproducibility, 6, 7
Residual functions, 232, 233–234
Residual functions for evaluated models, 272–273
Riazi and Whitson (1993) equation, 623
Rice and Teja (1982) method, 667
Riedel (1954) method, 161, 168–171, 179*t*, 665
Rigid rotator-harmonic oscillator (RRHO) approximation, 124, 126, 138, 141
RKPR model, 397
Robustness, 228
Rouse scaling, 609
Rowley (1988) method, 589–591, 591, 595

RRHO approximation. *See* Rigid rotator-harmonic oscillator (RRHO) approximation
Ruzicka and Domalski (1993) method, 175–177, 179*t*

SAC10, 445
Sadus (2018) method, 206
SAFT, 4, 337, 347, 410
SAFT models:
  Aspen-TDE-SAFT, 274, 275*t*, 277*t*, 278*t*, 279*t*, 280*t*
  defined, 410
  EGC-PC-SAFT. *See* EGC-PC-SAFT
  EGC-PC-SAFT(T_b), 254, 275*t*, 277*t*, 278*t*, 279*t*, 280*t*
  GC-PPC-SAFT, 247, 296
  GC-PPC-SAFT(IFP), 274, 275*t*, 277*t*, 278*t*, 279*t*, 280*t*
  HR-SAFT, 249
  long-range multipolar forces, 248
  parameters, 4
  PC-SAFT. *See* PC-SAFT
  PCC-SAFT, 310
  polar PC-SAFT (PCC-SAFT), 310
  SAFT EOS family, 225, 243
  SAFT for branched interactions, 266
  SAFT-VR, 253–256
  SAFT-VR-Mie model, 253–256
  SAFTγ, 259–266
  SAFTγ-CG-Mie, 260–261
  SAFTγ-WCA, 261–266
  Soft-SAFT, 253
SAFT-VR, 253–256, 346
SAFT-VR-Mie model, 253–256
SAFTγ, 259–266
SAFTγ-CG-Mie, 260–261
SAFTγ-WCA, 261–266
Salts, diffusion of, 648–652
Sample impurity, 19
Sample purity, 15–16
Sastri and Rao method:
  pure liquid surface tension, 666, 667, 668*t*, 669*t*
  low-temperature liquid viscosity, 508–513, 517–518
  thermal conductivities of pure liquids, 576–577, 577*t*, 578*t*, 582, 582*t*, 583*t*
Saturated liquid density, 147–155
  discussion and recommendations, 154–155
  Elbro et al. (1991) method, 152–154, 155*t*
  Rackett equation, 148–152, 155*t*
Scatchard (1931)-Hildebrand (1929) contribution. *See* SH
SCED. *See* Separation of cohesive energy density (SCED)
Schmidt number, 626
Schrödinger's equation, 5
Second cross virial coefficients, 319–322
Second virial coefficients, 201–207
Segmental activity coefficient (SAC), 352, 387
"segments," 350
Self-diffusion coefficient, 607, 608*f*
Self-diffusivity, 608–613
Semiempirical EOS, 270
Sensitivity coefficient, 15
Separation of cohesive energy density (SCED), 361–362, 374
SH:
  defined, 347, 410
  FH theory, 374
  IDACs, 418
  liquid-liquid equilibria (LLE), 428, 429*f*
  SCED, 361
  van Laar equation, 418
Shifted-rotated plot, 162–164
Shoulder SW (ShSW) model, 258*f*
Simplification *a*, 371
Simplification *b*, 371
Simulated annealing, 128
Simulation surrogates, 271–272
Single-property regression and visualization, 25
SLE. *See* Solid-liquid equilibria (SLE)
SLeDbPGL6ed.txt, 448
Small segment QCT (SS-QCT), 352–354, 362, 408, 410
SMILES, 92
Soares, R. D. P., 324
Soave EOS, 149, 225, 252, 674
Soave-Redlich-Kwong (SRK), 4
  mixtures, 323*t*
  pure gases and liquids, 235, 272*t*, 275*t*, 277*t*, 278*t*, 279*t*, 280*t*
Soft-SAFT, 253
Solid-liquid equilibria (SLE), 30, 337, 436–443
  accuracy, 450
  evaluation and discussion, 449–450
  purity of solid phase, 417
  SLE calculations with UNIQUAC model, 438–441
  SLE computations for aqueous amino acids, 441–443
  solid-phase composition, 437*f*
Solid-phase composition, 437*f*
Solid vapor pressure, 199–201
Solubilities of gases in liquids, 391, 391*t*, 395
Solubility parameter ratio, 299*f*
Solubility plot, 440*t*, 443*f*, 444*f*
"Solution of concepts" concept, 288
Solution-of-groups model, 375
Solvatochromic method, 420–422
SPACE, 362, 423
SPEADMD:
  binary VLE models, 405*t*
  bubble pressure calculation, 401–402
  compressed liquid density deviations, 278*t*
  critical region density deviations, 279*t*
  defined, 410
  group-contribution SAFT model, 337
  liquid-liquid equilibria (LLE), 448, 448*t*
  measures of compressibility deviations, 279*t*
  measures of thermal properties deviations, 280*t*
  methanol+benzene, 335*t*
  mixtures, 300, 306–308, 324, 326*t*, 327*t*, 328*t*
  perturbation models, 306–308
  pure gases and liquids, 257–259
  repulsive mixing rule, 346
  residual functions for dispersion interactions, 272*t*, 273, 323*t*
  saturated liquid density deviations, 277*t*
  simulation surrogates, 272
  solid-liquid equilibria (SLE), 449, 449*t*
  vapor density deviations, 278*t*
  vapor pressure deviations, 275*t*
Specialized phase behavior in mixtures, 417–454
  concluding remarks, 450–451
  discussion and recommendations, 443–450
  IDACs. *See* Infinite dilution activity coefficients (IDACs)
  LLE. *See* Liquid-liquid equilibria (LLE)
  notation (acronyms/symbols/Greek), 451–452
  solubilities of solids in liquids, 436–443. *See also* Solid-liquid equilibria (SLE)
Speed of sound, 230, 271, 274, 279*t*
Square gradient theory, 659
Square well spheres (SWS) model, 221, 222*f*
SRK. *See* Soave-Redlich-Kwong (SRK)
SS-QCT. *See* Small segment QCT (SS-QCT)
Standard enthalpy of formation, 97
Standard enthalpy of reaction, 98
Standard reference materials, 17
Standard state enthalpy of formation, 117–124, 131–133

Standard-state fugacity, 373
Standard state heat of combustion, 131
Standard uncertainty, 15
State properties, 231
Statistical associating fluid theory, 4, 337, 347, 410. *See also* SAFT models
Statistical mechanics, 2
Steam oxidation of propane, 97
Stefan-Maxwell equation, 627
Stein and Brown (1994) method, 81–85
Stereoisomerism, 105
Stewart (1989) method, 91
Stiel and Thodos (1964) method:
  pressure and thermal conductivities of gases, 556
  thermal conductivity of high-pressure gas mixtures, 567–568, 574
  thermal conductivity of polyatomic gases, 548, 551, 552, 552*t*
Stockmayer potential, 615
Stokes, 456
Stokes-Einstein relation, 629, 638
Strömsöe et al (1970) experimental data, 127, 127*t*
Strong solvation, 347–349, 354–357
Structure and bonding, 5
Suarez et al. (1989) method, 679–680, 681
Sublimation pressure, 199–201
Sugar solutions, 384–386
Support Vector Machines (SVM) regression, 25
Surface tension, 659–686
  defined, 659
  mixtures. *See* Surface tensions of mixtures
  notation (acronyms/symbols/Greek), 682–685
  pure liquids. *See* Pure liquid surface tension
  units of measure, 659
Surface tensions of mixtures, 671–682
  aqueous systems, 675–678
  discussion and recommendations, 681
  Gasem et al. (1989) method, 673, 674
  Hugill and van Welsenes (1986) method, 667, 672, 675
  interfacial tensions in liquid-liquid binary systems, 681–682
  Li and Fu (1991) method, 681–682
  Macleod-Sugden correlation, 674, 681
  Meissner and Michaels (1949) method, 675–676, 676*t*, 681
  parachor-based approach, 671–675
  Suarez et al. (1989) method, 679–680, 681
  Tamura et al. (1955) method, 676–678, 681
  thermodynamic-based relations, 678–680
  Zuo and Stenby (1997), 672, 674, 675, 681
SW12-TDE, 274, 275*t*, 277*t*, 278*t*, 279*t*, 280*t*
Symmetry entropy, 104
Szyszkowski equation, 675, 676*t*, 681

T_b. *See* Boiling point (T_b)
T_c. *See* Critical temperature (T_c)
T_m. *See* Melting point (T_m)
T-y-x diagram, 365, 365*f*
Tabular uncertainty budget, 16
Takahashi and Hongo (1982) method, 623
Takahashi (1974) correlation, 620, 623*f*
TAMie. *See* Transferable anisotropic Mie (TAMie)
Tamura et al. (1955) method, 676–678, 681
Tarakad and Danner (1977) method, 207, 207*t*
tcPR:
  defined, 236, 238, 410
  evaluation of, 406
  mixtures, 314–315, 323*t*
  pure gases and liquids, 272, 275*t*, 277*t*, 278*t*, 279*t*, 280*t*
tcPR-GE:
  binary VLE models, 405*t*
  bubble pressure calculation, 402
  defined, 334, 410
tcPR-GE (*Cont.*):
  evaluation of, 406
  liquid-liquid equilibria (LLE), 448*t*
  methanol+benzene, 335*t*
  mixtures, 326*t*, 327*t*, 328*t*
  solid-liquid equilibria (SLE), 449, 449*t*
  VLE *K*-factor, 400
tcPRq:
  binary VLE models, 405*t*
  defined, 334, 410
  methanol+benzene, 335*t*
  quadratic mixing, 405
  solid-liquid equilibria (SLE), 449*t*
TDE. *See* ThermoData Engine (TDE)
Teja and Rice interpolation procedure, 580*f*, 581
Teja and Rice method, 528–530, 530–531, 531
Temperature:
  bubble-point, 374
  consolute, 428
  critical temperature (T_c), 86–87
  gas diffusion, 626–627
  high-temperature liquid viscosity, 518–520
  Lennard-Jones viscosity collision integral, 459*f*
  liquid diffusion, 644–645
  liquid viscosity, 505–506
  low-pressure thermal conductivities of gases, 553, 553*f*
  low-pressure VLE, 369
  low temperature. *See* Low-temperature liquid viscosity
  modeling of temperature effects, 338
  pure liquid surface tension, 669, 670*f*, 671*f*
  thermal conductivity of liquids, 583–584
  thermodynamic properties. *See* Pure fluid thermodynamic properties - temperature
  triple point, 29*f*, 60
  viscosities of various liquids, 501*f*
Ternary D-Glucose(1)-Sucrose(2)-Water(3), 440*f*440*t*
Ternary mixtures, 406–407
Ternary systems, 403–404, 430
Ternary VLE, 380–384, 407*t*
Test database, 35, 85. *See also Properties of Gases and Liquids* (Elliott et al.)
TFF. *See* Transferable force field (TFF)
TFF-SPEADMD, 259, 326*t*, 327*t*, 328*t*
Theory of gas transport properties, 456–457. *See also* Gas viscosity
Theory of mixture modeling, 291–294
Thermal conductivity, 545–604
  defined, 545
  gas. *See* Thermal conductivity of gases
  liquid. *See* Thermal conductivity of liquids
  notation (acronyms/symbols/Greek), 596–600
  theory, 545–546
  units of measure, 545
Thermal conductivity of gases, 547–574
  high-pressure gas mixtures, 567–574
  low-pressure gas mixtures, 562–567
  polyatomic gases, 547–553
  pressure, 553–562
  temperature, 553, 553*f*
Thermal conductivity of high-pressure gas mixtures, 567–574
  Chung et al. method, 570–571, 574
  discussion and recommendations, 574
  Stiel and Thodos (1964) method, 567–568, 574
  TRAPP method, 571–573, 574
Thermal conductivity of liquid mixtures, 588–595
  Baroncini et al. method, 589, 591, 594–595
  discussion and recommendations, 591–593
  examples, 594–595
  Filippov method, 588–589, 591, 592*f*, 594

Thermal conductivity of liquid mixtures (*Cont.*):
  Jamieson et al. (1975) method, 589, 594
  power law method, 591
  Rowley (1988) method, 589–591, 591, 595
  thermal conductivities of several binaries, 588, 588*f*
Thermal conductivity of liquids, 574–595
  liquid mixtures, 588–595
  pressure, 585–587
  pure liquids, 575–583
  temperature, 583–584
Thermal conductivity of low-pressure gas mixtures, 562–567
  Chung et al. method, 564, 566–567
  corresponding states methods, 564
  discussion and recommendations, 564–567
  Mason and Saxena approximation, 563–564, 565, 566
  typical gas-mixtures thermal conductivities, 562–563, 562*f*
  Wassiljewa equation, 563, 565
Thermal conductivity of polyatomic gases, 547–553
  Chung et al. method, 550, 551, 551–552, 552, 552*t*
  discussion and recommendations, 551–553
  Eucken method, 547, 551, 552, 552*t*
  Misic and Thodos (1961) method, 548–550, 551, 552, 552*t*
  modified Eucken method, 547–548, 551, 551–552, 552, 552*t*
  Stiel and Thodos (1964) method, 548, 551, 552, 552*t*
Thermal conductivity of pure liquids, 575–583
  Danner and Daubert (1997) method, 579–580, 582, 582*t*, 583*t*
  discussion and recommendations, 582–583
  Latini et al. method, 576, 576*t*, 579, 582, 582*t*, 583*t*
  Missenard method, 575, 578–579, 582, 582*t*, 583*t*
  other techniques, 580–582
  Sastri and Rao method, 576–577, 577*t*, 578*t*, 582, 582*t*, 583*t*
  Teja and Rice interpolation procedure, 580*f*, 581
ThermoData Engine (TDE), 24, 28 92, 141, 419, 458, 500, 552
Thermodynamic consistency, 25–30, 294
Thermodynamic correction factor α, 642
Thermodynamic perturbation theory (TPT), 223–226
Thermodynamic properties of ideal gases, 97–146
  availability of data and computer software, 141
  Benson method, 102–107
  discussion and recommendations, 133–141
  Domalski and Hearing (1993, 1994) method, 107–111
  enthalpy, 97–99
  enthalpy of combustion, 131–133
  entropy, 99, 100*t*
  ideal gas absolute entropy, 137–139
  ideal gas heat capacity, 112–114, 139–141
  ideal gas standard state Gibbs energy of formation, 139
  modified Joback method, 112–114
  notation (acronyms/symbols/Greek), 141–144
  prediction methods evaluated, 133–141
  QM methods. *See* Quantum mechanical methods
  standard state enthalpy of formation, 131–133
Thermodynamic properties of mixtures, 287–332. *See also* Mixtures
Thermodynamic properties of pure gases and liquids, 217–286.
    See also Equations of state (EOS)
Thermodynamics Research Center (TRC), 6, 7, 13, 24, 458, 500, 552.
    See also NIST/TRC
ThermoLit, 395
Thermophysical properties, 14–16
Third cross virial coefficients, 322
Third virial coefficients, 207–208
Thol et al. EOS, 271
Thomas (1946) method, 517–518
Three-parameter Margules. *See* Margules three-parameter model
    (Margules3)
Three-parameter Wilson equation, 344n
"Tight" accuracy of local correlation calculations (lcorthr=tight), 123
Tihic et al. method, 256
Tihic method, 256
Titration, 22
Total excess Gibbs energy (G^E), 342, 370
TPT. *See* Thermodynamic perturbation theory (TPT)
TPT/virial series, 223
TPT1:
  activity modeling, 362
  bonding energy, 423
  chemical-physical interactions, 302–304, 337
  defined, 244, 410
  evaluation of, 407
  graphing, 337
  heteronuclear chain, 263
  IDACs, 447
  LFHB, compared, 354
  oligomers, 263
  reactionary stoichiometry, 357
  strong solvation, 356*f*
  water-methanol mixture, 399
Tracer diffusion coefficient, 607, 608*f*
Traditional estimation methods, 2–5
Transferable anisotropic Mie (TAMie), 271
Transferable force field (TFF), 255, 259, 275*t*, 277*t*, 278*t*, 279*t*, 280*t*
Transferable intermolecular potentials, 1
"Transferable" potential models, 5
Translated consistent Peng-Robinson. *See* tcPR
Transport property prediction model. *See* TRAPP method
TRAPP method, 3
  high-pressure gas mixture viscosities, 496–499
  MBWR, 240–241
  pressure and thermal conductivities of gases, 558–561
  pressure and viscosity of pure gases, 490–491*t*, 492–494, 495
  thermal conductivity of high-pressure gas mixtures, 571–573, 574
TraPPE-UA model, 271
TRC. *See* Thermodynamics Research Center (TRC)
TRC database. *See* NIST/TRC
Triple point temperature, 29*f*, 60
True value, 22, 23*f*
Tsonopoulos method, 202–204, 207*t*
Turner et al. (1998) method, 91
Two-body intermolecular interactions, 371
Two-parameter and three-parameter CSP, 316
Two-parameter Margules. *See* Margules two-parameter model
    (Margules2)
Twu (1985, 1986) equation, 531
Tyn-Calus (1975) method, 630–632, 634, 639
Tyn-Calus (1975a) method, 642
Type A/type B, 14
Type VI behavior, 364

UA model. *See* United atom (UA) model
UFRGS. *See* University Federal Rio Grande Sul (UFRGS) module
Uncertainty, 11–33
  assessment procedures. *See* Modern uncertainty assessment procedures
  budget, 16
  combined standard, 15
  essential nature, 12
  example (binary mixture), 20–22
  example (binodal composition determined by titration), 22
  example (calibrant property values), 20
  example (sample impurity), 19
  expanded, 15
  grossly optimistic uncertainty assessments, 18
  GUM, 13, 14, 17, 31
  historical background, 13
  importance, 11–13
  in-depth knowledge required for critical assessment, 19
  IUPAC guidelines, 17, 23

Uncertainty (*Cont.*):
level of confidence, 12, 12*f*
NIST Guide for the Expression of Uncertainty (Technical Note 1297), 13–14, 15
NIST Uncertainty Machine, 17
peer review, 18
propagation procedures, 17
publication, 17–18
rate of accumulation of thermophysical-property data, 18, 18*f*
reference materials, 16
reliability, 12
sample purity, 15–16
standard, 15
standard reference materials, 17
tabular uncertainty budget, 16
thermophysical properties, 14–16
type A/type B, 14
updated metadata, 17
VIM 3, 14
Uncertainty budget, 16
UNIFAC:
Do-mod UNIFAC, 371, 376, 377, 448*t*, 450*t*
IDACs, 445*t*, 446*t*
KT-modified-UNIFAC, 384
KT-UNIFAC, 375
Larsen's UNIFAC model, 678
liquid-liquid equilibria (LLE), 448*t*
NIST-KT-UNIFAC. *See* NIST-KT-UNIFAC
NIST-mod-UNIFAC. *See* NIST-mod-UNIFAC
original UNIFAC method. *See* UNIFAC-Original
predicting activity coefficients, 375–386
ternary VLE at 45°C, 380–382
ternary VLE at 760 mmHg, 382–384
UNIFAC-Original, 376, 380, 405*t*, 425, 445*t*, 446*t*
UNIFAC predictions, 24, 42
UNIFAC-VISCO method, 523–527, 530, 531
UNIQUAC, 292–294
advantages, 344
binary VLE models, 405*t*
consistently reliable, 364
defined, 357, 410
ethanol(1)+water(2), 367*f*
experimental solubility (polar sold organic solutes), 438
g^E, 336*t*
limitations, 420
liquid-liquid equilibria (LLE), 431, 447, 448, 448*t*
local compositions, 358
Mertl (1972) experimental data, 366
methanol+benzene, 335*t*
multicomponent g^E expressions, 342
multicomponent isobaric VLE data, 372–374
solid-liquid equilibria (SLE), 438–441, 449*t*
ternary VLE, 407
VLE and LLE, 344
United atom (UA) model, 270, 271
United explicit atom (EA) model, 270
Universal gas constant *(R)*, 218, 218*t*
Universal quasichemical theory. *See* UNIQUAC
University Federal Rio Grande Sul (UFRGS) module, 324
Updated metadata, 17
Upper consolute temperature, 428
User added model for process simulation, 302

V_c. *See* Critical volume (V_c)
Valence connectivity index, 91
Van der Waals activity models, 292
van der Waals attraction (dispersion), 92
van der Waals EOS, 219, 251, 292, 296
van Laar equation:
g^E, 335*t*
nonideal carbons, 344
Scatchard (1931)-Hildebrand (1929) (SH) model, 418
UNIQUAC model, 358
Van Ness test, 30
van Velzen et al. (1972) method, 517–518
Vanillin, 28, 29*f*
Vapor and supercritical fugacities, 333
Vapor fugacity, 396
Vapor-liquid critical properties, 36–66
Ambrose (1979) method, 85–92
Constantinou and Gani (1994) method, 42–48
Emami et al. (2009) method, 59–60, 61–62*t*
estimation methods, compared, 85–92
Fedors (1979) method, 85–92
Joback (1984) method, 40–42
Lydersen (1955) method, 37–39
Nannoolal et al. (2007) method, 48–59
Wilson and Jasperson (1996) method, 60–66
Vapor-liquid equilibria in mixtures, 333–416
activity coefficients, 341–343, 374–389
activity modeling. *See* Activity modeling
basic principles (guidelines), 408–409
binary mixtures, 405–406
bubble pressure. *See* Bubble pressure calculation
correlating low-pressure binary VLE, 363–369
COSMO-RS/SAC, 386–389
discussion and recommendations, 404–408
excess Gibbs energy (g^E), 335–336*t*, 342, 343*t*
fugacity of a pure liquid, 339–340
Gibbs-Duhem equation, 341–343
Henry's law, 390–396
isobaric VLE data, 365–367, 372–374
isothermal VLE data, 364–365
lattice fluid hydrogen bonding (LFHB) theory, 354
linear free energy relations (LFER), 362
local composition theory, 357–361, 362
local surface Guggenheim (LSG) model, 359
methanol+benzene, 334*f*, 335*t*
modeling of temperature effects, 338
multicomponent VLE at low pressure, 370–372
notation (acronyms/symbols/Greek), 409–412
QCT. *See* Quasichemical theory (QCT)
Raoult's law/modified Raoult's law, 340–341, 342
separation of cohesive energy density (SCED), 361–362
simplifications in VLE relation, 340–341
sugar solutions, 384–386
temperature, effect of, on low-pressure VLE, 369
ternary mixtures, 406–407
ternary VLE, 380–384
thermodynamics of VLE, 338–339
total excess Gibbs energy (G^E), 342
UNIFAC. *See* UNIFAC
VLE *K*-factor, 398–399
VLE with equations of state, 396–404
Vapor-liquid equilibrium (VLE) consistency tests, 28–30
Vapor-liquid equilibrium (VLE) measurements, 25
Vapor-liquid equilibrium (VLE) models, 8
Vapor-liquid equilibrium (VLE) pressures (P_{VLE}), 21, 21*f*
Vapor pressure. *See* Liquid vapor pressure
Vapor pressure correlations, 158–164
vdW1-fluid behavior, 297, 300
Vetere (1991) method, 169–171, 179*t*
Veytsmann's formula, 354
Vignes (1966) method, 640, 641*f*, 642, 642*f*
VIM 3. *See* International vocabulary of metrology—Basic and general concepts and associated terms (VIM 3)

Vinograd and McBain (1941) method, 651–652
Virial coefficients, 201–208
  cross coefficients, 318
  hard-sphere (HS) potential, 205
  higher order, 208
  Iglesias-Silva and Hall (2001) method, 204–205, 207*t*
  Lennard-Jones (LJ) potential, 206
  Mie potential, 206
  molecular-based methods, 205–206
  second, 201–207
  second cross, 319–322
  third, 207–208
  third cross, 322
  Tsonopoulos method, 202–204, 207*t*
Virial equations of state, 318–322
Viscometer, 20
Viscosity, 455–544
  commercial viscometers, 20
  defined, 456
  diffusion, 635, 638
  function of thermodynamic state of the fluid, 455
  gas. *See* Gas viscosity
  IFACs, 635–639
  liquid. *See* Liquid viscosity
  Newtonian fluids, 455
  nonequilibrium property, 455
  notation (acronyms/symbols/Greek), 532–538
  units of measure, 456
Visual analysis, 25
VLE consistency tests. *See* Vapor-liquid equilibrium (VLE) consistency tests
VLE *K*-factor, 398–399
VLE measurements. *See* Vapor-liquid equilibrium (VLE) measurements
VLE models. *See* Vapor-liquid equilibrium (VLE) models
VLE pressures. *See* Vapor-liquid equilibrium (VLE) pressures (P_{VLE})
Vogel equation, 505
Volume-Translated Peng-Robinson, 26
Volume-translated-PR (vt-PR), 275*t*, 277*t*, 278*t*, 279*t*, 280*t*
Volume translation, 237–238
vt-PR. *See* Volume-translated-PR (vt-PR)

W/(m·K), 545
W (Weizmann) methods, 119
Wagner equation, 161–162
Wagner models (EOS), 239, 241–242
Waring constraint, 28
Wassiljewa equation, 563, 565
water+EGBE, 364
Watson relation, 164
WCA. *See* Weeks-Chandler-Anderson (WCA)
WebMO Basic, 117, 125, 126*t*
Weeks-Chandler-Anderson (WCA), 261
Weighting nonrandom factors, 352
Weinaug-Katz (1943) equation, 672, 673
Wertheim's blister potential model, 245
Wertheim's chain term, 248
Wertheim's (1984) first-order thermodynamic perturbation theory, 302–304. *See also* TPT1
Wertheim's steric hindrance condition, 355
Wertheim's theory of association, 225, 226, 244–248
WG 3. *See* Working Group 3 (WG 3)
Wilke and Chang (1955) method, 629–630, 644, 646
Wilke and Lee (1955) method, 618, 619, 620, 627
Wilke (1950) method, 469–470*t*, 471–472, 479
Wilson:
  binary VLE models, 405*t*
  bubble pressure calculation, 402
  consistently reliable, 364
  g^E, 335*t*
  limitations, 420
  liquid-liquid equilibria (LLE), 435
  local compositions, 358
  mean mixture model, 371
  methanol+benzene, 335*t*
  miscibility gap, 344
  multicomponent g^E expressions, 342
  solid-liquid equilibria (SLE), 449, 449*t*
  ternary mixtures, 406, 407*t*
  three-parameter Wilson equation, 344n
  UNIQUAC model, 358
Wilson and Jasperson (1996) method, 60–66
Wilson equation. *See* Wilson
Wilson model. *See* Wilson
Wohl's approximation (simplification *b*), 371
Wong-Sandler mixing rules, 312–313, 313–314, 403
Working Group 3 (WG 3), 31n3

Yamada and Gunn (1973) method, 148, 150, 151, 151*f*, 152, 155*t*
Yaws correlations, 424
Yaws databases, 425
Yoon and Thodos (1970) method, 463–464

Z. *See* Compressibility factor *(Z)*
Zabransky and Rubiczka (2004, 2005) amendment, 175
Zero-point energy (ZPE), 115, 116
Zero-point vibration energy, 115
Zuo and Stenby (1997) method:
  pure liquid surface tension, 665–666, 666–667, 668*t*, 669*t*
  surface tensions of mixtures, 672, 674, 675, 681